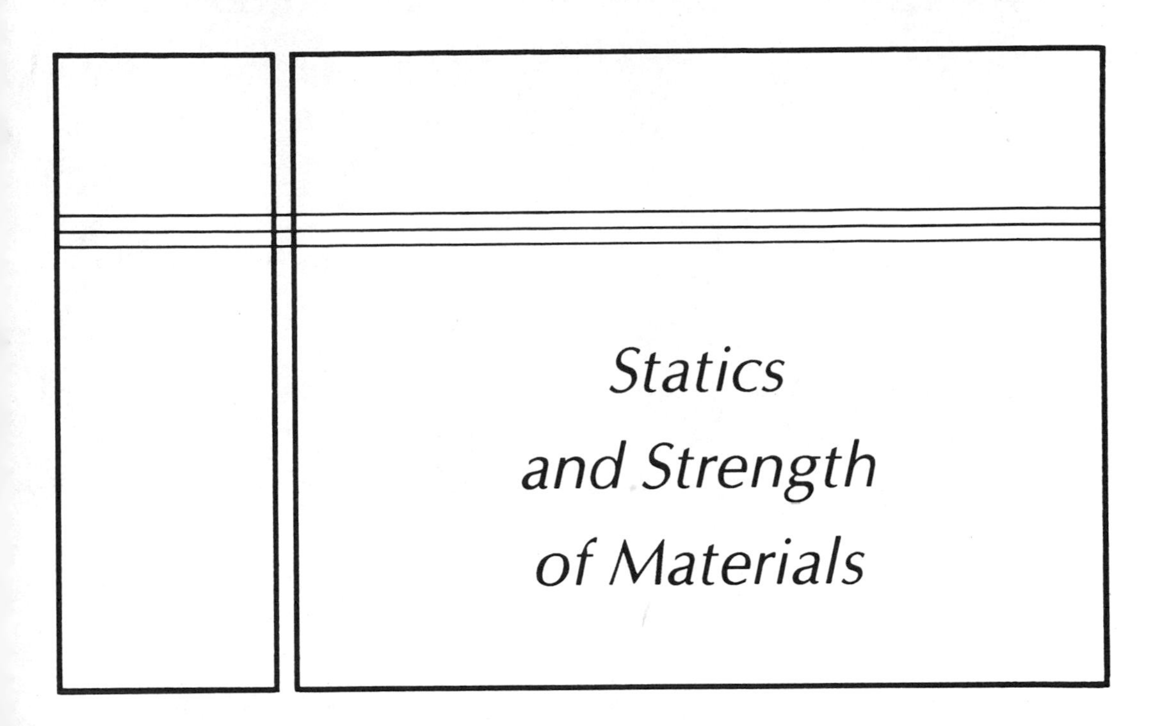
Statics
and Strength
of Materials

Second Edition

Statics and Strength of Materials

Karl K. Stevens

Department of Mechanical Engineering
Florida Atlantic University

Prentice-Hall, Inc., Englewood Cliffs, New Jersey 07632

Library of Congress Cataloging-in-Publication Data

Stevens, Karl K. (date)
Statics and strength of materials.

Includes index.
1. Statics. 2. Strength of materials.
I. Title.
TA351.S73 1987 620.1′12 86-22498
ISBN 0-13-844671-7

Editorial/production supervision
and interior design: Sophie Papanikolaou
Cover design: Lundgren Graphics
Manufacturing buyer: Rhett Conklin

Printed in the United States of America
10 9 8 7 6 5 4 3

ISBN 0-13-844671-7 025

Prentice-Hall International (UK) Limited, *London*
Prentice-Hall of Australia Pty. Limited, *Sydney*
Prentice-Hall Canada Inc., *Toronto*
Prentice-Hall Hispanoamericana, S.A., *Mexico*
Prentice-Hall of India Private Limited, *New Delhi*
Prentice-Hall of Japan, Inc., *Tokyo*
Prentice-Hall of Southeast Asia Pte. Ltd., *Singapore*
Editora Prentice-Hall do Brasil, Ltda., *Rio de Janeiro*

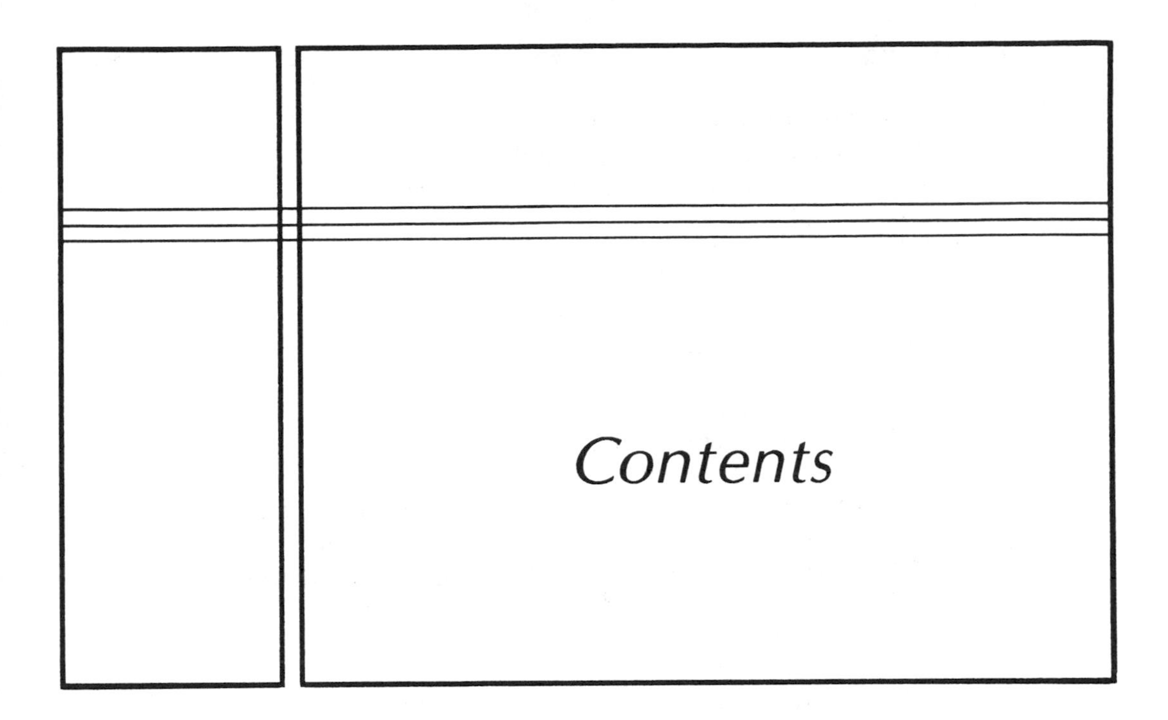

Contents

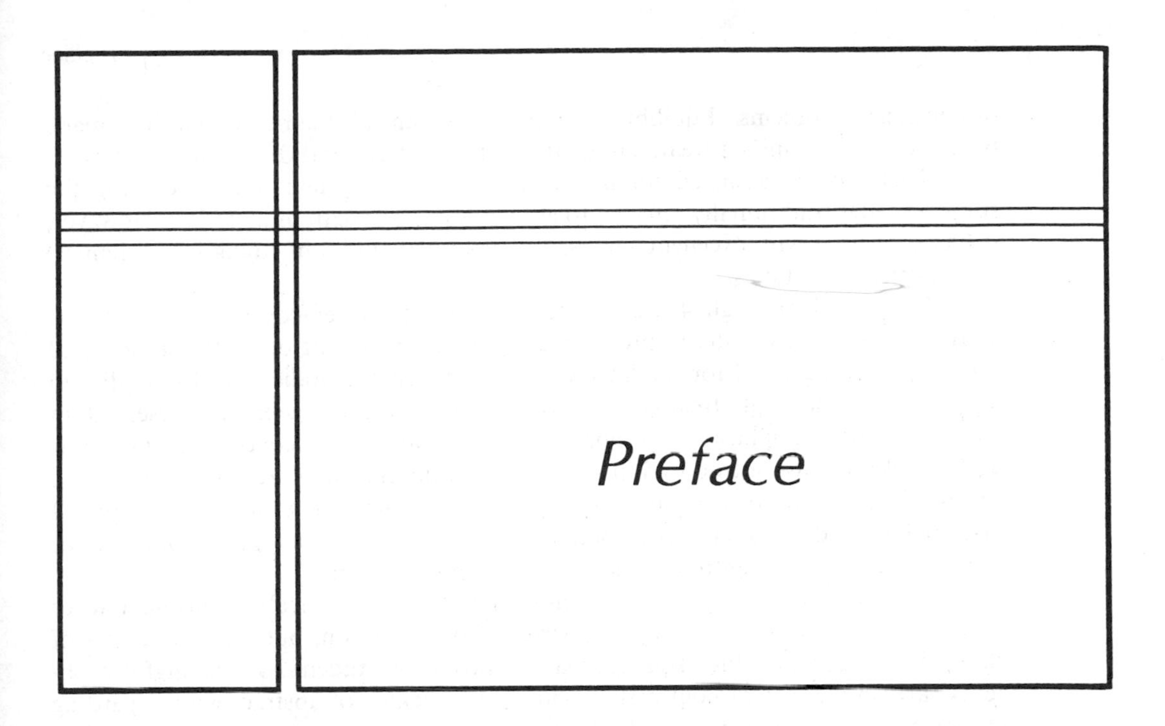

Preface

Published in 1979, the first edition of this text has been very well received, but the need for revision has become apparent. Foremost, this second edition contains an expanded set of exercise problems, most of which are completely new. There are now over 900 problems for student solution, an increase of some 15 percent over the number in the first edition. A section on the use of singularity functions for determination of beam deflections has been added, and some of the material on general states of stress and the treatment of the generalized Hooke's law have been moved forward to improve continuity of the presentation and to provide increased emphasis on the three-dimensional nature of stresses and strains. Graphical procedures, such as Mohr's circle, were used extensively in the first edition. Due to the current widespread use of computers for solution of engineering problems, increased emphasis has been placed upon analytical procedures suitable for programming. The sign conventions for shear stresses and shear strains for Mohr's circle have been changed to be consistent with those used for the stress and strain transformation equations. This should eliminate a major source of confusion. The engineering sign convention for shear forces and stresses in beams has been retained.

Material has been updated where necessary, and some has been rewritten in an attempt to improve clarity; several errors that persisted in the first edition have been corrected. Otherwise, the treatment remains much the same. This implies a rigorous, yet practical orientation of the text, with a physical approach to the development of the subject matter. The development is gradual and vector-based, although scalar-geometric methods are used for most two-dimensional and for some simple three-dimensional problems. Equilibrium remains a central theme, which has made possible a highly unified treatment of the statics of rigid and deformable bodies.

This book is designed for use in a first undergraduate course covering the subject matter traditionally referred to as statics and strength of materials. However, it has been used with excellent success for separate courses in strength (mechanics) of materials and statics.

Chapters 1 through 4 are devoted to subject matter covered in traditional statics courses. This material provides a complete treatment of vector statics, and serves as a background for the later work on deformable bodies and for studies in dynamics. A thorough treatment of vectors and vector algebra is presented in Chapter 2, with emphasis upon applications to statics. The concepts of moments and couples are also introduced in this chapter. Equilibrium is treated in Chapter 3, followed by a study of statically equivalent force systems in Chapter 4. Chapter 4 also includes a discussion of distributed forces and the related concepts of center of gravity, mass center, centroids, and area moments of inertia.

Consideration of equilibrium before the study of statically equivalent force systems and resultants is a major departure from tradition, but the advantages of doing so are considerable. This approach exposes the student to meaningful problems early in the text without first building up a backlog of abstract work regarding the properties of force systems. Early work with forces and moments as they relate to meaningful physical problems gives the student a better feeling for these quantities and their physical effects. Once the principles of equilibrium have been presented, the study of statically equivalent force systems follows as a natural consequence of the need to know how to handle distributed forces. This places the importance of static equivalence in better perspective than when it is considered before equilibrium. Furthermore, statically equivalent force systems can then be defined in terms of their effect upon equilibrium. This very clearly brings out the limited nature of static equivalence, which is important for the work in later chapters involving deformable bodies.

The work on deformable bodies is contained in Chapters 5 through 11. With equilibrium being the common thread throughout, the difference between problems involving rigid and deformable bodies lies fundamentally in the degree of sophistication required in the mathematical modeling of the problem. Accordingly, the determination of stresses and deformations is approached from the viewpoint that these are indeterminate problems for which the material properties and geometry of the deformations must be taken into account in order to obtain a solution.

The concepts of stress and strain are introduced in Chapter 5, along with a discussion of material properties and the generalized Hooke's law. Problems involving average, normal, shear, and bearing stresses are also considered in this chapter. Chapters 6 through 10 deal with axially-loaded members, torsion, bending, combined loadings, and buckling, respectively. Both elastic and inelastic responses are considered. A brief, but complete, discussion of the common methods of experimental strain and stress analysis is given in Chapter 11.

Every attempt has been made to present the material in a consistent and orderly fashion. The fundamentals are presented as succinctly as possible, consistent with good understanding. On the other hand, considerable discussion has been devoted to the meaning and interpretation of the fundamentals and the procedures for applying them to engineering problems. However, individual instructors will find that they have considerable latitude to expand upon subjects and to add their own flavoring to the presentation.

This book contains more than enough material for a five-semester-hour course. For courses of fewer credits, it will be necessary to delete certain material. Consequently, the presentation has been organized in such a way that topics can be easily deleted to fit course content. In particular, discussions of the inelastic response of members is placed in separate sections, and the exercise problems are placed at the end of the appropriate section instead of at the end of the chapters. Aside from the emphasis upon equilibrium, the work on rigid and deformable bodies has not been intermixed. The traditional US–British system of units and the Système Internationale d'Unités (SI) are used to an approximately equal extent.

Completion of a project such as this is an evolutionary process. I am deeply indebted to the hundreds of students who, through their questions, pointed out areas where clarity of the presentation could be improved. I am also indebted to students and teachers who took the time to point out errors and to pass along their comments and criticisms of the text. In this regard, I would particularly like to thank Professor James Wilson of Duke University. Mr. Sueming Shen and Mr. Mike Carling provided invaluable assistance with the solutions of the exercise problems. As usual, the Prentice-Hall staff has beem most helpful and supportive, and its substantial contributions are gratefully recognized. Finally, I want to express my deepest gratitude to Jo, Scamp, Pee Wee, and Pedro for their love, support, and inspiration during completion of this project.

Karl K. Stevens
Boca Raton, Florida

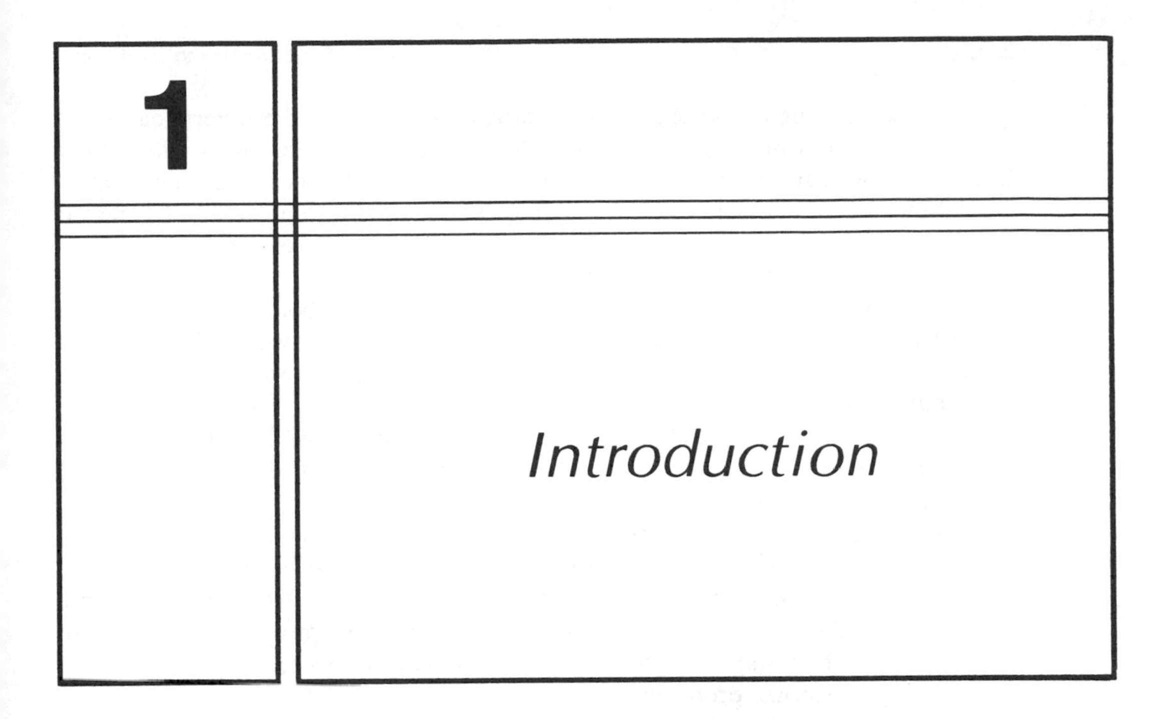

1.1 SCOPE

Let us begin with a brief overview of where we are going and what we hope to accomplish in the following chapters.

In this text we shall be concerned with the determination of the forces acting upon material bodies and their mechanical response to these forces. By mechanical response, we shall generally mean the deformation of a body and the stresses produced within it. Our interest in such problems stems from the fact that they pervade science and technology and are of direct concern to engineers, designers, and scientists.

For instance, structural engineers must be able to determine the loads acting upon a building and to design the building so that it responds to these loads in an acceptable manner. The same is true for other structures such as nuclear reactors, bridges, highways, machine parts, airplanes, pipelines, storage tanks, and electrical transmission lines. Designers of electrical machinery must account for forces produced by the magnetic and electrical fields present, and orthopedic surgeons require knowledge of the forces to which the human body is subjected in order to better repair bone fractures and to design artificial limbs and replacement parts for bones and joints. Mechanical engineers must determine the forces acting upon and within machines in order to properly design gears, bearings, and other machine elements. Agricultural engineers are faced with the problem of determining the effects of the forces to which fruits, vegetables, and grains are subjected during harvesting, handling, and storage.

These are but a few examples of problems involving the determination of forces acting upon physical systems and their response to these forces. How many other examples can you think of? Obviously, problems like these are innumerable.

Many of the problems we have mentioned are complex and cannot be treated in all generality in an introductory text such as this. Consequently, it will be necessary to restrict the types of problems to be considered.

First, we shall consider only static problems involving bodies at rest, or more precisely, bodies in equilibrium. A large number of the problems encountered in practice fall into this category. The behavior of bodies in motion is treated in texts on dynamics and vibrations.

Second, we shall consider only solid bodies. Gasses and liquids will not be considered per se, although we shall consider the loadings they impart to a solid body, such as a container or dam.

Third, although actual structures and machines are usually complex, they often consist of simple component parts such as rods, shafts, beams, and columns. We shall consider only the response of these basic structural elements and simple combinations thereof. This will enable us to solve many problems that are important in their own right and will provide the fundamentals necessary for consideration of more complex problems.

We have now identified in general terms the class of problems to be considered in this text. But what information will be needed to solve these problems? This question can probably best be answered by considering the steps involved in the solution of an engineering problem.

The first step in any engineering analysis is to clearly identify and focus attention upon the particular system or subsystem of interest. For example, suppose that we wish to find the forces that a driver exerts upon the seat of an automobile. Should we consider the driver? The seat? The automobile? The combination of driver, seat, and automobile? Or just what? A clear definition and statement of the problem usually help resolve this question.

After identifying the system of interest, the next step is to formulate an idealized model of the problem. The idea is to simplify the problem as much as possible by considering only those factors that have an important bearing upon the quantities to be determined, while neglecting those which do not. For instance, the weight of a floor beam might be only a very small fraction of the total load it supports and could, therefore, be neglected when determining the deflection (sag) of the beam.

Once a model has been decided upon, it is analyzed mathematically by using whatever physical laws are applicable. The results obtained are then examined to see if they make good sense and correspond to physical reality. A further check of the validity of the model may be obtained by performing critical experiments, or by comparing the results with those obtained from analyses based on more sophisticated models. If the model is found to be deficient, it is modified and the process repeated until satisfactory results are obtained.

What does the preceding discussion tell us about the things we must know in order to solve statics problems? Obviously, we must know how to construct appropriate models of problems, we must know the physical laws that apply, and we must know how to describe the quantities of interest in a form amenable to mathematical analysis.

The remaining sections of this chapter are devoted to a brief discussion of the basic concepts and physical laws that apply to statics problems. This will be followed in Chapter 2 by a study of how to describe certain quantities of interest, forces in particular, in a convenient mathematical form.

Once this background material has been mastered, we shall be in a position to solve some engineering problems. There will be two phases to our work. The first involves the determination of all the forces acting upon the system of interest; the second involves the determination of the response to these forces. Chapters 3 and 4 are devoted to the first phase. Here we shall develop and learn to apply the mathematical conditions for equilibrium and, at the same time, begin to develop some insight into problem modeling. Equilibrium is the single most important topic to be considered in this text, and the results presented in these chapters will be used throughout the remainder of our work.

Chapters 5–11 are devoted to the determination of the response of load-carrying members to applied loadings. In particular, we shall consider procedures for analyzing a member to see if it can fulfill its intended purpose and for designing a member so that it can safely support a given loading. The mechanical properties of common engineering materials will also be discussed, insofar as they relate to the problems of interest.

1.2 BASIC CONCEPTS AND PRINCIPLES

The study of the forces acting upon physical systems and their response to these forces forms the branch of science known as *mechanics*. Mechanics is one of the oldest branches of science, and some of its principles, such as the principle of the lever, probably came into play during man's earliest attempts to cope with his environment.

Although simple principles of mechanics were undoubtedly used in prehistoric times, it is unlikely that they were understood. The study of mechanics dates to the time of Aristotle (384–322 B.C.) and Archimedes (287–212 B.C.). Despite this early start, a satisfactory formulation of the principles of mechanics was not available until Sir Isaac Newton postulated the three laws of motion which are now the basis of most engineering applications of mechanics. These laws were presented in his *Philosophiae Naturalis Principia Mathematica* (The Mathematical Principles of Natural Science) published in 1686. The modern theories of relativity and quantum mechanics show Newton's laws to be inexact. However, the innumerable problems that have been solved successfully using these laws testify to the fact that their precision is extremely high, except in problems involving subatomic particles or velocities approaching the speed of light.

It will not be one of our objectives to trace in detail the history of the developments in mechanics. Those who are interested will find a number of good books available on this important and fascinating subject.

There are three main divisions of mechanics: *statics*, *kinematics*, and *dynamics*. Statics is concerned with systems in equilibrium and with the force interactions that operate to establish equilibrium. Kinematics is the study of the geometry of motion. Dynamics is concerned with the relationships between force interactions and the motions they produce. As we have already mentioned, we shall be concerned only with the statics of solid bodies.

Basic Concepts

Mechanics is based upon the concepts of force, mass, length, and time. Although each of us has a degree of familiarity with these concepts as a result of our physical intuition and experiences, these concepts cannot truly be defined. Rather, we give a rough indication of their physical significance and leave their actual description to be determined from the laws and postulates that describe the relationships between them. Thus, we may speak of the mass of an object as being a measure of the quantity of matter in it, but this statement can hardly be construed as being a precise definition of mass.

Our intuitive concepts of mass, length, and time will be adequate for problems to be considered in this text, particularly since time does not enter into statics problems and mass enters only indirectly. Force, however, plays a dominant role in statics and merits further comment.

The Concept of Force

The concept of force in mechanics provides a simple and convenient way to describe the very complex physical interactions between systems that alter or tend to alter the motion or state of rest of the systems. The key word here is *interaction*, which implies that two systems always participate in the creation of a force.

For example, when we stretch a rubber band, there is a force interaction between the band and our hands. This interaction causes the band to stretch, and it also produces the effect that we experience as a resistance to this stretching. The situation is illustrated schematically in Figure 1.1, where the arrows $\mathbf{F}_1$ represent the effect of the force interaction on our hands and the arrows $\mathbf{F}_2$ represent its effect on the band. According to Newton's third law, force interactions have an equal and opposite effect on the interacting systems. Thus, the effects $\mathbf{F}_1$ and $\mathbf{F}_2$ shown in Figure 1.1 are of equal magnitude and opposite direction. In other words, whatever tendency there is for our hands to stretch the band, the band has the same tendency to resist the stretching and hold our hands together.

Although the word force is properly used to denote an interaction between systems, it is more commonly used to denote the individual effects of the interaction. Thus, the individual effects $\mathbf{F}_1$ and $\mathbf{F}_2$ shown in Figure 1.1 would each be called a force. We shall also use the word force in this context, but in so doing it is essential to remember that we are talking about only one aspect of an interaction between two systems.

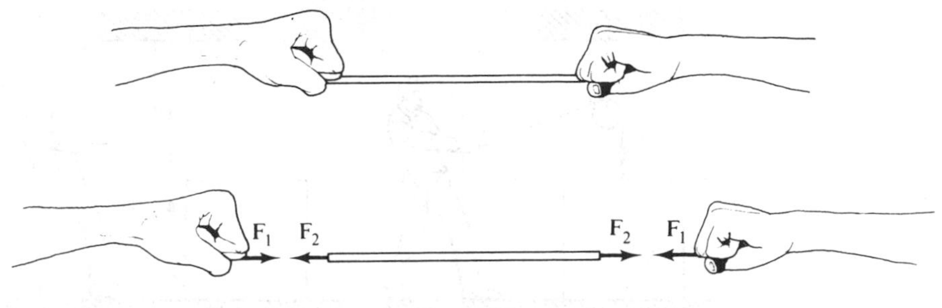

Figure 1.1

Force interactions arise when bodies come into direct physical contact, but they can also occur between systems that are separated from one another. Gravitational, electrical, and magnetic force interactions are of this latter type. Forces may act essentially at a point, as in the case of a ladder resting against a wall [Figure 1.2(a)], or they may be distributed over an area, as in the case of the wind loading on the side of a building [Figure 1.2(b)]. Some forces, such as the gravitational attraction between an object and the earth [Figure 1.2(c)], are distributed throughout a volume.

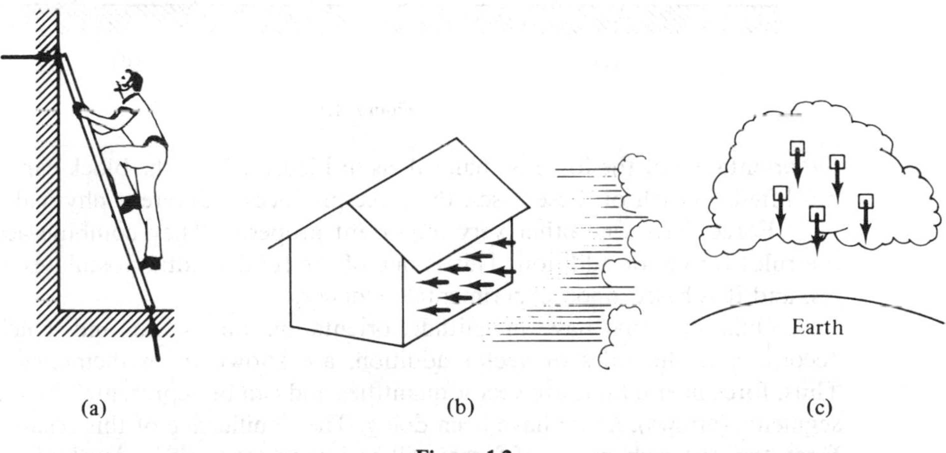

Figure 1.2

The physical effect of a force depends upon its *magnitude*, *orientation*, *sense*, and *point of application*, and these four factors must be specified in order to completely describe the force. For example, the behavior of the block shown in Figure 1.3(a) clearly depends upon the magnitude of the force exerted upon it by the man via the rope and pulley. If the force is too small, the block will remain at rest, but if the man pulls hard enough, the block can be lifted. The block can also be lifted if the rope is attached to a different point, as in Figure 1.3(b), but the block will rotate as it lifts. If the sense of the force is reversed [Figure 1.3(c)], the block will only be pressed more firmly against the surface upon which it rests. Finally, if

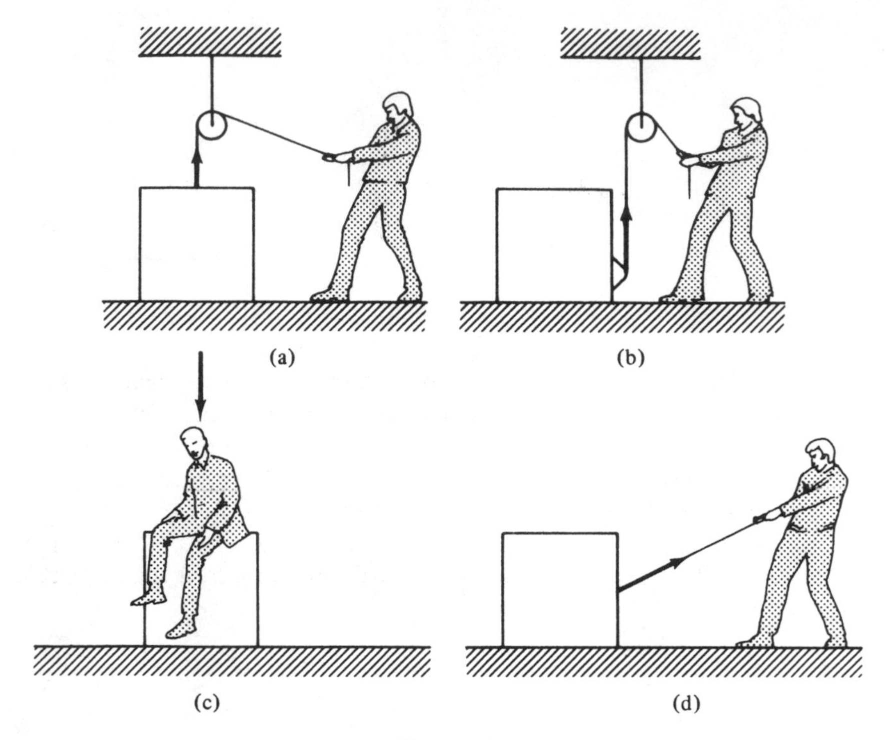

Figure 1.3

the orientation of the force is changed, as in Figure 1.3(d), the block can be slid, but not lifted. In each of these cases, the force produces a different physical effect.

Forces have one other very important property: They combine according to the rules for vector addition. This is one of the fundamental postulates of mechanics, and it is based upon experimental evidence.

Quantities that have magnitude, orientation, and sense, and which combine according to the rules of vector addition, are known in mathematics as *vectors*. Thus, force interactions are vector quantities and can be represented by directed line segments (arrows), as we have been doing. The significance of this result is twofold. First, the vector character of forces will be important to us in developing a physical feeling for them and the actions they produce. Second, all the rules of vector algebra can be used in mathematical operations involving forces. This will be helpful in the derivation of basic equations and in the solution of problems. Vectors and vector algebra will be discussed in Chapter 2.

Weight and Mass

The force exerted on a body because of its gravitational interaction with the earth is called the *weight* of the body. As will be shown in Chapter 4, the magnitude W of the weight of a body with mass m situated in a uniform gravitational field is given

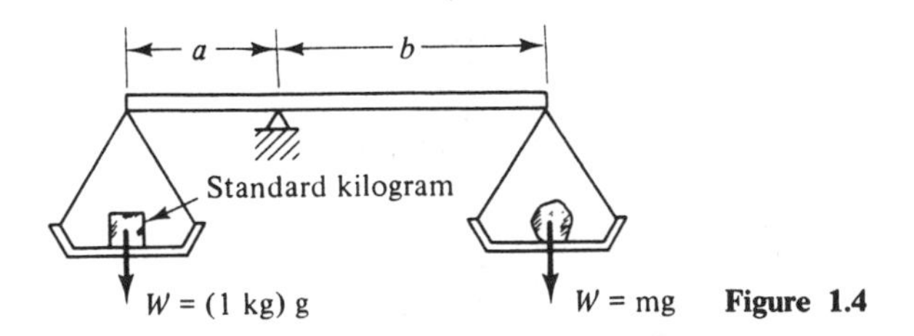

Figure 1.4

by the relation

$$W = mg \tag{1.1}$$

where g is the so-called *acceleration of gravity*. The value of g at or near the surface of the earth is 32.2 feet/second^2 (9.81 meters/second^2).

The distinction between weight and mass is widely misunderstood, mostly because of the common usage versus the standard technical usage of these terms. For example, grocery store scales in the United States are calibrated in units of force, usually pounds and ounces, and we speak of 5 pounds of potatoes as being a certain quantity of the vegetable. The true meaning, however, is the mass of potatoes which experiences a force of magnitude 5 pounds due to the gravitational attraction of the earth.

The mass of an object can be determined by balancing it against the standard unit of mass, the kilogram, on a balance scale (Figure 1.4). (The kilogram will be discussed in Section 1.3.) From Eq. (1.1), we find that the weight of the standard kilogram is $W = (1 \text{ kg})\ g$ and the weight of the mass m is $W = mg$. When the scale is balanced, the tendency of each of these forces to rotate the scale arm about the fulcrum will be equal, and we have $(1 \text{ kg})\ ga = mgb$, or $m = a/b$ kilograms. Since g does not appear in this result, the scale actually measures mass instead of weight. This is why balance scales are used in commerce.

Even though a balance scale measures mass, it can be calibrated to read in force units. The grocery scale is an example. Herein lies the source of much of the confusion between mass and weight. From the balance condition for the scale, we find that the weight W of the object is equal to $g(a/b)$ units of force. Since g varies from place to place, the calibration of the scale in force units would be valid only at a particular location. Any change in the weight of the object resulting from a change in g would not be detected by the scale because it actually measures mass.

Force can be measured with a spring scale (Figure 1.5). Experiments show that the elongation of a properly constructed spring is proportional to the force applied to it. The spring scale is based on this principle. If a standard kilogram is suspended from the scale, the elongation of the spring will correspond to g units of force.

Basic Principles

The relationship between the concepts of force, mass, length, and time are defined by several fundamental postulates based upon experimental evidence. These are the

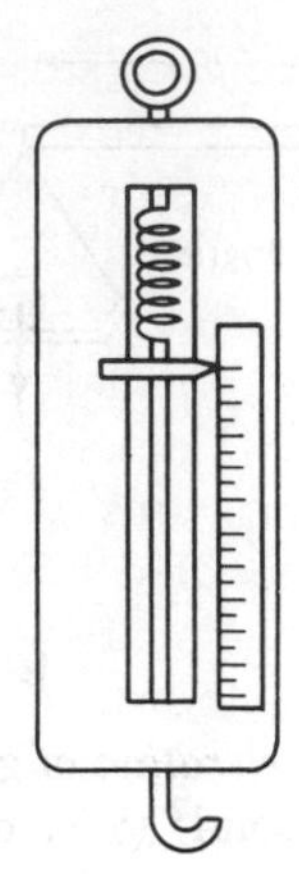

Figure 1.5

law of vector combination of forces mentioned previously, Newton's laws of motion, and laws, such as Newton's law of gravitation and Coulomb's law, which describe the nature of certain types of force interactions. These laws are sufficient to determine the response of a body to the forces acting upon it, if supplemented with information concerning the geometry of the deformations of the body and the resistive behavior of the material of which it is comprised. Further discussion of these basic laws and principles will be deferred until the chapters in which they are first used.

1.3 UNITS

In order to carry out calculations and measurements, one must first have a system of units and measures. Measurement systems are devised by choosing arbitrary standard units for a certain set of fundamental physical quantities such as mass, length, and time. For instance, the ancient Egyptians divided the period from sunrise to sunset into 12 hours of day and used the hour as their standard unit of time. The standard unit of length used by the Romans was the foot, which was derived originally from the length of a man's foot. When first introduced by the French Republican Government in 1793, the standard unit of length called the meter was defined as a specific fraction of the earth's circumference. The meter was later redefined as the distance between two marks on a certain platinum-iridium bar when at the temperature of melting ice.

Standards

Obviously, "standard" units of length and time based upon the length of a human foot and the period from sunrise to sunset are subject to variation and are not really standard. In order to eliminate ambiguities, standard units are now defined in terms of natural physical constants whenever possible.

The *meter*, which is the standard unit of length, is currently defined in terms of the wavelength of the orange–red line of the krypton–86 atom. The frequency of the radiation emitted by the cesium–133 atom in a transition between two of its fundamental hyperfine energy levels is used to define the standard unit of time, the *second*. The *kilogram* is the standard unit of mass. It is a platinum–iridium cylinder, 39 mm high and 39 mm in diameter, preserved at the International Bureau of Weights and Measures near Paris. The *degree kelvin* is the standard unit of temperature and is defined in terms of the triple point of water.

Once an acceptable standard unit for a quantity has been established, measurements can be made by comparison with the standard. Other units, which are multiples or submultiples of the standard unit, may be introduced as a matter of convenience. The Roman system of units, for example, included the inch, which was defined as one-twelfth of a foot, and the mile, which was equal to 5,000 feet.

US – British System of Units and SI

Numerous systems of units have been devised and used, but over the years most countries have come to use a metric system of measurement. A modernized version of this system, called the *Systéme Internationale d'Unités* (abbreviated SI), was established by international agreement in 1960. Both SI and the traditional US–British system of units are used in the United States. (The United Kingdom now uses SI almost exclusively.)

In the US–British system of units, force, length, time, temperature, electrical current, and luminous intensity are taken to be fundamental quantities, and the corresponding basic units are the pound, foot, second, degree fahrenheit, ampere, and candela. The units of all other physical quantities are derived in terms of these six basic units. For example, the unit of mass, as derived from Newton's second law, is the $\text{lb} \cdot \text{s}^2/\text{ft}$, sometimes called the *slug*.

The fundamental quantities in SI are mass, length, time, temperature, electrical current, and luminous intensity. The corresponding basic units are the kilogram, meter, second, degree kelvin (usually converted into degree celsius in common use), ampere, and candela. As in the US–British system, the units of all other physical quantities are expressed in terms of these basic units. Thus, the unit of force, as derived from Newton's second law, is the $\text{kg} \cdot \text{m/s}^2$. This unit is called the *newton*.

Multiples and submultiples of the basic units are used in both the US–British system and SI. For example, in the US–British system we find the inch, yard, and mile as units of length and the kip (1 kip = 1,000 pounds) as a unit of force. In SI we find the millimeter (10^{-3} meters) and kilometer (10^3 meters) as units of length and the ton (1 ton = 1,000 kilograms) as a unit of mass.

One of the major advantages of SI over the US–British system is that multiples and submultiples of the basic units are related by powers of 10 (the time units are an exception). This greatly simplifies the conversion between units because our number system is also based upon powers of 10. For example, since there are 100 centimeters in 1 meter, meters can be converted to centimeters simply by

TABLE 1.1 BASIC UNITS

Physical quantity	*Name of SI unit*	*SI symbol*	*Name of US–British unit*	*US–British symbol*
length	meter	m	foot	ft
force	newton	N	pound	lb
mass	kilogram	kg	slug	$lb \cdot s^2/ft$
time	second	s	second	s
temperature	degree celsius	°C	degree fahrenheit	°F

TABLE 1.2 EQUIVALENTS OF MEASURE

length	1 m = 3.281 ft	1 ft = 3.048×10^{-1} m
force	1 N = 2.248×10^{-1} lb	1 lb = 4.448 N
mass	1 kg = 6.854×10^{-2} slugs	1 slug = 1.459×10 kg
temperature	t_C(°C) and t_F(°F) are related by $t_C = 5/9(t_F - 32)$	

shifting the decimal point. In contrast, since there are 36 inches in a yard, conversion from yards to inches requires a multiplication by 36.

In view of the need for many engineers to be conversant with both SI and the US–British system of units, the following scheme is used in this text. Physical constants, material properties, and similar data are stated in US–British units, with the equivalent in SI given in parentheses. Approximately one-half of the example and exercise problems involve US–British units, with SI used in the remainder. Table 1.1 is a list of some of the basic units in both systems. Various equivalents of measure are given in Table 1.2.

In SI, it is recommended that numerical values be kept between 0.1 and 1,000 by the use of appropriate multipliers involving powers of 10. These multipliers are denoted by certain prefixes. The most common prefixes are listed in Table 1.3. Prefixes that shift the decimal point fewer than three places, such as centi (10^{-2}), are generally to be avoided.

TABLE 1.3 SI PREFIXES

Factor	*Prefix*	*SI symbol*
10^9	giga	G
10^6	mega	M
10^3	kilo	k
10^{-3}	milli	m
10^{-6}	micro	μ
10^{-9}	nano	n

1.4 COMPUTATIONS

In this section we shall consider several factors that frequently arise in computations.

Conversion of Units

It often is necessary to convert from one set of units to another. This may be accomplished as follows. Suppose that the magnitude of the moment of a force is given as 60 lb · ft and we wish to express it in the units N · m. We first write

$$M = 60 \text{ lb} \cdot \text{ft}$$

Since we want to replace feet by meters, we need to multiply by an expression containing meters in the numerator and feet in the denominator, and similarly for the conversion from pounds to newtons. From Table 1.2 we find that

$$1 \text{ m} = 3.281 \text{ ft} \qquad \text{or} \qquad \frac{1 \text{ m}}{3.281 \text{ ft}} = 1$$

and

$$1 \text{ N} = 0.2248 \text{ lb} \qquad \text{or} \qquad \frac{1 \text{ N}}{0.2248 \text{ lb}} = 1$$

Multiplying the value of M by these ratios, we have

$$M = (60 \text{ lb} \cdot \text{ft})\left(\frac{1 \text{ m}}{3.281 \text{ ft}}\right)\left(\frac{1 \text{ N}}{0.2248 \text{ lb}}\right)$$

Canceling out the units that appear in both the numerator and denominator and performing the numerical computations, we obtain

$$M = 60(1.356) = 81.35 \text{ N} \cdot \text{m}$$

Note that the value of M has not been changed because each of the ratios by which it was multiplied is equal to unity.

The factor 1.356 by which the moment expressed in lb · ft is multiplied to obtain the moment expressed in N · m is called a *conversion factor*. Conversion factors for a number of quantities are given in Table A.14 of the Appendix.

Dimensional Homogeneity

The principle of dimensional homogeneity states that basic equations representing physical phenomena, and all results derived from them, must be valid for all systems of units. This principle is of considerable theoretical and practical importance. It implies that all quantities must be expressed in the same units before their numerical values can be added or subtracted. Also, the units of quantities on the left side of an equality sign must be the same as those on the right side. If these conditions cannot be satisfied after performing any necessary conversion of units, the results in question are in error. Note, however, that satisfaction of the requirements of dimensional homogeneity does not guarantee that the results are correct. For example, the results could be off by a numerical factor, which would not be revealed in a check of the units.

Accuracy of Computations

With the use of computers and electronic calculators, it is possible to perform computations with a high degree of precision. Much of this precision is an illusion, however. Generally speaking, the computed value of a quantity can be no more accurate than the factors involved in it. This is a very important observation, particularly since the data in engineering problems often are not known to a high degree of precision. It is not uncommon, for example, for the values of certain material properties to be known only to within $\pm 10\%$, and the degree of uncertainty is often higher. Clearly, it would be nonsense to list to five or six decimal places a result involving a factor with a possible error of this magnitude. Furthermore, any precision that exceeds that which can be reasonably obtained when measuring the quantity in question has little or no practical significance.

We shall not be concerned in this text with a detailed analysis of errors in computations. However, all numerical results should at least be examined to see if the implied degree of precision is within reason.

1.5 CLOSURE

In this chapter we have discussed in general terms the basic concepts and principles upon which statics is based. At this point these ideas undoubtedly still seem a bit vague, and they must remain so until we see them used over and over again in the following chapters. It is only through repeated application that the basic principles can be truly understood and appreciated.

PROBLEMS

1.1. At points on or above the surface of the earth, the acceleration of gravity is inversely proportional to the square of the distance from the earth's center. At what altitude will the weight of an object be three-fourths that on the earth's surface? The radius of the earth is approximately 6,370 km.

1.2. An object that weighs 14.00 lb on the earth's surface weighs 2.31 lb on the surface of the moon. What is the local acceleration of gravity on the moon's surface?

1.3. Show that if the mass of an object is measured in kilograms and the acceleration of gravity is measured in meters per second squared, its weight will be in newtons.

1.4. The acceleration of gravity is approximately 0.236 percent less in Key West, Florida, than in Amsterdam, Netherlands. If diamonds were valued by weight instead of mass, what would be the loss in value of a million-dollar shipment from Amsterdam to Key West?

1.5. If 1 ft^3 of steel has a weight of 490 lb on the surface of the earth, what is its density (mass per unit volume) in $slugs/ft^3$? In kg/m^3?

1.6. If sawdust has a density (mass per unit volume) of 0.38 $slugs/ft^3$ and concrete a density of 2,400 kg/m^3, which has the greater weight, 1 m^3 of sawdust or 1 ft^3 of concrete?

1.7. Express your weight in newtons, your height in meters, and your waistline in millimeters. What is your mass in slugs? In kilograms? In tons?

1.8. A cruise ship is cruising at a speed of 18 knots (1 knot = 1 nmi/hr). If 1 nmi = 1.152 mi (statute), what is the speed in ft/s? In m/s?

1.9. Competition-size swimming pools commonly are 50 m long and 25 yd wide. What is the length of such a pool in yards? In feet? How many gallons of water are required to fill such a pool to a uniform depth of 2 m (1 U.S. gal = 3.785×10^{-3} m^3)?

1.10. A weather report gives the temperature in city A as 17°C, while another report gives the temperature in city B as 59°F. Which city has the cooler temperature?

1.11. At high altitude, water is observed to boil at a temperature of 98.69°C. What is the boiling point in °F?

1.12. What is the conversion factor by which a pressure (force per unit area) expressed in the units $lb/in.^2$ must be multiplied to obtain the pressure in N/m^2?

1.13. Determine the conversion factor by which the density of an object, measured in $slugs/ft^3$, must be multiplied to obtain the density in kg/m^3.

1.14. What is the conversion factor by which the advertised fuel consumption rate of an automobile, expressed in l/km, must be multiplied to obtain the rate in gal/mi? (1 U.S. gal = 3.785 l)

2

Vectors and Vector Quantities

2.1 INTRODUCTION

Before attempting to apply the basic ideas and principles of Chapter 1 to the solution of problems, we first need to know how to describe forces and related vector quantities in an appropriate and convenient mathematical form. We also need to know how to perform certain basic operations with these quantities, such as addition and subtraction. These topics will be considered in this chapter.

The basic properties of vectors will be discussed, along with vector algebra and its application to statics problems. Only those aspects of vector analysis for which we shall actually have use will be considered, and primary emphasis will be placed upon applications to physical problems. A more detailed and rigorous treatment of vector analysis may be found in the many available texts on this subject.

2.2 VECTORS

Definition and Basic Properties of Vectors

Quantities such as mass, length, time, and temperature have only a magnitude and are called *scalars*. Scalars are real numbers and obey the rules of elementary algebra. In contrast, vector quantities have both a *magnitude* and *direction* and combine according to the parallelogram law (this law will be discussed in Section

2.3). The direction of a vector includes both its *orientation* in space and its *sense*. We have already discussed in Chapter 1 the fact that forces are vector quantities. Other examples of vector quantities that occur in physical problems are velocity, acceleration, electric field intensity, and magnetic induction. Vectors can also be used to define the position in space of one point relative to another.

A vector can be represented graphically by a directed line segment (arrow). For example, Figure 2.1(a) shows the representation of a 50-lb push exerted on the crate by the man. The length of the arrow represents the magnitude of the vector, which is defined by a positive number indicating the number of units of measure involved. The force vector shown in Figure 2.1(a) has a magnitude of 50 lb. The orientation of the vector is defined by the angle between the line along which it lies, or *line of action*, and some convenient reference line. In Figure 2.1(a) the vector is oriented 30° above the horizontal. The sense of the vector is indicated by the arrowhead. Figure 2.1(b) shows the vector representation of the velocity of an aircraft in level flight traveling northeast at 800 km/hr.

Since vectors and scalars combine according to different rules, it is essential to distinguish between them. In this text, vector quantities will be denoted by letters in boldface type. The magnitude of a vector, which is a scalar, will be represented by a letter in lightface italic type or by bars enclosing the boldface vector symbol. Thus, A or $|\mathbf{A}|$ denotes the magnitude of the vector $\mathbf{A}$.

If two vectors have the same line of action, they are said to be *collinear*. Vectors that lie in the same plane are called *coplanar*, while those whose lines of action intersect at a common point are referred to as *concurrent* vectors. Two vectors $\mathbf{A}$ and $\mathbf{B}$ are *equal* ($\mathbf{A} = \mathbf{B}$) if they have the same magnitude, orientation, and sense; they need not be collinear.

It is important to note that equal vectors are not necessarily *equivalent* in the sense that they correspond to the same physical effect. For example, in Section 1.2 we found that the effect of a force depends not only upon its magnitude and direction, but also upon its point of application. Thus, two equal forces that act at different points on a body may produce entirely different responses. For example, a

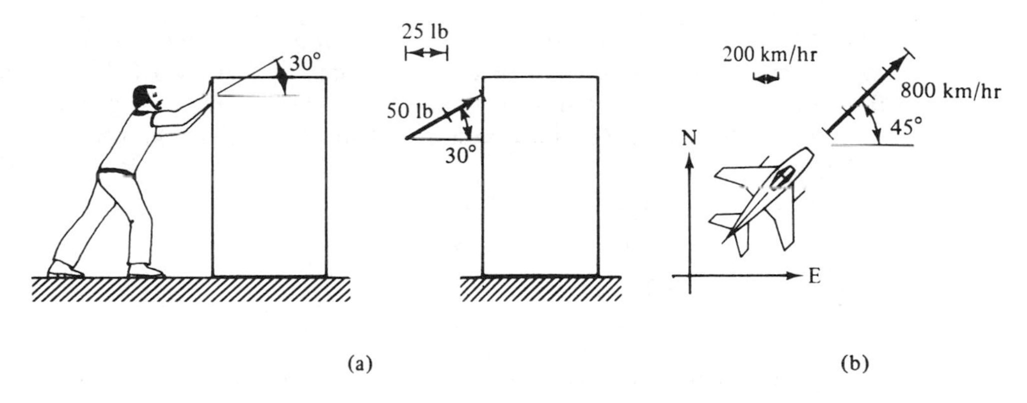

Figure 2.1

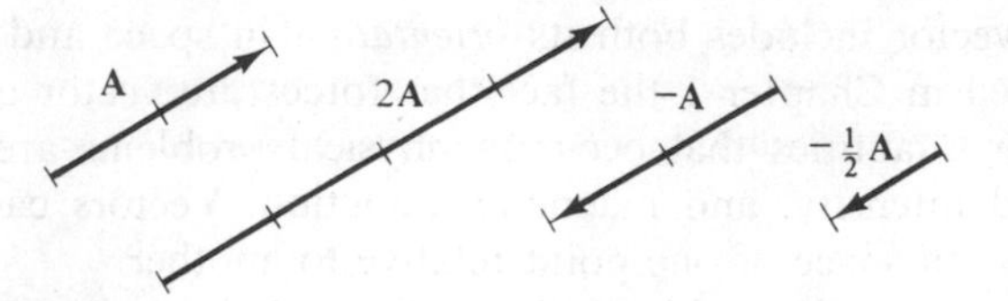

Figure 2.2

diving board will deflect considerably more if one stands at the tip instead of at the base, even though the force exerted upon the board has the same magnitude and direction in each instance. This distinction between the equality and equivalence of two forces is very important, and will be discussed in detail in Chapter 4.

Multiplication of a Vector by a Scalar

If a vector **A** is multiplied by a scalar (number) h, the product $h\mathbf{A}$ is defined as a vector whose magnitude is $|h|$ times the magnitude of **A** and whose sense is equal or opposite to that of **A**, depending upon whether h is positive or negative (Figure 2.2). Note that multiplication of a vector by (-1) merely reverses its sense. The rule for multiplication of a vector and a scalar also covers division of a vector by a scalar h, since this operation is equivalent to multiplication by the factor $1/h$, provided $h \neq 0$. Multiplication of a vector by a scalar obeys the same laws as the multiplication of two scalars.

Unit Vectors

Vectors with a magnitude of unity are called *unit vectors* and will be denoted by the boldface letter **e**. Any vector **A** can be turned into a unit vector simply by dividing by its magnitude. Thus,

$$\mathbf{e}_A = \frac{\mathbf{A}}{|\mathbf{A}|} \tag{2.1}$$

where the subscript denotes the vector with which the unit vector is associated. The set of units used to describe a vector is the same as that used for its magnitude; therefore, the unit vector will be without units, or *dimensionless*. A special set of unit vectors, directed along coordinate axes and denoted by **i**, **j**, and **k**, will be introduced in Section 2.5.

Equation (2.1) can be rewritten as

$$\mathbf{A} = |\mathbf{A}|\mathbf{e}_A$$

or, in a more general form, as

$$\mathbf{A} = \pm A\mathbf{e} \tag{2.2}$$

where **e** is a unit vector parallel to **A**. This relation indicates that the magnitude, orientation, and sense of a vector can be expressed separately. The scalar A (or $|\mathbf{A}|$)

defines the magnitude of **A**, the dimensionless unit vector **e** defines its orientation in space, and the $\pm$ sign defines its sense; a plus sign indicates that **A** has the same sense as **e**, while a minus sign indicates they have opposite senses.

Zero Vector

The *zero*, or *null*, *vector* **0** has zero magnitude by definition; its direction is undefined. It will be convenient to think of **0** as having all possible directions, so that it is parallel or perpendicular to any other vector.

2.3 ADDITION, SUBTRACTION AND RESOLUTION OF VECTORS

Vector Addition and Subtraction

The sum of two vectors **A** and **B** is defined by the *parallelogram law*, which is illustrated in Figure 2.3(a). According to this law, the vectors are placed tail to tail and a parallelogram is constructed by using them as two of the sides. The sum, **S** = **A** + **B**, corresponds to the diagonal of the parallelogram. The magnitude and orientation of **S** can be obtained graphically or by using trigonometry; its sense is determined by the construction.

It is evident from Figure 2.3(a) that the sum can also be obtained by laying out the two vectors tip to tail and constructing the vector, with tail at the starting point, necessary to form a closed triangle. This construction, which is illustrated in Figures 2.3(b) and 2.3(c), is known as the *triangle rule* for vector addition. Note that the order in which the vectors are added is immaterial.

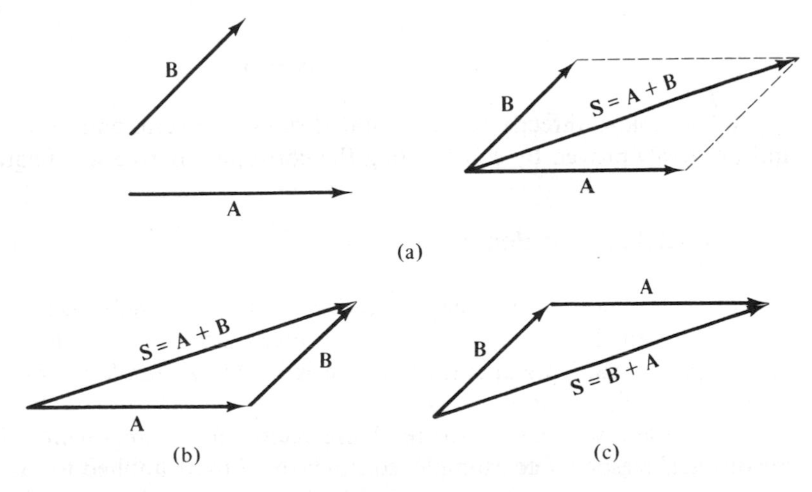

Figure 2.3

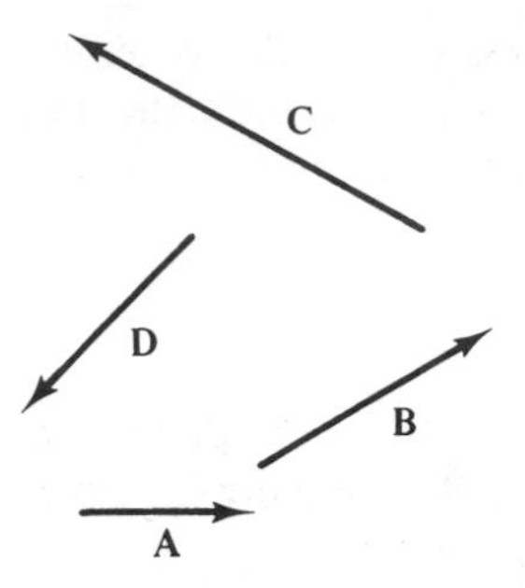

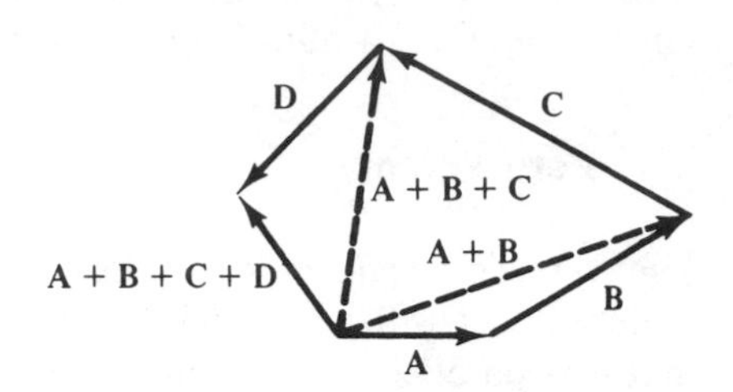

Figure 2.4

The subtraction of a vector can be defined as the addition of the corresponding negative vector. Thus, the difference of two vectors, $\mathbf{A} - \mathbf{B}$, can be rewritten as $\mathbf{A} + (-\mathbf{B})$ and the rules of addition applied.

If several vectors are to be added or subtracted, the triangle rule or parallelogram law can be applied to them two at a time. The procedure for addition using the triangle rule is illustrated in Figure 2.4. The resulting diagram reveals that the sum can also be obtained by laying out all the vectors tip to tail and constructing the vector, with tail at the starting point, necessary to close the loop. This is called the *polygon rule* for vector addition because a polygon is formed when the vectors all lie in the same plane. Although this rule applies for vectors with any orientation, it is of practical use only when the vectors are coplanar.

Addition and subtraction of vectors obey the same laws as addition and subtraction of scalars:

$$\begin{aligned} \mathbf{A} + \mathbf{B} &= \mathbf{B} + \mathbf{A} \quad \text{(commutative law)} \\ \mathbf{A} + (\mathbf{B} + \mathbf{C}) &= (\mathbf{A} + \mathbf{B}) + \mathbf{C} \quad \text{(associative law)} \\ \mathbf{A} - \mathbf{B} &= \mathbf{A} + (-\mathbf{B}) \\ \mathbf{A} + \mathbf{0} &= \mathbf{A}, \qquad \mathbf{A} - \mathbf{A} = \mathbf{0} \end{aligned} \tag{2.3}$$

These laws follow directly from the stated rules for vector addition and subtraction and are easily proved by constructing the corresponding vector diagrams.

Resolution of a Vector

Just as two or more vectors can be added to obtain a single vector, a single vector can be broken down into two or more constituent parts called *components*, or *component vectors*. This operation, which is known as *resolution* of a vector, is the reverse of vector addition.

There are two reasons for resolving vectors into components. First, it is done for physical reasons. For example, consider the force $\mathbf{F}$ applied to the wrench shown in Figure 2.5. Only the component of $\mathbf{F}$ that is perpendicular to the handle of the

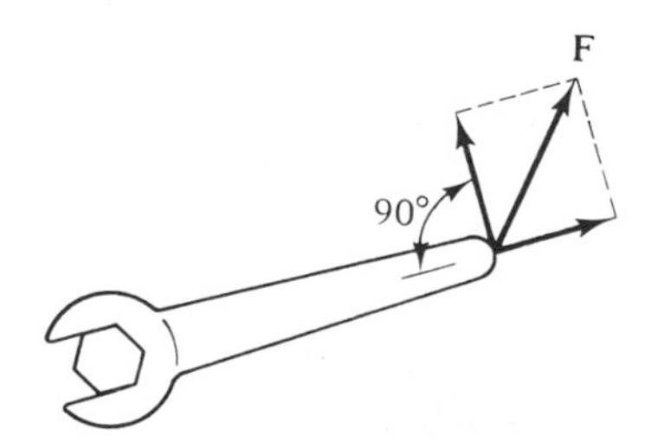

Figure 2.5

wrench is effective in turning it; therefore, we would need to know this component when determining the tendency of **F** to rotate the bolt. Second, the computations involved in vector algebra can often be simplified by first resolving the vectors into components that are mutually perpendicular, as in Figure 2.5. Such components are called *rectangular*, or *orthogonal*, *components*. The advantages of using them will be demonstrated in Section 2.5.

Resolution of a vector into components is accomplished by using the rules of vector addition and the fact that the sum of the components must equal the original vector. Naturally, the exact procedure depends upon what is known about the components. Two typical cases are considered in Examples 2.3 and 2.4.

Example 2.1. Addition of Vectors. Find the sum **S** of the two forces **P** and **Q** acting upon the bracket in Figure 2.6(a).

Trigonometric solution. We first sketch the vector triangle (or parallelogram) as shown in Figure 2.6(b). It need not be accurately constructed, but it should be sketched approximately to scale so that the figure isn't misleading. The sense of **S** is determined by the sketch. It remains to find its magnitude and orientation.

Two sides of the triangle and the included angle are known. Using the law of cosines, we have

$$S^2 = P^2 + Q^2 - 2PQ\cos B$$
$$S^2 = (5\text{ lb})^2 + (3\text{ lb})^2 - 2(5\text{ lb})(3\text{ lb})\cos 75°$$
$$S = 5.12\text{ lb}$$

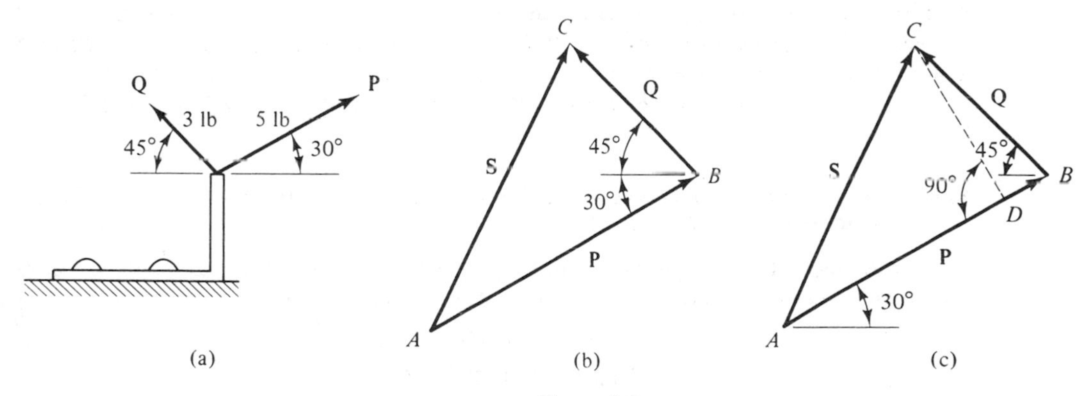

Figure 2.6

Applying the law of sines, we write

$$\frac{\sin A}{Q} = \frac{\sin 75°}{S}$$

from which

$$\sin A = \left(\frac{3 \text{ lb}}{5.12 \text{ lb}}\right)(\sin 75°)$$

$$\sin A = 0.566$$

$$A = 34.5°$$

Thus,

$$\mathbf{S} = 5.12 \text{ lb} \quad \angle 64.5° \qquad \textbf{Answer}$$

Alternate trigonometric solution. Instead of using the law of sines and cosines, we can proceed as follows. We form right triangles by dropping a perpendicular from C to AB, as shown in Figure 2.6(c). For triangle BCD, we have

$$DB = Q \cos 75° = (3 \text{ lb}) \cos 75° = 0.78 \text{ lb}$$

$$CD = Q \sin 75° = (3 \text{ lb}) \sin 75° = 2.90 \text{ lb}$$

so

$$AD = AB - DB = 5.00 \text{ lb} - 0.78 \text{ lb} = 4.22 \text{ lb}$$

From triangle ACD, we get

$$S^2 = (CD)^2 + (AD)^2 = (2.90 \text{ lb})^2 + (4.22 \text{ lb})^2$$

$$S = 5.12 \text{ lb}$$

$$\sin A = \frac{CD}{S} = \frac{2.90 \text{ lb}}{5.12 \text{ lb}} = 0.566$$

$$A = 34.5°$$

This is the same result obtained previously. The advantage of this approach is that it involves only right triangles. Note that the angle A could also have been determined from expressions for $\cos A$ or $\tan A$.

Graphical solution. We lay out the vector triangle to a convenient scale, say 2 in. = 1 lb. The scale is arbitrary, but the larger the triangle, the greater the accuracy attainable. The triangle will not be reconstructed here. Referring to our previous triangle, Figure 2.6(b), we would measure the length AC: $AC = 10.24$ in. Thus, $S = 10.24$ in. (1 lb/2 in.) = 5.12 lb. The angle that **S** makes with the horizontal, as measured with a protractor, is found to be 64.5°.

Example 2.2. Addition and Subtraction of Vectors. Given the three vectors in Figure 2.7(a), find the vector $\mathbf{T} = \mathbf{P} + \mathbf{Q} - \mathbf{R}$.

Solution. Starting with the vector **P**, we sketch the vector polygon as shown in Figure 2.7(b). Note that **R** is subtracted by adding $-\mathbf{R}$. From the geometry of the polygon, we find

$$Q = P \cos 60° + T \cos \theta$$

$$T \sin \theta = P \sin 60° + R$$

where θ defines the orientation of **T**. Rearranging these equations and dividing the second

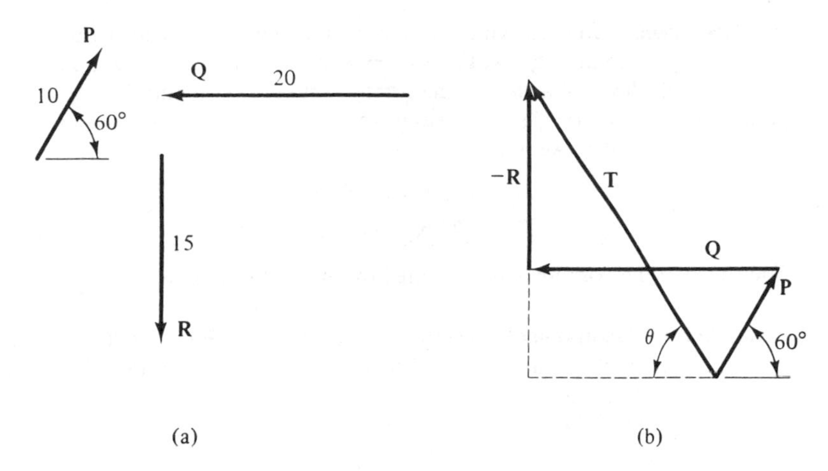

Figure 2.7

equation by the first, we obtain

$$\frac{T \sin \theta}{T \cos \theta} = \tan \theta = \frac{P \sin 60° + R}{Q - P \cos 60°}$$

$$\tan \theta = \frac{10 \sin 60° + 15}{20 - 10 \cos 60°} = 1.577$$

$$\theta = 57.6°$$

The magnitude of **T** can now be determined from either of our first two equations. Using the first equation, we get

$$T = \frac{Q - P \cos 60°}{\cos \theta} = \frac{20 - 10 \cos 60°}{\cos 57.6°} = 28.0$$

Thus,

T = 28.0 ⦪57.6° **Answer**

Of course, the solution could also be obtained graphically.

Example 2.3. Resolution of a Vector. Resolve the force **F** in Figure 2.8(a) into rectangular components that are horizontal and vertical.

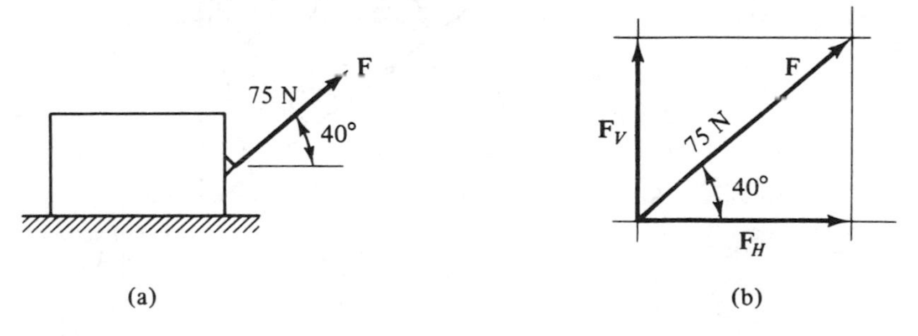

Figure 2.8

Solution. Let $\mathbf{F}_H$ and $\mathbf{F}_V$ denote the horizontal and vertical components of **F**, respectively. Recognizing that $\mathbf{F}_H + \mathbf{F}_V = \mathbf{F}$, we lay out lines at the tip and tail of **F** which are parallel to the lines of action of the component vectors [Figure 2.8(b)]. This forms the vector parallelogram (a rectangle in this case) from which the components can be determined. Using simple trigonometry, we have

$$F_H = (75 \text{ N})\cos 40° = 57.5 \text{ N}$$
$$F_V = (75 \text{ N})\sin 40° = 48.2 \text{ N}$$
Answer

The sense of the components is defined by the vector diagram.

Example 2.4. Resolution of a Vector. Two tug boats [Figure 2.9(a)] are assisting a freighter out of a harbor. (a) How hard should each tug push so that their combined effort is a 100,000-lb force in the direction the ship is heading? (b) If tug A remains oriented as shown, how should tug B be oriented so that the force it must exert is a minimum? Assume that the forces exerted by the tugs are parallel to the direction they are heading.

Solution. (a) The problem requires that we find the magnitudes of the component vectors of the 100,000-lb force with lines of action oriented in the directions the tugs push. Let **A** and **B** be the forces exerted by tugs *A* and *B*, respectively, and let **F** be their sum (F = 100,000 lb). We lay out lines at the tip and tail of **F** with the orientation of **A** and **B** to form a vector parallelogram [Figure 2.9(b)]. Considering one of the triangles formed, we have

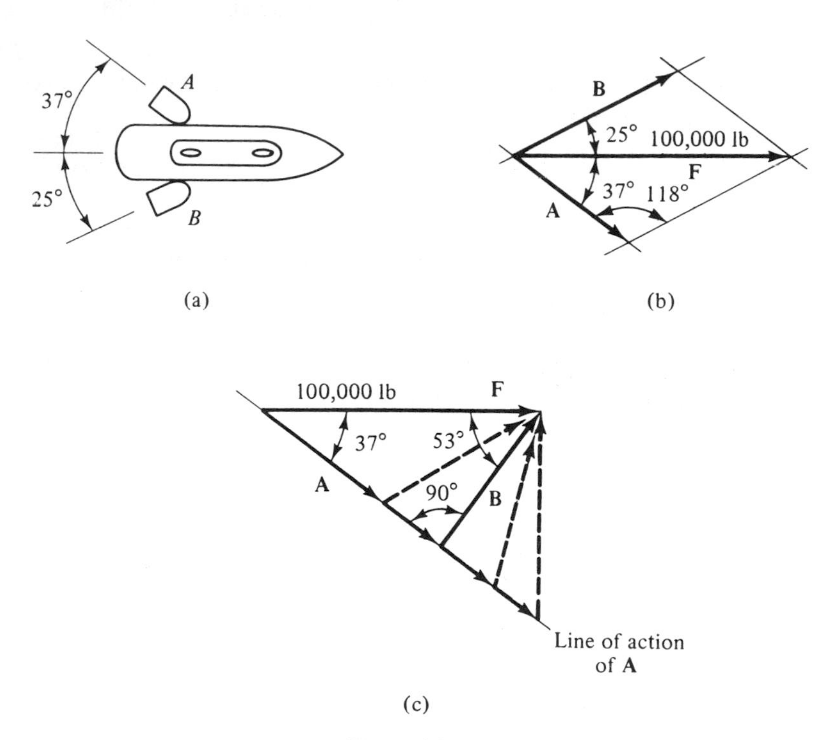

Figure 2.9

from the law of sines

$$\frac{\sin 37^\circ}{B} = \frac{\sin 118^\circ}{F} = \frac{\sin 25^\circ}{A}$$

Solving for A and B, we get

$$A = 48{,}000 \text{ lb}$$
$$B = 68{,}000 \text{ lb} \qquad \textbf{Answer}$$

(b) Here we use the triangle rule. We know that **A** must lie along a line that makes an angle of 37° with **F** [Figure 2.9(c)]. We also know that the tail of **B** coincides with the tip of **A** and the tip of **B** coincides with the tip of **F**, since **A** + **B** = **F**. However, we do not know where **A** ends and **B** begins. Several possibilities are shown. Clearly, B is smallest when **B** is perpendicular to **A**. Thus, tug B should be oriented at an angle of 53° with respect to the longitudinal axis of the ship. The magnitudes of the forces that the tugs must exert in this case are easily determined from the vector triangle: A = 80,000 lb and B = 60,000 lb.

PROBLEMS

2.1. to 2.4 Find the sum of the forces shown (a) graphically and (b) analytically.

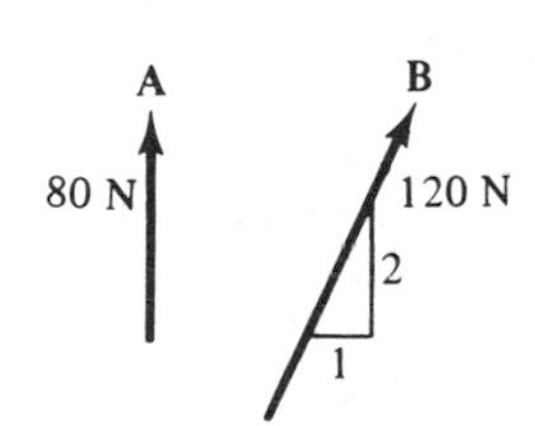

Problem 2.1

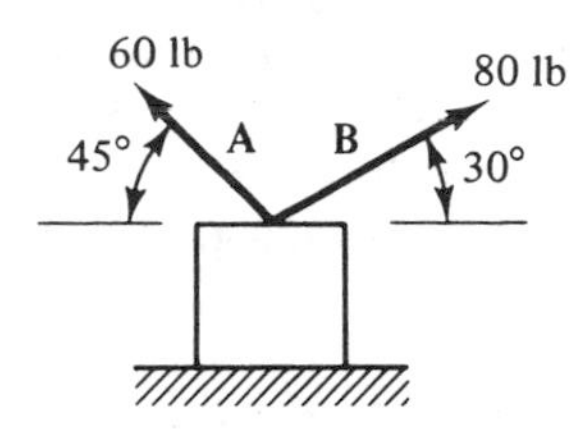

Problem 2.2

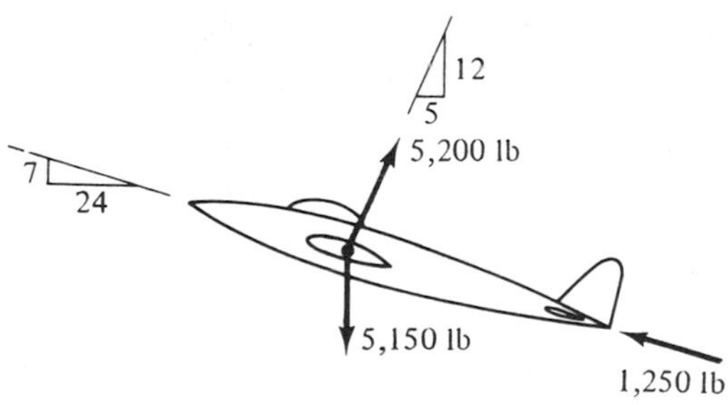

Problem 2.3

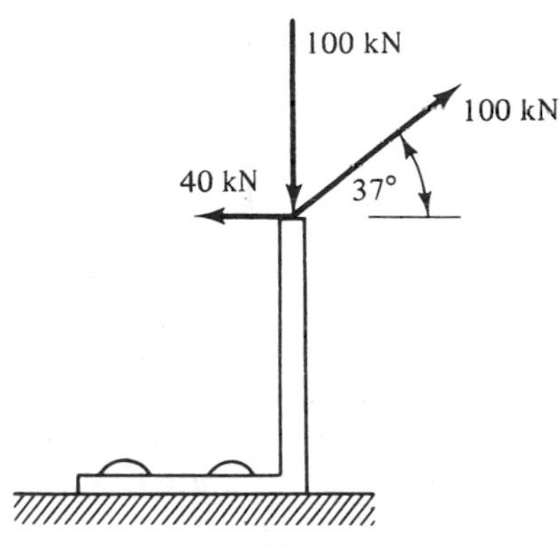

Problem 2.4

2.5. Find **A** − **B** and **B** − **A** for the forces in Problem 2.1

2.6. For the forces in Problem 2.2, find **A** − **B** and **B** − **A**.

2.7. Two ships are spotted on a radar screen. Ship A is 50 km due east of the radar location and ship B is 30 km due west and 40 km due north of it. Represent the positions of the ships by vectors. Determine the vector that represents the position of ship A relative to ship B.

2.8. Two tow trucks pull on a disabled truck as shown. If Q = 50 kN and θ = 40°, determine the magnitude of the force **P** so that the total force **F** acting lies along the

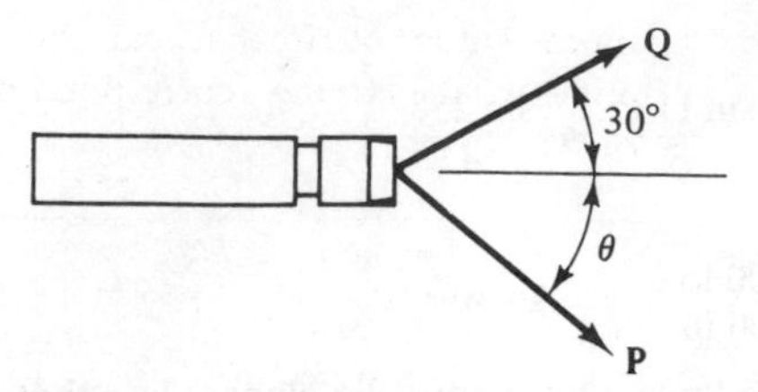

Problem 2.8

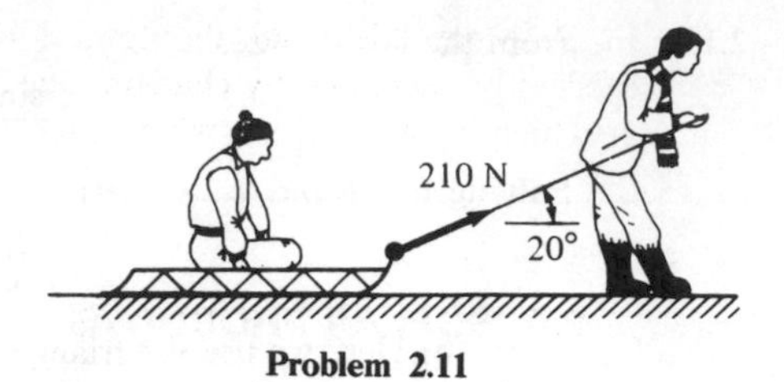

Problem 2.11

longitudinal axis of the truck. What is the magnitude of **F**?

2.9. If the sum of the four forces shown is zero, determine the magnitudes of forces **A** and **B**.

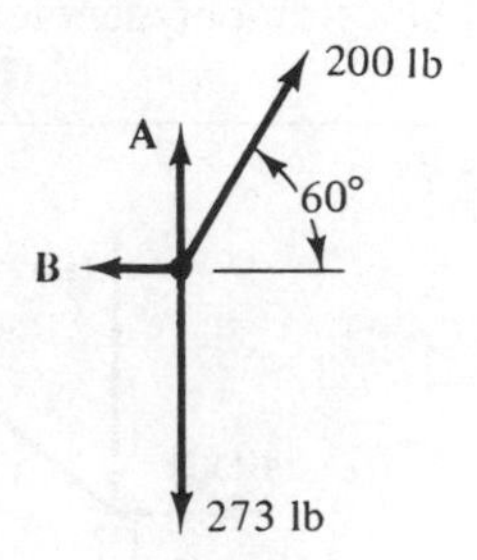

Problem 2.9

2.10. A 200-lb block is supported by two ropes, as shown. It is known that the sum of the forces acting at point B is **0**. Determine the magnitudes of the tensions $\mathbf{T}_1$ and $\mathbf{T}_2$ on the ropes.

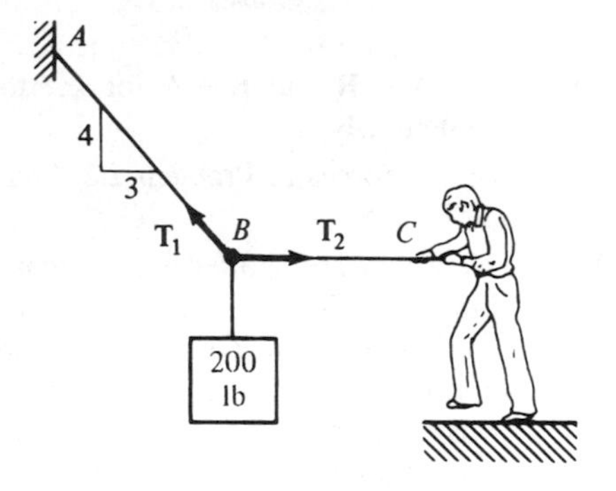

Problem 2.10

2.11. A man is pulling on a sled with a force of 210 N, as shown. Determine the horizontal and vertical components of the force.

2.12. Determine the horizontal and vertical components of the forces **A** and **B** in Problem 2.2, and resolve **A** into a vertical component and a component collinear with **B**.

2.13. If the total force exerted by the two tow trucks in Problem 2.8 has a magnitude of 40 kN and is directed straight ahead, at what angle θ should the one truck pull so that the force **P** it must exert is a minimum? What are the corresponding magnitudes of **P** and **Q**?

2.14. A freighter driven aground during a storm is to be pulled off the beach by two cables anchored offshore and connected to winches on board. The tensions on the cables are adjusted to provide a total force **F** acting perpendicular to the center line of the vessel. If $\theta = 40°$ and the maximum allowable tensions on the cables are $T_1 = 100$ kips and $T_2 = 120$ kips, what is the maximum force **F** that can be generated? What are the corresponding values of T_1 and T_2?

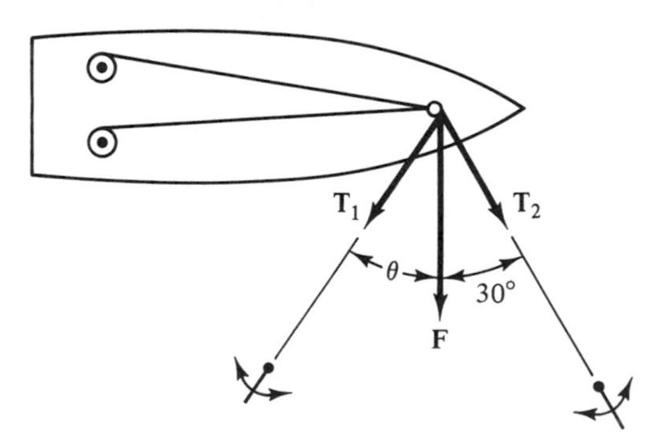

Problem 2.14

2.15. In Problem 2.14, can the total force **F** exerted be increased by changing the angle at which $\mathbf{T}_1$ acts? If so, what is the maximum value of **F** that can be achieved and what is the corresponding value of θ?

2.4 SCALAR PRODUCTS

There are two forms of vector multiplication. One is the scalar product, which we shall consider in this section, and the other is the vector product, which we shall consider in Section 2.7. The scalar product is closely associated with the concept of the projection of a vector onto a line or axis, which enters into some of the most important and useful results in vector algebra.

The *scalar product* of two vectors **A** and **B** is a number denoted by $\mathbf{A} \cdot \mathbf{B}$ and defined by the relation

$$\mathbf{A} \cdot \mathbf{B} = AB \cos \theta \tag{2.4}$$

where A and B are the respective magnitudes of the vectors and θ is the angle between them. The angle between two vectors is defined as the smaller of the two angles between their positive directions, so $0 \leqslant \theta \leqslant 180°$. The scalar product is also called the *dot* or *inner product*.

Interpreted physically, the quantity $B \cos \theta$ in Eq. (2.4) represents the *projection* of **B** onto **A** (Figure 2.10), which we write as

$$\text{Proj}_{\mathbf{A}}\, \mathbf{B} = B \cos \theta$$

Similarly, the quantity $A \cos \theta$ can be interpreted as the projection of **A** onto **B**:

$$\text{Proj}_{\mathbf{B}}\, \mathbf{A} = A \cos \theta$$

The projections will be either positive or negative, depending upon whether **A** and **B** form an acute angle ($\theta < 90°$) or obtuse angle ($\theta > 90°$). If $\theta = 90°$, the projections are zero and so is the dot product $\mathbf{A} \cdot \mathbf{B}$. The preceding results indicate that the dot product of two vectors can also be expressed as

$$\mathbf{A} \cdot \mathbf{B} = A\, \text{Proj}_{\mathbf{A}}\, \mathbf{B} = B\, \text{Proj}_{\mathbf{B}}\, \mathbf{A} \tag{2.5}$$

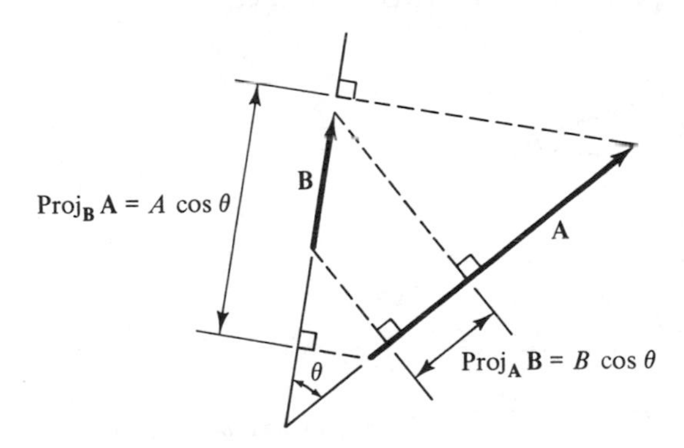

Figure 2.10

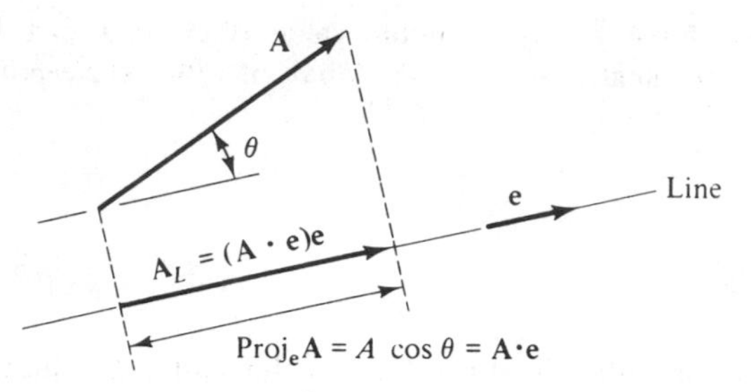

Figure 2.11

If one of the vectors is a unit vector **e** along a given line or axis, the dot product gives the projection of the second vector onto that line or axis (Figure 2.11), since

$$\mathbf{A} \cdot \mathbf{e} = (1)\text{Proj}_{\mathbf{e}}\,\mathbf{A} = A\cos\theta$$

Using Eq. (2.2), the component $\mathbf{A}_L$ of the vector **A** along the line can then be written as

$$\mathbf{A}_L = (\mathbf{A} \cdot \mathbf{e})\mathbf{e} \tag{2.6}$$

where the projection $\mathbf{A} \cdot \mathbf{e}$ defines its magnitude and sense and the unit vector **e** defines its orientation in space. This relationship is useful in three-dimensional problems for determining the component of a vector in a given direction.

The scalar product obeys the same laws as the product of real numbers:

$$\begin{aligned}
&\mathbf{A} \cdot \mathbf{B} = \mathbf{B} \cdot \mathbf{A} \quad \text{(commutative law)}\\
&\mathbf{A} \cdot (\mathbf{B} + \mathbf{C}) = \mathbf{A} \cdot \mathbf{B} + \mathbf{A} \cdot \mathbf{C} \quad \text{(distributive law)}\\
&(h\mathbf{A}) \cdot \mathbf{B} = \mathbf{A} \cdot (h\mathbf{B}) = h(\mathbf{A} \cdot \mathbf{B});\ h \text{ is any scalar}\\
&\mathbf{A} \cdot \mathbf{A} = A^2\\
&\mathbf{A} \cdot \mathbf{B} = 0 \text{ when } \mathbf{A} \text{ is perpendicular to } \mathbf{B}
\end{aligned} \tag{2.7}$$

All but the second of these laws follow directly from the definition of the scalar product, Eq. (2.4). To prove the second law, we note from Eq. (2.5) that

$$\mathbf{A} \cdot (\mathbf{B} + \mathbf{C}) = A\,\text{Proj}_{\mathbf{A}}\,(\mathbf{B} + \mathbf{C})$$

From Figure 2.12, we see that

$$\text{Proj}_{\mathbf{A}}\,(\mathbf{B} + \mathbf{C}) = \text{Proj}_{\mathbf{A}}\,\mathbf{B} + \text{Proj}_{\mathbf{A}}\,\mathbf{C}$$

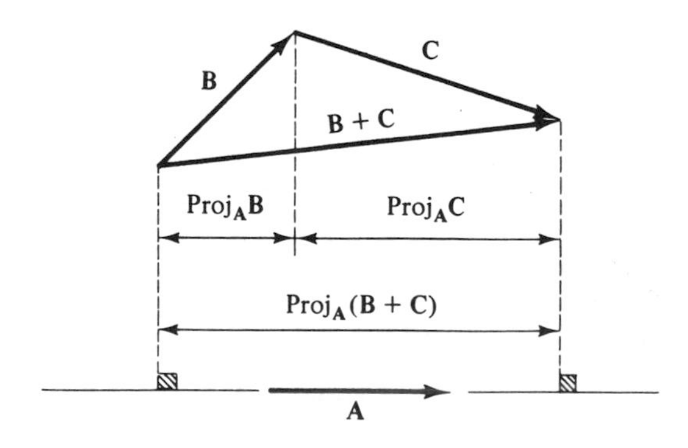

Figure 2.12

so

$$\mathbf{A}\cdot(\mathbf{B}+\mathbf{C}) = A\,\text{Proj}_{\mathbf{A}}\,\mathbf{B} + A\,\text{Proj}_{\mathbf{A}}\mathbf{C} = \mathbf{A}\cdot\mathbf{B}+\mathbf{A}\cdot\mathbf{C}$$

Example 2.5. Projections of a Vector. The cables shown in Figure 2.13(a) exert forces **P** and **Q** on the block, as indicated. Determine (a) the projection of **Q** onto **P**, (b) the projection of **P** onto **Q**, (c) the dot product of **P** and **Q**, (d) the projection of **P** onto the horizontal, and (e) the horizontal component of **P**.

Solution. (a) The angle between the two vectors is $\theta = 118°$; so

$$\text{Proj}_{\mathbf{P}}\,\mathbf{Q} = Q\cos\theta = (100\text{ kN})\cos 118° = -47.0\text{ kN} \qquad \textbf{Answer}$$

(b)

$$\text{Proj}_{\mathbf{Q}}\,\mathbf{P} = P\cos\theta = (113\text{ kN})\cos 118° = -53.1\text{ kN} \qquad \textbf{Answer}$$

The two projections are shown in Figure 2.13(b). They are both negative because the angle between P and Q is greater than 90°.

(c) From Eq. (2.4) we have

$$\mathbf{P}\cdot\mathbf{Q} = PQ\cos\theta = (113\times10^3\text{ N})(13\times10^3\text{ N})\cos 118°$$
$$= -5.31\times10^9\text{ N}^2 \text{ or } -5.31\text{ GN}^2 \qquad \textbf{Answer}$$

(d) Let the positive direction along the horizontal be to the right, as indicated by the unit vector **e** in Figure 2.13(c). The angle between **P** and **e** is $\theta = 143°$; so

$$\text{Proj}_{\mathbf{e}}\,\mathbf{P} = \mathbf{P}\cdot\mathbf{e} = \mathbf{P}\cos\theta = (113\text{ kN})\cos 143° = -90.2\text{ kN} \qquad \textbf{Answer}$$

If the positive direction along the horizontal is taken to be to the left, the sign of the projection will be reversed.

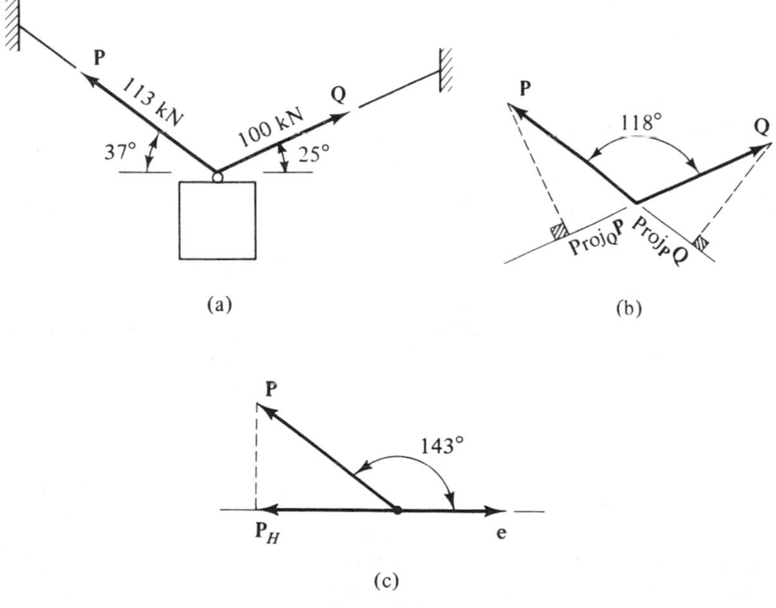

Figure 2.13

(e) If we denote the horizontal component of **P** by $\mathbf{P}_H$, we have from Eq. (2.6) and the results of part (d)

$$\mathbf{P}_H = (\mathbf{P} \cdot \mathbf{e})\mathbf{e} = -90.2\mathbf{e} \text{ kN} \qquad \textbf{Answer}$$

The minus sign indicates that the sense of $\mathbf{P}_H$ is opposite to that of **e**; i.e., $\mathbf{P}_H$ is directed to the left, as is obvious from Figure 2.13(c).

PROBLEMS

2.16. to 2.18. For the vectors **A** and **B** shown find (a) the projection of **A** onto **B**, (b) the projection of **B** onto **A**, (c) the dot product **A** · **B**, and (d) the projections of **A** and **B** onto the horizontal.

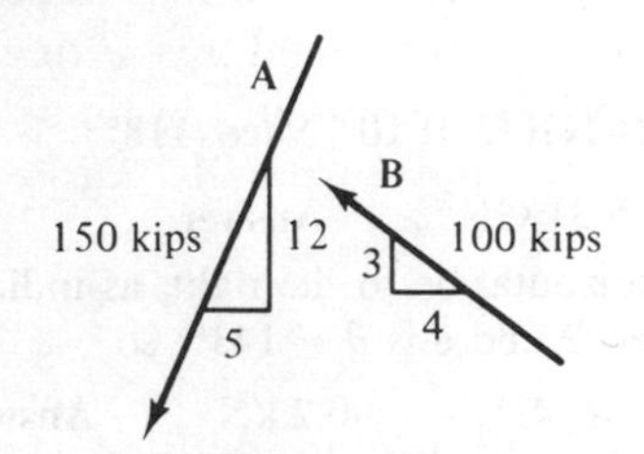

Problem 2.16

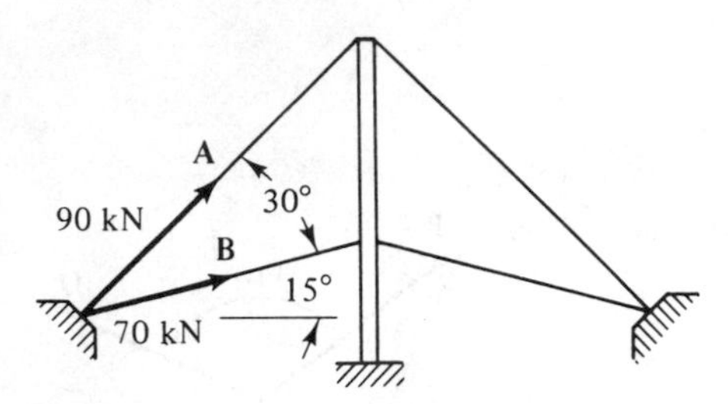

Problem 2.17

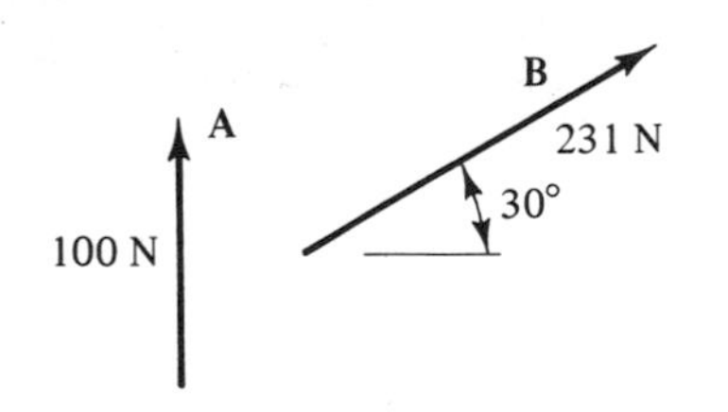

Problem 2.18

2.19. The dot product of two vectors **A** and **B** is **A** · **B** = 52.0, where **A** is as shown and **B** lies along line *b-b*. If $B = 15$, find **B** and the component of **A** along line *b-b*.

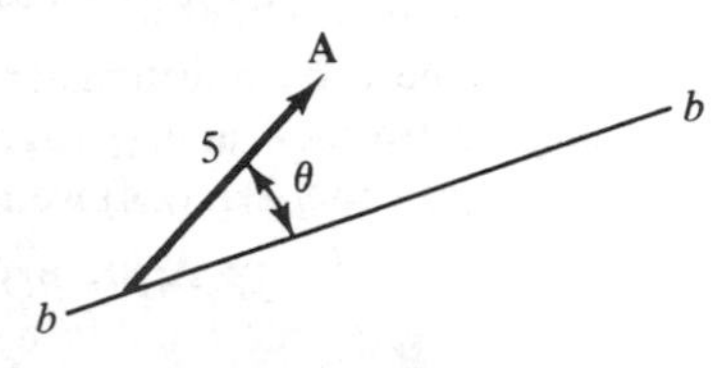

Problem 2.19

2.20. Same as Problem 2.19, except that $\theta = 30°$ and B is unknown.

2.21. The vector **S** shown is the sum of two vectors **A** and **B**, where **A** lies along line *a-a*. If **A** · **S** = 160.0 m², find **A** and **B**.

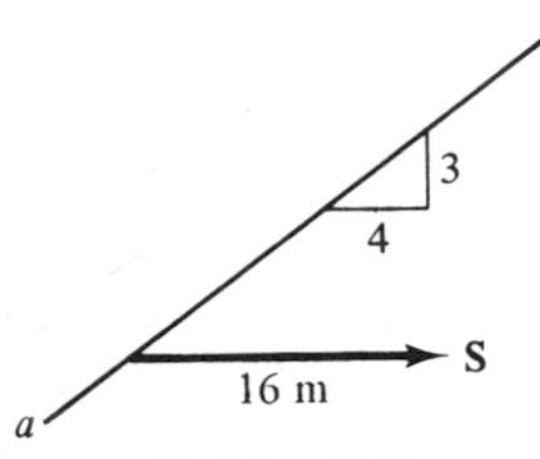

Problem 2.21

2.22. Same as Problem 2.21, except that **A** · **S** = −192.0 m².

2.23. If **S** is the sum of the vectors **A** and **B** shown and **A** · **S** = 9 ft² and **B** · **S** = 16 ft², find **S** and **B**.

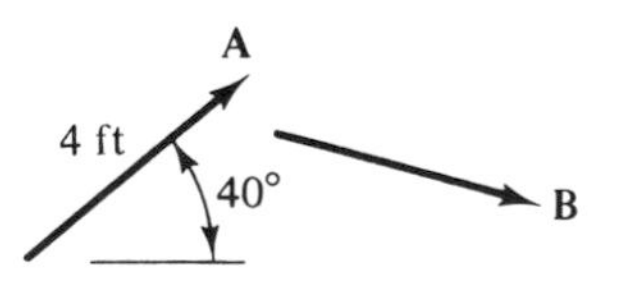

Problem 2.23

2.5 VECTORS REFERRED TO RECTANGULAR COORDINATE SYSTEMS

Up to this point, we have represented vectors by directed line segments in space. In this section we shall consider an alternate representation that is usually more convenient to use in computations.

Vectors in Component Form

Consider a rectangular Cartesian coordinate system with axes x, y, and z (Figure 2.14). Now let us introduce a set of unit vectors **i**, **j**, and **k** which define the positive directions along the coordinate axes. These vectors will be called the *base vectors*. The special notation is used to distinguish them from other unit vectors, **e**. In this text we shall use only *right-handed coordinate systems* for which x, y, and z lie along the thumb, index, and middle fingers of the right hand, respectively.

Any vector **A** located within a given coordinate system can be written as the sum of its three rectangular components $\mathbf{A}_x$, $\mathbf{A}_y$, and $\mathbf{A}_z$ parallel to the coordinate axes (Figure 2.14):

$$\mathbf{A} = \mathbf{A}_x + \mathbf{A}_y + \mathbf{A}_z$$

This is easily verified by using the polygon rule for vector addition, as indicated in the figure. From the geometry of the figure, or from Eq. (2.6), we have

$$\begin{aligned} \mathbf{A}_x &= (\mathbf{A} \cdot \mathbf{i})\mathbf{i} = A_x\mathbf{i} \\ \mathbf{A}_y &= (\mathbf{A} \cdot \mathbf{j})\mathbf{j} = A_y\mathbf{j} \\ \mathbf{A}_z &= (\mathbf{A} \cdot \mathbf{k})\mathbf{k} = A_z\mathbf{k} \end{aligned} \tag{2.8}$$

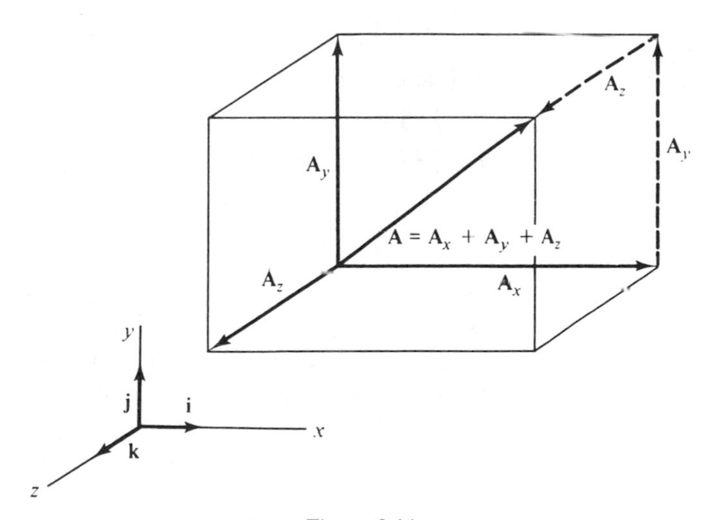

Figure 2.14

Thus, any vector **A** can be expressed in the *component form*

$$\mathbf{A} = A_x\mathbf{i} + A_y\mathbf{j} + A_z\mathbf{k} \tag{2.9}$$

where A_x, A_y, and A_z are the projections of **A** onto the coordinate axes (the *rectangular projections* of the vector). The projections are shown as being positive in Figure 2.14, but, of course, they can be negative. Equations of the same form as Eq. (2.9) hold for other orthogonal coordinate systems, such as cylindrical or spherical coordinates, but with different sets of base vectors.

We now have two ways to represent a vector. It can be represented graphically by a directed line segment in space, or, according to Eq. (2.9), it can be represented by a set of three numbers (scalars) which are its projections onto the axes of a rectangular coordinate system. As we shall see in the following, the latter representation is usually the most convenient to use in computations.

For brevity, we shall write the units for vectors given in component form only at the end of the complete vector expression. For example, we write a vector $\mathbf{F} = -(3 \text{ lb})\mathbf{i} + (4 \text{ lb})\mathbf{j}$ as $\mathbf{F} = -3\mathbf{i} + 4\mathbf{j}$ lb, where it is understood that the units apply to each component of the vector.

Scalar Products

The scalar product of two vectors **A** and **B** expressed in component form can be computed by using the distributive property stated in Eq. (2.7) of Section 2.4. We have

$$\begin{aligned}
\mathbf{A} \cdot \mathbf{B} &= (A_x\mathbf{i} + A_y\mathbf{j} + A_z\mathbf{k}) \cdot (B_x\mathbf{i} + B_y\mathbf{j} + B_z\mathbf{k}) \\
&= A_xB_x\mathbf{i} \cdot \mathbf{i} + A_xB_y\mathbf{i} \cdot \mathbf{j} + A_xB_z\mathbf{i} \cdot \mathbf{k} \\
&\quad + A_yB_x\mathbf{j} \cdot \mathbf{i} + A_yB_y\mathbf{j} \cdot \mathbf{j} + A_yB_z\mathbf{j} \cdot \mathbf{k} \\
&\quad + A_zB_x\mathbf{k} \cdot \mathbf{i} + A_zB_y\mathbf{k} \cdot \mathbf{j} + A_zB_z\mathbf{k} \cdot \mathbf{k}
\end{aligned}$$

Now, the base vectors **i**, **j**, and **k** are mutually perpendicular; so

$$\begin{aligned}
&\mathbf{i} \cdot \mathbf{i} = \mathbf{j} \cdot \mathbf{j} = \mathbf{k} \cdot \mathbf{k} = 1 \\
&\mathbf{i} \cdot \mathbf{j} = \mathbf{j} \cdot \mathbf{i} = \mathbf{i} \cdot \mathbf{k} = \mathbf{k} \cdot \mathbf{i} = \mathbf{j} \cdot \mathbf{k} = \mathbf{k} \cdot \mathbf{j} = 0
\end{aligned}$$

Thus,

$$\mathbf{A} \cdot \mathbf{B} = A_xB_x + A_yB_y + A_zB_z \tag{2.10}$$

This form of the dot product is particularly convenient when the angle between the two vectors is unknown, as is often the case. In fact, this angle can be determined by computing $\mathbf{A} \cdot \mathbf{B}$ both from Eq. (2.10) and from Eq. (2.4) and equating the two expressions. These steps yield the result

$$\cos\theta = \frac{\mathbf{A} \cdot \mathbf{B}}{AB} = \frac{A_xB_x + A_yB_y + A_zB_z}{AB} \tag{2.11}$$

where θ is the angle between **A** and **B**.

Magnitude and Orientation of a Vector in Space

The magnitude of a vector **A** expressed in component form can be computed by using the scalar product. From Eqs. (2.4) and (2.10), we have

$$\mathbf{A} \cdot \mathbf{A} = A^2 = A_x^2 + A_y^2 + A_z^2$$

or

$$A = \sqrt{A_x^2 + A_y^2 + A_z^2} \tag{2.12}$$

This result is just a statement of the Pythagorean theorem in three dimensions.

The orientation of a vector **A** in space is defined by the *direction angles* θ_x, θ_y, and θ_z between it and the positive coordinate axes (Figure 2.15). By definition, the values of these angles are restricted to the interval 0° to 180°, inclusive. The direction angles of a vector are directly related to its rectangular projections, as can be shown from Eq. (2.8) and the definition of the scalar product. We have

$$\begin{aligned} A_x &= \mathbf{A} \cdot \mathbf{i} = A \cos \theta_x \\ A_y &= \mathbf{A} \cdot \mathbf{j} = A \cos \theta_y \\ A_z &= \mathbf{A} \cdot \mathbf{k} = A \cos \theta_z \end{aligned} \tag{2.13}$$

The quantities $\cos \theta_x$, $\cos \theta_y$, and $\cos \theta_z$ are called the *direction cosines* of the vector. Using Eqs. (2.13), (2.2), and (2.9), we can write

$$\mathbf{A} = A(\cos \theta_x \mathbf{i} + \cos \theta_y \mathbf{j} + \cos \theta_z \mathbf{k}) = A\mathbf{e}$$

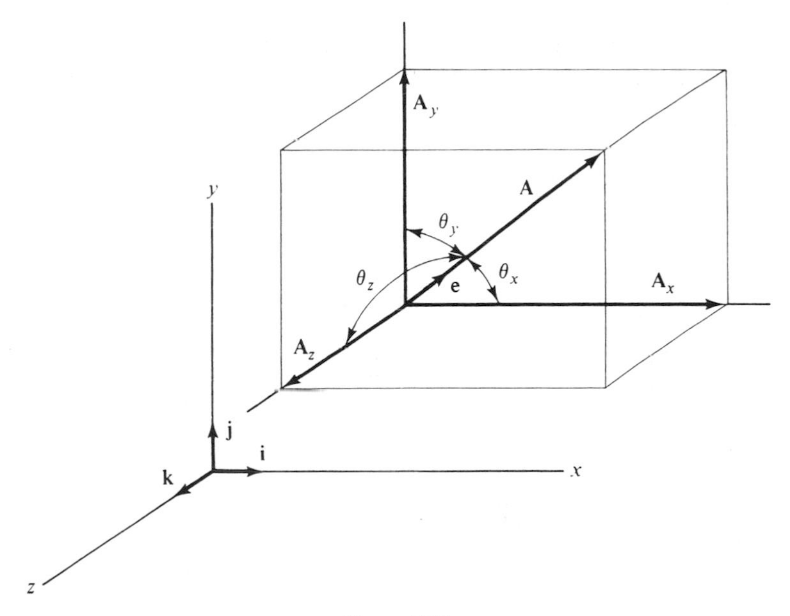

Figure 2.15

or

$$\mathbf{e} = \cos\theta_x\mathbf{i} + \cos\theta_y\mathbf{j} + \cos\theta_z\mathbf{k} \tag{2.14}$$

where **e** is a unit vector along **A** (Figure 2.15). From Eq. (2.14), we see that the rectangular projections of a unit vector along a given line are just the direction cosines of that line. Now

$$\mathbf{e}\cdot\mathbf{e} = \cos^2\theta_x + \cos^2\theta_y + \cos^2\theta_z = 1 \tag{2.15}$$

which indicates that the direction angles are not independent. If any two of them are given, the third must be such that Eq. (2.15) is satisfied.

Vector Sums

Vector addition and subtraction is greatly simplified when the vectors are expressed in component form. We have already shown that the projection of the sum of two vectors is equal to the sum of their projections (see Figure 2.12 in Section 2.4). Thus, to add vectors, we need only add their corresponding rectangular projections:

$$\mathbf{A} + \mathbf{B} = (A_x + B_x)\mathbf{i} + (A_y + B_y)\mathbf{j} + (A_z + B_z)\mathbf{k} \tag{2.16}$$

Similarly,

$$\mathbf{A} - \mathbf{B} = (A_x - B_x)\mathbf{i} + (A_y - B_y)\mathbf{j} + (A_z - B_z)\mathbf{k} \tag{2.17}$$

These methods of addition and subtraction hold for any number of vectors, and they are particularly convenient for three-dimensional problems. Notice that the results are obtained in component form, which is usually preferable. However, the magnitude and orientation of the resulting vectors can always be obtained by using Eqs. (2.12) and (2.13), if desired.

Example 2.6. Addition of Vectors. Find the sum **S** of the two forces **P** and **Q** acting upon the bracket in Figure 2.16(a) and determine its magnitude and orientation. Use rectangular projections.

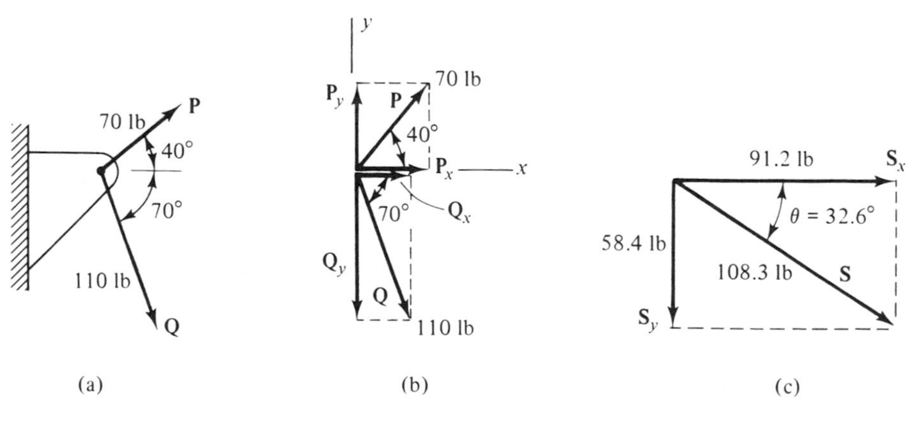

Figure 2.16

Solution. We first establish a coordinate system and sketch the component vectors [Figure 2.16(b)]. The rectangular projections of the vectors can be computed by using Eq. (2.13), but it is simpler to determine them directly from the geometry of the figure:

$$P_x = (70 \text{ lb})\cos 40° = 53.6 \text{ lb}$$
$$P_y = (70 \text{ lb})\sin 40° = 45.0 \text{ lb}$$
$$Q_x = (110 \text{ lb})\cos 70° = 37.6 \text{ lb}$$
$$Q_y = -(110 \text{ lb})\sin 70° = -103.4 \text{ lb}$$

Note that Q_y is negative because the component vector points in the negative coordinate direction. Now,

$$S_x = P_x + Q_x = 53.6 \text{ lb} + 37.6 \text{ lb} = 91.2 \text{ lb}$$
$$S_y = P_y + Q_y = 45.0 \text{ lb} + (-103.4 \text{ lb}) = -58.4 \text{ lb}$$

so

$$\mathbf{S} = S_x\mathbf{i} + S_y\mathbf{j} = 91.2\mathbf{i} - 58.4\mathbf{j} \text{ lb} \qquad \textbf{Answer}$$

The components of **S** are shown in Figure 2.16(c). Referring to the figure, or using Eq. (2.12), we have

$$S = \sqrt{S_x^2 + S_y^2} = \sqrt{(91.2)^2 + (-58.4)^2} \text{ lb} = 108.3 \text{ lb} \qquad \textbf{Answer}$$

and

$$\tan\theta = \frac{58.4 \text{ lb}}{91.2 \text{ lb}} = 0.64$$
$$\theta = 32.6° \qquad \textbf{Answer}$$

The orientation of **S** can also be determined using Eqs. (2.13).

Example 2.7. Vector Addition and Dot Product. Given the three vectors **A** = 3**i** + 2**j** − 6**k**, **B** = 2**i** + 4**j** + 7**k**, and **C** = −2**i** + 9**j** + 3**k**, find (a) the vector **D** = **A** + **B** − **C** and its magnitude and direction angles and (b) the dot product of the vectors **A** and **B** and the angle between them.

Solution. (a)

$$D_x = A_x + B_x - C_x = 3 + 2 - (-2) = 7$$
$$D_y = A_y + B_y - C_y = 2 + 4 - 9 = -3$$
$$D_z = A_z + B_z - C_z = -6 + 7 - 3 = -2$$
$$\mathbf{D} = 7\mathbf{i} - 3\mathbf{j} - 2\mathbf{k} \qquad \textbf{Answer}$$

From Eq. (2.12), we have

$$D = \sqrt{(7)^2 + (-3)^2 + (-2)^2} = 7.87 \qquad \textbf{Answer}$$

The direction angles of **D** are determined from Eq. (2.13):

$$\cos\theta_x = \frac{D_x}{D} = \frac{7}{7.87} = 0.89 \qquad \theta_x = 27.2°$$

$$\cos\theta_y = \frac{D_y}{D} = \frac{-3}{7.87} = -0.38 \qquad \theta_y = 112.4° \qquad \textbf{Answer}$$

$$\cos\theta_z = \frac{D_z}{D} = \frac{-2}{7.87} = -0.25 \qquad \theta_z = 104.7°$$

(b) Using Eq. (2.10), we find

$$\mathbf{A}\cdot\mathbf{B} = A_xB_x + A_yB_y + A_zB_z = 3(2) + 2(4) + (-6)(7) = -28 \qquad \textbf{Answer}$$

Also,

$$A = \sqrt{(3)^2 + (2)^2 + (-6)^2} = 7$$

$$B = \sqrt{(2)^2 + (4)^2 + (7)^2} = 8.31$$

The angle between the two vectors is given by Eq. (2.11):

$$\cos\theta = \frac{\mathbf{A}\cdot\mathbf{B}}{AB} = \frac{-28}{7(8.31)} = -0.48 \qquad \theta = 118.8° \qquad \textbf{Answer}$$

PROBLEMS

2.24. to 2.26. For the given set of vectors **A** and **B**, find **A** + **B**, **A** − **B**, 2**A** + 3**B** and **A** • **B** by using rectangular projections. Also determine the angle between **A** and **B**.

2.24. **A** = 2**i** − 4**j**   **B** = 2**i** − **j** + 3**k**

2.25. **A** = −2**i** + 4**j** + 16**k**   **B** = 6**i** − 2**j** + 3**k**

2.26. **A** = 8**i** + 6**j** − 4**k**   **B** = 2**i** − 4**j** − 2**k**

2.27. Determine whether or not the vectors **A** = 6**i** − 2**j** + 3**k** and **B** = −2**i** + 4**j** + 12**k** are perpendicular.

2.28. Determine the value of z for which the vector **A** = 2**i** − **j** + 3**k** will be perpendicular to the vector **B** = 2**i** − 4**j** + z**k**.

2.29. Rework Problem 2.3 using rectangular projections.

2.30. Rework Problem 2.4 using rectangular projections.

2.31. to 2.34. Express the vector shown in component form and determine its direction angles.

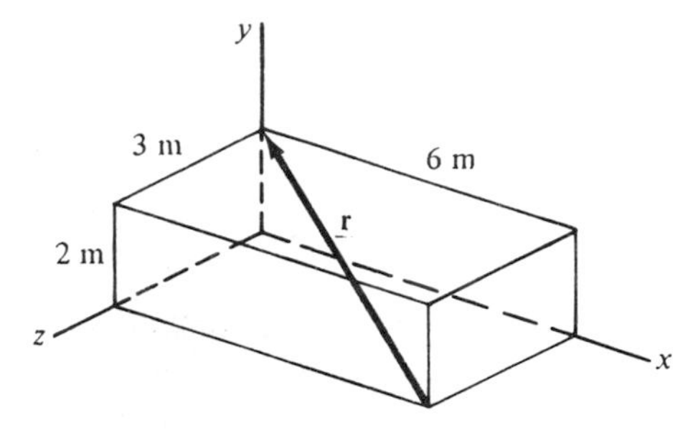

Problem 2.31

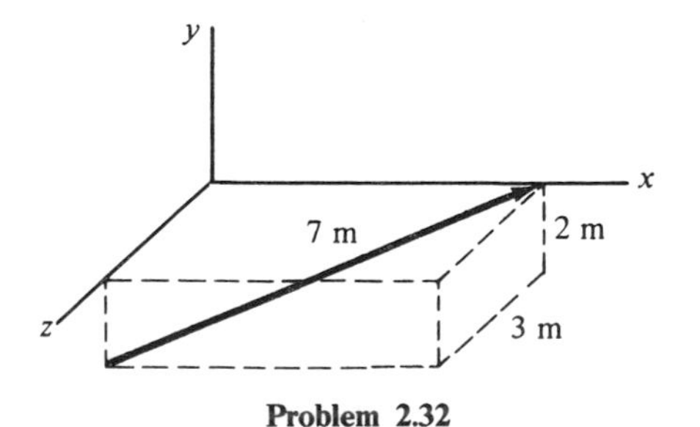

Problem 2.32

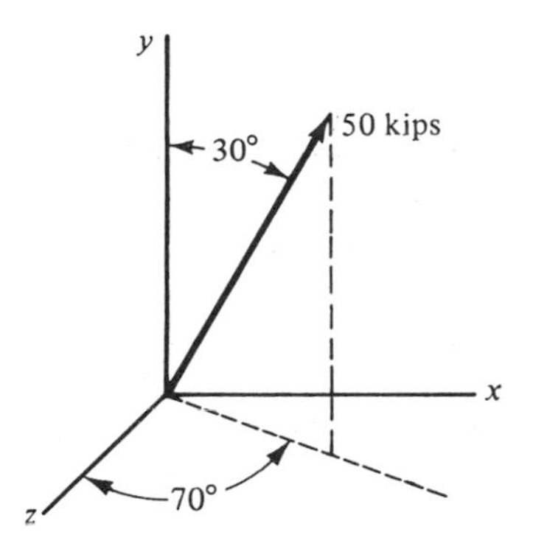

Problem 2.33

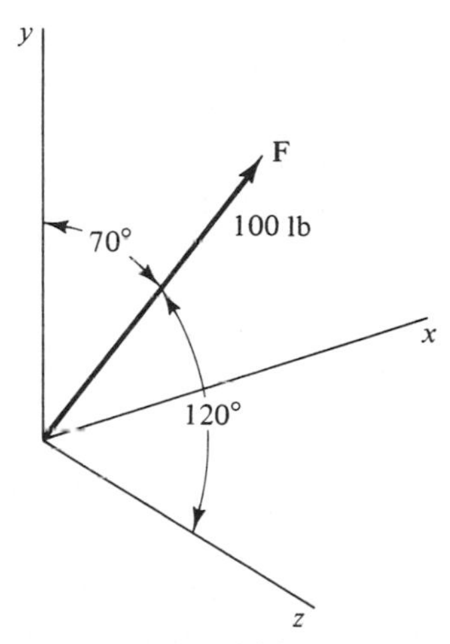

Problem 2.34

2.35. and 2.36. For the vectors shown, find (a) their sum **S**, (b) their dot product, and (c) the angle between them. Also determine the magnitude and direction angles of **S**. Use rectangular components.

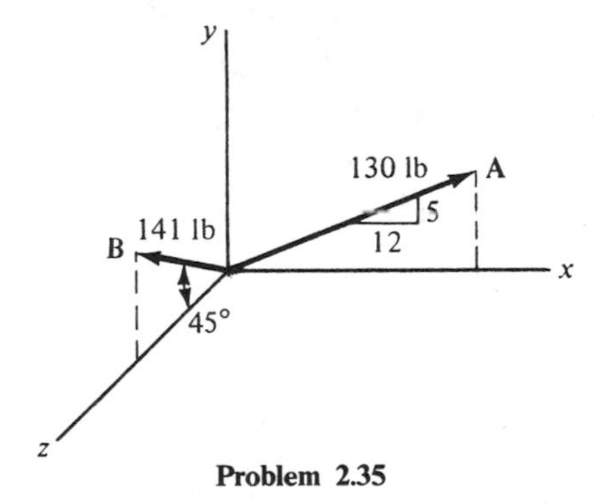

Problem 2.35

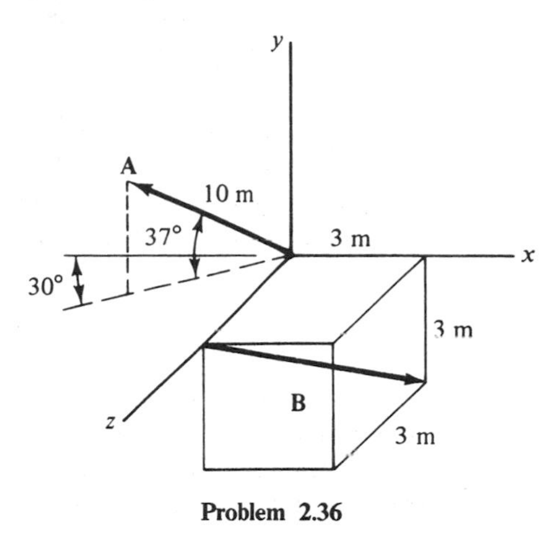

Problem 2.36

2.37. Knowing that the sum of the forces acting at point A on the lamp shown must be zero, determine the magnitudes of the tensions $\mathbf{T}_1$ and $\mathbf{T}_2$ on the cords for $\theta = 45°$. Use rectangular components.

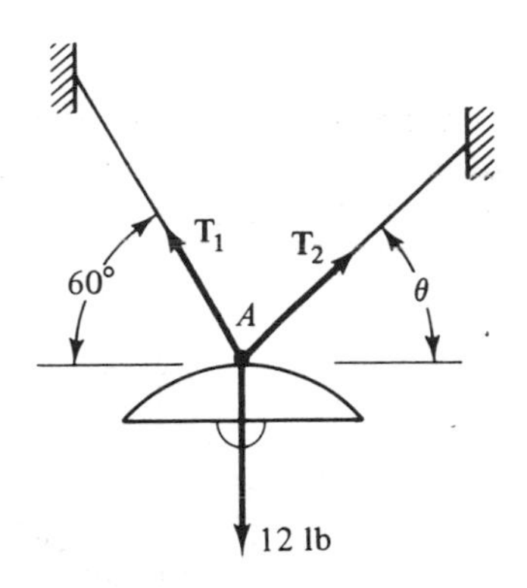

Problem 2.37

2.38. In Problem 2.37, for what angle θ will the tension $\mathbf{T}_2$ be minimum? What are the corresponding magnitudes of $\mathbf{T}_1$ and $\mathbf{T}_2$? Use rectangular components.

2.6 POSITION VECTORS

In this section we shall consider the use of vectors to describe the position of a point in space or the position of one point relative to another. This representation will be useful in a number of applications, and it is used extensively in the study of kinematics (the geometry of motion).

The position of a point P relative to the origin O of a given coordinate system can be defined by a *position vector* $\mathbf{r}_{OP}$, which is directed from O to P and whose rectangular projections are equal to the coordinates of the point [Figure 2.17(a)]:

$$\mathbf{r}_{OP} = x_P\mathbf{i} + y_P\mathbf{j} + z_P\mathbf{k} \tag{2.18}$$

Similarly, the position of one point relative to another can be defined by a position vector between them. For example, the vector $\mathbf{r}_{QP}$ shown in Figure 2.17(b) goes from point Q to point P and defines the position of P relative to Q. Similarly, the vector $\mathbf{r}_{PQ}$ would go from point P to point Q and would define the position of Q with respect to P. Note that the first and second letters in the subscript on $\mathbf{r}$ denote the points at its tail and tip, respectively.

Referring to Figure 2.17(b), we see that

$$\mathbf{r}_{OQ} + \mathbf{r}_{QP} = \mathbf{r}_{OP}$$

or

$$\mathbf{r}_{QP} = \mathbf{r}_{OP} - \mathbf{r}_{OQ} = (x_P - x_Q)\mathbf{i} + (y_P - y_Q)\mathbf{j} + (z_P - z_Q)\mathbf{k} \tag{2.19}$$

According to this equation, the rectangular projections of a vector between two points in space are equal to the corresponding coordinates of its tip minus those of its tail. In many cases, the projections can also be obtained by inspection by

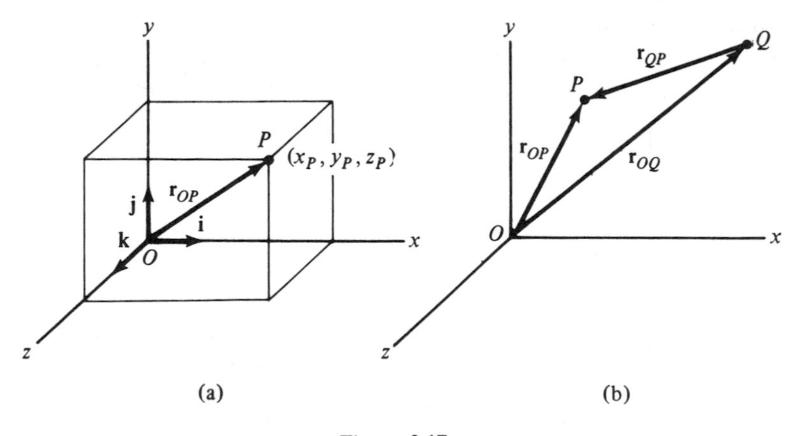

Figure 2.17

determining how far, and in which direction, one must move parallel to the coordinate axes to get from the tail of the vector to the tip (see Example 2.8).

Position vectors are particularly useful for finding unit vectors along a line that passes through two known points. All that is required is to form the position vector between the points and then divide by its magnitude. Once a unit vector along a line has been established, any other vector **A** along the line can be expressed in component form by using the relationship $\mathbf{A} = \pm A\mathbf{e}$ from Eq. (2.2).

Example 2.8. Determination of a Position Vector. Determine the position vector $\mathbf{r}_{AB}$ between points A and B on the wire shape shown in Figure 2.18.

Solution. It is helpful to first sketch the vector. The rectangular projections of $\mathbf{r}_{AB}$ are the coordinates of its tip minus those of its tail; so

$$\mathbf{r}_{AB} = (0 - 60)\mathbf{i} + (40 - 10)\mathbf{j} + [30 - (-20)]\mathbf{k}\ \text{mm}$$

$$= -60\mathbf{i} + 30\mathbf{j} + 50\mathbf{k}\ \text{mm} \qquad \textbf{Answer}$$

The position vector can also be obtained by inspection. It is clear from Figure 2.18 that we can get from A to B by going 60 mm in the negative x direction, a total of 30 mm in the positive y direction, and a total of 50 mm in the positive z direction. Thus,

$$\mathbf{r}_{AB} = -60\mathbf{i} + 30\mathbf{j} + 50\mathbf{k}\ \text{mm}$$

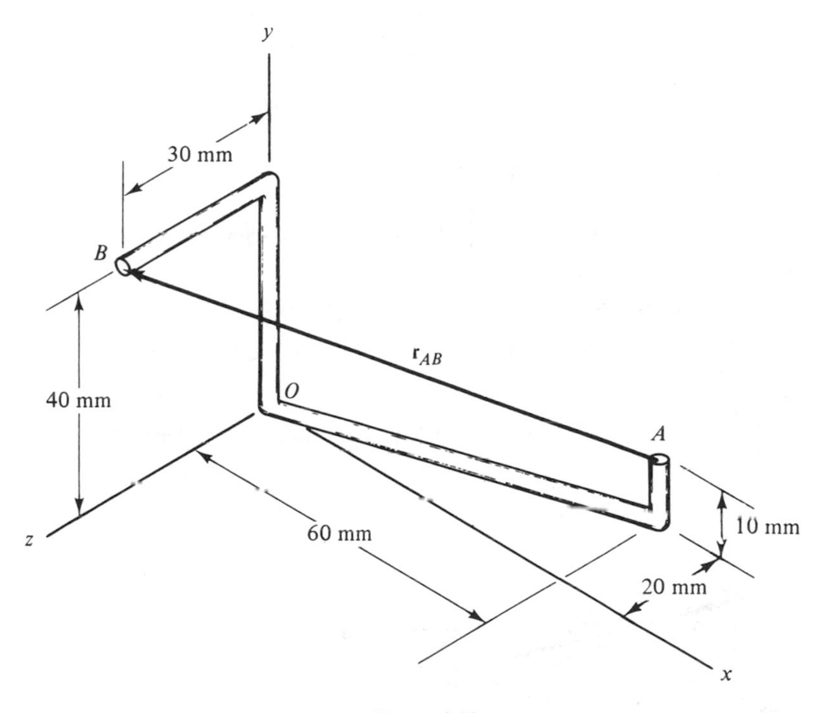

Figure 2.18

Example 2.9. Representation of a Vector in Space. A stone slab (Figure 2.19) is partially supported by a cable AB that exerts a 10,000-lb force **T** on it as shown. (a) Express **T** in component form and (b) determine its component $\mathbf{T}_{OB}$ along diagonal OB of the slab.

Solution. (a) We first determine the unit vector $\mathbf{e}_{BA}$ along the line of action of **T**. Points B and A have coordinates $0, 3, 6$ ft and $7, 0, 4$ ft, respectively; so

$$\mathbf{r}_{BA} = 7\mathbf{i} - 3\mathbf{j} - 2\mathbf{k} \text{ ft}$$

$$r_{BA} = \sqrt{(7)^2 + (-3)^2 + (-2)^2} = \sqrt{62} \text{ ft}$$

$$\mathbf{e}_{BA} = \frac{\mathbf{r}_{BA}}{r_{BA}} = \frac{1}{\sqrt{62}}(7\mathbf{i} - 3\mathbf{j} - 2\mathbf{k})$$

Now **T** has the same sense as $\mathbf{e}_{BA}$; so

$$\mathbf{T} = T\mathbf{e}_{BA} = \frac{10{,}000 \text{ lb}}{\sqrt{62}}(7\mathbf{i} - 3\mathbf{j} - 2\mathbf{k})$$

$$= 8{,}890\mathbf{i} - 3{,}810\mathbf{j} - 2{,}540\mathbf{k} \text{ lb} \qquad \textbf{Answer}$$

(b) From Eq. (2.6), the component $\mathbf{T}_{OB}$ of **T** along OB is equal to the projection of **T** onto the line multiplied by a unit vector along it. Either of the unit vectors $\mathbf{e}_{OB}$ or $\mathbf{e}_{BO}$ can be used. Suppose we use $\mathbf{e}_{OB}$. Referring to Figure 2.19, we have

$$\mathbf{r}_{OB} = 3\mathbf{j} + 6\mathbf{k} \text{ ft}$$

$$r_{OB} = \sqrt{(3)^2 + (6)^2} = \sqrt{45} \text{ ft}$$

$$\mathbf{e}_{OB} = \frac{1}{\sqrt{45}}(3\mathbf{j} + 6\mathbf{k})$$

The projection of **T** onto line OB is

$$\mathbf{T} \cdot \mathbf{e}_{OB} = \frac{1}{\sqrt{45}}[-3{,}810(3) - 2{,}540(6)] = -3{,}980 \text{ lb}$$

The vector component $\mathbf{T}_{OB}$ can now be written as

$$\mathbf{T}_{OB} = (\mathbf{T} \cdot \mathbf{e}_{OB})\mathbf{e}_{OB} = -3{,}980\mathbf{e}_{OB} \text{ lb}$$

or

$$\mathbf{T}_{OB} = -1{,}780\mathbf{j} - 3{,}560\mathbf{k} \text{ lb} \qquad \textbf{Answer}$$

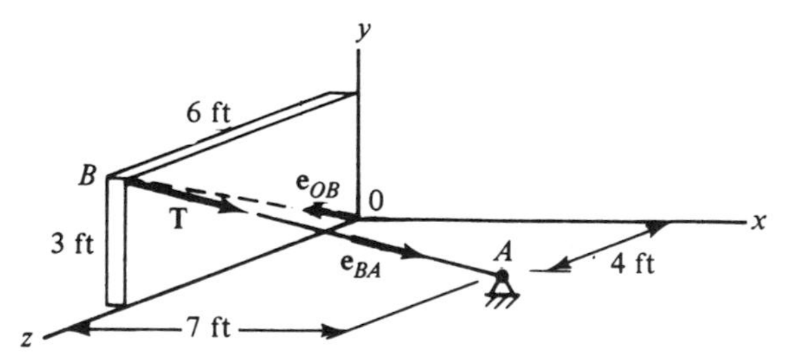

Figure 2.19

PROBLEMS

2.39. and 2.40. For the given set of points P and Q, determine (a) the position vector of each point relative to the origin, (b) the position vector of point Q relative to point P, and (c) the unit vector directed from P toward Q. Show each vector on a sketch.

2.39.
P: 6 m, -2 m, -3 m Q: 4 m, -1 m, 3 m

2.40.
P: 4 ft, 0 ft, 2 ft Q: 6 ft, -3 ft, 2 ft

2.41. According to Newton's law of gravitation, two particles attract each other with equal and opposite forces that lie along a line connecting their centers. The magnitude of these forces is proportional to the product of the masses of the particles and is inversely proportional to the square of the distance between them. The constant of proportionality is $G = 6.658 \times 10^{-11}$ $m^3/(kg \cdot s^2)$. Using this law, express in component form the force exerted on a 2-kg particle located at 4 m, 4 m, -2 m by a 1.5-kg particle located at the origin.

2.42. Same as Problem 2.41, except that the 1.5-kg particle is located at 2 m, -1 m, 3 m.

2.43. The 200-lb force shown lies along line AB. Determine the projection of the force onto line OC and find its vector component along this line. What is the angle between the force and OC?

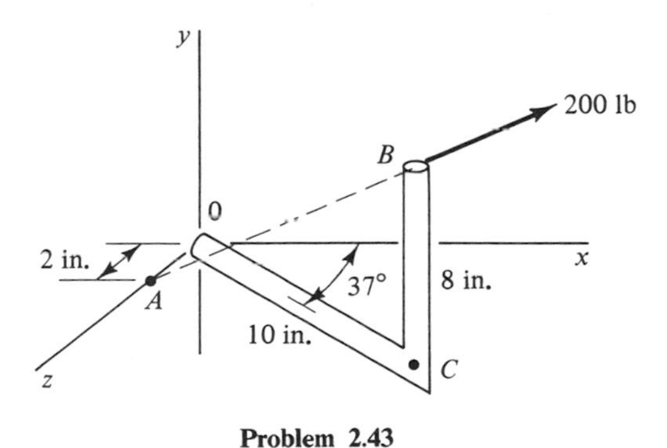

Problem 2.43

2.44. Same as Problem 2.43, except that line OC is replaced by line OB.

2.45. The magnitude of the tension on support cable AD of the tower shown is known to be 280 kN. Express the force $\mathbf{T}_1$ exerted on the anchor at D in component form and determine its direction angles.

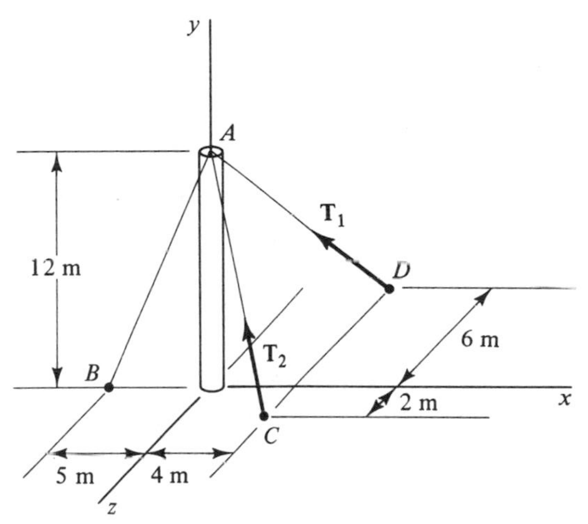

Problem 2.45

2.46. For the tower in Problem 2.45, express the force $\mathbf{T}_2$ exerted on the anchor at C in component form and determine its direction angles. The magnitude of the tension on cable AC is 350 kN.

2.47. What is the angle between support cables AB and AC on the tower in Problem 2.45?

2.7 VECTOR PRODUCTS

The *vector product* of two vectors **A** and **B** is a vector denoted by $\mathbf{A} \times \mathbf{B}$ and whose magnitude is given by the relation

$$|\mathbf{A} \times \mathbf{B}| = AB \sin \theta \tag{2.20}$$

In this equation, A and B are the respective magnitudes of the vectors and θ in the angle between them (Figure 2.20). The product $\mathbf{A} \times \mathbf{B}$, which is also called the *cross*, or *outer*, *product*, is perpendicular to the plane formed by **A** and **B**. Its sense is defined by the so-called *right-hand screw rule*. According to this rule, if the fingers of the right hand are placed along the first vector, **A**, and then rotated toward the second vector, **B**, the thumb points in the direction of their cross product $\mathbf{A} \times \mathbf{B}$.

The vector product obeys the following laws:

$$\begin{aligned} \mathbf{A} \times \mathbf{B} &= -(\mathbf{B} \times \mathbf{A}) \quad \text{(anticommutative law)} \\ \mathbf{A} \times (\mathbf{B} + \mathbf{C}) &= \mathbf{A} \times \mathbf{B} + \mathbf{A} \times \mathbf{C} \quad \text{(distributive law)} \\ (h\mathbf{A}) \times \mathbf{B} &= \mathbf{A} \times (h\mathbf{B}) = h(\mathbf{A} \times \mathbf{B}); \ h \text{ is any scalar} \\ \mathbf{A} \times \mathbf{A} &= \mathbf{0} \\ \mathbf{A} \times \mathbf{B} &= \mathbf{0} \text{ when } \mathbf{A} \text{ is parallel to } \mathbf{B} \end{aligned} \tag{2.21}$$

The first law follows directly from the right-hand screw rule and indicates that the order of the factors in the vector product cannot be changed without altering its sign. This is the only deviation of the rules of vector algebra from those of ordinary algebra that we shall encounter. All the other laws follow from the definition of the vector product. Proof of the distributive law is left as an exercise in Problem 2.59.

Using the distributive property of the cross product, we have for vectors expressed in component form

$$\begin{aligned} \mathbf{A} \times \mathbf{B} &= (A_x\mathbf{i} + A_y\mathbf{j} + A_z\mathbf{k}) \times (B_x\mathbf{i} + B_y\mathbf{j} + B_z\mathbf{k}) \\ &= A_xB_x(\mathbf{i} \times \mathbf{i}) + A_xB_y(\mathbf{i} \times \mathbf{j}) + A_xB_z(\mathbf{i} \times \mathbf{k}) \\ &\quad + A_yB_x(\mathbf{j} \times \mathbf{i}) + A_yB_y(\mathbf{j} \times \mathbf{j}) + A_yB_z(\mathbf{j} \times \mathbf{k}) \\ &\quad + A_zB_x(\mathbf{k} \times \mathbf{i}) + A_zB_y(\mathbf{k} \times \mathbf{j}) + A_zB_z(\mathbf{k} \times \mathbf{k}) \end{aligned}$$

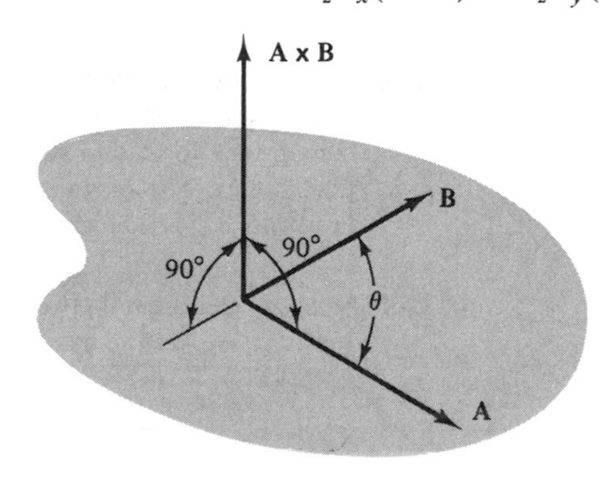

Figure 2.20

Since the base vectors are mutually perpendicular and have unit magnitude, the cross product of any two of them is just the positive or negative of the third base vector, with the sense readily determined from the right-hand screw rule. Thus, $\mathbf{i} \times \mathbf{j} = \mathbf{k}, \mathbf{k} \times \mathbf{j} = -\mathbf{i}, \mathbf{i} \times \mathbf{k} = -\mathbf{j}$, etc. Substituting these results into the preceding expression and noting that the cross product of a vector with itself is zero, we obtain

$$\mathbf{A} \times \mathbf{B} = (A_y B_z - A_z B_y)\mathbf{i} - (A_x B_z - A_z B_x)\mathbf{j} + (A_x B_y - A_y B_x)\mathbf{k} \qquad (2.22)$$

Equation (2.22) can also be expressed as a determinant with the base vectors as the elements of the first row and the rectangular projections of the first and second vectors in the cross product as the elements of the second and third row, respectively:

$$\mathbf{A} \times \mathbf{B} = \begin{vmatrix} \mathbf{i} & \mathbf{j} & \mathbf{k} \\ A_x & A_y & A_z \\ B_x & B_y & B_z \end{vmatrix} \qquad (2.23)$$

Expansion of this determinant in terms of elements of the first row confirms that Eqs. (2.22) and (2.23) are the same:

$$\mathbf{A} \times \mathbf{B} = \mathbf{i}\begin{vmatrix} A_y & A_z \\ B_y & B_z \end{vmatrix} - \mathbf{j}\begin{vmatrix} A_x & A_z \\ B_x & B_z \end{vmatrix} + \mathbf{k}\begin{vmatrix} A_x & A_y \\ B_x & B_y \end{vmatrix}$$

or

$$\mathbf{A} \times \mathbf{B} = (A_y B_z - A_z B_y)\mathbf{i} - (A_x B_z - A_z B_x)\mathbf{j} + (A_x B_y - A_y B_x)\mathbf{k}$$

If one or both of the vectors have several zero components, it is usually simpler to compute the cross product term by term by using the distributive property (see Example 2.10).

Example 2.10. Vector Product of Two Vectors. Determine the vector product $\mathbf{A} \times \mathbf{B}$ of the two vectors $\mathbf{A} = 2\mathbf{i} + 4\mathbf{j}$ and $\mathbf{B} = -3\mathbf{i} + \mathbf{k}$.

Solution. Using the determinant form of the vector product, we have

$$\mathbf{A} \times \mathbf{B} = \begin{vmatrix} \mathbf{i} & \mathbf{j} & \mathbf{k} \\ 2 & 4 & 0 \\ -3 & 0 & 1 \end{vmatrix}$$

$$\begin{aligned} \mathbf{A} \times \mathbf{B} &= \mathbf{i}[(4)(1) - (0)(0)] \quad \mathbf{j}[(2)(1) - (0)(-3)] \\ &\quad + \mathbf{k}[(2)(0) - (4)(-3)] \\ &= 4\mathbf{i} - 2\mathbf{j} + 12\mathbf{k} \qquad \textbf{Answer} \end{aligned}$$

The same result is obtained by computing the vector product term by term:

$$\begin{aligned} \mathbf{A} \times \mathbf{B} &= (2\mathbf{i} + 4\mathbf{j}) \times (-3\mathbf{i} + \mathbf{k}) \\ &= -6(\mathbf{i} \times \mathbf{i}) + 2(\mathbf{i} \times \mathbf{k}) - 12(\mathbf{j} \times \mathbf{i}) + 4(\mathbf{j} \times \mathbf{k}) \\ &= 4\mathbf{i} - 2\mathbf{j} + 12\mathbf{k} \end{aligned}$$

PROBLEMS

2.48. to 2.50. For the given set of vectors **A** and **B**, determine the cross product $\mathbf{A} \times \mathbf{B}$.

2.48. $\mathbf{A} = 8\mathbf{i} + 2\mathbf{j}$ $\quad \mathbf{B} = 10\mathbf{i} + 5\mathbf{j} - 2\mathbf{k}$
2.49. $\mathbf{A} = 2\mathbf{i} + 3\mathbf{j} - \mathbf{k}$ $\quad \mathbf{B} = 5\mathbf{i} - 9\mathbf{j} + 8\mathbf{k}$
2.50. $\mathbf{A} = 6\mathbf{i} - 2\mathbf{j} + 3\mathbf{k}$ $\quad \mathbf{B} = 8\mathbf{i} + 6\mathbf{j} - 4\mathbf{k}$

2.51. Determine the cross product of vectors **A** and **B** in Problem 2.35 and find a unit vector normal to the plane formed by **A** and **B**.

2.52. Same as Problem 2.51, except use the vectors **A** and **B** in Problem 2.36.

2.53. Determine whether or not the two vectors $\mathbf{A} = 2\mathbf{i} + 3\mathbf{j} - \mathbf{k}$ and $\mathbf{B} = 5\mathbf{i} - 9\mathbf{j} + 8\mathbf{k}$ are parallel.

2.54. Determine the values of y and z for which the vector $\mathbf{B} = 4\mathbf{i} + y\mathbf{j} + z\mathbf{k}$ will be parallel to the vector $\mathbf{A} = -2\mathbf{i} + \mathbf{j} - \mathbf{k}$.

2.55. Same as Problem 2.54, except that $\mathbf{A} = 2\mathbf{i} + 3\mathbf{j} - \mathbf{k}$.

2.56. Show that the vectors $\mathbf{A} = -2\mathbf{i} + \mathbf{j} - \mathbf{k}$, $\mathbf{B} = 4\mathbf{i} - 3\mathbf{j} + 2\mathbf{k}$ and $\mathbf{C} = 2\mathbf{i} + 3\mathbf{j} + \mathbf{k}$ all lie in the same plane. Write an expression for a unit vector normal to this plane.

2.57. Same as Problem 2.56, except that $\mathbf{C} = -4\mathbf{i} + \mathbf{j} - 2\mathbf{k}$.

2.58. Show that the magnitude of the cross product $\mathbf{A} \times \mathbf{B}$ can be interpreted geometrically as the area of a parallelogram with **A** and **B** as two of its sides.

2.59. Use the definition of the vector product given in Eq. (2.20) to construct a geometric proof of the distributive law $\mathbf{A} \times (\mathbf{B} + \mathbf{C}) = (\mathbf{A} \times \mathbf{B}) + (\mathbf{A} \times \mathbf{C})$.

2.8 MOMENT OF A FORCE ABOUT A POINT

When a force is applied to an object, the object may tend to rotate as well as translate. For example, the friction force developed between the front wheel of an automobile and the road causes the wheel to rotate instead of simply sliding along the surface. A measure of the tendency of a force to rotate the body upon which it acts is given by the *moment* of the force. As we shall see in Chapter 3, moments play a fundamental role in the conditions for equilibrium of a body. They are also encountered in dynamics problems involving rotational motions. Moments about a point will be considered in this section, and moments about a line, or axis, will be discussed in Section 2.9.

The *moment about a point P* of a force **F** is

$$\mathbf{M}_P = \mathbf{r} \times \mathbf{F} \tag{2.24}$$

where **r** is a position vector from P to any point on the line of action of **F** (Figure 2.21). Note that the moment is a vector quantity. It is perpendicular to the plane formed by **r** and **F**, and its sense is determined by the right-hand screw rule for vector products. We shall represent moments by two-headed arrows, as in Figure 2.21, to distinguish them from forces.

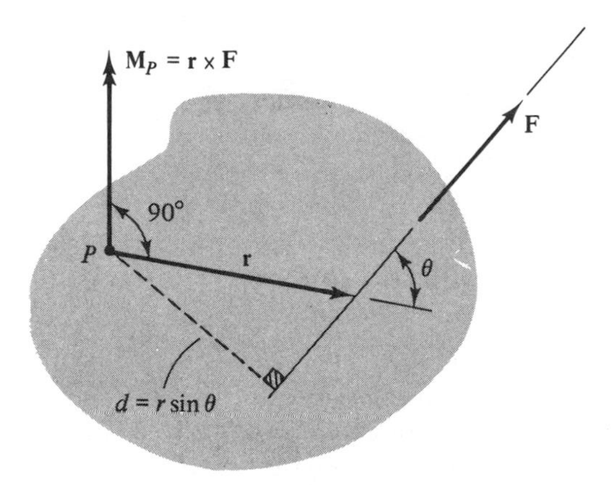

Figure 2.21

The magnitude of the moment is given by the definition of the cross product, Eq. (2.20):

$$M_P = |\mathbf{r} \times \mathbf{F}| = rF \sin\theta \tag{2.25}$$

From Figure 2.21, the quantity $r \sin\theta$ is recognized as the perpendicular distance d from the point P to the line of action of the force. Similarly, the quantity $F \sin\theta$ is seen to be the magnitude F_N of the component of the force normal to $\mathbf{r}$. Thus, the magnitude of the moment can also be expressed as

$$M_P = Fd \qquad \text{or} \qquad M_P = F_N r \tag{2.26}$$

The distance d is called the *moment arm* of the force. Since $M_P = 0$ when $d = 0$, we conclude that any force whose line of action passes through a point P will have zero moment about that point. We also conclude from Eq. (2.26) that only the component of force perpendicular to the position vector contributes to the moment; thus, any force parallel to $\mathbf{r}$ will have zero moment. Equation (2.26) indicates that moments have units of force times distance. Common units for moments are lb · ft, lb · in., and N · m (newton-meters).

If a force $\boldsymbol{F}$ is expressed in terms of its component vectors $\mathbf{F}_1, \mathbf{F}_2, \mathbf{F}_3, \ldots,$ its moment is

$$\mathbf{M}_P = \mathbf{r} \times \mathbf{F} = \mathbf{r} \times (\mathbf{F}_1 + \mathbf{F}_2 + \mathbf{F}_3 + \cdots)$$

or

$$\mathbf{M}_P = (\mathbf{r} \times \mathbf{F}_1) + (\mathbf{r} \times \mathbf{F}_2) + (\mathbf{r} \times \mathbf{F}_3) + \cdots = \mathbf{M}_{P1} + \mathbf{M}_{P2} + \mathbf{M}_{P3} + \cdots \tag{2.27}$$

Thus, the moment of a force is equal to the sum of the moments of each of its components. This result, which follows directly from the distributive property of the vector product, is very useful in applications. It was derived originally (long before vector algebra was invented) by P. Varignon (1654–1722), and is known as *Varignon's theorem*.

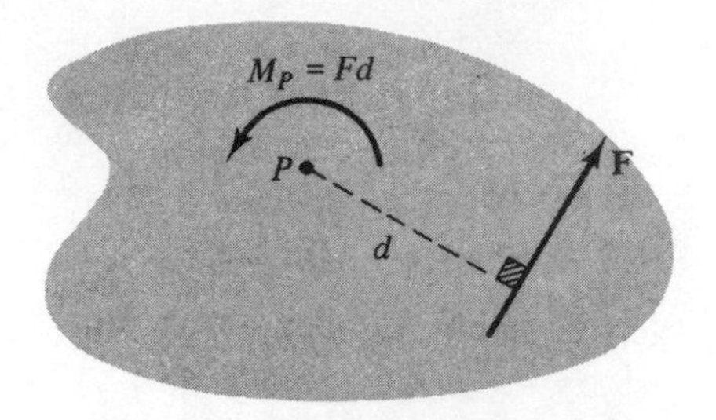

Figure 2.22

In two-dimensional problems the moment vector is perpendicular to the page, which makes it difficult to show on sketches. An alternative is to represent the moment vector by a curved arrow in the plane of the forces, as in Figure 2.22. This arrow can be thought of as indicating the direction of rotation of the fingers of the right hand when applying the right-hand screw rule. The sign convention for moments represented by curved arrows is arbitrary; in this text it will be denoted by a curved arrow with a plus sign over the moment symbol. Thus, $\overset{+}{\curvearrowleft}M$ signifies that counterclockwise moments are considered positive.

Although the moment of a force can always be determined by using cross products (*vector approach*), it is often simpler to compute its magnitude from Eq. (2.26) and determine its orientation and sense by inspection (*physical approach*). This is particularly true for two-dimensional problems. The curved arrow representing the moment will lie in the plane of the forces in this case, and the sense of the moment can be easily determined by noting the direction in which the force appears to be moving around the point about which the moment is computed. Each of these approaches is illustrated in the following examples.

Example 2.11. Moment About a Point. The door in Figure 2.23(a) is subjected to a 2-kN force, as shown. Determine the moment of the force about point A.

Solution. Since the problem is two-dimensional, there are several ways to proceed.

(a) ***Vector Approach***. We first set up a coordinate system, express the force in component form, and determine the position vector **r**; **r** can go from A to any point on the line of action of **F**. Point B is chosen because its coordinates are known. Using the coordinate system shown in Fig. 2.23(a), we have

$$\mathbf{F} = 1.73\mathbf{i} + 1.00\mathbf{j} \text{ kN}$$

$$\mathbf{r}_{AB} = -2\mathbf{i} + 3\mathbf{j} \text{ m}$$

$$\mathbf{M}_A = \mathbf{r}_{AB} \times \mathbf{F} = (-2\mathbf{i} + 3\mathbf{j}) \times (1.73\mathbf{i} + 1.00\mathbf{j}) \text{ kN} \cdot \text{m}$$

$$= -2(1.00)\mathbf{k} - 3(1.73)\mathbf{k} = -7.2\mathbf{k} \text{ kN} \cdot \text{m} \qquad \textbf{Answer}$$

The determinant form of the cross product can also be used. $\mathbf{M}_A$ lies along the negative z axis, as shown.

(b) ***Physical Approach***. The magnitude of the moment is $M_A = Fd$, where d is the perpendicular distance from A to the line of action of **F** [Eq. (2.26)]. From Figure 2.23(b), we

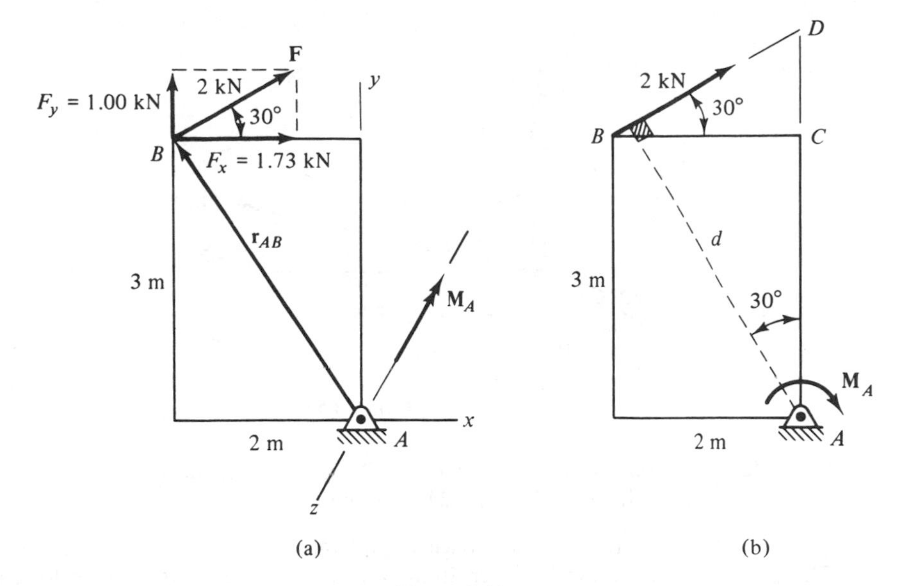

Figure 2.23

find

$$CD = 2 \tan 30° \text{ m} = 1.15 \text{ m}$$
$$AD = AC + CD = 4.15 \text{ m}$$
$$d = AD \cos 30° = 3.59 \text{ m}$$

so

$$M_A = Fd = (2 \text{ kN})(3.59 \text{ m}) = 7.2 \text{ kN} \cdot \text{m} \qquad \textbf{Answer}$$

The sense of the moment is determined by inspection. Since the force appears to be moving clockwise about point A, the moment is clockwise [Figure 2.23(b)].

(c) ***Physical Approach Using Components***. Instead of working with the perpendicular distance d, as in (b), we can make use of the fact that the moment of the force is equal to the sum of the moments of its components (Varignon's theorem). Referring to Figure 2.23(a) and taking counterclockwise moments to be positive, we have

$$\overset{+}{\curvearrowleft} M_A = -(1.73 \text{ kN})(3 \text{ m}) - (1.00 \text{ kN})(2 \text{ m})$$

$$= -7.2 \text{ kN} \cdot \text{m} \qquad \textbf{Answer}$$

The negative answer means that $\mathbf{M}_A$ is actually clockwise, as in Figure 2.23(b).

All of the procedures used here lead to the same result. The vector approach is a bit longer than the others; the physical approach using components is the simplest. The choice of procedure in problems like this is primarily a matter of convenience and personal preference.

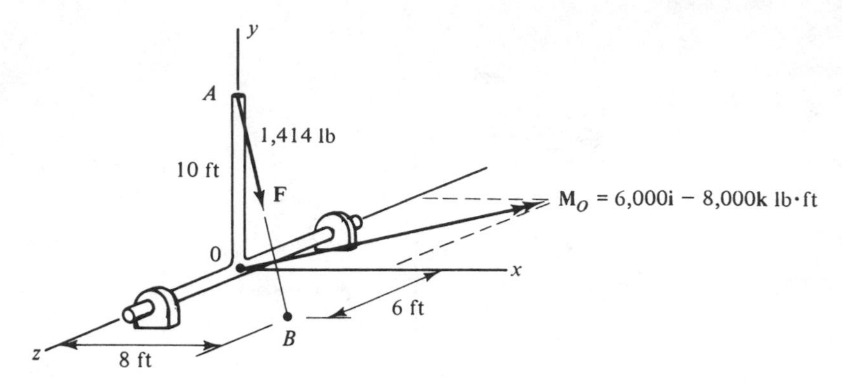

Figure 2.24

Example 2.12. Moment About a Point. A T-bar is subjected to a force **F** with magnitude 1,414 lb, as shown in Figure 2.24. Determine the moment about the origin *O*.

Solution. The vector approach is preferable because of the complicated geometry. The position vector **r** can be taken from *O* to either *A* or *B*. It is taken to *A* because this results in the simplest expression. In component form (details not given here), we have

$$\mathbf{F} = 800\mathbf{i} - 1{,}000\mathbf{j} + 600\mathbf{k}\ \text{lb}$$

$$\mathbf{r}_{OA} = 10\mathbf{j}\ \text{ft}$$

so

$$\mathbf{M}_O = \mathbf{r}_{OA} \times \mathbf{F} = 10\mathbf{j} \times (800\mathbf{i} - 1{,}000\mathbf{j} - 600\mathbf{k})\ \text{lb} \cdot \text{ft}$$

$$\mathbf{M}_O = 6{,}000\mathbf{i} - 8{,}000\mathbf{k}\ \text{lb} \cdot \text{ft} \qquad \textbf{Answer}$$

The moment vector is as shown.

PROBLEMS

2.60. to 2.63. For the situation shown, determine the moment of the force about point *P*.

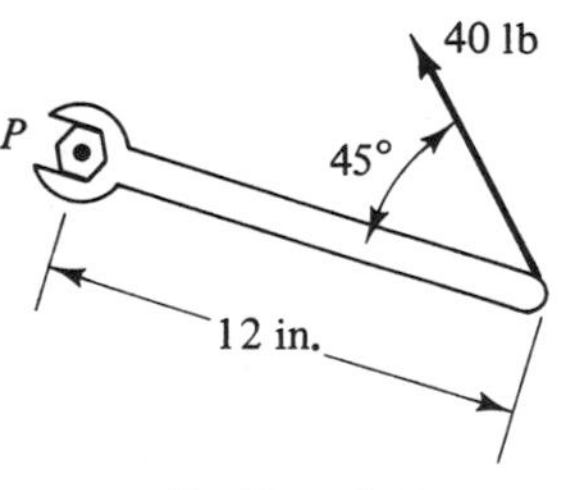

Problem 2.60

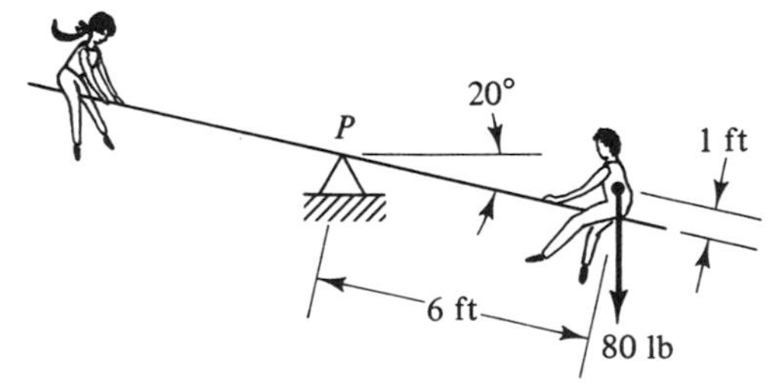

Problem 2.61

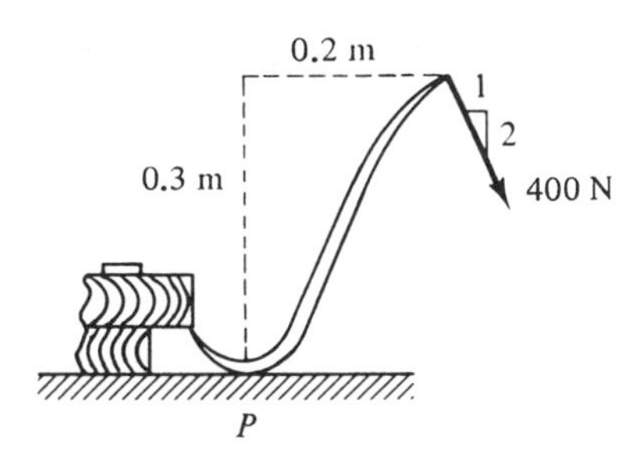

Problem 2.62

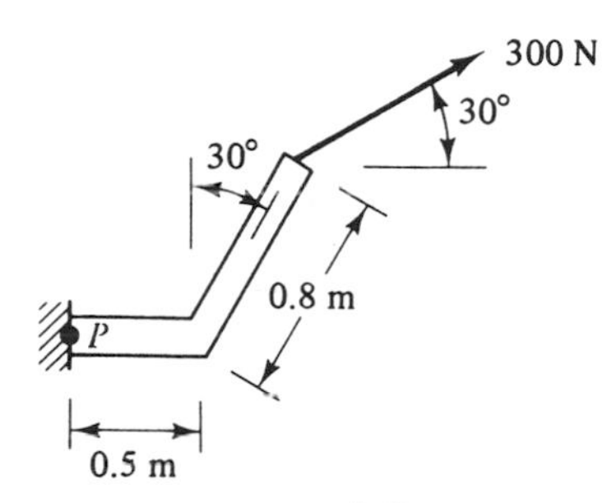

Problem 2.63

2.64. A door is hinged at one corner and held in position by a force **F**, as shown. For what magnitude of **F** will the total moment about hinge A be zero?

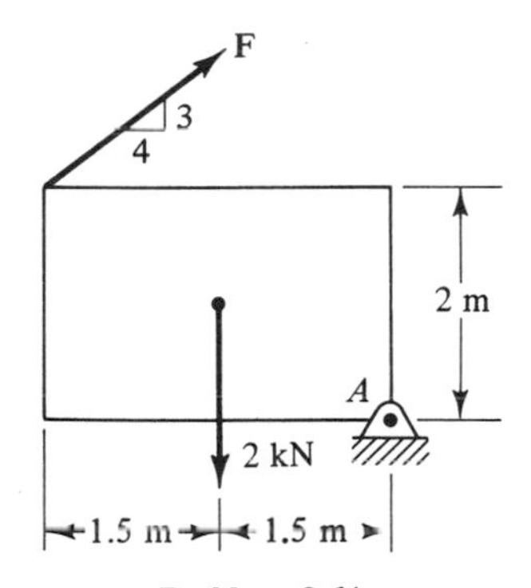

Problem 2.64

2.65. The forces exerted on a retaining wall by the soil are as shown. Determine the distance d for which the total moment about corner A of the wall will be zero.

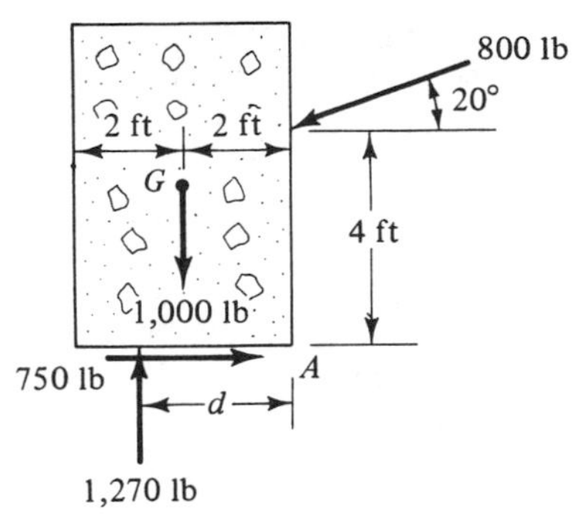

Problem 2.65

2.66. A force **F** is applied to the end of a semicircular slab by means of a rope attached at point A as shown. Determine the moment of the force about point B for $\theta = 70°$.

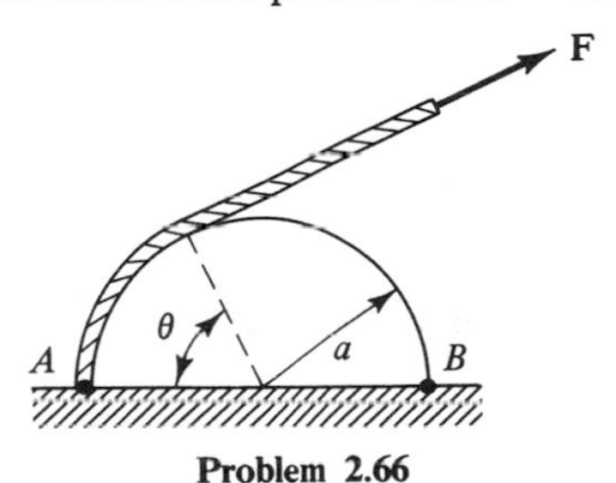

Problem 2.66

2.67. In Problem 2.66, for what value of θ will the moment about point B be (a) maximum and (b) minimum? What are the corresponding values of the moment?

2.68. A 400-N force is applied to a bracket as shown. For $\theta = 30°$, determine (a) the mo-

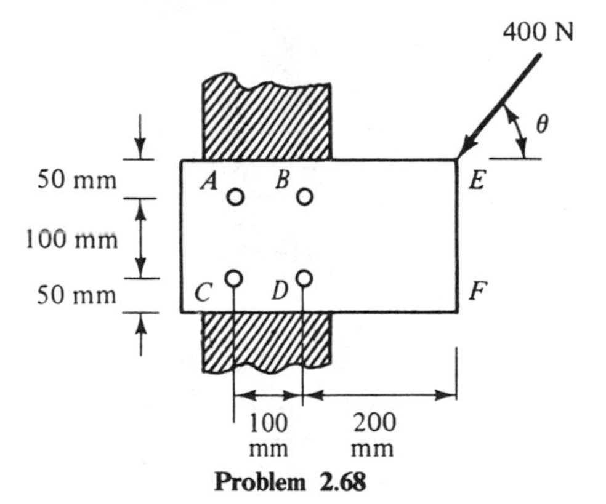

Problem 2.68

ment of the 400-N force about point D, (b) the vertical force applied at E that would produce this same moment about D, and (c) the smallest force applied at E that would produce this same moment about D.

2.69. For what orientation θ of the 400-N force in Problem 2.68 will the moment of this force about point C be maximum? For what orientation will it be minimum? What are these maximum and minimum values of the moment?

2.70. and 2.71. For the force given, determine (a) the moment about the origin O and (b) the perpendicular distance from O to the line of action of the force.

2.70. $\mathbf{F} = 400\mathbf{i} - 600\mathbf{j} + 300\mathbf{k}$ lb acting at 4, −2, 4 ft

2.71. $\mathbf{F} = -2\mathbf{i} - 3\mathbf{j} + 5\mathbf{k}$ kN acting at 6, 2, −3 m

2.72. A 550-lb force lies along a line passing through points A and B as shown. Determine the moment of the force about points A, B, C, and D.

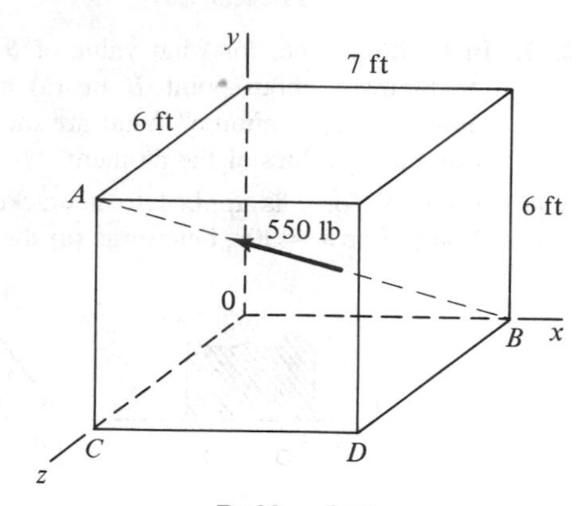

Problem 2.72

2.73. Tubs of parts are delivered from a store room to the work area by the curved chute shown. Determine the moment about point A when a 50-lb tub is located at end B of the chute.

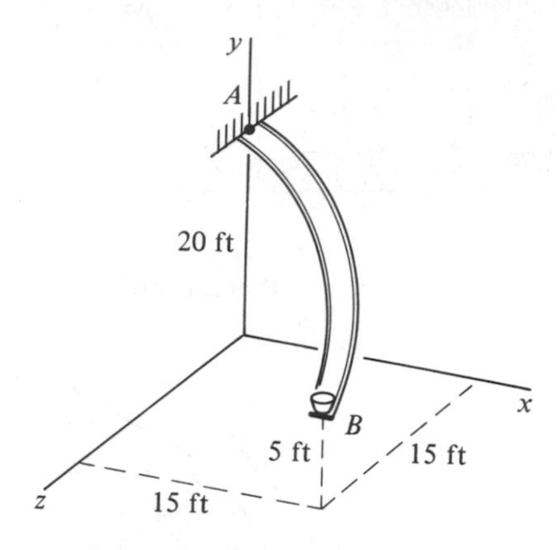

Problem 2.73

2.74. The spray of water exiting a showerhead exerts a 5-N force on the head in the direction shown. Determine the moment of this force about point A. What is the perpendicular distance from A to the line of action of the force? The curved pipe lies in the x-y plane and is in the shape of a circular arc with a radius of 100 mm.

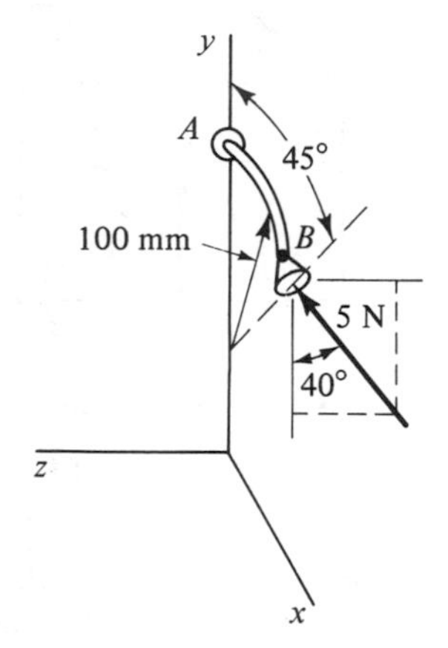

Problem 2.74

2.75. An adjustable work light with a mass of 25 kg is subjected to a 50-N force acting parallel to the x axis, as shown. Determine the total moment about point A when $\theta = 45°$. Neglect the mass of the supporting members.

2.76. Same as Problem 2.75, except that $\theta = 30°$.

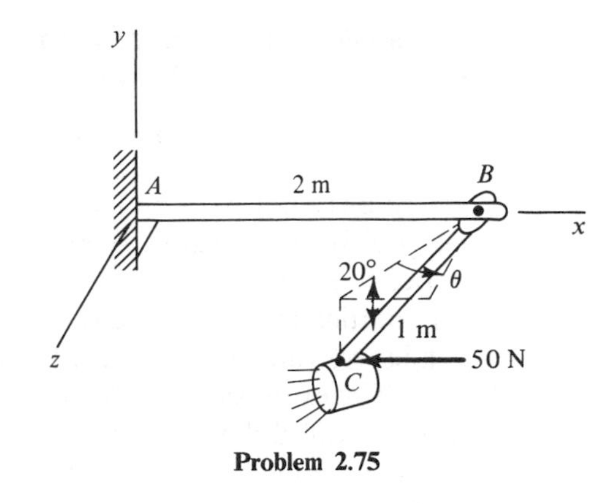

Problem 2.75

2.9 MOMENT OF A FORCE ABOUT A LINE

In many applications the object of interest is constrained to rotate about a particular line or axis. Bodies supported by bearings or hinges fall into this category; typical examples are a door, the crankshaft in an automobile engine, and the bit in a drill. The tendency of a force to produce rotation about a specific line is given by the *moment about the line*.

Consider a line L, with $\mathbf{M}_P$ the moment about some point P along it (Figure 2.25). The moment about the line, $\mathbf{M}_L$, is simply the component of $\mathbf{M}_P$ along L. If

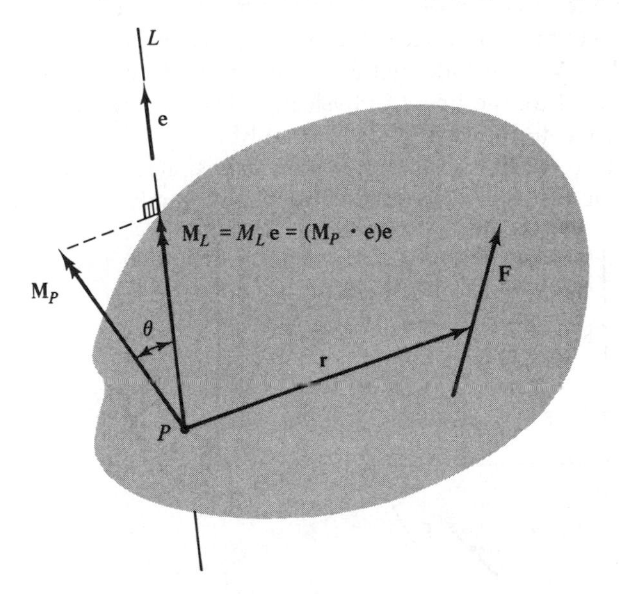

Figure 2.25

the angle θ between $\mathbf{M}_P$ and L is known, $\mathbf{M}_L$ can be determined directly ($M_L = M_P \cos\theta$). Otherwise, it is convenient to use the scalar product [Eq. (2.6)]:

$$\mathbf{M}_L = M_L\mathbf{e} = (\mathbf{M}_P \cdot \mathbf{e})\mathbf{e} = [(\mathbf{r} \times \mathbf{F}) \cdot \mathbf{e}]\mathbf{e} \tag{2.28}$$

where $\mathbf{e}$ is a unit vector along the line and $M_L = \text{Proj}_{\mathbf{e}}\,\mathbf{M}_P$. The projection M_L will be positive or negative, depending upon whether $\mathbf{M}_L$ and $\mathbf{e}$ have the same or opposite sense. In the way of physical interpretation, the direction of the tendency for rotation about the line is given by the fingers of the right hand when the thumb is aligned with $\mathbf{M}_L$ (the right-hand screw rule).

The expression for the projection M_L in Eq. (2.28) is of the form $\mathbf{A} \cdot (\mathbf{B} \times \mathbf{C})$, which is called the *scalar triple product*. This product can be determined in two steps by first performing the cross product and then the dot product, or it can be computed in one step by using the relationship

$$\mathbf{A} \cdot (\mathbf{B} \times \mathbf{C}) = \begin{vmatrix} A_x & A_y & A_z \\ B_x & B_y & B_z \\ C_x & C_y & C_z \end{vmatrix} \tag{2.29}$$

Equation (2.29) can be easily verified by performing the indicated vector operations and comparing the result with the expanded form of the determinant.

From Eq. (2.28), we see that $\mathbf{M}_L$ will be zero if $\mathbf{M}_P = \mathbf{0}$ or is perpendicular to $\mathbf{e}$. Now $\mathbf{M}_P$ is perpendicular to the plane formed by the vectors $\mathbf{r}$ and $\mathbf{F}$; so the latter condition implies that $\mathbf{r}$, $\mathbf{F}$, and the line L all lie in the same plane. Consequently, L and the line of action of the force either intersect or are parallel. The condition $\mathbf{M}_P = \mathbf{0}$ also implies an intersection of these two lines, since $\mathbf{F}$ passes through P in this case. Thus, we conclude that $\mathbf{M}_L = \mathbf{0}$ for any force whose line of action intersects, or is parallel to, the line L. Physically, this implies that only the component of $\mathbf{F}$ perpendicular to the line contributes to $\mathbf{M}_L$.

If the force is normal to the line and a perpendicular distance d away from it (Figure 2.26), the magnitude of the moment about the line is simply the magnitude

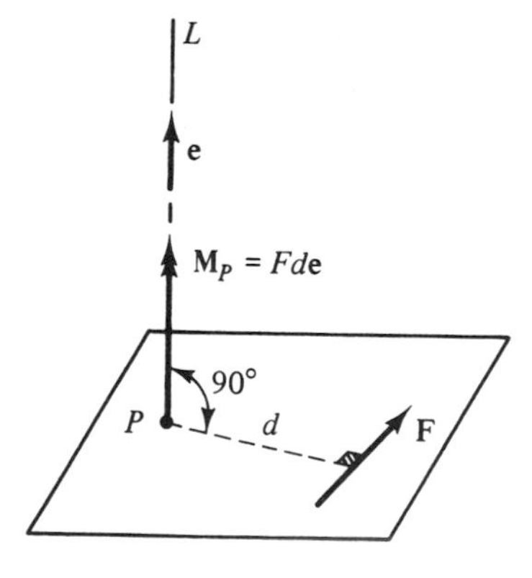

Figure 2.26

of the force times the distance d:

$$|\mathbf{M}_L| = Fd \qquad \mathbf{F} \perp L \tag{2.30}$$

To show this, we choose the point P, which is arbitrary, such that it lies in the same plane as $\mathbf{F}$. Then $\mathbf{M}_P = Fd\mathbf{e}$ and $|\mathbf{M}_L| = \mathbf{M}_P \cdot \mathbf{e} = Fd$, which confirms Eq. (2.30).

Moments about a line often can be used to advantage when determining moments about a point. In this regard, we note that the components of the moment about a point P, $\mathbf{M}_P = M_{Px}\mathbf{i} + M_{Py}\mathbf{j} + M_{Pz}\mathbf{k}$, are the moments about lines through P parallel to the coordinate axes. This is easy to show. The unit vector along a line through P parallel to the x axis is $\mathbf{i}$; so the moment about this line is $(\mathbf{M}_P \cdot \mathbf{i})\mathbf{i} = M_{Px}\mathbf{i}$. Similarly, $M_{Py}\mathbf{j}$ and $M_{Pz}\mathbf{k}$ can be shown to be moments about lines through P parallel to the y and z axes, respectively. Thus, the moment about a point can be determined term by term by finding the moments about lines through the point parallel to the coordinate axes. This procedure is illustrated in Example 2.15.

Example 2.13. Moment About a Line. The stone slab in Figure 2.27 is partially supported by a cable AB that exerts a 10,000-lb force $\mathbf{T}$ on it as shown. Determine the moment about (a) the z axis, (b) the diagonal OB, and (c) the diagonal CD. Note that the dimensions and loading are the same as in Example 2.9.

Solution. (a) First, we determine the moment about some point on the z axis. There are two logical choices: point C and the origin. Suppose we pick point C and take the position vector from C to B:

$$\mathbf{r}_{CB} = 3\mathbf{j} \text{ ft}$$

In Example 2.9 we found that

$$\mathbf{T} = 8{,}890\mathbf{i} - 3{,}810\mathbf{j} - 2{,}540\mathbf{k} \text{ lb}$$

so

$$\begin{aligned}\mathbf{M}_C = \mathbf{r}_{CB} \times \mathbf{T} &= 3\mathbf{j} \times (8{,}890\mathbf{i} - 3{,}810\mathbf{j} - 2{,}540\mathbf{k}) \text{ lb} \cdot \text{ft} \\ &= -7{,}620\mathbf{i} - 26{,}670\mathbf{k} \text{ lb} \cdot \text{ft}\end{aligned}$$

The moment about the z axis is just the z component of $\mathbf{M}_C$:

$$\mathbf{M}_{Oz} = (\mathbf{M}_C \cdot \mathbf{k})\mathbf{k} = -26{,}670\mathbf{k} \text{ lb} \cdot \text{ft} \qquad \textbf{Answer}$$

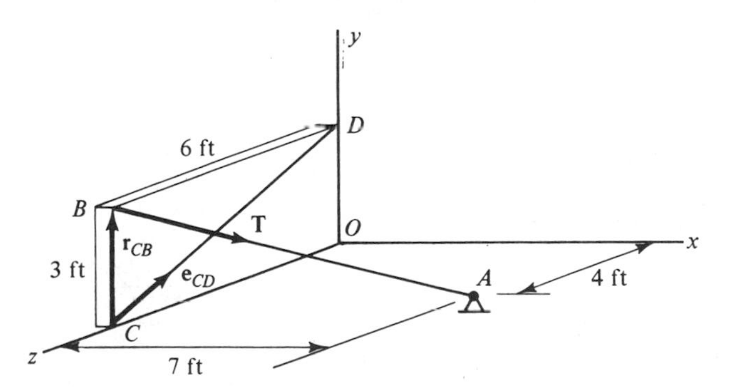

Figure 2.27

(b) The line of action of $\mathbf{T}$ intersects OB; therefore, the moment about this line is zero.

(c) Because we already know the moment about point C on line CD, it only remains to find a unit vector along this line. Since the sense of the unit vector is arbitrary, we take it from C to D:

$$\mathbf{r}_{CD} = 3\mathbf{j} - 6\mathbf{k} \text{ ft}$$

$$r_{CD} = \sqrt{(3)^2 + (-6)^2} = \sqrt{45} \text{ ft}$$

$$\mathbf{e}_{CD} = \frac{\mathbf{r}_{CD}}{r_{CD}} = \frac{1}{\sqrt{45}}(3\mathbf{j} - 6\mathbf{k})$$

From Eq. (2.28), we have

$$\mathbf{M}_{CD} = (\mathbf{M}_C \cdot \mathbf{e}_{CD})\mathbf{e}_{CD} = \left[(-7{,}620\mathbf{i} - 26{,}670\mathbf{k}) \cdot \frac{1}{\sqrt{45}}(3\mathbf{j} - 6\mathbf{k})\right]\mathbf{e}_{CD}$$

$$= \frac{6}{\sqrt{45}}(26{,}670)\mathbf{e}_{CD} = 23{,}850\mathbf{e}_{CD} \text{ lb} \cdot \text{ft}$$

or

$$\mathbf{M}_{CD} = 10{,}670\mathbf{j} - 21{,}340\mathbf{k} \text{ lb} \cdot \text{ft} \qquad \textbf{Answer}$$

Alternate solution: The solution could also have been obtained by using the determinant form of the scalar triple product, Eq. (2.29). For part (c) of the problem, for example,

$$M_{CD} = \mathbf{e}_{CD} \cdot (\mathbf{r}_{CB} \times \mathbf{T}) = \begin{vmatrix} 0 & \frac{3}{\sqrt{45}} & \frac{-6}{\sqrt{45}} \\ 0 & 3 & 0 \\ 8{,}890 & -3{,}810 & -2{,}540 \end{vmatrix}$$

Expanding the determinant, we get

$$M_{CD} = 8{,}890(3)\frac{6}{\sqrt{45}} = 23{,}850 \text{ lb} \cdot \text{ft}$$

so

$$\mathbf{M}_{CD} = 23{,}850\mathbf{e}_{CD} \text{ lb} \cdot \text{ft}$$

This is the same result obtained previously.

Example 2.14. Moment About a Line. A 180-N force is exerted on a wrench as shown in Figure 2.28. What is the moment tending to unscrew the bolt?

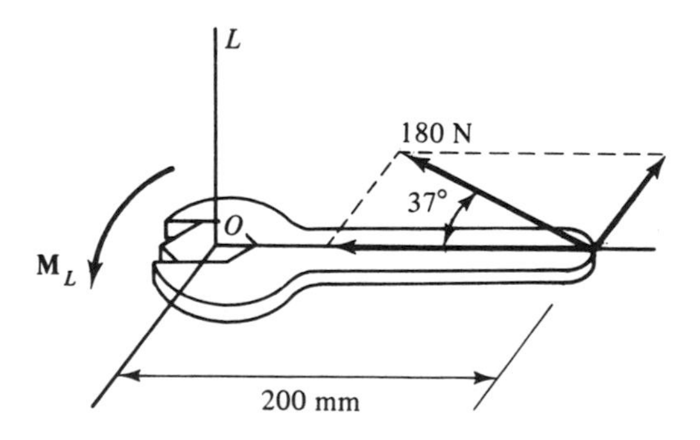

Figure 2.28

Solution. The quantity desired is the moment about the axis L of the bolt. Resolving the force into components, we see that the component along the handle will not contribute to the moment because its line of action intersects the bolt axis. The moment due to the other component is readily determined by inspection. From Eq. (2.30), we have

$$M_L = Fd = (180 \sin 37° \text{ N})(0.2 \text{ m}) = 21.7 \text{ N} \cdot \text{m} \qquad \textbf{Answer}$$

Since the force appears to be moving counterclockwise about the bolt axis when viewed from above, $\mathbf{M}_L$ can be represented either by a counterclockwise curved arrow in the plane of the wrench, as shown, or by a vector pointing upward along the bolt axis.

Example 2.15. Moment About a Point. The force $\mathbf{F}$ in Figure 2.29 is parallel to the y axis. Find its moment about the origin by computing moments about lines.

Solution. The problem is three-dimensional, but the moment can be determined by inspection by computing the moments about the coordinate axes. Since the force is perpendicular to the x and z axes and is parallel to the y axis, we have

$$|\mathbf{M}_{Ox}| = Fb \qquad |\mathbf{M}_{Oz}| = Fa \qquad |\mathbf{M}_{Oy}| = 0$$

The sense of the moments is obtained by the right-hand screw rule. If we curve the fingers of the right hand in the direction of rotation about the x axis, we see that our thumb points in the positive x direction. Thus,

$$\mathbf{M}_{Ox} = Fb\mathbf{i}$$

Similarly,

$$\mathbf{M}_{Oz} = -Fa\mathbf{k}$$

The moments about the coordinate axes are the vector components of the moment about the origin; so

$$\mathbf{M}_O = Fb\mathbf{i} - Fa\mathbf{k} \qquad \textbf{Answer}$$

It is left as an exercise to show that the same result is obtained by using the cross product.

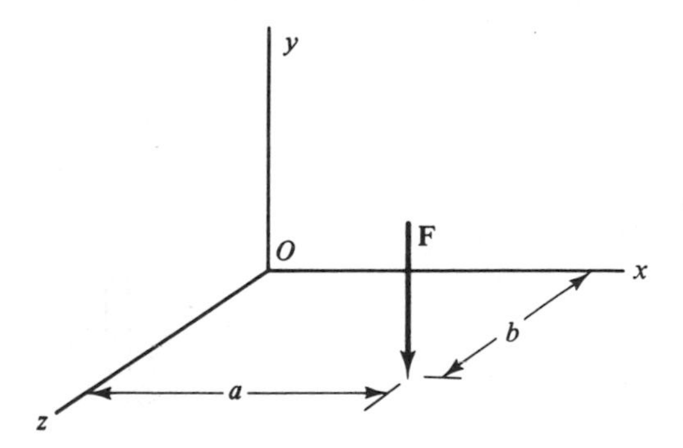

Figure 2.29

PROBLEMS

2.77. and 2.78. For the given set of vectors, determine the scalar triple product $\mathbf{A} \cdot (\mathbf{B} \times \mathbf{C})$.

2.77. $\mathbf{A} = 2\mathbf{i} + 2\mathbf{j} - \mathbf{k}$ $\mathbf{B} = \mathbf{i} + \mathbf{j}$
$\mathbf{C} = 4\mathbf{i} - 2\mathbf{j} - 6\mathbf{k}$

2.78. $\mathbf{A} = -3\mathbf{i} + 6\mathbf{j} - 2\mathbf{k}$ $\mathbf{B} = 4\mathbf{i} + 2\mathbf{j} - \mathbf{k}$
$\mathbf{C} = 2\mathbf{i} + 3\mathbf{j} - 3\mathbf{k}$

2.79. Show that the scalar triple product of any three vectors that lie in same plane is zero.

2.80. If the tensions on a belt are as shown, determine the total moment tending to rotate the shaft about its longitudinal axis.

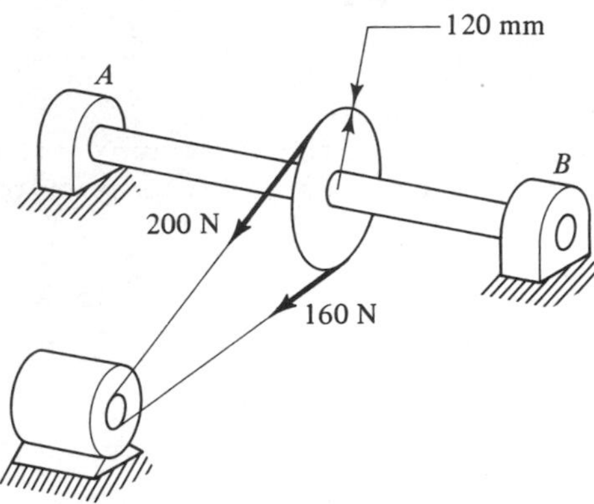

Problem 2.80

2.81. A 50-N force is applied to a door, as shown. What is the moment tending to rotate the door about its hinges? The force is oriented 45° above the horizontal and lies in a plane perpendicular to the door.

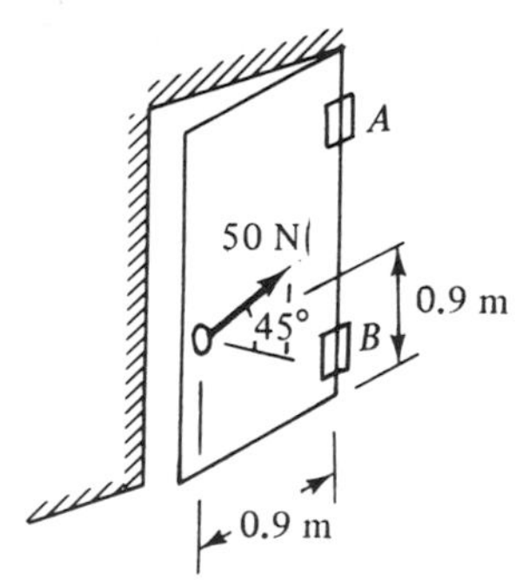

Problem 2.81

2.82. A 900-lb reel of electrical cable is supported as shown. If the tension on the cable has a magnitude of 260 lb, determine the total moment due to the reel weight and cable tension about a line through supports C and D. Will the assembly tend to tip about this line? Each of the support members is 4 ft long and has negligible weight; the radius to the outer layer of cable on the reel is 1 ft.

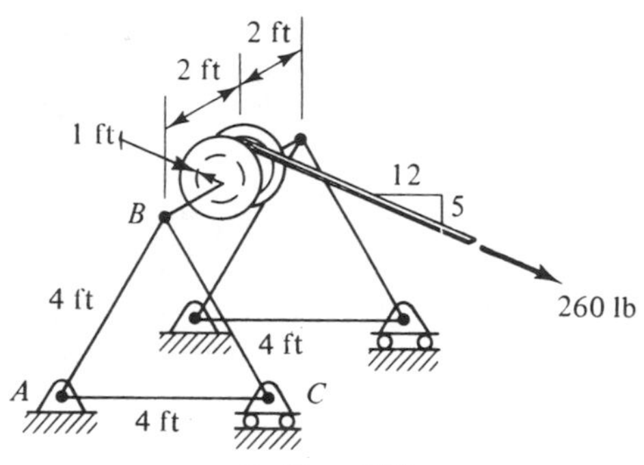

Problem 2.82

2.83. In Problem 2.72 (Section 2.8), determine the moment of the force about (a) the x axis, (b) line CD, and (c) diagonal OD.

2.84. The tension on a telephone wire has a magnitude of 800 N and is directed as shown. Determine the moment tending to twist the pole about its longitudinal axis.

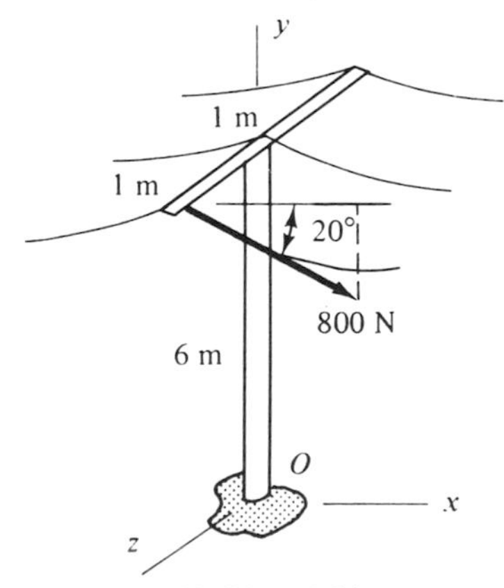

Problem 2.84

2.85. The tension on the cable of a jib crane has a magnitude of 20 kN and is directed as shown. Determine the moment tending to rotate the crane about its vertical axis. What is the moment at O tending to bend the vertical crane member about the x axis? About the z axis?

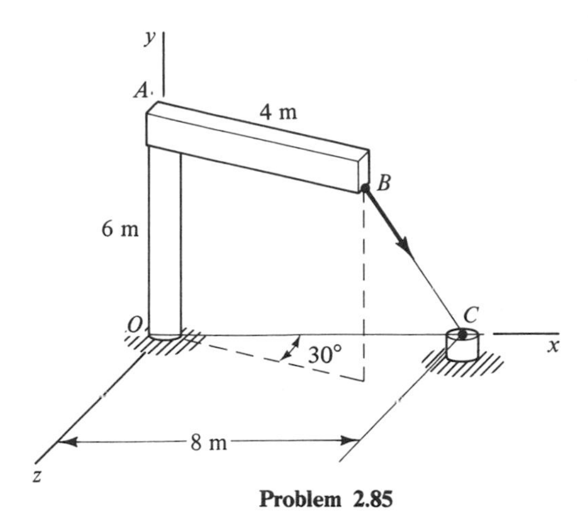

Problem 2.85

2.86. Same as Problem 2.85, except that the object being moved by the crane is located along the positive z axis, 6 m from the origin.

2.87. Rework Problem 2.73 by considering moments about lines through point A parallel to the coordinate axes.

2.88. Rework Problem 2.74 by considering moments about lines through point A parallel to the coordinate axes.

2.10 COUPLES

Figure 2.30 illustrates a situation that is not uncommon in physical problems. As shown in the figure, the fluid jets of the lawn sprinkler produce equal and opposite forces acting some distance apart. The sum of these forces is zero, but the sum of their moments about point O is nonzero. Thus, the sole physical effect of the forces is to tend to rotate the arm of the sprinkler. This type of force system, with zero total force and nonzero total moment, is called a *couple*. Couples are extremely important in the study of mechanics because they represent a pure rotational effect.

A couple, which we shall denote by $\mathbf{C}$, can consist of any number of forces and is represented by the total moment of the force system of which it is comprised. However, unlike the moment of a force, the moment of a couple is the same about every point. To prove this, we consider the total moment of a general force system,

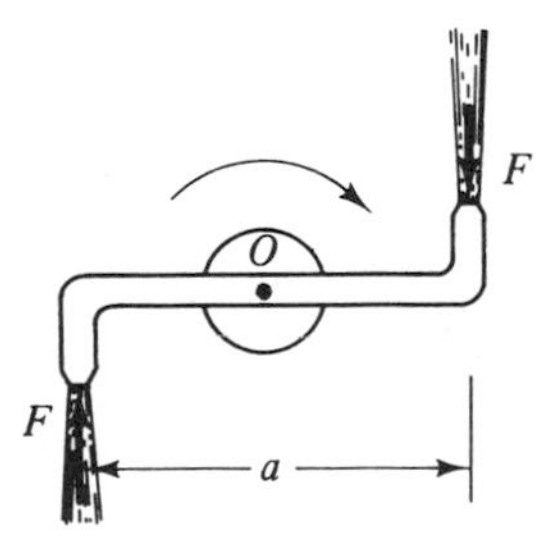

Figure 2.30

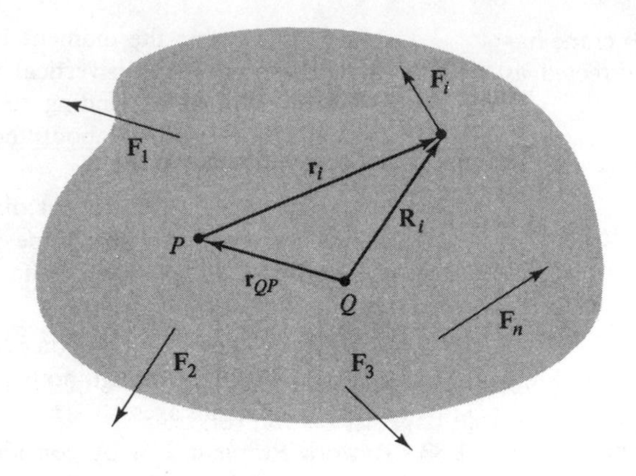

Figure 2.31

such as shown in Figure 2.31, about two different points P and Q. Referring to the figure, we have

$$\sum \mathbf{M}_P = \sum (\mathbf{r}_i \times \mathbf{F}_i)$$
$$\sum \mathbf{M}_Q = \sum (\mathbf{R}_i \times \mathbf{F}_i)$$

Now

$$\mathbf{R}_i = \mathbf{r}_i + \mathbf{r}_{QP}$$

so

$$\sum \mathbf{M}_Q = \sum \left[(\mathbf{r}_i + \mathbf{r}_{QP}) \times \mathbf{F}_i\right] = \sum \mathbf{M}_P + \mathbf{r}_{QP} \times \left(\sum \mathbf{F}_i\right) \tag{2.31}$$

But $\Sigma \mathbf{F}_i = \mathbf{0}$ for a couple, in which case $\Sigma \mathbf{M}_Q = \Sigma \mathbf{M}_P$. Since points P and Q are arbitrary, it follows that the total moment of a couple is the same about every point.

In the case of a simple couple consisting of two equal and opposite forces spaced a distance apart (Figure 2.32), we have for the couple vector

$$\mathbf{C} = \sum \mathbf{M}_P = \mathbf{r}_1 \times \mathbf{F} + \mathbf{r}_2 \times (-\mathbf{F}) = (\mathbf{r}_1 - \mathbf{r}_2) \times \mathbf{F} \tag{2.32}$$

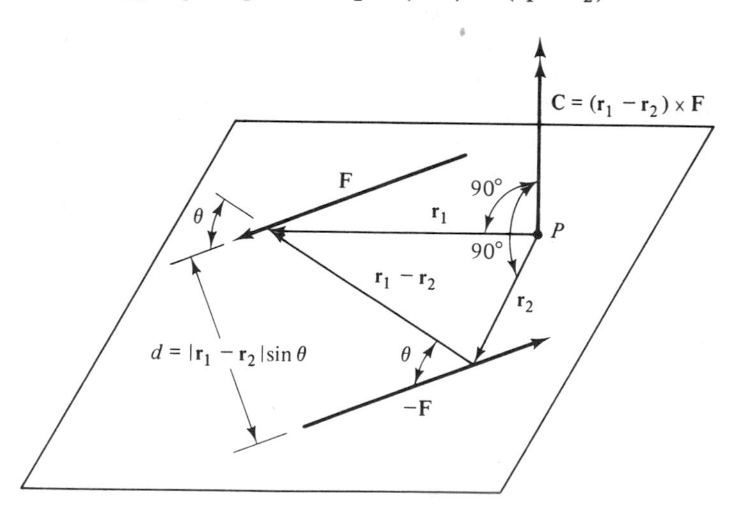

Figure 2.32

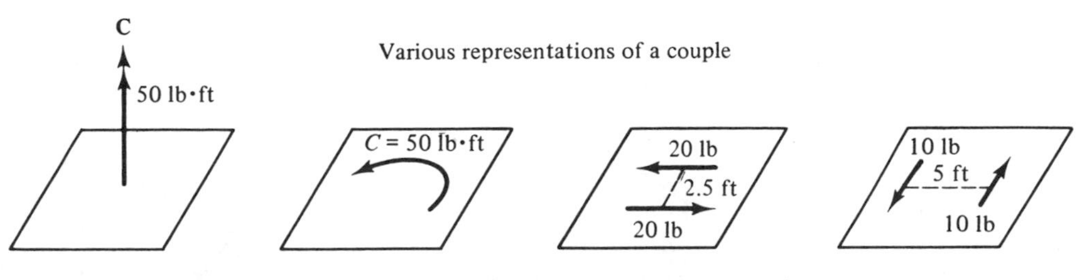

Figure 2.33

where $\mathbf{r}_1$ and $\mathbf{r}_2$ are the position vectors from P to the lines of action of $\mathbf{F}$ and $-\mathbf{F}$, respectively. The magnitude of the couple is given by the relation

$$C = |\mathbf{r}_1 - \mathbf{r}_2| F \sin\theta = Fd \tag{2.33}$$

where d is the perpendicular distance between the two forces. Since the vector $(\mathbf{r}_1 - \mathbf{r}_2)$ lies in the plane of the forces, the couple is perpendicular to this plane. Its sense can be determined by inspection or by the right-hand screw rule.

Since the moment of a couple is the same about every point, the couple vector $\mathbf{C}$ has no definite point of application. It can be located anywhere on the body upon which it acts. In Figure 2.32, for example, it was arbitrarily placed at point P. For two-dimensional problems, couples can also be represented by curved arrows in the plane of the forces or by two equal and opposite parallel forces spaced an appropriate distance apart (Figure 2.33). Finally, we note that a couple, being a moment vector, can be combined directly with the moment of a force.

Example 2.16. Moment of a Couple. If it takes a 40-N · m couple to turn the steering wheel of a bus, what forces acting as shown in Figure 2.34 must be applied to the wheel?

Solution. The moment of the couple is equal to the magnitude of the forces times the perpendicular distance between them [Eq. (2.33)]. Thus,

$$F = \frac{C}{d} = \frac{40\ \text{N}\cdot\text{m}}{0.45\ \text{m}} = 88.9\ \text{N} \qquad \textbf{Answer}$$

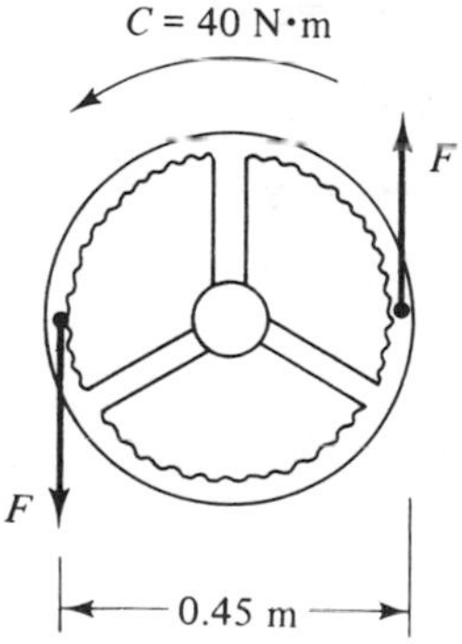

Figure 2.34

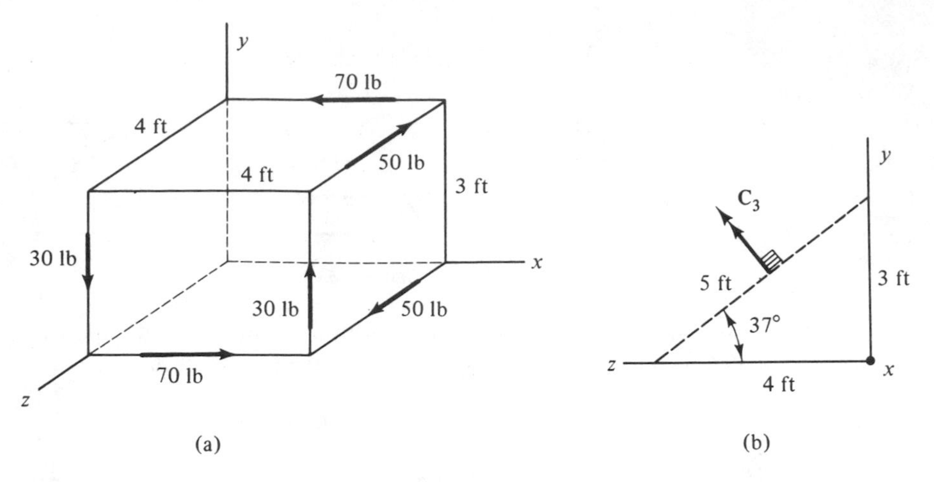

Figure 2.35

Example 2.17. Addition of Couples. Express the couples acting upon the block in Figure 2.35(a) in vector form and determine their sum.

Solution. Each pair of forces forms a couple with magnitude Fd [Eq. (2.33)] acting perpendicular to the plane in which the forces lie.

The 50-lb forces form a couple of magnitude

$$C_1 = Fd = 50 \text{ lb } (3 \text{ ft}) = 150 \text{ lb} \cdot \text{ft}$$

parallel to the x axis and pointing in the negative x direction. Thus,

$$\mathbf{C}_1 = -150\mathbf{i} \text{ lb} \cdot \text{ft} \qquad \textbf{Answer}$$

Similarly, the couple formed by the 30-lb forces is

$$\mathbf{C}_2 = 120\mathbf{k} \text{ lb} \cdot \text{ft} \qquad \textbf{Answer}$$

The couple formed by the 70-lb forces is directed as shown in Figure 2.35(b) and has magnitude

$$C_3 = Fd = 70 \text{ lb } (5 \text{ ft}) = 350 \text{ lb} \cdot \text{ft}$$

Resolving $\mathbf{C}_3$ into components, we have

$$\begin{aligned} \mathbf{C}_3 &= 350 \cos 37°\mathbf{j} + 350 \sin 37°\mathbf{k} \text{ lb} \cdot \text{ft} \\ &= 280\mathbf{j} + 210\mathbf{k} \text{ lb} \cdot \text{ft} \qquad \textbf{Answer} \end{aligned}$$

The sum of the couples is

$$\mathbf{C} = \mathbf{C}_1 + \mathbf{C}_2 + \mathbf{C}_3 = -150\mathbf{i} + 280\mathbf{j} + 330\mathbf{k} \text{ lb} \cdot \text{ft} \qquad \textbf{Answer}$$

and can be shown as acting anyplace upon the block.

PROBLEMS

2.89. to 2.91. Determine whether or not the force system shown forms a couple. If it does, determine the couple and show it on a sketch. If it doesn't, what additional force acting at point *B* is necessary for the system to be a couple, and what is this couple?

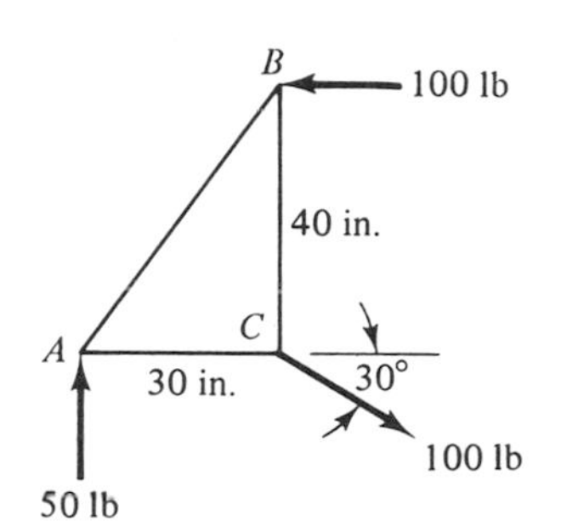

Problem 2.89

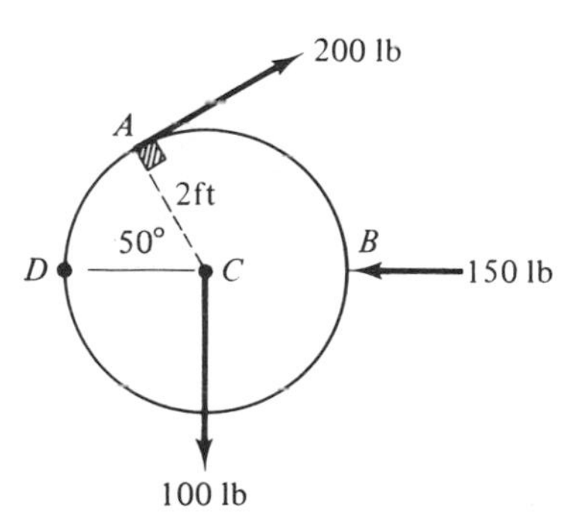

Problem 2.90

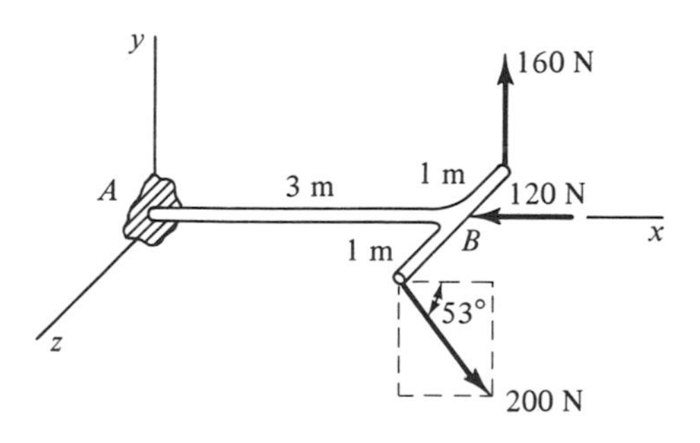

Problem 2.91

2.92. A large pipe is screwed into a threaded connection by attaching two chains and pulling on them as shown. Determine the couple tending to turn the pipe.

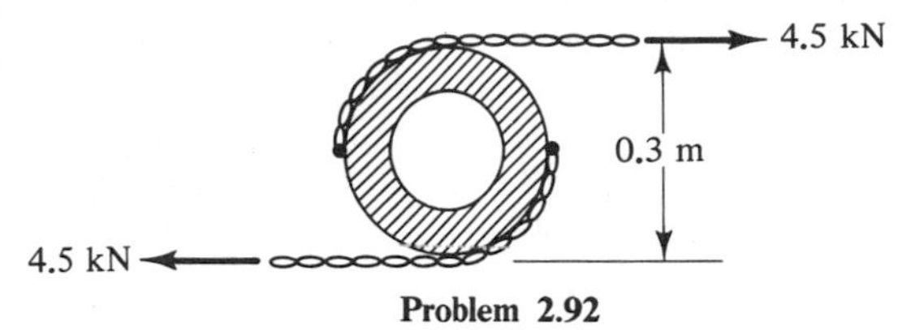

Problem 2.92

2.93. It takes a 50-lb · ft couple to close a certain pipeline valve. Determine the forces acting upon the handle as shown necessary to generate this couple.

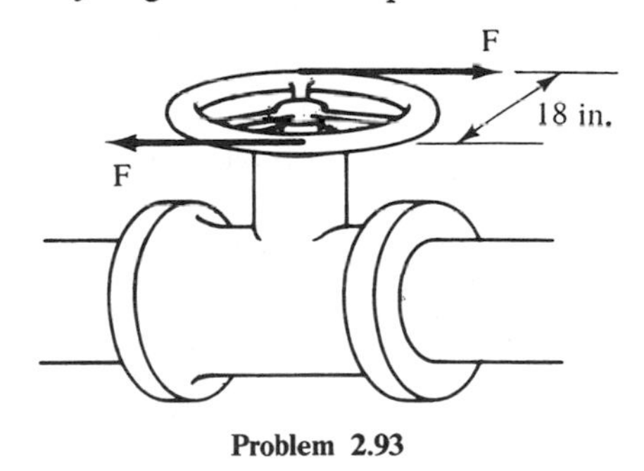

Problem 2.93

2.94. A 60-lb · ft couple is required to loosen

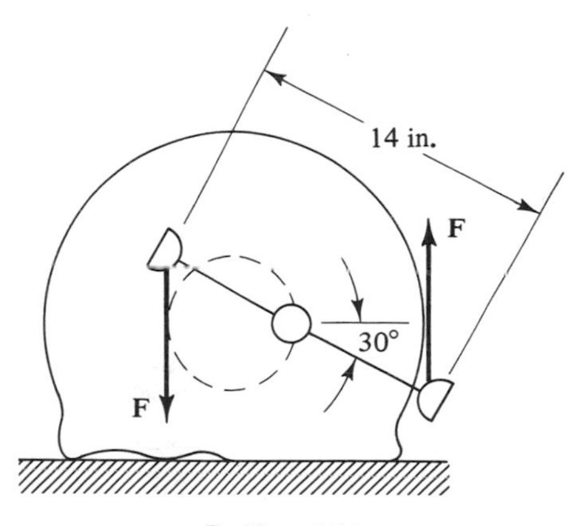

Problem 2.94

magnitude of vertical force **F** must a motorist exert to generate this couple if the lug wrench is used as shown? Would a larger or smaller vertical force be required if the wrench were positioned horizontally?

2.95. The three forces acting on the plate shown form a couple. Determine (a) the vertical forces acting at points B and D that would produce the same couple, and (b) the smallest forces acting at points B and D that would produce the same couple.

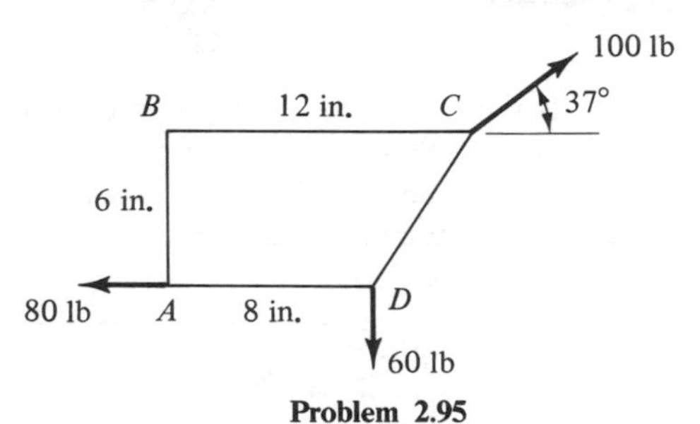

Problem 2.95

2.96. The three forces in Problem 2.95 are to be replaced by a pair of forces that produces the same couple. What are the smallest forces that can be used and where on the rectangular plate should they be applied? Show the results on a sketch.

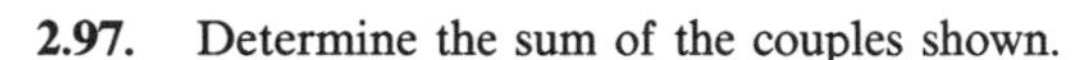

2.97. Determine the sum of the couples shown.

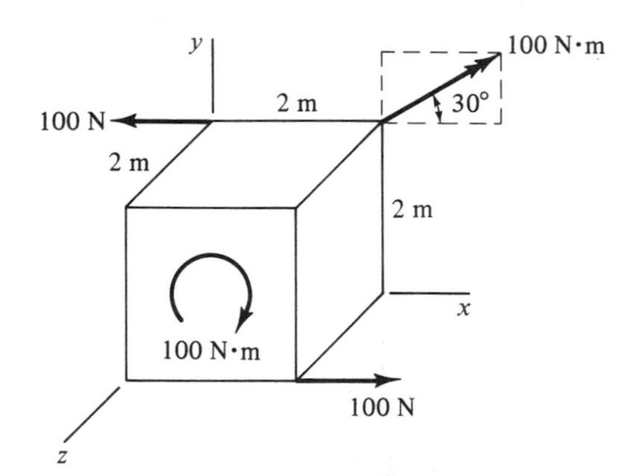

Problem 2.97

2.98. Determine the couple **C** for which the total couple acting upon the plate shown is zero.

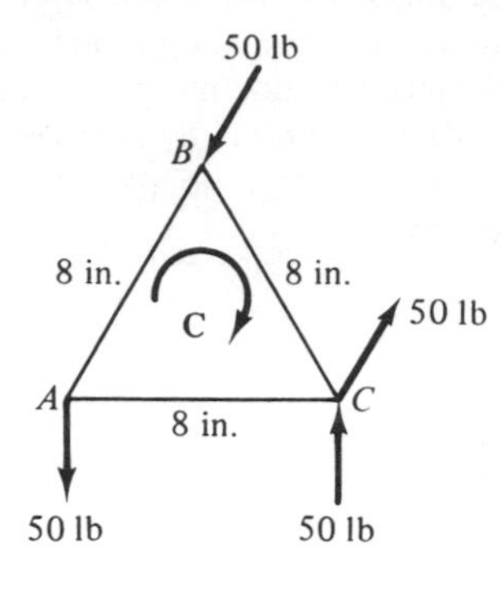

Problem 2.98

2.99. Two forces and a couple act on a plate as shown. Determine (a) the total moment about the origin, and (b) the total moment about the y axis.

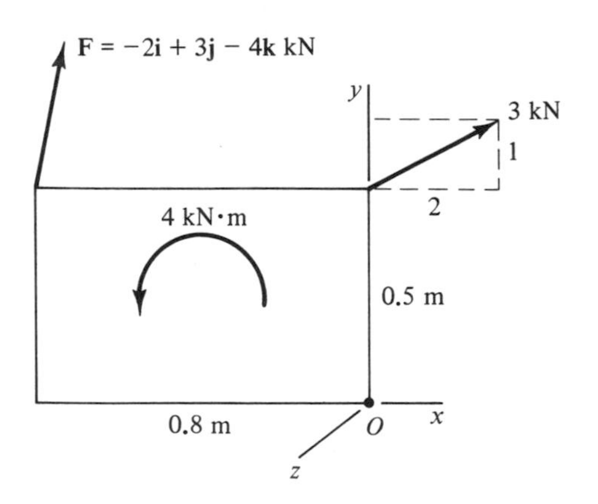

Problem 2.99

2.100. The block shown is subjected to two forces and a couple. Determine (a) the total moment about point A, and (b) the total moment about the z axis.

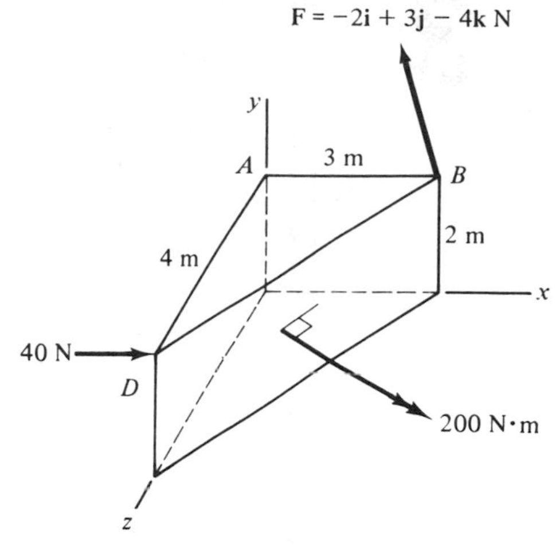

Problem 2.100

2.11 CLOSURE

In this chapter we have considered various vector operations and their application in determining certain quantities of interest. The vector approach to the solution of problems is very elegant and powerful, but as illustrated in a number of examples, it often is simpler to proceed from the basic definitions and simple geometric concepts. Hence, it is important that one learn to choose mathematical tools to fit the problem. In our work to follow we shall use the vector approach whenever it is warranted, but we shall not hesitate to drop it when it isn't.

As should be evident by now, vector algebra is primarily a convenient way of handling the geometry of a problem. The more complicated the geometry, the more useful vector algebra becomes. Also, the use of vectors and vector algebra makes it possible to state basic concepts, definitions, and equations in a general but compact form.

3

Equilibrium

3.1 INTRODUCTION

A body that is initially at rest and that remains at rest when acted upon by forces and couples is said to be in a state of *equilibrium*. For such a state to exist, the forces and couples acting must satisfy certain conditions. When these conditions are expressed in mathematical form, they can be used to determine information about these forces and couples that may not be known beforehand. The determination of such information, which we shall refer to as a *force analysis*, is a prerequisite for determining the response of a body to applied forces, or in designing it to withstand some prescribed loading.

Our goal in this chapter is to develop the equilibrium conditions and to learn how to use them in a force analysis. We begin with a discussion of Newton's laws of motion, upon which the equilibrium conditions are based.

3.2 NEWTON'S LAWS OF MOTION

Newton's Third Law

Newton's third law concerns the nature of force interactions. It states that if one body exerts a force on another, the second body exerts a force on the first which is equal in magnitude, opposite in sense, and which has the same line of action. Stated

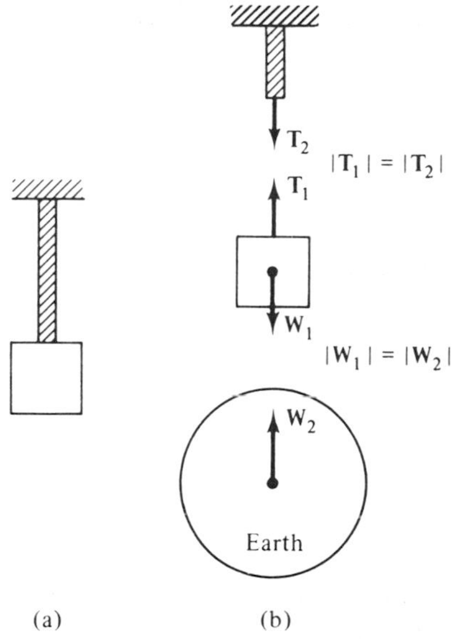

Figure 3.1

another way, we may say that for every force effect there is an equal, opposite, and collinear countereffect.

To illustrate the third law, let us consider the simple example of a block suspended from a cord [Figure 3.1(a)]. The cord prevents the block from falling down and, therefore, exerts an upward force on it, represented by $\mathbf{T}_1$ in Figure 3.1(b). The countereffect to this force is a downward force $\mathbf{T}_2$ of equal magnitude exerted on the cord by the block. There is also a gravitational interaction between the block and the earth. The force exerted on the block (its weight) is denoted by $\mathbf{W}_1$ in Figure 3.1(b). The countereffect to $\mathbf{W}_1$ is an upward force $\mathbf{W}_2$ of equal magnitude exerted on the earth by the block.

Newton's third law often seems confusing at first exposure, but it is actually very simple to apply. As indicated in Chapter 1 (Section 1.2), forces originate from an interaction between two bodies. Thus, to find the countereffect to any force, it is only necessary to determine the other body involved in that particular force interaction. In other words, we simply consider the physical origin of the force.

Most force interactions obey Newton's third law, but there are some that do not. For instance, the forces between two moving charges and between a moving charge and a magnetic field are equal and opposite, but they are not collinear. Hence, they violate the third law, at least in the form in which we have stated it.

Newton's First Law

Newton's first law states that if a body is at rest or is moving in a straight line with constant speed, the sum of the forces acting upon the body must be zero.

As an illustration of the first law, let us reconsider the block shown in Figure 3.1(b). Since the block is at rest, the sum of the forces acting upon it must be zero.

Thus, $T_1 - W_1 = 0$, or $T_1 = W_1$. In other words, the cord pulls up on the block with a force which is equal in magnitude to the force with which the earth pulls down on it. It is important to note that $\mathbf{T}_1$ is not the countereffect of $\mathbf{W}_1$ referred to in Newton's third law, even though the two forces are equal and opposite. As mentioned previously, the countereffect to $\mathbf{T}_1$ is the force $\mathbf{T}_2$ shown in Figure 3.1(b), and the countereffect to $\mathbf{W}_1$ is the force $\mathbf{W}_2$.

Newton's first law is a special case of his second law, which states that the acceleration of a body is proportional to the sum of the forces acting upon it. A body at rest or moving in a straight line with constant speed has no acceleration; thus, according to the second law, the sum of the forces acting upon the body must be zero. This is the same result stated in the first law. Since we shall not be dealing with accelerating bodies, we shall have no direct need for Newton's second law.

Newtonian Reference Frames

Newton's laws of motion are valid only in certain reference frames (coordinate systems) called *Newtonian reference frames*. Strictly speaking, coordinate systems fixed to the earth do not qualify because of the rotation of the earth about its axis and its motion about the sun. However, these effects are small and can be neglected in most instances. Hence, coordinate systems fixed to the earth are very close approximations to Newtonian reference frames, and we shall use them exclusively in this text.

PROBLEMS

3.1. Identify the forces acting upon a tennis racket when the ball is struck and show these forces on a sketch of the racket. What are the countereffects to these forces?

3.2. Identify the forces acting upon log A in the stack of firewood shown and identify the countereffects to each. Show the forces on a sketch of the log.

Problem 3.2

3.3. A bundle of pipe is held aloft by an overhead crane with the sling arrangement shown. Identify the forces acting upon the crane hook and the countereffects to each. Assume that the hook has negligible mass. If the pipe bundle weighs 10,000 lb, what is the magnitude of the tension on the crane cable?

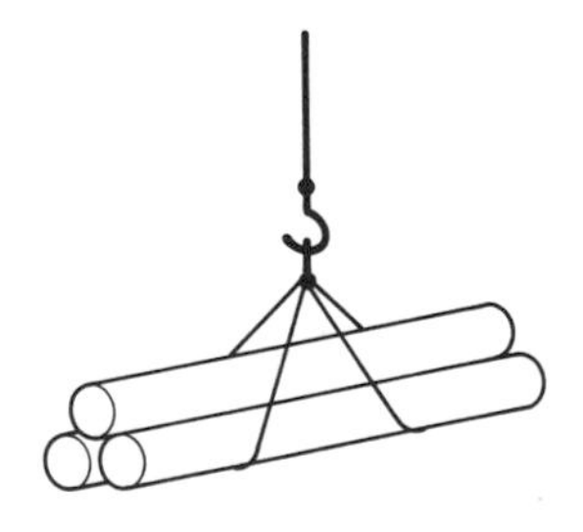

Problem 3.3

3.4. A child with a mass of 32 kg is sitting at rest in a swing, as shown. Identify the forces acting upon the child and upon the seat of the swing, and determine the countereffects to each. What is the magnitude of the force exerted upon the seat of the swing by the child?

Problem 3.4

3.3 GENERAL EQUILIBRIUM CONDITIONS

In this section, we shall consider the conditions that the forces and couples acting upon a body must satisfy in order for it to be in equilibrium.

According to Newton's first law, the sum of the forces exerted on a body at rest must be zero. Notice, however, that this law says nothing about the moments, or rotational effects, of the forces. Clearly, the total moment must also be zero, else the body would rotate.

The fundamental problem here is that Newton's first law (and second law), as originally stated, applies only for very small bodies, or *particles*, with negligible dimensions and nonzero mass. However, it can be extended to bodies of finite size as follows.

Consider a system consisting of two particles, and let $\mathbf{f}_1$ and $\mathbf{f}_2$ be the forces due to the interaction between them (Figure 3.2). These forces are called *internal*

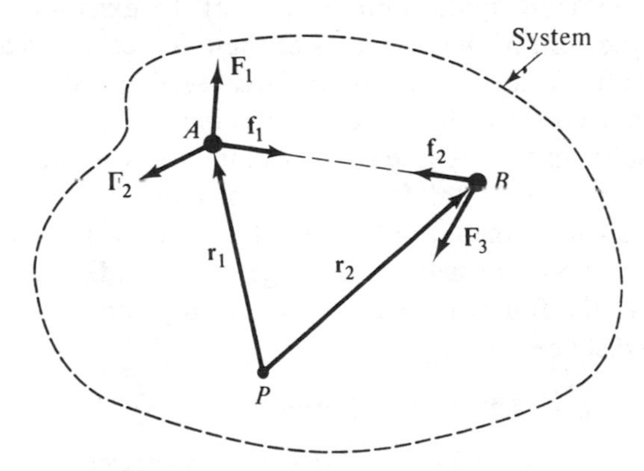

Figure 3.2

forces, since they are due to interactions between bodies *within* the system. Assuming that the internal forces obey Newton's third law, we have $\mathbf{f}_1 = -\mathbf{f}_2$. Suppose that there are also forces, such as $\mathbf{F}_1$, $\mathbf{F}_2$, and $\mathbf{F}_3$, exerted on the particles due to interactions with bodies outside the system. Such forces are called *external forces*. Clearly, all the forces acting upon a particular particle must have the same point of application because a particle has negligible dimensions.

We shall say that the system is in equilibrium if each particle within it is in equilibrium. In this case, by Newton's first law, the sum of the forces acting upon each particle must be zero. For particle A we have

$$\sum \mathbf{F}_A = \mathbf{F}_1 + \mathbf{F}_2 + \mathbf{f}_1 = \mathbf{0}$$

and for particle B

$$\sum \mathbf{F}_B = \mathbf{f}_2 + \mathbf{F}_3 = \mathbf{0}$$

The total force acting upon the system is

$$\sum \mathbf{F} = \sum \mathbf{F}_A + \sum \mathbf{F}_B = \mathbf{F}_1 + \mathbf{F}_2 + \mathbf{F}_3 + \mathbf{f}_1 + \mathbf{f}_2 = \mathbf{0}$$

Now let us consider the total moment of these forces about some point P. Referring to Figure 3.2, we have

$$\sum \mathbf{M}_P = \mathbf{r}_1 \times \left(\sum \mathbf{F}_A\right) + \mathbf{r}_2 \times \left(\sum \mathbf{F}_B\right)$$

But $\Sigma \mathbf{F}_A = \Sigma \mathbf{F}_B = \mathbf{0}$; so the total moment must also be zero, as stated previously.

Since the forces $\mathbf{f}_1$ and $\mathbf{f}_2$ have the same line of action, the moment condition can be rewritten as

$$\sum \mathbf{M}_P = \mathbf{r}_1 \times (\mathbf{F}_1 + \mathbf{F}_2 + \mathbf{f}_1 + \mathbf{f}_2) + \mathbf{r}_2 \times \mathbf{F}_3 = \mathbf{0}$$

But $\mathbf{f}_1 = -\mathbf{f}_2$; so the conditions on the forces and moments reduce to

$$\sum \mathbf{F} = \mathbf{F}_1 + \mathbf{F}_2 + \mathbf{F}_3 = \mathbf{0}$$

and

$$\sum \mathbf{M}_P = (\mathbf{r}_1 \times \mathbf{F}_1) + (\mathbf{r}_1 \times \mathbf{F}_2) + (\mathbf{r}_2 \times \mathbf{F}_3) = \mathbf{0}$$

In other words, if the system is in equilibrium, the sum of the *external* forces acting upon it is zero and so is the sum of the moments of these forces about an arbitrary point. The internal forces need not be considered because their effects cancel out.

Although we shall not go through the details, it should not be too difficult to see that the preceding results hold for a system consisting of any number of particles acted upon by any number of external forces, provided the internal forces obey Newton's third law. In particular, these results apply to bodies of finite extent, since such bodies can be thought of as consisting of a large number of very small pieces, or particles. Thus, we have the following general equilibrium conditions:

If a system is in equilibrium, then

$$\sum \mathbf{F} = \mathbf{0} \quad \text{and} \quad \sum \mathbf{M}_P = \mathbf{0} \tag{3.1}$$

where $\Sigma \mathbf{F}$ is the sum of the external forces acting upon the system and $\Sigma \mathbf{M}_P$ is the total *moment of these forces about an arbitrary point, including the moments of any couples which may be acting.*

Equations (3.1) are *necessary* conditions for equilibrium; i.e., if the system is in equilibrium, these equations must be satisfied. They are not, in general, *sufficient* conditions for equilibrium; satisfaction of these equations does not necessarily guarantee that the system will be in equilibrium. This presents no difficulties, however, for we shall be dealing only with systems known to be in equilibrium. Equations (3.1) are both necessary and sufficient conditions for equilibrium of a rigid body. Proof that they are sufficient requires use of Newton's second law and other knowledge beyond the level of this text.

It is important to note that Eqs. (3.1) hold for any system in equilibrium, regardless of the material of which it is comprised. For example, they hold for a mass of fluid at rest, as well as for solid bodies. They also apply to moving systems under certain conditions, since Newton's first law, upon which they are based, applies to particles moving with constant velocity as well as to particles at rest. For instance, Eqs. (3.1) hold for bodies that move in a straight line at constant speed without rotation and for bodies that rotate at a constant rate about a fixed axis through their mass center. Typical examples are an airplane in straight, level flight at constant speed and the pulley on an electric motor rotating at constant speed. However, problems involving motion of any kind are usually relegated to texts on dynamics.

When expressed in component form, Eqs. (3.1) yield the six scalar equations:

$$\begin{aligned} &\sum F_x = 0 \qquad \sum F_y = 0 \qquad \sum F_z = 0 \\ &\sum M_{Px} = 0 \qquad \sum M_{Py} = 0 \qquad \sum M_{Pz} = 0 \end{aligned} \tag{3.2}$$

These equations can be used in a force analysis of a system to solve for unknown information concerning the external forces and couples acting. Since there are six equations, we can generally solve for six unknowns. If all of the unknowns concerning the external forces and couples can be determined from the equilibrium equations, the problem is said to be *statically determinate*. If not, it is said to be *statically indeterminate*. For the present, we shall consider only statically determinate problems. Statically indeterminate problems will be considered in later chapters.

When there are more unknowns than equations of equilibrium in a problem, it is tempting to try to obtain additional equations by considering moments about more than one point. Unfortunately, this procedure does not work. According to Eq. (2.31), the relationship between the total moment about two points P and Q is

$$\sum \mathbf{M}_Q = \sum \mathbf{M}_P + \mathbf{r}_{QP} \times \sum \mathbf{F}_i$$

But $\Sigma \mathbf{F}_i = \mathbf{0}$ and $\Sigma \mathbf{M}_P = \mathbf{0}$ for a system in equilibrium. Thus, the condition $\Sigma \mathbf{M}_Q = \mathbf{0}$ will be satisfied automatically, and no new information can be obtained from it. Consequently, we conclude that there are no more than six independent equations of equilibrium for a given system.

3.4 MODELING OF PROBLEMS—FREE-BODY DIAGRAMS

The equilibrium equations are relatively simple to apply when the system of interest is given and the nature of the forces and couples acting upon it are known. In fact, there is little more to it than the addition of forces and calculation of moments considered in Chapter 2.

Unfortunately, the situation is not so simple in practice. We usually are confronted with a physical system, such as a machine, structure, or tool, along with some applied loadings that are either known or can be estimated. It is left up to us to (1) select the specific system of interest, (2) determine the nature of all unknown external forces and couples exerted on the system by its supports, connections, gravitational attraction, etc., and (3) make whatever assumptions are reasonable to facilitate the solution of the problem. In other words, we must select the system and model the problem before we apply the equations of equilibrium.

The specific system of interest can usually be determined from a clearly stated problem definition; some of the aspects of problem modeling are discussed in the following. Throughout our discussion, keep in mind that the basic idea is to simplify the problem as much as possible by considering only factors that have an important bearing upon the quantities to be determined and neglecting those that do not.

Free-Body Concept

Once the system to be analyzed has been selected, we imagine that it is completely isolated, or freed, from its surroundings, including any other bodies with which it may be in contact. We then draw a separate sketch showing only the system of interest and the external forces and couples acting upon it. These forces and couples include the applied loadings, the weight, and any other loadings exerted upon the system by its surroundings. All the forces and couples acting are represented by vectors, with symbols used to denote unknown quantities. The resulting sketch is called a *free-body diagram*, or FBD. This diagram focuses attention upon the particular system to be investigated and clearly indicates the forces and couples involved. In other words, the free-body diagram is the basis for the equilibrium equations.

A simple example of a free-body diagram is given in Figure 3.3. Figure 3.3(a) shows a man standing on a plank across a stream, and Figure 3.3(b) shows the corresponding free-body diagram for the plank. In this diagram the vectors $\mathbf{F}_A$ and $\mathbf{F}_B$ represent the upward forces that the banks of the stream exert on the member and $\mathbf{W}$ denotes its weight. The force $\mathbf{N}$ is exerted by the man's feet.

Since the loads applied to a body are usually known, it is no problem to include them in the free-body diagram. The major difficulty lies in deciding what other forces and couples of significance may be acting. This aspect of construction of the free-body diagram will be discussed next.

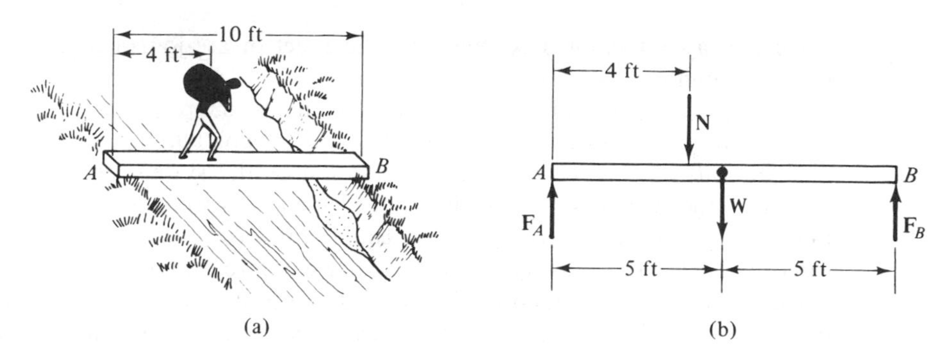

Figure 3.3

Force Interactions at a Distance-Weight

As indicated in Chapter 1, the weight of a body is the force exerted on it due to the gravitational attraction of the earth. If the mass of the body is the quantity specified, its weight can be computed from the relation $W = mg$ given in Eq. (1.1). Note that the weight will be in pounds if the mass is measured in slugs (lb $\cdot$ s^2/ft) and the acceleration of gravity in feet per second squared. If the mass is given in kilograms and the acceleration of gravity in meters per second squared, the weight will be in newtons.

Insofar as the equilibrium of a body is concerned, its weight can be considered to be a single force acting downward toward the center of the earth with point of application at the *center of gravity* G of the body. The location of the center of gravity is often obvious from the body's geometry. If the body is symmetrical and uniform throughout, its center of gravity will lie at the intersection of the planes of symmetry. For example, the center of gravity of a uniform cylinder coincides with its geometric center. The procedure for locating the center of gravity of irregular shaped bodies will be discussed in Chapter 4. In many instances, the weight of a body is small in comparison to the applied loads and can be neglected in the force analysis.

A body is actually subjected to forces due to gravitational interactions with every other body in the universe, not just with the earth. However, as long as we are considering systems located at or near the surface of the earth, these other bodies are usually so small or so far away that their gravitational effects are negligible. There may also be force interactions at a distance due to the presence of electrical or magnetic fields, but they will not be considered here.

Contact Forces–Reactions

Contact forces arise whenever a body is connected to or in contact with another body. Such forces are prevalent in engineering problems. For instance, structures and machines are held in place by various kinds of supports, and they usually consist of a number of parts that are in contact or are joined together by various types of connections.

The forces exerted on a system by contacts, supports, and connections are often distributed over an area, as in the case of a book lying on a table. These force distributions prevent translation and rotation of the system and hold it in equilibrium under the influence of the applied loads. As we shall show in Chapter 4, these effects can be represented by an equivalent force system consisting of a single force and/or a single couple. These equivalent forces and couples are called the *reactions*, and are the quantities determined in a force analysis. Their relationship to the actual distribution of force that may be acting will be examined in detail in Chapter 4.

The nature of the reactions exerted on a body by most types of contacts, connections, and supports is either intuitively obvious or can be determined with a little physical reasoning. We shall attempt to illustrate this fact in the following examples.

Suppose we wish to determine the nature of the forces that the floor exerts on the block shown in Figure 3.4(a). Since the floor prevents the block from falling through space, it obviously must exert an upward force on it. If we attempt to slide the block, there will normally be some resistance because of the friction between it and the floor. Thus, the floor can exert forces on the block which are perpendicular and parallel to the plane of contact [Figure 3.4(b)]. These forces are usually referred to as the *normal* and *friction force*, respectively. The orientation and sense of these forces are known, but their magnitudes and the location of the normal force are unknown. If the friction force is small enough to be considered negligible, we say that the bodies or surfaces in contact are *smooth*, or *frictionless*. This is an idealization; there is no such thing as a perfectly frictionless surface.

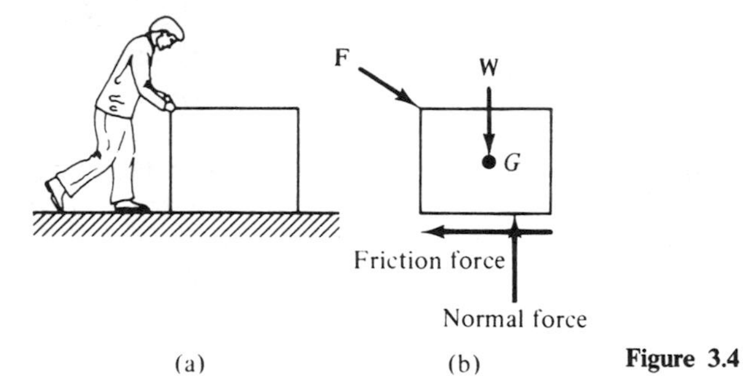

Figure 3.4

As a second example, let us consider the slender bar shown in Figure 3.5(a). The left end of the bar is connected to the wall by means of a *smooth pin*, or *ideal hinge*. Again, the concept of a frictionless hinge is an idealization. The right end of the bar is supported by a wire *BC*. We wish to determine the nature of the reactions on the bar at the supports.

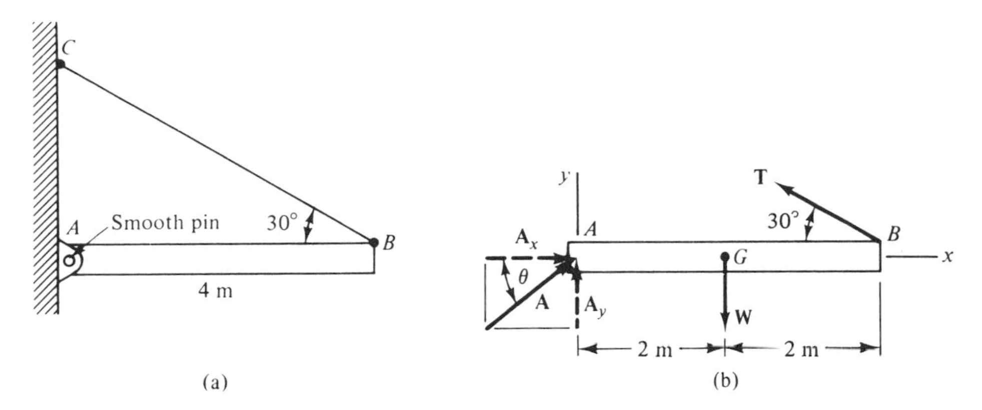

Figure 3.5

Flexible members, such as wires, ropes, strings, and cables, can only "pull" on the bodies to which they are attached. The force exerted lies along the axis of the member and is equal in magnitude to the tension on it. Thus, the wire exerts a force **T** on the bar of unknown magnitude and directed from *B* toward *C* [Figure 3.5(b)].

Since the pin at *A* will resist translation of the bar in any direction within the plane of the figure, it can exert a force **A** on the bar of unknown magnitude and direction [Figure 3.5(b)]. Alternatively, we can say that the pin exerts two unknown components of force on the bar, as indicated by the dashed arrows in the figure. If we attempt to rotate the bar about the pin, the pin will offer no resistance because it is frictionless. Thus, there is no couple exerted on the bar. The free-body diagram is completed by adding the weight **W** of the member and the dimensions. It should be noted, however, that additional reactions will be present if there are loads acting that tend to move the bar out of the plane of the page.

Our final example concerns the reactions on a post embedded in a concrete slab [Figure 3.6(a)]. Obviously, the slab will resist translation and rotation of the post in any direction. Thus, it can exert a force and couple of unknown magnitude and direction on the post. Alternatively, we can say that the slab can exert three components of force and three components of couple [Figure 3.6(b)]. In either case, there are six unknowns at the supports. A support that prevents both translation and rotation of a body is called a *fixed*, or *clamped*, *support*.

Generalizing the results obtained for the preceding examples, we can say that if a support or connection prevents or restricts the translation of a body in a given direction, it can exert a force on the body in that direction. If it prevents or restricts the rotation of the body about a particular axis, it can exert a couple on the body about that axis.

Since a force is completely defined by its three rectangular components, the nature of the force exerted on a body can be determined by pretending to attempt to move the body in three mutually perpendicular directions and observing whether

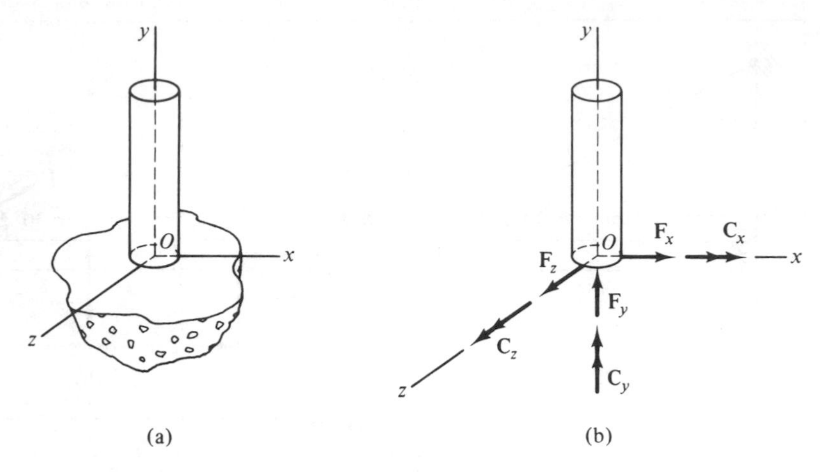

Figure 3.6

or not the support or connection would resist such motions. For two-dimensional problems, all forces lie in one plane, and only two perpendicular directions within this plane need be considered.

Similarly, the nature of the couple exerted on the body can be determined by pretending to rotate the body about three mutually perpendicular axes and observing whether or not the support or connection would resist such rotations. In two-dimensional problems, all couple vectors are perpendicular to the plane of the problem, and only rotation about an axis perpendicular to this plane need be considered.

The reactions exerted by some common types of supports and connections are shown in Figure 3.7. Study these carefully and see if you can verify the results presented by considering the motions that each support will or will not resist. These results should not be memorized. The point we have been trying to make is that the nature of the reactions can be deduced from simple physical reasoning.

Many of the supports and connections encountered in practice do not correspond exactly to any one of the idealizations shown in Figure 3.7. For example, Figure 3.8 shows a structure comprised of three boards nailed together. The connections at A and B will resist, but not completely prevent, rotation. Thus, they are not exactly equivalent to fixed supports, nor are they equivalent to pinned joints. It is a matter of engineering judgment as to how these connections should be modeled. Assuming that only a few nails are used at each connection, the resistance to rotation will likely be small. In this case, it would be reasonable to treat the connections as pinned joints.

Rigid and Deformable Bodies

When loads are applied to a body, the body will deform. Strictly speaking, the equilibrium equations apply only after all deformation has ceased and every part of

Support or connection	Reactions	
	In-plane loadings	Three-dimensional loadings
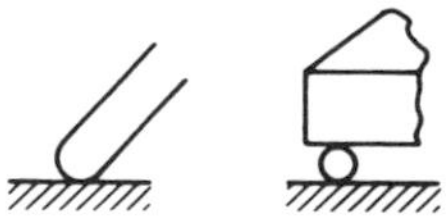 Smooth surface or ball	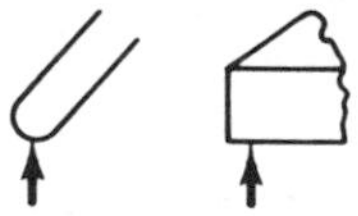 Single force normal to surface (1 unknown)	Same as in-plane case
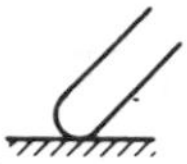 Rough surface	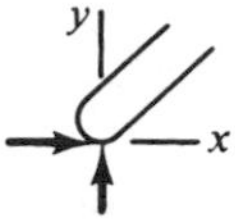 Normal force and friction force (2 unknowns)	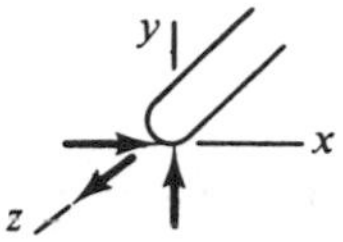 Normal force and two comp. friction (3 unknowns)
 Roller or rocker	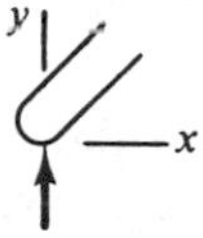 Single force normal to surface (1 unknown)	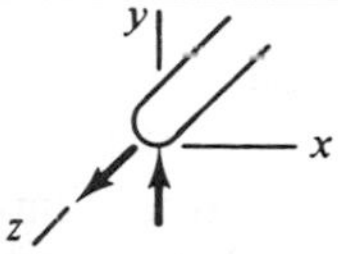 Normal force and friction force (2 unknowns)
 Cable, string, wire, thin rod, etc.	 Single force along cable (1 unknown)	Same as in-plane case
 Smooth pin or hinge	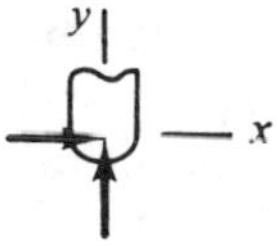 Two components of force (2 unknowns)	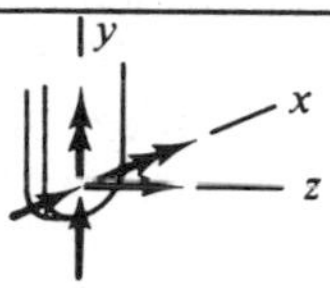 3 components of force and 2 of couple (5 unknowns)
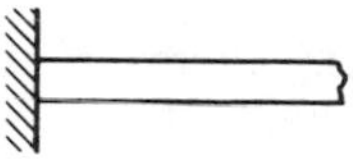 Fixed support	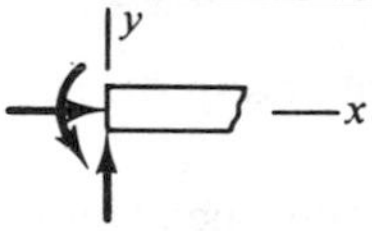 2 components of force and 1 of couple (3 unknowns)	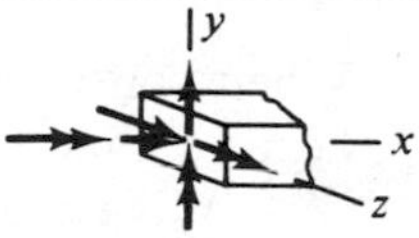 3 components of force and 3 of couple (6 unknowns)
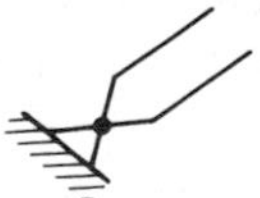 Ball and socket	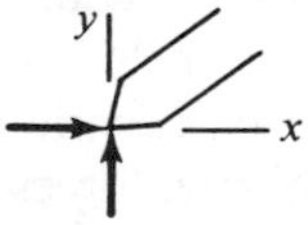 2 components of force (2 unknowns)	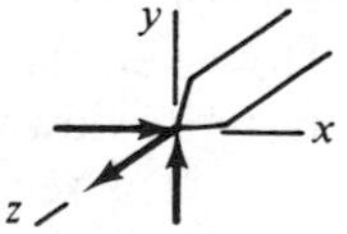 3 components of force (3 unknowns)

Figure 3.7

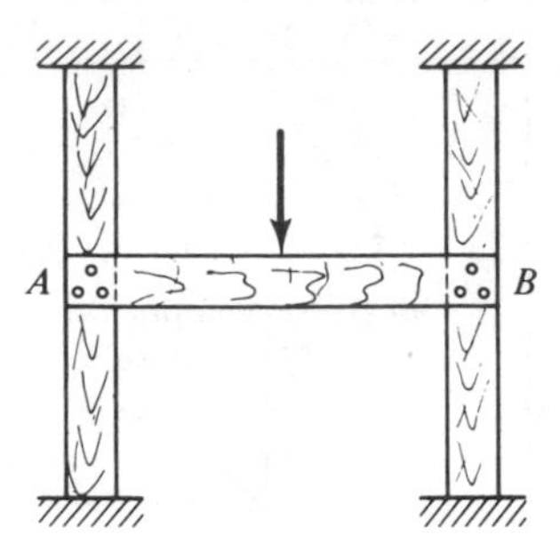

Figure 3.8

the body has come to rest. Hence, the force analysis should be based upon the dimensions and geometry of the deformed body. In most problems, however, the deformations are very small compared to the original dimensions of the body, and little error is introduced if the force analysis is based upon the undeformed geometry.

It is no coincidence that the deformations of most engineering structures are relatively small. These structures are purposely designed this way. For example, a highway bridge that sags several feet when vehicles cross it would be of little value, even though it may be in no danger of breaking.

If the deformations of a body are neglected when performing a force analysis, this is the same as assuming the body is rigid. The concept of a *rigid body* is widely used in mechanics, but it is an idealization. All bodies deform to some extent when loaded.

Most of the problems considered in this text are such that the rigid-body assumption can be used in the force analyses. There are, however, problems for which it is essential to base the equilibrium equations upon the deformed geometry of the body. These problems will be considered in Chapter 10. Due to its importance, the rigid-body assumption will be considered in more detail in Section 3.11, after we have had a chance to make use of it in the solution of some problems.

Free-Body Diagrams

When constructing the free-body diagram of a system, there are several points to keep in mind.

It is important that only the external forces and couples exerted on the system be shown on the free-body diagram. Forces and couples that are internal to the system, that the system exerts on other bodies, or that can be considered negligible do not enter into the equilibrium equations for the system and should not be shown. However, the diagram should include dimensions and any other information needed to write down the equilibrium equations.

The sense of any unknown forces and couples can be chosen arbitrarily on the free-body diagram. If a positive value is obtained for these quantities, their sense is as assumed. If their values turn out to be negative, their sense is opposite to that assumed. In contrast, all known forces and couples must be shown on the free-body diagram with their proper sense. Otherwise, a sign error will result.

There are two ways of representing forces and couples on the free-body diagram. They can be shown as vectors, which are then defined by appropriate mathematical expressions, or they can be represented by their projections, with their orientation and sense defined by the figure of the free-body diagram (as in Example 3.1). The representation in terms of projections is usually the most convenient, except in some three-dimensional problems. We shall use it in most instances.

The ideas presented in this section regarding problem modeling and construction of the free-body diagram are further illustrated in the following examples.

Example 3.1. Free-Body Diagram of a Ladder. The ladder shown in Figure 3.9(a) has a mass of 25 kg and is held in place by a rope attached to the wall. The wall and floor are smooth. Draw an FBD of the ladder.

Solution. We first isolate the ladder [Figure 3.9(b)]. The forces acting upon the ladder are its weight, the reactions at the points of contact with the wall and floor, and the force due to the rope. The smooth wall and floor exert forces normal to the plane of contact, which we denote by B_x and A_y, respectively. The rope exerts a force along its axis, which we denote by T. From Eq. (1.1), the weight of the ladder is

$$W = mg = (25 \text{ kg})(9.81 \text{ m/s}^2) = 245 \text{ N}$$

We assume that the ladder is uniform; therefore, its center of gravity is located at its midpoint. The complete FBD is as shown.

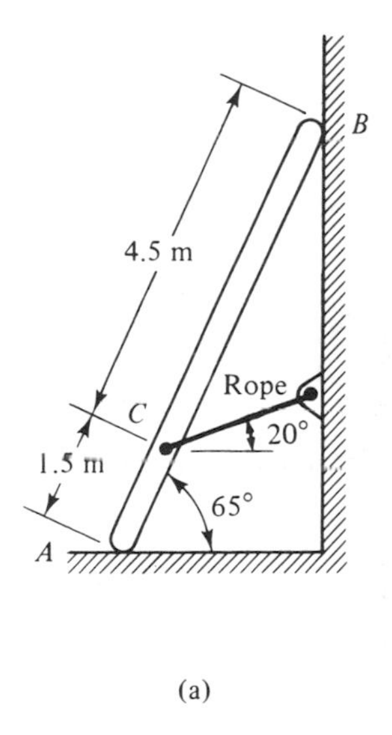

(a)

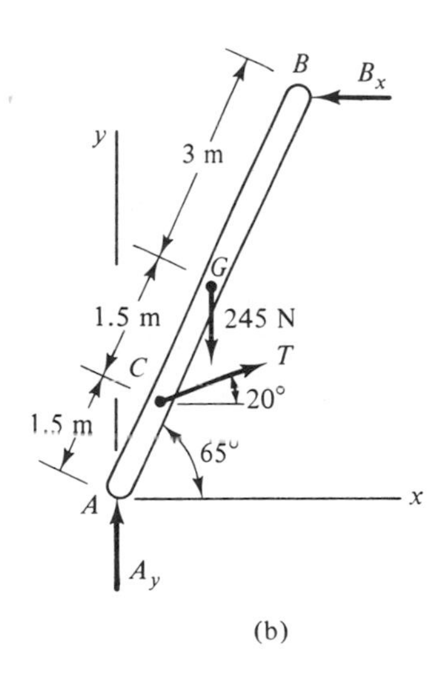

(b)

Figure 3.9

Example 3.2. Free-Body Diagrams of a Pulley and Support. A block is suspended from a rope passing over a pulley of radius r, as shown in Figure 3.10(a). The bearing at O is frictionless, and the pulley, rope, and support have negligible weight. Draw an FBD of the support OA and of the system consisting of the pulley and rope. Show that the tension on the rope has the same magnitude on both sides of the pulley.

Solution. First, we isolate the rope and pulley [Figure 3.10(b)]. The support at O prevents horizontal and vertical motion of the pulley and, therefore, can exert two components of force on it. Since these forces are unknowns, their sense can be chosen arbitrarily. There is no couple acting on the pulley at O because the frictionless bearing offers no resistance to rotation. The tensions on the rope lie along the axis of the member. The resulting FBD of the system is as shown.

Next, we isolate member OA [Figure 3.10(c)]. The fixed support at A prevents horizontal and vertical motion of the member, as well as rotation. Consequently, it can exert two components of force and one component of couple. Again, the sense of these unknown reactions has been chosen arbitrarily. The forces at O are obtained from the FBD of the pulley and Newton's third law. Since the forces O_x and O_y acting upon the pulley are exerted by the support, the pulley will exert equal and opposite forces upon the support at this location. Consequently, the FBD of the support is as shown. Note carefully the use of

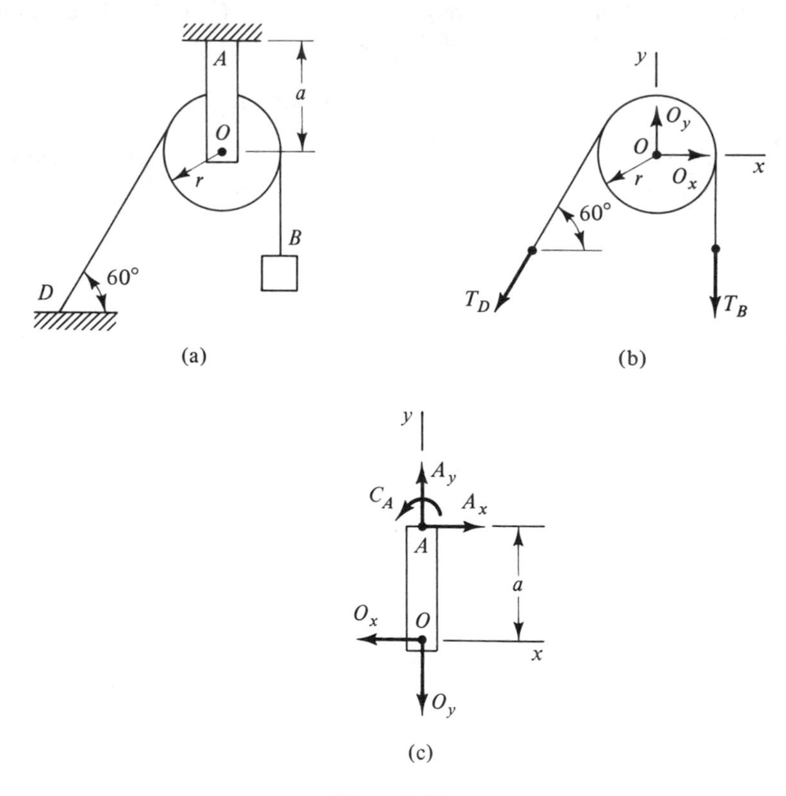

Figure 3.10

Newton's third law at point O. The importance of this law in constructing free-body diagrams cannot be overstated.

The equilibrium requirement that $\Sigma \mathbf{M}_O = \mathbf{0}$ for the FBD in Figure 3.10(b) yields the result

$$\sum \overset{+}{\curvearrowleft} M_O = T_D r - T_B r = 0$$

or

$$T_D = T_B \qquad \textbf{Answer}$$

Thus, the magnitude of the tension on the rope is the same on both sides of the pulley. This same result is obtained in every case of a rope or other flexible member of negligible weight passing over a smooth surface or a pulley with a frictionless support.

PROBLEMS

3.5. to 3.9. Draw a complete free-body diagram for the body, member, or structure shown. If no value is given for the mass or weight, assume that it is negligible.

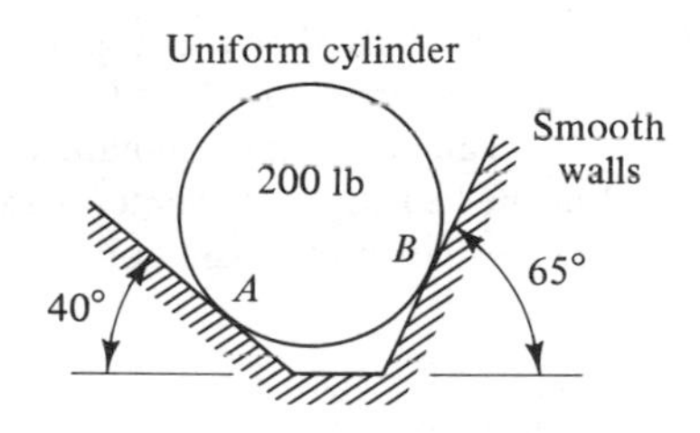

Problem 3.5

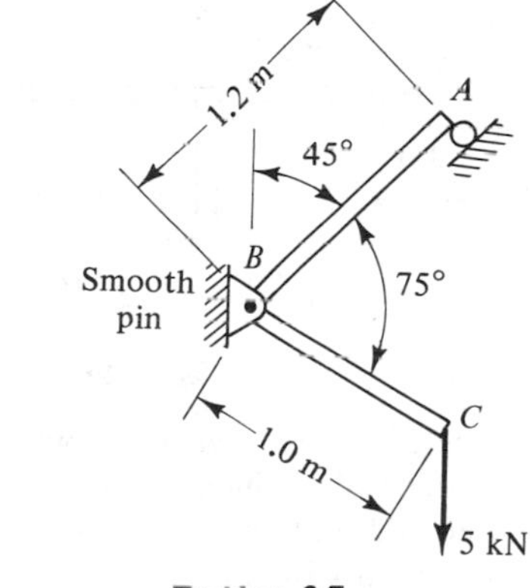

Problem 3.7

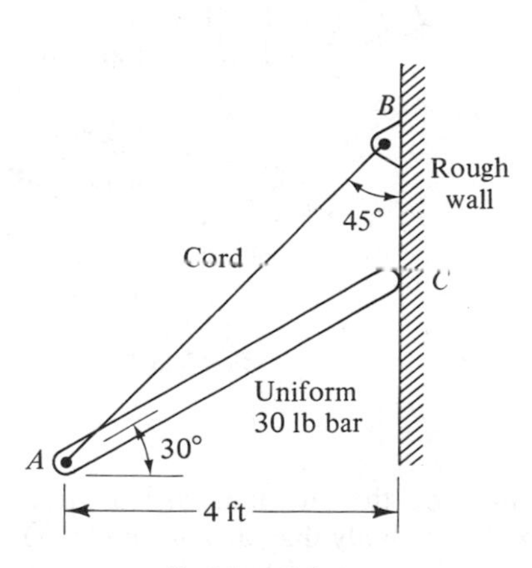

Problem 3.6

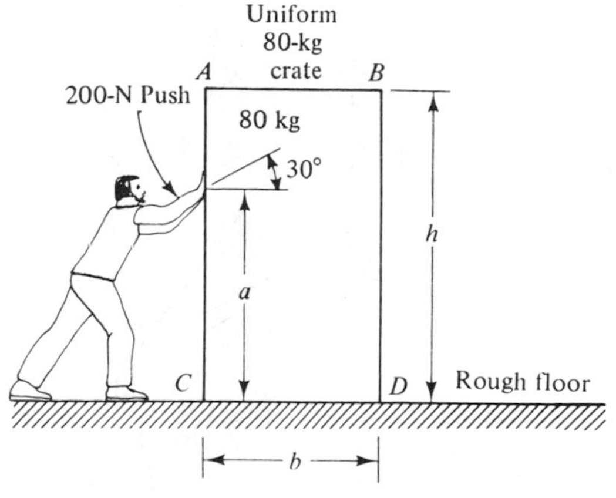

Problem 3.8

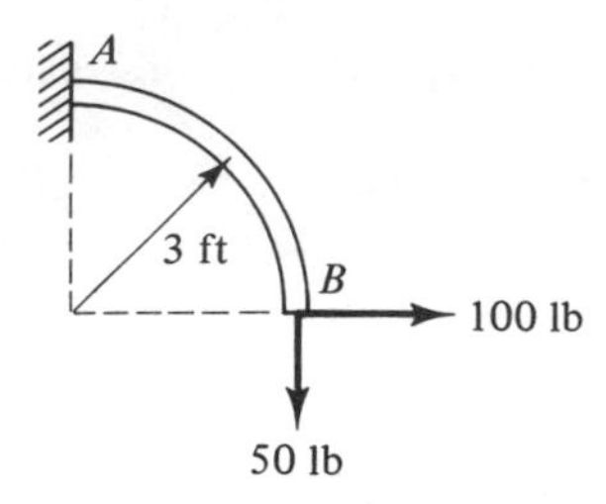

Problem 3.9

3.10. A frame with negligible weight is supported by a smooth pin at A and a rope that passes over a frictionless pulley at C, as shown. Draw an FBD of the frame.

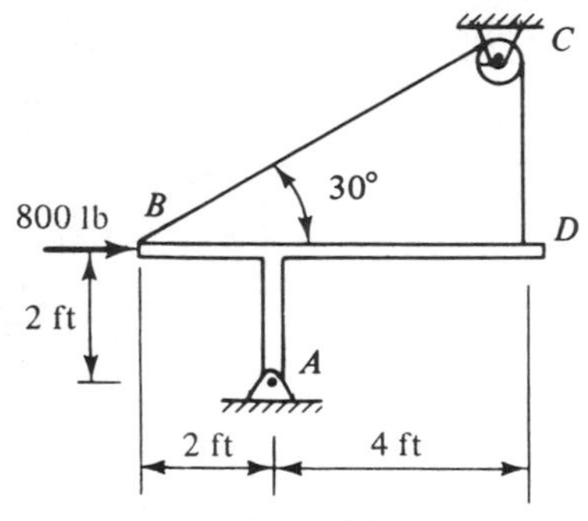

Problem 3.10

3.11. The 20-kg uniform bar shown is held in a horizontal position by a smooth hinge at one end and an inclined wire at the other. Draw an FBD for the bar.

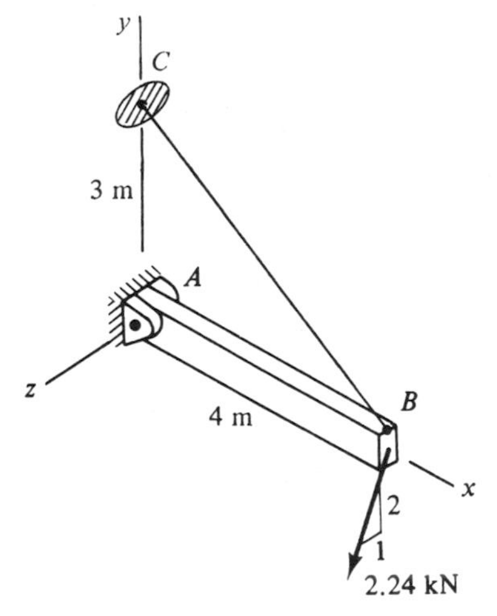

Problem 3.11

3.12. The uniform rod AB shown has a mass of 50 kg. It is supported by a ball and socket joint at B and a string DC attached to its midpoint C. The wall at A is smooth. Draw an FBD of the rod.

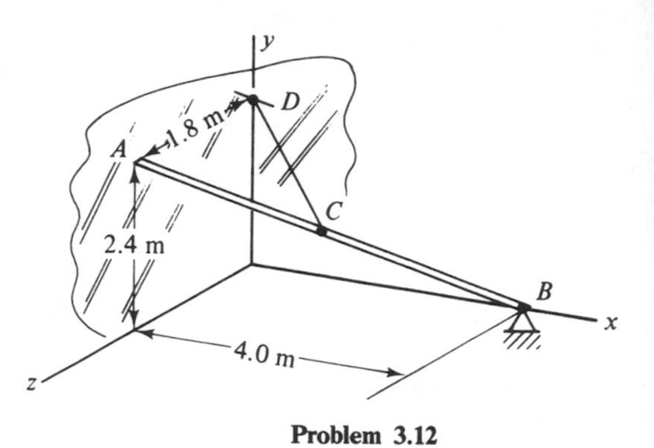

Problem 3.12

3.13. The line shaft shown is supported by frictionless bearings at A and B. The bearing at A can resist axial motion of the shaft, but the bearing at B cannot. Draw an FBD of the shaft-pulley assembly.

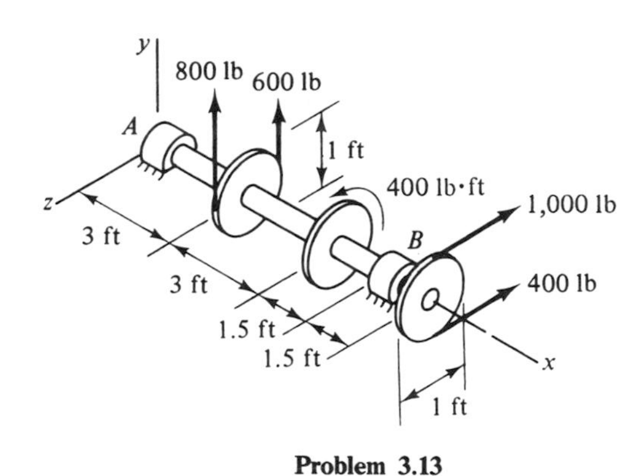

Problem 3.13

3.14. to 3.16. For the situation shown, draw complete free-body diagrams for body ① and body ②. If no value is given for the mass or weight, it may be assumed to be negligible.

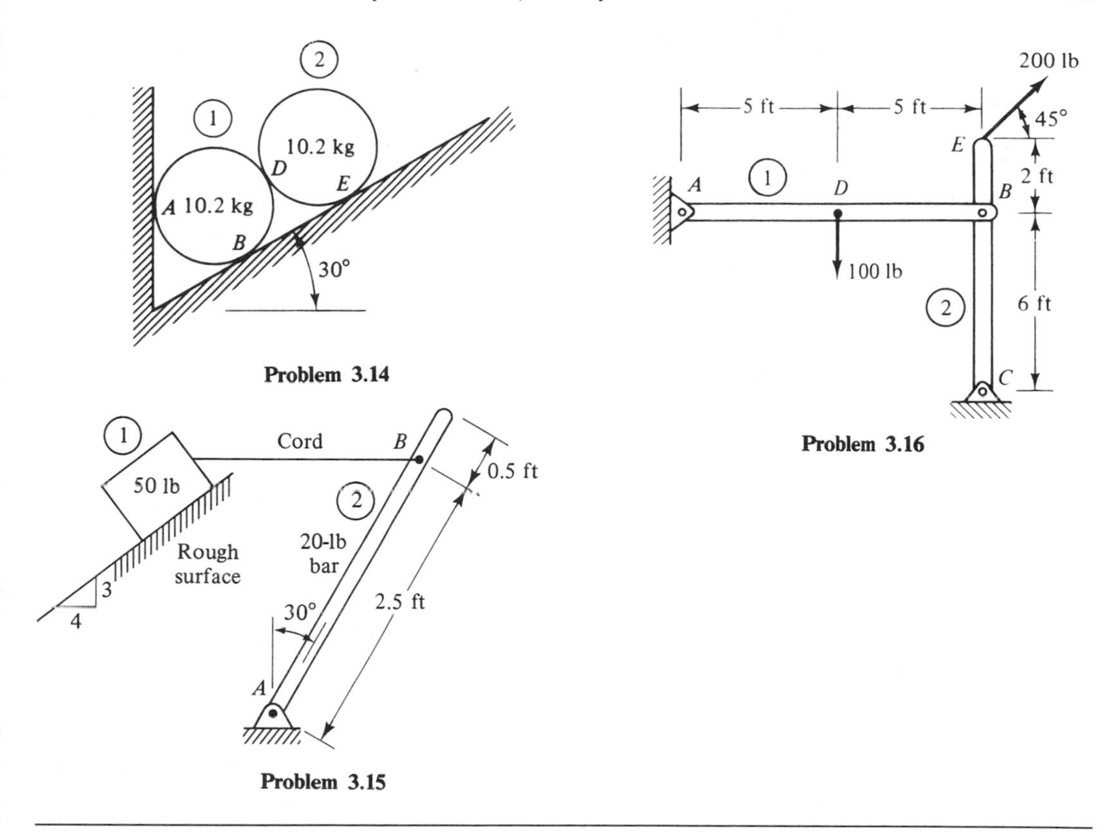

Problem 3.14

Problem 3.15

Problem 3.16

3.5 FORCE ANALYSIS OF SOME SPECIAL SYSTEMS

As indicated in Section 3.3, there are no more than six independent equations of equilibrium for a given system. There are, however, cases in which there are fewer than six independent equations because some of them are satisfied automatically. Recognition of these cases usually will simplify the force analysis. Several of the special types of systems commonly encountered are considered in this section, and the procedures for their force analysis are illustrated in example problems. First, however, a word of caution.

It often is tempting to try to take shortcuts in a force analysis by omitting the free-body diagram and writing down the equations of equilibrium directly from the figure in the problem statement. Do not get into this habit! Although this procedure may work for simple problems, it invariably results in confusion and errors when applied to more complex problems. A well-organized and systematic plan of attack is the key to success in a force analysis. Since the organization is provided by the free-body diagram, its importance cannot be overstated.

Concurrent Force Systems

Recall from Section 2.2 that the lines of action of concurrent forces all intersect at a common point. In order for a system of such forces to be in equilibrium, it is only necessary that $\Sigma\mathbf{F} = \mathbf{0}$. The moment equation of equilibrium, $\Sigma\mathbf{M}_p = \mathbf{0}$, is satisfied automatically. This can be readily seen by considering moments about the point of concurrency. Thus, there are at most three independent equations of equilibrium for concurrent force systems:

$$\sum F_x = 0 \qquad \sum F_y = 0 \qquad \sum F_z = 0 \tag{3.3}$$

Coplanar Force Systems

Coplanar forces all lie in the same plane, which is the same as saying that the problem is two dimensional. Let us arbitrarily choose coordinates such that the forces lie in the *x*-*y* plane. Since the forces have no *z* component, the equilibrium equation $\Sigma F_z = 0$ is satisfied automatically. Moments about a point *P* in the *x*-*y* plane will have only a *z* component; so the total moment is $\Sigma\mathbf{M}_P = \Sigma M_{Pz}\mathbf{k}$. Thus, the equations $\Sigma M_{Px} = 0$ and $\Sigma M_{Py} = 0$ are also satisfied, and the equation $\Sigma M_{Pz} = 0$ can be written simply as $\Sigma M_P = 0$. This leaves three independent equations of equilibrium:

$$\sum F_x = 0 \qquad \sum F_y = 0 \qquad \sum M_P = 0 \tag{3.4}$$

These equations correspond to the sum of forces in two perpendicular directions within the plane equal to zero and the sum of moments about some point in the plane equal to zero.

It sometimes is convenient to replace one or both of the force equations of equilibrium in Eqs. (3.4) with additional moment equations. This can be done, provided we still end up with three independent equations of equilibrium.

For instance, suppose that in Eqs. (3.4) we wish to replace the condition $\Sigma F_y = 0$ with $\Sigma M_Q = 0$, where *Q* is a second point in the plane of the forces. This new set of equations will be valid if it can be shown to be equivalent to the original set.

From Eq. (2.31), we have

$$\sum \mathbf{M}_Q = \sum \mathbf{M}_P + \mathbf{r}_{QP} \times \left[\left(\sum F_x\right)\mathbf{i} + \left(\sum F_y\right)\mathbf{j}\right]$$

Since $\Sigma\mathbf{M}_P = \mathbf{0}$, $\Sigma\mathbf{M}_Q = \mathbf{0}$, and $\Sigma F_x = 0$, this expression reduces to

$$\mathbf{r}_{QP} \times \left(\sum F_y\right)\mathbf{j} = \mathbf{0}$$

which is equivalent to $\Sigma F_y = 0$ if $\mathbf{r}_{QP}$ and $\mathbf{j}$ are not parallel. Interpreted geometrically, this means that points *P* and *Q* cannot lie on a line that is perpendicular to the *x* direction. If this condition is satisfied, Eqs. (3.4) can be replaced with the equations

$$\sum F_x = 0 \qquad \sum M_P = 0 \qquad \sum M_Q = 0 \tag{3.5}$$

where P and Q are two points in the plane of the forces. In applying these results, it is important to remember that the x direction can be chosen arbitrarily.

Equations (3.4) can also be replaced by three moment equations. Let P, Q, and S be points in the plane of the forces. From Eq. (2.31), we have

$$\sum \mathbf{M}_Q = \sum \mathbf{M}_P + \mathbf{r}_{QP} \times \sum \mathbf{F} = \mathbf{0}$$

$$\sum \mathbf{M}_S = \sum \mathbf{M}_P + \mathbf{r}_{SP} \times \sum \mathbf{F} = \mathbf{0}$$

If we take $\Sigma \mathbf{M}_P = \mathbf{0}$, the first of these equations implies that either $\mathbf{r}_{QP}$ is parallel to $\Sigma \mathbf{F}$ or $\Sigma \mathbf{F} = \mathbf{0}$. Similarly, the second equation implies that either $\mathbf{r}_{SP}$ is parallel to $\Sigma \mathbf{F}$ or $\Sigma \mathbf{F} = \mathbf{0}$. These two equations are, therefore, equivalent to $\Sigma F_x = 0$ and $\Sigma F_y = 0$ if $\mathbf{r}_{QP}$ and $\mathbf{r}_{SP}$ are not both parallel to $\Sigma \mathbf{F}$. Interpreted geometrically, this means that the points P, Q, and S cannot all lie on the same line. Under this condition, Eqs. (3.4) can be replaced by the equations

$$\sum M_P = 0 \qquad \sum M_Q = 0 \qquad \sum M_S = 0 \tag{3.6}$$

where P, Q, and S are three points in the plane of the forces.

Equations (3.4), (3.5), and (3.6) are equivalent under the conditions stated, and the particular set of equations used in any given problem is strictly a matter of convenience and personal preference.

Two-Force Systems

A *two-force system*, or *member*, is one that is subjected to external forces at only two points (and no couples). Cables, ropes, wires, and rods that are loaded at their ends and that have negligible weight are examples of two-force members commonly encountered in practice.

In order for a two-force system to be in equilibrium, the forces acting upon it must be equal and opposite and lie along a line connecting their points of application (Figure 3.11). Obviously, the forces must be equal and opposite in order to have $\Sigma \mathbf{F} = \mathbf{0}$. Furthermore, they must have the same line of action in order to satisfy the condition $\Sigma \mathbf{M}_P = \mathbf{0}$. This can be readily seen from Figure 3.11 by

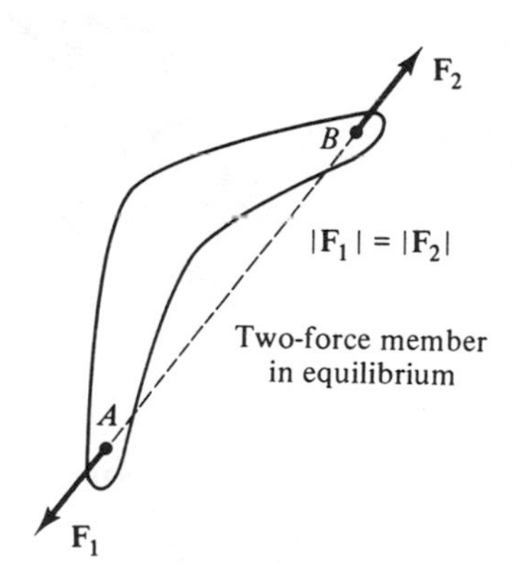

Figure 3.11

considering moments about the point of application of one of the forces. If there are several forces acting at each of the points of application, they can be added to obtain a single force at each point. The preceding statements concerning the equilibrium of the system then apply.

If the forces acting upon a straight two-force member tend to stretch it, we say that the member is in *tension*. If the forces tend to compress it, we say that the member is in *compression*.

Three-Force Systems

A *three-force system*, or *member*, is one that is subjected to external forces at three different points. In order for such a system to be in equilibrium, the forces must be coplanar and either concurrent or parallel.

Consider a system subjected to three forces, $\mathbf{F}_1$, $\mathbf{F}_2$, and $\mathbf{F}_3$. For equilibrium, $\mathbf{F}_1 + \mathbf{F}_2 + \mathbf{F}_3 = \mathbf{0}$, which implies that the vector polygon for the forces is a closed triangle. Interpreted geometrically, this means that the three forces must lie in the same plane. If we assume that the forces are not all parallel, the lines of action of two of them will intersect at some point P [Figure 3.12(a)]. If we take moments about this point, it is clear that the line of action of the third force must also pass through this point in order to satisfy the equilibrium condition $\Sigma\mathbf{M}_P = \mathbf{0}$. Thus, the forces must be concurrent. The only exception is when all three of the forces are parallel, as in Figure 3.12(b). If there are several forces acting at each of the points of application, they can be added to obtain a single force at each point. The preceding statements concerning the equilibrium of the system then apply.

Once it is recognized that the forces acting upon a three-force system must be coplanar and either concurrent or parallel, we are left with two independent

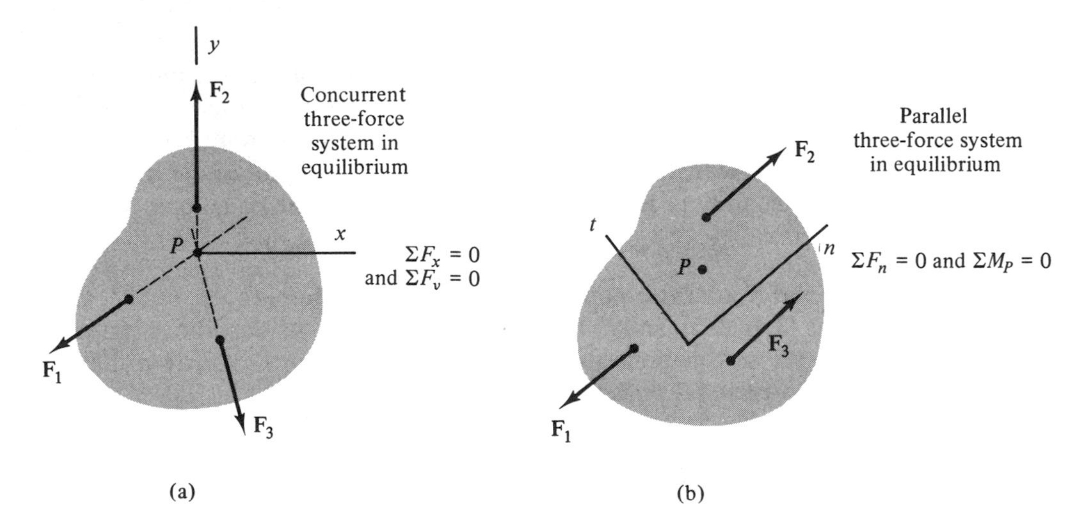

Figure 3.12

equations of equilibrium. If the forces are concurrent, these equations correspond to summation of forces in two perpendicular directions within the plane equal to zero. [Figure 3.12(a)]. If the forces are parallel, we have summation of forces in one direction within the plane and summation of moments about some point in the plane equal to zero [Figure 3.12(b)]. In many problems, however, it is just as convenient to make use only of the fact that the forces must be coplanar. In this case, there are three independent equations of equilibrium, and Eqs. (3.4), (3.5), or (3.6) apply.

Example 3.3. Force Analysis of Ropes. A block with weight W is supported by two ropes, as shown in Figure 3.13(a). Determine the tension on the ropes.

Solution. We take the block as our system, but there are other possibilities. For example, we could choose a system consisting of the block and both ropes. The FBD of the block is as shown in Figure 3.13(b). The forces T_{AB} and T_{BC} shown are those which the ropes exert on the block; by Newton's third law, they are equal and opposite to the forces the block exerts on the ropes.

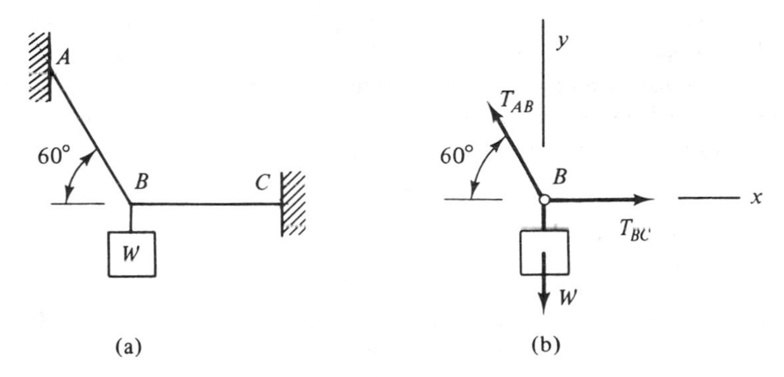

Figure 3.13

Since the forces on the block are coplanar and concurrent, there are two independent equations of equilibrium. Taking coordinates as shown, we have

$$\sum F_x = T_{BC} - T_{AB}\cos 60° = 0$$

$$\sum F_y = T_{AB}\sin 60° - W = 0$$

Solving these equations, we get

$$T_{AB} = 1.15W \qquad T_{BC} = 0.58W \qquad \textbf{Answer}$$

The orientation and sense of the tensions are defined by the FBD. This problem can also be solved by constructing the vector polygon and determining the magnitudes of the tensions graphically or by using trigonometry.

Example 3.4. Force Analysis of a Lawn Roller. The lawn roller shown in Figure 3.14(a) has a weight of 200 lb with center of gravity at point O. Determine the pull P that the man must exert parallel to the handle to prevent the roller from rolling down the incline.

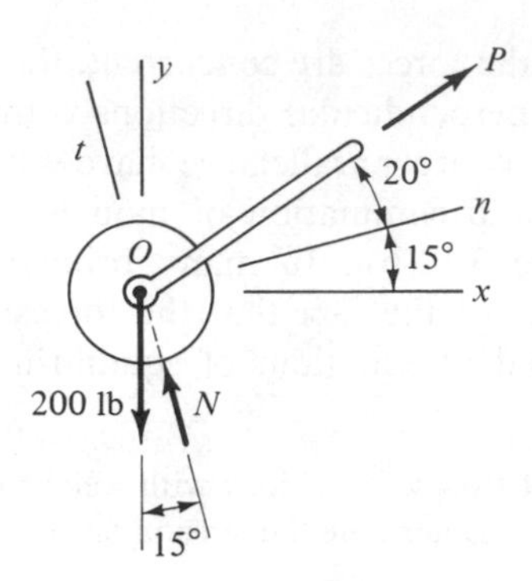

Figure 3.14

Solution. The FBD of the roller is as shown in Figure 3.14. Note that the roller is a three-force member; therefore, the force N exerted by the incline must be oriented such that its line of action passes through point O. Since the forces acting are concurrent and coplanar, there are two equilibrium equations, $\sum F_x = 0$ and $\sum F_y = 0$, from which to determine the two unknowns P and N.

The coordinate system used when applying the equilibrium equations is arbitrary, but a judicious choice of axes can often simplify the computations. To illustrate this, let us consider one set of axes, n and t, which are parallel and perpendicular to the plane, and another set, x and y, which are horizontal and vertical.

For the first set of axes, we have

$$\sum F_n = P\cos 20° - (200\text{ lb})\sin 15° = 0$$

from which we obtain

$$P = 55\text{ lb} \qquad \textbf{Answer}$$

The second equilibrium equation, $\sum F_t = 0$, is not needed in this problem.

For the second set of axes, we have

$$\sum F_x = P\cos 35° - N\sin 15° = 0$$

$$\sum F_y = P\sin 35° + N\cos 15° - 200\text{ lb} = 0$$

These two equations yield the same result as our first equation, but their solution requires more work.

Example 3.5. Force Analysis of a Door. The uniform door shown in Figure 3.15(a) has a weight of 60 lb and is supported by hinges at points A and B. The construction of the hinges is such that only the bottom one can exert a force in the vertical direction. Determine the reactions on the door at the hinges.

Solution. A FBD of the door is shown in Figure 3.15(b). Although the hinges can exert couples about axes parallel to the x and z axes, there is very little tendency for the door to rotate about such axes under the given loading. Thus, any couples exerted by the hinges will be negligible, and they have been omitted from the FBD. Similarly, there is no tendency for the door to move normal to the plane of the page; so the z components of the hinge forces have also been omitted.

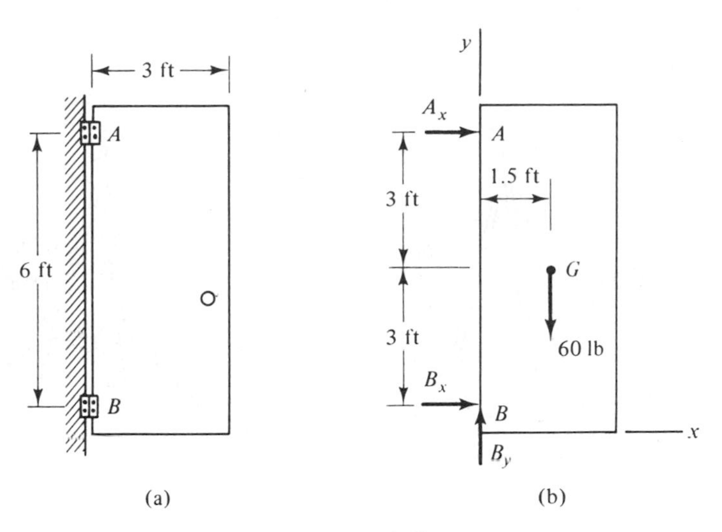

Figure 3.15

Since the forces acting upon the door are coplanar, there are three independent equations of equilibrium from which to determine the three unknowns A_x, B_x, and B_y. We have

$$\sum F_x = B_x + A_x = 0$$

$$\sum F_y = B_y - 60 \text{ lb} = 0$$

$$\sum \overset{+}{\curvearrowleft} M_A = B_x(6 \text{ ft}) - (60 \text{ lb})(1.5 \text{ ft}) = 0$$

Solving these equations, we get

$$B_x = 15 \text{ lb} \quad B_y = 60 \text{ lb} \quad A_x = -15 \text{ lb} \qquad \textbf{Answer}$$

The negative value of A_x means that the horizontal force on the door at A has a sense opposite to that shown in the figure. The point about which moments are computed is, of course, arbitrary. The idea is to try to select the point that most simplifies the resulting calculations.

This problem can also be solved by summing forces in one direction, say the y direction, and summing moments about two points, such as A and B, which do not lie along a line perpendicular to the direction in which the forces are summed [see Eqs. (3.5)]. A third approach is to sum moments about three points, such as A, B, and G, which do not all lie on the same line [see Eqs. (3.6)]. It is left an exercise to show that these approaches lead to the same results obtained previously.

Example 3.6. Force Analysis of a Tripod. The tripod shown in Figure 3.16(a) supports a 20-kg surveying instrument. The legs of the tripod are equally spaced and hinged at the top. If the mass of the tripod is negligible compared to that of the instrument, determine the reactions on the legs at the ground.

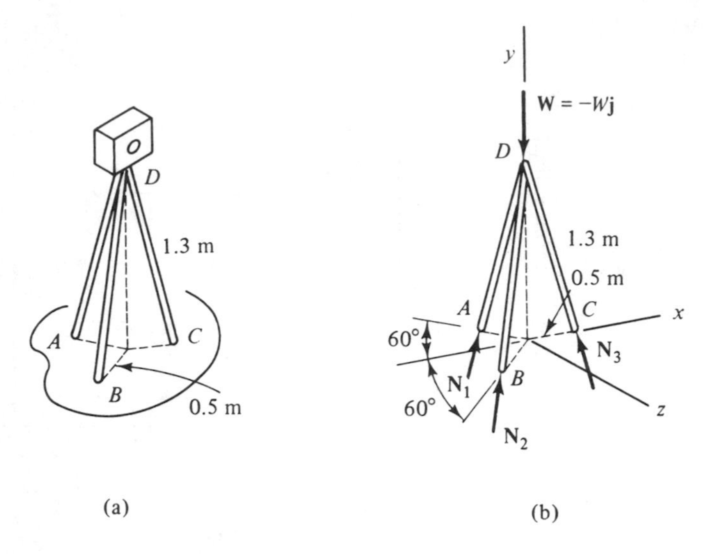

Figure 3.16

Solution. The legs of the tripod are two-force members because they have negligible weight and are loaded only at the top and bottom. Thus, the forces exerted on the legs by the ground must lie along the axes of the legs, as indicated in the FBD of Figure 3.16(b).

Taking coordinates as shown, we find from the geometry of the FBD

$$\mathbf{r}_{AD} = 0.25\mathbf{i} + 1.20\mathbf{j} + 0.43\mathbf{k}$$

and

$$\mathbf{N}_1 = N_1 \frac{\mathbf{r}_{AD}}{r_{AD}} = N_1 \frac{(0.25\mathbf{i} + 1.20\mathbf{j} + 0.43\mathbf{k})}{1.30}$$

Similarly,

$$\mathbf{N}_2 = N_2 \frac{(0.25\mathbf{i} + 1.20\mathbf{j} - 0.43\mathbf{k})}{1.30}$$

$$\mathbf{N}_3 = N_3 \frac{(-0.50\mathbf{i} + 1.20\mathbf{j})}{1.30}$$

Since the forces on the tripod are concurrent, the equilibrium equation is

$$\sum \mathbf{F} = \mathbf{N}_1 + \mathbf{N}_2 + \mathbf{N}_3 + \mathbf{W} = \mathbf{0}$$

or, in component form,

$$\sum F_x = 0.25N_1 + 0.25N_2 - 0.50N_3 = 0$$

$$\sum F_y = 1.20N_1 + 1.20N_2 + 1.20N_3 - 1.30W = 0$$

$$\sum F_z = 0.43N_1 - 0.43N_2 = 0$$

The weight of the instrument is

$$W = mg = (20 \text{ kg})(9.81 \text{ m/s}^2) = 196.2 \text{ N}$$

Substituting this value into the preceding equations and solving, we obtain

$$N_1 = N_2 = N_3 = 70.9 \text{ N} \qquad \textbf{Answer}$$

Even though this problem is three-dimensional, the geometry is sufficiently simple that a solution can be easily obtained without use of vector algebra. This can be done by first resolving the forces into horizontal and vertical components and then considering side and top views of the tripod. Application of this procedure is left as an exercise.

PROBLEMS

3.17. A 12-ft length of steel pipe is held aloft by a crane, as shown. Determine the tension on the cable sling ACB if the pipe weighs 600 lb and the cable length is (a) 16 ft and (b) 20 ft.

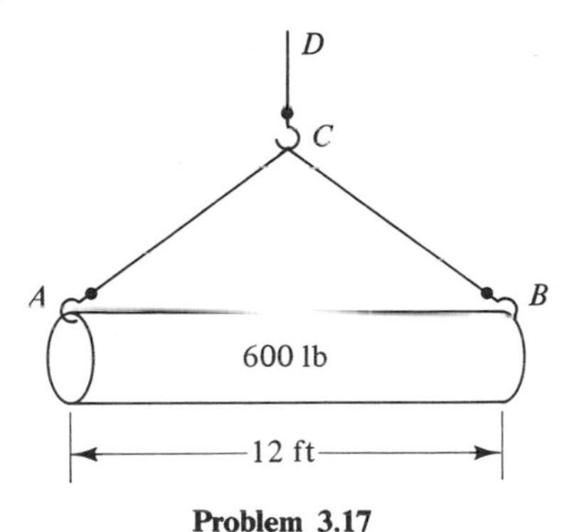

Problem 3.17

3.18. A heavy object is suspended above the floor by the arrangement of ropes shown. Determine the force P required for equilibrium if $W = 3$ kips and $\theta = 40°$.

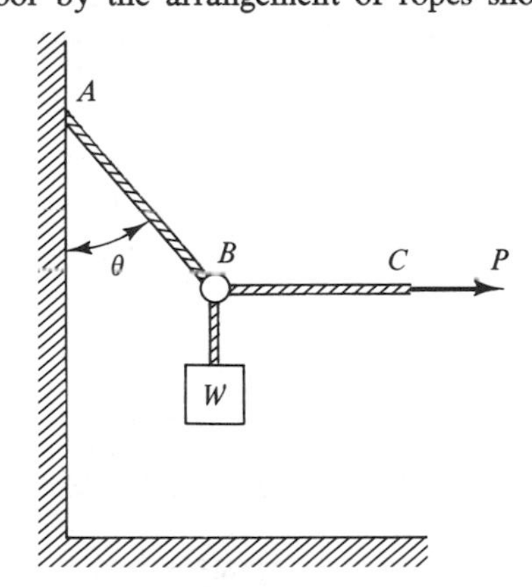

Problem 3.18

3.19. In Problem 3.18, determine the relationship between the force P, the weight W, and the angle θ for equilibrium. Show how the angle varies with increasing force by plotting θ versus the ratio P/W.

3.20. Determine the forces exerted on the uniform 50-kg cylinder shown by the smooth support surfaces.

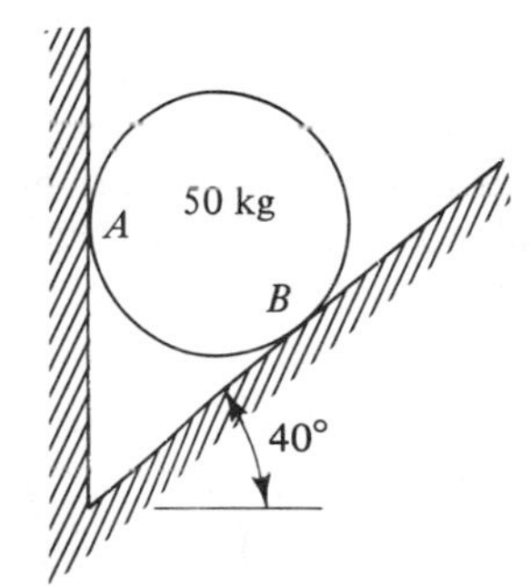

Problem 3.20

3.21. Determine the reactions on the member shown at the fixed support.

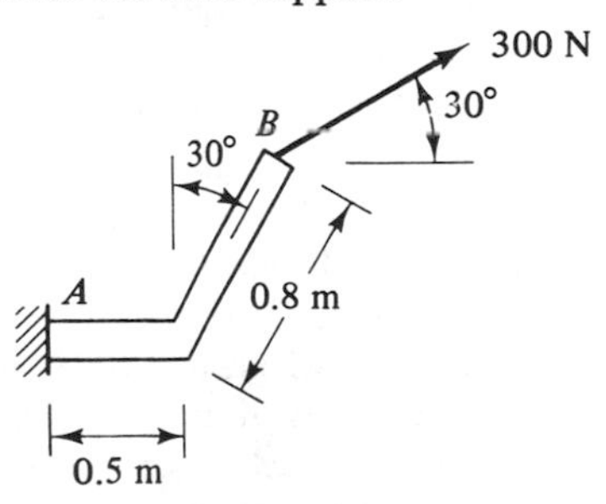

Problem 3.21

3.22. The caster on a movable platform has negligible weight and experiences an upward force of 1,200 lb at the point of contact with the floor, as shown. Determine the reactions on the caster at point A.

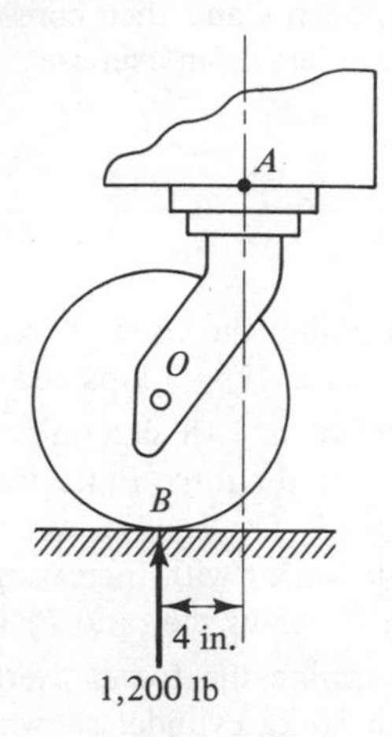

Problem 3.22

3.23. A machine part with negligible weight is supported by a smooth pin at A and a roller at B, as shown. Determine the reactions at each support.

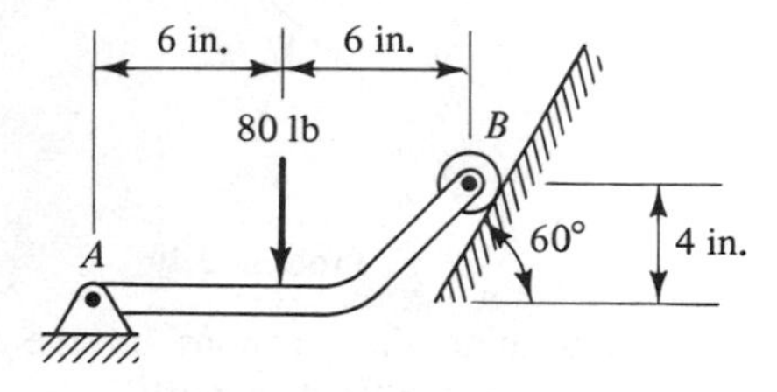

Problem 3.23

3.24. A device for lifting coils of sheet metal consists of a heavy frame, one end of which is inserted through the center of the coil as shown. If the frame has a mass of 200 kg, with center of gravity at G, determine the distance a such that the frame will not rotate when picking up a 1,500-kg coil.

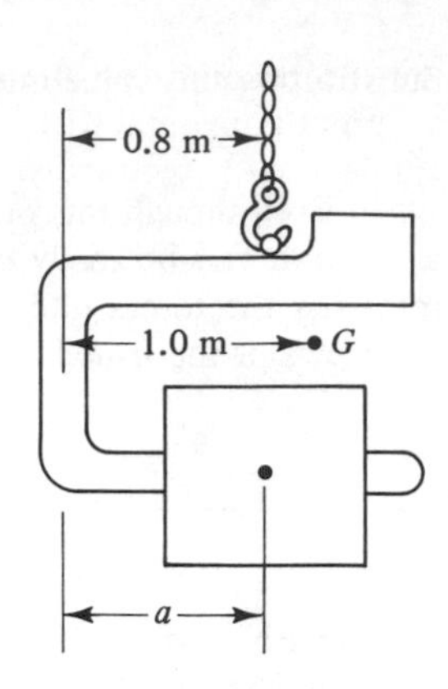

Problem 3.24

3.25. For the structure shown, determine the tension on the cable and the reactions at D.

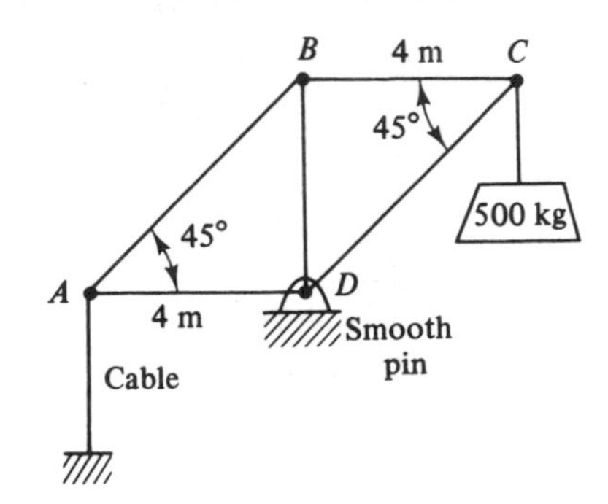

Problem 3.25

3.26. The tilt-cart shown and its contents weigh 800 lb, with center of gravity at G. De-

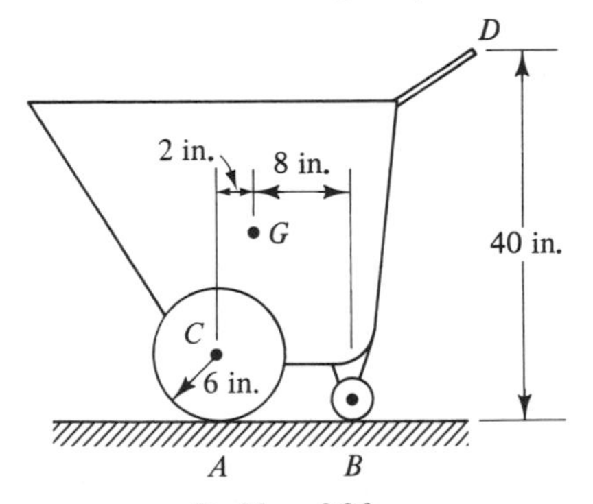

Problem 3.26

termine the reactions on the wheels at A and B. What horizontal force P applied to the handle at D is required to tip the cart? Assume that the front wheels are blocked so that the cart cannot move forward.

3.27. Determine the reactions on the uniform sphere supported as shown.

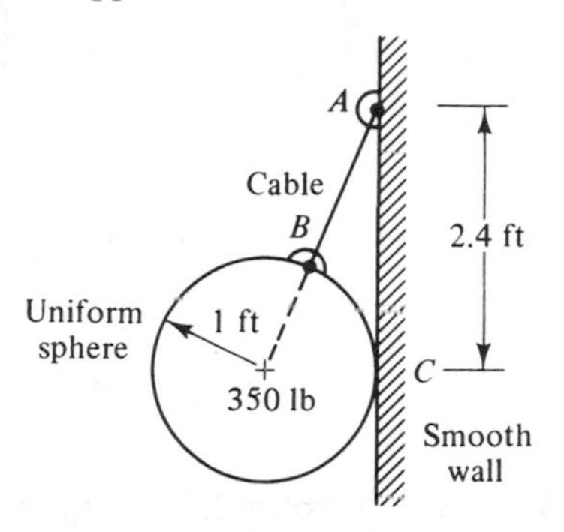

Problem 3.27

3.28. A uniform slender bar with one end resting on a rough surface and the other suspended by a string is in equilibrium in the position shown. Determine the tension on the string and the reaction at point A. Express the results in terms of the weight W of the bar.

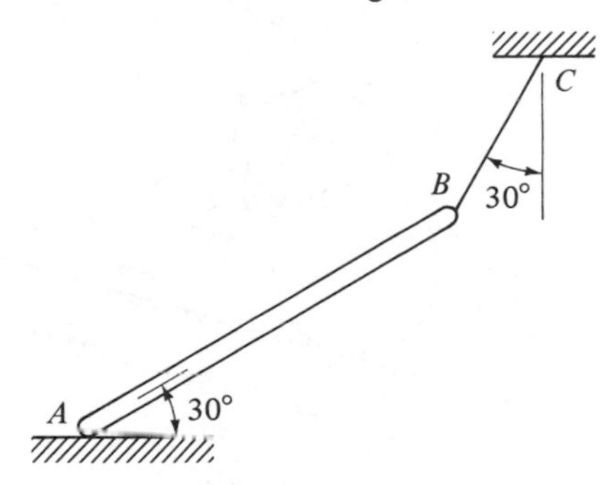

Problem 3.28

3.29. The stepped-pulley shown has negligible mass and is supported by a frictionless bearing at O. If $T = 100$ N, determine the couple C required for equilibrium and the corresponding reactions at O.

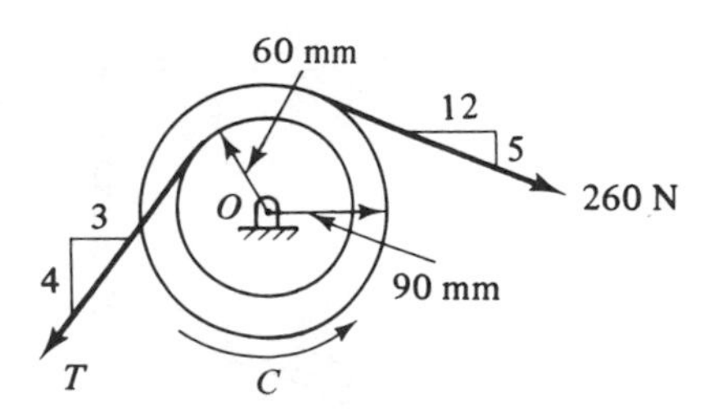

Problem 3.29

3.30. In Problem 3.29, determine the tension T for equilibrium and the corresponding reactions at O if $C = 10$ N · m.

3.31. What couple C must an electric motor apply to the drum of the winch shown to lift the 800-lb object? The applied couple is multiplied by a factor of 12 through a system of gears.

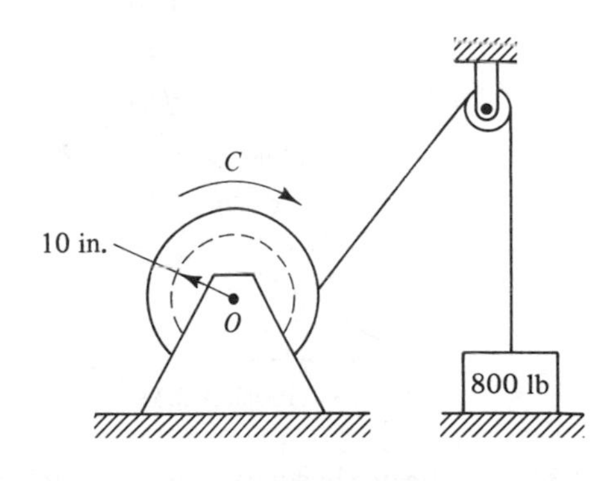

Problem 3.31

3.32. A barrel filled with a powdered material is lifted by wrapping a strap around it near the top and pulling upward on the free end, as shown. Determine the angle θ at which the barrel will hang.

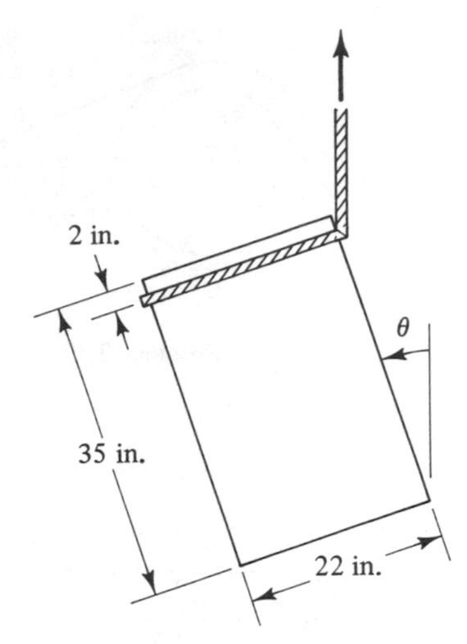

Problem 3.32

3.33. A 60-lb box is held on a smooth incline by a rope passing over a smooth peg as shown. Find (a) the tension on the rope and (b) the magnitude and location of the normal force exerted on the box by the plane.

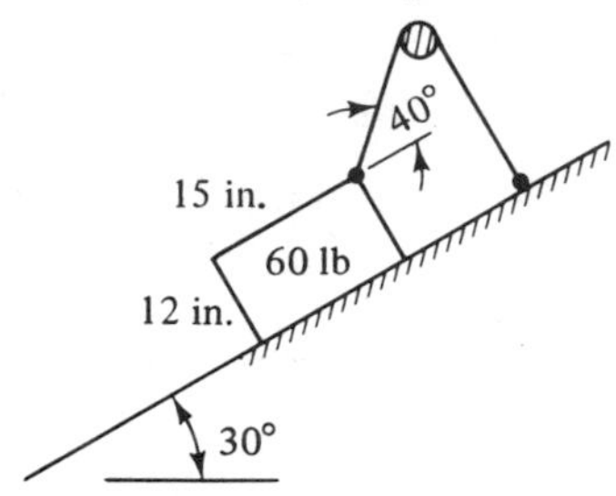

Problem 3.33

3.34. The beam shown has a mass per unit length of 25.5 kg/m and is supported by smooth pegs at A and B. Find (a) the reactions at the supports for $a = 0.5$ m and (b) the maximum value for a for which the beam can be held in equilibrium.

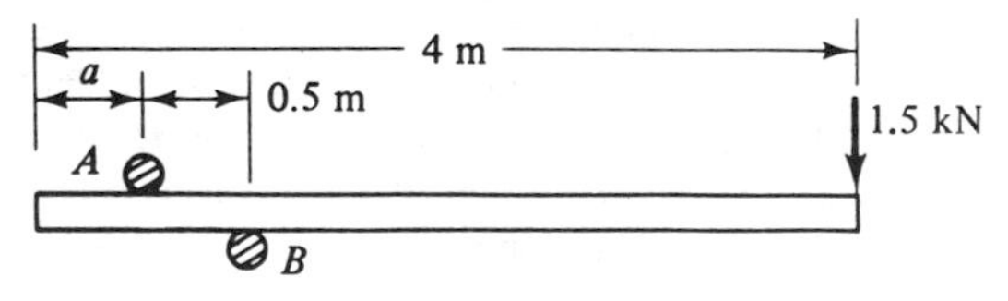

Problem 3.34

3.35. What is the maximum force F that can be applied to the bar shown if the magnitude of the reaction at the pin O is not to exceed 2 kN? The bar has negligible mass, and the pin is smooth.

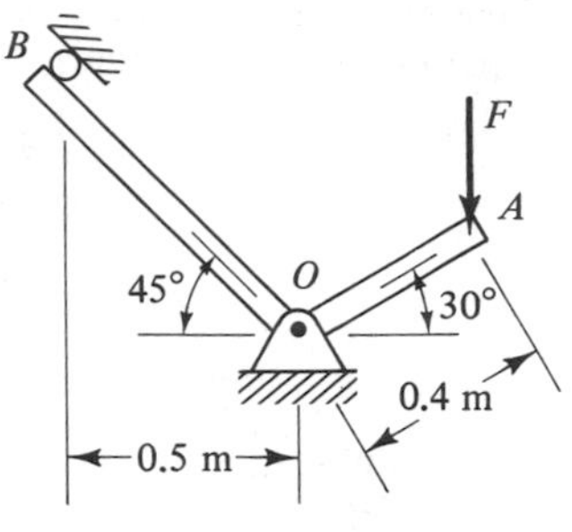

Problem 3.35

3.36. Same as Problem 3.35, except that the force F is replaced by a clockwise couple C.

3.37. The riding lawnmower shown and driver have a combined weight of 1,200 lb, with center of gravity at G. Assuming that the mower travels at constant speed, what is the steepest incline that can be negotiated without tipping over?

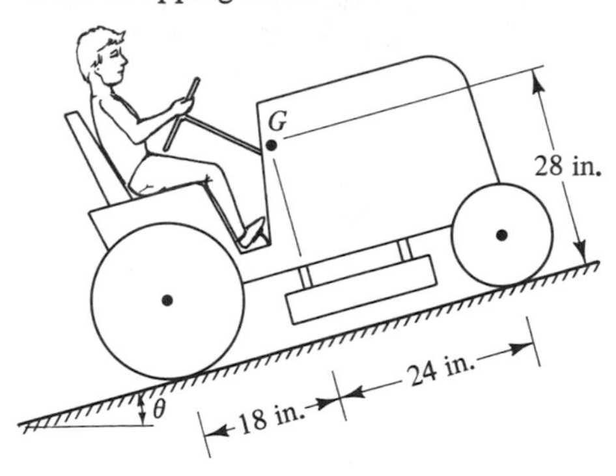

Problem 3.37

3.38. A movable partition is stabilized by a base at each end, as shown. Standards for office furniture require that the panel be able to tilt up to 10° from the vertical without falling over. Determine the minimum base length a for which this particular panel will meet the standard.

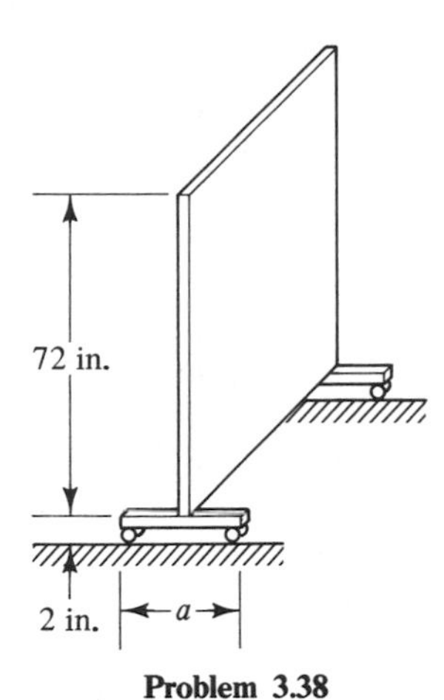

Problem 3.38

3.39. Fresh concrete is moved to an overhead construction site in a circular bucket connected to a crane hook by three chains equally spaced around the periphery, as shown. Determine the tension on each chain if the bucket and contents have a mass of 1,500 kg.

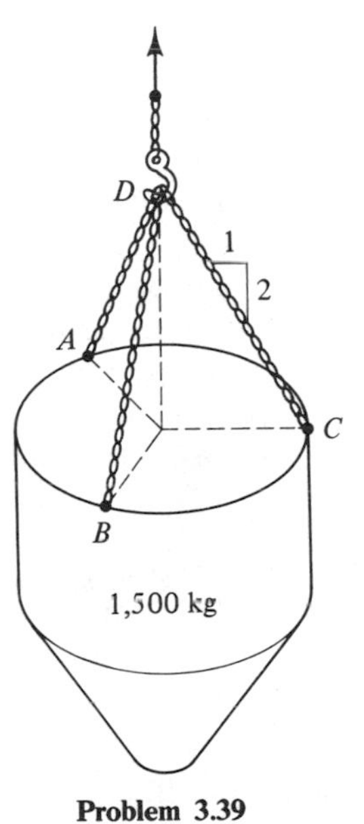

Problem 3.39

3.40. The rotor from a large turbine has weight W and is supported by a sling consisting of a continuous loop of flexible material placed around the member as shown. If the tensions on each of the four segments of the loop have the same magnitude, what is the tension T on the sling?

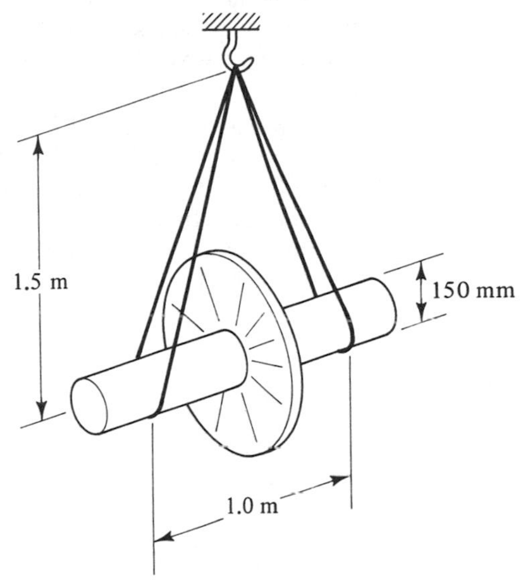

Problem 3.40

3.41. Determine the force acting on each leg of the tripod shown. The legs are supported

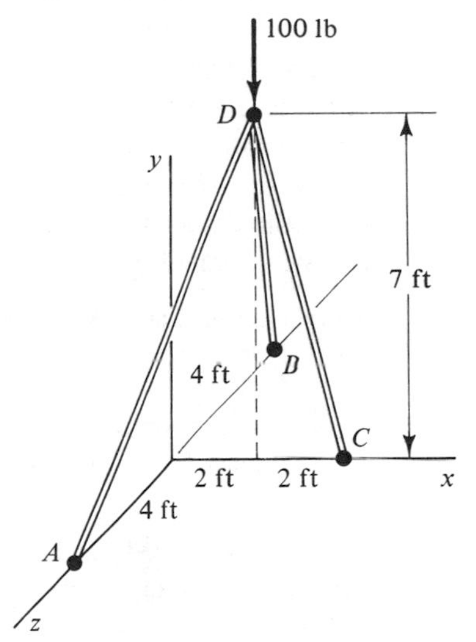

Problem 3.41

by ball-and-socket connections at top and bottom and have negligible weight.

3.42. For the structure in Problem 2.45 (Section 2.6), determine the support cable tensions for which the total loading on the tower at A will be a compressive force of 10 kips.

3.43. A 400-kg container is suspended from two cables, as shown. Find (a) the force F acting parallel to the x-axis required to hold the container (point C) 2 m away from the wall, and (b) the corresponding tensions on the two cables.

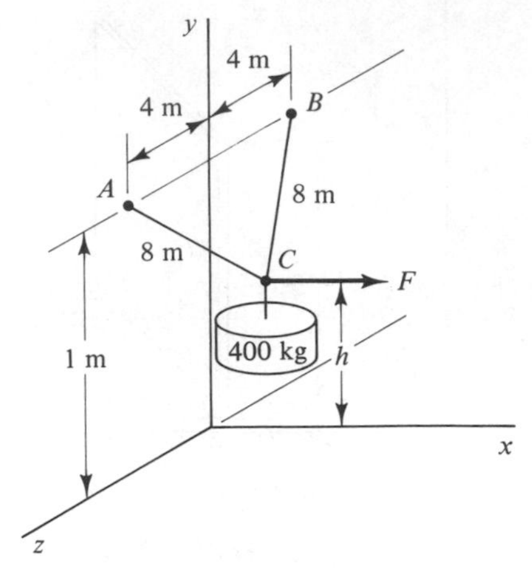

Problem 3.43

3.44. In Problem 3.43, determine the cable tensions and the height h as functions of the applied force F. Show the variations in these quantities by plotting them versus F.

3.6 FORCE ANALYSIS OF GENERAL THREE-DIMENSIONAL SYSTEMS

Force analyses of general three-dimensional systems involve no principles not already encountered in our study of two-dimensional and other special force systems. The only difference is that all six equilibrium equations must usually be considered in the general three-dimensional case. Some of the equilibrium equations may be satisfied automatically, but the situation differs from problem to problem. The following examples illustrate the steps involved in the force analysis. Note that it is convenient to use a formal vector approach if the geometry of the problem is at all complicated.

Example 3.7. Force Analysis of a Boom. A safe with a weight of 10,000 lb is suspended from a boom of negligible weight in preparation for loading onto a truck [Figure 3.17(a)]. The boom is supported by a smooth pin at O and a cable at A. What are the reactions on the boom and the tension on the cable? Could the pin connection at O be replaced by a ball and socket connection?

Solution. Since the pin at O can resist translation in all three coordinate directions, the force **R** exerted on the boom at O will have three unknown components. Similarly, the pin can resist rotation about the x and y axes; therefore, the couple **C** exerted on the boom will have two unknown components. Thus, the FBD of the boom is as shown in Figure

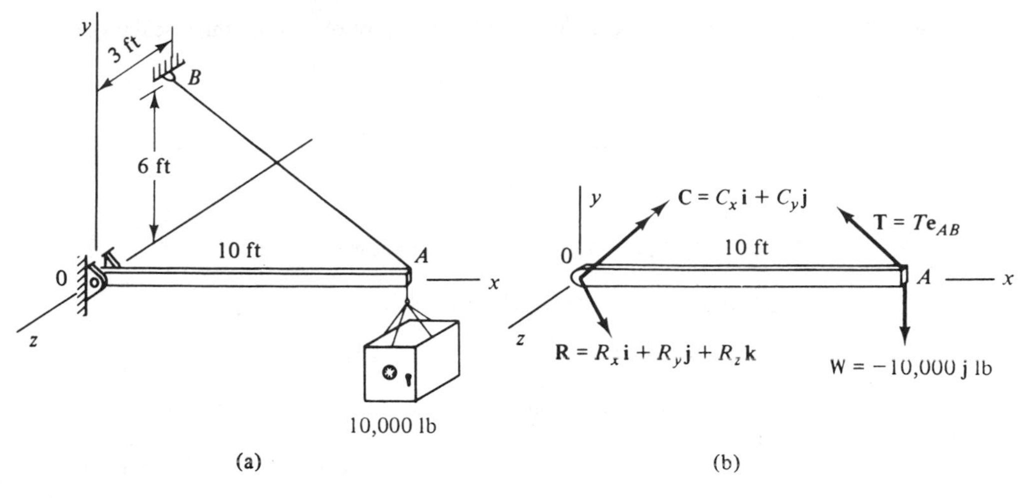

Figure 3.17

3.17(b), where **T** is the force exerted by the cable. From the geometry of the problem, we find

$$\mathbf{r}_{AB} = -10\mathbf{i} + 6\mathbf{j} - 3\mathbf{k} \text{ ft}$$

$$r_{AB} = \sqrt{(-10)^2 + (6)^2 + (-3)^2} = 12.04 \text{ ft}$$

$$\mathbf{T} = T\mathbf{e}_{AB} = T\frac{\mathbf{r}_{AB}}{r_{AB}} = \frac{T}{12.04}(-10\mathbf{i} + 6\mathbf{j} - 3\mathbf{k})$$

For equilibrium

$$\sum \mathbf{F} = \mathbf{R} + \mathbf{T} + \mathbf{W} = \mathbf{0}$$

or

$$\sum F_x = R_x - \frac{10}{12.04}T = 0$$

$$\sum F_y = R_y + \frac{6}{12.04}T - 10{,}000 \text{ lb} = 0$$

$$\sum F_z = R_z - \frac{3}{12.04}T = 0$$

Taking moments about point A, we have

$$\sum \mathbf{M}_A = \mathbf{C} + \mathbf{r}_{AO} \times \mathbf{R} = \mathbf{0}$$

where

$$\mathbf{r}_{AO} = -10\mathbf{i} \text{ ft}$$

Thus,

$$\sum \mathbf{M}_A = C_x\mathbf{i} + \mathbf{C}_y\mathbf{j} - 10\mathbf{i} \times (R_x\mathbf{i} + R_y\mathbf{j} + R_z\mathbf{k}) = \mathbf{0}$$

Performing the cross products and collecting like terms, we obtain the three scalar equations

$$\sum M_{Ax} = C_x = 0$$
$$\sum M_{Ay} = C_y + 10R_z = 0$$
$$\sum M_{Az} = -10R_y = 0$$

Solving the six equations of equilibrium for the six unknowns, we have

$$R_x = 16{,}670 \text{ lb} \quad R_y = 0 \quad R_z = 5{,}000 \text{ lb}$$
$$T = 20{,}070 \text{ lb} \quad C_x = 0 \quad C_y = -50{,}000 \text{ lb} \cdot \text{ft}$$

Answer

If the pin at O is replaced with a ball and socket connection $\mathbf{C} = \mathbf{0}$, because a ball and socket can't resist rotation. It is impossible to satisfy all the equilibrium equations in this case, which means that the boom cannot be held in equilibrium in the position shown. In order to maintain equilibrium, it would be necessary to add an additional constraint. Another cable attached to the boom would do, provided it was anchored to the wall at an appropriate point.

Example 3.8. Force Analysis of a Platform. Figure 3.18(a) shows a platform with weight W_P which is held in a horizontal position by three vertical wires. A man with weight W_M stands on the platform. Where should he stand so that the tension on each wire is the same?

Solution. An FBD of the platform is shown in Figure 3.18(b). The position of the man is denoted by the coordinates x_M and z_M. The problem is three-dimensional, but a formal vector approach isn't necessary. The orientation of all the forces is readily defined on the FBD, and moments can be determined by inspection.

Setting the sum of forces in the vertical direction equal to zero and the sum of moments about the coordinate axes equal to zero (which is the same as summing moments about the origin and setting each component of the sum equal to zero), we obtain the equilibrium

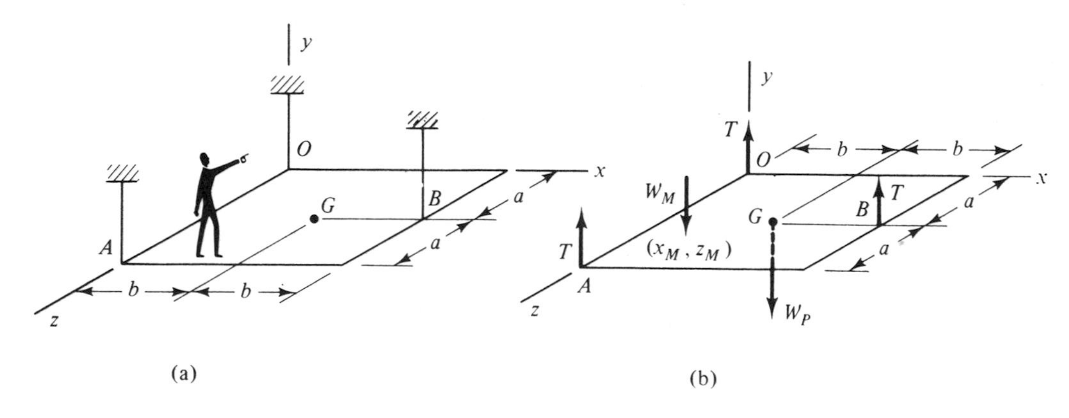

Figure 3.18

equations

$$\sum F_y = 3T - W_P - W_M = 0$$

$$\sum \overset{+}{\curvearrowleft} M_{0x} = W_M z_M + W_P a - Ta - T(2a) = 0$$

$$\sum \overset{+}{\curvearrowleft} M_{0z} = -W_P b - W_M x_M + T(2b) = 0$$

All other equilibrium equations are satisfied automatically. The sign convention used for moments is that of the right-hand screw rule; a moment is positive if it appears to be counterclockwise when viewed from the positive coordinate direction.

Solving for the unknowns, we have

$$T = \frac{W_P + W_M}{3} \qquad z_M = a \qquad x_M = \frac{b}{3}\left(2 - \frac{W_P}{W_M}\right) \qquad \textbf{Answer}$$

Note that $x_M < 0$ if $W_P > 2W_M$. In this case there is no place the man can stand on the platform and have the tensions on the wires be the same.

Example 3.9. Force Analysis of a T-Bar. The T-bar shown in Figure 3.19(a) has a mass of 32.6 kg and is supported by smooth bearings at A and B and a smooth wall at C. Determine the force that the wall exerts on the bar.

Solution. Instead of summing forces and considering moments about some point, we can obtain the solution more directly if we recognize that the sum of the moments about the

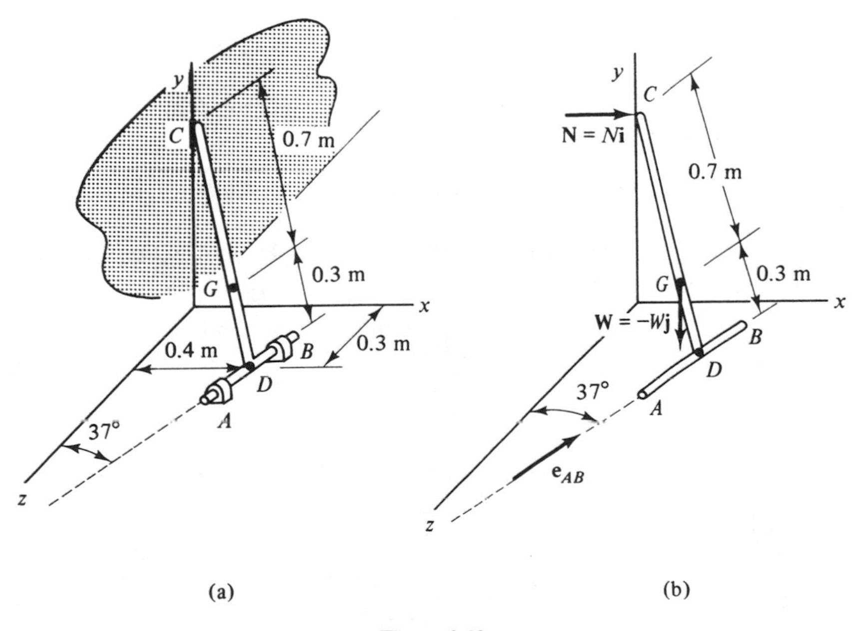

Figure 3.19

bearing axis (line AB) must be zero. The only forces that have a moment about this axis are the normal force due to the wall and the weight. These forces are shown in Figure 3.19(b). Note that this figure is not an FBD because we have not shown all of the reactions on the bar.

The magnitude of the moment about line AB can be expressed as [Eq. (2.28)]

$$M_{AB} = \mathbf{M}_D \cdot \mathbf{e}_{AB} = (\mathbf{r}_{DG} \times \mathbf{W} + \mathbf{r}_{DC} \times \mathbf{N}) \cdot \mathbf{e}_{AB}$$

From the given dimensions we see that

$$\mathbf{r}_{DG} = 0.3\mathbf{r}_{DC}$$

so

$$M_{AB} = [\mathbf{r}_{DC} \times (0.3\mathbf{W} + \mathbf{N})] \cdot \mathbf{e}_{AB}$$

The normal force is perpendicular to the wall (y-z plane) and the weight is downward; so

$$\mathbf{N} = N\mathbf{i} \qquad \mathbf{W} = -(32.6 \text{ kg})(9.81 \text{ m/s}^2)\mathbf{j} = -320\mathbf{j} \text{ N}$$

The vertical coordinate of point C is not given, but it can be determined from the known distance DC. We have

$$\mathbf{r}_{DC} = -0.4\mathbf{i} + y_C\mathbf{j} - 0.3\mathbf{k} \text{ m}$$

$$r_{DC}^2 = (-0.4)^2 + y_C^2 + (-0.3)^2 = (1 \text{ m})^2$$

from which we find

$$y_C = \sqrt{0.75} = 0.87 \text{ m}$$

The unit vector $\mathbf{e}_{AB}$ can be expressed in component form simply by resolving it into components along the coordinate axes:

$$\mathbf{e}_{AB} = \sin 37°\mathbf{i} - \cos 37°\mathbf{k} = 0.6\mathbf{i} - 0.8\mathbf{k}$$

Using the determinant form of the scalar triple product [Eq. (2.29)] for M_{AB}, we obtain the equilibrium equation

$$M_{AB} = \begin{vmatrix} 0.6 & 0 & -0.8 \\ -0.4 & y_C & -0.3 \\ N & -0.3W & 0 \end{vmatrix} = 0$$

or

$$-0.6(0.3)(0.3W) - 0.8[0.4(0.3W) - Ny_C] = 0$$

Solving for N, we get

$$N = \frac{0.15W}{0.8y_C} = \frac{0.15(320)\text{N} \cdot \text{m}}{0.8(0.87 \text{ m})} = 69 \text{ N} \qquad \textbf{Answer}$$

PROBLEMS

3.45. Two beams with a weight per unit length of 30 lb/ft are welded together to form the shape shown. Find the tension on each of the vertical support wires. The beams lie in a horizontal plane.

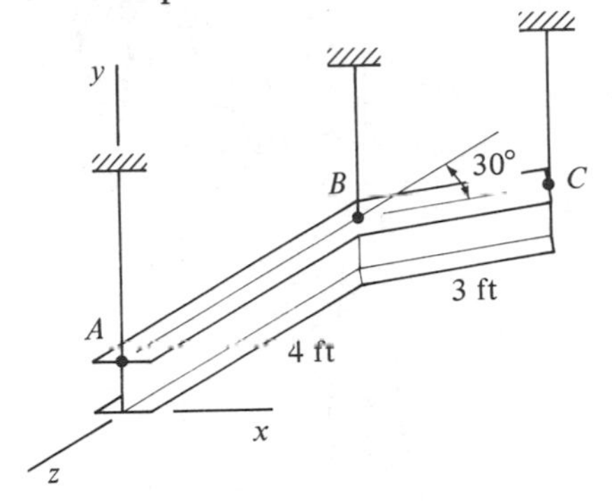

Problem 3.45

3.46. A uniform rectangular plate with a weight of 1,500 lb is held in a horizontal position, as shown. Determine the tension on each of the vertical support cables.

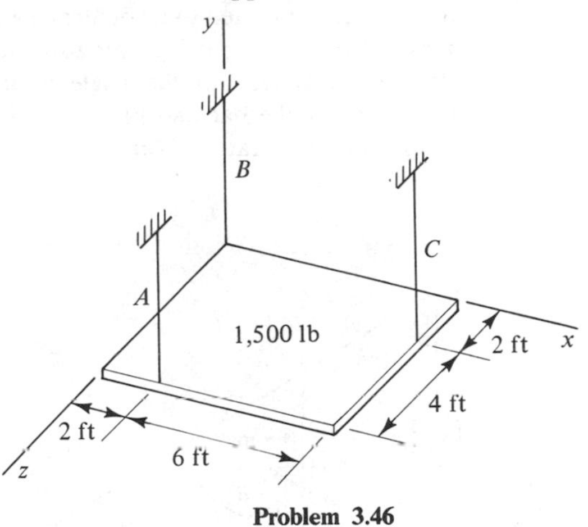

Problem 3.46

3.47. In Problem 3.46, it is proposed to equalize the tension on the cables by placing a 300-lb block at an appropriate location on the plate. Will this procedure work? If so, where on the plate should the block be placed, and what is the corresponding cable tension?

3.48. A slender pole with negligible mass is supported by two smooth eye-bolts and is loaded as shown. Determine the reactions at A and B.

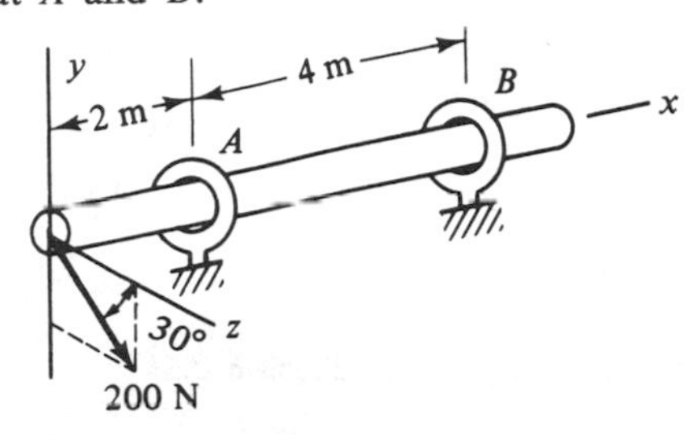

Problem 3.48

3.49. The rectangular plate shown has a mass of 673 kg and is supported by two slender rods and a ball and socket joint at O. Determine the reactions at O and the tensions on the rods.

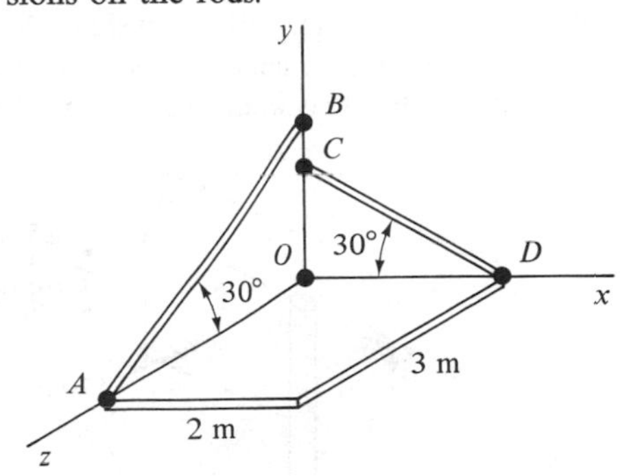

Problem 3.49

3.50. Same as Problem 3.49, except that point C coincides with point B and rod CD is no longer inclined at 30° from the horizontal.

3.51. A pipe bent into the shape of quarter circle is fixed at end A and loaded at end B, as shown. Determine the reactions at A if (a) the pipe has negligible weight and (b) the pipe has a weight per unit length of 5.8 lb/ft.

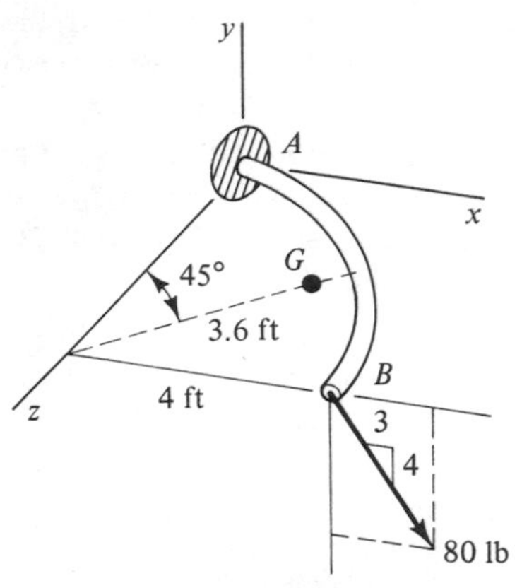

Problem 3.51

3.52. For the bar in Problem 3.11 (Section 3.4), determine the reactions at the hinge and the tension on the support wire. Neglect the weight of the bar.

3.53. Determine the reactions at the bearings for the line shaft in Problem 3.13 (Section 3.4).

3.54. A uniform sign, which weighs 40 lb, is supported by a uniform pole with a weight of 120 lb. The wind exerts a net force of 170 lb normal to the sign and acting at its center, as shown. Determine the reactions at the base O.

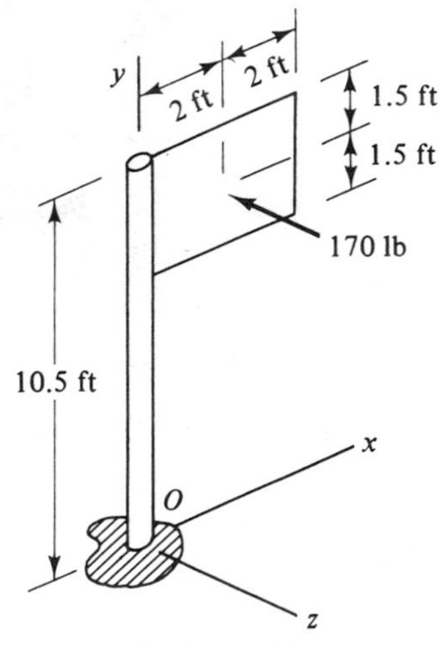

Problem 3.54

3.55. Determine expressions for the reactions at the base O of the robotic manipulator shown due to a payload of mass m, for $\theta = 0°$. Arm AB lies in the x-y plane.

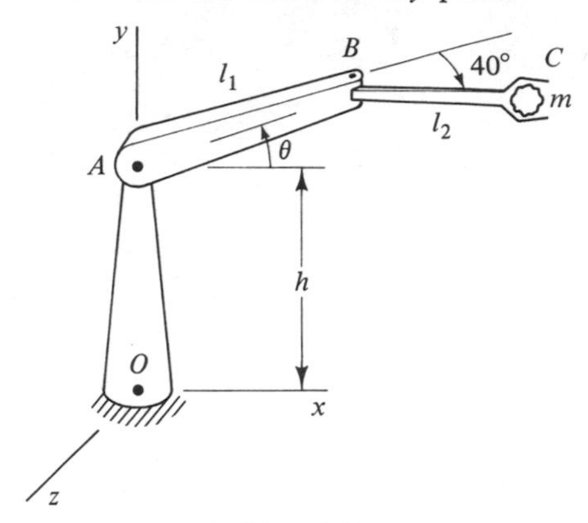

Problem 3.55

3.56. Same as Problem 3.55, except that $\theta = 30°$.

3.57. The T-bar shown has mass m, with center of gravity G. The bar is hinged at A and B, and end C is connected to a counterweight of mass M by a rope passing over a smooth pulley at D. Determine the angle of inclination, θ, of the bar and plot its variation with the mass ratio M/m.

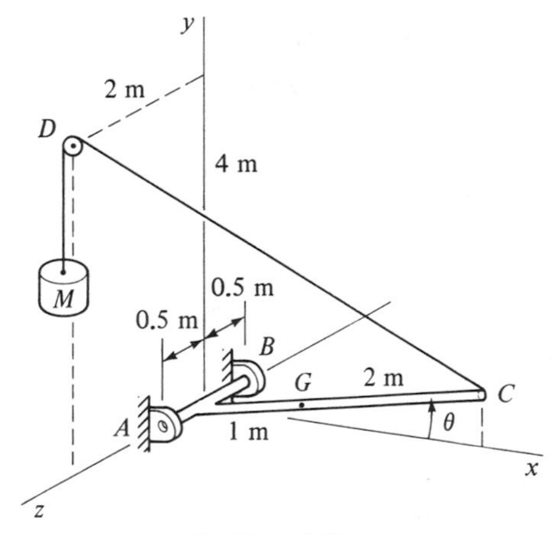

Problem 3.57

3.58. Determine the mass M of the counterweight required to hold the T-bar in Problem 3.57 in a horizontal position if $m = 75$ kg. What are the reactions on the bar at A and B and what is the tension on the rope for this case? Ignore any components of the reactions believed to be insignificant.

3.59. It is proposed that the uniform triangular plate shown be held in a horizontal position by placing the tip of the plate in a smooth socket at O and connecting corners A and B to a cord passing over a smooth pulley at C. The plate has weight W. Is this arrangement sufficient to hold the member in equilibrium? If so, what are the reactions at O and the tension on the cord? If not, what changes could be made in the supports to correct the situation?

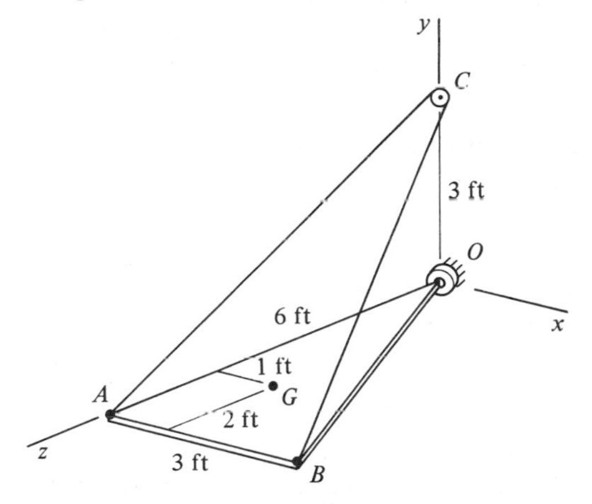

Problem 3.59

3.60. The curved pipe shown has negligible weight and lies in a horizontal plane. It is supported by a smooth sleeve bearing at A and by a vertical wire at B. Is this arrangement of supports sufficient to hold the pipe in equilibrium when the couple C is applied to the free end? If so, what are the reactions at A and the tension on the wire at B? If not, how could the supports be changed to correct the situation?

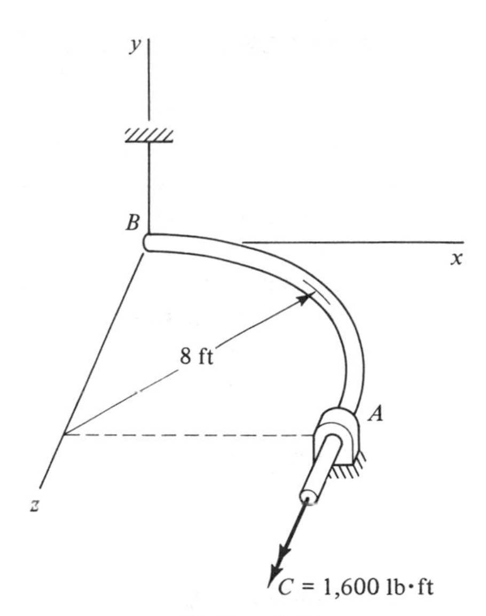

Problem 3.60

3.61. The lamp shown has a mass of 1.02 kg and is supported by tubular arms, which pass through holes in a thin metal bracket at A and B. A collar on the arm at B keeps the assembly from sliding downward. Assuming that the arms have negligible mass, determine the reactions on the lamp assembly at A and B. Neglect any components of the reactions believed to be insignificant. Arm ABC lies in the x-y plane.

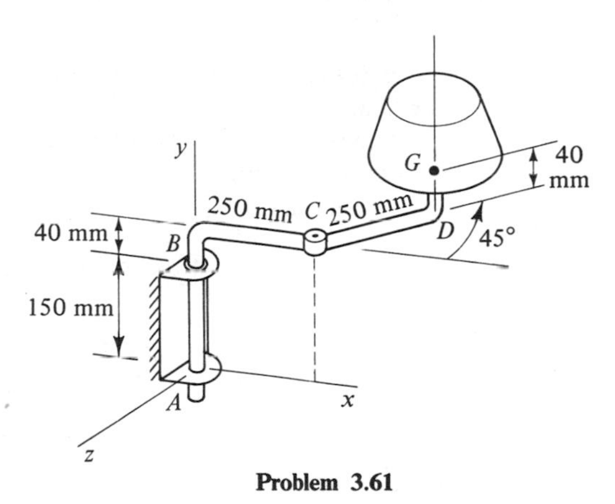

Problem 3.61

3.7 FORCE ANALYSIS OF COMPLEX SYSTEMS

Up to this point, all the force analyses encountered have been relatively simple in that only one free-body diagram had to be considered. This was because the systems consisted of only one major element. However, engineering structures and machines often consist of a number of members. The force analysis of such systems and their various elements involves no new principles, but the large number of free-body diagrams and equilibrium equations that are often involved does make it essential to have a systematic plan of attack.

Although there is no one procedure that is "best" for all problems, the following steps represent a reasonable approach in many instances:

1. Determine the reactions by considering the equilibrium of the entire structure.
2. Dismember the structure and draw FBDs of its various elements. Be sure to apply Newton's third law when considering the forces that one member exerts on another, and take advantage of any two-force members that may be present. FBDs of all the members may not be needed in order to determine the quantities of interest, but it usually does not require much extra effort to go ahead and draw them. There is less likelihood of making an error if all FBDs are drawn at one time.
3. Consider the equilibrium of as many individual members as necessary in order to determine all the desired unknowns. Try to avoid solving for unknowns that are not of interest, but this will not always be possible.

With experience, you will be able to modify these steps to best fit each different situation as it arises. These ideas are illustrated in the following examples.

Example 3.10. Force Analysis of a Bolt Cutter. What force P must be applied to the handles of the bolt cutter shown in Figure 3.20(a) if it takes a force of 75 lb to cut the bolt?

Solution. Consideration of the equilibrium of the entire cutter gives no information other than the fact that the force P on each handle must be the same. Thus, we consider the individual parts. The FBDs of one of the handles and jaws are shown in Figure 3.20(b). Note that AD is a two-force member; therefore, the orientation of force F_{AD} is known.

For the jaw

$$\overset{+}{\curvearrowleft}\sum M_A = B_y(3 \text{ in.}) - 75 \text{ lb}\,(1 \text{ in.}) - B_x d = 0$$

$$\sum F_x = B_x = 0$$

from which we get

$$B_y = 25 \text{ lb} \qquad B_x = 0$$

For the handle

$$\overset{+}{\curvearrowleft}\sum M_C = B_y(\tfrac{1}{2} \text{ in.}) + B_x(1 \text{ in.}) - P(10 \text{ in.}) = 0$$

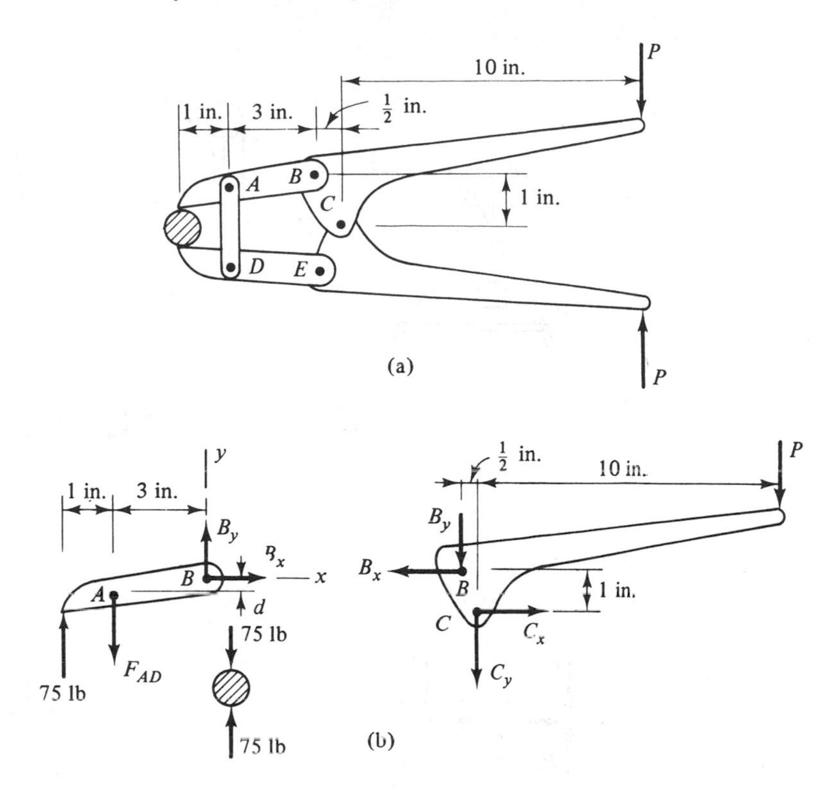

Figure 3.20

so

$$P = 1.25 \text{ lb} \qquad \textbf{Answer}$$

Example 3.11. Force Analysis of a Frame. Determine the forces acting upon member AC of the pin-connected frame shown in Figure 3.21(a).

Solution. We shall start by considering the equilibrium of the whole frame, even though we are interested only in the forces on member AC. It is convenient to first remove the pulley so that attention can be focused on the basic structure [Figure 3.21(b)].

Referring to the FBD of the pulley [Figure 3.21(b)], we have

$$\overset{+}{\curvearrowleft} \sum M_C = Wr - Tr = 0$$

$$\sum F_x = C_x + T = 0$$

$$\sum F_y = C_y - W = 0$$

from which we obtain

$$T = W \qquad C_x = -W \qquad C_y = W$$

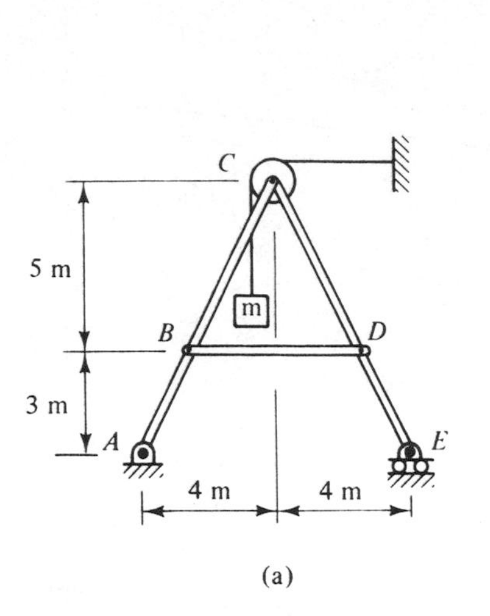

(a)

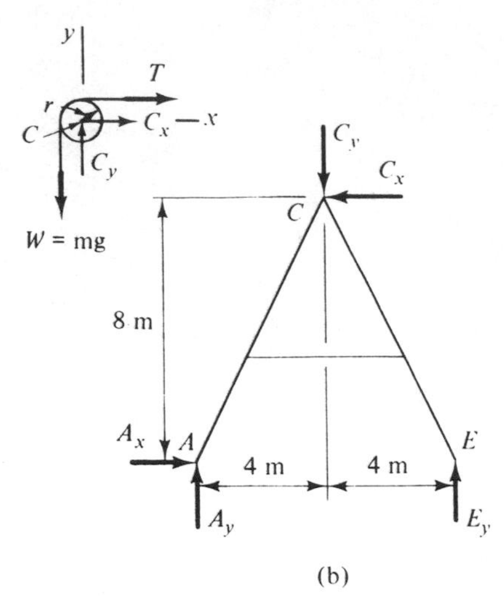

(b)

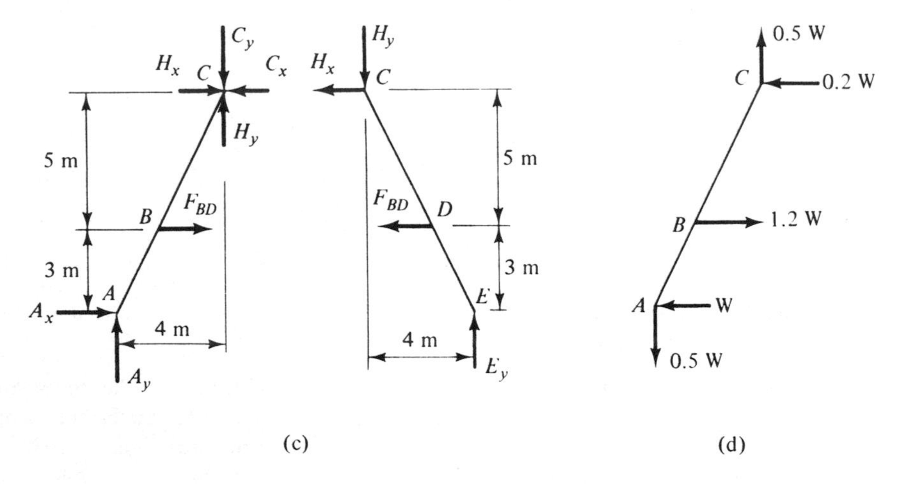

(c) (d)

Figure 3.21

For equilibrium of the frame

$$\sum \overset{+}{\curvearrowleft} M_A = C_x(8\text{ m}) - C_y(4\text{ m}) + E_y(8\text{ m}) = 0$$

$$\sum F_y = A_y + E_y - C_y = 0$$

$$\sum F_x = A_x - C_x = 0$$

Solving these equations, we obtain for the reactions

$$A_x = -W \qquad A_y = -0.5W \qquad E_y = 1.5W \qquad \textbf{Answer}$$

We now draw the FBDs of the individual members, but in doing so we find ourselves in a bit of a predicament at point C. Do we show the forces C_x and C_y as acting upon member AC or upon member EC? Actually, the members share these loads. However, the fraction of the loading that each member carries cannot be determined beforehand, and it is not necessary to do so. We need only show the forces as acting on one member or the other. The members also exert forces on each other, and the values of these forces are automatically adjusted in the analysis so that the correct total force on each member is obtained. This same situation will be encountered whenever there are three or more members connected at a common point or there are two or more members connected at a point where forces are also applied.

The FBDs of members AC and EC are shown in Figure 3.21(c), where C_x and C_y have been shown as acting upon AC. Note that we have made use of the fact that BD is a two-force member. The remaining unknowns can be obtained by considering the equilibrium of either AC or EC. We choose EC because it involves fewer forces:

$$\overset{+}{\curvearrowleft} \sum M_C = E_y(4 \text{ m}) - F_{BD}(5 \text{ m}) = 0$$

$$\sum F_x = -H_x - F_{BD} = 0$$

$$\sum F_y = -H_y + E_y = 0$$

Solving these equations, we get

$$F_{BD} = 1.2W \qquad H_x = -1.2W \qquad H_y = 1.5W \qquad \textbf{Answer}$$

The net forces acting upon member AC are summarized in Figure 3.21(d).

The equilibrium equations for member AC can be used as a check. This step will be left as an exercise. It will also be left as an exercise to show that the same results would be obtained if the forces C_x and C_y were considered to act on member EC. The values of H_x and H_y will be different, but the net forces on each of the bars at C will be the same.

PROBLEMS

3.62. A 50-kg block is held in the position shown by a series of ropes. Determine the forces P and Q and the tension on rope BC.

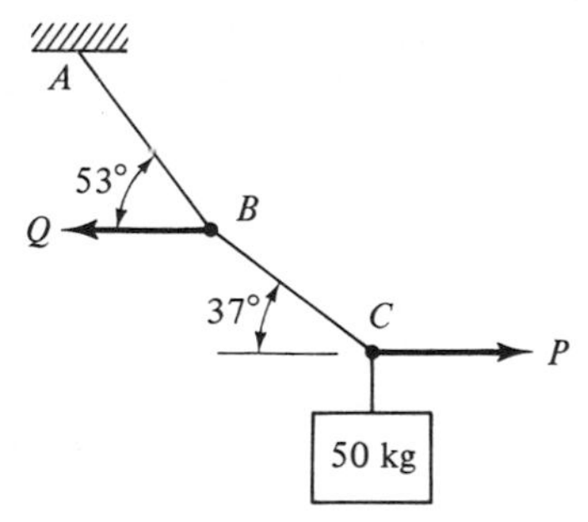

Problem 3.62

3.63. Two identical 100-kg cylinders are supported as shown. Determine all of the forces acting upon each cylinder. All surfaces are smooth.

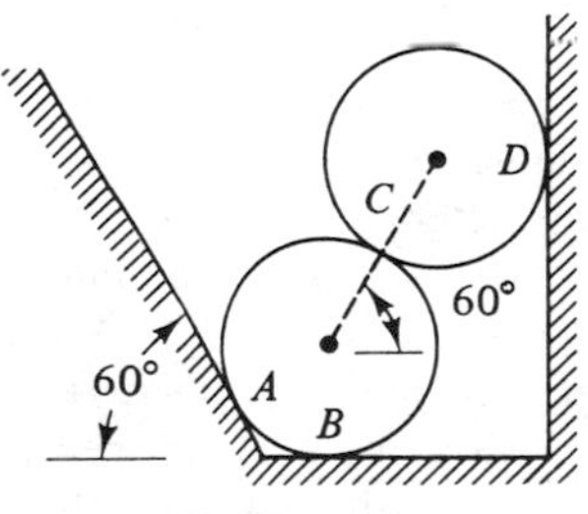

Problem 3.63

3.64. and 3.65. For the rope and pulley arrangement shown, determine the pull P necessary for equilibrium.

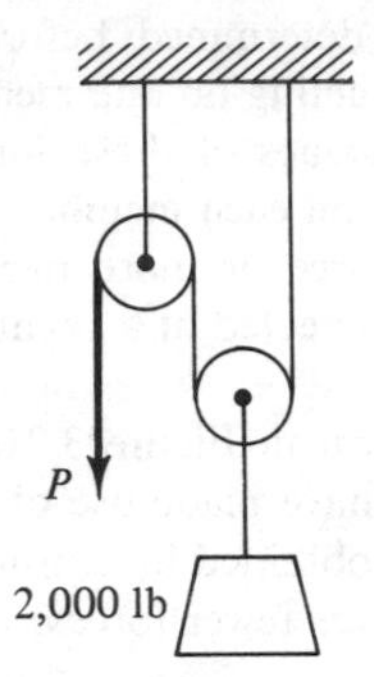

Problem 3.64

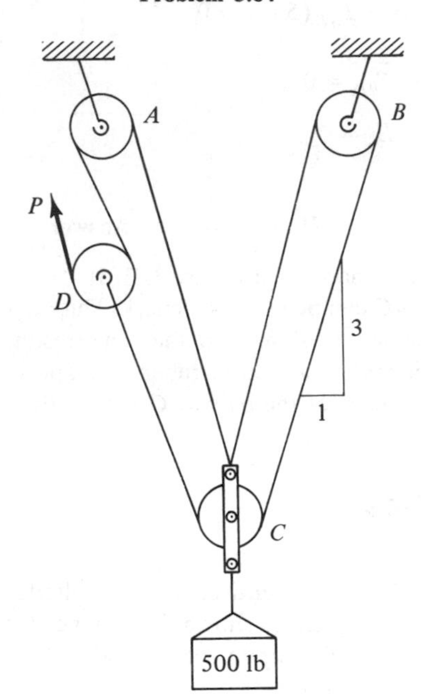

Problem 3.65

3.66. The top unit of the differential chain hoist shown consists of two grooved pulleys with diameters d_1 and d_2 keyed together to rotate as a unit. The diameter of the lower pulley is $\frac{1}{2}(d_1 + d_2)$. A continuous chain passes over the pulleys as shown. Determine the force F necessary to lift an object with weight W if $d_1 = 0.8d_2$. Assume that there is no tension on the slack side of the chain.

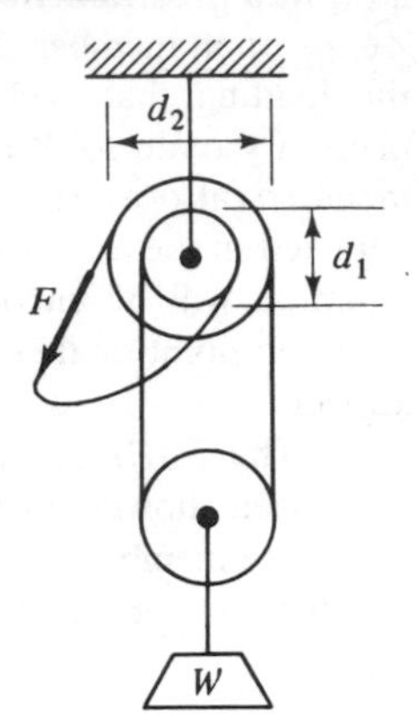

Problem 3.66

3.67. The uniform beam shown weighs 48 lb and is supported by vertical ropes at A and B. The pulleys are frictionless. Determine the tension on each of the support ropes.

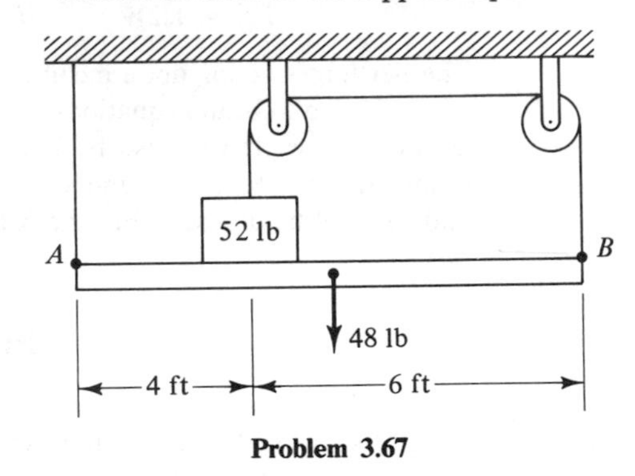

Problem 3.67

3.68. A uniform 50-kg bar is supported as shown. The pulley is frictionless and has a mass of 20 kg. Determine the horizontal and verti-

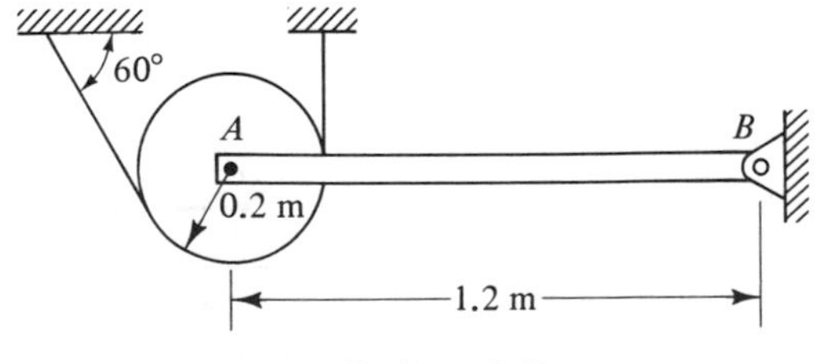

Problem 3.68

cal components of the forces acting on the bar at A and B.

3.69. For the pliers shown, determine the ratio of the gripping force P exerted on the bolt at B and the applied force F. (The ratio P/F is called the *mechanical advantage*.)

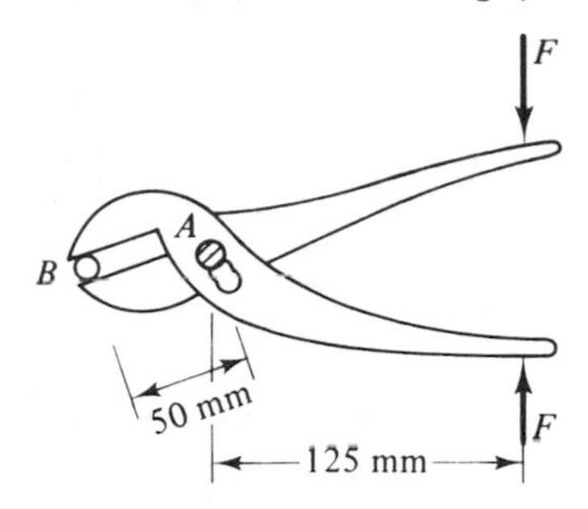

Problem 3.69

3.70. The device shown is used for punching holes in leather. What is the punching force P exerted on the leather if $F = 8$ lb? What is the mechanical advantage of this device (the ratio P/F)?

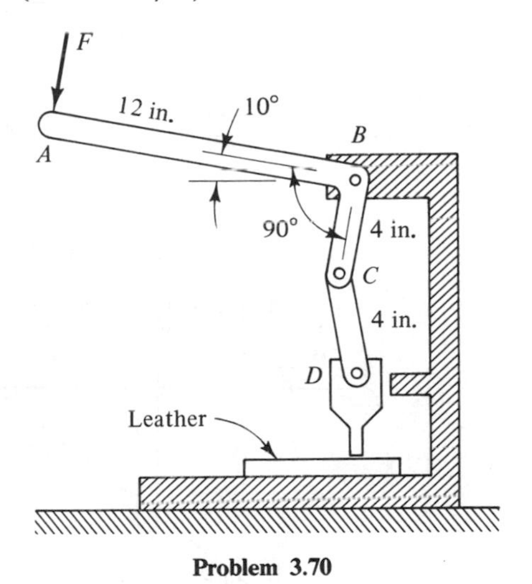

Problem 3.70

3.71. The safety hook shown closes and locks automatically when a load is applied. Determine the forces acting on link OAB when lifting an object with a mass of 3 ton.

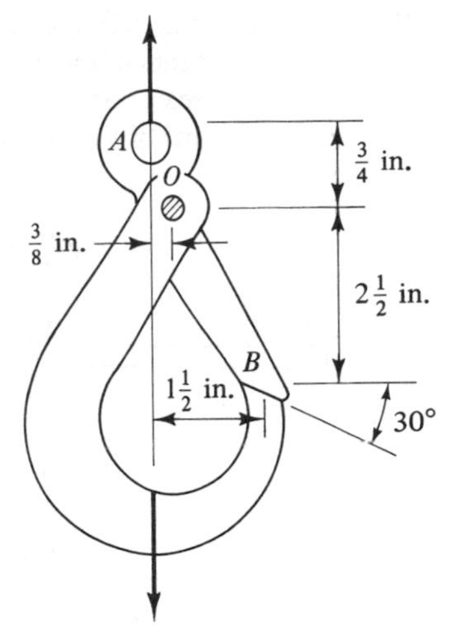

Problem 3.71

3.72. The essential features of a device for lifting heavy barrels of material are as shown. Determine the horizontal and vertical components of the forces acting on member ABC as a fraction of the weight W of the barrel and contents. Ignore friction and the weights of the members.

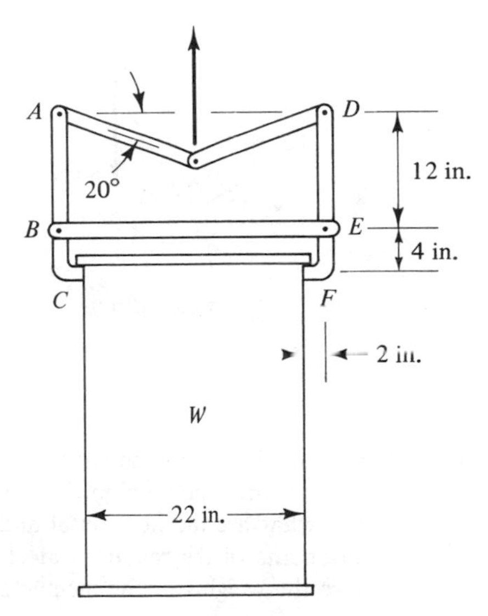

Problem 3.72

3.73. and 3.74. For the pin-connected frame shown, determine the horizontal and vertical components of the reactions at the supports. Ignore the weights of the members.

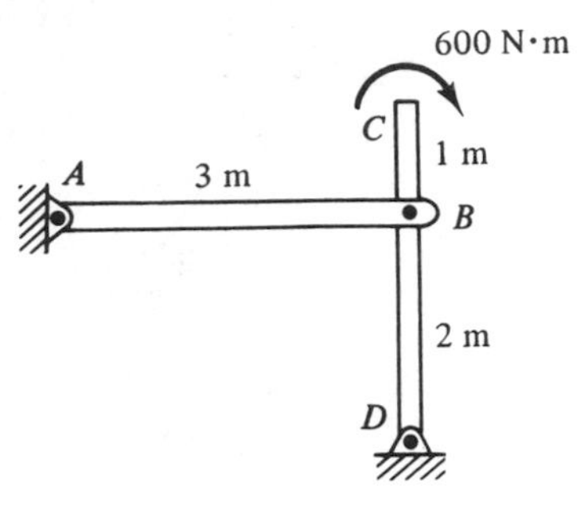

Problem 3.73

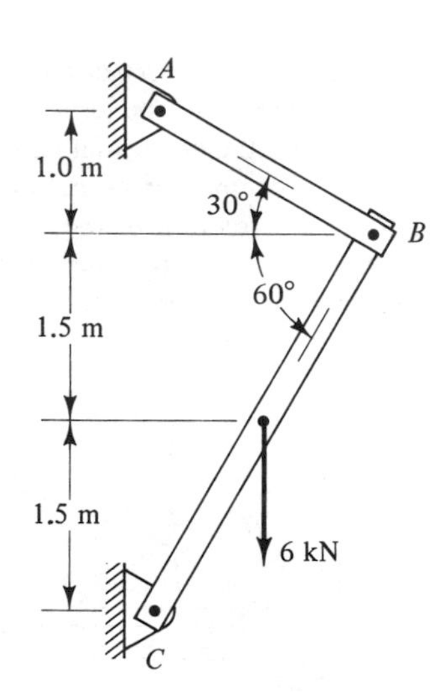

Problem 3.74

3.75. If the turn-buckle in the structure shown is tightened until the tension on cable EC is 8 kN, what are the horizontal and vertical components of the reactions at A and D? Ignore the weights of the members.

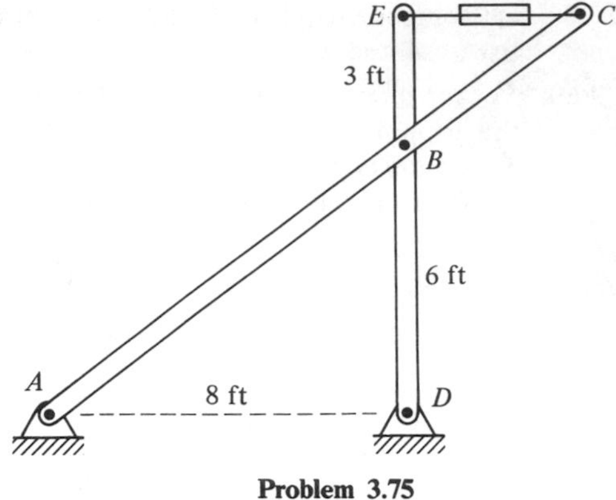

Problem 3.75

3.76. and 3.77. Determine the horizontal and vertical components of the forces acting on member ABC of the frame shown and show the results on a sketch of the member.

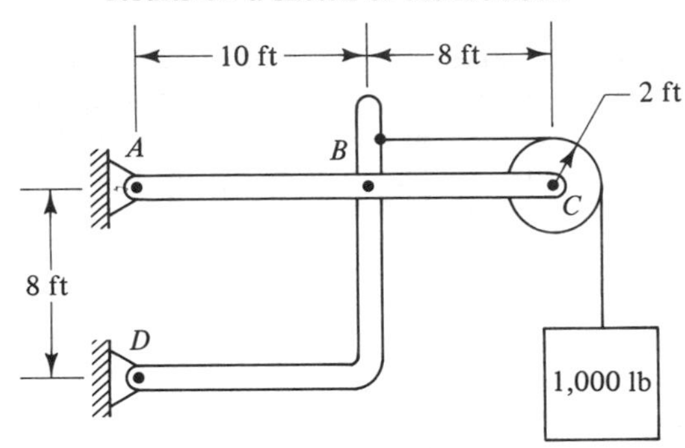

Problem 3.76

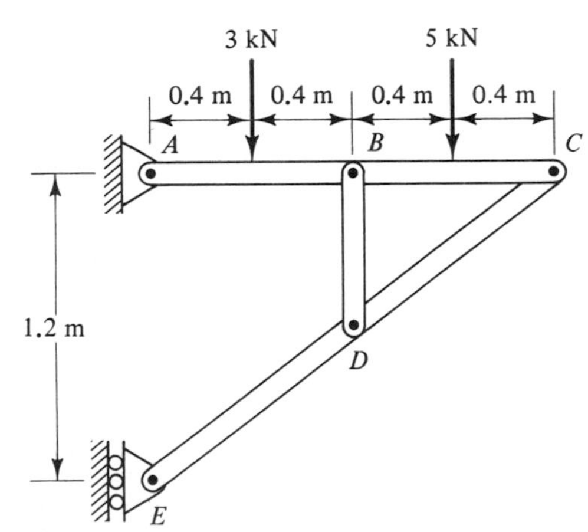

Problem 3.77

3.78. and 3.79. For the pin-connected frame shown, determine the horizontal and vertical components of the forces acting on each member. Assume that the members have negligible mass.

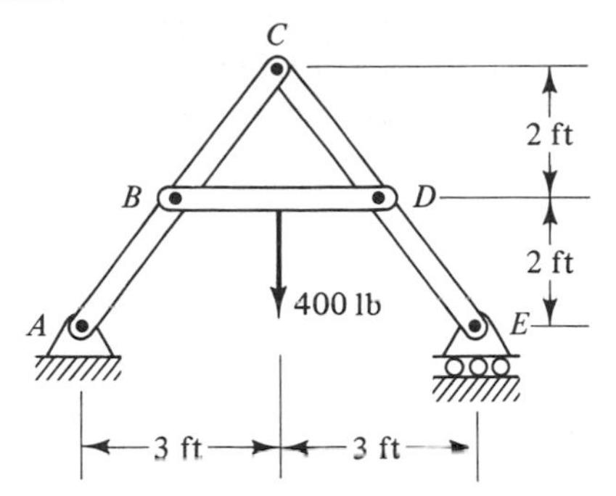

Problem 3.78

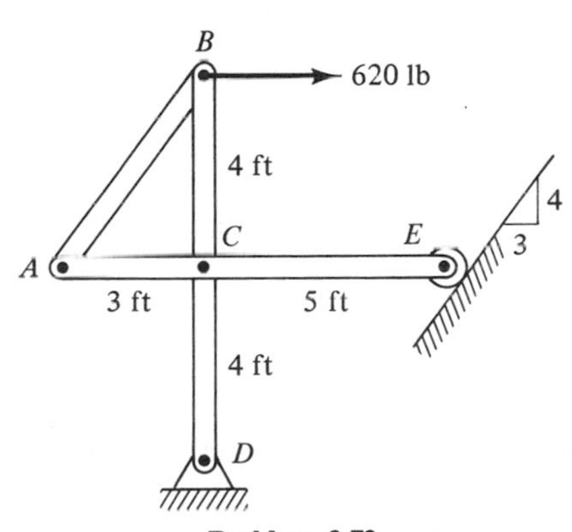

Problem 3.79

3.80. Find the reactions at each axle of the tractor-trailer unit shown. Assume that the reactions at the rear axles C and D are of equal magnitude. The tractor weighs 8,000 lb and the trailer and contents weigh 40,000 lb. The "fifth-wheel" at E connecting the units behaves like a hinged joint.

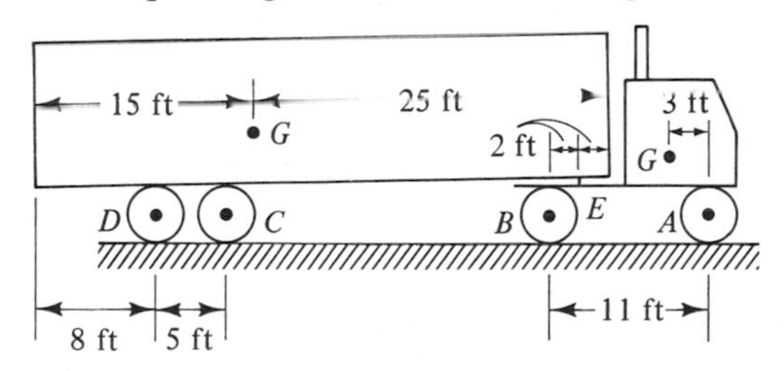

Problem 3.80

3.81. The essential features of the operating mechanism of a paper stapler are as shown. When the handle is depressed, link AB moves downward, deflecting the leaf spring BC. If a force of 5 N/mm is required to deflect the spring, what is the minimum force P needed to bring points D and E into contact? What are the corresponding reactions at the base C of the spring and at the pin F? Ignore the weights of the members.

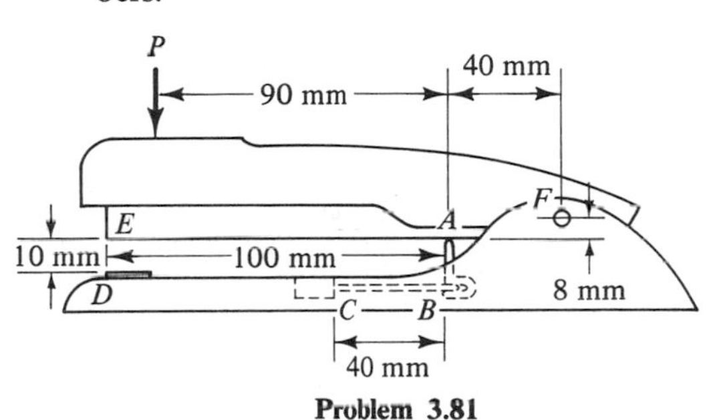

Problem 3.81

3.82. If $\theta = \phi = 30°$, what couple C must be exerted on arm OD of the simple robot

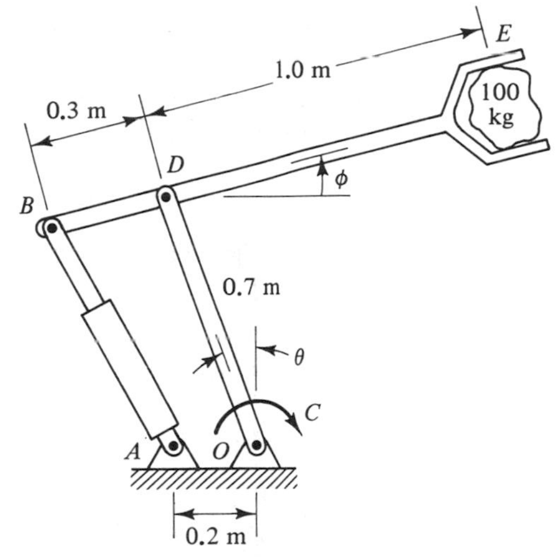

Problem 3.82

shown to maintain equilibrium while supporting a 100-kg payload? What is the corresponding force exerted on arm *BDE* by the hydraulic cylinder *AB*? Ignore friction and the mass of the members.

3.83. Same as Problem 3.82, except that $\theta = 45°$ and $\phi = 30°$.

3.84. The rack-and-pinion arrangement shown is used to position a heavy object with weight *W*. What force *F* is required to maintain equilibrium for the position indicated?

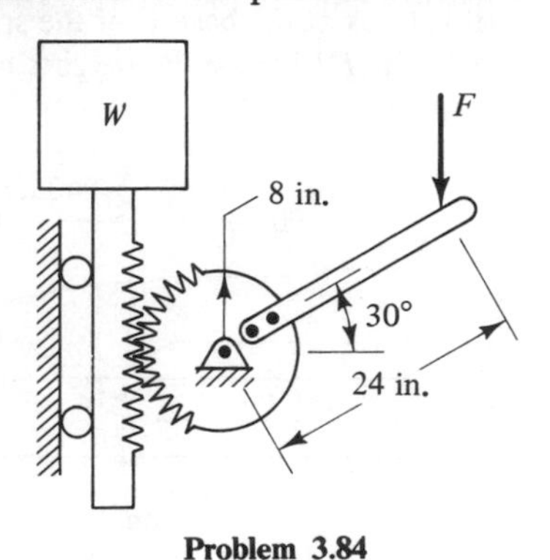

Problem 3.84

3.85. A schematic of a large gear train used to lift ore at a mine is as shown. Assuming a constant speed of rotation, what couple *C* must be applied to the input shaft to provide a torque (couple) of 100,000 lb · ft at the output shaft? Gears *A* and *D* have pitch diameters of 6 in. and gears *B* and *E* have pitch diameters of 48 in.

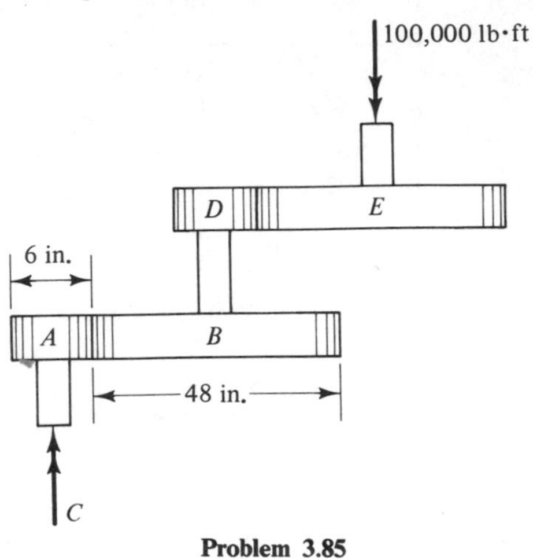

Problem 3.85

3.8 FORCE ANALYSIS OF TRUSSES

A *truss* is a structure consisting of a number of straight, slender members connected at their ends to form a rigid unit. By rigid, we mean that the structure will not collapse under a small applied load. For example, the structure shown in Figure 3.22(a) is nonrigid, and it will collapse if given a slight push. However, it can be

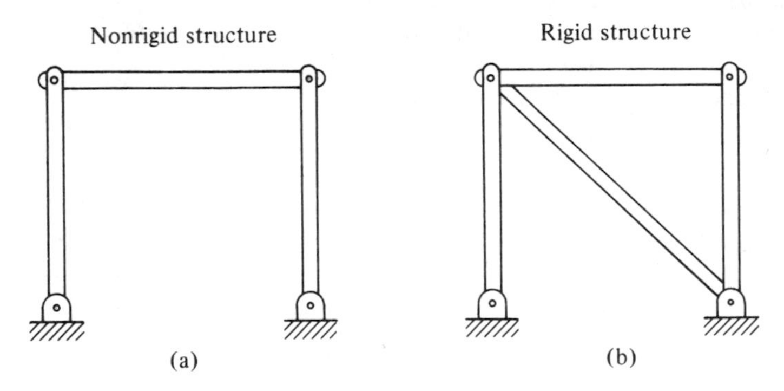

Figure 3.22

made rigid by adding a brace, as in Figure 3.22(b). Note that the resulting structure forms two connected triangles. The triangle is the simplest rigid unit, and most trusses consist of a number of connected triangular elements.

Trusses are efficient structures because they are lightweight and can support relatively large loads. They are commonly used in bridges, in the roofs of buildings, and in numerous other applications. Schematics of two common types of trusses are shown in Figure 3.23. These trusses are examples of so-called *plane trusses*, in which the individual members all lie in one plane. Trusses whose members do not all lie in one plane are called *space trusses*.

The force analysis of trusses is based upon the assumption that all members can be treated as two-force members. This feature distinguishes trusses from the frames and other complex systems considered in Section 3.7. The assumption of two-force members is reasonable if:

1. All connections are the equivalent of smooth pins (or ball and socket joints in the case of space trusses).
2. The weights of the members are negligible compared to the applied loads.

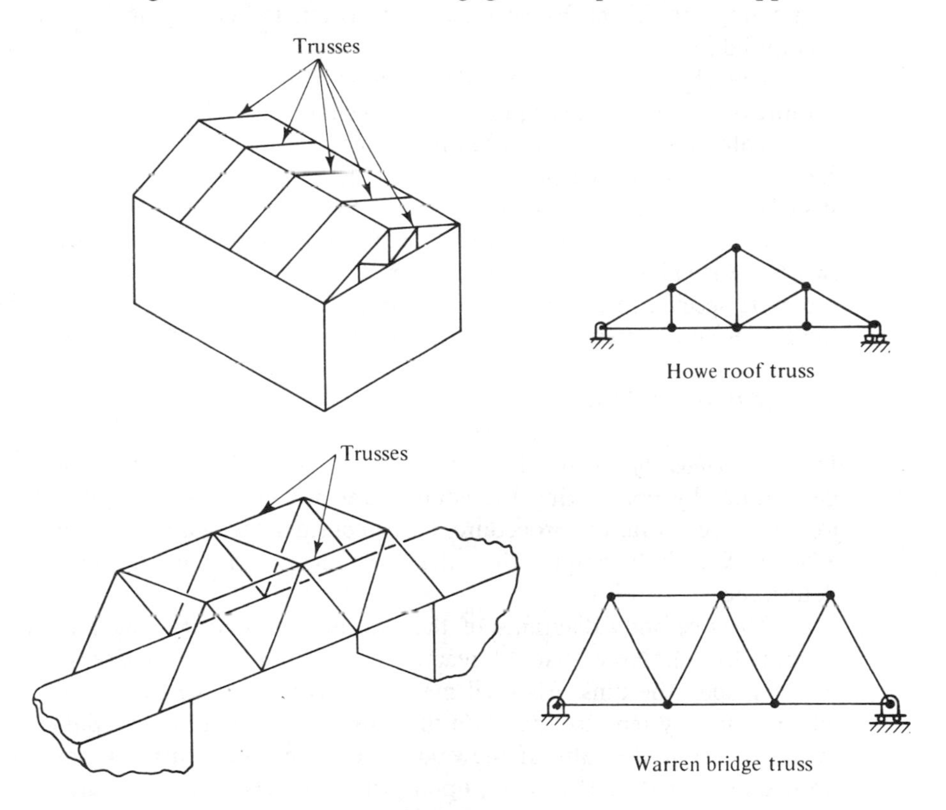

Figure 3.23

3. All loads and reactions consist only of forces (no couples) and act only at the joints.

Most trusses are designed so that the preceding conditions are met. For example, bridge decks are attached to the supportive side trusses only at the joints; consequently, the applied loads are transmitted to the trusses only at these locations. The connections in a truss are usually arranged so that the centerlines of the members joined together are concurrent, or nearly so. In this case, the joints behave like pinned connections to a good degree of approximation, even though the members may actually be joined by bolts, rivets, welds, nails, adhesives, or other types of connectors. If the centerlines of the members are not concurrent at the connections, couples of appreciable magnitude can be exerted on their ends and they can no longer be treated as two-force members.

If the weight of a truss member is significant, it can be accounted for approximately by replacing the weight W with loads equal to $W/2$ acting at each end of the member. This procedure makes it possible to continue to treat the members as two-force members, and it gives the correct average tension or compression acting upon them. However, any effects due to bending of the members are not accounted for.

The object of a force analysis of a truss is to determine the tensile or compressive forces acting upon the various members. The procedure for doing this is basically the same as that used for frames and other complex systems in Section 3.7. We first determine the reactions by considering the equilibrium of the entire structure (not always necessary), and then we dismember the structure and determine the forces on the individual members. However, since trusses consist only of two-force members, the procedures for carrying out the second of these two steps are more specific than in the case of frames. Two common methods for determining the forces on the members of a truss are discussed in the following.

Method of Joints

In the *method of joints*, the forces acting upon the members of the truss are determined by considering the equilibrium of the connecting pins at each of the joints. To illustrate the procedure, let us consider the simple truss shown in Figure 3.24(a). We shall assume that the reactions at A and D have already been determined.

The free-body diagrams of the members and connecting pins are shown in Figure 3.24(b). Note that all reactions and applied forces are shown as acting directly upon the pins. Also, all members have been assumed to be in tension, in which case they tend to "pull" on the pins. This assumption simplifies the interpretation of the final results. If the value of a force turns out to be positive, we know immediately that the member upon which it acts is in tension; if the force is negative, the member is in compression.

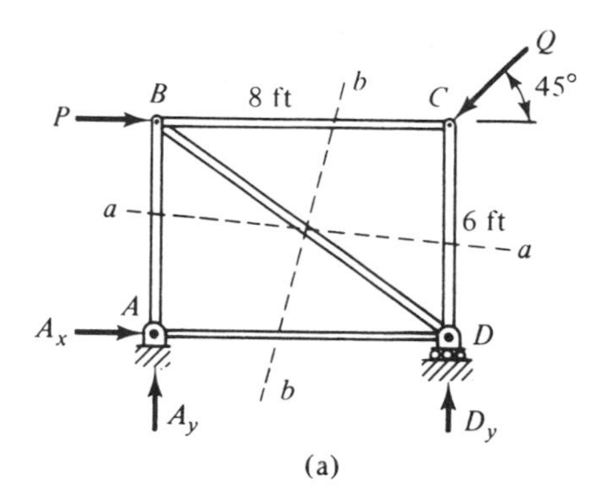

(a)

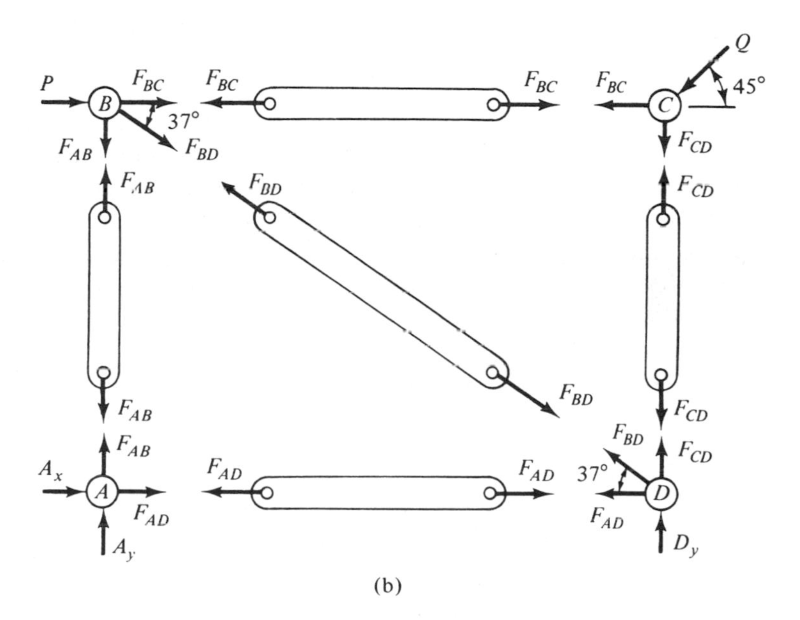

(b)

Figure 3.24

From the free-body diagrams of Figure 3.24(b), it is clear that the forces acting upon the members can be determined from the equations of equilibrium ($\Sigma F_x = 0$ and $\Sigma F_y = 0$) for the pins. (For space trusses we also have $\Sigma F_z = 0$.) We start at a pin at which there are no more than two unknowns, such as pins A or C in Figure 3.24(b). Once we have determined the unknowns at the first pin, we proceed to another. The process is repeated until all the unknowns have been determined. The order in which the pins are considered is immaterial; the only restriction is that the pin under consideration not involve more unknowns than equations of equilibrium.

For example, suppose we start at pin A and determine F_{AB} and F_{AD}. Once F_{AB} and F_{AD} are known, we can go to pin B or pin D, or we can go to pin C. Suppose we go to pin B and solve for F_{BC} and F_{BD} and then go to pin C and solve for the remaining unknown F_{CD}. The equilibrium equations for pin D are not needed; they can be used as a check.

It is not necessary to draw free-body diagrams of the truss members when using the method of joints. These diagrams were included here only to illustrate more clearly the force interactions between the members and the connecting pins. Only the free-body diagrams of the pins are needed.

Method of Sections

In the *method of sections*, the forces on the members of the truss are determined by imagining that the truss is cut into sections and then considering the equilibrium of the resulting pieces. To illustrate the procedure, let us again consider the truss of Figure 3.24(a).

If the truss is sectioned along line *a-a*, the free-body diagrams of the resulting pieces are as shown in Figure 3.25. Again, it is convenient to assume that all members are in tension. Since the forces acting upon each section of the truss are coplanar, there are three equilibrium equations for each section (there are six equations for space trusses). Thus, the forces F_{AB}, F_{BD}, and F_{CD} can be determined by considering the equilibrium of either the top or bottom piece of the truss.

In order to determine the forces on the rest of the members, it is necessary to section the truss in a different way. For example, the remaining unknowns F_{BC} and F_{AD} can be determined by sectioning the truss along line *b-b* [Figure 3.24(a)] and considering the equilibrium of either the left or right portions.

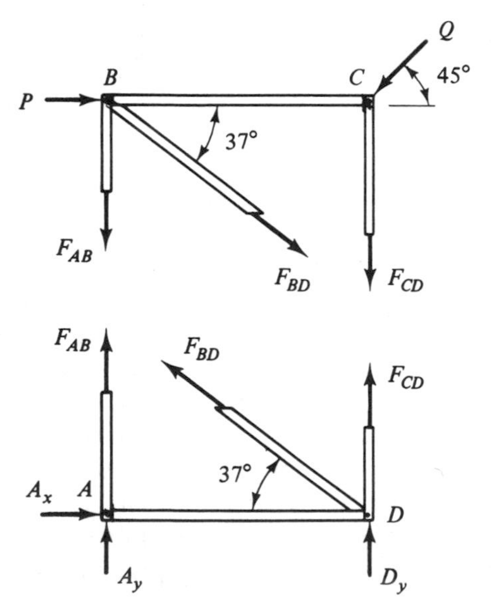

Figure 3.25

There is no definite procedure for determining where or how to section a truss, but the preceding example does reveal two general guidelines: (1) The section must pass through the member on which the force is to be determined and (2) the resulting free-body diagram should not involve more unknowns than equations of equilibrium. The first condition guarantees that the force of interest will appear on the free-body diagram and enter into the equilibrium equations, and the second condition guarantees that the force can be determined from these equations.

As a rule, the method of sections involves less work than the method of joints, but it requires more ingenuity to apply. The method of sections is particularly convenient when the forces on only a few members in the interior of a truss are to be determined. In many problems it is advantageous to use a combination of the method of sections and method of joints.

Zero-Force Members

Under any one given loading, the forces on some members of a truss may be zero. Member BD of the truss shown in Figure 3.26 is an example of such a *zero-force member*, as can be seen from the free-body diagram of pin D. Of course, a zero-force member under one loading may not be a zero-force member under a different loading. For instance, member BD would not be a zero-force member if the load were applied at joint D instead of joint B.

Even though they support no load, zero-force members have an important function. They serve to stabilize a truss and to hold in position the members that do carry a load. For example, if the zero-force member BD were removed from the truss shown in Figure 3.26, joint D would be free to move up or down, and any small lateral load or disturbance applied to member AD or DC would cause the truss to fold.

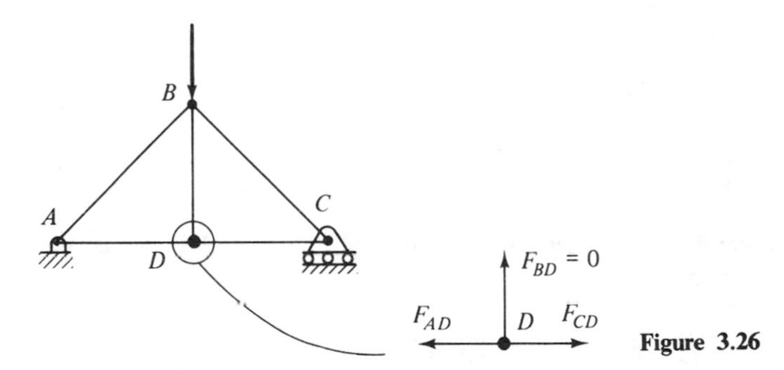

Figure 3.26

Example 3.12. Force Analysis of a Truss. Determine the forces on members BC and DE of the truss shown in Figure 3.27(a) by using (a) the method of joints and (b) the method of sections.

Solution. (a) The forces on members BC and CE can be determined by considering joint C; the force on member DE can then be obtained by considering joint E. If it is

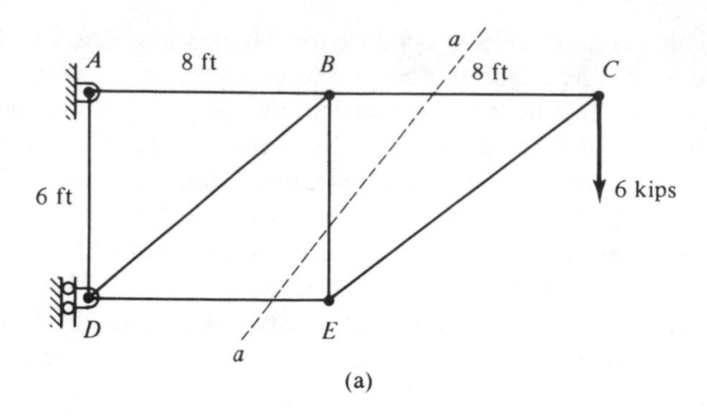

(a)

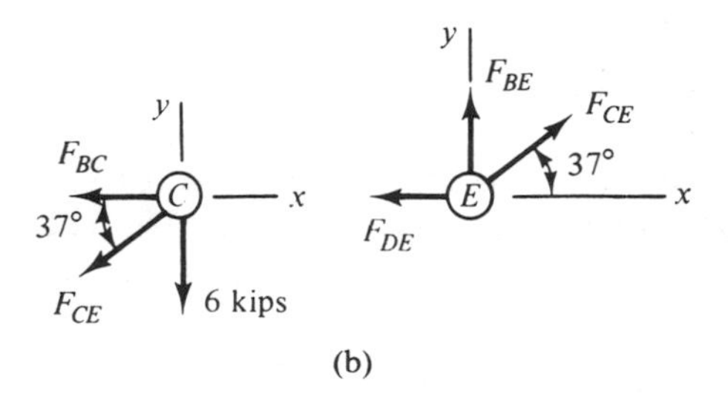

(b)

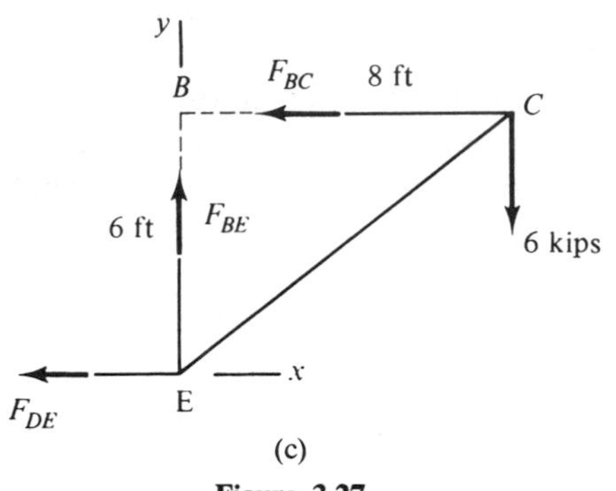

(c)

Figure 3.27

assumed that all members are in tension, the FBDs of pins C and D are as shown in Figure 3.27(b). For equilibrium, we have:

Pin C:

$$\sum F_x = -F_{BC} - F_{CE}\cos 37° = 0$$

$$\sum F_y = -6 \text{ kips} - F_{CE}\sin 37° = 0$$

$$F_{CE} = -10 \text{ kips or } 10 \text{ kips (compression)}$$ **Answer**

$$F_{BC} = +8 \text{ kips or } 8 \text{ kips (tension)}$$

Pin E:

$$\sum F_x = F_{CE}\cos 37° - F_{DE} = 0$$

$$F_{DE} = -8 \text{ kips or } 8 \text{ kips (compression)}$$ **Answer**

(b) Sectioning the truss along line *a-a* and considering the portion to the right, we obtain the FBD shown in Figure 3.27(c). Again, all members are assumed to be in tension. For equilibrium

$$\sum \overset{+}{\curvearrowleft} M_E = F_{BC}(6\text{ ft}) - 6\text{ kips}(8\text{ ft}) = 0$$

$$\sum F_x = -F_{DE} - F_{BC} = 0$$

from which we obtain the same results as before:

$$F_{BC} = 8\text{ kips (tension)}$$

$$F_{DE} = 8\text{ kips (compression)}$$

Note that the reactions did not enter into the computations in this problem; thus, there was no need to determine them. This is not usually the case, however.

PROBLEMS

3.86. The members of the truss shown are of equal length. Using the method of joints, determine the force acting on each member.

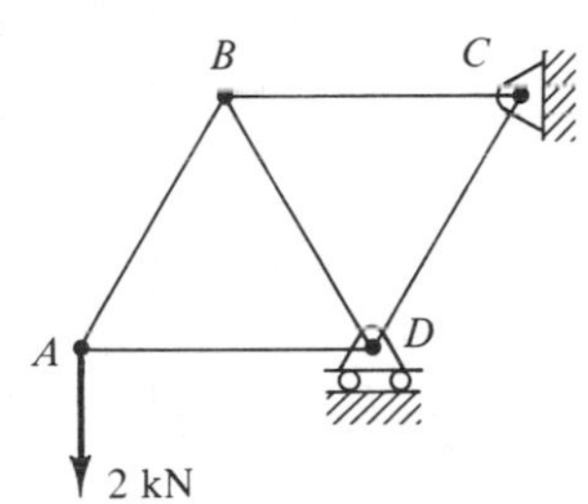

Problem 3.86

3.87. Each member of the truss shown is a uniform bar weighing 40 lb/ft. Determine the average tension or compression on each member due to the weights of the members. Use the method of joints.

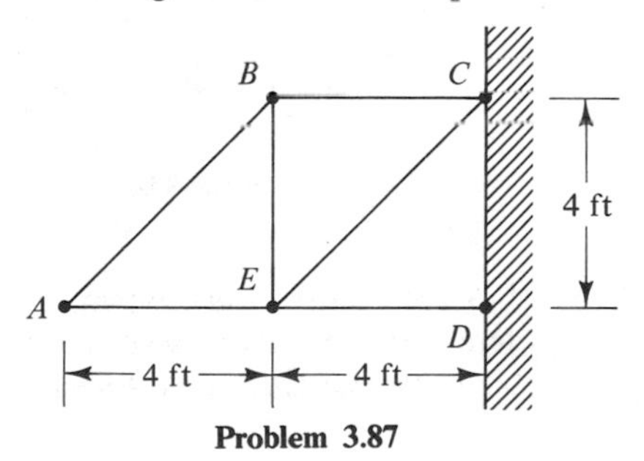

Problem 3.87

3.88. A truss is fabricated from five timbers connected as shown. Each member weighs 2.4 lb/ft. Determine the average tensile or compressive force on each member. If the weights of the members are neglected, what are the percentage errors in the resulting values of the forces?

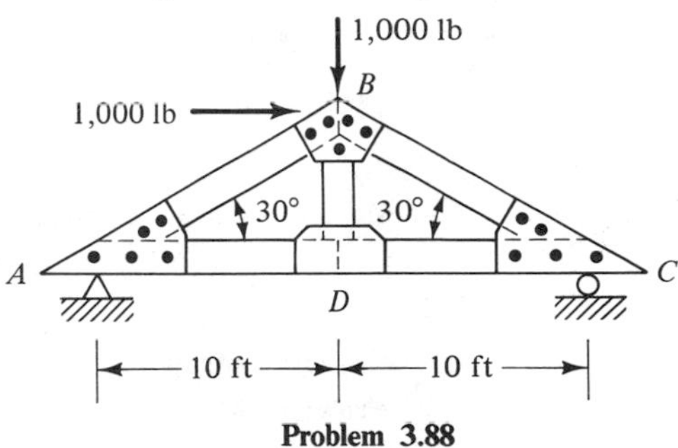

Problem 3.88

3.89. to 3.91. For the truss shown, determine the forces acting upon the members indicated by using the method of sections.

3.89. Members BC and CF.

3.90. Members DF, EC, and EG.

3.91. Members CD, CG, and FG.

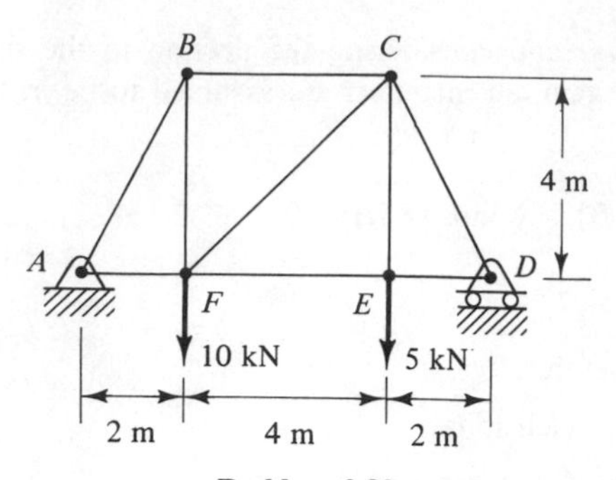

Problem 3.89

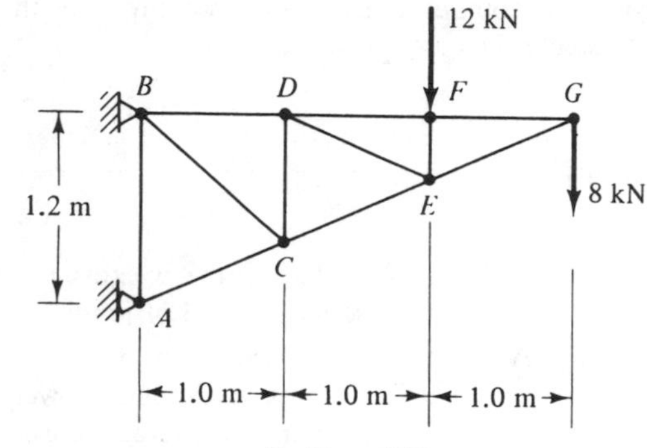

Problem 3.90

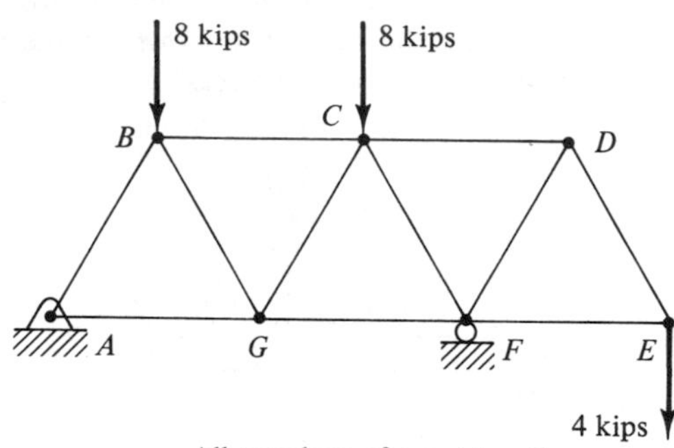

All members of equal length

Problem 3.91

3.92. For the truss in Problem 3.91, determine the forces on the four members which intersect at joint F. Use the method(s) of your choice.

3.93. Determine the forces acting on members BC, CE, and BF of the truss shown.

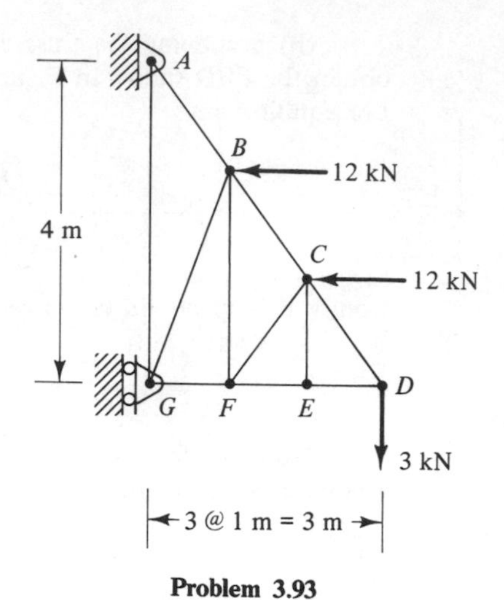

Problem 3.93

3.94. Identify the zero-force members in the truss shown and determine the forces on members FH and IJ.

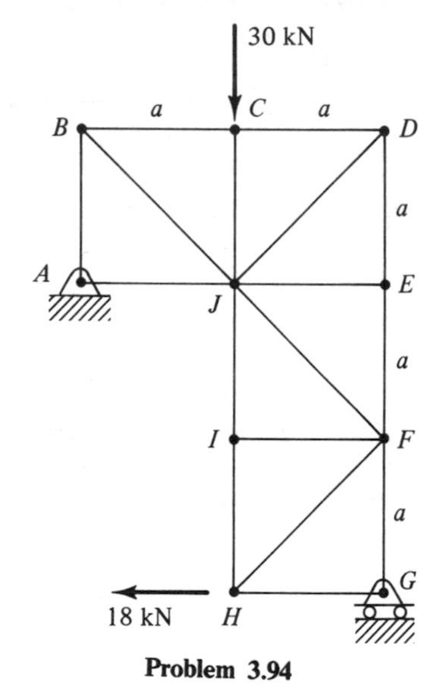

Problem 3.94

3.95. For the stadium truss shown, identify the zero-force members and determine the forces on members DE, EK, and KL.

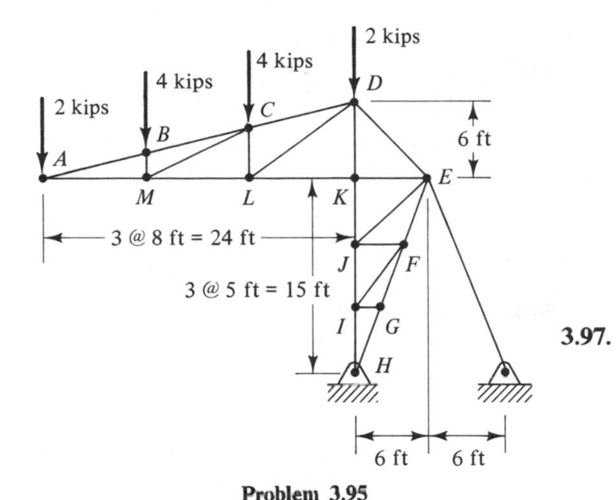

Problem 3.95

3.96. A drive-in theater screen is supported by four trusses, as shown. The screen is attached to joints D, E, and F of the first truss and to the corresponding joints of the adjoining trusses. If the screen has a weight of 96 kN and the wind exerts a net horizontal force of 72 kN against it, what are the forces acting on the members which intersect at joints D, E, and F? Assume that equal shares of the loads act at each of the points of attachment of the screen.

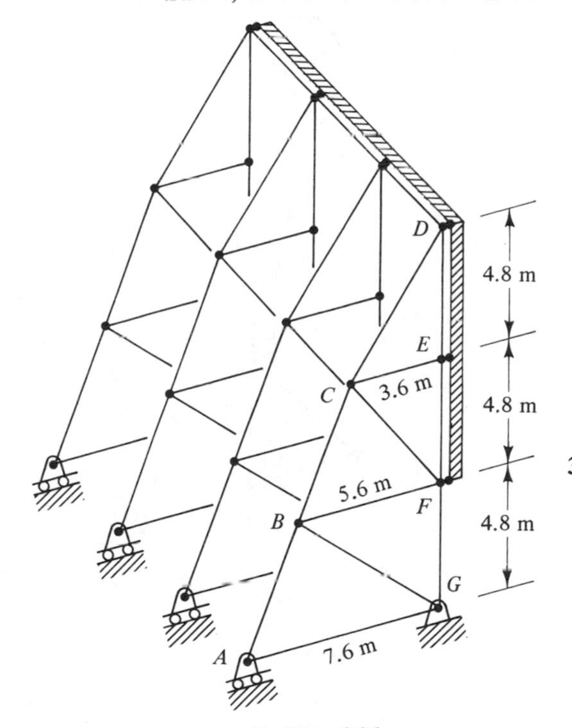

Problem 3.96

3.97. The boom CDE is hinged at C and is held in a horizontal position by a cable which passes over small pulleys at A and F, as shown. Determine the forces acting on the members of the supporting truss. Ignore the weight of the boom.

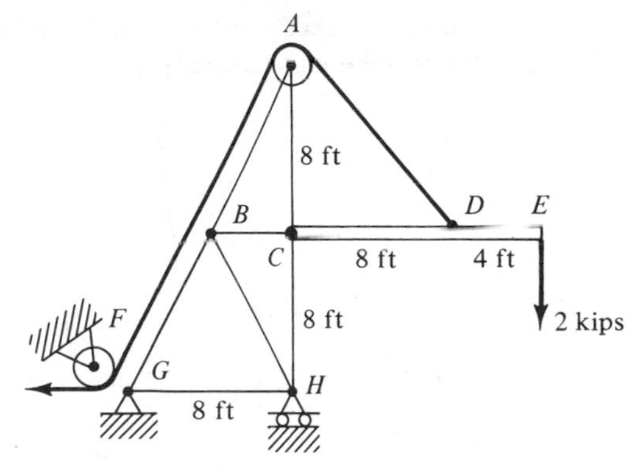

Problem 3.97

3.98. A conveyor belt moves across a 20-ft span on a series of rollers supported on each side by a truss made of two horizontal pipes connected by a series of braces welded in place as shown. If each roller weighs 60 lb and the belt and payload together weigh 120 lb/ft, what are the forces acting on members CD, CJ, DJ, and KJ? Assume that the tension on the belt is uniform along its length.

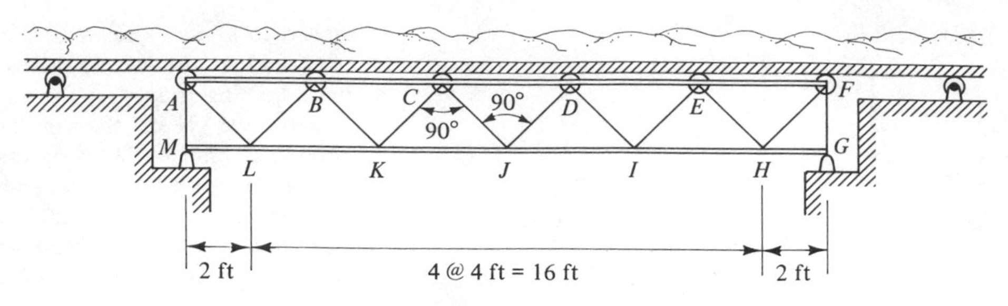

Problem 3.98

3.99. An object with a mass of 1 ton is suspended from a tripod, as shown. Determine the force acting on each of the members. All connections are equivalent to ball-and-socket joints, and the weights of the members are negligible.

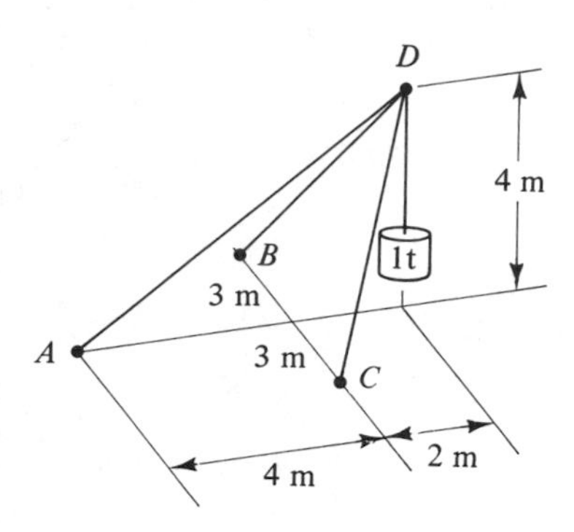

Problem 3.99

3.100. A space truss is used as the support tower for a windmill, as shown. (For clarity, not all members of the truss are depicted.) The windmill and tower have a combined weight of 12,000 lb, with center of gravity located along the centerline of the tower. Determine the forces on the five members that intersect at joint G. Ignore the weights of the individual members involved and assume that all connections are equivalent to ball-and-socket joints.

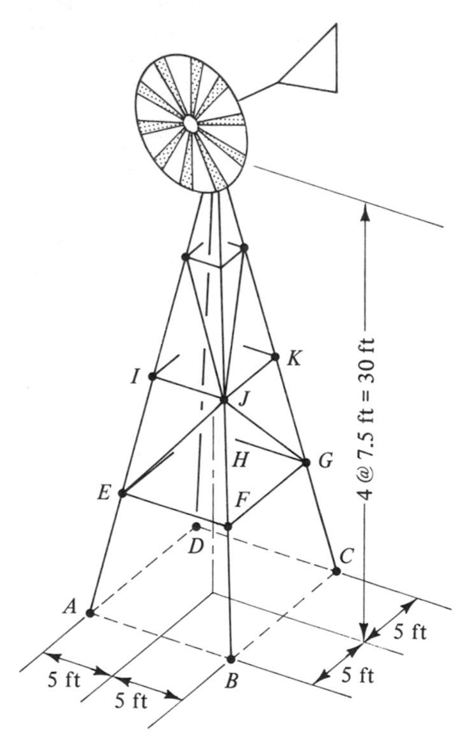

Problem 3.100

3.101. A section of the boom for a crane consists of a space truss constructed in the form of a cube with six diagonal members. Corner A is supported by a ball-and-socket joint and corner B is attached to a link parallel to the x axis; corners G and H rest on spherical rollers. For the loading and support system shown, determine the forces on all members that intersect at joint B.

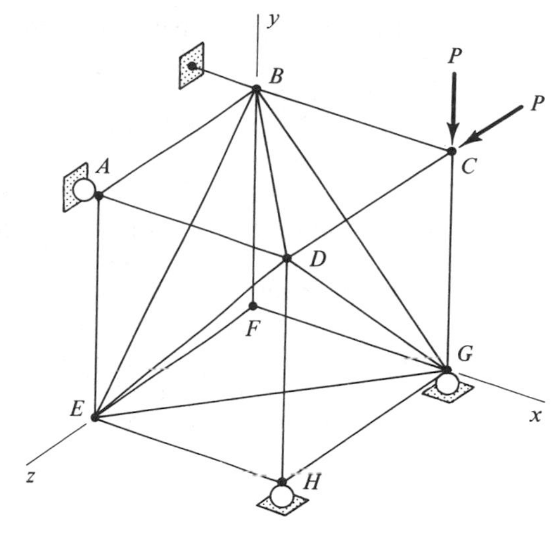

Problem 3.101

3.9 FORCE ANALYSES INVOLVING FRICTION

As we have already mentioned in Section 3.4, there is a friction force developed between rough surfaces in contact when one surface tends to slide relative to the other. This force is tangent to the plane of contact, and it always tends to oppose the relative motion of the body upon which it acts. In this section we shall consider the friction force in more detail.

Friction is somewhat of a paradox because it is both a tremendous asset and a liability. For instance, friction contributes to the wear in the engine of our automobile, but the vehicle could not move were it not for the friction between the tires and the road. There are several types of friction, but we shall consider only the so-called *dry*, or *Coulomb*, *friction* that occurs between nonlubricated surfaces.

The unique thing about the friction force is that there is a definite limit to its magnitude. Consider, for example, a block sitting on a rough surface and subjected to a horizontal force F [(Figure 3.28(a)]. A free-body diagram of the block is shown

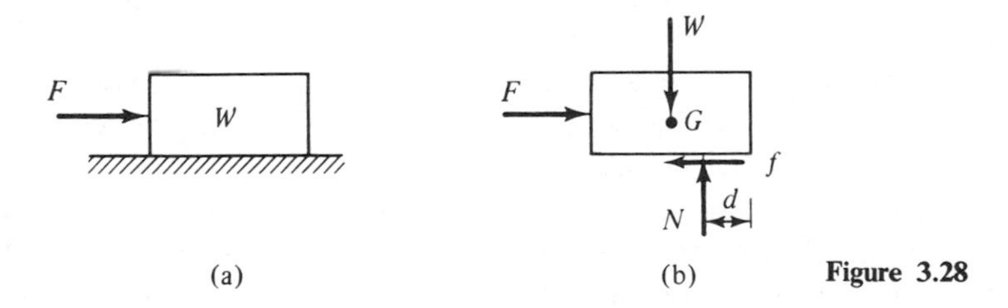

Figure 3.28

in Figure 3.28(b). Experiments show that (a) the block will remain in equilibrium if F is sufficiently small ($F = f$ in this case), (b) the block will start to move if F is increased to a certain value, and (c) the force required to keep the block in motion is less than that required to start it moving. The existence of these various types of responses is easily demonstrated. For example, try pushing a book or other fairly heavy object over a desk or table top. You will be able to sense with your hand the relative changes in F as you push harder upon the object and it starts to move.

The behavior of the block shown in Figure 3.28(a) is conveniently displayed by plotting the magnitude f of the friction force versus the applied force F, as in Figure 3.29. As F increases, f also increases until it reaches a maximum value f_{max}. At this instant the block is in a state of *impending motion*; it is still in equilibrium, but it is on the verge of moving. If F is at all increased beyond this point, the friction force will suddenly decrease. As a result, F will be greater than f and the block will move.

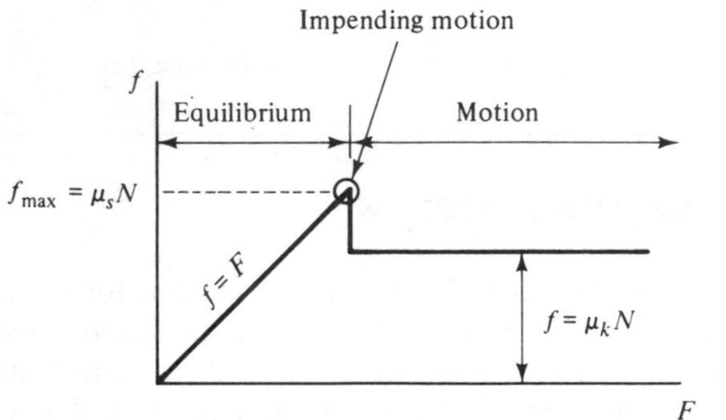

Figure 3.29

The maximum friction force that can be developed between any two given surfaces depends upon a number of factors, many of which are not entirely understood. However, experiments show that f_{max} is proportional to the normal force N between the contacting surfaces. Thus, we can write

$$f_{max} = \mu_s N \tag{3.7}$$

The constant of proportionality μ_s is called the *coefficient of static friction*. The friction force is also proportional to the normal force when there is relative motion between the contacting surfaces. In this case,

$$f = \mu_k N \tag{3.8}$$

where μ_k is the *coefficient of kinetic friction*. The laws of friction expressed in Eqs. (3.7) and (3.8) are based upon the experiments of the French physicist C. A. Coulomb, the results of which were published in 1781.

The coefficients of static and kinetic friction vary greatly for different materials and for different conditions of their surfaces. They also depend upon environmental factors such as temperature and humidity. Representative values of the

TABLE 3.1 COEFFICIENTS OF STATIC AND KINETIC FRICTION

Materials	μ_s	μ_k
Metal on metal	0.7	0.5
Metal on wood	0.6	0.4
Metal on stone	0.7	0.4
Metal on ice	0.03	0.02
Wood on wood	0.6	0.5
Rubber tires on dry pavement	0.9	0.8

coefficients of friction for various materials are given in Table 3.1. Note that $\mu_k < \mu_s$, which is in keeping with the fact that the friction force decreases when motion occurs.

The frictional characteristics of two contacting surfaces can also be expressed in terms of the angle of inclination ϕ of the resultant contact force R (Figure 3.30). From this figure, we have

$$\tan \phi = \frac{f}{N}$$

The angle ϕ is zero for frictionless surfaces, and it attains its maximum value ϕ_s when motion is impending.

For impending motion, $f = \mu_s N$; so

$$\tan \phi_s = \mu_s \tag{3.9}$$

When there is relative motion between the two surfaces, Eq. (3.8) applies, and we have

$$\tan \phi_k = \mu_k \tag{3.10}$$

The angles ϕ_s and ϕ_k are called the *angles of static* and *kinetic friction*, respectively. In some fields of engineering it is common practice to describe the friction forces in terms of these angles instead of the coefficients of friction.

The procedure for handling problems involving friction depends upon what is known about the problem. If it's known that motion is impending, then $f = f_{max} = \mu_s N$. In this case, the friction force is a known quantity (once N is determined) and,

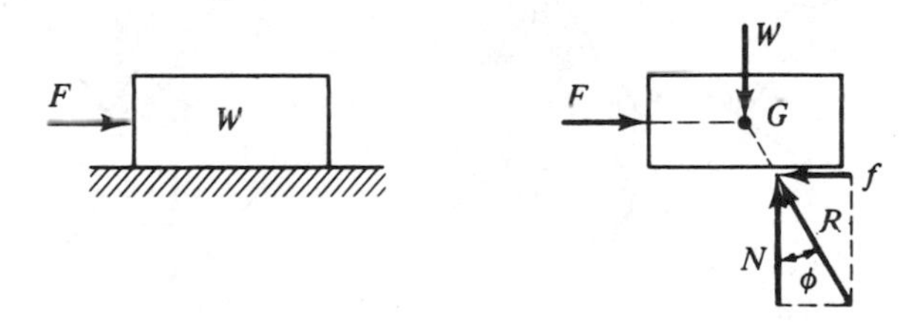

(a) (b) **Figure 3.30**

therefore, it must be shown on the free-body diagram with the proper sense. The friction force on a body always points in the direction opposite to that in which the body tends to move relative to the member with which it is in contact. The preceding comments also apply when the body is in motion, except that in this case $f = \mu_k N$.

If the system is in equilibrium and motion is not impending, the friction force is an unknown that must be determined from the equations of equilibrium. In this case, the friction force is just like any other unknown force, and its sense can be chosen arbitrarily on the free-body diagram.

In some problems it is not known beforehand whether the system will move or remain in equilibrium under the action of the applied forces. The following procedure is convenient for problems of this type:

1. Assume that the system is in equilibrium and solve for **f**.
2. Compare $|\mathbf{f}|$, which is the magnitude of the friction force required to maintain equilibrium, with $f_{max} = \mu_s N$, which is the maximum friction force available.
3. If $|\mathbf{f}| \leqslant f_{max}$, the system is in equilibrium as assumed, and the friction force is as determined in step 1. In other words, the surfaces can generate a friction force of sufficient magnitude to maintain equilibrium.
4. If $|\mathbf{f}| > f_{max}$, the surfaces cannot generate a friction force large enough to maintain equilibrium. The system will move, and $|\mathbf{f}| = \mu_k N$. The sense of **f** is as determined in step 1, and the direction of relative motion is opposite to the sense of **f**.

Example 3.13. Force Analysis of a Block. The metal block shown in Figure 3.31(a) has a weight of 70 lb, and the incline is made of wood. Determine the force F required to (a) prevent the block from sliding down the incline and (b) start it moving up the incline.

Solution. (a) When the block is on the verge of sliding down the incline, the FBD of the block is as shown in Figure 3.31(b). The friction force has magnitude $\mu_s N$ and in directed up the plane, opposite the direction in which the block tends to move relative to the fixed

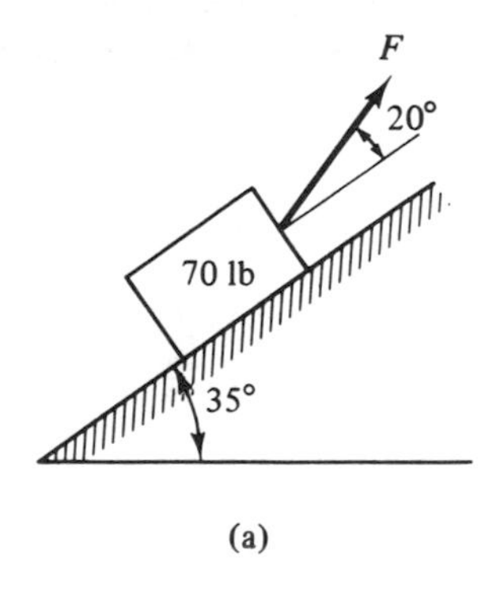

(a)

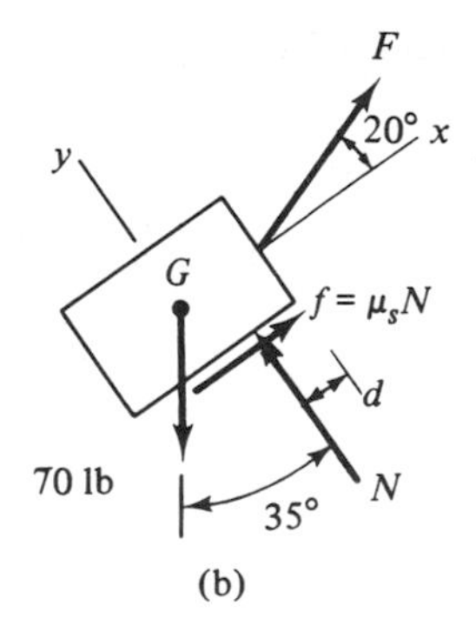

(b)

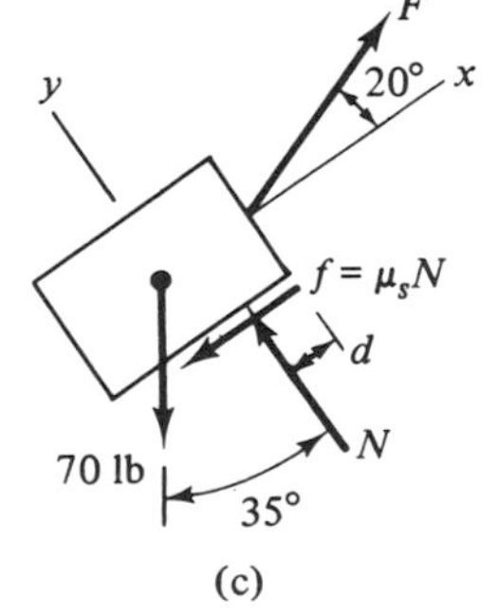

(c)

Figure 3.31

incline. For equilibrium,

$$\sum F_x = F\cos 20° + \mu_s N - (70 \text{ lb})\sin 35° = 0$$

$$\sum F_y = F\sin 20° + N - (70 \text{ lb})\cos 35° = 0$$

Solving these equations by using the value $\mu_s = 0.6$ obtained from Table 3.1 for metal on wood, we get

$$N = 54.7 \text{ lb} \qquad F = 7.8 \text{ lb} \qquad \textbf{Answer}$$

(b) When motion is impending up the incline, the FBD of the block is as shown in Figure 3.31(c). The only difference from the previous case is that the sense of the friction force is reversed. For equilibrium,

$$\sum F_x = F\cos 20° - \mu_s N - (70 \text{ lb})\sin 35° = 0$$

$$\sum F_y = F\sin 20° + N - (70 \text{ lb})\cos 35° = 0$$

from which wc obtain

$$N = 35.1 \text{ lb} \qquad F = 65.1 \text{ lb} \qquad \textbf{Answer}$$

To summarize, the block will slide down the incline if $F < 7.8$ lb, it will slide up the incline if $F > 65.1$ lb, and for $7.8 \text{ lb} \leqslant F \leqslant 65.1$ lb it will remain at rest.

Example 3.14. Determination of Friction Force. Determine the braking (friction) force acting upon the wheel shown in Figure 3.32(a) if the force F developed in the hydraulic

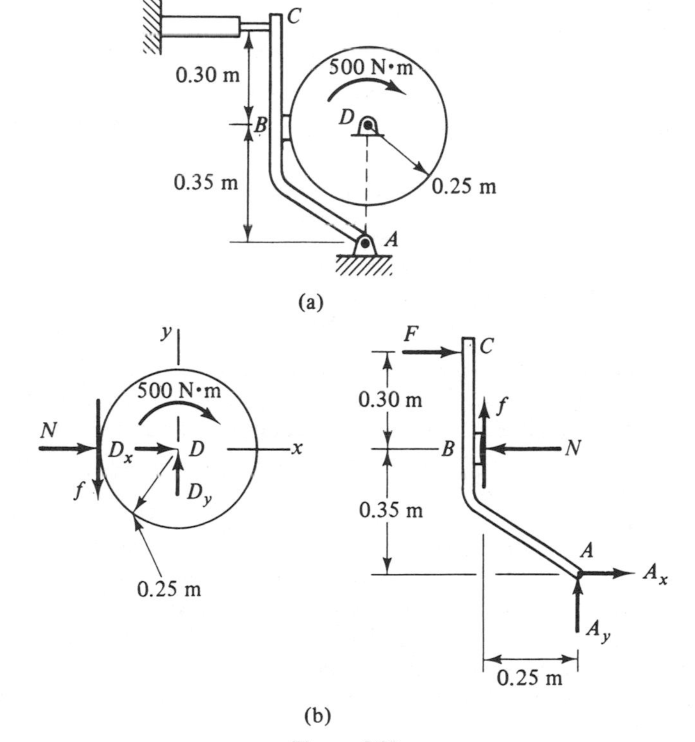

Figure 3.32

cylinder that actuates the brake is (a) 2 kN and (b) 1 kN. The coefficients of friction are $\mu_s = 0.4$ and $\mu_k = 0.3$.

Solution. Since we do not know beforehand whether the wheel will rotate or remain at rest under the action of the applied couple, we assume it is in equilibrium, In this case, the friction force **f** is an unknown. Referring to the FBDs of Figure 3.32(b), we have for the wheel

$$\sum \overset{+}{\curvearrowleft} M_D = f(0.25 \text{ m}) - 500 \text{ N} \cdot \text{m} = 0$$

For the brake arm,

$$\sum \overset{+}{\curvearrowleft} M_A = N(0.35 \text{ m}) - F(0.65 \text{ m}) - f(0.25 \text{ m}) = 0$$

(a) Solving the equilibrium equations with $F = 2$ kN, we obtain

$$f = 2.00 \text{ kN} \qquad N = 5.14 \text{ kN} \qquad f_{\max} = \mu_s N = 2.06 \text{ kN}$$

Since the friction force needed to maintain equilibrium is smaller than the maximum available friction force ($f < f_{\max}$), the wheel is in equilibrium as assumed. Thus, the friction force is

$$f = 2.00 \text{ kN} \qquad \textbf{Answer}$$

The sense of **f** is as shown on the FBDs.

(b) For $F = 1$ kN, we find

$$f = 2.00 \text{ kN} \qquad N = 3.29 \text{ kN} \qquad f_{\max} = \mu_s N = 1.32 \text{ kN}$$

Since $f > f_{\max}$, the surfaces cannot generate a friction force large enough to maintain equilibrium. Thus, the wheel will rotate clockwise (the direction of relative motion is opposite to **f**), and

$$f = \mu_k N = 0.3(3.29 \text{ kN}) = 0.99 \text{ kN} \qquad \textbf{Answer}$$

Example 3.15. Force Analysis of Blocks. In Figure 3.33(a) the uniform slab A weighs 5,000 lb and block B weighs 1,000 lb. What minimum force F is required to disturb equilibrium of the system? The coefficients of friction are as indicated in the figure.

Solution. This is an impending motion problem, but there is more than one way the system can start to move: (a) B can start to slide while A remains at rest or (b) B and A can both start to slide as a unit. The most straightforward way to proceed is to treat each of these cases as a separate problem and solve for F. The case that gives the smallest value of F is the one that actually occurs.

The FBD's of the blocks are shown in Figure 3.33(b). Note carefully the sense of the friction forces. Friction forces $\mathbf{f}_2$ and $\mathbf{f}_3$ are obviously directed to the left because both blocks tend to move to the right relative to the fixed surface. However, the situation may not be so obvious at the point of contact between the two blocks. Block A tends to move to the right relative to block B, while block B tends to move to the left relative to block A. Thus, $\mathbf{f}_1$ is directed as shown. Notice that once the direction of the friction force on one of the blocks has been established, the friction force on the other must be in accordance with Newton's third law.

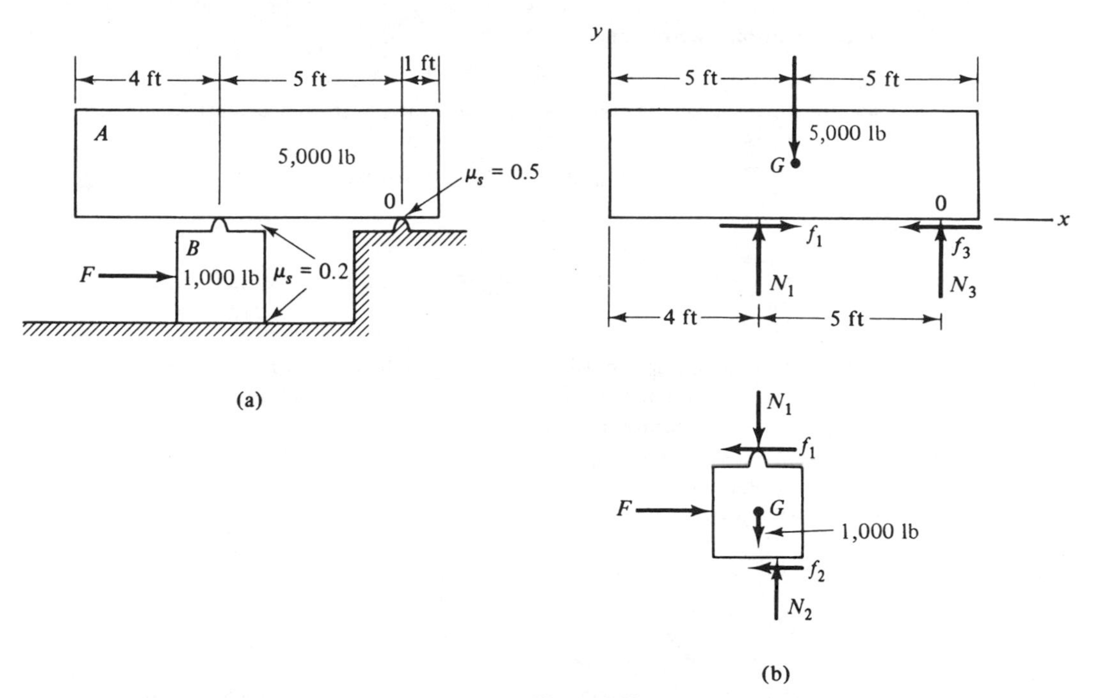

Figure 3.33

Case a. For this case, $f_1 = 0.2N_1$ and $f_2 = 0.2N_2$, while the friction force f_3 is unknown. For block B, we have

$$\sum F_x = F - 0.2N_1 - 0.2N_2 = 0$$

$$\sum F_y = N_2 - N_1 - 1{,}000 \text{ lb} = 0$$

For slab A,

$$\sum \overset{+}{\curvearrowleft} M_0 = 5{,}000 \text{ lb}(4 \text{ ft}) - N_1(5 \text{ ft}) = 0$$

Solving these equations, we get

$$F = 1{,}800 \text{ lb}$$

Case b. The only difference from the previous case is that now $f_3 = 0.5N_3$, while f_1 is unknown.

For block B, we have

$$\sum F_x = F - f_1 - 0.2N_2 = 0$$

$$\sum F_y = N_2 - N_1 - 1{,}000 \text{ lb} = 0$$

For slab A,

$$\sum \overset{+}{\curvearrowleft} M_0 = 5{,}000 \text{ lb}(4 \text{ ft}) - N_1(5 \text{ ft}) = 0$$

$$\sum F_y = N_1 + N_3 - 5{,}000 \text{ lb} = 0$$

$$\sum F_x = f_1 - 0.5N_3 = 0$$

Solving these equations, we obtain

$$F = 1{,}500 \text{ lb}$$

Case (b) gives the smallest value of F. Therefore, both blocks start to slide as a unit, and

$$F_{\min} = 1{,}500 \text{ lb} \qquad \textbf{Answer}$$

PROBLEMS

3.102. The log shown is to be pulled up a hillside by a chain attached to a tractor. If the mass of the log is 1.5 ton, what force F is required to (a) start the log moving up the hill, and (b) keep it moving up the hill? Assume that $\mu_s = 0.8$ and $\mu_k = 0.6$.

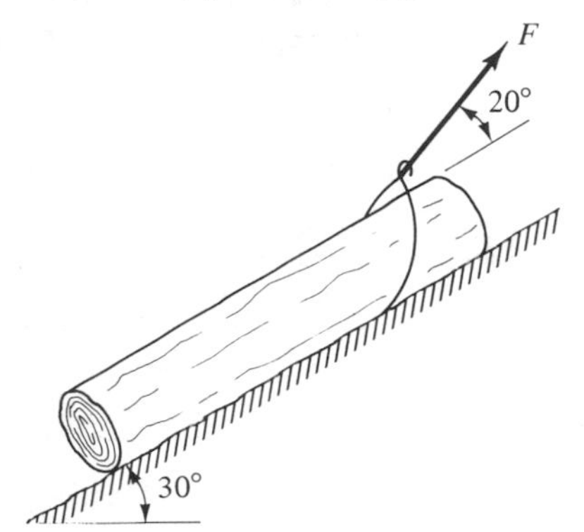

Problem 3.102

3.103. What is the steepest angle of inclination θ (the *angle of repose*) for which the block shown will remain in equilibrium if the static coefficient of friction is μ_s? How does this angle compare with the angle of static friction ϕ_s?

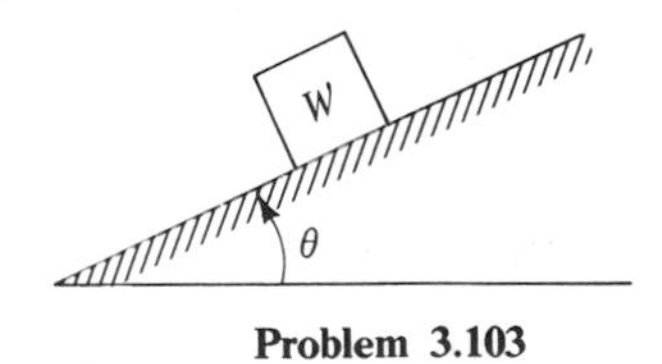

Problem 3.103

3.104. Determine the range of values of the weight W for which the system shown will remain in equilibrium; $\mu_s = 1/4$.

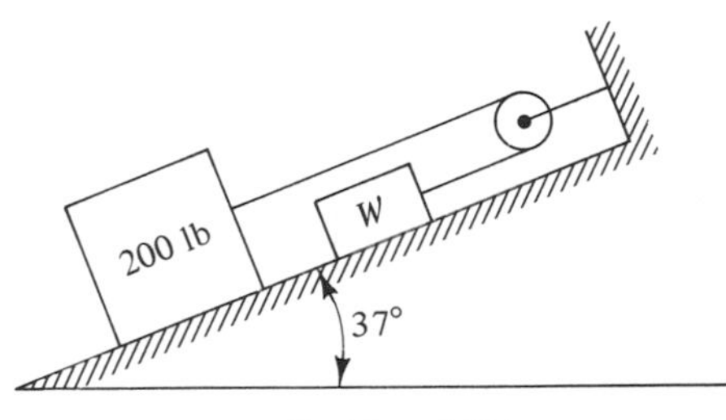

Problem 3.104

3.105. For the brake arrangement shown, determine the minimum force F required to prevent the block from descending. Assume that $\mu_s = 0.5$.

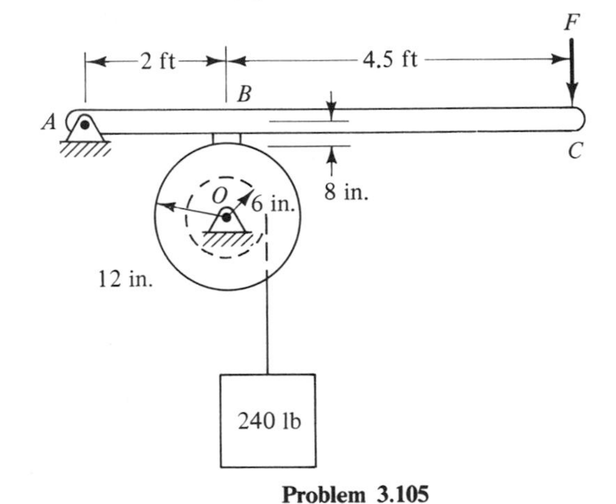

Problem 3.105

3.106. A cylinder with weight W rests in a horizontal V-block with included angle 2α, as shown. If the coefficient of friction is μ_s, determine an expression for the force F required to cause motion of the cylinder to impend.

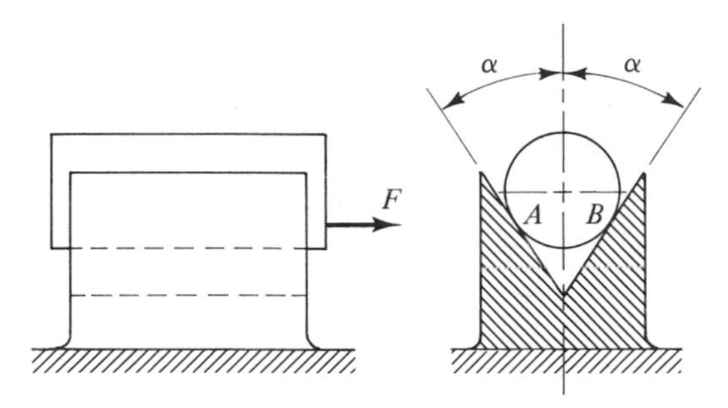

Problem 3.106

3.107. What force F is required to pull the plank shown from the stack of lumber if each plank has a mass of 10 kg? (See Table 3.1.)

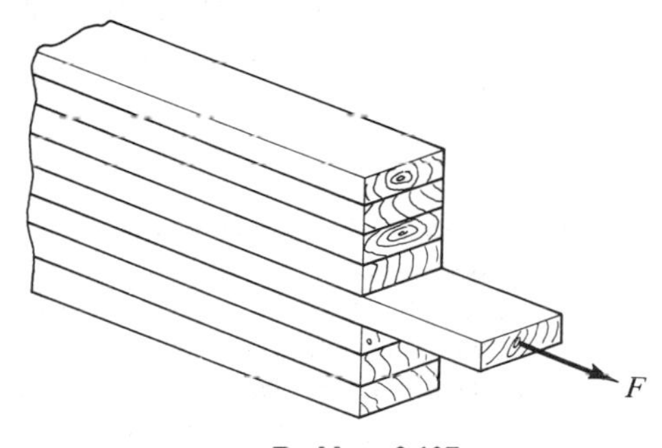

Problem 3.107

3.108. The steel drum shown is stored on its side with the label facing downward. If the drum and contents have a combined mass of 275 kg and the wall and floor are wood, estimate the couple C required to rotate the drum so that the label can be read. (See Table 3.1.)

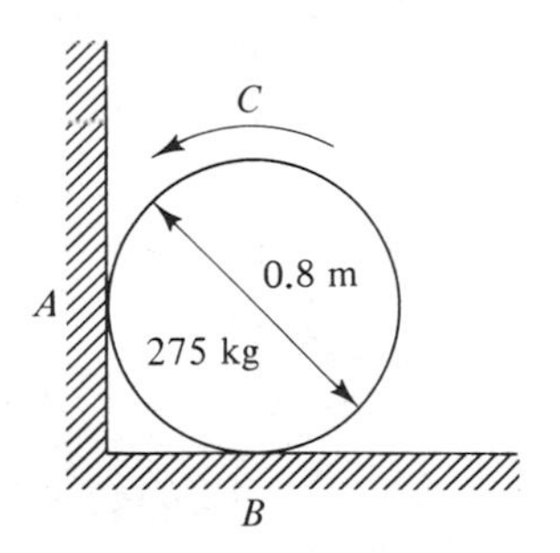

Problem 3.108

3.109. The 24-ft ladder shown weighs 40 lb and the worker weighs 180 lb. Determine the maximum distance d the base of the ladder can be placed from the wall if the ladder is not to slip when the worker climbs to the position indicated.

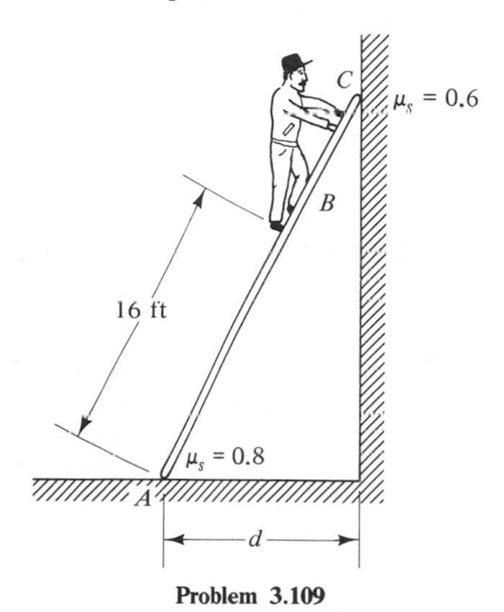

Problem 3.109

3.110. What is the angle of inclination θ of the steepest incline the vehicle shown can start up without slipping the driving wheels if it has (a) rear-wheel drive, (b) front-wheel drive, and (c) four-wheel drive? Assume rubber tires and dry pavement (see Table 3.1).

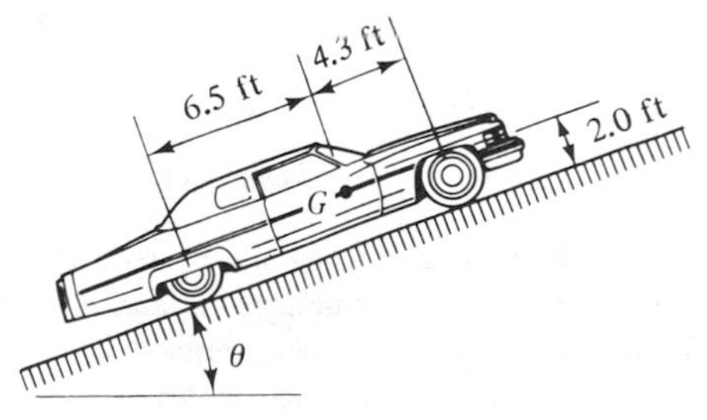

Problem 3.110

3.111. One end of a heavy machine with center of gravity at G is to be raised with a wedge, as shown. The machine has a mass of 1.85 ton and is made of metal. The wedge and floor are made of wood. Estimate the force F required to raise the machine (see Table 3.1 for values of μ_s). Surface AB remains very nearly horizontal, and a board nailed to the floor prevents slippage at A.

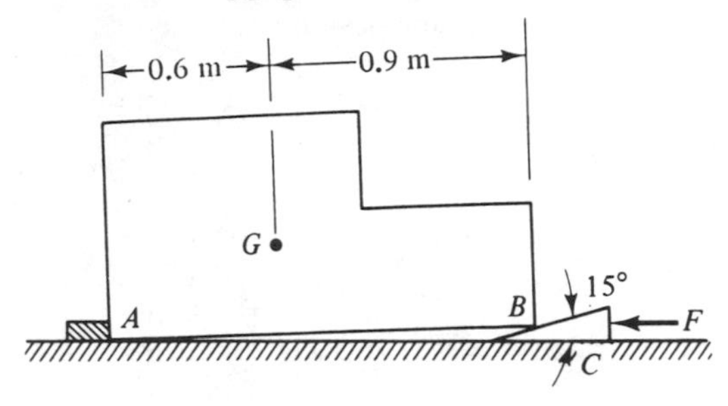

Problem 3.111

3.112. What is the minimum coefficient of static friction μ_s for which the bracket shown will remain in place regardless of the magnitude F of the applied force? The bracket has negligible mass and contacts the supporting rod at points A and B.

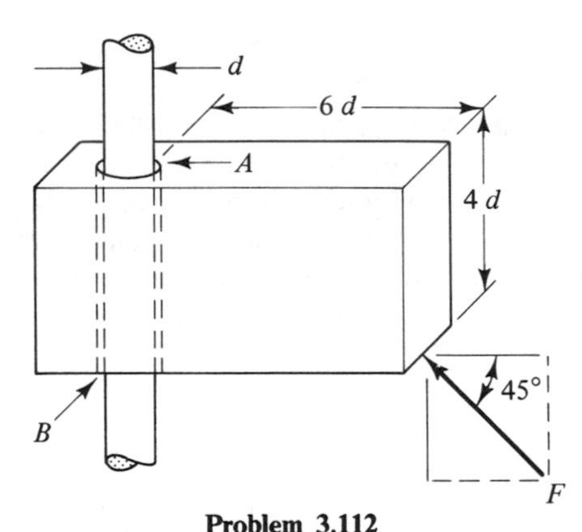

Problem 3.112

3.113. and 3.114. Determine whether or not the system shown is in equilibrium, and find the friction forces at all contact surfaces. The coefficients of friction are as indicated in the figure.

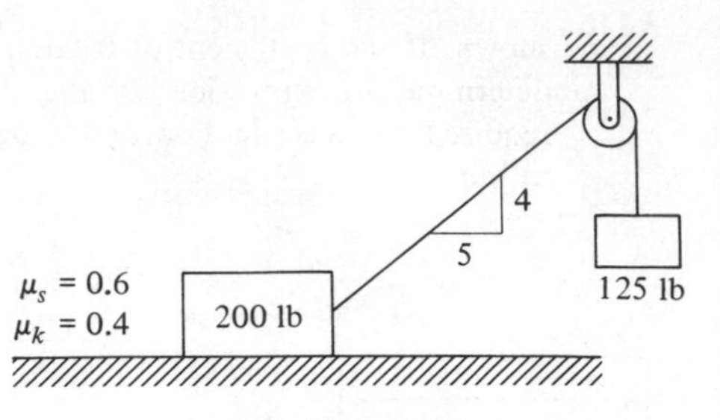

Problem 3.113

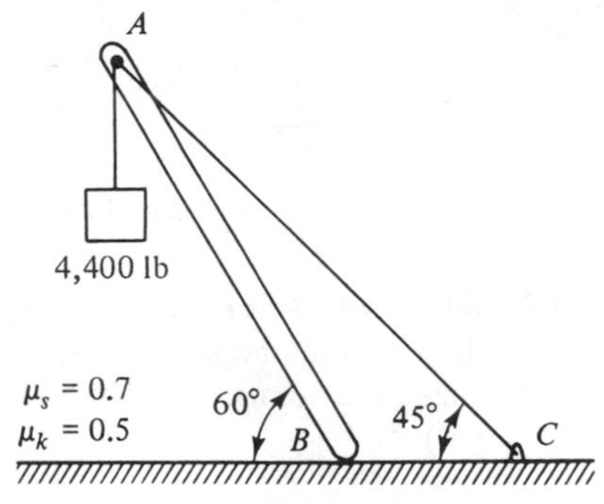

Problem 3.114

3.115. A worker attempts to move a heavy roll of material standing on end by putting a rope around it and pulling as shown. Determine the smallest force F required to cause the roll to move if (a) $h = d$ and (b) $h = 2d$; $\mu_s = 0.4$. Does the roll slide or tip?

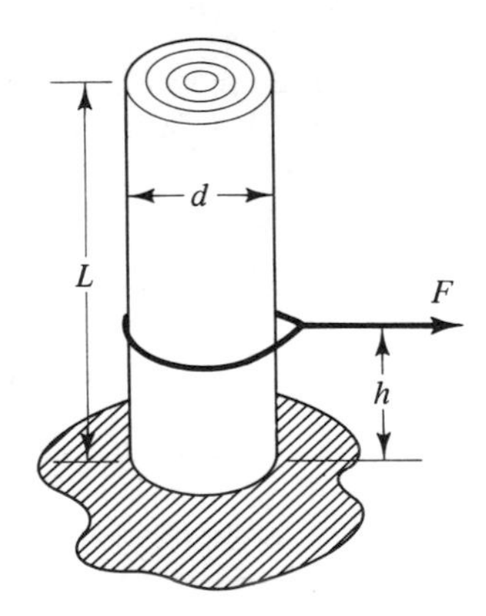

Problem 3.115

3.116. and 3.117. Determine the smallest force F that will disturb equilibrium of the system shown and describe clearly the manner in which it starts to move. The coefficients of friction are as indicated in the figure.

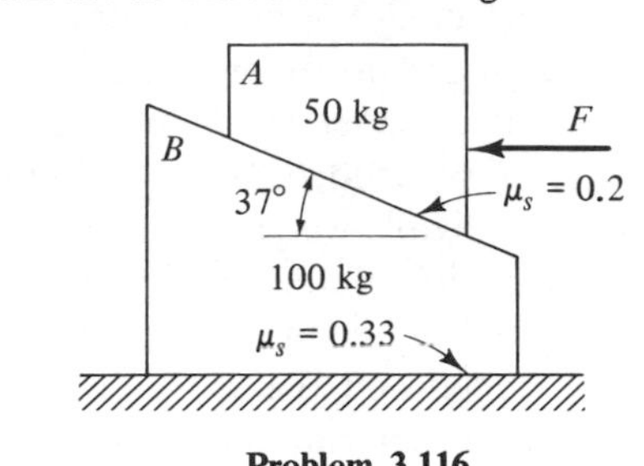

Problem 3.116

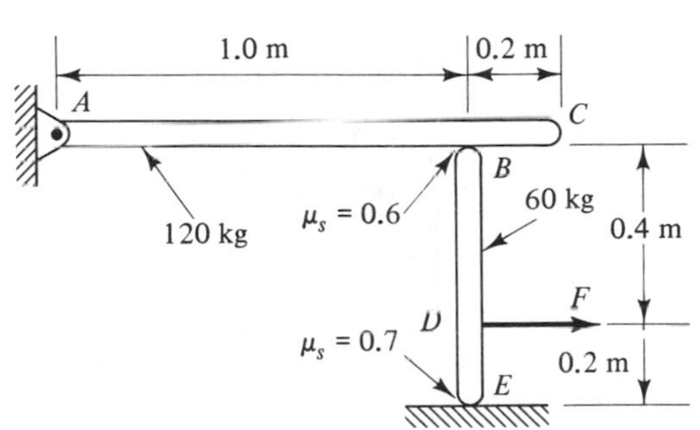

Problem 3.117

3.118. A small block is placed on an inclined surface, as shown. What is the minimum value of the static coefficient of friction for which the block will not slip when released?

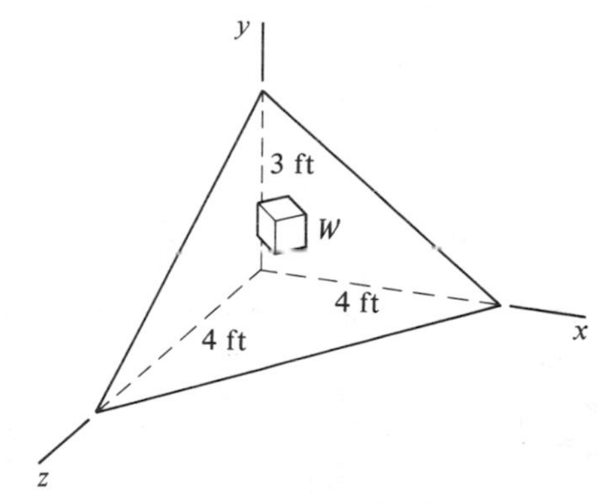

Problem 3.118

3.119. Determine the force F required to slide the lever shown along the square guide-rail if $\mu_s = 0.3$. The lever weighs 8 lb and fits loosely over the rail; the applied force acts parallel to the rail.

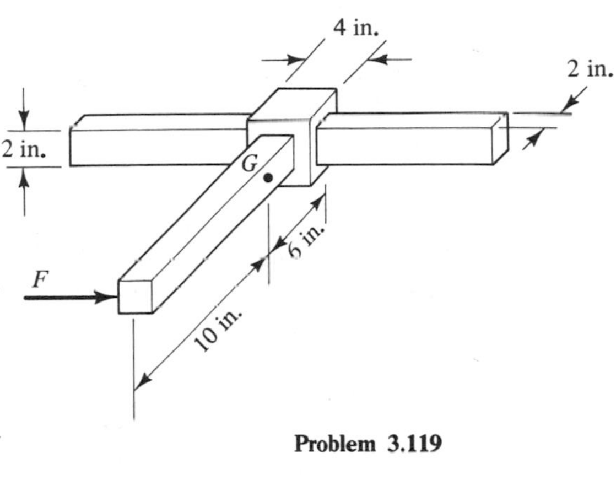

Problem 3.119

3.120. The uniform rod and block shown both have the same weight, and the coefficient of static friction between the block and surface is μ_s. If both cords and the rod are of length L, what is the maximum value of x for which equilibrium can be maintained?

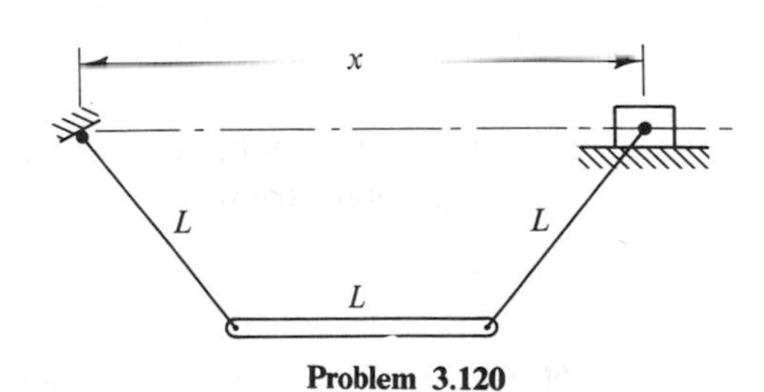

Problem 3.120

3.10 THE RIGID-BODY APPROXIMATION

Now that we have used the rigid-body assumption in the solution of a number of problems, let us consider the nature of this approximation in a bit more detail. In particular, we may ask: What do we gain by neglecting the deformations of a body when performing a force analysis, and how large are the errors introduced by this approximation? To answer these questions, let us consider the following problem.

Figure 3.34(a) shows a block with weight W suspended from the end of a rigid arm attached to a disc. The disc and arm have negligible weight, and the assembly is supported by a smooth bearing at A and a spring BC. Before the block is attached, the arm is horizontal and the spring is unstretched. After the block is attached and the system has come to rest, the spring is stretched and the arm is inclined at an angle θ from the horizontal, as shown. We wish to determine the tension T on the spring. The dimensions L and R and the weight W of the block are assumed known.

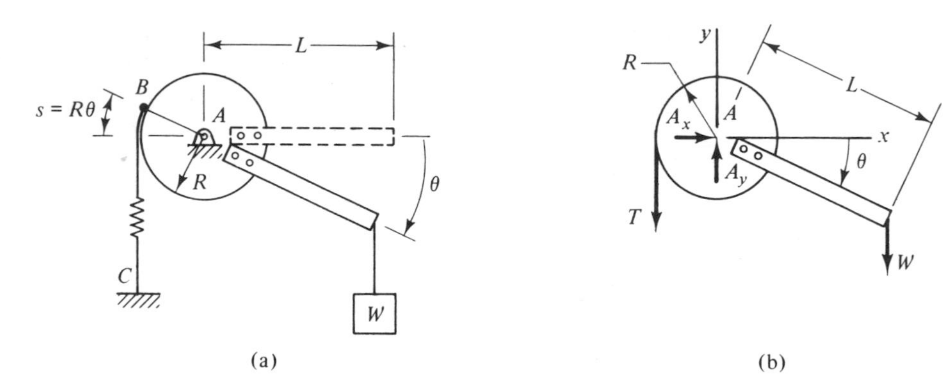

Figure 3.34

From the free-body diagram of figure 3.34(b), we have

$$\sum \overset{+}{\curvearrowleft} M_A = TR - WL\cos\theta = 0$$

or

$$T = \frac{WL}{R}\cos\theta \tag{3.11}$$

If the spring stretches only a small amount, θ will be approximately equal to zero (the rigid-body approximation), and we get

$$T_{\text{rigid}} = \frac{WL}{R} \tag{3.12}$$

However, if the spring stretches an appreciable amount, the assumption $\theta \approx 0$ is unreasonable. In this case, both θ and T are unknowns, and the problem cannot be solved from the equilibrium equations alone; it is statically indeterminate.

The additional information needed to solve the indeterminate problem is the resistive behavior of the spring. Suppose that the spring is tested by subjecting it to different tensions and measuring how much it stretches. The results of such an experiment performed on a typical coil spring are shown in Figure 3.35. Note that the amount of stretch s is proportional to the applied tension T. Thus, the data can be represented by the equation

$$T = ks \tag{3.13}$$

where k is the so-called *spring constant*. A spring that behaves according to Eq. (3.13) is said to be *linear*. For the spring of Figure 3.35, $k = 50$ lb/in.

From the geometry of Figure 3.34(a), we see that the spring will stretch an amount

$$s = R\theta \tag{3.14}$$

when the arm rotates through an angle θ. Combining this result with Eqs. (3.11) and (3.13), we obtain

$$kR^2\theta = WL\cos\theta \tag{3.15}$$

This equation can be solved for θ, provided the values of the other quantities are known. The solution can be obtained either by trial and error or by plotting both sides of the equation versus θ and determining the points of intersection of the two resulting curves. Once θ has been determined, the tension T can be found from Eq. (3.11).

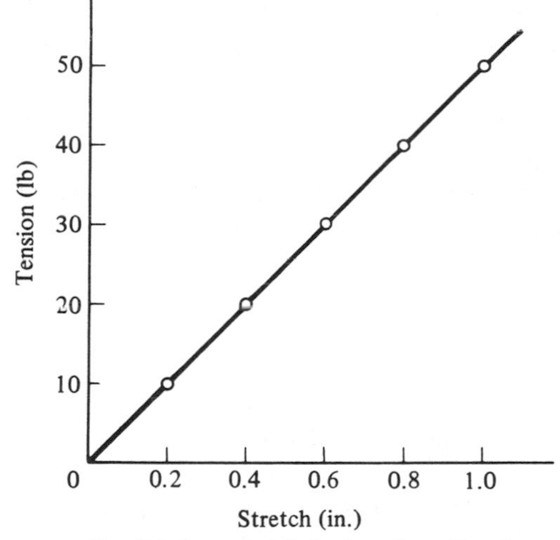

Load-deformation behavior of a coil spring **Figure 3.35**

The advantages of the rigid-body approximation should now be obvious. The exact solution for the tension not only requires a knowledge of the spring characteristics, but it also involves considerably more work than does the solution for T_{rigid}.

The error introduced by the rigid-body approximation can best be demonstrated by considering a specific case. A trial-and-error solution of Eq. (3.15) with $W = 50$ lb, $k = 50$ lb/in., $L = 3$ ft, and $R = 1$ ft gives $\theta = 13.904°$. From Eq. (3.11), we then find $T = 145.61$ lb. The value of T_{rigid} obtained from Eq. (3.12) is 150.00 lb, which exceeds the true value of T by 3%. An error of this magnitude would be acceptable for most engineering purposes. In fact, values of the parameters W, L, R, and k may not be known to this degree of accuracy.

An increase in the stiffness of the system decreases the angle of rotation of the arm and improves the accuracy of the rigid-body approximation. For example, if the spring is replaced by a $\frac{1}{16}$-in. diameter steel wire 3 feet long ($k = 2{,}560$ lb/in.), the error in the tension is only 0.001%! Clearly, the rigid-body approximation gives accurate results when the deformations are small, as they are in most typical engineering structures.

In later chapters we shall also be concerned with the deformations of structures; therefore, before leaving this section let us illustrate how the rigid-body approximation can be used to advantage in problems of this type.

Suppose that we wish to determine the stretch s of the spring, which can be computed from Eq. (3.11) once the tension is known. As we have seen, there is considerable work involved in obtaining the exact value of T. However, an approximate value of s can be easily determined by using the value of T_{rigid}: $s = T/k$ and $s_{\text{approx.}} = T_{\text{rigid}}/k$. For the particular system considered here, the error in the value of s is the same as the error in the value of the tension (3% for $k = 50$ lb/in.).

The preceding result suggests that we can assume a body is rigid while performing the force analysis and then use the values obtained for the forces and couples in finding the deformations of the body. This procedure greatly reduces the amount of work involved in determining the deformations, and we shall make considerable use of it in later chapters.

To summarize, the rigid-body approximation greatly reduces the amount of work involved in the the force analysis, and it introduces little error if the deformations are small. It can be used for all the problems we shall consider, with the exception of the buckling problems discussed in Chapter 10.

3.11 CLOSURE

Our work in this chapter has been concerned with equilibrium—the single most important topic we shall consider, and one of the most important in all science and engineering.

The general equations $\Sigma \mathbf{F} = \mathbf{0}$ and $\Sigma \mathbf{M}_P = \mathbf{0}$ are a mathematical statement of the physical requirement that the total force and moment acting upon a system must be zero if the system is to remain in equilibrium. These equations are used in force analyses to determine the unknown forces and couples that may be acting upon a body. This step is a prerequisite for determining the response of the body to applied loads or for properly designing it to withstand a given loading.

As we have seen, the key to a successful force analysis is the construction of proper free-body diagrams. For most problems, the force analysis can be based upon the dimensions of the system before the loads are applied (the rigid-body approximation). This enables us to separate the problem of determining the forces and couples acting upon the body from the problem of determining the deformations they produce, which greatly simplifies the force analysis.

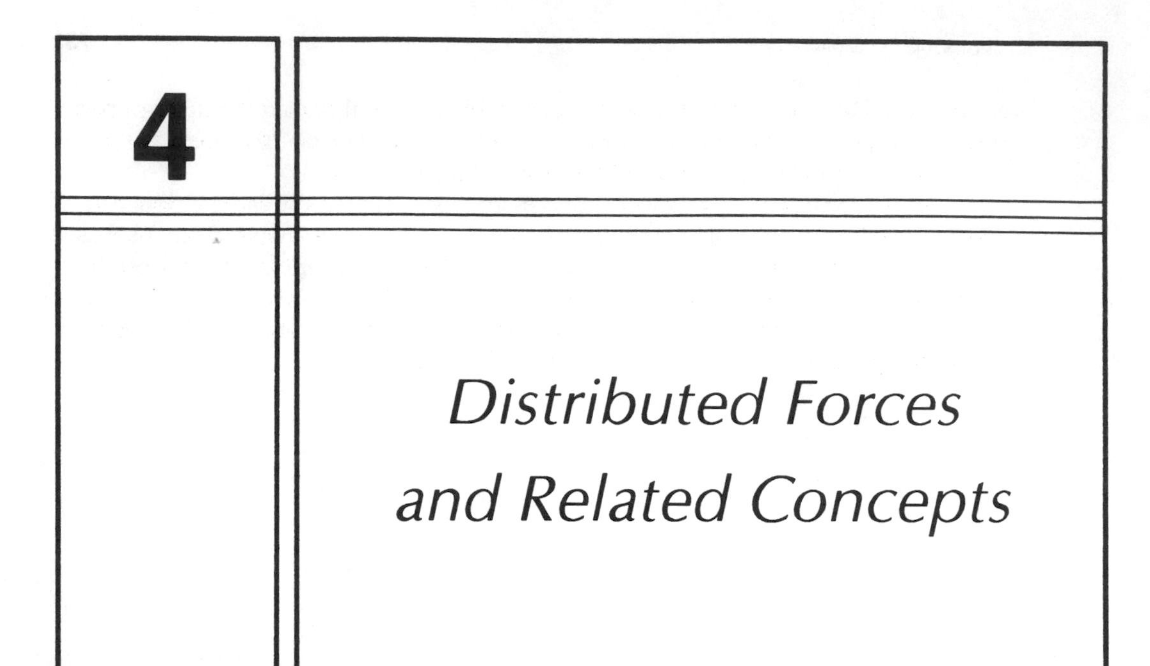

4.1 INTRODUCTION

In the preceding chapters we dealt only with *concentrated forces* (also called *discrete* or *point forces*) acting at a specific point. However, the concept of a concentrated force is an idealization. In reality, a force never acts at a point; it always exists as a *distributed force* acting over some area or volume. For example, if two spheres are pressed together, they will flatten slightly, and the contact force between them will be distributed over a small area. Similarly, the tension on a string is distributed over its cross section, and the gravitational attraction of the earth is distributed throughout the volume of the body upon which it acts. This being the case, what is our justification for working with concentrated forces?

It is reasonable to treat a force as being concentrated if the area or volume over which it is distributed is small enough compared to the other dimensions of the body to be considered a point. Typical examples are the force exerted on a telephone pole by a support cable and the reactions at the ends of a ladder due to the support surfaces. However, this argument does not justify the use of concentrated forces in every instance. It does not explain, for example, why the weight of a body can be treated as a concentrated force acting at the center of gravity or why the force distributed over the bottom surface of a block sitting on a smooth surface can be treated as a concentrated normal force.

The justification for working with concentrated forces in the aforementioned cases lies in the fact that distributed forces can be replaced by "equivalent" concentrated forces for the purpose of performing a force analysis. Our goal in this chapter is to determine the nature of this equivalence and to learn how to use it to advantage in the solution of problems. We shall begin by considering systems of discrete forces, with which we are already familiar. The results obtained will then be applied to distributed force systems. Certain geometric quantities associated with distributed force problems will also be considered.

4.2 STATICALLY EQUIVALENT FORCE SYSTEMS

Consider a body subjected to two different systems of forces, I and II (Figure 4.1). If the systems have the same total force and the same total moment about a common point, they will have the same effect upon the equilibrium of the body, and one system can be replaced by the other when performing a force analysis. Such force systems are said to be *statically equivalent*, or *equipollent*, a condition which we shall denote by the symbol $\sim$.

In the case of discrete forces, the conditions for static equivalence can be written as

$$\begin{aligned}\left(\sum \mathbf{F}\right)_{\mathrm{I}} &= \left(\sum \mathbf{F}\right)_{\mathrm{II}} \\ \left(\sum \mathbf{M}_P\right)_{\mathrm{I}} &= \left(\sum \mathbf{M}_P\right)_{\mathrm{II}}\end{aligned} \tag{4.1}$$

where the point P about which moments are computed is arbitrary. These conditions also hold if there are distributed forces. The only differences are the procedures for computing the total force and moment, which will be discussed in later sections.

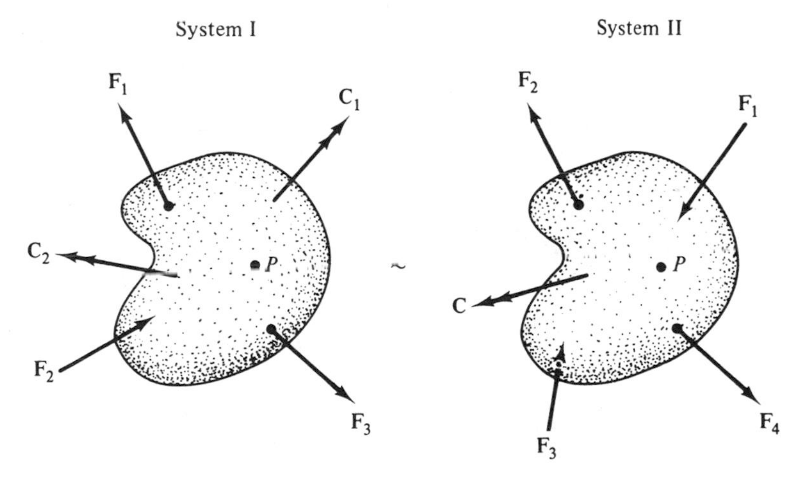

Figure 4.1

When expressed in component form, Eqs. (4.1) yield six scalar equations:

$$
\begin{aligned}
(\sum F_x)_{\mathrm{I}} &= (\sum F_x)_{\mathrm{II}} & (\sum F_y)_{\mathrm{I}} &= (\sum F_y)_{\mathrm{II}} & (\sum F_z)_{\mathrm{I}} &= (\sum F_z)_{\mathrm{II}} \\
(\sum M_{Px})_{\mathrm{I}} &= (\sum M_{Px})_{\mathrm{II}} & (\sum M_{Py})_{\mathrm{I}} &= (\sum M_{Py})_{\mathrm{II}} & (\sum M_{Pz})_{\mathrm{I}} &= (\sum M_{Pz})_{\mathrm{II}}
\end{aligned}
\tag{4.2}
$$

All six of these equations must be satisfied in order to have static equivalence. However, for two-dimensional and other special force systems, certain of these conditions will be satisfied automatically.

Several general results of practical interest can be deduced from Eqs. (4.1). First, we see that two couples are statically equivalent if they produce equal moments. Since the moment of a couple is the same about every point, it makes no difference where the couple is applied to a body insofar as static equivalence is concerned.

Second, two forces are statically equivalent if they are equal and have the same line of action. They need not have the same point of application. Two equal forces with different lines of action are not statically equivalent. In order to satisfy the moment condition in Eqs. (4.1), a couple with the appropriate moment must be added to one force system or the other, as in Figures 4.2(a) and 4.2(b).

It is essential to recognize that equipollent force systems are equivalent only in the sense that they have the same effect upon the equilibrium of a body. (In dynamics you will learn that equipollent force systems are also dynamically equivalent in the sense that they produce the same motion when applied to a given rigid

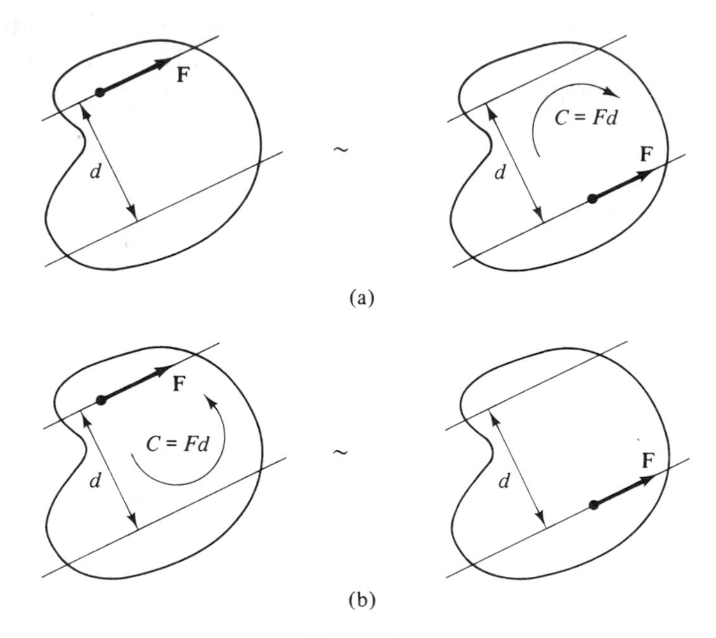

Figure 4.2

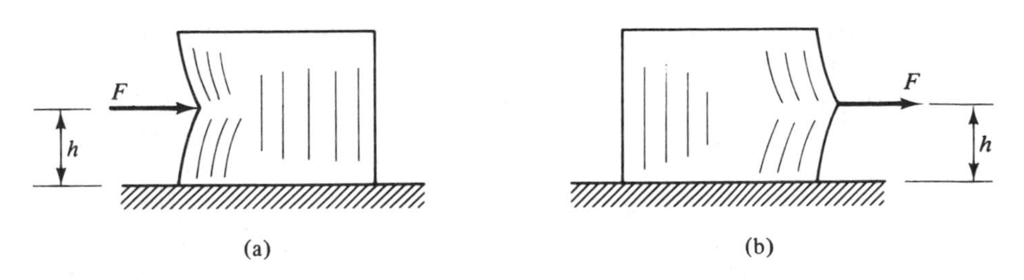

Figure 4.3

body.) Equipollent force systems are not equivalent in every sense. For example, Figure 4.3 shows a rubber block subjected to two different forces. The two forces are statically equivalent, but they produce completely different deformations of the body.

As a second example, let us consider the truss shown in Figure 4.4. The equilibrium of the truss is not affected if the load Q is applied at joint F instead of at joint B, since the two loadings are equipollent. However, the forces acting upon the individual members of the truss will be different. The force on member BF, for example, is zero when the load is applied at B, and it is a tension with magnitude Q when the load is applied at F.

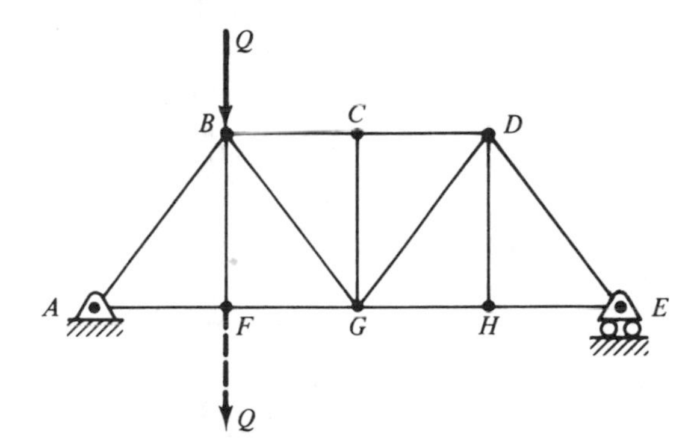

Figure 4.4

A still different type of situation is illustrated in Figure 4.5. Insofar as equilibrium is concerned, it makes no difference whether the stick shown is suspended from a cord or balanced on the end of our finger, since the forces exerted on the stick in each case are statically equivalent ($\mathbf{T} \sim \mathbf{N}$). However, the nature of the equilibrium is completely different. When the stick is suspended from a cord, the equilibrium position is stable. If the stick is disturbed slightly, it will return to its original position after the resulting oscillations have died out. The stick is in an unstable equilibrium position when it is balanced on our finger. Any slight bump will cause it to fall off, and it will not return to its original equilibrium position. (Stability will be discussed in Chapter 10.)

To summarize, replacement of a system of forces by an equipollent system does not affect the equilibrium of a body, but the deformations and forces

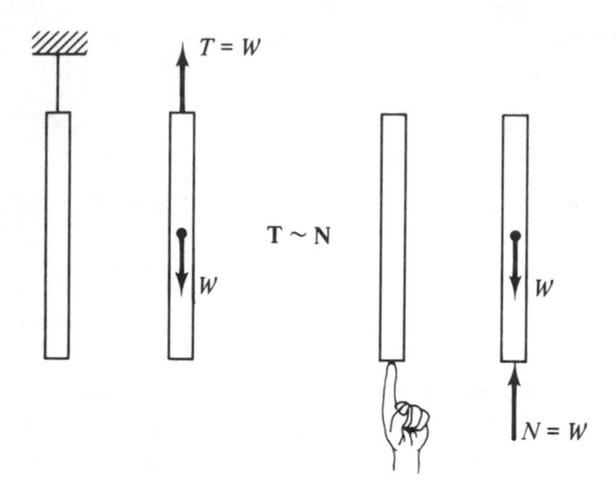

Figure 4.5

developed within it may be altered. Furthermore, the stability of the equilibrium can be affected. We shall need to keep these characteristics of equipollent force systems in mind when determining the response of bodies to applied loadings in later chapters.

Example 4.1. Statically Equivalent Couples. Replace the forces acting upon the block shown in Figure 4.6(a) by another statically equivalent force system.

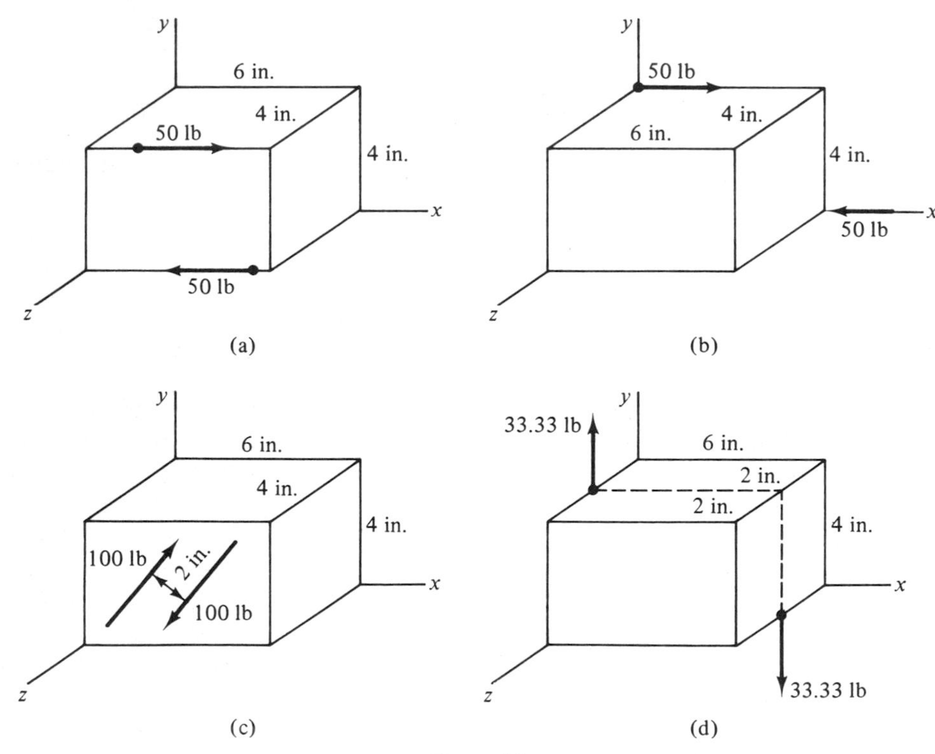

Figure 4.6

Solution. The two forces form a couple

$$\mathbf{C} = -200\mathbf{k}\ \text{lb} \cdot \text{in.}$$

The conditions for static equivalence, Eqs. (4.1), will be satisfied if this couple is replaced by any other couple with the same moment. There are an infinite number of ways to form such a couple. Several possibilities are shown in Figures 4.6(b)–4.6(d). What other possibilities can you think of?

Example 4.2. Statically Equivalent Force Systems. Replace the system of forces and couples acting upon the plate shown in Figure 4.7(a) by a statically equivalent system consisting of a vertical force acting along side AC and a horizontal force.

Solution. Let I denote the given force system and II the desired equipollent system. We now draw a sketch of system II showing all the known information about the forces and all the information that must be determined [Figure 4.7(b)]. The location of the line of action of the horizontal force is not known; therefore, it is shown an unknown distance d from the bottom of the plate. The sense of the unknown forces is chosen arbitrarily.

Applying the conditions for static equivalence, Eqs. (4.2), we obtain

$$\left(\sum F_x\right)_{\text{I}} = \left(\sum F_x\right)_{\text{II}}: 300\ \text{N} - 100\ \text{N} = F_H$$

$$\left(\sum F_y\right)_{\text{I}} = \left(\sum F_y\right)_{\text{II}}: -600\ \text{N} = -F_V$$

$$\left(\overset{+}{\curvearrowleft}\sum M_C\right)_{\text{I}} = \left(\overset{+}{\curvearrowleft}\sum M_C\right)_{\text{II}}: -300\ \text{N}\ (0.4\ \text{m}) - 600\ \text{N}\ (0.6\ \text{m}) + 100\ \text{N}\ (1.2\ \text{m})$$

$$+200\ \text{N} \cdot \text{m} = -F_H d$$

Solving these equations, we get

$$F_H = 200\ \text{N} \qquad F_V = 600\ \text{N} \qquad d = 0.8\ \text{m} \qquad \textbf{Answer}$$

Since F_H and F_V are positive, their sense is as assumed in Figure 4.7(b). Note that these forces do not have a definite point of application; they can be acting anywhere along their

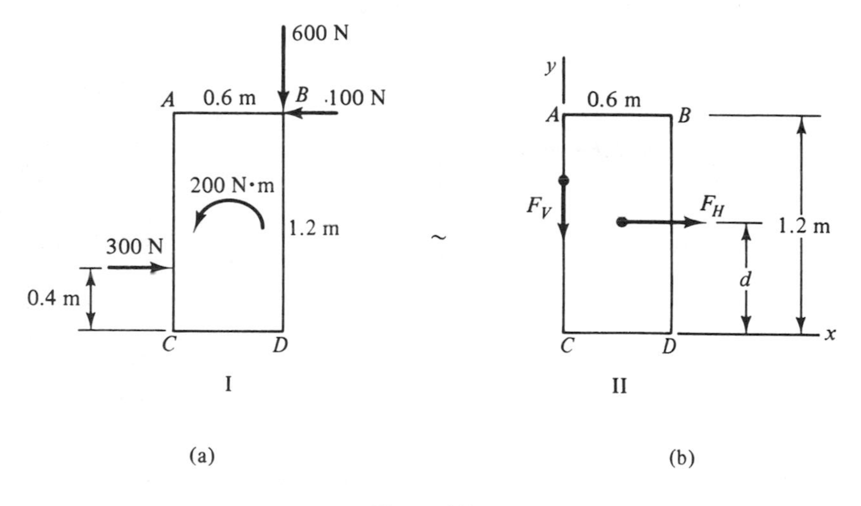

Figure 4.7

line of action. The fact that d is positive indicates that the line of action of F_H lies above side CD of the plate, as assumed.

Example 4.3. Statically Equivalent Force Systems. Determine the horizontal force acting at D that must be applied to the plate shown in Figure 4.8(a) in order for the total force system to be statically equivalent to a 40-lb upward force at B and a couple.

Solution. Neither force system is known completely. The unknown horizontal force at D is shown in Figure 4.8(a), and force system II is shown in Figure 4.8(b). The sense of the unknown force and couple is chosen arbitrarily.

For static equivalence, we have

$$\left(\sum F_x\right)_{\text{I}} = \left(\sum F_x\right)_{\text{II}}: \quad (100\text{ lb})\cos 37^\circ - F_H = 0$$

$$\left(\sum F_y\right)_{\text{I}} = \left(\sum F_y\right)_{\text{II}}: \quad -(100\text{ lb})\sin 37^\circ + 100\text{ lb} = 40\text{ lb}$$

$$\left(\sum \overset{+}{\curvearrowleft} M_A\right)_{\text{I}} = \left(\sum \overset{+}{\curvearrowleft} M_A\right)_{\text{II}}: \quad F_H(4\text{ in.}) + 100\text{ lb}\,(6\text{ in.}) = -C + 40\text{ lb}\,(6\text{ in.})$$

Since the second equation is satisfied identically, it yields no information; it serves only as a check. If this equation were not satisfied, it would be impossible to make the two force systems statically equivalent in the form stated. Solving for F_H and C from the first and third equations, we get

$$F_H = 80\text{ lb} \qquad C = -680\text{ lb}\cdot\text{in.} \qquad \textbf{Answer}$$

The negative value for C indicates that the sense of the couple is opposite to that assumed in Figure 4.8(b). Note that the couple can be shown as acting anywhere upon the plate because its moment is the same about any point. Also, the couple need not consist of any particular set of forces. Only the moment of the couple matters here.

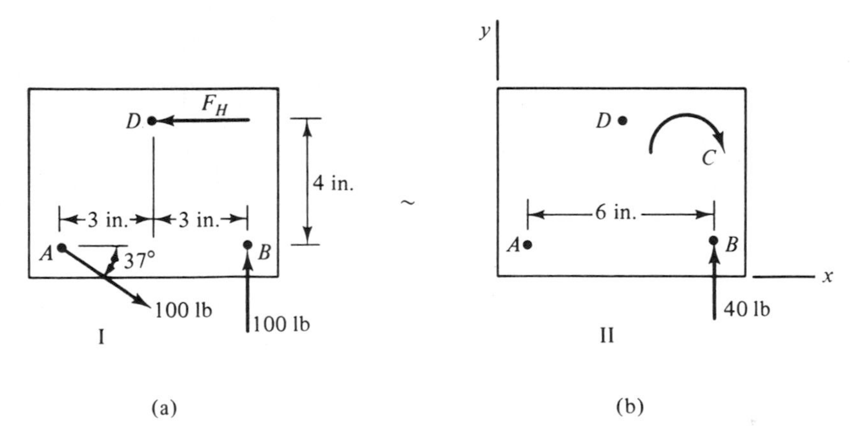

Figure 4.8

PROBLEMS

4.1. to 4.3. Determine whether or not the force systems I and II shown are statically equivalent. If not, can they be made equivalent by adding to system I (a) a force, (b) a couple, or (c) a force and a couple?

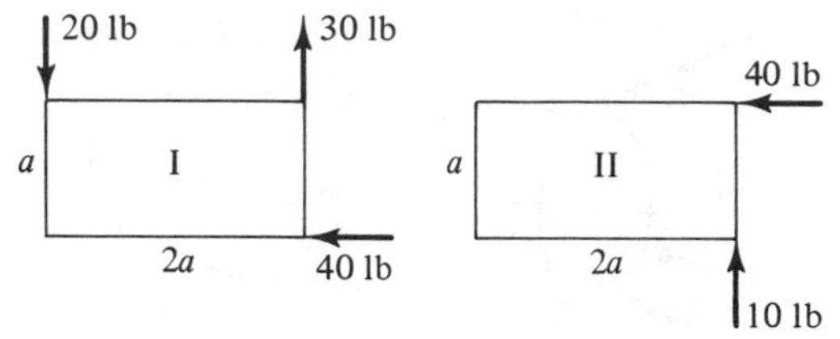

Problem 4.1

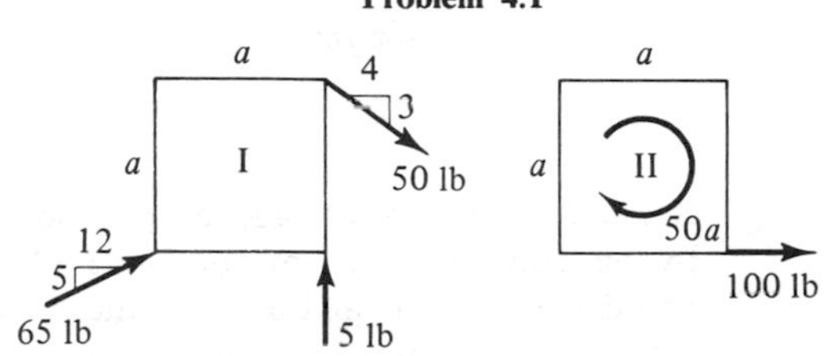

Problem 4.2

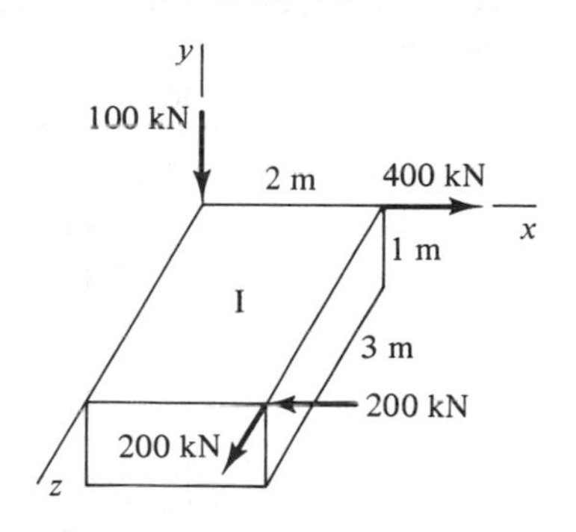

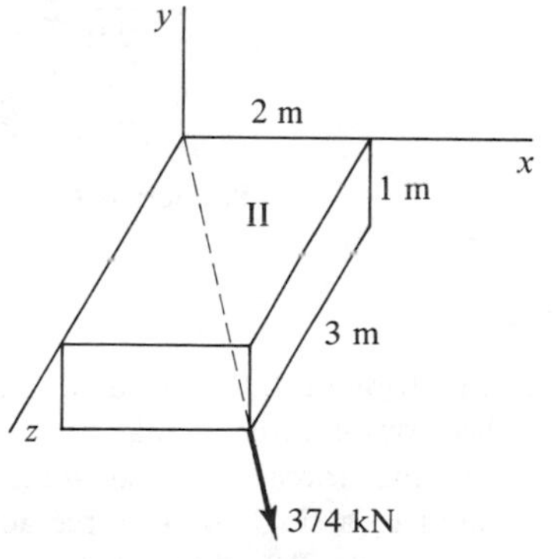

Problem 4.3

4.4. Replace the forces acting on the beam shown by an equivalent system consisting of a single force at A and a couple.

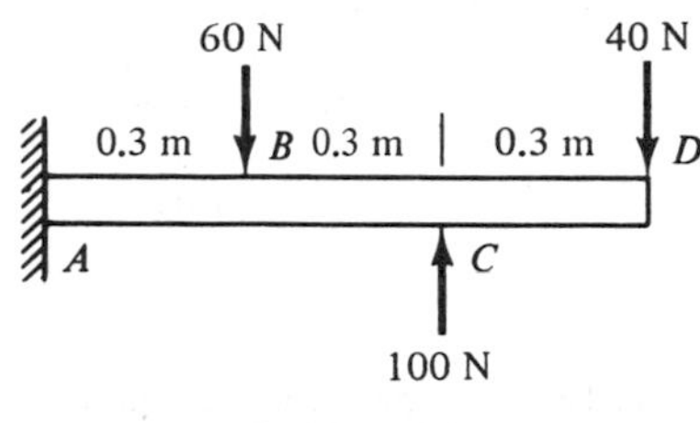

Problem 4.4

4.5. Replace the forces acting on the stepped pulley shown by an equivalent system consisting of a force at O and a couple.

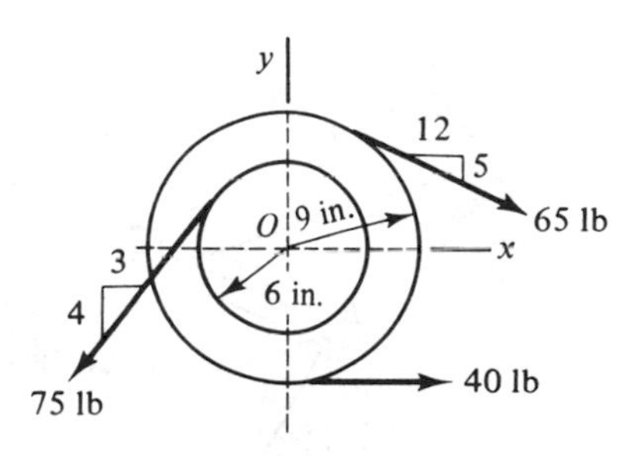

Problem 4.5

4.6. Replace the force acting upon the column shown by an equivalent system consisting of a force at the center O of the cross section and a couple. What is the couple tending to bend the column about (a) the x axis and (b) the y axis? Is there any tendency for the column to twist about the z axis?

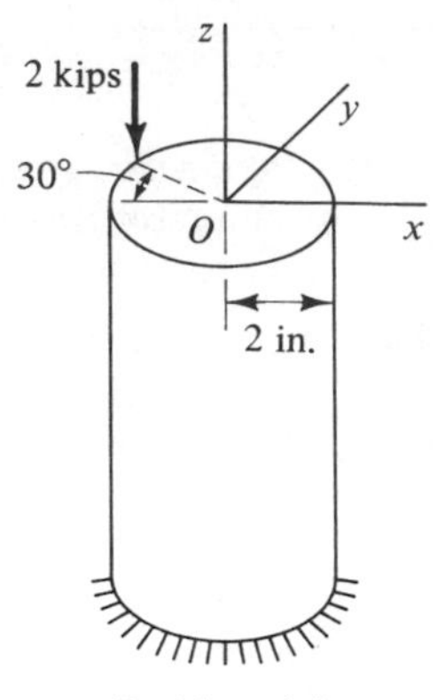

Problem 4.6

4.7. A force is applied at the end of a rigid arm attached to a circular shaft as shown. Replace the force by an equivalent system consisting of a force at O and a couple. What is the couple tending to twist the shaft about its longitudinal axis? Is there any tendency for the shaft to bend about the y or z axes?

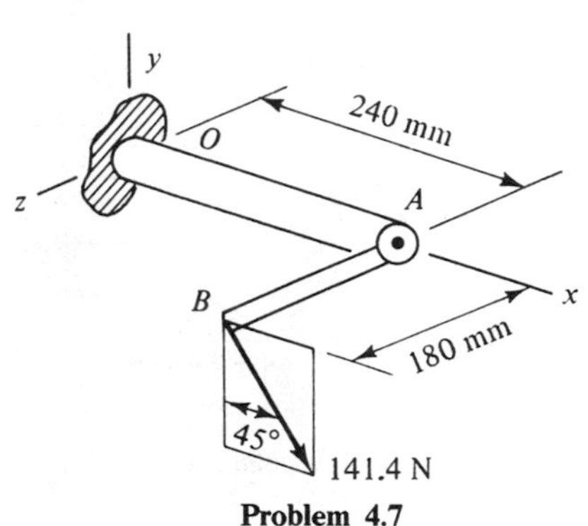

Problem 4.7

4.8. Three control rockets are equally spaced around the periphery of a space vehicle as shown. Rockets A and D produce a thrust T, while rocket B produces a thrust of $1.1T$. Each of the thrust vectors lies in a plane passing through the vehicle centerline and is inclined at an angle of 30° from this axis. Determine an equivalent force system consisting of a single force acting at the mass center G and a couple. Is there any tendency for the vehicle to rotate about the x, y, or z axes (roll, yaw, or pitch)?

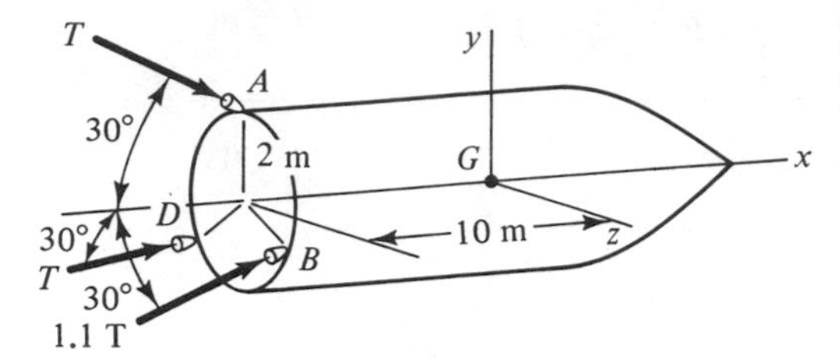

Problem 4.8

4.9. Replace the two forces and couple shown by an equivalent system consisting of a vertical force at E and a horizontal force acting at a point along side BD of the triangular plate. Show your results on a sketch.

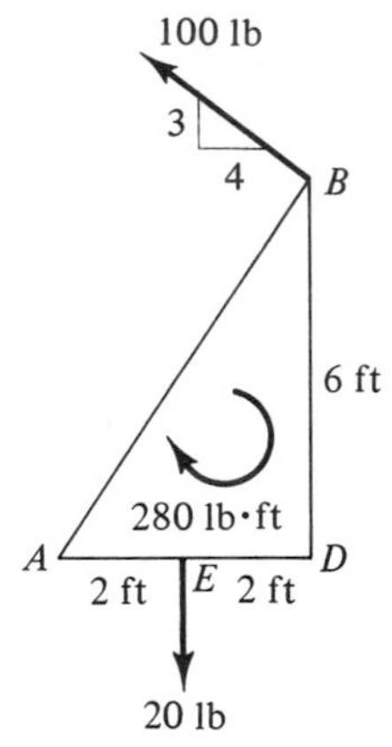

Problem 4.9

4.10. In Problem 4.9, determine completely the horizontal force **F** that must be added if the total force system shown is to be equivalent to a single 40-lb force acting upward along side BD of the plate.

4.11. A force $\mathbf{F} = 20\mathbf{i} + 30\mathbf{j} - 50\mathbf{k}$ kN is applied at end D of the bracket shown. What force $\mathbf{P}$ must be applied at point A if the total force system is to be equivalent to a horizontal and vertical force at O and a couple parallel to the z axis. What are the forces at O and the couple?

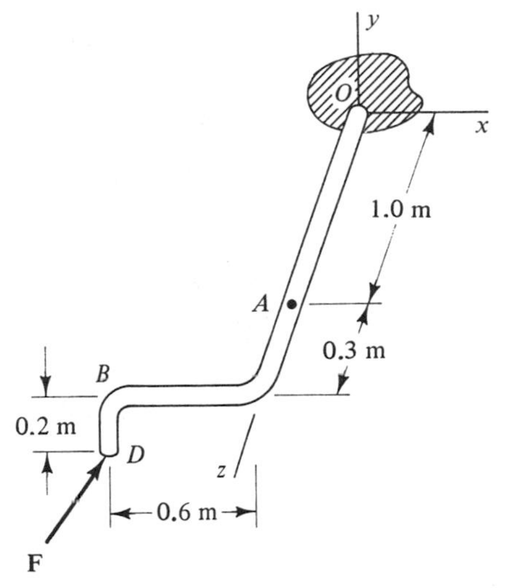

Problem 4.11

4.3 RESULTANTS OF FORCE SYSTEMS

When replacing one force system by another, it is usually desired to make the new system as simple as possible. The simplest force system equipollent to a given system is called the *resultant* of that system, and it can be determined from the conditions for static equivalence if we have some idea of its form. As we shall show in the following, the resultant is generally a force, a couple, or a force and a couple. We shall also develop guidelines to help determine a priori just which of these forms the resultant will take.

We begin by noting that any system of forces can be reduced to an equipollent system consisting of a single force $\mathbf{R}$ acting at an arbitrary point P and a single couple $\mathbf{C}^R$ (Figure 4.9). We shall call $\mathbf{R}$ the *resultant force* and $\mathbf{C}^R$ the *resultant couple*. Applying the conditions for static equivalence to the force systems of Figure 4.9, we have

$$\begin{aligned} \left(\sum \mathbf{F}\right)_{\mathrm{I}} = \left(\sum \mathbf{F}\right)_{\mathrm{II}}: &\qquad \left(\sum \mathbf{F}\right)_{\mathrm{I}} = \mathbf{R} \\ \left(\sum \mathbf{M}_P\right)_{\mathrm{I}} = \left(\sum \mathbf{M}_P\right)_{\mathrm{II}}: &\qquad \left(\sum \mathbf{M}_P\right)_{\mathrm{I}} = \mathbf{C}^R \end{aligned} \tag{4.3}$$

Thus, the resultant force $\mathbf{R}$ is simply the total force of the original system, and the resultant couple $\mathbf{C}^R$ is equal to the total moment of the system about point P. The

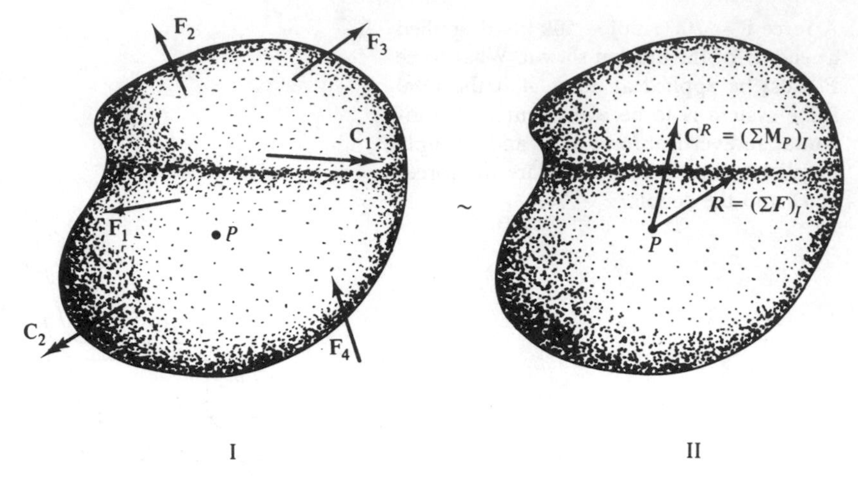

Figure 4.9

couple can be shown as acting at any point, but it is common practice to place it at the point of application of **R**, as in Figure 4.9.

For any one particular force system, **R**, $\mathbf{C}^R$, or both may be zero. If **R** is zero, the resultant is a couple; if $\mathbf{C}^R$ is zero, it is a single force. If both **R** and $\mathbf{C}^R$ are zero, the force system has a *null resultant*. When all the forces and couples acting upon a body have a null resultant, the body is in equilibrium.

If the resultant force and couple are perpendicular, and $\mathbf{R} \neq \mathbf{0}$, the force system can be further reduced to a single force. As illustrated in Figure 4.10, this reduction requires that **R** be located such that it provides the same moment about point P as does the couple $\mathbf{C}^R$, thus assuring that static equivalence is maintained.

With the preceding general results in mind, let us now consider some specific types of force systems.

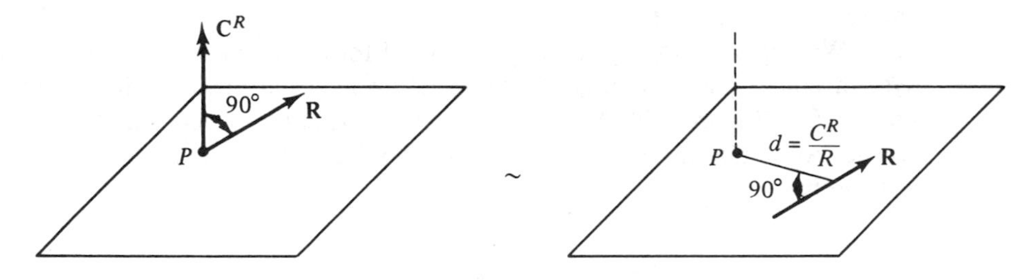

Figure 4.10

Concurrent Force Systems

If the forces are concurrent, their lines of action will intersect at some point P (Figure 4.11). Taking moments about P, we see that the resultant will be a single

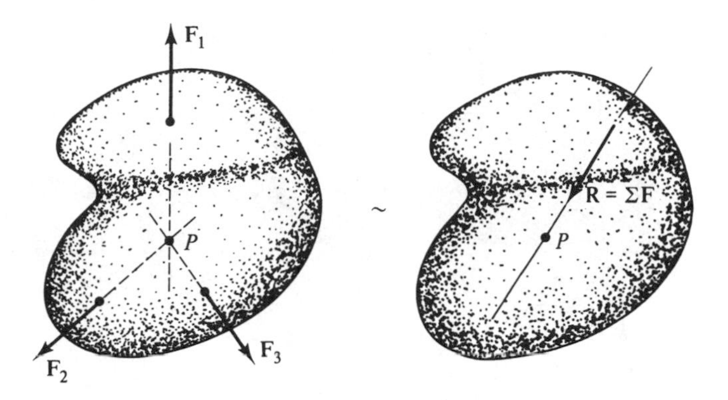

Figure 4.11

force **R** whose line of action passes through the point of concurrency. If $\mathbf{R} = \mathbf{0}$, the system has a null resultant.

Coplanar and Parallel Force Systems

Consider a system of coplanar forces (Figure 4.12). Using Eqs. (4.3) and taking moments about a point P in the plane of the forces, we see that $\mathbf{C}^R$ will be perpendicular to **R**. Thus, the system can be further reduced to a single force, provided $\mathbf{R} \neq \mathbf{0}$. This single force is the resultant of the system, and its line of action can be located by using the moment condition for static equivalence. If $\mathbf{R} = \mathbf{0}$, either the resultant will be a couple or the system will have a null resultant.

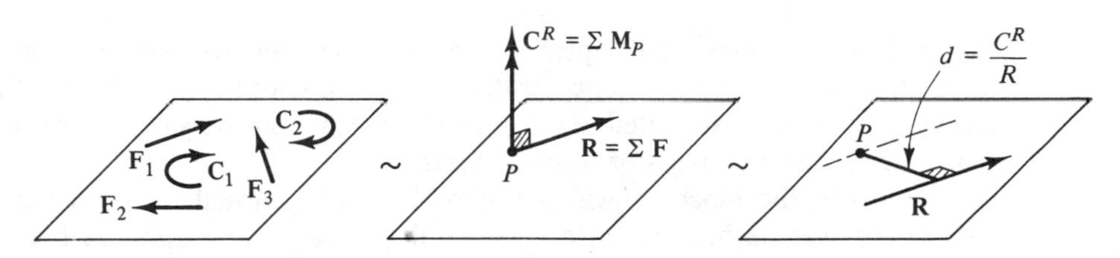

Figure 4.12

The preceding results also hold for systems of parallel forces (Figure 4.13). Since the moment of each force about a point P is perpendicular to the force, **R** and $\mathbf{C}^R$ will be perpendicular. Thus, the resultant will have the same form as in the case of coplanar forces. It is a single force **R** parallel to the original forces, provided $\mathbf{R} \neq \mathbf{0}$. For $\mathbf{R} = \mathbf{0}$, either the resultant is a couple or the system has a null resultant.

When determining the resultant of systems of parallel or coplanar forces, it is not necessary to first reduce the system to a force at a point and a couple. This was done here only for clarity. We can proceed directly from the given force system to the resultant, as illustrated in Examples 4.5 and 4.6 at the end of this section.

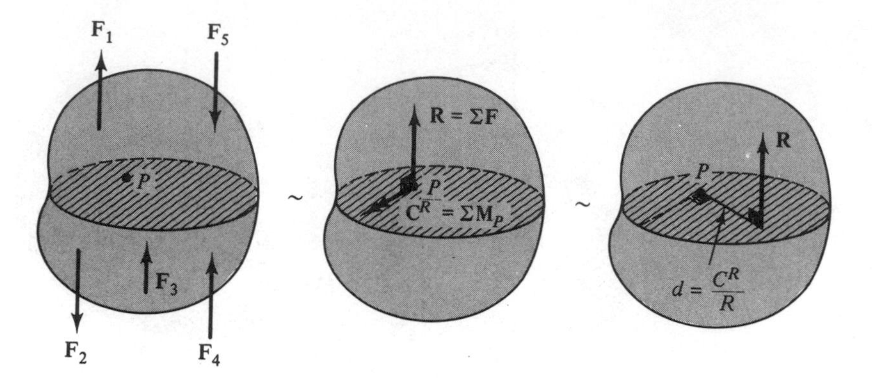

Figure 4.13

General Force Systems

For a general three-dimensional force system, it is not possible to determine a priori just what form the resultant will take. It will usually be a force and a couple. The component of the couple perpendicular to the force can be eliminated by moving the force to an appropriate point, as explained earlier. The resulting force system, consisting of a single force **R** and a couple parallel to it, is called a *wrench*. Reduction of a force system to a wrench has little practical application and will not be discussed further. It is usually most convenient to reduce the system to a single force and a couple and to stop at that point.

Application to Reactions

In the force analyses of Chapter 3 the reactions due to various supports and connections were treated as concentrated forces and couples, even though the forces acting were often distributed. This procedure can now be justified by using the properties of the resultants of force systems.

Consider the block shown in Figure 4.14. The normal force on the block is distributed over its bottom surface. Since the forces acting upon each element of

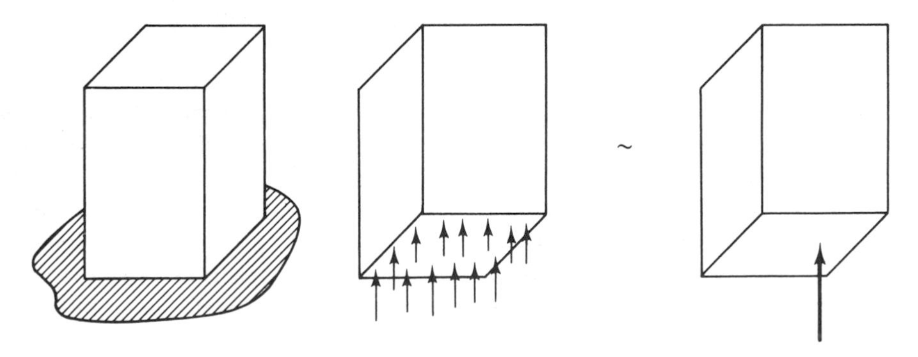

Figure 4.14

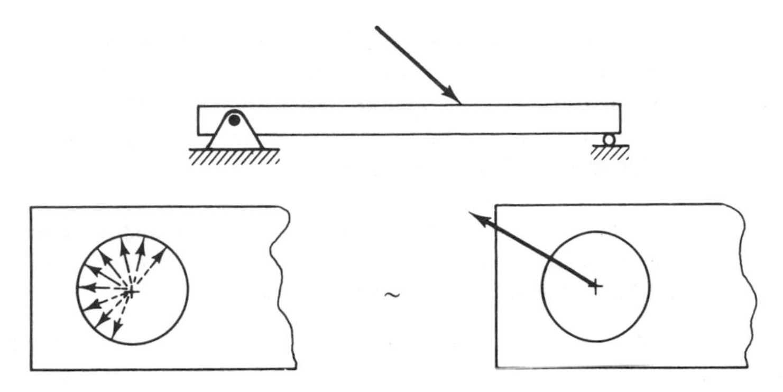

Figure 4.15

area are parallel, their resultant will be a single vertical force. Thus, the distributed normal force can be replaced by a statically equivalent concentrated force, as indicated.

As a second example, consider a beam supported at one end by a pinned connection (Figure 4.15). Assuming that the contact surface between the pin and the beam is smooth and circular, it follows that the forces exerted on each element of the surface by the pin must lie along radii of the hole in the beam. Since these forces are concurrent, they can be replaced by a single equivalent force acting at the center of the hole.

Our final example involves a beam with a fixed support (Figure 4.16). The wall exerts some distribution of force on the end of the beam, the details of which are generally unknown. However, any system of forces is equipollent to a force and a couple. Thus, the reactions due to the wall can be treated as a concentrated force and a couple, as shown.

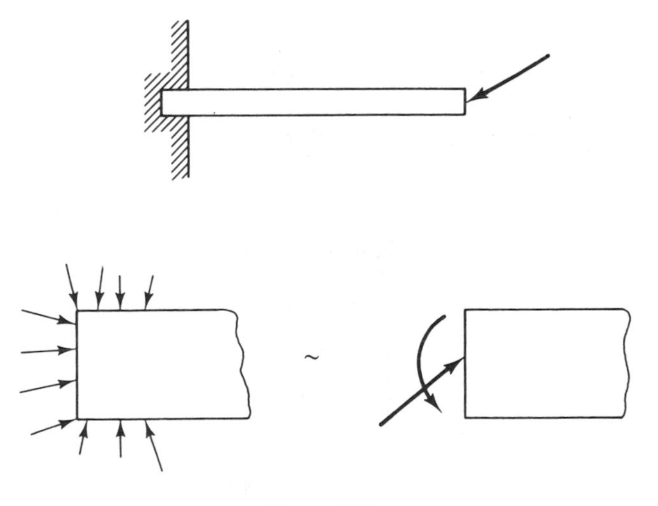

Figure 4.16

It is clear from the preceding examples that the quantities that we called the reactions in Chapter 3 are actually the resultant force and resultant couple of whatever system of forces is acting at the support or connection. It is these resultants that are determined in a force analysis; the actual force distribution cannot be determined from statics alone.

Example 4.4. Reduction of a Force System to a Force and a Couple. Reduce the force system shown in Figure 4.17(a) to a force **R** at the origin and a couple $\mathbf{C}^R$. Can the system be further reduced to a single force?

Solution. We first sketch the desired equipollent system [Figure 4.17(b)]. Applying the conditions for static equivalence, Eqs. (4.1), and determining the moments of the forces by inspection, we obtain

$$\left(\sum \mathbf{F}\right)_{\mathrm{I}} = \left(\sum \mathbf{F}_{\mathrm{II}}\right): \quad -50\mathbf{i} + 40\mathbf{j} + 60\mathbf{j}\ \mathrm{N} = \mathbf{R}$$

$$\mathbf{R} = -50\mathbf{i} + 100\mathbf{j}\ \mathrm{N} \qquad \textbf{Answer}$$

and

$$\left(\sum \mathbf{M}_0\right)_{\mathrm{I}} = \left(\sum \mathbf{M}_0\right)_{\mathrm{II}}: \quad (40\ \mathrm{N})(1.0\ \mathrm{m})\mathbf{k} - 60\ \mathrm{N}\,(0.6\ \mathrm{m})\mathbf{i} + 60\ \mathrm{N}\,(1.0\ \mathrm{m})\mathbf{k}$$

$$-50\ \mathrm{N}\,(0.6\ \mathrm{m})\mathbf{j} - 40\mathbf{j}\ \mathrm{N\cdot m} = \mathbf{C}^R$$

$$\mathbf{C}^R = -36\mathbf{i} - 70\mathbf{j} + 100\mathbf{k}\ \mathrm{N\cdot m} \qquad \textbf{Answer}$$

The force system can be further reduced to a single force only if $\mathbf{C}^R$ is perpendicular to **R**. Taking the dot product of the two vectors, we have

$$\mathbf{C}^R \cdot \mathbf{R} = (-36)(-50) + (-70)(100) = -5{,}200$$

Hence, the two vectors are not perpendicular, and the force system cannot be reduced to a single force.

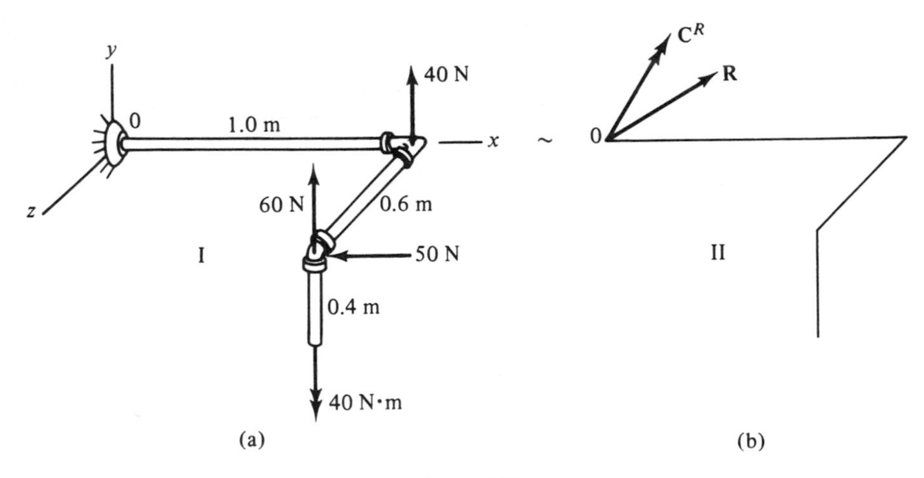

Figure 4.17

Example 4.5. Resultant of a Coplanar Force System. Determine the resultant of the forces acting upon the bracket shown in Figure 4.18(a) if (a) $F = 100$ lb and (b) $F = 200$ lb.

Solution. Since the forces are coplanar, we try a force **R** as the resultant [Figure 4.18(b)]. The magnitude and direction of **R** and the location of its line of action are unknown and are chosen arbitrarily in the figure. The resultant is placed at point E, although it can be acting at any other point along its line of action. The location of the line of action is defined by the distance d, but it can be defined in other ways. For example, it could be defined by the perpendicular distance between it and some known point, such as a corner of the bracket. For convenience, we work with **R** in component form. Applying the conditions for static equivalence, Eqs. (4.2), we have

$$\left(\sum F_x\right)_{\text{I}} = \left(\sum F_x\right)_{\text{II}}: \qquad -120 \text{ lb} + \qquad F\cos 53° = R_x$$

$$\left(\sum F_y\right)_{\text{I}} = \left(\sum F_y\right)_{\text{II}}: \qquad F\sin 53° - 40 \text{ lb} - 120 \text{ lb} = R_y$$

$$\left(\sum \overset{+}{\curvearrowleft} M_B\right)_{\text{I}} = \left(\sum \overset{+}{\curvearrowleft} M_B\right)_{\text{II}}: \qquad 40 \text{ lb } (10 \text{ in.}) = -R_y d$$

(a) Solving the preceding equations for $F = 100$ lb, we obtain

$$R_x = -60 \text{ lb} \qquad R_y = -80 \text{ lb} \qquad d = 5.0 \text{ in.} \qquad \textbf{Answer}$$

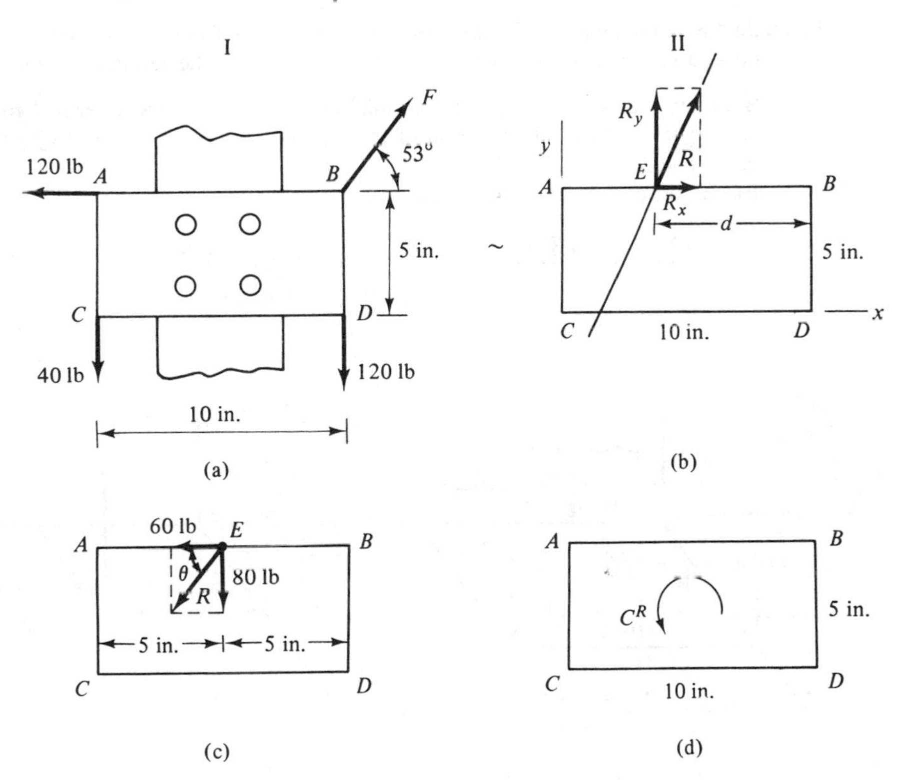

Figure 4.18

Since R_x and R_y are negative, the resultant is as shown in Figure 4.18(c). It is usually sufficient to leave the resultant in component form, but its magnitude and orientation can be easily determined. From Figure 4.18(c), we have

$$R = \sqrt{(60\text{ lb})^2 + (80\text{ lb})^2} = 100\text{ lb}$$

$$\tan\theta = \frac{80}{60} = 1.33$$

$$\theta = 53°$$

(b) Solving the force equations for static equivalence with $F = 200$ lb, we get

$$R_x = 0 \qquad R_y = 0$$

Thus, the resultant will not be a force; therefore, we try a couple [Figure 4.18(d)]. From the moment condition for static equivalence, we have

$$\left(\sum \overset{+}{\curvearrowleft} M_B\right)_I = \left(\sum \overset{+}{\curvearrowleft} M_B\right)_{II}: \qquad 40\text{ lb }(10\text{ in.}) = C^R$$

$$C^R = 400\text{ lb}\cdot\text{in.} \qquad \textbf{Answer}$$

Example 4.6. Resultant of a Parallel Force System. The forces exerted on a raft by its four occupants are as shown in Figure 4.19(a). Determine the resultant of these forces.

Solution. Since the forces are parallel and vertical, we try a vertical force **R** as the resultant [Figure 4.19(b)]. The location of the line of action of **R** is defined by the unknown distances a and b.

From the conditions for static equivalence, we have

$$\left(\sum F_y\right)_I = \left(\sum F_y\right)_{II}: \quad -150\text{ lb} - 100\text{ lb} - 110\text{ lb} - 200\text{ lb} = -R$$

$$R = 560\text{ lb} \qquad \textbf{Answer}$$

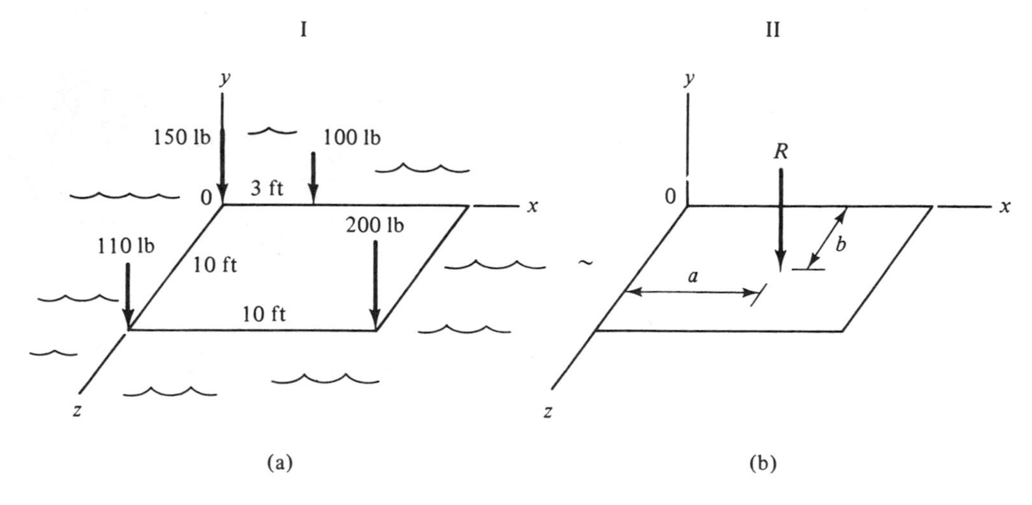

Figure 4.19

and

$$\left(\sum \mathbf{M}_0\right)_{\mathrm{I}} = \left(\sum \mathbf{M}_0\right)_{\mathrm{II}}: \quad -100 \text{ lb } (3 \text{ ft})\mathbf{k} + 200 \text{ lb } (10 \text{ ft})\mathbf{i} - 200 \text{ lb } (10 \text{ ft})\mathbf{k} + 110 \text{ lb } (10 \text{ ft})\mathbf{i} = Rb\mathbf{i} - Ra\mathbf{k}$$

or

$$3{,}100\mathbf{i} - 2{,}300\mathbf{k} \text{ lb} \cdot \text{ft} = Rb\mathbf{i} - Ra\mathbf{k}$$

where the moments have been determined by inspection. Solving for a and b from the moment equation, we obtain

$$a = \frac{2{,}300 \text{ lb} \cdot \text{ft}}{R} = \frac{2{,}300 \text{ lb} \cdot \text{ft}}{560 \text{ lb}} = 4.11 \text{ ft}$$

$$b = \frac{3{,}100 \text{ lb} \cdot \text{ft}}{R} = \frac{3{,}100 \text{ lb} \cdot \text{ft}}{560 \text{ lb}} = 5.54 \text{ ft}$$

Answer

PROBLEMS

4.12. and 4.13. Reduce the force system shown to a single force **R** acting at the origin and a couple $\mathbf{C}^R$. Can the system be further reduced to a single force?

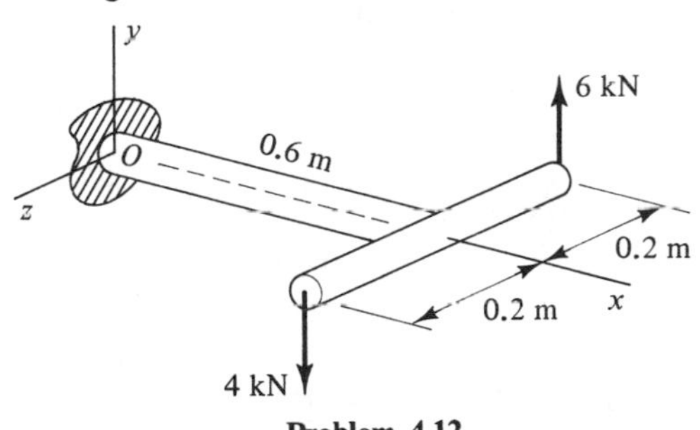

Problem 4.12

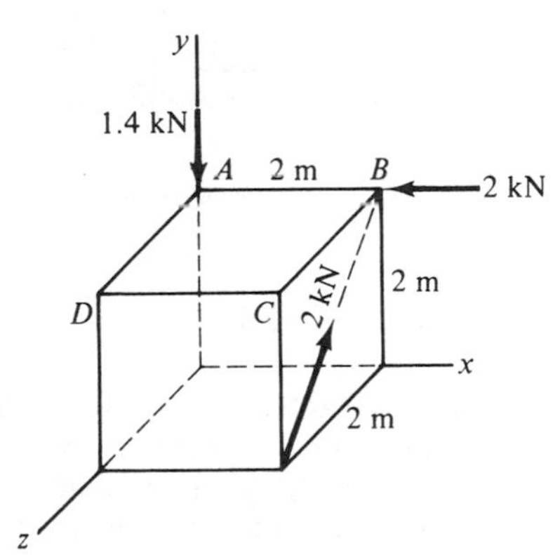

Problem 4.13

4.14. to 4.19. Determine completely and show on a sketch the resultant of the force system shown.

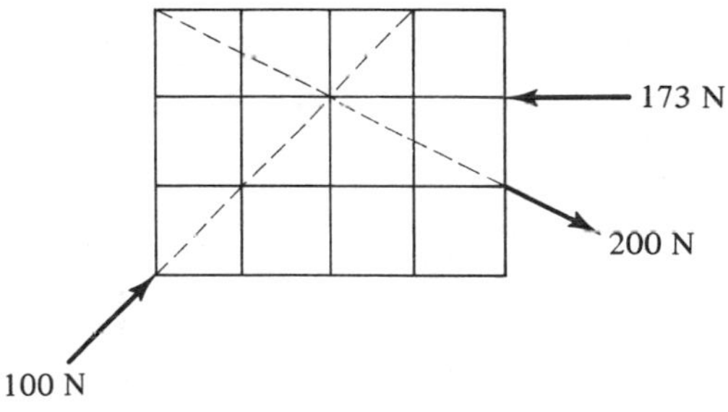

Problem 4.14

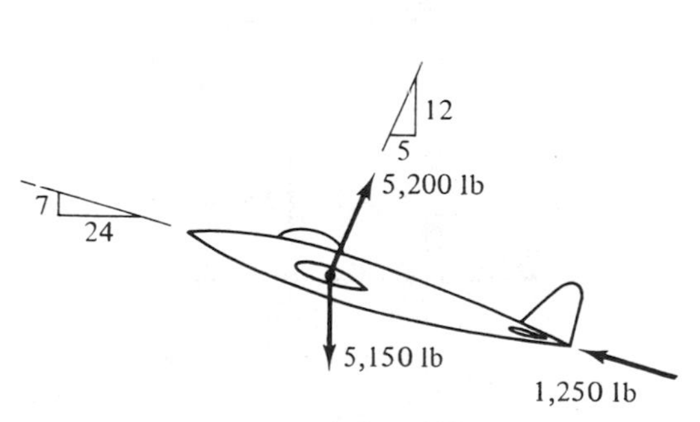

Problem 4.15

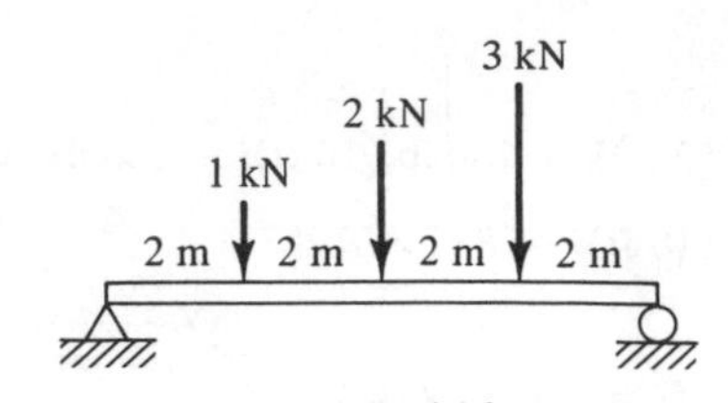

Problem 4.16

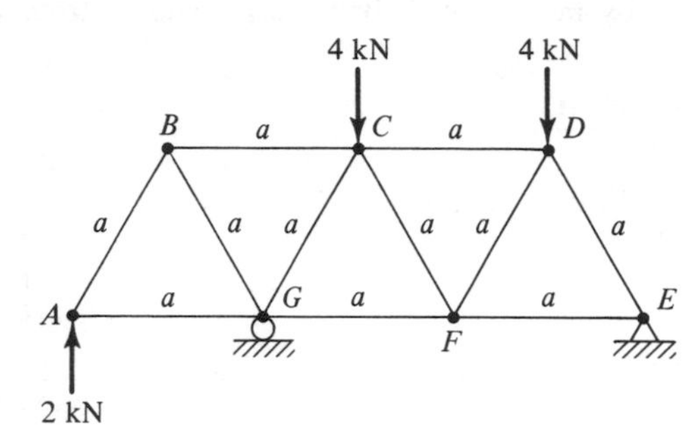

Problem 4.17

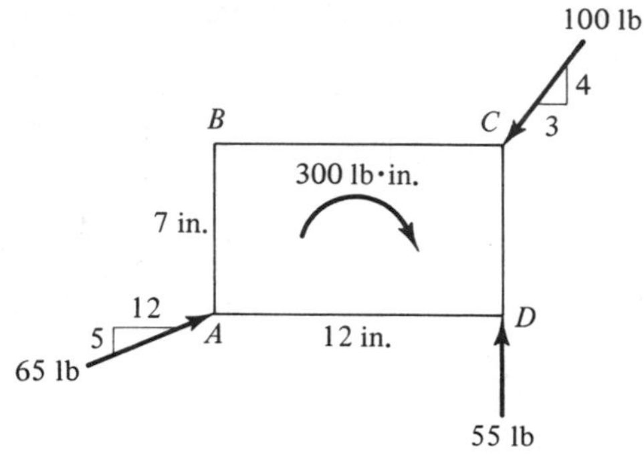

Problem 4.18

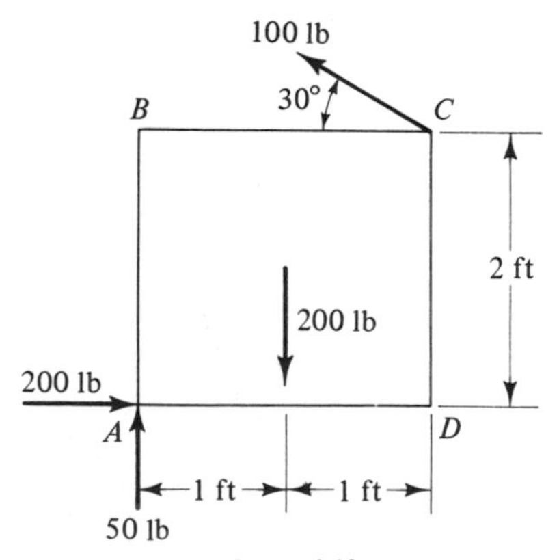

Problem 4.19

4.20. Determine the values of F and θ for which the resultant of the forces shown will be a couple. What is the magnitude and direction of this couple?

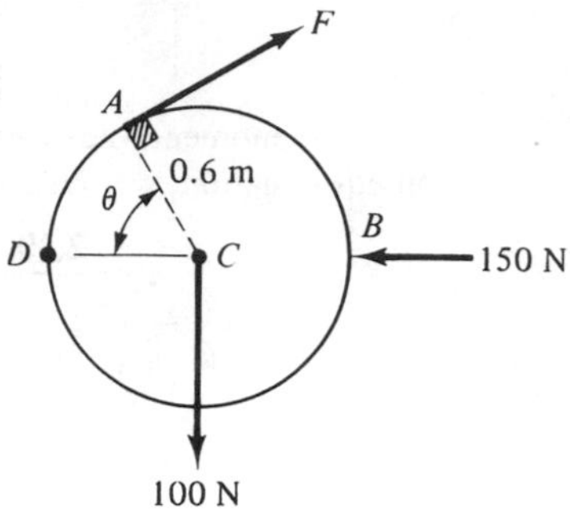

Problem 4.20

4.21. The resultant of the forces acting on the slab shown is known to be a single force passing through point A. Determine F and the resultant.

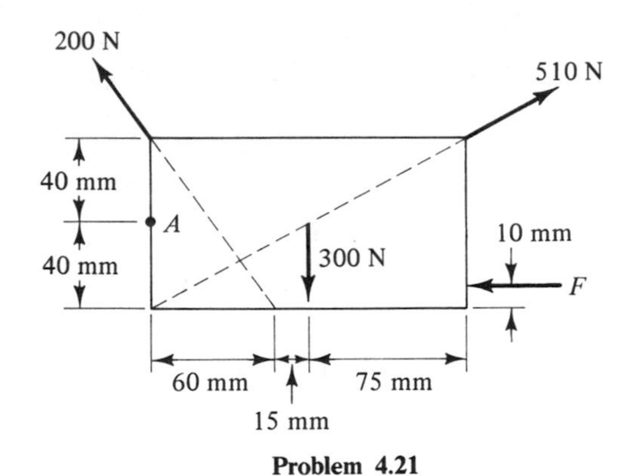

Problem 4.21

4.22. The resultant of the five forces shown is a couple with a magnitude of 22 kip · ft. Determine the magnitudes of the three unknown forces and the sense of the resultant couple.

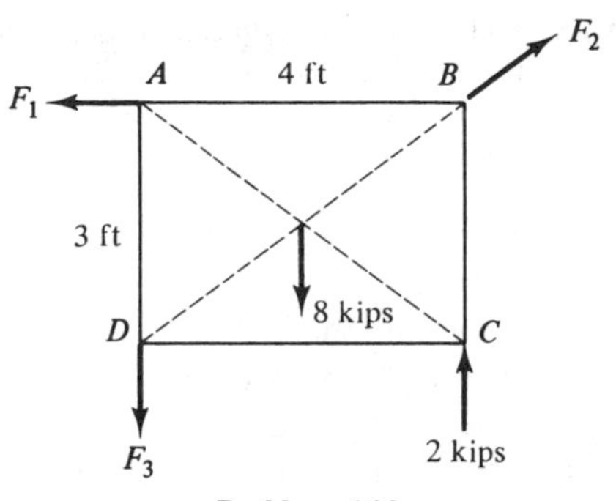

Problem 4.22

4.23. The support columns in a building exert the forces shown on the floor slab. Determine completely the resultant load on the slab.

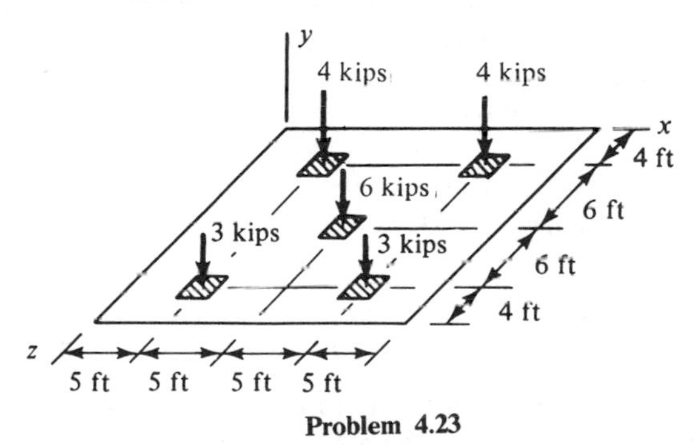

Problem 4.23

4.24. Determine the resultant of the forces shown and show the result on a sketch.

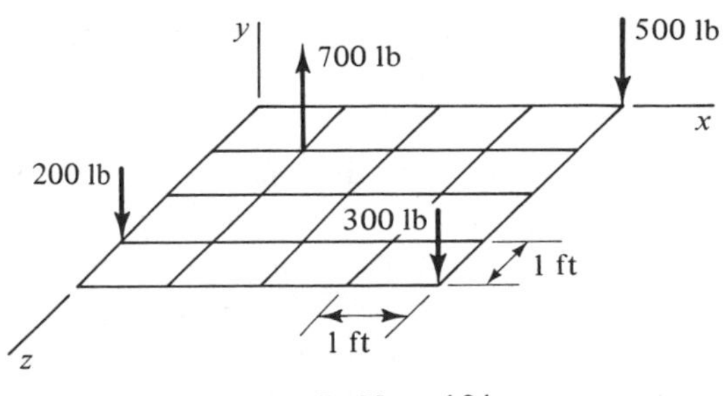

Problem 4.24

4.25. Two pieces of machinery, each with a mass of 4 tons, are mounted in the positions shown on a uniform platform supported by identical springs at each of its four corners. Where should a third 3-ton item be located if the platform is to be horizontal after the installation?

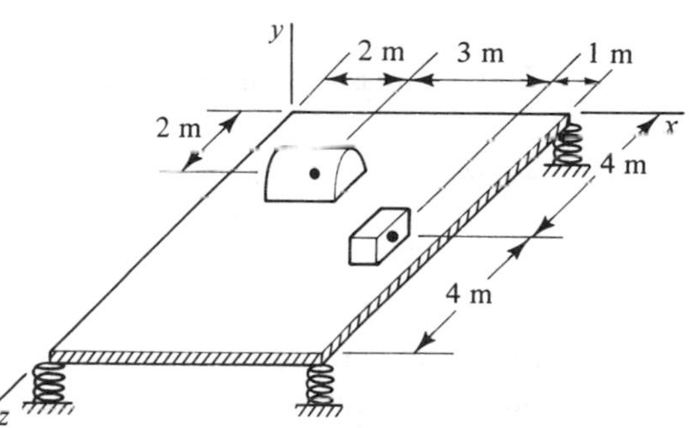

Problem 4.25

4.4 WEIGHT, CENTER OF GRAVITY, AND CENTER OF MASS

In our previous work, the gravitational attraction of the earth on an object has been treated as a single force (the weight) acting at a point called the center of gravity. We are now in a position to justify this step and to develop procedures for locating the center of gravity of bodies of arbitrary shape.

Resultant Gravitational Forces

Consider two bodies with mass m and M separated a distance r. According to *Newton's law of gravitation*, the bodies attract each other with equal and opposite forces that lie along a line connecting their centers and that have magnitude

$$F = \frac{GMm}{r^2} \tag{4.4}$$

where G is the *universal constant of gravitation*; $G = 3.442 \times 10^{-8}$ ft^4/(lb · s^4) or $g = 6.658 \times 10^{-11}$ m^3/(kg · s^2). This law, which is analogous to Coulomb's law of attraction between electric charges, was formulated by Sir Isaac Newton during his studies of the motions of the planets. Strictly speaking, the law of gravitation applies only for particles. However, it has also been shown to hold for spherical bodies of finite size whose density (mass per unit volume) is constant or varies only in the radial direction. The earth, being very nearly a stratified sphere, satisfies these conditions to a reasonable degree of approximation.

The attractive forces of the earth on a body of finite size can be determined by considering the body to consist of an infinite number of infinitesimal elements, or particles, with volume dV and mass $dm = \rho\, dV$, where ρ is the density. According to the law of gravitation, each of these elements is subjected to a force directed toward the earth's center and with magnitude defined by Eq. (4.4). Denoting the magnitude of the force acting upon a typical element by dW, we have

$$dW = \left(\frac{GM_{\text{earth}}}{r^2}\right)\rho\, dV$$

This expression can be rewritten as

$$dW = g\, dm = g\rho\, dV = \gamma\, dV$$

where

$$g = \frac{GM_{\text{earth}}}{r^2} \tag{4.5}$$

is the *acceleration of gravity* and $\gamma = \rho g$ is the weight per unit volume, or *specific weight*, of the body. (The acceleration of gravity due to the attraction of the moon or other celestial bodies can also be obtained from Eq. (4.5) by replacing M_{earth} with the mass of the particular body of interest.)

Unless a body is exceptionally large, the forces acting upon its various elements can be assumed to be parallel. Thus, the earth's gravitational attraction is statically equivalent to a single force. The magnitude of the resultant attractive force is known as the *weight* of the body and its point of application is called the *center of gravity*, or c.g. for short. Both the weight and location of the center of gravity can be determined from the conditions for static equivalence, as we shall now show.

Let $dW = \gamma\, dV$ denote the weight of an infinitesimal element of a body and $(\bar{x}_{\text{el}}, \bar{y}_{\text{el}}, \bar{z}_{\text{el}})$ denote the coordinates of its center of gravity G_{el} [Figure 4.20(a)]. Similarly, let $(\bar{x}, \bar{y}, \bar{z})$ denote the coordinates of the center of gravity G of the entire object and W denote its total weight [Figure 4.20(b)]. The locations of the centers of gravity G_{el} and G can also be defined by their respective position vectors $\mathbf{r}_{\text{el}}$ and $\mathbf{r}$.

If $\mathbf{e}$ is a unit vector in the direction of the local vertical, the force condition for static equivalence can be expressed as

$$\left(\sum \mathbf{F}\right)_{\text{I}} = \left(\sum \mathbf{F}\right)_{\text{II}}: \qquad \int \gamma\, dV\, \mathbf{e} = W\mathbf{e}$$

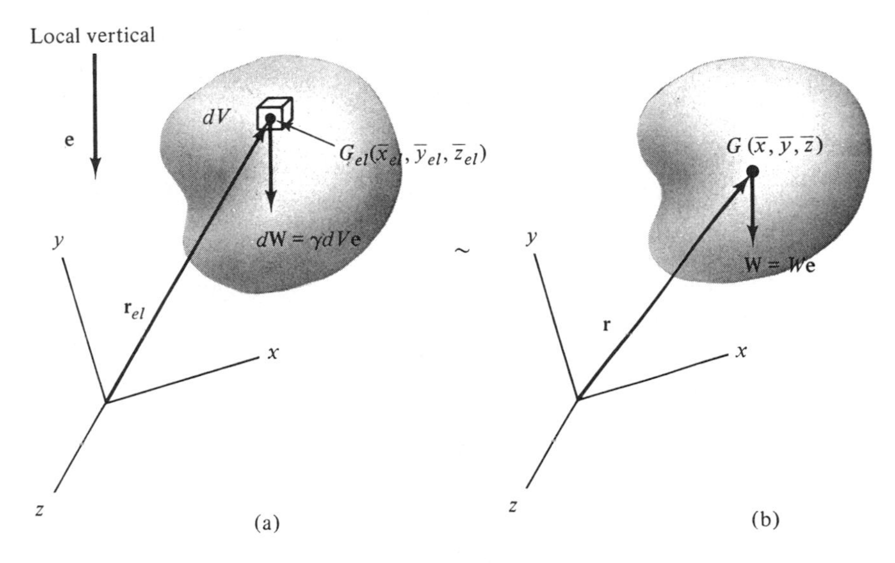

Figure 4.20

or

$$W = \int \gamma \, dV = \int g \, dm \tag{4.6}$$

Note that for system I, the summation becomes an integration since the forces are infinitesimal and infinite in number. For constant g, Eq. (4.6) reduces to

$$W = mg \tag{1.1}$$

where m is the total mass. This is the same result presented in Eq. (1.1). If the density also is constant, we have

$$W = \gamma V \tag{4.7}$$

where V is the total volume of the object.

The moment condition for static equivalence is

$$\left(\sum \mathbf{M}_0\right)_{\mathrm{I}} = \left(\sum \mathbf{M}_0\right)_{\mathrm{II}}: \qquad \int [\mathbf{r}_{\mathrm{el}} \times (\gamma \, dV \mathbf{e})] = \mathbf{r} \times (W\mathbf{e})$$

Using the distributive property of the cross product, we can rewrite this expression as

$$\left(\int \mathbf{r}_{\mathrm{el}} \gamma \, dV\right) \times \mathbf{e} = W\mathbf{r} \times \mathbf{e}$$

Equating the terms on the left side of the cross products and introducing Eq. (4.6), we obtain for the position vector of the center of gravity

$$\mathbf{r} = \frac{\int \mathbf{r}_{\mathrm{el}} \gamma \, dV}{\int \gamma \, dV} \tag{4.8}$$

Expressions for the coordinates of the center of gravity can be obtained from Eq. (4.8) by expressing the various position vectors in component form and equating the coefficients of the base vectors:

$$\bar{x} = \frac{\int \bar{x}_{el}\gamma\, dV}{\int \gamma\, dV} \qquad \bar{y} = \frac{\int \bar{y}_{el}\gamma\, dV}{\int \gamma\, dV} \qquad \bar{z} = \frac{\int \bar{z}_{el}\gamma\, dV}{\int \gamma\, dV} \tag{4.9}$$

In these equations, and in Eqs. (4.6) and (4.8), the integration is to be carried out over the entire volume of the body.

Equations (4.8) and (4.9) can also be expressed in terms of the density by introducing the relation $\gamma = \rho g$. For a uniform gravitational field, g will be a constant in both the numerator and denominator of these equations and it will cancel. The resulting expression for the position of the center of gravity,

$$\mathbf{r} = \frac{\int \mathbf{r}_{el}\rho\, dV}{\int \rho\, dV} \tag{4.10}$$

also defines the position of a point called the *center of mass*.

Since the center of gravity and the center of mass coincide for constant g, the two terms are often used interchangeably. However, the term center of gravity is properly associated with the effect of gravitational forces on a body, and center of mass with the manner in which the mass of a body is distributed. The mass center is an important concept in dynamics.

The evaluation of the integrals in the preceding equations will be considered in detail in Section 4.5. However, if the weight distribution of the body is symmetrical, one or more coordinates of the center of gravity can be determined by inspection.

Symmetry Considerations

The weight distribution of a body is said to have a *plane of symmetry* if for every element located a certain perpendicular distance to one side of the plane there is an element with equal weight located the same perpendicular distance to the other side. This condition is illustrated in Figure 4.21, where the y-z plane is the plane of symmetry. Expressing the first of Eqs. (4.9) in terms of weights, we have for the x coordinate of the center of gravity of the body

$$\bar{x} = \frac{\int \bar{x}_{el}\, dW}{W} = \frac{dW_1(a) + dW_1(-a) + dW_2(b) + dW_2(-b) + \cdots}{W} = 0$$

Since $\bar{x} = 0$, it follows that the center of gravity lies in the plane of symmetry. This is true in general, not just for this specific example.

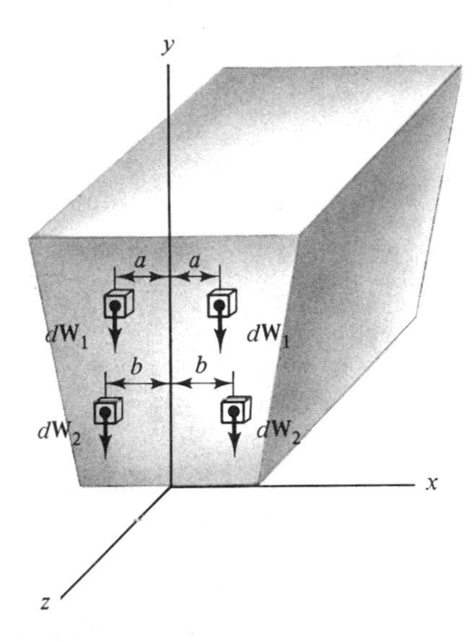

Figure 4.21

If a body has two planes of symmetry, its center of gravity must lie in both planes. Consequently, it must lie along their line of intersection. For example, the center of gravity of a homogeneous right circular cone lies along its axis. If there are three planes of symmetry that intersect at a point, the center of gravity coincides with their point of intersection. Thus, the center of gravity of many uniform bodies with simple shapes, such as spheres, cylinders, ellipsoids, and rectangular parallelepipeds, can be located completely by inspection.

The weight distribution of a body is said to have an *axis of symmetry* if for every element located a certain perpendicular distance to one side of the axis there is an element with equal weight located the same perpendicular distance to the other side. For example, the z axis is an axis of symmetry for the Z-shaped wire shown in Figure 4.22. If a body has an axis of symmetry, the center of gravity must lie along that axis. This can be readily shown by using the same procedure as in the case of a plane of symmetry.

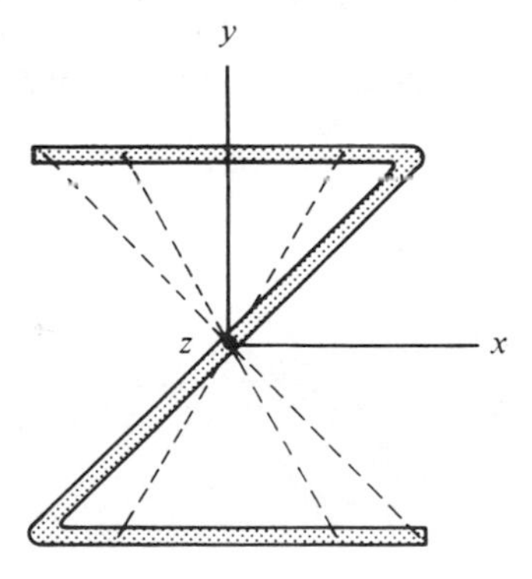

Figure 4.22

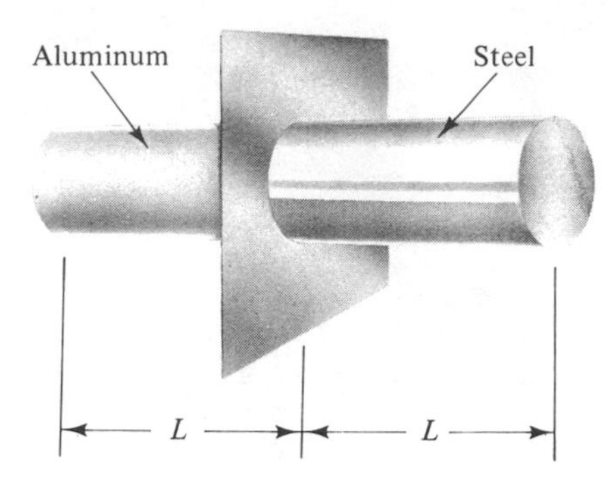

Figure 4.23

When using symmetry conditions to locate the center of gravity, it is essential to remember that it is symmetry of the weight distribution that is required. This is not to be confused with geometric symmetry. Geometric symmetry implies symmetry of the weight distribution only if the body is homogeneous and is located in a uniform gravitational field. For example, the composite shaft shown in Figure 4.23 has geometric symmetry with respect to the plane through its center, but it does not have weight symmetry. Symmetry of the mass distribution is required for location of the mass center by inspection.

Experimental Determination

In concluding this section, we note that the location of the center of gravity can also be determined experimentally. Usually, this requires that the body be supported in such a way that the reactions can be measured. The weight and location of the center of gravity are then determined from the equilibrium equations, as illustrated in Example 4.8.

Example 4.7. Location of the Mass Center. Locate the mass center of the homogeneous body shown in Figure 4.24.

Solution. The x-y plane is a plane of symmetry and the z axis is an axis of symmetry. Therefore, the mass center is at the origin of the coordinate system shown; $\bar{x} = \bar{y} = \bar{z} = 0$. Note that only geometric symmetry is required because the body is homogeneous.

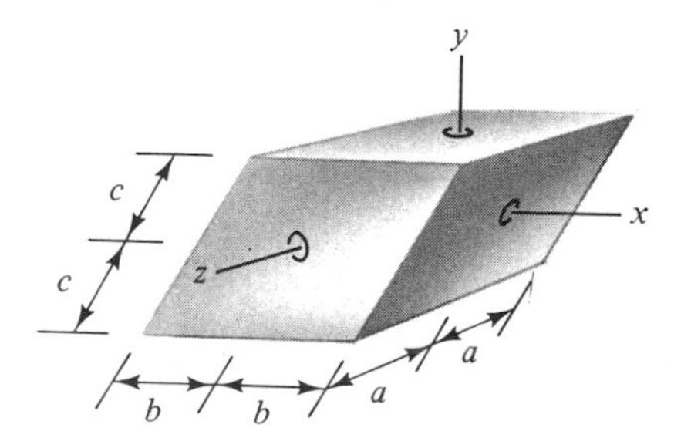

Figure 4.24

Example 4.8. Experimental Determination of Center of Gravity. A connecting rod from an engine is suspended from wires attached to a spring scale at each end, as shown in Figure 4.25(a). When the rod is horizontal, scale A reads 12 N and scale B reads 6 N. Determine the weight and location of the center of gravity of the connecting rod (assume that it is homogeneous).

Solution. By symmetry, $\bar{y} = \bar{z} = 0$. From the FBD of the rod [Figure 4.25(b)], we have

$$\sum F_y = 12\text{ N} + 6\text{ N} - W = 0$$

$$\overset{+}{\curvearrowleft}\sum M_0 = 6\text{ N}\,(0.3\text{ m}) - W\bar{x} = 0$$

from which we get

$$W = 18\text{ N} \qquad \bar{x} = 0.1\text{ m} \qquad \textbf{Answer}$$

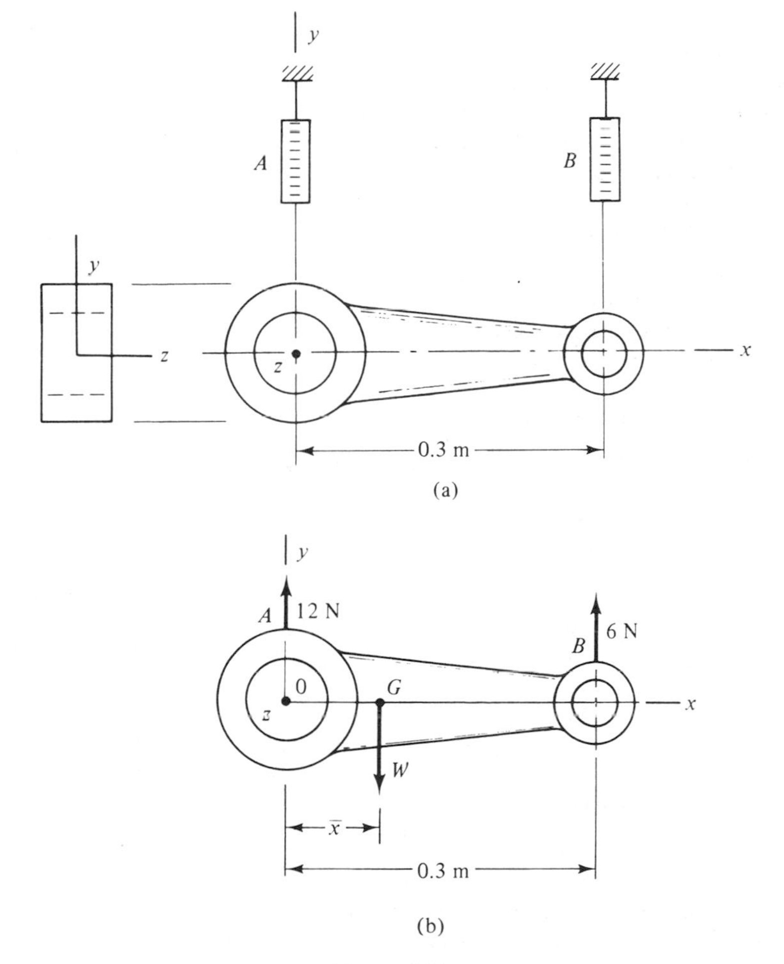

Figure 4.25

PROBLEMS

4.26. The mass of the earth is 4.08×10^{23} lb · s^2/ft and its average radius is 3,960 mi. Determine the acceleration of gravity at (a) the earth's surface and (b) 200 mi above the surface.

4.27. If the earth has a mass of 5.983×10^{24} kg and is 149.5×10^6 km away from the sun, which has a mass of 1.971×10^{30} kg, what is the magnitude of the attractive force between them?

4.28. What is the force of mutual attraction between two 100-mm diameter solid copper spheres placed such that they are just touching? How does this force compare in magnitude with the gravitational attraction of the earth on each sphere? The density of copper is 8,930 kg/m^3.

4.29. Three identical spheres are placed at the corners of an equilateral triangle with sides of length l, as shown. Determine the coordinates (a, b) of the location at which a fourth sphere will experience the same force of attraction from each of the other three.

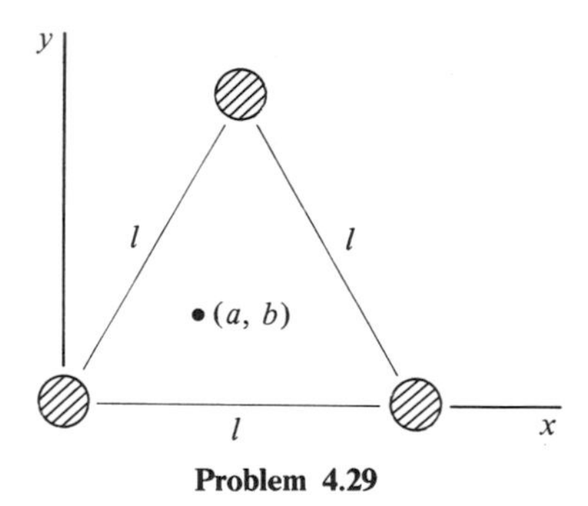

Problem 4.29

4.30. to 4.33. Locate the center of gravity of the uniform bodies shown by using symmetry conditions.

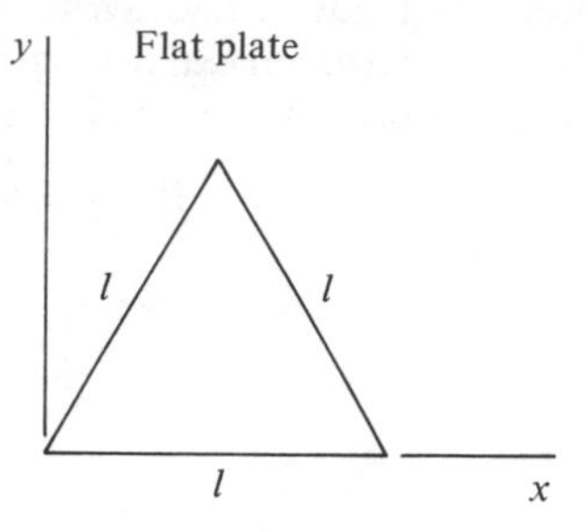

Problem 4.30

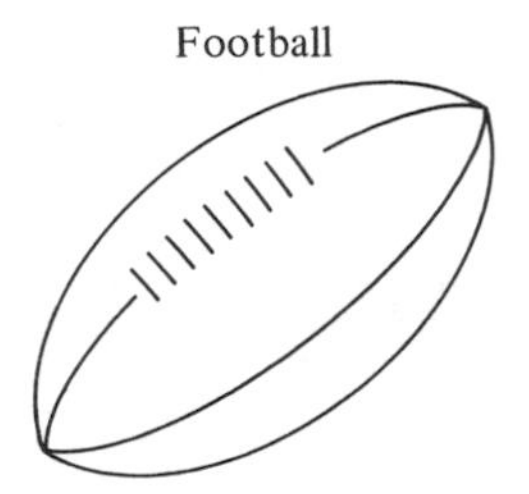

Problem 4.31

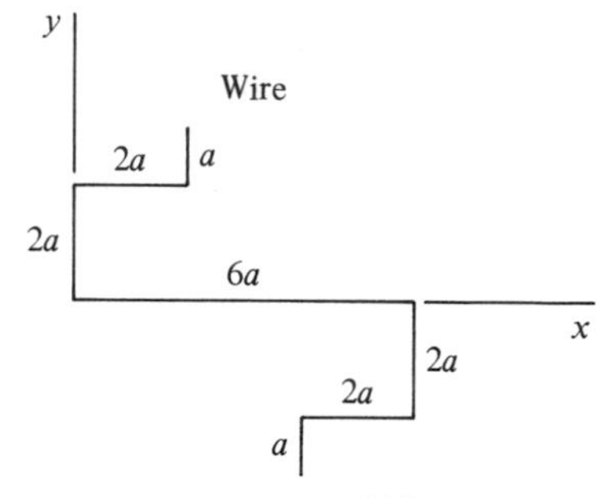

Problem 4.32

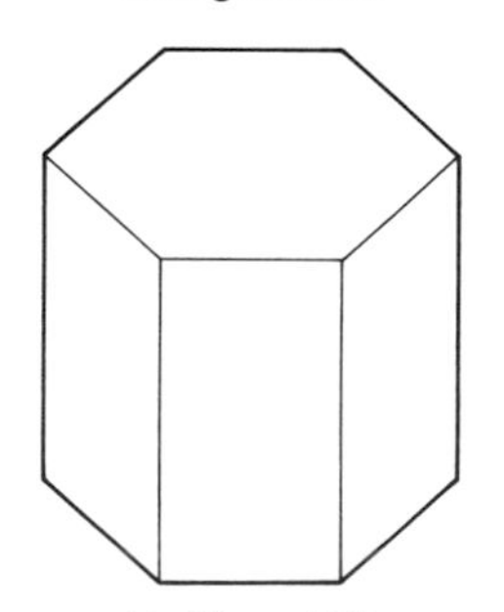

Problem 4.33

4.34. A uniform pulley is to be attached to a flange on a shaft by three identical bolts, as shown. Determine the angle θ for which the mass center of the assembly will be at the center of the pulley.

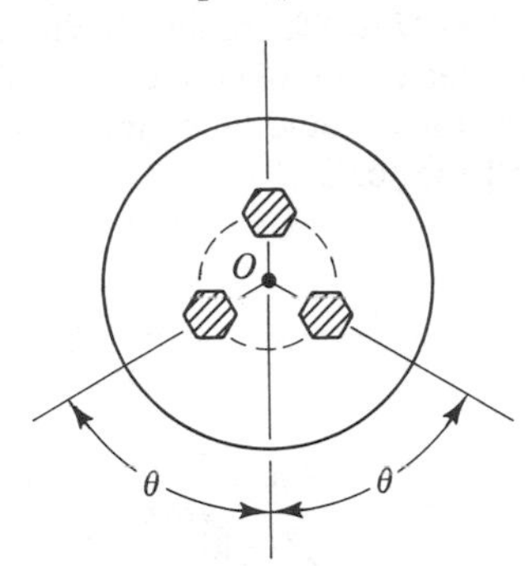

Problem 4.34

4.35. Show that if a body is suspended from a cord or frictionless pivot, its center of gravity will lie directly below the point of suspension. Thus, show that the center of gravity of an irregular-shaped plate can be found by suspending it from two or more different points and locating the point of intersection of vertical lines drawn through the points of suspension. Cut an irregular shape from cardboard and use this procedure to locate its center of gravity.

4.36. When the front wheels of a truck are placed on a scale, it registers 27 kN. When the rear wheels are placed on the scale, it registers 45 kN. What is the mass of the truck and how far back from the front axle is its mass center located? The wheelbase is 3.2 m.

4.37. The rotor assembly in a turbine is held in a horizontal position by vertical cables at each end, as shown. If the tensions on cables A and B are 35 kN and 20 kN, respectively, what is the weight and location of the center of gravity of the assembly?

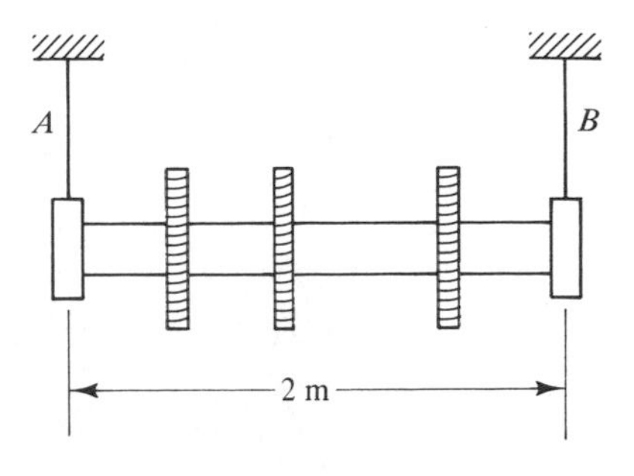

Problem 4.37

4.38. A 460-lb fiberglass automobile body rests on a flat horizontal surface, as shown. If it takes a vertical force of 300 lb to lift the nose off the surface and a vertical force of 290 lb to hold it 1 ft above the surface, what are the coordinates $\bar{x}$ and $\bar{y}$ of the center of gravity?

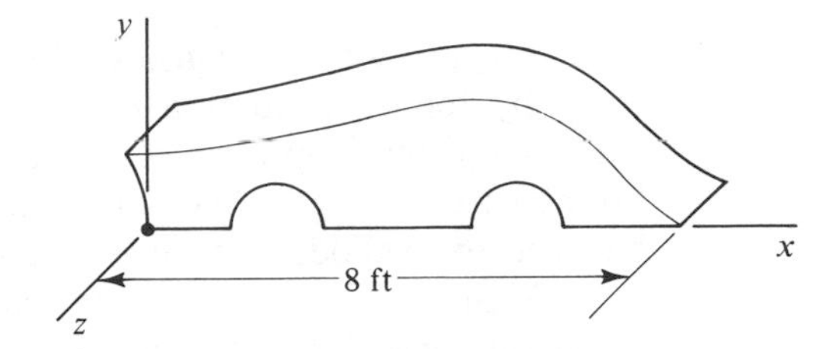

Problem 4.38

4.39. A framework made of three pieces of 10-in.-deep I-beam welded together is held in a horizontal position by three wires. If the tensions on the wires are as indicated, what are the coordinates of the center of gravity?

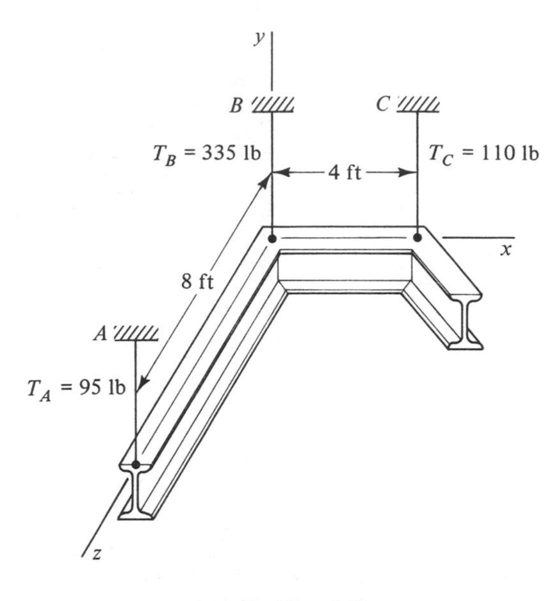

Problem 4.39

4.40. A triangular plate with an irregularly shaped cutout rests on a horizontal surface, as shown. The weight of the plate is 400 N. It takes a vertical force of 120 N to lift corner B off the surface and a vertical force of 90 N to lift corner C. Where is the mass center of the plate located and what vertical force would be required to lift corner A off the surface?

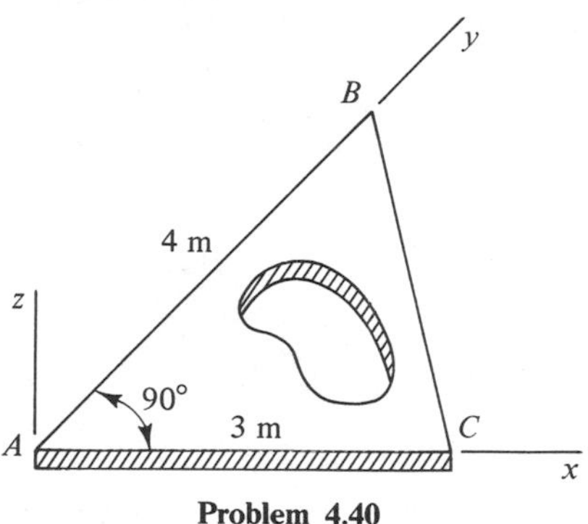

Problem 4.40

4.5 CENTERS OF GRAVITY BY INTEGRATION—CENTROIDS

In this section we shall consider the procedure for locating the center of gravity of a body by formal evaluation of the integrals in Eqs. (4.9). From a computational standpoint, this approach is practical only for bodies with simple shapes whose boundaries can be defined in equation form. More complicated shapes can be handled by methods of approximation or by evaluating the integrals numerically.

Many bodies are uniform throughout, at least to a reasonable degree of approximation. For such homogeneous bodies, γ is a constant, and Eqs. (4.9) reduce to

$$\bar{x} = \frac{\int \bar{x}_{\text{el}}\, dV}{V} \qquad \bar{y} = \frac{\int \bar{y}_{\text{el}}\, dV}{V} \qquad \bar{z} = \frac{\int \bar{z}_{\text{el}}\, dV}{V} \tag{4.11}$$

where

$$V = \int dV$$

is the total volume of the body. These equations for the coordinates of the center of gravity also define the position of a point called the *centroid C of a volume*.

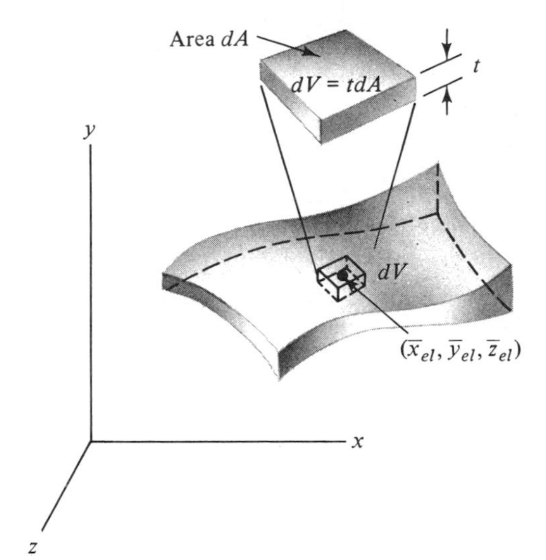

Figure 4.26

Although the centroid and center of gravity coincide for homogeneous bodies, they involve completely different physical concepts. The center of gravity is associated with the distribution of gravitational forces on a body. The centroid is associated with geometric shapes, and is a measure of the manner in which the volume, area, or length of the shape is distributed throughout space. We shall use $(\bar{x}, \bar{y}, \bar{z})$ to denote the coordinates of both centers of gravity and centroids. This should present no problem if we keep in mind the physical concepts associated with each. The symmetry conditions of Section 4.4 also apply to the location of centroids. However, only geometric symmetry is required in this case.

For thin, shell-like shapes that approximate a surface area (Figure 4.26), the volume of an element can be expressed as $dV = t\,dA$, where dA is the area of the element and t is its thickness. If t is a constant, Eqs. (4.11) become

$$\bar{x} = \frac{\int \bar{x}_{\text{el}}\, dA}{A} \qquad \bar{y} = \frac{\int \bar{y}_{\text{el}}\, dA}{A} \qquad \bar{z} = \frac{\int \bar{z}_{\text{el}}\, dA}{A} \tag{4.12}$$

where

$$A = \int dA$$

is the total surface area and $(\bar{x}_{\text{el}}, \bar{y}_{\text{el}}, \bar{z}_{\text{el}})$ are the coordinates of the centroid of the element of area dA. These equations define the *centroid of an area*.

The volume of an element of a thin, wire-like shape that approximates a line (Figure 4.27) can be expressed as $dV = A\,dL$, where A is the cross-sectional area of the element and dL is its length. If A is a constant, Eqs. (4.11) can be written as

$$\bar{x} = \frac{\int \bar{x}_{\text{el}}\, dL}{L} \qquad \bar{y} = \frac{\int \bar{y}_{\text{el}}\, dL}{L} \qquad \bar{z} = \frac{\int \bar{z}_{\text{el}}\, dL}{L} \tag{4.13}$$

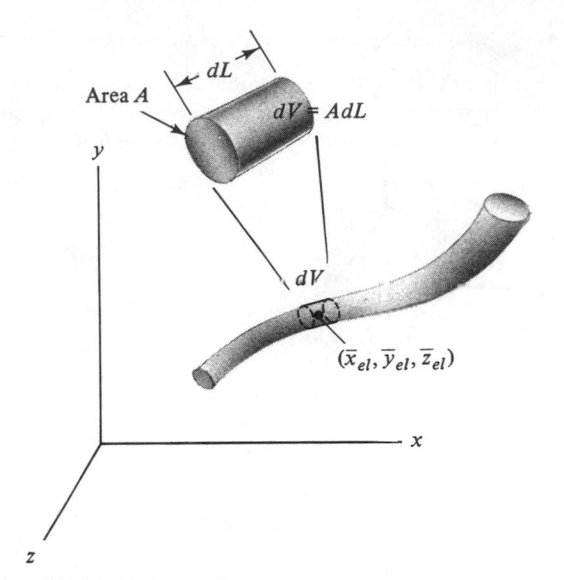

Figure 4.27

where

$$L = \int dL$$

is the total length of the shape and $(\bar{x}_{\mathrm{el}}, \bar{y}_{\mathrm{el}}, \bar{z}_{\mathrm{el}})$ are the coordinates of the centroid of the element of length dL. These equations define the *centroid of a line*.

At this point, it is important to note that the preceding equations define the location of the centroid of a shape in terms of the locations of the centroids of its constituent elements. In other words, $(\bar{x}_{\mathrm{el}}, \bar{y}_{\mathrm{el}}, \bar{z}_{\mathrm{el}})$ must be known before we can determine $(\bar{x}, \bar{y}, \bar{z})$. This presents no problem. The elements can always be chosen with simple shapes such that the locations of their centroids are obvious or can be determined from symmetry conditions.

Once an infinitesimal element has been selected, expressions for its area, length, or volume and location of its centroid can be readily determined from simple geometry, and the necessary integrals can be evaluated to locate the centroid of the shape. For mass centers and centers of gravity, the variation of the density ρ and specific weight γ throughout the body must also be known. Most problems can be handled conveniently by using single integrals and first-order elements (infinitesimal in one direction only). However, double and triple integrals can be used in many instances, if desired. An appropriate choice of the coordinate system can also simplify the computations. For example, polar coordinates are usually most convenient for problems involving circular shapes.

The locations of the centroids of many shapes are tabulated in engineering and mathematical handbooks. For reference purposes, results for several common shapes are listed in Table A.1 of the Appendix.

Example 4.9. Centroid of an Area. Locate the centroid of the area shown in Figure 4.28(a).

Solution. We take a vertical strip of width dx located an arbitrary distance x from the origin as the element of area. The height y of the element is given by the equation of the boundary, $y = x^2$. By symmetry, the centroid of the element is at its geometric center. Thus,

$$\bar{x}_{el} = x \qquad \bar{y}_{el} = \frac{y}{2} = \frac{x^2}{2} \qquad dA = y\,dx = x^2\,dx$$

To cover the entire area, all elements that lie between $x = 0$ and $x = 2$ must be considered. This establishes the limits of integration.

Substituting the expressions for $\bar{x}_{el}$, $\bar{y}_{el}$, and dA into Eqs. (4.12) and integrating, we obtain

$$A = \int dA = \int_0^2 x^2\,dx = \frac{x^3}{3}\bigg|_0^2 = \frac{8}{3}$$

$$A\bar{x} = \int \bar{x}_{el}\,dA = \int_0^2 (x)\,x^2\,dx = \frac{x^4}{4}\bigg|_0^2 = 4$$

$$A\bar{y} = \int \bar{y}_{el}\,dA = \int_0^2 \left(\frac{x^2}{2}\right) x^2\,dx = \frac{x^5}{10}\bigg|_0^2 = \frac{16}{5}$$

Solving for $\bar{x}$ and $\bar{y}$, we have

$$\bar{x} = \frac{4}{(8/3)} = 1.5 \qquad \bar{y} = \frac{(16/5)}{(8/3)} = 1.2 \qquad \textbf{Answer}$$

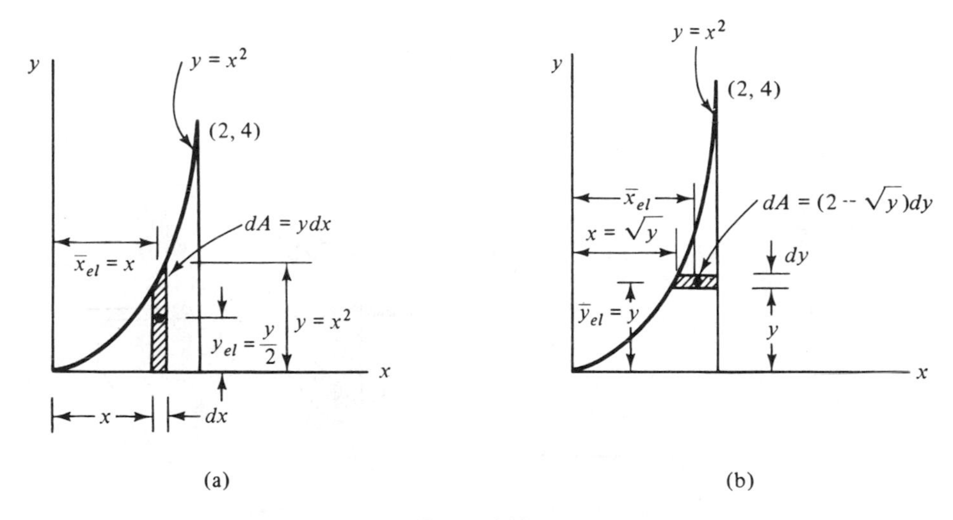

Figure 4.28

Alternate Solution. A horizontal strip can also be used as the element of area [Figure 4.28(b)]. The element has width dy and is located an arbitrary distance y from the origin. Since the centroid of the element lies at its midpoint, $\bar{x}_{el}$ is equal to the average of the horizontal coordinates of the two ends of the strip. The horizontal coordinate x of the left end is obtained from the equation of the boundary ($x = \sqrt{y}$), and the coordinate of the right end is given as $x = 2$. Thus,

$$\bar{x}_{el} = \frac{2 + \sqrt{y}}{2} \qquad \bar{y}_{el} = y \qquad dA = (2 - \sqrt{y})\,dy$$

Substituting these expressions into Eqs. (4.12) and integrating from $y = 0$ to $y = 4$, we get the same results as before.

The computations with the horizontal element are a bit lengthier than those with the vertical element. (The reverse may be true for other problems.) This is because the horizontal element runs from one curve ($x = \sqrt{y}$) to another ($x = 2$), while the vertical element runs from a coordinate axis ($y = 0$) to a curve ($y = x^2$). It is usually convenient to choose the element with one end on a coordinate axis, if possible.

Example 4.10. Centroid of a Volume. Locate the centroid of the pyramidal volume shown in Figure 4.29. The pyramid has height h and a square base with side b.

Solution. By symmetry, the centroid must lie along the x axis. Thus, $\bar{y} = \bar{z} = 0$, and only $\bar{x}$ remains to be determined. The element of volume is taken to be a square slab of thickness dx located an arbitrary distance x from the origin. By similar triangles, the length s of a side of the element is $s = (b/h)x$. Thus,

$$\bar{x}_{el} = x \qquad dV = s^2\,dx = (b/h)^2 x^2\,dx$$

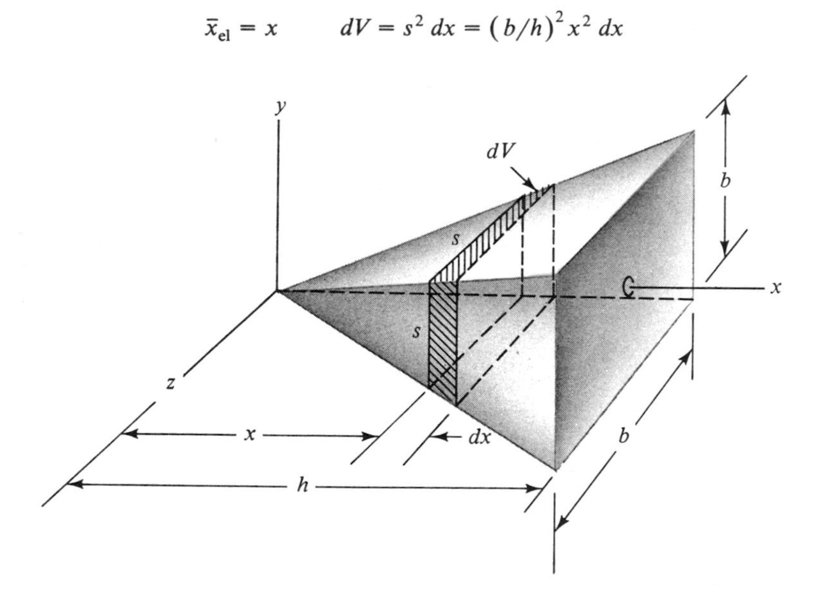

Figure 4.29

The integration is with respect to x and ranges from $x = 0$ to $x = h$. From Eqs. (4.11), we obtain

$$V = \int dV = \int_0^h \left(\frac{b}{h}\right)^2 x^2\,dx = \left(\frac{b}{h}\right)^2 \frac{x^3}{3}\Big|_0^h = \frac{b^2h}{3}$$

$$V\bar{x} = \int \bar{x}_{\text{el}}\,dV = \int_0^h x\left(\frac{b}{h}\right)^2 x^2\,dx = \left(\frac{b}{h}\right)^2 \frac{x^4}{4}\Big|_0^h = \frac{b^2h^2}{4}$$

$$\bar{x} = \frac{3}{4}h \qquad \bar{y} = 0 \qquad \bar{z} = 0 \qquad \textbf{Answer}$$

Example 4.11. Center of Gravity of a Wire. A thin, homogeneous wire is bent into the shape of a circular arc (Figure 4.30). Locate its center of gravity.

Solution. Since the wire is homogeneous, the center of gravity coincides with the centroid. By symmetry, $\bar{y} = 0$. Using polar coordinates and choosing an element of length dL, we have from the geometry of the figure

$$\bar{x}_{\text{el}} = R\cos\theta \qquad dL = R\,d\theta$$

The integration is with respect to θ and ranges from $\theta = -\alpha$ to $\theta = +\alpha$. Thus, we have from Eqs. (4.13)

$$L = \int dL = \int_{-\alpha}^{\alpha} R\,d\theta = R\theta\Big|_{-\alpha}^{\alpha} = 2R\alpha$$

$$L\bar{x} = \int \bar{x}_{\text{el}}\,dL = \int_{-\alpha}^{\alpha} (R\cos\theta)\,R\,d\theta = R^2\sin\theta\Big|_{-\alpha}^{\alpha} = 2R^2\sin\alpha$$

$$\bar{x} = \frac{R\sin\alpha}{\alpha} \qquad \bar{y} = 0 \qquad \textbf{Answer}$$

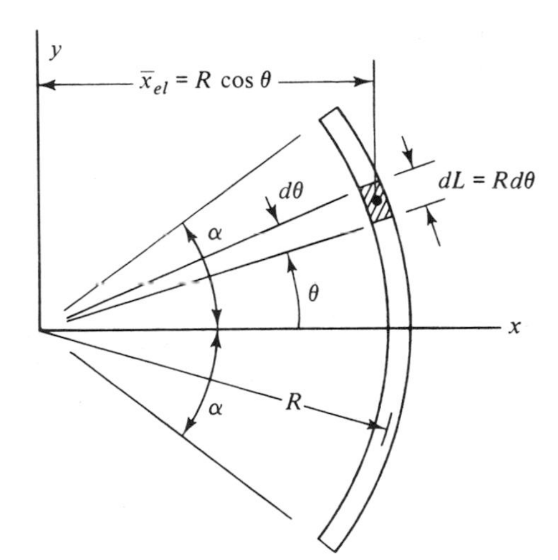

Figure 4.30

Example 4.12. Volume and Surface Area of Bodies of Revolution. For a complete or partial body of revolution, show that (a) the surface area is equal to the length of the generating curve multiplied by the distance its centroid travels during generation of the surface, and (b) the volume is equal to the generating area times the distance its centroid travels. (These results are known as the Pappus-Guldinus theorems.) (c) Use these theorems to determine the surface area and volume of a sphere of radius R.

Solution. (a) Consider the surface generated by revolving the curve shown in Figure 4.31(a) through an angle θ (measured in radians) about the x axis. The element of surface area dA associated with an element of the curve of length dL located a distance y from the axis of revolution is

$$dA = y\theta\, dL$$

Since θ is a constant for any one particular shape, the total surface area is

$$A = \theta \int y\, dL$$

From Eqs. (4.13), we see that

$$\int \bar{y}_{\text{el}}\, dL = \int y\, dL = \bar{y}L$$

Thus,

$$A = L\theta\bar{y} \qquad \textbf{Answer}$$

where $\theta\bar{y}$ is the distance travelled by the centroid of the generating curve.

(b) Consider the volume formed by rotating the area shown in Figure 4.31(b) through an angle θ about the x axis. The volume associated with an element of the area dA located a

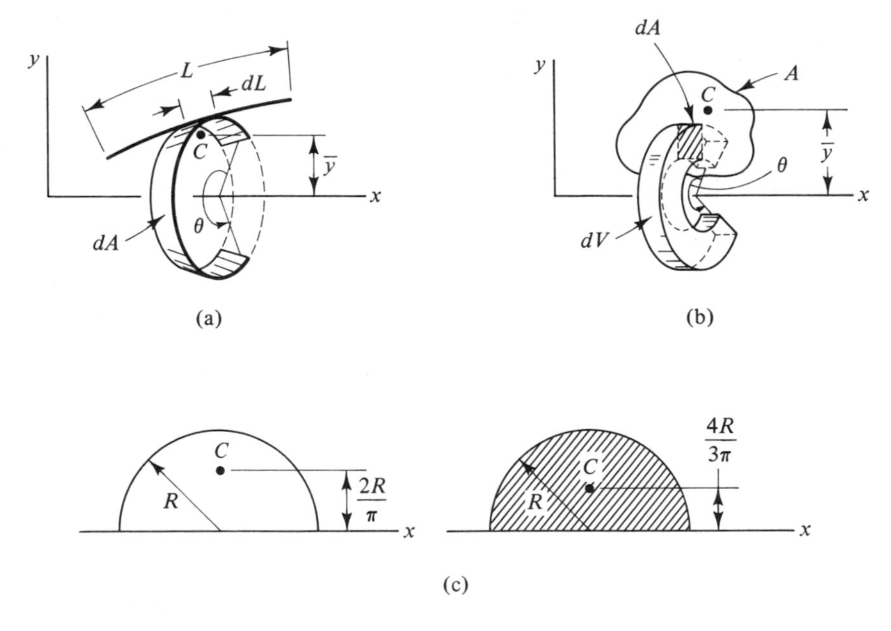

Figure 4.31

distance y from the x axis is

$$dV = y\theta\, dA$$

and the total volume is

$$V = \theta \int y\, dA$$

But

$$\int y\, dA = \bar{y}A$$

so

$$V = A\theta\bar{y} \qquad \textbf{Answer}$$

where $\theta\bar{y}$ is the distance travelled by the centroid of the generating area.

(c) The surface area and volume of a sphere can be generated by rotating the semicircular arc and area shown in Figure 4.31(c) through one complete revolution about the x axis. From Example 4.11, we have for the semicircular arc

$$\bar{y} = \frac{2R}{\pi}$$

Thus,

$$A = 2\pi\bar{y}L = 2\pi\left(\frac{2R}{\pi}\right)\pi R = 4\pi R^2 \qquad \textbf{Answer}$$

The location of the centroid of the semicircular area can be determined by integration or from Table A.1 of the Appendix. From the table, we find

$$\bar{y} = \frac{4R}{3\pi}$$

Thus,

$$V = 2\pi A\bar{y} = 2\pi\left(\frac{\pi R^2}{2}\right)\left(\frac{4R}{3\pi}\right) = \frac{4}{3}\pi R^3 \qquad \textbf{Answer}$$

PROBLEMS

4.41. to 4.48. Determine by integration the coordinates $\bar{x}$ and $\bar{y}$ of the centroid of the area shown.

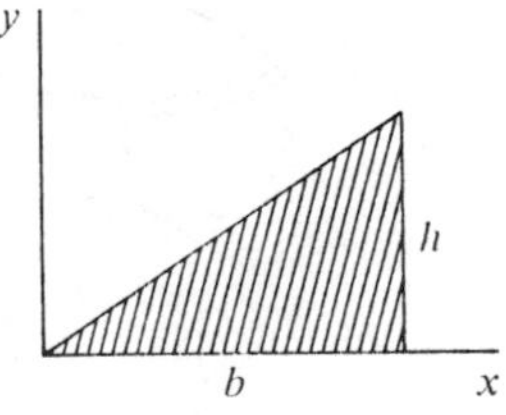

Problem 4.41

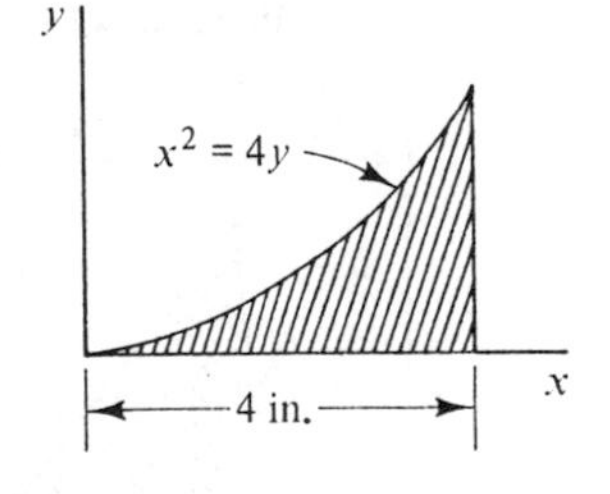

Problem 4.42

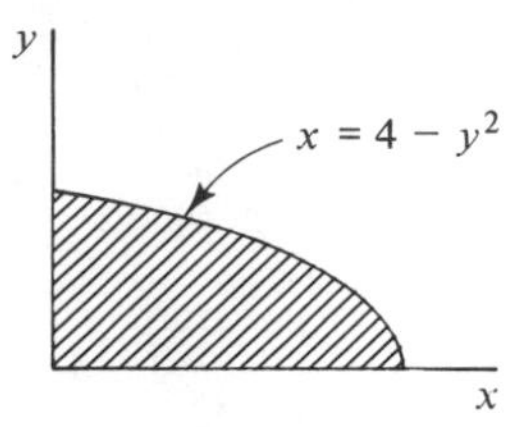

Problem 4.43

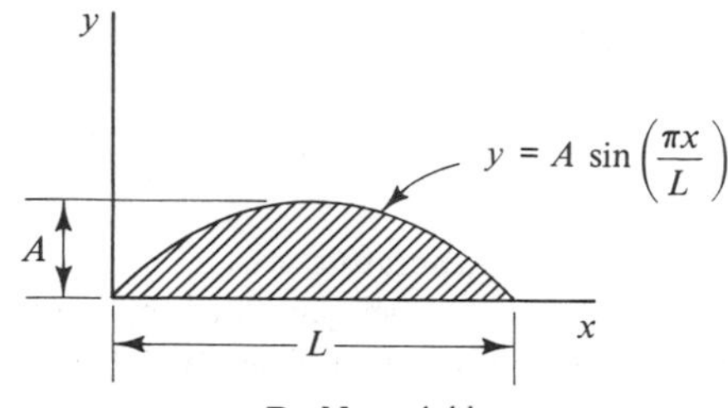

Problem 4.44

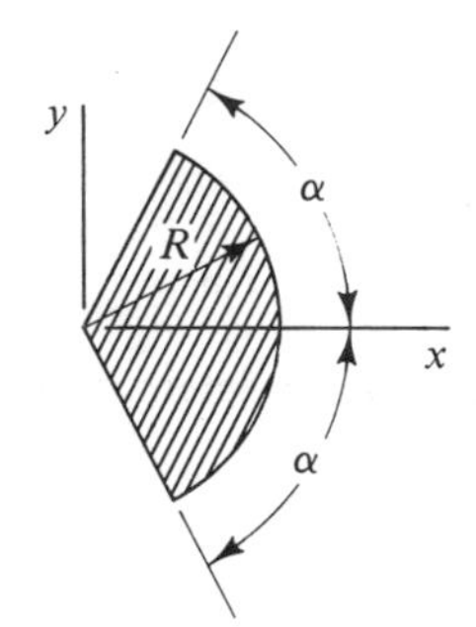

Problem 4.45

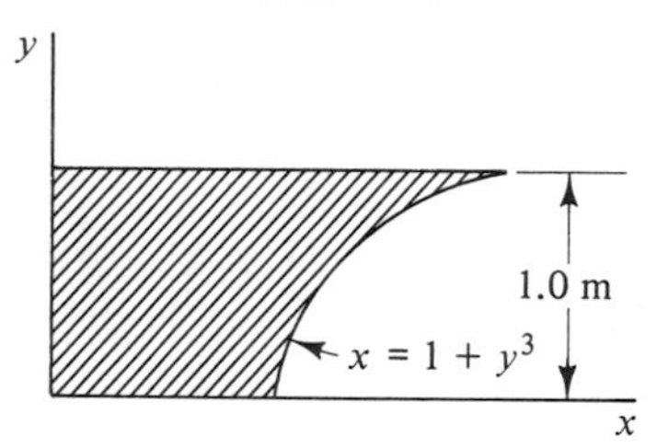

Problem 4.46

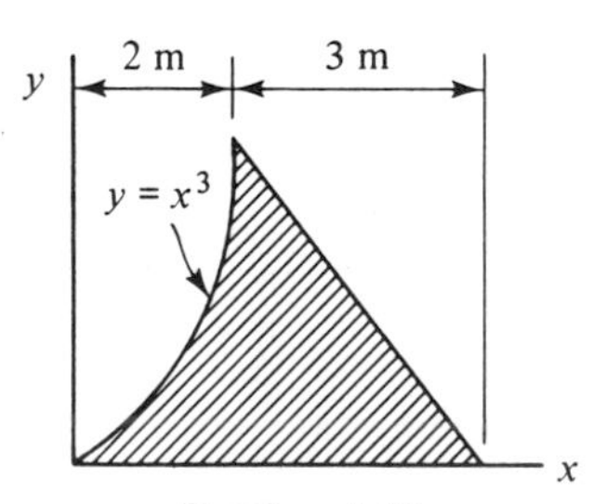

Problem 4.47

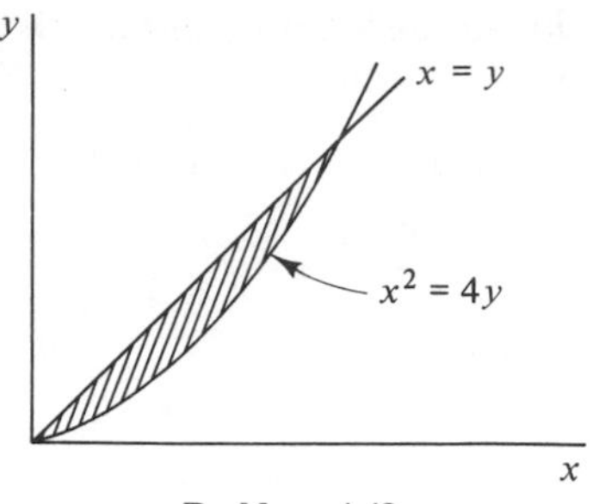

Problem 4.48

4.49. and 4.50. For the line shown, determine the coordinates $\bar{x}$ and $\bar{y}$ of the centroid. Note that from the Pythagorean theorem, $dL = [(dy/dx)^2 + 1]^{1/2} dx = [(dx/dy)^2 + 1]^{1/2} dy$.

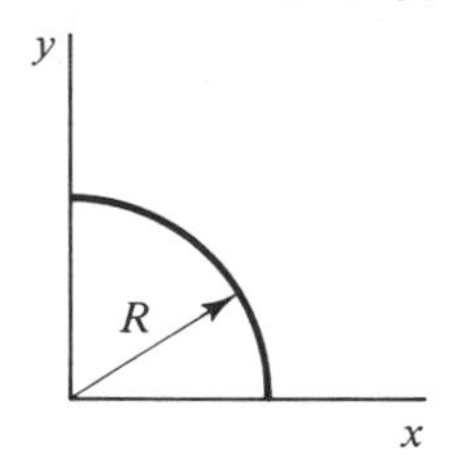

Problem 4.49

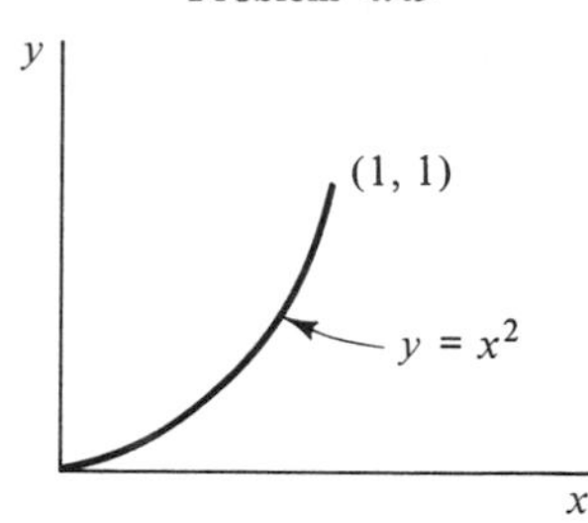

Problem 4.50

4.51. Locate the mass center of the uniform thin-walled spherical cap shown.

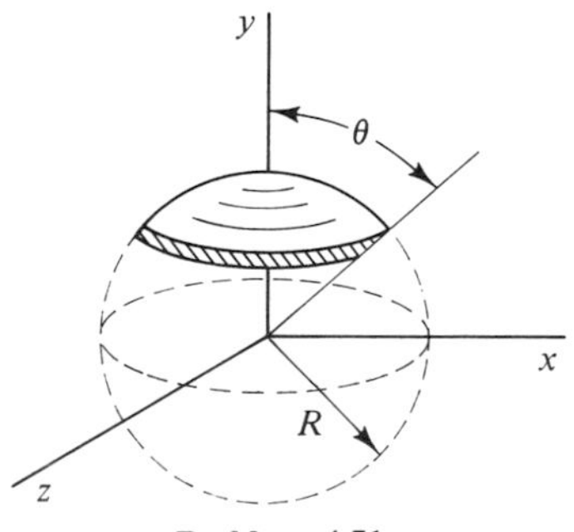

Problem 4.51

4.52. Locate the center of gravity of a thin conical shell with height h and a circular base with radius R.

4.53. Locate the center of gravity of a solid hemispherical object of radius R. Assume that the object is homogeneous.

4.54. Locate the centroid of the volume formed by rotating the area in Problem 4.42 through an angle of 360° about the x axis.

4.55. Same as Problem 4.54, except that the area is rotated through an angle of 180°.

4.56. and 4.57. Locate the mass center of the solid uniform body shown.

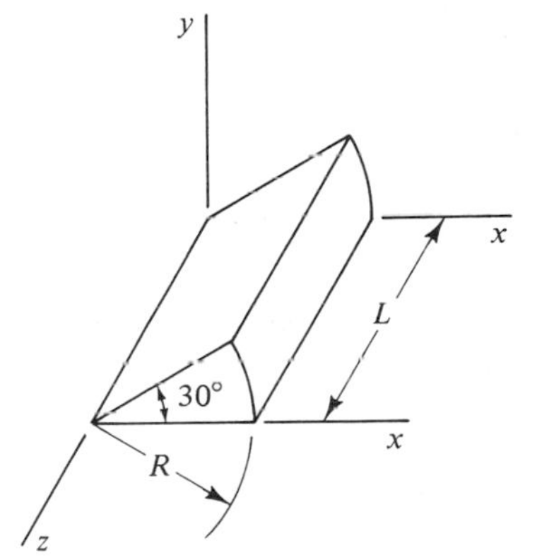

Problem 4.56

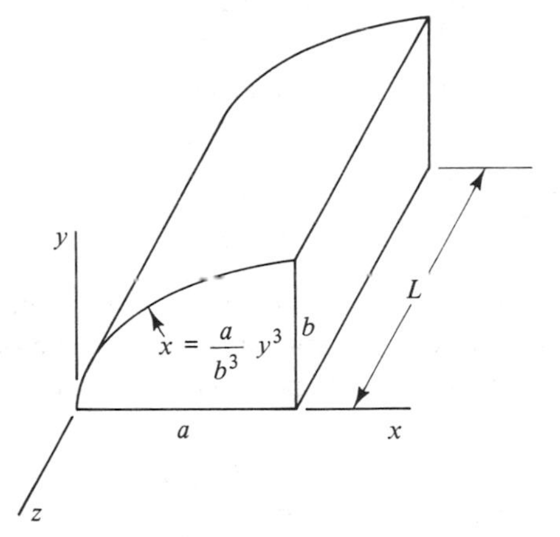

Problem 4.57

4.58. Use the Pappus-Guldinus theorems (see Example 4.12) to compute the surface area and volume of a solid cone with height h and a circular base with radius R.

4.59. Use the Pappus-Guldinus theorems (see Example 4.12) to compute the surface area and volume of the half-torus shown.

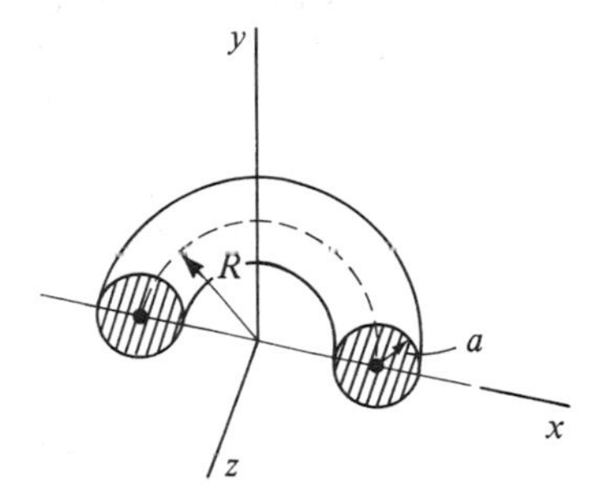

Problem 4.59

4.60. A lamp shade has the shape and dimensions shown. How many square feet of fabric are required to cover the shade?

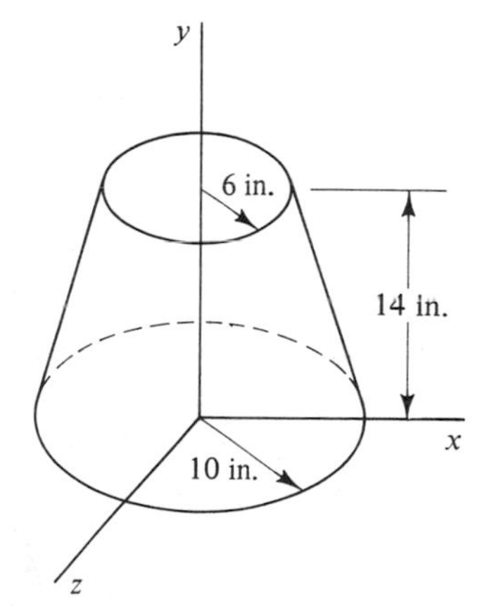

Problem 4.60

4.6 CENTERS OF GRAVITY AND CENTROIDS OF COMPOSITE BODIES AND SHAPES

Many objects can be divided into finite-sized pieces of simple shape whose centers of gravity can be easily located either from symmetry, integration, or from tables. For such composite bodies, the center of gravity can be located with a minimum of effort.

Consider a composite body consisting of pieces with volumes $V_1, V_2, \ldots, V_n$. Integrals over the total volume can be expressed as the sum of the integrals over each of these subvolumes. In particular, Eq. (4.8) for the position vector of the center of gravity can be written as

$$\mathbf{r} = \frac{\int_V \mathbf{r}_{\text{el}} \gamma \, dV}{\int_V \gamma \, dV} = \frac{\int_{V_1} \mathbf{r}_{\text{el}} \gamma \, dV + \int_{V_2} \mathbf{r}_{\text{el}} \gamma \, dV + \cdots + \int_{V_n} \mathbf{r}_{\text{el}} \gamma \, dV}{\int_{V_1} \gamma \, dV + \int_{V_2} \gamma \, dV + \cdots + \int_{V_n} \gamma \, dV}$$

From Eqs. (4.6) and (4.8), we recognize that

$$\int_{V_i} \gamma \, dV = W_i \qquad \int_{V_i} \mathbf{r}_{\text{el}} \gamma \, dV = W_i \mathbf{r}_i$$

where W_i is the weight of the ith piece of the body and $\mathbf{r}_i = \bar{x}_i \mathbf{i} + \bar{y}_i \mathbf{j} + \bar{z}_i \mathbf{k}$ is the position vector of its center of gravity. Thus, we have

$$\mathbf{r} = \frac{W_1 \mathbf{r}_1 + W_2 \mathbf{r}_2 + \cdots + W_n \mathbf{r}_n}{W_1 + W_2 + \cdots + W_n} = \frac{\sum W_i \mathbf{r}_i}{\sum W_i} \tag{4.14}$$

or in component form,

$$\bar{x} = \frac{\sum W_i \bar{x}_i}{\sum W_i} \qquad \bar{y} = \frac{\sum W_i \bar{y}_i}{\sum W_i} \qquad \bar{z} = \frac{\sum W_i \bar{z}_i}{\sum W_i} \tag{4.15}$$

This same procedure yields for the coordinates of the centroid of a composite volume

$$\bar{x} = \frac{\sum V_i \bar{x}_i}{\sum V_i} \qquad \bar{y} = \frac{\sum V_i \bar{y}_i}{\sum V_i} \qquad \bar{z} = \frac{\sum V_i \bar{z}_i}{\sum V_i} \tag{4.16}$$

where V_i is the volume of the ith piece and $(\bar{x}_i, \bar{y}_i, \bar{z}_i)$ are the coordinates of its centroid. Similar relations hold for composite areas and lines and can be obtained from Eqs. (4.16) by replacing the Vs by As and Ls, respectively. For example, for a composite area,

$$\bar{x} = \frac{\sum A_i \bar{x}_i}{\sum A_i} \qquad \bar{y} = \frac{\sum A_i \bar{y}_i}{\sum A_i} \qquad \bar{z} = \frac{\sum A_i \bar{z}_i}{\sum A_i} \tag{4.17}$$

where A_i is the area of the ith piece and $(\bar{x}_i, \bar{y}_i, \bar{z}_i)$ are the coordinates of its centroid.

Bodies with holes or cavities can be thought of as solid bodies minus one or more pieces. In this case, the weights of the pieces removed are considered to be negative quantities in Eqs. (4.14) and (4.15). The same procedure applies for centroids. The volume, area, or length of a piece removed from a shape is simply considered to be negative.

Example 4.13. Center of Gravity of a Shaft. Locate the center of gravity of the composite shaft shown in Figure 4.32. The specific weight of steel is 77 kN/m^3 and the specific weight of aluminum is 27 kN/m^3. Each portion of the shaft is homogeneous.

Solution. We choose coordinates and number the pieces as shown. By symmetry, $\bar{y} = \bar{z} = 0$, and $\bar{x}_1 = -L/2$ and $\bar{x}_2 = L/2$. The weights of the pieces are $W_1 = \gamma_1 V_1$ and $W_2 = \gamma_2 V_2$. Since the pieces have the same volume,

$$W_1 = \left(\frac{\gamma_1}{\gamma_2}\right) W_2 = \frac{77 \text{ kN/m}^3}{27 \text{ kN/m}^3} W_2 = 2.85 W_2$$

Applying the first of Eqs. (4.15), we have

$$\bar{x} = \frac{W_1 \bar{x}_1 + W_2 \bar{x}_2}{W_1 + W_2} = \frac{2.85 W_2(-L/2) + W_2(L/2)}{2.85 W_2 + W_2} = -0.24L \qquad \textbf{Answer}$$

Thus, the center of gravity lies along the centerline a distance $0.24L$ to the left of the center of the shaft. Note that it was not necessary to calculate the actual weights of the pieces in this example.

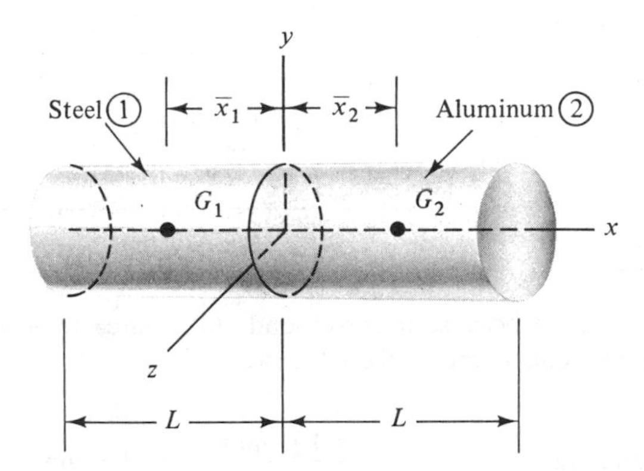

Figure 4.32

Example 4.14. Center of Gravity of a Sheet-Metal Part. The sheet-metal part shown in Figure 4.33 is homogeneous and has constant thickness. Locate its center of gravity.

Solution. Since the body is homogeneous, its center of gravity coincides with the centroid of the area. We choose coordinates and divide the shape into pieces as indicated. Piece 2 is a solid rectangle and piece 3 is the circular area removed from it to form the hole. By symmetry, $\bar{z} = 0$. The location of the centroids of the various pieces can be determined from symmetry and from Table A.1 of the Appendix. Since there are a number of pieces, it is

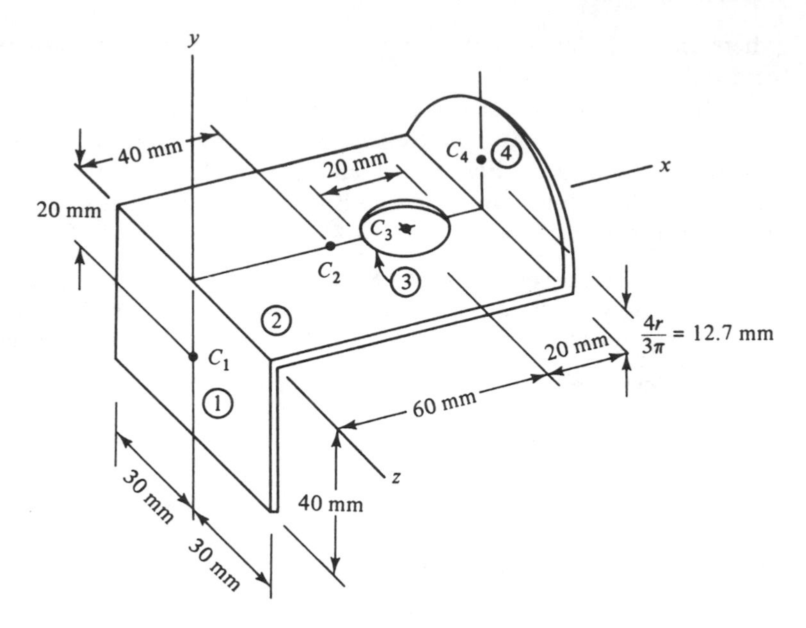

Figure 4.33

convenient to tabulate the data necessary for making the computations indicated in Eqs. (4.17):

No.	A_i (mm^2)	$\bar{x}_i$ (mm)	$\bar{y}_i$ (mm)	$A_i\bar{x}_i$ (mm^3)	$A_i\bar{y}_i$ (mm^3)
1	2.40×10^3	0.0	-20.0	0.0	-48.0×10^3
2	4.80×10^3	40.0	0.0	192.0×10^3	0.0
3	-0.31×10^3	60.0	0.0	-18.6×10^3	0.0
4	1.41×10^3	80.0	12.7	112.8×10^3	18.0×10^3
Σ	8.30×10^3			286.2×10^3	-30.0×10^3

Note that A_3 is considered negative because it corresponds to an area removed from the shape. Using Eqs. (4.17) and the sums found in the table, we have

$$\bar{x} = \frac{286.2 \times 10^3 \text{ mm}^3}{8.3 \times 10^3 \text{ mm}^2} = 34.5 \text{ mm} \quad \bar{y} = \frac{-30.0 \times 10^3 \text{ mm}^3}{8.3 \times 10^3 \text{ mm}^2} = -3.6 \text{ mm}$$

$$\bar{z} = 0$$

Answer

Example 4.15. Centroid of an Area. Locate the centroid of the cross-sectional area of the structural member shown in Figure 4.34(a).

Solution. We choose coordinates and divide the area into elements as shown. By symmetry, $\bar{x} = 0$. The geometric properties of the angle sections are obtained from Table A.5

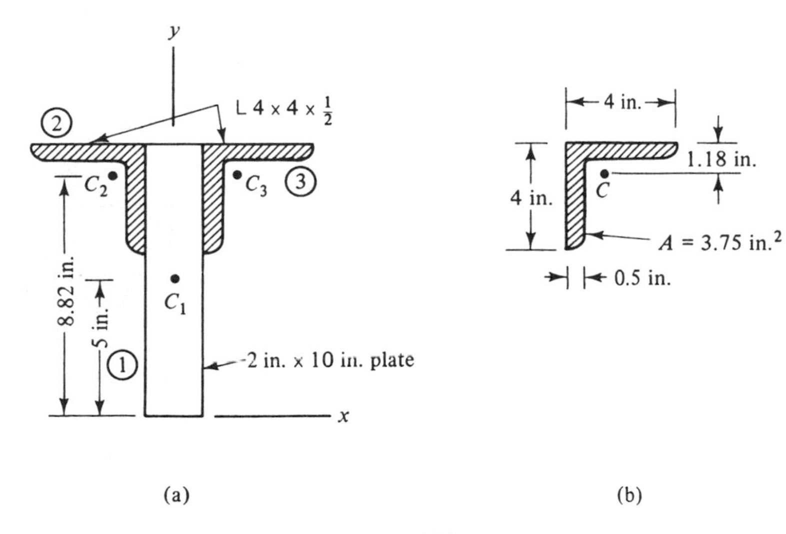

Figure 4.34

of the Appendix. The data are listed in Figure 4.34(b). Using a tabular format and Eqs. (4.17), we have

No.	A_i (in.^{2})	$\bar{y}_i$ (in.)	$A_i\bar{y}_i$ (in.^{3})
1	20.00	5.00	100.00
2	3.75	8.82	33.08
3	3.75	8.82	33.08
Σ	27.50		166.16

$$\bar{y} = \frac{166.16 \text{ in.}^3}{27.50 \text{ in.}^2} = 6.04 \text{ in.} \qquad \bar{x} = 0 \qquad \textbf{Answer}$$

PROBLEMS

4.61. to 4.70. Locate the centroid of the shape shown.

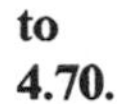

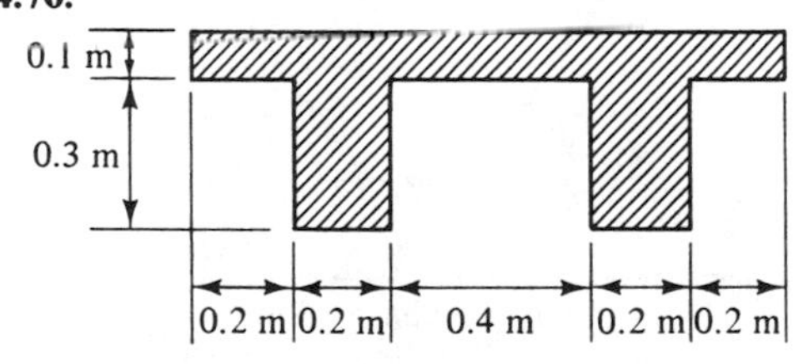

Problem 4.61

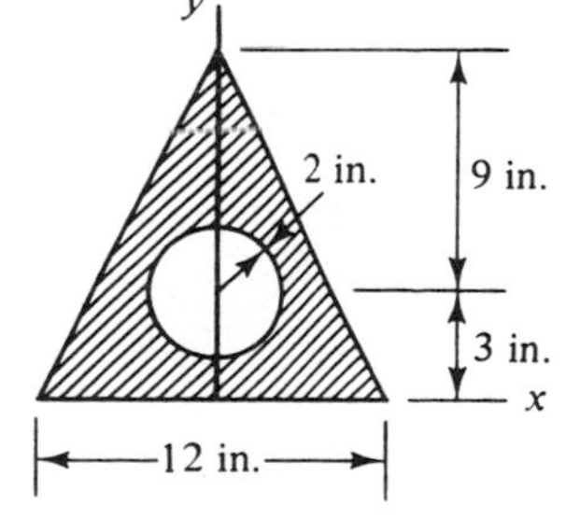

Problem 4.62

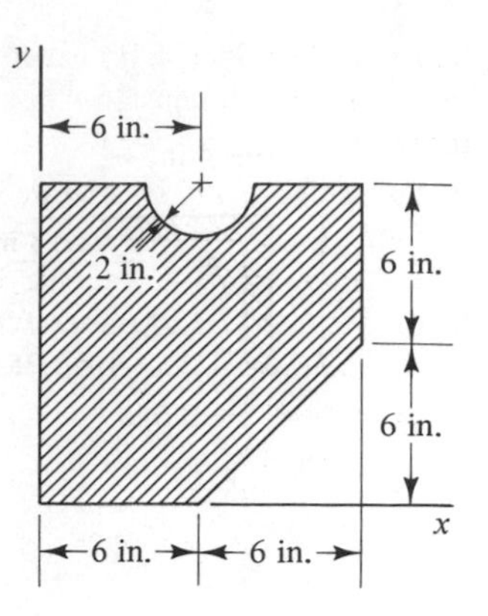

Problem 4.63

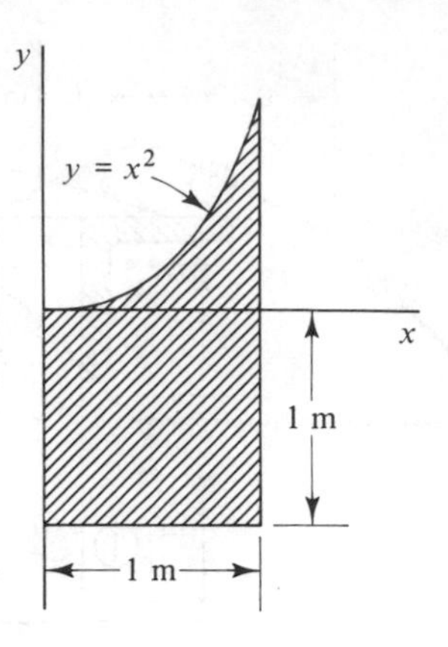

Problem 4.66

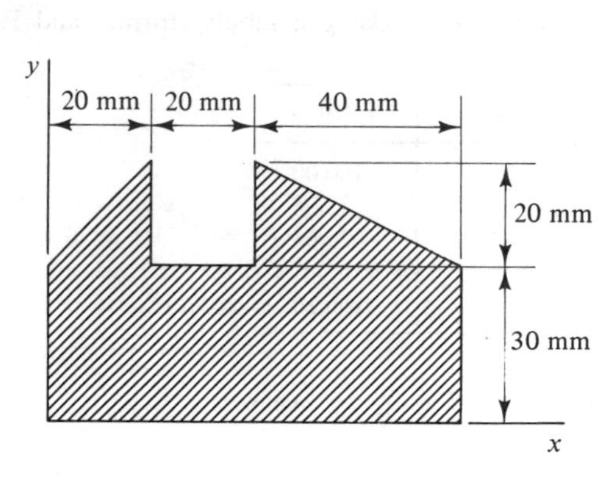

Problem 4.64

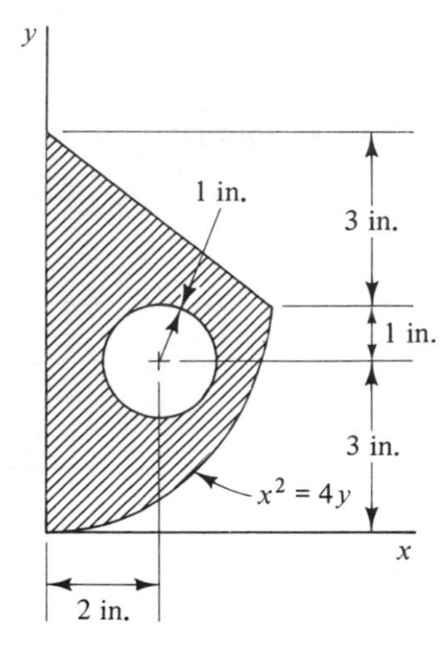

Problem 4.67

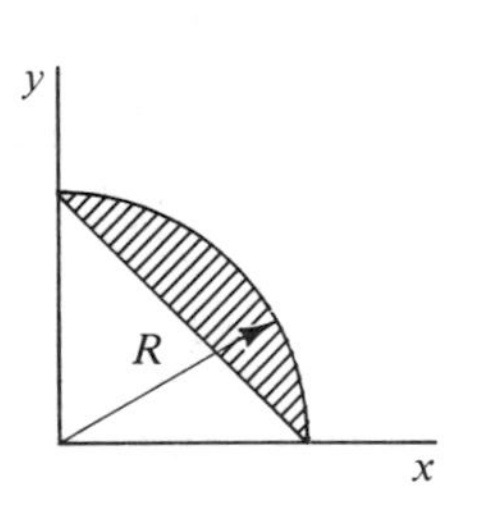

Problem 4.65

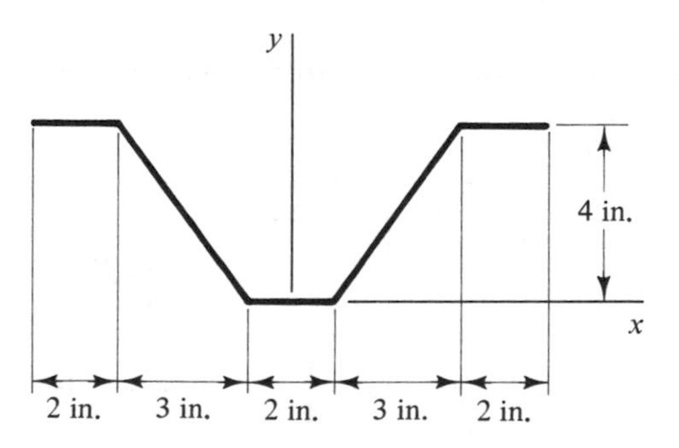

Problem 4.68

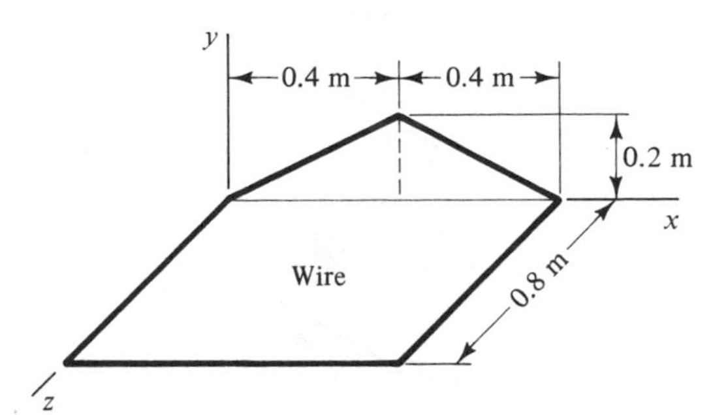

Problem 4.69

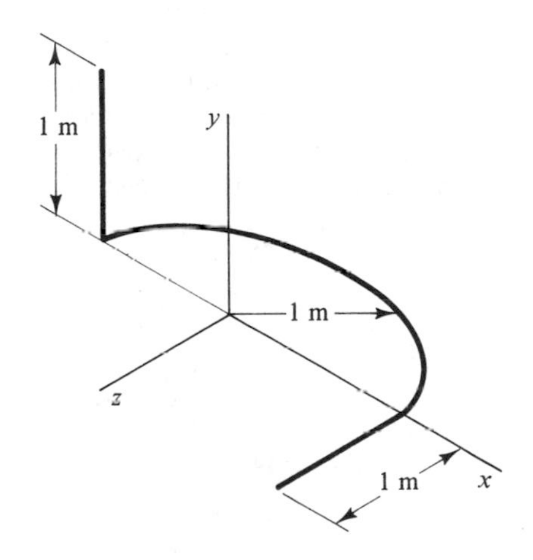

Problem 4.70

4.71. If $a = 2$ in., determine the dimension b for which the centroid of the symmetrical area shown will be located 5 in. above the x-axis.

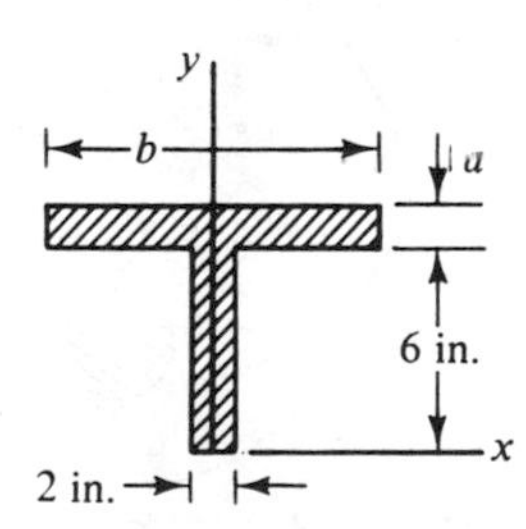

Problem 4.71

4.72. Same as Problem 4.71, except that $b = 8$ in. and the dimension a is to be determined.

4.73. A flat steel plate is attached to the top surface of a 254 × 102 mm Universal beam with a mass of 28 kg/m (see Table A.8) to form a structural member with the cross section shown. Locate the centroid of the cross section.

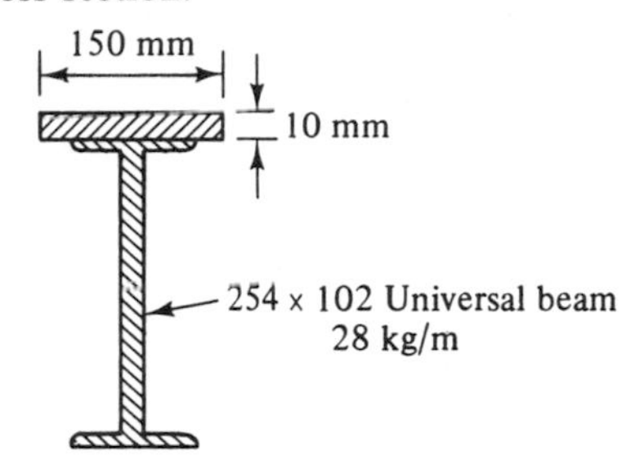

Problem 4.73

4.74. Three timbers are nailed together to form a symmetrical section with the nominal dimensions shown. Locate the centroid of the section by using the geometric properties of timbers given in Table A.12 of the Appendix.

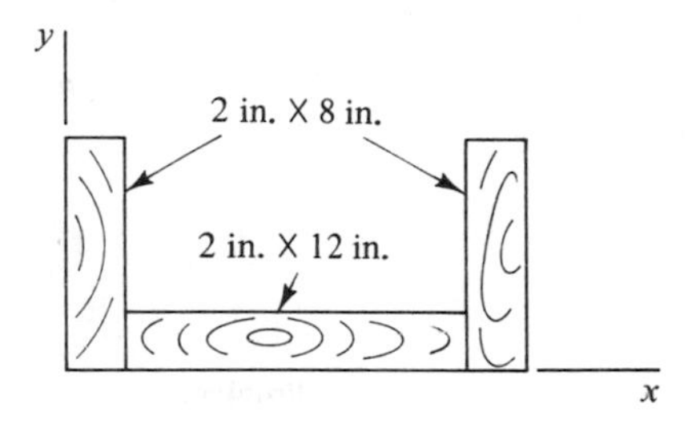

Problem 4.74

4.75. Locate the mass center of a cylindrical can with radius R and height $h = 4R$ if it is open at the top.

4.76. A sheet-metal part has the shape shown. Locate its mass center.

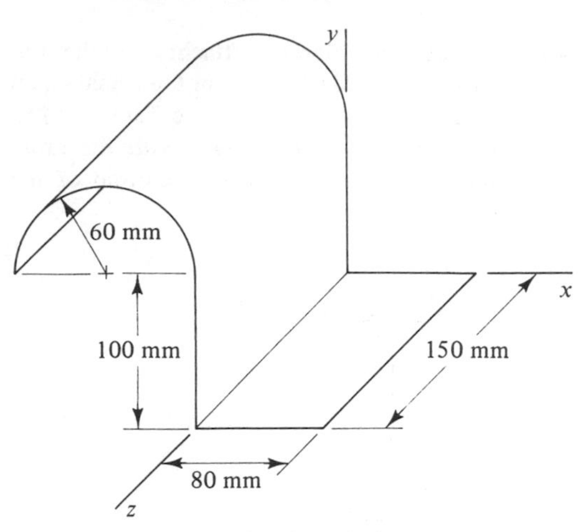

Problem 4.76

4.77. A solid hemisphere is joined to a solid cone, as shown. Locate the centroid of the combined volume.

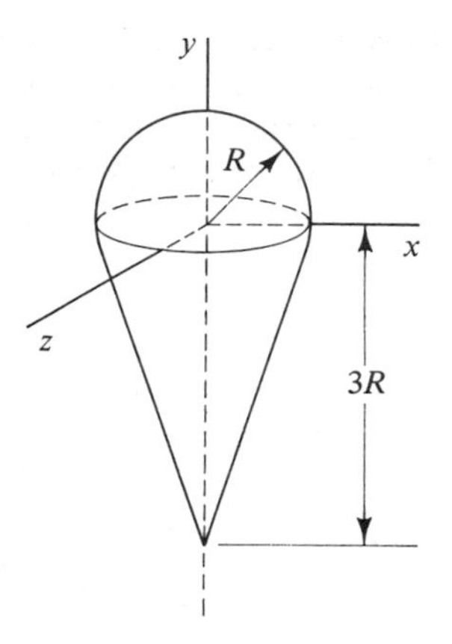

Problem 4.77

4.78. A section of steel pipe with a nominal diameter of 6 in. is joined to an aluminum plate, as shown. Locate the center of gravity of the assembly and the centroid of the volume that it occupies. See Table A.7 of the Appendix for the properties of the pipe. The specific weight of aluminum is 170 lb/ft^3.

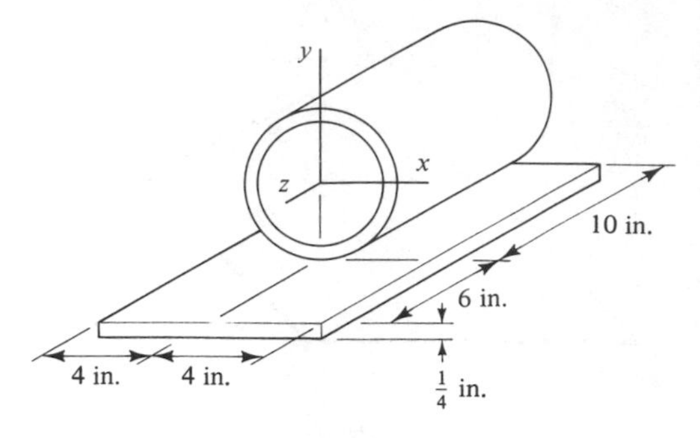

Problem 4.78

4.79. Locate the center of gravity of the trough shown. The ends are made of wood with a specific weight of 30 lb/ft^3 and the central portion is made of $\frac{1}{8}$-in. thick steel that weighs 490 lb/ft^3. The actual dimensions of the wood are as indicated.

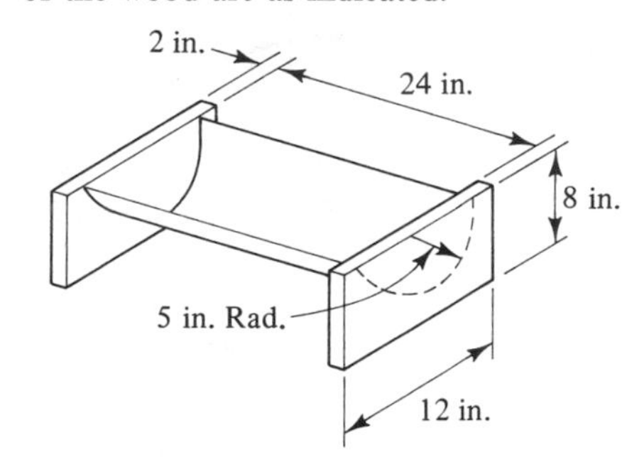

Problem 4.79

4.80. The mass center of a thin aluminum plate with a mass of 100 kg is to be adjusted

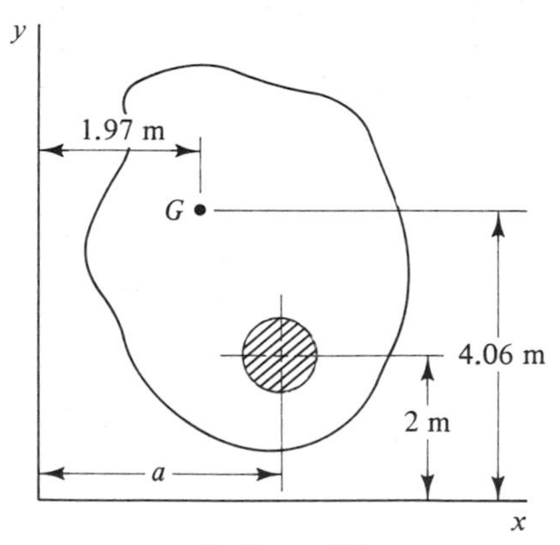

Problem 4.80

from its current position G to the point $\bar{x} = 2$ m, $\bar{y} = 4$ m by addition of a steel disc at the location shown. What mass of steel is needed and what is the required dimension a?

4.81. A piece of rigid polyvinyl chloride (PVC) pipe with an inside diameter of 150 mm and length L is stood on end and is partially filled with concrete. What is the required depth d of concrete if the mass center of the assembly is to be located a distance $0.4L$ above the base? The pipe has a mass per unit length of 6 kg/m and the density of concrete is 2.4 ton/m^3.

4.82. In Problem 4.81, for what depth d of concrete will the distance from the base to the center of mass be a minimum?

4.83. A cast iron flywheel with an outer radius of 6 in. has the cross section shown. If the specific weight of the cast iron is 450 lb/ft^3, what is the weight of the rim? (See Example 4.12 in Section 4.5.)

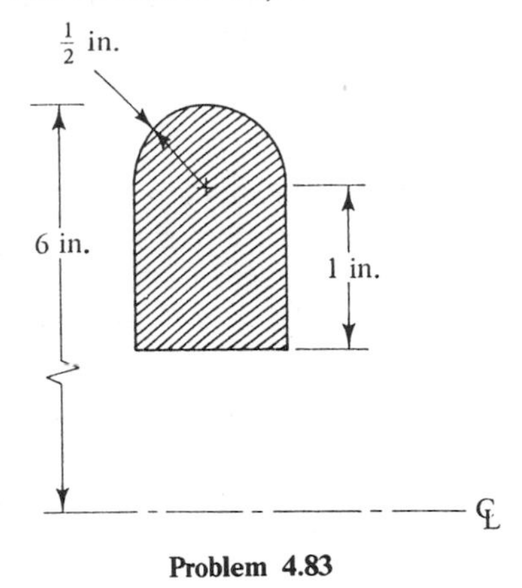

Problem 4.83

4.84. In Problem 4.83, how many gallons of paint would be required to paint 10,000 of the rims if each gallon covers 250 ft^2 of area?

4.7 SURFACE FORCES

Forces that are distributed over a surface area are called *surface forces*. Common examples are the forces between solid bodies in contact and the pressure loadings exerted on a solid body by a fluid. Some common types of surface forces will be considered in this section, along with the procedures for determining their resultants.

Surface and Line Forces

Surface forces are defined in terms of the force acting per unit of area. If the force acting upon an element of area ΔA is $\Delta \mathbf{F}$ (Figure 4.35), the average force per unit

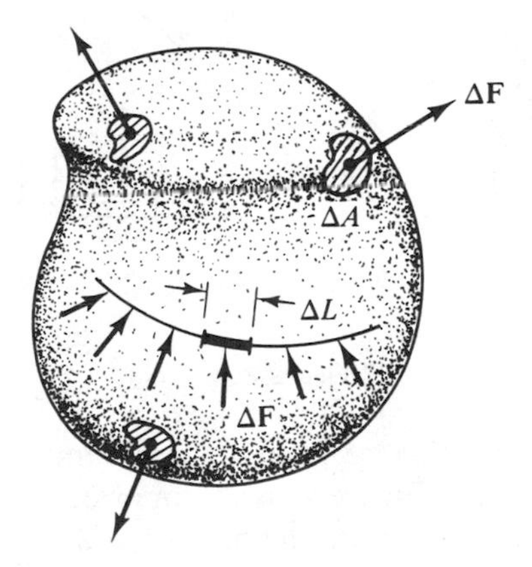

Figure 4.35

area is $\Delta\mathbf{F}/\Delta A$. The *surface force at a point*, which we shall denote by **s**, is defined as the limiting value of this ratio as the area surrounding the point approaches zero. Stated mathematically.

$$\mathbf{s} = \lim_{A \to 0} \frac{\Delta \mathbf{F}}{\Delta A} = \frac{d\mathbf{F}}{dA} \tag{4.18}$$

Both the magnitude and direction of **s** can vary from point to point over the surface. Common units for the magnitude of a surface force are lb/ft^2 and lb/in.^2 (psi). In SI, the units are N/m^2, or pascals.

If the surface area is long and narrow, the distributed force can usually be assumed to vary only along its length. Such forces are called *line forces*. An example would be loading that material stacked on a floor would impart to the top surface of a floor beam. Line forces are expressed in terms of the force acting per unit of length of the line. Thus, the *line force at a point*, **q**, is defined as

$$\mathbf{q} = \lim_{\Delta L \to 0} \frac{\Delta \mathbf{F}}{\Delta L} = \frac{d\mathbf{F}}{dL} \tag{4.19}$$

where $\Delta\mathbf{F}$ is the force acting upon a line element of length ΔL (Figure 4.35). Common units for the magnitude of a line force are lb/ft, lb/in., and N/m.

Surface forces can often be represented graphically by means of a *loading diagram*, which is a plot of the variation of **s** over the surface. For example, if identical cartons measuring 1 ft on a side and weighing 50 lb are stacked four deep on a warehouse floor (Figure 4.36), each unit of area of the floor is subjected to the same surface force of $200\ \text{lb/ft}^2$. The corresponding loading diagram would be as shown. The height of the diagram represents the magnitude of the surface force, and the arrows indicate its direction. If the cartons are stacked to a variable depth, the loading diagram would have a variable height.

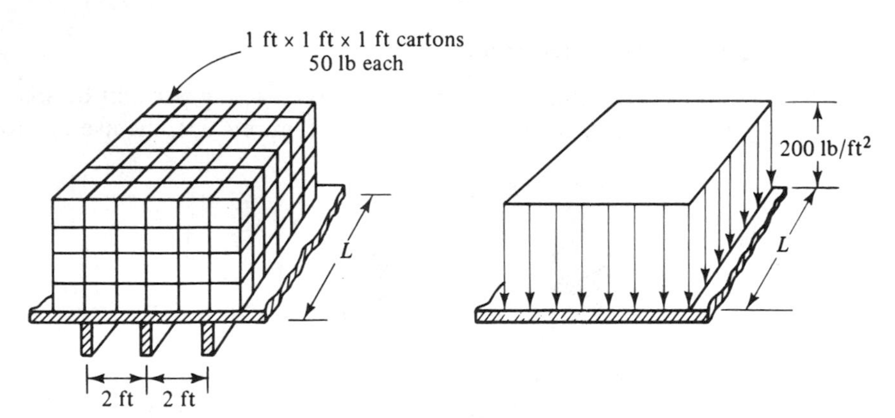

Figure 4.36

Loading diagrams can also be used to represent line forces. For example, suppose that the floor shown in Figure 4.36 is supported by beams spaced 2 ft apart. If we assume that each beam supports the load on a 2-ft-wide strip of the floor, or

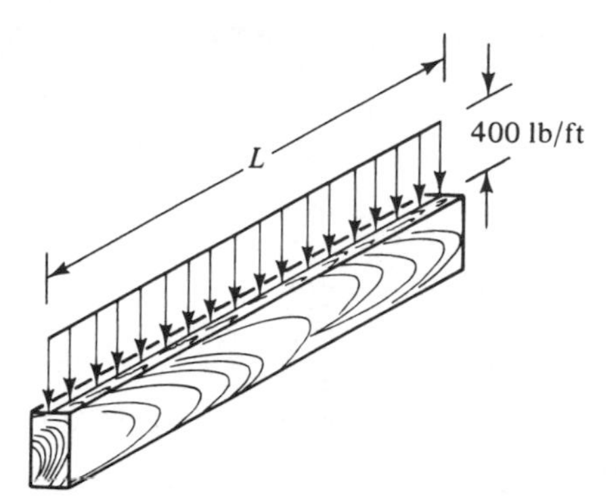

Figure 4.37

400 lb/ft of length, the corresponding loading diagram is as shown in Figure 4.37. The height of the diagram represents the magnitude of the line force. Note that the loading diagram forms an area for line forces and a volume for surface forces.

Pressure Loadings

The magnitude of the surface force exerted by a fluid is called *pressure*. Pressure loadings are characterized by the fact that they are always directed toward the surface upon which they act and are perpendicular to it. Also, the pressure at any point in a fluid is the same in all directions.

The pressure in a liquid at rest, the so-called *hydrostatic pressure*, is due to the weight of the liquid lying above the particular point of interest. Consider a point situated at a depth h (Figure 4.38). The weight of a column of liquid with cross-sectional area dA above this point is $dW = \gamma h\, dA$, where γ is the specific weight of the liquid. The pressure at the point, which is the force per unit area, is $p = dF/dA = dW/dA$, or

$$p = \gamma h \tag{4.20}$$

Thus, the hydrostatic pressure increases linearly with depth. This relationship can also be used to estimate the pressure loadings due to loose, granular materials, such

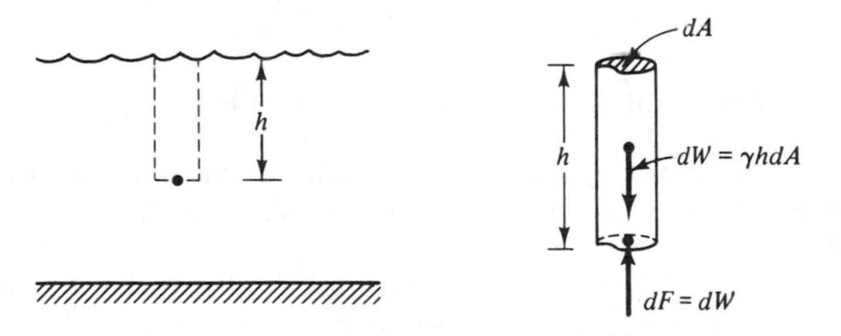

Figure 4.38

TABLE 4.1 SPECIFIC WEIGHTS AND DENSITIES OF SOME COMMON LIQUIDS

Materials	γ (lb/ft^3)	ρ (kg/m^3)
Water (fresh)	62.4	1,000
Water (sea)	64.0	1,025
Motor oil	55.6	890
Olive oil	57.4	920
Turpentine	54.3	870
Ethyl alcohol	49.3	790
Mercury	846	13,550

as dry sand and grain. The specific weights and densities of some common liquids are given in Table 4.1.

Gasses usually have negligible weight. Thus, they exert a uniform pressure over the surface upon which they act. For example, if the can shown in Figure 4.39(a) is filled with air at pressure p, the surface force on the end of the can is as shown in Figure 4.39(b). Figure 4.39(c) shows the loading if the can is filled with a liquid with specific weight γ.

All bodies situated at the earth's surface are subjected to an atmospheric pressure of approximately 14.7 lb/in.^{2} (10.1×10^4 N/m^2) at sea level due to the weight of the overlying atmosphere. If all surfaces of the body are exposed to the atmosphere, this loading has a null resultant and need not be considered in a force analysis. Consequently, it is usually the *gage pressure*, or difference between the total (absolute) pressure and atmospheric pressure, which is of interest. Equation 4.20 defines the gage pressure in a liquid at rest.

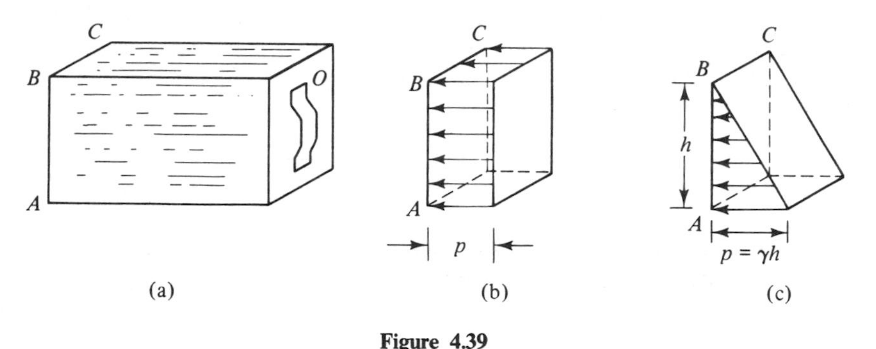

Figure 4.39

Resultants of Parallel Surface and Line Forces

Distributed loadings often have the same orientation over the entire surface or line. In this case they form a system of parallel forces, and the resultant can be readily obtained.

Figure 4.40(a) shows a loading diagram for a general system of parallel surface forces. According to Eq. (4.18), each element of area dA is subjected to a force $d\mathbf{F} = s\,dA\,\mathbf{e}$, where $\mathbf{e}$ is a unit vector in the direction of the surface force. Since these

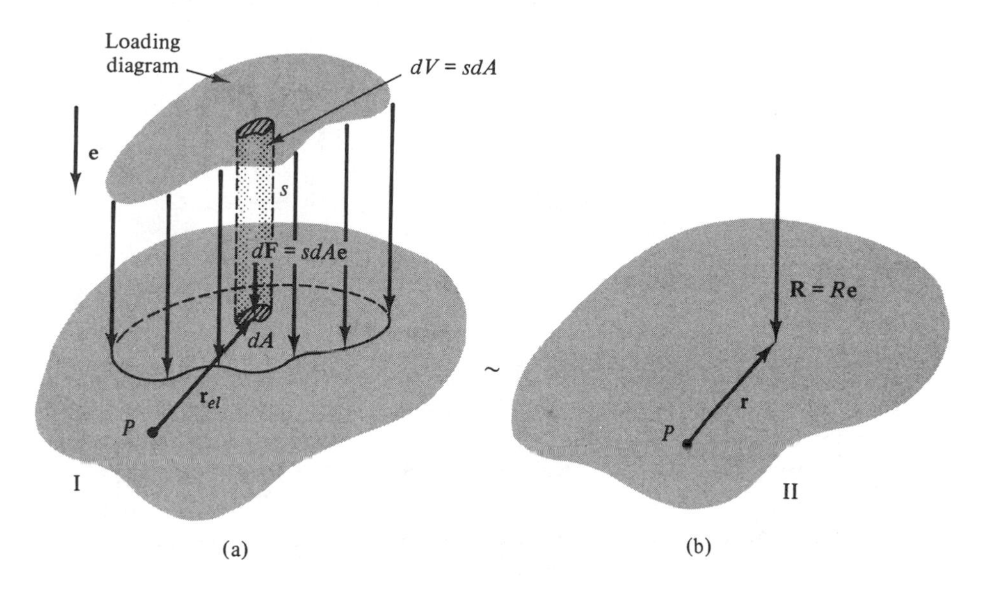

Figure 4.40

forces are all parallel, their resultant will be a single force **R** [Figure 4.40(b)]. From the conditions for static equivalence, we have

$$\left(\sum \mathbf{F}\right)_{\mathrm{I}} = \left(\sum \mathbf{F}\right)_{\mathrm{II}}: \qquad \int dF\mathbf{e} = R\mathbf{e}$$

$$\left(\sum \mathbf{M}_P\right)_{\mathrm{I}} = \left(\sum \mathbf{M}_P\right)_{\mathrm{II}}: \quad \int (\mathbf{r}_{\mathrm{el}} \times dF\mathbf{e}) = \mathbf{r} \times R\mathbf{e}$$

where $\mathbf{r}$ and $\mathbf{r}_{\mathrm{el}}$ are the respective position vectors of **R** and $d\mathbf{F}$ relative to an arbitrary point P. Rewriting the moment condition as

$$\left(\int \mathbf{r}_{\mathrm{el}}\, dF\right) \times \mathbf{e} = (\mathbf{r}R) \times \mathbf{e}$$

and solving for R and $\mathbf{r}$, we obtain

$$R = \int dF = \int s\, dA \tag{4.21}$$

$$\mathbf{r} = \frac{\int \mathbf{r}_{\mathrm{el}}\, dF}{\int dF} = \frac{\int \mathbf{r}_{\mathrm{el}} s\, dA}{\int s\, dA} \tag{4.22}$$

If the variation of s over the surface is known, these integrals can be evaluated to determine the resultant force and its location. However, the integrals have a geometric interpretation that makes their formal evaluation unnecessary in many cases.

Referring to Figure 4.40(a), we see that $s\,dA$ is equal to the volume dV of a column under the loading diagram with cross-sectional area dA. If the quantity $s\,dA$

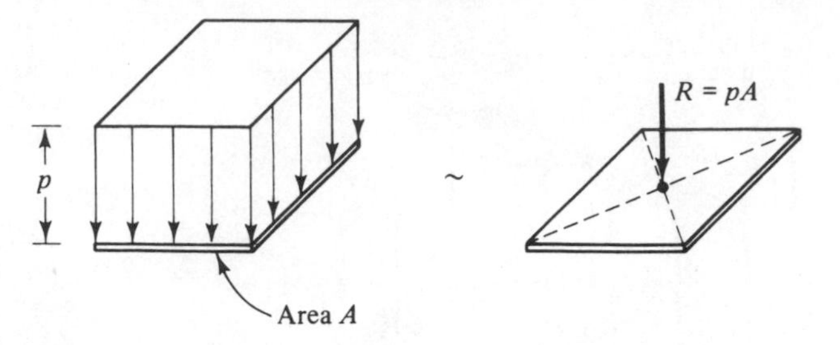

Figure 4.41

is replaced by dV in Eqs. (4.21) and (4.22), they correspond to the equations for the total volume of a shape and the position vector of its centroid (see Section 4.5). Interpreted physically, this means that the resultant of a parallel surface force has magnitude equal to the total volume under the loading diagram and its line of action passes through the centroid of that volume. This result makes it possible to determine the resultant of many simple loadings without actually performing an integration. For example, we can immediately say that the resultant of the uniform pressure loading shown in Figure 4.41 has magnitude $R = pA$ and acts at the center of the plate.

Equations (4.21) and (4.22) can also be used to determine the resultants of parallel line forces. For a line force, $dF = q\,dL$ [Eq. (4.19)]. Now $q\,dL$ is the area dA of an element with width dL under the loading diagram [Figure 4.42(a)]; therefore, Eqs. (4.21) and (4.22) correspond to the equations for the total area of a shape and the position vector of its centroid. Thus, the resultant of a parallel line force has magnitude equal to the area under the loading diagram and its line of action passes through the centroid of that area [Figure 4.42(b)].

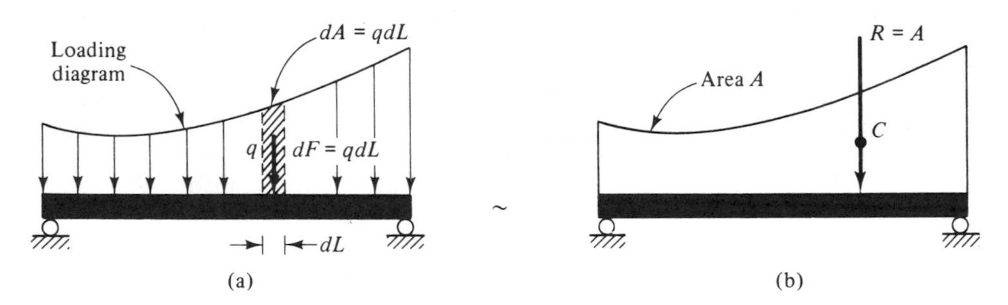

Figure 4.42

Pressure Loadings on Curved Surfaces

Figure 4.43(a) shows an end view of the pressure distribution on the curved upstream face of a dam. If we attempt to determine the resultant force on the dam directly from the surface force distribution, we find that the integrals involved are

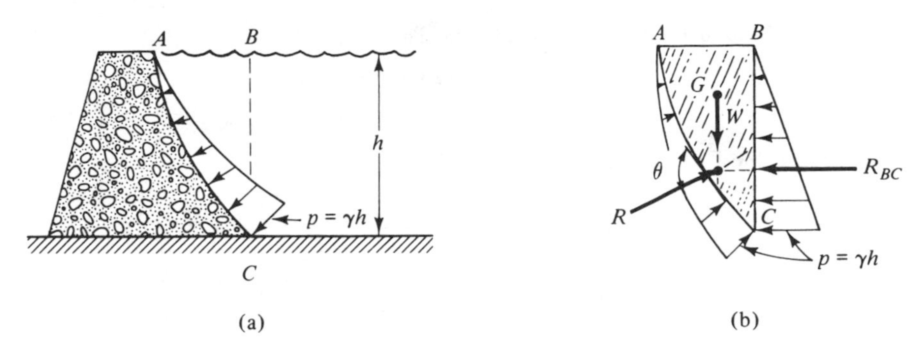

Figure 4.43

complicated because both the magnitude and the orientation of the loading vary over the surface. This complication can be avoided by using an alternate procedure wherein the resultant is determined indirectly from a force analysis of a portion of the fluid.

For example, suppose we wish to determine the resultant force on the face of the dam. To do this, we consider the equilibrium of the volume of fluid bounded by the dam and the vertical plane BC. The free-body diagram is shown in Figure 4.43(b). Since the pressure loading on plane BC involves parallel surface forces, its resultant $\mathbf{R}_{BC}$ can be easily determined. Once the weight of the fluid has been computed and its center of gravity located, the only remaining unknown is the resultant $\mathbf{R}$ of the pressure acting upon the curved surface AC of the fluid volume. This force can be determined from the equations of equilibrium, with the resultant force on the dam obtained from Newton's third law. Uniform pressure loadings due to gasses can be handled in the same way.

Example 4.16 Line Loads on a Beam. A beam supports the distributed loads shown in Figure 4.44(a). Determine the reactions at the supports.

Solution. We first replace the distributed loads by their resultants $\mathbf{R}_1$ and $\mathbf{R}_2$. Each resultant has a magnitude equal to the area under the corresponding loading diagram and

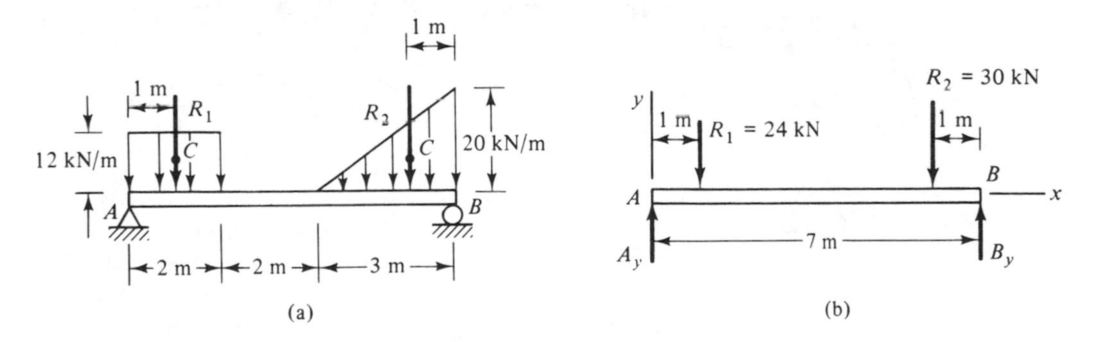

Figure 4.44

acts through the centroid of that area. Thus,

$$R_1 = (12 \text{ kN/m})(2 \text{ m}) = 24 \text{ kN}$$

$$R_2 = \tfrac{1}{2}(20 \text{ kN/m})(3 \text{ m}) = 30 \text{ kN}$$

From the FBD of Figure 4.45(b), we have

$$\overset{+}{\curvearrowleft}\sum M_A = B_y(7 \text{ m}) - 24 \text{ kN}(1 \text{ m}) - 30 \text{ kN}(6 \text{ m}) = 0$$

$$\sum F_y = A_y + B_y - 24 \text{ kN} - 30 \text{ kN} = 0$$

or

$$B_y = 29.1 \text{ kN} \qquad A_y = 24.9 \text{ kN} \qquad \textbf{Answer}$$

Example 4.17 Resultant Force on a Dam. Determine the resultant force **R** acting upon the upstream face of the dam shown in Figure 4.45(a). The dam is 30 ft long.

Solution. We consider the equilibrium of the volume of water bounded by the curved surface AB and the plane AC. Since the pressure varies linearly with depth, the FBD is as shown in Figure 4.45(b). We now replace all distributed forces by their resultants. None of the loadings varies along the length of the dam; so all the resultants lie in a vertical plane located at its midpoint. The weight and location of the center of gravity of the volume of fluid can be determined by integration or by using the results in Table A.1 of the Appendix. These calculations are not given here. The resultant of the distributed forces on plane AC is equal to the volume under the corresponding loading diagram and acts through the centroid of that

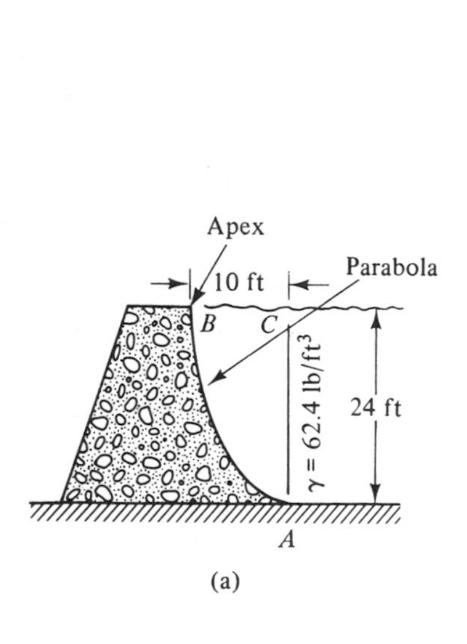

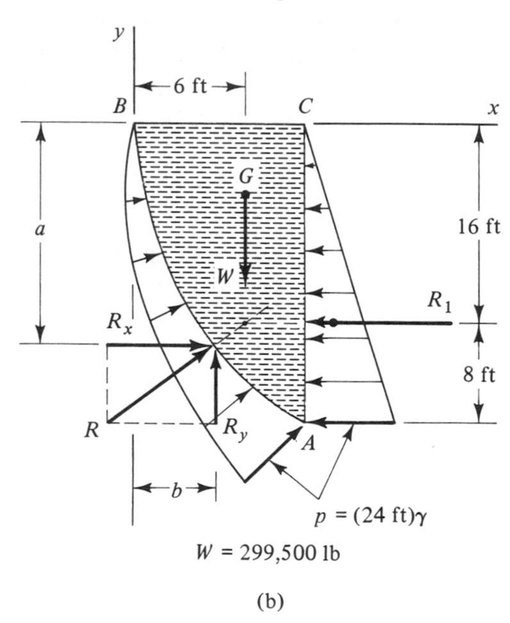

Figure 4.45

volume. Carrying out the indicated computations, we have

$$R_1 = \tfrac{1}{2}(24\text{ ft})(62.4\text{ lb/ft}^3)(24\text{ ft})(30\text{ ft}) = 539{,}100\text{ lb}$$

For equilibrium

$$\sum F_x = R_x - R_1 = 0$$

$$\sum F_y = R_y - W = 0$$

from which we obtain

$$R_x = 539{,}100\text{ lb} \qquad R_y = 299{,}500\text{ lb}$$

$$R = \sqrt{R_x^2 + R_y^2} = 616{,}700\text{ lb} \qquad \textbf{Answer}$$

Several different approaches can be used to locate **R**. One is to recognize that the volume of fluid is a three-force body, for which the forces acting must pass through a common point for equilibrium. Another approach is to use the moment equation of equilibrium. We shall use this approach, as it is the more general of the two. We have

$$\overset{+}{\curvearrowleft}\sum M_B = R_x a + R_y b - W(6\text{ ft}) - R_1(16\text{ ft}) = 0$$

For the coordinates shown in Figure 4.45(b), the equation of the parabolic surface of the dam is

$$x = \frac{10\text{ ft}}{(24\text{ ft})^2}y^2$$

Thus

$$b = \frac{10\text{ ft}}{(24\text{ ft})^2}a^2$$

Combining this result with the moment equation of equilibrium and solving, we get

$$a^2 + 103.7a - 2{,}004 = 0$$

$$a = 16.7\text{ ft}, \qquad b = 0.3\text{ ft} \qquad \textbf{Answer}$$

By Newton's third law, the force on the dam is equal and opposite to **R**. It is interesting to note that **R** is not perpendicular to the curved surface, although the pressure is; **R** will be normal to the surface only if it is flat or circular.

Example 4.18. Resultant Force on a Gate. Determine the resultant water force on the triangular gate in the irrigation dam shown in Figure 4.46(a).

Solution. The resultant force on the gate is equal to the volume under the loading diagram [Figure 4.46(b)] and acts through its centroid. However, the geometry of the diagram is sufficiently complicated in this case that it is simpler to start from first principles. Since we do know from the symmetry of the diagram that the resultant acts through a point on the centerline of the gate, only its vertical location need be determined.

We select a horizontal strip with width dy located a distance y down from the top of the gate as the element of the area [Figure 4.46(c)]. The force acting upon this element is

$$dF = p\,dA = \gamma y\,dA$$

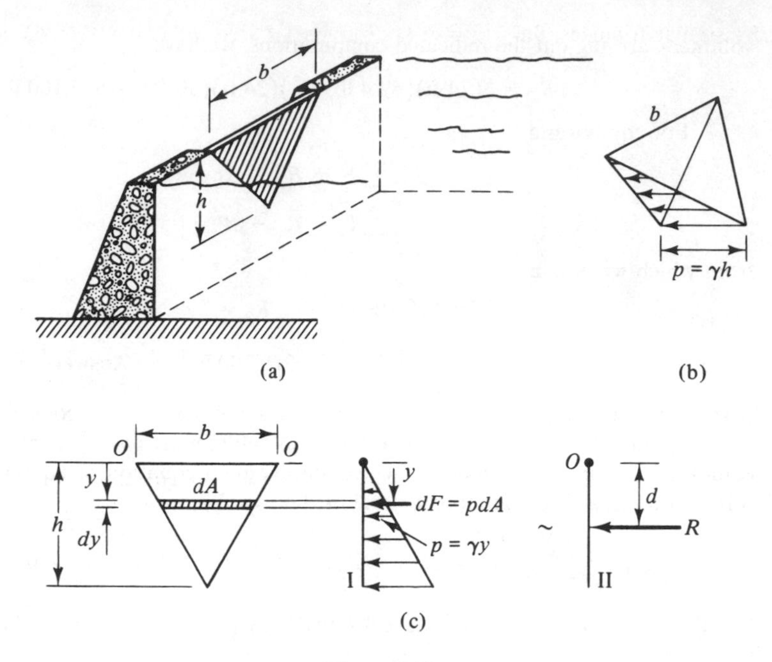

Figure 4.46

Denoting the resultant of the distributed forces by **R** and applying the conditions for static equivalence, we have

$$\left(\sum \mathbf{F}\right)_{\mathrm{I}} = \left(\sum \mathbf{F}\right)_{\mathrm{II}}: \qquad \int \gamma y \, dA = R$$

$$\left(\sum M_0\right)_{\mathrm{I}} = \left(\sum M_0\right)_{\mathrm{II}}: \qquad -\int y(\gamma y \, dA) = -Rd$$

From the definition of the centroid of an area, Eqs. (4.12), we see that

$$\int y \, dA = A\bar{y}$$

where A is the area of the gate and $\bar{y}$ is the distance to its centroid. Thus,

$$R = \gamma A\bar{y} = \gamma\left(\frac{1}{2}bh\right)\left(\frac{h}{3}\right) = \frac{1}{6}\gamma bh^2 \qquad \textbf{Answer}$$

Solving for the distance d, we get

$$d = \frac{\gamma \int y^2 \, dA}{R}$$

By similar triangles, the width of the element of area dA is $b(1 - y/h)$; so

$$dA = b(1 - y/h)\,dy$$

$$\int y^2\,dA = \int_0^h y^2 b(1 - y/h)\,dy = \frac{1}{12}bh^3$$

$$d = \frac{\gamma\left(\dfrac{bh^3}{12}\right)}{\gamma\dfrac{bh^2}{6}} = \frac{h}{2} \qquad \textbf{Answer}$$

The integral $\int y^2\,dA$ in the numerator of the expression for d is called the moment of inertia of the gate area. Moments of inertia arise in a number of problems, and will be discussed further in Section 4.8.

Example 4.19. Buoyant Forces. Show that the buoyant force on an object floating or submerged in a homogeneous fluid has magnitude equal to the weight of the volume of displaced fluid and that its line of action passes through the centroid of that volume (Archimedes' law).

Solution. The pressures acting upon the surface of a submerged or floating object depend only upon the depth and are the same as those acting upon a volume of fluid occupying the same region as the submerged portion of the body. Thus, the resultant forces on the surface will also be the same, and can be obtained from a force analysis of the displaced volume of fluid. From the FBDs of either Figure 4.47(a) or 4.47(b), we have

$$\sum F_y = F_B - \gamma V = 0$$

$$F_B = \gamma V \qquad \textbf{Answer}$$

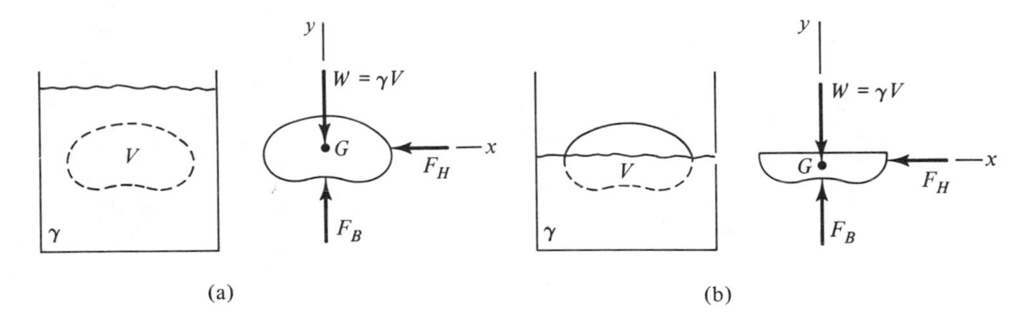

Figure 4.47

Thus, the net upward force (buoyant force) is equal to the weight of the displaced fluid; the net horizontal force on the object is zero. For moment equilibrium, F_B must pass through G. For a homogeneous fluid, the c.g. coincides with the centroid of the volume occupied. Consequently, the bouyant force passes through the centroid of the volume of displaced fluid.

PROBLEMS

4.85. to 4.89. Determine the reactions at the supports for the beam shown.

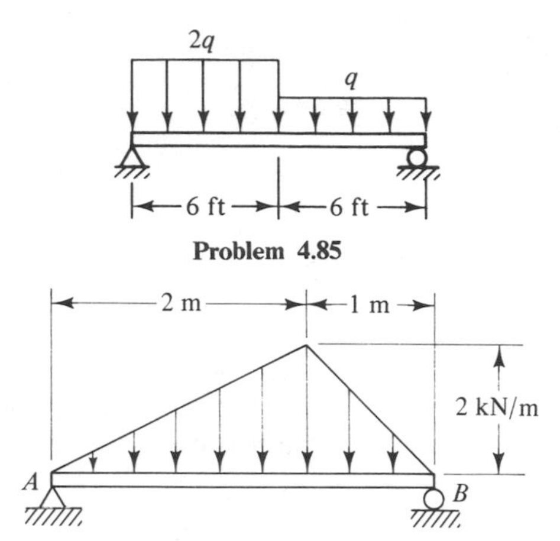

Problem 4.85

Problem 4.86

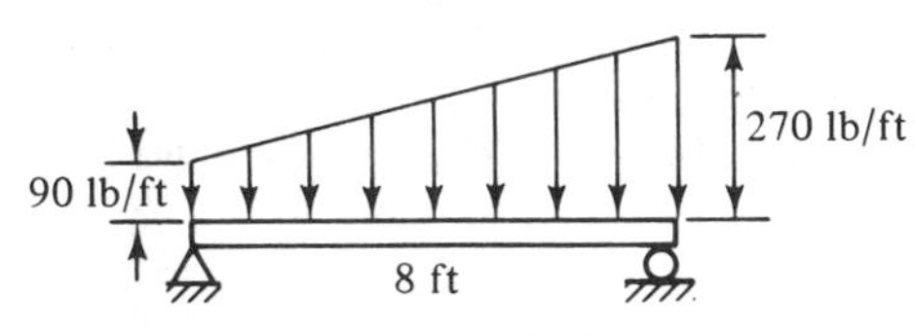

Problem 4.87

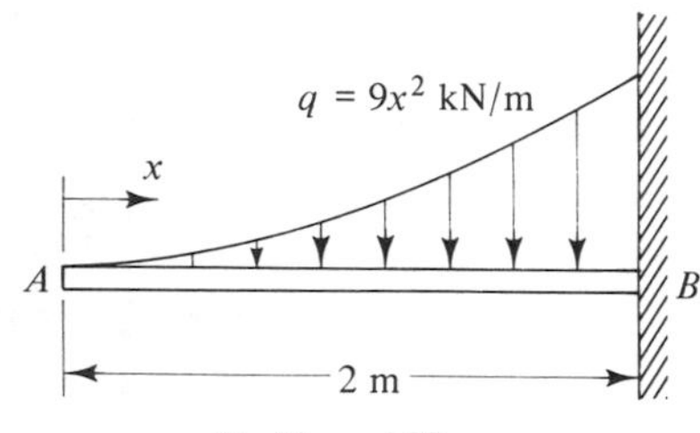

Problem 4.88

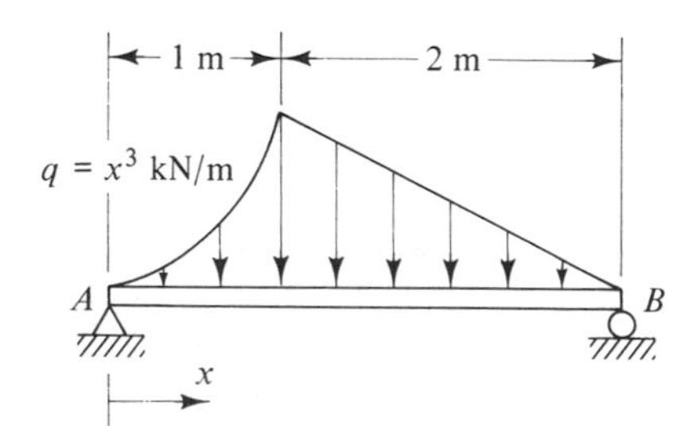

Problem 4.89

4.90. Determine the horizontal and vertical components of the reactions at A and D for the frame shown and determine the force on member AC.

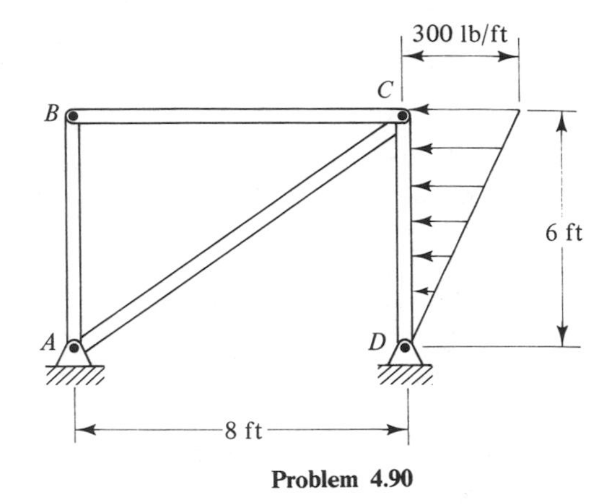

Problem 4.90

4.91. Determine the resultant of the force system shown.

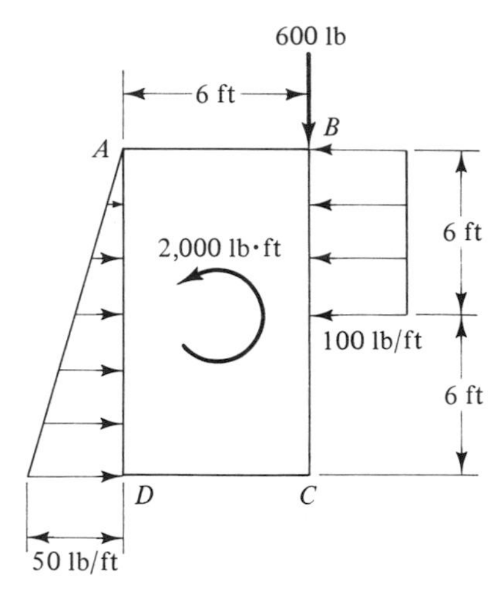

Problem 4.91

4.92. The 140-kg steel plate shown is lifted by two electromagnets. If the diameter of each magnet is 0.2 m, what are the magnitudes of the average surface forces exerted on the plate?

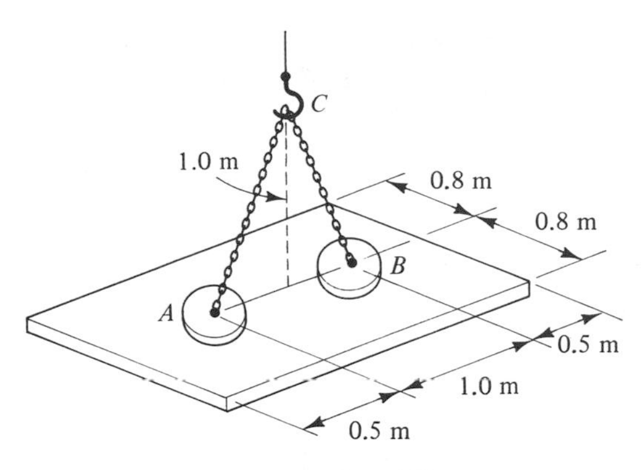

Problem 4.92

4.93. The gravity dam shown is 6 m long and is made of concrete with a density of 2,400 kg/m^3. If $h = 1$ m, what is the resultant force acting on the upstream face of the dam? The density of the water is 1,000 kg/m^3.

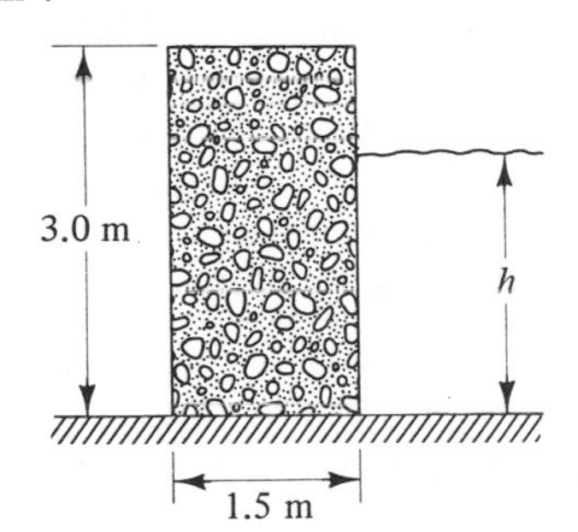

Problem 4.93

4.94. In Problem 4.93, will the dam first tip or overflow as the water depth increases?

4.95. A vertical window in an underwater observation room is 3 m long and 2 m high. If the top of the window is 2 m below the surface, what is the resultant force on the window and where is it located? The density of sea water is 1,025 kg/m^3.

4.96. A rectangular pan with vertical sides contains a 2-in.-deep layer of motor oil on top of a 6-in.-deep layer of water (see Table 4.1). Determine completely the resultant force per unit of length acting on one end of the container.

4.97. Determine the magnitude and location of the resultant force acting on the inclined side CD of the eave trough shown when it is filled with water to a depth of 3 in. Consider a 1-ft-long section of the trough.

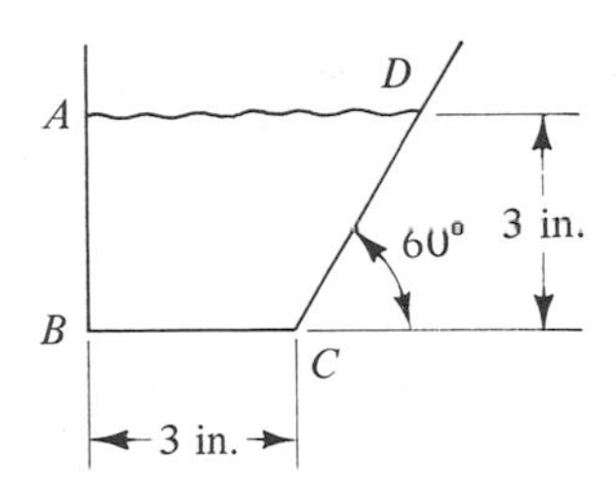

Problem 4.97

4.98. An access hole in the inclined side of a tank is covered with a rectangular plate as shown. Obtain an expression for the magnitude R of the resultant force acting upon the plate and for the distance d that defines its location.

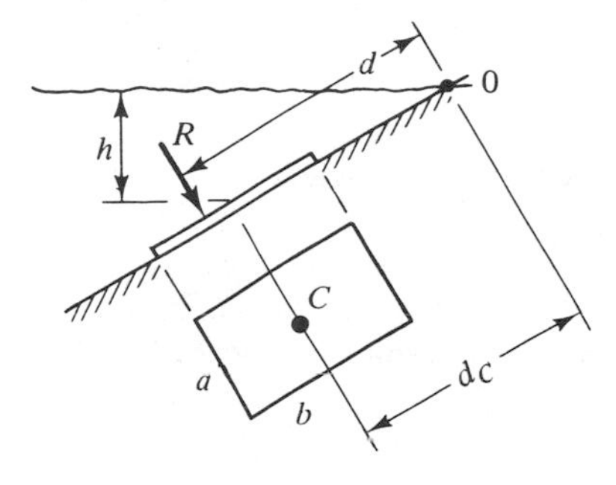

Problem 4.98

4.99. Determine the magnitude and location of the resultant force per unit of length acting upon the curved portion AC of the pool wall shown. The specific weight of the water is 62.4 lb/ft^3.

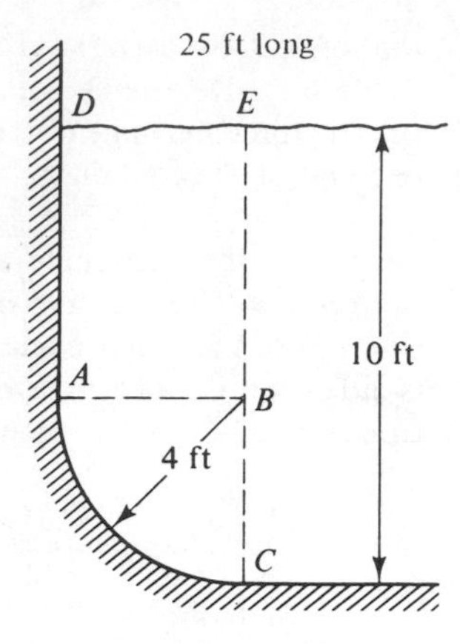

Problem 4.99

4.100. Determine the magnitude and location of the resultant force per unit of length acting upon the upstream face of the dam shown.

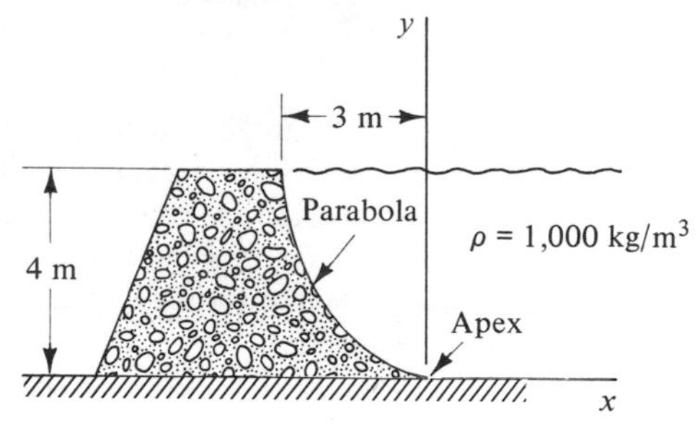

Problem 4.100

4.101. A large log is blocking the flow in a small stream as shown. Determine the magnitude and location of the resultant force exerted on the log by the water. Consider a segment of the log one foot long.

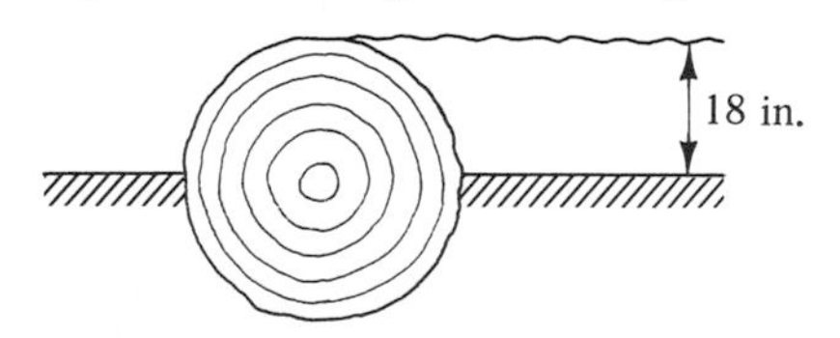

Problem 4.101

4.102. Sewage sludge ($\gamma = 90\ \text{lb/ft}^3$) is held in a 6-ft-wide rectangular channel by a heavy gate hinged at the top as shown. What is the required weight W of the gate if it is to lift and allow the sludge to flow out when it reaches a depth of 4 ft?

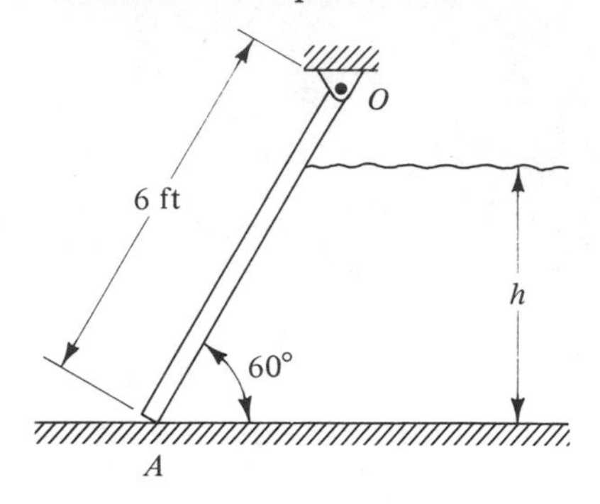

Problem 4.102

4.103. Determine the magnitude and location of the resultant force acting on the vertical end of the trough in Problem 4.97.

4.104. The streambed behind a dam has the parabolic profile shown. Determine the resultant force exerted on the vertical face of the dam. Express the results in terms of the specific weight γ of the water and the dimensions h and a.

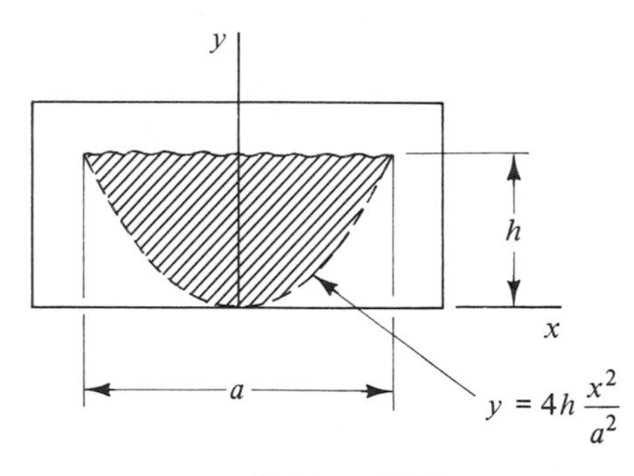

Problem 4.104

4.105. Grain with a density of 770 kg/m^3 is stored in a bin to the depth shown. The depth is uniform in the direction normal to the figure. What is the magnitude and location of the resultant force on end $ABCD$ of the bin?

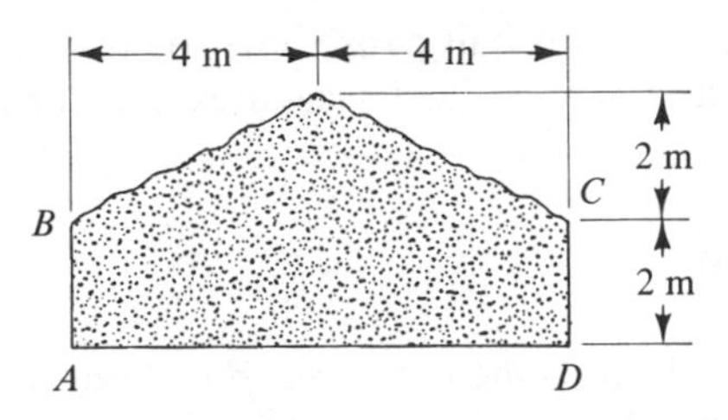

Problem 4.105

4.106. A fishing float with negligible mass is made of two conical segments joined as shown. What force is required to pull the float beneath the surface?

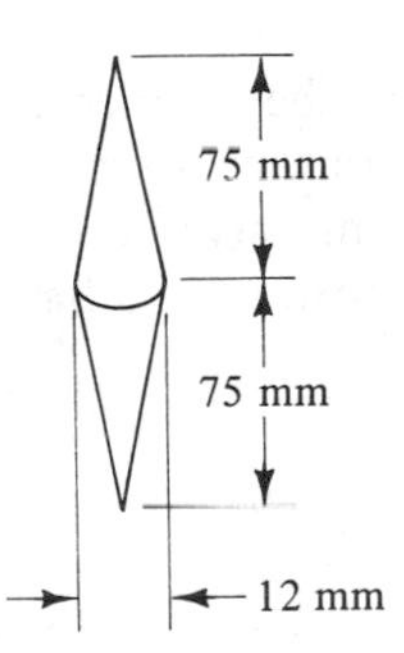

Problem 4.106

4.107. An instrument package is suspended beneath the surface of the sea by a conical buoy with the dimensions shown. The package weighs 562 lb and has a volume of 7.7 ft^3; the float weighs 250 lb. What is the depth of immersion d of the buoy?

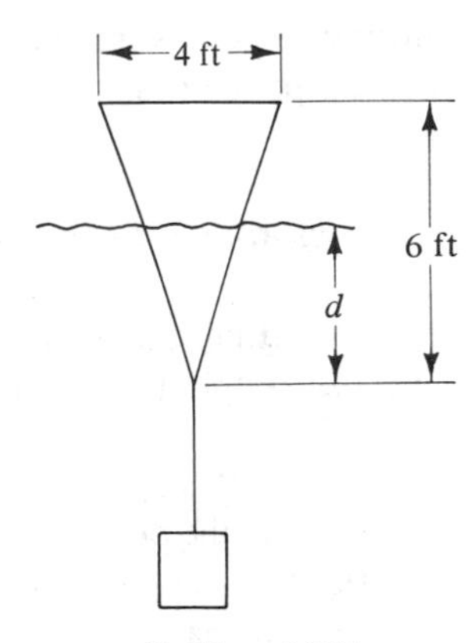

Problem 4.107

4.108. The uniform triangular slab shown is made of wood with a specific weight half that of the liquid in which it is immersed. The slab is hinged at A and connected to a chain of negligible weight at C. Determine the ratio of the tension T on the chain and the weight W of the slab as a function of the fractional depth of immersion d/a. Plot T/W versus d/a and explain the results.

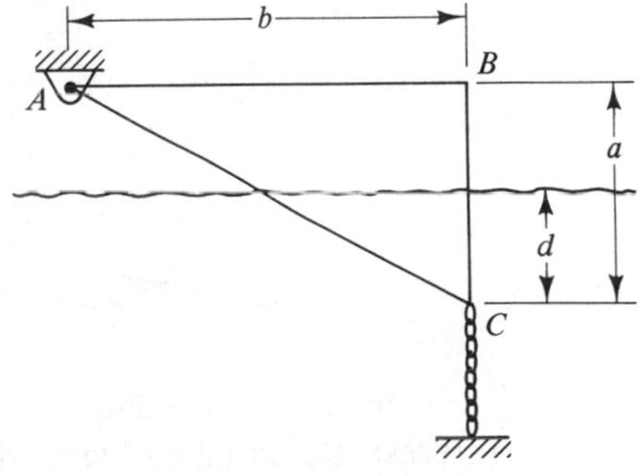

Problem 4.108

4.8 AREA MOMENTS OF INERTIA

When locating the resultant of the pressure loading in Example 4.18, we encountered an integral quantity called the *area moment of inertia*. Integrals of this type arise whenever the magnitude of the surface force varies linearly with distance, as in the case of a pressure loading. Since these surface forces will be encountered on numerous occasions in the following chapters, we shall pause here to consider some useful properties of area moments of inertia and the procedures for computing them.

Moments of Inertia by Integration

By definition, the moment of inertia dI_a of an element of area dA about a line or axis aa within the plane of the area [Figure 4.48(a)] is

$$dI_A = d^2(dA) \tag{4.23}$$

where d is the perpendicular distance from the axis to the element. The moment of inertia for the total area is the sum of the moments of inertia for each of its constituent parts:

$$I_a = \int dI_a = \int d^2(dA) \tag{4.24}$$

The moment of inertia is a relative measure of the manner in which an area is distributed with respect to the axis of interest. The further the area is away from the axis, the larger its moment of inertia. From the definitions in Eqs. (4.23) and (4.24), it is clear that area moments of inertia are always positive and have the dimensions of length to the fourth power.

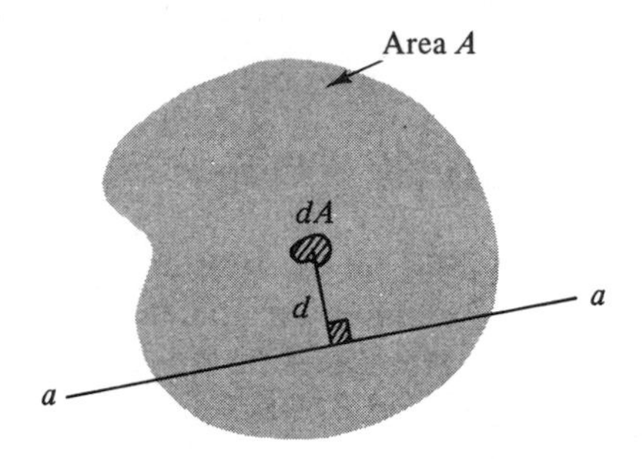

Figure 4.48

In many problems it is the value of I about the coordinate axes that are of interest. Referring to Figure 4.49 and applying the definition in Eq. (4.24), we have

$$I_x = \int y^2\,dA$$
$$I_y = \int x^2\,dA \tag{4.25}$$

The moment of inertia about an axis perpendicular to the area is called the *polar moment of inertia* and is commonly denoted by J. Thus, the polar moment of inertia about the z axis for the area shown in Figure 4.49 is

$$J_z = \int r^2\,dA = \int (x^2 + y^2)\,dA$$

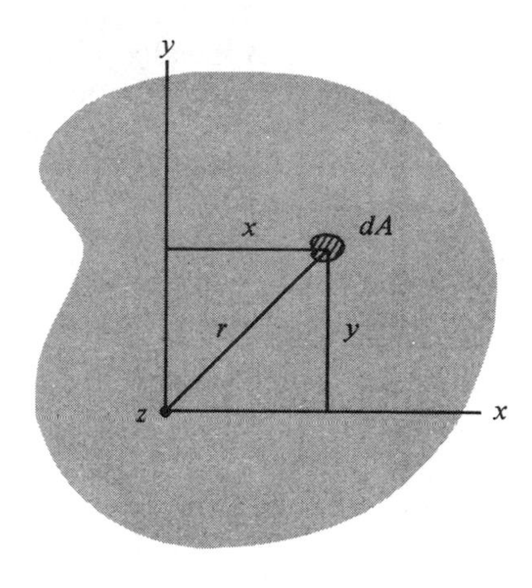

Figure 4.49

or

$$J_z = I_x + I_y \tag{4.26}$$

This equation states that the sum of the moments of inertia about two perpendicular axes within the plane of the area is equal to the moment of inertia about a third concurrent axis perpendicular to the plane. If the moments of inertia about any two of the coordinate axes are known, the value about the third axis can be determined directly from Eq. (4.26). Moments of inertia about axes through the centroid are of particular significance, as we shall see a bit later. Values of I and J about centroidal axes will be denoted by $\bar{I}$ and $\bar{J}$.

The distribution of an area with respect to a given axis can also be described in terms of the *radius of gyration*, r_a. By definition,

$$I_a \text{ (or } J_a) = r_a^2 A \tag{4.27}$$

where A is the total area of the shape. The radius of gyration has no particular physical significance. It is a fictitious dimension, which, when squared and multiplied by the area, yields the moment of inertia. Equations are sometimes expressed in terms of the radius of gyration instead of the moment of inertia as a matter of convenience. We shall do this in Chapter 10, for example.

As in the case of centroids, the greatest difficulty in determining moments of inertia by integration lies in the formulation of the integrals. Again, the first, and most important, step is the selection of the element of area. When using single integration involving rectangular elements of area, the element should be selected parallel to the axis of interest, if possible. Otherwise, each portion of the element is a different distance from the axis, and the definitions in Eqs. (4.23) and (4.24) cannot be used directly (see Example 4.20). One alternative is to use double integration. The other is to use the parallel axis theorem, which will be discussed later in this section.

Values of the moment of inertia and radius of gyration of many common shapes have been computed and tabulated in engineering and mathematical handbooks. Results for some common geometrical and structural shapes are given in Tables A.1 and A.3 through A.12 of the Appendix. Usually, only the values of I and J about centroidal axes are given. However, if these values are known, the values about any other parallel axis can be readily determined, as we shall now show.

Parallel Axis Theorem

Consider the area shown in Figure 4.50. Let $a'a'$ be an axis through the centroid and aa be any other axis parallel to it. The distance between the axes is d. From the definition of moment of inertia, Eq. (4.24), we have

$$I_a = \int d_a^2(dA) = \int (d_{a'} + d)^2\, dA$$

$$= \int d_{a'}^2(dA) + 2d\int d_{a'}(dA) + d^2\int dA$$

where d_a is the distance from axis aa to the element of area and $d_{a'}$ is the distance to the element from the centroidal axis $a'a'$.

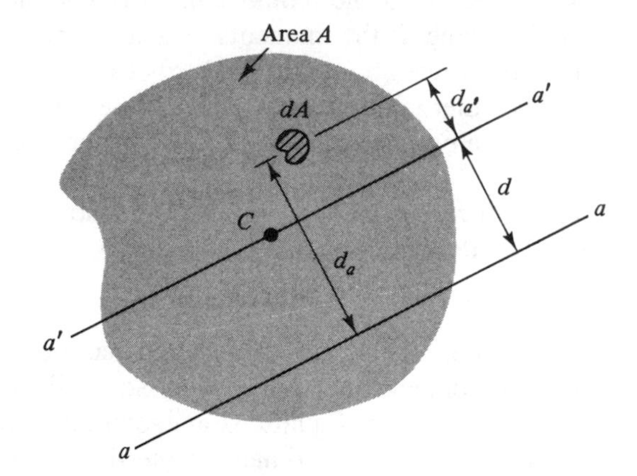

Figure 4.50

The first term in the preceding expression is recognized to be the moment of inertia, $\bar{I}_{a'}$, about the centroidal axis. Using the definition of the centroid of an area, Eqs. (4.12), the integral in the second term can be expressed as

$$\int d_a'(dA) = A\bar{d}$$

where $\bar{d}$ is the distance from the centroidal axis to the centroid. Since this distance is zero, the integral expression must also be zero. Thus, we have

$$I_a = \bar{I}_{a'} + Ad^2 \tag{4.28}$$

This result, which is called the *parallel axis theorem*, relates the moment of inertia about a centroidal axis to that about any other parallel axis. The theorem also applies for polar moments of inertia. It is widely used both for direct determination of moments of inertia and for formulation of integral expressions for area moments. In the latter case, the parallel axis theorem is used to determine dI for the element about the desired axis, and the resulting expression integrated to determine I for the entire area (see Example 4.21).

Composite Shapes

An approach similar to that used for centroids can be used to compute the moment of inertia of composite areas. The values of the moments of inertia of simple shapes about centroidal axes can be determined from tables, and the parallel axis theorem can be used to transfer them to the desired axis. The moments of inertia of each element are then summed to determine I for the whole area. Cutouts and holes are handled by considering their moments of inertia and areas to be negative.

Example 4.20. Moment of Inertia of a Rectangle. Determine the moments of inertia of a rectangle with height h and base b about axes parallel to its sides and passing through its centroid [Figure 4.51(a)].

Solution. To compute $\bar{I}_x$, we select a horizontal strip as the element of area. Referring to Figure 4.51(a), we have from Eqs. (4.25)

$$\bar{I}_x = \int y^2\, dA = \int_{-h/2}^{h/2} y^2 (b\, dy) = \frac{bh^3}{12} \qquad \textbf{Answer}$$

To compute $\bar{I}_y$, we select a vertical element of area [Figure 4.51(b)]. The horizontal element cannot be used directly because each portion of it lies at a different distance from the y axis. We have

$$\bar{I}_y = \int x^2\, dA = \int_{-b/2}^{b/2} x^2 (h\, dx) = \frac{hb^3}{12} \qquad \textbf{Answer}$$

This result can also be obtained from the expression for $\bar{I}_x$ simply by interchanging the x and y dimensions.

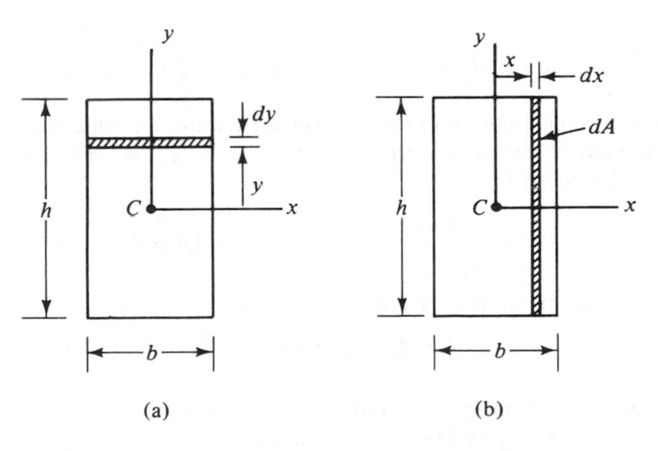

Figure 4.51

Example 4.21. Moment of Inertia of an Area. Determine the moments of inertia I_x, I_y, and J_z for the area shown in Figure 4.52.

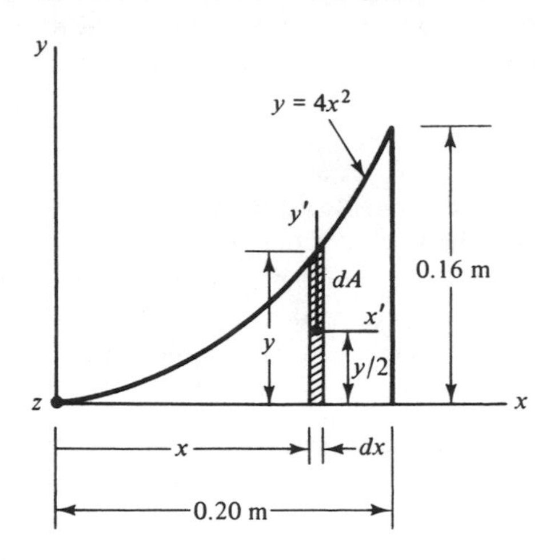

Figure 4.52

Solution. Taking a vertical strip as the element of area, we have

$$dA = y\,dx = 4x^2\,dx$$

Since the entire element is the same distance from the y axis, I_y can be computed directly from Eqs. (4.25):

$$I_y = \int x^2\,dA = \int_0^{0.2} x^2(4x^2)\,dx = \left.\frac{4x^5}{5}\right|_0^{0.2} = 25.6 \times 10^{-5}\ \text{m}^4 \qquad \textbf{Answer}$$

To compute I_x, we must either select another element parallel to the x axis or use the parallel axis theorem. Using the latter approach, and the results of Example 4.20, we have for the rectangular element

$$d\bar{I}_{x'} = \frac{bh^3}{12} = \frac{1}{12}y^3\,dx$$

$$dI_x = d\bar{I}_{x'} + Ad^2 = \frac{1}{12}y^3\,dx + y\,dx\left(\frac{y}{2}\right)^2 = \frac{1}{3}y^3\,dx = \frac{64}{3}x^6\,dx$$

(This result could also have been obtained by using the equation $bh^3/3$ from Table A.1 for the moment of inertia of a rectangle about its base.) We now integrate dI_x over the entire area to determine I_x:

$$I_x = \int_0^{0.2} \frac{64}{3}x^6\,dx = \left.\frac{64}{21}x^7\right|_0^{0.2} = 3.9 \times 10^{-5}\ \text{m}^4 \qquad \textbf{Answer}$$

From Eq. (4.26), the polar moment of inertia is

$$J_z = I_x + I_y = 29.5 \times 10^{-5}\ \text{m}^4 \qquad \textbf{Answer}$$

Example 4.22. Moment of Inertia of a Composite Area. A triangular bracket with a circular hole is welded to an angle section to form the shape shown in Figure 4.53(a). Determine the moment of inertia and radius of gyration about axis aa.

Solution. The area is considered to consist of three elements [Figure 4.53(b)]. Their centroidal moments of inertia, obtained from Tables A.1 and A.5 of the Appendix, are as indicated. The parallel axis theorem is then used to determine their moments of inertia about axis aa. Carrying out the indicated computations in tabular form, we have

No.	$\bar{I}_i$ (in.^{4})	A_i (in.^{2})	d_i (in.)	$A_i d_i^2$ (in.^{4})
1	89.0	15.00	2.37	84.3
2	36.0	18.00	4.00	288.0
3	−0.8	−3.14	4.00	−50.3
Σ	124.2	29.86		322.0

Note that $\bar{I}$ and A for the circle are considered negative because this piece is removed to form the hole. Using the sums found in the table, we obtain

$$I_a = \sum \bar{I}_i + \sum A_i d_i^2 = 124.2 + 322.0 = 446.2 \text{ in.}^4$$

Answer

$$r_a = \sqrt{\frac{I_a}{A}} = \sqrt{\frac{446.2 \text{ in.}^4}{29.9 \text{ in.}^2}} = 3.9 \text{ in.}$$

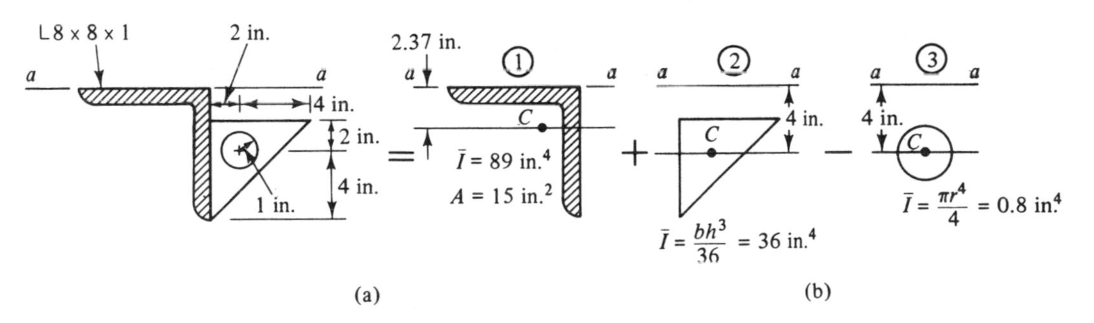

Figure 4.53

PROBLEMS

4.109. to 4.112. Determine by integration the moments of inertia I_x and I_y for the area shown. Also determine J_z.

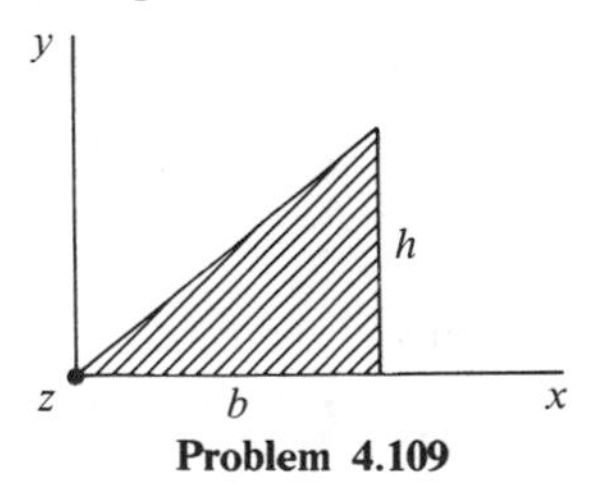

Problem 4.109

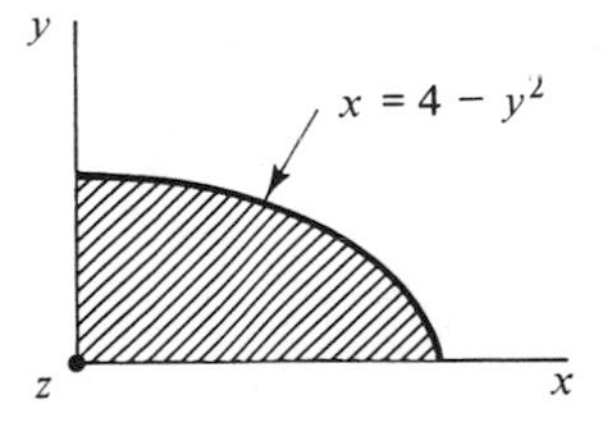

Problem 4.110

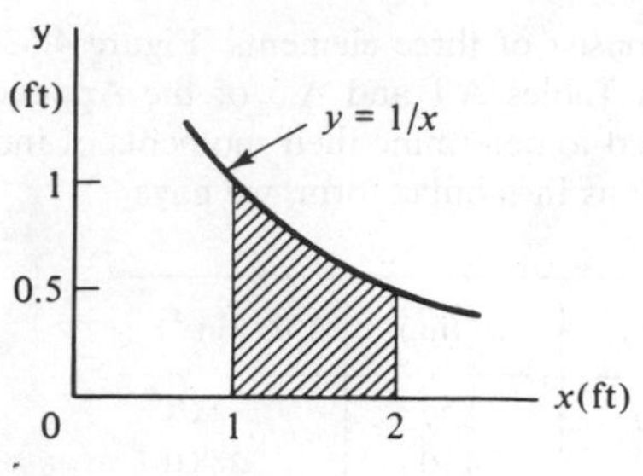

Problem 4.111

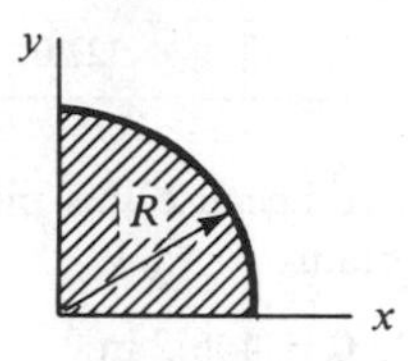

Problem 4.112

4.113. Determine I_x, I_y, and J_z for the rectangular area shown. Also determine the radii of gyration about the x, y, and z axes.

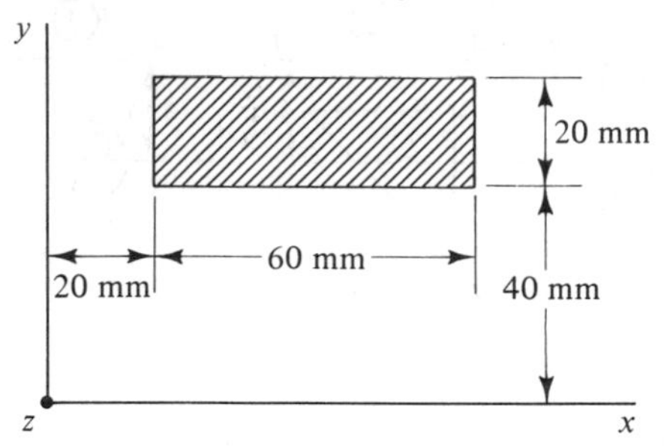

Problem 4.113

4.114. If $I_b = 216 \text{ mm}^4$ for the area shown, find I_a and the dimension b.

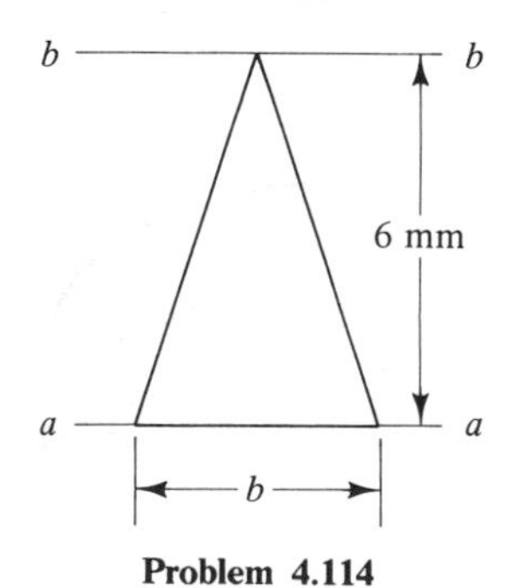

Problem 4.114

4.115. and 4.116. Determine the moments of inertia of the area shown about the x and y axes.

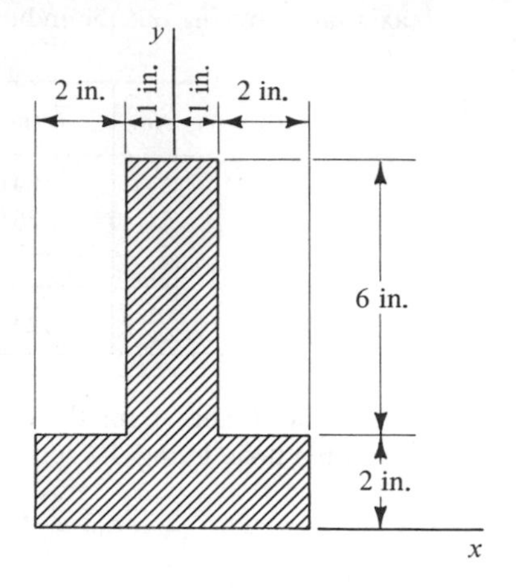

Problem 4.115

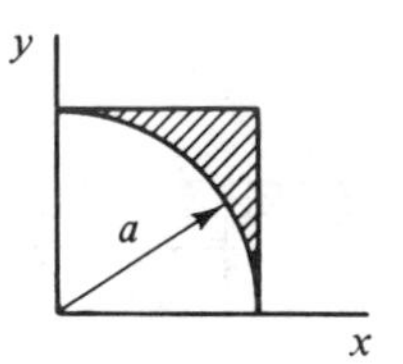

Problem 4.116

4.117. Determine the radius of gyration of the area shown about the z axis.

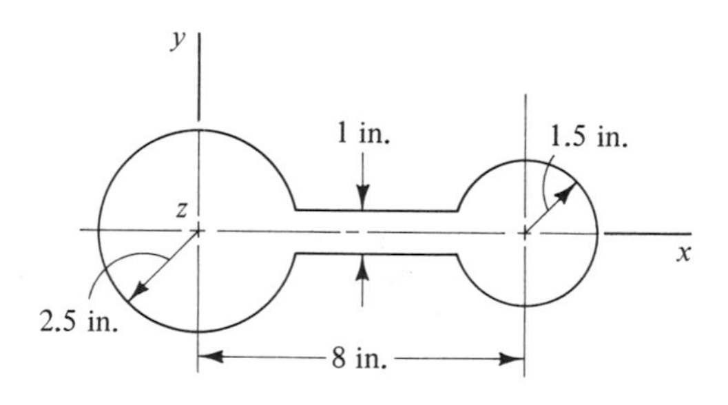

Problem 4.117

4.118. If $I_x = 1{,}000 \text{ in.}^4$ for the T-section shown, determine $\bar{y}$ and find the moment of inertia about a horizontal axis through the centroid.

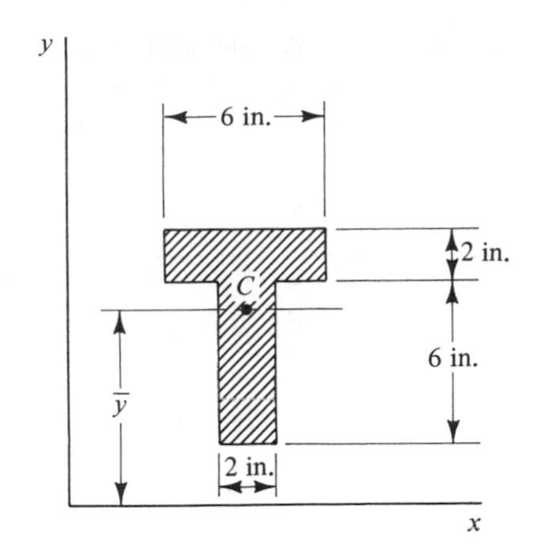

Problem 4.118

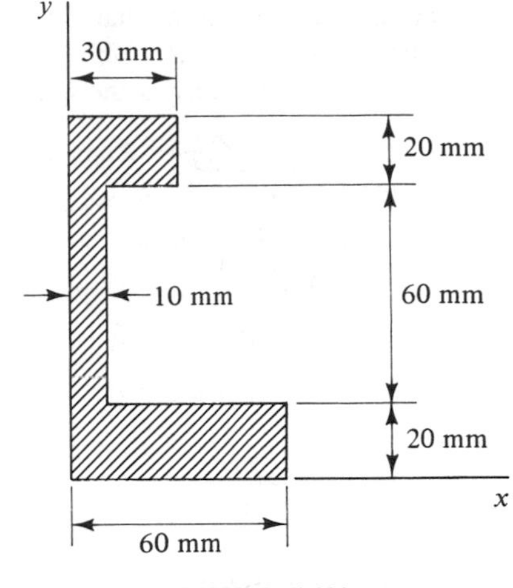

Problem 4.120

4.119. Determine the spacing d between the two rectangular areas shown for which $I_x = I_y$. What are the corresponding values of the moments of inertia?

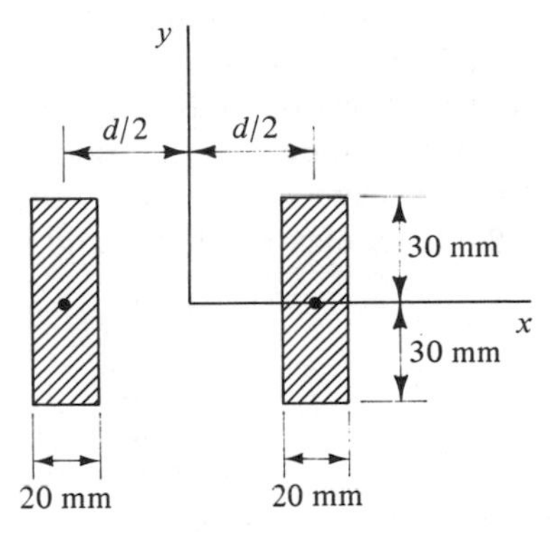

Problem 4.119

4.120. An aluminum extrusion has the cross section shown. Determine the moment of inertia about a horizontal axis through the centroid of the section.

4.121. A channel section is connected to a W-shape beam to form a structural member with the symmetrical cross section shown. Locate the centroid of the cross section by using the geometric properties for the members given in Tables A.3 and A.4 of the Appendix, and determine the moments of inertia about horizontal and vertical centroidal axes.

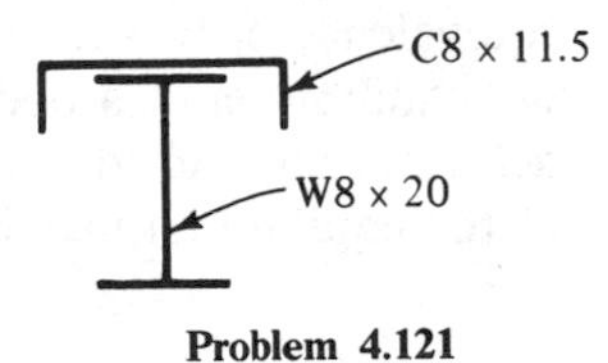

Problem 4.121

4.122. Three timbers are nailed together to form a symmetrical section with the nominal dimensions shown. Determine the moments of inertia of the section about horizontal and vertical centroidal axes. See Table A.12 of the Appendix for the geometric properties of the timbers.

4.123. Determine the polar moment of inertia about the longitudinal axis of a hollow tube with inside radius R_i and outside radius R_o. What is the radius of gyration about this axis? Show that if the walls are thin, $J \approx 2\pi R_{avg}^3 t$, where R_{avg} is the average radius and t is the wall thickness.

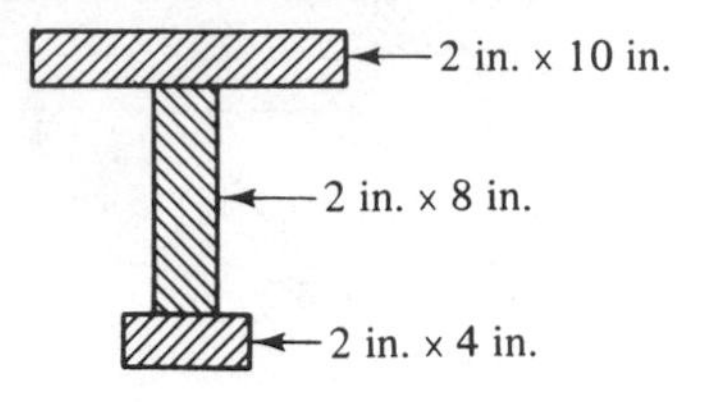

Problem 4.122

4.124. A panel of corrugated sheet metal with thickness t has the cross section shown. Determine I_x for the section.

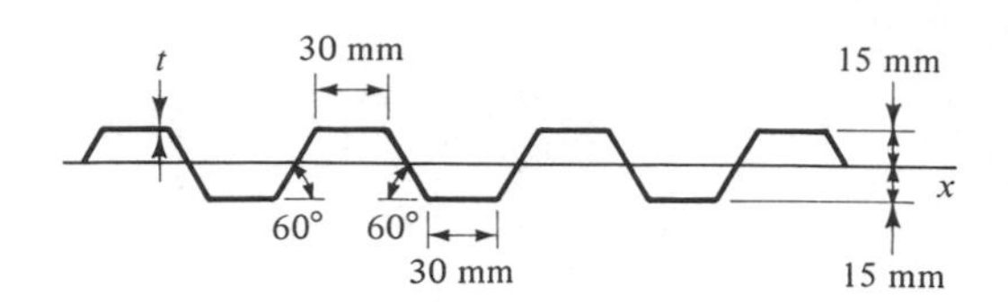

Problem 4.124

4.9 CLOSURE

In this chapter we have considered the conditions for the static equivalence of two force systems. These conditions apply to force systems in general, but they are particularly useful when dealing with distributed forces. They were used in Section 4.3 to show that the reactions due to various supports and connections can be treated as concentrated forces and couples in a force analysis. They were also used to locate the center of gravity of a body and to develop procedures for handling distributed applied loads. We shall use these conditions again in later chapters when determining the stresses developed within load-carrying members.

It is essential that we not forget the limitations on static equivalence. Static equivalence of two force systems implies only that they have the same effect upon the equilibrium of a body. If one of the force systems is replaced by the other, the deformations and forces developed within the body may be altered and the stability of the equilibrium may be affected.

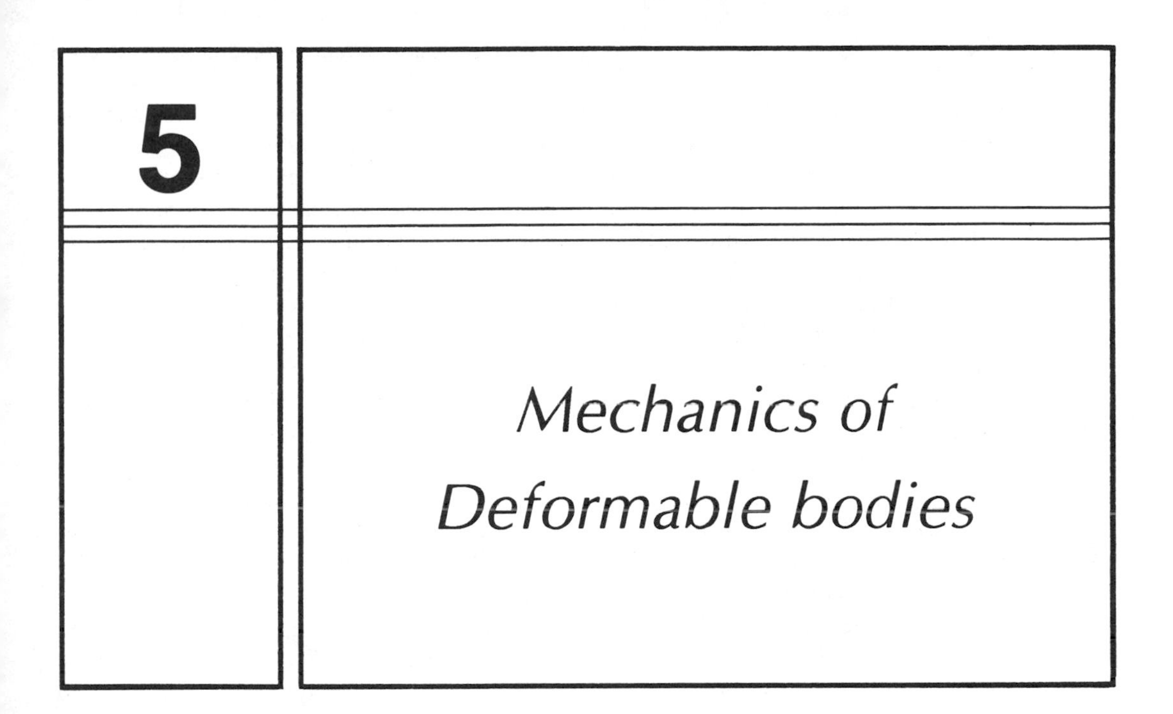

5.1 INTRODUCTION

Our discussions in the previous chapters have been concerned primarily with the problem of determining the forces and couples acting upon various bodies loaded and supported in different ways. This was done without regard to the response of the bodies to the applied loadings, other than assuming that the resulting deformations are small (the rigid-body approximation).

Clearly, the ability of a load-carrying member to fulfill its intended purpose requires that the member not break under the applied loadings. The "strength" of a member depends in large measure upon the stresses, or intensity of the forces developed within it. Deformations of a member can also dictate the limits of its usefulness. A floor that sags an excessive amount is a "failure," even though it may be in no danger of breaking.

The effects of forces and couples upon load-carrying members will be investigated in this chapter and in those to follow. Our goal is to develop relationships between the applied loadings and the stresses and deformations they produce. These relationships are needed for analyzing members to see if they can adequately perform their job and for designing (sizing) members to carry a given loading.

Once the force analysis of a body has been completed, the rigid-body approximation must be abandoned if its response is to be determined. Obviously, we cannot determine the deformations of a body if we insist on assuming that it is rigid. This is one of the major points of departure from our work in earlier chapters. Thus, we shall now be considering the *mechanics of deformable bodies*.

The mechanics of deformable bodies is more involved than the mechanics of rigid bodies, but not unduly so. Actually, we have already performed an analysis of a deformable system; namely, the example of the rigid arm supported by a spring considered in Section 3.10. A review of this example shows that, in order to obtain a solution, we had to consider equilibrium, the geometry of the deformations, and the resistive properties of the material (spring). These three items are the basic ingredients in the mechanics of deformable bodies.

We have already considered equilibrium. The geometry of deformations and resistive properties of materials will be discussed in this chapter, along with the application of all three concepts to some simple, but important, problems. Structural elements such as rods, shafts, and beams under simple loadings will be considered in Chapters 6 through 8. These results will be combined in Chapter 9 in a study of more complicated members and loadings. The buckling of columns under compressive loads will be considered in Chapter 10.

It was mentioned in Section 3.4 that most engineering structures are designed such that the deformations are small. Accordingly, *small deformations will be assumed throughout* all our work. As a result, the rigid-body approximation can still be used in force analyses, even though the values of the forces and couples so obtained are then used to determine the deformations. The justification for this procedure, which greatly simplifies the analysis, was given in Section 3.10. This approach can be used for all but buckling problems, for which the equilibrium equations must be based upon the deformed geometry of the body.

In the way of historical perspective, we note that the systematic study of the response of load-carrying members dates back at least to Leonardo da Vinci (1452–1519), who performed experiments on the strength of wires and beams. The advent of railroads in the late 1800's provided the impetus for much of the basic work in this area, and problems associated with the design of aircraft, space vehicles, and nuclear reactors have led to extensive studies of the more advanced aspects of the subject.

5.2 STRESSES AND STRESS RESULTANTS

Loadings applied to a body are transmitted through it to the supports, which provide the reactions necessary to hold the entire member in equilibrium. In the process, forces are developed within the body that resist the applied loads and maintain equilibrium of each of its parts. These resistive forces are distributed over surfaces within the body, and their magnitude per unit area is called *stress*. The resultants of these forces are called *stress resultants* (also called internal reactions or internal forces).

As an illustration, consider a bar made of several blocks glued together end to end [Figure 5.1(a)]. If we pull on the end of the bar with a small force **F**, it is clear

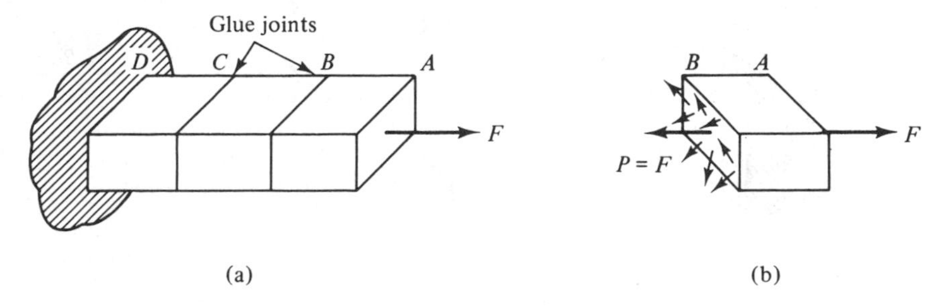

Figure 5.1

that the glue joints will resist the tendency for the blocks to be pulled apart. The joints provide resistive forces that hold the various pieces of the bar together in equilibrium. They also serve to transmit the applied load from block to block to the wall, which supports the entire member.

The resultant force transmitted by a joint can be determined by passing an imaginary cutting plane through it and then considering the equilibrium of one of the resulting pieces of the bar. (Note that this is the method of sections used in Chapter 3 to determine the forces in the members of a truss.) For example, sectioning the bar along joint B and considering the equilibrium of the piece to the right [Figure 5.1(b)], we see that the stress resultant, denoted by $\mathbf{P}$, must be equal and opposite to the applied force $\mathbf{F}$. If we push on the bar instead of pull on it, the sense of the stress resultant is reversed. In both cases, the resultant is a force perpendicular to the cut surface. This type of stress resultant is called a *normal force*, and is associated with the tendency of a member to elongate or shorten.

A different situation is illustrated in Figure 5.2. In this case, the stress resultant is a force parallel to the joint and is associated with the tendency of one part of the member to slide with respect to the other. This tendency is called *shear*, and the stress resultant is called a *shear force*. Shear forces will be denoted by $\mathbf{V}$.

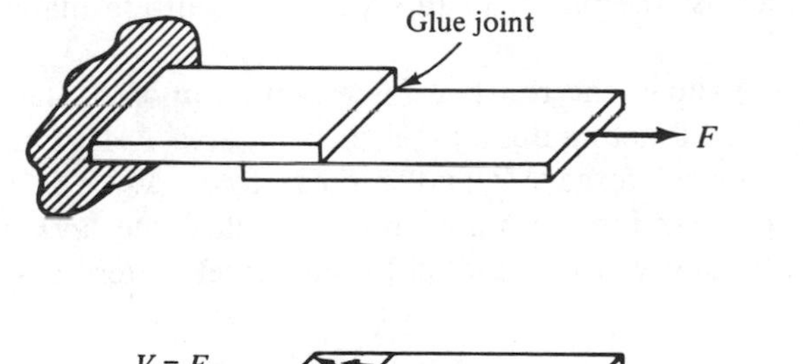

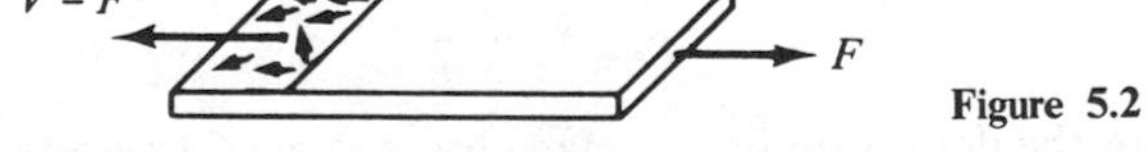

Figure 5.2

The stress resultant can also be a couple. If the couple is perpendicular to the cutting plane [Figure 5.3(a)], the stress resultant, denoted by $\mathbf{T}$, is called a *torque*. Torques are associated with the tendency of a member to twist. Couples that lie in

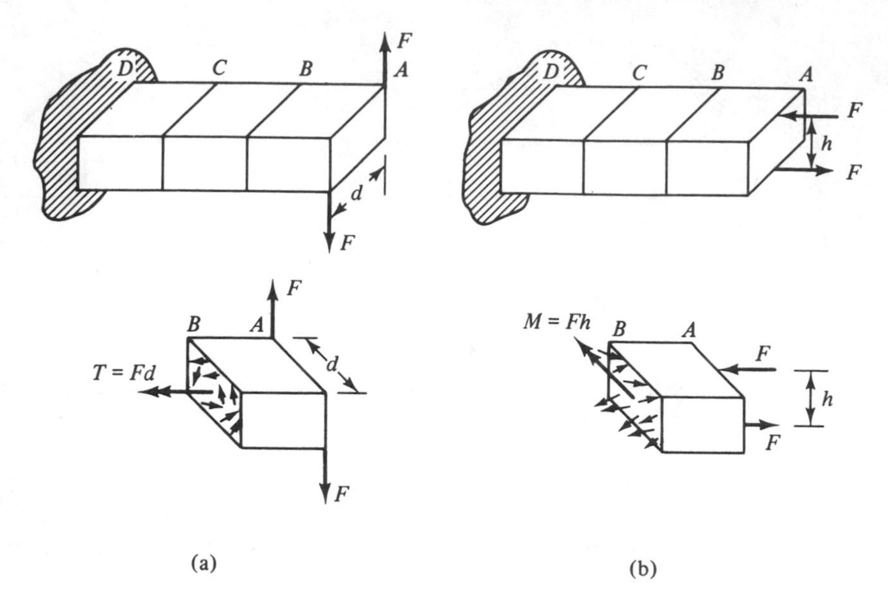

Figure 5.3

the cutting plane [Figure 5.3(b)] are associated with the tendency of a member to bend and are called *bending moments*. Bending moments will be denoted by **M**. Under more complex loading conditions, the various types of stress resultants can occur simultaneously.

The preceding ideas apply directly to bodies of one piece. The only difference is that they are held together by interatomic forces instead of glue joints. Any body can be thought of as consisting of two pieces, one on either side of an imaginary plane [Figure 5.4(a)]. Each piece exerts forces on the other piece, just as the adjoining blocks in our previous examples exert forces on each other. These forces are transmitted across the plane. In this sense, the cutting plane is analogous to a glue joint.

Figure 5.4(b) shows the resistive forces acting upon a plane through a body. The resultant force $\Delta\mathbf{R}$ acting upon an element of area ΔA within the plane can be resolved into a normal force $\Delta\mathbf{P}$ and a shear force $\Delta\mathbf{V}$. The magnitudes of the normal force and shear force per unit area are called the *normal stress* and *shear stress*, respectively, and will be denoted by the Greek letters σ and τ. The *average stresses* are

$$\sigma_{\text{avg}} = \frac{\Delta P}{\Delta A} \qquad \tau_{\text{avg}} = \frac{\Delta V}{\Delta A} \tag{5.1}$$

and, from the definition of a surface force [Eq. (4.18)], the *normal* and *shear stresses at a point* in the plane are

$$\sigma = \lim_{\Delta A \to 0} \frac{\Delta P}{\Delta A} = \frac{dP}{dA} \qquad \tau = \lim_{\Delta A \to 0} \frac{\Delta V}{\Delta A} = \frac{dV}{dA} \tag{5.2}$$

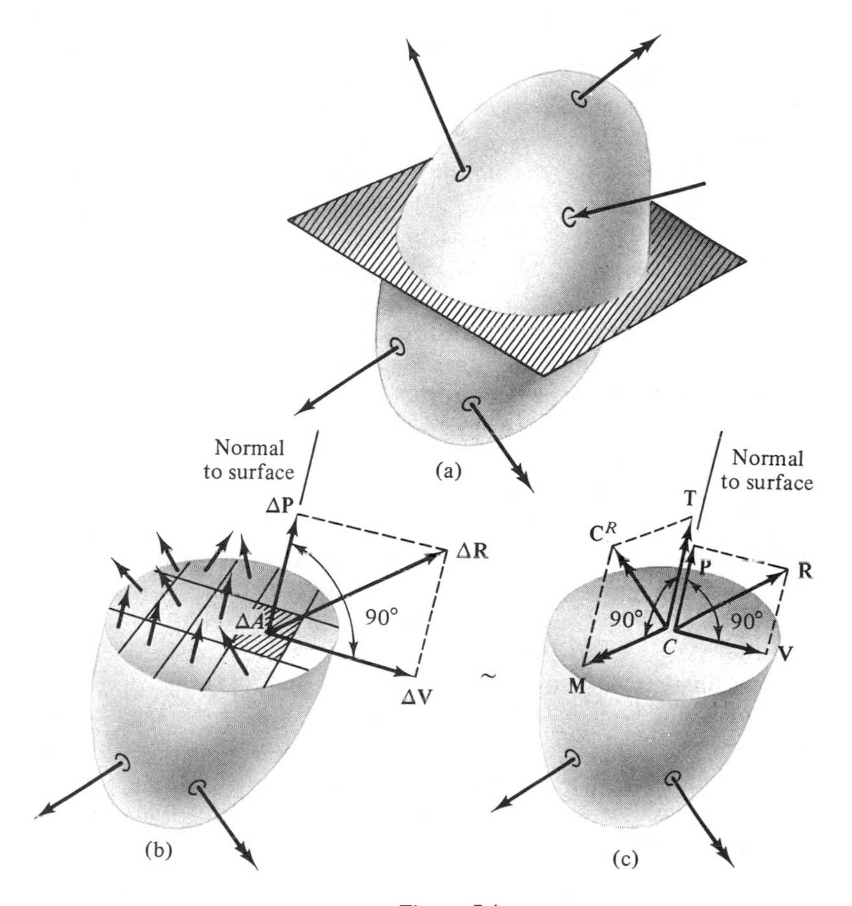

Figure 5.4

If the stresses are uniformly distributed over the plane, the average stresses and the stresses at a point are the same. Common units for stress are lb/in^2. (psi) or N/m^2 (pascals). Another common unit is ksi (1 ksi = 1,000 psi).

Normal stresses (and forces) directed away from the cut section are called *tensile* and will be considered positive. When directed toward the section, they are called *compressive* and will be considered negative. It is important to note that this sign convention relates to the response of the member and is separate from any sign convention we may use in a force analysis. There is no universal sign convention for shear stress. In many instances, only its magnitude is of importance.

Under general loading conditions, the stress resultant acting upon any given plane within a body will be a force and a couple [Figure 5.4(c)]. For convenience, the resultant force will always be placed at the centroid of the cut section. (Recall from Section 4.3 that the resultant couple depends upon the location of the resultant force.) The resultant force can be resolved into a normal force **P** perpendicular to the plane and a shear force **V** parallel to the plane. Similarly, the resultant couple can be resolved into a torque **T** normal to the plane and a bending moment **M** parallel to the plane. If desired, the shear force and bending moment can be further resolved into components within the cut section.

Both the stresses and stress resultants generally vary throughout the body. They also depend upon the orientation of the plane on which they act. Thus, the orientation of the cutting plane must always be specified in order for stated values of the stresses and stress resultants to have meaning.

The stresses are related to the stress resultants via the conditions for static equivalence. As shown earlier, the stress resultants can be determined by using the method of sections. Unfortunately, the stresses cannot be determined so easily. Since there are any number of possible stress distributions with the same resultant, additional information is needed for a unique determination of the stresses. But is it necessary to know the stresses and not just the stress resultants? Yes, and for two very important reasons.

First, the strength of a material is best expressed in terms of stresses instead of forces. For example, the force that a glue joint in the bar shown in Figure 5.1(a) can support is not a good measure of the strength of the glue. Obviously, joints with a larger area can support a larger force, even though the properties of the glue remain the same. A better measure of the material capabilities is the force per unit area, or stress, that it can withstand.

Second, the deformations produced by the applied loads are intimately related to the stress distribution. This is illustrated in Figure 5.5, which shows two different distributions of normal stress acting upon identical rubber blocks. Each stress distribution has the same resultant, but it produces a different pattern of deformation. Clearly, it is not enough to know only the stress resultants if the deformations are to be determined.

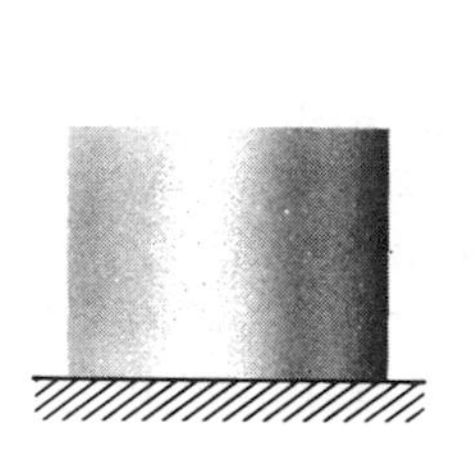
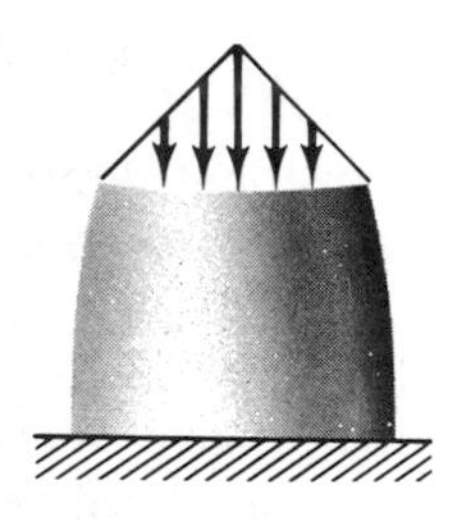
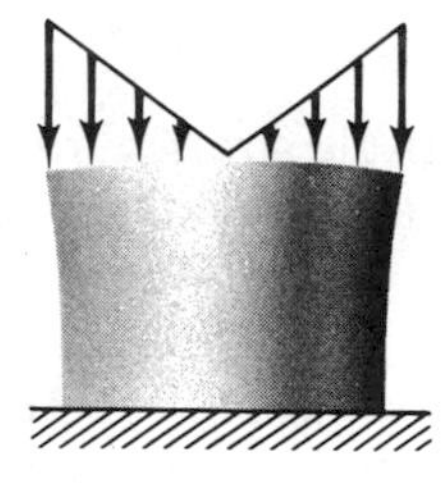

Figure 5.5

At this point, we are in no position to outline a general procedure for determining the stresses in a body. This must await our study of the geometry of the deformations and the resistive properties of materials. However, there are many problems for which average stresses have significance and can be readily computed. Some of these problems will be considered in Sections 5.3 and 5.4. The following examples illustrate the procedures for determining the stress resultants.

Example 5.1. Stress Resultants in a Beam. Determine the normal and shear forces and bending moment at sections *a-a* and *b-b* in the beam shown in Figure 5.6(a).

Solution. We first determine the reactions at A and B. From the FBD [Figure 5.6(b)], we have

$$\sum F_x = A_x + B_x = 0$$

$$\sum F_y = B_y - 6{,}000 \text{ lb} = 0$$

$$\overset{+}{\curvearrowleft}\sum M_B = 6{,}000 \text{ lb } (5 \text{ ft}) - A_x(4 \text{ ft}) = 0$$

$$A_x = 7{,}500 \text{ lb} \qquad B_x = -7{,}500 \text{ lb} \qquad B_y = 6{,}000 \text{ lb}$$

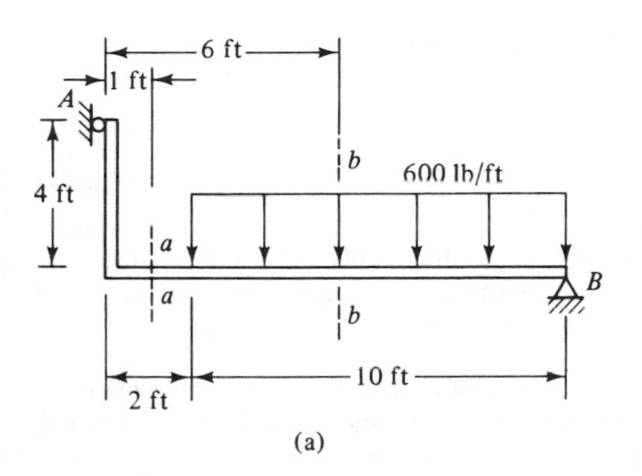

(a)

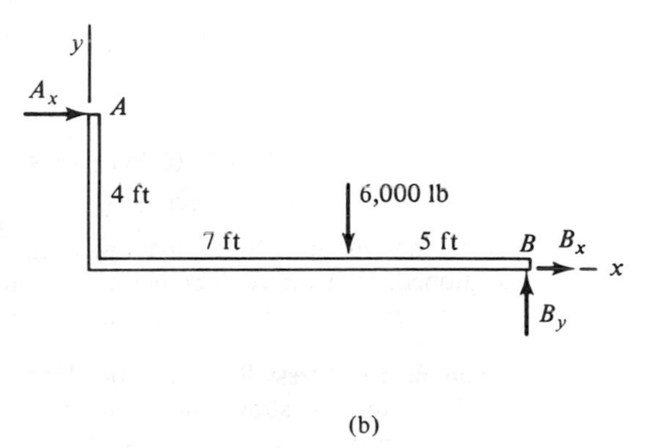

(b)

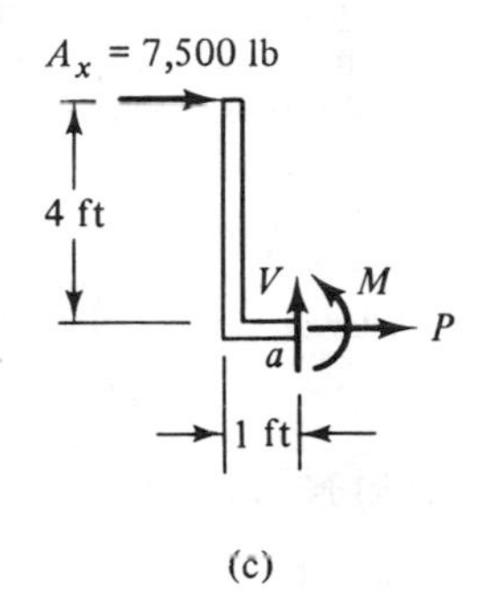

(c)

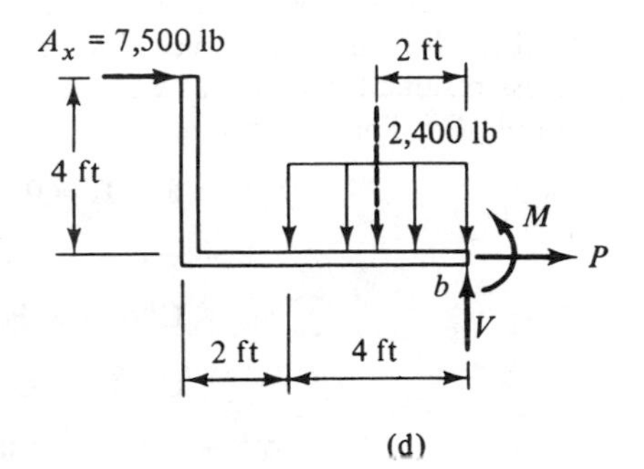

(d)

Figure 5.6

Section *a-a*. Sectioning the beam at *a-a* and considering the piece to the left, we obtain the FBD shown in Figure 5.6(c). For equilibrium,

$$\sum F_x = P + 7{,}500 \text{ lb} = 0$$

$$\sum F_y = V = 0$$

$$\overset{+}{\curvearrowleft}\sum M_a = M - 7{,}500 \text{ lb } (4 \text{ ft}) = 0$$

so

$$P = -7{,}500 \text{ lb or } 7{,}500 \text{ lb (compression)} \qquad V = 0$$
$$M = 30{,}000 \text{ lb} \cdot \text{ft}$$ **Answer**

The portion of the beam to the right of *a-a* could also have been used. The stress resultants on this segment are equal and opposite to those on the left-hand segment, by Newton's third law.

Section *b-b*. Sectioning through the beam and the distributed load at *b-b* and proceeding as before, we have from Figure 5.6(d)

$$\sum F_x = P + 7{,}500 \text{ lb} = 0$$

$$\sum F_y = V - 2{,}400 \text{ lb} = 0$$

$$\overset{+}{\curvearrowleft} \sum M_b = M + 2{,}400 \text{ lb (2 ft)} - 7{,}500 \text{ lb (4 ft)} = 0$$

or

$$P = -7{,}500 \text{ lb or } 7{,}500 \text{ lb (compression)} \qquad V = 2{,}400 \text{ lb}$$
$$M = 25{,}200 \text{ lb} \cdot \text{ft}$$ **Answer**

Note that the distributed load was replaced by its resultant only after the member was sectioned. If this were done before sectioning the member, the forces within the body could be altered (see Section 4.2), leading to incorrect values for the stress resultants.

Example 5.2. Stress Resultants In a Rod. Find the stress resultants acting upon section *a-a* of the rod shown in Figure 5.7(a). Identify the shear and normal forces, bending moments, and torque. Describe the responses associated with each.

Solution. We section the bar at *a-a* and consider the portion to the right [Figure 5.7(b)]. Note that this choice eliminates the need to determine the reactions at the wall. The stress resultant is shown as a single force **R** acting at the centroid of the cross section and a couple $\mathbf{C}^R$. For equilibrium:

$$\sum \mathbf{F} = \mathbf{F} + \mathbf{R} = \mathbf{0}$$

$$\mathbf{R} = -\mathbf{F} = -40\mathbf{i} + 60\mathbf{j} - 20\mathbf{k} \text{ N}$$ **Answer**

$$\sum \mathbf{M}_a = \mathbf{C}^R + \mathbf{r} \times \mathbf{F} = \mathbf{0}$$

$$\mathbf{C}^R + (2\mathbf{i} + 2\mathbf{k}) \times (40\mathbf{i} - 60\mathbf{j} + 20\mathbf{k}) \text{ N} \cdot \text{m} = \mathbf{0}$$

$$\mathbf{C}^R = -120\mathbf{i} - 40\mathbf{j} + 120\mathbf{k} \text{ N} \cdot \text{m}$$ **Answer**

The components of **R** are shown in Figure 5.7(c). The normal force (perpendicular to the cut section) is 40 N (tension). It is associated with a stretching of portion AB of the member. The 60-N and 20-N forces are the components of the shear force within the plane. They are associated with the tendency of part aBC of the bar to move downward and in the positive z direction relative to the piece attached to the wall. The components of $\mathbf{C}^R$ are shown in Figure 5.7(d). The torque (normal to the cut section) is associated with a tendency of portion AB of the member to twist about the x axis. The other two components are bending moments. The 40-N · m moment is associated with a bending of the bar in the x-z plane and the 120-N · m moment with bending in the x-y plane.

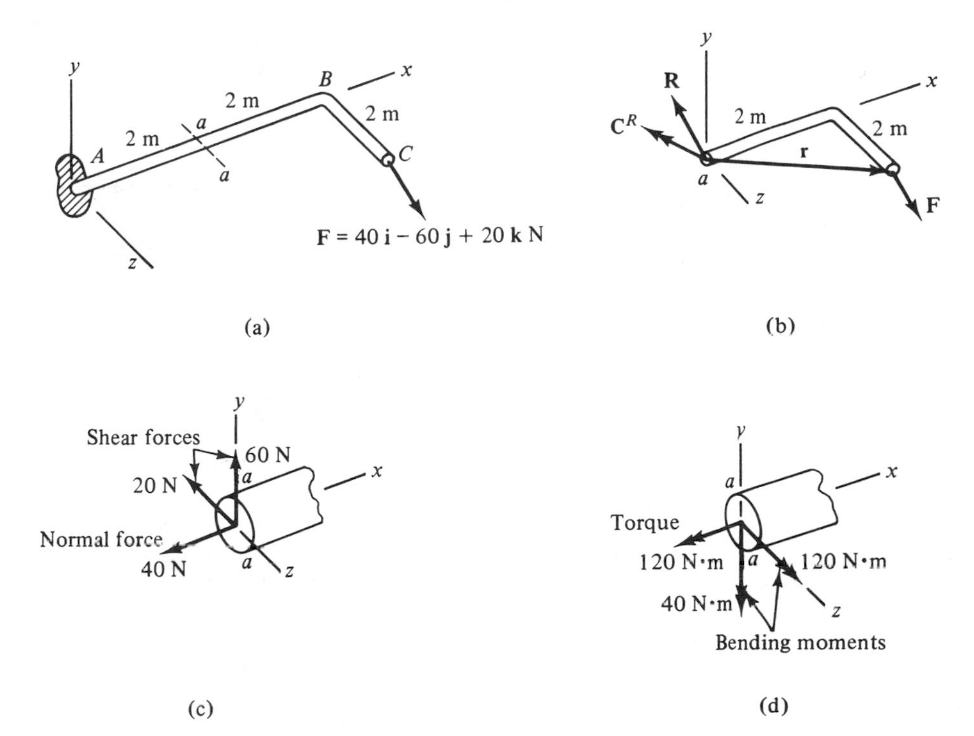

Figure 5.7

PROBLEMS

5.1. to 5.11. For the member or structure shown, determine the stress resultants on the sections indicated. Identify the normal and shear forces, bending moments, and torques, where applicable.

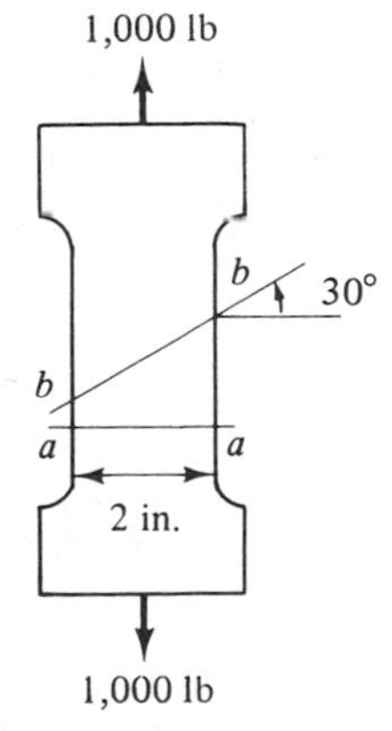

Problem 5.1

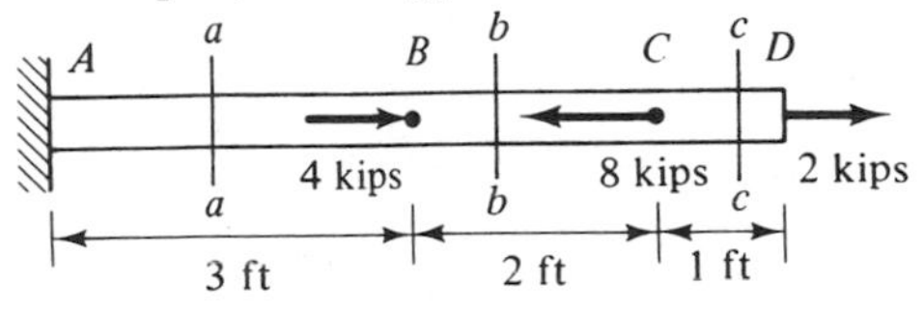

Problem 5.2

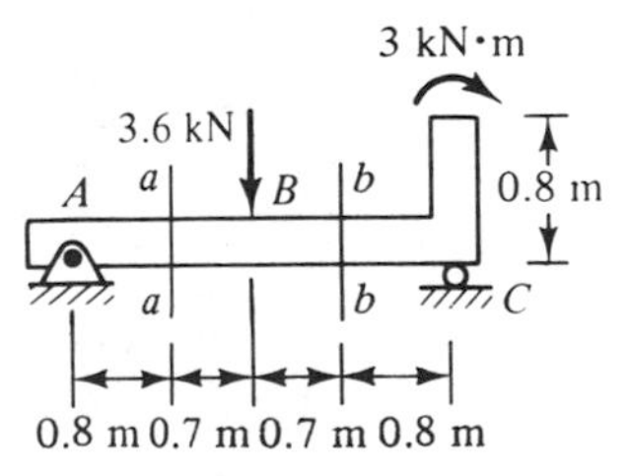

Problem 5.3

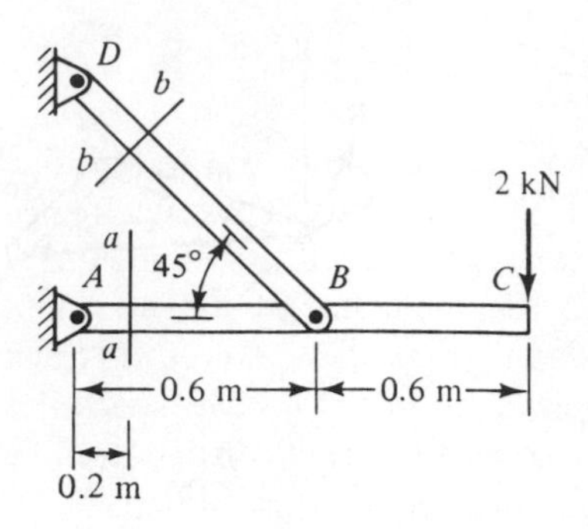

Problem 5.4

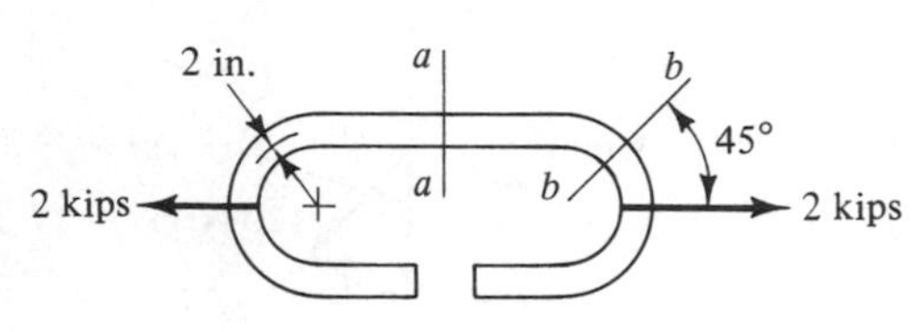

Problem 5.5

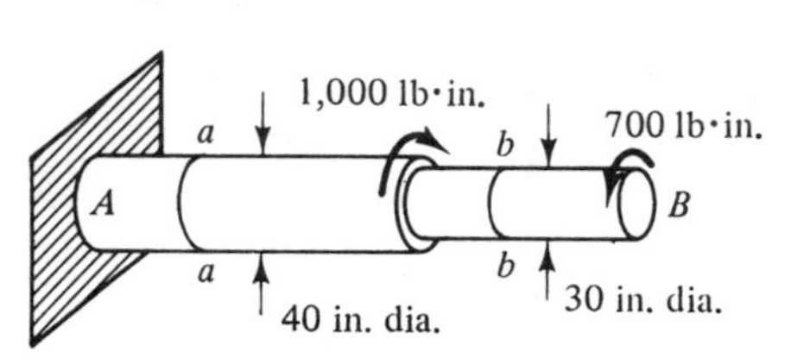

Problem 5.6

2 in.
4 in.
b
b
20°
a
a
5 in.
5 in.
4 lb

Problem 5.7

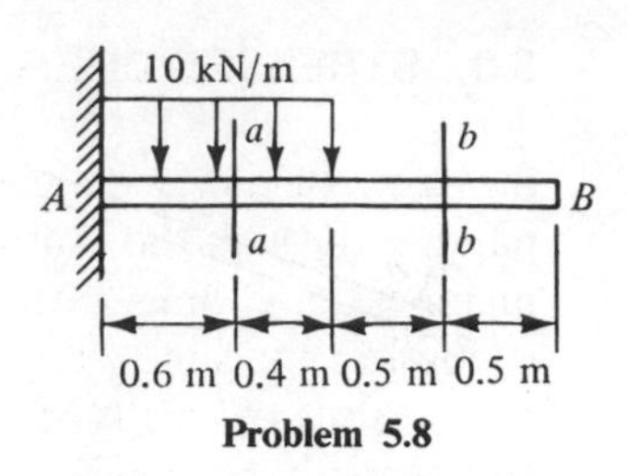

Problem 5.8

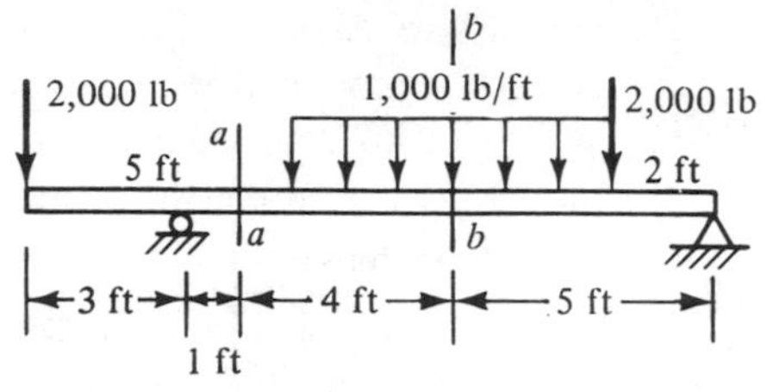

Problem 5.9

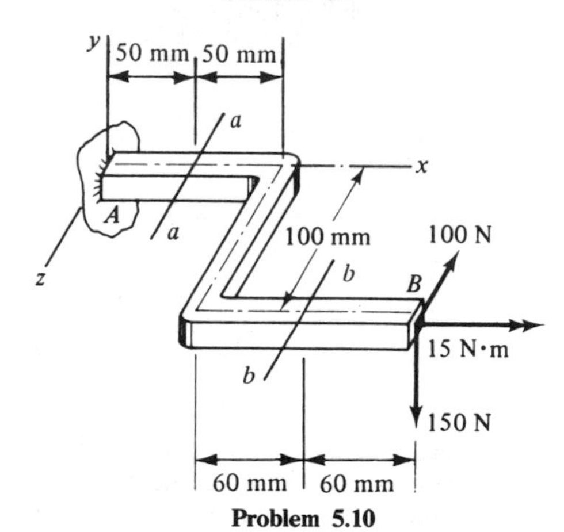

Problem 5.10

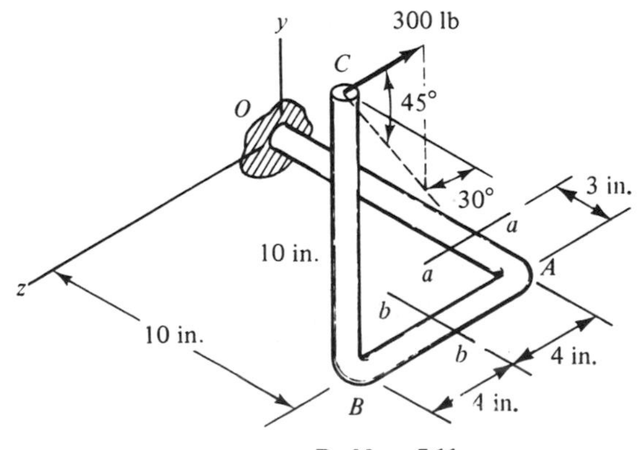

Problem 5.11

5.3. STRESSES DUE TO AXIAL LOADING

Straight, slender members subjected to forces that lie along the longitudinal axis passing through the centroid of the cross section are said to be *axially loaded*. Such members are one of the more common types of structural elements. Typical examples are support wires and cables and the members of a truss. The stress resultant in an axially loaded member consists only of a force; therefore, the average stresses can be computed directly from their definitions. As we shall show, the actual stresses in axially loaded members are often uniformly distributed, in which case they are equal to the average stresses and can also be determined.

Consider an axially loaded member with arbitrary cross section in equilibrium under the action of two or more forces [Figure 5.8(a)]. The normal and shear forces on a plane oriented at an arbitrary angle θ, measured as shown, are

$$P = R\cos\theta \qquad V = R\sin\theta$$

where R is the magnitude of the stress resultant. The area A_P of the plane over which these forces act is

$$A_P = \frac{A}{\cos\theta}$$

where A is the cross-sectional area of the member. This result is easily derived, as shown in Figure 5.8(b).

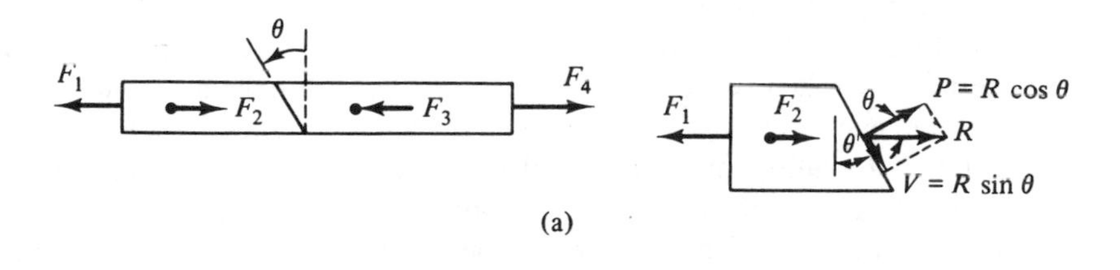

(a)

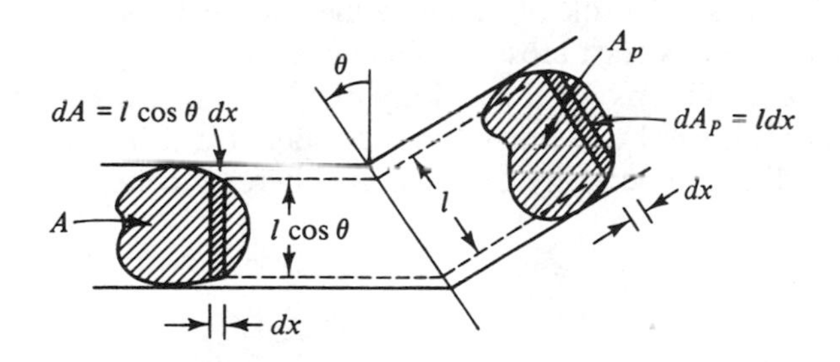

$A = \int l\cos\theta dx = \int \cos\theta dA_P$
$= A_P\cos\theta$

(b)

Figure 5.8

We now make the assumption that the actual stresses are uniformly distributed, in which case they are equal to the average stresses. From the definitions of the average stresses, Eqs. (5.1), we have for the normal stress σ_θ on the plane

$$\sigma_\theta = \sigma_{avg} = \frac{P}{A_P} = \frac{R}{A}\cos^2\theta \tag{5.3}$$

and for the shear stress τ_θ

$$\tau_\theta = \tau_{avg} = \frac{V}{A_P} = \frac{R}{A}\sin\theta\cos\theta = \frac{R}{2A}\sin 2\theta \tag{5.4}$$

These equations relate the stresses to the stress resultant, which, in turn, is related to the applied loads.

From Eq. (5.3), we see that the normal stress is a maximum on planes perpendicular to the axis of the member, for which $\theta = 0$ and $\cos\theta = 1$ (Figure 5.9). Since $P = R$ on these planes, the maximum value of normal stress is simply the normal force divided by the cross-sectional area of the member:

$$\sigma_{max} = \frac{R}{A} = \frac{P}{A} \tag{5.5}$$

The maximum shear stress occurs on planes oriented at $\pm 45°$ with respect to the longitudinal axis (Figure 5.9), as can be seen from Eq. (5.4). Thus,

$$|\tau_{max}| = \frac{1}{2}\frac{|R|}{A} = \frac{1}{2}|\sigma_{max}| \tag{5.6}$$

The fact that the magnitude of the maximum shear stress is only one-half that of the maximum normal stress does not mean that shear stresses are unimportant. Some materials are much weaker in shear than in tension or compression.

The assumption of uniformly distributed stresses is correct, and Eqs. (5.3) through (5.6) valid, for axially loaded members made of a homogeneous material, except near points of load application and sudden changes in the cross section, such as notches or holes. The member may be tapered slightly, and the applied loads may vary along its length. For compressive loads, the member must not be so slender that it buckles. As a rule of thumb, a member will not buckle if its length is no more than ten times its least cross-sectional dimension.

The conditions for the validity of the load–stress relations given in Eqs. (5.3) through (5.6) will be examined in more detail in Chapter 6. However, the need for axial loading can be easily demonstrated here by considering the stresses on a plane

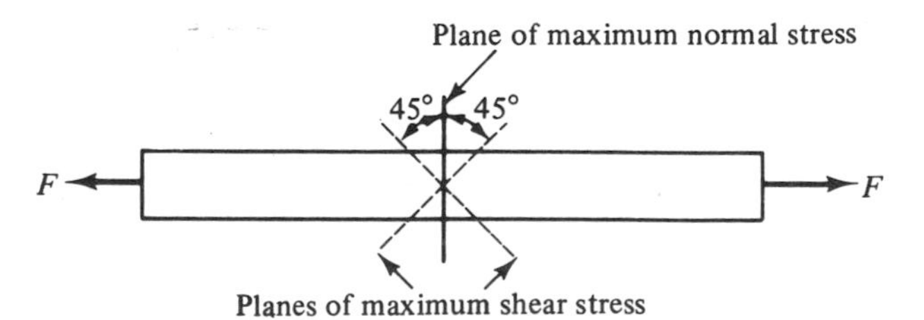

Figure 5.9

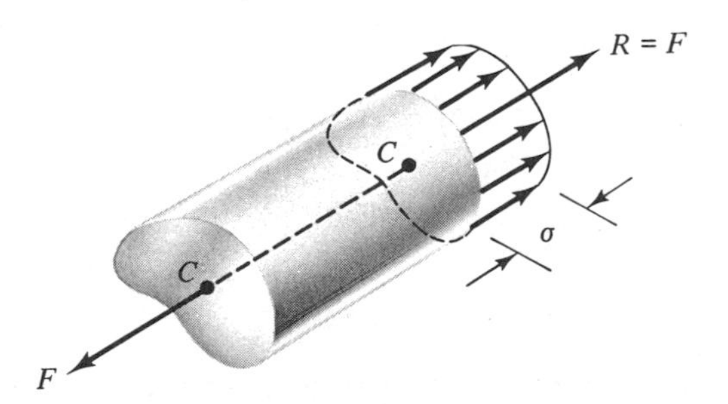

Figure 5.10

perpendicular to the axis of the member. If the stress distribution is uniform, as has been assumed, the stress resultant will lie along the longitudinal axis passing through the centroid of the cross section (Figure 5.10). (Recall from Section 4.7 that the resultant of a surface force passes through the centroid of the volume under the loading diagram.) Consequently, the applied loads must also lie along the centroidal axis in order to maintain equilibrium. If they don't, the stresses cannot be uniformly distributed, and Eqs. (5.3) through (5.6) do not apply.

There is no need to memorize any of the equations presented in this section. The important thing to remember is that the stresses on any given plane are simply the normal and shear forces acting upon that plane divided by the area over which they act.

Example 5.3. Stresses In a Plate. A broken plate is welded back together, as shown in Figure 5.11(a). (a) What is the maximum load F that can be applied if the tensile stress in the weld cannot exceed 5,000 psi? (b) What are the maximum normal and shear stresses in the plate for this value of load and on what planes do they occur? Assume axial loading.

Solution. (a) Sectioning the plate along the weld and considering the piece to the left [Figure 5.11(b)], we find that the tensile force on the weld is

$$P = F \sin 30° = \frac{F}{2}$$

The area over which this force acts is

$$A_P = \frac{A}{\sin 30°} = 2\left(\frac{1}{2}\text{ in.}\right)(4\text{ in.}) = 4\text{ in.}^2$$

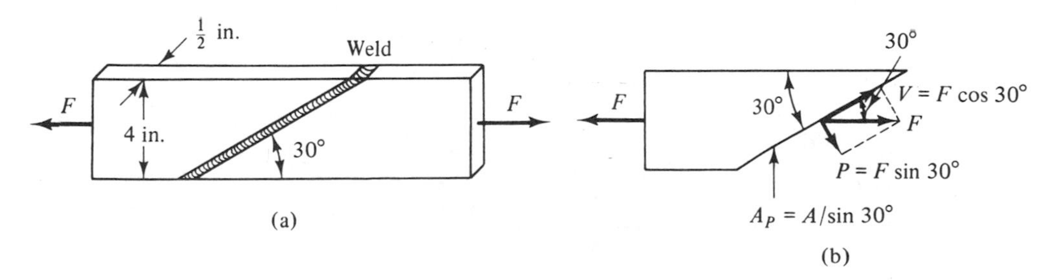

Figure 5.11

Thus, the condition on the tensile stress is

$$\sigma = \frac{P}{A_P} = \frac{(F/2)}{4\ \text{in.}^2} \leqslant 5{,}000\frac{\text{lb}}{\text{in.}^2}$$

from which we obtain for the allowable load

$$F \leqslant 40{,}000\ \text{lb} \qquad \textbf{Answer}$$

(b) The maximum normal stress occurs on planes perpendicular to the axis. On these planes, $P = F$; so

$$\sigma_{\text{max}} = \frac{40{,}000\ \text{lb}}{(\frac{1}{2}\text{in.})(4\ \text{in.})} = 20{,}000\ \text{psi(tension)} \qquad \textbf{Answer}$$

The maximum shear stress is one-half the maximum normal stress [Eq. (5.6)]:

$$|\tau_{\text{max}}| = \tfrac{1}{2}|\sigma_{\text{max}}| = 10{,}000\ \text{psi} \qquad \textbf{Answer}$$

It occurs on planes oriented $\pm 45°$ from the axis of the member.

Example 5.4. Design of a Column. A column in a building supports the loads shown in Figure 5.12(a). What is the required cross-sectional area of the member if the maximum allowable compressive stress is 20 MN/m^2? Neglect the weight of the member and assume that it does not buckle.

Solution. From the FBDs of Figure 5.12(b), we see that the largest compressive force occurs in portion AB of the member and is equal to 80 kN. Thus, the condition on the compressive stress is

$$\sigma_{\text{max}} = \frac{P}{A} = \frac{80 \times 10^3\ \text{N}}{A} \leqslant 20 \times 10^6\ \text{N/m}^2$$

from which we obtain

$$A \geqslant 4.0 \times 10^{-3}\ \text{m}^2 \qquad \textbf{Answer}$$

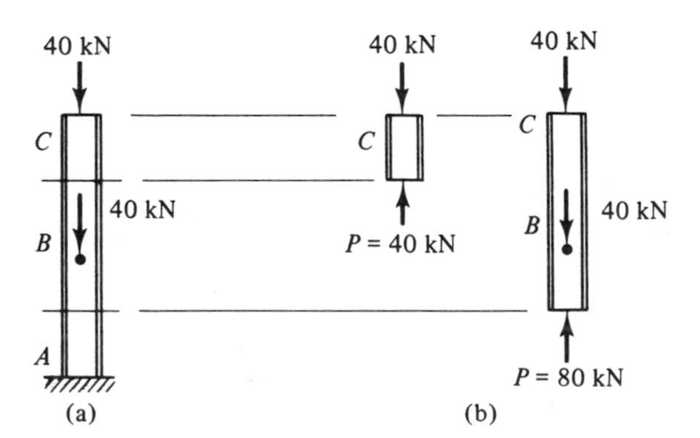

Figure 5.12

PROBLEMS

5.12. A speedboat tows a water skier and kite at constant speed and elevation, as shown. The tension on the towline is 900 N. What is the maximum tensile stress in the 5-mm diameter line?

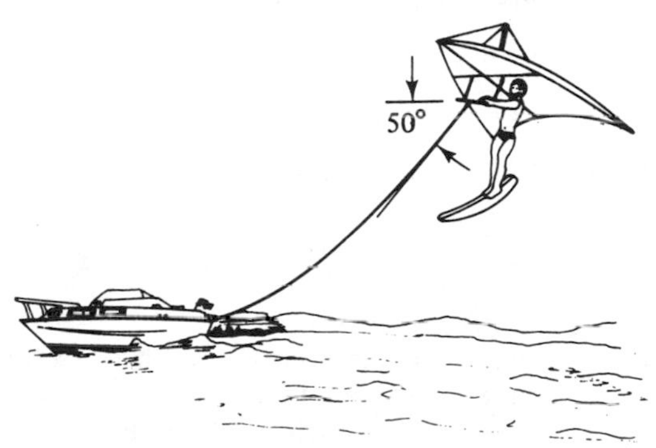

Problem 5.12

5.13. A 500-kg sign is to be supported as shown, by two links made of steel rod. What is the minimum required diameter of the rods if the allowable tensile stress is 60 MN/m^2?

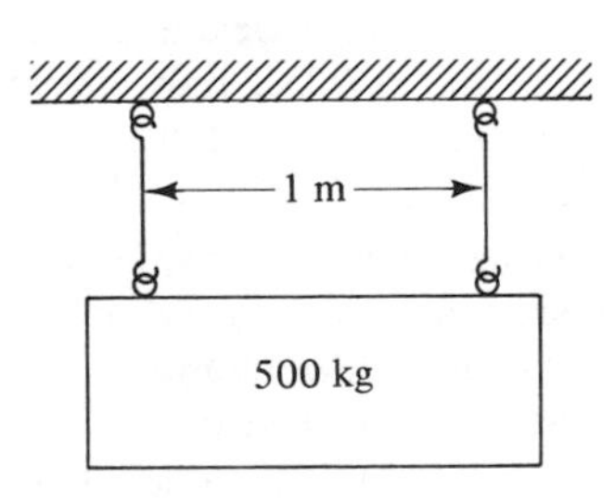

Problem 5.13

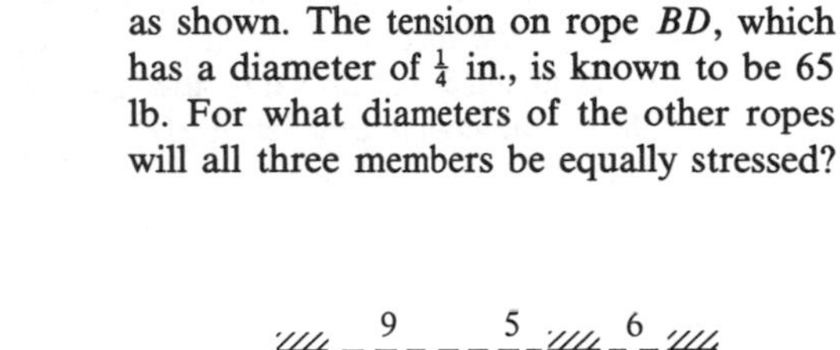

5.14. A 160-lb block is supported by three ropes, as shown. The tension on rope BD, which has a diameter of $\frac{1}{4}$ in., is known to be 65 lb. For what diameters of the other ropes will all three members be equally stressed?

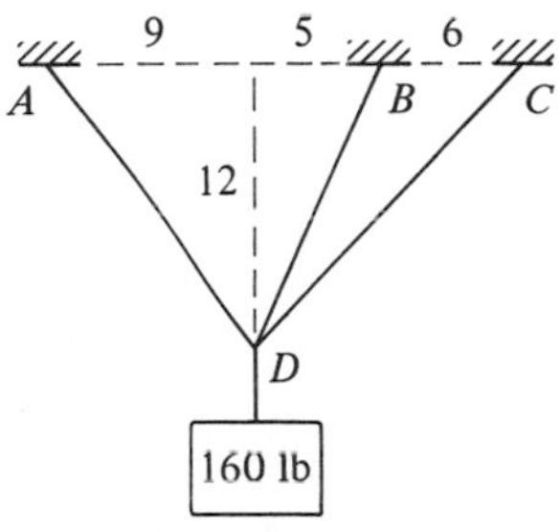

Problem 5.14

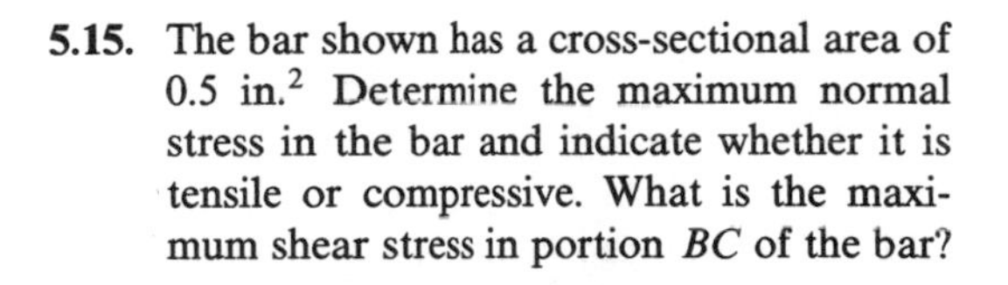

5.15. The bar shown has a cross-sectional area of 0.5 in.^{2} Determine the maximum normal stress in the bar and indicate whether it is tensile or compressive. What is the maximum shear stress in portion BC of the bar?

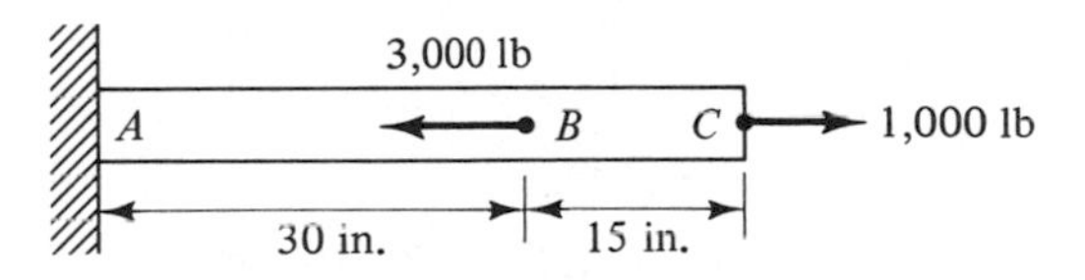

Problem 5.15

5.16. A 4 in. × 4 in. (nominal) wood post (see Table A.12) resting on a concrete base as shown is used to support a roof. The allowable compressive stresses in the wood and concrete are 1,800 psi and 1,250 psi, respectively. What is the maximum force F that can be supported if (a) the weights of the members are neglected, and (b) the weights of the members are included. The post weighs 30 lb/ft^3 and the specific weight of the concrete is 150 lb/ft^3.

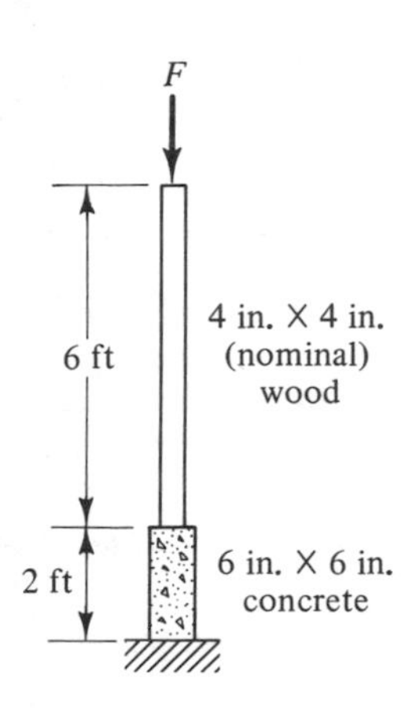

Problem 5.16

5.17. The bar shown has a square cross section 25 mm on a side. Determine the normal and shear stresses on (a) section *a-a* and (b) section *b-b*.

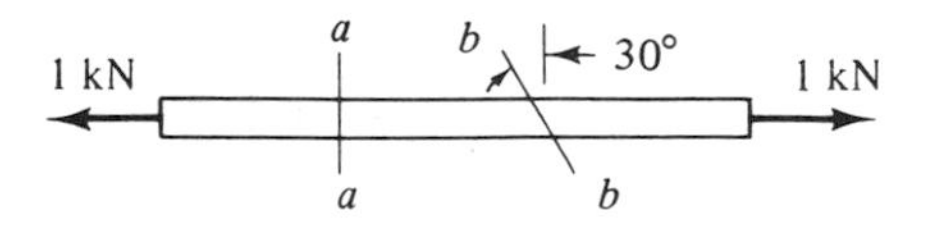

Problem 5.17

5.18. For the wood block shown, the allowable shear stress parallel to the grain is 1 MN/m^2 and the maximum allowable compressive stress in any direction is 4 MN/m^2. What is the maximum compressive load F the block can support?

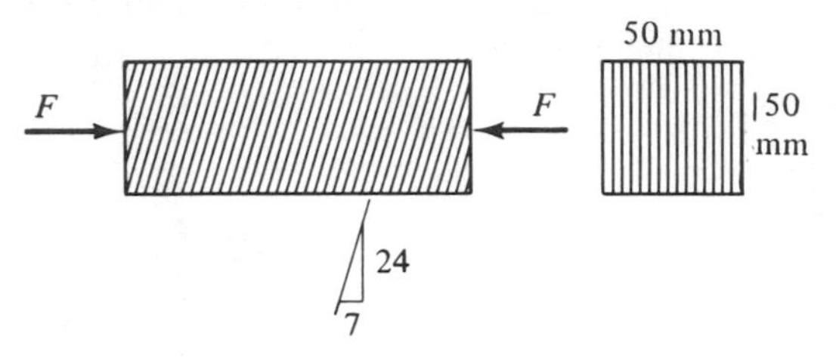

Problem 5.18

5.19. Determine the normal and shear stresses on the weld in the member shown as a function of the joint angle θ. Plot the results for $0° \leqslant \theta \leqslant 90°$.

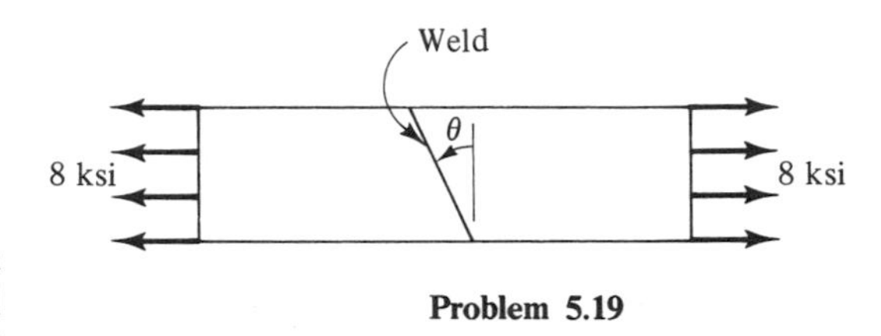

Problem 5.19

5.20. If the maximum allowable normal and shear stresses on the weld in Problem 5.19 are 6.0 ksi and 3.2 ksi, respectively, what is the minimum value of the joint angle θ that can be used?

5.21. Determine the required cross-sectional areas of members AB and AI of the truss shown if the allowable stresses are 140 MN/m^2 in tension and 70 MN/m^2 in compression. Assume that compression members do not buckle.

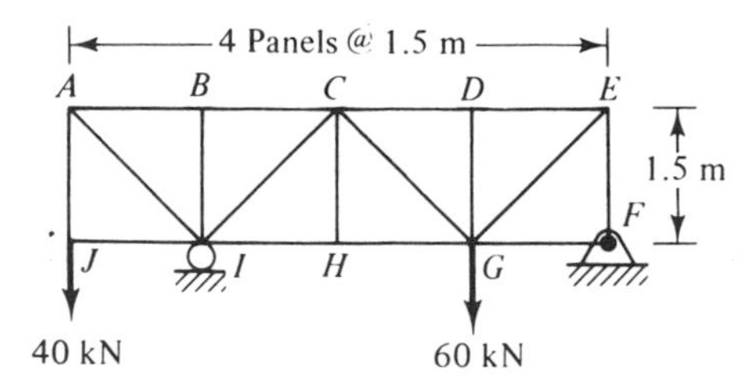

Problem 5.21

5.22. Same as Problem 5.21, except that the members of interest are CG and GH.

5.23. What is the maximum weight W that can be supported by the arrangement shown if the allowable tensile stress in cable AB is 8,000 psi and the allowable compressive stress in boom BC is 1,000 psi? The cable has a diameter of $\frac{1}{4}$ in. and the cross-sectional area of the boom is 2 in.^{2}

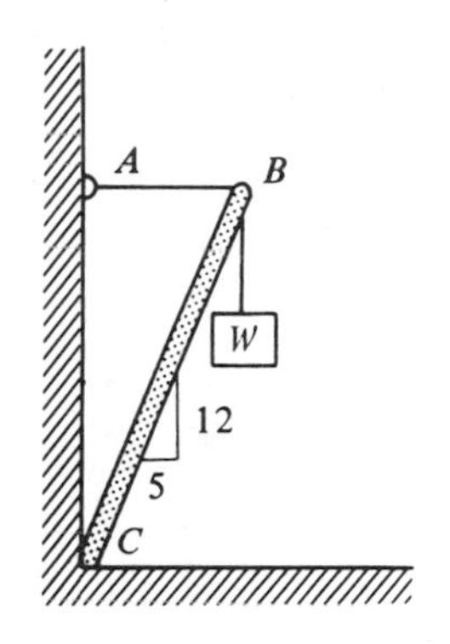

Problem 5.23

5.24. In Problem 5.23, determine the cross-sectional area of the boom for which both it and the cable will be simultaneously stressed to their limits. If the boom is to be made of pipe, what nominal diameter should be used (see Table A.7)?

5.25. Member BD of the frame shown has a cross-sectional area of 1.5 in.^{2} If the allowable stresses are 18 ksi in tension and 12 ksi in compression, what is the intensity q of the maximum distributed loading that can be supported? Assume that the other members are sufficiently strong and that the weights of all members are negligible.

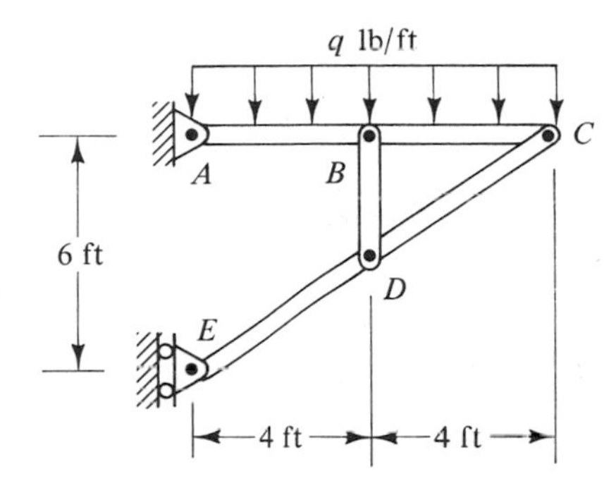

Problem 5.25

5.26. The uniform rod AB shown has a mass of 50 kg. It is supported by a ball and socket joint at B and a cord DC attached to its midpoint C. The wall at A is smooth. Three different cords with respective breaking strengths of 300 N, 600 N, and 1 kN are available. Which of the cords can be used?

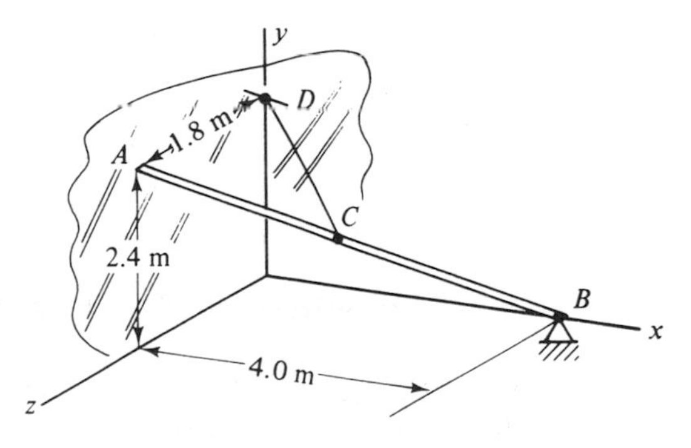

Problem 5.26

5.27. The uniform sign shown weighs 270 lb and is supported by a ball and socket joint at A and two slender rods BG and EF with negligible weight. If the rods have a diameter of $\frac{1}{8}$ in., to what tensile stresses are they subjected?

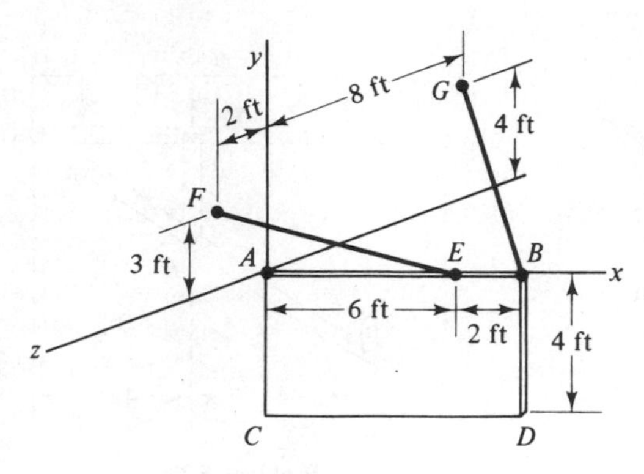

Problem 5.27

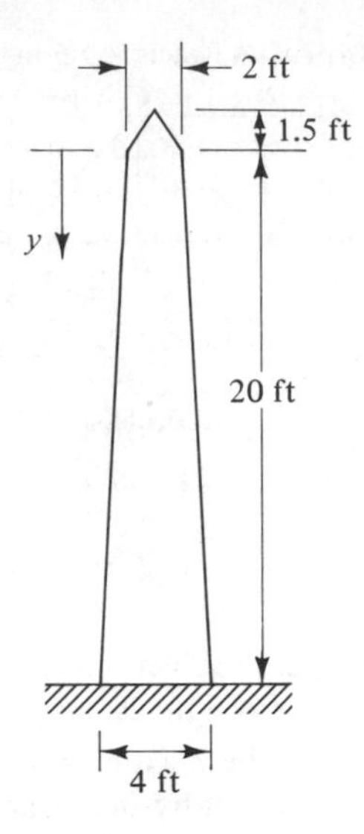

Problem 5.28

5.28. The granite obelisk shown has a square cross section and a pyramidal top. Determine the compressive stress over the cross section and plot its variation with the distance y from the top. Granite has a specific weight of approximately 165 lb/ft^3.

5.29. A body of revolution supports a compressive load F as shown. If the radius of the top surface is r_0 and the specific weight of the material is γ, how should the radius r vary with the distance y in order that the compressive stress at all cross sections be the same? Include the effect of the weight of the member.

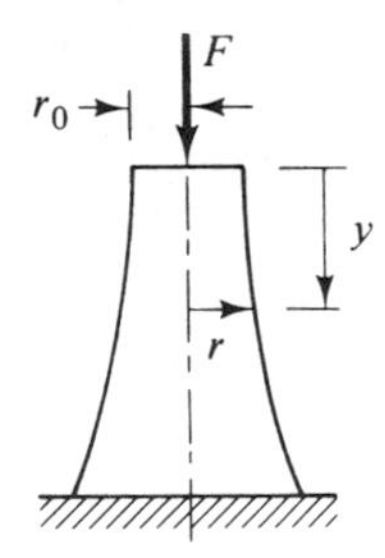

Problem 5.29

5.4 AVERAGE NORMAL, SHEAR, AND BEARING STRESSES

In many problems the actual stresses are difficult, if not impossible, to determine; therefore, it is necessary to work with average stresses. This situation is commonly encountered in structural joints where loads are transferred from one member to another. Average stresses do not necessarily provide a close estimate of the true stresses in such problems, but this can be accounted for by reducing the maximum allowable loads in order to compensate for inaccuracies in the analysis. This point will be considered in more detail in Section 5.10. The procedure for determining the average stresses is the same as that used for axially loaded members, as we shall show in the following.

For example, suppose that we wish to determine the significant stresses in the various members of the assembly shown in Figure 5.13(a). The assembly consists of two plates connected by a bolt and loaded in tension. The most important step in the analysis of problems of this type is to draw free-body diagrams of each of the members so that the forces acting upon them can be clearly identified. Figure 5.13(b) shows the major forces acting upon the plates and upon the bolt. Other forces that may be acting, such as those due to friction or tightening of the bolt, are usually of secondary importance and can be neglected. This is fortunate, for they are difficult to determine accurately.

Once the forces acting upon the various members have been determined, the next step is to decide just which stresses are of interest and on what planes. Obviously, the stresses that are likely to cause a failure of the joint are of primary concern. For example, since there is a possibility of shearing the bolt off along the interface between the two plates (plane *a-a*), the shear stress in the bolt on this plane is of interest.

Sectioning the bolt along plane *a-a* and considering the equilibrium of the top piece [Figure 5.13(c)], we obtain the shear force $V = F$. Thus, the average shear stress in the bolt on this plane is

$$\tau_{\text{avg}} = \frac{V}{A_{aa}} = \frac{F}{A_B}$$

where A_B is the cross-sectional area of the bolt.

In addition to shearing off the bolt, there are several other ways the connection can fail. For example, the plates may break because of the tensile stresses developed. These stresses are largest on plane *b-b* because the area available to carry the load is smallest there. Sectioning the plate along this plane [Figure 5.13(d)], we find that the average tensile stress is

$$\sigma_{\text{avg}} = \frac{P}{A_{bb}} = \frac{F}{t(w - d)}$$

where the dimensions t, w, and d are as defined in the figure. Since both plates in this example have the same dimensions, the stress will be the same in each.

It is also possible for a piece of a plate to tear out (shear) along planes *c-c* and *d-d* [Figure 5.13(e)], which would allow the plates to separate. Considering the piece removed, we see that the average shear stress on these planes is

$$\tau_{\text{avg}} = \frac{V}{A_{cc}} = \frac{V}{A_{dd}} = \frac{(F/2)}{ts}$$

where s is the distance from the center of the hole to the end of the plate.

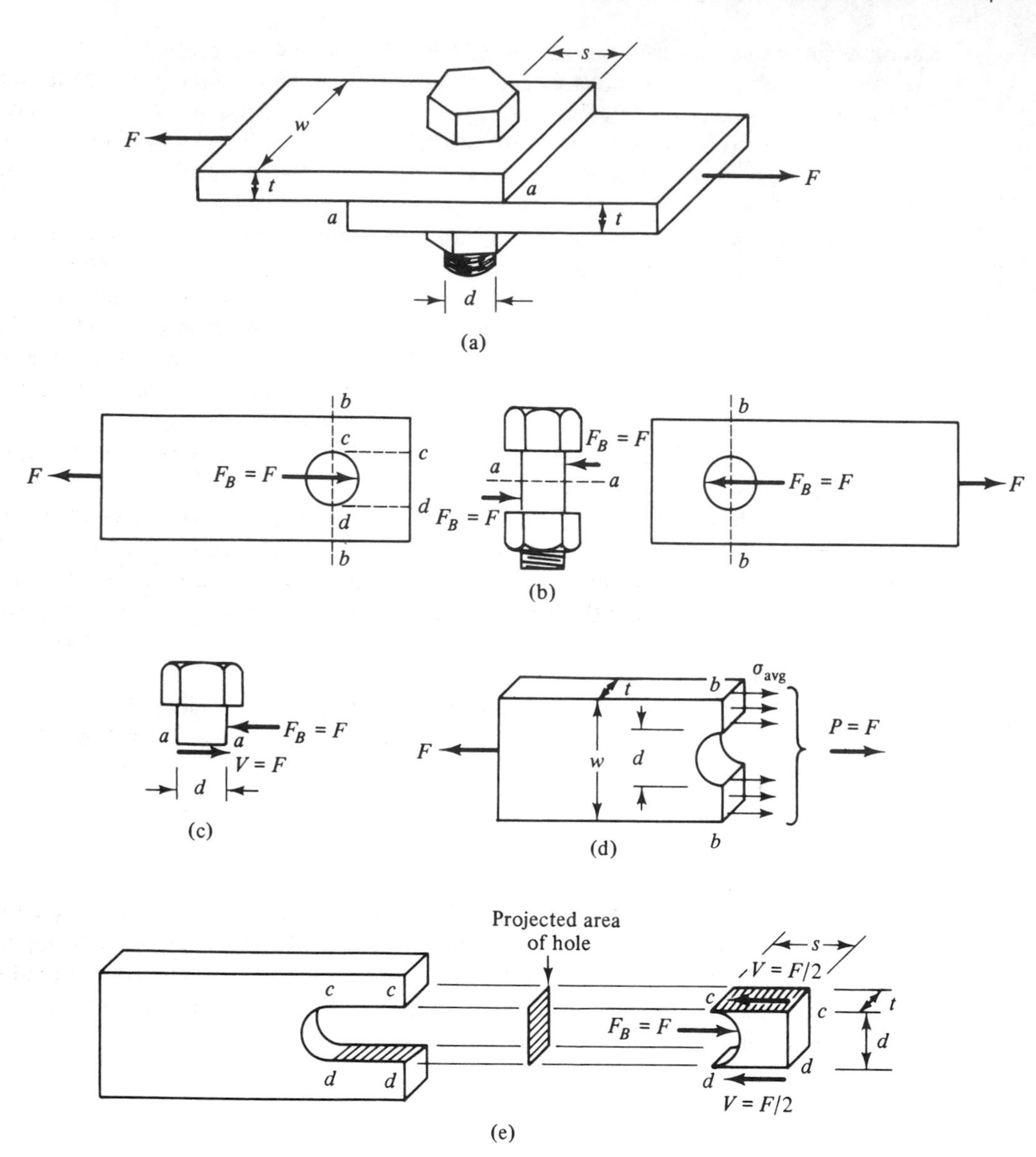

Figure 5.13

Another, and less obvious, possible mode of failure is excessive deformation of the plates due to the contact pressure between them and the bolt. This causes the holes to elongate, and it can result in a loose joint. In some applications this would not matter so long as the joint remained intact, but in others it would. For example, if this were a rivet joint in a pressure vessel, elongation of the holes could cause the vessel to leak.

The contact pressure generated when one solid body bears against another is called the *bearing stress*, which we shall denote by σ_B. The average bearing stress is the bearing force divided by the area over which it acts:

$$\sigma_{B_{\text{avg}}} = \frac{F_{\text{Bearing}}}{A_{\text{Bearing}}} \tag{5.7}$$

The actual bearing stresses in a connection depend upon many factors, including how snugly the bolt or rivet fits the hole. Thus, it is common practice to work with an average value of bearing stress obtained by dividing the bearing force by the projected area of the hole. Referring to Figure 5.13(e), we have

$$\sigma_{B_{\text{avg}}} = \frac{F_B}{A_{\text{proj}}} = \frac{F}{td}$$

where d is the diameter of the hole and t is the thickness of the plate.

This completes our analysis of the connection. The equations obtained relate the various stresses to the applied load and dimensions of the members. Note carefully the procedure used, for it applies to all average stress problems. If the stresses and planes of interest are not specified, they can usually be identified by examining the likely modes of failure. The procedure is further illustrated in the following examples.

In closing, we note that proper design and analysis of connections is a very important, but complex, subject. Various design codes and specifications have been developed to assist with these tasks. Here, we have considered only some of the basis concepts involved in joint analysis and design, as illustrations of the procedures for computing average stresses.

Example 5.5. Stresses in a Glue Joint. Three 20-mm by 100-mm boards are glued together to form the member shown in Figure 5.14(a). For an applied load of 15 kN, determine

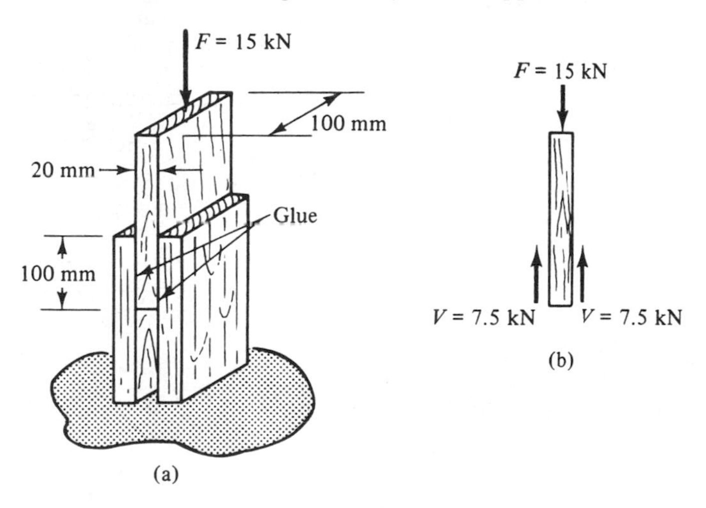

Figure 5.14

(a) the average shear stress in the glue and (b) the average bearing stress between the member and the floor.

Solution. (a) From the FBD of the top board [Figure 5.14(b)], we see that the shear force in each joint is $V = 7.5$ kN, acting over a square area 100 mm on a side. Thus, the average shear stress in the glue is

$$\tau_{avg} = \frac{V}{A} = \frac{7.5 \text{ kN}}{(0.1 \text{ m})^2} = 750 \text{ kN/m}^2 \qquad \textbf{Answer}$$

The average shear stress can also be computed by dividing the total shear force acting upon both joints, $2V$, by the total joint area, $2A$.

(b) A force analysis of the entire member shows that the total contact force on the floor is 15 kN acting over an area equal to the cross-sectional area of both bottom boards. Thus,

$$\sigma_{B_{avg}} = \frac{15 \text{ kN}}{2(0.02 \text{ m})(0.10 \text{ m})} = 3.75 \times 10^3 \text{ kN/m}^2 \text{ or } 3.75 \text{ MN/m}^2 \qquad \textbf{Answer}$$

Example 5.6. Stresses In a Connection. The steel truss member shown in Figure 5.15(a) supports a tension of 20 kips and is connected to a gusset plate by two rivets. (a) What

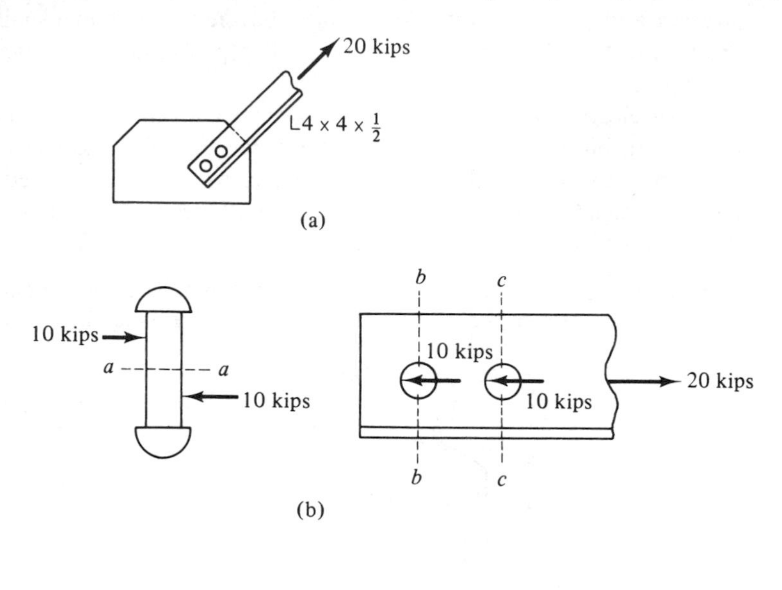

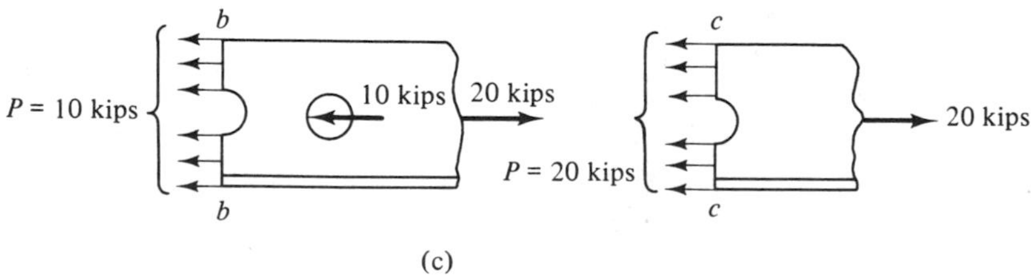

Figure 5.15

diameter rivets should be used if the allowable average shear and bearing stresses are 15 ksi and 27 ksi, respectively? (b) What is the maximum average tensile stress in the member for this size rivet?

Solution. (a) When there is more than one rivet or other connecting element, as in this example, one common procedure is to assume that the shear stress is uniformly distributed over the total area in shear. This implies that each of the connecting elements supports a shear force proportional to its cross-sectional area. Thus, the forces acting upon the angle and the rivets are as shown in Figure 5.15(b).

The rivets tend to shear off at the interface between the two members (plane *a-a*). The average shear stress in the rivets on this plane is

$$\tau_{avg} = \frac{V}{A_{rivet}} = \frac{10{,}000 \text{ lb}}{(\pi d^2/4) \text{ in.}^2} \leqslant 15{,}000 \text{ lb/in.}^2$$

from which we obtain for the diameter of the rivets

$$d \geqslant 0.92 \text{ in.}$$

The average bearing stress is the bearing force divided by the projected area of the rivet hole, which is the hole diameter times the thickness of the angle:

$$\sigma_{B_{avg}} = \frac{10{,}000 \text{ lb}}{(0.5 \text{ in.})\, d} \leqslant 27{,}000 \text{ lb/in.}^2$$

$$d \geqslant 0.74 \text{ in.}$$

Since both stress conditions must be satisfied, the required diameter is the larger of these two values (0.92 in.). Since rivets do not come in odd sizes, a 1-in. diameter would be used.

(b) The maximum tensile stress will occur at one or the other of the rivet holes, since the cross-sectional area is smallest there. The normal force is largest at the first hole [Figure 5.15(c)]; therefore, that is the location of the largest stress. From Table A.5, the cross-sectional area of the solid angle is 3.75 in.^{2} and its thickness is 0.5 in. Thus, the net area is

$$A_{net} = 3.75 \text{ in.}^2 - (1 \text{ in.})(0.5 \text{ in.}) = 3.25 \text{ in.}^2$$

and

$$\sigma_{avg} = \frac{P}{A_{net}} = \frac{20{,}000 \text{ lb}}{3.25 \text{ in.}^2} = 6{,}150 \text{ psi(tension)} \qquad \textbf{Answer}$$

PROBLEMS

5.30. A large picture with a mass of 50 kg is hung from a single nail driven into a wall as shown. What is the average shear stress in the nail if it has a diameter of 4 mm?

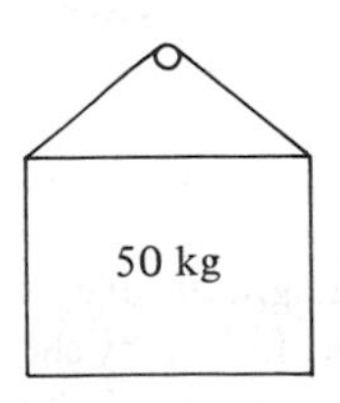

Problem 5.30

5.31. Two boards 100-mm wide and 15-mm thick are joined as shown with an adhesive with a shear strength of 500 kN/m^2. What is the required overlap L if the joint is to be as strong as the individual members, for which the allowable compressive stress is 20 MN/m^2?

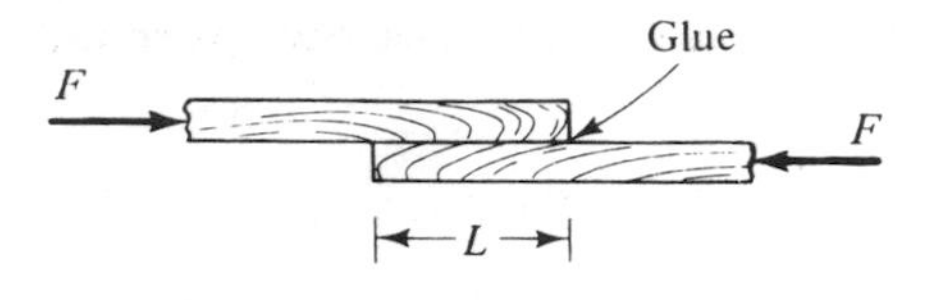

Problem 5.31

5.32. An aluminum pulley is keyed to the shaft of an engine by a square steel key, as shown. What is the minimum key size d required to transmit a torque of 240 lb · in. if the allowable shear stress in the key is 8 ksi and the allowable bearing stress on the aluminum is 12 ksi? Neglect any friction between the pulley and the shaft. The pulley hub is $\frac{3}{4}$ in. thick and the key is $\frac{3}{4}$ in. long.

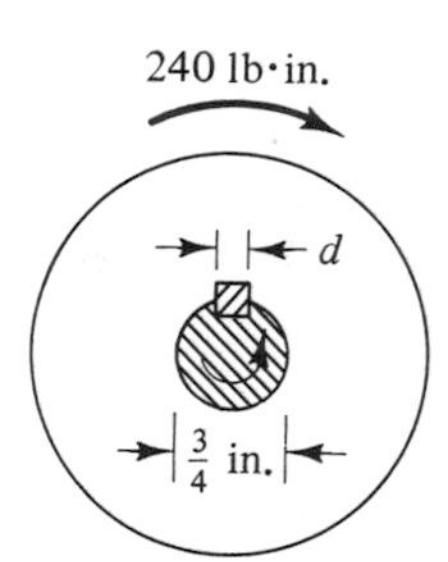

Problem 5.32

5.33. The design of fillet welds is based upon assumed failure by shear over the minimum section (throat) of the weld, as shown. Thus, show that a fillet weld can support a force per unit of length of $q = 0.707t\tau_a$, where t is the weld size and τ_a is the allowable shear stress. Determine q for a $\frac{1}{2}$-in. weld with $\tau_a = 18$ ksi.

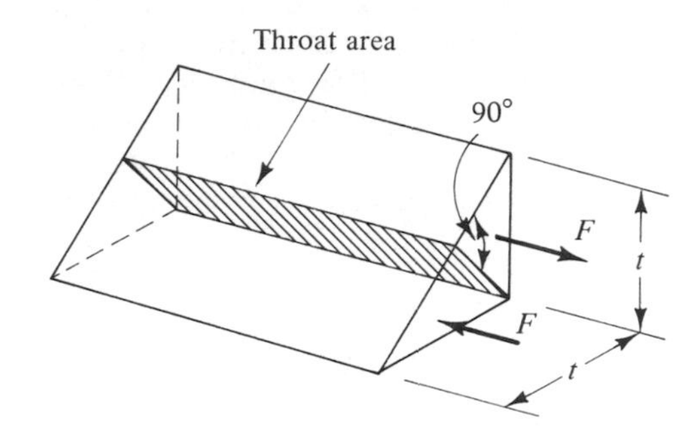

Problem 5.33

5.34. A simple truss is made of 150 mm × 150 mm timbers, as shown. If the allowable bearing and shear stresses are 6.0 MN/m^2 and 1.2 MN/m^2, respectively, what are the minimum required dimensions a and b? Assume the joint at B behaves like a pinned connection.

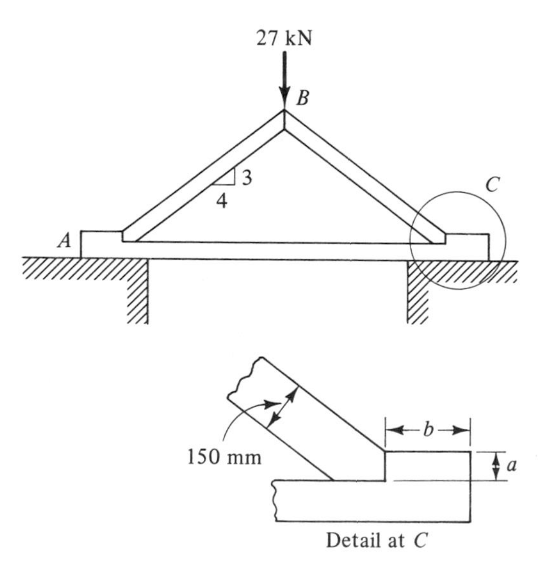

Problem 5.34

5.35. The shear strength of an adhesive is tested by using it to connect five 100-mm wide by 25-mm thick pieces of wood as shown and loading the assembly to failure. If the joints fail in shear when $F = 26.9$ kN, what is the average shear strength of the adhesive? What is the bearing stress between the test assembly and the top and bottom loading plates at the failure load?

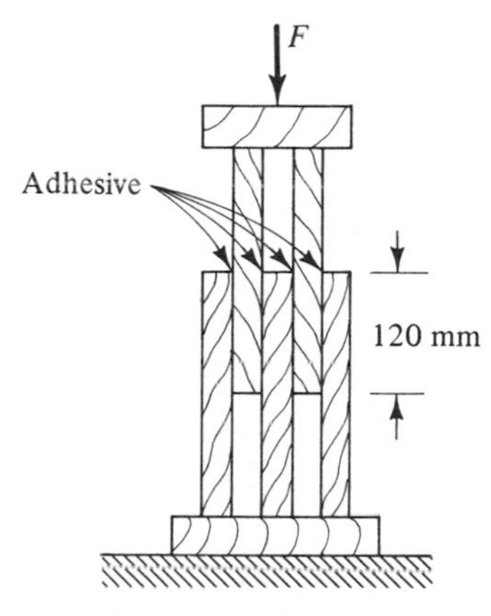

Problem 5.35

5.36. A buoy is connected to the buoy chain by a shackle and $\frac{3}{4}$-in. diameter shackle pin, as shown. The maximum tension on the buoy chain (during storms) is estimated to be 10 kips. What is the average shear stress in the shackle pin at this load and what is the bearing stress on the attachment lug? If the pin diameter decreases by 0.02 in. per year because of corrosion and wear, what is the predicted life of the connection? Assume an allowable shear stress of 16 ksi.

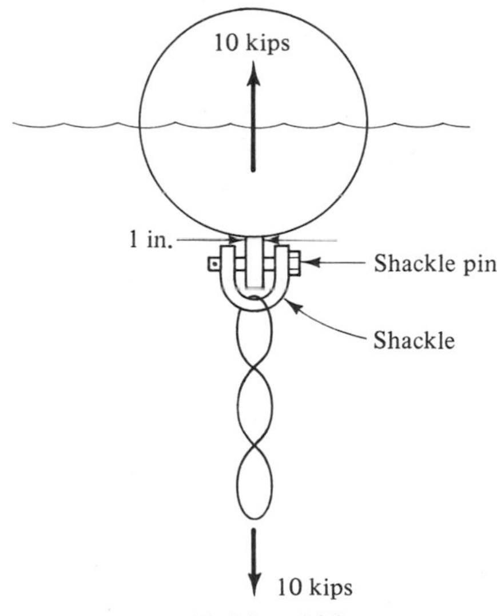

Problem 5.36

5.37. Two plates 3 in. wide and $\frac{1}{4}$ in. thick are connected by two $\frac{1}{2}$-in. diameter rivets. For the loading shown, determine (a) the shear stress in the rivets, (b) the maximum tensile stress in the plates, and (c) the bearing stress between the plates and the rivets.

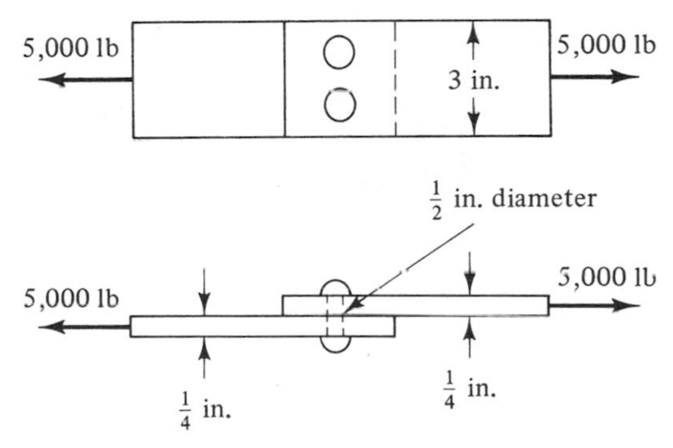

Problem 5.37

5.38. Same as Problem 5.37, except that the rivets are arranged in a longitudinal row.

5.39. A heavy trap door with a mass of 200 kg is supported by two door hinges. Determine

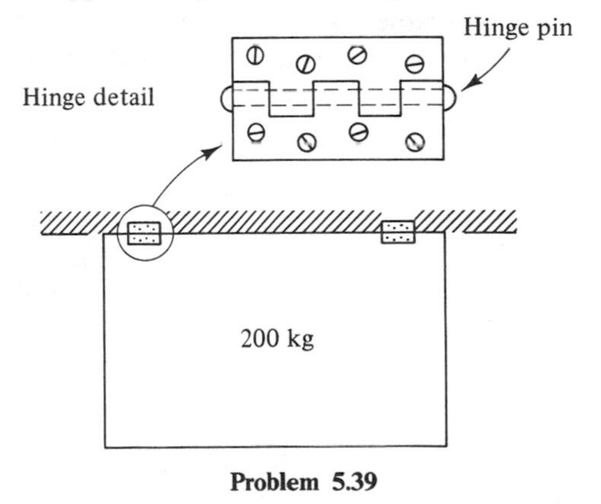

Problem 5.39

the average shear stress in the 4-mm diameter hinge pins when the door is hanging in the vertical position shown. What is the average shear stress in the screws used to attach the hinges if they each have a net cross-sectional area of 5 mm^2?

5.40. A bracket made of a piece of 89-mm × 89-mm angle section with a mass of 10.58 kg/m is attached to the flange of a 305-mm × 102-mm Universal beam with a mass of 33 kg/m by two 16-mm diameter bolts as shown. If the allowable shear and bearing stresses are 85 MN/m^2 and 255 MN/m^2, respectively, what is the maximum allowable load P? (See Tables A.8 and A.10.)

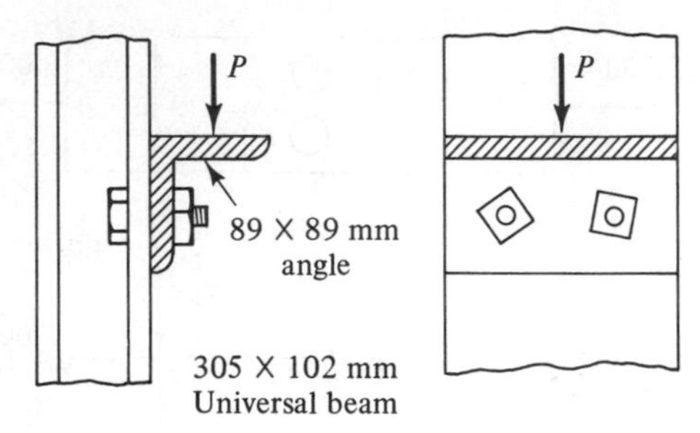

Problem 5.40

5.41. Three members, each 3 in. wide and $\frac{3}{4}$ in. thick, are to be joined as shown. The designer is debating whether to use one 1-in. diameter bolt or two $\frac{3}{4}$-in. diameter bolts arranged side by side for the connection. If the allowable tensile stress is 1.5 times the allowable shear stress and half the allowable bearing stress, which design will give the strongest connection? What is the minimum dimension s required to prevent tear-out in each case?

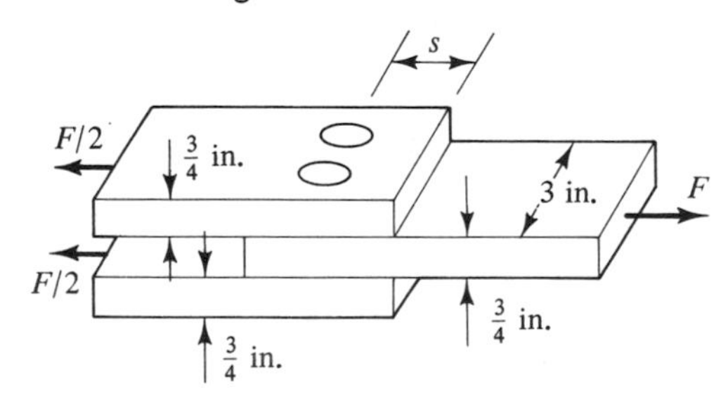

Problem 5.41

5.42. A bar is attached to a vertical support by two $\frac{1}{2}$-in. fillet welds (see Problem 5.33), as shown. By what percentage could the strength of the connection be increased by adding a transverse weld of the same size across the end of the bar? Assume that transverse welds have the same strength as longitudinal welds of the same size.

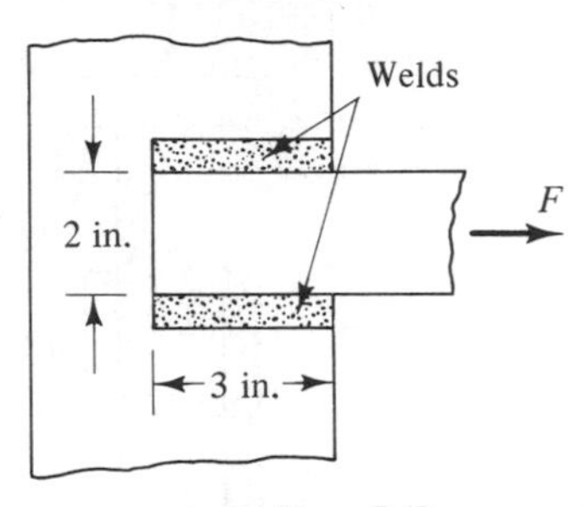

Problem 5.42

5.43. The plate shown is attached to the bracket with fillet welds for which q = 3,750 lb/in. (see Problem 5.33). In a proposed redesign, the welds are to be replaced with $\frac{1}{2}$-in. diameter bolts with an allowable shear stress of 15 ksi. How many bolts should be used if the capacity of the connection is not to be reduced?

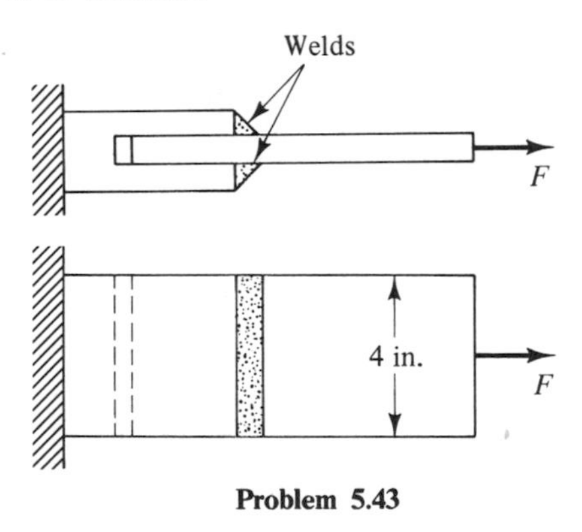

Problem 5.43

5.44. Same as Problem 5.43, except that the plate is 3 in. wide.

5.5 STATES OF STRESS AT A POINT

Inherent to the definition of normal and shear stresses is the fact that their values depend on the orientation of the planes upon which they act. This raises a crucial question, which we have so far carefully avoided. Since the possible cutting planes through any one point in a body are infinite in number, how can we ever hope to determine, describe, and depict the stresses present at that point? We shall consider this question in this section, along with some other important properties of stresses. First, however, do we need complete knowledge of the stresses existing at a point? Sometimes yes and sometimes no. For example, if we are interested only in the maximum normal stress in an axially loaded bar, we need consider only planes perpendicular to the axis of the member; any other stresses that may be acting on other planes are of no consequence. On the other hand, if we wish to determine the maximum pressure to which a steel tank can be subjected without failure, knowledge of all the stresses is required. Stresses on certain planes can be determined using the load-stress relationships that we shall develop for various types of members in the chapters to follow.

Stresses on a Material Element

If the stresses on three mutually perpendicular planes through a point are known, it can be shown that the stresses on any other plane through that point can be determined from them. A proof of this statement for two-dimensional problems will be given in Chapter 9. Proof for the three-dimensional case is beyond the scope of this text, but can be found in any number of more advanced books on mechanics of deformable bodies.

Consider three orthogonal planes through a point in a body [Figure 5.16(a)]. Let us assume that these planes correspond to x, y, z coordinate planes, although this need not be the case. It is convenient to show the various stresses existing at the point as acting upon an infinitesimal element of material surrounding the point and with sides parallel to the three cutting planes. This representation is illustrated in Figure 5.16(b), where the element is shown greatly enlarged for clarity.

Subscripts are used to distinguish between the various stresses. For example, σ_x, σ_y, and σ_z denote the normal stresses on planes perpendicular to the x, y, and z axes, respectively. For brevity, we shall refer to these planes as the x, y, and z faces of the element. The first subscript on the shear stress denotes the face of the element upon which it acts and the second subscript indicates the direction of the stress. Thus, τ_{xy} and τ_{xz} are the shear stresses acting upon the x face in the y and z directions, and so forth.

In Figure 5.16(b), only the stress components on three of the six faces of the element are shown. However, for equilibrium of the element, components of stress of equal magnitude, but opposite sense, must exist on the other three opposing faces. Actually, the magnitudes of the stresses may, and usually do, vary slightly

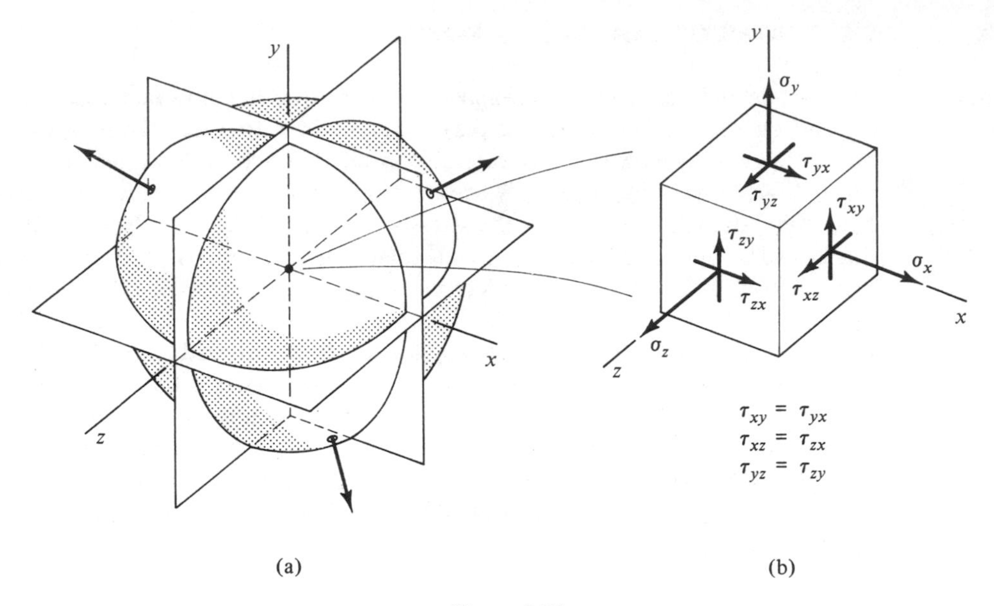

Figure 5.16

from one face of the element to another. These slight variations can be neglected for the types of problems considered in this text.

Normal stresses are considered positive if they are tensile and negative if they are compressive. A shear stress is considered to be positive if it and the outward normal of the plane on which it acts both point in either a positive or negative coordinate direction. Otherwise, the shear stress is taken to be negative. Notice that this sign convention for shear stress has meaning only after a coordinate system has been established. According to the conventions stated here, all the stresses shown on the element in Figure 5.16(b) are positive.

The stresses acting upon a material element define what is known as the *state of stress* at that point in the body. The most general situation possible involves a *triaxial state of stress*, wherein there are stresses on the element acting in all three

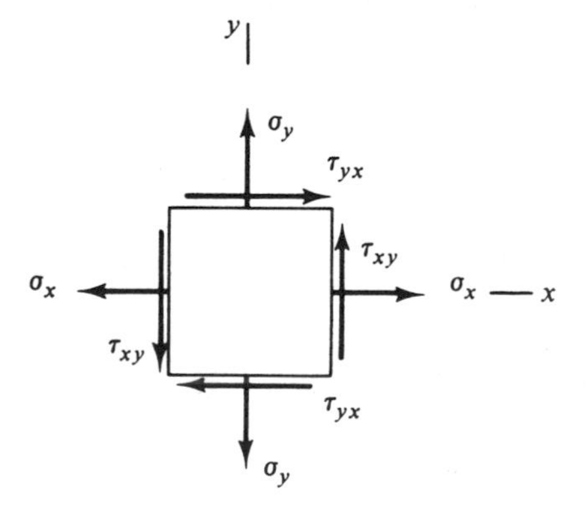

Figure 5.17

directions, as in Figure 5.16(b). Situations in which there are stresses acting in only two directions are referred to as *biaxial states of stress*. If there are stresses acting in one direction only, the state of stress is said to be *uniaxial*. The last two cases fall into a more general category known as a state of *plane stress*, wherein all the stresses acting lie in one plane. Most of the problems we shall consider fall into the plane stress category. In this case, a two-dimensional view of the element is sufficient to display the stresses [Figure 5.17].

Symmetry of the Shear Stresses

The shear stresses acting on a material element are not independent; rather, there is a definite relationship between them as a consequence of the moment conditions for equilibrium.

Suppose the element in Figure 5.16(b) has dimensions dx, dy, and dz. If we view the element from along the positive z axis and consider only those forces which contribute to the moment about that axis, we have the situation shown in Figure 5.18. Referring to the figure, we have

$$\overset{+}{\curvearrowleft}\sum M_z = (\tau_{xy}\, dy\, dz)\, dx - (\tau_{yx}\, dx\, dz)\, dy = 0$$

or

$$\tau_{xy} = \tau_{yx}$$

Similarly, if we consider moments about the other two coordinate axes, we find that $\tau_{xz} = \tau_{zx}$ and $\tau_{yz} = \tau_{zy}$.

The general result

$$\tau_{yx} = \tau_{xy} \qquad \tau_{zx} = \tau_{xz} \qquad \tau_{zy} = \tau_{yz} \tag{5.8}$$

implies that the shear stresses acting around an element in any one plane must be equal in magnitude. Furthermore, their sense must be such that they meet tip to tip or tail to tail. It also implies that shear stresses cannot exist alone. They are always accompanied by other shear stresses acting on orthogonal planes.

As a result of the symmetry conditions in Eq. (5.8), six components of stress $(\sigma_x, \sigma_y, \sigma_z, \tau_{xy}, \tau_{xz}, \tau_{yz})$ are required to completely define a general state of stress.

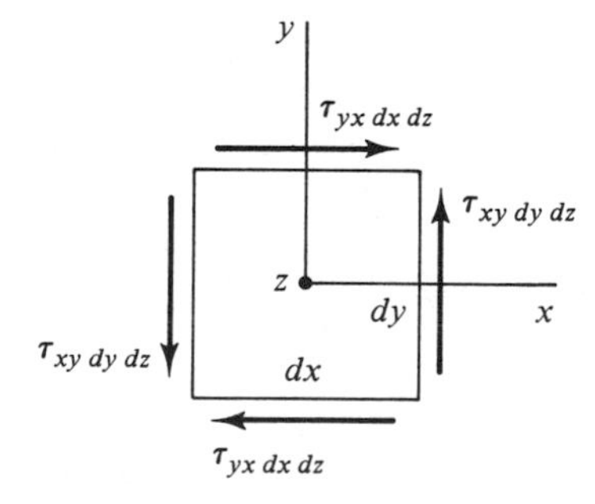

Figure 5.18

For plane stress, say in the *x-y* plane, only three components are required $(\sigma_x, \sigma_y, \tau_{xy})$.

5.6 DEFORMATION AND STRAIN

Now that we have developed some feel for stresses and their properties, we turn our attention to the *deformations* produced by the applied loads. By deformation, we mean a change in the geometry of a body. This can be a change in size, a change in shape, or both.

All bodies deform when loaded: A rubber band elongates when pulled, a diving board deflects downward when we stand at its end, a rubber ball flattens when squeezed, and a floor sags when we walk across it. The degree of deformation depends upon the resistive properties of the material and the magnitude and distribution of the applied loads. The deformations may be relatively large and visible to the naked eye, as in the case of a rubber band, or they may be very small, as in the case of a floor. For members made of very stiff materials like steel or concrete, the deformations may be so small that they can be detected only with sensitive instruments. As stated in the introduction to this chapter, we shall consider only those problems for which the deformations are small.

Normal and Shear Strains

We begin our discussion by considering a rod subjected to an axial load. Under a tensile load, the member will elongate, and under a compressive load, it will shorten (Figure 5.19). The amount of elongation or shortening is not really the best measure of the tendency of the applied load to produce deformation, however, because it depends upon the original length of the member. A long rod will elongate more than a short rod of the same size and material under the same loading, as can be easily demonstrated by using rubber bands of different lengths. A better measure of the tendency of the force to produce deformation is the change in length it causes per unit of length of the rod. This ratio, which is called the *average normal strain*, is independent of the length of the member.

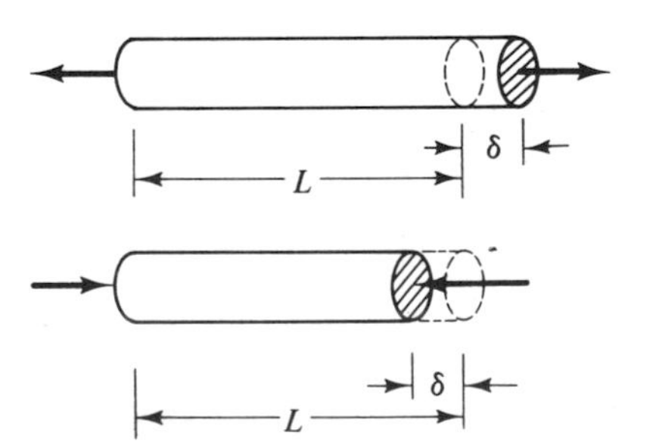

Figure 5.19

Denoting the elongation and normal strain by the Greek letters δ and ε, respectively, we have

$$\varepsilon_{avg} = \frac{\delta}{L} = \frac{L_{final} - L_{initial}}{L_{initial}} \tag{5.9}$$

where L is the length of the member. If the member increases in length, ε is positive and is called a *tensile strain*; if the member shortens, ε is negative and is called a *compressive strain*. This sign convention is consistent with that stated earlier for normal stresses, in that a tensile stress is associated with a tensile strain and a compressive stress is associated with a compressive strain. Physically, normal strains provide a measure of the change in size of a body when it is loaded.

Now let us consider the deformation due to a shear force. The assembly shown in Figure 5.20(a) consists of a metal plate bonded between two rubber pads attached to fixed surfaces. This arrangement is frequently used in machinery and engine mounts. When a horizontal force is applied to the plate, a shear force is exerted on the pads and the pads deform approximately as indicated.

The distance δ_s that one surface of a pad moves relative to the other [Figure 5.20(b)] is called the *shear deformation*. As in the case of the axially loaded rod, the amount of deformation depends upon the dimensions of the body. In this particular example, δ_s depends upon the dimension L. However, the angle of inclination of the side of the pad is independent of L and, therefore, provides a better measure of the tendency of the shear force to produce shear deformation.

Accordingly, we define the *average shear strain* as the change in angle between the two originally perpendicular sides of the pad, measured in radians. Denoting the shear strain by the Greek letter γ, we have

$$\gamma_{avg} = \frac{\pi}{2} - \alpha \tag{5.10}$$

where α is the angle between the two sides after loading [Figure 5.20(b)]. The average shear strain can also be defined in terms of the shear deformation. Referring to Figure 5.20(b) and recognizing that the change in angle will be small for small

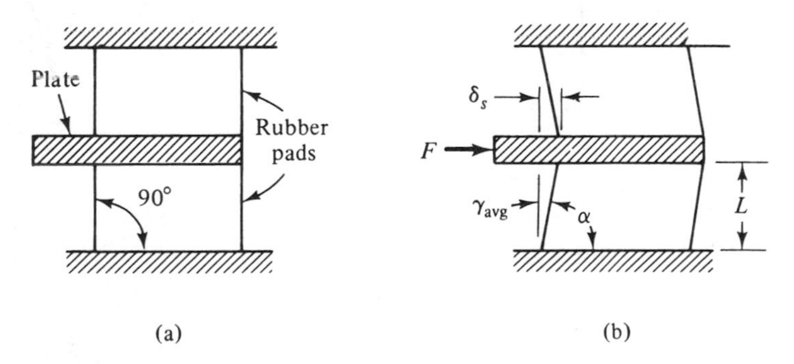

Figure 5.20

deformations, we obtain

$$\gamma_{avg} \approx \tan(\gamma_{avg}) = \frac{\delta_s}{L} \tag{5.11}$$

Shear strains are associated with shear stresses, and they are a measure of the change in shape, or distortion, of a body due to the applied loads. They are considered positive if the angle between the two perpendicular sides decreases and negative if it increases. This convention is consistent with that stated earlier for shear stresses, in that a positive shear stress corresponds to a positive shear strain.

The concepts of normal and shear strain developed from the preceding examples apply to any member. To show this, consider two perpendicular directions n and t at a point P within a body before it is loaded [Figure 5.21(a)]. Let PQ and PR be short line segments along these directions with lengths ΔL_n and ΔL_t, respectively. The individual segments are analogous to the rod in our previous example (Figure 5.19), and the pair of perpendicular segments is analogous to the sides of the shear pad shown in Figure 5.20(a).

Upon loading, the body will deform and points P, Q, and R will move to new positions P', Q', and R' [Figure 5.21(b)]. In general, the length of the line segments will increase or decrease and the angle between them will change. The deformations that may occur can be easily demonstrated by drawing lines on a balloon and then inflating it. Notice that the original straight lines may become curved, but segments $P'Q'$ and $P'R'$ can still be considered straight because they are short. The deformations have been shown greatly exaggerated in the figure.

If we let $\Delta\delta_n$ be the change in length of line segment PQ, the average normal strain at the point P associated with the direction n is

$$\varepsilon_n(P)_{avg} = \frac{P'Q' - PQ}{PQ} = \frac{\Delta\delta_n}{\Delta L_n} \tag{5.12}$$

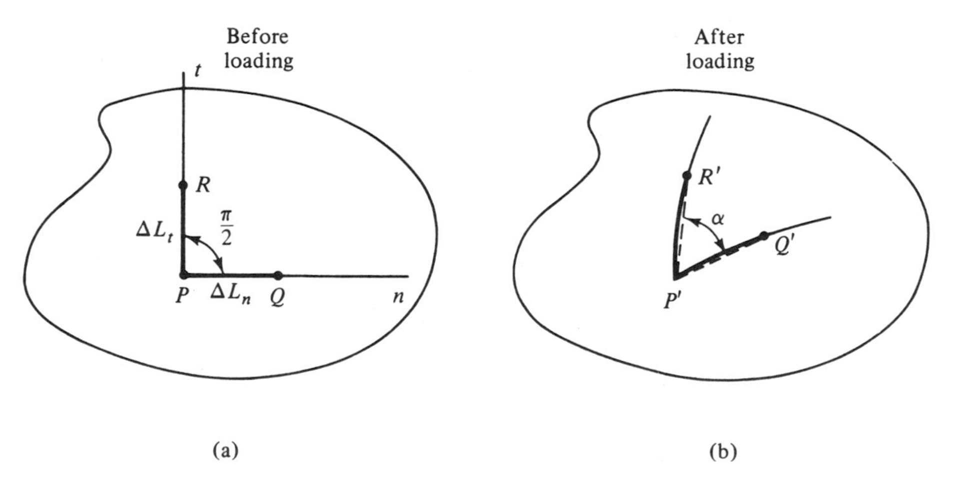

Figure 5.21

The *normal strain at a point* in the original direction n is defined as the limit of this ratio as ΔL_n tends to zero:

$$\varepsilon_n(P) = \lim_{\Delta L_n \to 0} \frac{\Delta \delta_n}{\Delta L_n} = \frac{d\delta_n}{dL_n} \tag{5.13}$$

The average normal strain is the same as the strain at a point, if the latter does not vary over the length of the line segment.

The average shear strain at P associated with the perpendicular directions n and t is

$$\gamma_{nt}(P)_{\text{avg}} = \frac{\pi}{2} - \alpha \tag{5.14}$$

where α is the angle between the chords $P'Q'$ and $P'R'$ [Figure 5.21b)]. The *shear strain at a point* is then defined as

$$\gamma_{nt}(P) = \frac{\pi}{2} - \lim_{\substack{\Delta L_n \to 0 \\ \Delta L_t \to 0}} \alpha \tag{5.15}$$

Average shear strains are the same as shear strains at a point if straight lines before loading remain straight after loading. In this case, the angle α in Eqs. (5.14) and (5.15) is independent of the length of the line segments considered.

It is important to note that normal strains are always measured in a particular direction and shear strains are always measured between two perpendicular directions. Accordingly, the values of the strains at a point depend upon the directions considered, just as the values of the stresses at a point depend upon the orientation of the plane considered. Of course, the strains may also vary from point to point within the body. The directions associated with the strains are indicated by n and t, or other appropriate subscripts. For example, ε_{AB} signifies the strain of a particular line element AB.

Strains are dimensionless quantities and have the same value no matter what units of length are used. Since strains are usually quite small in magnitude, it is customary to express them in terms of *micro strain*. One unit of micro strain, which we shall denote by the symbols $\mu\varepsilon$, is equal to a strain of 10^{-6}. For example, 250 $\mu\varepsilon$ represents a strain of 250×10^{-6}. Strains can also be expressed as a percentage.

State of Strain at a Point

Equations (5.12)–(5.15) hold for any two perpendicular directions, since the choice of the directions n and t is arbitrary. If we apply them to the x, y, and z coordinate directions at a point in a body, the resulting six components of strain $(\varepsilon_x, \varepsilon_y, \varepsilon_z, \gamma_{xy}, \gamma_{xz}, \gamma_{yz})$ define the *state of strain* at that point, just as the six stress components define the state of stress. If all the deformations occur in one plane, the strains associated with the third direction will be zero. This case is referred to as *plane strain*. For example, $\varepsilon_z = \gamma_{xz} = \gamma_{yz} = 0$ for plane strain in the xy plane, in which case the state of strain is completely defined by the three strain components ε_x, ε_y and γ_{xy}.

The state of strain at a point in a body can be displayed by showing the associated deformations of an infinitesimal element of material located at that point. Figure 5.22 shows the representation for positive and negative normal and shear strains.

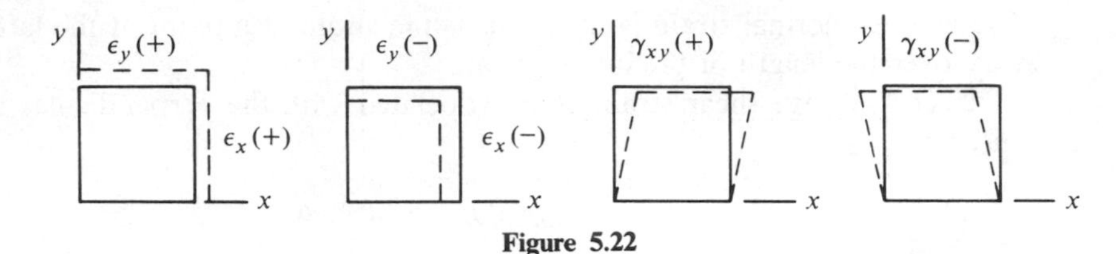

Figure 5.22

Example 5.7. Normal Strain of a Wire. The frame shown in Figure 5.23(a) consists of three rigid bars hinged together at the ends and braced with an elastic cord AC. Upon loading, member BC moves 50 mm to the right. Determine the average normal strain of the cord.

Solution. Referring to Figure 5.23(b), we have for the final length of the cord

$$A'C' = \sqrt{(1{,}550 \text{ mm})^2 + (1{,}500 \text{ mm})^2} = 2{,}157 \text{ mm}$$

In this calculation the vertical coordinate of point C' is taken equal to the length of bar DC. This is a good approximation, since the angle of inclination θ of the bar is very small. Of course, the actual coordinate of C' can be computed; it is 1,499.2 mm. The initial length of the cord is

$$AC = 1{,}500\sqrt{2} \text{ mm} = 2{,}121 \text{ mm}$$

so the average normal strain is

$$\varepsilon_{AC} = \frac{A'C' - AC}{AC} = \frac{2{,}157 \text{ mm} - 2{,}121 \text{ mm}}{2{,}121 \text{ mm}}$$
$$= 0.017 \text{ or } 17{,}000\ \mu\varepsilon \text{ (tension)} \qquad \textbf{Answer}$$

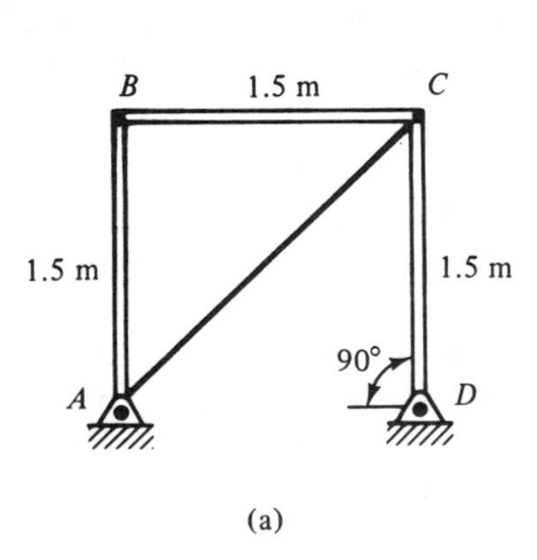

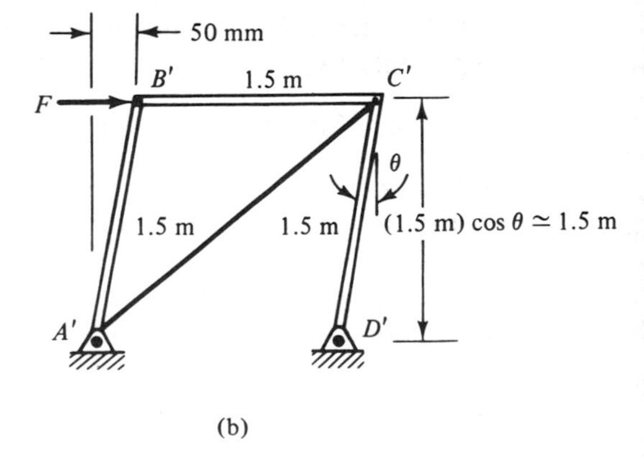

Figure 5.23

Example 5.8. Shear Strain in a Plate. A rectangular plate [Figure 5.24(a)] distorts into a parallelogram [Figure 5.24(b)] when loaded. Determine the shear strain γ_{xy} at points A and C. Are the values computed average shear strains or the actual shear strains at these points?

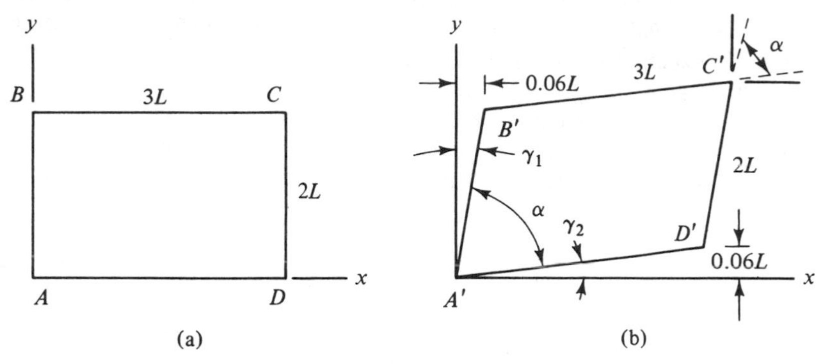

Figure 5.24

Solution. $\gamma_{xy}(A)$ is the change in angle between the positive x and y directions at A, or

$$\gamma_{xy}(A) = \gamma_1 + \gamma_2$$

Since the deformations are small,

$$\gamma_1 \approx \sin \gamma_1 = \frac{0.06L}{2L} = 0.03$$

$$\gamma_2 \approx \sin \gamma_2 = \frac{0.06L}{3L} = 0.02$$

and

$$\gamma_{xy}(A) = 0.05 \text{ or } 5\%. \qquad \textbf{Answer}$$

The angle between the x and y axes decreases; so γ_{xy} is positive.

The shear strain at C is the change in angle between the positive x and y directions there. From Figure 5.24(b), it can be seen that

$$\gamma_{xy}(C) = \gamma_{xy}(A) \qquad \textbf{Answer}$$

In fact, γ_{xy} has the same value at all points in the plate.

Since the sides of the plate remain straight, the angles γ_1 and γ_2 do not depend upon the lengths of the line segments used to compute them. Thus, the values computed are the actual shear strains.

PROBLEMS

5.45. A 10-ft-long tow rope stretches 6 in. when used to move a disabled vehicle. What is the average strain of the rope?

5.46. A rubber band with a total length of 4 in. is stretched over a rolled newspaper with a diameter of 2 in. What is the average normal strain of the band?

5.47. What is the change in length of a bar originally 2 m long if the average longitudinal strain is (a) 2,500 $\mu\varepsilon$ and (b) -320 $\mu\varepsilon$?

5.48. A metal sphere is heated until the circumferential strain is 1,500 $\mu\varepsilon$. What is the percentage change in volume of the sphere?

5.49. A rigid horizontal bar hinged at one end is supported by a vertical wire, as shown. When the bar is loaded it is observed to rotate downward through an angle of 1°. What is the normal strain of the wire?

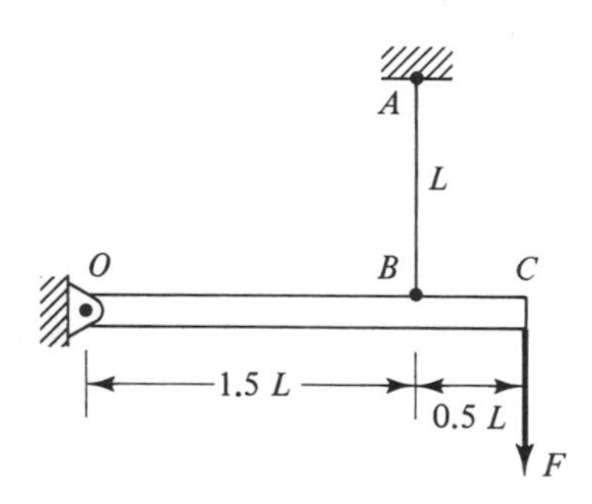

Problem 5.49

5.50. A rigid bar AB is supported by an elastic cord AC, as shown. What is the normal strain of the cord if the bar rotates downward through an angle of 20° when it is released?

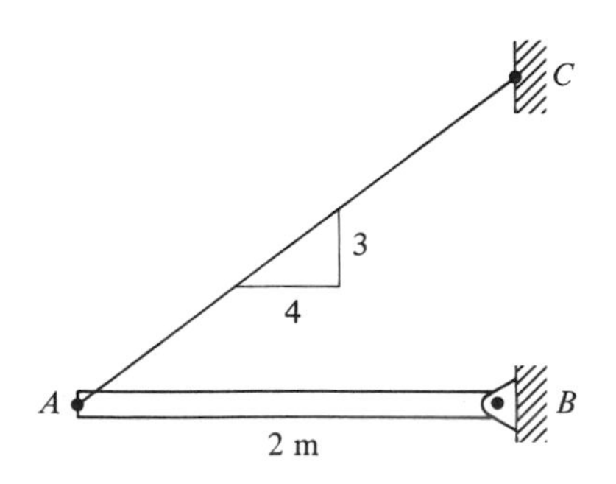

Problem 5.50

5.51. When a block is suspended from the three cables shown, point D is observed to move downward a distance of $0.01L$, where L is the length of the center cable. What is the ratio of the normal strains of the outer cables and that of the middle cable?

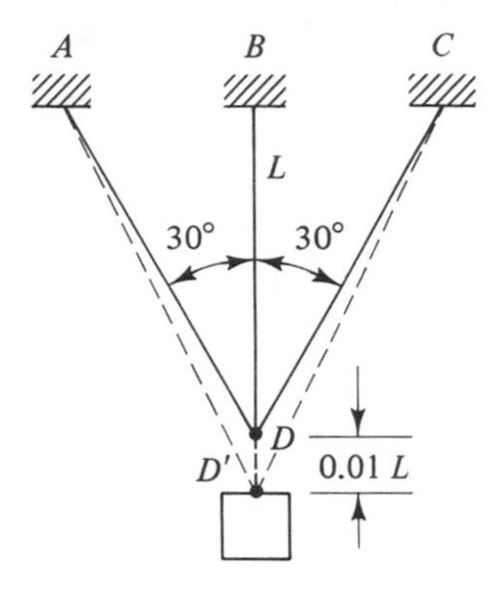

Problem 5.51

5.52. When the load is applied to the member shown, point B is observed to move downward 1 mm. What are the average normal strains in portions AB and BC of the member? What is the normal strain of the entire member ABC?

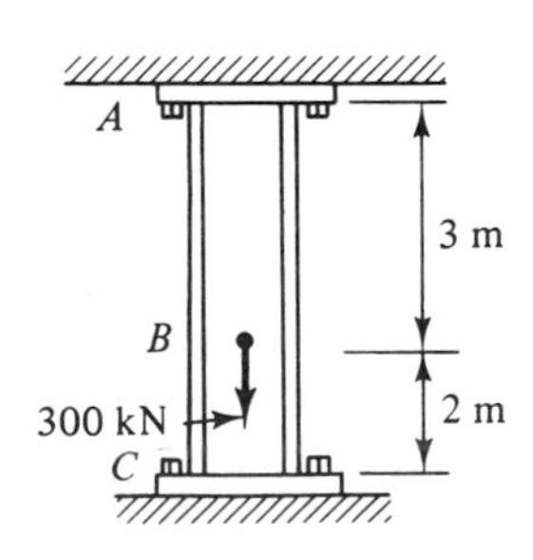

Problem 5.52

5.53. A vertical wire BD is attached to the mid-point of a stretched horizontal cable AC, as shown. When a block is suspended from end D of the wire, B moves to B' and D moves to D'. What is the displacement DD' of the block if the strain of the horizontal cable is 5,000 $\mu\varepsilon$ and the strain of the vertical wire is 1.0%?

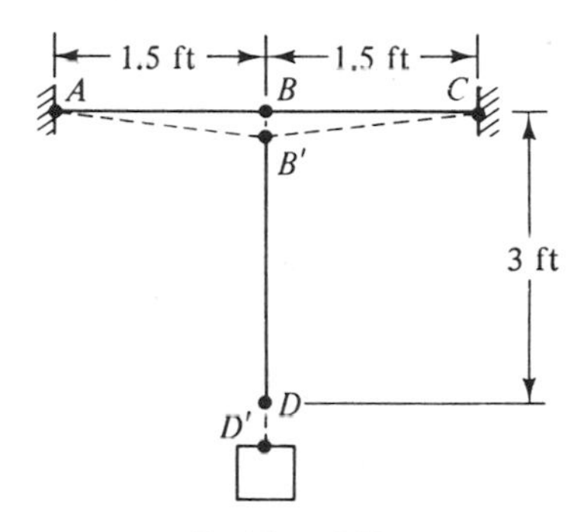

Problem 5.53

5.54. A small square marked on a shaft before loading deforms as shown when the member is twisted. What is the shear strain γ_{xy}?

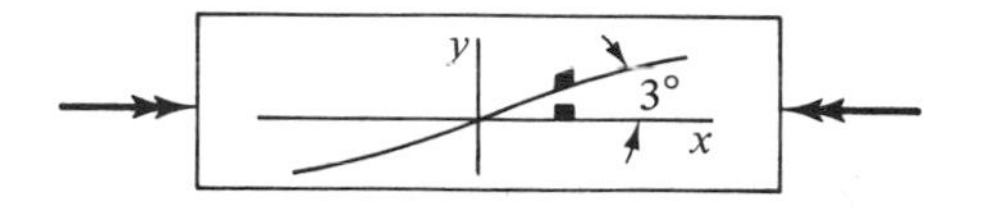

Problem 5.54

5.55. A motor mount made of a rectangular pad of rubber deforms as shown when loaded. What is the shear strain γ_{xy}?

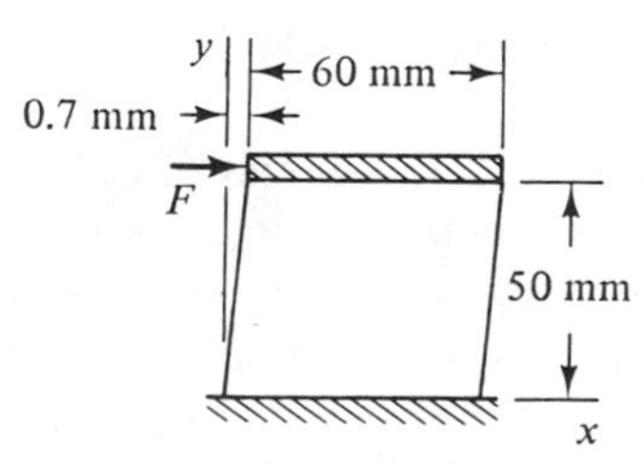

Problem 5.55

5.56. When a rectangular block is subjected to a shear force, the originally straight sides become curves described by the equation $x = ay^3/h^3$, where a and h are as defined in the figure. Determine the shear strain γ_{xy} at (a) any point x, y in the block, (b) the origin $x = y = 0$, and (c) the center of the block $x = b/2$, $y = h/2$.

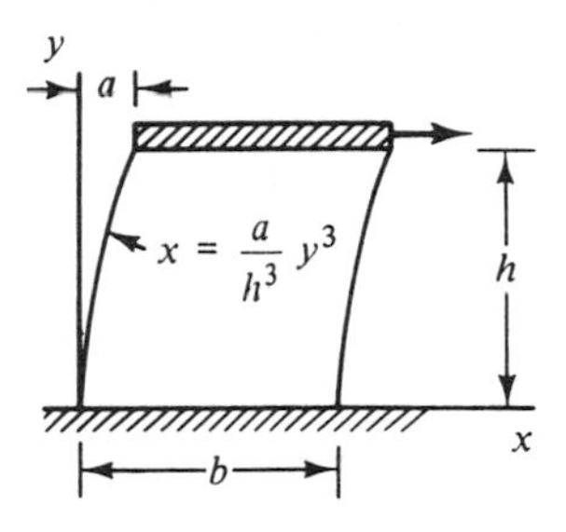

Problem 5.56

5.57. A square marked on a plastic bar deforms as shown when the member is subjected to a tensile loading. If diagonals AC and BD were originally 2.0 in. long, determine the strains ε_x, ε_y, and γ_{xy}.

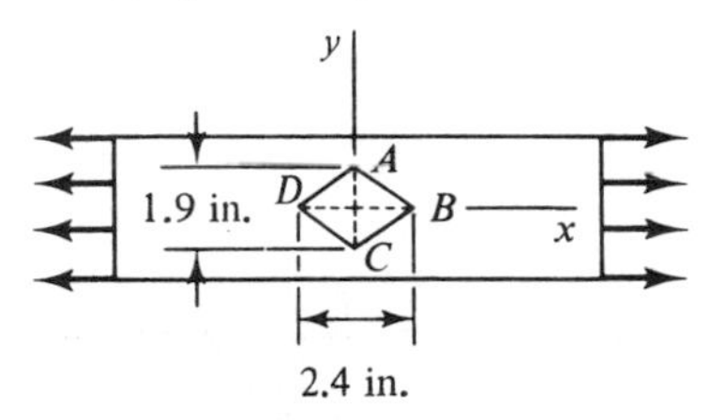

Problem 5.57

5.58. When deformed, the rectangle shown distorts into a parallelogram. Sides AB and CD elongate 0.02 mm and rotate 0.002

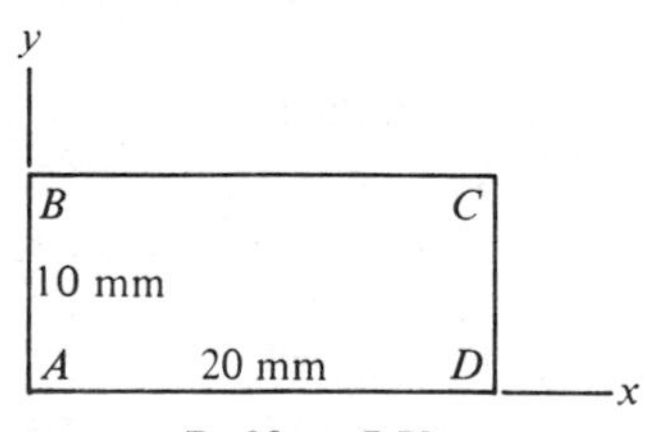

Problem 5.58

radian clockwise; sides AD and BC elongate 0.02 mm and rotate 0.003 radian clockwise. Determine the strains ε_x, ε_y, and γ_{xy}.

5.59. The deformed shape of rectangle $OABC$ shown is indicated by the dashed lines. If the normal strains are $\varepsilon_x = -200\ \mu\varepsilon$ and $\varepsilon_y = 400\ \mu\varepsilon$, what are the dimensions a and b and the shear strain γ_{xy}?

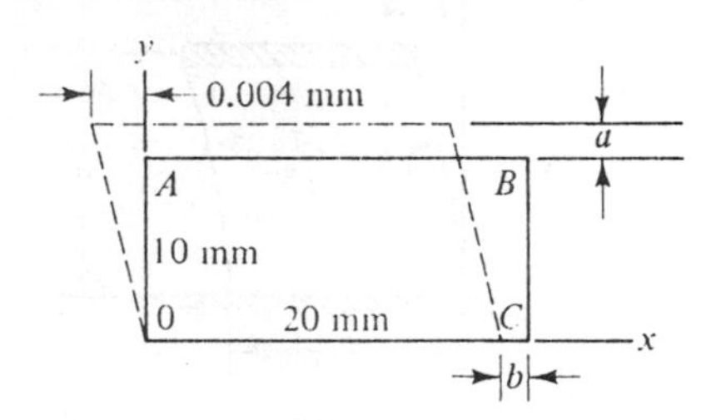

Problem 5.59

5.60. A thin triangular plate deforms as shown when loaded. Determine the shear strain γ_{nt} and the normal strains ε_n and ε_t.

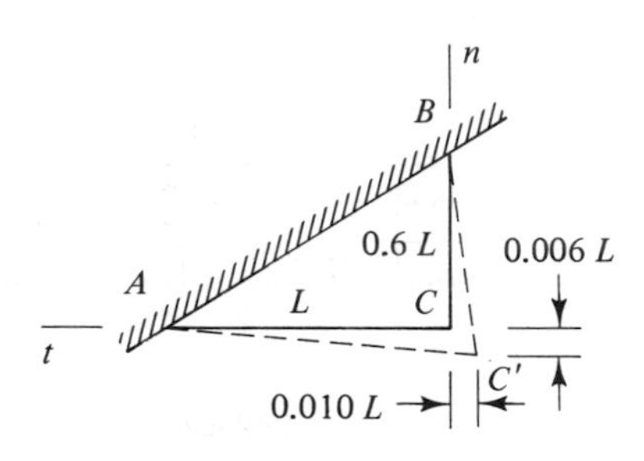

Problem 5.60

5.61. The isosceles triangle ABC shown deforms into triangle $A'B'C'$ when loaded, with point D' at the midpoint of side $A'C'$. Determine the normal strains ε_n and ε_t and the shear strain γ_{nt} at point D. The displacements are shown greatly exaggerated for clarity.

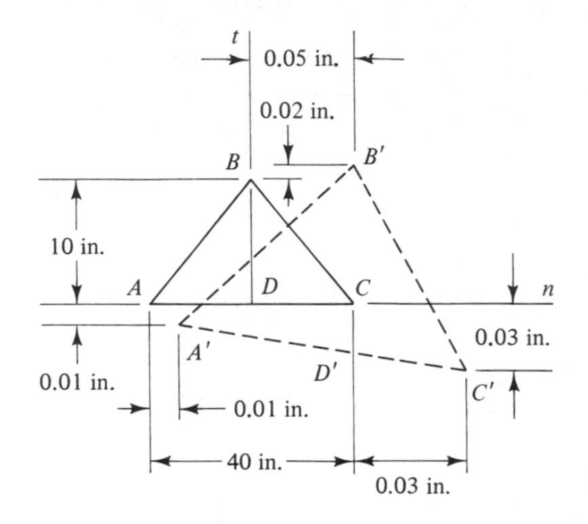

Problem 5.61

5.7 MATERIAL TESTS AND DATA REPRESENTATION

The remaining basic ingredient in our study of the mechanics of deformable bodies is the resistive properties of materials. These properties relate the stresses to the strains and can only be determined by experiment. In this and the following two sections, we shall consider the behavior of some common engineering materials under tensile, compressive, and shear loadings and the tests for determining this behavior. Our primary concern will be the description of observed material responses. The metallurgical and other physical and chemical factors associated with these responses will not be considered in any detail. They are discussed in texts on materials science.

Tensile and Compressive Tests

One of the simplest tests for determining mechanical properties of a material is the so-called *tensile test*. In this test, a load is applied along the longitudinal axis of a rod or bar, usually with rectangular or circular cross section. The applied load and the resulting elongation of the member are measured. The load is then increased and the process repeated until the desired load levels are reached or the member breaks. The procedure used in *compressive tests* is basically the same, except that the member is loaded in compression.

There are many different testing machines available for performing tensile and compressive tests. All of them involve a fixed and a movable surface between which the specimen can be stretched or compressed. The testing machine shown in Figure 5.25 is typical. The top and bottom crossheads are driven upward as a unit by a

Figure 5.25 Tinius Olsen 60,000 lb Standard Super "L" Universal Testing Machine. (*Courtesy* Tinius Olsen Testing Machine Co., Inc.)

hydraulic cylinder, while the middle crosshead remains stationary. Thus, a specimen gripped between the top and middle heads will be loaded in tension; one placed between the central and bottom heads will be compressed. The load applied to the specimen is determined by components internal to the machine and is displayed on a load dial. Many of the testing machines in use today are electronically controlled and can subject a specimen to various prescribed loadings or deformation histories. Automatic data recording systems are also widely used.

Standards governing specimen dimensions and test procedures have been developed by the American Society for Testing Materials (ASTM) and other agencies to assure some degree of uniformity between data obtained in different laboratories. A typical tensile specimen is shown in Figure 5.26. It has an enlarged portion on each end for gripping in the test machine and a uniform central test section of smaller size. The test section is designed as an axially loaded member; therefore, the stresses and strains within it can be easily computed. Before the specimen is loaded, a known distance called the *gage length* is marked off in the center of the test section. It is the elongation or shortening of this gage length which is measured. One-inch and 2-inch gage lengths are common for metal specimens, but other lengths can be used.

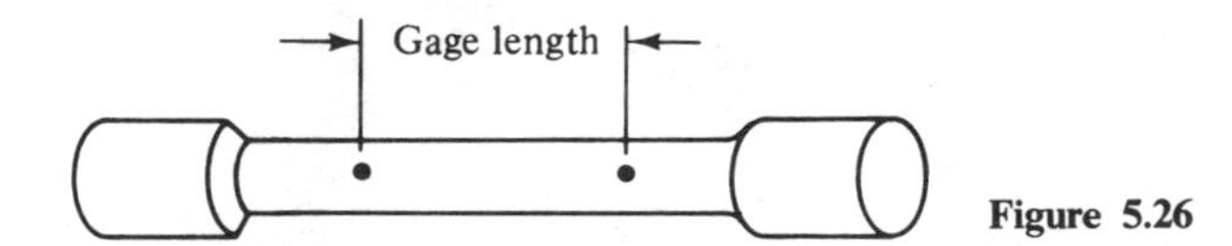

Figure 5.26

The deformations are usually very small, particularly at low load levels, and must be magnified in order to be measured accurately. There are numerous instruments available for making these measurements. The instrument shown in Figure 5.27 is held in place by leaf springs, and it has two knife edges 2 in. apart that establish the gage length. One knife edge remains fixed and the other moves along with the specimen as it deforms. The deformations are magnified electronically. The movement of the knife edge moves the core of the differential transformer, which creates an electrical signal proportional to the deformation. This signal can be used for automatic recording of the load-deformation data. Other devices and techniques for measuring deformations will be discussed in Chapter 11.

Stress-Strain Diagrams

Load-deformation data obtained from tensile or compressive tests do not give a direct indication of the material behavior, because they also depend upon the specimen geometry. The load required to produce a certain amount of deformation obviously will depend upon the size of the member. Similarly, the change in the gage length at a given load will depend upon the gage length used; the longer the gage length, the greater the change. This dependence upon specimen geometry is evident

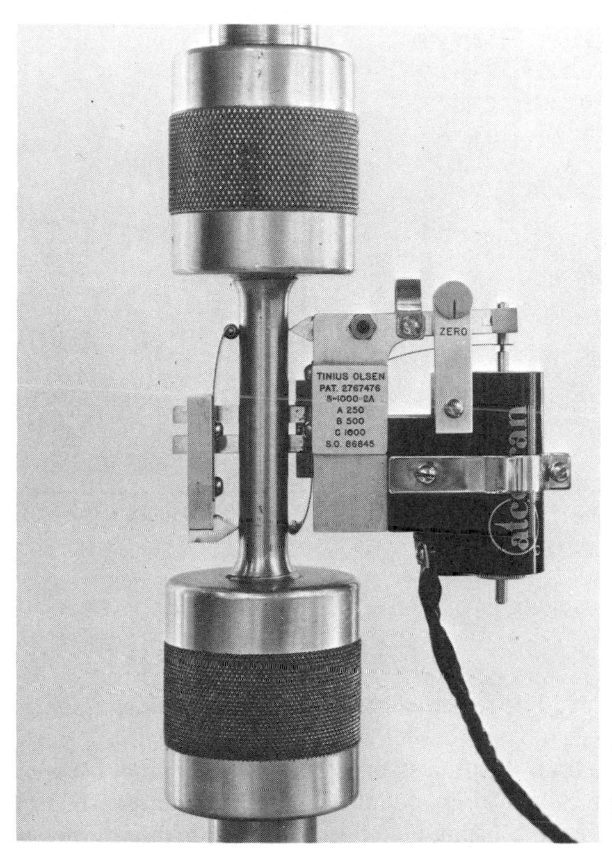

Figure 5.27 Tinius Olsen Model S-1000 Averaging Type Extensometer shown in place on a standard specimen. (*Courtesy* Tinius Olsen Testing Machine Co., Inc.)

in Figure 5.28(a), which shows a portion of the load-deformation curves for aluminum tensile specimens with different cross-sectional areas and gage lengths. All specimens were taken from the same bar of material.

Figure 5.28(b) shows the same data with the loads and deformations converted to stresses and strains via the relationships

$$\sigma = \frac{F}{A} \qquad \varepsilon = \frac{\delta}{L}$$

Here, σ is the normal stress on a plane perpendicular to the longitudinal axis of the specimen, ε is the normal strain in the longitudinal direction, F is the applied load, A is the original cross-sectional area, δ is the change in the gage length, and L is the original gage length.

The fact that the data now fall along a single curve indicates that this conversion eliminates the effects of specimen size and gage length. Consequently, the resulting *stress-strain curve*, or *diagram*, gives a direct indication of the material properties. This result is of no little consequence. It confirms our earlier statements that force per unit area (stress) and deformation per unit length (strain) are the

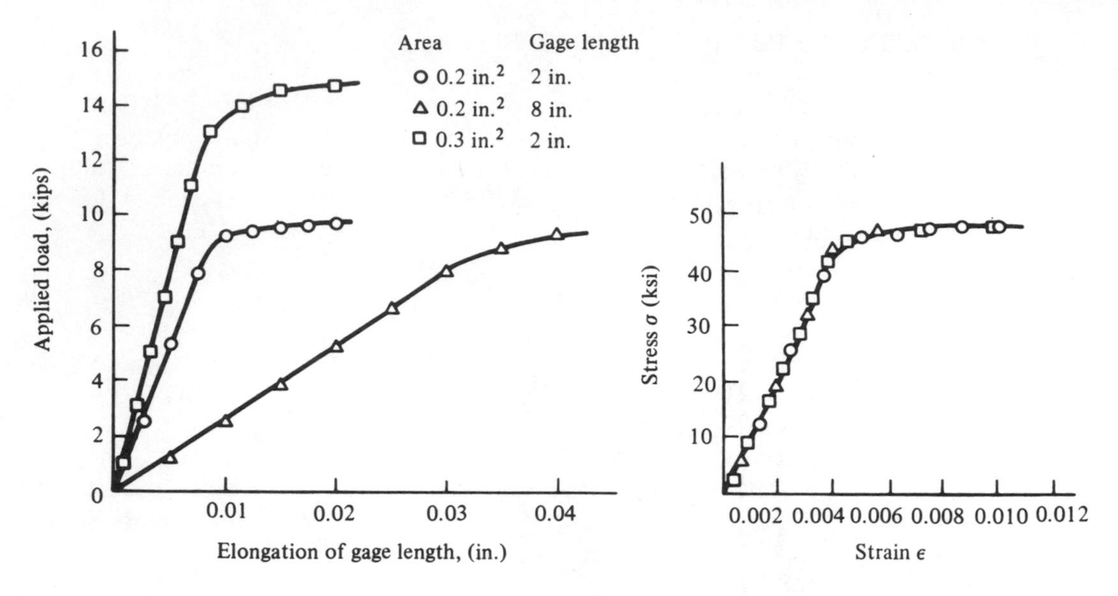

Figure 5.28 Load-Deformation Relations (a) and Stress-Strain Relation (b) for 2017-T451 Aluminum Alloy.

appropriate measures for expressing the effects of applied loadings upon material bodies.

Note that the stress-strain diagram is based upon the original cross-sectional area and gage length, even though these quantities change continuously as the test progresses. These changes have a negligible effect, except possibly during the final stages of the test before the specimen breaks. Moreover, since the job of design engineers is to determine the necessary original dimensions of members, material data based upon the *engineering stress* and *engineering strain* computed by using the original specimen dimensions are of the most use to them. Material scientists concerned with the detailed responses of materials need to know the precise values of stress and strain to which the material is subjected at each instant. Accordingly, they usually work with the *true stress* and *true strain* based upon the instantaneous values of area and gage length. All of the problems considered in this text involve only engineering stress and strain.

Shear Tests

Material behavior in shear can be determined by twisting a specimen in the form of a thin-walled tube. By measuring the applied couple and the resulting angle of twist, the shear stresses and strains can be determined and a shear stress-shear strain diagram can be constructed. A more detailed discussion of this test must await our study of torsion in Chapter 7.

5.8 RESISTIVE BEHAVIOR OF MATERIALS

Different materials exhibit different responses, and these responses are reflected in the shapes of their stress-strain diagrams. The responses typical of common structural materials subjected to slowly applied tensile, compressive, and shear loadings are discussed in this section. Emphasis will be placed upon tensile behavior, for the behavior in compression and shear is similar in many respects.

Tensile Behavior

Figure 5.29 shows complete tensile stress-strain curves for structural steel, Plexiglas, and cast iron. These curves, and those for most other structural materials loaded in tension, fall into one of the three general categories shown in Figure 5.30.

Diagram OA in Figure 5.30 is very nearly linear up to the stress at which the specimen breaks (denoted by $\times$). Glass and certain plastics have stress-strain curves like this one. Diagram OBC is typical of metals such as steel, aluminum, and brass. The stress is proportional to strain up to point B, at which the material is said to *yield*. Beyond this point, the strain increases much more rapidly than the stress. Materials such as cast iron and concrete (in compression) have curves with the shape ODE. The stress-strain diagrams for these materials are characterized by the fact that they have no well-defined linear region.

There are two distinct ranges of material response: the *elastic range* and the *inelastic*, or *plastic*, *range*. In the elastic range (OA, OB, or OD in Figure 5.30), the transmission of force from point to point within the material is reversible. Ideally speaking, if a material is loaded within the elastic range and then unloaded, each atom will regain its original equilibrium position and the entire member will return to its original dimensions. As can be seen from Figure 5.29(a), the elastic range corresponds to relatively small strains (about 1,200 $\mu\varepsilon$ for structural steel). Nevertheless, most members are designed so that this range is not exceeded under normal circumstances.

The inelastic range (BC or DE in Figure 5.30) is caused by a disruption of the bonds between atoms in some localized regions within the material. When this happens, shear deformations and slip between atoms occur, and some new bonds are established. This phenomenon is not completely reversible, and the member will not return to its original dimensions when unloaded. This behavior is clearly illustrated in Figure 5.31, which shows the loading and unloading behavior of an aluminum alloy in tension. The strain that remains after the load is removed is called the *permanent strain*. Experiments show that for most metals the stress-strain curve for unloading is approximately linear and parallel to the initial straight line portion of the curve for loading, although not always to the degree indicated in Figure 5.31.

If the load on the specimen is continuously increased, the phenomenon of disruption and reestablishment of atomic bonds is eventually propagated throughout the member. A limit of inelastic deformation is reached when no new bonds are

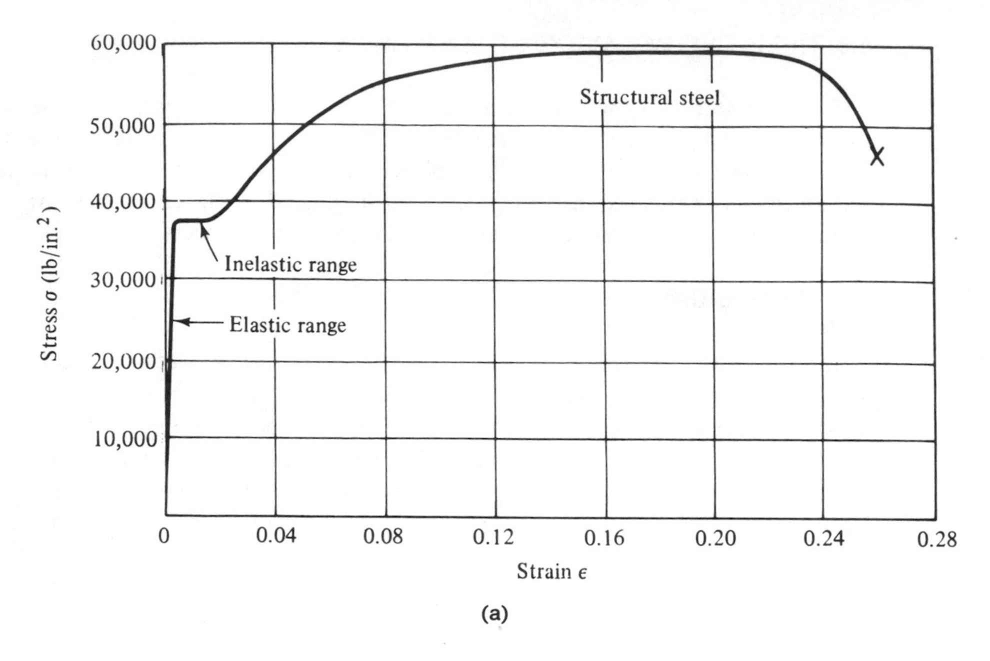

(a)

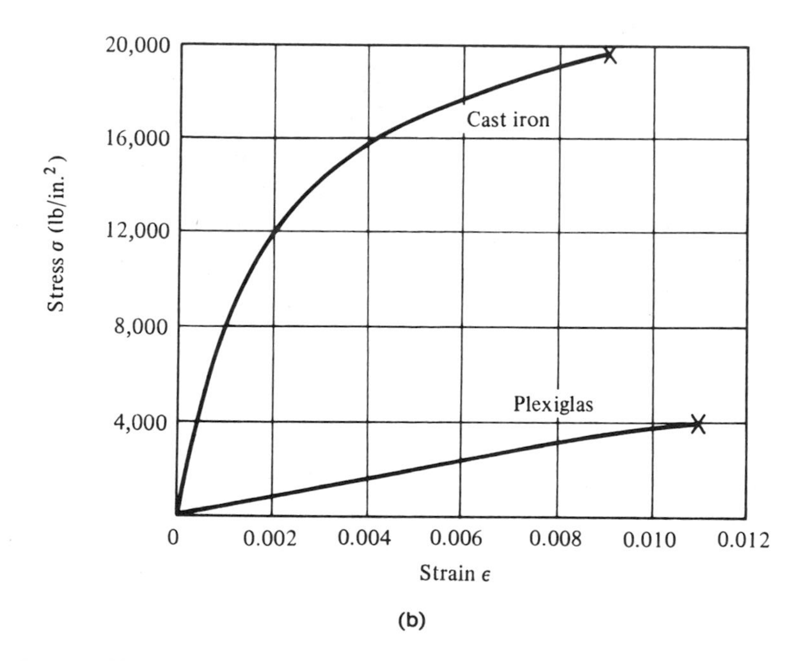

(b)

Figure 5.29 Tensile Behavior of (a) Structural Steel and (b) Plexiglas and Cast Iron.

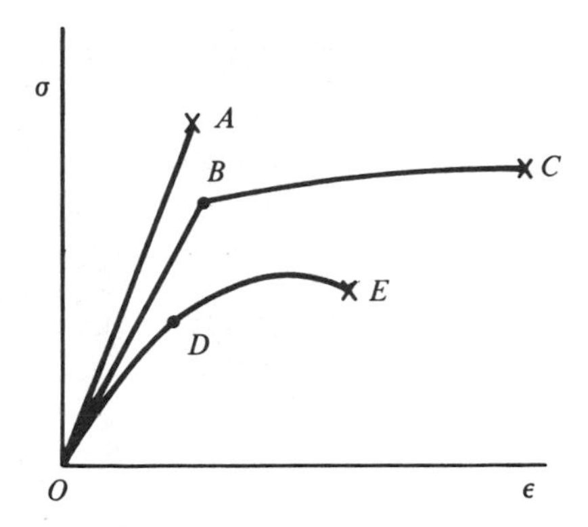

Figure 5.30

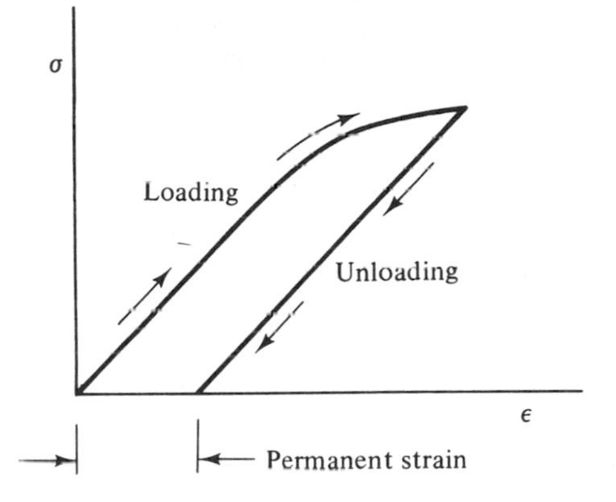

Figure 5.31 Loading-Unloading Behavior of an Aluminum Alloy in Tension.

established, with the result that the specimen fractures or separates. These limits are shown as points A, C, or E in Figure 5.30.

Material behavior can also be categorized according to the amount of inelastic deformation that occurs prior to fracture. Materials such as glass, plaster, and cast iron which exhibit very little inelastic deformation are said to be *brittle*. Steel, aluminum, and other materials which undergo comparatively large inelastic deformations are called *ductile*.

Brittle materials tend to fail suddenly with little warning when they are overloaded. In contrast, ductile materials yield and continue to support the load while undergoing inelastic deformation. This ability to deform inelastically helps prevent catastrophic accidents, and it is one of the reasons metals such as steel and aluminum are so outstanding as construction materials. Ductility is also an important factor in metalforming operations.

The fracture of brittle materials is associated with tensile stresses, but ductile materials tend to fail as a result of the shear stresses that develop. This is illustrated in Figure 5.32, which shows the fracture patterns for a brittle and a ductile material. The brittle material [Figure 5.32(a)] breaks normal to the axis of the specimen because this is the plane of maximum tensile stress. Ductile materials usually exhibit

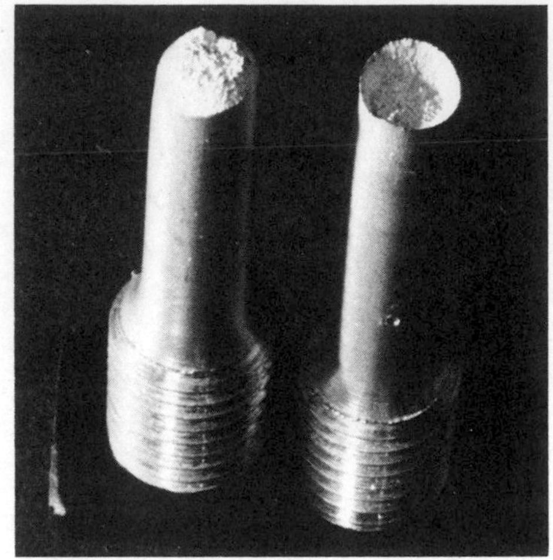

Figure 5.32 Typical Fracture Patterns of (a) Brittle and (b) Ductile Materials in Tension.

a cup-and-cone type fracture [Figure 5.32(b)]. The sides of the cup and cone portions are inclined at approximately 45° to the specimen axis, which corresponds to the planes of maximum shear stress.

The reduced cross section of the specimen shown in Figure 5.32(b) in the vicinity of the fracture is due to a phenomenon called *necking*. This localized reduction in area is responsible for the decrease in stress beyond the maximum point on the stress-strain diagram [see Figure 5.29(a), for example]. The load required to continue the deformation decreases as the specimen necks down because of the smaller cross-sectional area. Since the stress-strain curve is based upon the original area of the specimen, it also reflects this decrease.

Compressive Behavior

Materials that behave in a ductile fashion in tension usually exhibit approximately the same response in compression, well into the inelastic range, but the later stages of the tensile and compressive curves differ significantly. Since compression specimens expand instead of necking down, the compressive stress-strain curve continues to rise instead of reaching a maximum and then dropping off. Materials that are brittle in tension usually exhibit ductile behavior in compression. Most structural materials, brittle as well as ductile, behave approximately the same in tension and in compression over the elastic range.

Behavior in Shear

The shear stress-shear strain curve for most materials closely resembles the tensile curve. Of course, the magnitudes of the stresses and strains involved are different.

For example, a material that yields at a certain stress in tension will yield at approximately one-half that stress in shear.

Poisson Effect

Materials loaded in one direction undergo deformations and strains both in that direction and in directions perpendicular to it. This phenomenon is called the *Poisson effect* in honor of S. D. Poisson (1781–1840), who studied it analytically. This effect can be easily demonstrated by stretching a large rubber band or squeezing a rubber ball. The ratio of the magnitudes of the strain in a direction transverse to the loading and in the direction of the loading (longitudinal direction) is known as *Poisson's ratio*, which we shall denote by the Greek letter ν. Thus,

$$\nu = \frac{|\varepsilon_{\text{transverse}}|}{|\varepsilon_{\text{longitudinal}}|} \tag{5.16}$$

The sense of the transverse strain resulting from the Poisson effect is always opposite to that of the strain in the direction of loading. For example, tension members contract laterally when loaded, while compression members expand.

Environmental and Other Factors

In concluding this section, it is important that our comments on material behavior be put in proper perspective. Material behavior is not absolute; it depends upon many factors. Temperature, rate of loading, number of repetitions of the load, and the state of stress all influence material behavior. Other possible factors are irradiation or exposure to corrosive environments.

Materials may also exhibit certain directional effects. For example, the behavior of wood is different when loaded parallel to the grain than when loaded normal to the grain. Materials whose properties depend upon direction are said to be *anisotropic*; materials whose properties are the same in all directions are called *isotropic*.

Most materials exhibit directional effects if the specimens tested are sufficiently small. For example, the individual crystals of a metal are anisotropic. However, these effects tend to cancel in bodies of larger size because of the large number of randomly oriented crystals involved. As a result, the material may behave anisotropically on a microscopic scale and isotropically on a macroscopic scale. Anisotropy can also be produced by fabrication processes such as forging and rolling, which tend to give a preferred alignment to the grains of metals. Anisotropy is often introduced intentionally to enhance the material performance, as in fiber reinforced composite materials and steel reinforced concrete.

Nonhomogeneous materials have properties that vary from point to point. Most structural materials are nonhomogeneous on a microscopic scale, but they are reasonably homogeneous on a macroscopic scale.

Although it is important to be aware of the existence of factors such as those mentioned, their incorporation into the analysis of the response of load-carrying members is beyond the scope of this book. Unless stated otherwise, we shall assume that (1) the materials considered are homogeneous and isotropic, at least to a reasonable degree of approximation; (2) all loads are slowly applied and nonrepetitive; and (3) only moderate temperatures are involved. The results presented will apply in a large percentage of cases. Material behavior under general states of stress will be discussed briefly in Section 5.10 and in Chapter 9.

5.9 MATERIAL PROPERTIES AND STRESS-STRAIN RELATIONS

Material responses can be described quantitatively in terms of certain mechanical properties and by equations that represent the stress-strain relations. This quantitative description, which is necessary for computational purposes, is discussed in this section.

Material Properties

Most of the mechanical properties of interest can be determined directly from the stress-strain diagram. In discussing these properties we shall refer to Figure 5.33, which shows stress-strain curves typical of those for mild steel and aluminum in tension. Except where noted, the definitions given also apply for compression and shear. The symbols used to denote the various quantities, if any, are given in parentheses after the name of the property.

Among the properties of interest are the following:

1. *Proportional Limit* (σ_{PL} or τ_{PL}). The stress above which stress is no longer proportional to strain.

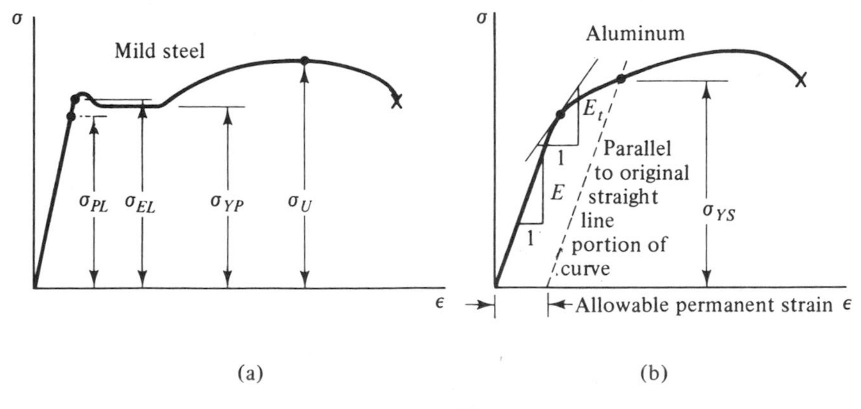

Figure 5.33

2. *Modulus of Elasticity* (E). Slope of the initial linear portion of the stress-strain diagram. This quantity has the same units as stress and is a measure of the material stiffness; the larger the value of E, the larger the stress required to produce a given strain. For the common structural metals, the value of E ranges from approximately 10×10^6 psi (69×10^9 N/m^2) for aluminum to 30×10^6 psi (207×10^9 N/m^2) for steel.
3. *Tangent Modulus* (E_t). Slope of the stress-strain curve beyond the proportional limit. There is no one single value of E_t; it varies with the amount of strain. The value of E_t at the origin is used as the modulus of elasticity for materials such as cast iron whose stress-strain curves have no well-defined linear range.
4. *Shear Modulus* (G). Slope of the initial linear portion of the shear stress-shear strain curve. This quantity is a measure of the material stiffness in shear and has the same units as stress. Typical values of G are 4×10^6 psi (28×10^9 N/m^2) for aluminum and 12×10^6 psi (83×10^9 N/m^2) for steel.
5. *Elastic Limit* (σ_{EL} or τ_{EL}). The maximum stress that can be applied without resulting in permanent deformation upon unloading. The elastic limit is difficult to determine experimentally; the proportional limit is usually used in place of the elastic limit because they have very nearly the same values for many materials.
6. *Yield Point* (σ_{YP} or τ_{YP}). The stress at which there are large increases in strain for little or no increase in stress. Among the common structural materials, only steel exhibits this type of response.
7. *Yield Strength* (σ_{YS} or τ_{YS}). The maximum stress that can be applied without exceeding a specified value of permanent strain (usually 0.2%) upon unloading. The procedure for determining the yield strength is illustrated in Figure 5.33(b). The yield strength is not really a material property because it depends upon the amount of permanent strain allowed. It is used for materials that do not have a yield point to define a stress above which excessive inelastic deformations will occur.
8. *Ultimate Strength* (σ_U or τ_U). The maximum stress the material will withstand.
9. *Percent Elongation*. The strain at fracture in tension, expressed as a percentage. This is a measure of ductility. The larger the percent elongation, the more ductile the material. If the complete stress-strain curve is not available, the percent elongation can be computed from the definition of average strain:

$$\text{Percent elongation} = \frac{\text{final gage length} - \text{initial gage length}}{\text{initial gage length}} \times 100 \qquad (5.17)$$

The broken specimen must be pieced back together in order to measure the final gage length.

10. *Percent Reduction of Area*. The reduction in cross-sectional area of a tensile specimen at fracture, expressed as a percentage. This quantity is another

measure of ductility. It is defined by the relationship

$$\text{Percent reduction area} = \frac{A_{\text{initial}} - A_{\text{final}}}{A_{\text{initial}}} \times 100 \tag{5.18}$$

The final area is measured at the fracture site where the cross section is smallest. The percent reduction of area cannot be determined from the stress-strain diagram.

11. *Poisson's Ratio* (ν). Poisson's ratio can be determined from tensile or compressive tests, provided both the transverse and longitudinal strains are measured [Eq. (5.16)]. If the material is isotropic, the value of ν is independent of direction; so the change in any cross-sectional dimension of the test specimen can be used to determine the transverse strain. For stress levels below the elastic limit, ν is a constant. Values for the common structural metals range from about 0.27 to 0.35. Above the elastic limit, the value of ν depends on the stress level.

Relationship Between E, G, and ν

For isotropic materials, the modulus of elasticity, shear modulus, and Poisson's ratio are related via the expression

$$G = \frac{E}{2(1 + \nu)} \tag{5.19}$$

Proof of this relationship will be given in Chapter 9 (Section 9.7).

Statistical Nature of Material Properties

Supposedly identical tests do not necessarily yield the same values for the material properties. This may be because of natural variations in the material from specimen to specimen, differences in the loading or environmental conditions from test to test, inaccuracies introduced in reading values off the stress-strain curves, and a myriad other reasons. If there is only a small variation in the values obtained for a particular quantity, it is sufficient to use an average value for it based upon the results of several tests. However, if the values vary widely, a larger number of tests must be conducted and the data analyzed by using statistical methods.

Representative values for several important properties are listed for a number of materials in Table A.2 of the Appendix. More detailed information, and information on other materials, can be found in engineering handbooks, manufacturers' literature, and other references.

Stress-Strain Relations

Below the proportional limit the stress-strain curve is a straight line completely defined by its slope. Thus, we have the stress-strain relations

$$\sigma = E\varepsilon \qquad (\sigma \leqslant \sigma_{PL}) \tag{5.20}$$

for normal stresses and normal strains and

$$\tau = G\gamma \qquad (\tau \leqslant \tau_{PL}) \tag{5.21}$$

for shear stresses and shear strains, where E and G are the modulus of elasticity and the shear modulus, respectively. Materials that obey linear relations of this type are said to be *linearly elastic*. Equation (5.20) is commonly called *Hooke's law* in honor of Robert Hooke (1635–1703), who carried out experiments on wires and observed a linear relationship between the applied load and resulting elongation. Equation (5.21) is the counterpart of Hooke's law for shear.

Problems involving brittle materials are often such that an equation describing the complete stress-strain curve is required. Highly brittle materials, such as glass and Plexiglas in tension, obey Hooke's law to a good degree of approximation all the way to fracture [see Figure 5.29(b), for example]. Nonlinear stress-strain curves, such as that for cast iron in tension [Figure 5.29(b)], can usually be represented by a power law relationship of the form

$$\sigma = K\varepsilon^n \tag{5.22}$$

where K and n are constants obtained by fitting the equation to the data.

In most problems involving ductile materials, the strains are sufficiently small that only the elastic range and first part of the inelastic range need be described. For materials with a yield point, the resulting stress-strain relations are simple. Hooke's law holds up to the proportional limit, beyond which the stress has a constant value equal to the yield point of the material (Figure 5.34):

$$\begin{aligned} \sigma &= E\varepsilon \qquad (\sigma \leqslant \sigma_{YP}) \\ \sigma &= \sigma_{YP} \qquad (\sigma > \sigma_{YP}) \end{aligned} \tag{5.23}$$

Materials that exhibit responses described by Eqs. (5.23) are called *elastic-perfectly plastic*.

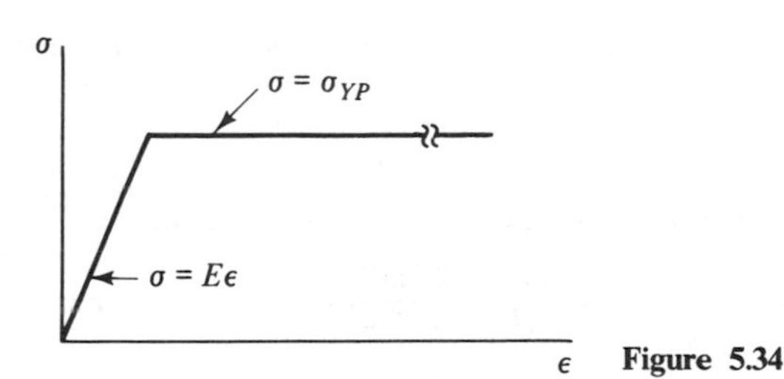

Figure 5.34

The behavior of ductile materials without a yield point is more difficult to describe. However, it is usually sufficiently accurate to treat the material as being elastic-perfectly plastic, with a yield point equal to the actual yield strength. The nature of this approximation is illustrated in Figure 5.35. It is widely used in structural engineering and machine design, and it will be used almost exclusively in this book. Note that, for this representation, the yield point, yield strength, and proportional limit all have the same values.

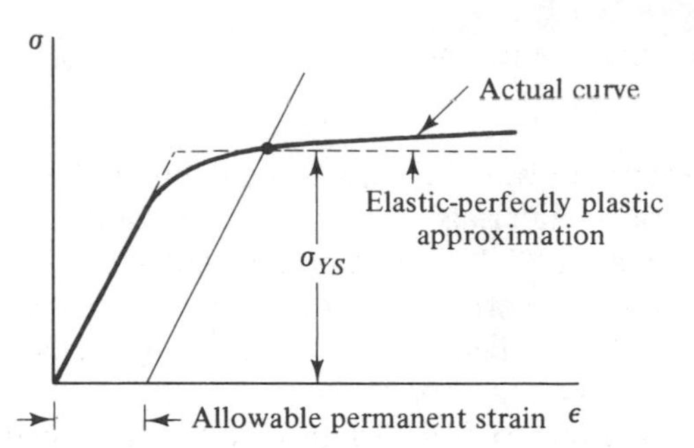

Figure 5.35

Obviously, more accurate representations of the stress-strain curves are possible, but only at the expense of increased mathematical complexity. After an appropriate change of symbols, all of the stress-strain relations discussed here also apply for shear.

Example 5.9 Tensile Test. During a tensile test the original 50.000-mm gage length is observed to have increased to 50.030 mm and the original diameter of 12.000 mm to have decreased to 11.998 mm. The applied load is 4.68 kN. Determine the modulus of elasticity and Poisson's ratio. Assume that the stress is below the proportional limit.

Solution. The stress is

$$\sigma = \frac{F}{A} = \frac{4.68 \times 10^3 \text{ N}}{(\pi/4)(12)^2(10^{-6}) \text{ m}^2} = 41.38 \times 10^6 \text{ N/m}^2$$

and the strains in the longitudinal and transverse directions are

$$\varepsilon_L = \frac{50.030 \text{ mm} - 50.000 \text{ mm}}{50.000 \text{ mm}} = 0.0006$$

$$\varepsilon_t = \frac{11.998 \text{ mm} - 12.000 \text{ mm}}{12.000 \text{ mm}} = -0.0002$$

Hooke's law applies because the stresses are below the proportional limit; therefore,

$$E = \frac{\sigma}{\varepsilon} = \frac{41.38 \times 10^6 \text{ N/m}^2}{6 \times 10^{-4}} = 69.0 \times 10^9 \text{ N/m}^2 \text{ or } 69.0 \text{ GN/m}^2 \qquad \textbf{Answer}$$

From the definition of Poisson's ratio, Eq. (5.16), we obtain

$$\nu = \frac{|\varepsilon_t|}{|\varepsilon_L|} = \frac{0.0002}{0.0006} = 0.33 \qquad \textbf{Answer}$$

PROBLEMS

5.62. A 0.500-in. diameter circle drawn on a compression member before it is loaded distorts into an ellipse after loading, as shown. The major and minor axes of the ellipse are 0.515 and 0.455 in. long, respectively. Determine the modulus of elasticity and Poisson's ratio for this material. Assume linearly elastic behavior.

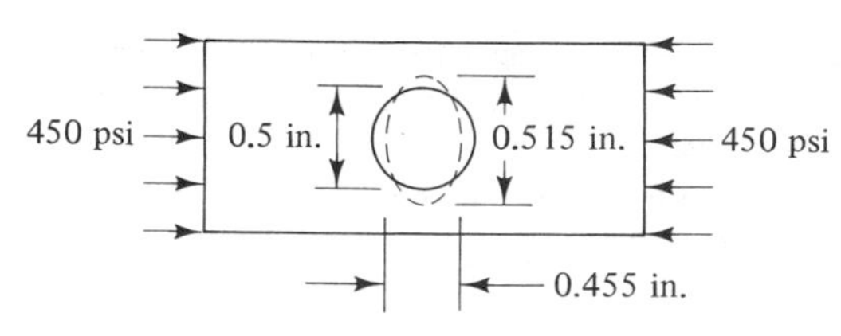

Problem 5.62

5.63. When a tensile specimen is subjected to a load of 5 kips, the original 2-in. gage length is observed to have increased by 0.0050 in. and the original 0.505-in. diameter to have decreased to 0.5046 in. Assuming that the stress is below the proportional limit, what are the values of E and ν for this material?

5.64. A tensile specimen is loaded until the transverse strain is $-450\ \mu\varepsilon$. If the material is elastic with $E = 12 \times 10^6$ psi and $\nu = 0.3$, what is the maximum tensile stress in the member?

5.65. What is the shear modulus G for an isotropic elastic material with $E = 110$ GN/m^2 and $\nu = 0.34$?

5.66. Same as Problem 5.65, except that $E = 72$ GN/m^2 and $\nu = 0.33$.

5.67. Compare the yield strengths for 0.2% allowable permanent strain for cast iron and aluminum. Use the stress-strain diagrams shown in Figures 5.28(b) and 5.29(b).

5.68. Compare the percent elongations for cast iron and structural steel. Use the stress-strain diagrams shown in Figure 5.29.

5.69. The following data are for a steel tensile specimen: initial diameter = 15.40 mm, final diameter = 10.98 mm, initial gage length = 200 mm, final gage length = 260 mm. Determine the percent reduction of area and percent elongation.

5.70. The tensile stress-strain diagram for a material is as shown. Determine (a) the proportional limit, (b) the modulus of elasticity, (c) the ultimate strength, and (d) the yield strength at 0.2% allowable permanent strain. If the specimen was originally 250-mm long, what is its final length at the instant of fracture? What is the load required to stretch a 2-m long rod made of this material by 12 mm if the diameter is 20 mm?

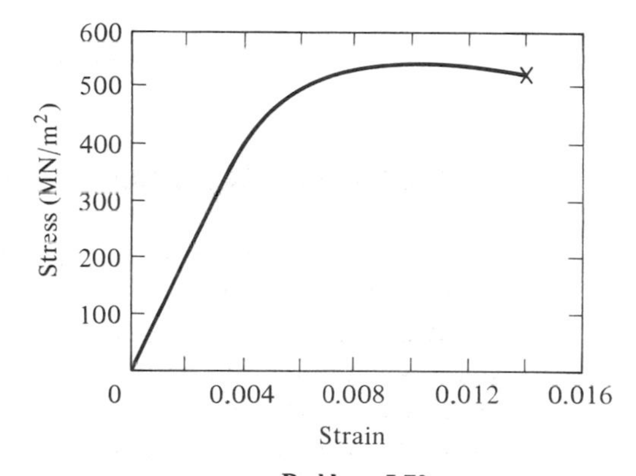

Problem 5.70

5.71. In Problem 5.70, what is the tangent modulus of the material at (a) $\varepsilon = 5{,}000\ \mu\varepsilon$ and (b) $\sigma = 500$ MN/m^2?

5.72. A tension member made of material with the stress-strain diagram given in Problem 5.70 is loaded to a strain of 8,000 $\mu\varepsilon$ and then unloaded. If the member was originally 24 in. long, what is its final length?

5.73. The following data are from a compression test on a specimen with a diameter of 1.000 in. The gage length is 2.000 in. and the extensometer used has a magnification factor of 10. Plot the stress-strain diagram and determine (a) the modulus of elasticity, (b) the yield strength at 0.2% allowable permanent strain and (c) the proportional limit.

Load (kips)	*Extensometer Reading (in.)*
0	0
8.4	0.020
10.2	0.040
11.5	0.060
12.5	0.080
13.3	0.100
14.0	0.120
14.7	0.140
15.3	0.160
⋮	⋮

5.74. Show that the stress-strain relationship in Eq. (5.22) is a straight line with slope n when plotted on a log-log scale. Use this result to find K and n for the material in Problem 5.73.

5.10 GENERALIZED HOOKE'S LAW

The stresses acting upon a material element at a point in a body for a general triaxial state of stress were shown in Figure 5.16(b). It is clear that the normal stresses acting upon the element will tend to stretch or compress it and the shear stresses will tend to distort it. Our goal in this section is to determine the relationships between the various stresses and strains. We shall consider only the case of small deformations of linearly elastic isotropic materials. As a result, the strains will be proportional to the stresses, and the principle of superposition will apply. That is, the total strain due to a combination of stresses is the sum of those due to each of the individual stresses. We shall also make use of the fact that, for isotropic materials, a normal stress produces only normal strains and a particular component of shear stress produces only the corresponding component of shear strain. This important result follows from symmetry arguments, but a proof is beyond the scope of this text.

Figure 5.36 shows the resulting deformations of a material element due to a normal stress σ_x. The original shape of the element is indicated by the solid lines and the deformed shape by the dashed lines. The normal strain in the direction of the stress is determined by Hooke's law [Eq. (5.20)], and the strains in the other two directions are related to it through the definition of Poisson's ratio [Eq. (5.16)].

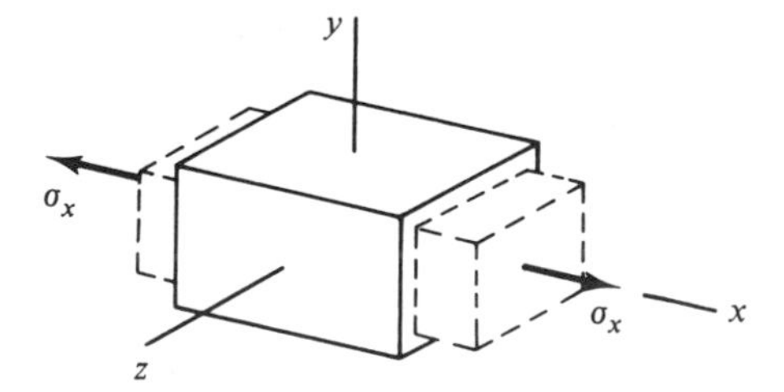

Figure 5.36

Thus, the stress σ_x produces a normal strain in the x direction of

$$\varepsilon_x^{(1)} = \frac{\sigma_x}{E}$$

and normal strains in the y and z directions of

$$\varepsilon_y^{(1)} = -\nu\varepsilon_x^{(1)} = -\frac{\nu\sigma_x}{E}$$

and

$$\varepsilon_z^{(1)} = -\nu\varepsilon_x^{(1)} = -\frac{\nu\sigma_x}{E}$$

Here E and ν are the modulus of elasticity and Poisson's ratio of the material, respectively. Similarly, stresses σ_y and σ_z produce strains of

$$\varepsilon_y^{(2)} = \frac{\sigma_y}{E} \qquad \varepsilon_x^{(2)} = \varepsilon_z^{(2)} = -\nu\varepsilon_y^{(2)} = -\frac{\nu\sigma_y}{E}$$

and

$$\varepsilon_z^{(3)} = \frac{\sigma_z}{E} \qquad \varepsilon_x^{(3)} = \varepsilon_y^{(3)} = -\nu\varepsilon_z^{(3)} = -\frac{\nu\sigma_z}{E}$$

The total normal strains in the three coordinate directions are

$$\begin{aligned}
\varepsilon_x &= \varepsilon_x^{(1)} + \varepsilon_x^{(2)} + \varepsilon_x^{(3)} = \frac{\sigma_x - \nu(\sigma_y + \sigma_z)}{E} \\
\varepsilon_y &= \varepsilon_y^{(1)} + \varepsilon_y^{(2)} + \varepsilon_y^{(3)} = \frac{\sigma_y - \nu(\sigma_x + \sigma_z)}{E} \\
\varepsilon_z &= \varepsilon_z^{(1)} + \varepsilon_z^{(2)} + \varepsilon_z^{(3)} = \frac{\sigma_z - \nu(\sigma_x + \sigma_y)}{E}
\end{aligned} \tag{5.24}$$

Since a component of shear stress produces only the corresponding component of shear strain, the relationships between the shear stresses and shear strains are

$$\gamma_{xy} = \frac{\tau_{xy}}{G} \qquad \gamma_{xz} = \frac{\tau_{xz}}{G} \qquad \gamma_{yz} = \frac{\tau_{yz}}{G} \tag{5.25}$$

where G is the shear modulus of the material.

Equations (5.24) and (5.25) are called the *generalized Hooke's law*. They define the stress-strain relations for a linearly elastic isotropic material subjected to a general triaxial state of stress. These relations apply for any set of mutually perpendicular axes, not just xyz.

It is sometimes convenient to invert Eqs. (5.24) so that the stresses are expressed in terms of the strains. Carrying out the necessary algebra, we obtain

$$\begin{aligned}
\sigma_x &= \frac{E}{1+\nu}\left[\varepsilon_x + \frac{\nu}{1-2\nu}(\varepsilon_x + \varepsilon_y + \varepsilon_z)\right] \\
\sigma_y &= \frac{E}{1+\nu}\left[\varepsilon_y + \frac{\nu}{1-2\nu}(\varepsilon_x + \varepsilon_y + \varepsilon_z)\right] \\
\sigma_z &= \frac{E}{1+\nu}\left[\varepsilon_z + \frac{\nu}{1-2\nu}(\varepsilon_x + \varepsilon_y + \varepsilon_z)\right]
\end{aligned} \tag{5.26}$$

For plane stress in the xy plane, $\sigma_z = \tau_{xz} = \tau_{yz} = 0$, and Eqs. (5.25) and (5.26) reduce to

$$\sigma_x = \frac{E}{(1-\nu^2)}(\varepsilon_x + \nu\varepsilon_y)$$
$$\sigma_y = \frac{E}{(1-\nu^2)}(\varepsilon_y + \nu\varepsilon_x) \tag{5.27}$$
$$\tau_{xy} = G\gamma_{xy}$$

Example 5.10. Deformations of a Plate. A structural steel plate with $E = 210\ \text{GN/m}^2$ and $\nu = 0.3$ has the dimensions shown in Figure 5.37 before loading. The plate is then subjected to a state of plane stress in the xy plane with $\sigma_x = 150\ \text{MN/m}^2$. For what value of the stress σ_y will the dimension Y of the plate remain unchanged? What are the final dimensions of the plate in the other two directions?

Solution. For plane stress, $\sigma_z = 0$. Also, $\varepsilon_y = 0$, since the dimension Y does not change. Thus, we have from the second of Eqs. (5.24)

$$\varepsilon_y = \frac{\sigma_y - \nu\sigma_x}{E} = 0$$

or

$$\sigma_y = \nu\sigma_x = 0.3(150\ \text{MN/m}^2) = 45\ \text{MN/m}^2 \qquad \textbf{Answer}$$

From the first and third of Eqs. (5.24), we now obtain

$$\varepsilon_x = \frac{\sigma_x - \nu\sigma_y}{E} = \frac{(150 \times 10^6\ \text{N/m}^2) - 0.3(45 \times 10^6\ \text{N/m}^2)}{210 \times 10^9\ \text{N/m}^2} = 0.00065$$

$$\varepsilon_z = \frac{-\nu(\sigma_x + \sigma_y)}{E} = \frac{-0.3(195 \times 10^6\ \text{N/m}^2)}{210 \times 10^9\ \text{N/m}^2} = -0.00028$$

The final dimensions X' and Z' of the plate are

$$X' = X(1 + \varepsilon_x) = 300\ \text{mm}(1.00065) = 300.195\ \text{mm}$$
$$Z' = Z(1 + \varepsilon_z) = 10\ \text{mm}(0.99972) = 9.997\ \text{mm} \qquad \textbf{Answer}$$

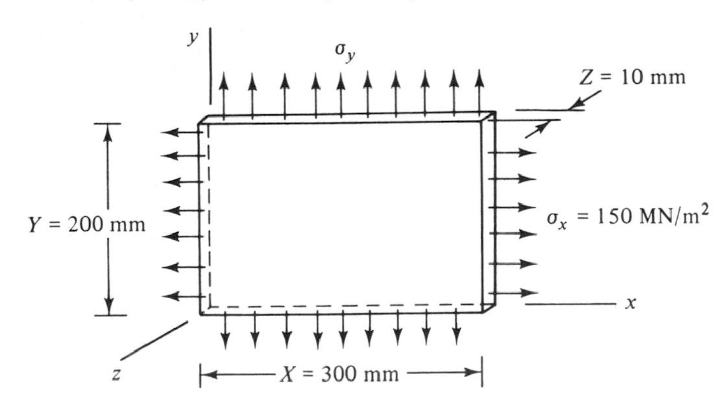

Figure 5.37

PROBLEMS

5.75. to 5.77. For the state of plane stress shown, determine the corresponding strains ε_x, ε_y, and γ_{xy} and sketch the deformed shape of the element. The material properties are as indicated.

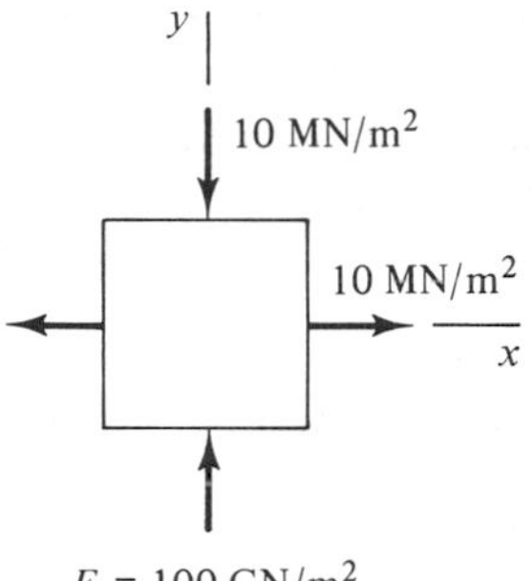

$E = 100\ GN/m^2$
$G = 42\ GN/m^2$
$\nu = 0.20$

Problem 5.75

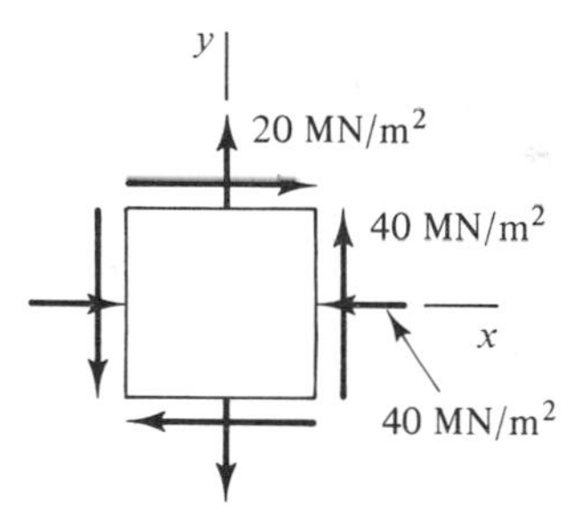

$E = 80\ GN/m^2$
$G = 30\ GN/m^2$
$\nu = 0.35$

Problem 5.76

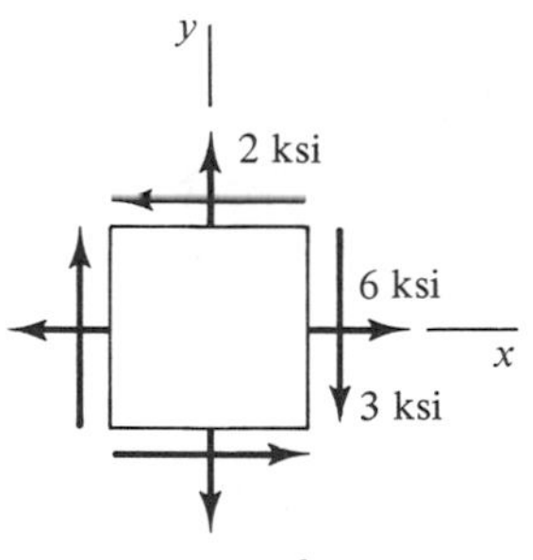

$E = 30 \times 10^6$ psi
$G = 12 \times 10^6$ psi
$\nu = 0.3$

Problem 5.77

5.78. A $\frac{1}{4}$-in. thick steel plate is subjected to the stresses shown. What is the change in thickness of the plate? $E = 30 \times 10^6$ psi and $\nu = 0.3$.

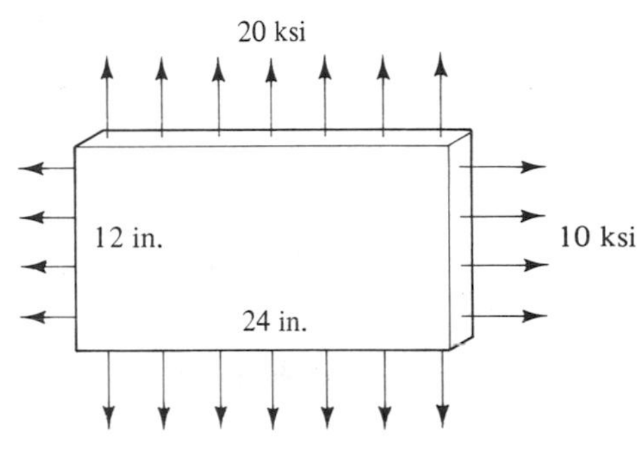

Problem 5.78

5.79. When a rectangular plate 300-mm long in the x-direction and 200-mm long in the y-direction is subjected to uniform stresses σ_x and σ_y ($\sigma_z = 0$), the 300-mm length becomes 300.20 mm and the 200-mm length becomes 199.92 mm. Determine σ_x and σ_y; $E = 68\ GN/m^2$ and $\nu = 0.33$.

5.80. For what stress ratio σ_x/σ_y will the flat plate shown undergo equal changes in the x- and y-dimensions?

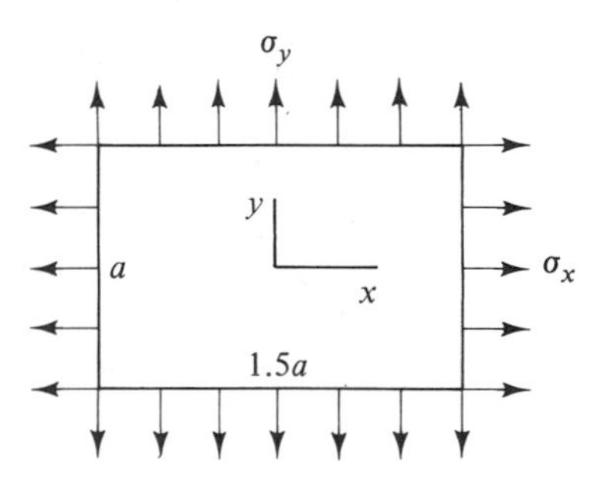

Problem 5.80

5.81. A rectangular block is subjected to uniformly distributed forces with the magnitudes shown. Determine the change in each dimension of the block and the change in its volume; $E = 72\ \text{GN/m}^2$ and $\nu = 0.33$.

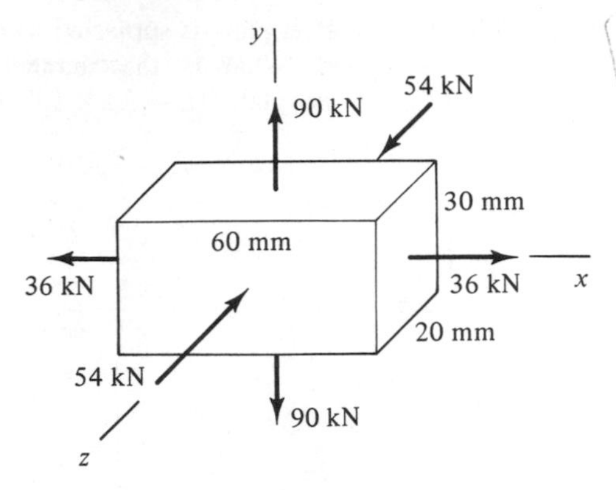

Problem 5.81

5.82. A rectangular block is subjected to uniformly distributed forces with the magnitudes shown. For what value of the force F will there be no change in length in the x direction? What are the changes in the other dimensions for this value of F? $E = 10 \times 10^6$ psi and $\nu = \frac{1}{4}$.

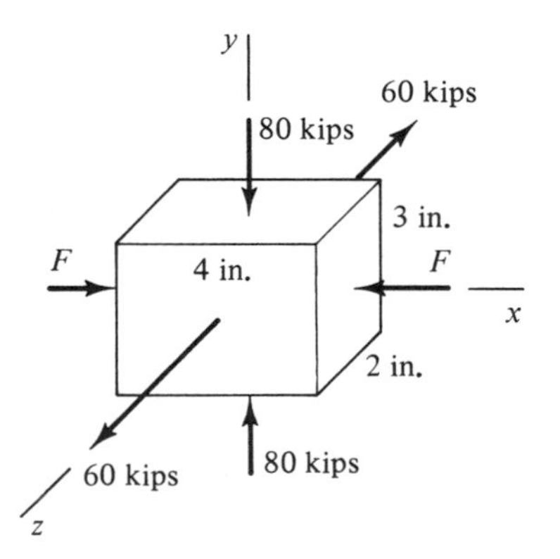

Problem 5.82

5.83. A solid foam plastic used to provide buoyancy has a compressive modulus of elasticity of 3 ksi and a Poisson's ratio of 0.2. What is the percentage loss in buoyancy of a cube of this material when it is lowered into the ocean to a depth where the pressure is 200 psi?

5.84. A thin plate with width w, thickness t, and length L is placed between two smooth rigid walls and then subjected to a uniform stress in the x-direction, as shown. Determine the stress in the y-direction and the change in thickness. Express the results in terms of the material properties E and ν and the applied stress σ.

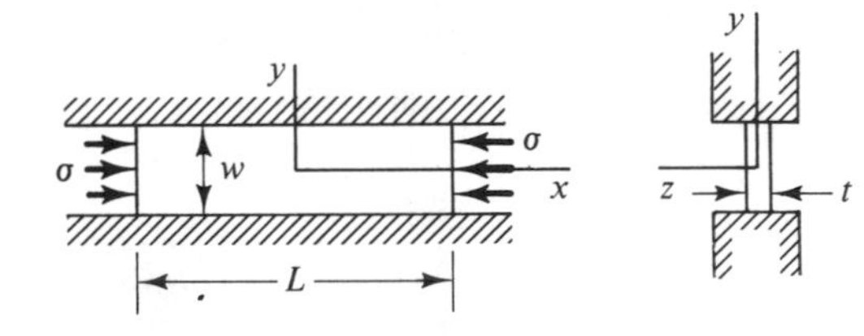

Problem 5.84

5.85. The 2 in. × 3 in. × 6 in. rubber block shown initially fits stress free into a cavity in a large steel block. What is the magnitude of

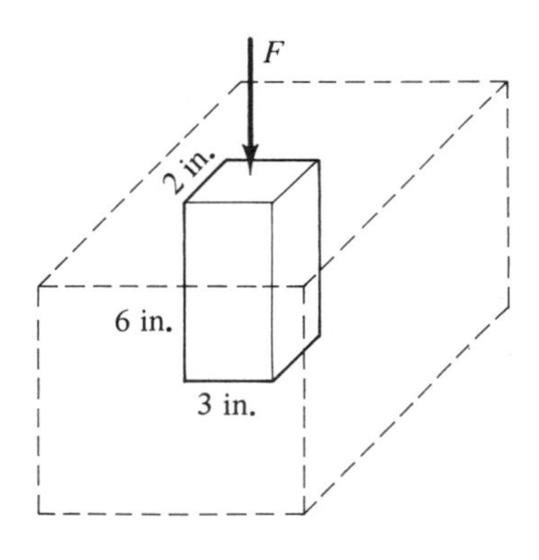

Problem 5.85

the applied load F if it causes the block to shorten by 0.30 in? Use $E = 52$ ksi and $\nu = 3/7$. Ignore any deformations of the steel block.

5.86. Two equal-sized cubes made of different materials are placed between rigid surfaces, as shown. What is the contact pressure between the cubes when the bottom cube is subjected to a uniform pressure p on all four exposed sides? Express the result in terms of p and the material properties; ignore the weight of the cubes.

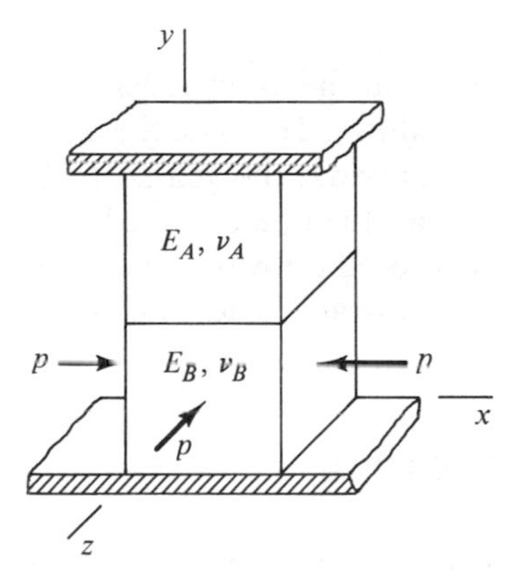

Problem 5.86

5.87. A rectangular block fits between two smooth rigid surfaces with a slight gap, as shown. Determine the stress σ_y and the change in length in the x-direction, assuming that the gap closes. What is the minimum value of σ_x required to close the gap? Express the results in terms of the dimensions given and the material properties E and ν.

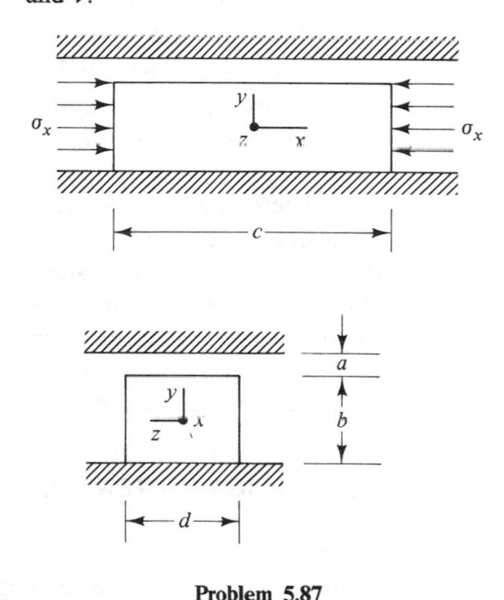

Problem 5.87

5.11 LOAD-STRESS AND LOAD-DEFORMATION RELATIONS

We are now in a position to outline a general procedure for obtaining equations that relate the applied loadings to the stresses and deformations they produce. Recall from Section 5.2 that the actual stresses cannot be determined from the equations of statics alone. Additional information about the manner in which the stresses are distributed is required.

One possible approach is to make assumptions about the stress distribution and then check to see if the results obtained are reasonable. A more systematic approach is to relate the stress distribution to some quantity that can be readily observed and/or measured. But what quantity would this be? For obvious reasons, quantities occurring within the body cannot be easily observed and measured, if at all. The one quantity that meets our requirements is the deformations occurring on the surface. Observation of the pattern of these deformations is the key to the determination of the stresses and strains throughout the body. The procedure is as follows:

1. Observe the geometry of the deformation pattern produced by the loading.
2. Obtain the distribution of strain from the equations relating strains to deformations (Section 5.6). At this point, only the manner in which the strains are distributed can be determined; their magnitudes remain unknown.
3. Obtain the stress distribution from the strain distribution by using the stress-strain relations (Sections 5.9 or 5.10).
4. Relate the stresses to the stress resultants by using the conditions for static equivalence (Section 4.7).
5. Relate the stress resultants to the applied loadings by using the method of sections (Section 5.2).

These steps, which need not be carried out in the exact order listed, yield the relation between the stresses and the applied loadings. To relate the deformations to the loads, we go back to steps 2 and 3. Step 3 provides the actual strains, since the stresses are now known. Actual deformations can then be obtained from step 2 by solving the strain-deformation relations. This procedure will be used extensively in the following chapters to determine the load-stress and load-deformation relations for common structural members.

5.12 DESIGN CRITERIA AND SAFETY FACTORS

Load-stress and load-deformation relations can be used to check to see if a given member will perform adequately under a given loading (analysis) or to determine the material and size of member necessary to support a particular load (design). The design criteria by which the performance of a member or adequacy of its design are judged depend, in large measure, upon the intended purpose of the member. As we have mentioned several times before, a member need not break in order to fail.

There are four general types of structural failure under static loadings: fracture, general yielding, excessive deformations, and buckling. The first three will be considered in this section, along with the significant material properties and design criterion associated with each. Buckling will be discussed in Chapter 10.

Fracture

When a member breaks, its usefulness as a load-carrying member obviously ceases. Most members made of brittle material fail in this manner. Unfortunately, since the deformations are usually very small, there is little warning of the impending failure. Fracture of materials is associated with the ultimate stress; therefore, the appropriate design criterion for this mode of failure is that the stresses be kept below the ultimate strength of the material. Members made of ductile materials also break, but they usually fail first because of general yielding or excessive deformations.

General Yielding

Members made of ductile materials can undergo very large inelastic deformations before breaking. As a result of these large deformations, the geometry of the structure can change so drastically that it literally collapses. The design criterion for this type of failure is obvious—the applied load must be kept below the load at which collapse occurs. The collapse load of simple members and structures will be considered in the following chapters. Suffice it to say at this point that the collapse load is intimately related to the yield strength of the material, for this is the stress at which the inelastic deformations start to become significant.

Excessive Deformations

This type of failure occurs when the deformations, even though they may be small, exceed some allowable value. For example, deformations in machine tools must be kept small so that small tolerances on the dimensions of the workpiece can be maintained. The amount of deformation that can be tolerated in any given problem is often a matter of engineering judgment. Excessive deformation is a result of insufficient "stiffness" of the member or structure. For many problems in which deformations are the limiting factor, the stresses are below the proportional limit. In these cases, the significant material parameters are the elastic modulus, E, or shear modulus, G.

Safety Factors

It should be apparent by now that there is a degree of inexactness involved in determining the response of load-carrying members and that it would be rather presumptuous to think that one could design right to the limit of the capability of the material without encountering difficulties. For example, various approximations are involved in the analysis, and the loads, dimensions, and material properties may not be known precisely. In addition, the structure may not be built according to specifications, or it may later be used in a manner not anticipated by the designer.

The necessary margin for uncertainty is usually provided by introducing a *safety factor* into the design. The safety factor is defined as the ratio of the load to produce failure (the design load) and the anticipated load to which the member will be subjected (the working load):

$$\text{Safety factor } (SF) = \frac{\text{failure load}}{\text{working load}} \tag{5.28}$$

For example, a safety factor of 3 means that the member is designed to withstand three times the expected applied load. A safety factor can also be introduced by reducing the allowable stresses or deformations. However, we shall use the definition given here.

The value of the safety factor is often set by building codes or other specifications, and is based upon factors such as the consequences of a failure, economics, and the degree of confidence in the accuracy of the analysis. For example, since failure of a large building would involve potential loss of life, a substantial safety factor would be used in its design. Aircraft also require a high degree of safety. However, an airplane could not get off the ground if it were designed with a large safety factor because of the extra material and weight involved. A small safety factor must be used, but this is compensated for by making the analysis highly accurate.

5.13 CLOSURE

The function of a load-carrying member is to support or transmit applied loads, and its success or failure is governed by the magnitudes of the resulting stresses and deformations. In this chapter we outlined the basic procedures for obtaining the load-stress and load-deformation relations and for using them in design and analysis. Some simple problems were considered, but most of the applications will come in the following chapters.

We have also introduced the concept of using a small material element to display the various stresses acting at a point within a body. This representation will be of great help in later chapters in visualizing and analyzing stresses due to complex loadings.

A review of this chapter will show that only a few new concepts have been introduced, a fact easily lost in the deluge of new definitions and terminology. Many of the points discussed undoubtedly remain a bit vague, but they should fall into focus when they are used over and over again in the applications to follow. There are several key points that apply to all our work and they are of sufficient importance to merit repeating here:

1. The three basic ingredients in the mechanics of deformable bodies are equilibrium, the geometry of the deformations, and the resistive properties of materials.
2. We consider only small deformation problems. Many, but not all, problems encountered in practice fall into this category.
3. If the deformations are small, the geometry of the body before loading can be used in all force analyses, except for buckling problems. This is an approximation without which few problems of practical interest could be solved. The accuracy of this approximation was illustrated in Section 3.10.

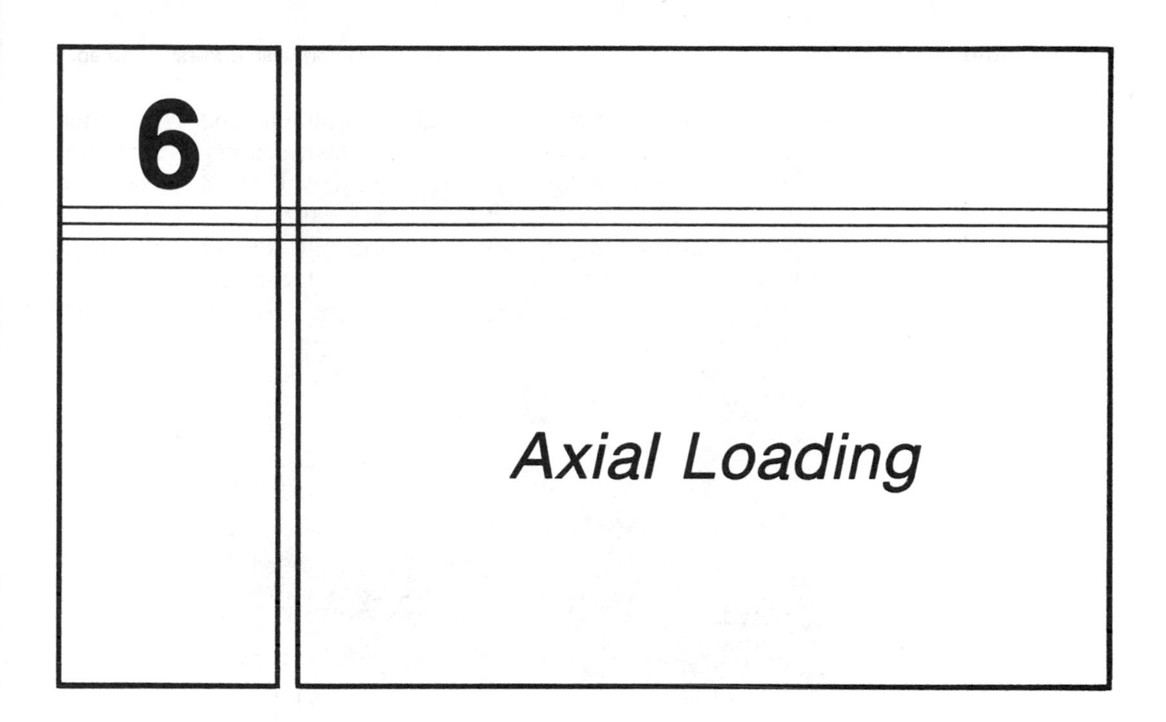

6.1 INTRODUCTION

The stresses in members subjected to simple axial loadings were discussed in Section 5.3. In this chapter we shall consider the stresses that result from more complex axial loadings, as well as the resulting deformations. We shall also consider some closely related problems concerning stresses and deformations in thin-walled pressure vessels. Our first step will be to examine more carefully the conditions under which the load-stress relations developed in Section 5.3 are valid.

6.2 VALIDITY OF LOAD-STRESS RELATIONS

Equations (5.3) through (5.6) derived in Section 5.3 define the actual stresses in an axially loaded member provided they are uniformly distributed over the section. Thus, it is only necessary to determine the conditions for a uniform stress distribution in order to establish the range of validity of these equations. This can be done by using the procedure outlined in Section 5.11.

Uniform axially loaded members deform in such a way that planes perpendicular to the axis before loading remain plane and perpendicular to the axis after loading. This is illustrated in Figure 6.1, which shows the deformation pattern in an axially loaded rubber sheet. Note that the deformed grids are distorted in the vicinity of the notches, but are uniform some distance away. Similar distortions in

(a) Before Loading

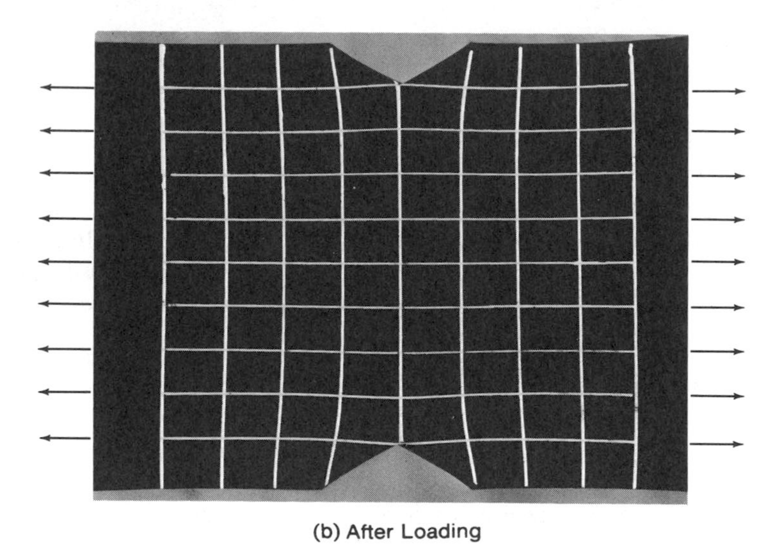

(b) After Loading

Figure 6.1 Deformation of an Axially Loaded Rubber Sheet as Indicated by Grid Patterns

the deformation pattern may also occur near supports, connections, and points of load application.

Now that we have observed the deformations in the sheet (step 1 in Section 5.11), we consider the strains associated with them. In doing so, it will be helpful to think of the member as consisting of a number of parallel fibers, or line elements, laid side by side.

Concentrating for the moment on the regions away from the notches, we have the situation illustrated in Figure 6.2(a). This figure shows the deformation of a typical row of squares across the sheet. Since each line element has the same original length, L, and elongates the same amount, δ, the strain is the same for each. Thus, the normal strain in the longitudinal direction is uniformly distributed across the sheet, as shown in Figure 6.2(b). This determines the strain distribution (step 2 in Section 5.11).

If the material is uniform across the sheet, each fiber will have the same properties (stress-strain diagram). Accordingly, each will experience the same stress because the strain is the same for each. Thus, the stress is also uniformly distributed across the sheet [Figure 6.2(c)]. This will not be the case near the notches, or at any other locations where the deformations are nonuniform. The stresses and deformations near the points of load application depend upon how the load is applied. For example, it is not too difficult to visualize that a load uniformly distributed over the end of the sheet would produce very different deformations locally than would a statically equivalent load applied as a concentrated force at its center. Members that have significant taper or are curved also exhibit nonuniform deformations.

The conditions for a uniform stress distribution and the validity of the load-stress relations given in Eqs. (5.3) through (5.6) can be summarized as follows:

1. The member must be axially loaded (this was shown in Section 5.3). Another way of saying this is that the stress resultant must consist only of a force

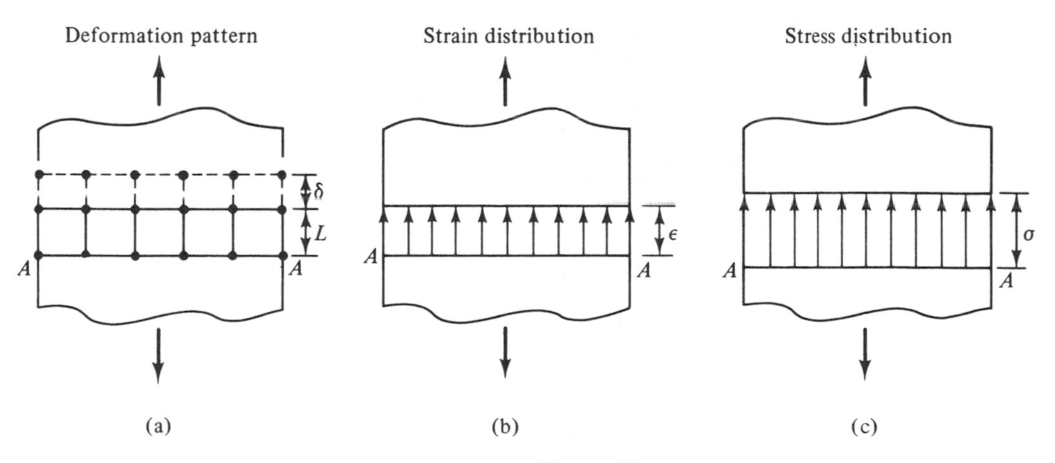

Figure 6.2

acting at the centroid of the cut section. If there is any torque or bending moment, the stresses cannot be uniformly distributed.

2. For compressive loads, the member must not be so slender that it buckles. Once the member bends, it is no longer axially loaded.
3. The member must be straight, or nearly so, for some distance on either side of the plane of interest.
4. The member must have constant cross-sectional area, or nearly so, for some distance on either side of the plane of interest.
5. The plane of interest must be some distance away from connections, supports, and points of load application.
6. The material must be uniform across the width of the member. Its properties can vary along the length, however, as in the case of two bars of different materials joined end to end to form a single composite member.

The regions of a member over which the load-stress relations are not strictly valid tend to be highly localized. As a rule of thumb, they extend only over a distance approximately equal to the width of the member. This is illustrated in Figure 6.3, which shows the experimentally determined strain distribution in an

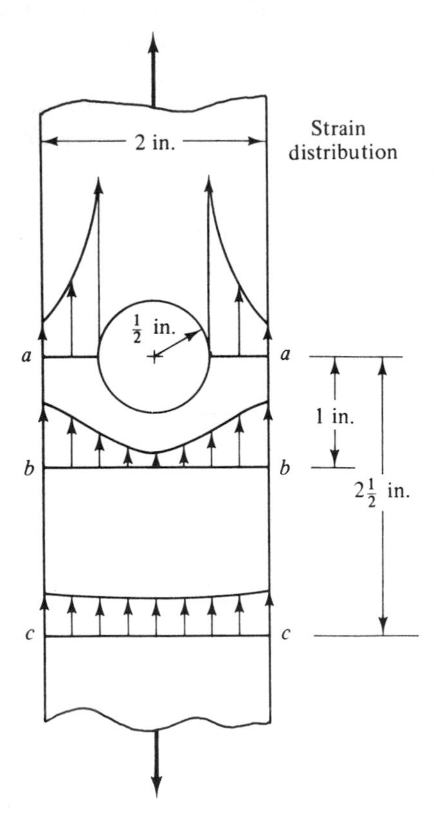

Figure 6.3

axially loaded bar containing a hole. The strains are significantly affected near the hole, but $1\frac{1}{4}$ bar-widths away at section *c-c* the strains are practically the same as if the hole weren't there.

This localizing effect, which is also evident near the notches in the sheet shown in Figure 6.1(b), occurs in all types of members under all types of loadings and is known as *St. Venant's principle*. This principle is very important, for without it there would be little hope of determining the stresses and deformations analytically. It makes it possible to develop simple load-stress and load-deformation relations that apply throughout most of the body. Special techniques can then be developed for handling the remaining regions involving the localized effects.

The maximum normal stress at holes, notches, or other geometrical discontinuities can be obtained by applying a correction factor to the values computed from the relationship $\sigma = P/A$. These correction factors, which are usually determined experimentally, will be discussed in Section 6.6. The situation near supports, connections, and points of load application is not so easily handled. There, the usual procedure is to work with the average stresses and to account for any errors by using a safety factor. In some cases, empirical methods of analysis have been devised.

PROBLEM

6.1. For each of the members shown, indicate whether or not the equation $\sigma = P/A$ is valid at the section indicated. If it is not valid, which of the necessary conditions are violated?

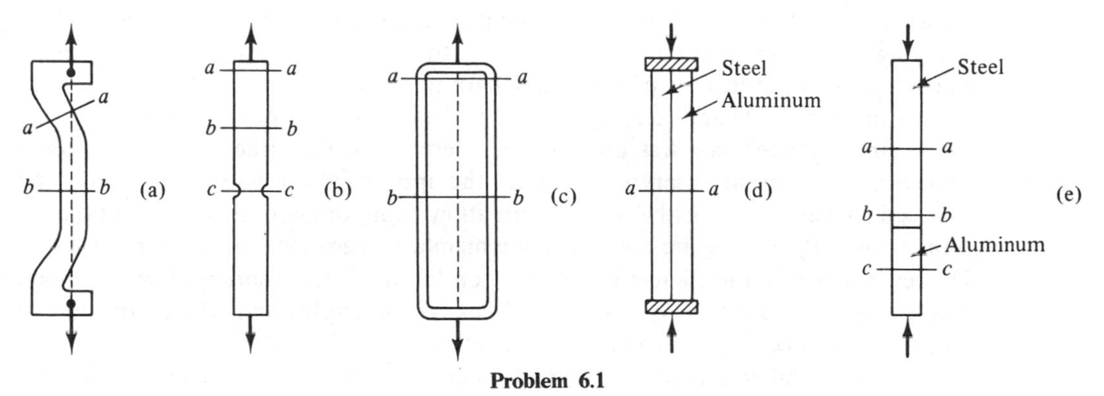

Problem 6.1

6.3 DEFORMATIONS

The elongation or shortening of an axially loaded member can be determined by combining the expression for the normal stress with the definition of normal strain and the stress-strain relation for the material. For the present, we shall assume that the material is linearly elastic.

Consider a line element of length dx located an arbitrary distance x from the end of the member (Figure 6.4). From the definition of normal strain, Eq. (5.13), the elongation of the element is

$$d\delta = \varepsilon\, dx$$

From Hooke's law [Eq. (5.20)] and the load-stress relation [Eq. (5.5)], we obtain for the strain

$$\varepsilon = \frac{\sigma}{E} = \frac{P}{AE}$$

where P is the normal force on planes perpendicular to the axis of the member, A is the cross-sectional area, and E is the modulus of elasticity. Combining these expressions and integrating over the length of the member, we obtain

$$\delta = \int_0^L \frac{P}{AE}\, dx \qquad (\sigma \leqslant \sigma_{PL}) \tag{6.1}$$

The quantities P, A, and E must be expressed as functions of x before the integral can be evaluated. If they are constant over the length of the member, Eq. (6.1) can be written as

$$\delta = \frac{PL}{AE} \qquad (\sigma \leqslant \sigma_{PL}) \tag{6.2}$$

The sign of δ is the same as that of P; it is positive if P is tensile and negative if it is compressive.

The preceding equations define the total deformation of the entire member. However, the deformation of any particular segment of it can be determined by carrying out the integration in Eq. (6.1) over the length of the segment or by using the segment length instead of the total length in Eq. (6.2).

Equations (6.1) and (6.2) apply only if the stresses are below the proportional limit, since Hooke's law was used in their derivation. For other types of material behavior, it is only necessary to introduce the appropriate stress-strain relation and proceed as before. Since the load-deformation equations are also based upon the relation $\sigma = P/A$ there are regions of the member where they are not strictly valid. However, this has a negligible influence upon the total deformation of the member. These regions are relatively short (St. Venant's principle) and, therefore, do not contribute appreciably to the value of the integral in Eq. (6.1).

If the member is loaded at several points along its length, the normal force P will vary from section to section. In this case, it is helpful to plot the variation of P along the length of the member so that its value at any location can be readily

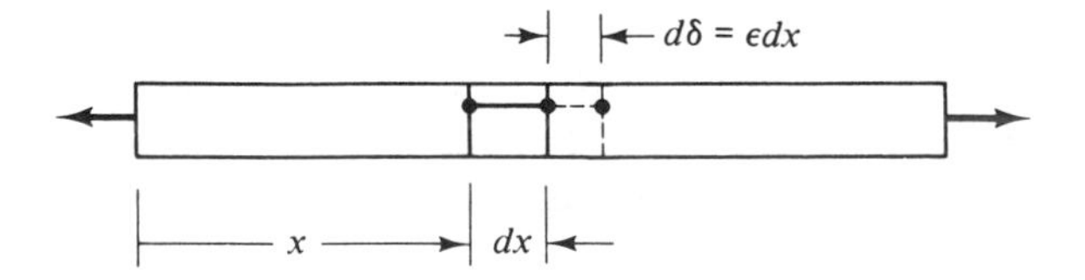

Figure 6.4

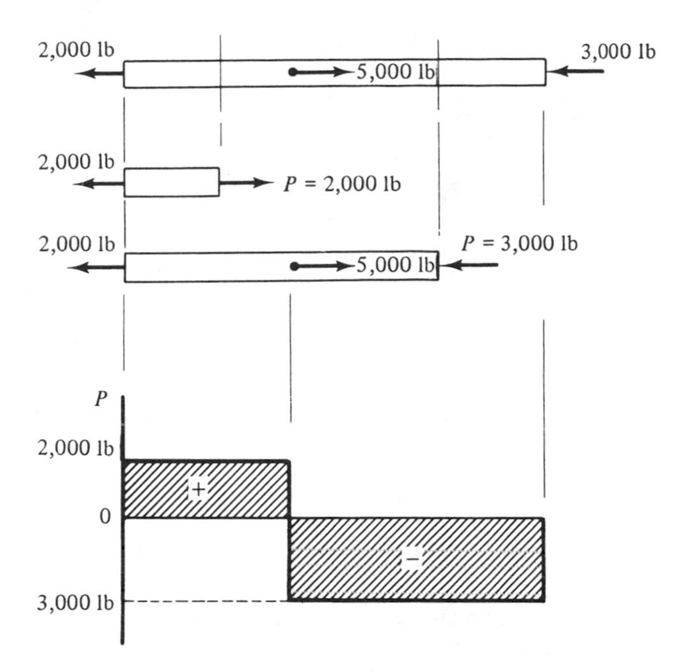

Figure 6.5

determined (Figure 6.5). This plot, called the *normal force diagram*, is useful in determining both the stresses and the deformations. In accordance with our sign convention, tensile forces are plotted as positive and compressive forces as negative.

Once the value of P in each portion of the member has been determined, the deformation of each segment can be computed and the results summed to determine the total deformation. In other words, the member is treated as a series of shorter members placed end to end, each with constant normal force. The deformation of the entire member is then equal to the sum of the deformations of its parts. This approach can also be used when different segments of the member have different cross-sectional areas or modulus of elasticity.

Example 6.1. Deformation of Truss Members. Members AB and AE of the truss shown in Figure 6.6(a) are made of structural steel and have cross-sectional areas of 2.4×10^{-3} m^2 and 1.2×10^{-3} m^2, respectively. How much will these members elongate or shorten as a result of the 400-kN applied load? Assume that compression members do not buckle.

Solution. The first step is to determine the forces on the members of interest. A force analysis of the entire truss (details not given here) shows that the reactions at A and D are 200-kN upward forces. From the FBD of joint A [Figure 6.6(b)], we then have

$$\begin{aligned}
\sum F_y &= F_{AB}\sin 53^\circ + 200\text{ kN} = 0 \\
\sum F_x &= F_{AE} + F_{AB}\cos 53^\circ = 0 \\
F_{AB} &= -250\text{ kN or } 250\text{ kN (compression)} \\
F_{AE} &= 150\text{ kN (tension)}
\end{aligned}$$

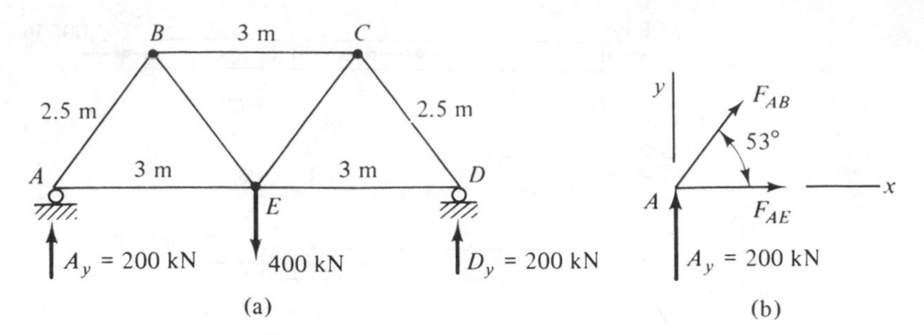

Figure 6.6

We next check the stresses to see if they are below the proportional limit:

$$\sigma_{AB} = \left(\frac{P}{A}\right)_{AB} = \frac{250 \times 10^3 \text{ N}}{2.4 \times 10^{-3} \text{ m}^2} = 104 \times 10^6 \text{ N/m}^2 \text{ (compression)}$$

$$\sigma_{AE} = \left(\frac{P}{A}\right)_{AE} = \frac{150 \times 10^3 \text{ N}}{1.2 \times 10^{-3} \text{ m}^2} = 125 \times 10^6 \text{ N/m}^2 \text{ (tension)}$$

Both of these values are well below the proportional limit of structural steel (see Table A.2); therefore, Eq. (6.2) applies:

$$\delta_{AB} = \left(\frac{PL}{AE}\right)_{AB} = \frac{(-250 \times 10^3 \text{ N})(2.5 \text{ m})}{(2.4 \times 10^{-3} \text{ m}^2)(207 \times 10^9 \text{ N/m}^2)}$$

$$= -1.26 \times 10^{-3} \text{ m or } 1.26 \text{ mm (shortening)} \qquad \textbf{Answer}$$

$$\delta_{AE} = \left(\frac{PL}{AE}\right)_{AE} = \frac{(150 \times 10^3 \text{ N})(3 \text{ m})}{(1.2 \times 10^{-3} \text{ m}^2)(207 \times 10^9 \text{ N/m}^2)}$$

$$= 1.81 \times 10^{-3} \text{ m or } 1.81 \text{ mm (elongation)} \qquad \textbf{Answer}$$

Example 6.2. Deformation of a Stepped Bar. (a) What load F can be applied to the aluminum bar shown in Figure 6.7(a) if the maximum normal stress cannot exceed 24 ksi? Use a safety factor of 3. (b) How much will the bar deform under this loading? Assume linearly elastic behavior.

Solution. (a) We first determine the reaction at A and the normal force diagram [Figure 6.7(b)]. It is clear from this diagram that the largest normal stress occurs in BC. (The normal force in BC is only one-half that in AB, but the area is six times smaller.) Thus, we have

$$\sigma_{BC} = \left(\frac{P}{A}\right)_{BC} = \frac{F}{0.5 \text{ in.}^2} \leqslant 24{,}000 \text{ lb/in.}^2$$

$$F \leqslant 12{,}000 \text{ lb}$$

This defines the failure load. The actual load that can be applied to the bar (the working load) is this value divided by the safety factor [Eq. (5.28)]:

$$F_{\text{working}} = \frac{F_{\text{failure}}}{SF} = \frac{12{,}000 \text{ lb}}{3} = 4{,}000 \text{ lb} \qquad \textbf{Answer}$$

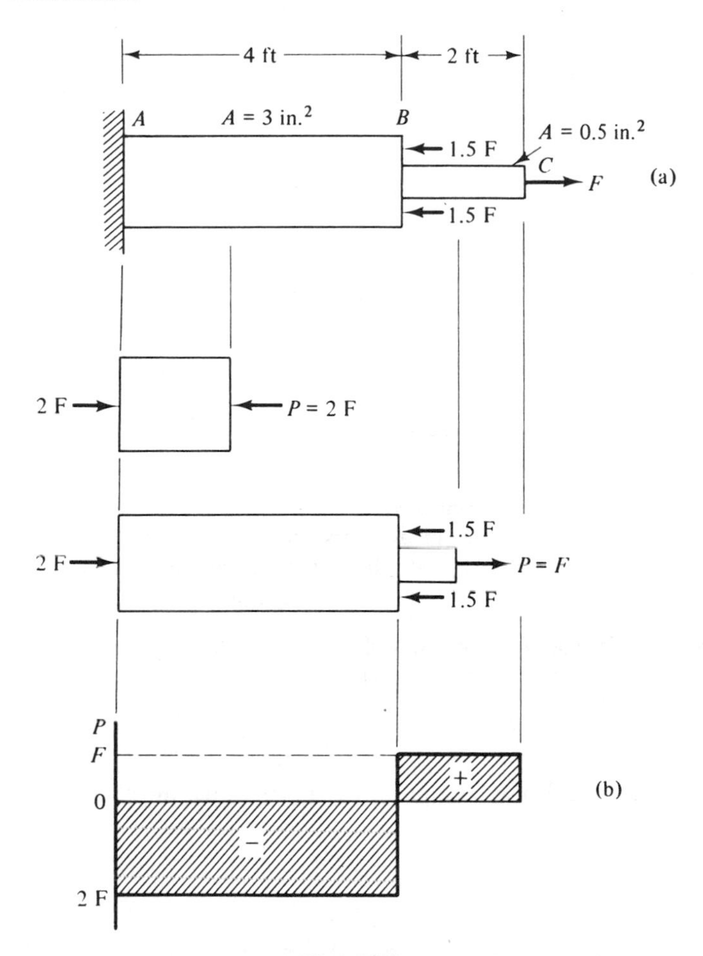

Figure 6.7

(b) The total deformation of the member is the sum of the deformations of each part. The working load is used in the computations because it is the load actually applied to the bar. We have

$$\delta = \delta_{AB} + \delta_{BC} = \left(\frac{PL}{AE}\right)_{AB} + \left(\frac{PL}{AE}\right)_{BC}$$

or

$$\delta = \frac{(-8{,}000\text{ lb})(48\text{ in.})}{(3\text{ in.}^2)(10 \times 10^6\text{ lb/in.}^2)} + \frac{(4{,}000\text{ lb})(24\text{ in.})}{(0.5\text{ in.}^2)(10 \times 10^6\text{ lb/in.}^2)}$$

$$= -0.0128\text{ in.} + 0.0192\text{ in.} = 0.0064\text{ in. (elongation)} \qquad \textbf{Answer}$$

Units must be watched carefully when computing δ because A and E are usually given in terms of square inches and pounds per square inch, respectively, while L is often given in feet.

Example 6.3. Elongation of a Bar Due to Its Own Weight. A bar with cross-sectional area A, specific weight γ, and length L is suspended vertically from one end [Figure 6.8(a)].

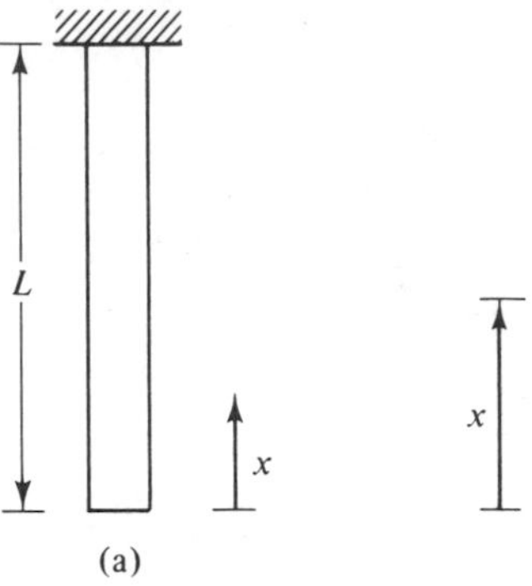

(a)

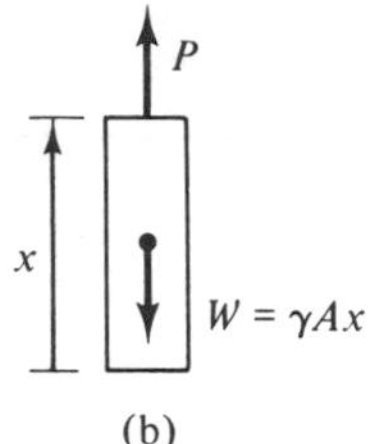

(b)

Figure 6.8

How much will the bar elongate under the action of its own weight? Assume that Hooke's law applies.

Solution. For convenience, we take the origin of coordinates at the free end of the bar. Sectioning the bar at an arbitrary location and considering the equilibrium of the resulting piece [Figure 6.8(b)], we obtain for the normal force

$$P = \gamma A x$$

Since P varies continuously along the length of the member, it is necessary to integrate in order to determine δ. From Eq. (6.1), we have

$$\delta = \int_0^L \frac{P}{AE}\,dx = \int_0^L \frac{\gamma x}{E}\,dx = \frac{\gamma L^2}{2E} \qquad \textbf{Answer}$$

The elongation can also be expressed in terms of the total weight of the member, which is $W = \gamma A L$:

$$\delta = \frac{WL}{2AE}$$

PROBLEMS

6.2. A $\frac{1}{2}$-in. diameter rod made of 2024-T3 aluminum elongates 0.35 in. when subjected to a tensile load of 2 kips. What is the original length of the rod?

6.3. A steel wire with a diameter of $\frac{1}{8}$ in. is attached to a rigid wall at one end and to an eye-bolt at the other end as shown. If the nut on the bolt is advanced one and one-half turns beyond that required to take up any slack, what is the tension induced in the wire? The bolt has eight threads per inch. Use $E = 30 \times 10^6$ psi.

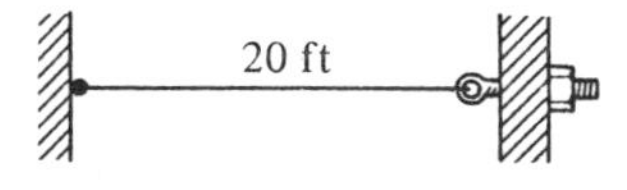

Problem 6.3

6.4. The uniform beam AB shown has a mass

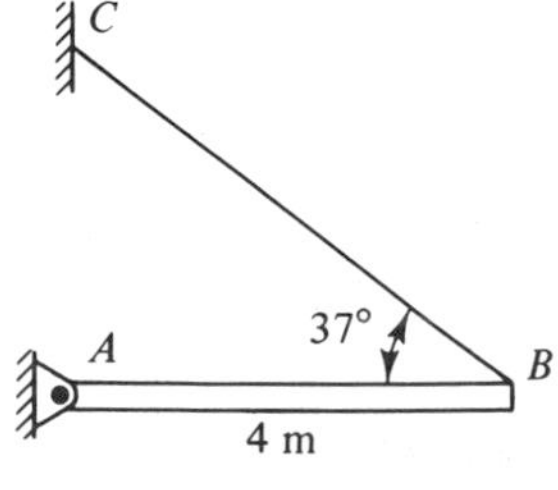

Problem 6.4

per unit length of 81.6 kg/m. By how much will the support cable BC elongate under this loading if it has a diameter of 10 mm? Use $E = 207\ GN/m^2$.

6.5. The members of the truss shown are made of 64-mm × 64-mm equal angles with a mass of 5.96 kg/m (see Table A.10). Determine the elongation or shortening of members BD and DE. Use $E = 207\ GN/m^2$ and neglect the weights of the members. Assume that compression members do not buckle.

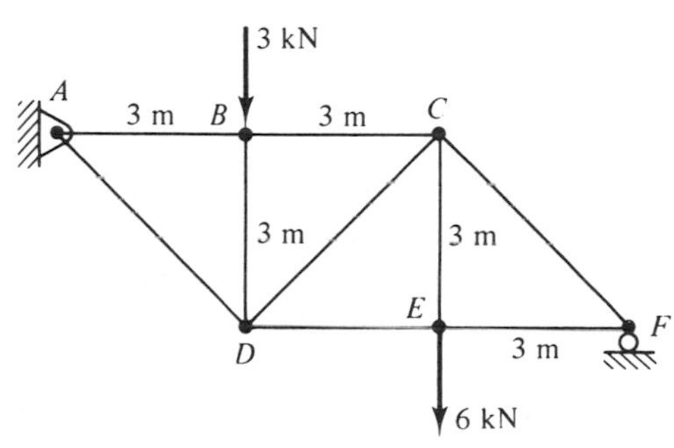

Problem 6.5

6.6. A steel member 6-ft long is subjected to a tensile load of 30 kips. What is the minimum required cross-sectional area of the member if it is not to yield and if the elongation is not to exceed 0.15 in? Use a safety factor of 2.

6.7. A steel ($E = 30 \times 10^6$ psi) bar with a cross-sectional area of 0.5 in.^{2} is loaded as shown. Determine the maximum normal stress in the bar and the total change in its length.

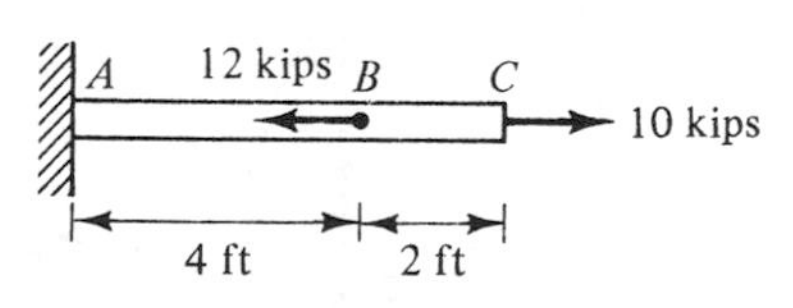

Problem 6.7

6.8. An aluminum ($E = 72\ GN/m^2$) bar with a cross-sectional area of 150 mm^2 is loaded as shown. Construct the normal force diagram for the member and determine its overall change in length. What is the maximum normal stress in the bar and in which portion does it occur?

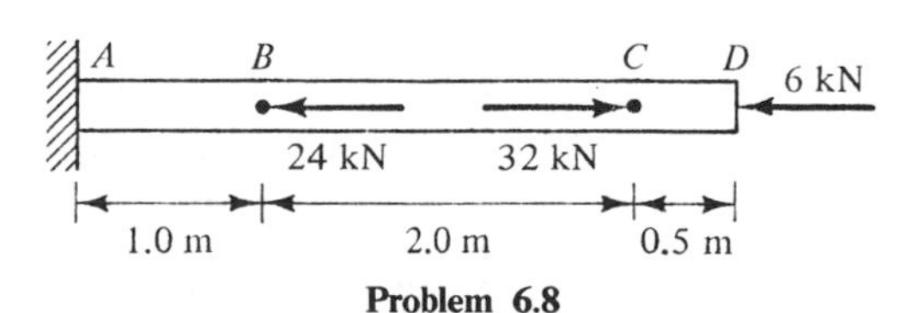

Problem 6.8

6.9. If the tensile stress in portion AB of the rod shown is not to exceed 140 MN/m^2, what is the minimum allowable value of the load F? For this value of F, determine the change in length of portion BC of the rod and the total change in its length. $A = 300$ mm^2 and $E = 207\ GN/m^2$.

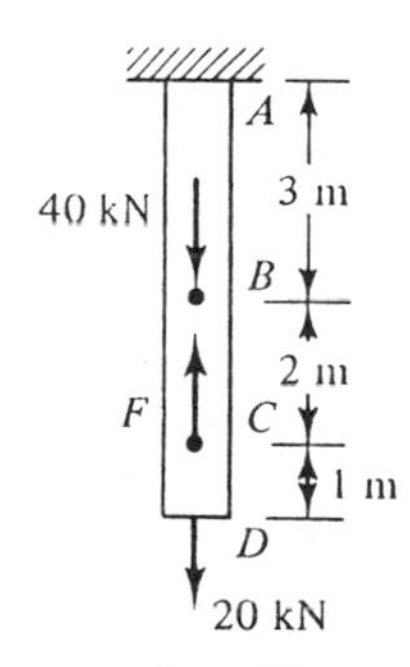

Problem 6.9

6.10. In Problem 6.9, what is the maximum allowable value of the load F if the maximum allowable stress in portion AB of the rod is 80 MN/m^2 compression? What is the overall change in length of the member for this value of F?

6.11. A short pedestal is made of a length of 3-in. nominal diameter steel pipe (see Table A.7) placed atop a length of 5-in. nominal diameter pipe as shown. If the allowable compressive stress is 15 ksi, what is the maximum allowable value of the load F? How far does the top of the pedestal settle under this loading and what is the bearing stress between the bottom pipe and the floor?

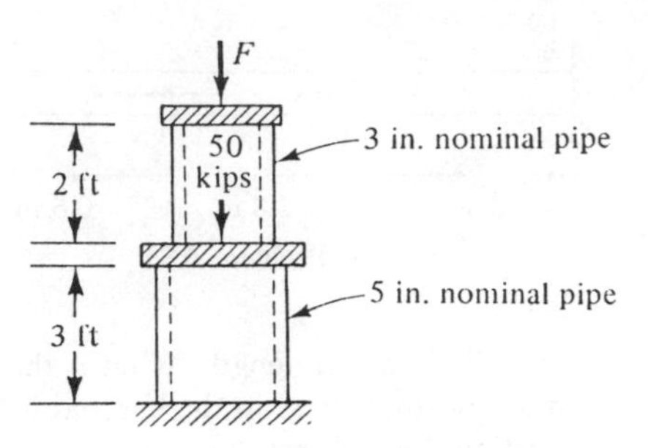

Problem 6.11

6.12. What is the overall change in length of the composite member shown?

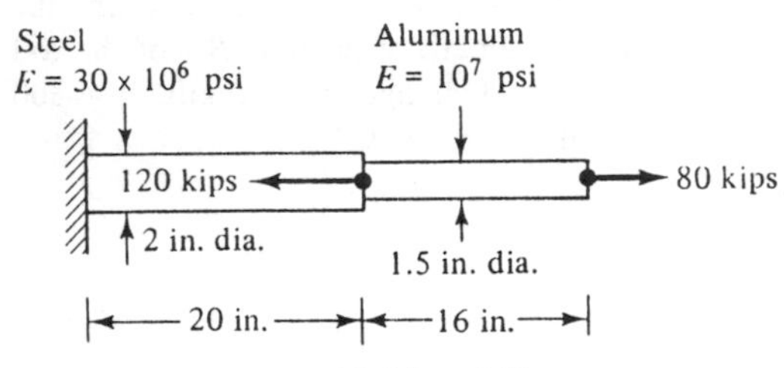

Problem 6.12

6.13. Select the lightest weight W-shape (see Table A.3) that will support the loads shown if the overall change in length is not to exceed 0.04 in. $E = 30 \times 10^6$ psi.

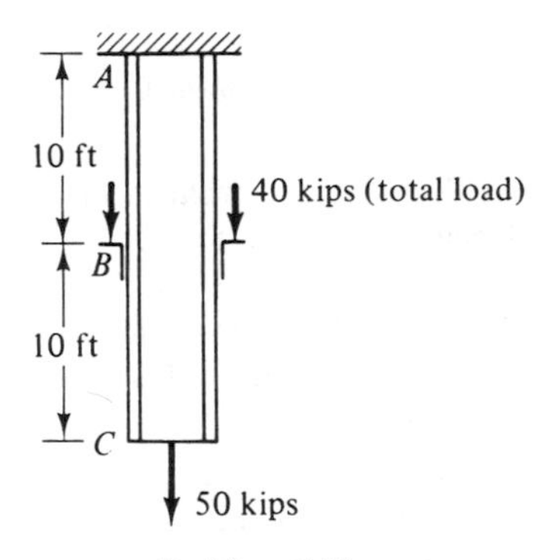

Problem 6.13

6.14. A square concrete pillar with a density of 2,400 kg/m^3 is loaded as shown. What is the overall shortening of the member if (a) its weight is neglected and (b) its weight is included? $E = 21$ GN/m^2.

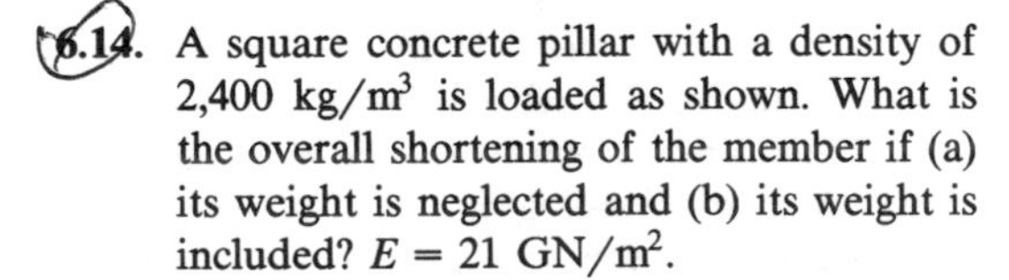

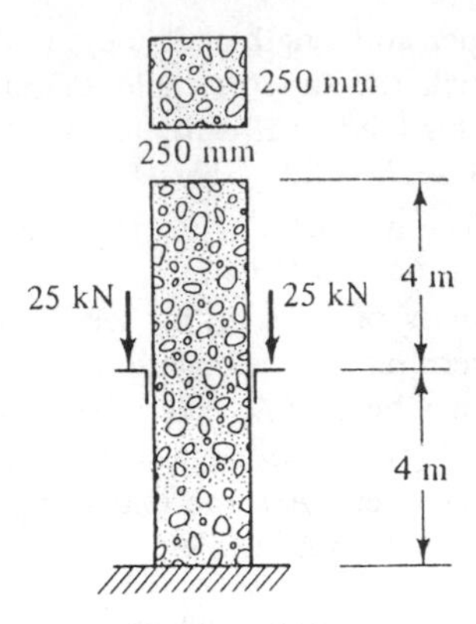

Problem 6.14

6.15. Determine the overall shortening of the obelisk in Problem 5.28 (Section 5.3) under the action of its own weight. Ignore any deformations of the pyramidal top.

6.16. Same as Problem 6.13, but with the effect of the weight of the member included.

6.17. A uniform bar 2 m long made of a material with a density of 6,000 kg/m^3 and whose stress-strain curve is described by the equation $\sigma(\text{MN/m}^2) = 560\ \varepsilon^{0.3}$, is suspended from one end. Determine the total elongation of the member due to its own weight.

6.18. A round bar with a cross-sectional area of 300 mm^2 is loaded as shown. The strain gage on the bar indicates an axial strain of 1,400 $\mu\varepsilon$, and the stress-strain curve for the material is as indicated. Determine F and the total elongation of the member.

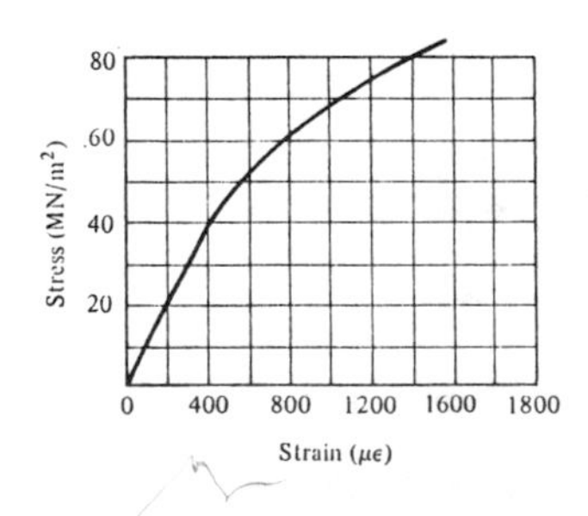

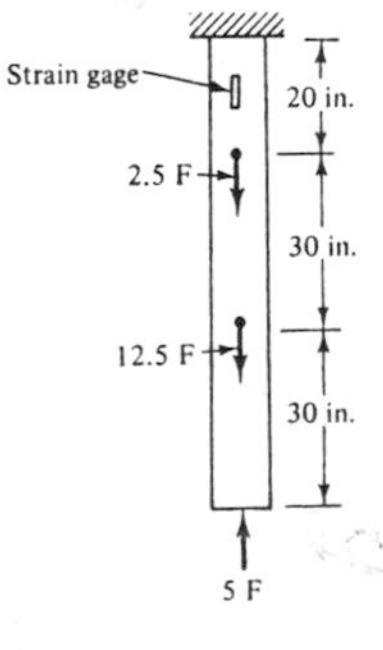

Problem 6.18

6.4 STATICALLY INDETERMINATE PROBLEMS

Consider the problem of determining the forces acting upon the rigid bar shown in Figure 6.9(a). The bar is hinged at the left end and supported by cables at B and C. From the free-body diagram [Figure 6.9(b)], it is evident that there are four unknowns and only three equations from which to determine them. Thus, the forces acting cannot be determined solely from the equations of equilibrium. This is an example of what was referred to in Section 3.3 as a *statically indeterminate* problem.

Indeterminate situations arise because there are more constraints than necessary to maintain equilibrium. In this problem, for instance, the bar could be held in place by a single cable. The additional constraints also place restrictions upon the deformations. When expressed analytically, these restrictions provide the additional equations necessary to obtain a solution.

Referring to Figure 6.9(c), we see that the cables cannot elongate independently of one another when the bar is loaded. Instead, their elongations must satisfy a definite relationship. From the geometry of the figure, we have

$$\frac{\delta_{BD}}{a} = \frac{\delta_{CE}}{b}$$

or

$$\delta_{BD} = \left(\frac{a}{b}\right)\delta_{CE}$$

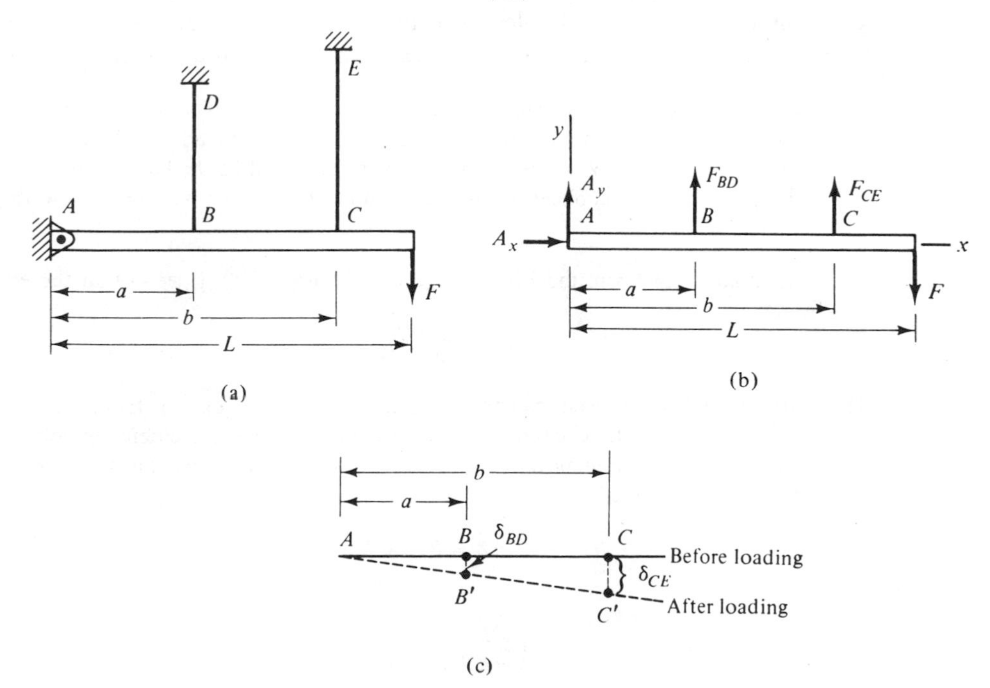

Figure 6.9

This relationship is called the *compatibility condition*, for it states the condition that the deformations must satisfy in order to be compatible with the physical situation.

The fourth equation needed for the force analysis of the bar is obtained by substituting the appropriate load-deformation relations into the compatibility condition. Assuming that the stresses in the cables are below the proportional limit, we obtain

$$\left(\frac{FL}{AE}\right)_{BD} = \left(\frac{a}{b}\right)\left(\frac{FL}{AE}\right)_{CE}$$

This equation and the three equations of equilibrium are sufficient for determining all the forces acting upon the bar and upon the cables. Once the forces on the cables have been determined, the assumption concerning the stresses within them can be checked.

The most important thing to notice about this example is that the deformations had to be considered in order to perform the force analysis, even though they were of no interest otherwise. This is characteristic of statically indeterminate problems.

The compatibility conditions are generally different for each different problem, but a clue to their form can usually be obtained from the problem statement. Sketches of the expected deformations, as in Figure 6.9(c), are an invaluable aid in determining the necessary relationships between them. Further illustrations are given in the following examples. The basic procedures outlined here will also be used in later chapters for statically indeterminate torsion and bending problems.

Example 6.4 Forces In a Statically Indeterminate Bar. A bar with cross-sectional area A and length L is attached to immovable walls at each end and loaded as shown in Figure 6.10(a). (a) What are the reactions at each end of the bar and (b) how far does the point of load application move? Assume that the stresses are below the proportional limit.

Solution. (a) From the FBD of the bar [Figure 6.10(b)], we obtain the equilibrium condition

$$F_A + F_C = F$$

There are two unknown reactions and only one equilibrium equation; therefore, the problem is statically indeterminate. Consequently, we must also consider the deformations.

Since the walls are immovable, the total elongation of the bar must be zero. Thus, the compatibility condition is

$$\delta = \delta_{AB} + \delta_{BC} = 0$$

or

$$\left(\frac{PL}{AE}\right)_{AB} + \left(\frac{PL}{AE}\right)_{BC} = 0$$

Using the values of P obtained from the normal force diagram [Figure 6.10(b)] and

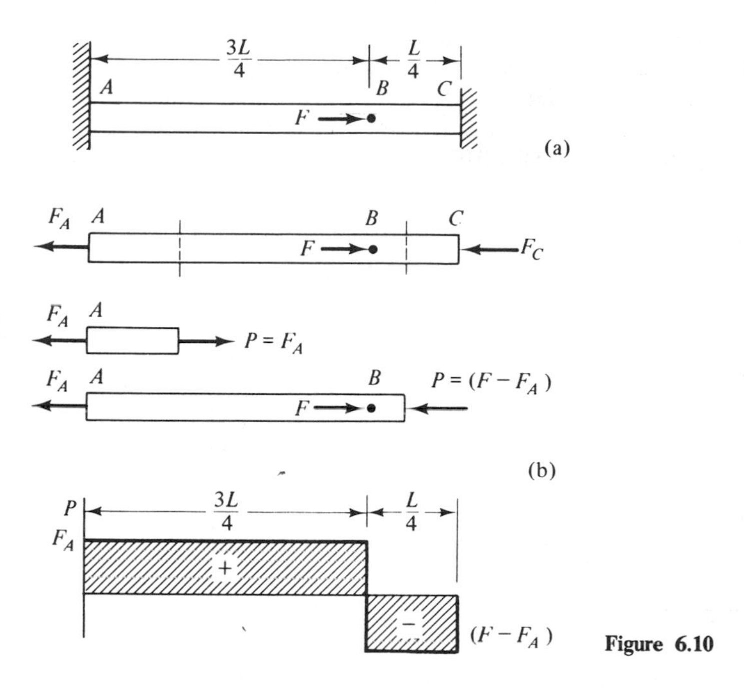

Figure 6.10

recognizing that the quantity AE cancels out, we have

$$F_A \frac{3L}{4} + \left[-(F - F_A) \frac{L}{4} \right] = 0$$

$$F_A = \frac{F}{4}$$

Thus, the reactions are

$$F_A = \frac{F}{4}$$

$$F_C = \frac{3F}{4}$$

Answer

This problem can also be approached from the viewpoint that the amount of elongation of portion AB of the bar must equal the amount of shortening of portion BC. In this case, the compatibility condition is written as

$$|\delta_{AB}| = |\delta_{BC}|$$

The results obtained are the same, but there is more chance of making sign errors by using this approach.

(b) The point of load application moves to the right a distance equal to the elongation of portion AB of the bar, which is

$$\delta_{AB} = \left(\frac{PL}{AE} \right)_{AB} = \frac{(F/4)(3L/4)}{AE} = \frac{3}{16} \frac{FL}{AE}$$ **Answer**

This distance is also equal to the amount of shortening of portion BC.

Example 6.5. Forces In a Composite Column. A column consisting of a 10-in. standard steel pipe filled with high-strength concrete supports a total load of 100 kips applied through a rigid loading plate [Figure 6.11(a)]. What portion of the load is carried by each material if (a) the concrete is initially level with the top of the pipe and (b) the level of the concrete is initially 0.01 in. below the top of the pipe? $E_S = 30 \times 10^6$ psi and $E_C = 3 \times 10^6$ psi.

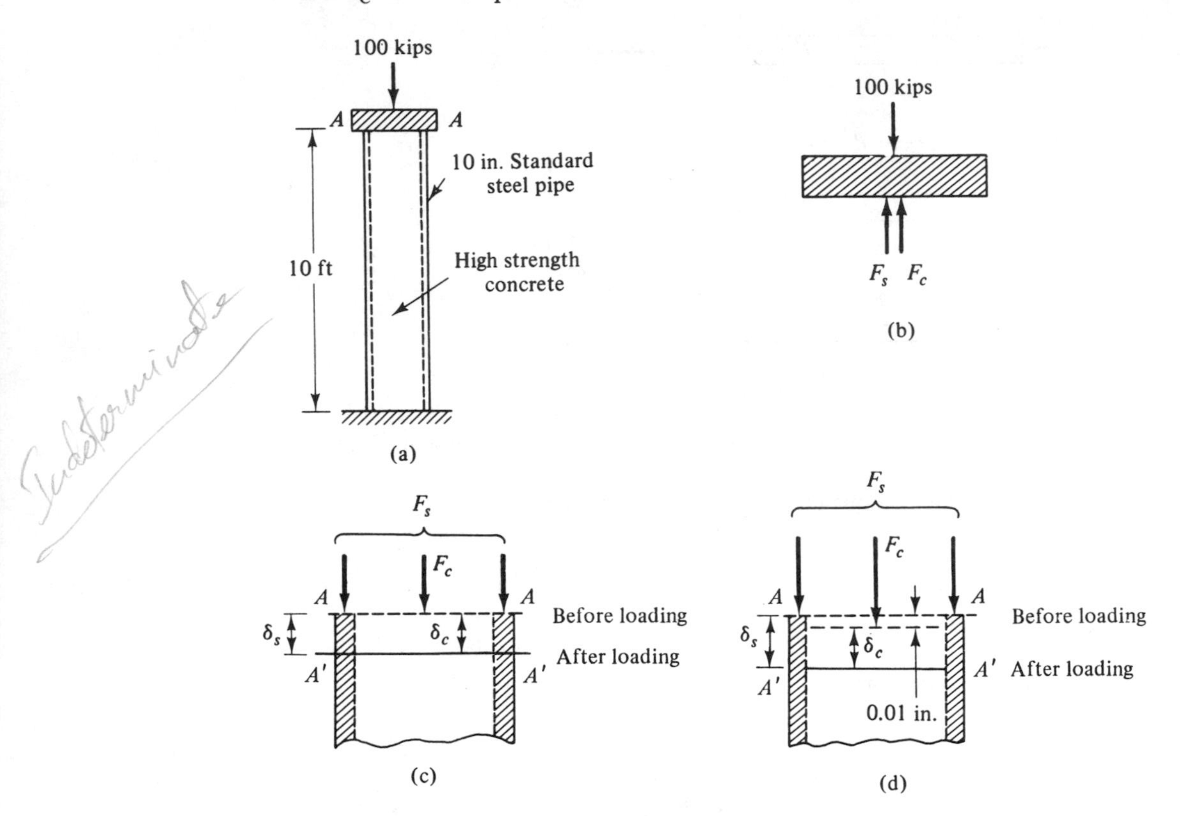

Figure 6.11

Solution. (a) Both parts of the member are obviously in compression and will shorten.

Let F_S and F_C denote the forces acting upon the steel and concrete, respectively. From the FBD of the loading plate [Figure 6.11(b)], we have

$$F_S + F_C = 100 \text{ kips}$$

Since the concrete is level with the top of the pipe, both materials must deform the same amount [Figure 6.11(c)]. Thus, the compatibility condition can be written as

$$|\delta_S| = |\delta_C|$$

or, assuming linearly elastic material behavior, as

$$\left(\frac{FL}{AE}\right)_S = \left(\frac{FL}{AE}\right)_C$$

From Table A.7, the cross-sectional area of the pipe is $A_S = 11.9 \text{ in.}^2$ and its inside diameter is 10.020 in. Thus,

$$A_C = \frac{\pi d^2}{4} = \frac{\pi(10.02 \text{ in.})^2}{4} = 78.9 \text{ in.}^2$$

Substituting the values of A and E into the compatibility condition and recognizing that the lengths cancel, we have

$$F_S = \frac{(EA)_S}{(EA)_C} F_C = \frac{(30 \times 10^6 \text{ lb/in.}^2)(11.9 \text{ in.}^2)}{(3 \times 10^6 \text{ lb/in.}^2)(78.9 \text{ in.}^2)} F_C = 1.5 F_C$$

Combining this expression with the equilibrium equation, we obtain

$$F_S = 60 \text{ kips (compression)}$$
$$F_C = 40 \text{ kips (compression)}$$

Answer

The positive values obtained for F_S and F_C indicate that they are directed as shown on the FBD.

We now check the assumption of linearly elastic material behavior. The maximum compressive stresses in the steel and in the concrete are

$$\sigma_S = \left(\frac{P}{A}\right)_S = \frac{60 \text{ kips}}{11.9 \text{ in.}^2} = 5.0 \text{ ksi}$$

$$\sigma_C = \left(\frac{P}{A}\right)_C = \frac{40 \text{ kips}}{78.9 \text{ in.}^2} = 0.5 \text{ ksi}$$

The stress-strain diagram for concrete does not have a definite linear range. However, the stress σ_C is well below the ultimate strength of 5 ksi (see Table A.2); therefore, it is reasonable to approximate the stress-strain curve by a straight line over this limited range of stress. Since the stress in the steel is also below the proportional limit, the solution is valid.

(b) The question here is, does the steel deform enough so that the loading plate comes into contact with the concrete? We assume that it does. If the load on the concrete turns out to be compressive, the assumption is confirmed; if it turns out to be tensile, which is physically impossible, the assumption is incorrect and all the load is carried by the pipe.

Referring to Figure 6.11(d), we see that the compatibility condition is

$$|\delta_S| = |\delta_C| + 0.01 \text{ in.}$$

Otherwise, the solution is the same as in part (a). We have

$$\left(\frac{FL}{AE}\right)_S = \left(\frac{FL}{AE}\right)_C + 0.01 \text{ in.}$$

$$F_S = \frac{(EA)_S}{(EA)_C} F_C + \frac{(AE)_S}{L_S}(0.01 \text{ in.})$$

The coefficient of F_C was computed in part (a). Thus,

$$F_S = 1.5 F_C + \frac{(11.9 \text{ in.}^2)(30 \times 10^6 \text{ lb/in.}^2)(0.01 \text{ in.})}{120 \text{ in.}}$$
$$= 1.5 F_C + 29{,}800 \text{ lb}$$

Combining this equation with the equilibrium equation, we obtain

$$F_C = 28 \text{ kips (compression)}$$

$$F_S = 72 \text{ kips (compression)}$$

Answer

The positive values obtained for F_C and F_S indicate that both forces are compressive; therefore, our assumption that the loading plate makes contact with the concrete is correct.

The stress in the steel is now

$$\sigma_S = \left(\frac{P}{A}\right)_S = \frac{72 \text{ kips}}{11.9 \text{ in.}^2} = 6.1 \text{ ksi}$$

which is still well below the proportional limit. Since the stress in the concrete is less than in part (a), it is also below the proportional limit. Thus, our solution is valid.

PROBLEMS

6.19. Determine the stresses in each of the wires supporting the rigid bar shown if $F = 20$ kN.

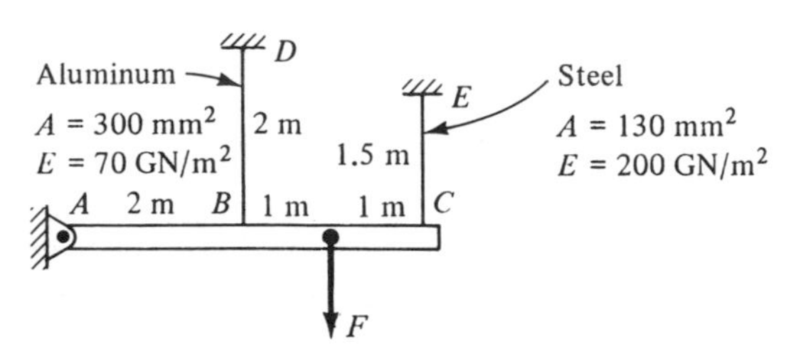

Problem 6.19

6.20. What is the maximum load F that can be applied to the rigid horizontal bar in Problem 6.19 if the structural steel wire is not to yield? Use a safety factor of 3.

6.21. Find the ratio of the cross-sectional areas of the aluminum and steel wires for which the rigid slab shown will hang in a horizontal position.

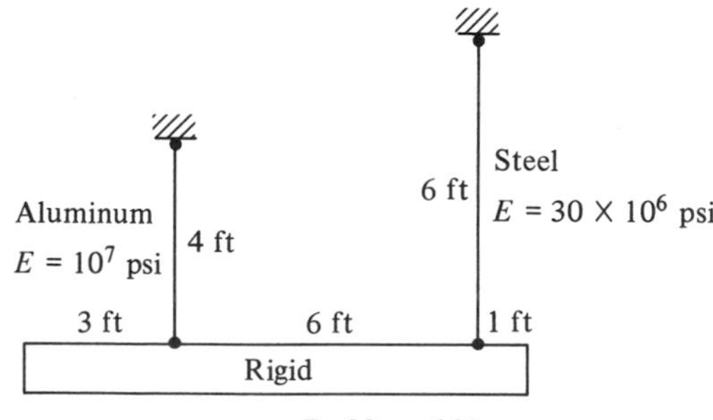

Problem 6.21

6.22. A rigid bar is supported by a steel rod with cross-sectional area A and by a wood post with cross-sectional area $10A$, as shown. What is the minimum required value of A if the bar is not to rotate more than 1° when the load is applied and if the stress in the post is not to exceed 2,000 psi? Assume that the post does not buckle. What diameter pin is required to connect the rod to the bar at C if the allowable shear stress in the pin is 5,000 psi?

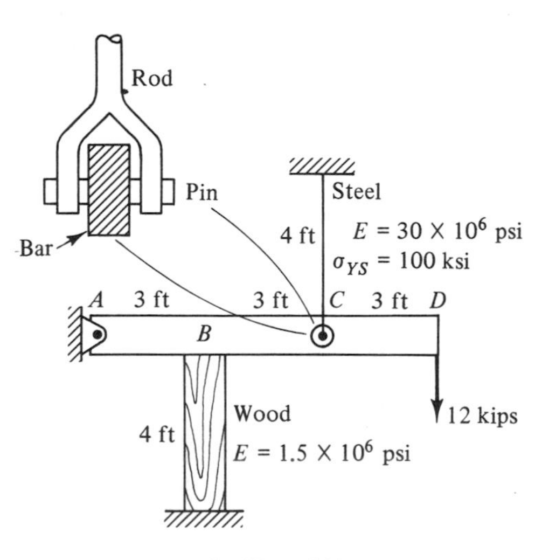

Problem 6.22

6.23. Two rectangular blocks made of steel ($E = 30 \times 10^6$ psi) and aluminum ($E = 10 \times 10^6$ psi) are compressed between rigid plates, as shown. Both blocks are 4-in. thick. Determine the location x of the applied load such that the rigid plates will remain horizontal when the load is applied. What are the corresponding stresses in the blocks?

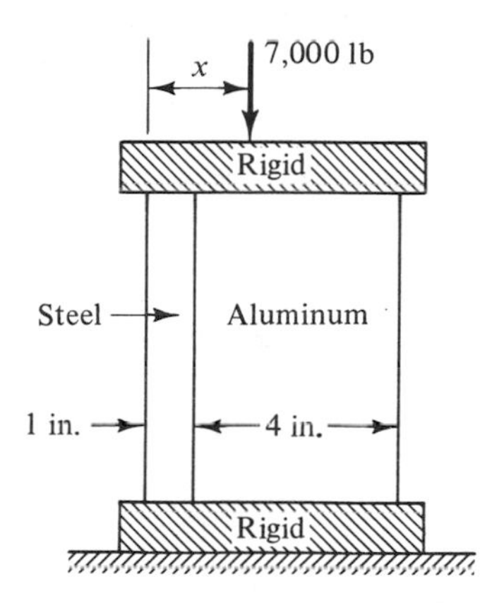

Problem 6.23

6.24. A rigid slab with mass m is supported by three columns, as shown. If $m = 12{,}000$ kg, what is the compressive force on each of the members? Assume that the columns do not buckle.

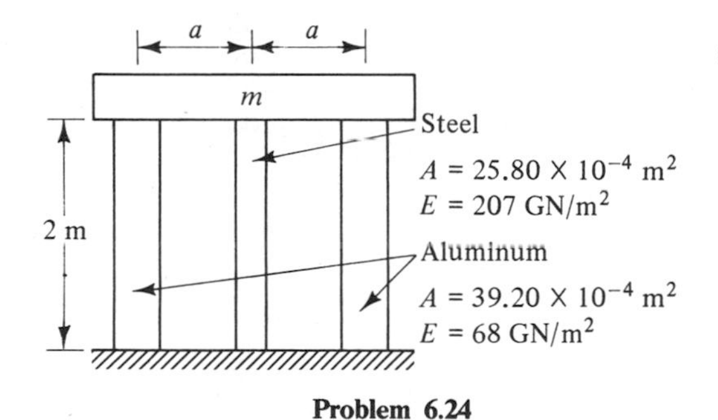

Problem 6.24

6.25. In Problem 6.24, for what mass m of the slab will the stress in the aluminum members be two-thirds that in the steel, if there initially is a gap of 1.5 mm between the top of the center member and the slab?

6.26. A 3,000-lb object is supported by three steel wires, as shown. Two of the wires are cut longer than the third one; thus, they are initially slack. Determine the tensile stress in each wire after the load is applied. Through what distance does the object move downward as a result of the elongations of the wires?

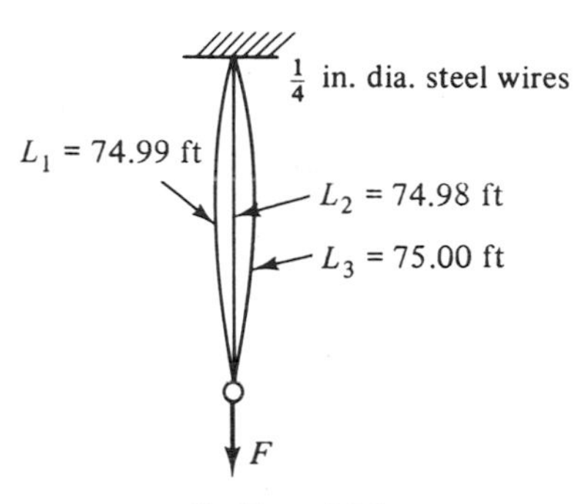

Problem 6.26

6.27. An electrical cable consists of a single 4-mm-diameter copper wire covered with a 1.5-mm thick plastic insulating material. How much will the cable elongate per kN of tensile load and per meter of length? What is the ratio of the tensile stress in the plastic and in the copper? Use $E_{copper} = 100$ GN/m^2 and $E_{plastic} = 10$ GN/m^2.

6.28. The pedestal shown consists of a 12-in.-diameter (nominal) steel pipe filled to the top with concrete. What is the maximum applied load F that can be supported if the allowable stresses are 30 ksi in the pipe and 2 ksi in the concrete? Use a safety factor of 3.

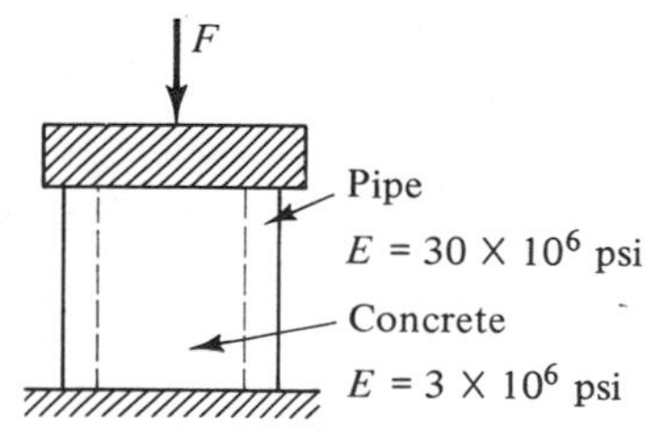

Problem 6.28

6.29. The concrete pedestal shown is reinforced by five steel rods with a diameter of $\frac{1}{2}$ in. and a yield strength of 80 ksi. The concrete has an ultimate compressive strength of 5 ksi. What is the largest load F that can be supported without yielding the steel or crushing the concrete? Use a safety factor of 2. How does this compare with the load that could be supported by the pedestal if it had no reinforcing? $E_S = 30 \times 10^6$ psi and $E_C = 3 \times 10^6$ psi.

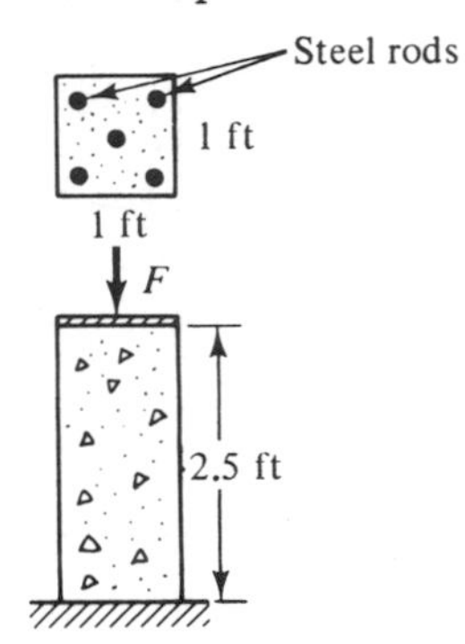

Problem 6.29

6.30. A brass sleeve 0.4-m long with 20-mm ID and 25-mm OD is slipped over a 12-mm diameter high-strength steel bolt and the nut is advanced one-eighth turn beyond the point where the ends of the members first come into contact. If the bolt has 0.6 threads/mm, what is the stress induced in each of the members?

6.31. A composite bar is firmly attached to unyielding supports at the ends and is subjected to the axial load F shown. If the aluminum is stressed to 70 MN/m², what is the stress in the steel? Assume elastic behavior.

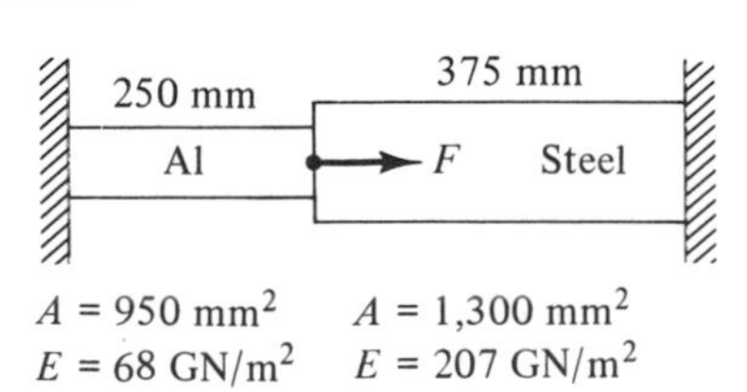

Problem 6.31

6.32. Same as Problem 6.31, except that initially there is a gap of 0.05 mm between the right end of the steel bar and the support.

6.33. In an attempt to increase the load-carrying capacity of a 10 in.× 10 in. (nominal) oak column, 4 in.× 4 in.× $\frac{1}{4}$ in. steel angle sections are attached at each of the four corners, as shown. By what percentage does the presence of the angle sections change the load-carrying capacity of the column? The allowable stresses are 20 ksi for the steel and 4 ksi for the oak; $E_s = 30 \times 10^6$ psi and $E_w = 1.8 \times 10^6$ psi.

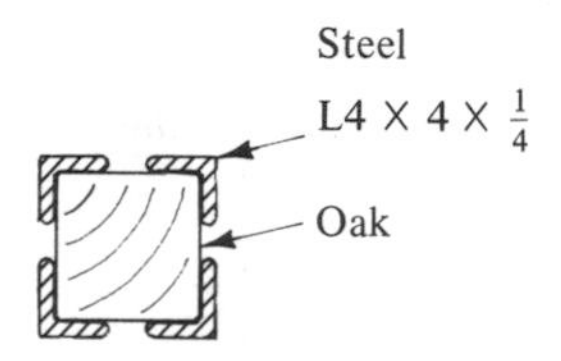

Problem 6.33

6.34. The bar shown has a cross-sectional area of 400 mm². The stress-strain diagram for the material is as indicated, and is the same in tension and in compression. If $F = 150$ kN, determine the normal stress in each portion of the bar.

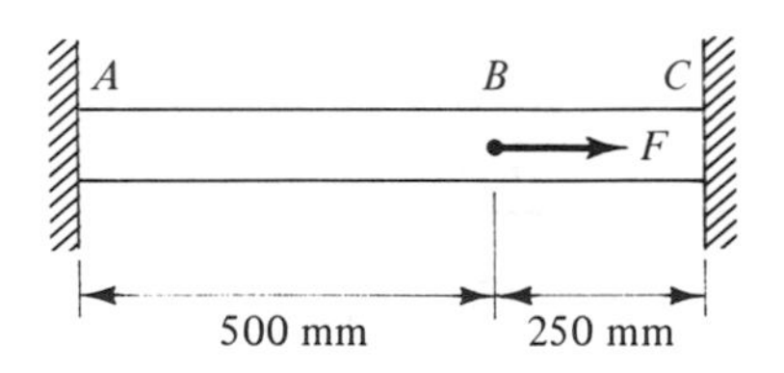

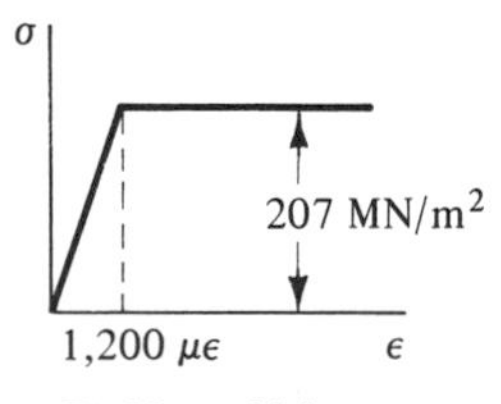

Problem 6.34

6.35. Same as Problem 6.34, except that $F = 100$ kN.

6.5 THERMAL STRAINS AND STRESSES

When a material is heated or cooled, it expands or contracts. For example, consider a dimension L_0 of a body at some reference temperature T_0. Under a uniform temperature change $\Delta T = T - T_0$, this dimension will change by an amount

$$\delta = L - L_0 = \alpha L_0 \Delta T \tag{6.3}$$

This relation is based upon experimental observations and is valid for moderate changes in temperature. The constant of proportionality, α, is called the *coefficient of thermal expansion*. It has the units of $(°\text{F})^{-1}$ or $(°\text{C})^{-1}$. Typical values of α for some common materials are given in Table A.2 of the Appendix.

Equation (6.3) applies to all linear dimensions of a body subjected to a uniform temperature change. For example, it applies to the change in diameter of a rod as well as to the change in length. If ΔT varies from point to point within the body, the situation is not so simple. In this case, Eq. (6.3) must be applied to infinitesimal line elements and the results integrated to determine the total dimensional change. Problems involving nonuniform temperature changes will not be considered in this text.

The thermally induced change in dimension gives rise to a normal strain

$$\varepsilon_T = \frac{\delta}{L_0} = \alpha \Delta T \tag{6.4}$$

This strain, denoted by ε_T, is called the *thermal strain* to distinguish it from the *mechanical strain* due to the applied loads. The total strain and deformation is the sum of those due to the applied loads and the temperature change. Note that the thermal strain and deformation are positive if the temperature increases and negative if it decreases.

Thermal strains due to a uniform temperature change differ from mechanical strains in that there is no stress associated with them unless the expansion or contraction of the body is restricted. Then, very large stresses can be generated. These so-called *thermal stresses* are responsible for the buckling of pavements on hot summer days. They are also a problem in large-scale structures, such as bridges, piping systems, and railroad tracks, where expansion joints or some other means of accommodating the thermal deformations must be provided. By their very nature, thermal stress problems are statically indeterminate.

Example 6.6. Stresses and Deformations In a Cooled Rod. A brass rod $[\alpha = 19 \times 10^{-6}$ $(°\text{C})^{-1}]$ has a length of 3 m and a cross-sectional area of 1.2×10^{-3} m^2 at 20°C. The rod is then subjected to an axial tensile loading of 24 kN and the temperature is decreased to 0°C. Determine (a) the maximum normal stress within the member and (b) the change in its length; $E = 83$ GN/m^2.

Solution. (a) Since the thermal contraction is not restricted in any way, the only stress is that due to the applied load. The change in cross-sectional area due to the temperature change is proportional to the square of α and is considered negligible. Thus,

$$\sigma = \frac{P}{A} = \frac{24 \times 10^3 \text{ N}}{1.2 \times 10^{-3} \text{ m}^2} = 20 \times 10^6 \text{ N/m}^2 \text{ or } 20 \text{ MN/m}^2 \text{ (tension)} \qquad \textbf{Answer}$$

(b) The total change in length is the sum of the changes due to the temperature decrease and the applied load. Since the stress in the rod is below the proportional limit (see Table A.2), we have

$$\begin{aligned}\delta_{\text{total}} &= \frac{PL}{AE} + \alpha L\,\Delta T \\ &= \frac{(24 \times 10^3 \text{ N})(3 \text{ m})}{(1.2 \times 10^{-3} \text{ m}^2)(83 \times 10^9 \text{ N/m}^2)} \\ &\quad + \left[19 \times 10^{-6}(^\circ\text{C})^{-1}\right](3 \text{ m})(0^\circ\text{C} - 20^\circ\text{C}) \\ &= -0.42 \times 10^{-3} \text{ m or } 0.42 \text{ mm (shortening)} \qquad \textbf{Answer}\end{aligned}$$

Here, the contraction due to the decrease in temperature exceeds the elongation due to the applied load.

Example 6.7. Thermal Stresses In a Bar. A 6-ft-long steel bar with a cross-sectional area of 3 in.^{2} is placed between two walls and the temperature is increased by 70°F [Figure 6.12(a)]. Assuming that the bar does not buckle, what is the maximum normal stress within it if (a) the walls are immovable and (b) the walls move apart 0.01 in.?

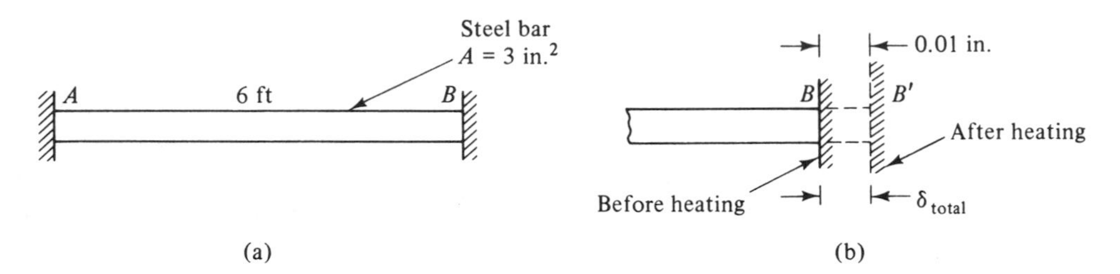

Figure 6.12

Solution. (a) As the bar tends to expand, the walls exert compressive forces of unknown magnitude F on it and prevent it from changing length. The compatibility condition is

$$\delta_{\text{total}} = \frac{-FL}{AE} + \alpha L\,\Delta T = 0$$

from which we obtain

$$\begin{aligned}F = \alpha AE\,\Delta T &= \left[6.5 \times 10^{-6}(^\circ\text{F})^{-1}\right](3 \text{ in.}^2)(30 \times 10^6 \text{ lb/in.}^2)(70^\circ\text{F}) \\ &= 40{,}950 \text{ lb}\end{aligned}$$

The maximum normal stress is

$$\sigma = \frac{P}{A} = \frac{40{,}950 \text{ lb}}{3 \text{ in.}^2} = 13{,}650 \text{ psi (compression)} \qquad \textbf{Answer}$$

Since this stress is well below the proportional limit of steel, the solution is valid.

(b) In this case, the bar can expand an amount equal to the distance the walls move apart [Figure 6.12(b)]. Thus, the compatibility condition is

$$\delta_{\text{total}} = \frac{-FL}{AE} + \alpha L \Delta T = 0.01 \text{ in.}$$

from which we obtain

$$F = \alpha AE \Delta T - \frac{AE}{L}(0.01 \text{ in.})$$

The first term on the right-hand side of the equation is simply the value of F computed in part (a). Thus,

$$F = 40{,}950 \text{ lb} - \frac{(3 \text{ in.}^2)(30 \times 10^6 \text{ lb/in.}^2)(0.01 \text{ in.})}{72 \text{ in.}} = 28{,}450 \text{ lb}$$

$$\sigma = \frac{P}{A} = \frac{28{,}450 \text{ lb}}{3 \text{ in.}^2} = 9{,}480 \text{ psi (compression)} \qquad \textbf{Answer}$$

A comparison of this result with that obtained in part (a) shows that the thermal stress can be reduced considerably by providing for even a relatively small amount of expansion or contraction.

This problem can also be approached in a slightly different way by imagining that one wall is removed so that the bar can expand freely and then computing the force F required to compress it back to its required length. It is left to the reader to show that this approach leads to the same results presented here.

PROBLEMS

6.36. Design drawings for an offshore drilling platform specify that a certain steel member be 30-m long, with a tolerance of ± 10 mm. If the member is cut to exact dimension in a shop where the temperature is 20°C, will the length still be within tolerance when the member reaches a temperature of 60°C while awaiting assembly in the field?

6.37. If the member in Problem 6.36 experiences temperatures ranging from -8°C to $+60$°C before assembly in the field, is there a dimension to which it can be cut in the shop at 20°C so that the length will always be within tolerance? If so, what is this length? Assume that the member can be cut to an exact dimension.

6.38. Vertical expansion joints are placed at regular intervals in the long concrete walls of a building to alleviate the effects of thermal expansion and contraction. What is the minimum required spacing of the joints if each can accommodate an expansion or contraction of $\frac{1}{8}$ in? The wall temperature is expected to range from -30°F to $+120$°F, and the joints were installed at 70°F. Use $\alpha = 6 \times 10^{-6} (°\text{F})^{-1}$.

6.39. An aluminum [$\alpha = 12.6 \times 10^{-6}$ $(°\text{F})^{-1}$] plate 3-ft long, 2-ft wide, and $\frac{3}{4}$-in thick contains a 1-ft diameter hole at the center. What are the dimensions of the plate and diameter of the hole if the temperature is raised 100°F? What is the percentage change in area of the hole?

6.40. A 15-mm diameter brass bolt 0.4-m long is subjected to a tensile loading of 18 kN while undergoing a temperature decrease of 60°C. Determine the changes in the length and diameter of the bolt. What is the maximum tensile stress in the bolt?

6.41. The two bars shown each have a cross-sectional area of 1,200 mm^2 and fit snugly between the immovable walls. If the temperature is increased by 60°C, what is the maximum normal stress in each bar, and how far and in which direction does the interface B between them move?

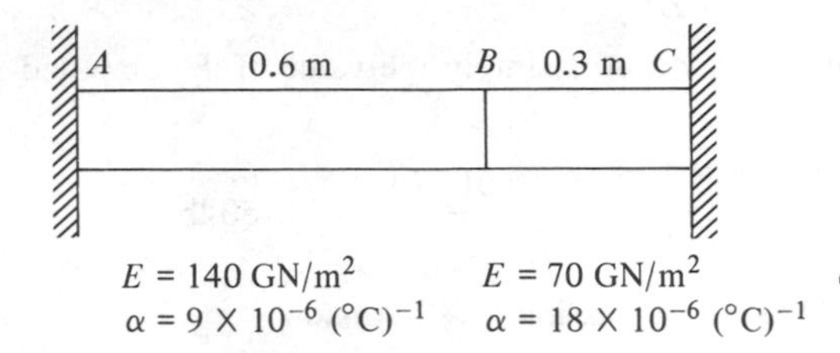

$E = 140\ GN/m^2$
$\alpha = 9 \times 10^{-6}\ (°C)^{-1}$

$E = 70\ GN/m^2$
$\alpha = 18 \times 10^{-6}\ (°C)^{-1}$

Problem 6.41

6.42. Same as Problem 6.41, except that the walls move apart 0.2 mm when the temperature is increased.

6.43. A stepped cylinder is attached to immovable surfaces at each end and then loaded as shown. For what change in temperature will the axial force in the larger portion be half that in the smaller portion? Express the results in terms of the material properties E and α, the area A, and the applied load F.

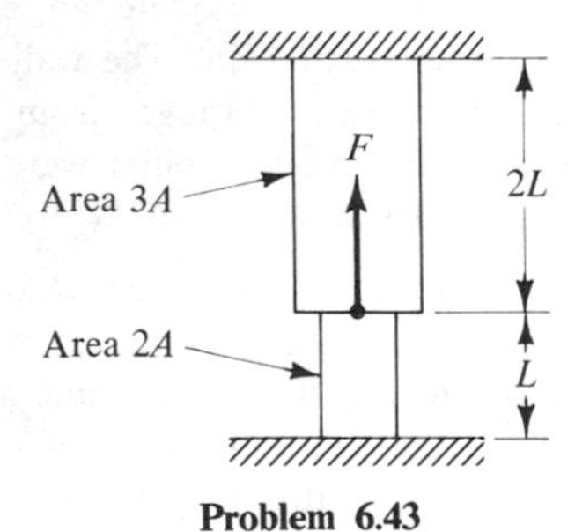

Problem 6.43

6.44. A 2024-T3 aluminum bar is attached to rigid walls at each end by smooth pins as shown. Determine the required diameter d of the pins if the assembly is to withstand a temperature decrease of 60°F. The allowable average normal stress in the bar is 20 ksi and the allowable average shear stress in the pins is 10 ksi.

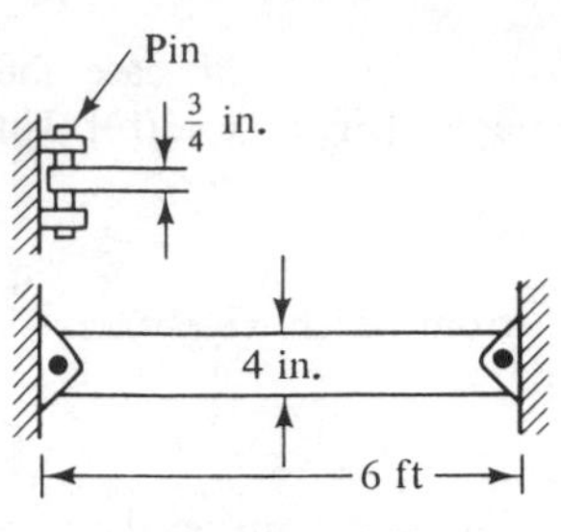

Problem 6.44

6.45. A 254-mm × 102-mm steel Universal beam with a mass of 22 kg/m (see Table A.8) is used as a column in a building. The member is attached to rigid supports at the top and bottom and is loaded as shown. What are the reactions at the supports after a temperature decrease of 30°C? Neglect the weight of the member.

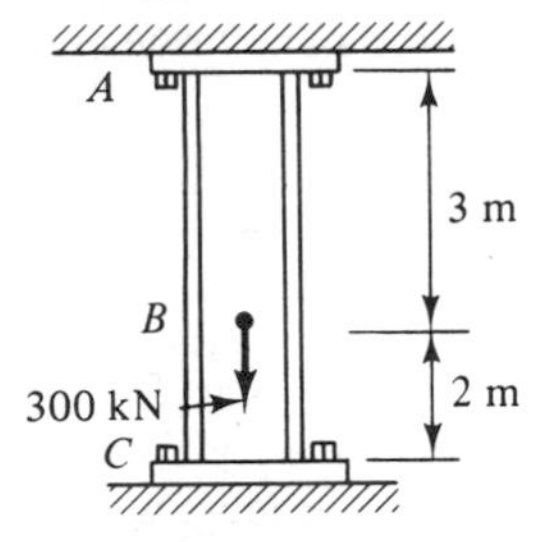

Problem 6.45

6.46. A 50-lb object is suspended by a galvanized (zinc-coated) steel wire at 70°F. The wire has a diameter of $\frac{1}{4}$ in. and the zinc coating is 0.006-in. thick. What is the maximum normal stress in the steel core and in the zinc coating at 30°F? $E_z = 16 \times 10^6$ psi and $\alpha_z = 17.2 \times 10^{-6} (°F)^{-1}$.

6.47. In Problem 6.46, for what change in temperature, if any, will the zinc coating be stress free?

6.48. A rigid floor slab with a mass of 3,200 kg rests on three columns, as shown. What is the compressive force on each of the members (a) at installation and (b) after a temperature decrease of 20°C?

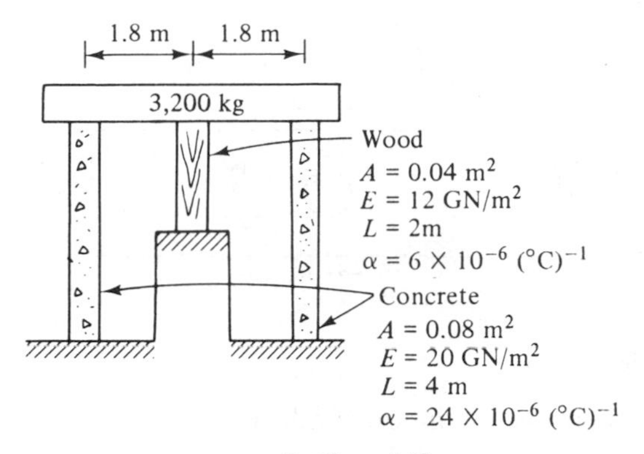

Problem 6.48

6.49. The rigid bar ABC shown is originally horizontal. Determine the vertical force F that must be applied at end A to return the bar to a horizontal position after a temperature rise of 100°F. What are the resulting normal stresses in support members BE and CD?

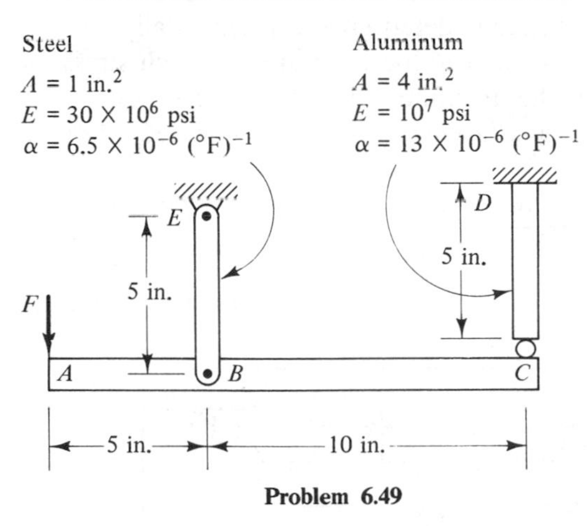

Problem 6.49

6.50. Three parallel bars 250-mm long, 12-mm thick, and 25-mm wide are riveted together at each end with a single 12-mm diameter rivet, as shown. The outer bars are steel and the center bar is aluminum. After the bars are riveted together, the temperature decreases by 40°C. Determine the maximum average normal stress in each bar and the average shear stress in the rivets.

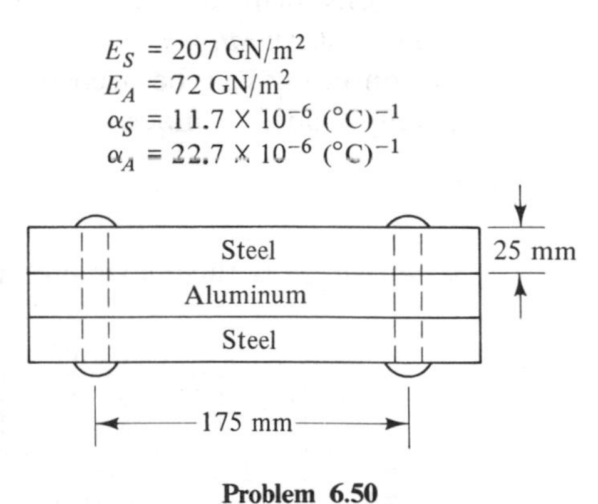

Problem 6.50

6.6 STRAIN AND STRESS CONCENTRATIONS

The strains in a member depend upon its geometry and the applied loads. Discontinuities in the geometry, such as holes and notches, give rise to localized areas of high strain. These effects, which are evident in Figures 6.1 and 6.3, are called *strain concentrations*. Strain concentrations may also occur at other discontinuities, such as voids and rigid inclusions in the material, and at points of load application.

The effect of geometrical discontinuities on the strains can be described in terms of a *strain concentration factor*, K_ε, defined as the ratio of the maximum strain with the discontinuity and the value the strain would have if the discontinuity were

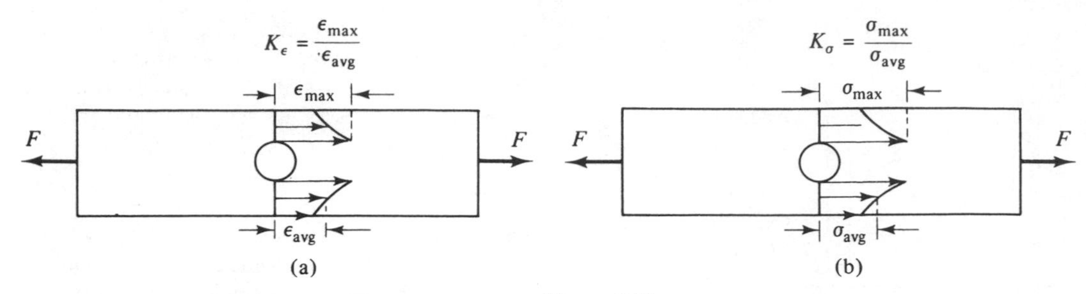

Figure 6.13

not present. For an axially loaded member, this definition can be expressed as

$$K_\varepsilon = \frac{\varepsilon_{max}}{\varepsilon_{avg}} \tag{6.5}$$

where ε_{avg} is the average strain over the cross section and ε_{max} is the maximum strain [Figure 6.13(a)]. The value of K_ε is usually determined experimentally.

Geometrical discontinuities can also result in localized areas of high stress, or *stress concentrations*. The increase in the maximum value of stress due to a discontinuity can be accounted for by a *stress concentration factor*, K_σ. For an axially loaded member,

$$K_\sigma = \frac{\sigma_{max}}{\sigma_{avg}} \tag{6.6}$$

where σ_{max} is the maximum stress and σ_{avg} is the average stress over the cross

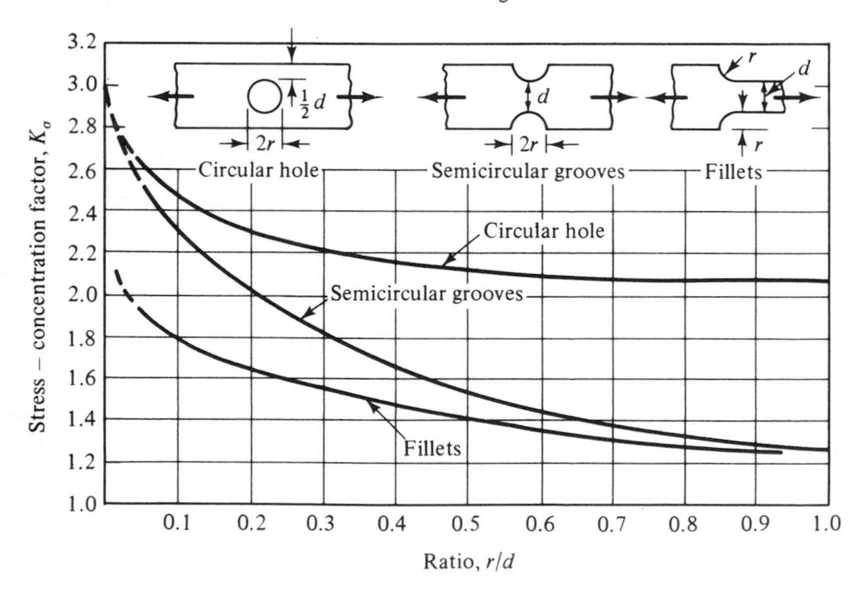

Figure 6.14 Stress Concentration Factors for Axially Loaded Flat Bars.

section [Figure 6.13(b)]. Thus, the maximum stress can be determined by multiplying the value obtained from the relation $\sigma = P/A$ by the factor K_σ.

The value of K_σ depends upon the strain distribution (i.e., upon K_ε) and the stress-strain relation for the material. For linearly elastic material behavior, the stress is directly proportional to strain, and $K_\sigma = K_\varepsilon$. The corresponding values of K_σ for axially loaded flat bars containing circular holes, semicircular grooves, and fillets are given in Figure 6.14. These results are based upon average stresses computed by using the net cross-sectional area of the member. For members with fillets, the area of the smaller portion is used. Note that the more drastic the change in cross section, the larger the value of K_σ. Stress concentration factors for other discontinuities, members, and loadings are available in engineering handbooks and other references.

Materials that exhibit brittle behavior usually have stress-strain diagrams that are very nearly linear all the way to fracture. Consequently, the values of K_σ given in Figure 6.14 can be used to determine the failure load for members made of such materials. The load to produce fracture is determined by setting the maximum stress at the discontinuity equal to the ultimate strength of the material.

If the material can undergo inelastic deformations, the stress distribution tends to become uniform once the yield strength is exceeded. As a result, the stress concentration factor is lower than the values given in Figure 6.14 ($K_\sigma < K_\varepsilon$). This situation is illustrated in Figure 6.15, which shows the stress distribution at a hole in a bar made of an elastic-perfectly plastic material for three different load levels. The conditions at locations a, b, and c on the cross section are denoted by the corresponding points on the stress-strain diagram.

As the load is increased from zero, the maximum strain increases and so does the maximum stress, until it is just equal to the yield strength of the material. This is the situation shown in Figure 6.15(a). Up to this load, $K_\sigma = K_\varepsilon$, and the values given in Figure 6.14 apply. If the load is increased further, the maximum strain increases, but the maximum stress remains equal to the yield strength. Strains at some distance away from the hole are now at, or above, the proportional limit strain; therefore, the stress distribution becomes more nearly uniform [Figure 6.15(b)]. This causes K_σ to decrease, since the average stress increases while the maximum stress remains the same. Thus, $K_\sigma < K_\varepsilon$.

If the load continues to increase, a situation is eventually reached where the strains are at, or above, the proportional limit strain everywhere over the cross section. The stresses are then uniformly distributed, as shown in Figure 6.15(c). This is referred to as the *fully plastic condition*, and the load at which it occurs is called the *fully plastic load*, F_{fp}. The maximum stress is equal to the average stress in this case; therefore, $K_\sigma = 1$. In other words, the discontinuity has no effect other than to decrease the cross-sectional area available to carry the load. All other effects are wiped out by the ability of the material to undergo inelastic deformations This will not be the case for all types of loadings, however. Materials that are ductile under static loadings tend to behave in a brittle fashion when subjected to cyclic loadings,

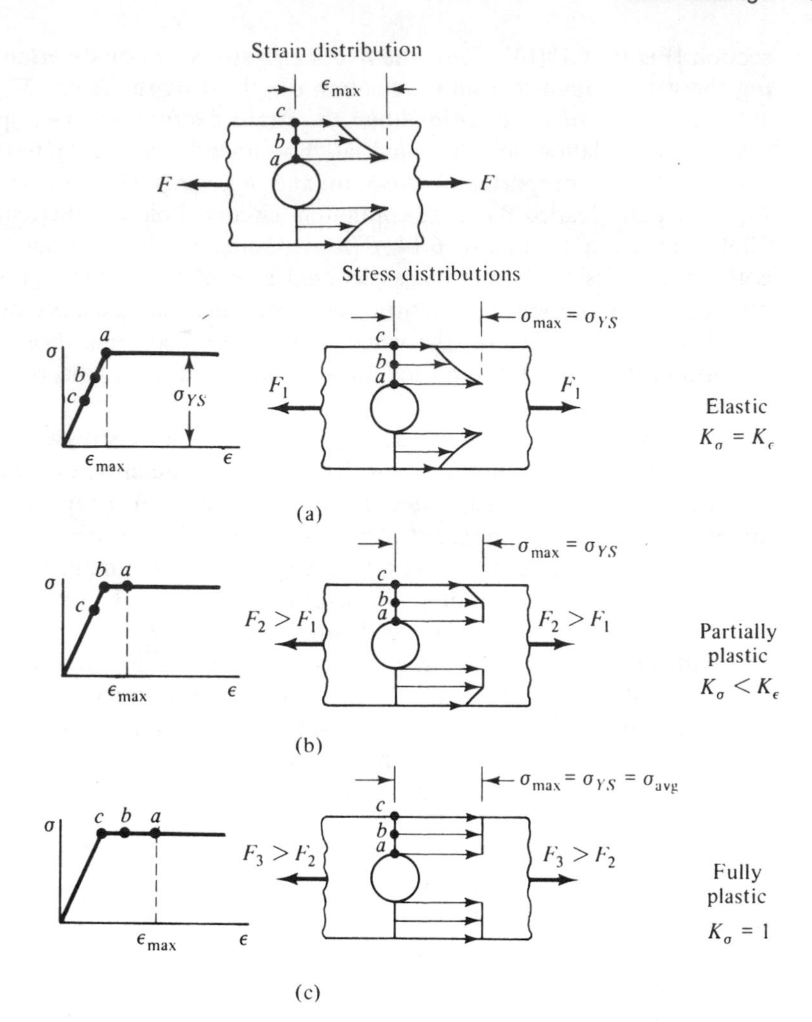

Figure 6.15

such as occur in moving machine parts or vibrating structures. For such loadings, the stress concentration remains and is an important consideration.

A ductile member does not necessarily fail when the fully plastic condition is reached. Experiments show that the region of large strains is still very much localized near the discontinuity; therefore, the overall deformations of the member will be relatively small. Gross yielding of the member and the accompanying large deformations do not result until the stresses at sections away from the discontinuity reach the yield strength of the material. However, the member may fracture at the reduced section before this happens. It is usually sufficiently accurate to take $K_\sigma = 1$ when computing the fracture load for ductile materials.

It should be clear from the preceding discussion that strain concentrations are most serious in members made of materials that behave in a brittle manner. The discontinuities result in greatly increased stresses, which can easily lead to fracture. For statically loaded members that behave in a ductile manner, the material can yield in the localized regions of high stress and continue to carry the load.

Example 6.8. Design of a Member with a Hole. What thickness t is required if the axially loaded member shown in Figure 6.16 is to support a load of 10 kips with a safety factor of 2 and (a) it is made of a brittle material with an ultimate strength of 40 ksi and (b) it is made of a ductile material with a yield strength of 36 ksi and an ultimate strength of 65 ksi? For satisfactory performance, the member must not break or undergo gross yielding.

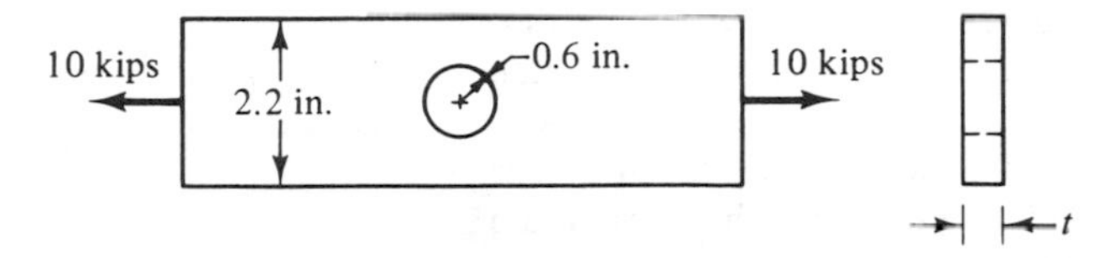

Figure 6.16

Solution. (a) The brittle material can only fail by fracture. The design (failure) load is equal to the safety factor times the working load, or 20 kips. Referring to Figure 6.14, we have $r/d = 0.6$ and $K_\sigma = 2.1$. Now

$$\sigma_{max} = K_\sigma \sigma_{avg} = K_\sigma \frac{P}{A}$$

where A is the net cross-sectional area at the hole. Fracture occurs when the maximum stress is equal to the ultimate strength of the material. Thus, we have the requirement that

$$\sigma_{max} = \frac{2.1(20{,}000 \text{ lb})}{(1 \text{ in.})t} \leqslant 40{,}000 \text{ lb/in.}^2$$

or

$$t \geqslant 1.05 \text{ in.} \qquad \textbf{Answer}$$

(b) Fracture at the hole occurs when the maximum stress there is equal to the ultimate strength of the material. We take $K_\sigma = 1$. Thus,

$$\sigma = \frac{P}{A} = \frac{20{,}000 \text{ lb}}{(1 \text{ in.})t} \leqslant 65{,}000 \text{ lb/in.}^2$$

or

$$t \geqslant 0.31 \text{ in.}$$

Gross yielding occurs when the stress away from the hole reaches the material yield strength:

$$\sigma = \frac{P}{A} = \frac{20{,}000 \text{ lb}}{(2.2 \text{ in.})t} \leqslant 36{,}000 \text{ lb/in.}^2$$

or

$$t \geqslant 0.25 \text{ in.}$$

Failure by fracture at the hole governs, and the minimum required thickness is 0.31 in.

PROBLEMS

6.51. For the member shown in Figure 6.3 (Section 6.2), estimate the values of the strain concentration factor K_ε at the hole and at the section 1 in. away from it. All strains are plotted to the same scale in the figure, and the maximum strain at the hole is 1,500 $\mu\varepsilon$.

6.52. The strain distribution on section *a-a* near a concentrated force applied to the end of an axially loaded bar is approximately as shown. Determine (a) the strain concentration factor K_ε, (b) the stress concentration factor K_σ if the material is linearly elastic, and (c) the stress distribution and the value of K_σ if the material has the stress-strain diagram shown.

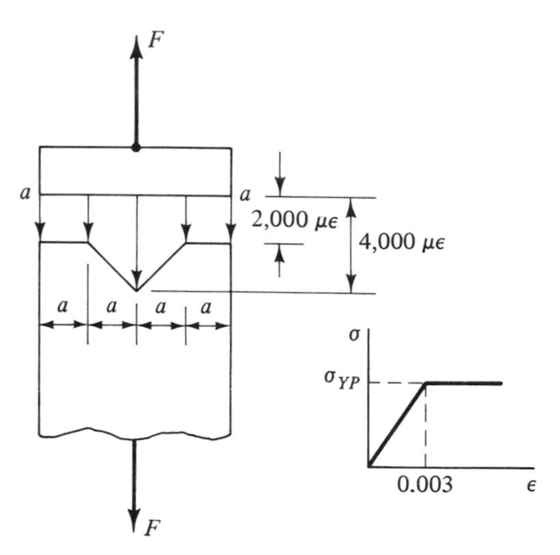

Problem 6.52

6.53. An axially loaded bar $\frac{1}{4}$-in. thick and 1-in. wide has a $\frac{1}{2}$-in. diameter hole at its center. The bar is subjected to a tensile load of 5 kips. Determine the maximum tensile stress in the bar if (a) the material exhibits brittle behavior and (b) the material exhibits ductile behavior, with a yield strength of 80 ksi and an ultimate strength of 125 ksi.

6.54. The member shown is $\frac{1}{2}$ in.-thick and is made of a material with a yield point of 35 ksi and an ultimate tensile strength of 60 ksi. Determine (a) the load F at which yielding first occurs, (b) the fully plastic load F_{fp}, and (c) the maximum load that can be supported with a safety factor of 3 without fracture or gross yielding.

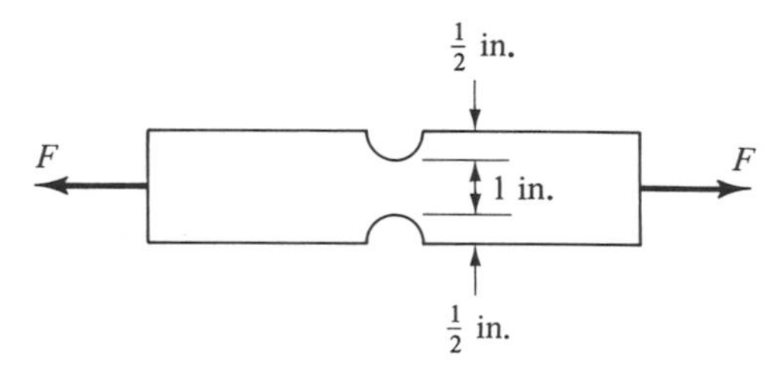

Problem 6.54

6.55. If the bar shown is 10-mm thick and is made of a ductile material with a yield strength of 300 MN/m² and an ultimate strength of 475 MN/m², determine (a) the load F at which yielding first occurs, (b) the fully plastic load F_{fp}, and (c) the maximum load F that can be supported without fracture or gross yielding.

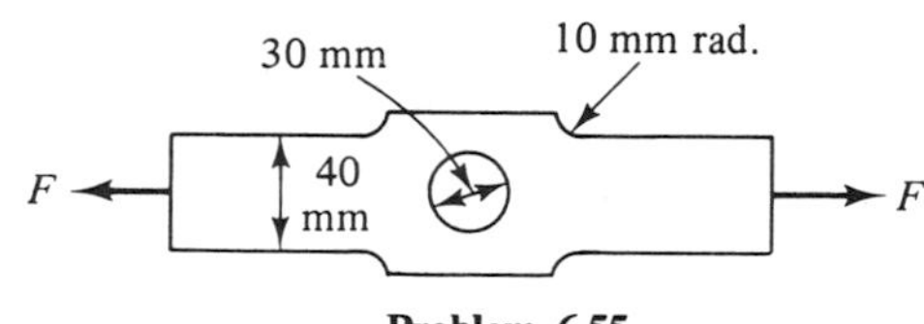

Problem 6.55

6.56. If the bar in Problem 6.55 is made of a brittle material with an ultimate tensile strength of 70 MN/m², what is its required thickness t if it is to support a tensile load 25 kN with a safety factor of 2?

6.57. The simple member shown, consisting of two identical rectangular structural steel bars connected by a single bolt, is to support a working load of 20 kips with a safety factor of 2. For satisfactory performance, the member must not break, fail in bearing, or undergo general yielding. Design the member by selecting the size of bolt and bars to be used and by determining the dimension s required to prevent tear-out. Assume that the properties of the bolt are the same as those for the bars and that the allowable stress in bearing is 1.3 times the yield strength. Bolts are available in diameters up to 1 in. in $\frac{1}{8}$-in. increments and from 1- to 4-in. diameters in increments of $\frac{1}{4}$ in. The bars are available with widths and thicknesses up to 1 in. in $\frac{1}{8}$-in. increments and in $\frac{1}{4}$-in. increments for sizes greater than 1 in.

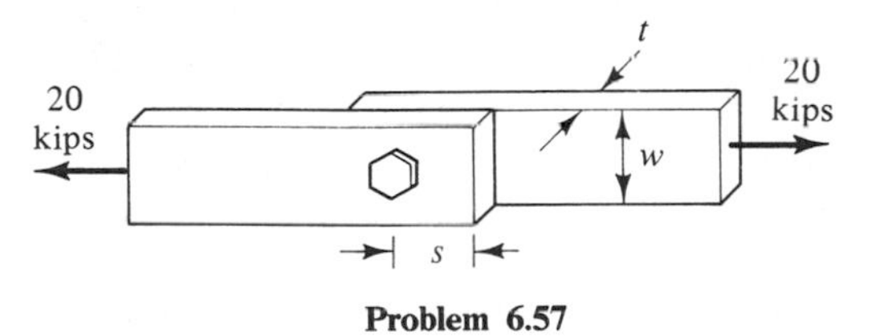

Problem 6.57

6.58. Same as Problem 6.57, except that the material is 2024-T3 aluminum.

6.59. The member shown is made of a brittle material with an ultimate strength of 180 MN/m^2. If the member is to support a tensile load of 30 kN, what is the maximum allowable hole diameter d?

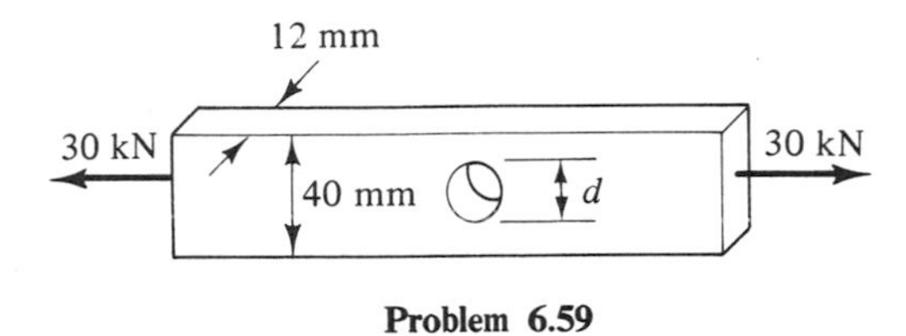

Problem 6.59

6.60. If the circular hole in the member in Problem 6.59 is replaced by semicircular grooves, what is the maximum allowable groove radius r?

6.7 INELASTIC BEHAVIOR OF STATICALLY INDETERMINATE STRUCTURES

When the stresses in an axially loaded member made of a ductile material reach the yield strength, failure by general yielding usually occurs for little or no increase in load. However, if the member is part of a statically indeterminate structure, its deformations are restricted by the other members and additional load can be supported until they, too, yield.

To illustrate this, let us consider the example of a rigid bar hinged at one end and supported by two cables [Figure 6.17(a)]. For simplicity, we assume that both cables have the same cross-sectional area and length and are made of the same material. We further assume that the material behavior is essentially elastic-perfectly plastic.

An elastic analysis shows that the loads on the cables are

$$F_{BD} = \frac{FLa}{(a^2 + b^2)} \qquad F_{CE} = \frac{FLb}{(a^2 + b^2)}$$

where F is the applied load and the dimensions a, b, and L are as defined in Figure

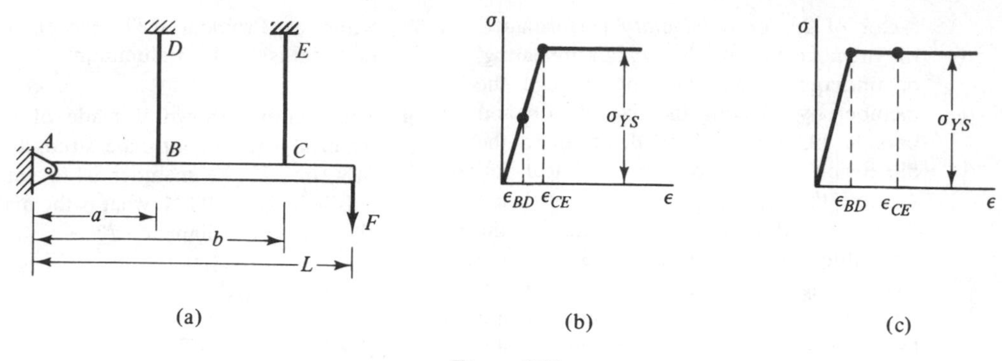

Figure 6.17

6.17(a). Cable *CE* is the most highly stressed because it has the same area as cable *BD* and supports a larger load.

As the applied load is increased from zero, cable *CE* will yield first. The stress in cable *BD* will still be within the elastic range when this happens [Figure 6.17(b)]. Consequently, its elongation will be small and the rotation of the bar will be limited. Any further increases in load must be supported entirely by cable *BD* because the stress in cable *CE* will remain equal to the yield strength. Eventually, the second cable will also yield [Figure 6.17(c)]. Both cables will be free to undergo large deformations when this happens, allowing the structure to literally collapse. The load at which this occurs is called the *limit load* for the structure.

Since the stresses in the cables are known to be equal to the yield strength at the limit load, the forces acting upon them can be determined by multiplying the stresses by the cross-sectional areas. Once these forces are known, the problem becomes statically determinate, and the limit load can be obtained directly from the equilibrium equations. Thus, a fully plastic analysis of an indeterminate structure is much simpler than an elastic analysis.

There may be more than one possible pattern of deformations that will lead to collapse of a structure. The various possibilities are called *collapse mechanisms*. Among the various possible mechanisms, the one with the smallest value of limit load is the one which actually occurs.

Example 6.9. Limit Load for a Simple Structure. A rigid bar is supported by three wires and is loaded as shown in Figure 6.18(a). The outer wires have a cross-sectional area of 1×10^{-3} m² and are made of a steel alloy with a yield strength of 690 MN/m². The center wire has a cross-sectional area of 2×10^{-3} m² and is made of aluminum ($\sigma_{YS} = 480$ MN/m²). What is the limit load for this structure?

Solution. The limit load is reached when all three wires have yielded. The force on each wire at this load is equal to the material yield strength times the cross-sectional area:

$$F_A = F_C = F_{\text{steel}} = (690 \times 10^6 \text{ N/m}^2)(10^{-3} \text{ m}^2) = 690 \times 10^3 \text{ N}$$

$$F_B = F_{\text{alum}} = (480 \times 10^6 \text{ N/m}^2)(2 \times 10^{-3} \text{ m}^2) = 960 \times 10^3 \text{ N}$$

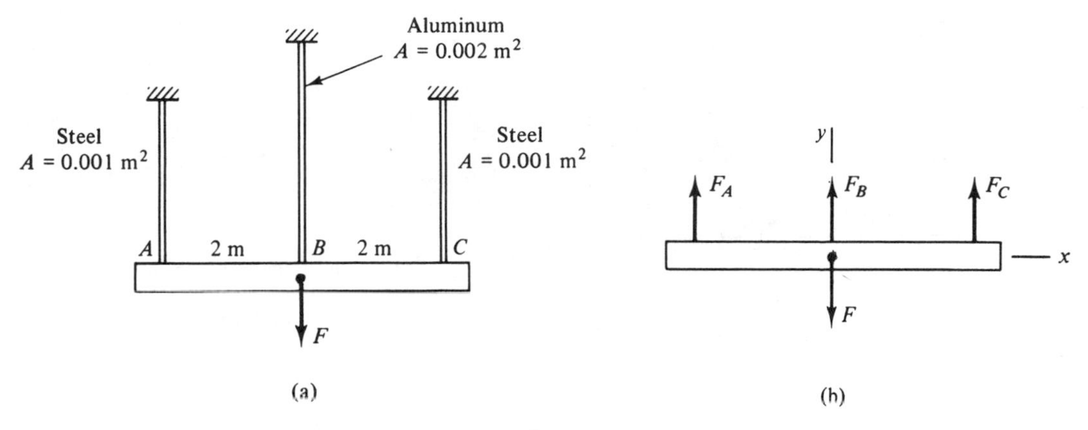

Figure 6.18

From the FBD of the bar [Figure 6.18(b)], we have

$$\sum F_y = F_A + F_B + F_C - F = 0$$

Thus, the fully plastic load is

$$F_{fp} = 2F_A + F_B = 2(690 \times 10^3 \text{ N}) + 960 \times 10^3 \text{ N}$$

$$= 2.34 \times 10^6 \text{ N or } 2.34 \text{ MN} \qquad \textbf{Answer}$$

Notice that the lengths of the wires have no bearing upon the limit load.

PROBLEMS

6.61. to 6.70. Determine the limit load for the structure shown. The cross-sectional dimensions and material properties are as indicated in the figure; otherwise, express the results in terms of the cross-sectional area A and yield strength σ_{YS}.

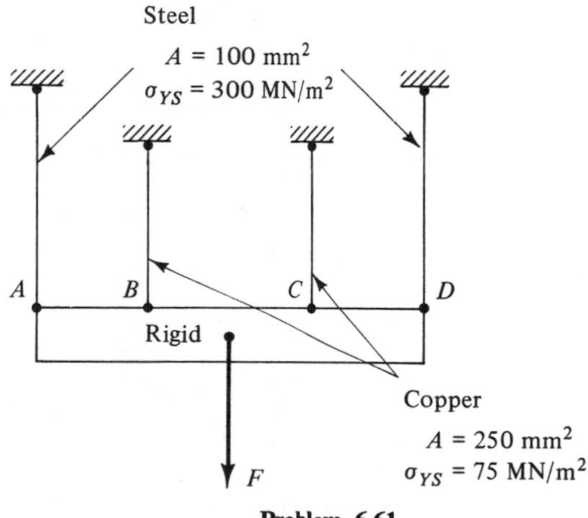

Problem 6.61

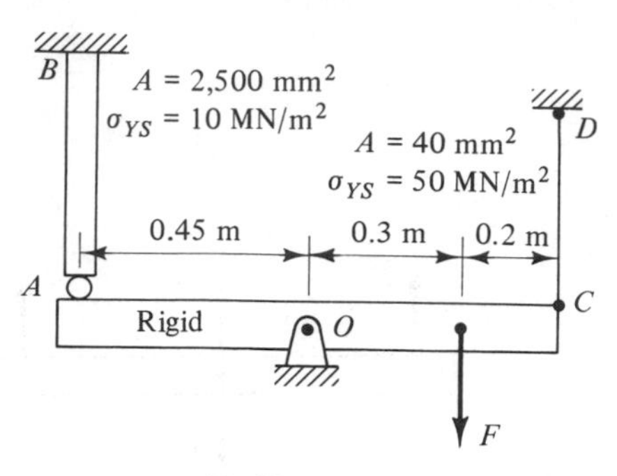

Problem 6.62

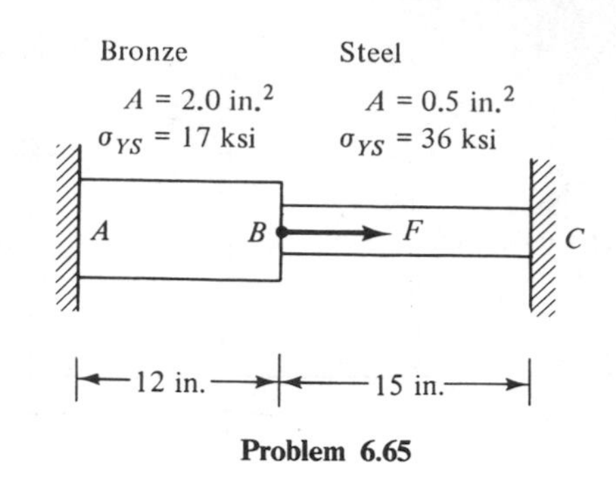

Problem 6.65

1 in. Rad.
2024–T3 Aluminum
F
4 in.
8 in.
8 in.
8 in.
2 in.

Problem 6.66

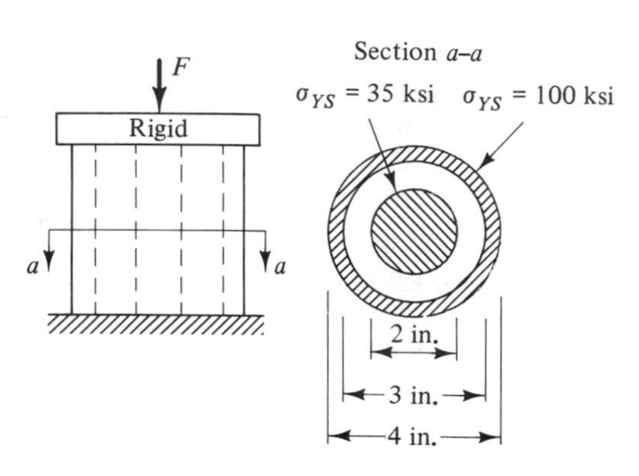

Problem 6.63

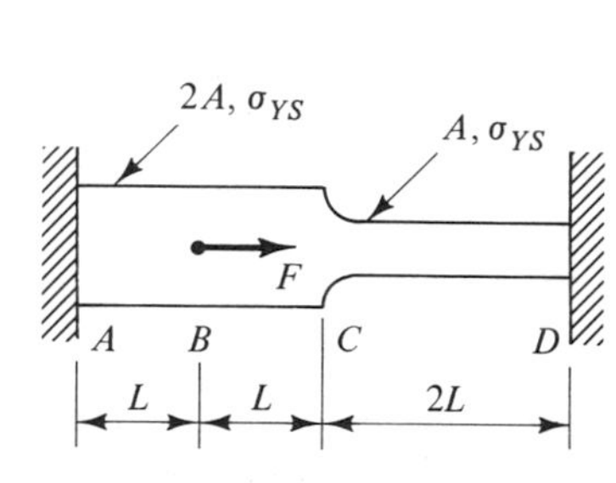

Problem 6.67

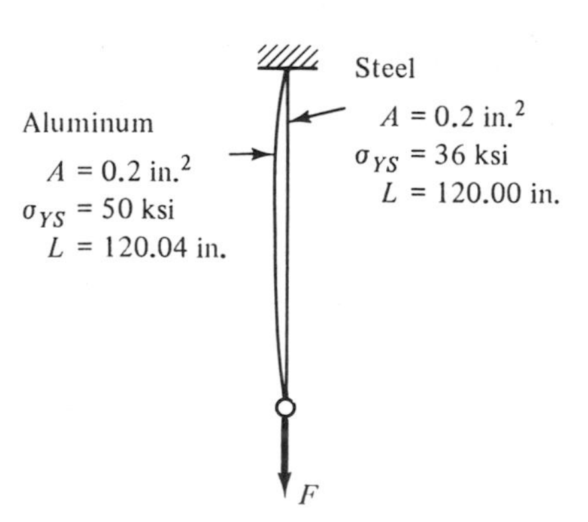

Problem 6.64

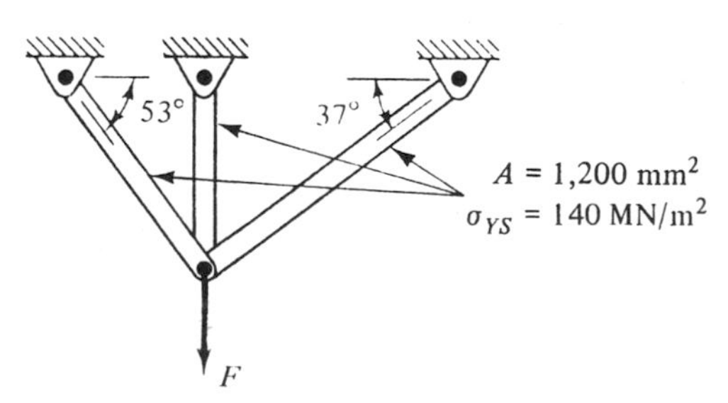

Problem 6.68

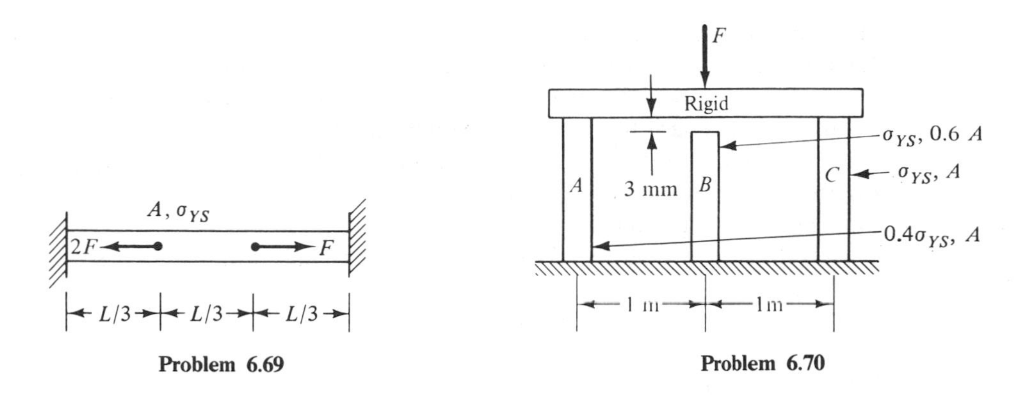

Problem 6.69

Problem 6.70

6.8 THIN-WALLED CYLINDERS AND SPHERES

Pressurized cylinders and spheres are commonly encountered in engineering practice, often in the form of piping systems and storage vessels. If the walls of the member are thin (as a rule of thumb, one-tenth of the radius or less), the stresses within it can be readily determined.

Consider a thin-walled circular cylinder with closed ends subjected to a uniform internal gage pressure p [Figure 6.19(a)]. The cylinder tends to elongate

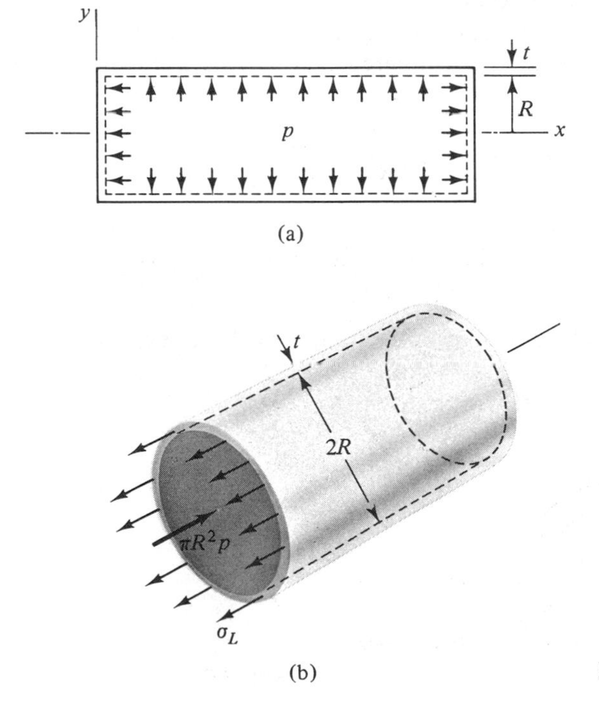

Figure 6.19

when pressurized, giving rise to normal stresses acting upon planes perpendicular to the axis. Sectioning the cylinder and the gas or fluid within it along such a plane, we obtain the free-body diagram shown in Figure 6.19(b). The resultant force due to the internal pressure lies along the centerline of the cylinder; therefore, the conditions for axial loading are met. Consequently, the normal stress σ_L acting in the longitudinal direction will be uniformly distributed over the cross section if the material is homogeneous.

Referring to Figure 6.19(b), we have for equilibrium of the segment of the cylinder

$$\sum F_x = \pi R^2 p - 2\pi R t \sigma_L = 0$$

Thus, the longitudinal stress is

$$\sigma_L = \frac{pR}{2t} \tag{6.7}$$

where p is the gage pressure and R and t are the radius and wall thickness of the cylinder, respectively. No distinction is made between the inside and outside radius because they are very nearly equal for thin walls.

Cylinders also tend to expand in the radial direction when pressurized, causing the walls to stretch circumferentially. This is illustrated in Figure 6.20, where the change in radius, ΔR, has been shown greatly exaggerated for clarity. As a result of the radial expansion, there are normal stresses σ_C acting in the circumferential direction on planes parallel to the cylinder axis. These stresses can be determined by using the procedure outlined in Section 5.11.

The normal strain in the circumferential direction is the change in circumference divided by the original circumference. Referring to Figure 6.20, we have at the inside of the cylinder

$$\varepsilon_{\text{inside}} = \frac{2\pi(R + \Delta R) - 2\pi R}{2\pi R} = \frac{\Delta R}{R}$$

while at the outside

$$\varepsilon_{\text{outside}} = \frac{2\pi(R + \Delta R + t) - 2\pi(R + t)}{2\pi(R + t)} = \frac{\Delta R}{R + t}$$

If the wall thickness is small compared to the radius, the latter expression reduces to

$$\varepsilon_{\text{outside}} \approx \frac{\Delta R}{R} = \varepsilon_{\text{inside}}$$

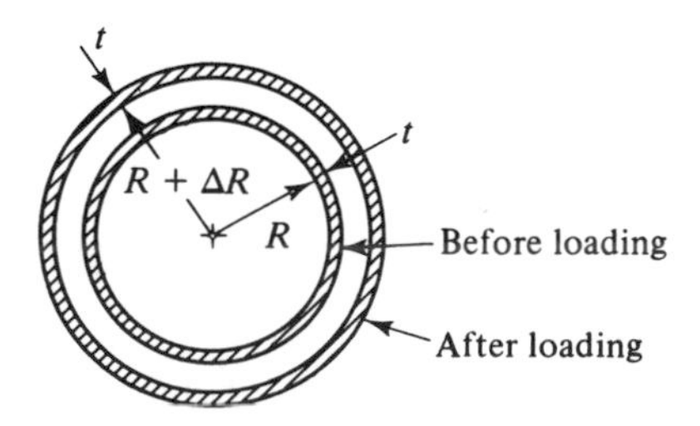

Figure 6.20

Thus, the circumferential strains are approximately the same at the inside and outside of the cylinder and, consequently, at all points in between. As a result, the circumferential stresses will be very nearly uniform through the wall thickness if the material is homogeneous.

The circumferential stresses can be related to the internal pressure by considering the equilibrium of a segment of the cylinder. Figure 6.21 shows the free-body diagram of a segment obtained by removing the ends and then sectioning the cylinder and the fluid within it lengthwise (longitudinal stresses and forces have been omitted for clarity). For equilibrium,

$$\sum F_y = 2Lt\sigma_C - 2RLp = 0$$

from which we obtain for the circumferential stress

$$\sigma_C = \frac{pR}{t} \tag{6.8}$$

A comparison of Eqs. (6.7) and (6.8) shows that the circumferential stresses are twice as large as the longitudinal stresses. Equation (6.8) also holds for thin-walled hoops, bands, and rings subjected to radial pressure.

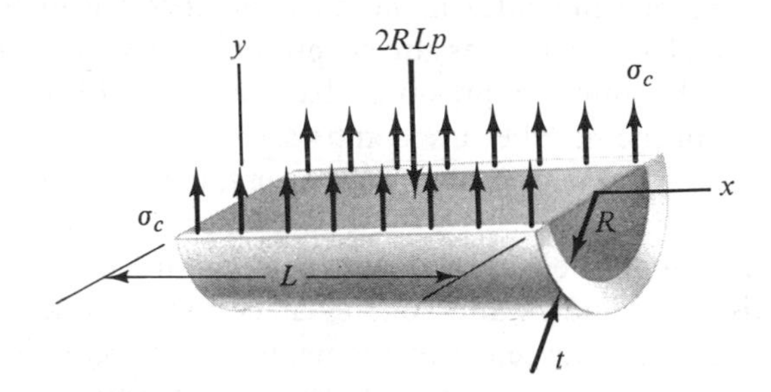

Figure 6.21

Spheres

A sphere subjected to a uniform internal pressure expands into a larger sphere. From this observation it can be shown that the stresses are uniformly distributed through the wall thickness if the walls are thin and the material is homogeneous. Since the procedure is the same as that used for the circumferential stresses in cylinders, it will not be repeated here.

Sectioning through the sphere and the fluid or gas within it (Figure 6.22), we obtain the equilibrium condition

$$\sum F_x = \pi R^2 p - 2\pi Rt\sigma = 0$$

or

$$\sigma = \frac{pR}{2t} \tag{6.9}$$

This is the same expression obtained for the longitudinal stress in a cylinder. Since

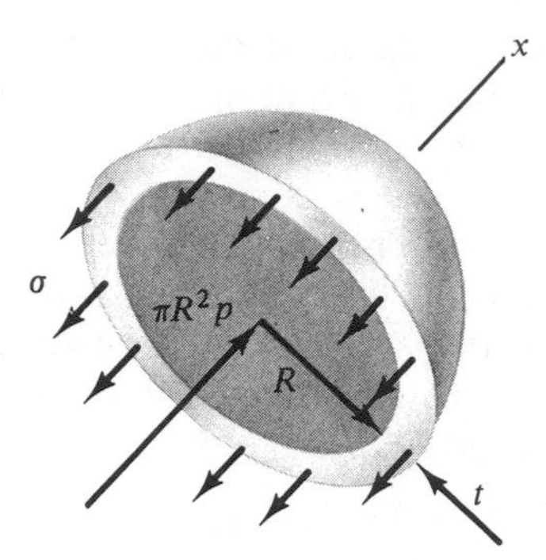

Figure 6.22

cutting planes with different orientation yield free-body diagrams identical to that shown in Figure 6.22, the stresses in a sphere are the same in all directions.

Validity of Pressure-Stress Relations

Although an internal pressure was assumed in the preceding derivations, Eqs. (6.7) through (6.9) also apply for external pressures if the member does not buckle and lose its cylindrical or spherical shape. The pressure to produce buckling is usually relatively small compared to the internal pressure that can be withstood. For external pressures, the stresses in the member are compressive.

Equations (6.7) and (6.8) do not ordinarily apply near the ends of a cylinder where the radial expansion may be restricted by the end elements. This restriction causes localized shearing and bending effects that are not accounted for in these equations. Similar effects arise in cylinders and spheres at points of attachment to other members and at the junction between wall segments of different thickness. Also, the pressure-stress relations do not apply near holes or other sources of stress concentration. The determination of the stresses in locations such as these is complicated and is beyond the scope of an introductory text such as this.

The assumption of a uniform pressure loading implies that the pressure is imparted by a gas of negligible weight. If imparted by a fluid, the pressure varies linearly with depth. In this case, the pressure-stress relations in Eqs. (6.7) through (6.9) are not strictly valid, but they can be used to obtain an estimate of the stresses. The accuracy of this estimate depends upon the nature of the particular problem of interest.

States of Stress

The various stresses acting at a point in a pressurized cylinder or sphere can be conveniently displayed on a small element of material, as described in Section 5.5. The stresses acting on an element at the outside surface of a cylinder and a sphere are shown in Figure 6.23. At the inside surface, there is also a compressive stress in the radial direction with magnitude equal to the internal pressure. This stress

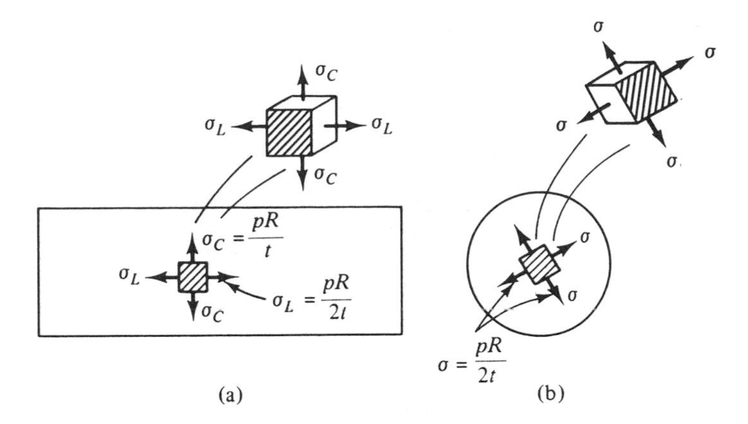

Figure 6.23

decreases through the wall thickness and is zero at the outside surface. For thin-walled pressure vessels, $R/t \gg 1$; therefore, $\sigma_C = 2\sigma_L \gg p$. Thus, the radial stress is much smaller than the longitudinal and circumferential stresses, and usually can be neglected.

Deformations

Deformations of cylinders and spheres are not as easily determined as deformations for axially loaded members because the material is simultaneously stretched or compressed in two directions. For linearly elastic material behavior, the deformations can be determined by using the generalized Hooke's law (see Example 6.10). For rings and similar members in which only a circumferential stress is acting, deformations and strains can be determined by using the simpler methods developed in Section 6.3 (see Example 6.11).

Example 6.10. Stresses In an Air Tank. A cylindrical steel tank with hemispherical ends contains air at a gage pressure of 120 psi [Figure 6.24(a)]. The radius of the tank is 8 in. and the wall thickness is $\frac{1}{4}$ in. Determine the stresses in each portion of the tank. What is the change in radius of the cylindrical portion when the tank is pressurized?

Solution. The longitudinal stresses in the cylindrical portion of the tank and the stresses in the hemispherical ends have the same values. From Eq. (6.7) or Eq. (6.9), we have

$$\sigma_L = \frac{pR}{2t} = \frac{(120 \text{ lb/in.}^2)(8 \text{ in.})}{2(\frac{1}{4} \text{ in.})} = 1{,}920 \text{ psi} \qquad \textbf{Answer}$$

The circumferential stress in the cylindrical portion of the tank is [Eq. (6.8)]

$$\sigma_C = \frac{pR}{t} = 2\sigma_L = 3{,}840 \text{ psi} \qquad \textbf{Answer}$$

These stresses are shown on the elements in Figure 6.24(b). The orientation of the element in the hemispherical ends is arbitrary, since the stresses there are the same in all directions.

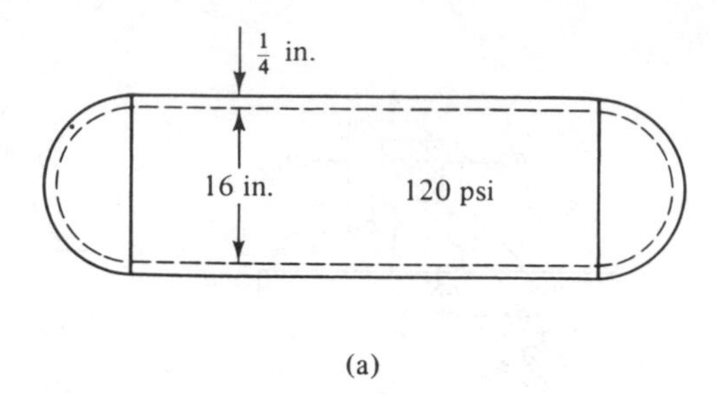

(a)

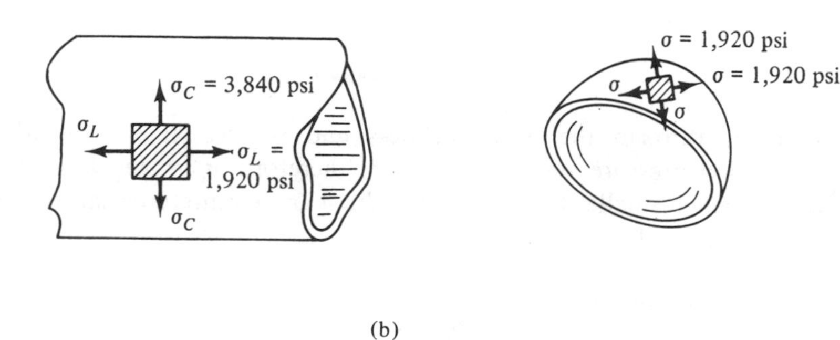

(b)

Figure 6.24

To compute the change in radius of the cylindrical portion, we take axes xyz with x in the longitudinal direction, y in the circumferential direction and z in the radial direction. Assuming linearly elastic material behavior and neglecting any radial stress ($\sigma_z = 0$), we obtain from Eqs. (5.24)

$$\varepsilon_y = \frac{\sigma_y - \nu\sigma_x}{E} = \frac{(3{,}840 \text{ lb/in.}^2) - 0.29(1{,}920 \text{ lb/in.}^2)}{30 \times 10^6 \text{ lb/in.}^2} = 109.4 \times 10^{-6}$$

This strain is also equal to the change in circumference of the cylinder divided by the original circumference. Letting ΔR denote the change in radius, we have

$$\varepsilon_y = \frac{2\pi(R + \Delta R) - 2\pi R}{2\pi R} = \frac{\Delta R}{R}$$

Equating the two expressions for ε_y, we obtain

$$\Delta R = R\varepsilon_y = (8 \text{ in.})(109.4 \times 10^{-6}) = 0.88 \times 10^{-3} \text{ in.} \qquad \textbf{Answer}$$

Example 6.11. Stresses In a Ring. A thin ring with an inside diameter of 500 mm and a wall thickness of 12 mm is fitted onto a shaft with a diameter of 503 mm by heating the ring until it slips over the shaft and then allowing it to cool to its original temperature. What is the resulting circumferential stress in the ring and what is the contact pressure between the ring and the shaft? Assume that the shaft deforms a negligible amount. The ring is made of cast iron with the stress-strain curve shown in Figure 5.29(b).

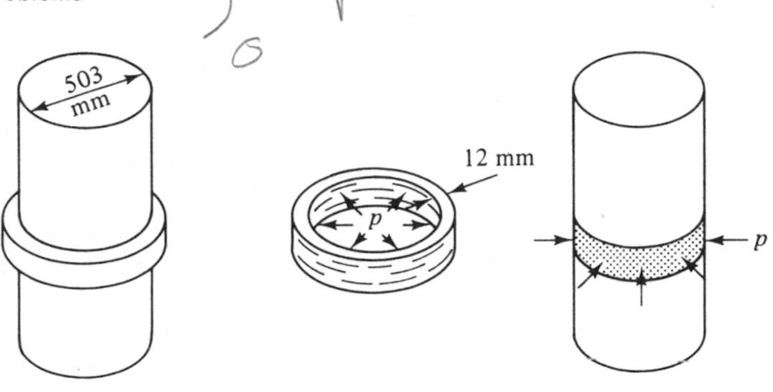

Figure 6.25

Solution. The shrinking of the ring onto the shaft has the same effect as subjecting it to a radial pressure sufficient to expand it from its original diameter of 500 mm to its final diameter of 503 mm (Figure 6.25). This pressure is exerted by the shaft and is assumed to be uniformly distributed over the contact surface.

The circumferential strain is equal to the change in circumference of the ring divided by the original circumference:

$$\varepsilon_C = \frac{\pi(503\text{ mm}) - \pi(500\text{ mm})}{\pi(500\text{ mm})} = 0.006$$

From the stress-strain curve in Figure 5.29, we find that the stress at this value of strain is approximately

$$\sigma_C = 18{,}000\text{ psi or }124\text{ MN/m}^2 \qquad \textbf{Answer}$$

The contact pressure is now determined from the equation for the circumferential stress, Eq. (6.8):

$$\sigma_C = \frac{pR}{t}$$

so

$$p = \frac{\sigma_C t}{R} = \frac{(124 \times 10^6\text{ N/m}^2)(12 \times 10^{-3}\text{ m})}{(0.503\text{ m}/2)}$$

$$= 5.92 \times 10^6\text{ N/m}^2\text{ or }5.92\text{ MN/m}^2 \qquad \textbf{Answer}$$

PROBLEMS

6.71. A cylindrical pressure vessel 20 in. in diameter with a wall thickness of $\frac{1}{4}$ in. is subjected to an internal gage pressure of 500 psi. Determine the longitudinal and circumferential stresses and show them acting upon a small material element.

6.72. The chamber of the hydraulic cylinder shown is made of a material for which the maximum allowable tensile stress is 20 ksi. What is the maximum force F the cylinder can support? Ignore the weight of the hydraulic fluid and any localized end effects.

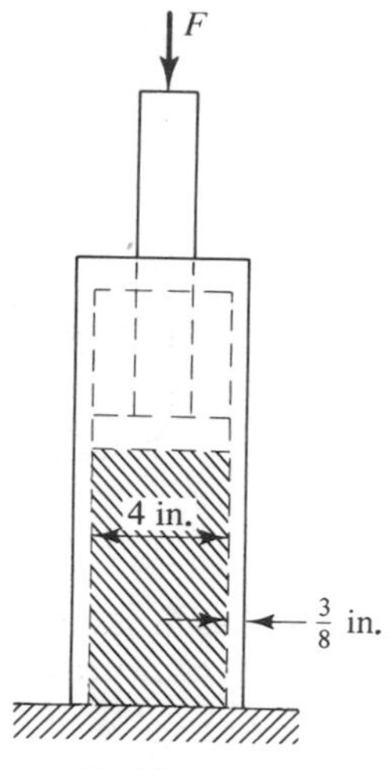

Problem 6.72

6.73. A cylindrical scuba tank 6.5 in. in diameter has a wall thickness of $\frac{1}{4}$ in. If the maximum allowable tensile stress in the main body of the tank is 80 ksi, what is the safety factor when the tank is filled with air at 2,250 psi?

6.74. A spherical vessel with a diameter of 6 m is to contain a corrosive gas at a pressure of 1.4 MN/m^2. The maximum allowable tensile stress in the tank wall is 85 MN/m^2. If the wall is expected to corrode at the rate of 0.2 mm/year, what is the minimum required wall thickness if the tank is to have a useful lifetime of at least ten years?

6.75. Derive an expression for the change in diameter of a thin-walled spherical pressure vessel when it is subjected to an internal pressure p. Assume linearly elastic material behavior.

6.76. A 300-mm diameter metal sphere with a coefficient of thermal expansion of $\alpha = 17 \times 10^{-6}(°C)^{-1}$ is covered with a 2.5-mm thick ceramic coating with $\alpha = 0.8 \times 10^{-6}(°C)^{-1}$, $E = 70$ GN/m^2, and $\nu = 0.2$. Experiments show that the coating will crack at a maximum tensile mechanical strain of 200 $\mu\varepsilon$. If the coating is stress free at 20°C, at what temperature can it be expected to crack? What is the contact pressure between the coating and the sphere at this temperature? Ignore any deformations of the sphere resulting from the pressure exerted upon it.

6.77. An inflatable building consists of thin sheet plastic held in position by a small internal pressure p maintained by an air blower. The density per square meter of the plastic is ρ_s. Derive an expression for the minimum pressure required to maintain the semicircular shape shown. Use a safety factor of 4.

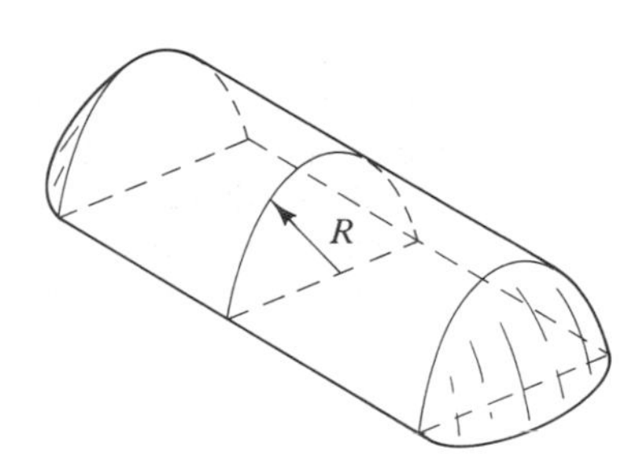

Problem 6.77

6.78. A certain ring made of 2024-T3 aluminum is 2 in. in diameter, $\frac{1}{4}$-in. wide, and $\frac{1}{8}$-in. thick. It is to be stretched by driving it onto a tapered rod, as shown. If the ring is driven a distance $d = \frac{1}{4}$ in. from the point where it first seats on the rod, what is the tensile stress in the ring and what is the contact pressure between it and the rod? Assume elastic-perfectly plastic material behavior and ignore any deformations of the rod.

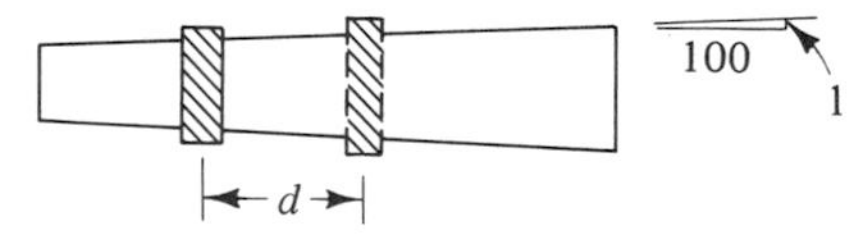

Problem 6.78

6.79. Same as Problem 6.78, except that the ring is driven 1 in. onto the rod.

6.80. A large rubber band 40-mm wide and 6-mm thick has a total peripheral length of 1.5 m. The band is used to hold a sheet plastic cover over the open end of a barrel with a diameter of 0.6 m, as shown. If the stress-strain diagram (based on the original cross-sectional area) for the band is as indicated, determine (a) the tension T on the band and (b) the contact pressure between the band and the barrel. Assume the barrel does not deform.

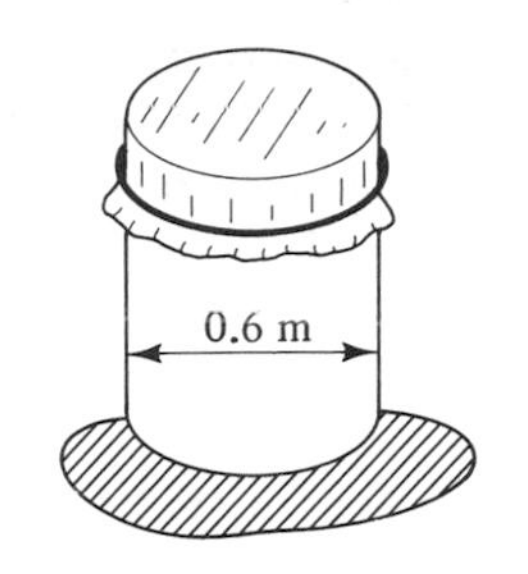

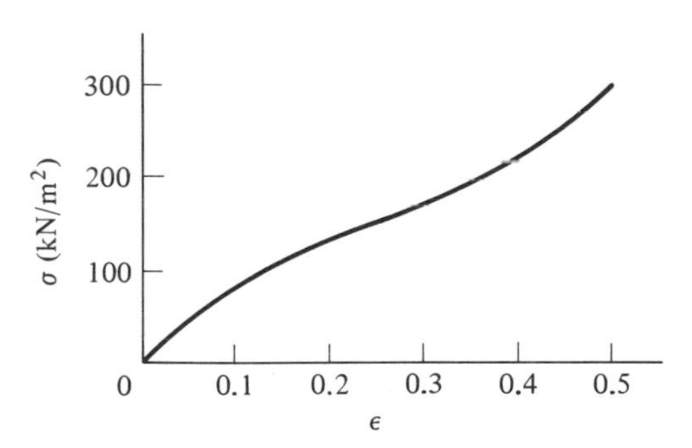

Problem 6.80

6.81. The rubber band described in Problem 6.80 is used to hold together four containers of frozen ice cream, as shown. Each container has a diameter of 300 mm. Assuming that the containers do not deform, determine (a) the tensile stress in the band and (b) the contact pressure between it and the containers.

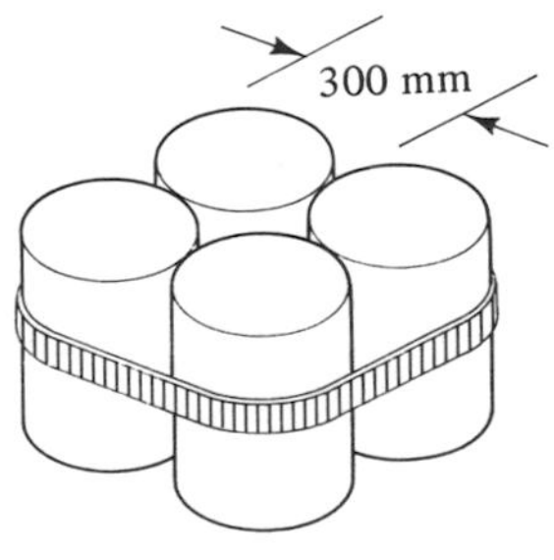

Problem 6.81

6.82. A 40-mm wide clamp consisting of two bands made of 1.5-mm thick sheet steel is placed around a 100-mm diameter pipe as shown and the screws at each side are tightened until they are subjected to a tensile force of 400 N. If the static coefficient of friction between the pipe and the band is 0.7, what is the maximum total vertical force F that can be applied to the band if it is not to slip? What is the tensile stress in the band? Assume that the pipe does not deform.

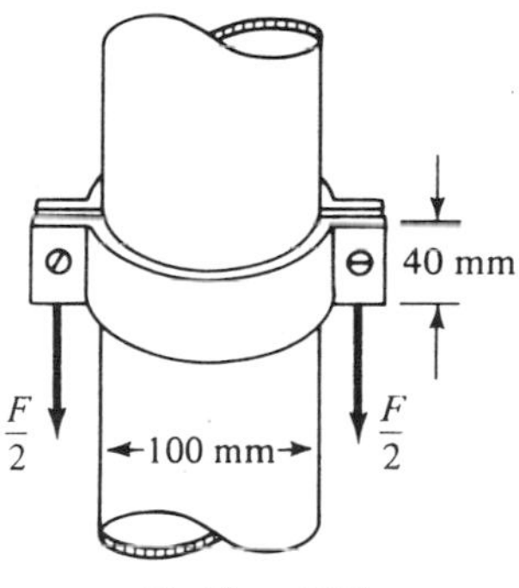

Problem 6.82

6.83. A thin-walled cylindrical tank with closed ends has radius R, wall thickness t, and length L. The tank just fits between rigid end walls when the internal pressure is zero. Calculate the forces exerted on the walls by the tank for an internal pressure p. Neglect the radial stresses in the tank.

6.84. A thin-walled tube with radius R, wall thickness t, and length L fits snugly into a smooth hole drilled into a rigid block. If the tube is subjected to a longitudinal compressive stress σ, what is the change in its length and what is the contact pressure p between it and the block? Ignore the radial stresses in the tube.

6.9 CLOSURE

In this chapter we have shown that the actual stresses in axially loaded members and in pressurized cylinders and spheres are generally uniformly distributed. Consequently, they have the same values as the average stresses and can be computed by dividing the normal and shear forces by the areas over which they act. The load-stress relations so obtained stem directly from the equations of equilibrium and, therefore, apply regardless of the material behavior. This property is unique to these types of members and loadings. In contrast, the load-deformation relations depend upon the material properties and have different forms for different ranges of stress.

7

Torsion

7.1 INTRODUCTION

In this chapter we shall consider the *torsion* of circular shafts. More precisely, we shall consider circular members loaded in such a way that the stress resultant is a couple that lies along the longitudinal axis and whose response consists of a twisting about that axis. Such members are commonly used as drive shafts in power transmission systems and in other mechanical and structural applications. Our goal is to determine the load-stress and load-deformation relations, both for the linearly elastic and inelastic cases, and to learn how to apply them.

As we shall see, there is a direct analogy between all aspects of the torsion problem and the analysis of axially loaded members. The only significant difference is that shear stresses and rotational effects are involved instead of normal stresses and elongations or contractions. Accordingly, the basic procedures used in Chapters 5 and 6 for axially loaded members still apply. This is true for both statically determinate and statically indeterminate problems.

Only members in equilibrium will be considered. Thus, the results obtained will apply to shafts at rest and to shafts rotating with constant angular velocity.

7.2 DEFORMATION PATTERN AND STRAINS

Following the procedure outlined in Section 5.11, we first consider the nature of the deformations in a twisted shaft. Experiments show that a circular shaft made of a homogeneous and isotropic material deforms in such a way that planes perpendicular to the axis before loading remain plane and perpendicular to the axis after loading, and radial lines in the cross section remain radial. Furthermore, the length does not change appreciably. In other words, the shaft behaves as a series of thin discs that rotate slightly with respect to one another when the shaft is twisted. As a result, lines originally parallel to the axis of the shaft distort into helixes. This pattern of deformation, which also follows from symmetry, is illustrated in Figure 7.1 and is confirmed by Figure 7.2.

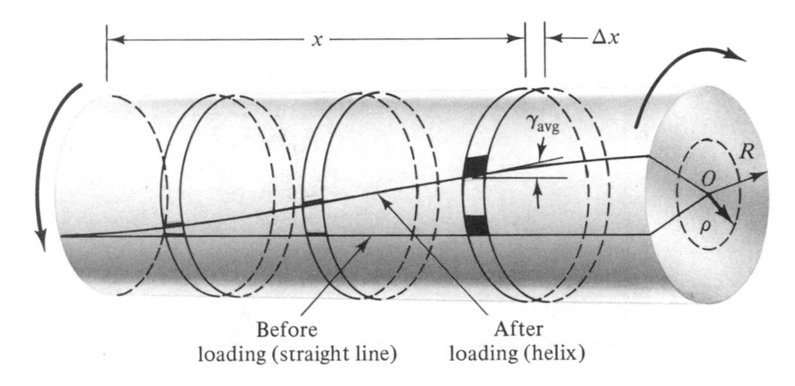

Figure 7.1

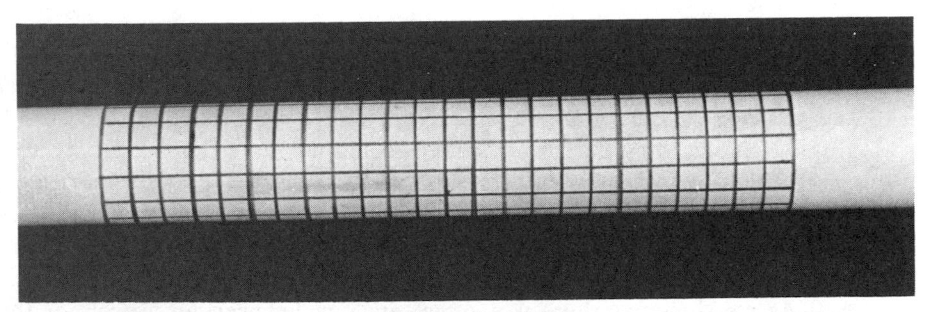

(a) Before Loading

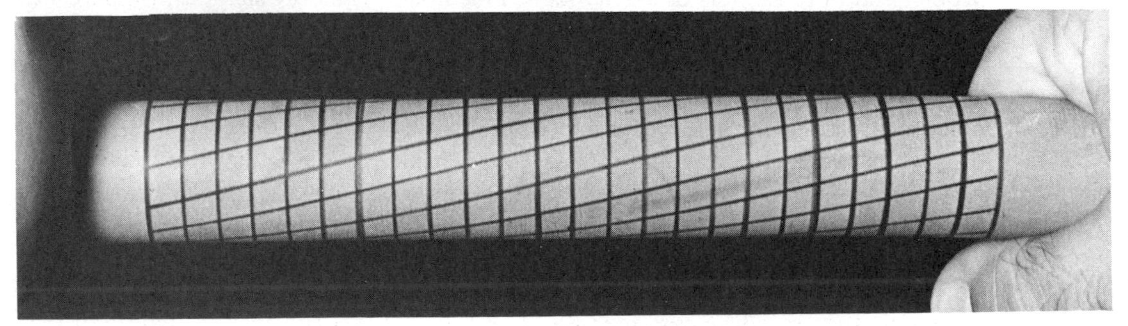

(b) After Loading

Figure 7.2 Deformation of a Rubber Shaft as Indicated by Grid Patterns.

The deformations can be related to the strains by considering a short segment of the shaft with length Δx (Figure 7.3). Lines AB and CD before loading become lines $A'B'$ and $C'D'$ after loading. They remain very nearly straight because they are short. The angle between the original and final line elements is the average shear strain, γ_{avg}. This is most easily seen from Figure 7.1. It is clear from Figure 7.3 that this angle varies with the radius ρ.

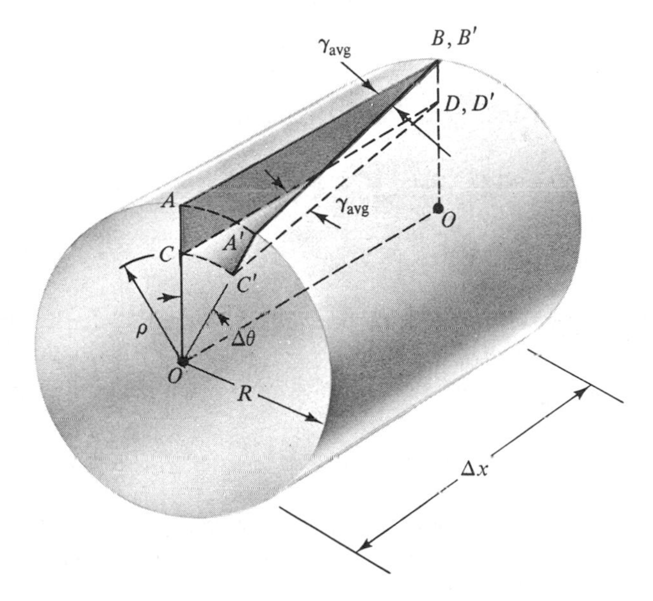

Figure 7.3

Denoting the amount of rotation of one end of the segment with respect to the other by $\Delta\theta$ and assuming that the deformations are small, we have from the geometry of Figure 7.3

$$\gamma_{avg} \approx \tan \gamma_{avg} = \frac{CC'}{\Delta x} = \frac{\rho\,\Delta\theta}{\Delta x}$$

The shear strain at a point is obtained by taking the limit as $\Delta x \to 0$:

$$\gamma = \lim_{\Delta x \to 0} \frac{\rho\,\Delta\theta}{\Delta x} = \frac{\rho\, d\theta}{dx} \tag{7.1}$$

Equation (7.1) shows that the shear strain varies linearly with the radius and is a maximum at the outside of the shaft where $\rho = R$. This is illustrated in Figure 7.4, where the height of the diagram denotes the relative magnitude of γ.

The strain distribution in a homogeneous and isotropic shaft depends only upon the geometry of the member and the manner in which it is loaded. It is independent of the resistive properties of the material. However, there are several conditions, in addition to those already mentioned, that must be satisfied before Eq.

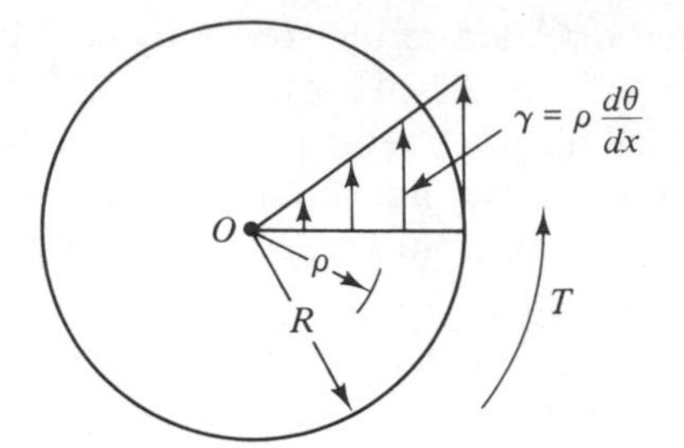

Figure 7.4

(7.1) is valid. For convenience, all of these requirements are summarized below:

1. The member must be straight and circular, or nearly so, for some distance on either side of the cross section of interest.
2. The loading must be such that the stress resultant is a couple that lies along the longitudinal axis.
3. The cross-sectional area must be constant, or nearly so, for some distance on either side of the section of interest.
4. The section of interest must be some distance away from connections, supports, and points of load application.
5. The material must be homogeneous and isotropic.

The fifth condition can be relaxed somewhat, but it is sufficiently general for our purposes. If the conditions on the geometry of the shaft are not met, there will be distortions in the deformation pattern and Eq. (7.1) does not apply. However, these effects tend to be localized (St. Venant's principle), and they generally die out over a distance approximately equal to the shaft diameter.

If the shaft is uniform and the torque doesn't vary along its length, then neither will the shear strain. In this case, Eq. (7.1) can be integrated directly to obtain

$$\theta = \int_0^L \frac{\gamma}{\rho}\, dx = \frac{\gamma L}{\rho} \tag{7.2}$$

where L is the length of the member and θ is the *angle of twist* (measured in radians) through which one end rotates relative to the other.

7.3 SHEAR STRESSES — LINEARLY ELASTIC CASE

Continuing with the procedure outlined in Section 5.11, we now determine the stress distribution and torque-stress relation for a shaft in torsion by combining the strain distribution with the stress-strain relation and the conditions for static equivalence. Linearly elastic material behavior will be assumed for the present. The inelastic behavior of shafts will be considered in Section 7.6.

Torque-Stress Relation

If the material is homogeneous and isotropic and the stresses are below the proportional limit, the stress-strain relation $\tau = G\gamma$ applies throughout the shaft. Combining this expression with the strain-deformation relation of Eq. (7.1), we obtain for the stress distribution

$$\tau = G\gamma = G\rho \frac{d\theta}{dx} \tag{7.3}$$

Thus, the shear stresses also vary linearly with the radius, ρ, as shown in Figure 7.5. They are oriented in the tangential direction and act over the entire cross section.

The shear stresses can be related to the stress resultant (torque), T, by using the conditions for static equivalence. Referring to Figure 7.5, we see that the force dF acting upon an element of area dA of the cross section has a moment about the center of the shaft of magnitude $dM = \rho\, dF = \rho\tau\, dA$. Thus, we have

$$\left(\overset{+}{\curvearrowleft}\sum M_0\right)_I = \left(\overset{+}{\curvearrowleft}\sum M_0\right)_{II} : \qquad \int \rho\tau\, dA = T \tag{7.4}$$

Substituting Eq. (7.3) into this relation and recognizing that G and $d\theta/dx$ do not vary over the cross section, we obtain

$$T = \int \rho\left(G\rho\frac{d\theta}{dx}\right) dA = G\frac{d\theta}{dx}\int \rho^2\, dA$$

The latter integral is the polar moment of inertia, J, of the cross-sectional area (see Section 4.8); therefore, this expression can be rewritten as

$$T = GJ\frac{d\theta}{dx} \tag{7.5}$$

Combining Eqs. (7.3) and (7.5), we obtain the torque-stress relation

$$\tau = \frac{T\rho}{J} \qquad (\tau \leqslant \tau_{PL}) \tag{7.6}$$

Here, T is the torque acting upon the particular cross section of interest and J is

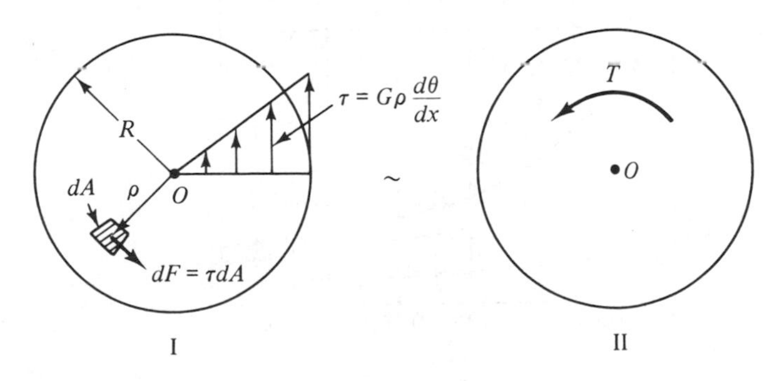

Figure 7.5

the polar moment of inertia of that cross section. For a shaft with radius R or diameter D, we obtain from Table A.1

$$J = \frac{\pi R^4}{2} = \frac{\pi D^4}{32} \tag{7.7}$$

Equation (7.6) defines the shear stresses acting upon planes perpendicular to the axis, which are also the maximum shear stresses in a shaft. Proof of this will be given in the following section. The sense of the shear stress is the same as that of the torque, as indicated in Figure 7.5.

The connection between the stresses and the applied loads is completed by relating the torque to the loads by using the method of sections, as discussed in Section 5.2. If the torque varies along the length of the shaft, as when couples are applied at intermediate points as well as at the ends, it is convenient to construct a *torque diagram* showing the value of the torque at every location along the member (see Example 7.2).

Validity of Torque-Stress Relation

The torque-stress relation given in Eq. (7.6) applies only if the stresses are below the proportional limit, since the relation $\tau = G\gamma$ was used in its derivation. For other types of material behavior, it is only necessary to use the appropriate stress-strain relation in Eq. (7.3) and then proceed as before. Equation (7.6) is also based upon the strain-deformation relation of Eq. (7.1); therefore, the five conditions stated in Section 7.2 concerning the geometry and loading of the shaft must also be met.

Many shafts contain oil holes, keyways, fillets, and other geometrical discontinuities, the effects of which can often be accounted for by multiplying the maximum value of shear stress obtained from Eq. (7.6) by a stress concentration

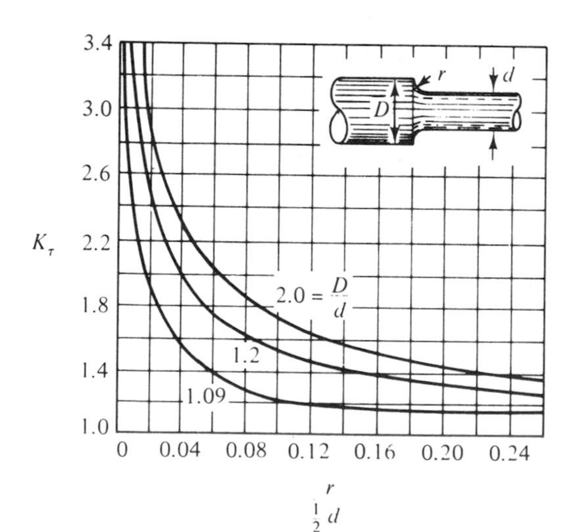

Figure 7.6 Torsional Stress-Concentration Factors for Stepped Circular Shafts.

factor, K_τ. Thus,

$$\tau_{max} = K_\tau \frac{TR}{J} \tag{7.8}$$

Values of K_τ for shafts with fillets are given in Figure 7.6 for the linearly elastic case. These values apply to the stresses in the smaller portion of the member.

As in the case of axially loaded members, geometrical discontinuities are most serious if the material behaves in a brittle manner. Shafts made of ductile materials can yield and continue to transmit the torque. The inelastic behavior of shafts will be considered in Section 7.6.

Hollow Shafts

From Figure 7.5, it can be seen that the shear stresses are smallest near the center of a shaft and that the moment arms of the forces due to these stresses are also small. Consequently, the material near the center contributes little to the torque-carrying capacity of a member, and considerable economy can be achieved by eliminating it. This is why hollow shafts are commonly used. For example, a hollow shaft with an inside radius equal to one-half the outside radius can transmit 94% of the torque carried by a solid shaft with the same outside radius, but it contains only 75% as much material.

The torque-stress relation of Eq. (7.6) also applies to hollow shafts, in which case J is the polar moment of inertia of the net cross-sectional area:

$$J = \frac{\pi}{2}\left(R_o^4 - R_i^4\right) = \frac{\pi}{32}\left(D_o^4 - D_i^4\right) \tag{7.9}$$

The subscripts i and o denote inside and outside, respectively. The stress distribution in a hollow shaft loaded within the elastic range is shown in Figure 7.7.

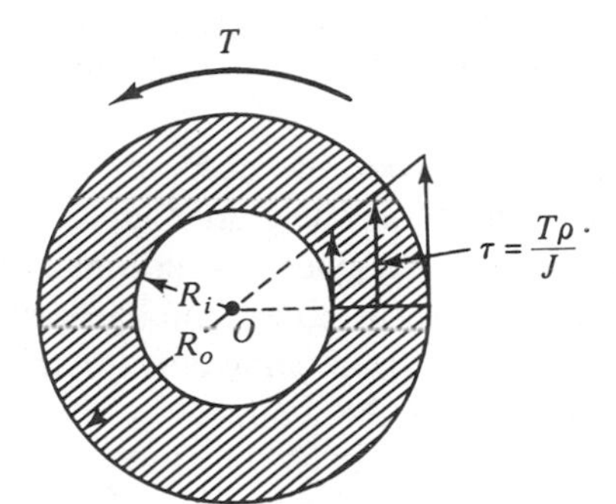

Figure 7.7

Thin-Walled Circular Shafts

For a hollow shaft with thin walls, the shear strain is very nearly uniform through the wall thickness [Figure 7.8(a)]. Accordingly, the stress distribution [Figure 7.8(b)], will also be nearly uniform. This is true regardless of the material properties.

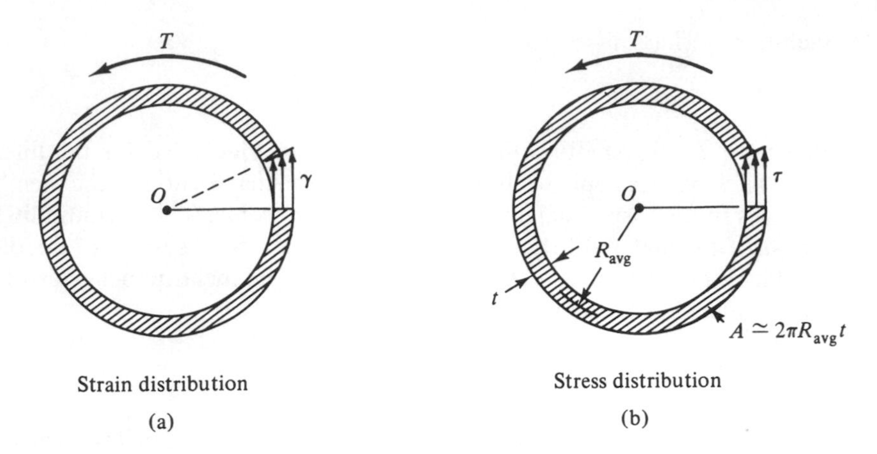

Figure 7.8

If we assume that τ does not vary over the cross section and that all of the material is located at approximately the same radius $\rho = R_{avg}$, the torque-stress relation of Eq. (7.4) becomes

$$T = \int \rho \tau \, dA \approx \tau R_{avg} \int dA = \tau R_{avg} A$$

where A is the cross-sectional area of the shaft and R_{avg} is the radius to the centerline of this area [Figure 7.8(b)]. Now $A \approx 2\pi R_{avg} t$, where t is the wall thickness; therefore, this expression can be rewritten as

$$\tau \approx \frac{T}{2\pi t R_{avg}^2} \tag{7.10}$$

The accuracy of this relation increases with decreasing wall thickness. However, shafts with very thin walls tend to buckle, or wrinkle, in which case Eq. (7.10) is no longer valid.

Equations (7.10) and (7.2) provide the basis for the experimental determination of the properties of a material in shear. If couples of known magnitude are applied to a thin-walled specimen with given dimensions and the resulting angles of twist measured, or vice versa, the shear stress and shear strain can be determined from these equations and the stress-strain curve constructed. This experiment is not easy to perform, however, because of the tendency of the member to buckle.

Power Transmission

Shafts are most commonly encountered as elements of rotating machinery. In this case, it often is convenient to express the response in terms of the power transmitted rather than the torque.

By definition, power is the time rate at which work is done. Accordingly, the power P associated with a constant torque (couple) T acting on a body rotating with constant angular velocity ω is

$$P = T\omega \tag{7.11}$$

In the US-British system of units, the unit of power is the *horsepower* (1 hp = 33,000 lb · ft/min); in SI, the *watt* is the unit of power (1 W = 1 N · m/s), although the kilowatt (kW) is a more convenient unit for many applications. The angular velocity is usually given in revolutions per minute (rpm), which we shall denote by n. Thus, Eq. (7.11) can be expressed in the more useful forms

$$P(\text{hp}) = \frac{T(\text{lb} \cdot \text{in.})\,n(\text{rpm})}{63{,}000} \tag{7.12}$$

and

$$P(\text{kW}) = \frac{T(\text{N} \cdot \text{m})\,n(\text{rpm})}{9{,}550} \tag{7.13}$$

Example 7.1. Shear Stress in a Stepped Shaft. A stepped shaft made of 2024-T3 aluminum alloy is loaded as shown in Figure 7.9(a). Determine the maximum shear stress in each section of the shaft and at the fillet.

Solution. The torque is the same in both portions of the shaft and is equal in magnitude to the applied couple [Figure 7.9(b)]. From the torque-stress relation of Eq. (7.6), we have

$$\tau_{\max} = \frac{TR}{J} = \frac{TR}{(\pi/2)\,R^4} = \frac{2T}{\pi R^3}$$

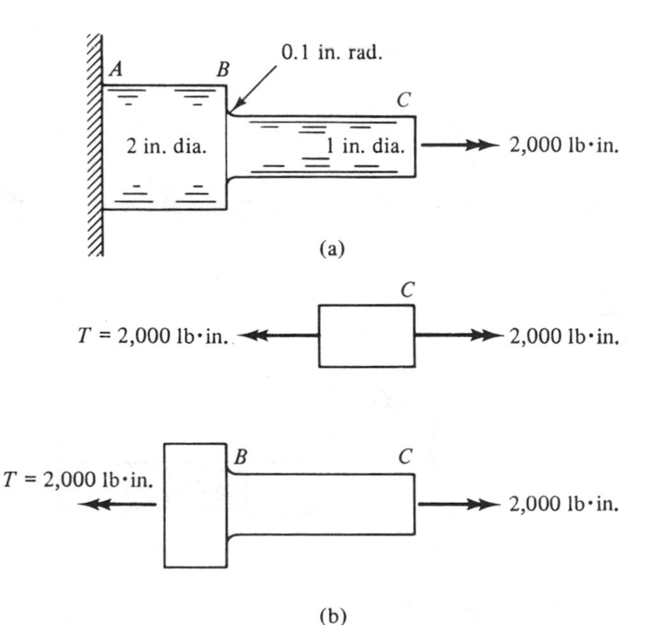

Figure 7.9

For section AB of the shaft,

$$\tau_{\max} = \frac{2(2{,}000 \text{ lb} \cdot \text{in.})}{\pi(1 \text{ in.})^3} = 1{,}270 \text{ psi} \qquad \textbf{Answer}$$

while for section BC,

$$\tau_{\max} = \frac{2(2{,}000 \text{ lb} \cdot \text{in.})}{\pi(0.5 \text{ in.})^3} = 10{,}190 \text{ psi} \qquad \textbf{Answer}$$

Referring to Figure 7.6, we have $r/(d/2) = 0.2$, $D/d = 2$, and $K_\tau = 1.45$. Applying this factor to the stress in the small portion of the shaft, we obtain for the stress at the fillet

$$\tau_{\max} = 1.45(10{,}190 \text{ psi}) = 14{,}780 \text{ psi} \qquad \textbf{Answer}$$

We now check to see if the stresses are below the proportional limit. If they aren't, Eq. (7.6) used in the computations isn't valid and the value $K_\tau = 1.45$ doesn't apply. From Table A.2, the yield strength of the material in shear is found to be 30 ksi. Since all stresses are well below this value, the results are valid.

Example 7.2. Allowable Power Input to a Drive Shaft. A motor rotating at 1,800 rpm delivers power to a drive shaft that distributes two-thirds of it to a machine at the left end and one-third of it to another machine at the right end [Figure 7.10(a)]. How much

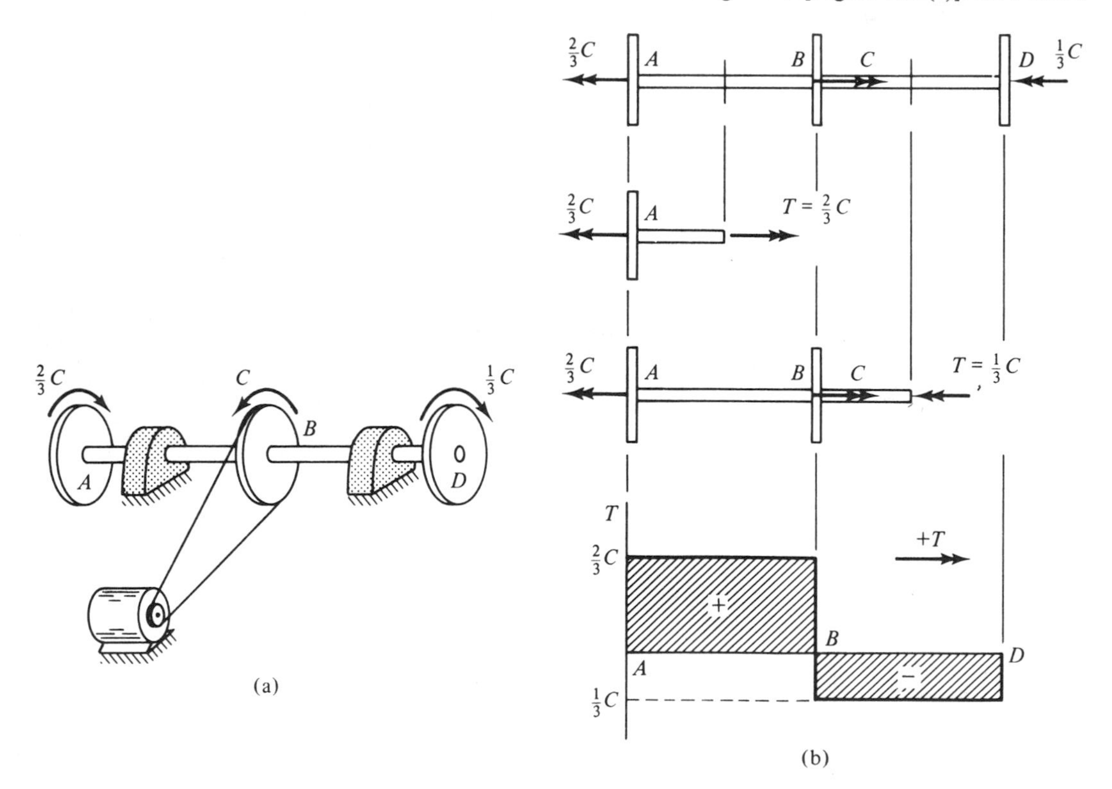

Figure 7.10

power can be input without exceeding the shear yield strength of 210 MN/m^2? The shaft is hollow with an outside diameter of 30 mm and an inside diameter of 20 mm. Use a safety factor of 1.5.

Solution. Let C be the couple exerted on the shaft by the motor. Since torque is proportional to power, the torque diagram is as shown in Figure 7.10(b). Here, we have arbitrarily chosen torque vectors directed to the right to be positive. Actually, the sign of the torque has little significance insofar as the shear stresses are concerned. Since the shaft is uniform, the stress will be largest where the torque is largest. The torque diagram indicates that this is in portion AB of the shaft, where $T = 2/3C$.

The polar moment of inertia of the shaft is

$$J = \frac{\pi}{32}\left(D_o^4 - D_i^4\right) = \frac{\pi}{32}\left[(30\text{ mm})^4 - (20\text{ mm})^4\right] = 6.38 \times 10^4\text{ mm}^4$$

From Eq. (7.6), we have

$$\tau_{\max} = \frac{TR}{J} = \frac{(2/3C)(15 \times 10^{-3}\text{ m})}{6.38 \times 10^{-8}\text{ m}^4} \leqslant 210 \times 10^6\text{ N/m}^2$$

or

$$C \leqslant 1.34\text{ kN} \cdot \text{m}$$

The working couple is obtained by dividing this value by the safety factor:

$$C_{\text{working}} = \frac{1.34 \times 10^3\text{ N} \cdot \text{m}}{1.5} = 893\text{ N} \cdot \text{m}$$

From Eq. (7.13), the allowable power input is

$$P = \frac{(893\text{ N} \cdot \text{m})(1{,}800\text{ rpm})}{9{,}550} = 168\text{ kW} \qquad \textbf{Answer}$$

PROBLEMS

7.1. What diameter solid shaft is required to transmit a torque of 180 lb · ft if the maximum allowable shear stress is 15 ksi?

7.2. A hollow shaft with the diameters shown has an allowable shear stress of 10 ksi.

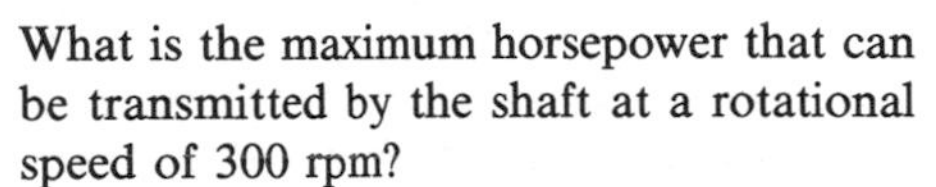

What is the maximum horsepower that can be transmitted by the shaft at a rotational speed of 300 rpm?

7.3. In an attempt to unscrew the $1\frac{1}{2}$-in. (nominal) diameter steel pipe shown, a plumber

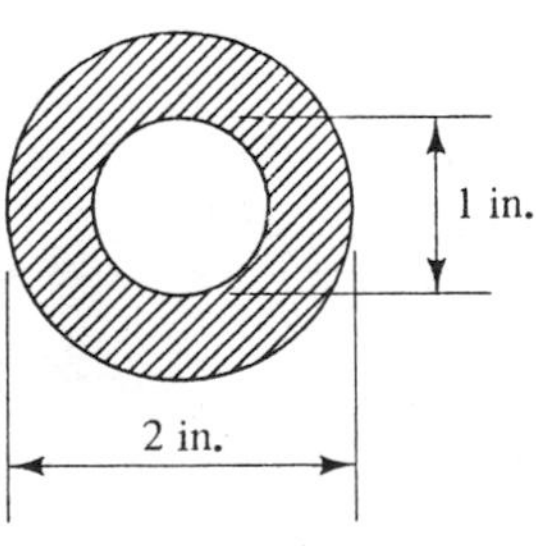

Problem 7.2

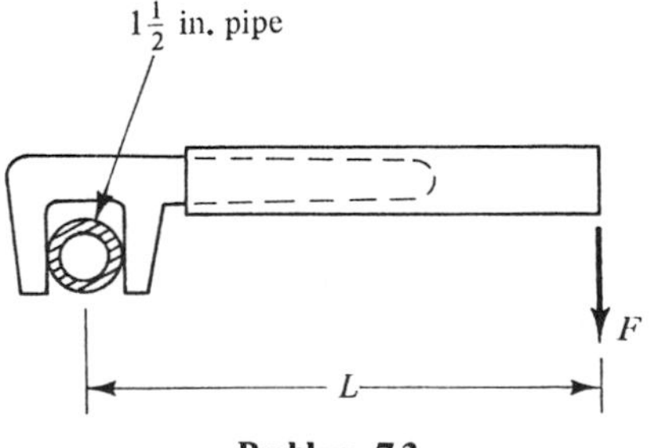

Problem 7.3

adds a "cheater" to the pipe wrench. What is the maximum allowable cheater length L if the torsional shear stress in the pipe is not to exceed 10 ksi? Assume the plumber can exert a maximum force $F = 120$ lb.

7.4. The T-bar shown has a solid cross section with a diameter of 30 mm. If $F = 200$ N, what is the maximum torsional shear stress in the member?

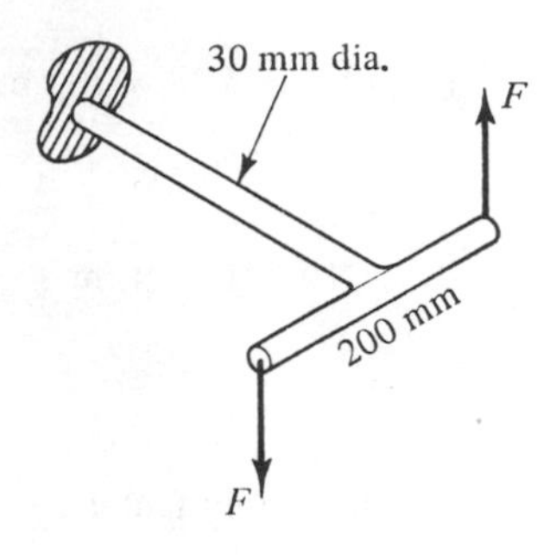

Problem 7.4

7.5. What is the maximum force F that can be applied to the T-bar in Problem 7.4 if the allowable shear stress is 125 MN/m^2? Use a safety factor of 2.5.

7.6. What is the largest couple C that can be applied to the stepped shaft shown if (a) stress concentrations are ignored and (b) the stress concentration at the fillet is included? The allowable shear stress is 100 MN/m^2.

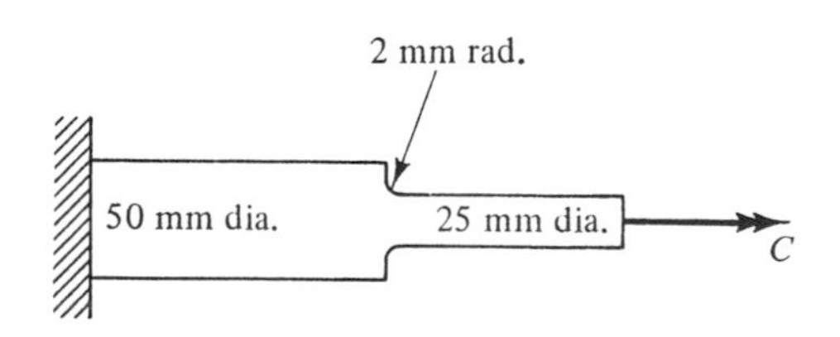

Problem 7.6

7.7. A solid shaft with a diameter of 2.5 in. transmits 400 hp at 600 rpm. Find a hollow shaft of the same material with an outside diameter of 3 in. that will transmit the same power with the same maximum stress level. What is the ratio of the weights of the hollow and solid shafts if they have the same length?

7.8. Determine the maximum shear stress in the stepped shaft shown and indicate clearly the section in which it occurs. Ignore any stress concentrations.

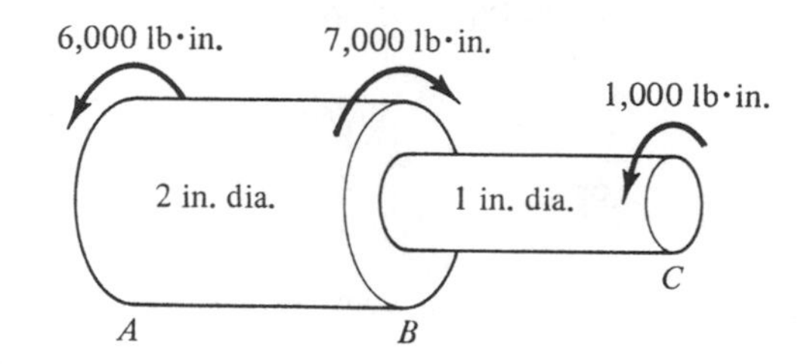

Problem 7.8

7.9. The hollow alloy steel shaft shown has an outside diameter of 100 mm and an inside diameter of 50 mm. Power P is input to the shaft at pulley B and is output in the fractional amounts indicated at pulleys A, C, and D. Determine the maximum shear stress in the shaft for a power input of 7.2 MW at 1,200 rpm and indicate clearly the section in which it occurs.

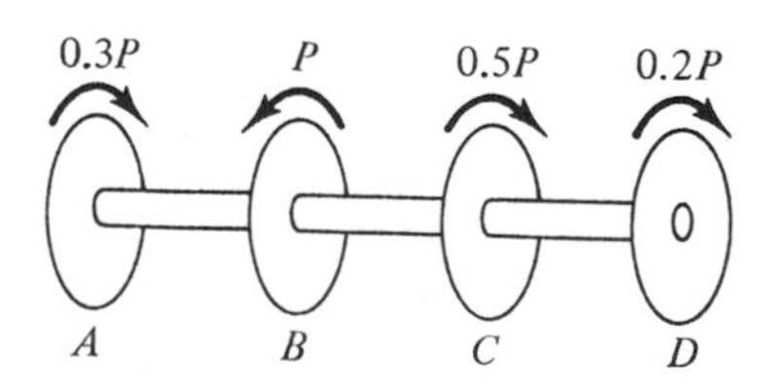

Problem 7.9

7.10. What is the maximum power P that can be input to the shaft in Problem 7.9 if the allowable shear stress is 70 MN/m^2?

7.11. A 4-in. (nominal) diameter structural steel pipe is attached to an immovable support by $\frac{1}{4}$-in. fillet welds, as shown. What is the total length of weld required if the connection is to be as strong as the pipe? The allowable shear stresses are 10 ksi for the pipe and 18 ksi for the weld (see Problem 5.33).

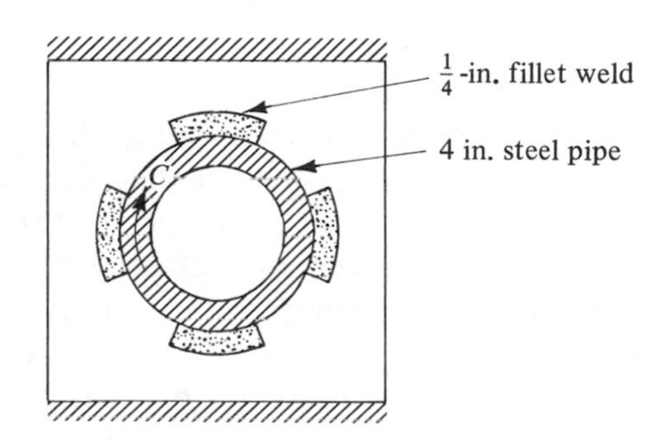

Problem 7.11

7.12. A hollow shaft made of structural steel is connected to a solid shaft made of 2024-T3 aluminum by a $\frac{1}{2}$-in. diameter structural steel bolt, as shown. What is the largest couple C that can be applied to the assembly without causing either shaft to yield or the bolt to shear off?

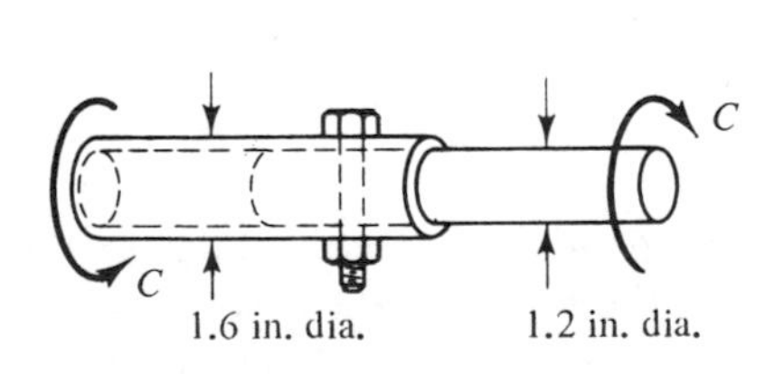

Problem 7.12

7.13. Two steel shafts are connected by gears, as shown. What is the largest couple C that can be applied to shaft DE if the maximum allowable torsional shear stress in each member is 20 ksi? Use a safety factor of 2.

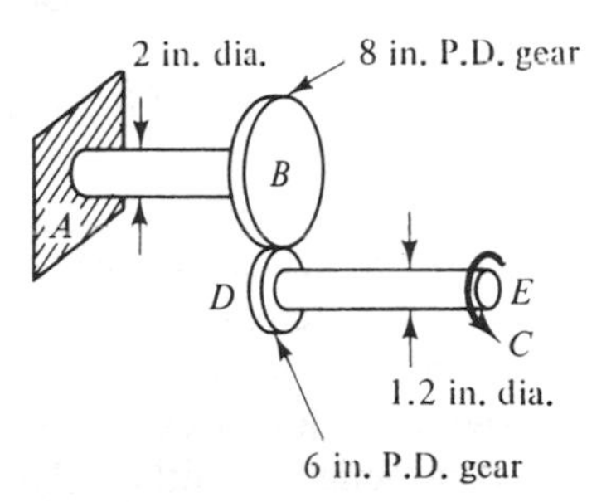

Problem 7.13

7.14. The drive assembly shown is to be made fail-safe against overloads by designing the coupling bolts such that they shear off before the shaft will yield. What diameter bolts should be used if a safety factor of 2.5 is required? Assume that the ultimate shear strength of the bolts is twice the shear yield strength for the shaft.

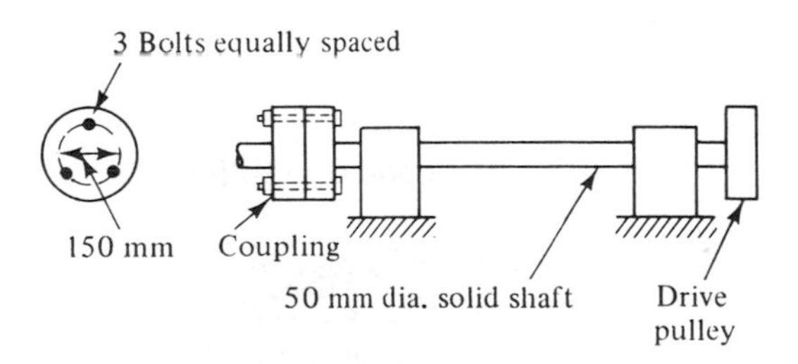

Problem 7.14

7.4 STRESSES ON OBLIQUE PLANES

Equation (7.6), or its equivalent for materials with a stress-strain relation other than $\tau = G\gamma$, defines the shear stresses acting upon planes perpendicular to the axis of a shaft. However, there are also other stresses acting that are of significance. In particular, there are longitudinal shear stresses acting upon planes parallel to the shaft axis and normal and shear stresses acting upon planes inclined to the axis. These stresses will be considered in this section.

Longitudinal Shear Stresses

Consider the stresses acting upon an element of material oriented with sides parallel and perpendicular to the shaft axis [Figure 7.11]. Since one side of the element lies in the shaft cross section, the shear stress acting upon it is given by Eq. (7.6), or its equivalent. This stress has the same sense as the torque, as indicated in the figure. However, shear stresses cannot exist alone. As shown in Section 5.5, they are always accompanied by other shear stresses of equal magnitude acting on orthogonal planes. Accordingly, there must also be shear stresses acting upon planes parallel to the shaft axis which have the same magnitude as those acting over the cross section and which vary with the radius in the same manner. This is illustrated in Figure 7.11 for the linearly elastic case. The longitudinal shear stresses are particularly important in wooden shafts. Since wood is relatively weak in shear parallel to the grain, these stresses tend to cause the shaft to split along its length when twisted.

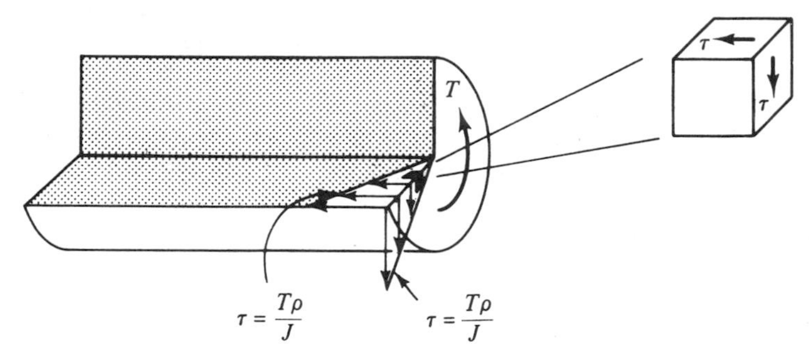

Figure 7.11

Stresses on Oblique Planes

The state of stress in a shaft on planes perpendicular and parallel to the axis is shown in Figure 7.12(a), where $\tau_{xy} = \tau_{yx}$ are the torsional shear stresses. Stresses on other planes can be determined by sectioning the element along the desired plane and considering the equilibrium of one of the resulting pieces.

Applying this procedure to the element in Figure 7.12(b) and then converting the stresses to forces, we obtain the free body diagram shown in Figure 7.12(c). In

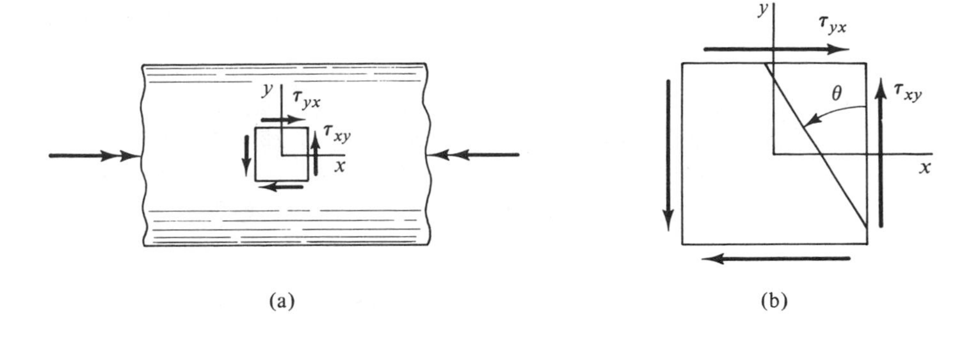

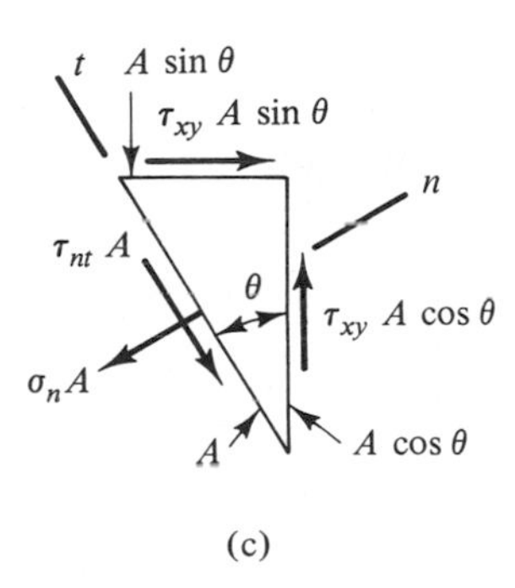

Figure 7.12

this figure, n and t are axes perpendicular and parallel to the plane of interest and σ_n and τ_{nt} are the normal and shear stresses acting on this plane (shown in their positive directions); θ is the angle of inclination of the plane with respect to the shaft cross section, and A is the area of the inclined plane. The equations of equilibrium are

$$\sum F_n = -\sigma_n A + (\tau_{xy} A \cos\theta)\sin\theta + (\tau_{xy} A \sin\theta)\cos\theta = 0$$

$$\sum F_t = -\tau_{nt} A + (\tau_{xy} A \cos\theta)\cos\theta - (\tau_{xy} A \sin\theta)\sin\theta = 0$$

from which we obtain

$$\sigma_n = 2\tau_{xy} \sin\theta\cos\theta = \tau_{xy}\sin 2\theta \tag{7.14}$$

and

$$\tau_{nt} = \tau_{xy}(\cos^2\theta - \sin^2\theta) = \tau_{xy}\cos 2\theta \tag{7.15}$$

These equations indicate that both normal and shear stresses exist in a twisted shaft.

From Eq. (7.14), it can be seen that τ_{nt} is largest when $\cos 2\theta = 1$, which corresponds to $\theta = 0°$ or $90°$. Thus, the maximum shear stress occurs on planes parallel and perpendicular to the axis of the shaft and is equal in magnitude to τ_{xy}:

$$|\tau_{nt}|_{\max} = |\tau_{xy}| \tag{7.16}$$

This confirms our earlier statement to this effect in Section 7.3.

The normal stress is a maximum when $\sin 2\theta = 1$, or $\theta = \pm 45°$. For $\theta = +45°$, σ_n is positive (tension), and for $\theta = -45°$, it is negative (compression). In both cases, it is equal in magnitude to τ_{xy}:

$$|\sigma_n|_{\max} = |\tau_{xy}| \tag{7.17}$$

These maximum shear and normal stresses and the planes upon which they act are shown in Figure 7.13. If the sense of the shear stress is reversed, the planes of maximum tension and compression also reverse. Notice that the shear stresses appear to pull the element in the direction of the maximum tension. This observation can be used to readily determine which of the maximum normal stresses is tension and which is compression.

Figure 7.13

The maximum tensile stress is usually the most significant stress in shafts made of brittle material because it is the one that tends to cause fracture. This is illustrated in Figure 7.14(a), which shows the fracture surface of a cast-iron shaft. The surface is oriented at 45° with respect to the axis, which corresponds to the plane of maximum tensile stress. This type of failure can be readily demonstrated by twisting a piece of chalk.

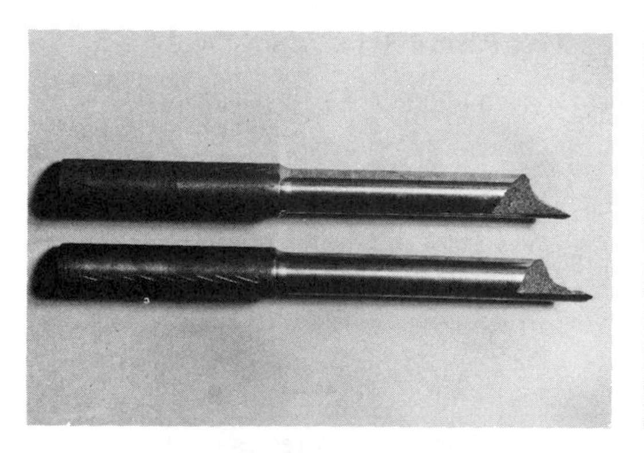

(a) Cast Iron

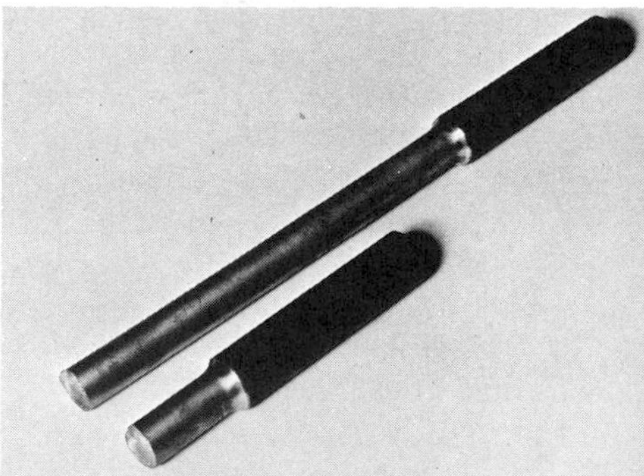

(b) Mild Steel

Figure 7.14 Fracture Patterns for (a) Brittle; and (b) Ductile Shafts.

Figure 7.14(b) shows the fracture surface for a steel shaft. As mentioned previously, ductile materials generally fail due to the shear stresses developed, which, in this case, are largest on planes perpendicular to the axis. Accordingly, the shaft breaks along one of these planes.

The compressive stresses developed can cause buckling, or wrinkling, of thin-walled hollow shafts. Experiments show that the wrinkle is oriented at 45° with respect to the axis, along the plane of maximum compressive stress.

Example 7.3. Design of a Cast-Iron Shaft. What diameter shaft is required to transmit a couple of 4,000 lb · in. with a safety factor of 2 if the material is cast iron with an ultimate tensile strength of 20 ksi?

Solution. The design couple is the safety factor times the applied couple, or 8,000 lb · in. The maximum tensile stress is equal in magnitude to the maximum torsional shear stress [Eq. (7.17)]:

$$\sigma_{max} = \tau_{max} = \frac{TR}{J} = \frac{16T}{\pi D^3}$$

Thus, we have

$$\sigma_{max} = \frac{16(8{,}000 \text{ lb} \cdot \text{in.})}{\pi D^3} \nless 20{,}000 \text{ lb/in}^2$$

or

$$D \geqslant 1.26 \text{ in.} \qquad \textbf{Answer}$$

PROBLEMS

7.15. A 30-mm diameter shaft is made of a material with allowable stresses of 30 MN/m^2 in shear and 35 MN/m^2 in tension. Determine the maximum torque T that can be transmitted by the shaft.

7.16. Which can support the larger torque, a 100-mm diameter wooden shaft with an allowable shear stress of 0.7 MN/m^2 parallel to the grain or a 10-mm diameter cast iron shaft with an ultimate tensile strength of 140 MN/m^2?

7.17. The state of stress in a twisted shaft is as shown. Determine the normal and shear stresses on a plane inclined at an angle θ with the shaft cross section and plot their variation with θ for $0 \leqslant \theta \leqslant 180°$.

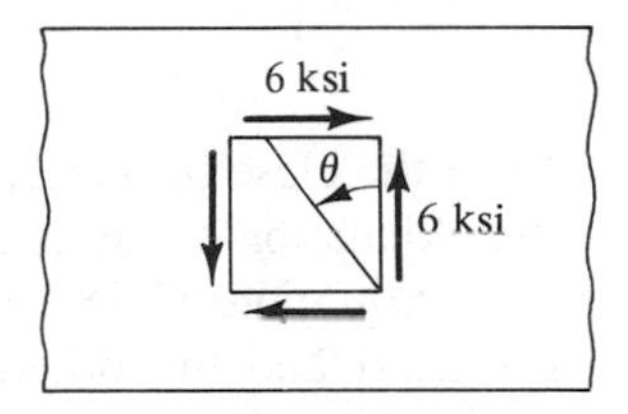

Problem 7.17

7.18. In Problem 7.17, what are the normal and shear stresses on a plane oriented 30° counterclockwise from the longitudinal axis of the shaft?

7.19. A thin-walled hollow shaft with an average diameter of 40 mm is to transmit 12 kW of power at 1,200 rpm. What is the minimum required wall thickness if the maximum compressive stress must be kept below 30 MN/m^2 to prevent buckling?

7.20. A broken wooden shaft is spliced together as shown. If the allowable tensile stress in the glue is 0.5 MN/m^2 and the allowable longitudinal shear stress in the wood is 0.7 MN/m^2, what should be the angle of inclination θ of the splice if the joint is to be as strong as the surrounding wood?

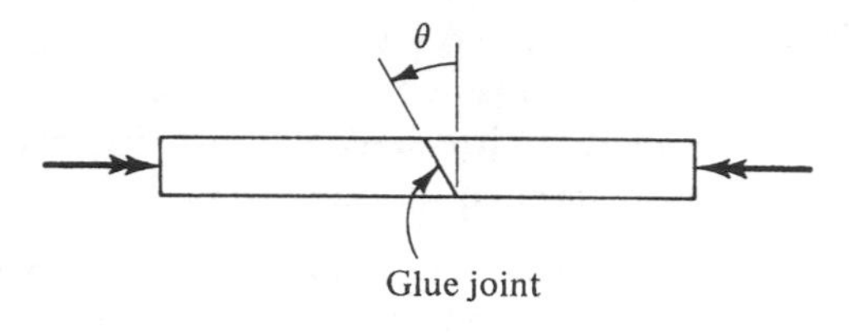

Problem 7.20

7.5 ANGLE OF TWIST

As indicated in Section 7.2, the angle θ through which one cross section of a shaft rotates with respect to another is called the *angle of twist*. If the stresses are below the proportional limit, this angle can be determined by integrating Eq. (7.5). We have

$$\theta = \int_0^L \frac{T}{JG}\,dx \qquad (\tau \leqslant \tau_{PL}) \tag{7.18}$$

where L is the length of the shaft and θ is measured in radians. The sense of θ is the same as that of the torque T. Accordingly, the shaft twists in the direction indicated by the fingers on the right hand when the thumb is placed along the torque vector.

The quantities T, J, and G must be expressed as functions of the distance x along the shaft before the preceding integral can be evaluated. If they are constant over the length of the member, Eq. (7.18) reduces to

$$\theta = \frac{TL}{JG} \qquad (\tau \leqslant \tau_{PL}) \tag{7.19}$$

Note the close similarity between this equation and the expression $\delta = PL/AE$ for the elongation of an axially loaded member.

Equations (7.18) and (7.19) define the total angle of twist between one end of the shaft and the other. To determine the twist of any particular portion of the shaft, it is only necessary to carry out the integration in Eq. (7.18) over the length of this portion or to use this length in place of the total length in Eq. (7.19).

Note that Eqs. (7.18) and (7.19) apply only if the shear stress is below the proportional limit because the stress-strain relation $\tau = G\gamma$ was used in their derivation. For other types of material behavior, it is only necessary to combine the appropriate stress-strain relation with the strain-deformation relation of Eq. (7.1) and the torque-stress relation of Eq. (7.4) and proceed as before.

There are regions of the shaft near the supports, points of load application, and geometrical discontinuities where Eq. (7.5) is not strictly valid. However, this has a negligible effect upon the total twist of the member. These regions are relatively short (St. Venant's principle) and, therefore, contribute little to the value of the integral in Eq. (7.18).

If the torque, polar moment of inertia, or shear modulus vary in steps along the length, the shaft can be treated as a series of shorter members placed end to end within which T, J, and G are constant. The twist of each portion can then be computed from Eq. (7.19) and the results summed to determine the twist of the entire member. In doing this, the sign convention for θ is the same as that for T, which can be chosen arbitrarily. The torque diagram is an invaluable aid in keeping track of the signs of T and θ in the various portions of the shaft.

Statically indeterminate torsion problems are handled in the same way as problems involving statically indeterminate axially loaded members. The only difference is that the compatibility condition involves angles of twist instead of elongations.

Our discussion of the deformations of shafts has purposely been brief because the procedures for determining them are basically the same as for axially loaded members. This is further illustrated in the following examples.

Example 7.4. Angle of Twist of a Shaft. Through what angle can a solid shaft with a diameter of 30 mm and a length of 2 m be twisted without exceeding the shear yield strength of 145 NM/m² if $G = 83$ GN/m²?

Solution. We first determine the torque from the torque-stress relation, Eq. (7.6):

$$T = \tau_{\max} \frac{J}{R}$$

Substituting this expression into the torque-twist relation of Eq. (7.19), we have

$$\theta = \frac{TL}{JG} = \tau_{\max} \frac{J}{R}\left(\frac{L}{JG}\right) = \frac{\tau_{\max} L}{RG}$$

or

$$\theta = \frac{(145 \times 10^6 \text{ N/m}^2)(2 \text{ m})}{(15 \times 10^{-3} \text{ m})(83 \times 10^9 \text{ N/m}^2)} = 0.23 \text{ rad or } 13.2^\circ \qquad \textbf{Answer}$$

Example 7.5. Twist of a Composite Shaft. A steel pipe is joined to a solid aluminum rod to form the composite shaft shown in Figure 7.15(a). Through what angle will the free end of the shaft rotate under the loading shown?

Solution. We first determine the reaction at the wall and the torque diagram [Figure 7.15(b)]. Here, we have used the sign convention that torque vectors directed to the right are positive. According to this convention, θ will be positive if the shaft twists in the counter-clockwise direction when viewed from the right end.

The material properties are obtained from Table A.2:

$$\text{Steel:} \quad G = 12 \times 10^6 \text{ psi} \qquad \text{Alum:} \quad G = 3.8 \times 10^6 \text{ psi}$$

$$\tau_{YS} = 60 \text{ ksi} \qquad \tau_{Ys} = 21 \text{ ksi}$$

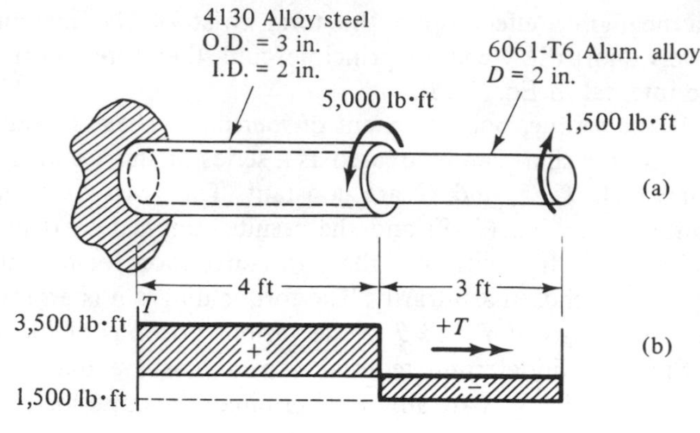

Figure 7.15

The polar moments of inertia are

$$J_S = \frac{\pi}{32}\left[(3\text{ in.})^4 - (2\text{ in.})^4\right] = 6.38\text{ in.}^4$$

$$J_A = \frac{\pi}{32}(2\text{ in.})^4 = 1.57\text{ in.}^4$$

We next check the stresses to see if they are below the proportional limit:

$$(\tau_{\max})_S = \left(\frac{TR}{J}\right)_S = \frac{(3{,}500\text{ lb}\cdot\text{ft})(12\text{ in./ft})(1.5\text{ in.})}{6.38\text{ in.}^4} = 9{,}870\text{ psi}$$

$$(\tau_{\max})_A = \left(\frac{TR}{J}\right)_A = \frac{(1{,}500\text{ lb}\cdot\text{ft})(12\text{ in./ft})(1\text{ in.})}{1.57\text{ in.}^4} = 11{,}460\text{ psi}$$

Since both of these values are well below the respective yield strengths, Eq. (7.19) applies.

The total angle of twist of the shaft is the sum of the angles of twist of each part:

$$\theta = \theta_S + \theta_A = \left(\frac{TL}{JG}\right)_S + \left(\frac{TL}{JG}\right)_A$$

or

$$\theta = \frac{(3{,}500\text{ lb}\cdot\text{ft})(12\text{ in./ft})(48\text{ in.})}{(6.38\text{ in.}^4)(12\times10^6\text{ lb/in.}^2)} + \frac{(-1{,}500\text{ lb}\cdot\text{ft})(12\text{ in./ft})(36\text{ in.})}{(1.57\text{ in.}^4)(3.8\times10^6\text{ lb/in.}^2)}$$

$$= 0.026 - 0.109 = -0.083\text{ radians}$$ **Answer**

Since θ_S is positive and θ_A is negative, the steel pipe twists in the counterclockwise direction (when viewed from the right end of the shaft) while the aluminum rod twists in the clockwise direction. The aluminum twists more than the steel; so the net twist of the free end is clockwise. Note that units must be watched carefully when computing θ.

Example 7.6. Torques in an Indeterminate Shaft. An undersized shaft is "strengthened" by slipping a hollow tube over it and connecting the two shafts at the ends so that they twist as a single unit [Figure 7.16(a)]. By what factors will this procedure reduce the shear stresses in the solid shaft and its angle of twist? Both shafts have the same length, and the quantity JG for the outer shaft is twice that of the inner shaft. Assume linearly elastic behavior.

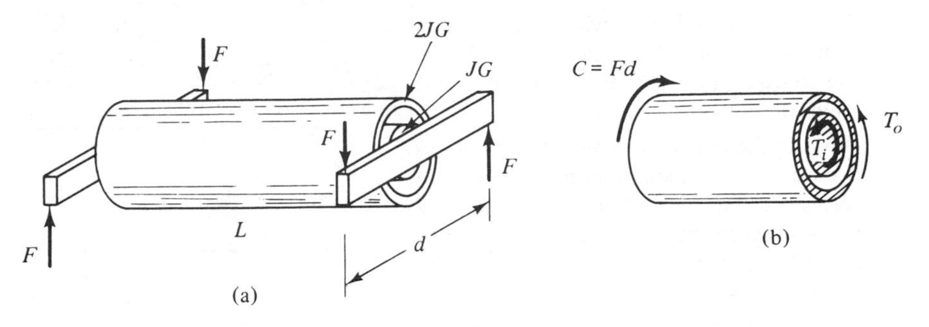

Figure 7.16

Solution. Let T_o and T_i denote the torques in the outer and inner shafts, respectively. Sectioning the composite shaft perpendicular to its axis and considering the equilibrium of the resulting piece [Figure 7.16(b)], we have

$$T_o + T_i = C$$

where $C = Fd$ is the magnitude of the applied couple. The problem is statically indeterminate because there are two unknowns and only one equilibrium equation from which to determine them.

The angle of twist of each shaft must be the same because they are connected at the ends. Thus, the compatibility condition is

$$\theta_o = \theta_i$$

Substituting the torque-twist relation of Eq. (7.19) into this expression, we obtain

$$\left(\frac{TL}{JG}\right)_o = \left(\frac{TL}{JG}\right)_i$$

or

$$T_o = \frac{2(JG)_i}{(JG)_i} T_i = 2T_i$$

Combining this result with the equilibrium equation, we get

$$T_i = \frac{C}{3}$$

Thus, the presence of the outer shaft reduces the torque in the inner shaft by two-thirds (the applied couple is assumed to be the same in each case). Since the shear stress and angle of twist are directly proportional to the torque, they are reduced by the same amount.

PROBLEMS

7.21. An aluminum ($G = 4 \times 10^6$ psi) wire 50 ft long has a diameter of $\frac{1}{4}$ in. What torque is required to twist the wire through two complete revolutions? What is the maximum shear stress in the wire for this value of torque?

7.22. A structural steel shaft with a diameter of 1 in. is 6 ft long. Through what angle can the shaft be twisted without exceeding the yield strength in shear?

7.23. What is the minimum required diameter of a solid shaft 10 ft long that must transmit a torque of 2 kip · ft with a safety factor of 2 without exceeding the allowable shear stress of 24 ksi or the allowable angle of twist of 5°? Use $G = 4 \times 10^6$ psi.

7.24. A solid shaft with a diameter of 60 mm transmits 300 kW at 600 rpm. Find a hollow shaft of the same material with an outside diameter of 75 mm that will transmit the same power with the same angle of twist per unit of length. What is the ratio of the weights of the hollow and solid shafts and what is the ratio of the maximum shear stress in each?

7.25. Determine the angle of twist (in degrees) of the free end of the steel shaft shown.

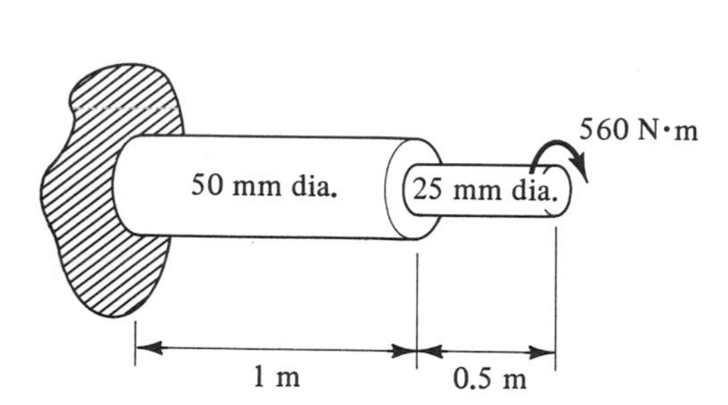

Problem 7.25

7.26. The composite shaft shown has a diameter of 2 in. Determine (a) the lengths L_1 and L_2 for which the angle of twist of the free end will be zero and (b) the maximum shear stress in the member.

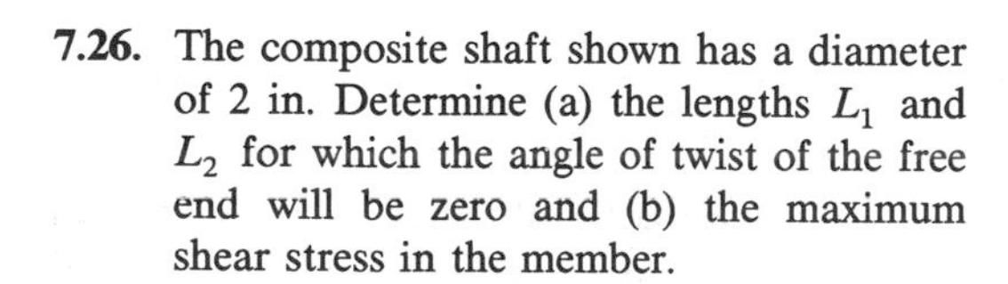

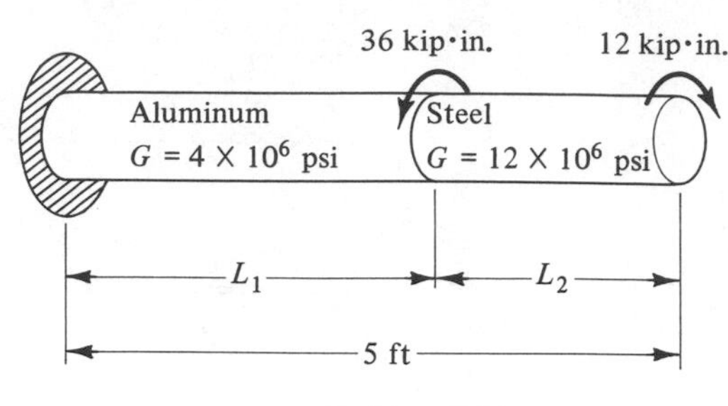

Problem 7.26

7.27. A stepped steel shaft is loaded as shown. Determine the magnitude of the couple C for which the angle of twist at the free end will be zero. What is the maximum shear stress in the shaft and through what angle does section AB twist?

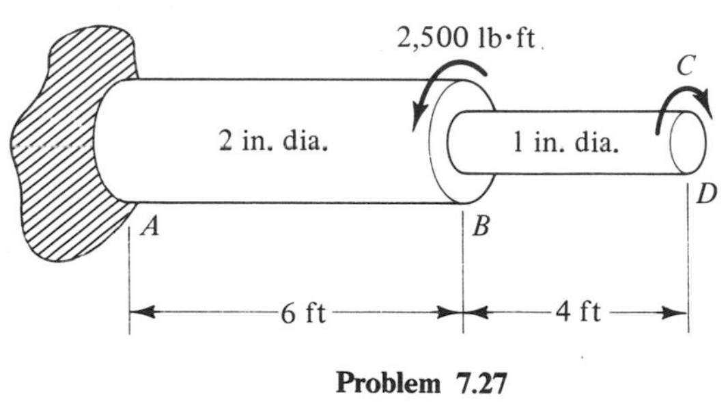

Problem 7.27

7.28. A hollow drive shaft with 100-mm OD and 50-mm ID is in equilibrium under the ac-

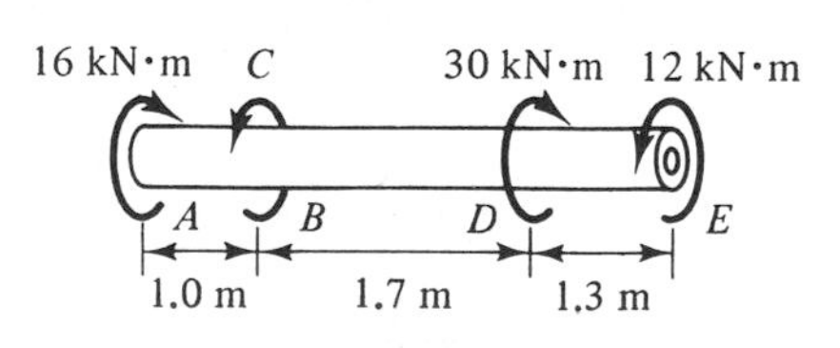

Problem 7.28

tion of the couples shown. Determine the relative angles of twist (a) between locations A and E and (b) between locations B and D. Use $G = 28\ \text{GN/m}^2$.

7.29. For the shaft in Problem 7.28, determine the angle of twist relative to the left end and plot its variation along the length of the member. What is the maximum angle of twist in the shaft and where does it occur?

7.30. When a steel shaft 50 mm in diameter and 2-m long transmits a torque of 1 kN · m, the twist is found to exceed the allowable value of 0.025 rad. The situation is to be corrected by slipping a hollow steel sleeve with 70-mm OD and 50-mm ID over the shaft and bonding it in place over its entire length. What is the required length L of the sleeve? Does it make any difference where the sleeve is located along the shaft?

7.31. A rod has a variable diameter, as shown. Derive an expression for the angle of twist of the free end when subjected to a torque T. Assume linearly elastic material behavior.

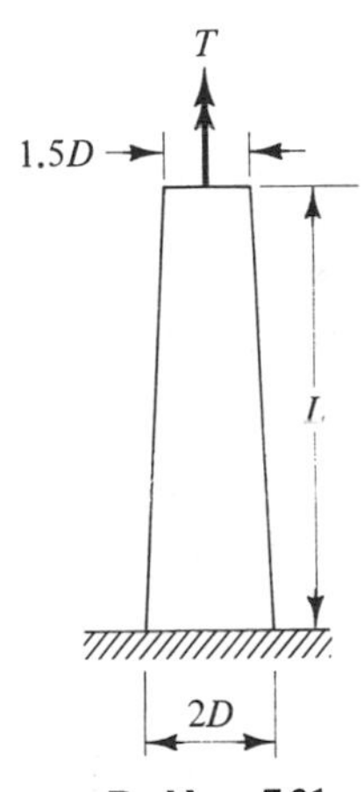

Problem 7.31

7.32. The ends of the uniform shaft shown are fixed so that they cannot rotate. Determine the reactions at the walls. Assume elastic behavior.

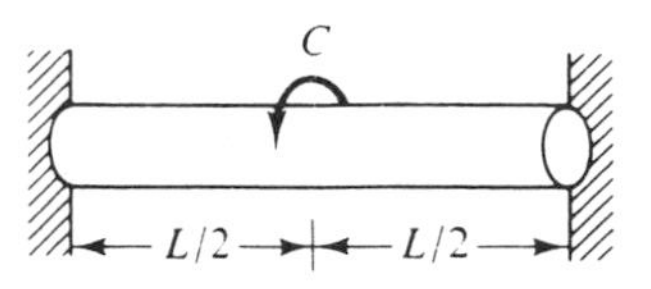

Problem 7.32

7.33. A composite shaft made of two different materials A and B bonded together end-to-end is fixed at each end, as shown. What is the largest couple C that can be applied at the junction between the two materials if the allowable shear stresses are $\tau_A = 15$ ksi and $\tau_B = 8$ ksi; $G_A = 10 \times 10^6$ psi and $G_B = 6 \times 10^6$ psi?

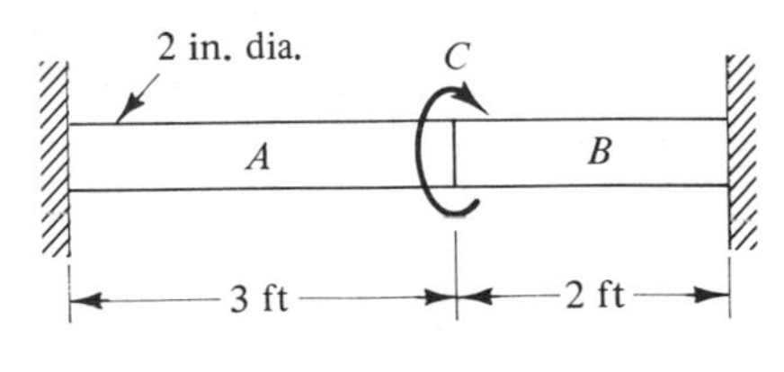

Problem 7.33

7.34. Same as Problem 7.33, except that the right end of the shaft rotates 1° when the couple is applied.

7.35. The assembly shown consists of two gears supported by bearings and connected to identical steel shafts ($\tau_{ys} = 200\ \text{MN/m}^2$). Through what angle will the rigid arm EG rotate when the load $F = 2$ kN is applied?

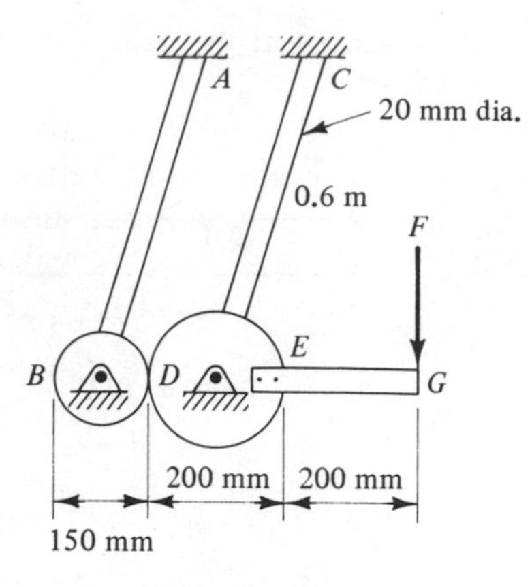

Problem 7.35

7.36. For the assembly in Problem 7.35, what is the maximum force F that can be applied if the shafts are not to yield. What are the angles of twist of the shafts for this value of F?

7.37. The free end of the shaft shown is attached to a rigid bar connected to two identical vertical wires. If the wires are stress free at 70°F, to what tensile stress will they be subjected at 30°F?

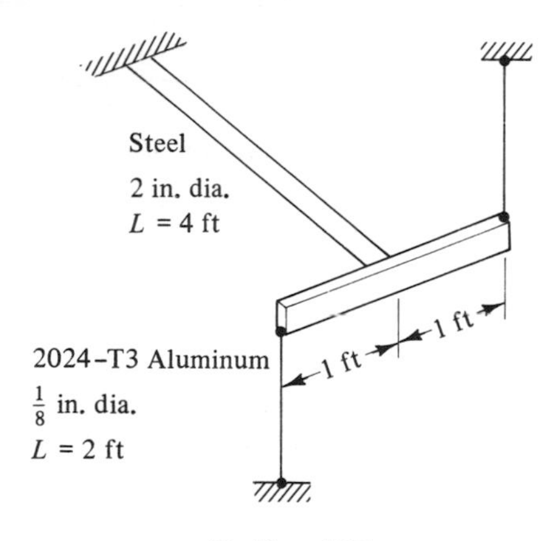

Problem 7.37

7.38. The 1-in. diameter steel torsion bar in an automobile is supported at an intermediate point by a fixed rubber mount, as shown. The bar and the mount are bonded together over their entire interface. Experiments show that the torque-twist relation for the mount is $T(\text{lb} \cdot \text{in.}) = 2 \times 10^4\, \theta(\text{rad.})$ for relatively small angles of rotation. By what percentage does the presence of the mount reduce the angle of twist at the free end of the torsion bar? What effect does the mount have upon the torsional stresses?

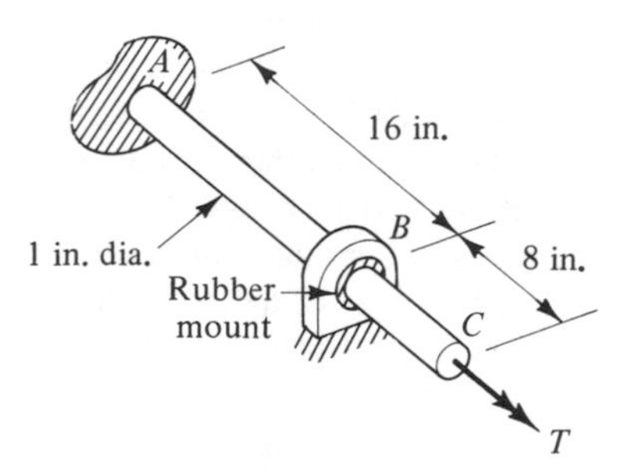

Problem 7.38

7.39. Derive the torque-stress and torque-twist relations for a circular shaft of radius R and length L made of a material whose shear stress-shear strain curve is described by the equation $\tau = k\gamma^n$, where k and n are material constants.

7.40. A composite shaft made of two different materials A and B bonded together over their entire length has the cross section shown. Determine the torque-stress and torque-twist relations for the member. Assume linearly elastic material behavior.

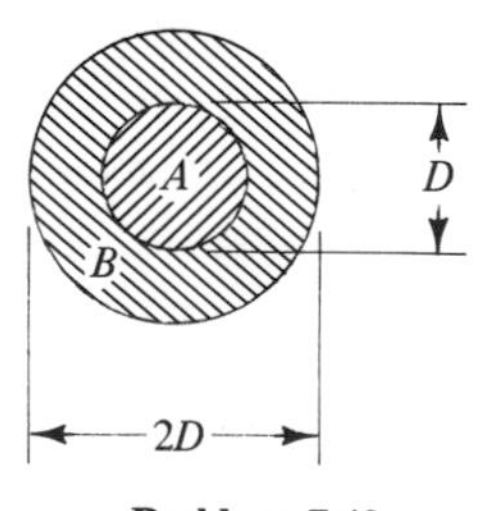

Problem 7.40

7.6 INELASTIC BEHAVIOR

In this section we shall consider the response of torsional members strained into the inelastic range. All of our preceding assumptions concerning the geometry and loading of the shaft still apply, plus the additional assumption that the material behavior is essentially elastic-perfectly plastic.

Stress Distribution

As the shaft undergoes inelastic deformations, the stress distribution tends to become uniform over the cross section, and the torque-stress and torque-twist relations derived in the preceding sections no longer apply. The situation is illustrated in Figure 7.17, which shows the shear stress distribution in a shaft made of an elastic-perfectly plastic material for three different angles of twist. The conditions at locations a, b, and c on the cross section are denoted by the corresponding points on the stress-strain diagram.

The shear strain in a shaft is defined by Eq. (7.1) and is proportional to the angle of twist regardless of the stress level. As the twist increases from zero, the maximum shear strain increases, and so does the maximum shear stress, until it is just equal to the yield strength of the material. This is the situation shown in Figure 7.17(a). Up to this point, the torque-stress and torque-twist relations given in the preceding sections are valid.

If the angle of twist is further increased, the maximum shear strain increases, but the maximum shear stress remains equal to the yield strength. Since strains at points in the interior of the shaft are now at or above the proportional limit strain, the stress distribution becomes more nearly uniform [Figure 7.17(b)]. This is referred to as the *partially plastic condition* because the strains are within the inelastic, or plastic, range only over a portion of the cross section.

As the angle of twist continues to increase, a situation is eventually reached in which the strains are at or above the proportional limit strain everywhere over the cross section, except near the center where the strain is always zero. The stress distribution is then very nearly uniform, as shown in Figure 7.17(c). This is referred to as the *fully plastic condition*, and the torque at which it occurs is called the *fully plastic torque*, T_{fp}.

Fully Plastic Torque

The torque-stress relation for the fully plastic case is obtained from Eq. (7.4):

$$T = \int \rho \tau \, dA$$

Referring to Figure 7.18, we take a ring with radius ρ and thickness $d\rho$ as the element of area dA. The shear stress is equal to the yield strength, τ_{YS}, and is assumed to be constant over the entire cross section. This assumption is reasonable,

Strain distribution

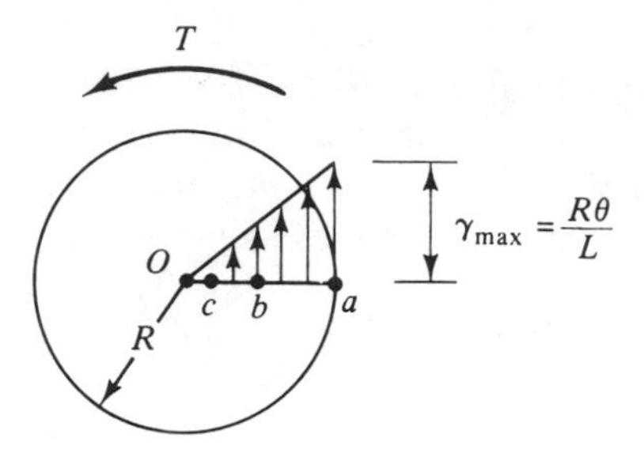

Stress distributions

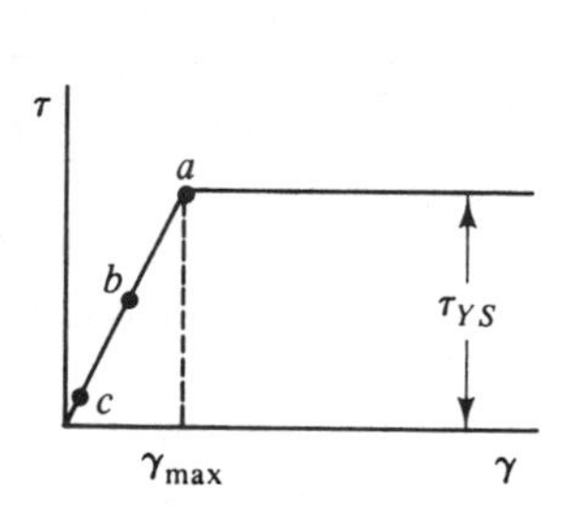

(a)

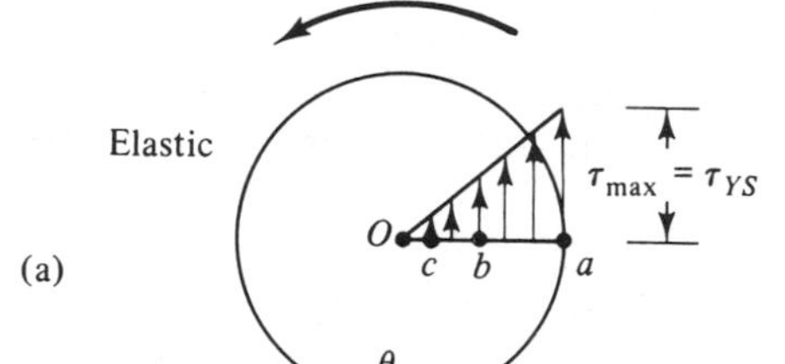

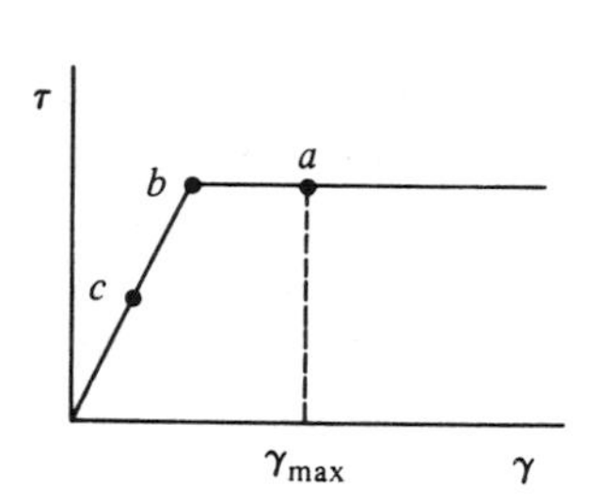

(b)

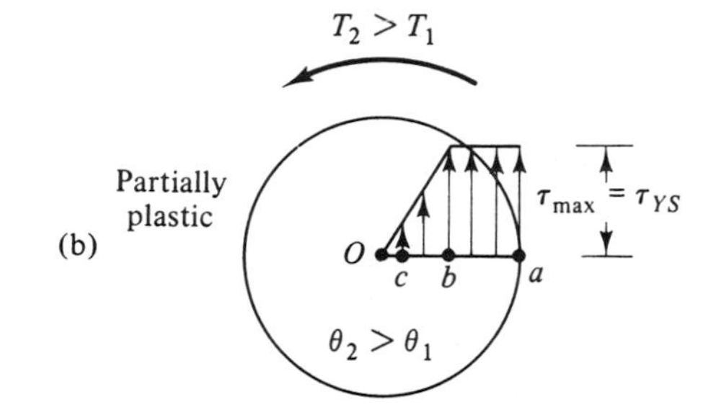

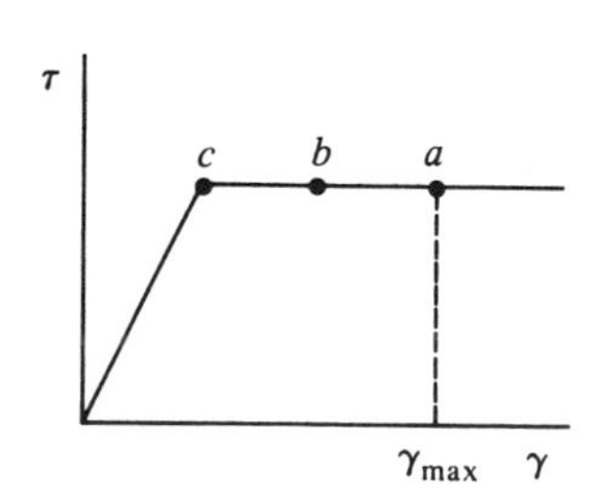

(c)

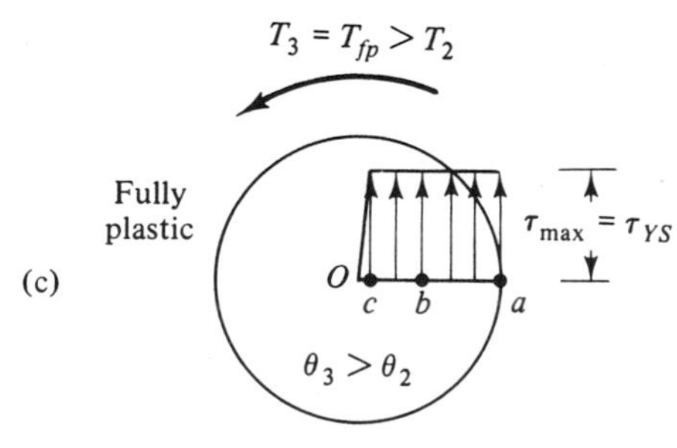

Figure 7.17

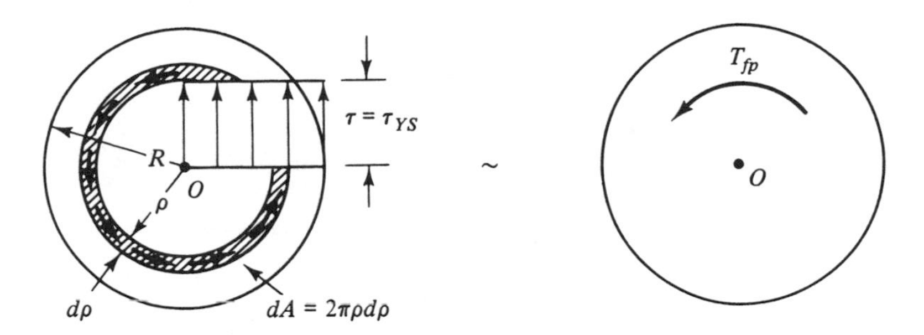

Figure 7.18

as the material at the center of the shaft contributes little to the torque. Thus, we have

$$T_{fp} = \int_0^R \rho \tau_{YS} (2\pi\rho \, d\rho) = \frac{2}{3}\pi\tau_{YS}R^3 \tag{7.20}$$

or, expressed in terms of the polar moment of inertia,

$$T_{fp} = \frac{4}{3}\frac{\tau_{YS}J}{R} \tag{7.21}$$

It is also informative to express the fully plastic torque in terms of the *yield torque*, T_{YS}, at which the maximum shear stress first reaches the yield strength. From Eq. (7.6), we have

$$T_{YS} = \frac{\tau_{YS}J}{R} \tag{7.22}$$

therefore, Eq. (7.21) can be written as

$$T_{fp} = \frac{4}{3}T_{YS} \tag{7.23}$$

Thus, the fully plastic torque is four-thirds of the torque at which the shaft starts to yield.

These various torques and the different ranges of behavior associated with them are illustrated in Figure 7.19, which shows a portion of the experimentally determined torque-twist curve for a mild steel shaft. The point at which the curve first starts to deviate from a straight line corresponds to the onset of yielding. This point is difficult to locate accurately because the deviation from linearity is initially very slight. The horizontal portion of the curve corresponds to the fully plastic condition. Points between these two extremes correspond to partially plastic states.

Notice that in the fully plastic range the angle of twist can increase to very large values with little or no increase in torque. Thus, the shaft usually fails by general yielding once the fully plastic condition is reached, unless the twisting is somehow restricted.

Figure 7.19 also shows the advantage of allowing some permanent deformations whenever possible. If no permanent deformations are allowed, the maximum

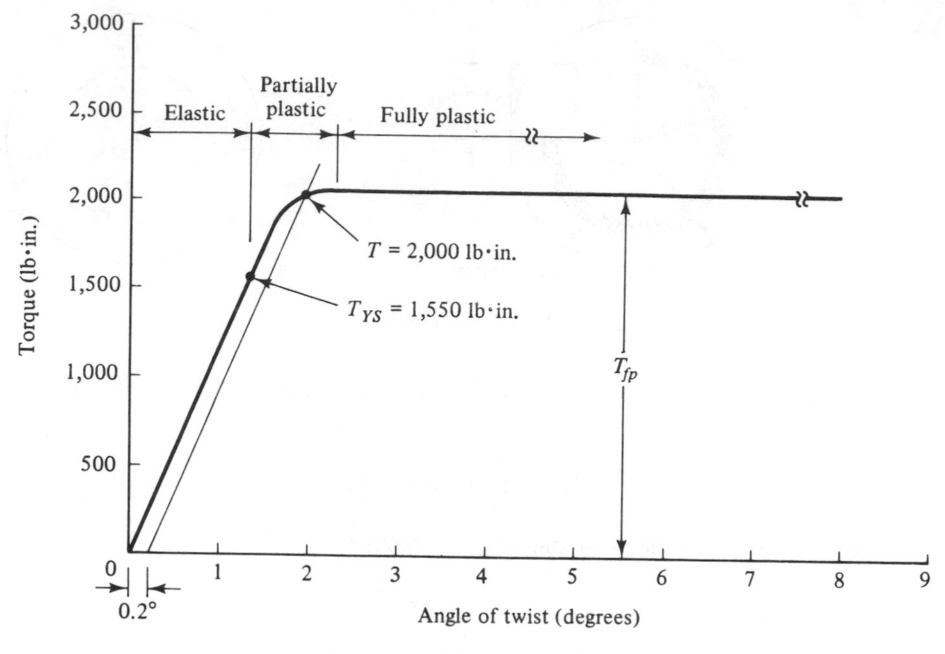

Figure 7.19

torque that can be supported is T_{YS}, which for this particular shaft is 1,550 lb · in. However, if a permanent twist of 0.2° is permitted, a torque of 2,000 lb · in. can be supported, which is an increase of 29%. As illustrated in the figure, this torque is determined in the same manner as the yield strength of a material.

Ultimate Torque

If the stress-strain diagram is reasonably flat in the vicinity of the ultimate strength, the stress distribution will be very nearly uniform, as in the fully plastic case. Thus, the torque-stress relation is the same as that given in Eq. (7.20), except that the yield strength is replaced by the ultimate strength, τ_U, and the fully plastic torque is replaced by the *ultimate torque*, T_U:

$$T_U = \frac{2}{3}\pi\tau_U R^3 \tag{7.24}$$

or

$$T_U = \frac{4}{3}\frac{\tau_U J}{R} \tag{7.25}$$

These equations define the maximum torque that can be applied to a shaft made of a ductile material without it breaking.

Hollow Shafts

The inelastic behavior of hollow shafts is the same as for solid shafts. However, the torque-stress relations cannot be obtained by simply replacing J in the various equations by that for a hollow shaft, as in the linearly elastic case. These relations must be derived independently. This can be done by carrying out the integration in Eq. (7.20) from the inside radius R_i of the shaft to the outside radius R_o. For the fully plastic torque, we obtain

$$T_{fp} = \frac{2}{3}\pi\tau_{YS}\left(R_o^3 - R_i^3\right) = \frac{4}{3}\tau_{YS}\left(\frac{J_o}{R_o} - \frac{J_i}{R_i}\right) \tag{7.26}$$

where J_o and J_i are the polar moments of inertia of areas with radius R_o and R_i, respectively. Similarly, the ultimate torque is

$$T_U = \frac{2}{3}\pi\tau_U\left(R_o^3 - R_i^3\right) = \frac{4}{3}\tau_U\left(\frac{J_o}{R_o} - \frac{J_i}{R_i}\right) \tag{7.27}$$

Example 7.7. Fully Plastic and Ultimate Torques for a Hollow Shaft. Determine the fully plastic and ultimate torques for a hollow shaft with an outside diameter of 80 mm and an inside diameter of 50 mm made of a ductile material for which $\tau_{YS} = 400\ \text{MN/m}^2$ and $\tau_U = 550\ \text{MN/m}^2$.

Solution. The fully plastic torque for a hollow shaft is defined by Eq. (7.26):

$$T_{fp} = \frac{2}{3}\pi\tau_{YS}\left(R_o^3 - R_i^3\right)$$

or

$$T_{fp} = \frac{2}{3}\pi\left(400 \times 10^6\ \text{N/m}^2\right)\left[(0.040\ \text{m})^3 - (0.025\ \text{m})^3\right]$$

$$= 40.5 \times 10^3\ \text{N}\cdot\text{m} \quad \text{or} \quad 40.5\ \text{kN}\cdot\text{m} \qquad \textbf{Answer}$$

The ultimate torque is given by the same expression, but with the yield strength replaced by the ultimate strength [Eq. (7.27)]:

$$T_U = \frac{2}{3}\pi\tau_U\left(R_o^3 - R_i^3\right) = \frac{\tau_U}{\tau_{YS}}T_{fp}$$

Thus,

$$T_U = \frac{\left(550\ \text{MN/m}^2\right)(40.5\ \text{kN}\cdot\text{m})}{\left(400\ \text{MN/m}^2\right)} = 55.7\ \text{kN}\cdot\text{m} \qquad \textbf{Answer}$$

Example 7.8. Inelastic Response of a Shaft. (a) What applied couple C is required to twist the shaft shown in Figure 7.20(a) through an angle of 11°? The stress-strain diagram for the material is given in Figure 7.20(b). (b) What is the maximum couple that can be applied without the shaft failing due to excessive inelastic deformations (general yielding)?

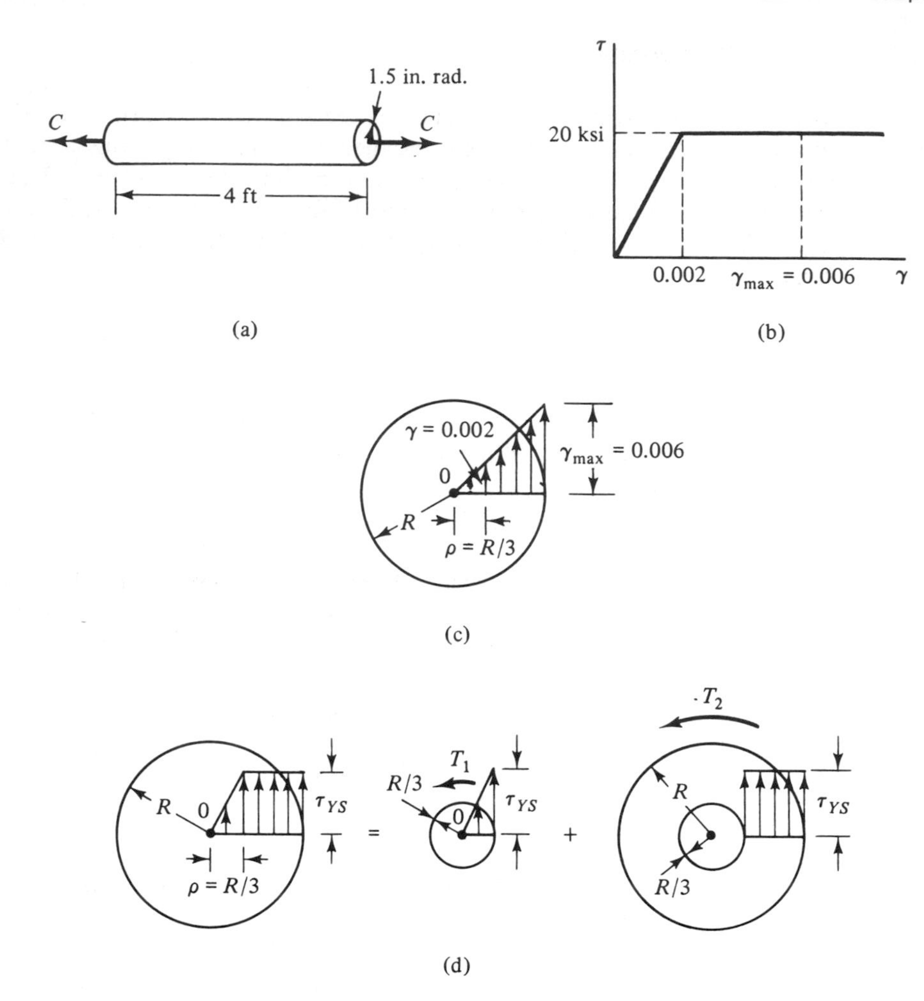

Figure 7.20

Solution. (a) We first check to see whether the strains are in the linearly elastic or inelastic range. From Eq. (7.2), the maximum strain is

$$\gamma_{\max} = \frac{\theta R}{L} = \frac{(11^\circ)(\pi\,\text{rad}/180^\circ)(1.5\text{ in.})}{48\text{ in.}} = 0.006$$

Since this value exceeds the strain at the proportional limit [Figure 7.20(b)], the shaft is strained into the inelastic range. The strain distribution is as shown in Figure 7.20(c). By similar triangles, the strain is found to be equal to the proportional limit strain at a radius of $\rho = R/3$.

The corresponding stress distribution is shown in Figure 7.20(d). As indicated in the figure, the shaft can be thought of as consisting of two parts: an inner elastic core with radius $R/3$ stressed to the yield strength and an outer fully plastic part with inner radius $R/3$ and outer radius R. The torques T_1 and T_2 carried by each of these parts can be obtained from

Eqs. (7.22) and (7.26), respectively:

$$T_1 = \frac{\tau_{YS}(\pi/2)(R/3)^4}{(R/3)} = \frac{\pi}{54}\tau_{YS}R^3$$

$$T_2 = \frac{2}{3}\pi\tau_{YS}\left[R^3 - (R/3)^3\right] = \frac{52}{81}\pi\tau_{YS}R^3$$

The total torque, which is equal in magnitude to the applied couple, is

$$T = T_1 + T_2 = \frac{53.5}{81}\pi\tau_{YS}R^3 \qquad \textbf{Answer}$$

or

$$T = \frac{53.5\pi}{81}(20{,}000 \text{ lb/in.}^2)(1.5 \text{ in.})^3 = 14.01 \times 10^4 \text{ lb} \cdot \text{in.} \qquad \textbf{Answer}$$

Note that only a small fraction of the torque is carried by the elastic core.

(b) General yielding occurs when the fully plastic condition is reached. Thus, the maximum couple that can be transmitted is equal in magnitude to the fully plastic torque. From Eq. (7.20), we have

$$T_{fp} = \frac{2}{3}\pi\tau_{YS}R^3 = \frac{2\pi}{3}(20{,}000 \text{ lb/in.}^2)(1.5 \text{ in.})^3 = 14.14 \times 10^4 \text{ lb} \cdot \text{in.} \qquad \textbf{Answer}$$

which is only slightly larger than the value obtained in part (a).

PROBLEMS

7.41. Compare the fully plastic and yield torques for a hollow shaft with an outside diameter of 50 mm and an inside diameter of 25 mm. The shear yield strength for the material is 140 MN/m^2.

7.42. Compare the fully plastic and ultimate torques for the hollow shaft of Problem 7.41 with those for a solid shaft with the same outside diameter. The ultimate shear strength is 300 MN/m^2.

7.43. Verify the expression given in Eq. (7.26) for the fully plastic torque for a hollow shaft with inner radius R_i and outer radius R_o. Show that this expression can also be obtained by subtracting the fully plastic torque for a solid shaft with radius R_i from that for a solid shaft with radius R_o.

7.44. A 1-in. diameter shaft contains a small keyway, for which the stress concentration factor is $K_\tau = 2.8$. what is the maximum torque that can be exerted on the shaft without causing it to (a) yield, (b) become fully plastic, and (c) twist off? $\tau_{YS} = 50$ ksi and $\tau_U = 75$ ksi.

7.45. A composite shaft is made by slipping a 3-in. nominal diameter steel pipe inside a 4-in. nominal diameter steel pipe and connecting them together at the ends. What is the maximum torque that can be transmitted without failure by general yielding if $\tau_{YS} = 20$ ksi?

7.46. A 0.8-m-long solid shaft with a radius of 40 mm is twisted until the maximum shear strain is 0.01. The shear modulus G of the material is 70 GN/m^2 and the yield point in shear is 175 MN/m^2. Sketch the distribution of shear strain and shear stress over the cross section and determine the angle of twist θ and the torque T.

7.47. Same as Problem 7.46, except that $\theta = 0.12$ rad and the maximum shear strain is an unknown to be determined.

7.48. For the shaft in Problem 7.45, what is the largest torque that can be transmitted without causing yielding in the inner member? Sketch the distribution of shear strain and shear stress over the cross section at this value of torque.

7.49. A solid circular shaft 3 in. in diameter and 4 ft long carries a torque of 15 kip · ft. If the material is 2024-T3 aluminum, what is the radius ρ_E of the elastic core and what is the angle of twist θ?

7.50. Same as Problem 7.49, except that the material is brass and $T = 14$ kip · ft.

7.51. The uniform structural steel shaft shown has a diameter of 60 mm and is attached to immovable supports at each end. If the angle of twist at the point of load application is 0.04 rad, what is the magnitude of the applied couple C?

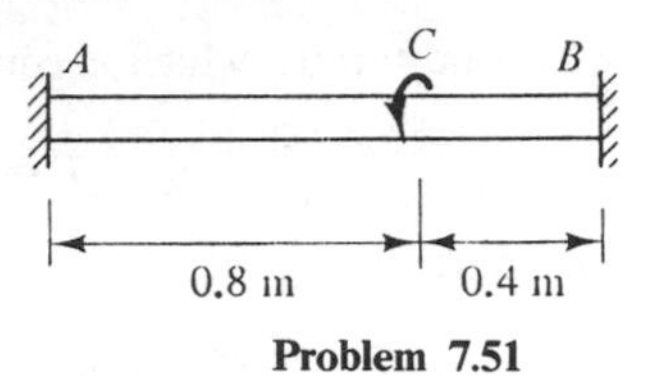

Problem 7.51

7.52. Same as Problem 7.51, except that the angle of twist is 0.06 rad.

7.7 LIMIT LOADS

When the fully plastic condition is reached in a shaft made of a ductile material, failure by general yielding and the associated large angles of twist usually occurs for little or no increase in load. If, however, the member is statically indeterminate, the angle of twist is restricted by the addition support conditions and further loading is possible before general yielding occurs.

To illustrate this, let us consider the example of a shaft attached to rigid walls at each end and subjected to a couple C applied at an intermediate point along its length [Figure 7.21(a)]. The material behavior is assumed to be elastic-perfectly plastic. An elastic analysis shows that the torques in the two portions of the shaft are

$$T_{AB} = \frac{C}{3} \qquad T_{BD} = \frac{2C}{3}$$

Thus, portion BD is the most highly stressed.

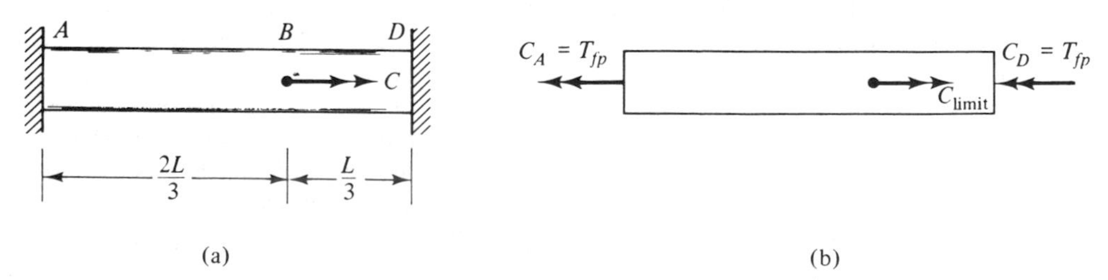

Figure 7.21

As the applied couple is increased from zero, the fully plastic condition is first reached in portion *BD* of the shaft. However, its angle of twist will be restricted by the adjoining portion, *AB*, which remains elastic, or possibly partially plastic. Any further increases in load must be supported entirely by portion *AB* because the torque in portion *BD* will remain equal to the fully plastic torque. Eventually, portion *AB* will also become fully plastic. When this happens, both parts of the shaft will be free to undergo large angles of twist, and its usefulness as a load-carrying member will generally be ended. As in the axial loading case, the load at which this occurs is called the *limit load*. Again, a structure may have several possible collapse mechanisms.

At the limit load, the torques are equal to the fully plastic torques, and the problem becomes statically determinate. Thus, the limit load can be determined solely from the equations of equilibrium, as is evident from the free-body diagram of the member [Figure 7.21(b)]. For this particular example, the limit load (couple) is

$$C_{\text{limit}} = 2T_{fp}$$

where the fully plastic torque is defined by Eq. (7.20).

Example 7.9. Limit Load of a Stepped Shaft. Derive an expression for the limit load, C_{limit}, for the stepped shaft shown in Figure 7.22(a). Express the result in terms of the material yield strength, τ_{YS}, and the diameter, D, of the smaller portion of the shaft.

Solution. General yielding cannot occur until both portions of the shaft become fully plastic. The free-body diagram for this case is shown in Figure 7.22(b). For equilibrium, the sum of the moments about the longitudinal axis must be zero, or

$$C_{\text{limit}} = (T_{fp})_{AB} + (T_{fp})_{BE}$$

From Eq. (7.20), we have

$$(T_{fp})_{BE} = \frac{2\pi}{3}\tau_{YS}\left(\frac{D}{2}\right)^3 = \frac{\pi}{12}\tau_{YS}D^3$$

$$(T_{fp})_{AB} = \frac{2}{3}\pi\tau_{YS}D^3$$

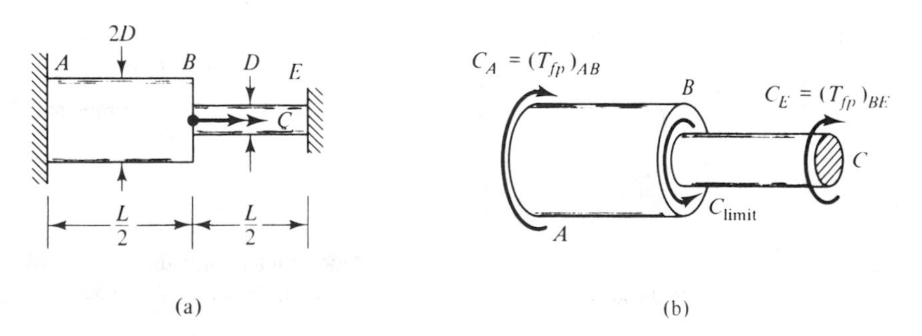

Figure 7.22

so

$$C_{\text{limit}} = \frac{3}{4}\pi\tau_{YS}D^3 \qquad \textbf{Answer}$$

Note that the lengths of the segments don't enter into the computations.

PROBLEMS

7.53. to 7.57. Determine the limit load (couple) for the member shown. The shaft diameter and material shear yield strength are as indicated; otherwise, express the results in terms of the fully plastic torque T_{fp} for the member.

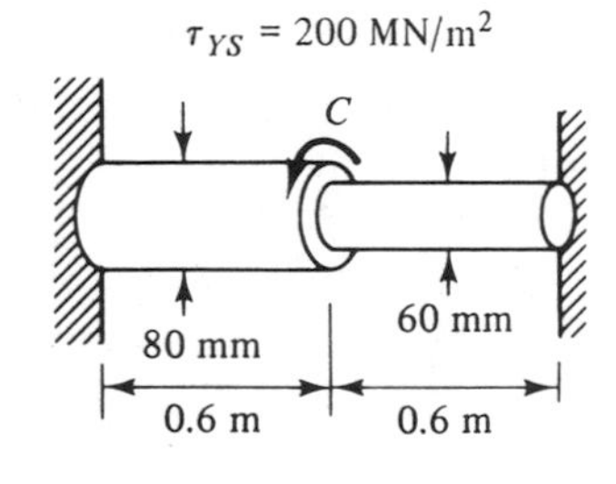

Problem 7.55

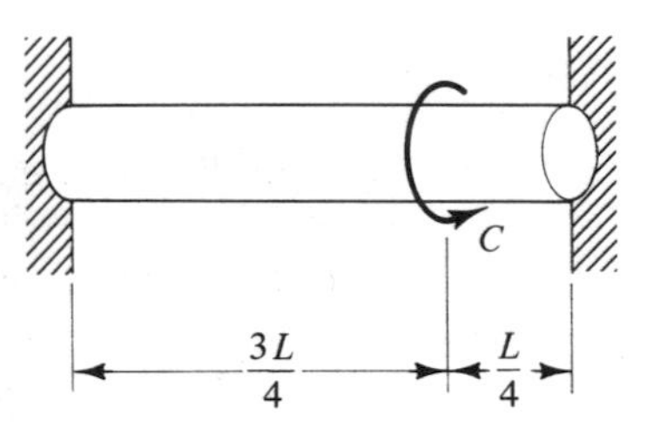

Problem 7.53

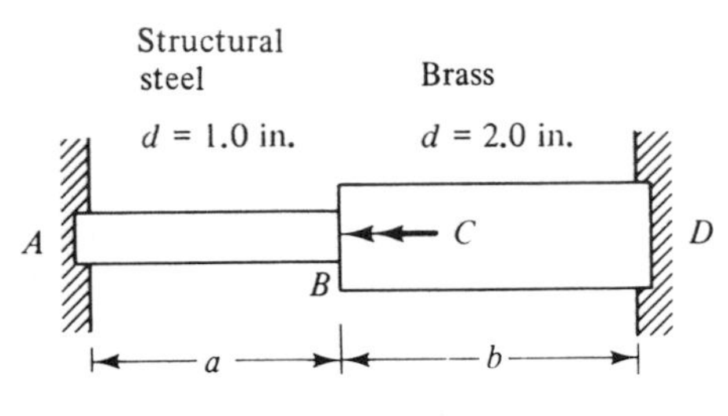

Problem 7.56

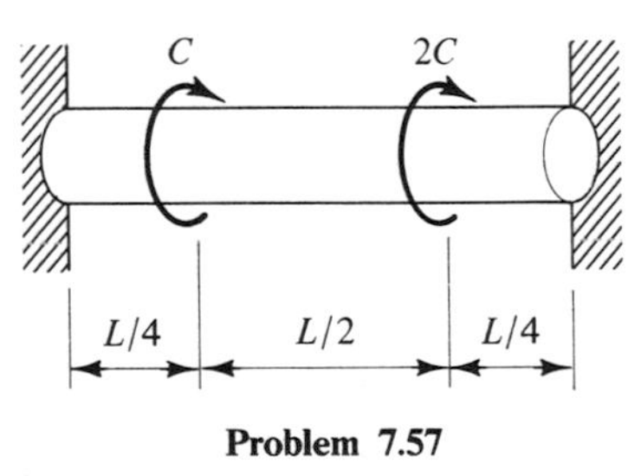

Problem 7.57

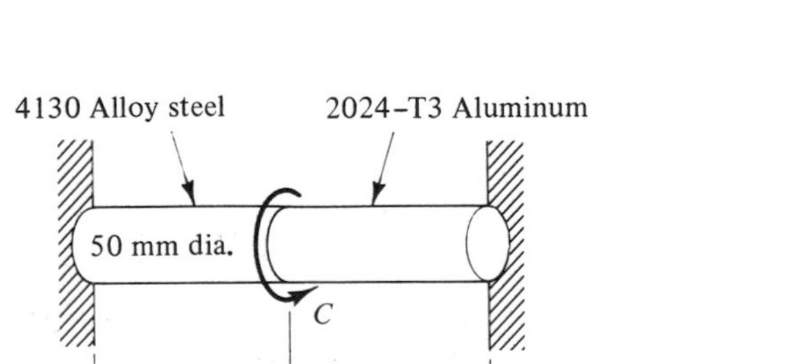

Problem 7.54

7.58. Determine the limit load for the assembly in Problem 7.35 (Section 7.5).

7.8 CLOSURE

This chapter was devoted to a study of the torsional response of solid and hollow circular shafts. The stress distribution in a shaft depends upon the stress-strain relation for the material and, consequently, so do both the torque-stress and torque-twist relations. Most torsional members are designed to operate within the linearly elastic range, but the inelastic response is also of importance. Accordingly, it was considered in some detail for the case of an elastic-perfectly plastic material.

The couples applied to a shaft are often referred to as torques, but here we used this term only to denote the axial component of the couple in the stress resultant. This was done in an attempt to avoid any confusion about which of these quantities is involved in the torque-stress and torque-twist relations. In many cases, but not in all, the torque is equal in magnitude to the applied couples. The distinction between the two should always be kept in mind, for many errors can be avoided by doing so.

As mentioned in Section 7.1, all the results presented in this chapter apply for circular shafts at rest and for circular shafts rotating with constant angular velocity. The torsion of shafts with noncircular cross sections is considerably more complicated than that of circular shafts and is treated in more advanced texts.

8

Bending

8.1 INTRODUCTION

To this point, we have considered the response of slender members to loadings acting along the longitudinal axis. These loadings were forces in the case of axially loaded members and couples in the case of torsional members. In this chapter we shall consider the response of straight, slender members to forces and couples acting perpendicular to the longitudinal axis. Such members are called *beams* and are probably the most common type of structural element.

The loadings applied to a beam cause its originally straight axis to deform into a curve (Figure 8.1). This type of deformation is referred to as *bending*, or *flexure*. Our goal in this chapter is to develop the load-stress and load-deformation relations for the bending response of beams and to learn how to apply them. Again, the general procedure outlined in Section 5.11 will be used, and the deformations will be assumed to be small.

Beams are usually classified according to their support conditions. For example, a beam supported by rollers, or the equivalent, at each end is called a *simply supported beam* or a *simple beam* [Figure 8.1(a)]. A *cantilever beam* is one that is fixed at one end and free at the other [Figure 8.1(b)]. If the free end is placed on a roller, as in Figure 8.1(c), it becomes a *propped cantilever*. Beams clamped on both ends are usually called *fixed-fixed beams* [Figure 8.1(d)], and those that extend beyond the supports are called *overhanging beams* [Figure 8.1(e)]. If a beam is

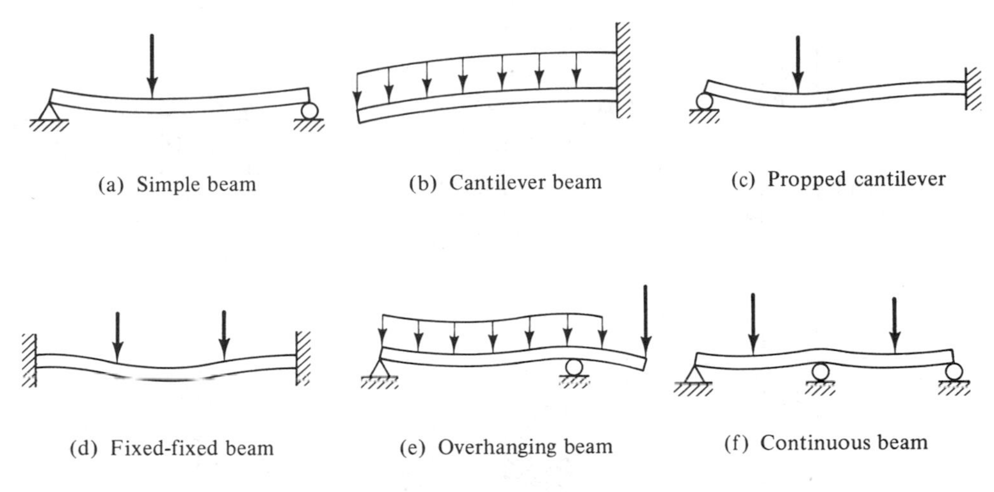

Figure 8.1

supported at more than two points [Figure 8.1(f)], it is referred to as a *continuous beam*. We shall use this nomenclature throughout.

Beams are also classified according to the shapes of their cross sections. For example, an I-beam has a cross section shaped like the letter I and a T-beam has a cross section shaped like the letter T.

8.2 SHEAR AND BENDING MOMENT DIAGRAMS

The loading on most beams is such that the stress resultant on planes perpendicular to the axis consists of a shear force, V, and a bending moment, M. In determining beam responses, it is highly convenient, if not essential, to first determine the *shear* and *bending moment diagrams* showing the values of these quantities along the entire length of the member. Because of the nature of the loadings on beams, these diagrams are usually more complicated than the normal force and torque diagrams considered previously. Thus, we shall consider them in some detail.

The basic procedure for determining the shear and bending moment diagrams is to determine the values of V and M at various locations along the member by using the method of sections, as discussed in Section 5.2, and then plot the results. As an illustration, let us consider the beam shown in Figure 8.2(a). Note that the reactions have already been determined. Ordinarily, this would be the first step taken.

We choose coordinates with the x axis to the right along the beam and the y axis vertically upward. The exact location of the x axis within the cross section will be left unspecified for the present. Sectioning the beam perpendicular to its axis at an arbitrary location between points A and B, we obtain the free-body diagram shown in Figure 8.2(b). The shear force will be considered positive if it is directed

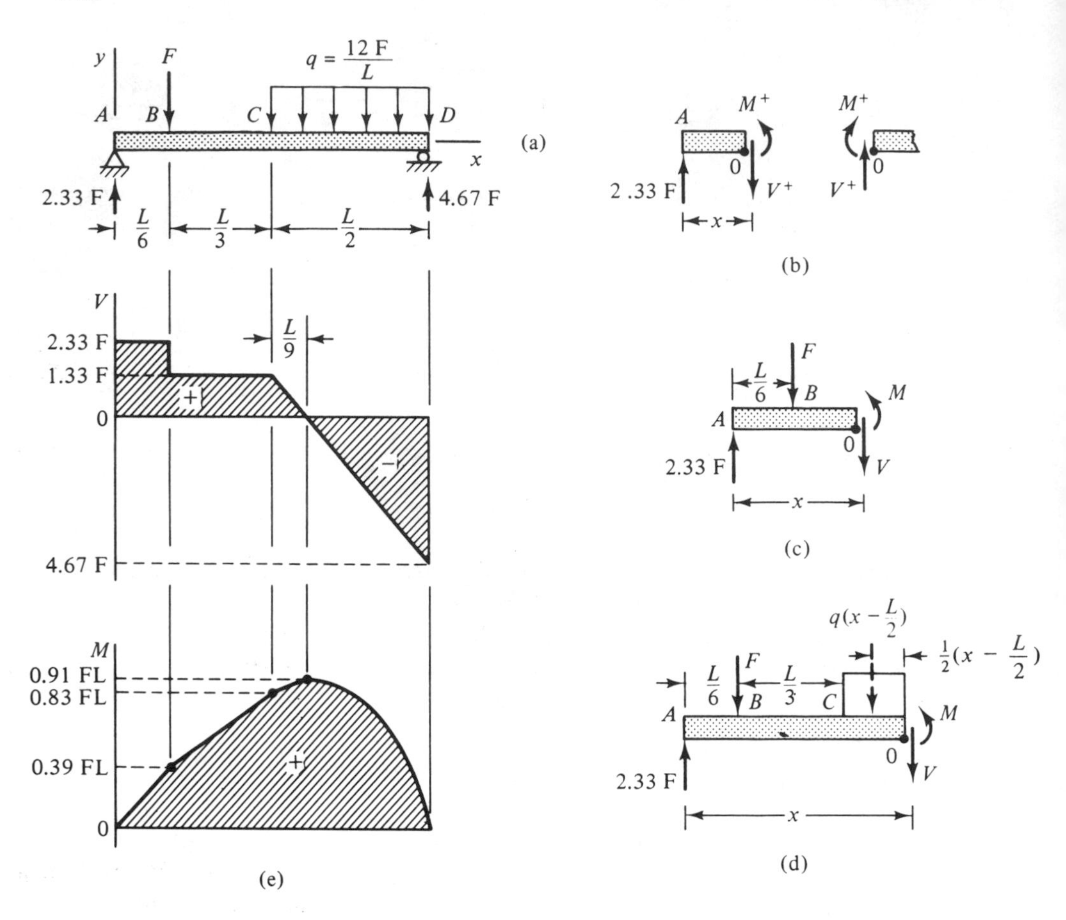

Figure 8.2

downward on the right-hand face of the beam segment, and the bending moment on this face will be considered positive if it is counterclockwise. Note that positive V and M have the opposite sense on the left-hand face of the adjoining segment, in accordance with Newton's third law. It is convenient to place V and M on the free-body diagram in their positive directions because this simplifies the interpretation of the signs on the answers.

The equilibrium equations for this segment of the beam are

$$\sum F_y = 2.33F - V = 0$$

$$\overset{+}{\curvearrowleft}\ \sum M_O = M - (2.33F)x = 0$$

from which we obtain

$$V = 2.33F \qquad M = 2.33Fx \qquad \left(0 \leqslant x \leqslant \frac{L}{6}\right)$$

These equations hold only over the left one-sixth of the beam because the free-body diagram is valid only over this range.

If the beam is sectioned between points B and C, the force F comes into play, and the free-body diagram is as shown in Figure 8.2(c). This diagram is valid over the central one-third of the beam. We have

$$\sum F_y = 2.33F - F - V = 0$$

$$\overset{+}{\curvearrowleft}\sum M_O = M - (2.33F)x + F\left(x - \frac{L}{6}\right) = 0$$

or

$$V = 1.33F \qquad M = \frac{FL}{6} + 1.33Fx \qquad \left(\frac{L}{6} \leqslant x \leqslant \frac{L}{2}\right)$$

Finally, if the beam is sectioned within portion CD, we obtain the free-body diagram shown in Figure 8.2(d). The equilibrium equations for this segment of the beam are

$$\sum F_y = 2.33F - F - q\left(x - \frac{L}{2}\right) - V = 0$$

$$\overset{+}{\curvearrowleft}\sum M_O = M - (2.33F)x + F\left(x - \frac{L}{6}\right) + \frac{q}{2}\left(x - \frac{L}{2}\right)\left(x - \frac{L}{2}\right) = 0$$

or

$$V = 7.33F - 12F\frac{x}{L} \qquad \left(\frac{L}{2} \leqslant x \leqslant L\right)$$

$$M = -1.33FL + 7.33Fx - 6F\frac{x^2}{L}$$

Notice that there is one set of equations for V and M for each portion of the beam and that each set applies until a change in loading is encountered. Plotting these equations, we obtain the shear and bending moment diagrams shown in Figure 8.2(e). It is convenient to plot the diagrams directly below the free-body diagram of the beam.

From Figure 8.2(e), we see that upward concentrated forces (A_y and D_y in this example) cause upward jumps in the shear diagram and downward concentrated forces (F in this example) cause downward jumps. The magnitudes of the jumps are equal to the magnitudes of the corresponding forces. Applied couples cause similar jumps in the moment diagram. Counterclockwise couples cause downward jumps and clockwise couples cause upward jumps (see Example 8.1).

The weight of a beam is often small compared to the applied loads, in which case the weight can be neglected with little error. It is a simple matter to include it, however. If the beam is uniform, its weight merely contributes an additional uniformly distributed loading. The weight per unit length of some common structural shapes is given in Tables A.3 through A.12 of the Appendix, and values for other members are available in handbooks.

Example 8.1. Shear and Bending Moment Diagrams. Determine the shear and bending moment diagrams for the cantilever beam shown in Figure 8.3(a).

Solution. The first step is to determine the reactions. From $\Sigma F_y = 0$ and $\Sigma M_A = 0$, the reactions at the wall are found to be a 1,000-lb upward force and a 14,000-lb · ft counterclockwise couple.

Sectioning the beam between A and B, we obtain the FBD shown in Figure 8.3(b). The equilibrium equations are

$$\sum F_y = 1{,}000 \text{ lb} - V = 0$$

$$\overset{+}{\curvearrowleft}\sum M_O = M + 14{,}000 \text{ lb} \cdot \text{ft} - (1{,}000 \text{ lb})\, x = 0$$

from which we obtain

$$V = 1{,}000 \text{ lb} \qquad M = -14{,}000 + 1{,}000x \text{ lb} \cdot \text{ft} \qquad (0 \leqslant x \leqslant 5)$$

The FBD obtained by sectioning the beam between B and C is shown in Figure 8.3(c). We have

$$\sum F_y = 1{,}000 \text{ lb} - V = 0$$

$$\overset{+}{\curvearrowleft}\sum M_O = M + 14{,}000 \text{ lb} \cdot \text{ft} - (1{,}000 \text{ lb})\, x - 2{,}000 \text{ lb}(2 \text{ ft}) = 0$$

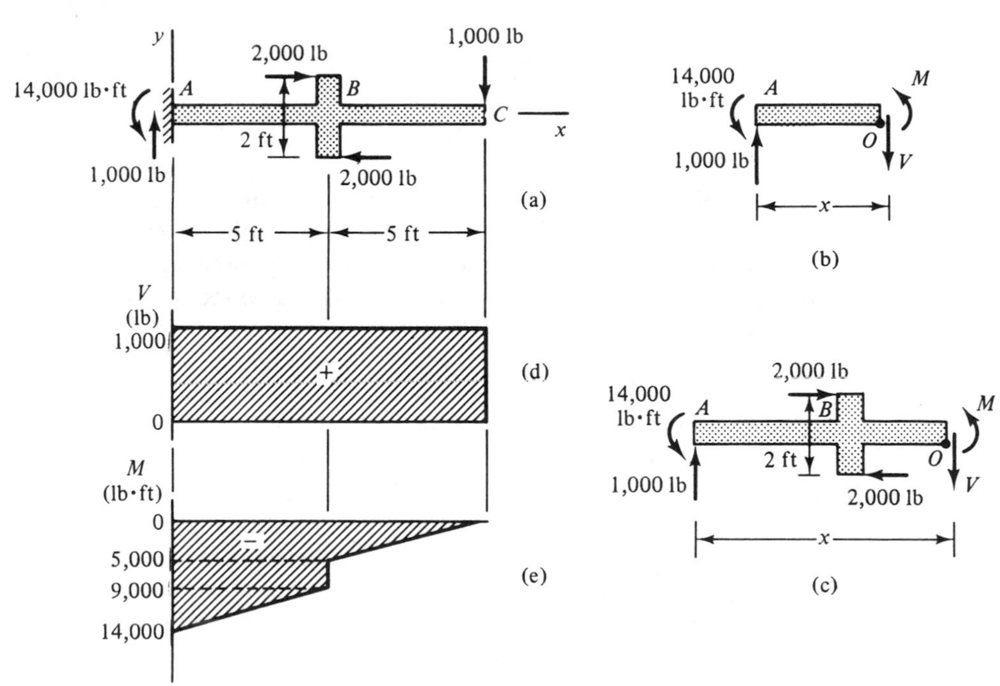

Figure 8.3

or

$$V = 1{,}000 \text{ lb} \qquad M = -10{,}000 + 1{,}000x \text{ lb} \cdot \text{ft} \qquad (5 \leqslant x \leqslant 10)$$

The corresponding shear and moment diagrams are shown in Figures 8.3(d) and 8.3(e), respectively.

Notice that the concentrated couples cause jumps in the moment diagram. The counterclockwise couple at the wall causes a downward jump, and the clockwise couple at the center of the beam causes an upward jump. The jumps are of the same magnitude as the corresponding couples.

PROBLEMS

8.1. to 8.8. For the beam shown, determine the equations for the shear force V and the bending moment M as a function of the distance x from the left end and construct the shear and moment diagrams. Neglect the weight of the member.

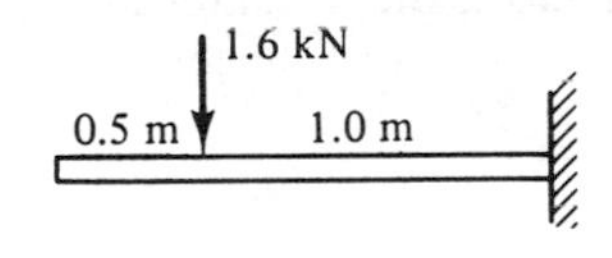

Problem 8.1

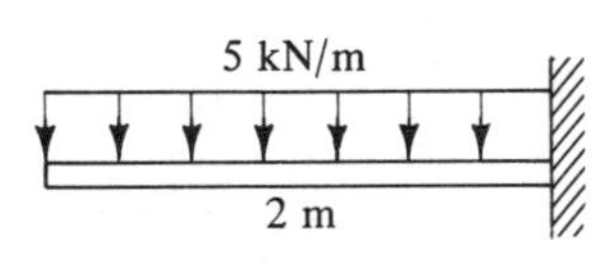

Problem 8.2

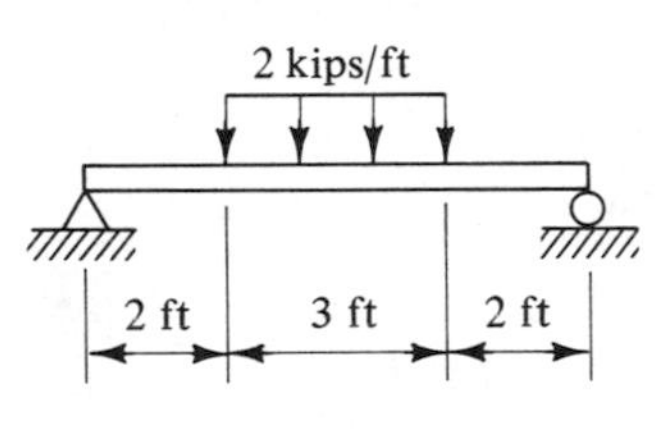

Problem 8.3

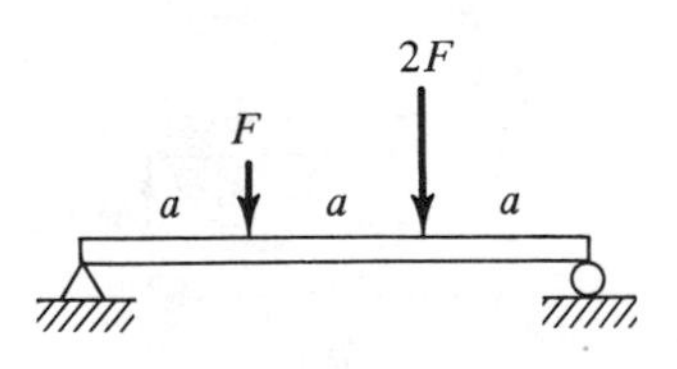

Problem 8.4

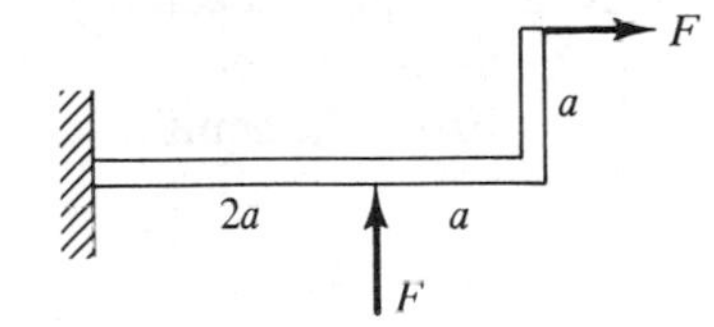

Problem 8.5

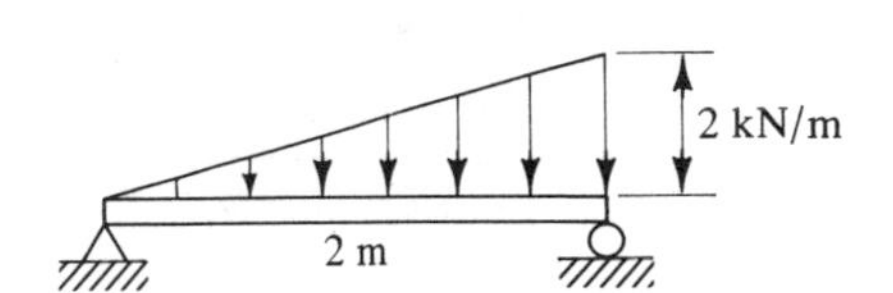

Problem 8.6

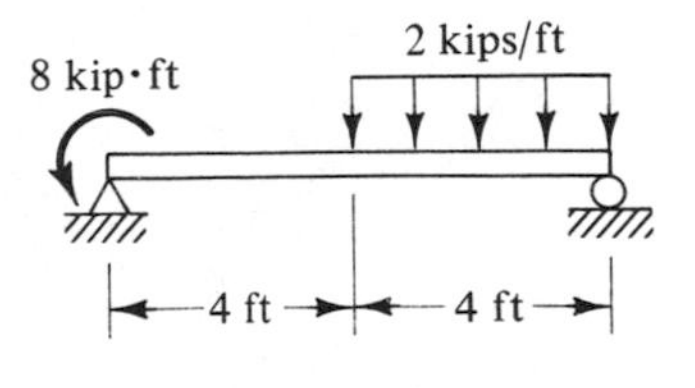

Problem 8.7

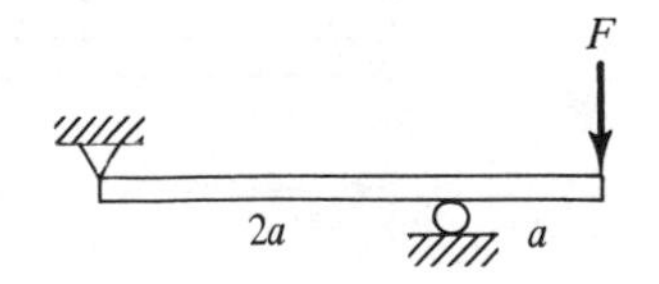

Problem 8.8

8.3 LOAD, SHEAR, AND MOMENT RELATIONS

Although the method of determining the shear and bending moment diagrams presented in the preceding section is straightforward, it is tedious to apply because it requires that the relations between the shear force, bending moment, and applied loads be derived for each different beam considered. An alternate approach is to derive relations between these quantities which apply to beams in general. As we shall see, this approach greatly facilitates the construction of the shear and bending moment diagrams, especially for simple loadings.

Figure 8.4(a) shows a beam subjected to a distributed load, $q(x)$, which will be considered positive if it acts in the positive y direction. We now consider the equilibrium of a short segment of the beam with length Δx located an arbitrary distance x from the end. The free-body diagram is as shown in Figure 8.4(b). The shear force and bending moment generally vary along the beam; so their values on the left- and right-hand faces of the segment are shown to differ by a small amount ΔV and ΔM, respectively. Since the segment is short, the loading can be considered to be uniform over its length.

From the equilibrium equations

$$\sum F_y = V + q\,\Delta x - (V + \Delta V) = 0$$

$$\overset{+}{\curvearrowleft}\sum M_O = (M + \Delta M) - q\,\Delta x\left(\frac{\Delta x}{2}\right) - M - V\Delta x = 0$$

we have

$$\Delta V = q\,\Delta x$$

$$\Delta M = V\Delta x + q\frac{(\Delta x)^2}{2}$$

Dividing these expressions by Δx and taking the limit as Δx tends to zero, we obtain

$$\lim_{\Delta x \to 0} \frac{\Delta V}{\Delta x} = \frac{dV}{dx} = q \tag{8.1}$$

$$\lim_{\Delta x \to 0} \frac{\Delta M}{\Delta x} = \frac{dM}{dx} = V \tag{8.2}$$

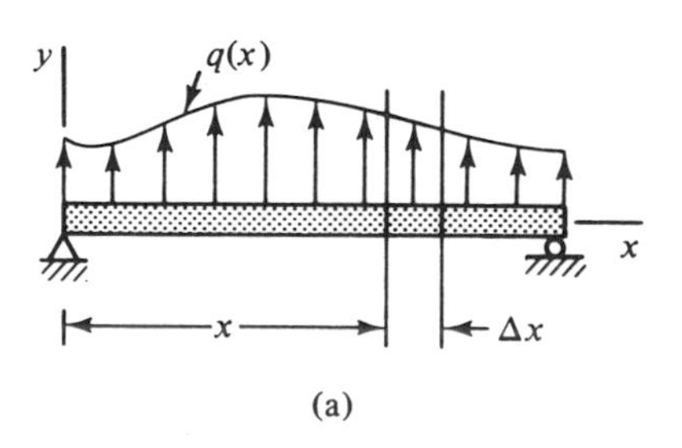

(a)

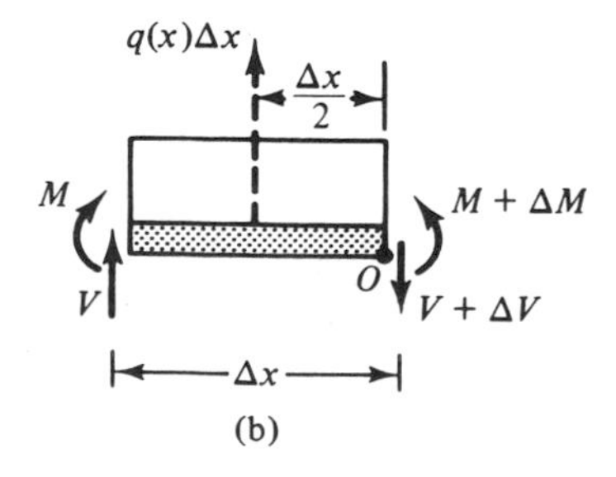

(b)

Figure 8.4

These equations define the relationship between the shear force, bending moment, and distributed load. Any concentrated forces or couples that may be acting are not accounted for in these relations. Consequently, they apply only over the regions between concentrated loads.

Equations (8.1) and (8.2) can be expressed in an alternate form by multiplying them by dx and integrating from a location $x = x_1$ where the shear and bending moment have values V_1 and M_1 to another location $x = x_2$ where they have values V_2 and M_2:

$$V_2 - V_1 = \int_{x_1}^{x_2} q\,dx \tag{8.3}$$

$$M_2 - M_1 = \int_{x_1}^{x_2} V\,dx \tag{8.4}$$

Interpreted geometrically, Eqs. (8.1) and (8.2) state that the slope of the shear diagram at any location is equal to the intensity of the distributed loading at that location, and that the slope of the moment diagram at any point is equal to the value of the shear force at that point. It is also evident from Eq. (8.2) that the relative maxima and minima on the moment diagram, which correspond to points of zero slope, occur where $V = 0$.

The integrals in Eqs. (8.3) and (8.4) represent areas under the loading and shear diagrams, respectively. Consequently, Eq. (8.3) indicates that the change in shear between any two locations along the beam is equal to the area under the loading diagram between these locations. Similarly, Eq. (8.4) indicates that the change in bending moment between any two points is equal to the area under the shear diagram between these points. These areas are considered to be positive if q and V are positive, respectively, and negative if they are negative.

For simple loadings, the areas under the loading and shear diagrams can be easily determined by using only geometry, and the shear and moment diagrams can be sketched by inspection by using the geometrical interpretations of Eqs. (8.1) through (8.4). The jumps in the diagrams due to concentrated forces and couples must be added in separately, however, for they are not accounted for in these equations. For more complex loadings, the areas under the loading and shear diagrams are not so easily computed, and it is usually necessary to determine V and M by integration of Eqs. (8.3) and (8.4).

When sketching the shear and moment diagrams, it is important that their accuracy be checked. This can be done by noting whether or not they close (end up at zero value). Closure of the shear diagram indicates that the sum of the vertical forces acting upon the beam is zero, as it must be for equilibrium. Similarly, closure of the moment diagram implies that the sum of the moments for the entire beam is zero. If either of the diagrams fails to close, there is an error in their construction or in the determination of the reactions.

Example 8.2. Shear and Moment Diagrams for a Simple Beam. Sketch the shear and bending moment diagrams for the simple beam shown in Figure 8.5(a).

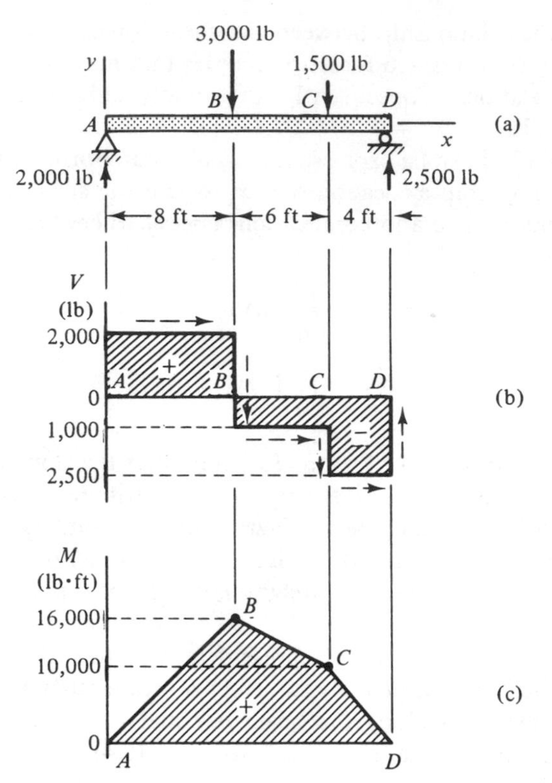

Figure 8.5

Solution. We first determine the reactions. The upward force at A is found to be 2,000 lb and the upward force at D is 2,500 lb.

We now sketch the shear diagram. From Eq. (8.1), we note that the diagram will consist of horizontal lines (zero slope) between the concentrated forces, since there is no distributed load ($q \equiv 0$). Working from left to right along the beam, we first encounter the upward concentrated force at A. Thus, the shear jumps up to a value of 2,000 lb at this point [Figure 8.5(b)]. It then remains constant up to point B, where it suddenly decreases by an amount equal to the magnitude of the downward force acting there. Continuing across the beam in this fashion, we obtain the complete diagram shown. The order of progression is indicated by the dashed arrows.

The moment diagram is now obtained by referring to the shear diagram. There will be no jumps in the value of M because there are no couples acting. Consequently, $M = 0$ at the left end of the beam.

The shear is positive and constant from A to B; therefore, according to Eq. (8.2), the moment diagram has constant positive slope over this interval [Figure 8.5(c)]. Furthermore, the change in moment in going from A to B is equal to the area under the shear diagram between these points [Eq. (8.4)]:

$$M_B - M_A = (2{,}000 \text{ lb})(8 \text{ ft}) = 16{,}000 \text{ lb} \cdot \text{ft}$$

This establishes the value of the moment at B. From B to C, the diagram has constant negative slope because of the constant negative shear. The change in moment over this interval is

$$M_C - M_B = (-1{,}000 \text{ lb})(6 \text{ ft}) = -6{,}000 \text{ lb} \cdot \text{ft}$$

for a total of 10,000 lb · ft at C. Continuing on across the beam in this fashion, we obtain the complete diagram shown.

Note that both diagrams close. This does not guarantee that they are correct, but it does provide a degree of confidence in them.

Example 8.3. Maximum Shear and Moment in a Beam. The overhanging beam shown in Figure 8.6(a) supports only its own weight. Determine the maximum shear force and bending moment in the beam and the locations at which they occur. The member is a 381-mm × 152-mm Universal beam with an actual depth of 388.6 mm.

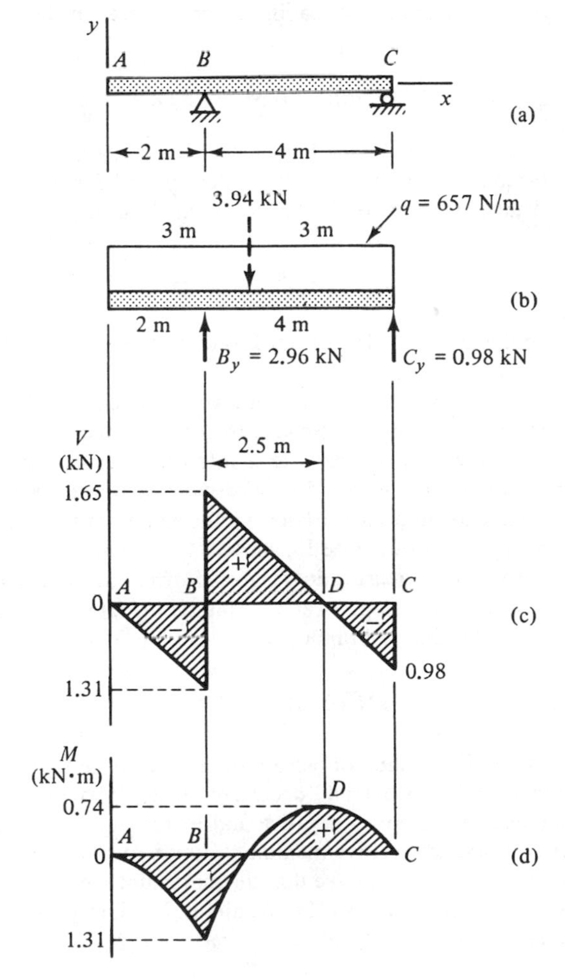

Figure 8.6

Solution. From Table A.8, we find that the beam has a mass of 67 kg/m. The weight per unit length is obtained by multiplying by the acceleration of gravity:

$$q = (67\ \text{kg/m})(9.81\ \text{m/s}^2) = 657\ \text{N/m}$$

From the FBD [Figure 8.6(b)], we have

$$\sum F_y = B_y + C_y - 3.94\ \text{kN} = 0$$

$$\overset{+}{\curvearrowleft}\sum M_C = -B_y(4\ \text{m}) + 3.94\ \text{kN}(3\ \text{m}) = 0$$

or

$$B_y = 2.96\ \text{kN} \qquad C_y = 0.98\ \text{kN}$$

Because of the uniform downward (negative) distributed loading, each portion of the shear diagram will consist of straight lines with the same negative slope. The change in shear between A and B is equal to the area under the loading diagram between these points [Eq. (8.3)]:

$$V_B - V_A = (-657\ \text{N/m})(2\ \text{m}) = -1.31\ \text{kN}$$

This establishes the value of the shear just to the left of B [Figure 8.6(c)]. At B, the shear suddenly increases an amount equal to the magnitude of the upward force acting there. It then decreases linearly with distance from B to C. The change in shear in going from B to C is

$$V_C - V_B = (-657\ \text{N/m})(4\ \text{m}) = -2.63\ \text{kN}$$

for a total of -0.98 kN just to the left of C. The upward force at C then brings the value of the shear back to zero.

The distance d from the left support to the point of zero shear can be determined from similar triangles. An equivalent procedure is to determine the distance required to reduce the shear from 1.65 kN to zero, given that it decreases at a rate of 657 N/m. Obviously, the required distance is $d = 1.65\ \text{kN}/(657\ \text{N/m}) = 2.5$ m. For more complex diagrams, it may be necessary to determine the points of zero shear by setting the equations for V equal to zero and solving for the corresponding values of x (see Example 8.4).

Since the shear is negative and increasing in magnitude from A to B, the moment diagram will have an increasing negative slope over this interval [Figure 8.6(d)]. The area under the shear diagram provides the change in moment:

$$M_B - M_A = \frac{1}{2}(-1.31\ \text{kN})(2\ \text{m}) = -1.31\ \text{kN} \cdot \text{m}$$

This establishes the moment at B. Continuing on across the beam, we obtain the complete diagram shown. Note that the slope is positive and decreasing from B to D because of the positive decreasing shear. Similarly, the slope is negative and increasing from D to C. Also note that the diagram closes and that the relative maximum occurs at a point of zero shear.

From the shear and moment diagrams, we see that the maximum shear force is 1.65 kN and occurs just to the right of the support at B. The maximum bending moment occurs at this same support and has a magnitude of 1.31 kN · m.

Example 8.4. Shear and Moment Diagrams for a Cantilever Beam. A cantilever beam supports the loading shown in Figure 8.7(a). Obtain the shear and bending moment diagrams and determine the magnitude and location of the maximum bending moment. The reactions are as indicated.

Solution. Since $q(x)$ varies linearly with x, V will be a function of x^2. Thus, the area under the shear diagram, which is a parabola, cannot be determined from simple geometry. V and M could be determined by using the procedure of Section 8.2, but we shall use integration in order to illustrate the steps involved.

At $x = 0$, the shear jumps up to a value $V(0) = q_0 L/10$ due to the concentrated force acting there. According to Eq. (8.3), the change in shear between this location and any other arbitrary location x is

$$V(x) - V(0) = \int_0^x q\,dx = -\int_0^x \frac{q_0 x}{L}\,dx$$

Thus,

$$V(x) = \frac{q_0 L}{10} - \frac{q_0 x^2}{2L}$$

A sketch of the shear diagram is shown in Figure 8.7(b). The point of zero shear is located by setting the equation for $V(x)$ equal to zero and solving for x. This gives $x = L/\sqrt{5}$.

The moment at the left end of the beam is zero because there is no couple acting there. From Eq. (8.4), the change in moment between this location and any other arbitrary location

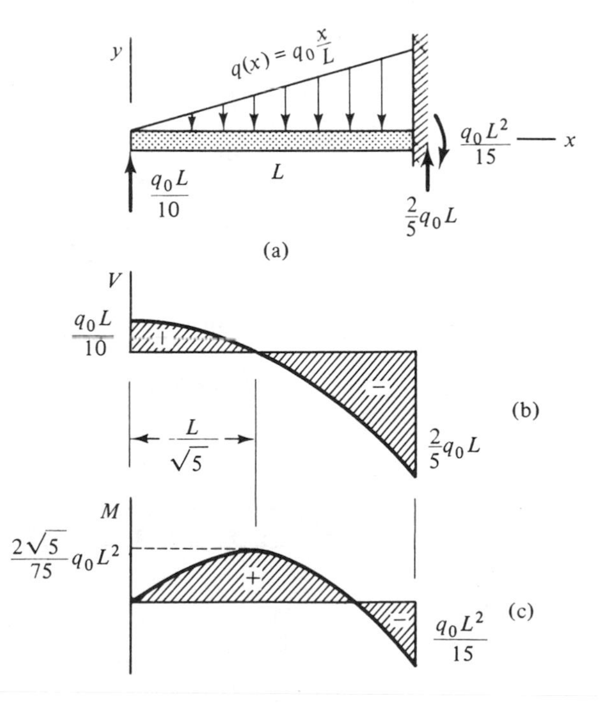

Figure 8.7

x is

$$M(x) - M(0) = \int_0^x V\,dx = \int_0^x \left(\frac{q_0 L}{10} - \frac{q_0 x^2}{2L} \right) dx$$

Thus,

$$M(x) = \frac{q_0 L x}{10} - \frac{q_0 x^3}{6L}$$

This curve is sketched in Figure 8.7(c). The relative maximum occurs at $x = L/\sqrt{5}$, where $V = 0$. The moment at this location is found to be $(2\sqrt{5}/75)q_0 L^2$. This is not the maximum moment in the beam, however. It occurs just to the left of the wall where $M = -q_0 L^2/15$. The clockwise couple acting at the wall then brings the diagram back to zero, as it should.

PROBLEMS

8.9. to 8.12. Sketch the shear and bending moment diagrams for the beams in Problems 8.5 through 8.8 (Section 8.2). Neglect the weights of the members.

8.13. to 8.21. Obtain the shear and bending moment diagrams for the beam shown and determine the magnitudes and locations of the maximum shear force and bending moment. Neglect the weight of the member.

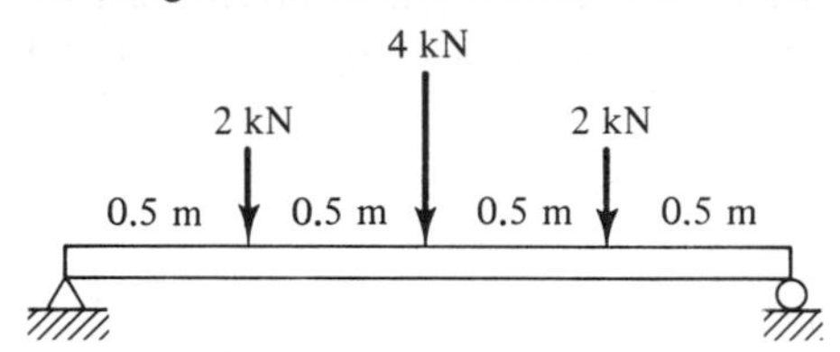

Problem 8.13

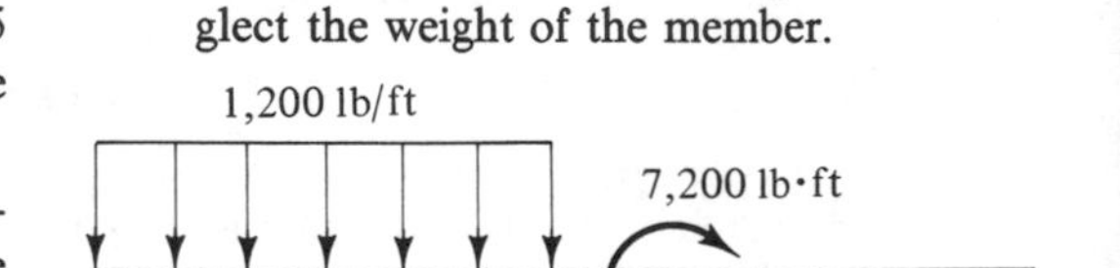

Problem 8.16

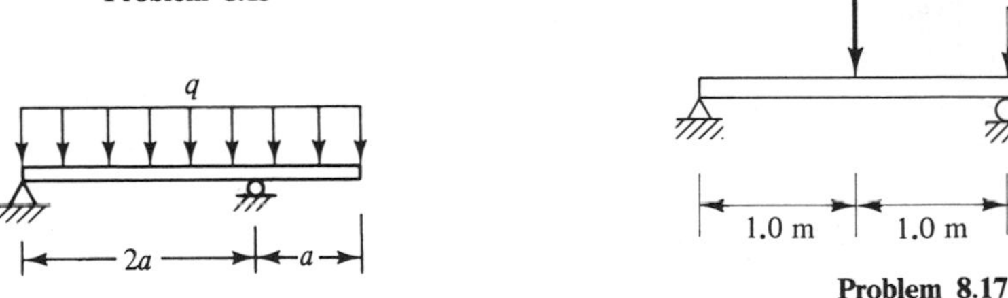

Problem 8.14

Problem 8.17

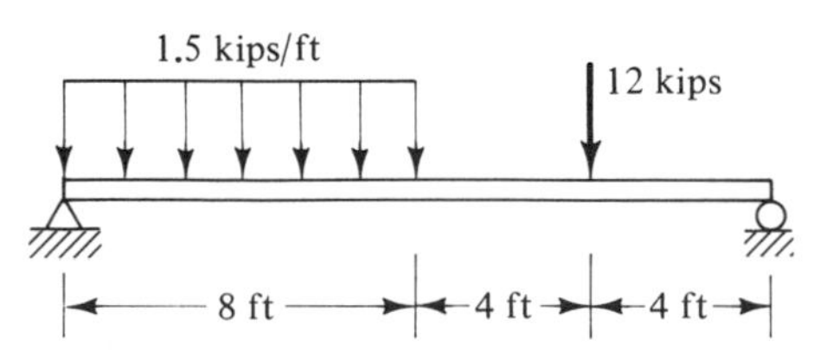

Problem 8.15

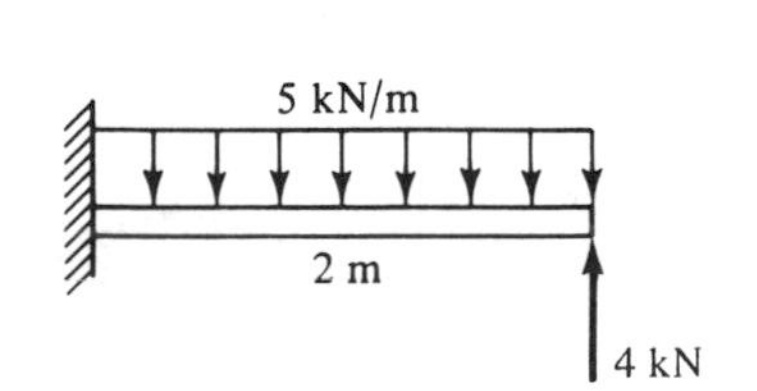

Problem 8.18

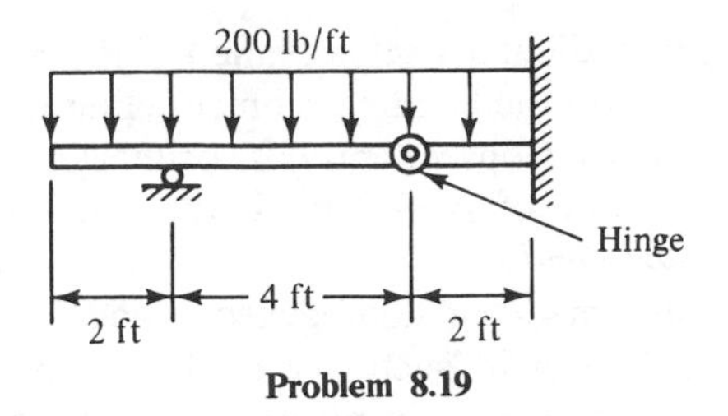

Problem 8.19

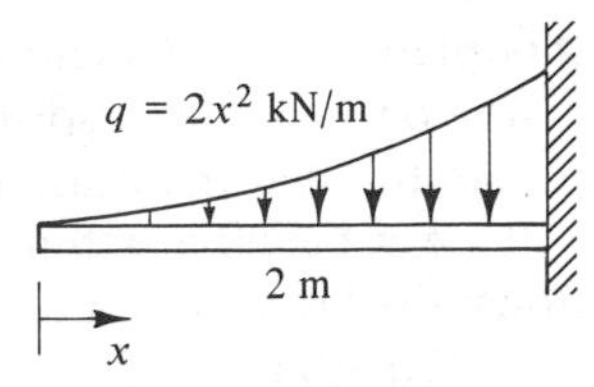

Problem 8.21

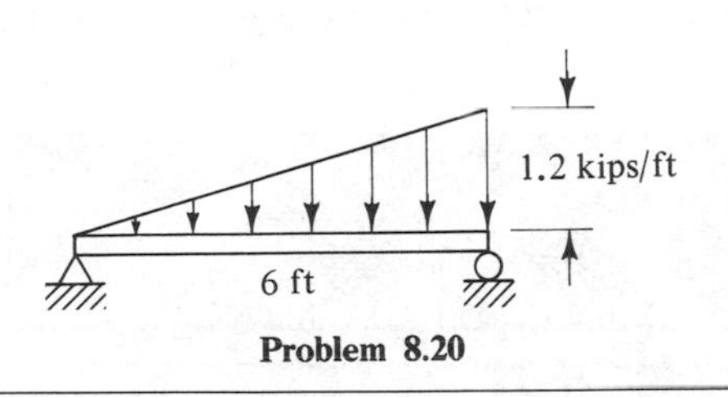

Problem 8.20

8.22. If the member in Problem 8.13 is a 254-mm × 146-mm Universal beam with a mass of 43 kg/m, what is the percent error in the value of the maximum bending moment due to neglecting the weight of the member?

8.23. The beam in Problem 8.19 has a weight of 67 lb/ft. What is the percent error in the value of the maximum bending moment if the weight of the member is neglected?

8.4 DEFORMATION PATTERN AND STRAINS

Now that procedures for determining the shear force and bending moment in a beam have been established, we shall turn our attention to the stresses and deformations associated with these quantities. We shall again follow the general procedure given in Section 5.11, the initial step of which is to consider the deformation pattern. First, however, we must be more specific about the type of members and loadings considered.

For the present, we assume that the loading is such that the stress resultant consists only of a bending moment and no shear force. In view of the relationship $dM/dx = V$ [Eq. (8.2)], this implies that the bending moment does not vary along the length of the member. This condition can be achieved by applying equal and opposite couples at the ends of the beam. It can also be achieved in other ways, at least over a portion of the length (see Figure 8.8).

We further assume that the beam has a longitudinal plane of symmetry, as most beams encountered in practice do, and that the bending moment lies within

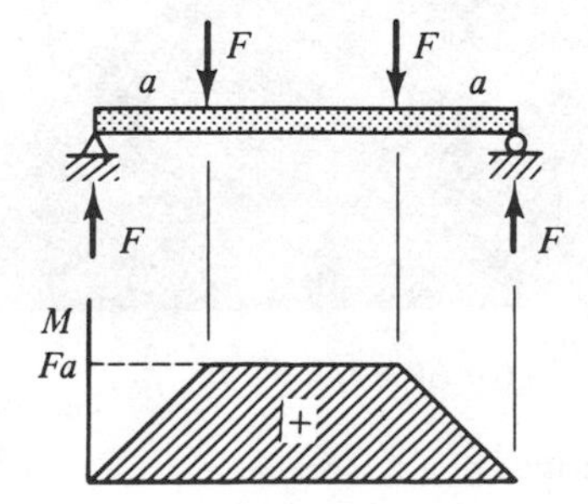

Figure 8.8

this plane. (The moment vector will be perpendicular to this plane.) If the material is homogeneous and isotropic, this restriction on the bending moment eliminates the possibility that the beam will twist as it deflects. Consequently, its primary response will be a bending within the longitudinal plane of symmetry. Under these conditions, the beam is said to be subjected to *pure bending*.

Experiments show that uniform beams made of homogeneous and isotropic materials and subjected to pure bending deform in such a way that planes perpendicular to the longitudinal axis before loading remain plane and perpendicular to the axis after loading. In other words, the beam cross sections rotate relative to one another when the beam deforms. This result, which also follows from symmetry, is illustrated in Figure 8.9.

(a) Before Loading

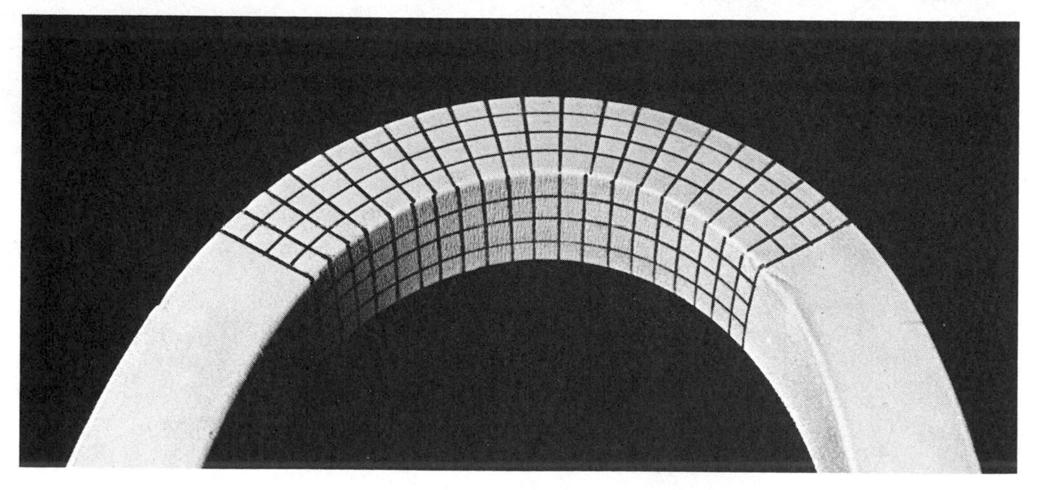

(b) After Loading

Figure 8.9 Deformation of a Rubber Beam as Indicated by Grid Patterns.

The deformations can be related to the strains by considering a small element of the beam of length Δx [Figure 8.10(a)]. In doing this, it will be helpful to think of the beam as consisting of a number of longitudinal fibers, or line elements, laid side by side.

If the bending moment is positive, as in Figure 8.10(b), a line element such as AB in the top portion of the beam will shorten while one in the bottom portion, such as CD, will elongate. At some point in between, there is a line element EF that does not change length. Actually, this line is the edge of an entire surface extending over the width and length of the beam, called the *neutral surface*. If the sense of the bending moment is reversed, the fibers at the top of the beam elongate while those at the bottom shorten. The deformations have been shown greatly exaggerated for clarity.

The intersection of the neutral surface with the longitudinal plane of symmetry is called the *neutral axis* of the beam, and its intersection with the beam cross section will be referred to as the *bending axis* [Figure 8.10(b)]. Notice that the beam cross sections rotate about the bending axis as the beam deforms. The exact location of the coordinate axes introduced in Section 8.2 can now be specified. The x axis is taken to coincide with the neutral axis in the undeformed beam, and the y axis is taken positive upward from it.

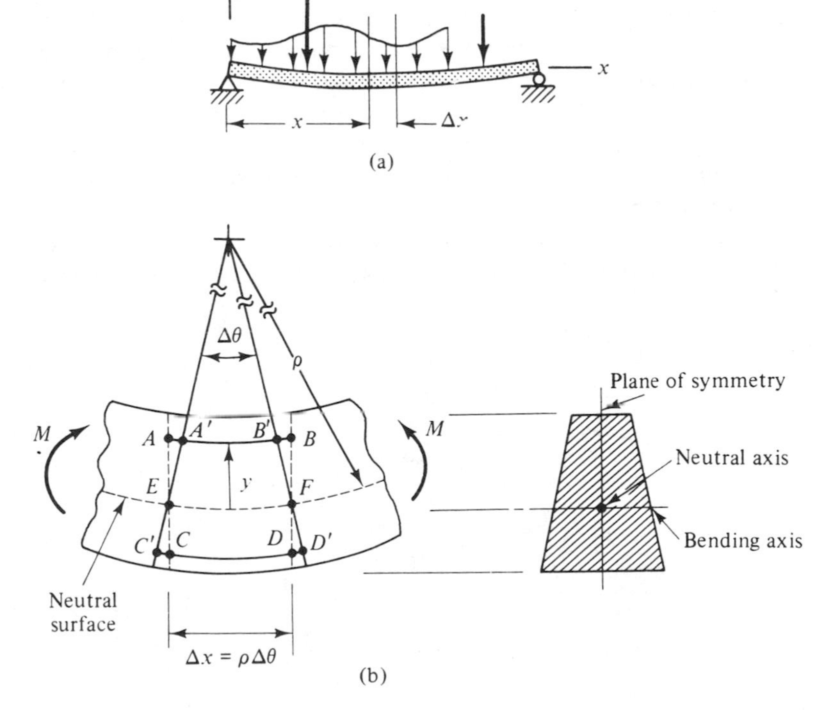

Figure 8.10

Now let us consider the strain of an arbitrary line element, such as AB, located a distance y from the neutral surface [Figure 8.10(b)]. From the geometry of the figure, the final length of the element is $A'B' = (\rho - y)\,\Delta\theta$, where ρ is the radius of curvature of the neutral axis in the deformed beam and $\Delta\theta$ is the included angle between adjacent cross sections. The initial length Δx of the element can be expressed in terms of $\Delta\theta$ by recognizing that line element EF does not change length; hence, $\Delta x = \rho\,\Delta\theta$. Thus, the normal strain in the longitudinal direction is

$$\varepsilon_{AB} = \frac{A'B' - AB}{AB} = \frac{(\rho - y)\,\Delta\theta - \rho\,\Delta\theta}{\rho\,\Delta\theta}$$

or

$$\varepsilon = -\frac{y}{\rho} \tag{8.5}$$

Equation (8.5) indicates that the normal strain varies linearly with the distance from the neutral surface and is largest at the top and bottom of the beam where the value of y is largest. The strain distribution is shown in Figure 8.11, where the height of the diagram denotes the relative magnitude of ε. Note that the strain does not vary across the width of the beam. Note also that the location of the neutral surface has not yet been established. This will be done in the following section.

The question now arises as to what effect the shear forces, which are present in most beams, have on the deformation pattern and the strains. Experiments and more advanced analyses show that the shear forces have a negligible effect on these quantities, unless the beam is exceptionally short (length of the same order of magnitude as the depth of the cross section). On the other hand, long beams with deep, narrow cross sections may buckle and undergo out-of-plane deformations, in which case Eq. (8.5) no longer applies. This behavior can be readily demonstrated by attempting to bend a strip of paper within the plane in which it lies. We shall assume that buckling does not occur.

In arriving at the strain distribution given in Eq. (8.5), we have invoked a number of assumptions and restrictions concerning the geometry and loading of the

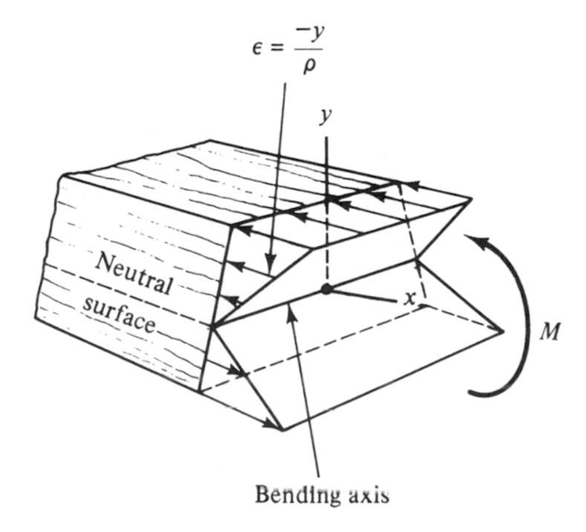

Figure 8.11

beam. There are also other conditions that must be met before Eq. (8.5) is valid. For convenience, all of these requirements are summarized below:

1. The beam must be straight and have constant cross-sectional area, or nearly so, for some distance on either side of the cross section of interest.
2. The beam must have a longitudinal plane of symmetry, and the bending moment must lie within this plane. Stated another way, the resultant applied loads must lie within the longitudinal plane of symmetry.
3. The member must not be so short that shearing effects are significant, nor so long and narrow that it buckles.
4. The section of interest must be some distance away from supports, connections, or points of load application.
5. The material must be homogeneous and isotropic.

Some of the preceding conditions can be relaxed somewhat, but they are sufficiently general for our purposes. The distortions in the strain distribution that occur near the supports, points of load application, and discontinuities in the cross section tend to be localized (St. Venant's principle) and generally die out over a distance approximately equal to the depth of the cross section. Note that the strain distribution depends only upon the geometry of the deformations and is completely independent of the stress-strain relations for the material.

8.5 NORMAL STRESSES — LINEARLY ELASTIC CASE

In this section we shall determine the distribution of normal stress and the moment-stress relation for a beam by combining the strain distribution with the stress-strain relation for the material and the conditions for static equivalence. The location of the neutral surface will also be determined in the process. Linearly elastic material behavior will be assumed for the present. The inelastic response of beams will be considered in Section 8.10.

Moment-Stress Relation

If the material is homogeneous and isotropic and the stresses are below the proportional limit, the stress-strain relation $\sigma = E\varepsilon$ applies throughout the beam. Combining this expression with the strain distribution of Eq. (8.5), we obtain for the stress distribution

$$\sigma = -\frac{Ey}{\rho} \tag{8.6}$$

Here, we have assumed that the modulus of elasticity, E, of the material is the same in tension and compression. As indicated in Chapter 5, this is a reasonable assumption for most common engineering materials. Equation (8.6) indicates that the normal stresses, like the normal strains, vary linearly with the distance y from the neutral surface [Figure 8.12(a)]. The normal stresses in beams are called *bending*, or *flexural*, *stresses* to distinguish them from those due to other types of loadings.

Now let us consider the moment-stress relation. Referring to Figure 8.12(b), we see that the force $dF = \sigma\, dA$ acting upon an element of area dA of the cross section has a moment about the bending axis (z axis) of magnitude $y\, dF = y\sigma\, dA$. The stresses are statically equivalent to the bending moment, M; therefore, we have

$$\left(\sum F_x\right)_{\text{I}} = \left(\sum F_x\right)_{\text{II}}: \qquad \int \sigma\, dA = 0 \tag{8.7}$$

$$\left(\sum M_z\right)_{\text{I}} = \left(\sum M_z\right)_{\text{II}}: \qquad -\int y\sigma\, dA = M \tag{8.8}$$

Equation (8.7) is a statement of the condition that the net normal force on the cross section must be zero.

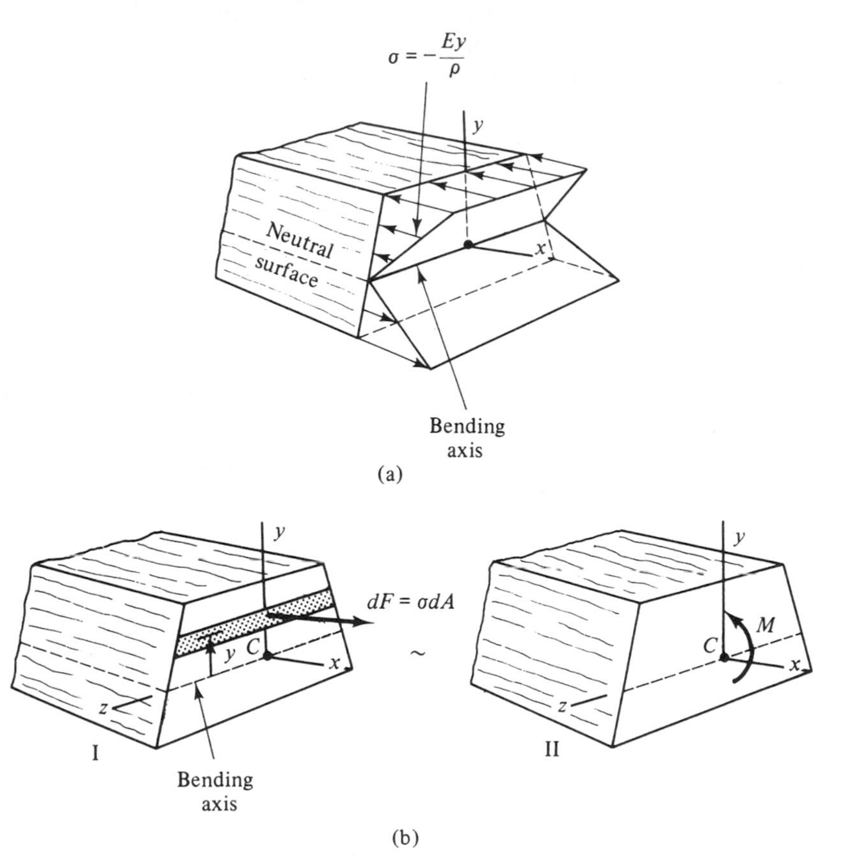

Figure 8.12

Substituting Eq. (8.6) into Eq. (8.7) and recognizing that E and ρ do not vary over the cross section, we have

$$-\frac{E}{\rho}\int y\,dA = -\frac{E}{\rho}A\bar{y} = 0$$

where A is the cross-sectional area and $\bar{y}$ is the distance from the bending axis to its centroid. This equation can be satisfied only if $\bar{y} = 0$, which implies that the bending axis passes through the centroid C of the cross section. This establishes the location of the neutral surface.

Combining Eqs. (8.6) and (8.8), we obtain

$$M = \int y\frac{(Ey)}{\rho}\,dA = \frac{E}{\rho}\int y^2\,dA$$

The latter integral is the moment of inertia, I, of the cross section about the bending axis [see Eq. (4.25)]; therefore, we have

$$M = \frac{EI}{\rho} \tag{8.9}$$

This equation relates the bending moment, M, to the curvature of the neutral axis, $1/\rho$, and is called the *moment-curvature relation*. We shall use this relation later to determine the deflection of beams.

The moment-stress relation is obtained by combining Eqs. (8.6) and (8.9):

$$\sigma = -\frac{My}{I} \qquad (\sigma \leqslant \sigma_{PL}) \tag{8.10}$$

Here, M is the bending moment acting upon the particular cross section of interest, I is the moment of inertia of that cross section about the bending axis, and y is the vertical distance from the bending axis to the particular point of interest within the cross section.

Note from Eq. (8.10) that a positive bending moment produces compression ($\sigma < 0$) in the top fibers of the beam where y is positive and tension ($\sigma > 0$) in the bottom fibers where y is negative. Conversely, a negative bending moment produces tension in the top fibers and compression in the bottom fibers. It is common practice to drop all signs in Eq. (8.10) and to use the expression only to compute the magnitude of the stress. The sense of the stress is then determined by inspection from the sense of the bending moment.

Equation (8.10) indicates that the larger the moment of inertia, I, of the cross section, the smaller the bending stresses. This explains why I-beams are so commonly used. With a cross section of this shape, it is possible to obtain a large moment of inertia with a minimum of material. Values of I for common structural shapes are given in handbooks; see Tables A.3 through A.12 of the Appendix, for example.

Design of Beams

The process of selecting a beam of appropriate size to support a given loading is complicated by the fact that the maximum stress depends upon two cross-sectional parameters, y and I. This difficulty can be eliminated by combining these two parameters into a single parameter, $S = I/y_{max}$, called the *section modulus*. Thus, the magnitude of the maximum stress can be expressed as

$$\sigma_{max} = \frac{M}{(I/y_{max})} = \frac{M}{S} \tag{8.11}$$

where S has the units of length cubed. Values of S for common structural shapes are given in handbooks and in Tables A.3 through A.12 of the Appendix.

Once the maximum bending moment and allowable stress have been determined, the required section modulus can be computed from Eq. (8.11). It is then a simple matter to select the appropriate size beam from tables of beam properties. There are usually several different sizes of beams with the required value of S. The one with the smallest weight per unit length involves the least amount of material, and, for economic reasons, it is the one that is usually used. Of course, the cross-sectional dimensions may also be a factor in the selection. For example, the depth of a floor beam must be compatible with the desired overall floor thickness. The procedure for "sizing" a beam will be illustrated in Example 8.7.

Validity of Moment-Stress Relation

The moment-stress relation, Eq. (8.10), is based upon Hooke's law and applies only if the stresses are below the proportional limit. Furthermore, the modulus of elasticity of the material must be the same in tension and compression. For other types of material behavior, it is only necessary to use the appropriate stress-strain relation in Eq. (8.6) and then proceed as before. It is important to note that the location of the neutral surface may or may not coincide with the centroid of the cross section in such cases. As indicated by Eq. (8.7), this surface must always be located such that there is no net normal force on the cross section. Since the strain-curvature relation, Eq. (8.5), was used in the derivation, the conditions given in Section 8.4 concerning the geometry and loading of the beam must also be satisfied.

The effects of geometrical discontinuities, such as holes, fillets, and notches, usually can be accounted for by multiplying the maximum value of stress obtained from Eq. (8.10) or Eq. (8.11) by a stress concentration factor, K_σ:

$$\sigma_{max} = K_\sigma\left(\frac{My_{max}}{I}\right) = K_\sigma\left(\frac{M}{S}\right) \tag{8.12}$$

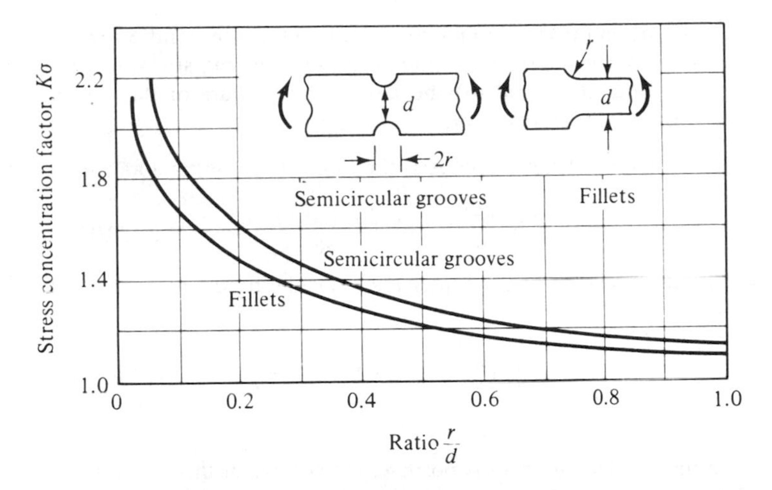

Figure 8.13 Stress-Concentration Factors for Beams of Rectangular Cross Section.

Values of K_σ for beams of rectangular cross section with fillets and semicircular grooves are given in Figure 8.13 for the linearly elastic case. These values apply to the stresses computed by using the dimensions of the minimum cross section. As has been mentioned several times before, geometrical discontinuities are most serious in members made of brittle materials.

Example 8.5. Flexural Stresses in a Cantilever Beam. A steel cantilever beam with a triangular cross section is loaded as shown in Figure 8.14(a). Determine at the center of

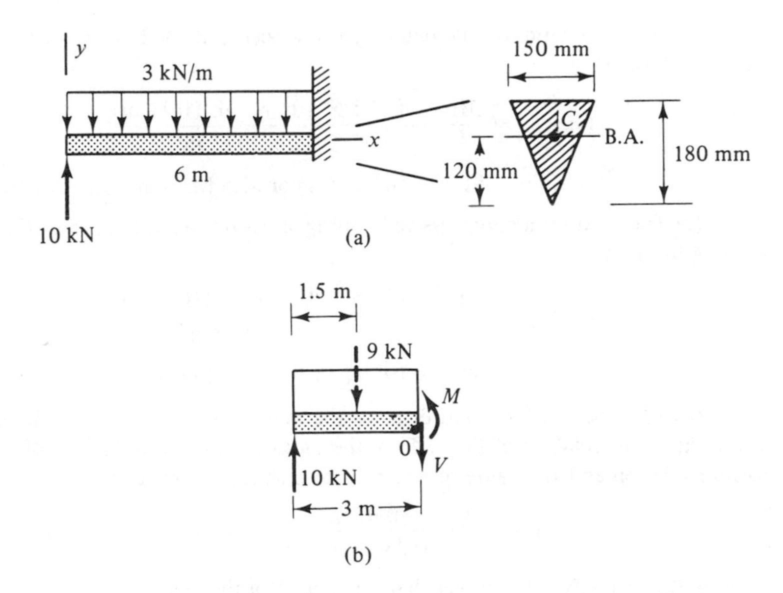

Figure 8.14

the beam (a) the normal stress and strain at a point 50 mm from the bottom of the cross section, (b) the maximum tensile bending stress, (c) the maximum compressive bending stress, and (d) the radius of curvature of the neutral axis. Assume linearly elastic material behavior.

Solution. From Table A.1, the moment of inertia about the bending axis is

$$I = \frac{bh^3}{36} = \frac{(0.15\text{ m})(0.18\text{ m})^3}{36} = 24.3 \times 10^{-6}\text{ m}^4$$

Sectioning the beam at the center and considering the piece to the left [Figure 8.14(b)], we have

$$\overset{+}{\curvearrowleft}\sum M_O = M + 9\text{ kN}(1.5\text{ m}) - 10\text{ kN}(3\text{ m}) = 0$$

$$M = 16.5\text{ kN} \cdot \text{m}$$

Since the bending moment is positive, the stresses at this location will be compressive above the bending axis and tensile below it.

(a) A point 50 mm from the bottom of the cross section is 70 mm below the bending axis. From Eq. (8.10), the bending stress at this point is

$$\sigma = \frac{My}{I} = \frac{(16.5 \times 10^3\text{ N} \cdot \text{m})(0.07\text{ m})}{24.3 \times 10^{-6}\text{ m}^4}$$

$$= 47.5 \times 10^6\text{ N/m}^2 \text{ or } 47.5\text{ MN/m}^2\text{ (tension)} \qquad \textbf{Answer}$$

Now, $E = 207\text{ GN/m}^2$ (Table A.2); therefore, the normal strain at this point is

$$\varepsilon = \frac{\sigma}{E} = \frac{47.5 \times 10^6\text{ N/m}^2}{207 \times 10^9\text{ N/m}^2} = 0.00023\text{ (tension)} \qquad \textbf{Answer}$$

(b) The maximum tensile bending stress occurs at the bottom of the cross section, and its magnitude is

$$(\sigma_t)_{\max} = \frac{My}{I} = \frac{(16.5 \times 10^3\text{ N} \cdot \text{m})(0.12\text{ m})}{24.3 \times 10^{-6}\text{ m}^4}$$

$$= 81.5 \times 10^6\text{ N/m}^2 \text{ or } 81.5\text{ MN/m}^2 \qquad \textbf{Answer}$$

(c) The maximum compressive bending stress occurs at the top of the cross section, and its magnitude is

$$(\sigma_c)_{\max} = \frac{My}{I} = \frac{(16.5 \times 10^3\text{ N} \cdot \text{m})(0.06\text{ m})}{24.3 \times 10^{-6}\text{ m}^4}$$

$$= 40.7 \times 10^6\text{ N/m}^2 \text{ or } 40.7\text{ MN/m}^2 \qquad \textbf{Answer}$$

(d) The radius of curvature, ρ, of the neutral axis can be determined either from the strain-curvature relation of Eq. (8.5) or the moment-curvature relation of Eq. (8.9). Using the former relation and the value of strain from part (a), we have

$$\rho = \left|\frac{y}{\varepsilon}\right| = \frac{0.07\text{ m}}{0.23 \times 10^{-3}} = 304.3\text{ m} \qquad \textbf{Answer}$$

Note that ρ is many times larger than the length of the beam. This will always be the case for small deformations.

Example 8.6. Maximum Flexural Stresses in a Beam. The T-beam shown in Figure 8.15(a) is made of structural steel and supports a uniformly distributed load of 10 kips/ft. Determine the maximum tensile and compressive bending stresses and the locations at which they occur.

Solution. The first step is to determine the reactions. The upward force at A is found to be 42 kips and the upward force at B is 98 kips. We now sketch the moment diagram [Figure 8.15(b)]. The location of the centroid of the cross section and the moment of inertia about the bending axis are determined next. Breaking the area into two rectangles [Figure 8.15(c)] and applying Eq. (4.17), we find that the distance $\bar{y}$ from the bottom of the cross section to the centroid is

$$\bar{y} = \frac{\Sigma A_i \bar{y}_i}{\Sigma A_i} = \frac{(40 \text{ in.}^2)(5 \text{ in.}) + (48 \text{ in.}^2)(12 \text{ in.})}{40 \text{ in.}^2 + 48 \text{ in.}^2} = 8.82 \text{ in.}$$

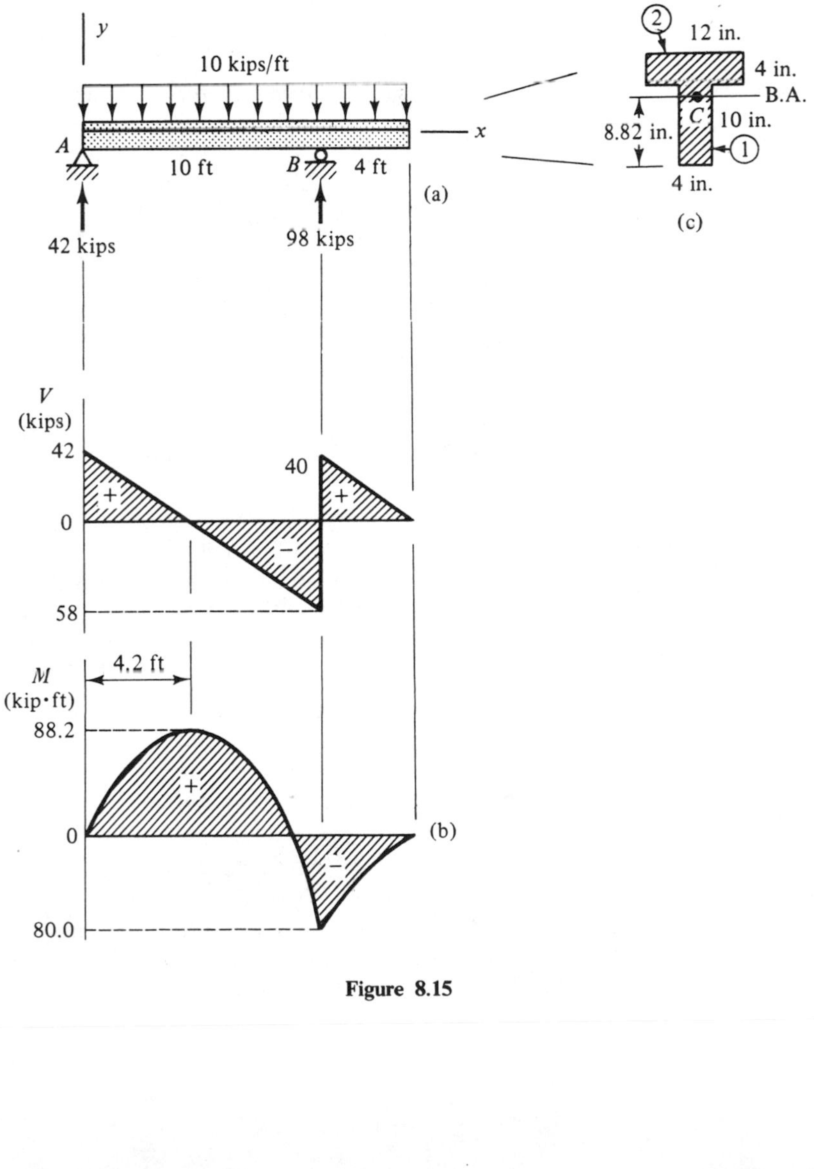

Figure 8.15

Using the parallel axis theorem [Eq. (4.28)], the moment of inertia about the bending axis is found to be

$$I = \sum(\bar{I} + Ad^2) = \frac{1}{12}(4\text{ in.})(10\text{ in.})^3 + (40\text{ in.}^2)(3.82\text{ in.})^2$$
$$+ \frac{1}{12}(12\text{ in.})(4\text{ in.})^3 + (48\text{ in.}^2)(3.18\text{ in.})^2$$
$$= 1{,}466\text{ in.}^4$$

The maximum stresses will occur at the location of one or the other peaks on the moment diagram. We consider each location separately. At 4.2 ft from the left end, the top fibers of the beam are in compression and the bottom fibers are in tension because M is positive. Thus, the maximum stresses there are

$$(\sigma_t)_{\max} = \frac{(88{,}200\text{ lb}\cdot\text{ft})(12\text{ in./ft})(8.82\text{ in.})}{1{,}466\text{ in.}^4} = 6{,}370\text{ psi}$$

$$(\sigma_c)_{\max} = \frac{(88{,}200\text{ lb}\cdot\text{ft})(12\text{ in./ft})(5.18\text{ in.})}{1{,}466\text{ in.}^4} = 3{,}740\text{ psi}$$

The moment is negative at the right support; therefore, the maximum stresses there are

$$(\sigma_t)_{\max} = \frac{(80{,}000\text{ lb}\cdot\text{ft})(12\text{ in./ft})(5.18\text{ in.})}{1{,}466\text{ in.}^4} = 3{,}390\text{ psi}$$

$$(\sigma_c)_{\max} = \frac{(80{,}000\text{ lb}\cdot\text{ft})(12\text{ in./ft})(8.82\text{ in.})}{1{,}466\text{ in.}^4} = 5{,}780\text{ psi}$$

A comparison of the two sets of values of stresses shows that the largest tensile stress is 6,370 psi and occurs at the bottom of the beam 4.2 ft from the left end. The largest compressive stress is 5,780 psi and occurs at the bottom of the beam over the right support. Note that the largest compressive stress does not occur where the bending moment is largest. The larger value of $y_{\max}$ for compressive stresses at the right support more than compensates for the smaller bending moment there. Since the maximum stresses are well below the yield strength of structural steel (see Table A.2), our results are valid.

Example 8.7. Design of a Beam. A simple beam with a span of 10 ft is to support a 10,000-lb concentrated load at its center [Figure 8.16(a)]. Determine the size of W-shape beam that will support this load and that would be most economical from a weight standpoint if (a) the weight of the beam is neglected and (b) the weight of the beam is included. The maximum allowable bending stress is 20,000 psi.

Solution. (a) We first determine the reactions and sketch the moment diagram [Figure 8.16(b)]. The maximum bending moment is found to be 25,000 lb · ft.

If we assume linearly elastic material behavior, we have from Eq. (8.11)

$$\sigma_{\max} = \frac{M}{S} \leqslant \sigma_{\text{allowable}}$$

or

$$S \geqslant \frac{(25{,}000\text{ lb}\cdot\text{ft})(12\text{ in./ft})}{20{,}000\text{ lb/in.}^2} = 15.00\text{ in.}^3$$

Turning to Table A.3 of the Appendix, we now look for those W-shapes with a value of S (about axis x-x) of 15 in.^{3} or greater. We start with the smallest beams at the bottom of the page and work upward. The first shape encountered that meets the requirements is a W

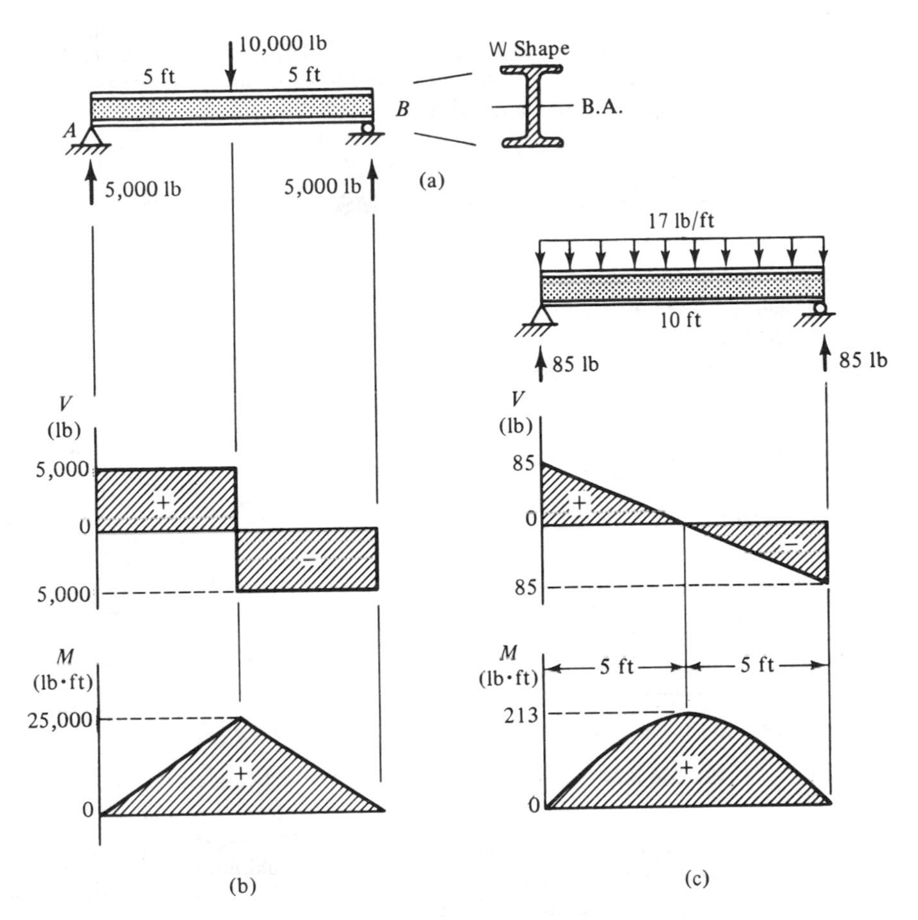

Figure 8.16

6 × 25 shape with $S = 16.7$ in.^{3}. (The second number in the designation, 25 in this case, denotes the nominal weight of the member per foot of length.) We continue on up the page to see if there are any lighter weight beams that can be used. We next find a W 8 × 20 shape with an S of 17.0 in.^{3}. All other 8-in. shapes are heavier than this one. In the 10-in. shapes, we find a W 10 × 17 with $S = 16.2$ in.^{3}. This is the lightest beam with the required value of S.

(b) We now check to see if the W 10 × 17 shape is still satisfactory when the weight of the member is included. This can be done either by computing the maximum stress and comparing it with the allowable value or by computing the new required value of S and comparing it with the actual value for the shape. The latter approach will be used here.

The additional bending moment due to the weight of the beam has a maximum value of 213 lb · ft at the center [Figure 8.16(c)], for a total moment of 25,213 lb · ft at that location. The required value of S is now

$$S \geqslant \frac{M}{\sigma_{\text{allowable}}} = \frac{(25{,}213 \text{ lb} \cdot \text{ft})(12 \text{ in./ft})}{20{,}000 \text{ lb/in.}^2} = 15.13 \text{ in.}^3$$

which is smaller than the actual value of 16.2 in.^{3}. Thus, the beam chosen by neglecting the weight is still satisfactory. This will often be the case, since the weight of the member is usually small compared to the applied loads.

PROBLEMS

8.24. A linearly elastic beam with the cross section shown supports a negative bending moment of 2 kN · m. What are the maximum tensile and compressive flexural stresses?

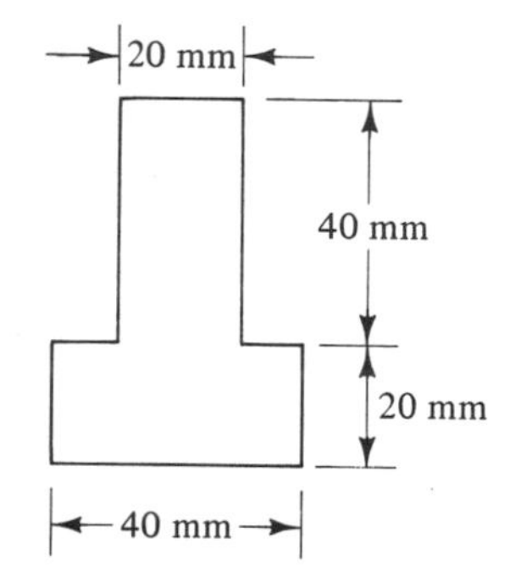

Problem 8.24

8.25. What is the maximum bending moment that can be supported by a 305-mm × 127-mm Universal beam with a mass of 42 kg/m (see Table A.8) if the maximum allowable bending stress is 140 MN/m^2?

8.26. Two 2 in. × 4 in. (nominal) timbers are glued together side by side to form a beam that is to support a vertical load. Assuming that the strength of the glue is not a factor, is the beam stronger with the glue joint oriented horizontally or vertically? Or is there any difference? Justify your answer.

8.27. Design specifications require that a simply supported wooden beam with a span of 16 ft support a uniformly distributed load of 100 lb/ft without exceeding the allowable bending stress of 2,500 psi. After a fire, the portion of the cross section which remains uncharred is found to approximate an ellipse with the dimensions shown. Assuming that the charred wood has zero strength, does the beam still meet specifications?

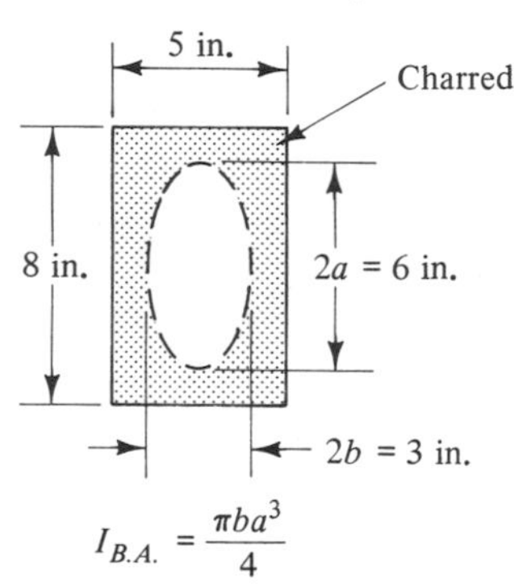

$$I_{B.A.} = \frac{\pi b a^3}{4}$$

Problem 8.27

8.28. While being cut to length, a 20-mm diameter steel rod is clamped to a table top, with a portion overhanging the table. What is the maximum allowable length L of overhang if the rod is not to yield under the action of its own weight? What is the radius of curvature of the neutral axis at the free end of the rod? Use $\sigma_{YS} = 36$ ksi.

8.29. A 203-mm × 133-mm Universal beam with a mass of 30 kg/m is to be strengthened by attaching a 15-mm thick plate of the same material to the top flange, as shown. By what percentage is the moment-carrying capacity of the beam increased if the maximum allowable stress is the same in each

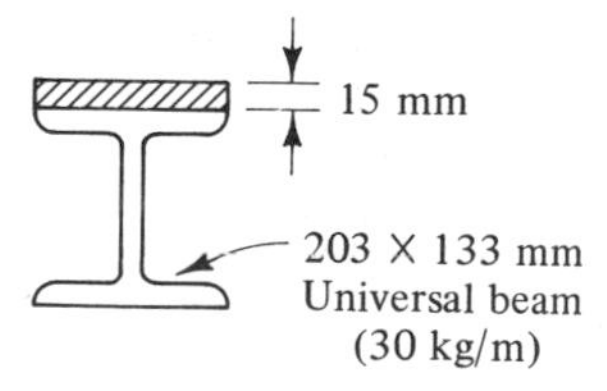

Problem 8.29

case? What would be the percentage increase if an identical plate were also added to the bottom flange?

8.30. The beam shown is made of a 4-in. × 6-in. (nominal) timber oriented with the larger dimension vertical. Determine the maximum tensile and compressive bending stresses and the radius of curvature of the neutral axis at midspan if (a) the weight of the member is neglected and (b) the weight of the member (4 lb/ft) is included. Use $E = 1.6 \times 10^6$ psi.

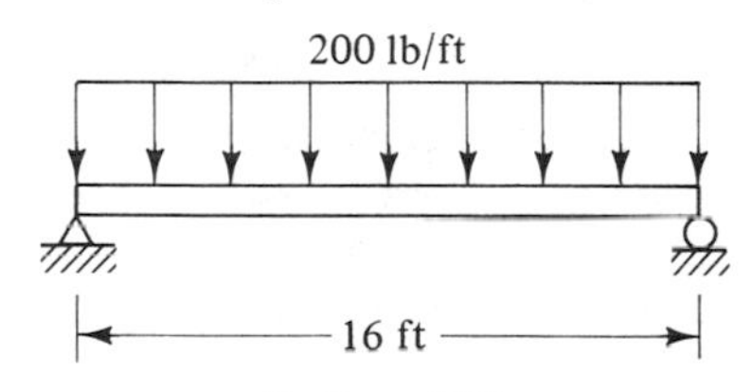

Problem 8.30

8.31. A length of 3/16-in. diameter brass wire is formed into a coil for shipping. What is the minimum allowable coil diameter if the wire is not to yield?

8.32. Determine the couple C required to bend a thin rectangular metal strip around a rigid cylinder, as shown. The strip has width b and thickness $2h$. The stress-strain relation for the material is $\sigma = k|\varepsilon|^n$, where k and n are material constants. What is the maximum bending stress in the strip?

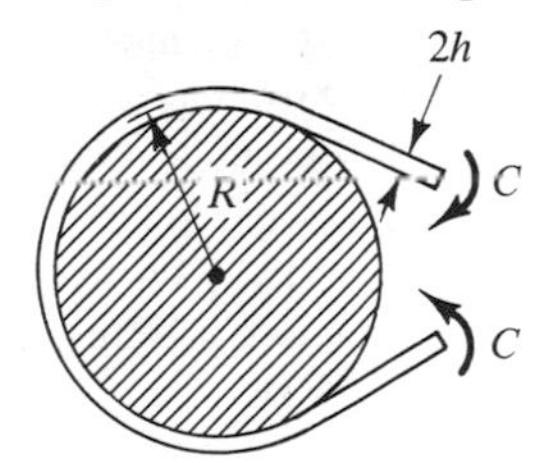

Problem 8.32

8.33. to 8.36. Determine the maximum tensile and compressive bending stresses in the beam shown and indicate where those stresses occur. Assume elastic behavior.

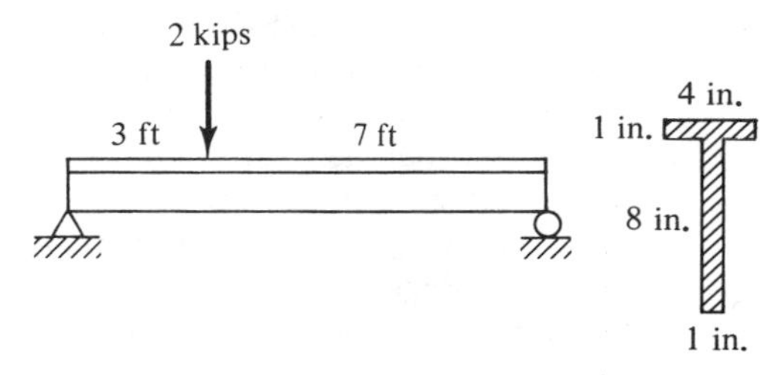

Problem 8.33

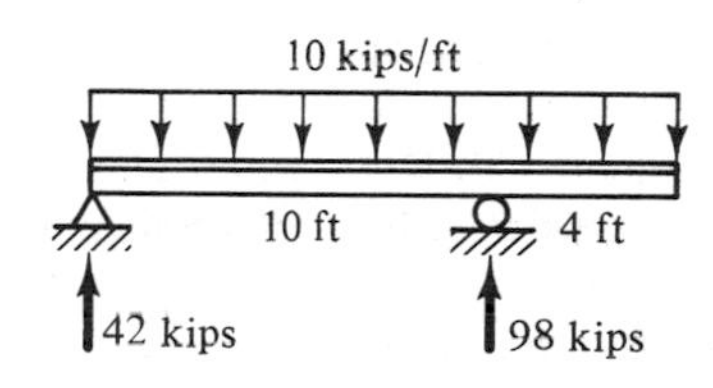

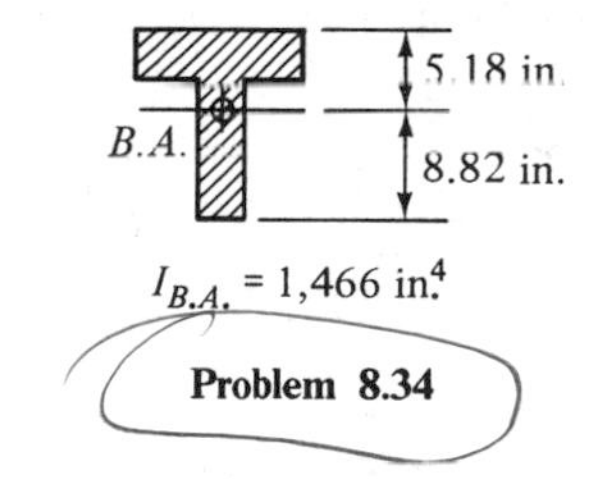

Problem 8.34

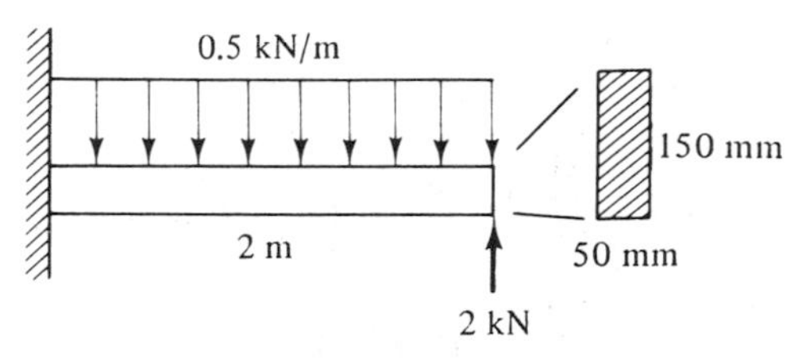

Problem 8.35

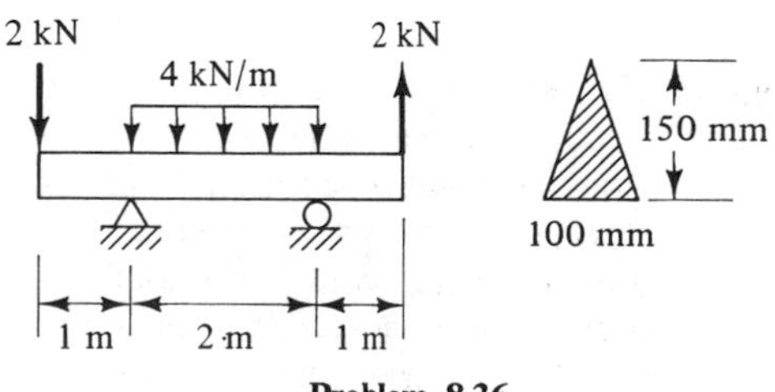

Problem 8.36

8.37. A 4-ft long structural steel cantilever beam is made of a $C\ 6 \times 13$ American Standard Channel oriented with the flanges down. What is the maximum vertical force F that can be applied at the free end without causing yielding if (a) the weight of the member is neglected and (b) the weight of the member is included? Use a safety factor of 2.

8.38. What is the lightest weight W-shape structural steel beam that will support a uniformly distributed load of 2 kips/ft over a simply supported span of 10 ft without yielding? Use a safety factor of 3 and include the effect of the weight of the member.

8.39. Determine the maximum allowable value of the load F for the beam shown if the allowable tensile and compressive stresses are 28 MN/m^2 and 70 MN/m^2, respectively.

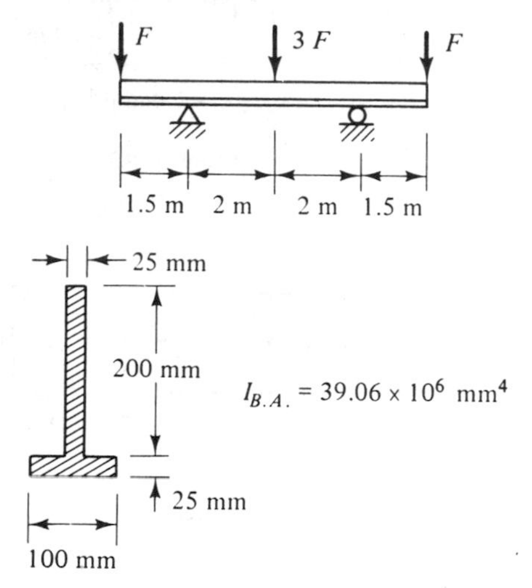

Problem 8.39

8.40. The arms on a material rack consist of two 3-mm thick parallel steel plates spaced 1 in. apart and connected at each end. For the dimensions shown, determine the maximum bending stress in the arm as a function of the distance x from the support and plot its variation over the length of the member. Assume a uniformly distributed downward load with magnitude q N/m.

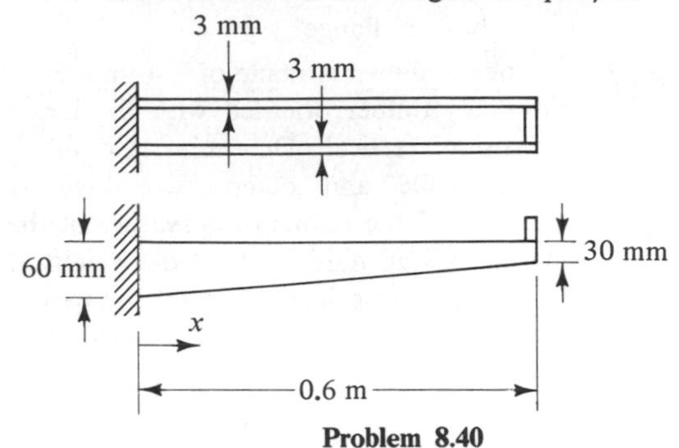

Problem 8.40

8.41. The gantry crane shown has a span of 14 ft and is to support a gross load of 16 kips. Select a W-shape structural steel beam for this application that will provide a safety factor of 1.5 against failure by the onset of yielding. Ignore the effect of the end braces.

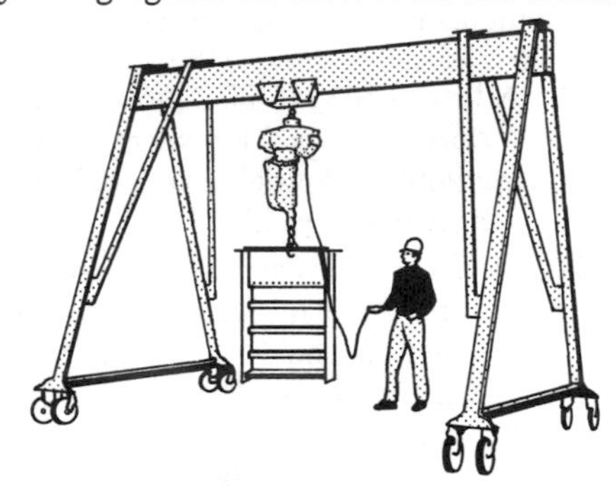

Problem 8.41

8.42. The horizontal members in the barrel rack shown are 8-ft long and are made of 2-in. square hollow structural steel tubes. Each barrel and contents has a maximum weight of 500 lb and the barrels are spaced 24 in. between centers. What is the tube wall

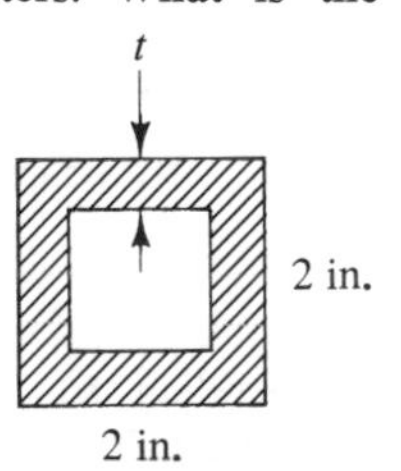

Problem 8.42

thickness t required to provide a safety factor of 2 against failure by the onset of yielding? Give your answer in 1/16ths of an inch.

8.43. By what percentage do the semicircular grooves reduce the load-carrying capacity of the beam shown? Assume linearly elastic material behavior.

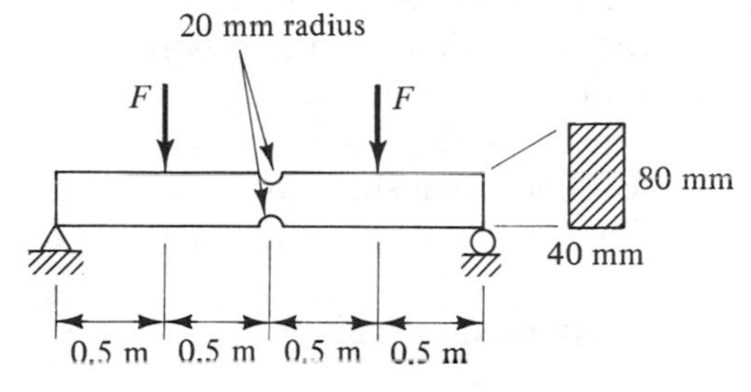

Problem 8.43

8.44. Determine the maximum flexural stress in the 20-mm thick member shown if (a) $a = 0.4$ m and (b) $a = 0.8$ m. Assume linearly elastic material behavior.

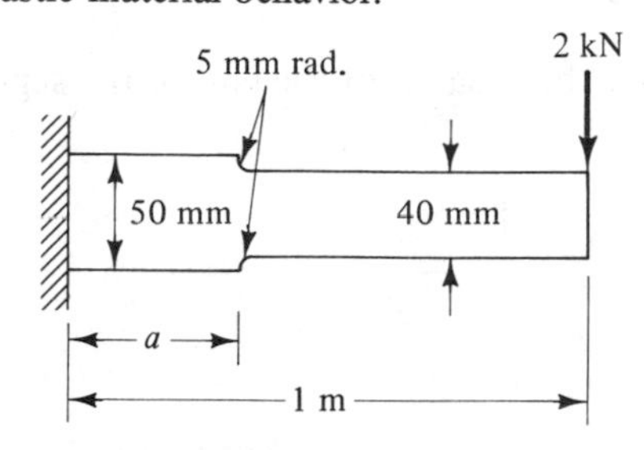

Problem 8.44

8.45. Determine the maximum bending stress in the member shown if (a) $a = 0.3$ m and (b) $a = 0.6$ m. Assume linearly elastic material behavior.

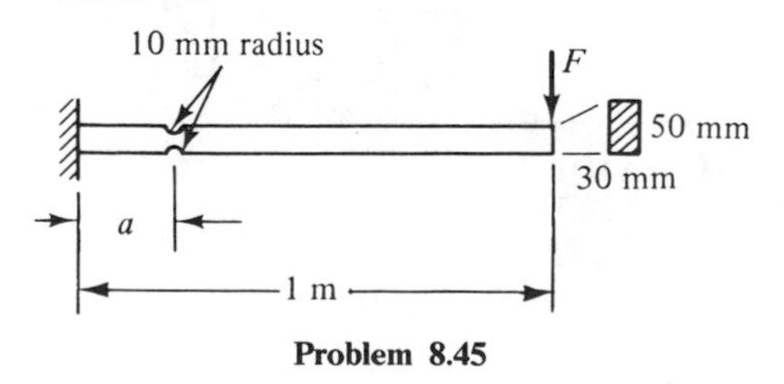

Problem 8.45

8.46. The composite beam shown is subjected to a bending moment of 10 kip · ft. Determine the maximum bending stresses in the steel and in the aluminum and sketch the variation in the bending stress over the cross section. *Hint*: The strains vary linearly with the distance from the bending axis regardless of the material properties.

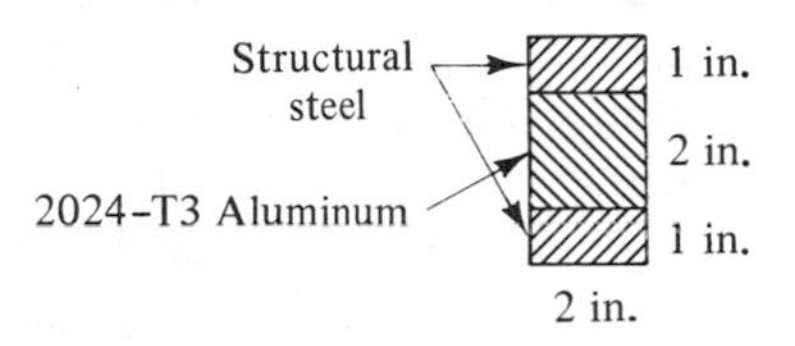

Problem 8.46

8.47. The reinforced concrete beam shown supports a positive bending moment of 240,000 lb · ft. If the steel reinforcing rods have a combined cross-sectional area of 8 in.^{2}, what is the maximum compressive stress in the concrete and the tensile stress in the steel? Assume the concrete has zero tensile strength and a modulus of elasticity equal to one-twelfth that of steel. *Hint*: The strains vary linearly with the distance from the bending axis regardless of the material properties.

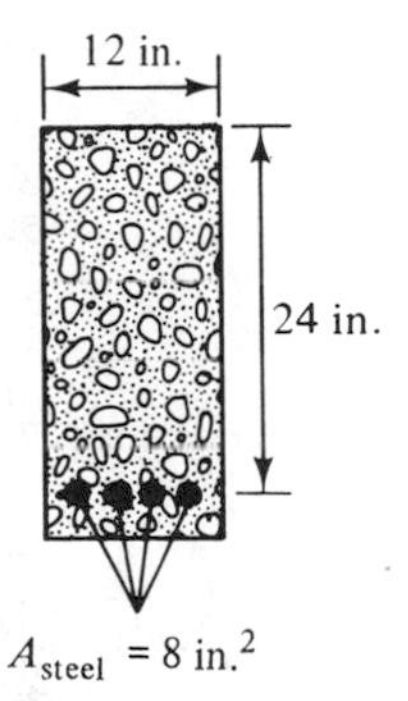

$A_{steel} = 8$ in.^{2}

Problem 8.47

8.48. In Problem 8.47, for what total area A of reinforcing rod will both the concrete and the steel be stressed to their respective limits of 2 ksi and 24 ksi?

8.6 SHEAR STRESSES IN BEAMS

Since the stress resultant in a beam usually includes a shear force, it follows that shear stresses must also be present. These stresses will be considered in this section for the case of linearly elastic material behavior.

Consider two cross sections AB and CD of a beam located a small distance Δx apart [Figure 8.17(a)]. Let M denote the value of the bending moment at section AB and $M + \Delta M$ the value at section CD. To determine the shear stress at a particular level in the beam, such as level EF, we section the beam horizontally along this level and between the two cross sections to obtain the segment $BEFD$ shown in Figure 8.17(b). Since the bending stresses acting upon the ends of the segment differ in magnitude, there must also be a horizontal shear force, ΔF, acting upon the top surface in order to maintain equilibrium. Shear forces also act over the ends of the segment.

Referring to Figure 8.17(b), we have the equilibrium condition

$$\sum F_x = \int (M + \Delta M)\frac{y}{I}\, dA - \Delta F - \int \frac{My}{I}\, dA = 0$$

or

$$\Delta F = \frac{\Delta M}{I} \int y\, dA$$

where the integration is to be carried out over the area A' of the end of the segment.

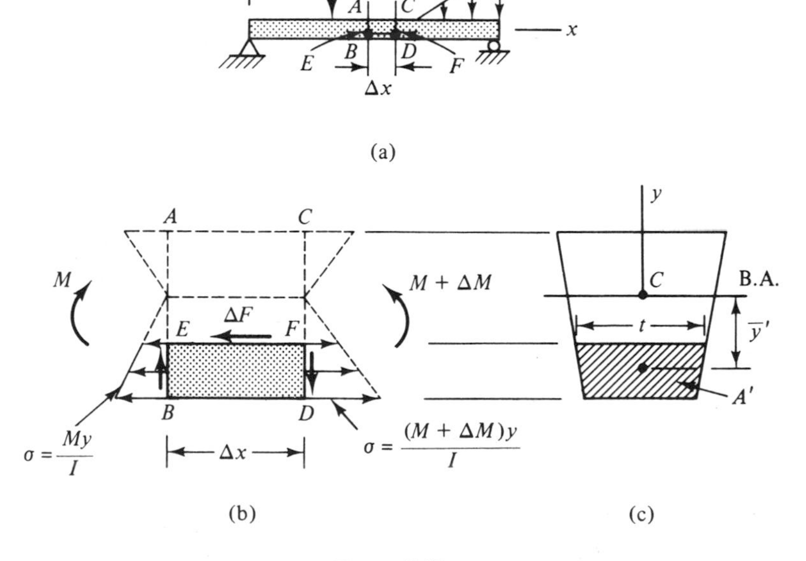

Figure 8.17

This expression can be rewritten as

$$\Delta F = \frac{\Delta M}{I} A' \bar{y}' \tag{8.13}$$

where I is the moment of inertia of the entire beam cross section about the bending axis and $\bar{y}'$ is the distance from this axis to the centroid of the area A' [Figure 8.17(c)]. This same result is obtained if we consider a beam segment above the level of interest instead of the segment below it. Consequently, we can use either the area above or below the level of interest as the area A'.

The shear force per unit of length of the beam, denoted by q_s and called the *shear flow*, is

$$q_s = \lim_{\Delta x \to 0} \frac{\Delta F}{\Delta x}$$

Using the expression for ΔF given in Eq. (8.13), we have

$$q_s = \lim_{\Delta x \to 0} \frac{\Delta M}{\Delta x} \frac{A' \bar{y}'}{I} = \frac{dM}{dx} \frac{A' \bar{y}'}{I}$$

But $dM/dx = V$, where V is the shear force at the cross section of interest; therefore,

$$q_s = \frac{V A' \bar{y}'}{I} \tag{8.14}$$

Shear flow is an important concept in the study of the torsion and bending of thin-walled members and in the study of the connections in built-up beam sections.

Since the actual distribution of shear stress across the thickness of a beam is difficult to determine, a uniform distribution is usually assumed. From Eq. (8.14), the average shear stress across the thickness is

$$\tau = \frac{q_s}{t} = \frac{V A' \bar{y}'}{It} \tag{8.15}$$

where t is the thickness of the beam at the level of interest [Figure 8.17(c)]. The shear stresses associated with the shear force V are often called *direct shear stresses* to distinguish them from those due to other types of loadings, such as torsion.

Except for rectangular sections, Eq. (8.15) gives an accurate estimate of the actual shear stresses only if t is small (thin-walled sections). It should also be noted that this equation and Eq. (8.14) for the shear flow are valid only when the bending stress formula is valid, since it was used in their derivation. In particular, Eqs. (8.14) and (8.15) apply only for linearly elastic material behavior.

Although we have considered only shear stresses on a plane parallel to the longitudinal axis, there are also vertical shear stresses acting over the cross section (Figure 8.18). The stresses on these two planes are equal in magnitude at any given point and, consequently, are both defined by Eq. (8.15). These results follow directly from the fact that if there is a shear stress acting upon one plane in a body, there must be a shear stress of equal magnitude acting upon a plane perpendicular to the first one, as shown in Section 5.5.

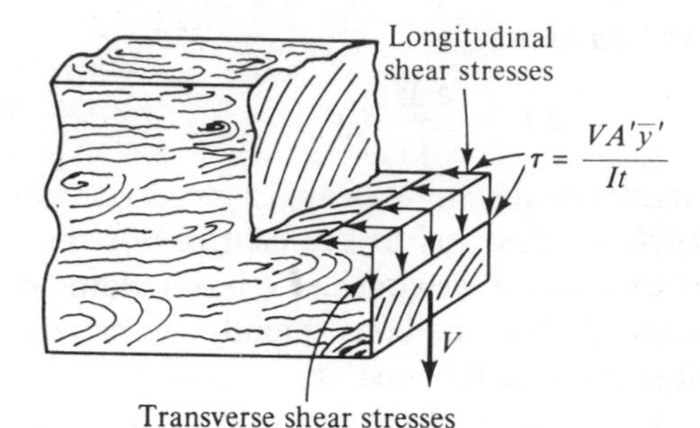

Figure 8.18

There are also horizontal shear stresses acting in the flanges of thin-walled members such as I-beams. Consideration of these stresses is beyond the scope of this text, but their existence can be easily demonstrated by considering the forces acting upon an element of the flange, as in Figure 8.19. Since the longitudinal forces due to the bending stresses differ by an amount ΔF when a shear force is present, it is clear that equilibrium can be maintained only if there are shear forces acting upon side $ABCD$ of the element and upon both ends. The latter forces give rise to the horizontal shear stresses.

It is evident from Eq. (8.15) that the maximum direct shear stress, τ_{max}, occurs on the cross section(s) of the beam where V is largest and at the level in the cross section where the quantity $A'\bar{y}'/t$ is a maximum. Little can be said in general about the location of τ_{max} for cross sections of arbitrary shape. For most common shapes, but not all, it occurs at the centroid of the section. We do know that the product $A'\bar{y}'$ is zero at the top and bottom of the cross section and increases as the level of interest moves toward the bending axis, at which point it attains its maximum value. This fact is usually sufficient to determine the location of τ_{max} for beams with cross

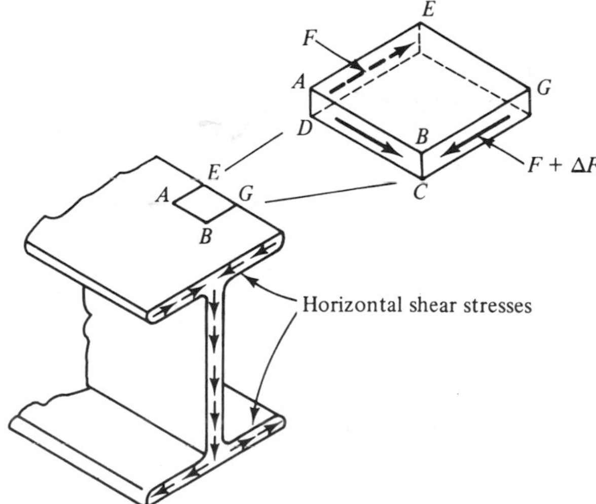

Figure 8.19

sections consisting of rectangular segments. The shear flow is always largest at the bending axis.

The maximum direct shear stress in a beam is usually small compared to the maximum bending stresses, unless the member is exceptionally short. Nevertheless, the direct shear stresses are often of significance, particularly in wooden beams. Since wood is relatively weak in shear parallel to the grain, such members tend to fail by splitting longitudinally due to the shear stresses developed. In fact, the original derivation of Eq. (8.15) resulted from a study of the horizontal cracks in wood ties on railroad bridges conducted by the Russian engineer D. I. Jouravsky in 1855.

Example 8.8. Shear Stresses In a Rectangular Beam. Determine the distribution of shear stress in a beam with rectangular cross section and subjected to a shear force V. What is the maximum value of the shear stress and where does it occur?

Solution. Selecting a point at an arbitrary distance y from the bending axis [Figure 8.20(a)], we find that the area above this point is $A' = b(h/2 - y)$ and the distance from the bending axis to its centroid is $\bar{y}' = (h/2 + y)/2$. Here, b is the width of the cross section and h is its height. At the point of interest, $t = b$. From Eq. (8.15), we obtain for the shear stress at this level

$$\tau = \frac{VA'\bar{y}'}{It} = \frac{Vb\left(\frac{h}{2} - y\right)\left(\frac{h}{2} + y\right)/2}{\left(\frac{1}{12}bh^3\right)b} = \frac{6V}{Ah^2}\left[\left(\frac{h}{2}\right)^2 - y^2\right] \qquad \textbf{Answer}$$

where $A = bh$ is the total cross-sectional area.

The magnitude of the shear stress varies parabolically over the depth of the beam. This distribution is shown in Figure 8.20(b), where the height of the diagram represents the relative magnitude of τ. Note that $\tau = 0$ at the top and bottom of the beam. This will always

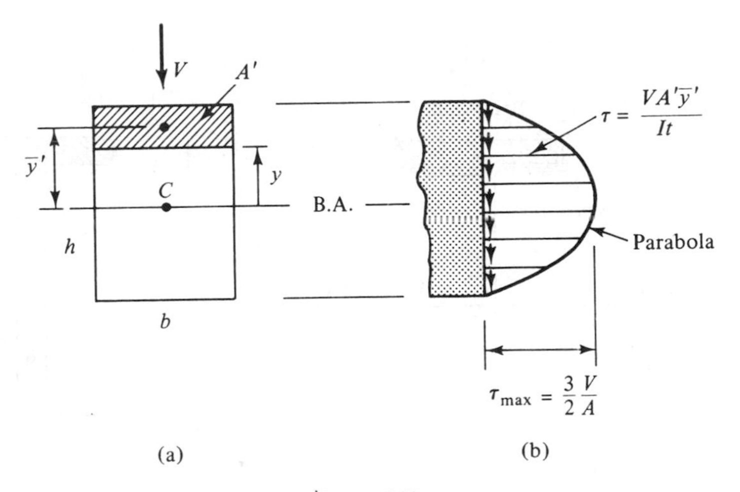

Figure 8.20

be the case regardless of the shape of the cross section because the area A' above or below these respective levels is zero. The maximum shear stress occurs at the bending axis where $y = 0$, and is equal to

$$\tau_{max} = \frac{3}{2}\frac{V}{A} \qquad \textbf{Answer}$$

Example 8.9. Shear Stresses In an I-Beam. A 254-mm × 102-mm Universal beam with a mass of 22 kg/m is subjected to a shear force of 5 kN. Determine the vertical shear stress at the three levels indicated in Figure 8.21(a) and sketch its distribution over the cross section.

Solution. The dimensions and geometric properties of the section obtained from Table A.8 are listed in Figure 8.21(a). Referring to Figure 8.21(b), we have at levels *a-a*

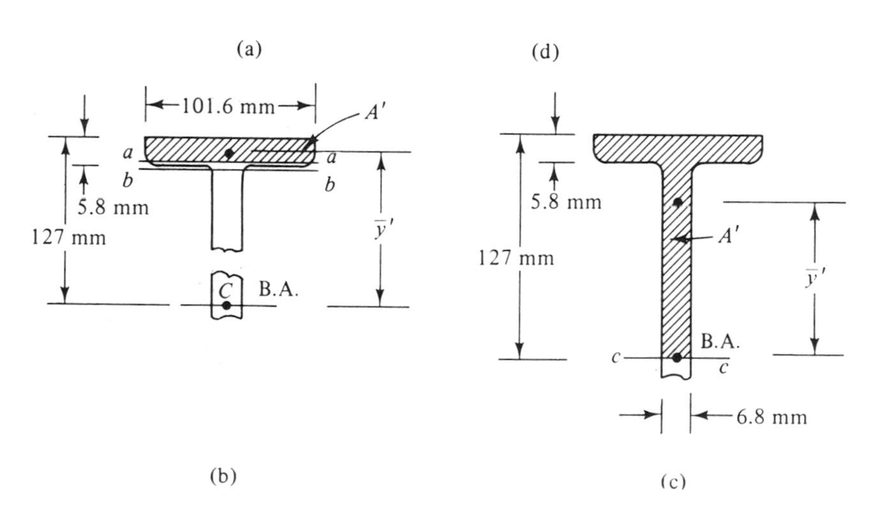

Figure 8.21

and *b-b*

$$A'\bar{y}' = (101.6 \text{ mm})(5.8 \text{ mm})(127 \text{ mm} - 2.9 \text{ mm}) = 73.1 \times 10^3 \text{ mm}^3$$

At level *c-c* [Figure 8.21(c)], we have

$$A'\bar{y}' = A'\bar{y}'(\text{flange}) + A'\bar{y}'\left(\frac{1}{2}\text{ web}\right)$$

$$= 73.1 \times 10^3 \text{ mm}^3 + \frac{(127 \text{ mm} - 5.8 \text{ mm})(6.8 \text{ mm})(127 \text{ mm} - 5.8 \text{ mm})}{2}$$

$$= 123.1 \times 10^3 \text{ mm}^3$$

From Eq. (8.15), the direct shear stresses at these levels are:

(*a-a*)
$$\tau = \frac{VA'\bar{y}'}{It} = \frac{(5 \times 10^3 \text{ N})(73.1 \times 10^{-6} \text{ m}^3)}{(28.63 \times 10^{-6} \text{ m}^4)(0.1016 \text{ m})}$$

$$= 0.13 \times 10^6 \text{ N/m}^2 \text{ or } 0.13 \text{ MN/m}^2$$ **Answer**

(*b-b*)
$$\tau = \frac{(5 \times 10^3 \text{ N})(73.1 \times 10^{-6} \text{ m}^3)}{(28.63 \times 10^{-6} \text{ m}^4)(0.0068 \text{ m})}$$

$$= 1.88 \times 10^6 \text{ N/m}^2 \text{ or } 1.88 \text{ MN/m}^2$$ **Answer**

(*c-c*)
$$\tau = \frac{(5 \times 10^3 \text{ N})(123.1 \times 10^{-6} \text{ m}^3)}{(28.63 \times 10^{-6} \text{ m}^4)(0.0068 \text{ m})}$$

$$= 3.16 \times 10^6 \text{ N/m}^2 \text{ or } 3.16 \text{ MN/m}^2$$ **Answer**

The stress distribution is sketched in Figure 8.21(d). The stress varies parabolically over the various portions of the cross section, which are very nearly rectangular, and is a maximum at the centroid. Note that the stresses in the web are much larger than those in the flanges. Thus, the web supports the majority of the shear load; any bending moment present is carried primarily by the flanges. Note that there are also horizontal shear stresses in the flanges, which we have not considered here (see Figure 8.19).

It is common practice to assume that the shear force in an I-beam is supported entirely by the web and that the shear stress is uniformly distributed over a web area computed by multiplying the overall depth of the beam by the web thickness. Using this procedure, we obtain for the average shear stress in the web

$$\tau = \frac{V}{A_{\text{web}}} = \frac{5 \times 10^3 \text{ N}}{(0.254 \text{ m})(0.0068 \text{ m})}$$

$$= 2.89 \times 10^6 \text{ N/m}^2 \text{ or } 2.89 \text{ MN/m}^2$$

This approach simplifies the computation of the shear stress and usually gives values in reasonable agreement with the maximum value obtained by using Eq. (8.15). For this particular example, the two values differ by 8.5%.

Example 8.10. Shear Stresses In a T-Beam. Two timbers are joined together with lag screws to form the beam shown in 8.22(a). (a) What is the maximum uniformly distributed load, q, that can be supported if the allowable direct shear stress in the wood is 95 psi? (b) What is the maximum allowable spacing between the lag screws for this loading if each screw can support a total shear force of 500 lb?

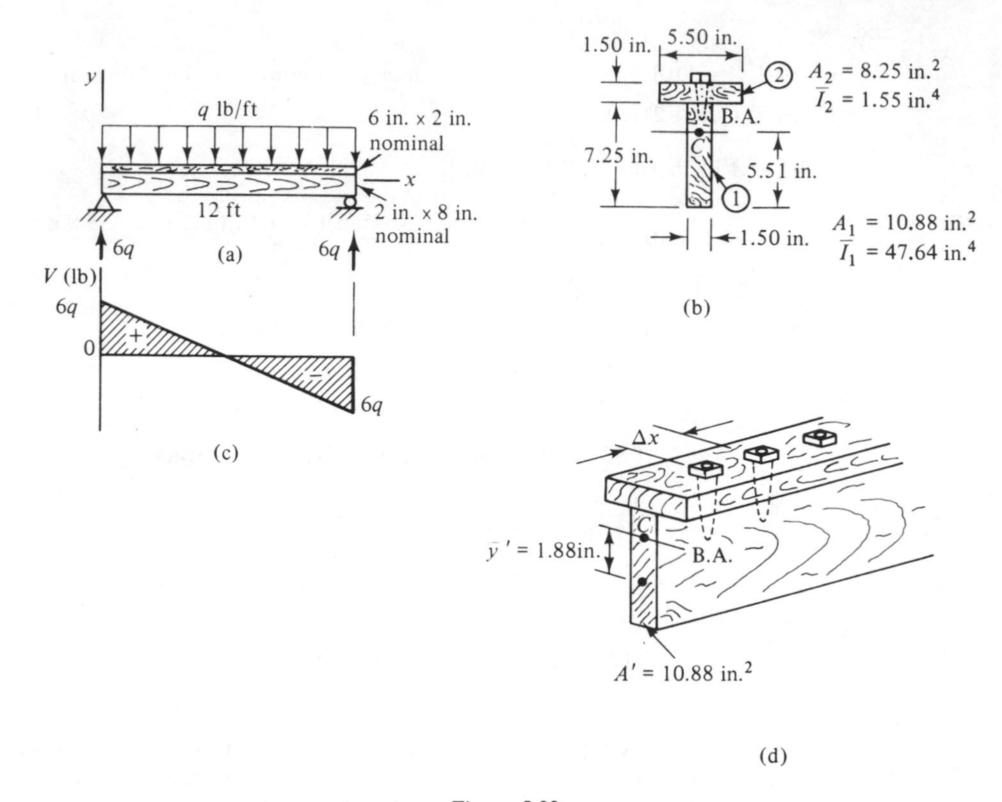

Figure 8.22

Solution. The actual dimensions and geometric properties of the timbers obtained from Table A.12 are listed in Figure 8.22(b). The centroid of the cross section is found to be located 5.51 in. above the bottom surface of the beam, and the moment of inertia of the cross section about the bending axis is 138.8 in.^{4}. From the shear diagram [Figure 8.22(c)], we see that the maximum shear force is $6q$ and occurs at the supports.

(a) The maximum direct shear stress occurs at the centroid of the section, since $A'\bar{y}'$ is largest there and t is smallest. Taking A' to be the area below this level, we obtain from Eq. (8.15)

$$\tau_{\max} = \frac{VA'\bar{y}'}{It} = \frac{(6q \text{ lb})(5.51 \text{ in.})(1.50 \text{ in.})(5.51 \text{ in.}/2)}{(138.8 \text{ in.}^4)(1.50 \text{ in.})} = 0.66q \text{ psi}$$

Thus, we have

$$0.66q \text{ psi} \leqslant 95 \text{ psi}$$

or

$$q \leqslant 144 \text{ lb/ft} \qquad \textbf{Answer}$$

(b) We assume that the lag screws are equally spaced a distance Δx apart [Figure 8.22(d)]. Thus, each screw must support the longitudinal shear force developed over a distance Δx at the intersection between the timbers. Taking A' to be the area below this level and using the maximum value of the shear force ($6q = 864$ lb), we obtain for the shear flow

$$q_s = \frac{(864 \text{ lb})(10.88 \text{ in.}^2)(1.88 \text{ in.})}{138.8 \text{ in.}^4} = 127 \text{ lb/in.}$$

The force supported by each screw is $q_s \, \Delta x$; therefore, we have

$$(127 \text{ lb/in.})(\Delta x \text{ in.}) \leqslant 500 \text{ lb}$$

$$\Delta x \leqslant 3.9 \text{ in.} \qquad \textbf{Answer}$$

This spacing is based on the maximum shear force at the supports. A larger spacing could be used away from these locations where V is smaller. However, a uniform spacing is usually used for convenience and for the added margin of safety it provides. Here, a uniform spacing of 4 in. would likely be used.

PROBLEMS

8.49. and 8.50 Beams with the cross sections shown support a positive shear force $V = 20$ kN. Determine the direct shear stress at the levels indicated and sketch its distribution over the cross section.

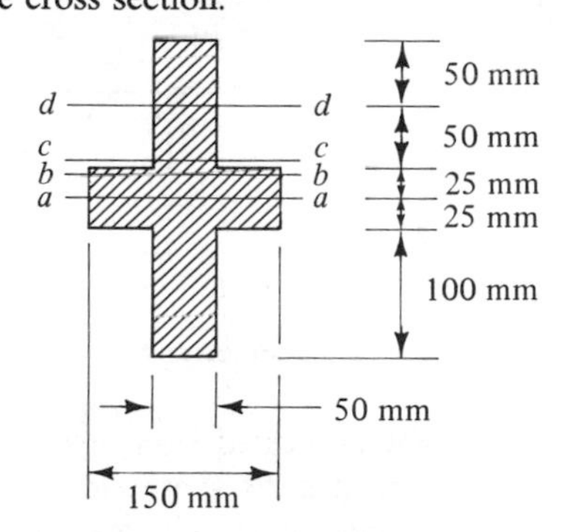

Problem 8.49

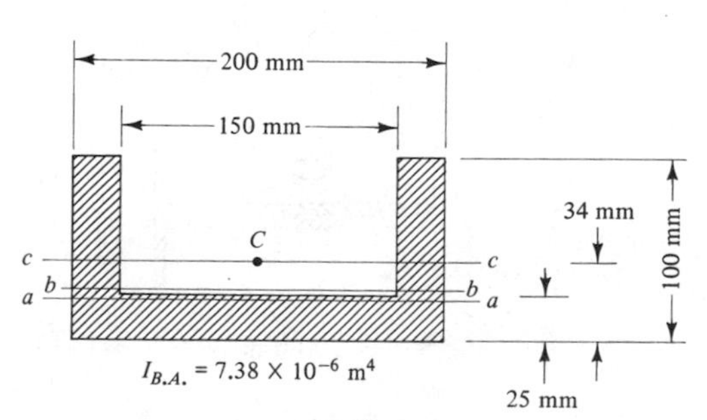

Problem 8.50

8.51. Compare the maximum direct shear stress in a beam with a square cross section with that in a beam with a circular cross section of the same area. Assume the shear force is the same in each case.

8.52. For the beam shown, determine the average horizontal shear stress at a point 2 ft from the wall and 3 in. up from the bottom of the member.

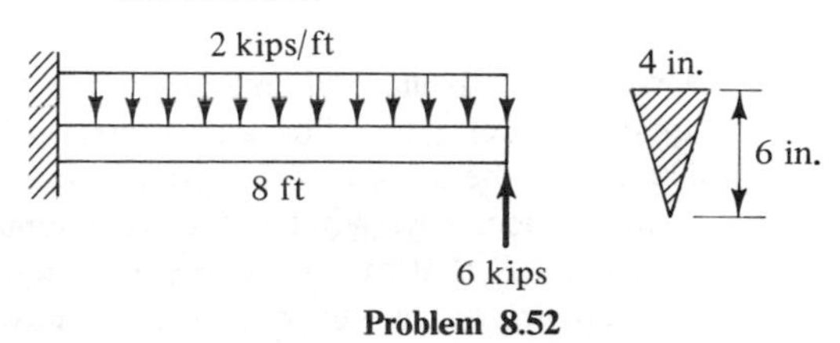

Problem 8.52

8.53. What is the maximum direct shear stress in the beam in Problem 8.52 and where does it occur?

8.54. Compute the ratio of the maximum bend-

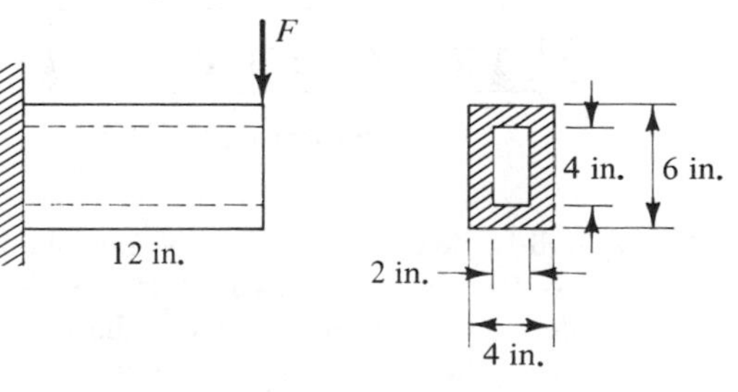

Problem 8.54

ing stress and the maximum direct shear stress for the beam shown.

8.55. A cantilever beam with a length of 2 m is made of a 203-mm × 133-mm Universal beam with a mass of 30 kg/m oriented with the web vertical. Compare the maximum direct shear stress and the maximum bending stress in the member for a uniformly distributed downward loading with magnitude q.

8.56. What diameter wooden rod is required to support the reel of electrical cable shown? The reel has a mass of 200 kg and is 0.8 m wide. The rod is 1.4 m long, with allowable stresses of 15 MN/m^2 in bending and 1.5 MN/m^2 in direct shear. Use a safety factor of 3. *Note*: The reel can be located anywhere along the rod, and the two are in contact only at the edges of the reel.

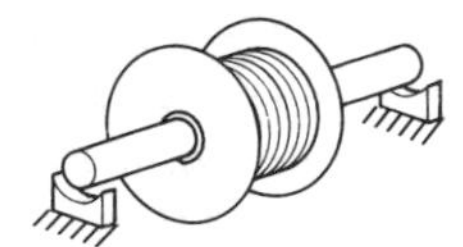

Problem 8.56

8.57. A railroad tie is made of an 8-in. (nominal) square timber with allowable stresses of 800 psi in bending and 80 psi in direct shear, respectively. What is the maximum total load F that can be supported if the reaction from the rail bed is uniformly distributed as shown?

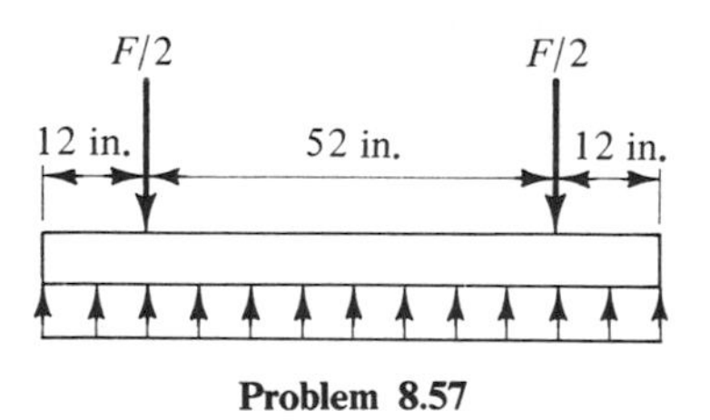

Problem 8.57

8.58. A hollow box beam is made of four planks glued together, as shown. If the beam is loaded until the shear stress in the glue is 75 psi, what are the corresponding maximum bending and direct shear stresses in the wood? Neglect the weight of the member.

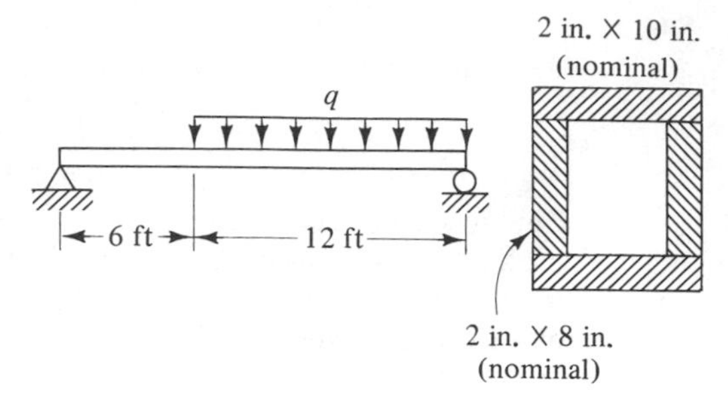

Problem 8.58

8.59. Two 3-in. × 2-in. steel plates are welded together to form a beam with the cross section shown. If each weld can withstand a shear force per unit length of 3 kips/in., what is the maximum shear force V that the beam can support?

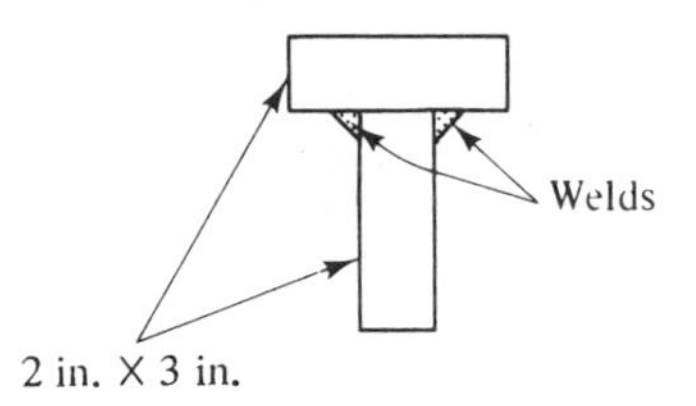

Problem 8.59

8.60. A beam fabricated from two 2-mm thick sheet metal parts spot-welded together as shown must support a maximum vertical

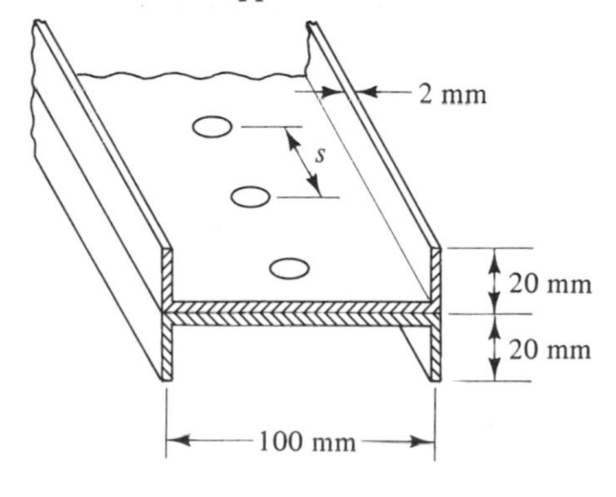

Problem 8.60

shear force of 100 N. What is the minimum required spacing s between the welds if each can support a shear force of 200 N?

8.61. Three 2-in. × 8-in. (nominal) planks oriented with the 8-in. dimension horizontal and connected with a single row of bolts are used as a simply supported beam with a span of 12 ft. What is the largest uniformly distributed load q that can be supported if the bending stress is not to exceed 1,200 psi? What is the maximum allowable bolt spacing for this loading if the $\frac{1}{2}$-in. diameter bolts have an allowable shear stress of 12 ksi?

8.62. A beam made of a 6 × 13 American Standard Channel is strengthened by attaching a 4-in. wide × $\frac{1}{2}$-in thick plate to its underside over the entire length of the member as shown. The plate is attached with a double row of machine screws, each of which can support a shear force of 300 lb. By what percentage does the addition of the plate increase the moment-carrying capacity of the member if the allowable stresses in tension and in compression are the same? What is the maximum allowable screw spacing s for a maximum vertical shear force of 1 kip?

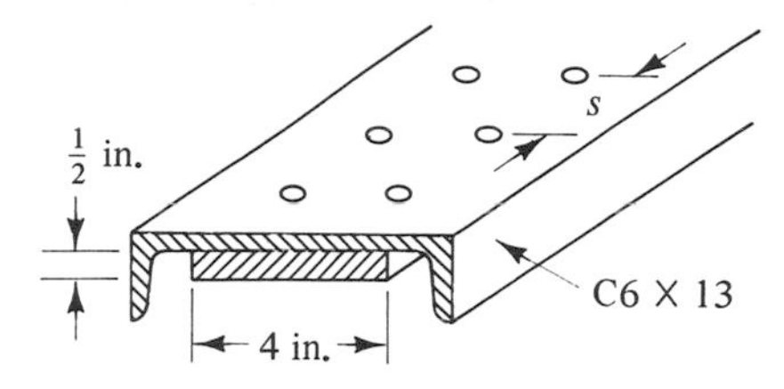

Problem 8.62

8.63. Same as Problem 8.62, except that the plate is $\frac{1}{4}$-in. thick.

8.7 DEFLECTION OF BEAMS

When a beam is loaded, the neutral axis, which is originally straight, deforms into a curve (Figure 8.23). We shall denote this so-called *elastic curve* by $v(x)$. The value of v is measured from the original position of the neutral axis (x axis) and is called the *deflection* of the beam. It is considered positive if the beam deflects upward in the positive y direction and negative if it deflects in the negative y direction. The slope of the deformed beam at any location x is $dv(x)/dx$ and will be denoted by $\theta(x)$.

The deflections of a beam are important because they can result in a "failure" of the member if they are excessive. For example, a floor beam that deflects an excessive amount would be of little value, even though it may be in no danger of

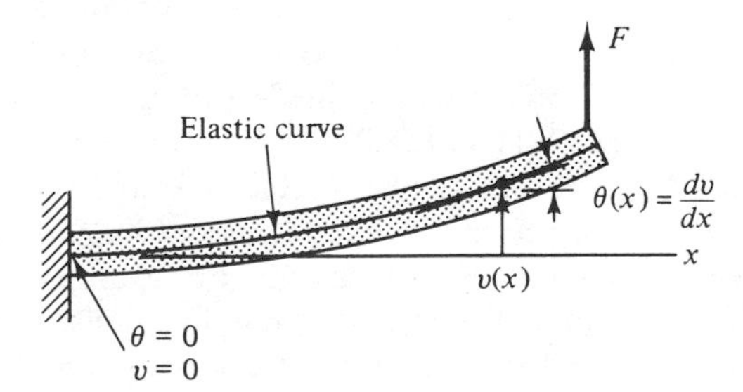

Figure 8.23

breaking. Consequently, limits are often placed upon the allowable deflections of a beam, as well as upon the stresses.

Our analysis of beam deflections will be based upon three fundamental assumptions. First, we assume that the material behavior is linearly elastic and that the beam geometry and loading conditions are such that the moment-curvature relation of Eq. (8.9) applies. Second, we assume that the deflections are small. And third, we assume that the contributions of the shear force, V, to the deflections are negligible. This last assumption is reasonable, unless the beam is exceptionally short.

The deflection, $v(x)$, of beam is related to its curvature via the relation

$$\frac{1}{\rho} = \frac{\dfrac{d^2v}{dx^2}}{\left[1 + \left(\dfrac{dv}{dx}\right)^2\right]^{3/2}} \tag{8.16}$$

The derivation of this equation, which applies for any plane curve, is given in texts on analytic geometry. In applying this equation, it is important to note that x is the coordinate of a point on the elastic curve of the deformed beam. However, we shall be considering only small deflections, in which case x can be taken to be the coordinate along the undeformed neutral axis, as we have been doing. Small deflections also imply that the slope dv/dx will be small, in which case the term $(dv/dx)^2$ is negligible compared to unity. Thus, Eq. (8.16) reduces to

$$\frac{1}{\rho} \approx \frac{d^2v}{dx^2} \tag{8.17}$$

Combining Eq. (8.17) with the moment-curvature relation of Eq. (8.9), we obtain as the governing differential equation for the deflections

$$\frac{d^2v}{dx^2} = \frac{M}{EI} \tag{8.18}$$

In this equation, M is the bending moment, E is the modulus of elasticity of the material, and I is the cross-sectional moment of inertia about the bending axis.

Once the bending moment and the so-called *flexural rigidity EI* are expressed in terms of x, Eq. (8.18) can be integrated once to obtain the slope of the beam, dv/dx, and a second time to determine the deflection, $v(x)$. In most problems, E and I are constants. For simple loadings, the moment equation can be readily determined by using the method of sections, as discussed in Section 8.2. For more complex loadings, the moment equation can be obtained by integrating the shear-load and moment-shear relations in Eqs. (8.1) and (8.2).

The integrated forms of Eq. (8.18) involve certain constants of integration, which are determined from the *boundary conditions* imposed upon the slope or deflection of the beam at the supports or other locations. For example, the supports of a simple beam prevent it from deflecting at these locations; therefore, the boundary conditions are that the deflection is zero at the ends $x = 0$ and $x = L$.

Stated mathematically, we have $v(0) = v(L) = 0$. A fixed support prevents both rotation and deflection. Thus, the boundary conditions for the cantilever beam shown in Figure 8.23 are that the slope and deflection are zero at the support, or $v(0) = \theta(0) = 0$. A sketch of the expected shape of the elastic curve is often helpful in determining the appropriate boundary conditions.

If the variation in the bending moment over the entire length of the beam is described by a single equation, the process of determining the slope and deflection is relatively straightforward (see Example 8.11). However, if there is more than one moment equation, as is often the case, the solution is more complex. In this case, there will be a different solution for $\theta(x)$ and $v(x)$ over each portion of the beam. These solutions must be pieced together by using the physical requirement that the neutral axis deform into a smooth curve without sudden jumps or sharp corners.

The procedure for matching the solutions is best illustrated by means of an example. Consider the beam shown in Figure 8.24(a). Since there is a different moment equation for each half of the member, there will be one set of solutions, $\theta_1(x)$ and $v_1(x)$, for the slope and deflection over the left half of the beam and another set, $\theta_2(x)$ and $v_2(x)$, for the corresponding quantities over the right half. Each set of solutions will involve two constants of integration, for a total of four.

Now, the elastic curve must be smooth, as indicated in Figure 8.24(b). From the figure, we see that this requires that both the slope and deflection have the same values at points immediately to the left and to the right of the center of the beam. Stated mathematically, we have for this example

$$\theta_1\left(\frac{L}{2}\right) = \theta_2\left(\frac{L}{2}\right)$$

$$v_1\left(\frac{L}{2}\right) = v_2\left(\frac{L}{2}\right)$$

These so-called *continuity conditions*, in addition to the two boundary conditions, are sufficient to determine the four constants of integration. This matching procedure will be further illustrated in Example 8.12.

Statically indeterminate problems can also be handled by using the basic procedures outlined in this section. In this case, there will be some reactions that

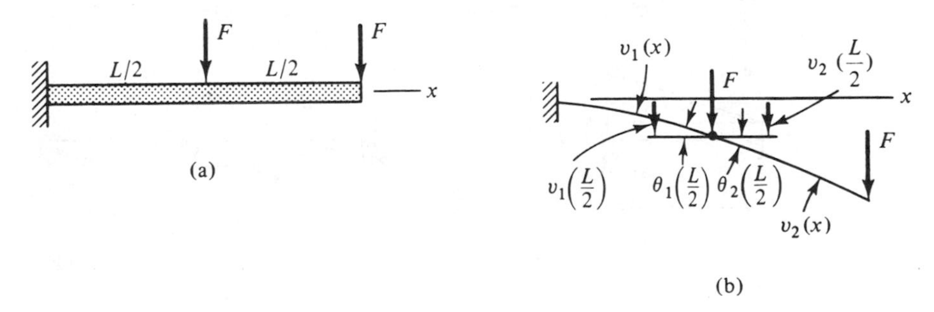

Figure 8.24

cannot be determined from a force analysis and that will remain as unknowns in the moment equation. However, the additional support conditions that make the problem indeterminate also provide enough additional boundary conditions so that both the constants of integration and the unknown reactions can be determined (see Example 8.13).

Many different techniques for determining the slope and deflection of a beam have been developed over the years. Some of these methods are graphical or semigraphical, but all are based upon Eq. (8.18). These various techniques will not be considered in this text.

Example 8.11. Deflection of a Simple Beam. For the beam shown in Figure 8.25(a), determine (a) the deflection equation, (b) the maximum deflection and the location at which it occurs, and (c) the slope at the left end. The flexural rigidity, EI, is constant.

Solution. (a) The reactions at A and B are found to be

$$A_y = \frac{q_0 L}{6} \qquad B_y = \frac{q_0 L}{3}$$

Sectioning the beam at an arbitrary location and considering the equilibrium of the piece to the left [Figure 8.25(b)], we have

$$\overset{+}{\curvearrowleft}\sum M_O = M + q(x)\left(\frac{x}{2}\right)\left(\frac{x}{3}\right) - \frac{q_0 L}{6}x = 0$$

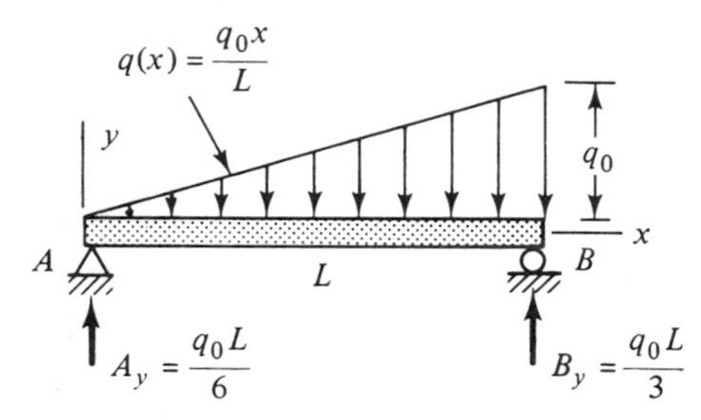

(a)

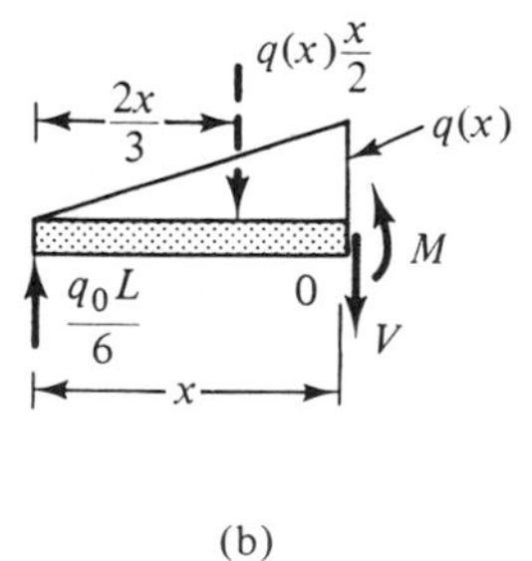

(b)

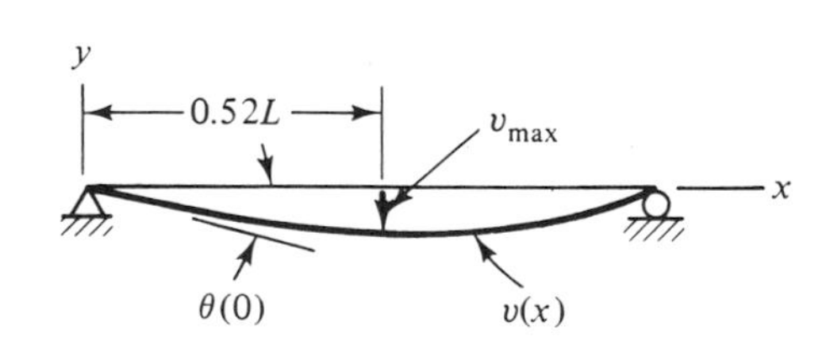

(c)

Figure 8.25

From this equation and Eq. (8.18), we obtain

$$EI\frac{d^2v}{dx^2} = M = q_0\left(\frac{Lx}{6} - \frac{x^3}{6L}\right)$$

Integrating both sides of this expression with respect to x, we have

$$EI\frac{dv}{dx} = EI\theta(x) = q_0\left(\frac{Lx^2}{12} - \frac{x^4}{24L}\right) + C_1$$

where C_1 is a constant of integration. A second integration yields

$$EIv(x) = q_0\left(\frac{Lx^3}{36} - \frac{x^5}{120L}\right) + C_1x + C_2$$

The boundary conditions are $v(0) = 0$ and $v(L) - 0$. Applying the first condition to the preceding equation, we find that $C_2 = 0$. The second condition requires that

$$0 = q_0\left(\frac{L^4}{36} - \frac{L^4}{120}\right) + C_1L$$

or

$$C_1 = -\frac{7}{360}q_0L^3$$

Thus, the deflection equation is

$$v(x) = \frac{q_0xL}{360EI}\left(10x^2 - 3\frac{x^4}{L^2} - 7L^2\right) \qquad \textbf{Answer}$$

(b) The maximum deflection occurs at a point of zero slope. The location of this point is determined by setting the equation for $\theta(x)$ equal to zero and solving for x. We have

$$EI\theta(x) = q_0\left(\frac{Lx^2}{12} - \frac{x^4}{24L}\right) - \frac{7}{360}q_0L^3 = 0$$

or

$$x^4 - 2L^2x^2 + \frac{7}{15}L^4 = 0$$

This is a quadratic equation for x^2. From the quadratic formula, we obtain

$$x^2 = L^2 \pm \sqrt{L^4 - \frac{7}{15}L^4}$$

$$= L^2 \pm 0.73L^2 = 0.27L^2 \quad \text{or} \quad 1.73L^2$$

The second solution is ruled out on the grounds that x cannot be greater than L; therefore,

$$x = \sqrt{0.27L^2} = 0.52L$$

Substituting this value of x into the deflection equation, we find that

$$v_{max} = -0.0065\frac{q_0L^4}{EI} \qquad \textbf{Answer}$$

(c) The slope at the left end of the beam is

$$\theta(0) = \frac{C_1}{EI} = -\frac{7q_0L^3}{360EI} \qquad \textbf{Answer}$$

The negative slope at this location is consistent with the downward deflection of the beam [Figure 8.25(c)].

Example 8.12. Deflection of a Simple Beam. A simple beam supports a single concentrated load F at a distance a from the left end [Figure 8.26(a)]. Determine the deflection equation if EI is a constant.

Solution. A force analysis of the entire beam reveals that the reactions are

$$A_y = \frac{bF}{L} \qquad B_y = \frac{aF}{L}$$

Sectioning the beam at a location between the left end and the applied load, we obtain the moment equation

$$M_1 = \frac{bFx}{L} \qquad (0 \leqslant x \leqslant a)$$

Similarly, the moment in the other portion of the beam [Figure 8.26(c)] is found to be

$$M_2 = \frac{bFx}{L} - F(x - a) \qquad (a \leqslant x \leqslant L)$$

Substituting the moment equations into the relation $EId^2v/dx^2 = M$ and integrating, we

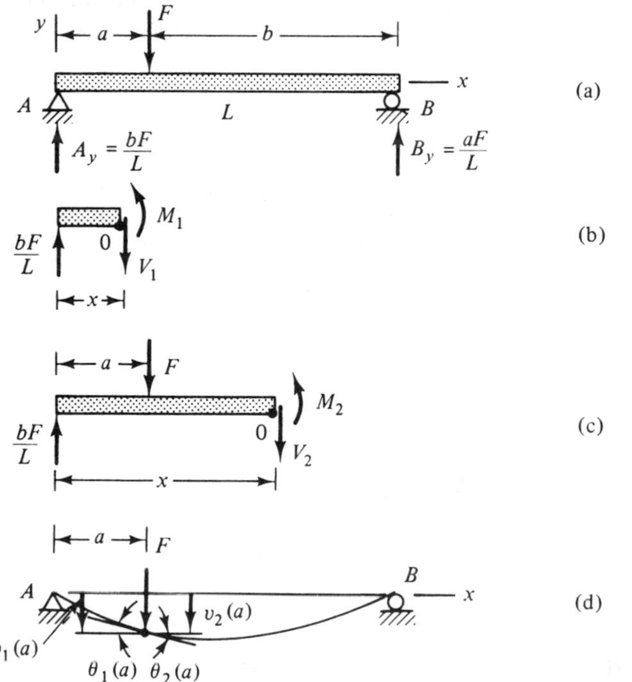

Figure 8.26

obtain for the slope and deflection in the first portion of the beam

$$EI\theta_1(x) = \frac{Fbx^2}{2L} + C_1$$

$$EIv_1(x) = \frac{Fbx^3}{6L} + C_1 x + C_2$$

and for the corresponding quantities in the second portion

$$EI\theta_2(x) = \frac{Fbx^2}{2L} - \frac{F(x-a)^2}{2} + C_3$$

$$EIv_2(x) = \frac{Fbx^3}{6L} - \frac{F(x-a)^3}{6} + C_3 x + C_4$$

Note that in determining θ_2 and v_2, the integral of terms of the form $(x-a)^n$ was written as $\dfrac{(x-a)^{(n+1)}}{(n+1)}$. Ordinarily, the term would first be expanded and then integrated. However, the two results differ only by a constant, which can be combined with the constant of integration. The algebra involved in determining the constants of integration is greatly reduced by using this procedure.

A sketch of the elastic curve is shown in Figure 8.26(d). It is clear from this sketch that the boundary conditions are $v_1(0) = 0$ and $v_2(L) = 0$ and that the continuity conditions between the two portions of the beam are $\theta_1(a) = \theta_2(a)$ and $v_1(a) = v_2(a)$. Applying these conditions to our solutions, we have:

$$v_1(0) = 0: \qquad 0 = C_2$$

$$v_2(L) = 0: \qquad 0 = \frac{FbL^2}{6} - \frac{Fb^3}{6} + C_3 L + C_4$$

$$\theta_1(a) = \theta_2(a): \quad \frac{Fba^2}{2L} + C_1 = \frac{Fba^2}{2L} + C_3$$

$$v_1(a) = v_2(a): \quad \frac{Fba^3}{6L} + C_1 a = \frac{Fba^3}{6L} + C_3 a + C_4$$

Solving these equations, we obtain

$$C_2 = C_4 = 0 \qquad C_1 = C_3 = \frac{Fb}{6L}(b^2 - L^2)$$

Thus, the deflection equations are, after rearrangement of terms,

$$v(x) = \frac{Fbx}{6EIL}(x^2 + b^2 - L^2) \qquad (x \leqslant a)$$

$$v(x) = \frac{Fb}{6EIL}\left[x^3 - \frac{L}{b}(x-a)^3 - x(L^2 - b^2)\right] \qquad (x \geqslant a)$$

Answer

Example 8.13. Analysis of an Indeterminate Beam. Determine the reactions and the deflection equation for the beam shown in Figure 8.27(a).

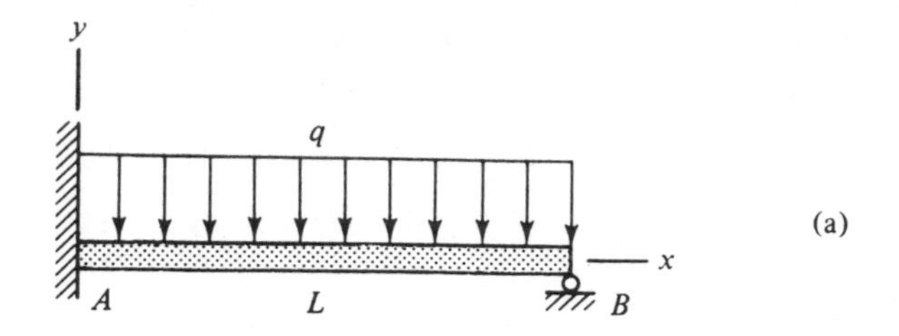

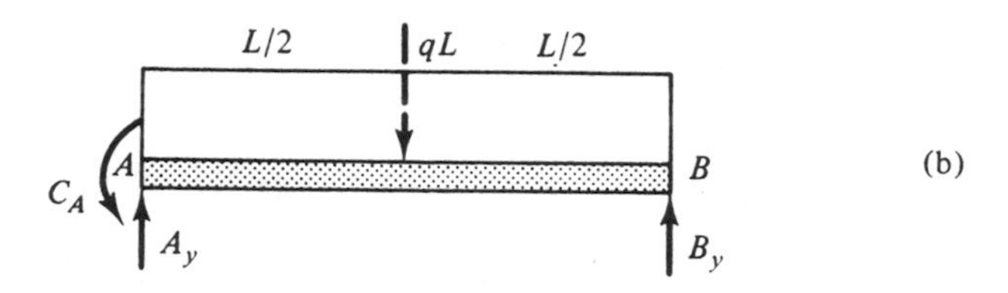

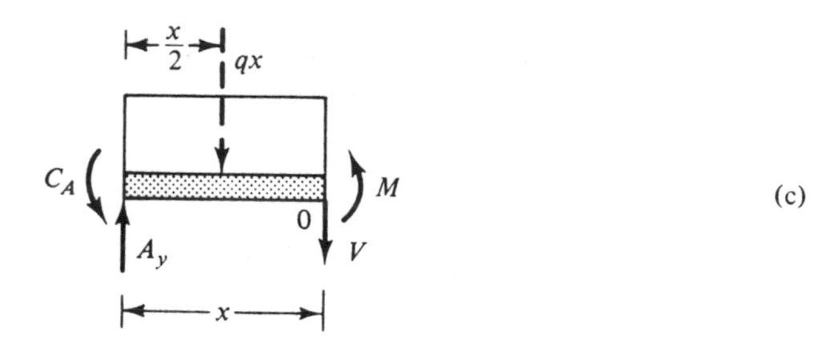

Figure 8.27

Solution. The FBD [Figure 8.27(b)] reveals that the problem is statically indeterminate because there are three unknown reactions and only two equations of equilibrium from which to determine them. These equations are

$$\sum F_y = A_y + B_y - qL = 0$$

$$\overset{+}{\curvearrowleft}\sum M_A = C_A + B_y L - \frac{qL^2}{2} = 0$$

To obtain a third equation, we must consider the deflections of the beam.

Sectioning the beam at an arbitrary distance x from the left end [Figure 8.27(c)], we have

$$\overset{+}{\sum} M_O = M + qx\left(\frac{x}{2}\right) + C_A - A_y x = 0$$

From this equation and Eq. (8.18), we obtain

$$EI\frac{d^2v}{dx^2} = M = A_y x - \frac{qx^2}{2} - C_A$$

Two integrations yield

$$EI\theta(x) = \frac{A_y x^2}{2} - \frac{qx^3}{6} - C_A x + C_1$$

$$EIv(x) = \frac{A_y x^3}{6} - \frac{qx^4}{24} - \frac{C_A x^2}{2} + C_1 x + C_2$$

The boundary conditions $\theta(0) = 0$ and $v(0) = 0$ imply that $C_1 = 0$ and $C_2 = 0$, respectively. However, there is a third boundary condition in this problem; namely, $v(L) = 0$. This condition provides the necessary third equation for determining the reactions:

$$\frac{A_y L^3}{6} - \frac{qL^4}{24} - \frac{C_A L^2}{2} = 0$$

From this equation and the two equilibrium equations, we obtain

$$A_y = \frac{5}{8}qL \qquad B_y = \frac{3}{8}qL \qquad C_A = \frac{qL^2}{8} \qquad \textbf{Answer}$$

Thus, the deflection equation is

$$v(x) = \frac{qx^2}{48EI}(5xL - 2x^2 - 3L^2) \qquad \textbf{Answer}$$

PROBLEMS

8.64. to 8.68 Derive the deflection equation for the uniform beam shown. Express the results in terms of the flexural rigidity EI, the length, and the applied loads.

Problem 8.64

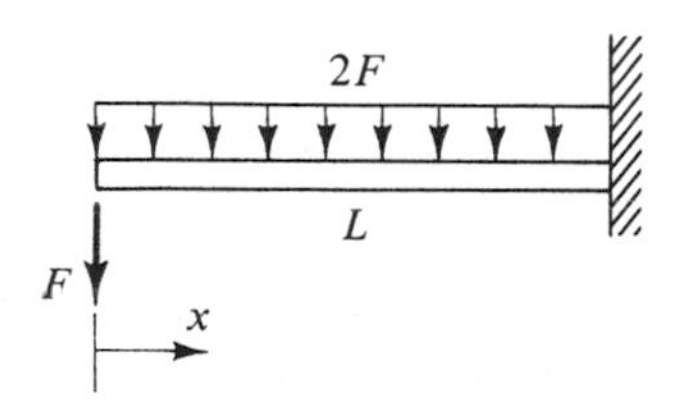

Problem 8.65

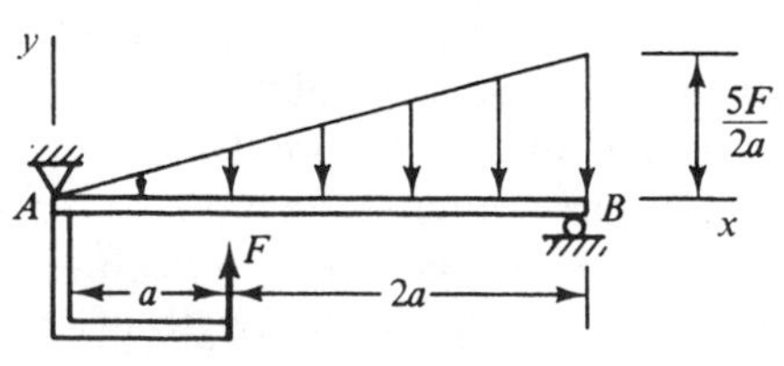

Problem 8.66

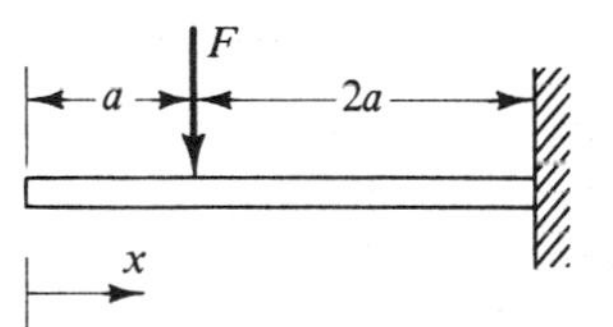

Problem 8.67

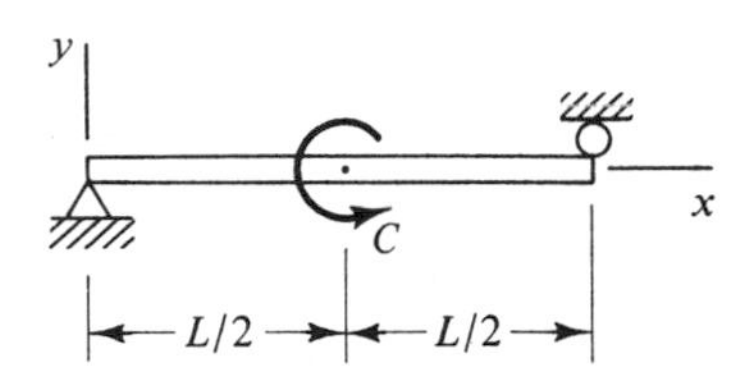

Problem 8.68

8.69. If the beam in Problem 8.21 (Section 8.3) is made of a 127-mm × 64-mm steel Channel oriented with the flanges down, what are the values of the slope and deflection at the free end?

8.70. For the beam in Problem 8.65, determine the maximum deflection and the location at which it occurs.

8.71. Repeat Problem 8.70 for the beam in Problem 8.66.

8.72. Determine the maximum deflection of the beam in Problem 8.68 and the location at which it occurs.

8.73. Determine the midspan deflection of the beam in Problem 8.3 (Section 8.2) if the member is made of 4-in. (nominal) diameter steel pipe. Ignore the weight of the member. Hint: Take advantage of the symmetry present.

8.74. An electrical contact in a switch is supported by a beryllium-copper wire, as shown. What diameter wire should be used if the member is not to yield when the switch is closed? Use a safety factor of 1.5; $E = 120\ \text{GN/m}^2$ and $\sigma_{YS} = 820\ \text{MN/m}^2$.

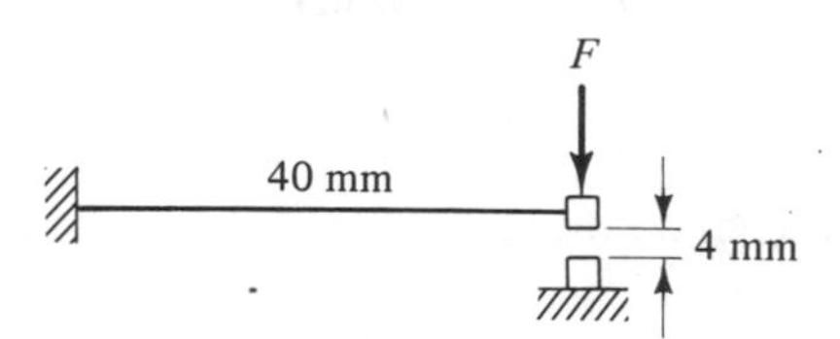

Problem 8.74

8.75. What is the maximum allowable length of a cantilevered section of 2-in. (nominal) steel pipe if the deflection under the action of its own weight is not to exceed 1/360 of its length?

8.76. Determine the lightest steel Universal beam that will support a uniformly distributed load of 8 kN/m over a simply supported span of 6 m if the allowable bending stress is 125 MN/m² and the maximum deflection is not to exceed 1/360 of the span length. Include the effect of the mass of the member.

8.77. Determine the reactions at the supports and the deflection equation for the beam shown. Compare the midspan deflection with that of an identical simply supported beam carrying the same load (see Example 8.11).

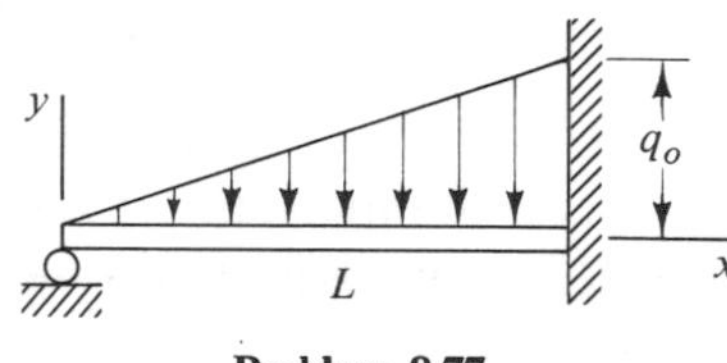

Problem 8.77

8.78. Because of improper soil preparation, the support at the left end of the beam in Problem 8.77 settles a distance $\Delta = 0.1$ in. relative to the support at the right end when the load is applied. If $q_0 = 1$ kip/ft, $L = 10$ ft, and $EI = 10^9\ \text{lb}\cdot\text{in.}^2$, what is the maximum bending moment in the beam and where does it occur?

8.8 SINGULARITY FUNCTIONS

It is an understatement to say that the integration method of determining beam deflections discussed in the preceding section becomes lengthy when the loadings are complex. For example, consider the work involved in Example 8.12 in determining the deflection of a simple beam with a single concentrated load. Just imagine what the situation would be if there were, say, three concentrated loads. There would be four different moment equations and four sets of solutions for the slope and deflection involving a total of eight constants of integration! These four sets of solutions would have to be matched together at three different locations. Clearly, the amount of work involved would be prohibitive. However, these difficulties can be avoided if the variation in the bending moment over the entire length of the beam can be described by a single equation. Fortunately, there is a way to do this.

Consider the two moment equations for the beam in Example 8.12:

$$M_1 = \frac{bFx}{L} \qquad (0 \leqslant x \leqslant a)$$

$$M_2 = \frac{bFx}{L} - F(x - a) \qquad (a \leqslant x \leqslant L)$$

We notice that the second equation would be valid for the entire beam if we could make the term $F(x - a)$ vanish for $x \leqslant a$. This can be done by multiplication by a unit step function, $U(x - a)$, which has the values

$$U(x - a) = \begin{cases} 0 & (x < a) \\ 1 & (x \geqslant a) \end{cases} \tag{8.19}$$

The result of this multiplication is illustrated in Figure 8.28, from which it can be seen that the unit step function acts as a "switch" to turn on the term $F(x - a)$ or to turn it off. Thus, if we write the moment equation as

$$M = \frac{bFx}{L} - F(x - a)U(x - a)$$

the result will be valid over the entire length of the member. A similar approach is used if there are other concentrated loads, or if there are distributed loads that do not act over the entire length of the beam.

The product $(x - a)U(x - a)$ is called a *singularity function*, which we shall distinguish from ordinary functions by enclosure in pointed brackets. Singularity functions have the property that

$$\langle x - a \rangle^n = \begin{cases} (x - a)^n & \text{if } x \geqslant a \\ 0 & \text{if } x < a \end{cases} \tag{8.20}$$

where n is a nonnegative integer ($n \geqslant 0$). The rules for integration and differentia-

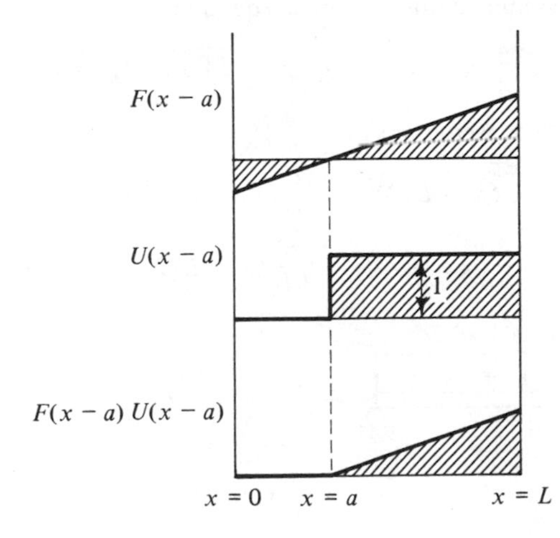

Figure 8.28

tion are

$$\int \langle x-a\rangle^n\,dx = \frac{1}{n+1}\langle x-a\rangle^{n+1} + C \qquad \text{for } n \geqslant 0 \tag{8.21}$$

$$\frac{d}{dx}\langle x-a\rangle^n = n\langle x-a\rangle^{n-1} \qquad \text{for } n > 0 \tag{8.22}$$

Singularity functions and their integrals and derivatives are also defined for negative values of the exponent n. These are needed to describe the loading function $q(x)$ when determining the shear and moment expressions by integration. However, we shall have no need for them here.

As we shall demonstrate in the following examples, the use of singularity functions makes it possible to describe the bending moment for the entire beam by a single equation. This procedure automatically accounts for the continuity relations between the various segments of the member and greatly simplifies the process of solving for slopes and deflections. Furthermore, the solutions are in a form ideal for implementation on a computer; the properties of the singularity function given in Eq. (8.20) can be represented by a simple IF statement in FORTRAN. The use of singularity functions to determine beam deflections is based upon a procedure developed by the German engineer A. Clebsch (1833–1872).

Example 8.14. Deflection of a Beam. For the beam shown in Figure 8.29, determine (a) the deflection equation and (b) the deflection at midspan. Use singularity functions.

Solution. (a) Solving for the reactions and using the method of sections, we find for the bending moments in the various segments of the beam

$$M_1 = \frac{5}{4}Fx \qquad (0 \leqslant x \leqslant L/4)$$

$$M_2 = \frac{5}{4}Fx - F(x - L/4) \qquad (L/4 \leqslant x \leqslant L/2)$$

$$M_3 = \frac{5}{4}Fx - F(x - L/4) - \frac{2F}{L}(x - L/2)^2 \qquad (L/2 \leqslant x \leqslant L)$$

These three equations can be combined into the single equation

$$M = EI\frac{d^2v}{dx} = \frac{5}{4}Fx - F\langle x - L/4\rangle - \frac{2F}{L}\langle x - L/2\rangle^2$$

Integrating twice with respect to x, we obtain

$$EI\theta(x) = \frac{5}{8}Fx^2 - \frac{F}{2}\langle x - L/4\rangle^2 - \frac{2F}{3L}\langle x - L/2\rangle^3 + C_1$$

$$EIv(x) = \frac{5}{24}Fx^3 - \frac{F}{6}\langle x - L/4\rangle^3 - \frac{F}{6L}\langle x - L/2\rangle^4 + C_1x + C_2$$

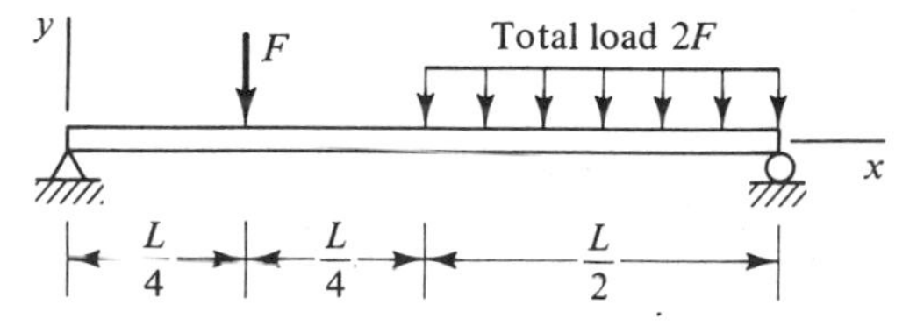

Figure 8.29

At $x = 0$, all the singularity functions have a value of zero. Thus, the boundary condition $v(0) = 0$ implies that $C_2 = 0$. The boundary condition $v(L) = 0$ at the right end of the beam, where the singularity functions are nonzero, requires that

$$\frac{5}{24}FL^3 - \frac{F}{6}(L - L/4)^3 - \frac{F}{6L}(L - L/2)^4 + C_1 L = 0$$

or

$$C_1 = -\frac{49}{384}FL^2$$

This completes our solution for the deflection equation, which can be written in the more compact form

$$v(x) = \frac{F}{384EI}\left[80x^3 - 64\langle x - L/4\rangle^3 - \frac{64}{L}\langle x - L/2\rangle^4 - 49L^2x\right] \qquad \textbf{Answer}$$

(b) At midspan, the first singularity function in the deflection equation is nonzero and the second one is zero. Thus,

$$v(L/2) = \frac{F}{384EI}\left[80(L/2)^3 - 64(L/2 - L/4)^3 - 49L^2(L/2)\right]$$

$$= -\frac{31FL^3}{768EI} \qquad \textbf{Answer}$$

Example 8.15. Moment Equation for a Beam. Use singularity functions to express the moment equation for the beam shown in Figure 8.30(a).

Solution. Let R_1 and R_2 denote the reactions at the left and right supports, respectively. Here, the distributed loading stops before the end of the beam is reached. A simple way to handle this situation is to extend the load to the right end of the beam and then add an equal and opposite loading over the right half of the member to cancel it out. Thus, we have the loading shown in Figure 8.30(b). Notice that both distributed loadings now extend to the right end of the member. Notice also that the loading has not been changed by this procedure.

The moment equation is

$$M(x) = R_1 x - \frac{q}{2}\langle x - L/4\rangle^2 + \frac{q}{2}\langle x - L/2\rangle^2 - C\langle x - 3L/4\rangle^0 \qquad \textbf{Answer}$$

Notice that the concentrated couple enters into the moment equation through a singularity function to the zero power, which is the same as the unit step function.

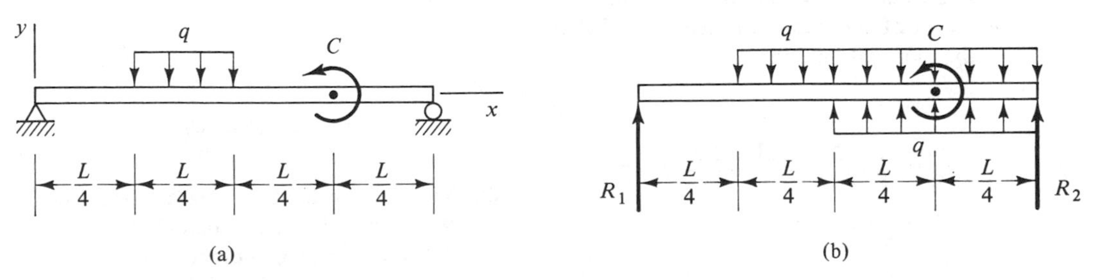

Figure 8.30

PROBLEMS

8.79. Solve Problem 8.67 (Section 8.7) using singularity functions.

8.80. Repeat Problem 8.79 for the beam in Problem 8.68 (Section 8.7).

8.81. Determine the deflection at the free end of the 50-mm square steel beam shown.

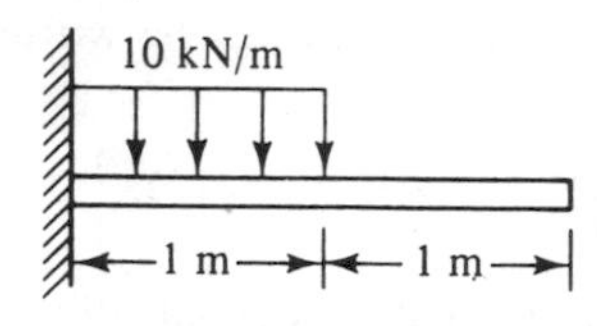

Problem 8.81

8.82. and 8.83 Obtain the deflection equation for the beam shown and determine the deflection at midspan.

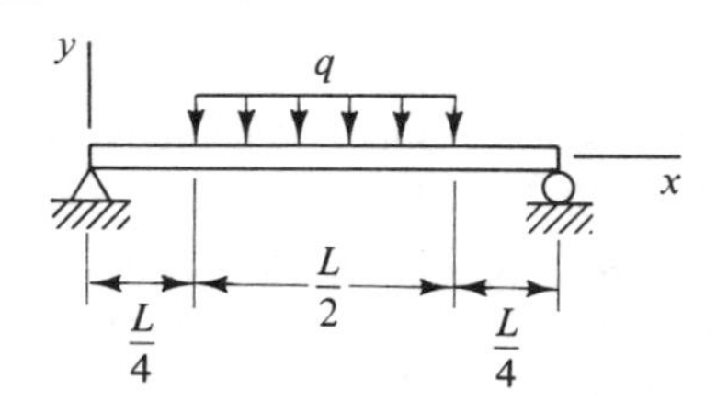

Problem 8.82

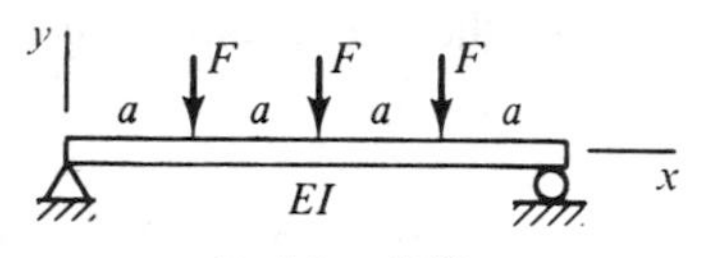

Problem 8.83

8.84. For the beam shown, derive the deflection equation and determine the magnitude and location of the maximum deflection. What is the slope at midspan?

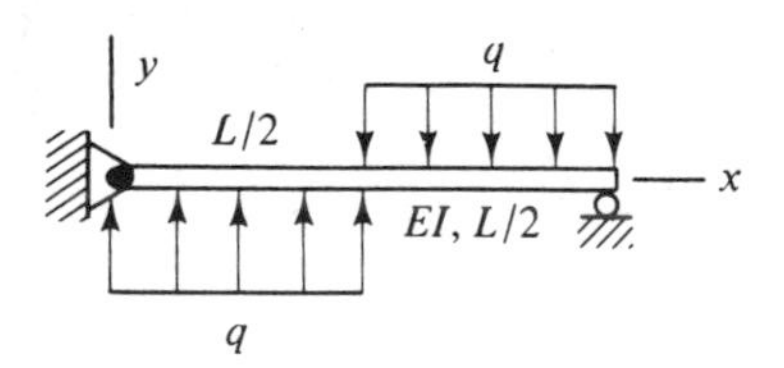

Problem 8.84

8.85. Derive the deflection equation for the beam shown and find the slope at the overhanging end. What is the maximum deflection and where does it occur?

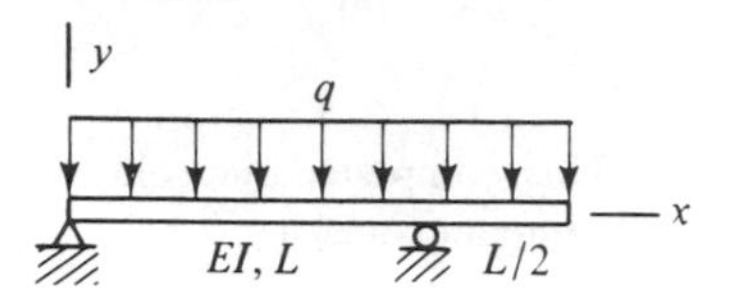

Problem 8.85

8.86. A 12-in. × 3-in. (nominal) plank is used as a diving board. What is the tip deflection and slope when the 160-lb diver stands at the end as shown? $E = 1.6 \times 10^6$ psi.

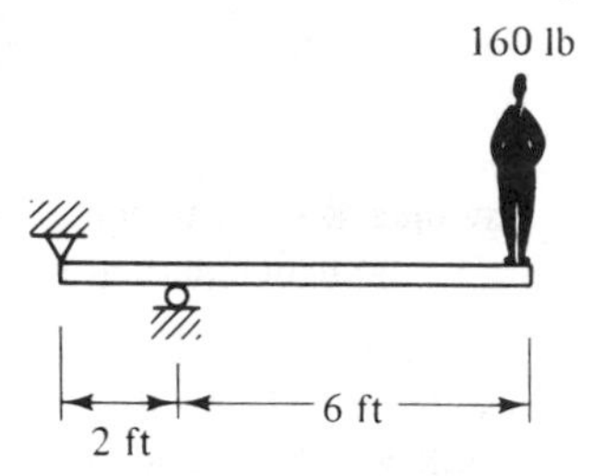

Problem 8.86

8.87. For what value of the couple C will the deflection at the free end of the beam shown be zero? What is the corresponding value of the deflection midway between the two supports?

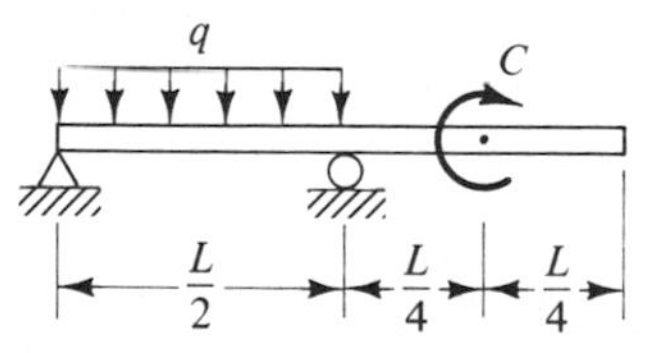

Problem 8.87

8.88. What is the lightest weight timber section that will support the loading shown if the deflection midway between the supports is not to exceed 1.5 in.? For stability, the depth of the cross section should not be more than three times its width. What is the deflection at the free end for the member selected? Use $E = 1.6 \times 10^6$ psi.

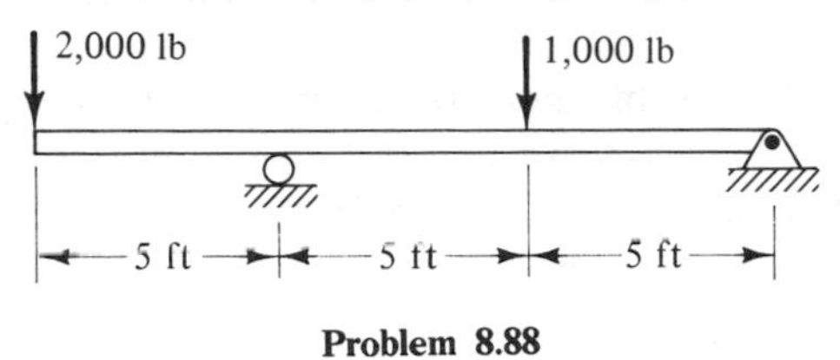

Problem 8.88

8.89. and 8.90. For the beam shown, determine the reactions at the supports and sketch the shear and moment diagrams.

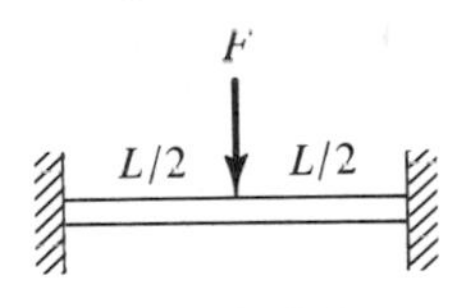

Problem 8.89

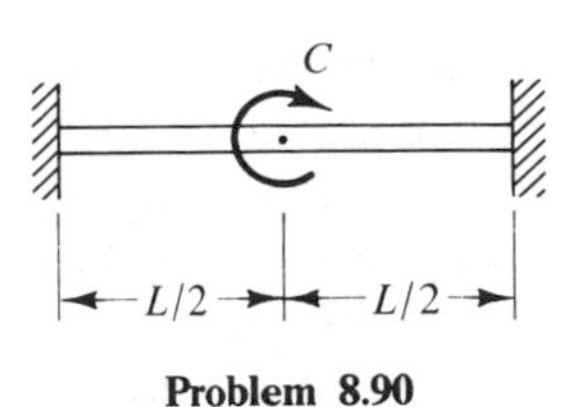

Problem 8.90

8.91. For the beam shown, determine the maximum deflection and bending moment and the locations at which they occur.

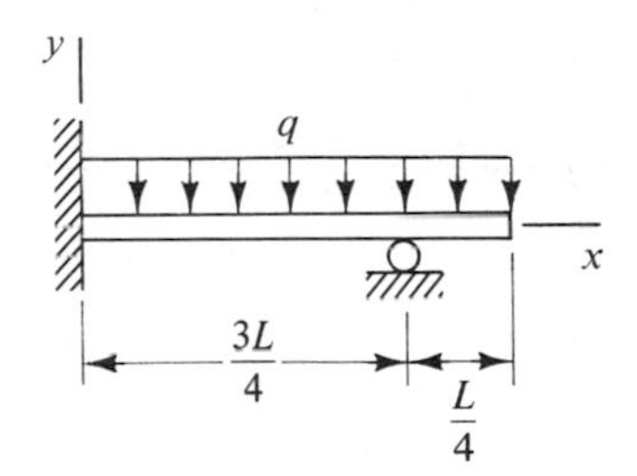

Problem 8.91

8.9 BEAM DEFLECTIONS BY SUPERPOSITION

The problems considered in the preceding sections show that the slope and deflection of a beam are linearly proportional to the applied loads. As indicated by Eq. (8.18), this will always be the case if the material is linearly elastic and the deflections are small. Consequently, the slope and deflection due to several loads are equal to the sum of those due to the individual loads. This result follows directly from the *principle of superposition*, which states that if an effect E_1 is linearly proportional to its cause C_1 and effect E_2 to its cause C_2, then C_1 and C_2 together will have an effect equivalent to E_1 and E_2 together. We have used this principle several times before, but without calling it by name.

The advantage of using superposition in beam deflection problems lies in the fact that solutions for the slope and deflection of common types of beams with simple loadings have been tabulated in engineering handbooks (see Table A.13 of the Appendix). Using superposition, we may combine these solutions to obtain the solution for more complicated loadings.

Superposition is also convenient to use for statically indeterminate problems. The procedure here is to remove enough supports to make the member statically determinate and to replace them by the reactions they exert. These reactions are then treated as part of the applied loading. From this point, the procedure is basically the same as that for statically determinate problems, except that the slopes and deflections due to the individual loadings must satisfy certain compatibility conditions so that the boundary conditions for the original problem are met. These compatibility conditions yield the additional equations necessary for determining the reactions.

For example, if we remove the roller from under the beam shown in Figure 8.31(a) and replace it by the force that it exerts, we obtain the loading shown in Figure 8.31(b). Decomposing this loading, as in Figure 8.31(c), we have as the compatibility condition between the deflections at B

$$v_B = (v_B)_1 + (v_B)_2 = 0$$

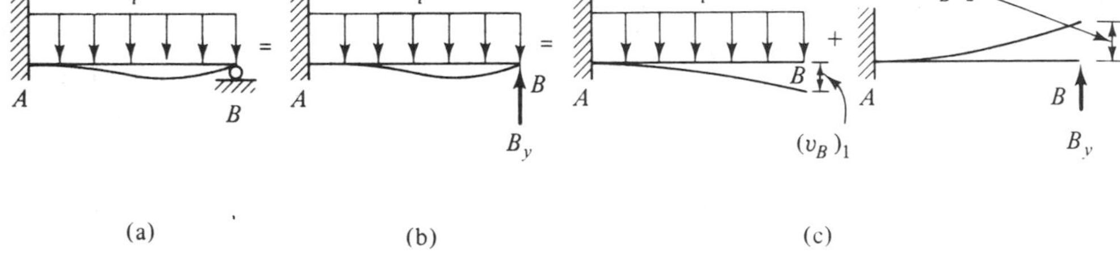

Figure 8.31

Both $(v_B)_1$ and $(v_B)_2$ are given in Table A.13 (cases 1 and 3). Now $(v_B)_1$ involves the applied loading and $(v_B)_2$ involves the unknown reaction B_y. Thus, B_y can be determined from the compatability condition. The other reactions can then be determined from the equations of equilibrium. Once the reactions are known, any other quantities of interest, such as bending moments, stresses, or deflections, can be found.

Example 8.16 Deflection of a Simple Beam. Determine the midspan deflection of the beam shown in Figure 8.32(a).

Solution. Decomposing the loading as shown in Figure 8.32(b), we have

$$v_C = (v_C)_1 - (v_C)_2$$

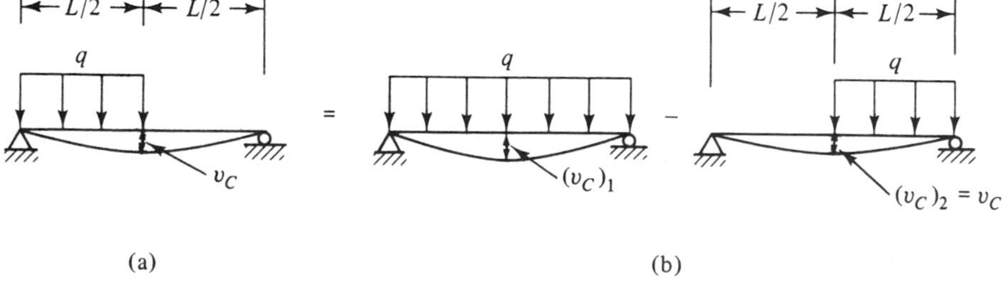

(a) (b)

Figure 8.32

But $(v_C)_2 = v_C$, since the first and last beams are the same except that one is turned end for end with respect to the other. Thus,

$$v_C = \frac{1}{2}(v_C)_1 = -\frac{5qL^4}{768EI} \qquad \textbf{Answer}$$

where the value of $(v_C)_1$ is obtained from Table A.13. The advantage of using superposition for this problem becomes obvious if we consider the amount of work involved in obtaining a solution by integration methods.

Example 8.17 Analysis of an Indeterminate Beam. Compare the maximum bending moment in the beam shown in Figure 8.33(a) with that in a simply supported beam with the same span and loading.

Solution. We remove the fixed support at the right end and replace it by the reactions it exerts [Figure 8.33(b)]. This loading is then broken down as shown in Figure 8.33(c). Referring to Table A.13 and using the conditions that the slope and deflection of the original beam must both be zero at the right end, we have

$$\theta_B - (\theta_B)_1 + (\theta_B)_2 + (\theta_B)_3 = \frac{-FL^2}{8EI} + \frac{B_y L^2}{2EI} - \frac{C_B L}{EI} = 0$$

or

$$-8C_B + L(4B_y - F) = 0$$

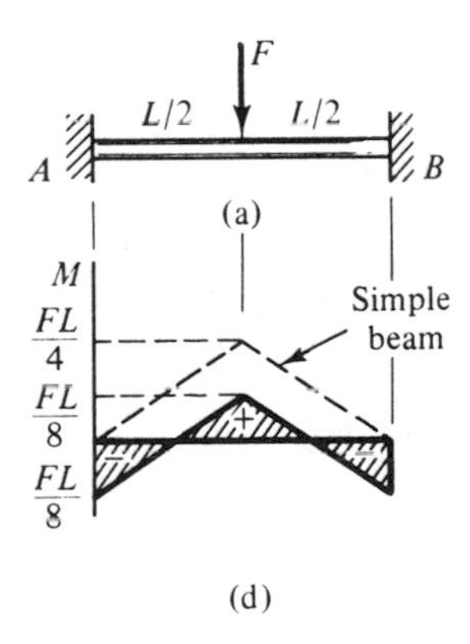

F L/2 L/2 C_B A B B_y = F L/2 L/2 A B + L B A B_y + L C_B A B

(b) (c)

Figure 8.33

and

$$v_B = (v_B)_1 + (v_B)_2 + (v_B)_3 = \frac{-5FL^3}{48EI} + \frac{B_y L^3}{3EI} - \frac{C_B L^2}{2EI} = 0$$

or

$$-24C_B + L(16B_y - 5F) = 0$$

Solving these equations, we obtain

$$B_y = \frac{F}{2} \qquad C_B = \frac{FL}{8}$$

A force analysis of the beam shows that the reactions at the left end have the same values. Thus, the moment diagram is as shown in Figure 8.33(d). The diagram for the corresponding simply supported beam is indicated by the dashed lines. These diagrams show that the maximum bending moment in the indeterminate beam is only one-half that in the simple beam. Also, since the moments in the indeterminate beam are more nearly uniform over the length, the material in the member is utilized more effectively. It is also of interest to note that the maximum deflection of the indeterminate beam is only one-fourth that of the corresponding simple beam [$-FL^3/(192EI)$ compared to $-FL^3/(48EI)$].

These results illustrate the reason for making a member indeterminate, which is to "stiffen" it and reduce the deformations and stresses. Stated another way, the load-carrying capacity of a member can be increased by making it statically indeterminate.

The advantages of using superposition for this problem are not so obvious. In fact, the solution could probably have been obtained with less effort by using singularity functions and integration. In such cases, the choice of solution method becomes a matter of personal preference.

PROBLEMS

8.92. A vertical post is subjected to two lateral forces, as shown. For what ratio F_2/F_1 of the forces will the deflection at the top of the post be zero? For what ratio of the forces will the slope be zero at the top?

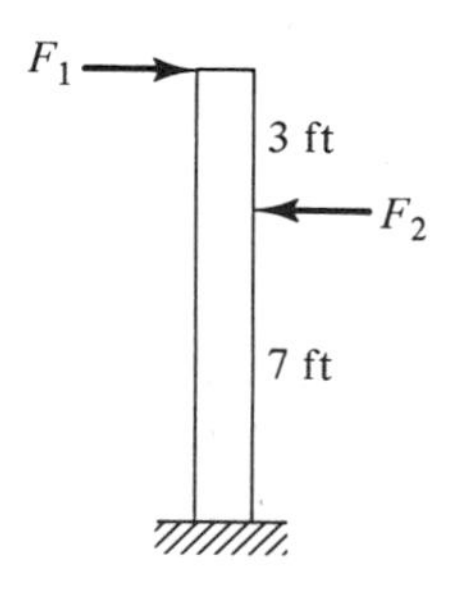

Problem 8.92

8.93. Determine the reactions at the supports for the beam shown and sketch the moment diagram. By what percentage does the presence of the support at B reduce the maximum bending moment in the member?

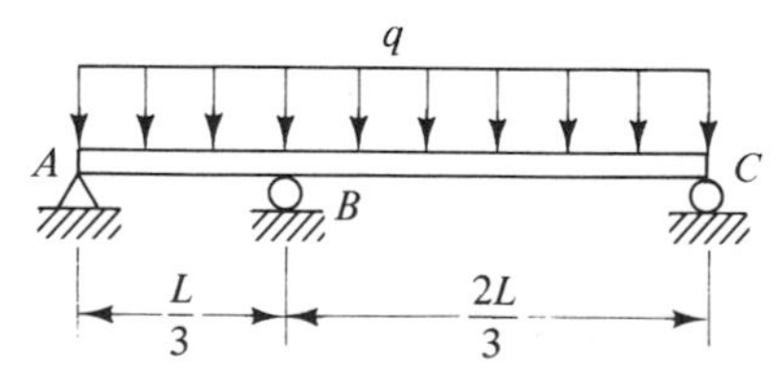

Problem 8.93

8.94. What is the lightest steel Universal beam that will support the loading shown if the allowable bending stress is 60 MN/m^2 and the maximum allowable deflection is 2 mm?

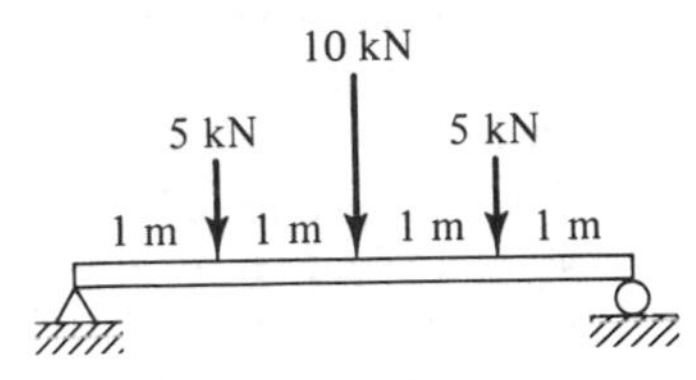

Problem 8.94

8.95. What size square timber is required to support the loading shown if the allowable stresses are 10 MN/m^2 in bending and 1 MN/m^2 in direct shear? How does this compare with the size member that would be required if the support at B were not present?

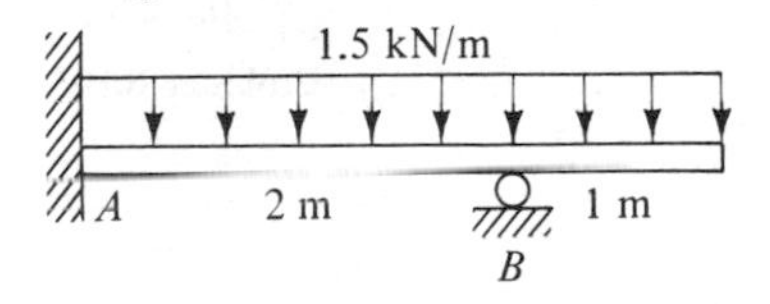

Problem 8.95

8.96. If the beam in Problem 8.95 is made of a 127-mm × 64-mm aluminum channel with the flanges vertical, what is the deflection at the free end? Use $E = 70$ GN/m^2.

8.97. A steel wire is attached to the end of a round steel cantilever beam as shown. What is the largest force F that can be applied to the beam if the maximum allowable tensile stress in the beam and in the wire is 20 ksi?

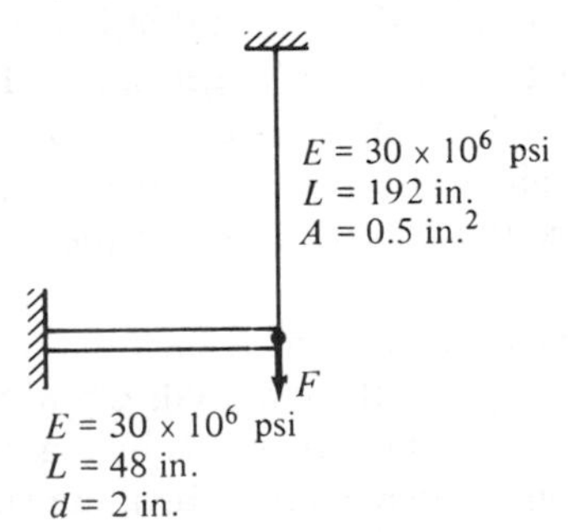

Problem 8.97

8.98. The end of a W 8 × 20 steel beam is connected to a rigid support by an aluminum wire, as shown. What is the maximum bending stress induced in the beam if the temperature of the wire is decreased by 50°F? What is the corresponding tensile stress in the wire?

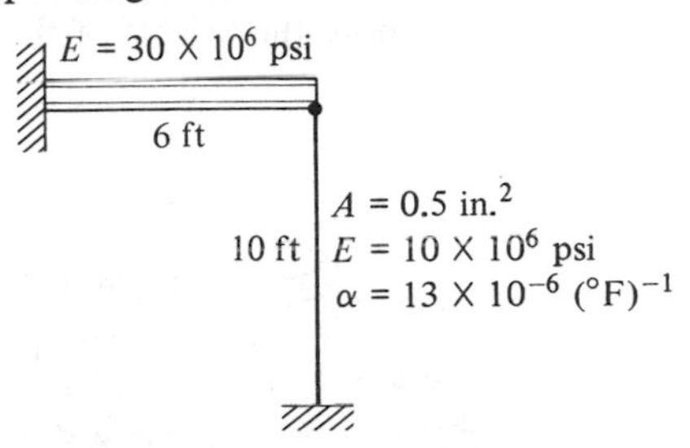

Problem 8.98

8.99. Two identical 20-mm square steel bars are connected at their centers by a 6-mm-diameter steel wire, as shown. A turnbuckle with 0.5 threads/mm is used to remove the slack in the wire. How many additional revolutions of the turnbuckle are possible before the onset of yielding? Which will yield first, the wire or the bars? Use $\sigma_{YP} = 300$ MN/m^2.

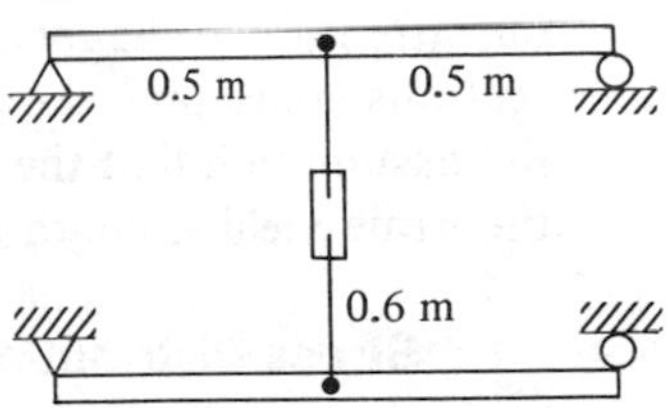

Problem 8.99

8.100. The two beams shown have the same value of EI and are just touching when un-

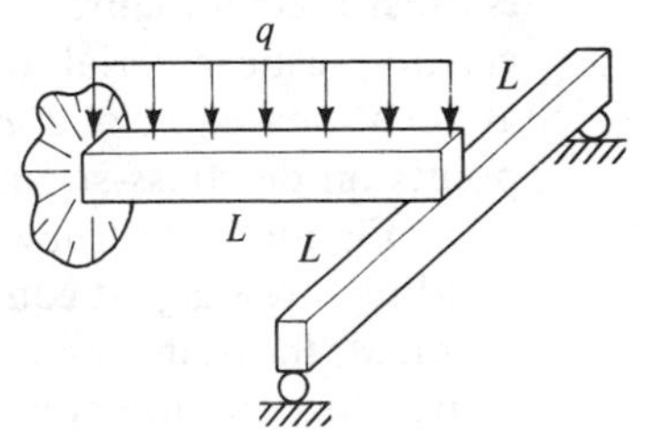

Problem 8.100

loaded. Determine the contact force between the beams and the deflection at the point of contact after the load is applied.

8.101. Three identical W 6 × 16 steel beams are arranged as shown. What is the deflection at the point of load application if $F = 10$ kips? Ignore the weights of the members.

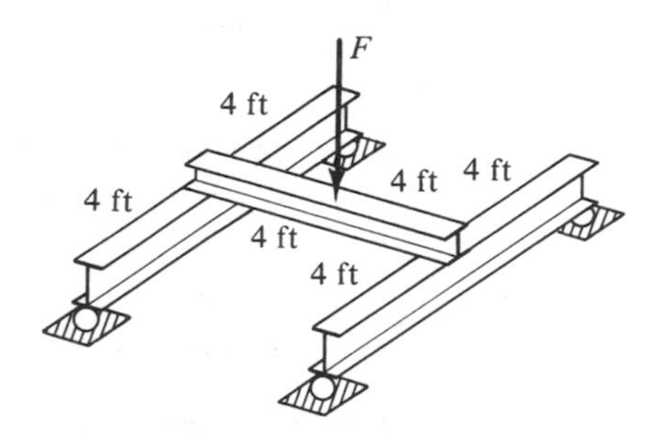

Problem 8.101

8.102. What is the maximum load F that can be applied to the arrangement of beams in Problem 8.101 if the maximum deflection is not to exceed 0.5 in.? Ignore the weights of the members and assume elastic behavior.

8.103. The L-shaped member shown is made of a 40-mm diameter steel rod and lies in a horizontal plane. Determine the vertical deflection at the free end.

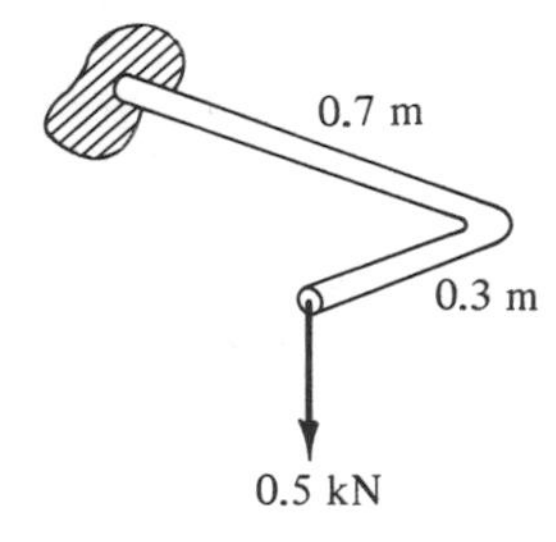

Problem 8.103

8.10 INELASTIC RESPONSE OF BEAMS

Now that we have considered the linearly elastic response of beams, we shall turn our attention to the inelastic response. All of the assumptions made in the preceding sections concerning the geometry and loading of the beam still apply, in addition to the assumption that the material behavior is essentially elastic-perfectly plastic with the same yield strength in tension and compression.

Stress Distribution

As the beam undergoes inelastic deformations, the bending stress distribution tends to become uniform over the cross section, and the moment-stress and moment-curvature relations derived in the preceding sections no longer apply. The situation is illustrated in Figure 8.34, which shows the bending stresses in a beam made of an elastic-plastic material for three different values of bending moment. The conditions at locations a, b, and c on the cross section are denoted by the corresponding points on the stress-strain diagram.

Figure 8.34(a) shows the situation when the moment is such that the maximum bending stress is just equal to the yield strength of the material. Up to this value of moment, the beam response is linearly elastic. If the moment is increased above this value, the curvature and strains increase, while the maximum bending stress remains equal to the yield strength. Since strains at points away from the top and bottom of

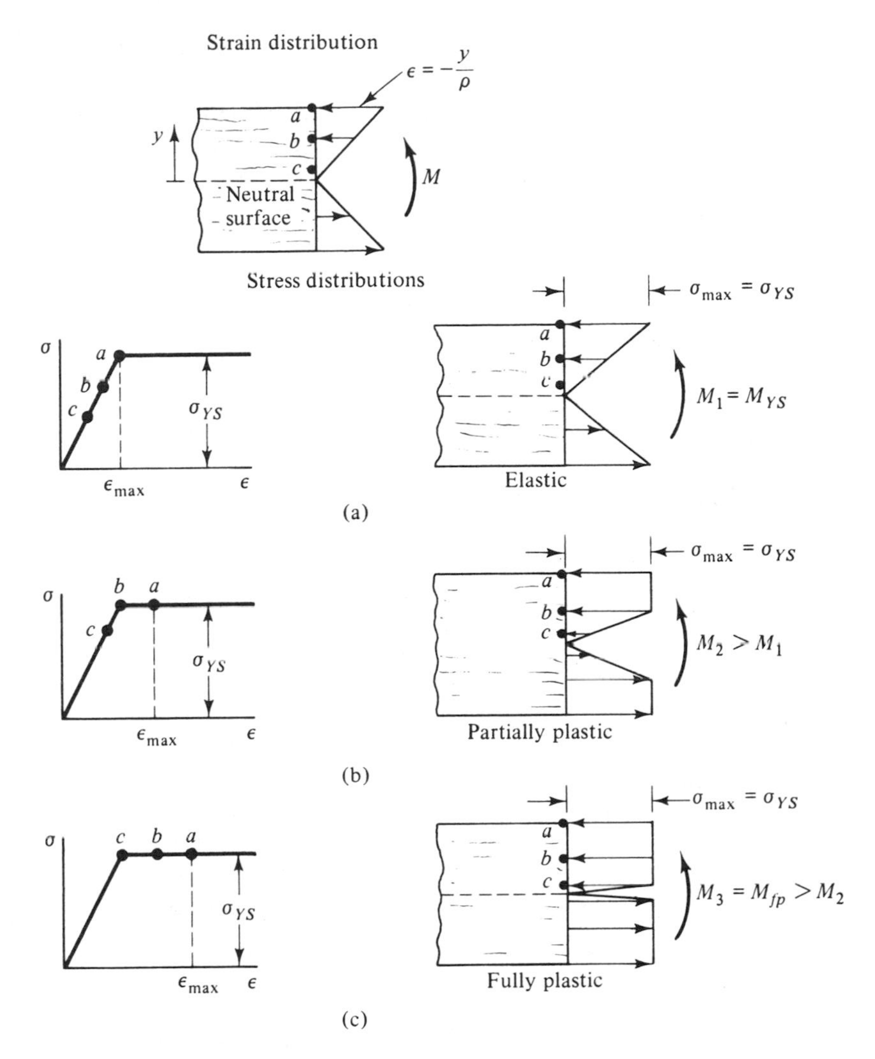

Figure 8.34

the cross section are now at or above the proportional limit strain, we have the partially plastic stress distribution shown in Figure 8.34(b). Note that the stresses in the elastic central portion of the beam remain linearly distributed.

If the bending moment continues to increase, a situation is eventually reached where the strains are at or above the proportional limit strain everywhere over the cross section, except near the bending axis where the strain is always zero. The stresses are then very nearly uniform over the entire cross section, as shown in Figure 8.34(c). The bending moment at which this fully plastic condition is reached is called the *fully plastic moment* and will be denoted by M_{fp}. It is customary to assume that the stresses are uniform over the entire cross section in the fully plastic case. This assumption is reasonable, as the stresses adjacent to the bending axis contribute little to the bending moment.

The location of the bending axis in the partially plastic and fully plastic cases is determined from the requirement that there be no net normal force on the cross section. In other words, the net tensile force acting upon one part of the cross section must balance the net compressive force acting upon the other part of the cross section. For the fully plastic case in which the stresses are uniformly distributed and the tensile stress is equal to the compressive stress, this condition implies that the area in tension must equal the area in compression. That is, the bending axis in the fully plastic case is located such that it divides the cross-sectional area in half. If the cross section has a horizontal axis of symmetry, as in the case of an I-beam or rectangular section, the bending axis will pass through the centroid. For unsymmetrical sections, such as a T-beam, it will not.

Fully Plastic Moment

Once the bending axis has been located, the fully plastic moment can be determined from the moment condition for static equivalence given in Eq. (8.8):

$$M = -\int y\sigma \, dA$$

However, it is simpler to determine it from first principles by simply computing the moment of the couple due to the resultant tensile and compressive forces acting upon each half of the cross section.

For example, consider a beam with a rectangular cross section (Figure 8.35). The magnitude of the net tensile and compressive forces acting is

$$F_T = F_C = \sigma_{YS}\left(\frac{bh}{2}\right)$$

and the distance between them is $h/2$. Thus,

$$M_{fp} = F_T\left(\frac{h}{2}\right) = \sigma_{YS}\left(\frac{bh^2}{4}\right) \tag{8.23}$$

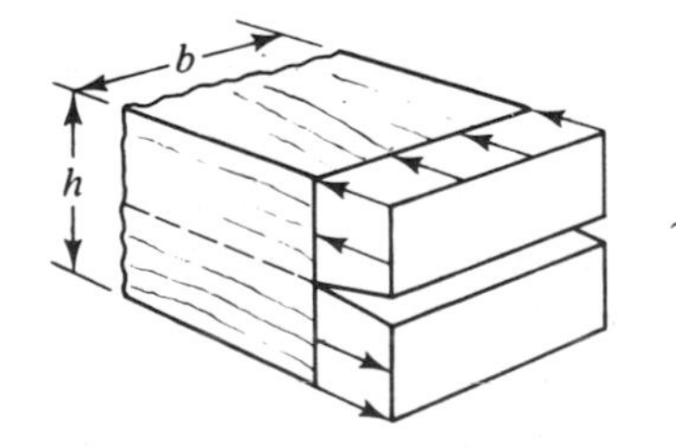

~
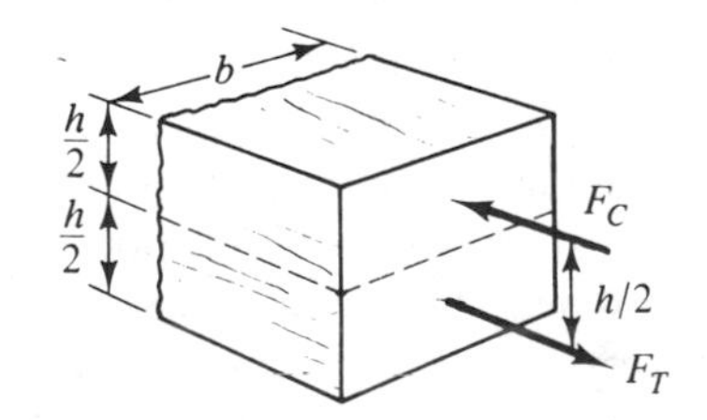

~
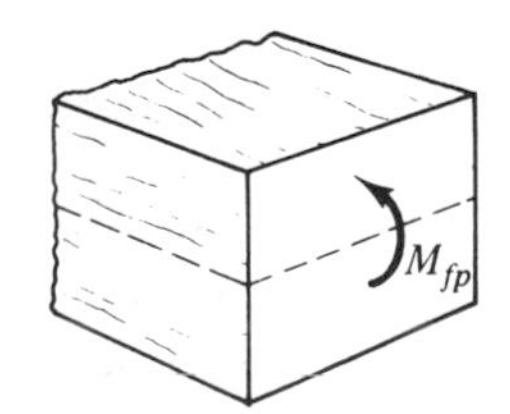

Figure 8.35

From Eq. (8.10), the moment M_{YS} at which yielding first occurs is

$$M_{YS} = \frac{\sigma_{YS} I}{(h/2)} = \sigma_{YS}\left(\frac{bh^2}{6}\right) \tag{8.24}$$

Comparing these equations, we find that for a rectangular section

$$M_{fp} = \frac{3}{2} M_{YS} \tag{8.25}$$

For cross sections with other shapes, the ratio between the two moments will be different.

These various bending moments and the different ranges of behavior associated with them are illustrated in Figure 8.36, which shows a portion of the experimentally determined moment-deflection curve for a mild steel beam. The point at which the curve first starts to deviate from a straight line corresponds to the onset of yielding. This point is difficult to locate accurately because the deviation from linearity is initially very slight. The horizontal portion of the curve corresponds to the fully plastic condition. Points in between these two extremes correspond to partially plastic states.

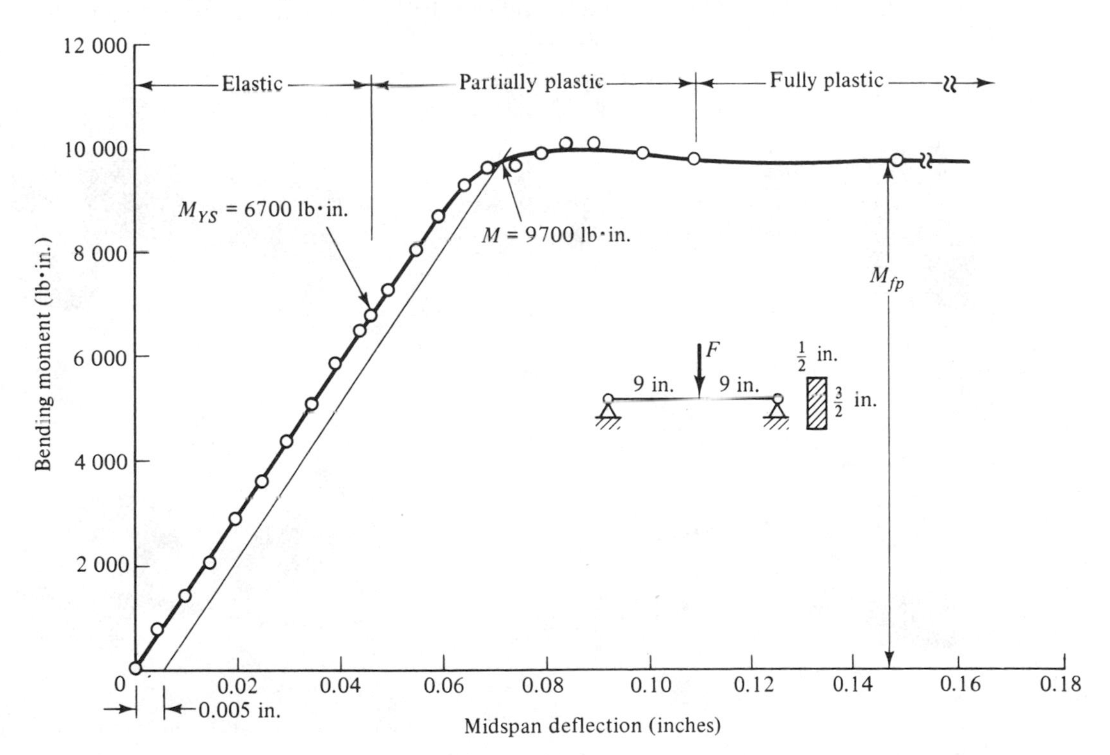

Figure 8.36 Moment-Deflection Relation for a Mild Steel Beam (Initial Portion Only).

The test specimen used to obtain this curve is shown in Figure 8.37(a). The flaking of the mill scale evident at its center indicates the regions over which the material has yielded. Note that even though the member was severely deformed, it did not break and could still support a load. Compare this with the behavior of a cast iron beam [Figure 8.37(b)], that suddenly breaks when the maximum tensile stress reaches the ultimate tensile strength of the material.

It is advantageous to allow some permanent deformation whenever possible, as can be seen from Figure 8.36. If no permanent deflection is allowed, the maximum bending moment that can be supported is M_{YS}, which for the particular beam tested is 6,700 lb · in. However, if a permanent midspan deflection of 0.005 in. is allowed, a moment of 9,700 lb · in. can be supported. This is an increase of 45%. The procedure for determining the latter value of moment is the same as that for finding the yield strength of a material.

There is, of course, a limit to the increase in load-carrying capacity that can be achieved by allowing permanent deformations. Once the fully plastic condition is reached, there is little or no additional resistance to bending at that section, and the beam behaves much as if it were hinged there. Consequently, the beam is said to have developed a *plastic hinge* at the location of the fully plastic moment. The formation of a plastic hinge can lead to large deflections with little or no increase in

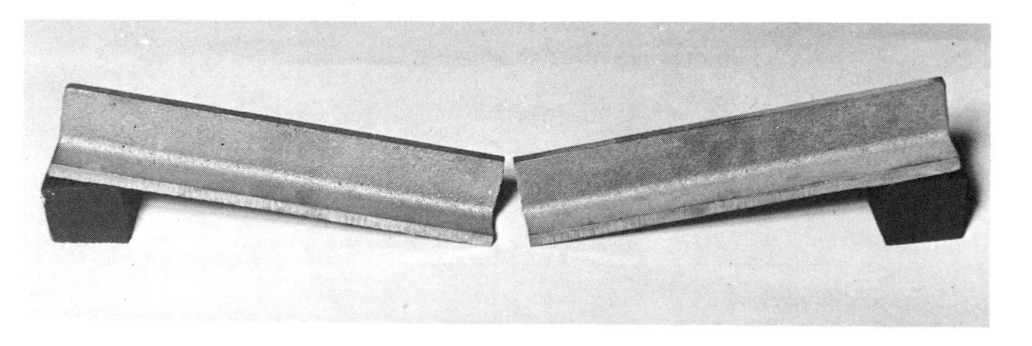

Figure 8.37 Response of (a) Mild Steel; and (b) Cast Iron Beams when Overloaded.

load, as indicated by the horizontal portion of the moment-deflection curve in Figure 8.36. Thus, a beam made of ductile material usually fails by general yielding once the fully plastic condition is reached, unless the deflection is somehow restricted.

Example 8.18. Fully Plastic Moment for a T-Beam. Determine the fully plastic moment for a beam made of 2024-T3 aluminum with the cross section shown in Figure 8.38(a).

Solution. We first locate the bending axis, which divides the cross-sectional area in half in the fully plastic case. The total area is

$$A = (0.05 \text{ m})(0.16 \text{ m}) + (0.05 \text{ m})(0.20 \text{ m}) = 0.018 \text{ m}^2$$

Therefore, $A/2 = 0.009 \text{ m}^2$. Thus, the bending axis is located 180 mm above the bottom of the section. The stress distribution is as shown in Figure 8.38(b). From Table A.2, the yield strength of the material is found to be 345 MN/m^2.

We now replace the stresses acting over each portion of the cross section by their resultants. This gives the three forces shown in Figure 8.38(c), where

$$F_1 = \sigma_{YS} A_1 = (345 \times 10^6 \text{ N/m}^2)(0.008 \text{ m}^2) = 2.76 \times 10^6 \text{ N}$$

$$F_2 = \sigma_{YS} A_2 = (345 \times 10^6 \text{ N/m}^2)(0.001 \text{ m}^2) = 0.35 \times 10^6 \text{ N}$$

$$F_3 = \sigma_{YS} A_3 = (345 \times 10^6 \text{ N/m}^2)(0.009 \text{ m}^2) = 3.11 \times 10^6 \text{ N}$$

These three forces are statically equivalent to the fully plastic moment. Taking moments about the bending axis (point O), we have from the condition $(\Sigma M_O)_{\text{I}} = (\Sigma M_O)_{\text{II}}$

$$M_{fp} = (2.76 \times 10^6 \text{ N})(0.045 \text{ m}) + (0.35 \times 10^6 \text{ N})(0.010 \text{ m})$$
$$+ (3.11 \times 10^6 \text{ N})(0.090 \text{ m})$$
$$= 408 \times 10^3 \text{ N} \cdot \text{m} \quad \text{or} \quad 408 \text{ kN} \cdot \text{m} \qquad \textbf{Answer}$$

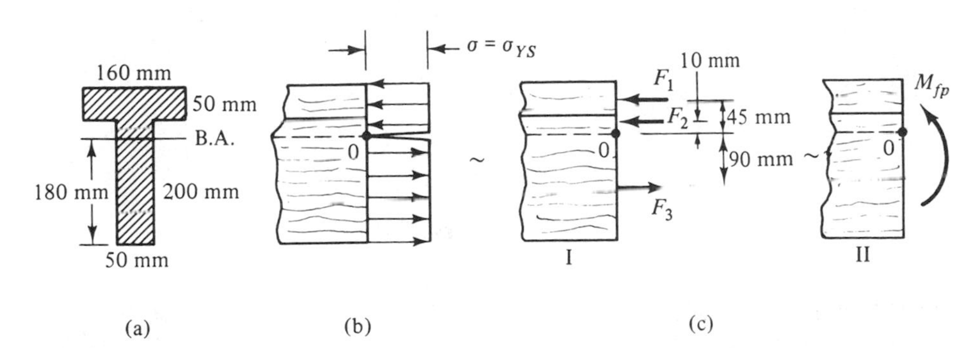

Figure 8.38

Example 8.19. Failure Load for a Cantilever Beam. The cantilever beam shown in Figure 8.39(a) is made of structural steel with a yield strength of 36 ksi. What is the maximum load F that can be supported with a safety factor of 2 against failure by general yielding?

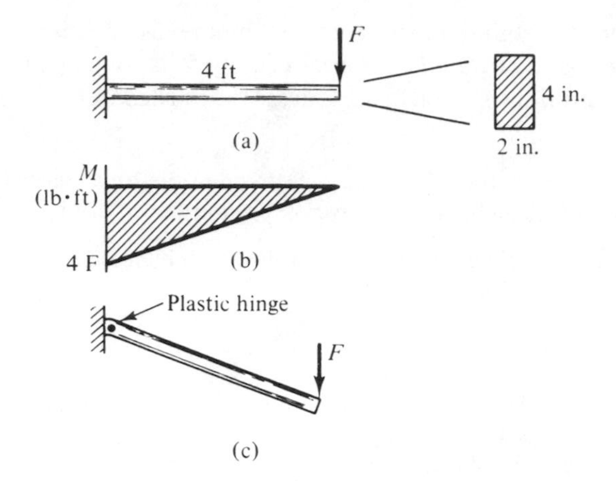

Figure 8.39

Solution. The beam will fail by general yielding once a fully plastic condition is reached somewhere along its length. From the moment diagram [Figure 8.39(b)], we see that the bending moment is largest at the wall. Thus, a plastic hinge will form there when the bending moment becomes equal to the fully plastic moment, allowing the beam to deflect approximately as indicated in Figure 8.39(c).

The fully plastic moment is computed from Eq. (8.23):

$$M_{fp} = \sigma_{YS}\left(\frac{bh^2}{4}\right) = \frac{(36 \times 10^3 \text{ lb/in.}^2)(2 \text{ in.})(4 \text{ in.})^2}{4} = 288 \times 10^3 \text{ lb} \cdot \text{in.}$$

Setting this moment equal to the bending moment at the wall, we obtain for the failure load

$$48F \text{ lb} \cdot \text{in.} = M_{fp} = 288 \times 10^3 \text{ lb} \cdot \text{in.}$$

$$F = 6{,}000 \text{ lb}$$

The working load is the failure load divided by the safety factor. Thus,

$$F_{\text{working}} = 3{,}000 \text{ lb} \qquad \textbf{Answer}$$

PROBLEMS

8.104. to 8.106. Determine the fully plastic moment M_{fp} for a beam with the cross section shown. Express the results in terms of the yield strength σ_{YS} and the dimensions given.

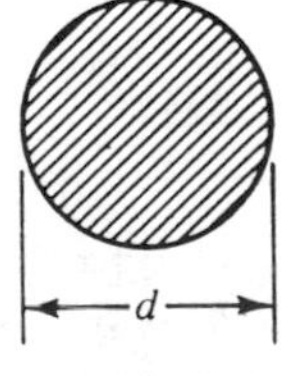

Problem 8.104

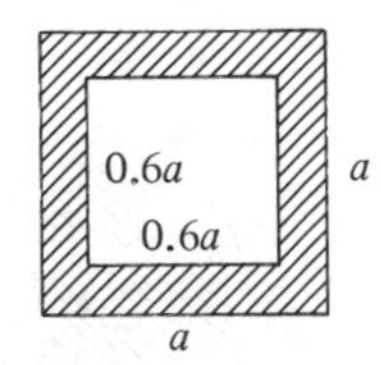

Problem 8.105

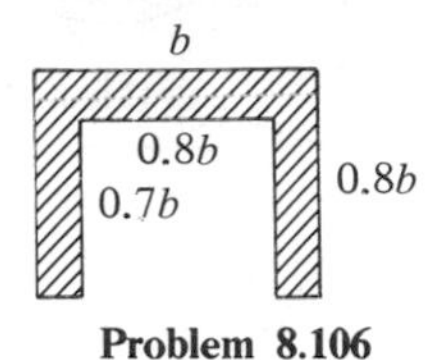

Problem 8.106

8.107. Determine the fully plastic moment for a beam with the cross section shown in Problem 8.46 (Section 8.5).

8.108. and 8.109 For the beam shown, determine the largest load that can be applied without causing collapse. The yield strength of the material and the required safety factor, if any, are as indicated in the figure.

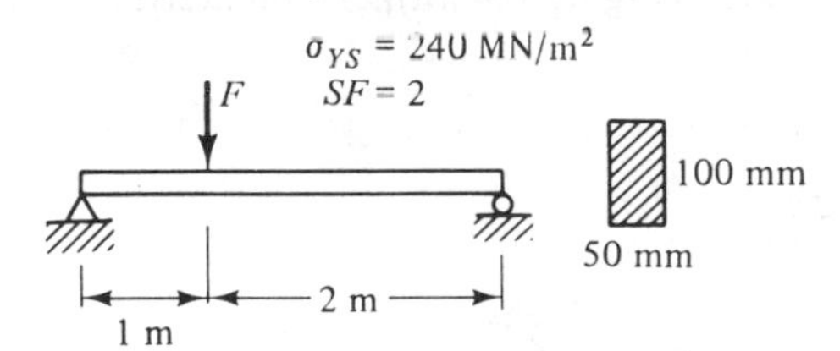

Problem 8.108

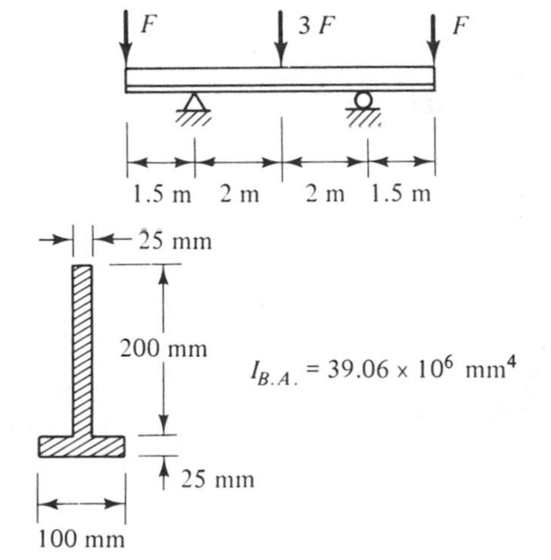

Problem 8.109

8.110. The solid circular beam shown is made of 2024-T3 aluminum. What diameter d of cross section is required to provide a safety factor of 2 against failure by collapse?

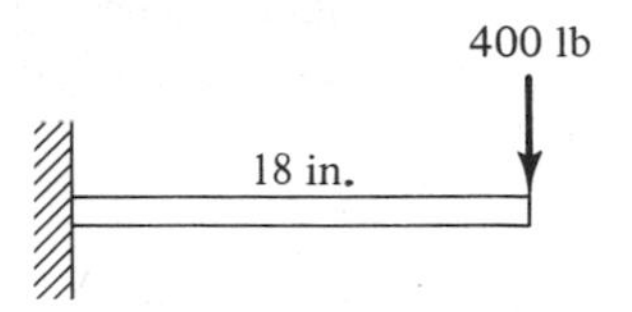

Problem 8.110

8.111. The beam shown is 2-in. wide by 5-in. deep. By what percentage does the presence of the semicircular grooves reduce the collapse load if (a) $d = L/4$ and (b) $d = 0.4L$? $\sigma_{YS} = 36$ ksi.

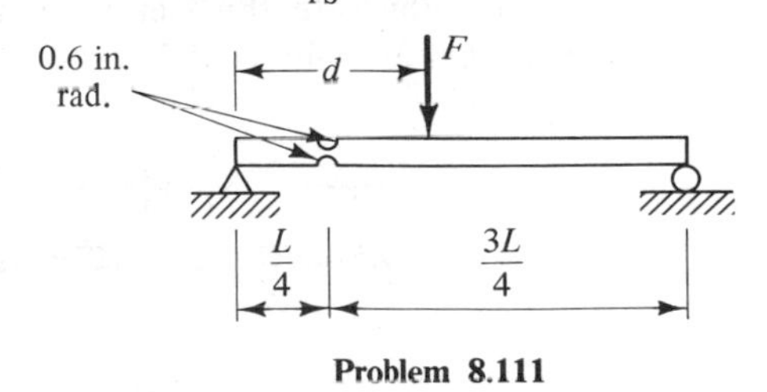

Problem 8.111

8.112. The beam shown is 1-in. thick and is made of structural steel. To lighten the member, the cross sectional area of the right half is reduced as shown. What is the minimum allowable dimension d if the collapse load for the beam is not to be reduced below its original value?

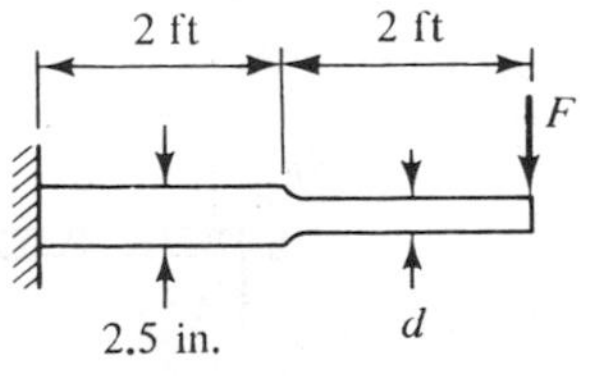

Problem 8.112

8.113. A rectangular beam with a cross section 2-in. wide and 4-in. deep is made of mild steel with a yield point of 30 ksi. The beam is subjected to a vertical loading until the maximum bending strain is 2,000 $\mu\varepsilon$. What is the value of the bending moment?

8.114. What magnitude of couple C is required to bend a 30-mm wide by 6-mm thick brass strip around a rigid cylinder, as shown?

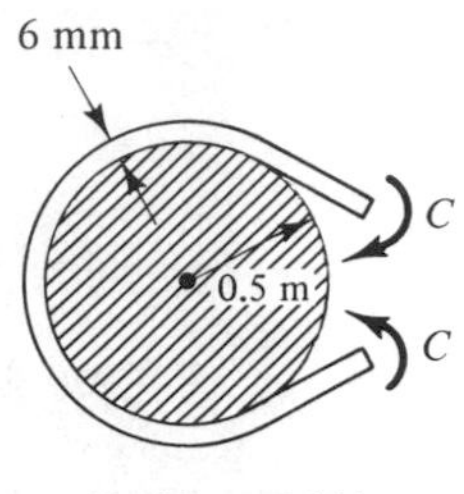

Problem 8.114

8.11 LIMIT LOADS

As indicated in the preceding section, a beam made of a ductile material usually fails by general yielding and the associated large deflections when a fully plastic condition is reached. If the member is statically indeterminate, however, the deflections are restricted by the additional support conditions, and further loading is possible.

To illustrate this, let us consider the example of a propped cantilever that supports a concentrated force F at midspan [Figure 8.40(a)]. The material behavior

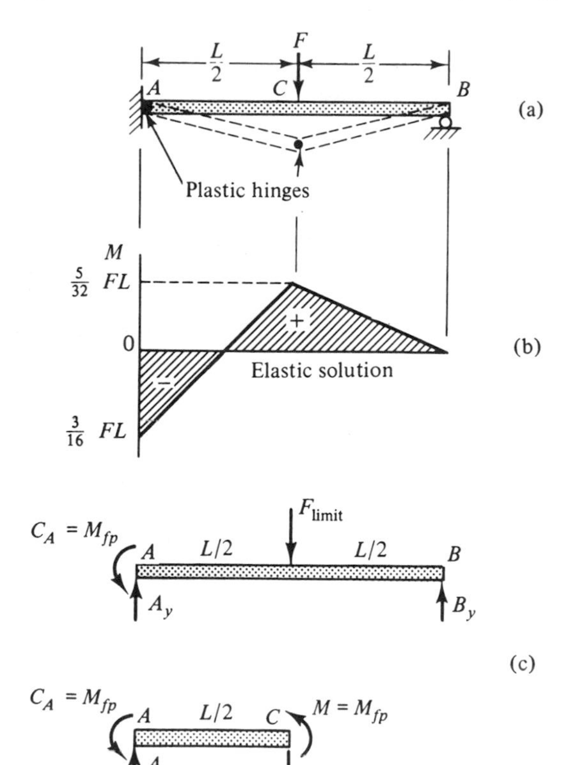

Figure 8.40

is assumed to be elastic-perfectly plastic. Now, plastic hinges form at the locations of maximum bending moment, which for a beam subjected to concentrated loads are always either at the loads or at the supports. Thus, plastic hinges can form at the wall and under the load.

An elastic analysis yields the moment diagram shown in Figure 8.40(b), which indicates that the largest bending moment occurs at the wall. Thus, the fully plastic condition will first be reached at this section. However, the formation of a plastic hinge at the wall does not lead to a collapse of the member because its deflection is restrained by the roller under the right end. Any further increases in load must be supported by sections away from the wall because the bending moment at the wall will remain equal to the fully plastic moment. Eventually, a second plastic hinge will form under the load, allowing the member to undergo large deflections. The resulting deformation pattern (collapse mechanism) is indicated by the dashed lines in Figure 8.40(a).

The limit load at which collapse occurs can be determined solely from equilibrium considerations because the bending moment at the plastic hinges is known to be equal to the fully plastic moment. This is evident from the free-body diagrams shown in Figure 8.40(c). The first diagram is for the entire beam and the second is for the portion obtained by sectioning the member just to the left of the applied load. There is one very important thing to note here—the moments at the plastic hinges are known quantities and, therefore, must be shown with the proper sense. The sense of these moments can usually be determined by inspection.

From the second diagram in Figure 8.40(c), we have

$$\overset{+}{\curvearrowleft}\sum M_C = M_{fp} + M_{fp} - A_y\left(\frac{L}{2}\right) = 0$$

and from the first diagram, we have

$$\overset{+}{\curvearrowleft}\sum M_B = F_{\text{limit}}\left(\frac{L}{2}\right) - A_y L + M_{fp} = 0$$

Solving these equations, we find that

$$F_{\text{limit}} = \frac{6}{L} M_{fp}$$

where M_{fp} is computed as explained in the preceding section.

This example points out the two steps involved in determining the limit load for beams and beam structures. One is an investigation of the geometry of the problem to determine the possible collapse mechanisms; the other is a force analysis of the beam and beam segments. If there is more than one possible collapse mechanism, the one that actually occurs is the one with the smallest limit load.

For distributed loads, the locations of the plastic hinges away from the supports are unknown. Furthermore, they cannot be determined from a moment diagram obtained from an elastic analysis of the deflections because the results become invalid once yielding occurs. However, we do know that the hinges form at points of maximum bending moment, which are also points of zero shear force for distributed loadings. (Recall from Section 8.3 that $dM/dx = V$.) This fact provides the additional information necessary to locate the plastic hinges. The procedure is illustrated in the following example.

Example 8.20 Limit Load for a Propped Cantilever Beam. Determine the limit load, q_{limit}, for the beam shown in Figure 8.41(a).

Solution. This problem is the same as that for the beam shown in Figure 8.40(a), except that the applied load is distributed instead of concentrated. The collapse mechanism will also be the same, but the location of the plastic hinge between the supports is unknown in this case.

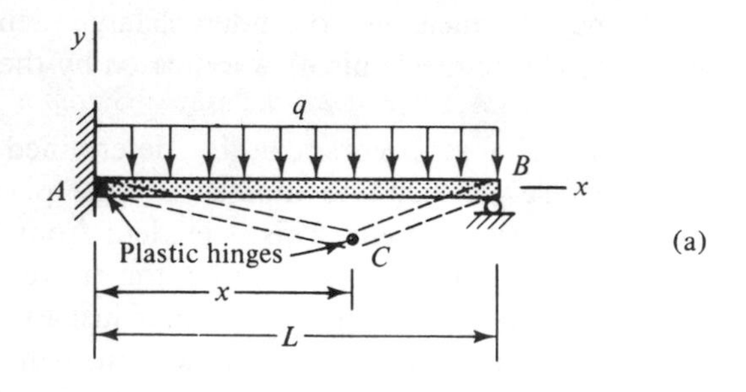

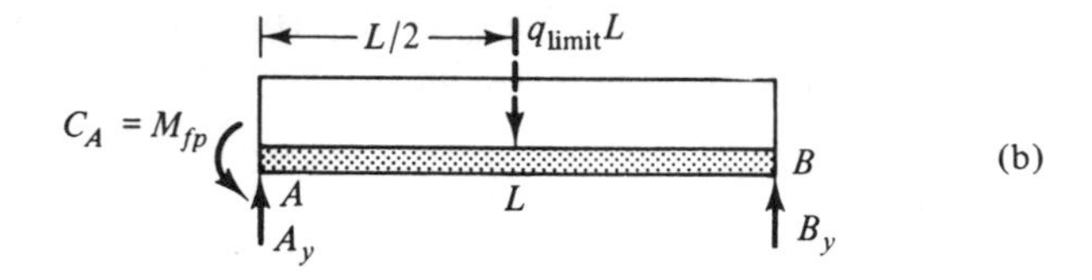

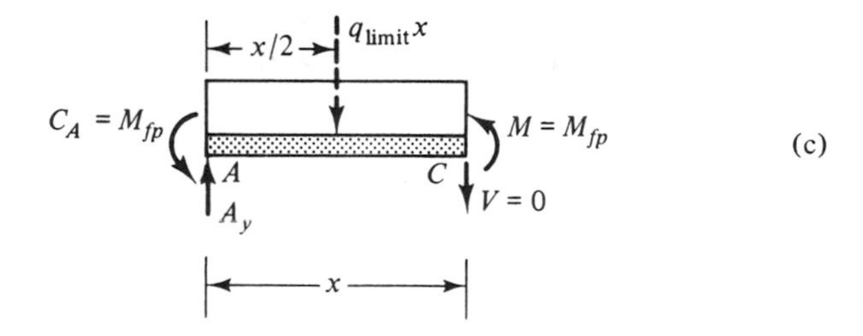

Figure 8.41

Figure 8.41(b) shows the FBD for the entire member. Assuming that the second plastic hinge occurs at a distance x from the wall, and recalling that $V = 0$ and $M = M_{fp}$ at the hinge, we obtain the FBD shown in Figure 8.41(c) for the beam segment to the left of this section. From the first diagram, we have

$$\overset{+}{\curvearrowleft}\sum M_B = q_{\text{limit}} L\left(\frac{L}{2}\right) - A_y L + M_{fp} = 0$$

and from the second diagram, we have

$$\sum F_y = A_y - q_{\text{limit}}x = 0$$

$$\overset{+}{\curvearrowleft}\sum M_C = M_{fp} + q_{\text{limit}}x\left(\frac{x}{2}\right) - A_y x + M_{fp} = 0$$

Solving these equations, we get

$$x = 0.59L \qquad q_{\text{limit}} = 11.65\frac{M_{fp}}{L^2} \qquad \textbf{Answer}$$

where M_{fp} is the value of the fully plastic moment for this particular beam.

Example 8.21 Limit Load for a Beam Structure. The beam shown in Figure 8.42(a) is hinged at the left end and supported at the right by a vertical rod. Determine the limit load for this structure. Both members are made of the same material and have a yield strength of 30 ksi.

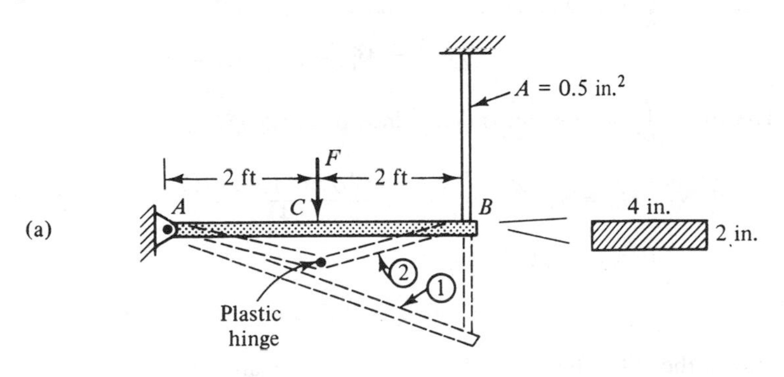

(b) A, 2 ft, C, F_{limit}, 2 ft, F_{fp}, B, A_y

(c) A, 2 ft, C, F_{limit}, 2 ft, F_B, B, A_y

A, 2 ft, C, $M = M_{fp}$, A_y, V

Figure 8.42

Solution. There are two possible collapse mechanisms, as indicated in Figure 8.42(a). The rod can yield, allowing the beam to rotate downward, or the rod can remain elastic with limited deformations and a plastic hinge can form in the beam under the load.

If the rod yields, the force on it will be

$$F_{fp} = \sigma_{YS} A_{\text{rod}} = (30{,}000 \text{ lb/in.}^2)(0.5 \text{ in.}^2) = 15{,}000 \text{ lb}$$

The FBD of the beam for this case is shown in Figure 8.42(b). We have

$$\overset{+}{\curvearrowleft} \sum M_A = -F_{\text{limit}}(2 \text{ ft}) + F_{fp}(4 \text{ ft}) = 0$$

$$F_{\text{limit}} = 2F_{fp} = 30{,}000 \text{ lb}$$

We now consider the second collapse mechanism. The FBDs of the beam and beam segment to the left of the load are shown in Figure 8.42(c). We have

$$\overset{+}{\curvearrowleft} \sum M_B = F_{\text{limit}}(2 \text{ ft}) - A_y(4 \text{ ft}) = 0$$

$$\overset{+}{\curvearrowleft} \sum M_C = M_{fp} - A_y(2 \text{ ft}) = 0$$

The fully plastic moment is determined from Eq. (8.23):

$$M_{fp} = \sigma_{YS}\frac{bh^2}{4} = \frac{(30{,}000 \text{ lb/in.}^2)(4 \text{ in.})(2 \text{ in.})^2}{4} = 120{,}000 \text{ lb} \cdot \text{in.}$$

Solving these equations, we get

$$F_{\text{limit}} = 10{,}000 \text{ lb} \qquad \textbf{Answer}$$

This is the actual limit load because it is the smaller of the two values. Thus, failure occurs due to formation of a plastic hinge under the load.

PROBLEMS

8.115. to 8.118 Identify and sketch the possible collapse mechanisms and determine the limit load for the beam or beam structure shown. The cross-sectional dimensions and yield strength of the material are as indicated; otherwise, express the results in terms of the fully plastic moment M_{fp} for the member.

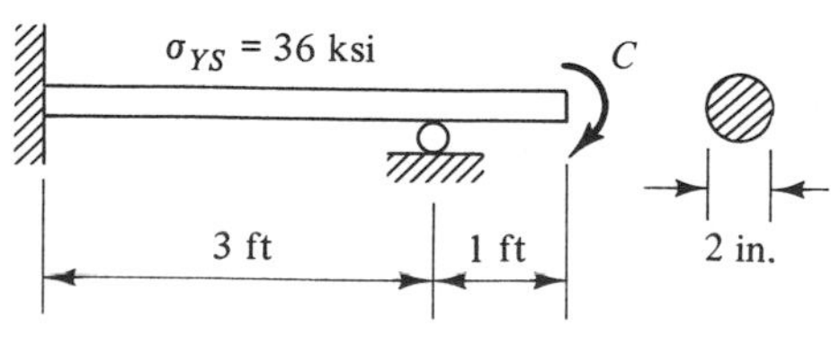

Problem 8.115

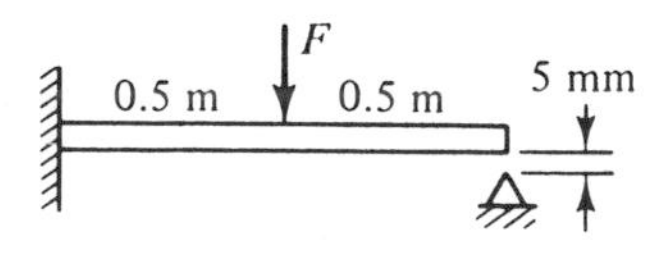

Problem 8.116

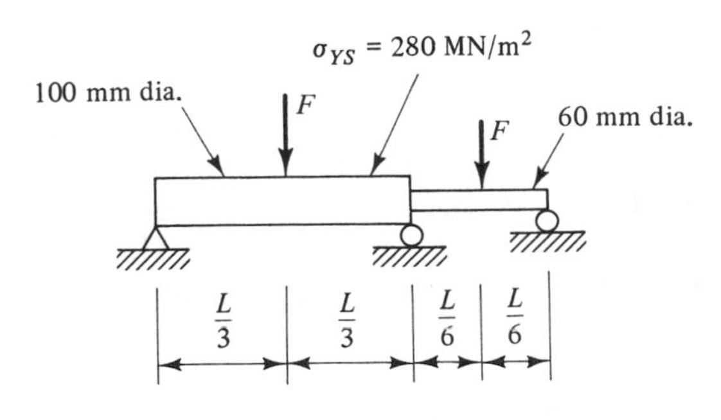

Problem 8.117

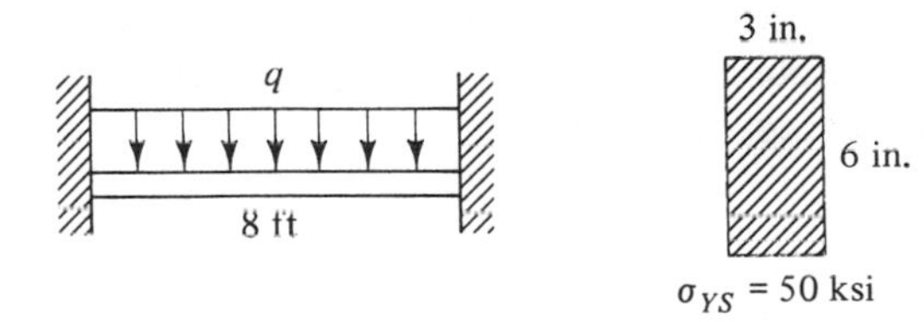

Problem 8.118

8.119. Identify and sketch the possible collapse mechanisms and determine the limit load for the structure in Problem 8.93 (Section 8.9). Express the results in terms of the fully plastic moment M_{fp} for the member.

8.120. Repeat Problem 8.119 for the beam in Problem 8.77 (Section 8.7).

8.121. A 2-in. square cantilever beam made of 2024-T3 aluminum is supported at its midpoint by a $\frac{1}{4}$-in. diameter structural steel wire, as shown. Identify the possible collapse mechanisms for the structure and determine the limit load.

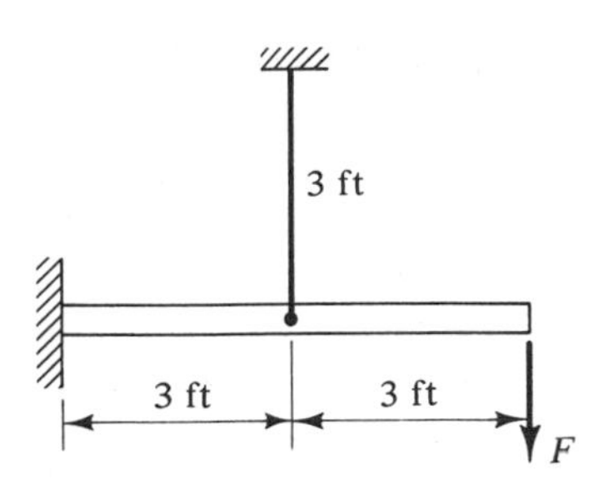

Problem 8.121

8.122. Two identical steel I-beams are arranged as shown. Identify the possible collapse mechanisms and determine the limit load. Express the results in terms of the fully plastic moment M_{fp} for the members.

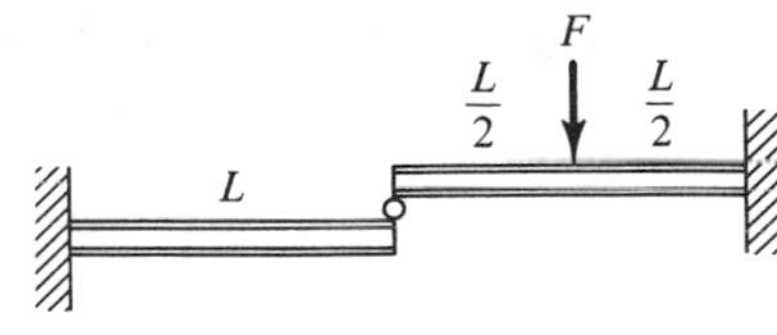

Problem 8.122

8.123. The right end of the steel beam shown rests on a roller and the left end is connected to two short pieces of 2-in. (nominal) steel pipe. The left end is prevented from deflecting vertically by a roller out of view beneath the beam. What is the limit load for this member? Use σ_{YS} = 60 ksi and τ_{YS} = 36 ksi.

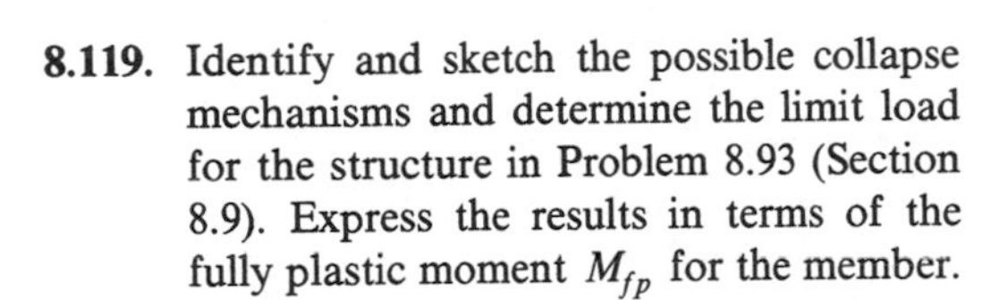

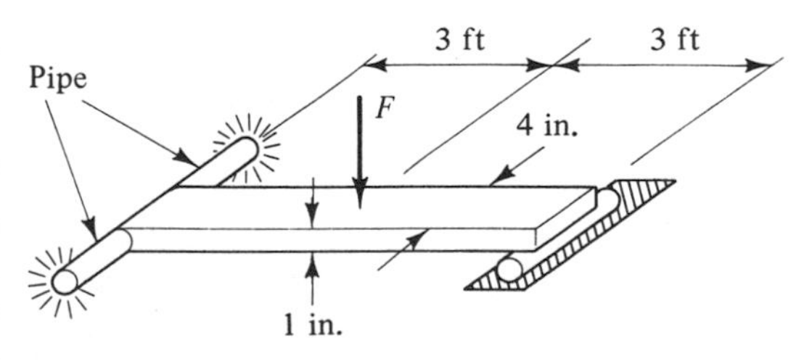

Problem 8.123

8.12 CLOSURE

In this chapter we considered the elastic and inelastic responses of beams, which are probably the most common type of load-carrying member. The analysis was more complicated than for axially loaded members or shafts, but for a very good reason. A review of the chapter shows that the loadings on beams are generally more complex and of greater variety than those for other members. Also, many different cross-sectional shapes were encountered. This is in contrast to shafts, where only circular members were considered.

Although we assumed that the beams are straight before loading, the results obtained also apply to members that are slightly curved. This is fortunate because there is no such thing as a perfectly straight beam. Under certain conditions, the results also apply to members with unsymmetrical cross sections. Furthermore, the basic concepts presented hold for composite members made of several different materials, the most common example of which is steel-reinforced concrete beams. However, a general treatment of composite members, or of members with significant initial curvature or unsymmetrical cross sections, is beyond the scope of this text. These topics are considered in more advanced works.

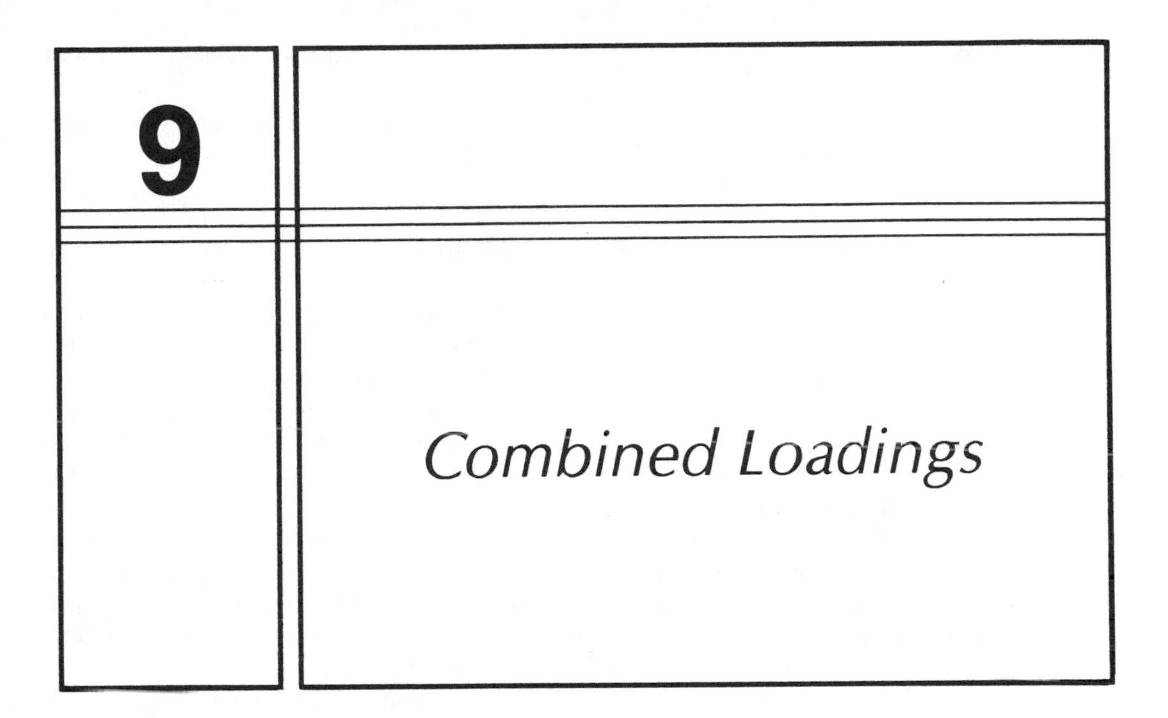

9.1 INTRODUCTION

To this point, we have considered the response of members subjected to the separate effects of axial loadings, torsion, bending, and uniform pressure loadings. Load-stress relations were derived that provide the stresses acting upon certain planes, usually parallel and perpendicular to the longitudinal axis of the member, and equations for computing the deformations were developed.

Although these results enabled us to solve many meaningful problems, several important questions remain unanswered. For example, how does a member respond when several types of loading, say torsion and bending, act simultaneously, as they often do in practice. How does one determine the maximum stresses in the member under such conditions, or how does one determine the stresses on planes other than those for which the basic equations apply? These and other related questions will be considered in this chapter.

Primary emphasis will be placed upon the stresses, as opposed to the deformations, since they are usually of the most concern in combined loading problems. For the most part, we shall consider problems involving only linearly elastic material behavior. An analysis of the inelastic response of members subjected to combined loadings is considerably more complicated than that for a single type of loading and is beyond the scope of an introductory text such as this. Nevertheless, some of the basic results to be presented are independent of the material properties, and, therefore, apply in both the elastic and inelastic cases. These will be pointed out where appropriate.

9.2 STRESSES DUE TO COMBINED LOADINGS

As long as the deformations are small and the material behavior is linearly elastic, the stresses in a member will be proportional to the applied loads. In this case, the stresses due to several different loadings acting simultaneously can be determined by using the principle of superposition (see Section 8.9). The stresses due to the combined loading is simply the combination of those due to each of the individual loadings. Note that this implies that the presence of one loading does not affect the stresses due to another.

It will be convenient to show the various stresses present at a given point in a body as acting upon an infinitesimal element of material surrounding the point, as we have done on several previous occasions. Recall from Section 5.5 that this procedure merely provides a means of displaying the stresses acting upon mutually perpendicular planes through the particular point of interest.

To illustrate the preceding ideas, let us consider a pressurized thin-walled cylinder that is also twisted and subjected to an axial load [Figure 9.1(a)]. The stresses on planes parallel and perpendicular to the longitudinal axis due to each of the loadings can be computed by using the load-stress relations derived in preceding chapters. These stresses are the longitudinal and circumferential normal stresses due to the internal pressure, the longitudinal normal stress due to the axial load, and the shear stresses due to the applied couple. Figure 9.1(b) shows these stresses acting

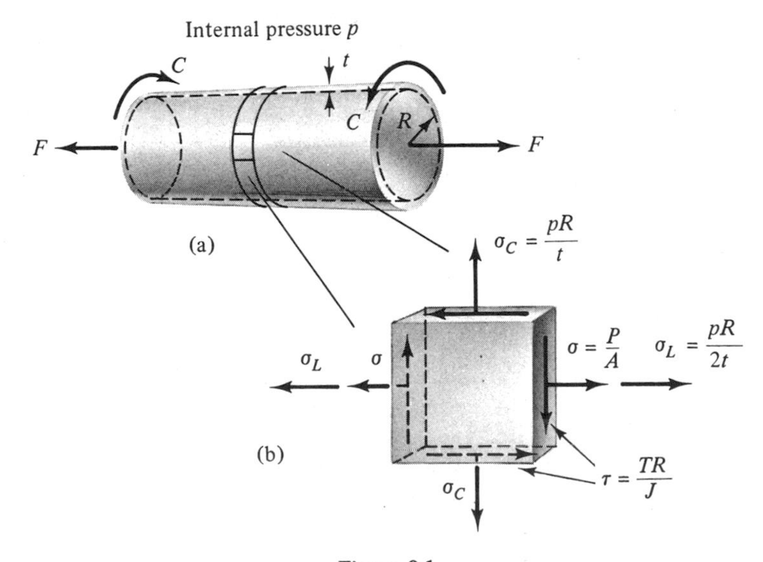

Figure 9.1

upon an element of material taken from the outside of the cylinder wall. The material behavior is assumed to be linearly elastic so that the torque-stress relation $\tau = T\rho/J$ applies.

The question now arises as to how the various stresses acting upon a material element combine. In general, special procedures are required because the stresses not only have magnitude and direction, but they also depend upon the orientation of the plane upon which they act. These procedures will be considered in the following sections. The one exception, which we shall consider in this section, involves like kinds of stresses (normal or shear) acting upon the same plane and in the same direction. In this case, the stresses can be combined algebraically. For example, the two normal stresses acting in the longitudinal direction in Figure 9.1(b) can be added to obtain the total normal stress in that direction.

Problems involving combined bending and axial loading or bending about two axes are probably the most common examples of situations in which the stresses can be added directly. Another common example involves the torsional and direct shear stresses in members subjected to simultaneous bending and torsion. The determination of the stresses in cases such as these involves no new principles, as can be seen from the following examples.

Example 9.1. Stresses In a Linkage. Determine the distribution of stress over the cross section in the central portion of the machine linkage shown in Figure 9.2(a). The member is 20 mm thick and is made of an alloy steel with a yield strength of 690 MN/m^2.

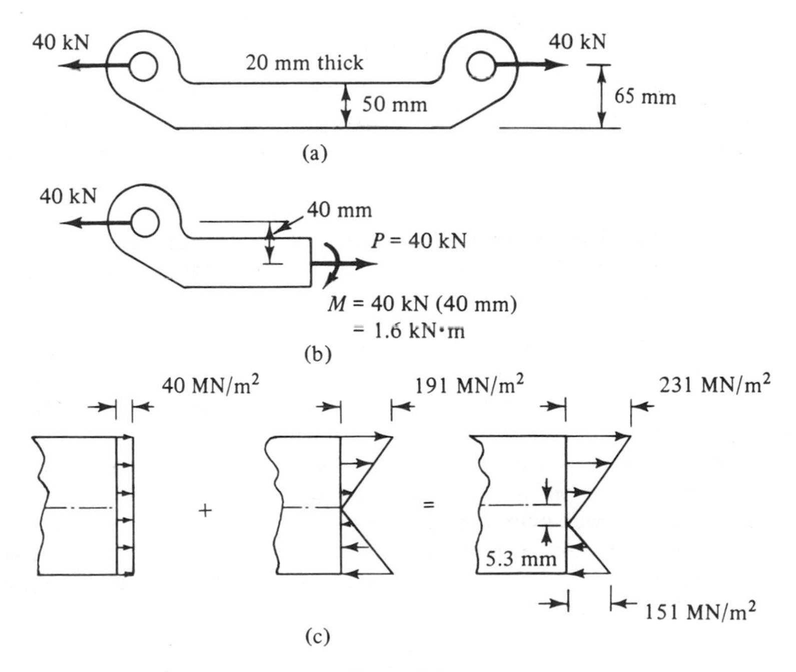

Figure 9.2

Solution. We first section the member and determine the stress resultants, as indicated in Figure 9.2(b). Note that the member is subjected to both axial loading and bending. The bending arises because of the offset between the line of action of the applied loads and the centroidal axis of the member. The bending moment and normal force are constant throughout the central portion of the linkage.

The stress due to the axial loading is

$$\sigma = \frac{P}{A} = \frac{40 \times 10^3 \text{ N}}{(0.05 \text{ m})(0.02 \text{ m})} = 40 \times 10^6 \text{ N/m}^2 \text{ or } 40 \text{ MN/m}^2$$

The moment of inertia about the bending axis is

$$I = \frac{1}{12} bh^3 = \frac{1}{12}(0.02 \text{ m})(0.05 \text{ m})^3 = 2.1 \times 10^{-7} \text{ m}^4$$

therefore, the maximum bending stress is

$$\sigma = \frac{My_{\max}}{I} = \frac{(1.6 \times 10^3 \text{ N} \cdot \text{m})(0.025 \text{ m})}{2.1 \times 10^{-7} \text{ m}^4} = 191 \times 10^6 \text{ N/m}^2 \text{ or } 191 \text{ MN/m}^2$$

Since both stresses act upon the same plane and in the same direction, they can be added directly [Figure 9.2(c)]. The maximum stress is below the yield strength of the material; therefore, all equations used are valid.

Note that the stresses still vary linearly over the depth of the cross section. The location of the level of zero stress can be obtained by setting the expression for the total stress equal to zero and solving for the corresponding value of y:

$$\sigma_{\text{total}} = \frac{P}{A} + \frac{My}{I} = 0$$

$$y = -\frac{PI}{MA} = \frac{-(40 \times 10^3 \text{ N})(2.1 \times 10^{-7} \text{ m}^4)}{(1.6 \times 10^3 \text{ N} \cdot \text{m})(10 \times 10^{-4} \text{ m}^4)} = -5.3 \times 10^{-3} \text{ m or } -5.3 \text{ mm}$$

Thus, the level of zero stress lies 5.3 mm below the centroid of the cross section. Its location can also be determined from Figure 9.2(c) by using similar triangles. The shift of the level of zero stress away from the centroid of the cross section is due to the presence of the normal force.

Example 9.2. Bending About Two Axes. A steel cantilever beam made of a W 8 × 31 shape is loaded as shown in Figure 9.3(a). Determine the maximum tensile and compressive stresses in the member.

Solution. Resolving the applied load into components [Figure 9.3(b)], we see that the vertical component will produce bending with xx as the bending axis and the horizontal component will produce bending with yy as the bending axis. Thus, the member is subjected to bending about two axes. The maximum bending moments occur at the wall and have the values indicated in Figure 9.3(c). The geometric properties of the cross section obtained from Table A.3 of the Appendix are listed in Figure 9.3(b).

Since the maximum bending stresses occur at the extremities of the cross section, the maximum tensile and compressive stresses will occur at one of the four corners of the cross section at the wall. Thus, we need consider only these four points. The maximum stress due to the bending moment M_1 is

$$\sigma = \frac{M_1}{S_{xx}} = \frac{(16{,}000 \text{ lb} \cdot \text{ft})(12 \text{ in./ft})}{27.4 \text{ in.}^3} = 7{,}000 \text{ psi}$$

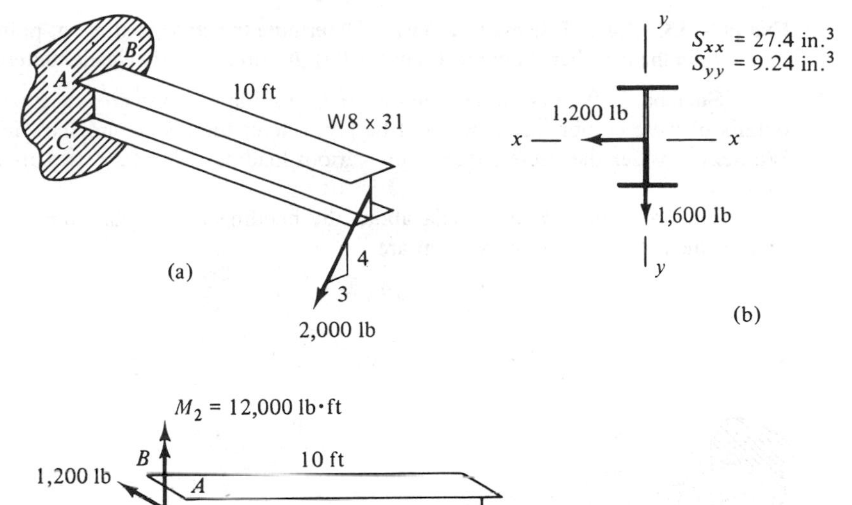

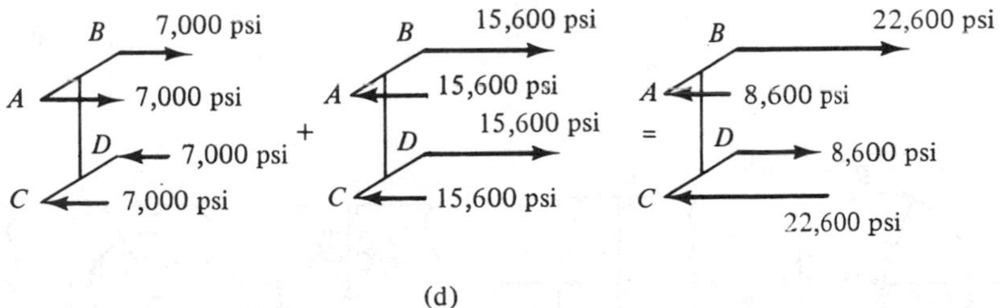

Figure 9.3

and is tensile at the top of the cross section and compressive at the bottom. The moment M_2 produces a tensile stress at points B and D and a compressive stress at points A and C of magnitude

$$\sigma = \frac{M_2}{S_{yy}} = \frac{(12{,}000 \text{ lb} \cdot \text{ft})(12 \text{ in./ft})}{9.24 \text{ in.}^3} = 15{,}600 \text{ psi}$$

Since both sets of stresses act upon the same plane and in the same direction, they can be added directly [Figure 9.3(d)]. The maximum tensile and compressive stresses are equal in magnitude and occur at points B and C, respectively:

$$(\sigma_t)_{\max} = (\sigma_c)_{\max} = 22{,}600 \text{ psi} \qquad \textbf{Answer}$$

Since the maximum stresses are below the yield strength of steel (see Table A.2), our results are valid.

Example 9.3. State of Stress In a Bar. Determine the state of stress at points A, B, C, and D in the member shown in Figure 9.4(a). Assume linearly elastic material behavior.

Solution. We first determine the stress resultants at the cross section of interest. The details of these computations will not be given here; the results are shown in Figure 9.4(b). We next consider the stresses due to the various loadings. These stresses are shown in Figure 9.4(c).

The area, moment of inertia about the bending axes, polar moment of inertia, and section modulus of the cross section are

$$A = \frac{\pi D^2}{4} = \pi \text{ in.}^2 \qquad J = \frac{\pi D^4}{32} = \frac{\pi}{2} \text{ in.}^4$$

$$I = \frac{\pi D^4}{64} = \frac{\pi}{4} \text{ in.}^4 \qquad S = \frac{I}{y_{\max}} = \frac{\pi}{4} \text{ in.}^3$$

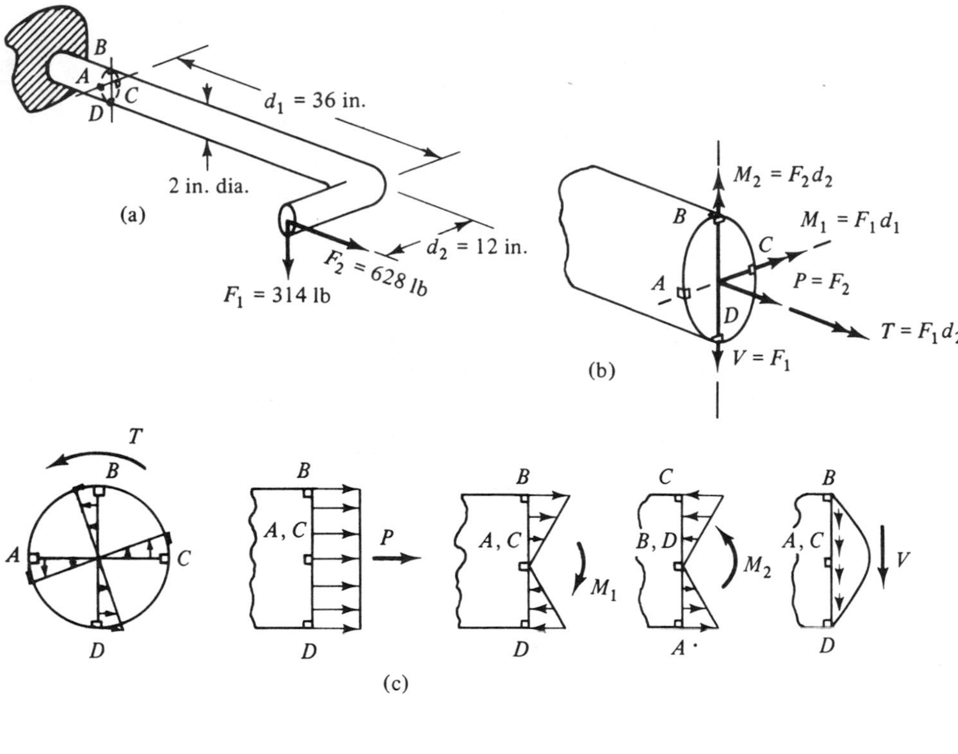

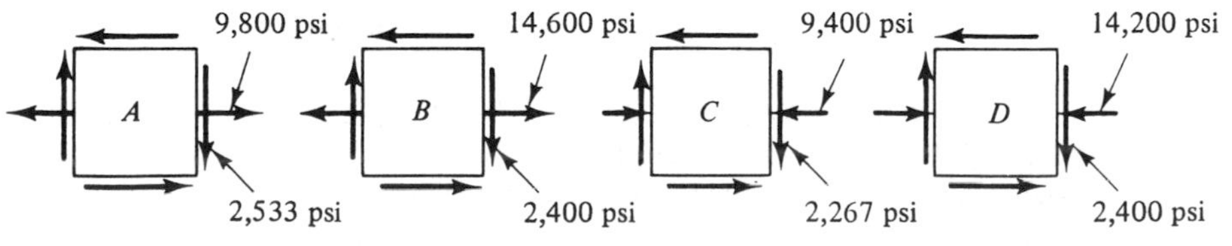

Figure 9.4

The torsional shear stress at the outside of the member is

$$\tau = \frac{TR}{J} = \frac{(12 \text{ in.})(314 \text{ lb})(1 \text{ in.})}{\pi/2 \text{ in.}^4} = 2{,}400 \text{ psi}$$

and the maximum bending stresses due to the moments M_1 and M_2 are

$$\sigma = \frac{M_1}{S} = \frac{(36 \text{ in.})(314 \text{ lb})}{\pi/4 \text{ in.}^3} = 14{,}400 \text{ psi}$$

and

$$\sigma = \frac{M_2}{S} = \frac{(12 \text{ in.})(628 \text{ lb})}{\pi/4 \text{ in.}^3} = 9{,}600 \text{ psi}$$

The normal stress due to the axial load is

$$\sigma = \frac{P}{A} = \frac{628 \text{ lb}}{\pi \text{ in.}^2} = 200 \text{ psi}$$

The direct shear stress is zero at points B and D. At points A and C it is

$$\tau = \frac{VA'\bar{y}'}{It} = \frac{(314 \text{ lb})(\pi/2 \text{ in.}^2)(4/3\pi \text{ in.})}{(\pi/4 \text{ in.}^4)(2 \text{ in.})} = 133 \text{ psi}$$

Referring to Figure 9.4(c) and noting the stresses acting at the various locations, we obtain the states of stress shown in Figure 9.4(d). The stresses at point C are as seen from the back side of the member, and those at D are as seen from below.

PROBLEMS

9.1. A large hook is loaded as shown. Determine the largest normal stress at the fixed end AB and indicate clearly whether it is tensile or compressive.

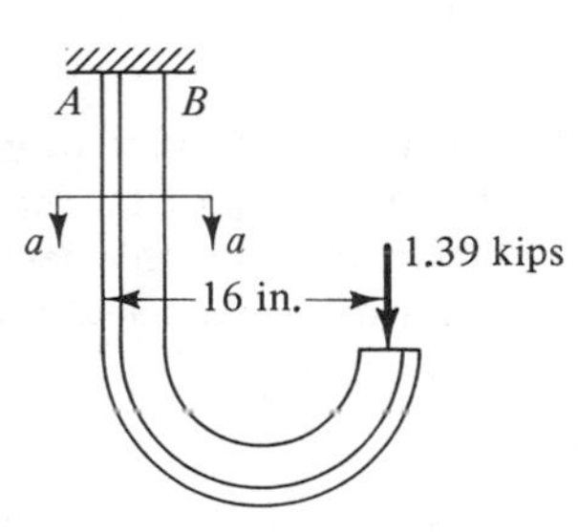

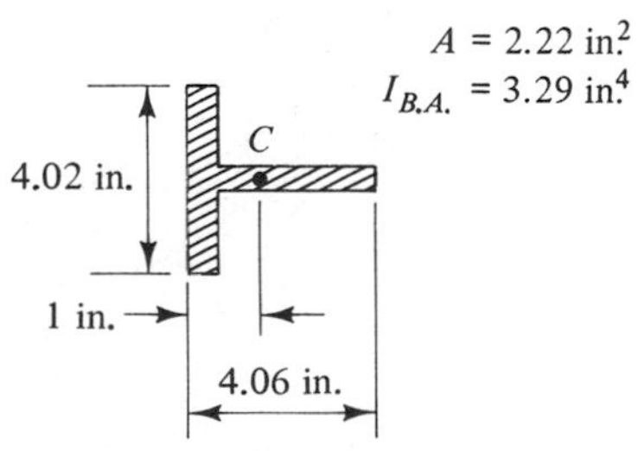

Problem 9.1

9.2. The support shown is made of 3-in. (nominal) diameter structural steel pipe. What is the largest load F that can be supported if the member is not to yield? Use a safety factor of 3.

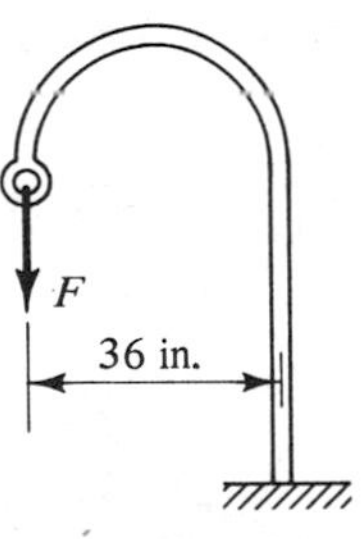

Problem 9.2

9.3. In Problem 9.2, what size pipe is required to support a load $F = 200$ lb?

9.4. The slotted link shown is made of steel with a yield point of 300 MN/m^2. What is the minimum required thickness t of the link if it is not to yield? Use a safety factor of 1.5 and ignore any stress concentrations.

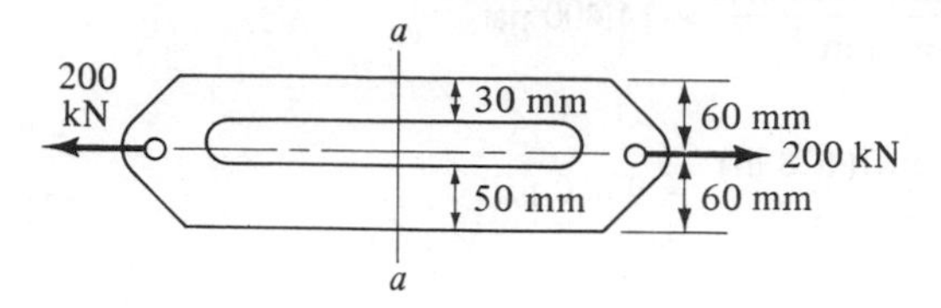

Problem 9.4

9.5. If the link in Problem 9.4 is 25 mm thick, determine and sketch the distribution of stress over the cross section at section a-a.

9.6. The 1-in. thick machine link shown is made of cast iron with an ultimate strength of 25 ksi in tension and 50 ksi in compression. What is the minimum allowable width w of the link? Use a safety factor of 2. Sketch the distribution of stress over the cross section at section a-a for this value of w.

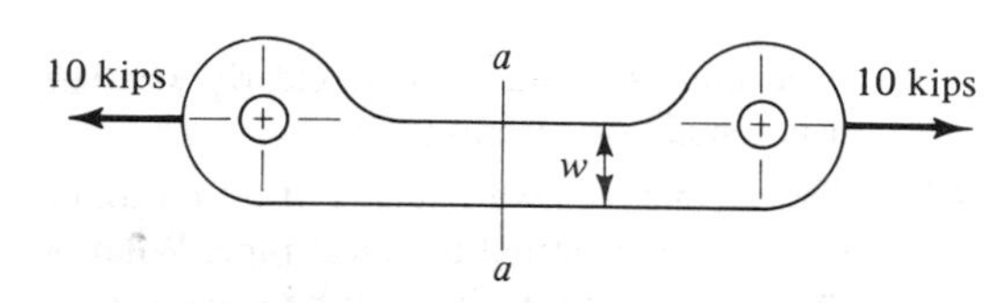

Problem 9.6

9.7. The inclined member shown has a solid circular cross section with a diameter of 2 in. Determine the maximum compressive stress in the member and indicate where it occurs.

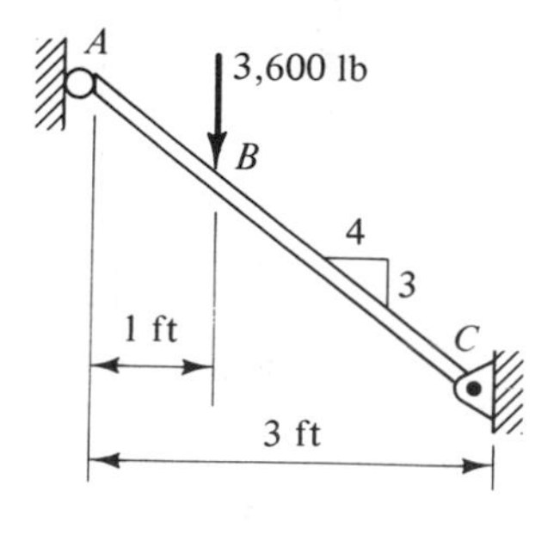

Problem 9.7

9.8. Determine the normal stresses at corners A, B, C, and D of the elastic block shown if $F = 40$ kN.

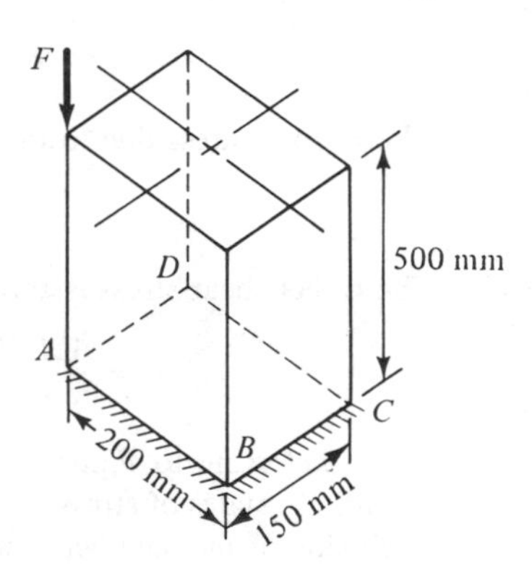

Problem 9.8

9.9. What is the largest force F that can be supported by the block in Problem 9.8 if the maximum allowable tensile stress is 80 MN/m^2?

9.10. What is the lightest weight W-shape beam that will support the loading shown if the maximum allowable tensile and compressive stresses are 15 ksi? Neglect the weight of the member.

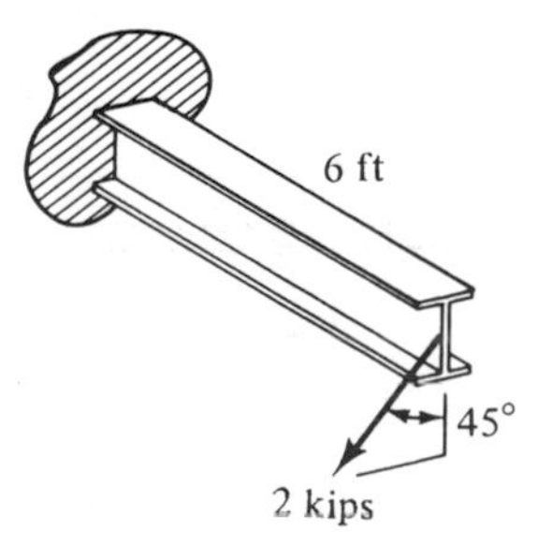

Problem 9.10

9.11. A 4-in. × 6-in. (nominal) timber oriented with the longer dimension vertical is supported at each end as shown. The member carries a distributed vertical loading of 100 lb/ft (including its weight) over its entire length. If the maximum allowable stresses are 1 ksi in tension and 3 ksi in compression, what is the largest side load F that can be applied at midspan?

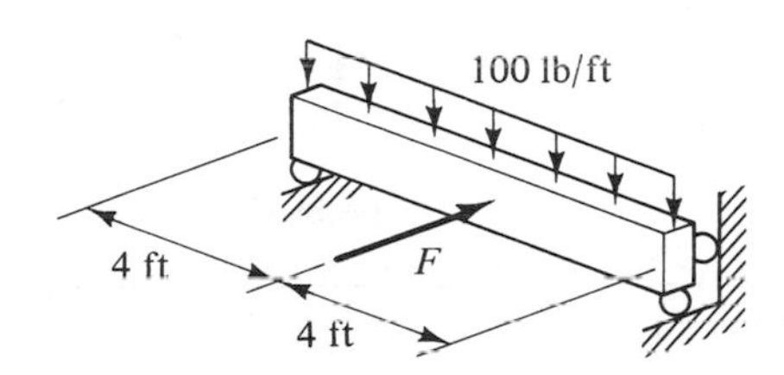

Problem 9.11

9.12. The pin-connected frame shown is to be made of structural steel Universal beams oriented with the webs normal to the plane of the figure. Select the required sizes of the members. Use a safety factor of 2.5 against failure by the onset of yielding and neglect the masses of the members.

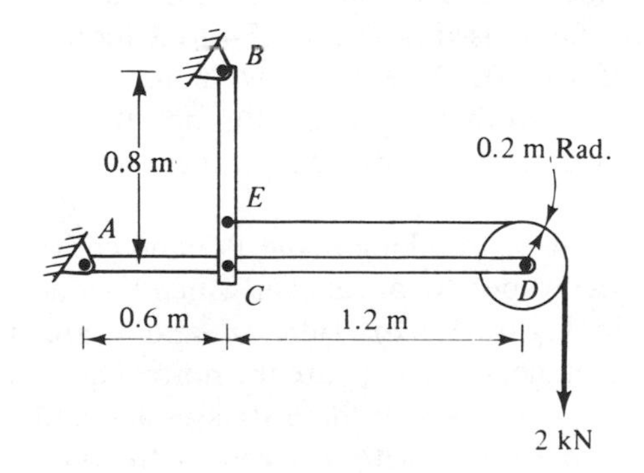

Problem 9.12

9.13. An elastic beam with a square cross section is loaded as shown. What is the smallest permissible dimension d if the maximum normal stress in the horizontal portion of the member is not to exceed 50 MN/m^2?

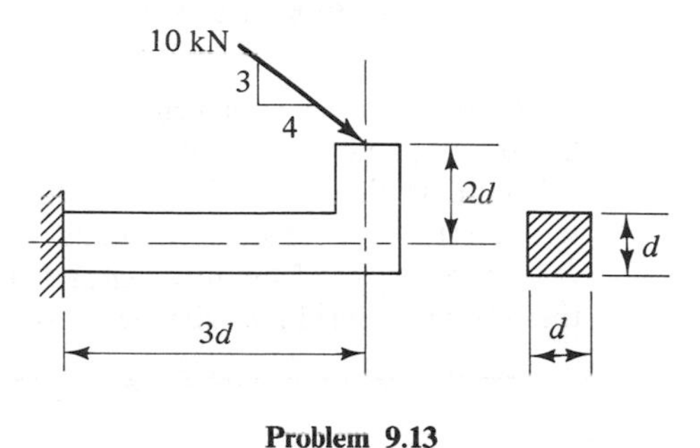

Problem 9.13

9.14. A cylindrical pressure vessel with an inside diameter of 20 in. and a wall thickness of $\frac{1}{2}$ in. is simply supported at each end and contains a gas under a pressure of 250 psi. The weight of the tank and gas combined is 600 lb per ft of length. To prevent buckling, the maximum longitudinal compressive stress in the cylinder wall must not exceed 500 psi. What is the maximum allowable length L for the vessel?

9.15. A cylindrical pressure vessel with an outside diameter of 4 ft and a wall thickness of $\frac{1}{4}$ in. is designed to hold a gas at a pressure of 120 psi. Because of faulty installation, the vessel is also subjected to a torque of 8,000 kip · in. Determine and show on a sketch the stresses acting on an element of material located (a) on the outside surface of the vessel and (b) on the inside surface.

9.16. Determine the states of stress at points A, B, and C on the beam shown and show the results on small material elements.

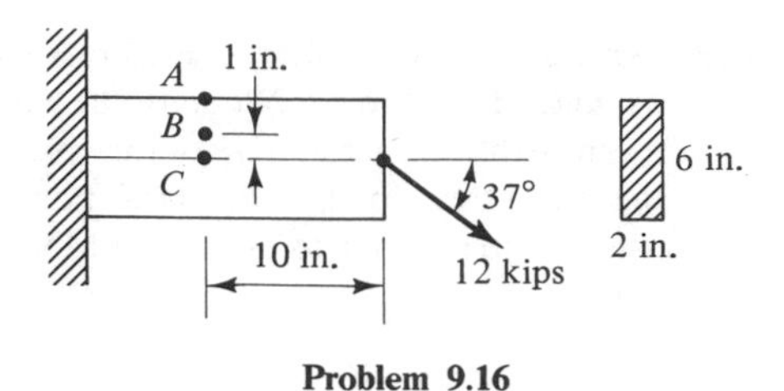

Problem 9.16

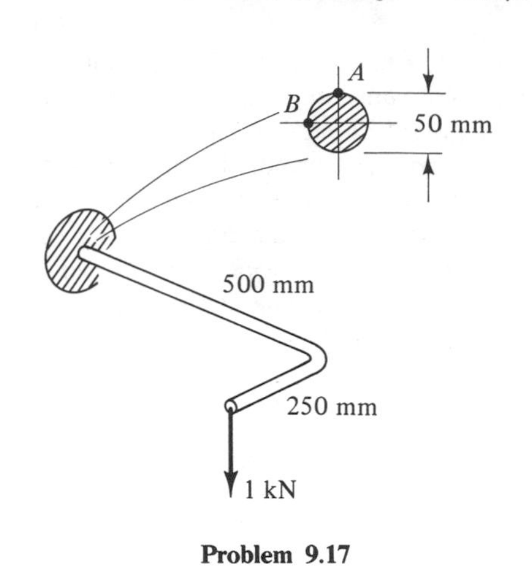

Problem 9.17

9.17. A 30-mm diameter L-shaped rod lies in a horizontal plane and is subjected to a vertical load at the free end, as shown. Determine the states of stress at points A and B on the surface of the member and show the results on small material elements.

9.3 STRESSES ON AN INCLINED PLANE

Once the state of stress at a point in a body is known, the stresses on any other plane through that point can be determined. The procedure for doing this for the case of plane stress will be discussed in this section.

Stress Transformation Equations

Figure 9.5(a) shows the stress components with respect to xy axes acting at a point in a body subjected to a general state of plane stress. All the stresses are shown acting in their positive directions. (At this point, it may be helpful to review the sign conventions for stresses established in Section 5.5.) To determine the stresses on some other plane inclined to the coordinate axes, such as plane PQ, we section the element along the plane of interest and consider the equilibrium of one of the resulting pieces. This is the same procedure used in Section 7.4 to determine the stresses on planes inclined to the axis of a shaft. Once again, we note that we are dealing with infinitesimal material elements, so that any plane through the element actually passes through the corresponding point in the body; the elements are shown enlarged for clarity only.

Sectioning the element shown in Figure 9.5(a) along plane PQ and converting the stresses to forces by multiplying by the respective areas over which they act, we obtain the free-body diagram shown in Figure 9.5(b). Here, n and t are axes perpendicular and parallel to the inclined plane, σ_n and τ_{nt} are the normal and shear stresses acting upon this plane, A is the area over which these stresses act, and θ is the angle through which the nt axes are rotated with respect to the xy axes, measured counterclockwise [Figure 9.5(a)].

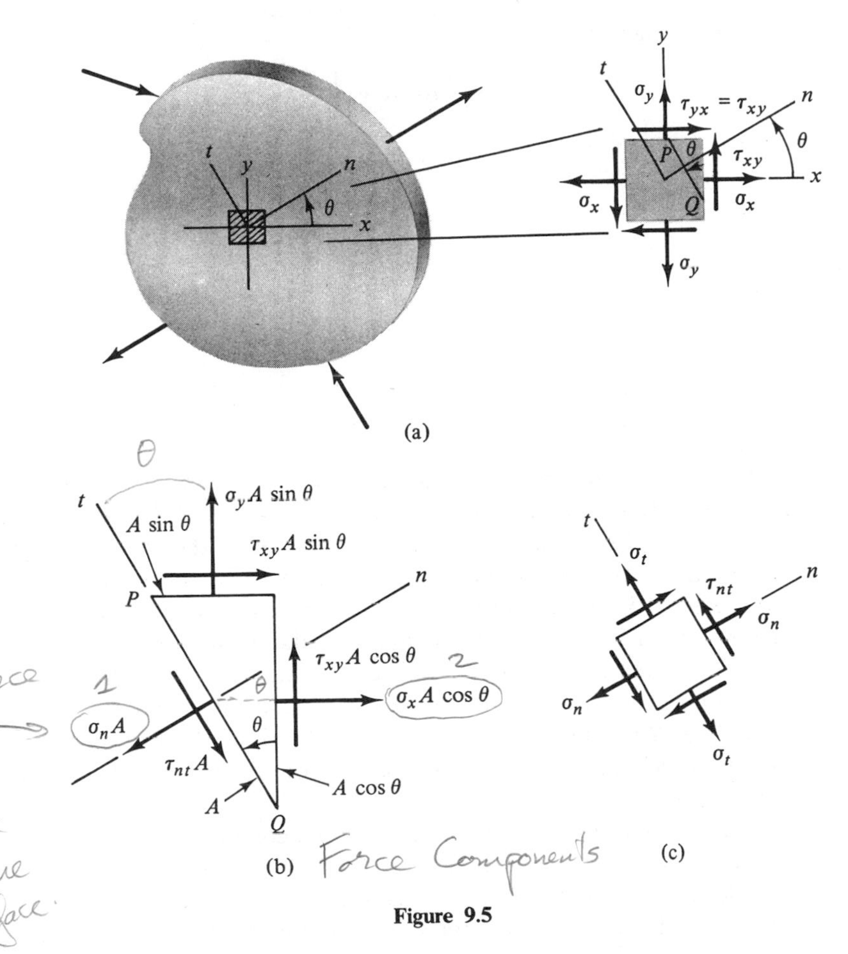

Figure 9.5

The equations of equilibrium are

$$\sum F_n = -\sigma_n A + (\sigma_x A \cos\theta)\cos\theta + (\tau_{xy} A \cos\theta)\sin\theta$$
$$+ (\tau_{xy} A \sin\theta)\cos\theta + (\sigma_y A \sin\theta)\sin\theta = 0$$
$$\sum F_t = -\tau_{nt} A - (\sigma_x A \cos\theta)\sin\theta + (\tau_{xy} A \cos\theta)\cos\theta$$
$$-(\tau_{xy} A \sin\theta)\sin\theta + (\sigma_y A \sin\theta)\cos\theta = 0$$

from which we obtain

$$\sigma_n = \sigma_x \cos^2\theta + \sigma_y \sin^2\theta + 2\tau_{xy} \sin\theta\cos\theta$$

and

$$\tau_{nt} = -(\sigma_x - \sigma_y)\sin\theta\cos\theta + \tau_{xy}(\cos^2\theta - \sin^2\theta)$$

These expressions can be put into a more convenient form by introducing the trigonometric identities

$$\cos^2\theta = \frac{1}{2} + \frac{1}{2}\cos 2\theta$$

$$\sin^2\theta = \frac{1}{2} - \frac{1}{2}\cos 2\theta$$

$$\sin\theta\cos\theta = \frac{1}{2}\sin 2\theta$$

We have

$$\sigma_n = \frac{\sigma_x + \sigma_y}{2} + \frac{\sigma_x - \sigma_y}{2}\cos 2\theta + \tau_{xy}\sin 2\theta \tag{9.1}$$

$$\tau_{nt} = -\frac{(\sigma_x - \sigma_y)}{2}\sin 2\theta + \tau_{xy}\cos 2\theta \tag{9.2}$$

These equations are known as the *stress transformation equations*. Given the values of the stress components with respect to xy axes, they determine the values with respect to the rotated axes nt. In other words, the transformation equations determine the stresses acting on a rotated element with sides aligned with the n and t axes [Figure 9.5(c)]. The normal stress, σ_t, on the t face of the rotated element is determined from Eq. (9.1) by replacing the value of θ by $\theta + \pi/2$.

It is of interest to note that Eqs. (5.3) and (5.4) for the normal and shear stresses on a plane inclined to the axis of a uniaxially loaded member are special cases of Eqs. (9.1) and (9.2) with $\sigma_y = \tau_{xy} = 0$. Similarly, Eqs. (7.14) and (7.15) for the stresses on planes inclined to the axis of a shaft are special cases of these equations with $\sigma_x = \sigma_y = 0$.

Considerable care must be taken to avoid computational errors when using the stress transformation equations. The signs on all quantities and the manner in which the angle θ is measured must be consistent with the conventions used in the derivations. All the stresses were shown in Figure 9.5 as acting in their positive directions. If any one of them has a sense opposite to that shown, it must be entered as a negative quantity in Eqs. (9.1) and (9.2). These equations are easily programmed on a computer, in which case the computations can be made rapidly and accurately. Since the stress transformation equations are based only upon equilibrium considerations, they apply regardless of the material behavior.

Representation of Stresses

The variation of the stresses σ_n and τ_{nt} with the angle θ can best be visualized from graphical representations of the stress transformation equations. Several types of plots are possible.

For example, consider the state of stress shown in Figure 9.6. Cartesian plots of the values of the stresses σ_n and τ_{nt} obtained from Eqs. (9.1) and (9.2) are shown in Figure 9.7. If this same information is plotted in a polar coordinate system, we obtain the polar plots shown in Figure 9.8. In these plots, the radial distance from

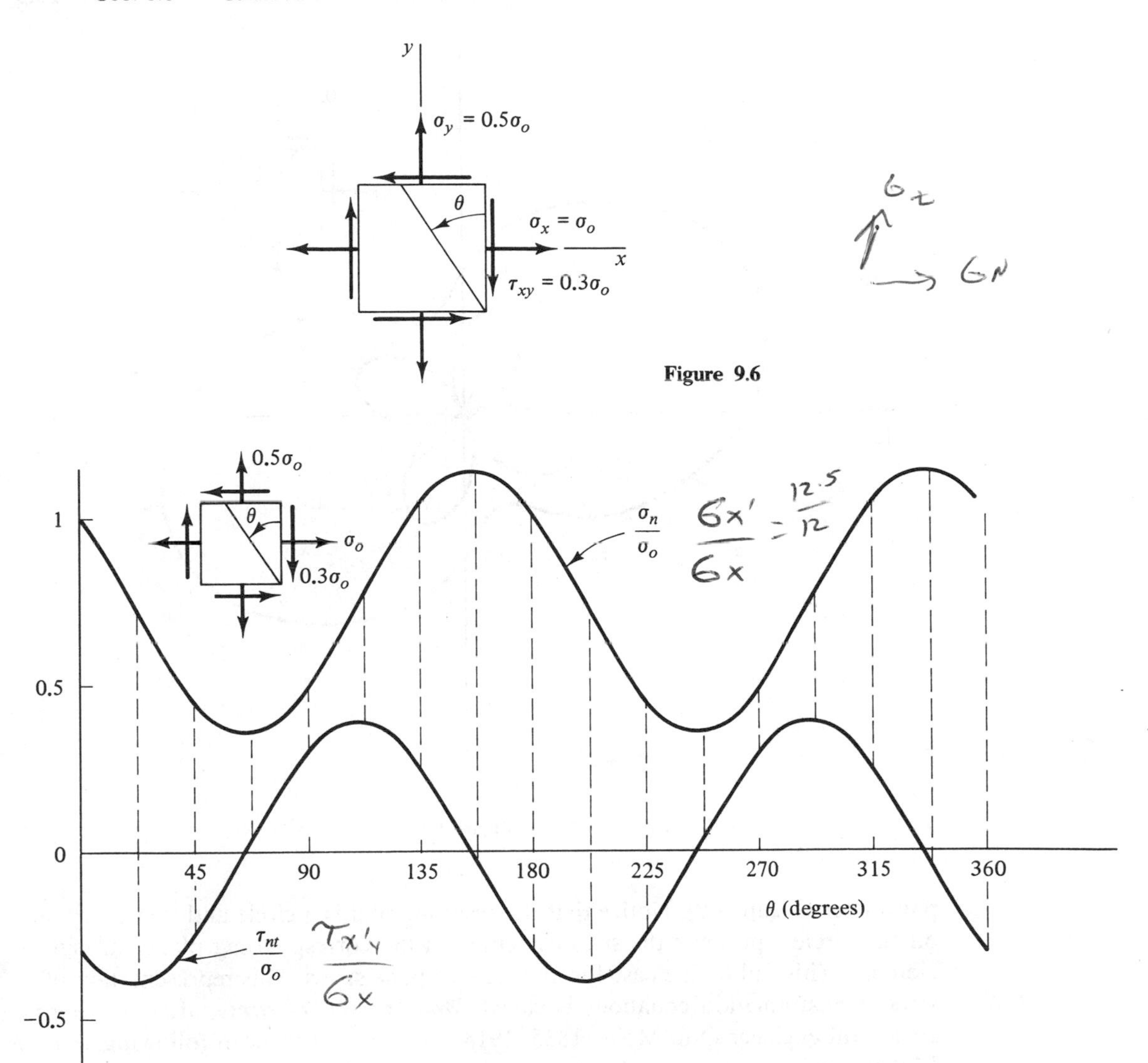

Figure 9.6

Figure 9.7

the origin to the curve represents the stress magnitude and the radial lines denote increments of θ equal to 22.5°. Both types of diagrams give a direct indication of the stresses acting on various planes. The polar plots give a more direct insight into the stress variations at the point, but the signs of the stresses and their sinusoidal nature are more obvious in the Cartesian plots. It should be noted that the overall appearance of both types of diagrams can vary slightly, depending upon the state of stress.

The stress transformation equations can be presented in a third form by plotting them in a plane with σ_n and τ_{nt} as coordinates and with the angle θ as a

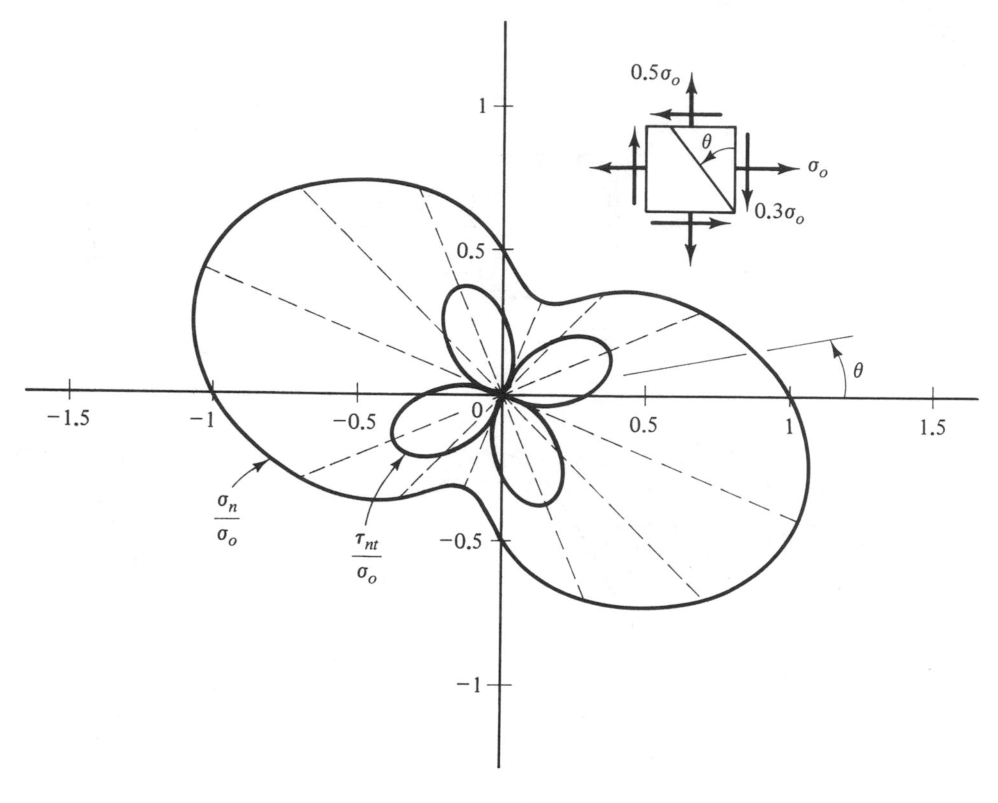

Figure 9.8

parameter (Figure 9.9). Notice that the resulting plot is a circle and that every point on the circle represents the stresses acting on the corresponding plane through the element. This will be the case for any state of pane stress. This representation of the stress transformation equations is called *Mohr's circle of stress*, after the German structural engineer Otto Mohr (1835–1918). As we shall show in following sections, Mohr's circle is a very useful concept.

Computer graphics programs which automatically plot the Mohr's circle and/or the polar and Cartesian diagrams are available, or can be written quite easily. All three types of diagrams reveal certain important features of the stress transformation equations. For example, it is obvious from the diagrams in Figures 9.7–9.9 that the values of the stresses are repeated twice over the interval $0 \leqslant \theta \leqslant 360°$. This is because of the double angle relationships in Eqs. (9.1) and (9.2). It can also be seen that the values of the normal and shear stresses exhibit relative maxima and minima at certain values of θ. Maximum stresses will be discussed in Section 9.5, and other important features revealed in the diagrams will be considered there and in Section 9.4.

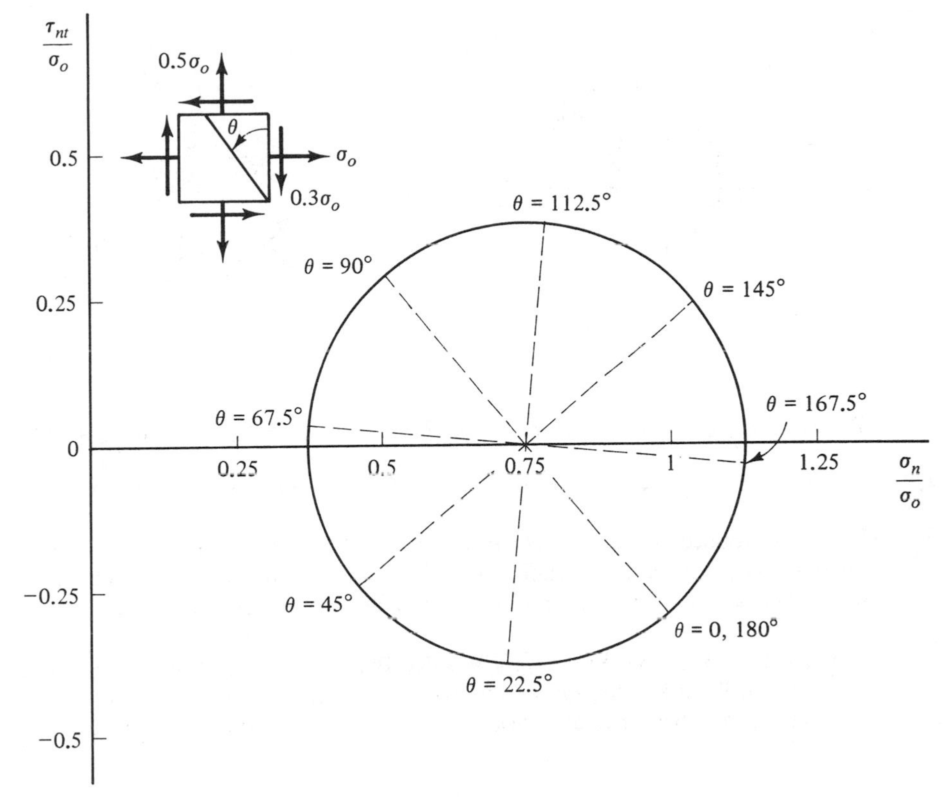

Figure 9.9

Example 9.4. Stress Components with Respect to Rotated Axes. Determine the stress components with respect to the *nt* axes for the state of plane stress given in Figure 9.10(a). Show the results on a rotated element with sides parallel to the *nt* axes.

Solution. According to our sign convention for stresses (Section 5.5) and for the angle θ, we have

$$\sigma_x = 10 \text{ MN/m}^2 \qquad \sigma_y = -6 \text{ MN/m}^2$$

$$\tau_{xy} = 4 \text{ MN/m}^2 \qquad \theta = 40°$$

Substituting these values into Eqs. (9.1) and (9.2), we obtain

$$\sigma_n = \tfrac{1}{2}[10 + (-6)] + \tfrac{1}{2}[10 - (-6)]\cos 80° + 4 \sin 80°$$

$$= 7.3 \text{ MN/m}^2 \qquad \textbf{Answer}$$

$$\tau_{nt} = -\tfrac{1}{2}[10 - (-6)]\sin 80° + 4 \cos 80°$$

$$= -7.2 \text{ MN/m}^2 \qquad \textbf{Answer}$$

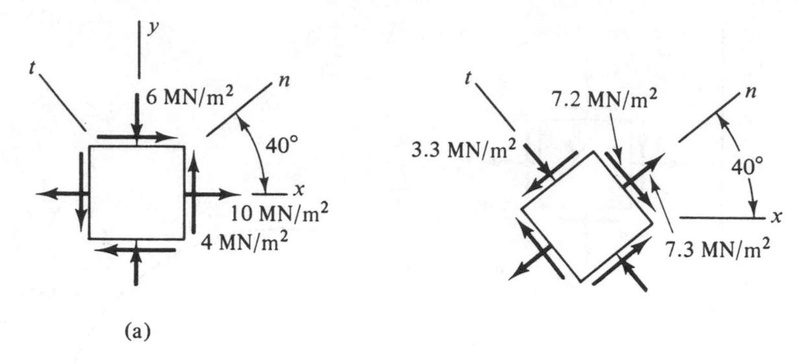

Figure 9.10

The value of σ_t is obtained by replacing θ by $\theta + \pi/2$ in Eq. (9.1). For $2\theta = 260°$, we get

$$\sigma_t = -3.3 \text{ MN/m}^2 \qquad \textbf{Answer}$$

The rotated element is shown in Figure 9.10(b). Note carefully the direction of the shear stress τ_{nt}. Since it is negative, it will act in a negative coordinate direction on planes whose outward normal is in a positive coordinate direction (see Section 5.5).

Example 9.5. Allowable Stress on a Wooden Block. The block of wood shown in Figure 9.11 will crack if the shear stress parallel to the grain exceeds 800 psi. If $\sigma_x = 500$ psi (tension), what range of values of σ_y can be applied without cracking the block?

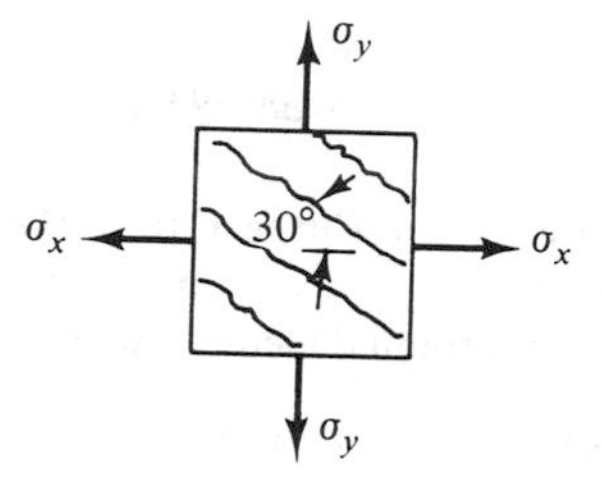

Figure 9.11

Solution. The grain of wood corresponds to an angle $\theta = 60°$ (or $-30°$). The condition on the shear stress parallel to the grain requires that $-800 \text{ psi} \leqslant \tau_{nt} \leqslant 800$ psi. Applying the first of these conditions to Eq. (9.2), we have

$$\tau_{nt} = -\tfrac{1}{2}(500 \text{ psi} - \sigma_y)\sin 120° \geqslant -800 \text{ psi}$$

or

$$\sigma_y \geqslant -1{,}350 \text{ psi}$$

Similarly, the condition $\tau_{nt} \leqslant 800$ psi requires that $\sigma_y \leqslant 2{,}350$ psi. Thus, the block will not crack if the value of σ_y lies in the range

$$-1{,}350 \text{ psi} \leqslant \sigma_y \leqslant 2{,}350 \text{ psi} \qquad \textbf{Answer}$$

PROBLEMS

9.18. to 9.20. Determine the stress components with respect to the *nt* axes for the state of stress shown. Show the results on a rotated element with sides parallel to the *nt* axes.

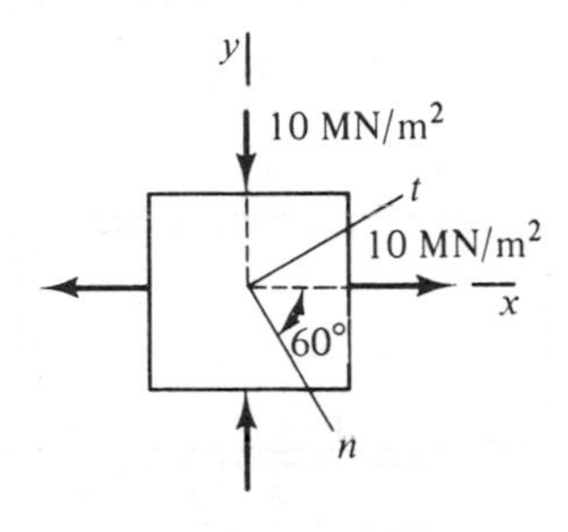

Problem 9.18

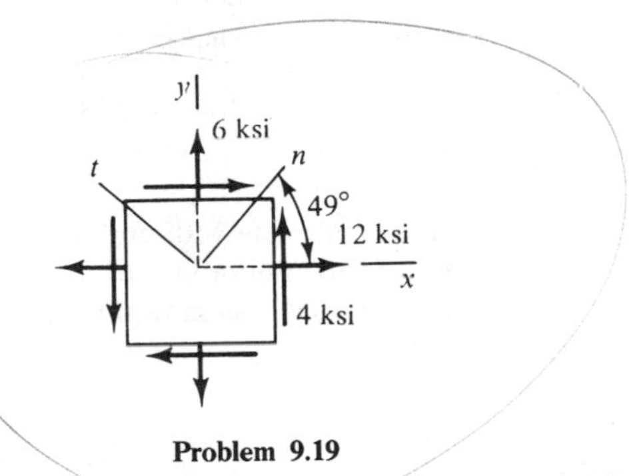

Problem 9.19

Problem 9.20

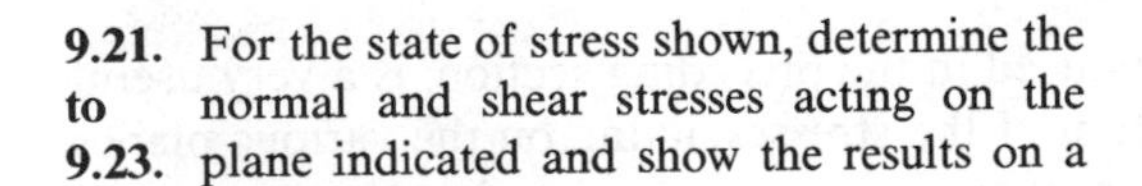

9.21. to 9.23. For the state of stress shown, determine the normal and shear stresses acting on the plane indicated and show the results on a sketch.

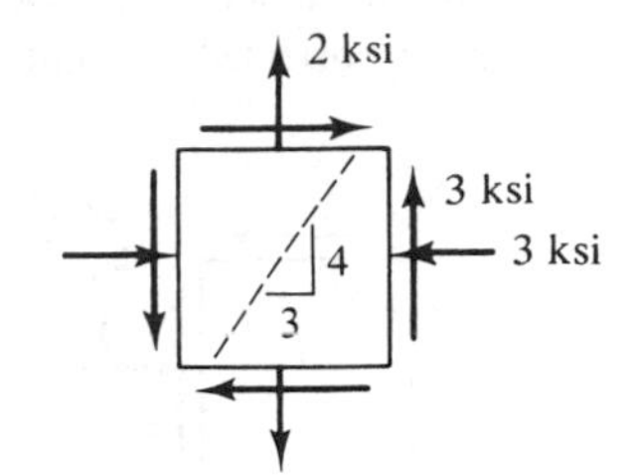

Problem 9.21

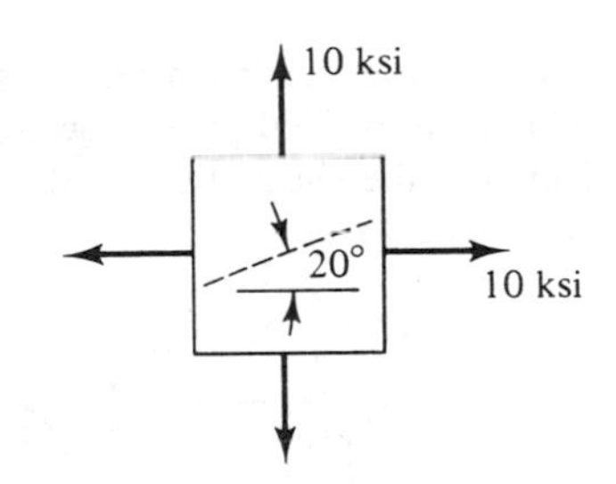

Problem 9.22

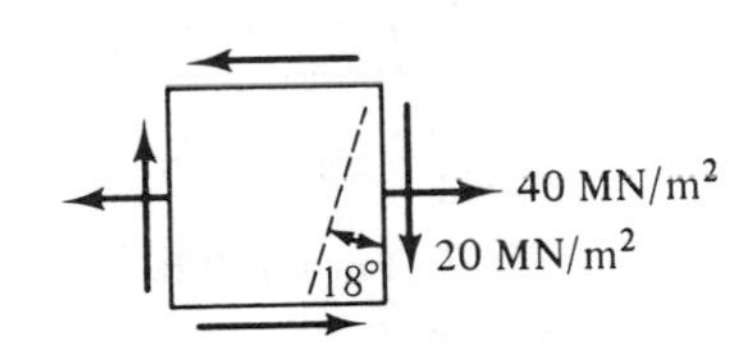

Problem 9.23

9.24. Two blocks are glued together as shown. If the normal stress on the joint is 15 MN/m^2, what is the value of σ_y?

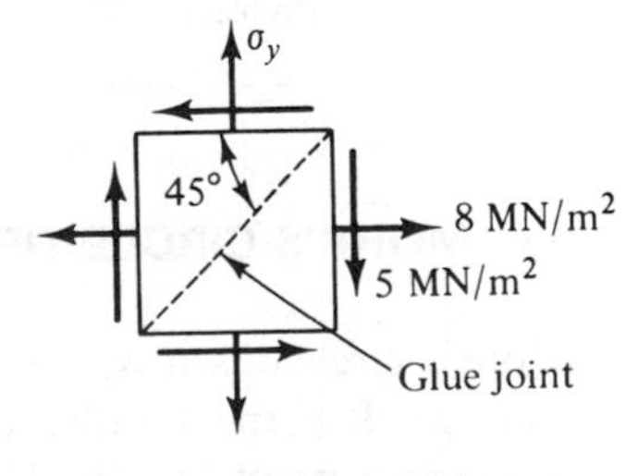

Problem 9.24

9.25. For the state of stress shown, it is known that $\sigma_n = -7.5$ ksi. Determine the angle θ and the stresses σ_t and τ_{nt}. Show the results on a properly oriented element.

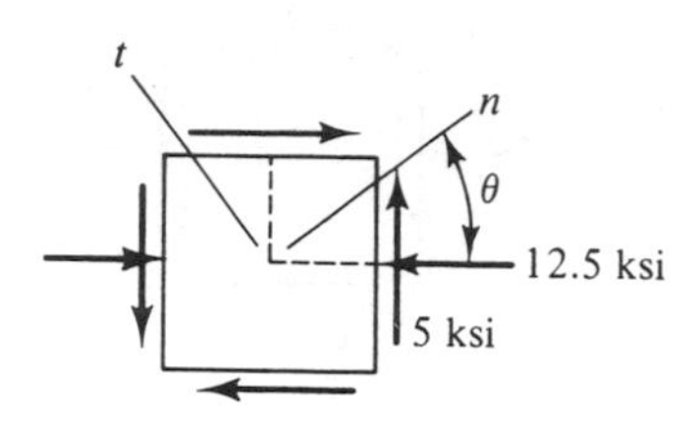

Problem 9.25

9.26. What is the maximum tensile stress σ_x to which the welded plate shown can be subjected if the shear stress in the weld is not to exceed 15 ksi?

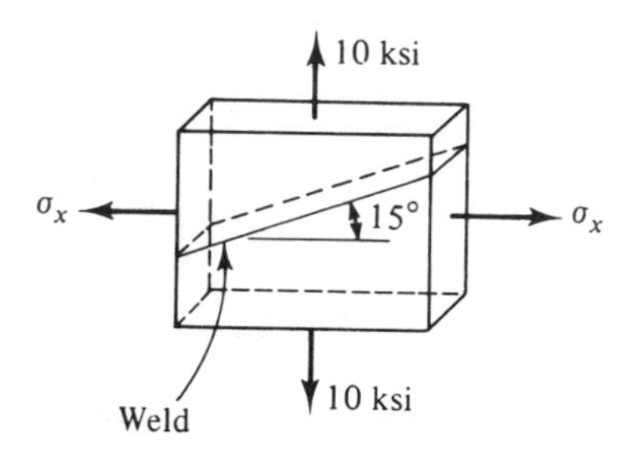

Problem 9.26

9.27. The stresses in the wall of a rocket motor casing are as shown, where p is the internal pressure. If the seam can withstand a tensile stress of 14 ksi, what pressure p can the casing withstand?

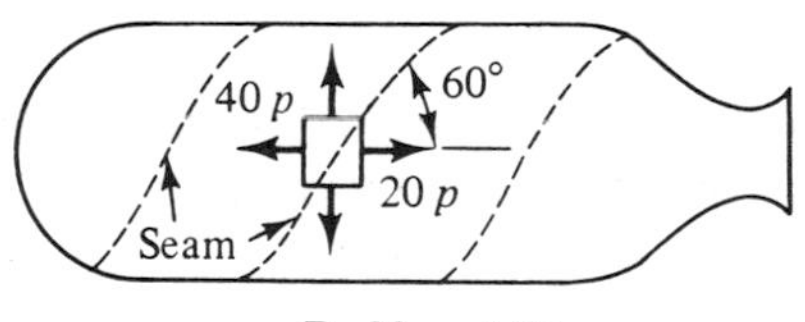

Problem 9.27

9.28. A steel plate $\frac{1}{2}$-in. thick × 5-in. wide × 10-in. long is subjected to uniformly distributed forces (not stresses) acting along its edges as shown. Determine the normal and shear stresses acting along plane *a-a*.

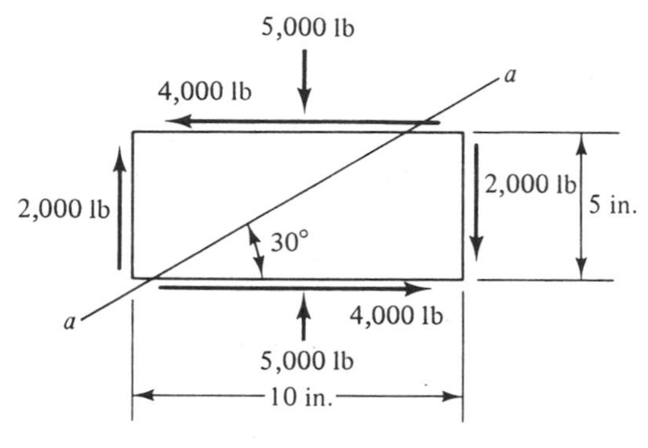

Problem 9.28

9.29. A skewed plate with unit thickness is subjected to uniformly distributed stresses along its sides as shown. Determine σ_x, σ_y, and τ_{xy}.

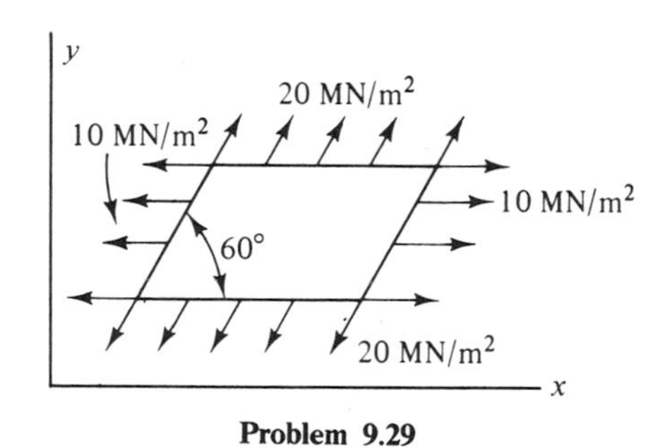

Problem 9.29

9.4 MOHR'S CIRCLE OF STRESS

Mohr's circle of stress, which we introduced in the preceding section, is a very useful concept. It permits a direct visualization of the stresses acting on the various planes through a point in a body and it is a convenient computational tool, as we shall show. We also shall show that the circle can be constructed directly from the given state of stress. First, however, we shall prove that the stress transformation equations do represent the equations of a circle in parametric form.

If we transpose the first term on the right-hand side of Eq. (9.1) to the left side and then square both it and Eq. (9.2), we obtain

$$\left[\sigma_n - \left(\frac{\sigma_x + \sigma_y}{2}\right)\right]^2 = \left[\left(\frac{\sigma_x - \sigma_y}{2}\right)\cos 2\theta + \tau_{xy}\sin 2\theta\right]^2$$

$$\tau_{nt}^2 = \left[-\left(\frac{\sigma_x - \sigma_y}{2}\right)\sin 2\theta + \tau_{xy}\cos 2\theta\right]^2$$

Expanding the right-hand sides of these expressions and then adding them, we have

$$\left[\sigma_n - \left(\frac{\sigma_x + \sigma_y}{2}\right)\right]^2 + \tau_{nt}^2 = \left(\frac{\sigma_x - \sigma_y}{2}\right)^2 + \tau_{xy}^2 \tag{9.3}$$

This equation defines a circle in a plane with σ_n and τ_{nt} as coordinates, as can be seen by comparison with the general equation of a circle in rectangular coordinates

$$(x - a)^2 + (y - b)^2 = R^2 \tag{9.4}$$

The significance of Mohr's circle lies in the fact that every point on it represents the stresses on some plane through the corresponding element. Thus, to determine the stresses on any given plane through the element, or with respect to any rotated set of axes, we need only find the coordinates of the corresponding points on the circle. As we shall see, this can be done by using only simple geometry.

Given a state of plane stress with respect to xy axes [Figure 9.12(a)], the step-by-step procedure for constructing Mohr's circle and for determining the stresses with respect to a rotated set of axes, or on any plane inclined to the xy axes, is as follows. These steps are illustrated in Figure 9.12(b):

1. Set up a coordinate system with normal stress as the horizontal coordinate and shear stress as the vertical coordinate.
2. Plot the stresses on the x and y faces of the element as points in this coordinate system. These points are denoted by (x) and (y), respectively, in Figure 9.12(b). Normal stresses are considered positive if tensile and negative if compressive. We have shown each of the normal stresses as being positive and with $\sigma_x > \sigma_y$, but this need not be the case. The shear stress τ_{xy} on the x face of the element is plotted in accordance with the sign convention established in Section 5.5. The shear stress τ_{yx} on the y face is plotted with the *opposite* sign. This reversal of the sign of τ_{yx} is necessary for proper construction of the circle. Thus, the shear stress on the x face of the element in Figure 9.12(a) is plotted as a positive quantity in Figure 9.12(b) and the shear stress on the y face is plotted as a negative quantity.
3. Connect the points (x) and (y). This establishes a diameter of the circle. The complete circle can then be constructed or sketched in. From the geometry of

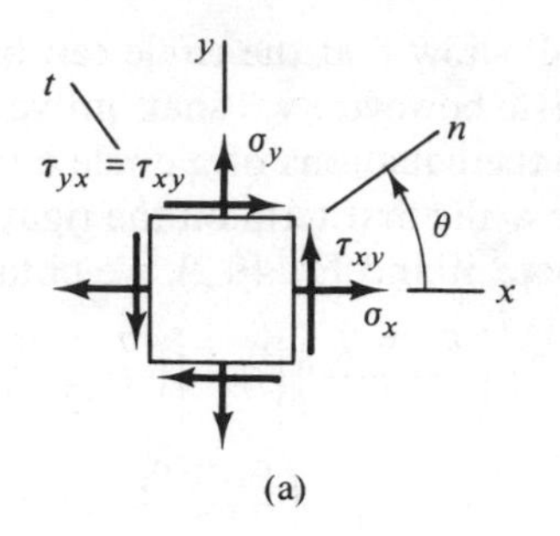

(a)

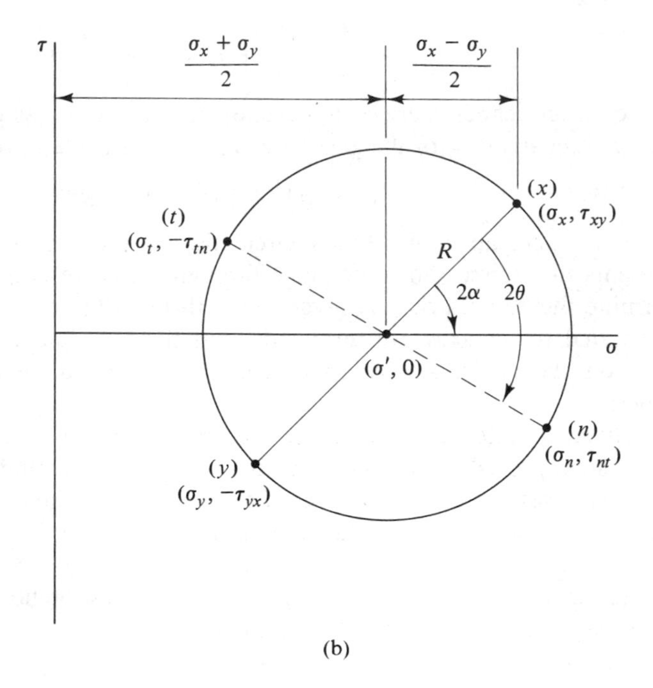

(b)

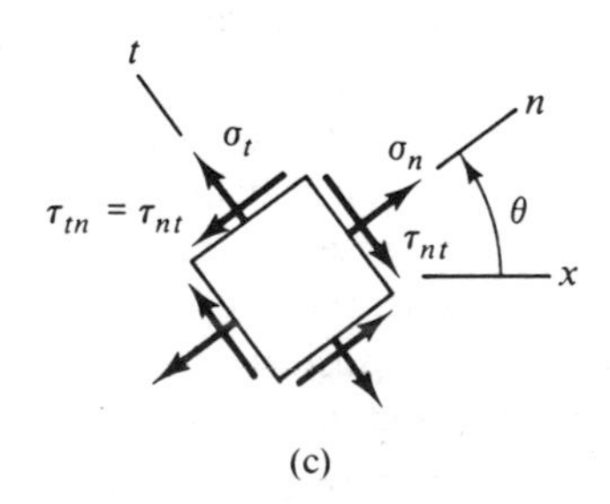

(c)

Figure 9.12

Figure 9.12(b), we see that the center of the circle will always lie on the horizontal axis. Furthermore, the coordinates $(\sigma', 0)$ of the center of the circle, its radius R, and the angle 2α between the diameter (x)-(y) and the horizontal axis can be determined from the known coordinates of points (x) and (y) by using only simple geometry. We have

$$\sigma' = \frac{\sigma_x + \sigma_y}{2} \qquad R = \sqrt{\left(\frac{\sigma_x - \sigma_y}{2}\right)^2 + \tau_{xy}^2} \qquad 2\alpha = \tan^{-1}\frac{\tau_{xy}}{\left(\dfrac{\sigma_x - \sigma_y}{2}\right)} \tag{9.5}$$

The first two of these relations can also be obtained from a direct comparison of Eqs. (9.3) and (9.4).

4. To locate the points (n) and (t) on the circle associated with the rotated nt axes, we rotate the diameter (x)-(y) of the circle in the direction *opposite* to that in which the axes are rotated, but through *twice* the angle. All angles on the circle are double those on the element because the stress transformation equations are functions of 2θ. The reversal of the direction of rotation from the element to the circle is a consequence of the sign convention used for shear stress, and is evident in Figure 9.9. Once points (n) and (t) have been located, their coordinates can be determined from the geometry of the circle. These coordinates provide the stresses on the n and t faces of the rotated element [Figure 9.12(c)]. The sense of these stresses is in accordance with the sign conventions given in step 2. This same procedure applies for determining the stresses on any given plane inclined to the xy axes. However, only the one point on the circle corresponding to the stresses on this plane need be considered in this case.

The procedures outlined here are further illustrated in the following examples. If Mohr's circle is constructed to scale, the stresses on any given plane can be determined graphically. However, the circle is usually used as a computational aid for determining these stresses analytically.

Example 9.6. Stress Components with Respect to Rotated Axes. Solve Example 9.4 using Mohr's circle.

Solution. We first redraw the element and construct Mohr's circle, as shown in Figure 9.13. According to our sign convention, the shear stress on the x face of the element is positive, while that on the y face is plotted as a negative quantity. Furthermore, the normal stress on the y face is considered negative because it is compressive.

From Figure 9.13(b), or from Eqs. (9.5), we find that the center of the circle is located a distance

$$\sigma' = \frac{10 + (-6)}{2} = 2 \text{ MN/m}^2$$

from the origin. Furthermore, the radius of the circle is

$$R = \sqrt{8^2 + 4^2} = 8.94 \text{ MN/m}^2$$

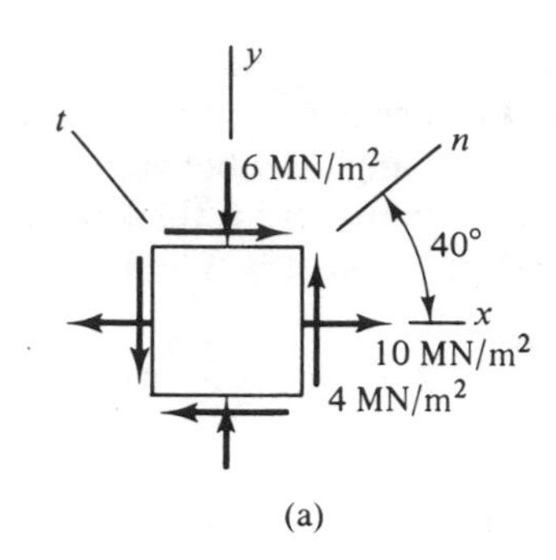

(a)

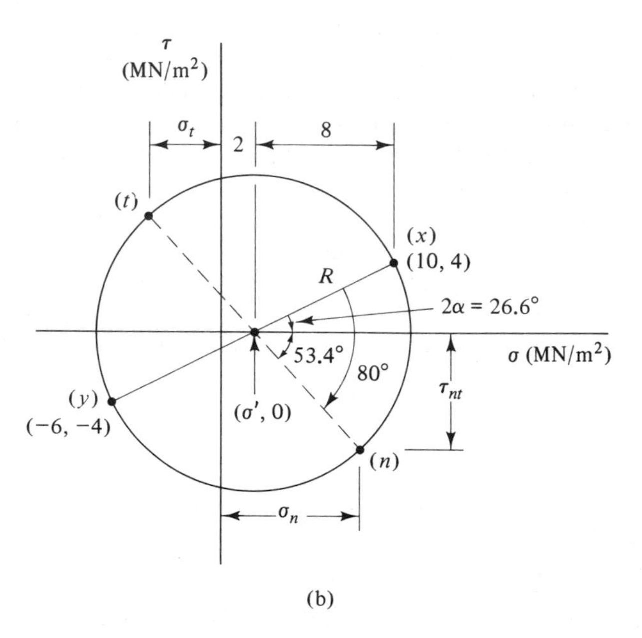

(b)

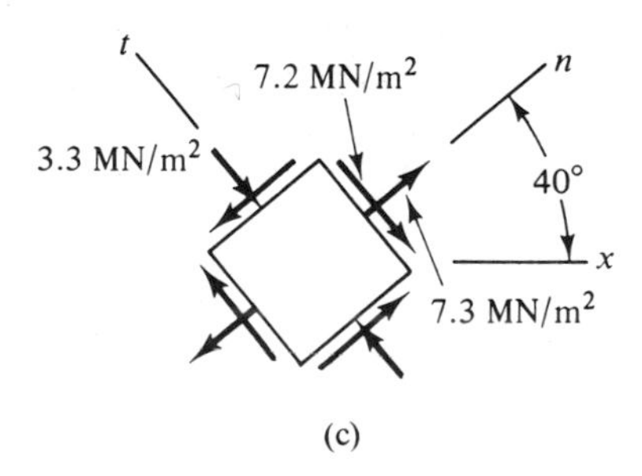

(c)

Figure 9.13

and

$$\tan 2\alpha = \frac{4}{8} \qquad 2\alpha = 26.6°$$

Since the nt axes are rotated 40° counterclockwise with respect to the xy axes, we rotate the diameter (x)-(y) of the circle 80° clockwise to determine the points (n) and (t) corresponding to the stresses associated with these axes. The angle between the diameter (n)-(t) and the horizontal axis is $80° - 2\alpha = 53.4°$. From the geometry of the circle, we then have

$$\sigma_n = \sigma' + R\cos 53.4° = 2 + 8.94\cos 53.4°$$

$$= 2 + 5.3 = 7.3 \text{ MN/m}^2$$

Answer

$$\sigma_t = \sigma' - R\cos 53.4° = 2 - 5.3 = -3.3 \text{ MN/m}^2$$

$$\tau_{nt} = -R\sin 53.4° = -8.94\sin 53.4° = -7.2 \text{ MN/m}^2$$

Figure 9.13(c) shows these stresses acting upon the rotated element. Note that we need determine the shear stress on only one face of the element because the values on the other faces are the same. The sense of the stresses is in accordance with the sign conventions used for the circle. These results are the same as those obtained in Example 9.4 using the stress transformation equations.

Example 9.7. Allowable Pressure in a Tank. A cylindrical tank with a radius of 10 in. and a wall thickness of 0.1 in. is welded along a helical seam that makes an angle of 55° with the longitudinal axis [Figure 9.14(a)]. What internal pressure, p, can the tank safely withstand if the tensile stress in the weld cannot exceed 30 ksi? Use a safety factor of 3. What is the shear stress in the weld at this pressure?

Solution. The longitudinal and circumferential stresses in the tank are (see Section 6.8)

$$\sigma_L = \frac{pR}{2t} = \frac{p(10 \text{ in.})}{2(0.1 \text{ in.})} = 50 \text{ p}$$

$$\sigma_C = \frac{pR}{t} = 2\sigma_L = 100 \text{ p}$$

Figure 9.14(b) shows these stresses acting upon a material element with sides oriented parallel and perpendicular to the longitudinal axis of the tank and containing a portion of the weld. Mohr's circle for this state of stress is shown in Figure 9.14(c).

We take the n axis perpendicular to the plane of interest. This axis is rotated 35° clockwise with respect to the x axis. Thus, the point (n) on the circle corresponding to the stresses on this plane lies 70° counterclockwise from point (x). From the geometry of the circle, we have for the tensile stress in the weld

$$\sigma_n = \sigma' - R\cos 70° = (75 \text{ p}) - (25 \text{ p})\cos 70° = 66.4 \text{ p} \leqslant 30{,}000 \text{ psi}$$

$$p \leqslant 450 \text{ psi}$$

The working pressure is obtained by dividing by the safety factor:

$$p_{\text{working}} = \frac{450 \text{ psi}}{3} = 150 \text{ psi}$$

Answer

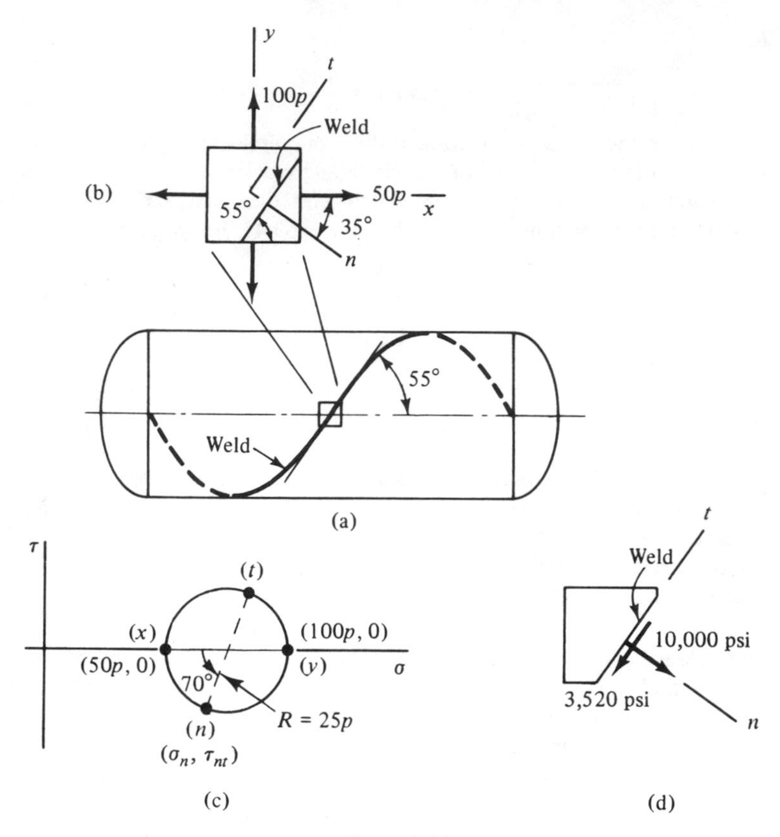

Figure 9.14

The shear stress in the weld at this pressure is

$$\tau_{nt} = -R \sin 70° = -(25 \text{ p}) \sin 70° = -25(150 \text{ psi}) \sin 70° = -3{,}520 \text{ psi} \qquad \textbf{Answer}$$

Figure 9.14(d) shows the stresses acting upon the plane of the weld at the working pressure. The sense of these stresses is in accordance with our established sign conventions.

It should be noted that our analysis accounts only for the stresses due to the internal pressure. Any stresses that may be present prior to the loading are not accounted for. Such initial stresses are often introduced during the fabrication process and are another reason why an adequate safety factor is required. Of course, initial stresses are a factor in most problems, not just pressurized cylinders.

PROBLEMS

9.30. to 9.35. Construct Mohr's circle for the state of stress shown. Determine the stress components with respect to the *nt* axes and show the results on a rotated element, or determine the normal and shear stresses on the plane indicated and show the results on a sketch, as required.

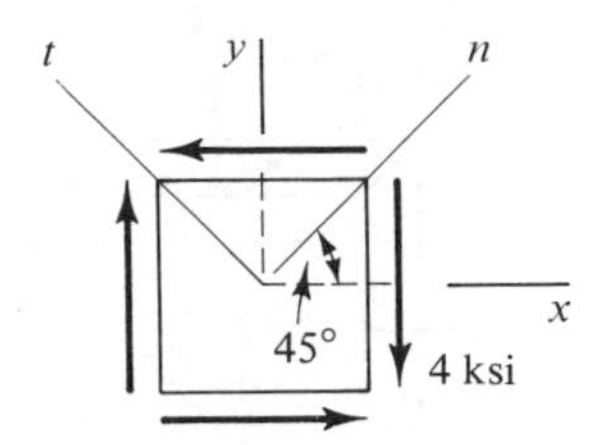

Problem 9.30

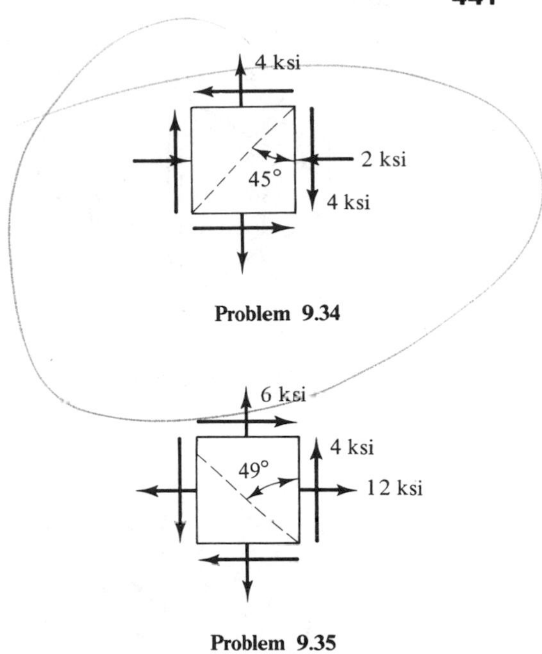

Problem 9.34

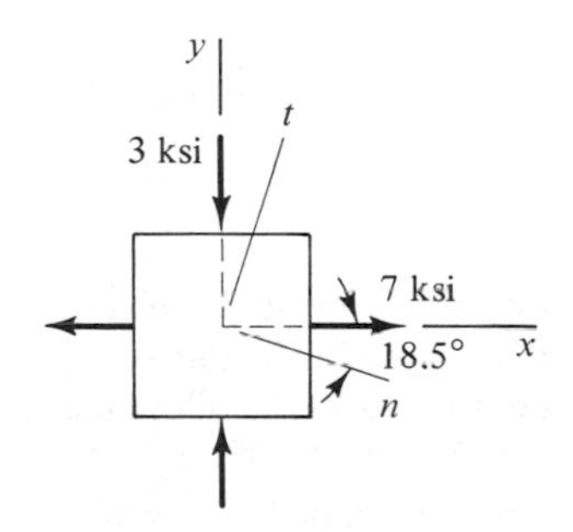

Problem 9.31

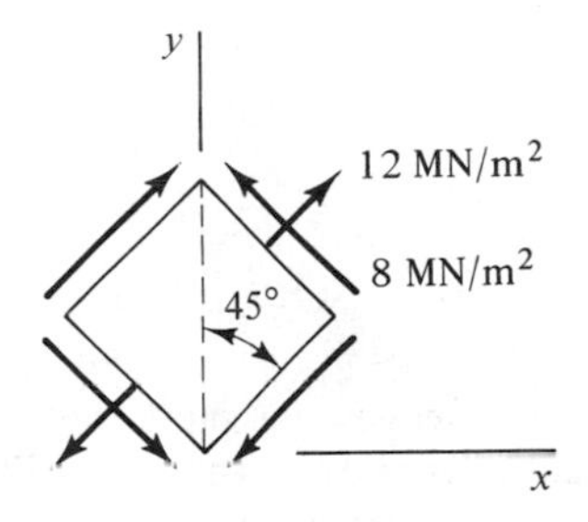

Problem 9.35

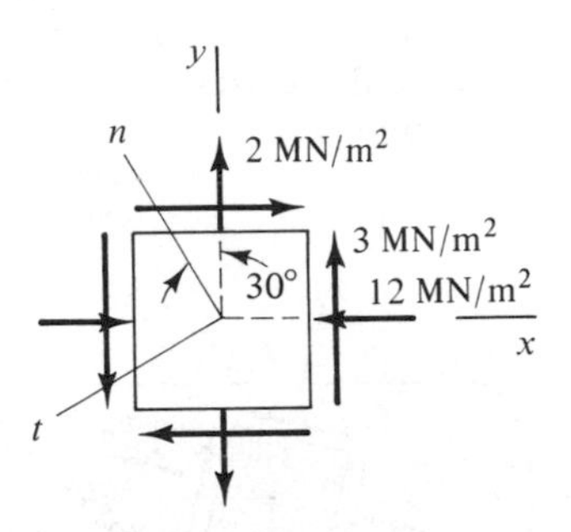

Problem 9.32

9.36. to 9.39. For the state of stress shown, determine σ_x, σ_y, and τ_{xy}. Use Mohr's circle.

Problem 9.36

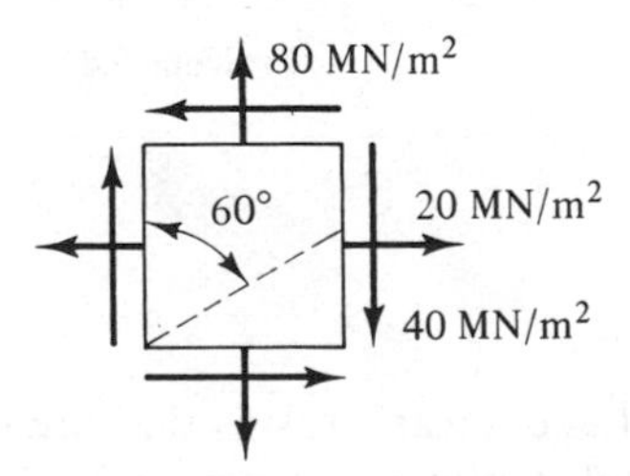

Problem 9.33

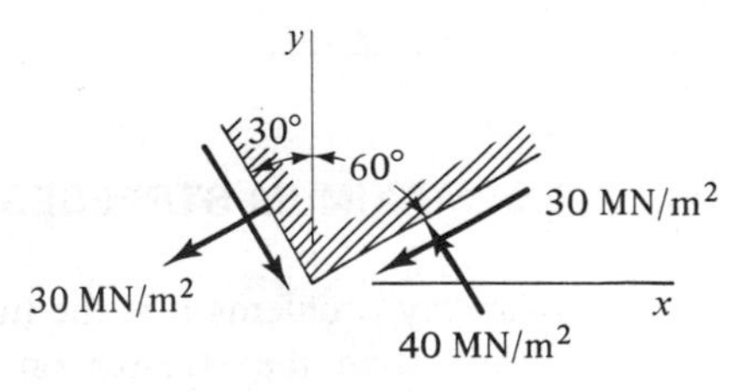

Problem 9.37

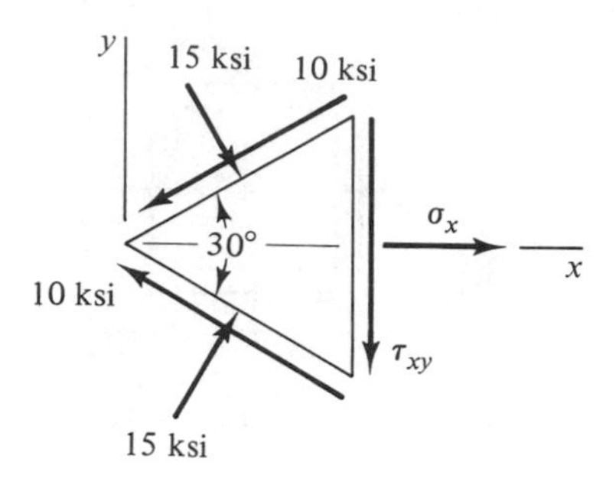

Problem 9.38

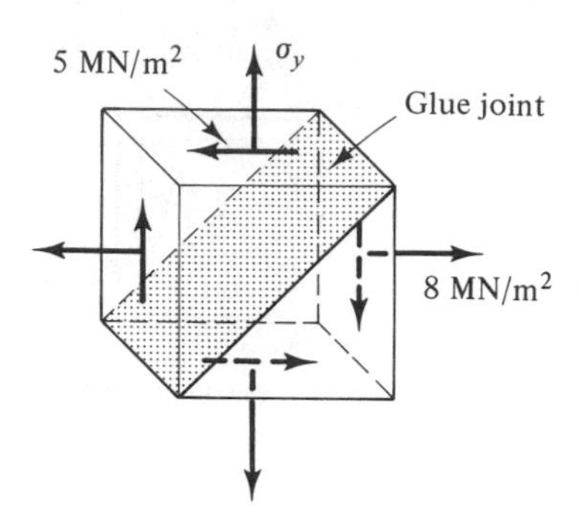

Problem 9.40

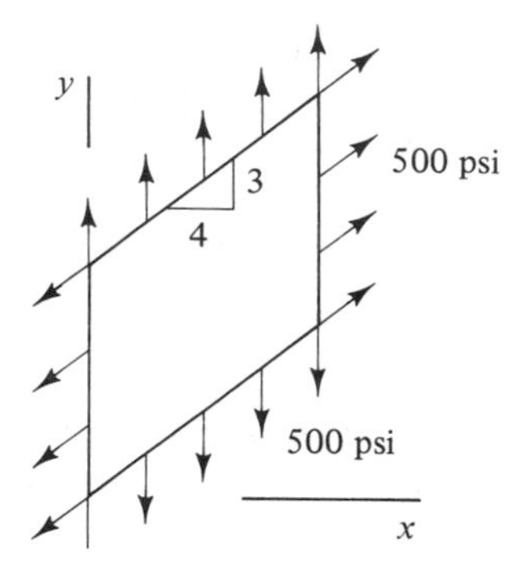

Problem 9.39

9.40. Two wedges are glued together to form a cube, as shown. The joint will break if the tensile stress in the glue exceeds 15 MN/m^2. What is the maximum allowable value of σ_y?

9.41. For the cube in Problem 9.40, determine the range of values of the stress σ_y that can be applied if the allowable shear stress in the glue is 10 MN/m^2.

9.42. An open-ended thin-walled cylinder with radius $r = 10$ in. and wall thickness $t = 0.1$ in. is subjected to an internal pressure p and an axial force F. Determine the values of p and F if $\sigma_n = 20$ ksi and $\sigma_t = 10$ ksi. What is the corresponding value of the shear stress τ_{nt}?

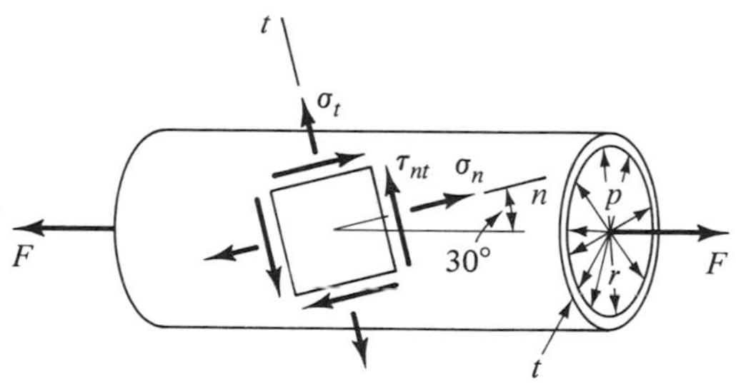

Problem 9.42

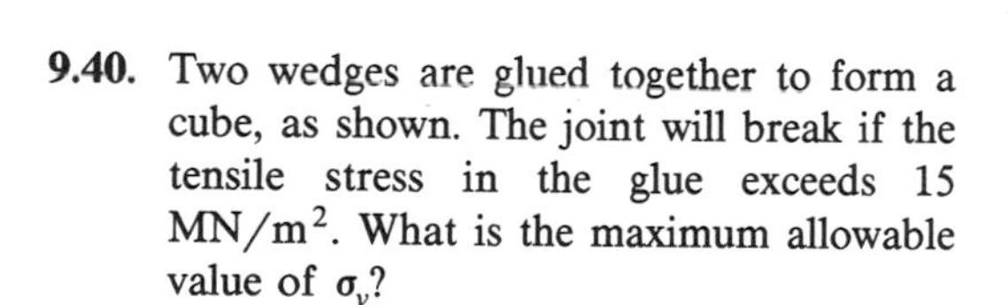

9.5 MAXIMUM STRESSES

In many problems it is the maximum normal and shear stresses that are of interest, rather than the stresses on any one particular plane. As we shall show in the following, these maximum stresses and the planes on which they occur can be readily determined by using Mohr's circle. They can also be determined directly from the stress transformation equations.

Principal Stresses and Maximum in-Plane Shear Stress

Consider a general state of plane stress at a point in a body [Figure 9.15(a)] and the corresponding Mohr's circle [Figure 9.15(b)]. It is clear from the latter figure that the algebraic largest and smallest normal stresses existing at this point correspond to the horizontal extremities of the circle. These stresses will be denoted by σ_1 and σ_2, respectively, and are called the *principal stresses*. The axes 1 and 2 associated with them are called the *principal axes of stress*, and the planes upon which they act are called the *principal planes*. There is a third principal axis normal to the plane of the figure. However, for a state of plane stress, the principal stress in this third direction is zero.

The points (a) and (b) at the extreme top and bottom of the circle correspond to the maximum and minimum in-plane shear stresses existing at this point in the body. These stresses differ only in sign, and both will be denoted by $(\tau_{nt})_{\max}$. There is no special name given the *ab* axes associated with them.

The principal stresses and the maximum in-plane shear stresses are easily determined. From the geometry of Mohr's circle [Figure 9.15(b)], we see that

$$\sigma_1 = \sigma' + R \qquad \sigma_2 = \sigma' - R \tag{9.6}$$

and

$$(\tau_{nt})_{\max} = \pm R = \pm \frac{(\sigma_1 - \sigma_2)}{2} \tag{9.7}$$

where σ' is the horizontal coordinate of the center of the circle and R is its radius. Both of these quantities can be determined from the known coordinates of points (x) and (y) on the circle, as was shown in the preceding section. The $\pm$ sign in Eq. (9.7) arises from the fact that the shear stresses are positive on one face of an element and negative on the adjoining face, according to the sign convention used for Mohr's circle. The principal stresses are shown in Figure 9.15(b) as being positive, but one or both may be negative.

The orientation of the principal axes is determined as follows. We rotate the diameter (x)-(y) of the circle in either direction until it is aligned with the diameter (1)-(2) which lies along the horizontal axis. The xy axes for the element are then rotated in the *opposite* direction, but through *one-half* the angle. This establishes the orientation of axes 1 and 2.

There is often some confusion as to which of the principal axes is axis 1 and which is axis 2. This question can be easily resolved by remembering that points on the circle correspond to axes on the element. Thus, if points (x) and (y) coincide with points (1) and (2), respectively, after rotation of the diameter of the circle, the rotated x axis is axis 1 and the rotated y axis is axis 2. This is the case shown in Figure 9.15(c). However, if point (x) coincides with point (2) after rotation, the

y
σy
τyx = τxy
τxy
x
σx
(a)
τ
σ1 = σ' + R
σ2 = σ' − R
(σ', (τnt)max)
(b)
(x)
(σx, τxy)
(τnt)max = R
R
2α
(2)
(σ', 0)
(1)
(σ2, 0)
(σ1, 0)
σ
2β = 2α + 90°
(y)
(σy, −τyx)
(a)
(σ', −(τnt)max)
(b)
2
α
y
σ2
σ1
1
α
x
σ1
σ2
(c)
y
a
β = α + 45°
b
β = α + 45°
x
σ = σ'
σ = σ'
(τnt)max
(d)

Figure 9.15

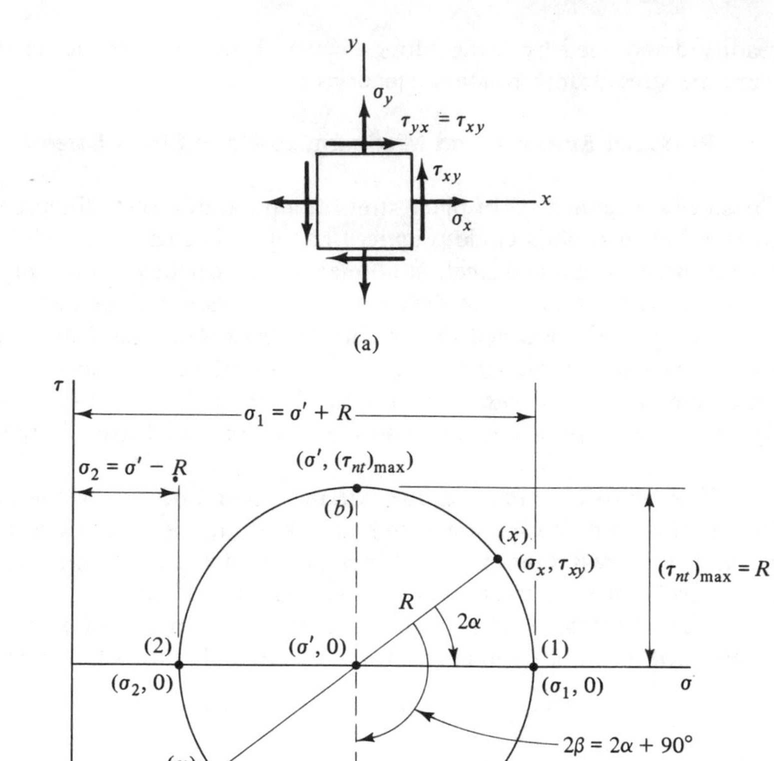

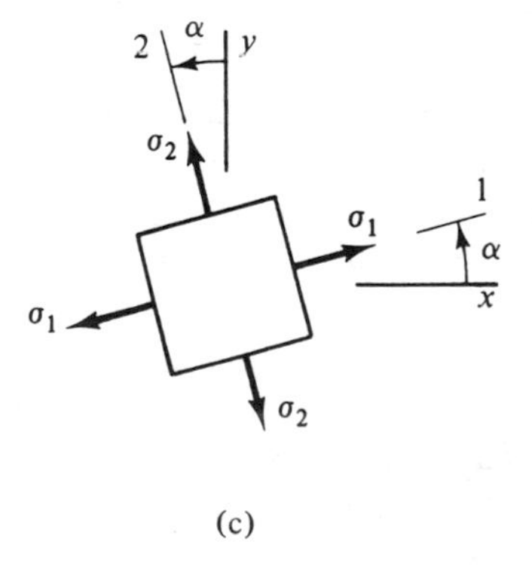

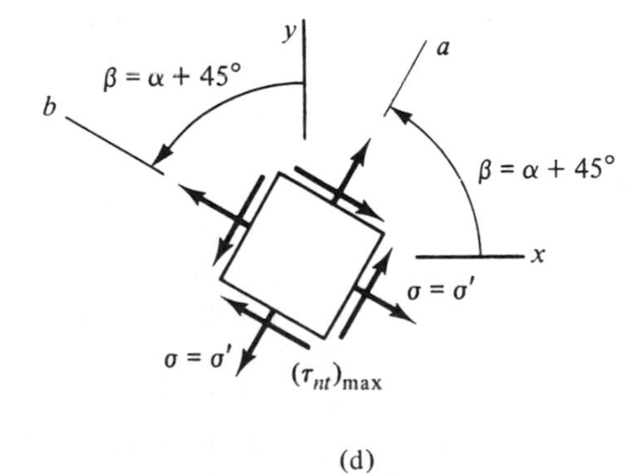

rotated x axis is axis 2 and the rotated y axis is axis 1. As we shall show in Example 9.8, the direction in which the diameter of the circle is rotated is immaterial. The results are the same in either case.

Once the principal axes have been determined, the principal stresses are shown on a rotated element with sides aligned with these axes, as in Figure 9.15(c). The faces of this element correspond to the planes upon which the principal stresses act. Note that there are no shear stresses on these planes (see also Figures 9.7 and 9.8). Thus, if a given state of stress involves only normal stresses, we can immediately say that they are the principal stresses.

This same procedure applies for determining the axes *ab* associated with the maximum in-plane shear stresses. In this case, the diameter (x)-(y) of Mohr's circle is rotated until it coincides with diameter (a)-(b), which is aligned vertically [Figure 9.15(b)]. Again, the direction of rotation is immaterial. Once the *ab* axes have been determined, the stresses associated with them are shown on a rotated element, as in Figure 9.15(d). In addition to the shear stresses, there are normal stresses acting upon all four faces of the element with magnitudes equal to the horizontal coordinate, σ', of the center of the circle. The sense of all the stresses is determined from the sign convention used for Mohr's circle. Note that the *ab* axes will always be 45° away from the principal axes because the corresponding diameters on Mohr's circle are 90° apart (see also Figures 9.7 and 9.8). In other words, the planes of maximum shear stress are 45° away from the principal planes. This result was obtained first for axially loaded members (Section 5.3) and later for shafts subjected to torsion (Section 7.4).

Principal stresses, maximum in-plane shear stresses, and the axes and planes associated with them can, of course, be determined without reference to Mohr's circle. If we combine Eqs. (9.5)–(9.7), we obtain

$$\sigma_{1,2} = \frac{\sigma_x + \sigma_y}{2} \pm \sqrt{\left(\frac{\sigma_x - \sigma_y}{2}\right)^2 + \tau_{xy}^2} \tag{9.8}$$

and

$$(\tau_{nt})_{\max} = \pm\sqrt{\left(\frac{\sigma_x - \sigma_y}{2}\right)^2 + \tau_{xy}^2} \tag{9.9}$$

In Eq. (9.8), the plus sign is associated with σ_1 and the minus sign with σ_2. With these equations, σ_1, σ_2, and $(\tau_{nt})_{\max}$ can be computed directly from the given state of stress. The normal stress acting upon the planes of maximum shear is equal to σ', and can be determined from Eqs.(9.5).

From Figure 9.15(b), we see that the orientation of the principal axes is defined by the angle 2α, which can be computed from Eqs. (9.5). This equation has two solutions, $2\alpha_1$ and $2\alpha_2$, which differ by 180°. The corresponding values of α_1 and α_2, which differ by 90°, define the orientation of the principal axes and planes. The question as to which of the angles α_1 and α_2 corresponds to the maximum value of the normal stress and which corresponds to the minimum value can be easily answered by substituting one of the angles into the stress transformation equation

[Eq. (9.1)] and observing whether the resulting value of the normal stress corresponds to σ_1 or to σ_2.

The axes a and b associated with the maximum in-plane shear stresses are defined by the angle 2β [Figure 9.15(b)]. From this figure, we have

$$2\beta = 2\alpha + 90°$$

so

$$\tan 2\beta = \tan(2\alpha + 90°) = -\cot 2\alpha \tag{9.10}$$

Combining this result with the expression for $\tan 2\alpha$ from Eqs. (9.5), we obtain

$$\tan 2\beta = -\frac{\left(\dfrac{\sigma_x - \sigma_y}{2}\right)}{\tau_{xy}} \tag{9.11}$$

The two roots of this equation, β_1 and β_2, define the orientation of the axes a and b.

The results in Eqs. (9.5)–(9.11) can also be derived directly from the stress transformation equations by using the mathematical conditions that $d\sigma_n/d\theta = 0$ and $d\tau_{nt}/d\theta = 0$ on the planes for which the values of σ_n and τ_{nt} have relative maxima or minima (see Problem 9.49).

Absolute Maximum Shear Stress

In determining the maximum shear stress at a point in a body, we must not lose sight of the fact that the material element is three-dimensional and that we have been considering only a two-dimensional view of it. It is entirely possible that a different view of the element will yield a value of shear stress larger than $(\tau_{nt})_{max}$.

To illustrate this, let us consider a material element aligned with the principal axes 1 and 2 [Figure 9.16(a)]. Let us also suppose that there is a principal stress, σ_3, acting in the third direction along axis 3 and that $\sigma_1 > \sigma_2 > \sigma_3 > 0$. The results obtained will also hold if one or more of the principal stresses is negative.

If the element is viewed from along the 3 axis, as we have done previously, the corresponding Mohr's circle is as shown in Figure 9.16(b). The largest shear stress observed is $(\tau_{nt})_{max}$, with magnitude $|\sigma_1 - \sigma_2|/2$. The Mohr's circles obtained by viewing the element from along the other two axes are also shown. Here, the largest shear stresses observed have magnitudes $|\sigma_1 - \sigma_3|/2$ and $|\sigma_2 - \sigma_3|/2$. The magnitude of the maximum shear stress existing at the point in question is the largest of the three values

$$\frac{|\sigma_1 - \sigma_2|}{2} \qquad \frac{|\sigma_1 - \sigma_3|}{2} \qquad \frac{|\sigma_2 - \sigma_3|}{2} \tag{9.12}$$

We shall refer to this stress as the *absolute maximum shear stress*, τ_{max}.

In many problems the principal stress σ_3 in the third direction is either zero or negligible compared to those in the other two directions. This does not necessarily

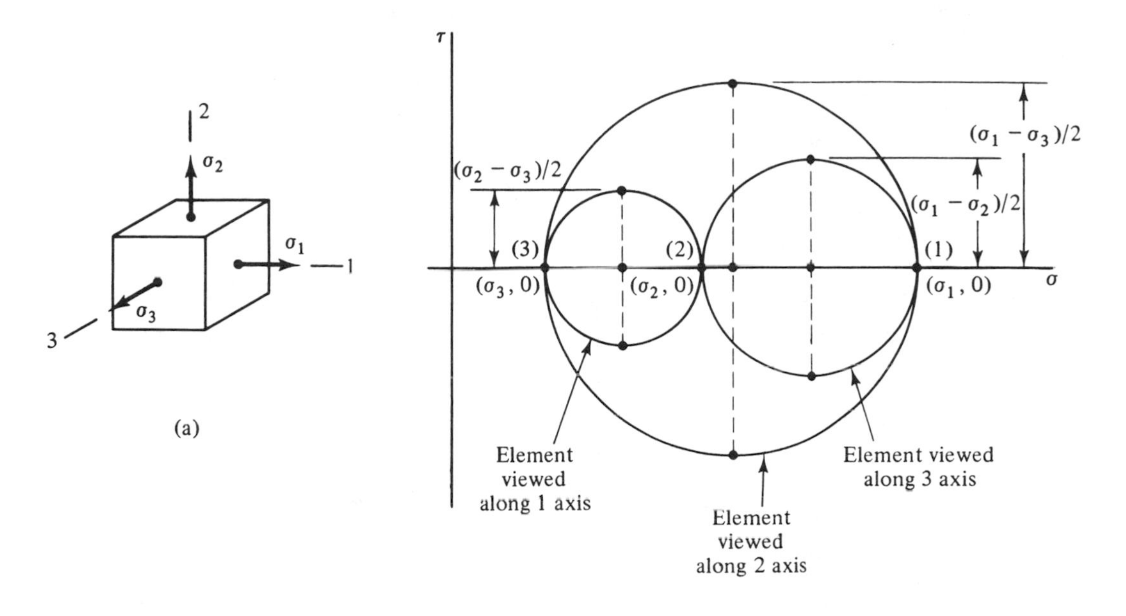

Figure 9.16

mean that the maximum in-plane shear stress will be the largest shear stress in the body, however. All three values in Eq. (9.12) must still be checked.

Example 9.8. Determination of Principal Stresses and Maximum Shear Stresses. For the state of stress shown in Figure 9.17(a), determine the principal stresses, the maximum in-plane shear stress, and the orientation of the corresponding axes (a) by using Mohr's circle and (b) by using Eqs. (9.5)–(9.11). Show the results on properly oriented elements. Also determine the absolute maximum shear stress.

Solution. (a) We first construct the Mohr's circle [Figure 9.17(b)]. The center of the circle has coordinates (8 ksi, 0) and its radius is

$$R = \sqrt{4^2 + 5^2} = 6.4 \text{ ksi}$$

Also,

$$\tan 2\alpha = -\frac{5}{4} \qquad 2\alpha = -51.3^\circ$$

From the geometry of the circle, we have

$$\sigma_1 = \sigma' + R = 8 + 6.4 = 14.4 \text{ ksi}$$

$$\sigma_2 = \sigma' - R = 8 - 6.4 = 1.6 \text{ ksi} \qquad \textbf{Answer}$$

$$|(\tau_{nt})_{\max}| = R = 6.4 \text{ ksi}$$

To locate the principal axes, we rotate the diameter (x)-(y) of the circle 51.3° counterclockwise until it is aligned with the horizontal diameter (1)-(2). The principal axes are

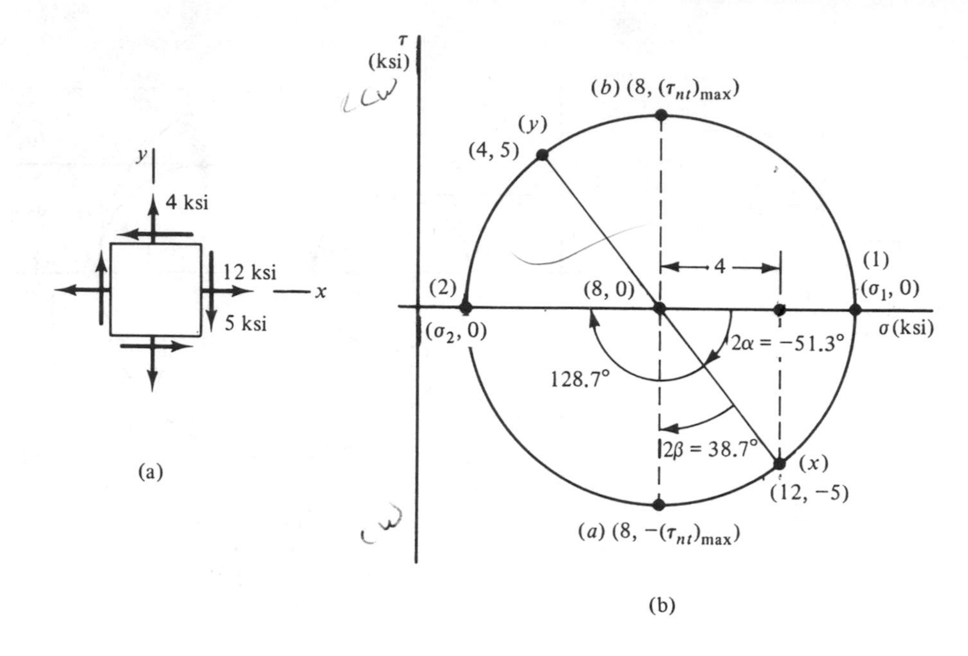

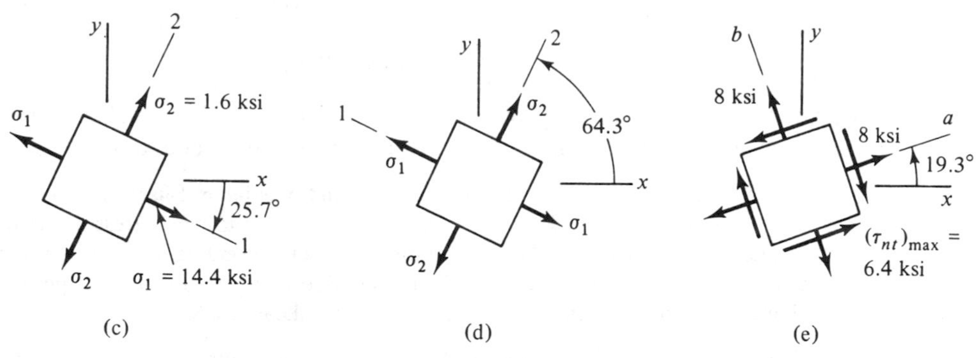

Figure 9.17

then obtained by rotating the xy axes in the opposite direction, but through one-half the angle. Since points (x) and (y) on the circle coincide with points (1) and (2), respectively, after rotation of the diameter, the rotated x axis is axis 1 and the rotated y axis is axis 2. Figure 9.17(c) shows the principal stresses acting upon the rotated element.

The principal axes could also have been determined by rotating diameter (x)-(y) of the circle clockwise through an angle of 128.7°. In this case, the rotated x axis becomes axis 2 and the rotated y axis becomes axis 1. The resulting element, shown in Figure 9.17(d), is seen to be identical to the element in Figure 9.17(c). This confirms our earlier statement that the direction of rotation is immaterial.

The axes associated with the maximum in-plane shear stress are obtained by rotating diameter (x)-(y) of the circle 38.7° clockwise until it is aligned with diameter (a)-(b). The a and b axes then correspond to the rotated x and y axes, respectively [Figure 9.17(e)]. The stresses acting upon the rotated element are as indicated. It is left as an exercise to show that the same results are obtained if the diameter of the circle is rotated counterclockwise. Note that the element in Figure 9.17(e) is rotated 45° with respect to the element in Figure 9.17(d). This is because principal planes and planes of maximum shear stress always are 45° apart.

The absolute maximum shear stress is the largest of the three values given in Eq. (9.12). Since $\sigma_3 = 0$ in this problem, we have

$$\left|\frac{\sigma_1 - \sigma_2}{2}\right| = |(\tau_{nt})_{\max}| = 6.4 \text{ ksi}$$

$$\left|\frac{\sigma_1 - \sigma_3}{2}\right| = \frac{14.4 - 0}{2} = 7.2 \text{ ksi}$$

$$\left|\frac{\sigma_2 - \sigma_3}{2}\right| = \frac{1.6 - 0}{2} = 0.8 \text{ ksi}$$

Thus,

$$\tau_{\max} = 7.2 \text{ ksi} \qquad \textbf{Answer}$$

(b) From Eqs. (9.5)–(9.7)

$$\sigma' = \frac{\sigma_x + \sigma_y}{2} = \frac{12 + 4}{2} = 8 \text{ ksi}$$

$$|(\tau_{nt})_{\max}| = R = \sqrt{\left(\frac{\sigma_x - \sigma_y}{2}\right)^2 + \tau_{xy}^2} = \sqrt{\left(\frac{12 - 4}{2}\right)^2 + (-5)^2}$$

$$= 6.4 \text{ ksi} \qquad \textbf{Answer}$$

$$\sigma_1 = \sigma' + R = 14.4 \text{ ksi} \qquad \sigma_2 = \sigma' - R = 1.6 \text{ ksi} \qquad \textbf{Answer}$$

$$\tan 2\alpha = \frac{\tau_{xy}}{\left(\frac{\sigma_x - \sigma_y}{2}\right)} = \frac{-5}{\left(\frac{12 - 4}{2}\right)} = -\frac{5}{4}$$

$$2\alpha_1 = -51.3° \qquad 2\alpha_2 = 2\alpha_1 + 180° = 128.7° \qquad \textbf{Answer}$$

If we substitute the value of $2\alpha_1$ into Eq. (9.1), we get $\sigma_n = 14.4$ ksi. Thus, the angle α_1 corresponds to σ_1, which means that principal axis 1 is located $-25.7°$ (or 25.7° CW) from the x axis [Figure 9.17(c)]. Since the principal axes are perpendicular, the orientation of axis 2 is also determined. From the relation $2\beta = 2\alpha + 90°$, or from Eq. (9.11), we have

$$2\beta_1 = 38.7° \qquad 2\beta_2 = 2\beta_1 + 180° = 218.7° \qquad \textbf{Answer}$$

Substituting the value of $2\beta_1$ into Eq. (9.2), we get $\tau_{nt} = -6.4$ ksi. Thus, the angle β_1 corresponds to the negative value of $(\tau_{nt})_{\max}$. In other words, the axis a associated with the minimum value of the shear stress is oriented $+19.3°$ (or 19.3° CCW) from the x axis [Figure 9.17(e)]. The angle β_2 can also be used to orient the element. It is left as an exercise to show that the results are the same as shown in Figure 9.17(e). *Hint*: Show the axes a and b for the rotated element and apply the sign convention for shear stress (Section 5.5).

The choice between use of Mohr's circle or the corresponding equations is a personal one. Both give identical results. For hand calculations, the Mohr's circle approach is faster and the signs of the stresses and the orientations of the various axes are more easily determined. However, the equations can be readily programmed on a computer. Regardless of the computational methods used, Mohr's circle can provide a fast and accurate check of the results.

PROBLEMS

9.43. to 9.45. For the state of stress shown, use Mohr's circle to determine the principal stresses, the maximum in-plane shear stress, and the orientation of the corresponding axes. Show the results on properly oriented elements. Also determine the absolute maximum shear stress.

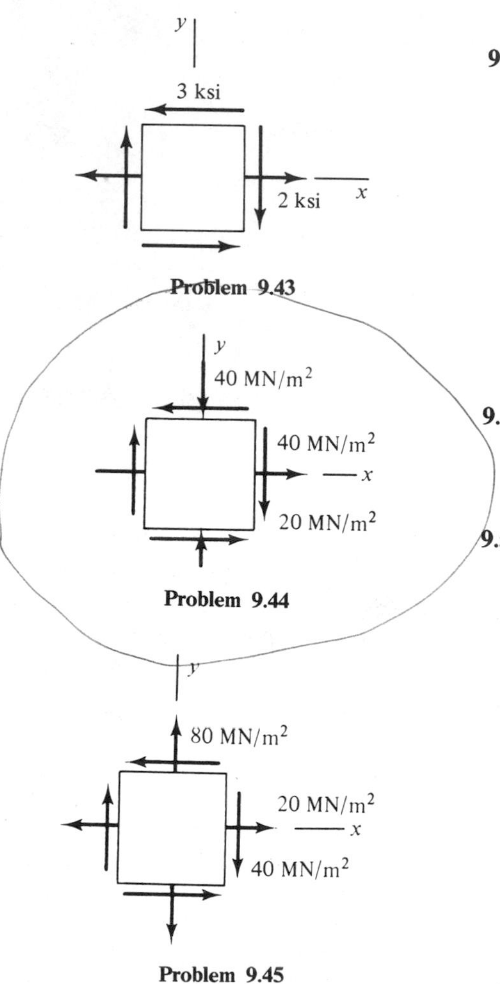

Problem 9.43

Problem 9.44

Problem 9.45

9.46. to 9.48. Solve Problems 9.43 to 9.45 using Eqs. (9.5)–(9.12) instead of Mohr's circle.

9.49. Derive Eqs. (9.8) and (9.9) for the principal stresses and the maximum in-plane shear stress by using the conditions that $d\sigma_n/d\theta = 0$ and $d\tau_{nt}/d\theta = 0$ on planes where the values of the stresses are relative maxima or minima. Also derive Eqs. (9.5) and (9.11) for the orientations of the principal planes and the planes of maximum in-plane shear stress.

9.50. If $\sigma_1 = 1{,}500$ psi for the state of stress shown, what is the value of σ_x?

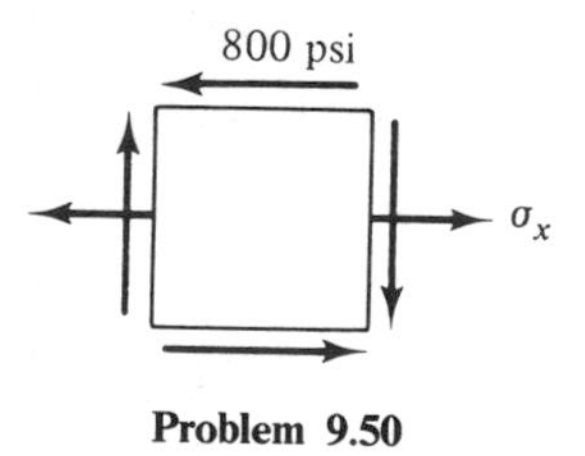

Problem 9.50

9.51. In Problem 9.50, what range of values of σ_x can be applied if the maximum tensile and compressive stresses are not to exceed 2,000 psi?

9.52. The uniform triangular plate shown is acted upon by a compressive stress 2σ on face AC and a tensile stress 2σ on face BC. Determine (a) the stresses σ_x and τ_{xy} acting

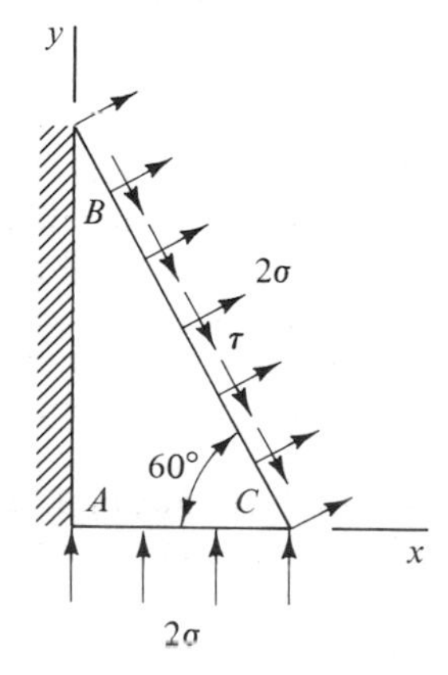

Problem 9.52

on face AB and the shear stress τ on face BC and (b) the principal stresses.

9.53. A solid circular shaft can transmit a torque of 3.56 kN · m without exceeding the maximum allowable tensile stress of 145 MN/m^2. What torque can be transmitted by the shaft if it is also subjected to a tensile force $F = 200$ kN?

9.54. A closed-end tube with an inside radius of 30 mm and a 3-mm wall thickness supports an axial tensile force $F = 10$ kN and an internal pressure p. By considering all the stresses acting upon an element of material at the inside surface of the tube, determine the maximum pressure it can withstand. The maximum allowable absolute shear stress is 53 MN/m^2. What is the maximum allowable pressure if the radial compressive stress is neglected?

9.55. A 2-in. diameter solid shaft is subjected to combined torsion and bending, as shown. If $M = 1{,}200$ lb · ft and $T = 1{,}500$ lb · ft, determine the principal stresses and the maximum in-plane shear stress at point A and show the results on properly oriented elements.

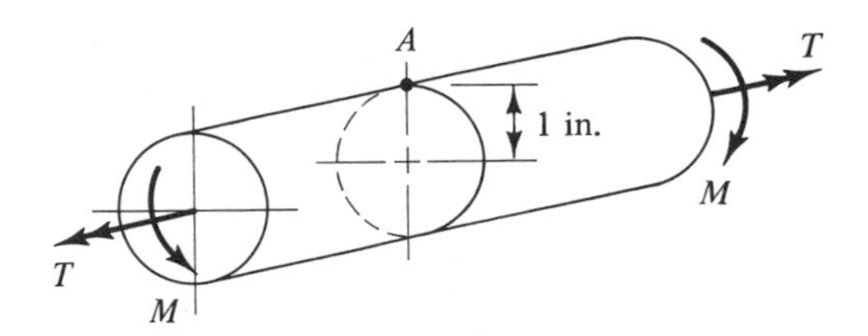

Problem 9.55

9.56. Determine the magnitudes of the bending moment M and torque T acting on the shaft in Problem 9.55 if the principal stresses are $\sigma_1 = 18$ ksi and $\sigma_2 = -2$ ksi.

9.57. Determine the principal stresses and maximum in-plane shear stresses at points A and B for the member in Problem 9.16 (Section 9.2) and show them on properly oriented elements.

9.58. Repeat Problem 9.57 for the member in Problem 9.17 (Section 9.2).

9.6 FAILURE DUE TO COMBINED LOADINGS

Now that we have considered the states of stress produced by combined loadings, we need to determine how they relate to the failure of a member. In particular, we shall need to be able to predict the conditions under which members made of brittle materials will fracture and those made of ductile materials will undergo inelastic deformations. The latter case is particularly significant because some of our basic load-stress relations do not apply in the inelastic range. The following example illustrates the basic problem at hand.

Consider a circular shaft made of a ductile material and subjected to an axial load, F, and a twisting couple, C. The loads at which inelastic action will occur if F and C act separately can be readily determined. We need only set the maximum normal stress due to F equal to the tensile yield strength of the material or the maximum shear stress due to C equal to the yield strength in shear. But what criterion do we use to determine the combination of F and C that produces yielding of the material?

Similarly, the individual values of F and C at which the shaft will fracture if made of a brittle material can be determined by setting the maximum tensile stresses developed equal to the tensile ultimate strength of the material. Furthermore, the plane of fracture can be predicted. The obvious question now is, what combination of F and C will cause fracture and on what plane will it likely occur?

Numerous criteria, or *theories of failure*, have been developed for predicting the conditions under which brittle materials will fracture or ductile materials will undergo inelastic deformations when subjected to general states of stress. Only two of these theories will be discussed here—one for brittle materials and one for ductile materials. Other theories of failure are discussed in more advanced texts.

The *maximum tensile stress theory* states that a brittle material will fracture when the maximum tensile stress reaches the ultimate strength of the material in tension. Fracture usually occurs along the plane on which the maximum tensile stress acts. The *maximum shear stress theory* states that a ductile material will yield and undergo inelastic deformations when the absolute maximum shear stress reaches the material yield strength in shear. Experiments show that these theories give reliable results for many materials, and they are widely used in the design and analysis of members subjected to combined loadings. The application of these theories is illustrated in the following example.

Example 9.9. Allowable Loads on a Shaft. The shaft shown in Figure 9.18(a) is subjected to combined axial loading and torsion. What is the largest tensile force F that can be applied without causing yielding or fracture if (a) the shaft is made of a ductile material with shear yield strength $\tau_{YS} = 145\ \text{MN/m}^2$ and (b) it is made of a brittle material with ultimate tensile strength $\sigma_U = 140\ \text{MN/m}^2$.

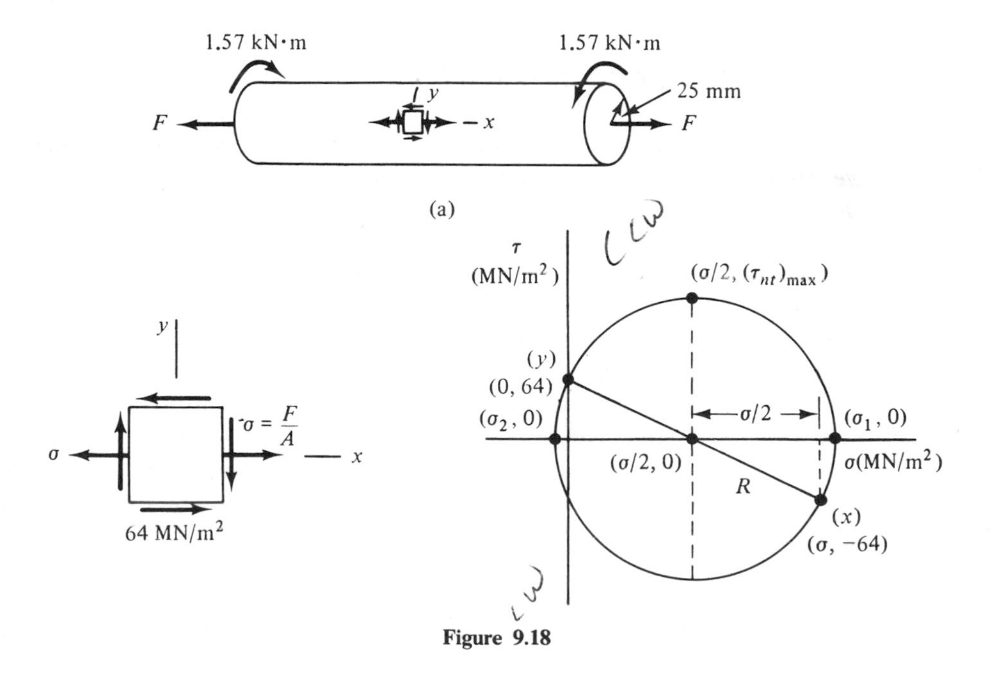

Figure 9.18

Solution. (a) The stresses are largest at the outside of the shaft, where the torsional shear stress is

$$\tau = \frac{TR}{J} = \frac{2T}{\pi R^3} = \frac{2(1.57 \times 10^3\ \text{N}\cdot\text{m})}{\pi(0.025\ \text{m})^3} = 64 \times 10^6\ \text{N/m}^2 \text{ or } 64\ \text{MN/m}^2$$

Thus, the state of stress in the shaft and the corresponding Mohr's circle are as shown in Figure 9.18(b), where $\sigma = F/A$ is the tensile stress due to the axial load. The cross-sectional area of the member is

$$A = \pi R^2 = \pi(0.025\ \text{m})^2 = 2 \times 10^{-3}\ \text{m}^2$$

The center of the Mohr's circle is located at $(\sigma/2, 0)$ and its radius is

$$R = \sqrt{\left(\frac{\sigma}{2}\right)^2 + (64)^2}\ \text{MN/m}^2$$

Since σ_1 is positive, σ_2 is negative, and $\sigma_3 = 0$, the absolute maximum shear stress will be equal to the maximum in-plane shear stress [see Eqs. (9.12)]. Thus,

$$\tau_{\max} = |(\tau_{nt})_{\max}| = R \leqslant 145\ \text{MN/m}^2$$

or

$$R^2 = (\sigma/2)^2 + (64)^2 \leqslant (145)^2$$

Solving for σ, we get

$$\sigma \leqslant 260\ \text{MN/m}^2$$

therefore,

$$F \leqslant (260 \times 10^6\ \text{N/m}^2)(2 \times 10^{-3}\ \text{m}^2)$$

$$F \leqslant 520\ \text{kN} \qquad \textbf{Answer}$$

This is the maximum force that can be applied without causing inelastic deformations.

(b) The maximum tensile stress in the shaft is

$$\sigma_1 = \frac{\sigma}{2} + R \leqslant 140\ \text{MN/m}^2$$

so we have

$$R^2 = \left(\frac{\sigma}{2}\right)^2 + (64)^2 \leqslant \left(140 - \frac{\sigma}{2}\right)^2$$

or

$$\sigma \leqslant 110\ \text{MN/m}^2$$

Thus,

$$F \leqslant (110 \times 10^6\ \text{N/m}^2)(2 \times 10^{-3}\ \text{m}^2)$$

$$F \leqslant 220\ \text{kN} \qquad \textbf{Answer}$$

This is the largest force that can be applied without causing fracture.

PROBLEMS

9.59. A solid circular shaft made of steel with shear yield strength $\tau_{YS} = 40$ ksi can transmit 25 hp at 1,800 rpm without yielding. What power can be transmitted at this same speed without yielding if the shaft is also subjected to a tensile force $F = 2{,}000$ lb?

9.60. Same as Problem 9.59, except that the shaft is subjected to a bending moment $M = 500$ lb · in. in addition to the tensile force F.

9.61. What is the largest load F that can be supported by the beam shown if (a) it is made of 6061-T6 aluminum and (b) it is made of grey cast iron? Use a safety factor of 2. For satisfactory performance, the member must not yield or fracture.

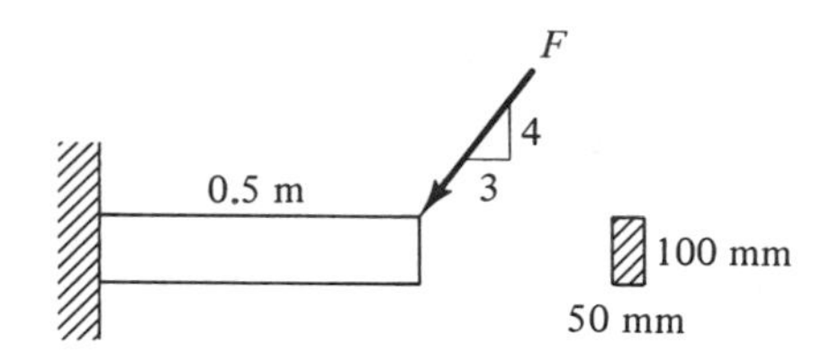

Problem 9.61

9.62. What is the minimum required cross-sectional dimension d for the member shown if (a) it is made of 2024-T3 aluminum and (b) it is made of titanium alloy? Use a safety factor of 1.5 against failure by the onset of yielding. What is the ratio of the weights of the titanium member and the member made of aluminum?

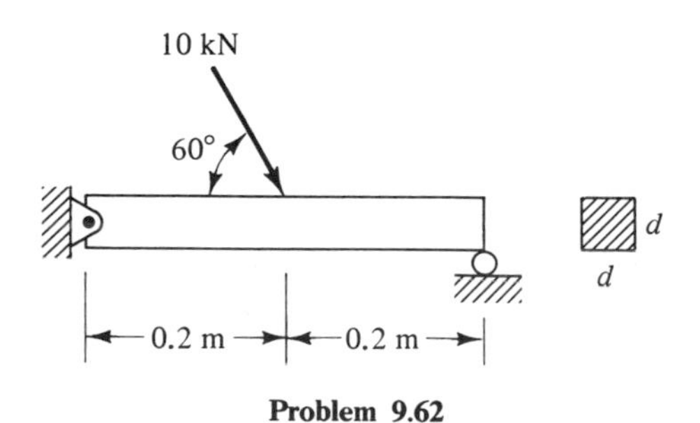

Problem 9.62

9.63. Design specifications for the 40-mm diameter steel shaft shown require that it transmit a torque of 800 N · m with a minimum safety factor of 2 against failure by yielding. Because of faulty installation, the center bearing is 6 mm lower than the others. Will the shaft still operate within specifications? Assume that the bearings act like simple supports with regard to bending; $\tau_{YS} = 145$ MN/m^2 and $E = 207$ GN/m^2.

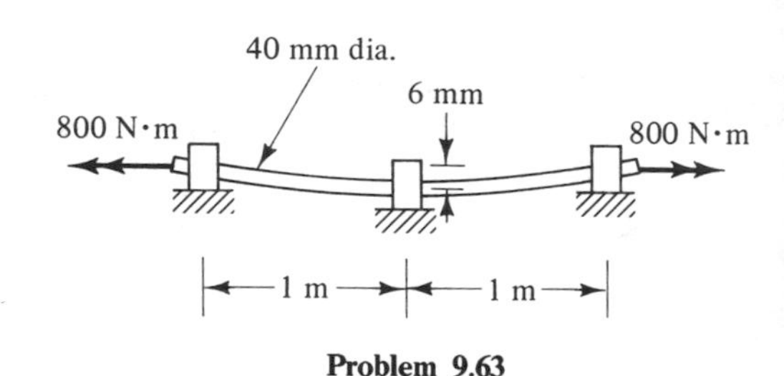

Problem 9.63

9.64. A 20-in. diameter cylindrical pressure vessel is designed to hold a gas at 200 psi with a safety factor of 2 against failure by the onset of yielding. Because of faulty installation, the cylinder is also subjected to a torque with a magnitude of 400 kip · in. By what percentage does the presence of the torque reduce the maximum pressure the cylinder can withstand?

9.65. The thin-walled cylinder shown is made of structural steel and contains a gas at 200 psi. To prevent buckling of the walls, the maximum compressive stress must not ex-

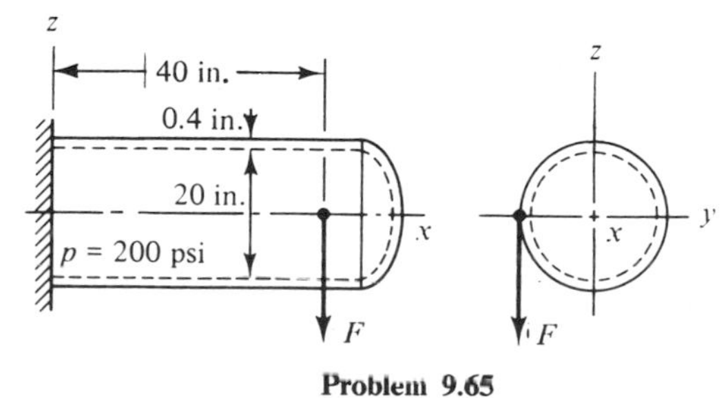

Problem 9.65

ceed 10 ksi. What is the maximum load F that can be applied to the cylinder without causing buckling?

9.66. A hydraulic cylinder with an internal diameter of 100 mm is to be made of a cast metal with maximum allowable stresses of 420 MN/m^2, 280 MN/m^2 and 180 MN/m^2 in tension, compression, and shear, respectively. What minimum wall thickness t is required to provide a safety factor of 2 at a working pressure of 8 MN/m^2?

9.67. An engineer suggests that the hydraulic cylinder described in Problem 9.66 can be made with thinner walls if it is also subjected to an axial compressive force. Is this correct? Justify your answer.

9.68. A large sign is supported on each side by a section of structural steel pipe, as shown. The sign weighs 800 lb and the maximum wind pressure on it is anticipated to be 0.4 psi. What nominal size pipe should be used if a safety factor of 1.5 against failure by the onset of yielding is required? Assume the connections between the pipes and the sign behave like pinned joints in bending.

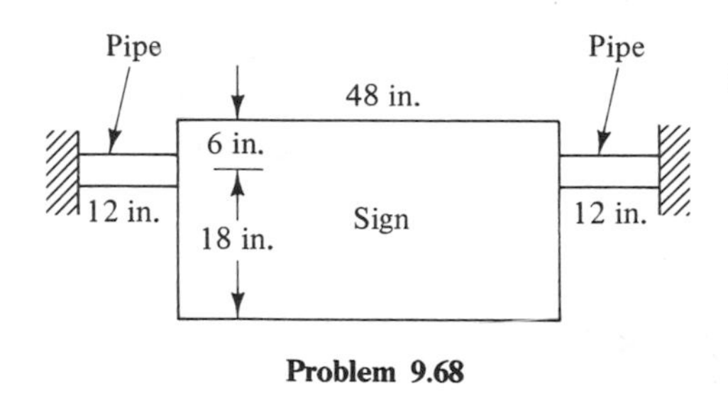

Problem 9.68

9.69. If 2-in. (nominal) diameter steel pipe is used to support the sign described in Problem 9.68, what is the safety factor against failure by the onset of yielding?

9.70. A 20-mm diameter acrylic rod with shear yield strength $\tau_{YS} = 20$ MN/m^2 and ultimate tensile strength $\sigma_U = 35$ MN/m^2 is loaded as shown. Will the member first yield or fracture as F is increased? What is the value of the failure load?

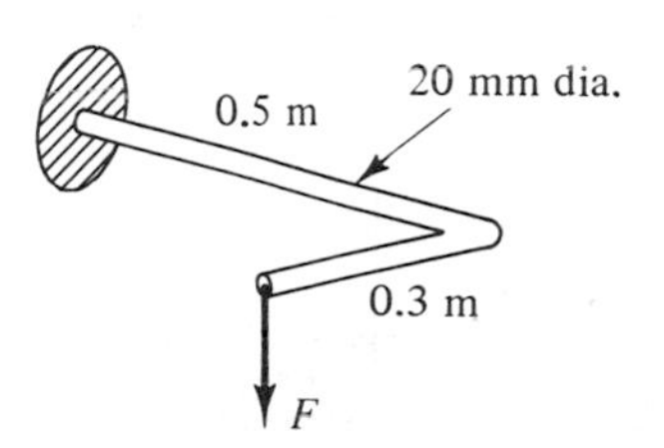

Problem 9.70

9.71. Derive an expression relating the torque T and bending moment M that a brittle circular shaft with ultimate tensile strength σ_U can support without fracture. Express the results in terms of T_U, the torque required to produce fracture if acting alone, and M_U, the moment required to produce fracture if acting alone.

9.72. Derive an expression relating the torque T and axial force P that a ductile circular shaft with shear yield strength τ_{YS} can support without yielding. Express the results in terms of the axial force P_{YS} which would produce yielding if acting alone and the torque T_{YS} which would produce yielding if acting alone.

9.7 TRANSFORMATION OF STRAINS

Once the strain components with respect to one set of axes are known at a point in a body, the strain components with respect to any other rotated set of axes can be determined. In particular, the maximum and minimum normal strains and the maximum shear strain existing at the point in question can be found, as well as the orientation of the corresponding axes. The necessary computational procedures are discussed in this section for the case of plane strain. These procedures are closely related to those used for stress transformation problems.

Strain Transformation Equations

Consider axes xy and nt at a point P in an undeformed body, with nt rotated with respect to xy through an angle θ, measured counterclockwise [Figure 9.19(a)]. Let PQ and PR be short line segments in the n and t directions, respectively. When the body deforms, points P, Q, and R will move to new positions P', Q', and R' [Figure 9.19(b)]. In general, the lengths of the line segments and the angle between

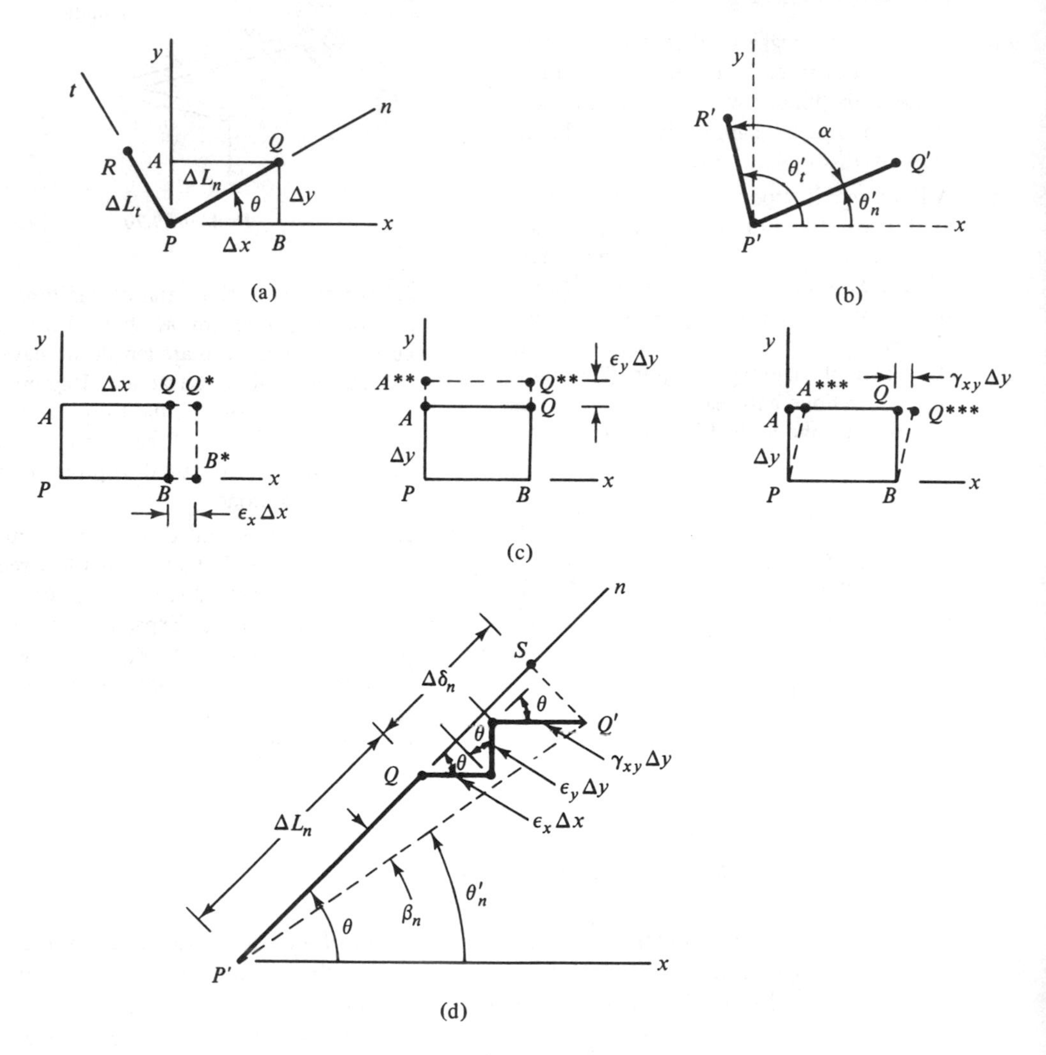

Figure 9.19

them will change. The problem is to express the resulting strain components $(\varepsilon_n, \varepsilon_t, \gamma_{nt})$ in terms of the strain components $(\varepsilon_x, \varepsilon_y, \gamma_{xy})$ and the angle θ.

To determine ε_n, we consider the rectangle $PAQB$ in Figure 9.19(a) with dimensions $\Delta x = \Delta L_n \cos\theta$ and $\Delta y = \Delta L_n \sin\theta$, where ΔL_n is the length of the diagonal PQ. Figure 9.19(c) shows the displacement of point Q relative to point P when this rectangle is subjected to each of the strains ε_x, ε_y, and γ_{xy}. Since the strains and the associated deformations are small, these displacements can be directly superimposed, as in Figure 9.19(d). The deformations are shown greatly exaggerated for clarity.

From the geometry of Figure 9.19(d), we find that the change in length of line segment PQ is

$$\Delta\delta_n = P'Q' - PQ = \varepsilon_x \Delta_x \cos\theta + \varepsilon_y \Delta_y \sin\theta + \gamma_{xy} \Delta_y \cos\theta$$

or, since $\Delta x = \Delta L_n \cos\theta$ and $\Delta y = \Delta L_n \sin\theta$,

$$\Delta\delta_n = \Delta L_n \left(\varepsilon_x \cos^2\theta + \varepsilon_y \sin^2\theta + \gamma_{xy} \sin\theta\cos\theta \right)$$

Substituting this expression into the definition of normal strain, Eq. (5.13), we obtain

$$\varepsilon_n = \lim_{\Delta L_n \to 0} \frac{\Delta\delta_n}{\Delta L_n} = \varepsilon_x \cos^2\theta + \varepsilon_y \sin^2\theta + \gamma_{xy} \sin\theta\cos\theta \tag{9.13}$$

Since the angle θ is arbitrary, this equation gives the normal strain in any direction at the point P. In particular, the normal strain ε_t can be obtained by replacing the angle θ by $\theta + \pi/2$.

Now let us consider the shear strain γ_{nt}. When the body deforms, line segments PQ and PR will rotate slightly and assume final orientations defined by the angles θ_n' and θ_t' in Figure 9.19(b). Let β_n and β_t be the respective angles of rotation. From Figure 9.19(d), we see that $\theta_n' = \theta - \beta_n$. Similarly, $\theta_t' = \theta + \pi/2 - \beta_t$. Thus, the final angle between the line segments is

$$\alpha = \theta_t' - \theta_n' = \frac{\pi}{2} - (\beta_t - \beta_n)$$

From the definition of shear strain, Eq. (5.15), we have

$$\gamma_{nt} = \frac{\pi}{2} - \lim_{\substack{\Delta L_n \to 0 \\ \Delta L_t \to 0}} \alpha - \lim_{\substack{\Delta L_n \to 0 \\ \Delta L_t \to 0}} (\beta_t - \beta_n) \tag{9.14}$$

The angle β_n can be obtained from the geometry of Figure 9.19(d). Since β_n is small, the arc length $\beta_n(\Delta L_n + \Delta\delta_n)$ will be approximately equal to the chord length $Q'S$. Thus, we have

$$\beta_n(\Delta L_n + \Delta\delta_n) = Q'S = \varepsilon_x \Delta x \sin\theta - \varepsilon_y \Delta_y \cos\theta + \gamma_{xy} \Delta_y \sin\theta$$

or, since $\Delta x = \Delta L_n \cos\theta$ and $\Delta y = \Delta L_n \sin\theta$,

$$\beta_n = \frac{1}{(1 + \Delta\delta_n/\Delta L_n)} \left[(\varepsilon_x - \varepsilon_y) \sin\theta\cos\theta + \gamma_{xy} \sin^2\theta \right]$$

Taking the limit as $\Delta L_n \to 0$ and noting that the quantity $\lim \Delta\delta_n/\Delta L_n = \varepsilon_n$ is very small compared to unity, we obtain

$$\lim_{\Delta L_n \to 0} \beta_n = (\varepsilon_x - \varepsilon_y)\sin\theta\cos\theta + \gamma_{xy}\sin^2\theta$$

The angle β_t can also be obtained from this expression by replacing θ by $\theta + \pi/2$. Making this substitution and recognizing that $\cos(\theta + \pi/2) = -\sin\theta$ and $\sin(\theta + \pi/2) = \cos\theta$, we have

$$\lim_{\Delta L_t \to 0} \beta_t = -(\varepsilon_x - \varepsilon_y)\sin\theta\cos\theta + \gamma_{xy}\cos^2\theta$$

The shear strain is obtained by substituting these results into Eq. (9.14):

$$\gamma_{nt} = -2(\varepsilon_x - \varepsilon_y)\sin\theta\cos\theta + \gamma_{xy}(\cos^2\theta - \sin^2\theta) \tag{9.15}$$

It is convenient to express Eqs. (9.13) and (9.15) in terms of 2θ by using the double-angle trigonometric identities listed in Section 9.3. We have

$$\varepsilon_n = \frac{\varepsilon_x + \varepsilon_y}{2} + \frac{\varepsilon_x - \varepsilon_y}{2}\cos 2\theta + \frac{\gamma_{xy}}{2}\sin 2\theta \tag{9.16}$$

$$\frac{\gamma_{nt}}{2} = -\frac{(\varepsilon_x - \varepsilon_y)}{2}\sin 2\theta + \frac{\gamma_{xy}}{2}\cos 2\theta \tag{9.17}$$

These equations are known as the *strain transformation equations*. Given the values of the strain components with respect to xy axes, they determine the values with respect to the rotated axes nt. These relations are based only upon geometry and, therefore, apply regardless of the material behavior.

The sign conventions for the strains are as indicated in Section 5.6. Normal strains are considered positive if tensile and negative if compressive. Shear strains are considered positive if the angle between the positive directions of the two associated axes decreases (see Figure 5.22). Recall that these sign conventions are consistent with those used for the stresses. That is, positive normal stresses correspond to positive normal strains and positive shear stresses correspond to positive shear strains. As described in Section 5.6, the state of strain associated with a particular set of axes can be displayed by showing the associated deformations of a small material element with sides aligned with these axes (see Figure 5.22 and Example 9.10). Strains can also be depicted by arrows, provided they are not confused with stresses.

Equations (9.16) and (9.17) are identical in form to the stress transformation equations, Eqs. (9.1) and (9.2). The only differences are that normal stresses are replaced by normal strains and the shear stresses are replaced by one-half the shear strains. Consequently, the procedures for determining the strain components with respect to a given set of axes are the same as for stresses. This is also true of the procedures for finding the maximum and minimum normal and shear strains and the orientations of the axes associated with them. Furthermore, the variations in the strains at a given point can be represented in the same ways as the stress variations. For example, they can be represented by Cartesian or polar plots similar to those shown in Figures 9.7 and 9.8. In particular, they can be represented by a *Mohr's circle of strain* in a plane with coordinates (ε, $\gamma/2$).

Mohr's Circle of Strain

Mohr's circle of strain (Figure 9.20) is laid out and used in exactly the same way as the circle of stress. In constructing the circle, the sign of the shear strain associated with point (x) corresponding to the x axis is taken in accordance with the sign convention for shear strains discussed earlier. However, the shear strain associated with point (y) corresponding to the y axis is plotted with the opposite sign. As explained in Section 9.4, this reversal of sign is necessary for proper construction of the circle. Use of Mohr's circle of strain will be demonstrated in Example 9.10.

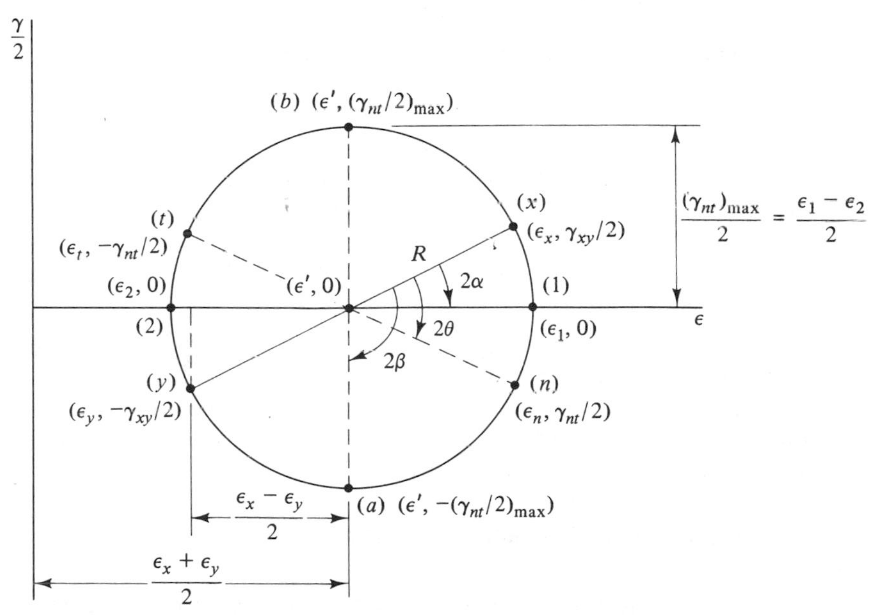

Figure 9.20

Principal strains and maximum shear strains

The algebraic maximum and minimum normal strains, denoted by ε_1 and ε_2, are called the *principal strains*. The associated axes 1 and 2 are called the *principal axes of strain*. Referring to Mohr's circle (Figure 9.20), we see that there are no shear strains associated with the principal axes. The extreme top and bottom of the circle correspond to one-half of the maximum shear strain, $(\gamma_{nt})_{max}$, which is accompanied by normal strains with magnitudes equal to the horizontal coordinate, ε', of the center of the circle. There is no special name for the axes ab corresponding to $(\gamma_{nt})_{max}$.

Equations for the principal strains, the maximum shear strain, and the orientations of the corresponding axes can be obtained from the strain transformation equations [Eqs. (9.16) and (9.17)]. They can also be obtained by analogy from Eqs. (9.5)–(9.11), or from the geometry of Mohr's circle of strain.

Using the latter approach, we find from Figure 9.20 that

$$\varepsilon' = \frac{\varepsilon_x + \varepsilon_y}{2} \tag{9.18}$$

and

$$\varepsilon_{1,2} = \frac{\varepsilon_x + \varepsilon_y}{2} \pm \sqrt{\left(\frac{\varepsilon_x - \varepsilon_y}{2}\right)^2 + \left(\frac{\gamma_{xy}}{2}\right)^2} \tag{9.19}$$

$$|(\gamma_{nt})_{\max}| = 2\sqrt{\left(\frac{\varepsilon_x - \varepsilon_y}{2}\right)^2 + \left(\frac{\gamma_{xy}}{2}\right)^2} = |\varepsilon_1 - \varepsilon_2| \tag{9.20}$$

The orientation of the principal axes is defined by the angle 2α, where

$$\tan 2\alpha = \frac{\gamma_{xy}}{\varepsilon_x - \varepsilon_y} \tag{9.21}$$

The angle 2β, which defines the orientation of the axes *ab* associated with the maximum shear strain, is defined by the relation

$$\tan 2\beta = -\frac{(\varepsilon_x - \varepsilon_y)}{\gamma_{xy}} \tag{9.22}$$

As can be seen from Figure 9.20, the principal axes and the axes associated with the maximum shear strain will always be 45° apart. For isotropic materials, the principal axes of strain coincide with the principal axes of stress. Furthermore, the axes associated with $(\gamma_{nt})_{\max}$ coincide with those for the maximum in-plane shear stress, $(\tau_{nt})_{\max}$ (see Problem 9.85).

Relationship Between E, G, and ν

We are now in a position to derive the relationship between the elastic constants E, G, and ν given in Eq. (5.19) of Section 5.9.

Consider an element of linearly elastic isotropic material subjected to principal stresses σ_1 and σ_2, with $\sigma_3 = 0$. From the generalized Hooke's law, Eqs. (5.24), we have for the principal strains

$$\varepsilon_1 = \frac{\sigma_1 - \nu\sigma_2}{E}$$

$$\varepsilon_2 = \frac{\sigma_2 - \nu\sigma_1}{E}$$

Subtracting the second of these equations from the first, we obtain

$$\varepsilon_1 - \varepsilon_2 = (\sigma_1 - \sigma_2)\frac{(1 + \nu)}{E}$$

Now $\sigma_1 - \sigma_2 = 2(\tau_{nt})_{\max}$ and $\varepsilon_1 - \varepsilon_2 = (\gamma_{nt})_{\max}$ from the corresponding Mohr's circles of stress and strain; therefore, the preceding expression can be written as

$$(\gamma_{nt})_{\max} = (\tau_{nt})_{\max} \frac{2(1+\nu)}{E}$$

Introducing the stress-strain relation $(\tau_{nt})_{\max} = G(\gamma_{nt})_{\max}$, we obtain

$$(\gamma_{nt})_{\max} = G(\gamma_{nt})_{\max} \frac{2(1+\nu)}{E}$$

or

$$G = \frac{E}{2(1+\nu)}$$

This is the relationship given in Eq. (5.19).

Example 9.10. Determination of Principal Strains and Maximum Shear Strain. The strain components at a point in a body subjected to plane strain are $\varepsilon_x = 750\ \mu\varepsilon$, $\varepsilon_y = -250\ \mu\varepsilon$ and $\gamma_{xy} = -600\ \mu\varepsilon$. Determine the principal strains, the maximum shear strain and the orientation of the corresponding axes. Sketch the original and deformed shapes of elements with sides aligned with these axes. Use Mohr's circle.

Solution. We first construct the Mohr's circle [Figure 9.21(a)]. The center of the circle has coordinates $(250, 0)\ \mu\varepsilon$ and its radius is

$$R = \sqrt{300^2 + 500^2} = 583\ \mu\varepsilon$$

For the angle 2α, we have

$$\tan 2\alpha = \frac{-300}{500} \qquad 2\alpha = -31^\circ$$

The principal strains are

$$\varepsilon_1 = \varepsilon' + R = 250 - 400583 = 833\ \mu\varepsilon$$

$$\varepsilon_2 = \varepsilon' - R = 250 - 583 = -333\ \mu\varepsilon$$

and the magnitude of the maximum shear strain is

$$|(\gamma_{nt})_{\max}| = 2R = 1{,}166\ \mu\varepsilon$$

The normal strains associated with $(\gamma_{nt})_{\max}$ are

$$\varepsilon_a = \varepsilon_b = \varepsilon' = 250\ \mu\varepsilon$$

Since diameter (1)-(2) of the circle lies 31° counterclockwise from diameter (x)-(y), the principal axes are oriented 15.5° clockwise from the xy axes [Figure 9.21(b)]. The corresponding principal strains are as shown. These strains represent a lengthening of the element in the 1 direction and a shortening in the 2 direction, as indicated by the dashed lines.

The axes ab associated with $(\gamma_{nt})_{\max}$ lie 29.5° counterclockwise from xy, since the corresponding diameters on the circle are 59° apart. The strains associated with these axes are shown in Figure 9.21(c). They represent a lengthening of the element in the a and b directions and an increase in the angle between the a and b axes (γ_{ab} is negative), as indicated by the dashed lines.

It is left as an exercise to show that these same results are obtained if Eqs. (9.18)–(9.22) are used instead of Mohr's circle.

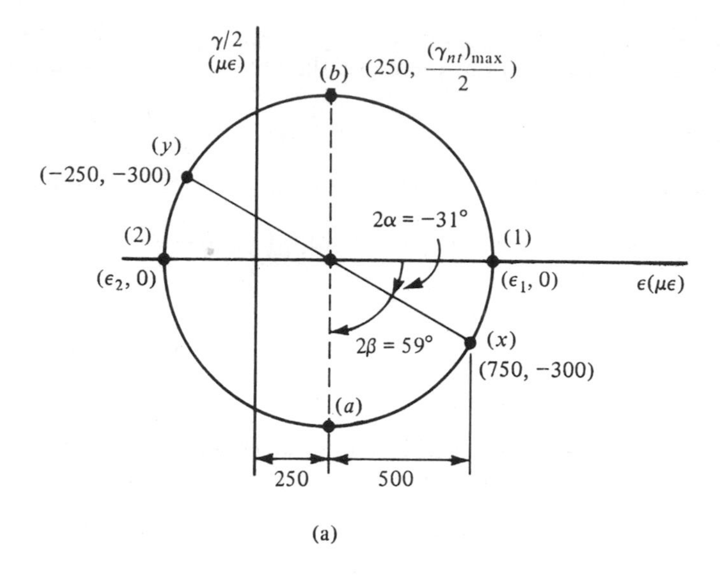

(a)

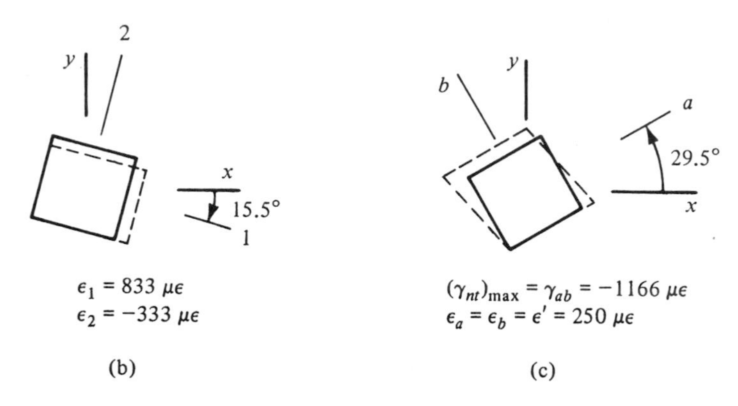

(b) (c)

Figure 9.21

PROBLEMS

9.73. to 9.75. For the following states of strain, determine the principal strains, the maximum shear strain, and the orientation of the corresponding axes. Sketch the original and deformed shapes of the corresponding elements. Use Mohr's circle.

9.73. $\varepsilon_x = 800\ \mu\varepsilon$ $\quad\varepsilon_y = -200\ \mu\varepsilon$

$\gamma_{xy} = 0$

9.74. $\varepsilon_x = 200\ \mu\varepsilon$ $\quad\varepsilon_y = 500\ \mu\varepsilon$

$\gamma_{xy} = -400\ \mu\varepsilon$

9.75. $\varepsilon_x = -800\ \mu\varepsilon$ $\quad\varepsilon_y = 500\ \mu\varepsilon$

$\gamma_{xy} = 600\ \mu\varepsilon$

9.76. to 9.78. Same as Problems 9.73 to 9.75, except that Eqs. (9.18)–(9.22) are to be used instead of Mohr's circle.

9.79. to 9.81. For the states of strain given in Problems 9.73 to 9.75, determine the strain components with respect to nt axes rotated 30° counterclockwise from the xy axes. Show the results on elements with sides aligned with these axes.

9.82. For the thin plate shown, $\varepsilon_x = -1{,}200\ \mu\varepsilon$, $\varepsilon_y = 0$, and $\gamma_{xy} = 500\ \mu\varepsilon$. If $\varepsilon_n = -750\ \mu\varepsilon$, determine the angle θ and the strains ε_t and γ_{nt}.

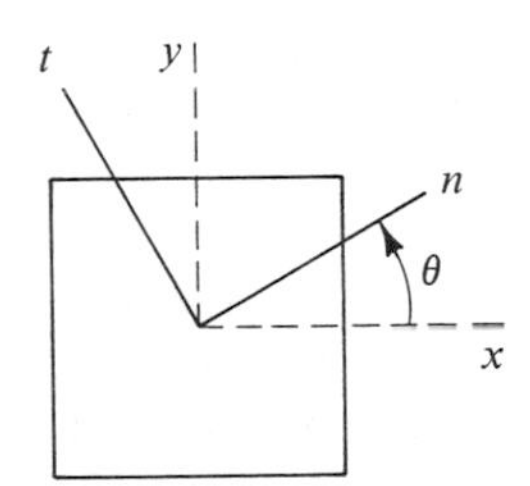

Problem 9.82

9.83. The deformed shape of rectangle $OABC$ shown is indicated by the dashed lines. It is known that the principal strains are $\varepsilon_1 = 800\ \mu\varepsilon$ and $\varepsilon_2 = -400\ \mu\varepsilon$. Determine the displacements a and b and the orientation of the principal axes.

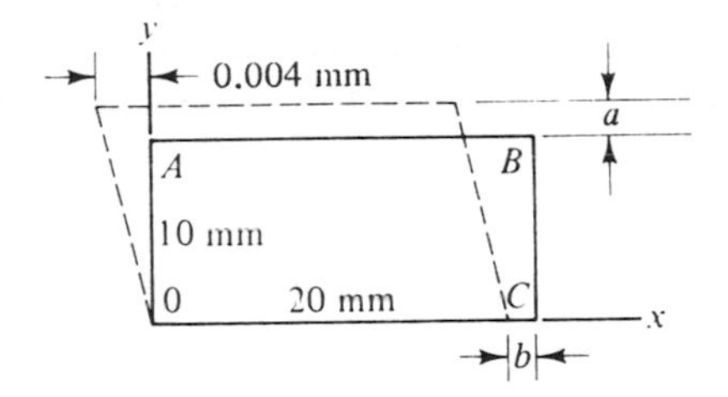

Problem 9.83

9.84. A 0.50-in. diameter circle drawn on a member before it is loaded distorts into an ellipse after loading, as shown. If the major and minor axes of the ellipse are 0.52 in. and 0.46 in. long, respectively, determine ε_x, ε_y, and γ_{xy}. What is the value of the maximum shear strain?

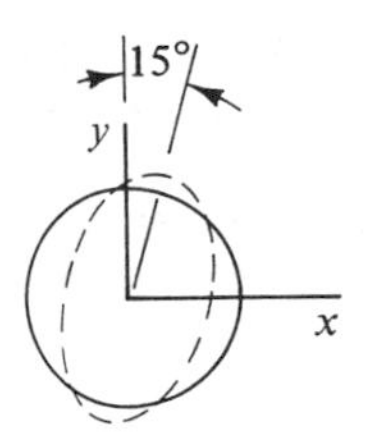

Problem 9.84

9.85. Show that Eqs. (9.21) and (9.22) reduce to Eqs. (9.5) and (9.11), respectively, when combined with the generalized Hooke's law for plane stress [Eqs. (5.27)]. Thus, show that the principal axes of stress coincide with the principal axes of strain and that the axes associated with the maximum shear strain coincide with the axes associated with the maximum in-plane shear stress for linearly elastic isotropic materials.

9.86. A member is loaded such that the strains at a point on the surface are $\varepsilon_x = -800\ \mu\varepsilon$, $\varepsilon_y = -200\ \mu\varepsilon$, and $\gamma_{xy} = 600\ \mu\varepsilon$. Determine the principal stresses and the orientation of the principal axes. The material is aluminum with $E = 10^7$ psi and $\nu = 0.33$.

9.87. Same as Problem 9.86, except that $\varepsilon_x = 200\ \mu\varepsilon$, $\varepsilon_y = -600\ \mu\varepsilon$, and $\gamma_{xy} = 800\ \mu\varepsilon$.

9.88. Determine the normal strain in direction A–A for the state of stress shown if $E = 200\ \text{GN/m}^2$ and $\nu = 0.3$.

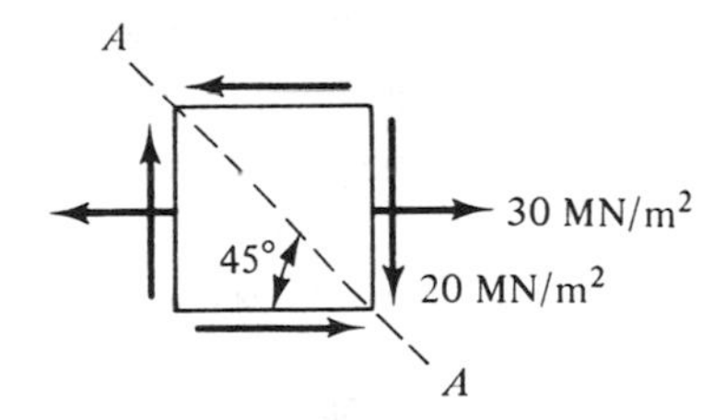

Problem 9.88

9.89. A thin-walled copper cylinder with closed ends and with radius to thickness ratio $R/t = 10$ contains a gas under a pressure of 600 psi. If a gage for measuring normal strains is attached to the outer surface of the pipe, what strain will the gage indicate if it is oriented (a) parallel to the pipe axis, (b) perpendicular to the pipe axis, and (c) at an angle of 25° from the pipe axis? Use $E = 12 \times 10^6$ psi and $\nu = \frac{1}{3}$.

9.90. A 40-mm diameter shaft is made of 2024-T3 aluminum. If the normal strain in a direction 20° from the shaft axis has a magnitude of 2,000 $\mu\varepsilon$, what is the magnitude of the applied torque?

9.8 STRAIN GAGE ROSETTES

Although it is possible to derive equations for computing the stresses in a variety of members, there are many problems in which the loading or geometry is so complicated that the only practical approach is to determine the stresses experimentally. It is usually the maximum stresses that are of interest, and, with few exceptions, these occur at the surface of a member. Consequently, the stresses can be determined by measuring the strains at the surface and then calculating the stresses from the stress-strain relations. Since a state of plane stress exists at a free surface, it is only necessary to know the stress components with respect to two perpendicular axes in the plane of the surface, say σ_x, σ_y, and τ_{xy}. From the generalized Hooke's law for plane stress, Eqs. (5.27), we see that these stresses can be determined if the two normal strains ε_x and ε_y and the shear strain γ_{xy} are known.

Normal strains are relatively easy to measure, and electrical resistance strain gages (discussed in Chapter 11) are widely used for this purpose. However, shear strains are difficult to measure accurately because they involve very small changes in angles. Fortunately, they can be determined indirectly by measuring the normal strain in a third direction at the point of interest. Thus, the state of strain at a point on the surface of a member can be determined by measuring the normal strains in three different directions at that point. Special assemblages of strain gages called *strain gage rosettes* are available for making such measurements. Two common configurations are the rectangular rosette [Figure 9.22(a)], in which the three gages

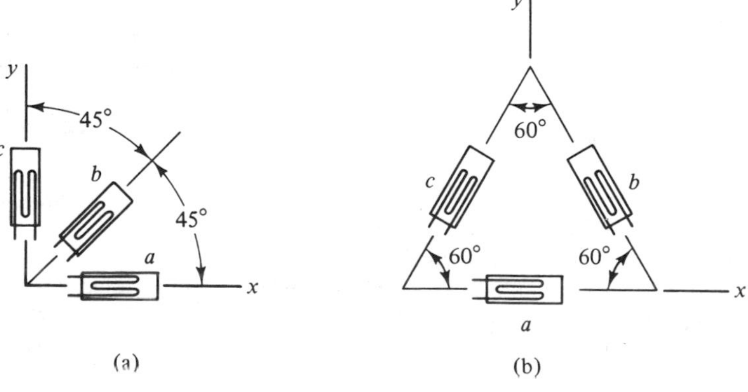

Figure 9.22

are 45° apart, and the delta rosette [Figure 9.22(b)] in which they are 60° apart. Each of the gages in a rosette detects the normal strain in the direction along which the gage is aligned. These three pieces of data are sufficient to determine the strain components ε_x, ε_y, and γ_{xy}.

For example, consider the rectangular rosette [Figure 9.22(a)]. Let ε_a, ε_b, and ε_c be the respective strain readings from the three gages. Since the choice of axes is arbitrary, we take axes xy as shown in the figure. Thus, the angles between the gages and the x axis are, respectively, $\theta_a = 0°$, $\theta_b = 45°$, and $\theta_c = 90°$. Applying the strain transformation equation, Eq. (9.13), in each of the directions a, b, and c, we obtain

$$\begin{aligned} \varepsilon_a &= \varepsilon_x \cos^2 0° + \varepsilon_y \sin^2 0° + \gamma_{xy} \sin 0° \cos 0° = \varepsilon_x \\ \varepsilon_b &= \varepsilon_x \cos^2 45° + \varepsilon_y \sin^2 45° + \gamma_{xy} \sin 45° \cos 45° = \tfrac{1}{2}(\varepsilon_x + \varepsilon_y + \gamma_{xy}) \qquad (9.23) \\ \varepsilon_c &= \varepsilon_x \cos^2 90° + \varepsilon_y \sin^2 90° + \gamma_{xy} \sin 90° \cos 90° = \varepsilon_y \end{aligned}$$

Thus, we have three equations from which to determine the three unknowns ε_x, ε_y, and γ_{xy}. Once these strains are known, the stress components can be determined from the generalized Hooke's law, provided the material is isotropic and linearly elastic.

Special equations for computing the principal stresses directly from the measured strains can be derived (see Problems 9.101 and 9.102). Such equations are convenient, and they form the basis for computerized methods of data reduction. However, we shall be concerned here only with the basic concepts and procedures involved in the data reduction process.

The experimental procedure outlined is widely used for determining the strains and stresses. However, a number of other procedures are available. Some of these are described in Chapter 11, and others are discussed in texts on experimental stress analysis.

Example 9.11. Experimental Determination of Principal Stresses. A delta rosette [Figure 9.23(a)] applied to the surface of a member made of 2024-T3 aluminum alloy gives the strain readings $\varepsilon_a = 1{,}224\ \mu\varepsilon$, $\varepsilon_b = -66\ \mu\varepsilon$, and $\varepsilon_c = 442\ \mu\varepsilon$. Determine the principal stresses at this location and the orientation of the principal axes.

Solution. We choose axes xy as shown in Figure 9.23(a). Thus, the angles between the gages and the x axis are, respectively, $\theta_a = 0°$, $\theta_b = 120°$, and $\theta_c = 60°$. Applying the strain transformation equation, Eq. (9.13), in each of the three directions a, b, and c, we have

$$\begin{aligned} \varepsilon_a &= \varepsilon_x \cos^2 0° + \varepsilon_y \sin^2 0° + \gamma_{xy} \sin 0° \cos 0° = \varepsilon_x \\ \varepsilon_b &= \varepsilon_x \cos^2 120° + \varepsilon_y \sin^2 120° + \gamma_{xy} \sin 120° \cos 120° = \frac{\varepsilon_x}{4} + \frac{3\varepsilon_y}{4} - \frac{\sqrt{3}}{4}\gamma_{xy} \\ \varepsilon_c &= \varepsilon_x \cos^2 60° + \varepsilon_y \sin^2 60° + \gamma_{xy} \sin 60° \cos 60° = \frac{\varepsilon_x}{4} + \frac{3\varepsilon_y}{4} + \frac{\sqrt{3}}{4}\gamma_{xy} \end{aligned}$$

Substituting the given values of ε_a, ε_b, and ε_c into these equations and solving, we obtain

$$\varepsilon_x = 1{,}224\ \mu\varepsilon \qquad \varepsilon_y = -157\ \mu\varepsilon \qquad \gamma_{xy} = 587\ \mu\varepsilon$$

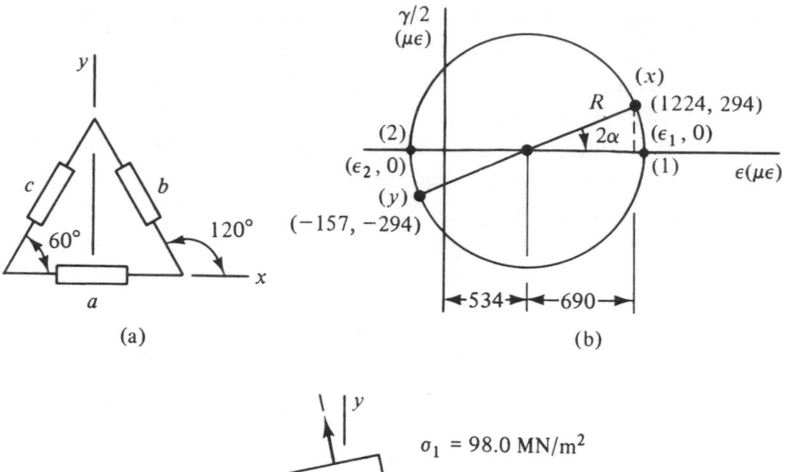

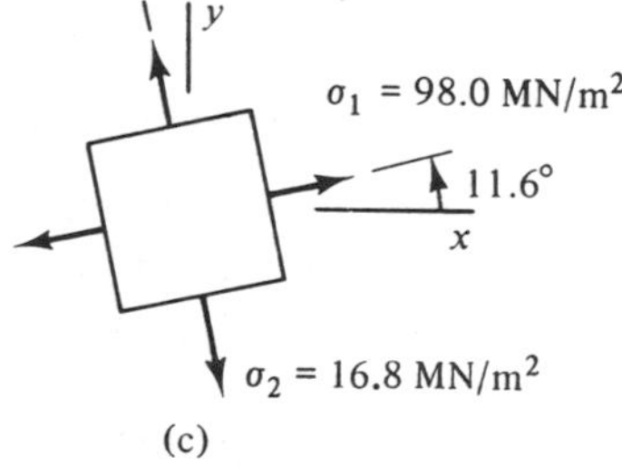

Figure 9.23

There are two ways to proceed from here. We can either determine σ_x, σ_y, and τ_{xy} from the generalized Hooke's law, Eqs. (5.27), and then find the principal stresses, or we can determine the principal strains ε_1 and ε_2 and then compute σ_1 and σ_2 from Eqs. (5.27). We shall use the latter approach here.

The Mohr's circle for strain is shown in Figure 9.23(b). The center of the circle is located at (534, 0) $\mu\varepsilon$ and its radius is

$$R = \sqrt{690^2 + 294^2} = 750\ \mu\varepsilon$$

Thus,

$$\varepsilon_1 = c + R = 534 + 750 = 1{,}284\ \mu\varepsilon$$

$$\varepsilon_2 = c - R = 534 - 750 = -216\ \mu\varepsilon$$

$$\tan 2\alpha = \frac{294}{690} \qquad 2\alpha = 23.1°$$

The principal stresses are computed from Eqs. (5.27) by using the values $E = 72$ GN/m^2 and $\nu = 0.33$ obtained from Table A.2:

$$\sigma_1 = \frac{E}{(1-\nu^2)}(\varepsilon_1 + \nu\varepsilon_2) = \frac{72 \times 10^9\ \text{N/m}^2}{(1 - 0.33^2)}[1{,}284 + 0.33(-216)]10^{-6}$$

$$= 98.0 \times 10^6\ \text{N/m}^2 \text{ or } 98.0\ \text{MN/m}^2 \qquad \textbf{Answer}$$

$$\sigma_2 = \frac{E}{(1-\nu^2)}(\varepsilon_2 + \nu\varepsilon_1) = \frac{72 \times 10^9\ \text{N/m}^2}{(1 - 0.33^2)}[-216 + 0.33(1{,}284)]10^{-6}$$

$$= 16.8 \times 10^6\ \text{N/m}^2 \text{ or } 16.8\ \text{MN/m}^2 \qquad \textbf{Answer}$$

For an isotropic material, the principal axes of stress coincide with the principal axes of strain, which are oriented 11.6° counterclockwise from the xy axes [Figure 9.23(c)].

PROBLEMS

9.91. and 9.92. The strains on the surface of a member were measured by using a delta rosette oriented as shown in Figure 9.22(b). Determine the strain components ε_x, ε_y and γ_{xy} for the gage readings indicated below.

9.91. $\varepsilon_a = 163\ \mu\varepsilon$ $\varepsilon_b = -222\ \mu\varepsilon$ $\varepsilon_c = -10\ \mu\varepsilon$

9.92. $\varepsilon_a = -410\ \mu\varepsilon$ $\varepsilon_b = 540\ \mu\varepsilon$ $\varepsilon_c = 2{,}170\ \mu\varepsilon$

9.93. and 9.94. Same as Problems 9.91 and 9.92, except that the strains were measured by using a rectangular rosette oriented as shown in Figure 9.22(a).

9.95. A rectangular rosette (Figure 9.22) is mounted on the outer surface of a thin-walled spherical pressure vessel. What is the anticipated relationship between the strain readings ε_a, ε_b, and ε_c?

9.96. A delta rosette (Figure 9.22) bonded to the surface of a member gives the following strain readings: $\varepsilon_a = 640\ \mu\varepsilon$, $\varepsilon_b = -336\ \mu\varepsilon$, and $\varepsilon_c = 498\ \mu\varepsilon$. Determine the principal strains and the orientation of the principal axes. Show the results on a properly oriented element.

9.97. A rectangular rosette mounted on the web of an I-beam as shown gives the following strain readings: $\varepsilon_a = 1{,}084\ \mu\varepsilon$, $\varepsilon_b = 264\ \mu\varepsilon$, and $\varepsilon_c = -200\ \mu\varepsilon$. Determine the principal strains, the maximum shear strain and the orientation of the corresponding axes. Show the results on properly oriented elements.

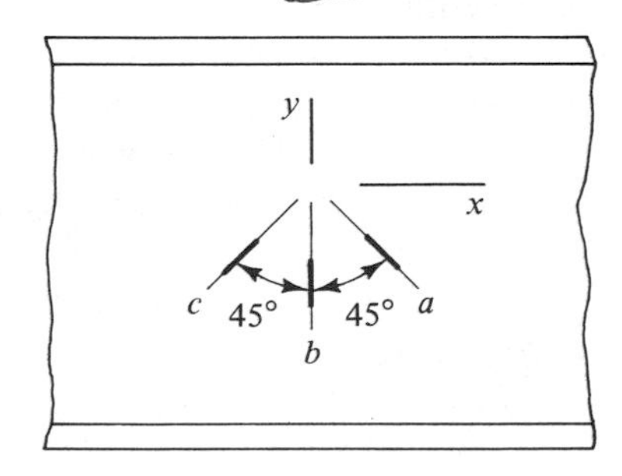

Problem 9.97

9.98. to 9.100. The strains on a free surface of a member were measured by using three strain gages oriented as shown. The strain readings and material properties are as indicated. Determine the principal stresses and maximum in-plane shear stress and show them on properly oriented elements. Also determine the absolute maximum shear stress.

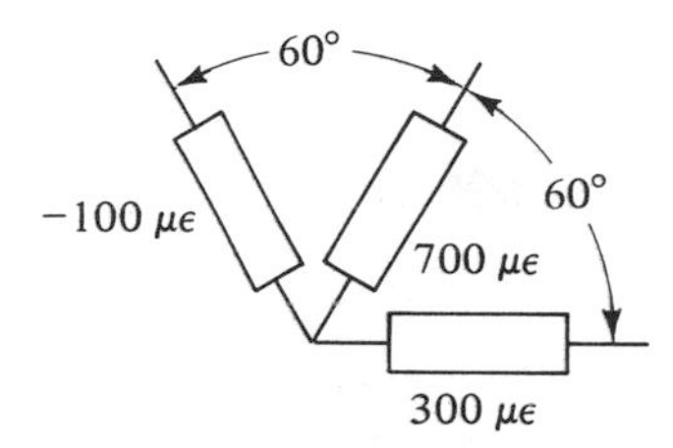

$E = 72\ \text{GN/m}^2$
$G = 27\ \text{GN/m}^2$
$\nu = 0.33$

Problem 9.98

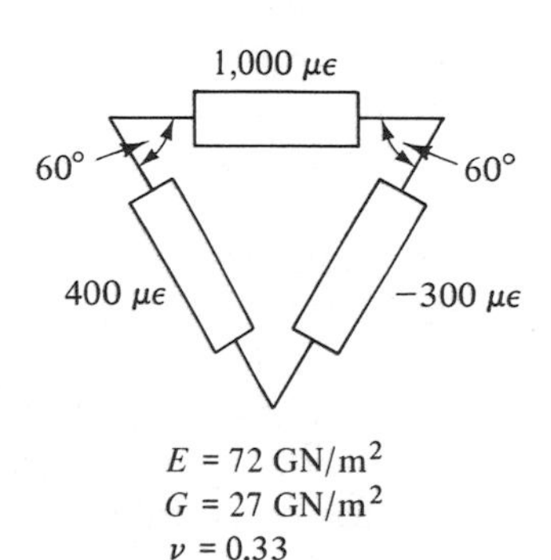

$E = 72\ \text{GN/m}^2$
$G = 27\ \text{GN/m}^2$
$\nu = 0.33$

Problem 9.99

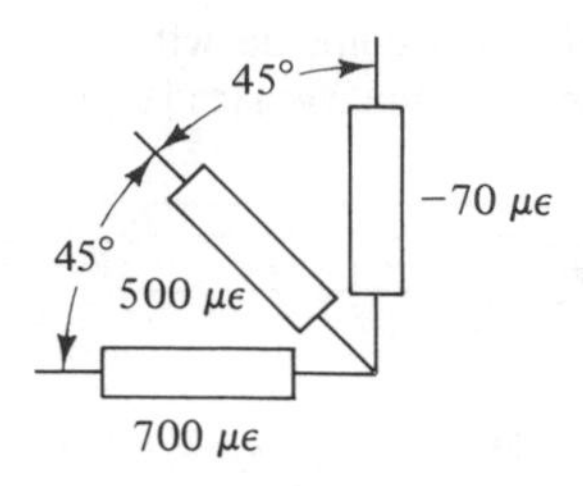

$E = 30 \times 10^6$ psi
$G = 12 \times 10^6$ psi
$\nu = 0.3$

Problem 9.100

9.101. Derive an expression for the principal stresses σ_1 and σ_2 in terms of the strain readings ε_a, ε_b, and ε_c from a rectangular rosette (Figure 9.22). Assume linearly elastic isotropic material behavior.

9.102. Repeat Problem 9.101 for a delta rosette (Figure 9.22).

9.103. A cylindrical pressure vessel made by capping the ends of a section of 100-mm diameter plastic pipe with a wall thickness of 3 mm is subjected to an internal pressure p and an axial force F. The strain readings from two gages mounted on the outside surface with the orientations shown are $\varepsilon_a = 700\ \mu\varepsilon$ and $\varepsilon_b = 500\ \mu\varepsilon$. If $E = 2\ \text{GN/m}^2$ and $\nu = 0.4$, determine p and F.

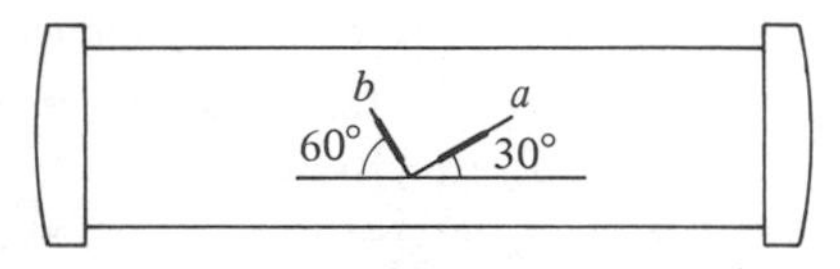

Problem 9.103

9.104. Same as Problem 9.103, except that $\varepsilon_a = 0$ and $\varepsilon_b = 400\ \mu\varepsilon$.

9.9 CLOSURE

In this chapter we have been concerned with procedures for determining the states of stress and strain in a body subjected to various types of combined loadings. Particular emphasis was placed upon the determination of the maximum stresses because they usually determine whether or not a member will fail. There are two steps involved in determining the maximum stresses. First, we must use our knowledge of the stress distribution in members obtained from the basic load-stress relations to determine the points in the body where the maximum stresses will likely occur. Second, we must determine the maximum stresses at these points and the planes on which they act. It is sometimes necessary to check more than one point before the actual maximum stresses can be determined.

The basic tools for transferring stress and strain components from one set of axes to another are the transformation equations and Mohr's circle. Both stresses and strains transfer in exactly the same way because they are both what is known in mathematics as *tensor* quantities, which transform in a specific way under a rotation of axes. Area moments of inertia are also tensor quantities.

Mohr's circle for stress and strain and the underlying transformation equations are based solely on equilibrium and geometric considerations. Consequently, they apply regardless of the material properties. However, the generalized Hooke's law used to relate the stresses and the strains holds only for linearly elastic isotropic materials.

10

Buckling

10.1 INTRODUCTION

A straight slender member subjected to an axial compressive load is called a *column*. Such members are commonly encountered in trusses and in the framework of buildings where they serve primarily to carry the weight of the structure and the applied loads to the foundation. They are also found as machine linkages, flagpoles, signposts, supports for highway overpasses, and numerous other structural and machine elements.

If a compression member is relatively short, it will remain straight when loaded, and the load-stress and load-deformation relations presented in Chapters 5 and 6 apply. However, a different type of behavior is observed for longer members. When the compressive load reaches a certain critical value, the column undergoes a bending action in which the lateral deflection becomes very large with little increase in load. This response is called *buckling* and usually leads to collapse of the member. Failures due to buckling are frequently catastrophic because they occur suddenly with little warning when the critical load is reached.

Our goal in this chapter is to determine the critical loads at which various types of columns will buckle. The analysis is much the same as for other load-carrying members, but there are some fundamental differences. The most significant of these is that the equilibrium equations must be based on the deformed geometry of the member instead of the undeformed geometry used in all previous cases. In other words, the rigid-body approximation cannot be used in deriving the equations governing the buckling of a column.

The concept of buckling is closely allied to the idea of stability of an equilibrium position introduced in Section 4.2. The latter concept is, in turn, related to the behavior of a system in equilibrium when subjected to the bumps, jolts, vibration, extraneous forces, and other disturbances, irregularities, and imperfections that are always present. Such factors have been ignored until now, for they had no significant influence on the response of the members considered. This is not the case for columns and other members subject to buckling. The relationship between buckling and the stability of an equilibrium position will be discussed in this chapter.

Buckling can occur with most types of members under the proper conditions. For example, we have already mentioned in Chapter 7 that a thin-walled tube tends to wrinkle (buckle) when twisted. Similarly, a thin-walled cylinder will buckle when subjected to an excessive external pressure, and a beam with a deep, narrow cross section can buckle by twisting and deflecting sideways if the span is too long. Problems like these are not as common as those involving columns, however, and will not be considered in this text.

10.2 BASIC CONCEPTS

Buckling of Perfect Systems

The concepts of stability of an equilibrium position and buckling can probably best be described by means of an example. To this end, let us consider the system shown in Figure 10.1(a). The system consists of a rigid bar supported at the bottom by a hinge and torsional spring and subjected to a force that always remains vertical. This system is a crude model of an elastic column, with the bending resistance of the member represented by the torsional spring.

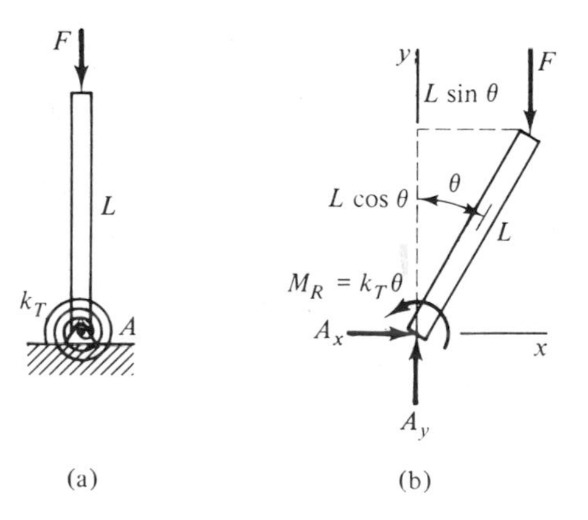

Figure 10.1

It is clear that the applied load will have no effect upon the system as long as the bar remains vertical and the load is aligned with its axis. However, if the system is disturbed so that the bar rotates through some angle θ [Figure 10.1(b)], there will be a moment of magnitude $FL \sin \theta$ tending to rotate the bar further in the same direction. This rotation is resisted by the moment exerted by the spring, which has magnitude $k_T\theta$. This moment tends to return the bar to its original position. We shall refer to these two moments as the *disturbing moment*, M_D, and the *restoring moment*, M_R, respectively.

Equilibrium of the bar requires that $M_D = M_R$, or

$$FL \sin \theta = k_T\theta \tag{10.1}$$

We wish to solve this equation for the angle θ through which the bar will rotate for a given load F. To do this, we rewrite the equation as

$$\sin \theta = \frac{\theta}{\beta} \tag{10.2}$$

where

$$\beta = \frac{FL}{k_T} \tag{10.3}$$

Equation (10.2) can be solved by choosing values of θ and determining the corresponding values of the load parameter β. However, it is more informative to solve it by plotting both sides of the equation and determining the values of θ at the intersection points [Figure 10.2(a)]. Note that the sine curve in Figure 10.2(a) represents the disturbing moment and the straight lines represent the restoring moment for various values of β. This procedure yields the solutions shown in Figure 10.2(b). There are other solutions corresponding to negative values of θ which are not shown.

The values of θ obtained from Eq. (10.2) define the equilibrium positions of the bar. Note that $\theta = 0$ is always a solution of this equation and is, therefore, always an equilibrium position. Furthermore, this is the only solution if $\beta < 1$ ($F < k_T/L$), as can be seen from Figure 10.2(a). For $\beta > 1$ ($F > k_T/L$), there is a second set of solutions for which θ is nonzero. The point at which the two solution curves cross is called the *bifurcation point* [Figure 10.2(b)].

Now let us interpret these solutions in terms of the mechanics of the problem. Suppose that the bar is disturbed some small amount $\Delta\theta$ from the equilibrium position $\theta = 0$. It is clear from Figure 10.2(a) that the restoring moment at this new position of the bar is larger than the disturbing moment for $\beta < 1$. Thus, the bar will return to its original position after the disturbance, in which case the equilibrium position $\theta = 0$ is said to be *stable*. There may be some oscillations, but they will eventually die out. In contrast, the disturbing moment is seen to be larger than the restoring moment for $\beta > 1$. The equilibrium position $\theta = 0$ is *unstable* in this case, and the bar will move away from this position if disturbed.

The dividing line between these two types of behavior corresponds to

$$\beta_{cr} = 1 \tag{10.4}$$

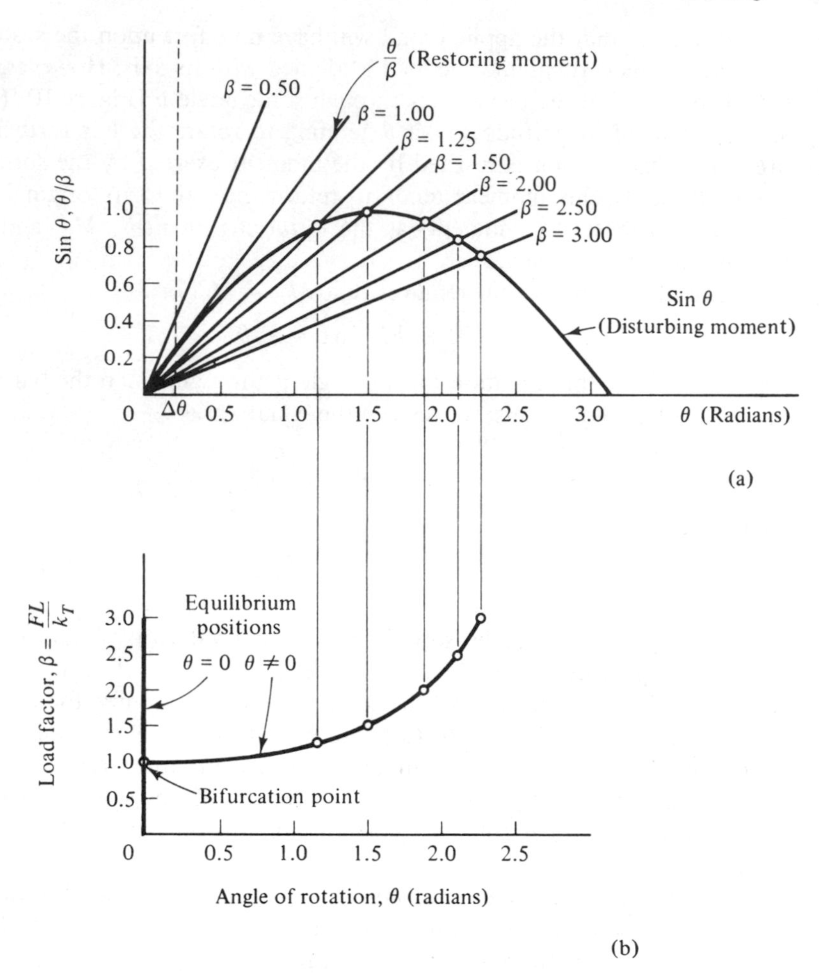

Figure 10.2

or

$$F_{cr} = \frac{k_T}{L} \tag{10.5}$$

This expression defines the *critical load*, F_{cr}, for the system.

Using this same procedure, it can be shown that the nonzero equilibrium positions, which exist only for $\beta > 1$, are stable. Thus, the stable and unstable equilibrium positions of the bar are as indicated in Figure 10.3.

Once the critical load is exceeded, any disturbance will cause the bar to undergo a relatively large rotation to the nonzero equilibrium position, which is now the stable one. It is this type of response, indicated by the arrows in Figure 10.3, that was referred to earlier as buckling. Notice that the angle of rotation of the bar increases very rapidly as the applied load is increased above the critical value.

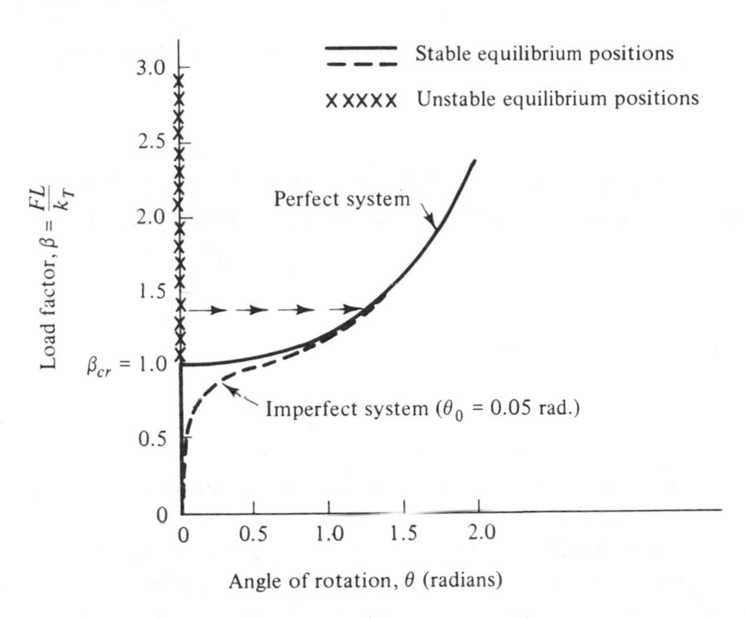

Figure 10.3

Systems with Imperfections

To this point, we have assumed a perfect system. This is not entirely realistic because there will always be some imperfections present. For example, the bar may not be exactly vertical, or the force may not be in perfect alignment with the axis of the bar. The question is, what effect do such imperfections have on the response of the system?

This question can be answered by considering an imperfection in the form of a small initial angle θ_0 between the bar and the vertical. The equilibrium equation in this case is

$$FL \sin(\theta + \theta_0) = k_T \theta \tag{10.6}$$

and its solution for $\theta_0 = 0.05$ radian (2.9°) is indicated by the dashed curve in Figure 10.3. Notice that, with the imperfection, there is some rotation of the bar for all nonzero values of load. Furthermore, all equilibrium positions are now stable, and there is no well-defined critical load. However, the angle of rotation increases rapidly and becomes very large in the vicinity of the critical load for the perfect system, so that load can still be taken as the buckling load. In other words, the buckling load for the system is not affected by small imperfections.

Determination of the Critical Load

If only the critical load is of interest, as is the case in most problems, it is not necessary to go through so complete an analysis. The desired information can be determined from a linear analysis, as we shall now show.

For small values of θ, $\sin\theta \cong \theta$, and Eq. (10.2) becomes

$$\theta = \frac{\theta}{\beta} \tag{10.7}$$

This equation has a nonzero solution only if $\beta = 1$, which corresponds to the critical load for the system. Similarly, Eq. (10.6) for the imperfect system reduces to

$$\theta + \theta_0 = \frac{\theta}{\beta} \tag{10.8}$$

which has the solution

$$\theta = \frac{\beta\theta_0}{1 - \beta} \tag{10.9}$$

This equation predicts an infinite angle of rotation for $\beta = 1$, which again corresponds to the critical load.

These results indicate that, for a perfect system, the critical load can be obtained from the condition that the linear equilibrium equation has a nonzero solution. In other words, we seek the load for which the system will remain in equilibrium in a slightly deflected position. For an imperfect system, the critical load is obtained from the condition that the displacements become infinite.

The results presented in this section were obtained for a specific example, but they also apply to the buckling of columns and many other common types of structural elements.

Example 10.1. Critical Load for a Bar-Spring System. Another simple model of a column consists of a rigid bar hinged at the bottom and supported at the top by a horizontal spring with spring constant k [Figure 10.4(a)]. The spring exerts zero force when the bar is vertical, and the applied force F always acts vertically downward. Determine the critical load for this system if (a) it is perfect and (b) there is an imperfection in the form of a small horizontal force F_H acting at the tip of the bar.

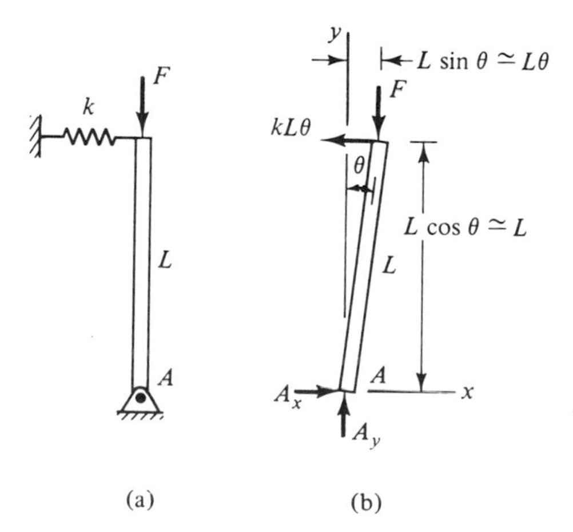

Figure 10.4

Solution. (a) Figure 10.4(b) shows the FBD of the bar when it is in a slightly deflected position. The spring exerts a force proportional to its elongation, which, for small values of θ, is $L\theta$. Summing moments about point A, we obtain

$$\overset{+}{\curvearrowleft}\sum M_A = kL^2\theta - FL\theta = 0$$

or

$$kL^2\theta = FL\theta$$

This equation yields a nonzero value of θ only if

$$F = F_{cr} = kL \qquad \textbf{Answer}$$

(b) Assuming that the horizontal force F_H acts toward the right, we have

$$\overset{+}{\curvearrowleft}\sum M_A = kL^2\theta - FL\theta - F_H L = 0$$

or

$$\theta = \frac{F_H}{kL - F}$$

The critical load is the load at which θ becomes infinite:

$$F_{cr} = kL \qquad \textbf{Answer}$$

This is the same result obtained in (a), which means that the presence of the horizontal force does not affect the buckling load for the system.

PROBLEMS

10.1. to 10.3. Determine the critical load for the system shown. The rigid bars have negligible weight, and all connections and supports are frictionless. Torsional springs exert a moment $M = k_T\theta$, where k_T is the torsional spring constant and θ is the relative angle through which the ends of the springs rotate. Assume the applied load always acts vertically downward.

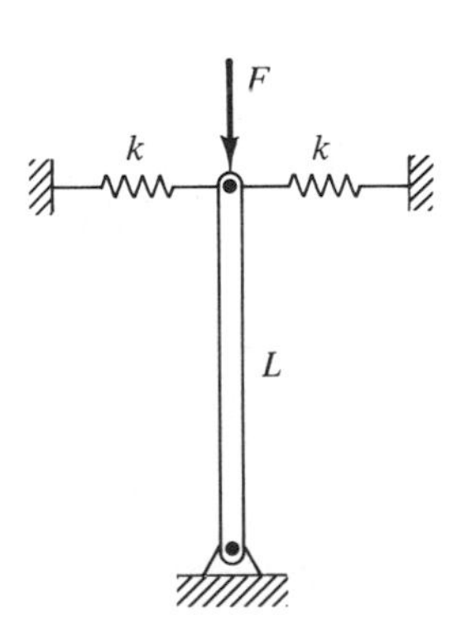

Problem 10.1

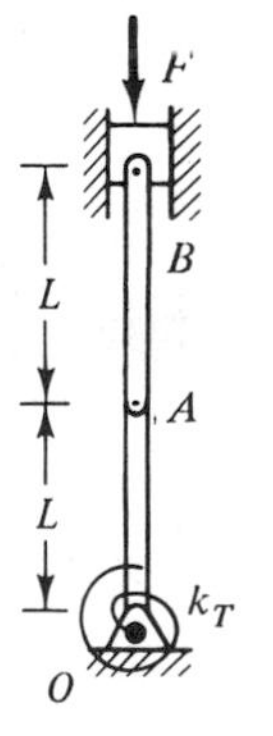

Problem 10.2

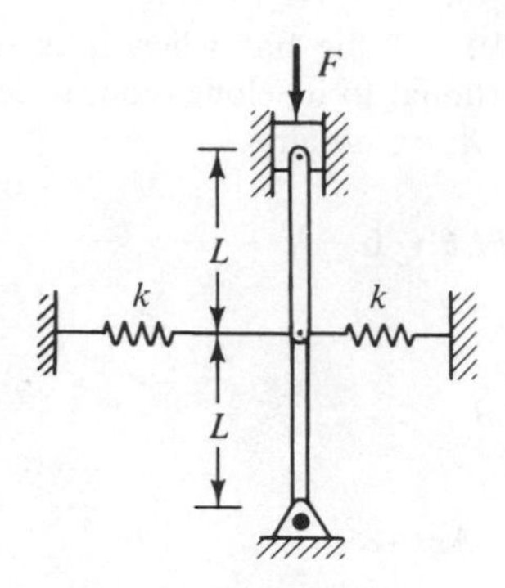

Problem 10.3

10.4. Same as Problem 10.1, except that the uniform bar has weight W.

10.5. Same as Problem 10.2, except that there is also a small horizontal force F_H acting at point A.

10.6. The rigid bar shown is supported by a frictionless pivot at B and is stabilized by a spring which is unstretched when the member is vertical. The bar is subjected to a load F applied by a rope connected to end A and passing through a smooth guide at O. Determine the critical value of F if (a) the weight of the member is neglected and (b) the weight W of member AB is included.

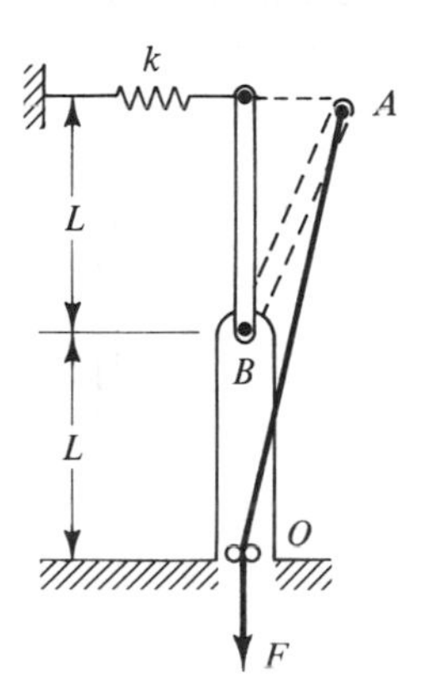

Problem 10.6

10.7. What is the maximum value of the mass M of the block for which the vertical equilibrium position of the system shown will be stable? The rigid supporting structure has negligible mass and is supported by a smooth hinge at O. The springs are undeformed for the position shown.

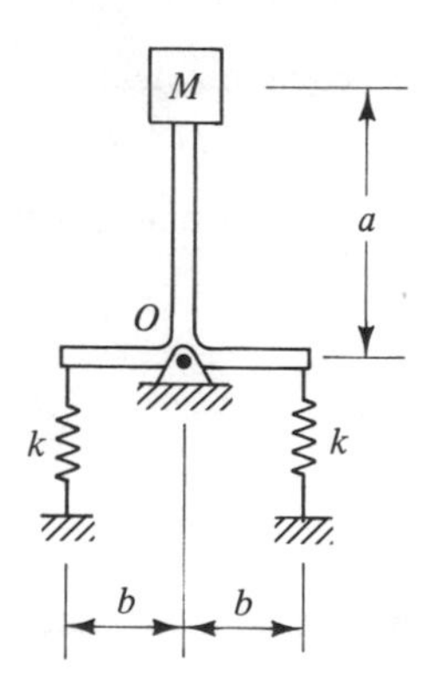

Problem 10.7

10.8. The bottom end of a uniform bar with weight W rests against a smooth floor and the top end is constrained to move vertically by a roller with negligible weight. Determine the range of values of the spring constant k of the lateral springs for which the system will be stable in the position shown.

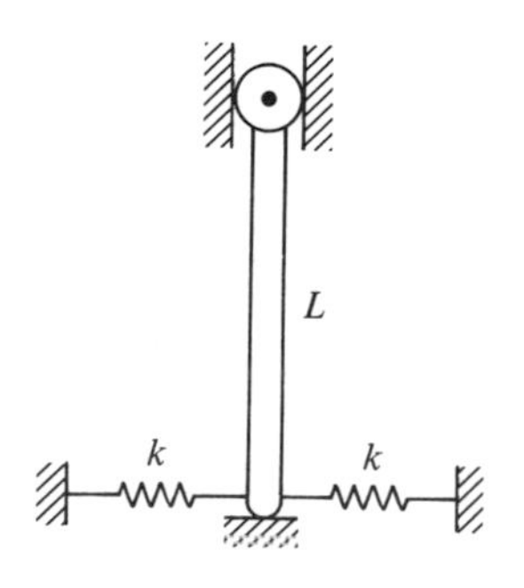

Problem 10.8

10.9. A rigid disc welded to the end of an elastic shaft is subjected to two equal and opposite forces that always remain vertical. The shaft is fixed at one end and supported by a bearing at the other end, so that it is free to twist. Determine the critical load for the system by considering the equilibrium of the disc when it is rotated through a small angle θ, as shown. Express the result in terms of the dimensions D and L and the torsional rigidity JG of the shaft.

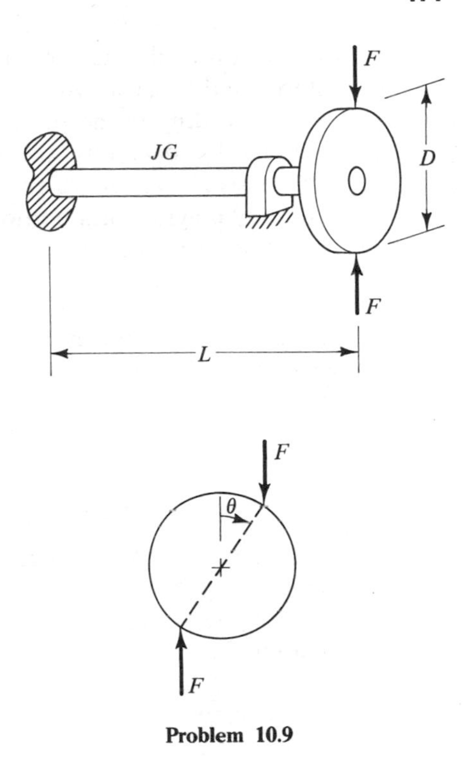

Problem 10.9

10.3 BUCKLING OF ELASTIC COLUMNS

Consider a hinge-ended column supported by smooth pins, or the equivalent, at each end and subjected to an axial compressive load F that always retains its original orientation [Figure 10.5(a)]. We assume that one end of the column is free to move longitudinally with respect to the other, so that the member must support the entire compressive load. It is further assumed that the material behavior is linearly elastic and that there are no imperfections, such as an initial deviation from

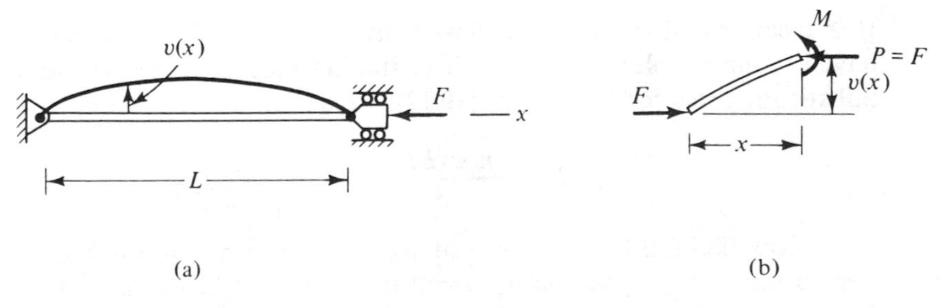

Figure 10.5

straightness of the member or an eccentricity of the applied load. The problem is to determine the load at which the column will buckle.

According to the procedure given in Section 10.2, the buckling load can be obtained by deflecting the column slightly and determining the value of the compressive load for which it will stay in this position. Figure 10.5(b) shows the free-body diagram of a portion of the column when it is deflected an amount $v(x)$. The bending moment is seen to be

$$M = -Fv$$

Substituting this expression into Eq. (8.18) for the small deflection of a linearly elastic beam, we obtain as the governing differential equation for the bending of the column

$$EI\frac{d^2v}{dx^2} = -Fv \tag{10.10}$$

or

$$\frac{d^2v}{dx^2} + k^2v = 0 \tag{10.11}$$

where

$$k^2 = \frac{F}{EI} \tag{10.12}$$

The term on the right-hand side of Eq. (10.10) represents the disturbing moment, which tends to deflect the column, and the term on the left-hand side represents the restoring moment, which tends to bring it back to its undeflected equilibrium position $v(x) = 0$.

The general solution to Eq. (10.11) is

$$v(x) = A\cos kx + B\sin kx \tag{10.13}$$

where A and B are constants. This can be readily confirmed by direct substitution. Applying the boundary conditions $v(0) = 0$ and $v(L) = 0$, we have

$$v(0) = 0: \qquad 0 = A$$

$$v(L) = 0: \qquad 0 = B\sin kL$$

It is clear from the second of these relations that $\sin kL$ must be zero in order to have a nonzero solution ($B \neq 0$). This implies that $kL = n\pi$, where n is an integer. Substituting this result into Eq. (10.12), we obtain

$$F_{cr} = \frac{n^2\pi^2EI}{L^2} \qquad n = 1, 2, 3, \ldots \tag{10.14}$$

Equation (10.14) has physical significance only for $n = 1$ because the undeflected equilibrium position $v(x) = 0$ is unstable at loads above the lowest critical load. The values of F_{cr} corresponding to $n > 1$ cannot be attained physically. Thus, the buckling load for the column is

$$F_{\text{cr}} = \frac{\pi^2 EI}{L^2} \tag{10.15}$$

This result is attributed to the Swiss mathematician Leonhard Euler (1707–1783). Accordingly, Eq. (10.15) is called *Euler's formula*, and the corresponding load is called the *Euler critical load*.

In Eq. (10.15), E is the modulus of elasticity of the material, L is thc length of the column, and I is the moment of inertia of the cross section about the bending axis. If the column is free to deflect in any direction, as in the case of ball and socket supports, it will tend to bend about the axis corresponding to the smallest value of I. This axis offers the least resistance to bending, and, as can be seen from Eq. (10.15), gives the smallest value of the critical load. Of course, if the column is constrained to bend about a certain axis, the value of I about that axis must be used.

According to Eqs. (10.13) and (10.15), the deflected shape of the column is

$$v(x) = B \sin kx = B \sin\left(\frac{\pi x}{L}\right) \tag{10.16}$$

That is, the column buckles into a sine wave. Note that the constant B, which defines the amplitude of the deflection, remains undetermined. It can be obtained only from a nonlinear analysis. This is of little concern because we are usually interested only in the critical load.

The load-deflection curve corresponding to Eq. (10.16) is shown in Figure 10.6, which is a plot of the applied load, F, versus the midspan deflection, v_0. For comparison purposes, the curve obtained from the exact equilibrium equations for large deflections is also shown. Both of these curves are based on the assumption of linearly elastic material behavior.

In actuality, the bending moment and stresses become large as the deflection increases, and the material will yield if it is ductile. Eventually, a plastic hinge will

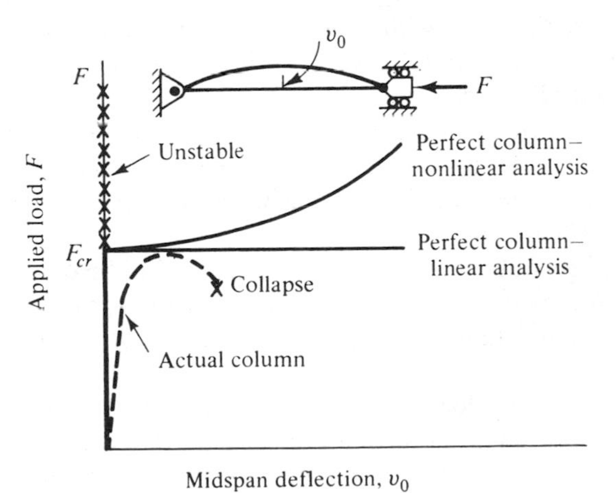

Figure 10.6

form at the most highly stressed section, and the member will collapse. Thus, the actual load-deflection behavior of a typical column made of a ductile material is as indicated by the dashed curve in Figure 10.6. Because of imperfections, the maximum load that can be supported is somewhat less than the Euler critical load. Consequently, a safety factor should always be used with Eq. (10.15). Imperfections are also responsible for there being some deflection at all values of load.

Euler's formula for the critical load is based upon the assumption of linearly elastic material behavior. Many of the columns encountered in practice are such that this condition is not met; so the validity of Eq. (10.15) should always be checked.

Since most columns remain very nearly straight prior to buckling, the average compressive stress at the critical load is $\sigma_{cr} = F_{cr}/A$. Writing the moment of inertia as $I = Ar^2$, where r is the radius of gyration, we have from Eq. (10.15)

$$\sigma_{cr} = \frac{F_{cr}}{A} = \frac{\pi^2 E}{(L/r)^2} \tag{10.17}$$

Thus, the condition for the validity of Euler's formula can be expressed as

$$\sigma_{cr} = \frac{\pi^2 E}{(L/r)^2} \leqslant \sigma_{PL} \tag{10.18}$$

or

$$(L/r)^2 \geqslant \frac{\pi^2 E}{\sigma_{PL}} \tag{10.19}$$

where σ_{PL} is the proportional limit of the material. The quantity L/r in these equations is called the *slenderness ratio* of the column.

The various ranges of column behavior are conveniently displayed on a plot of the critical stress versus the slenderness ratio (Figure 10.7). Columns that satisfy the condition given in Eq. (10.18) or Eq. (10.19), and for which Euler's formula applies, are designated *long columns*. For steel with $E = 30 \times 10^6$ psi and $\sigma_{PL} = 30$ ksi, a long column is one for which $L/r \geqslant 100$. Columns with $L/r > 200$ are seldom used because they can support very little load.

Compression members with a slenderness ratio of approximately 10 or less do not usually buckle. They fail due to general yielding when the compressive stress reaches the yield strength of the material. These *short columns*, or *compression blocks*, were considered in Chapters 5 and 6.

Columns that fall in between these two extremes are said to be of *intermediate length*. They fail by buckling, but Euler's formula does not apply because the stresses exceed the proportional limit. The inelastic buckling of columns will be considered in Section 10.5.

Euler's formula for the critical load for an elastic column is remarkable in that it contains no measure of the strength of the material. The failure load depends only upon the material stiffness and the geometry of the member. This is in contrast to

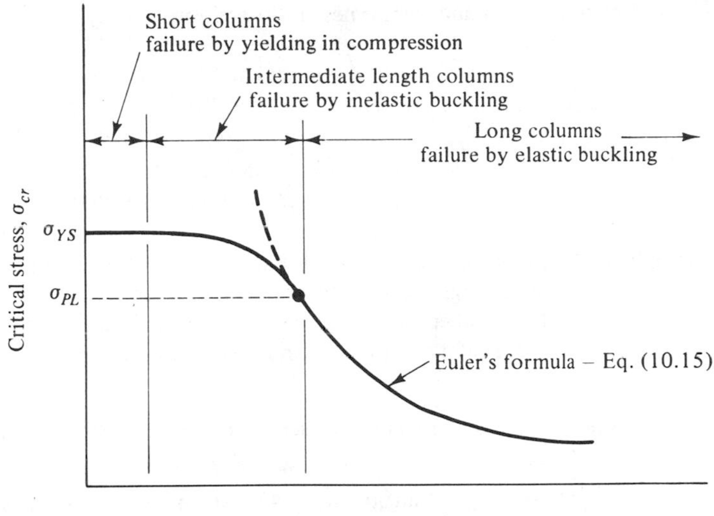

Figure 10.7

the members considered in earlier chapters, for which the failure load was usually found to depend upon the ultimate or yield strength of the material.

Example 10.2. Adequacy of a Column. It is proposed that 4-in. × 4-in. nominal timber columns with $E = 1.6 \times 10^6$ psi and $\sigma_{PL} = 5{,}400$ psi be used as intermediate supports under a floor beam in a building. Each column is 8-ft long and must support an estimated load of 10,000 lb. Are the proposed columns adequate, and, if so, what safety factor do they provide? What is the slenderness ratio of the columns? Assume hinged ends.

Solution. The geometric properties of the members are obtained from Table A.12: $I = 12.51$ in.^{4} and $A = 12.25$ in.^{2}. Assuming linearly elastic material behavior, we obtain from Euler's formula [Eq. (10.15)]

$$F_{cr} = \frac{\pi^2 EI}{L^2} = \pi^2 \frac{(1.6 \times 10^6 \text{ lb/in.}^2)(12.51 \text{ in.}^4)}{(96 \text{ in.})^2} = 21{,}400 \text{ lb}$$

The corresponding critical stress is

$$\sigma_{cr} = \frac{F_{cr}}{A} = \frac{21{,}400 \text{ lb}}{12.25 \text{ in.}^2} = 1{,}750 \text{ psi}$$

which is well below the proportional limit of 5,400 psi. Thus, the assumption of linearly elastic material behavior is correct, and Euler's formula is valid.

The columns are adequate because the applied load is less than the buckling load. The safety factor is

$$SF = \frac{\text{failure load}}{\text{working load}} = \frac{21{,}400 \text{ lb}}{10{,}000 \text{ lb}} = 2.14 \qquad \textbf{Answer}$$

The radius of gyration and slenderness ratio are, respectively,

$$r = \sqrt{\frac{I}{A}} = \sqrt{\frac{12.51 \text{ in.}^4}{12.25 \text{ in.}^2}} = 1.01 \text{ in.}$$

and

$$\frac{L}{r} = \frac{96 \text{ in.}}{1.01 \text{ in.}} = 95 \qquad \textbf{Answer}$$

Example 10.3. Sizing of a Column. What is the most economical (from a weight standpoint) unequal angle section to use for member *BC* of the truss shown in Figure 10.8(a)? The member is to be made of steel with $E = 207 \text{ GN/m}^2$ and $\sigma_{YS} = 210 \text{ MN/m}^2$. Use a safety factor of 2 and assume that the member is free to buckle in any direction.

Solution. Using the method of sections [Figure 10.8(b)], we have

$$\overset{+}{\curvearrowleft}\sum M_E = -T_{BC}(3 \sin 60° \text{ m}) - 45 \text{ kN}(3 \text{ m}) + 45 \text{ kN}(3 \cos 60° \text{ m}) = 0$$

$$T_{BC} = -26 \text{ kN} \quad \text{or} \quad 26 \text{ kN (compression)}$$

The buckling load must be at least as large as the design load, which is the safety factor times the actual load, or 52 kN:

$$F_{\text{cr}} = \frac{\pi^2 EI}{L^2} \geqslant 52 \text{ kN}$$

The member will tend to buckle about the axis of least moment of inertia; therefore,

$$I_{\min} \geqslant \frac{(52 \times 10^3 \text{ N})(3 \text{ m})^2}{\pi^2(207 \times 10^9 \text{ N/m}^2)} = 22.91 \times 10^{-8} \text{ m}^4 \quad \text{or} \quad 22.91 \text{ cm}^4$$

Turning to Table A.11, we seek the angle section with the smallest mass per unit length that has a minimum moment of inertia of 22.91 cm^4, or larger. (Note that I is minimum about axis VV.) Starting at the bottom of the table and working upward, we find that there

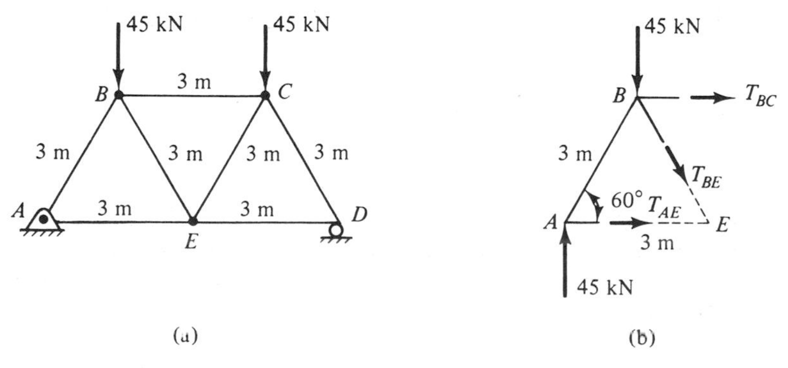

Figure 10.8

are several possibilities:

Nominal Size (mm)	I_{VV} (cm^4)	*Mass/Length (kg/m)*
76 × 64	24.4	11.17
89 × 64	24.6	10.57
89 × 76	25.5	7.89
102 × 64	23.2	9.69

There are other sections in each group in the table that have the required value of I, but they have a larger mass than those listed. The 89-mm × 76-mm section with a mass per unit length of 7.89 kg/m is seen to be the most economical.

Since Euler's formula was used in the computations, we must check to see if it is valid. From Table A.11, the area of the section selected is found to be 10.05 cm^2. Thus,

$$\sigma_{cr} = \frac{F_{cr}}{A} = \frac{52 \times 10^3 \text{ N}}{10.05 \times 10^{-4} \text{ m}^2} = 5.17 \times 10^7 \text{ N/m}^2 \quad \text{or} \quad 51.7 \text{ MN/m}^2$$

which is well below the yield strength.

PROBLEMS

10.10. Determine the minimum slenderness ratio for which Euler's formula applies for hinge-ended columns made of (a) structural steel, (b) 2024-T3 aluminum, (c) titanium alloy, and (d) brass.

10.11. Determine the buckling load of a brass rod 25 mm in diameter and 0.75-m long. What is the slenderness ratio of the member? Assume hinged ends.

10.12. Same as Problem 10.11, except that the rod is made of 6061-T6 aluminum.

10.13. A wooden yardstick is compressed by placing one end against the floor and pushing down on the other end with the flat of the hand. Estimate the load at which the stick will buckle if it is 1-in. wide and $\frac{1}{4}$-in. thick. Use $E = 1.6 \times 10^6$ psi.

10.14. A 1-in. diameter solid structural steel member is to support a compressive load of 1,000 lb with a safety factor of 3. What is the maximum allowable length for the member? Assume hinged ends.

10.15. A 2-in. × 4-in. (nominal) timber 8-ft long is used as a column in a construction project. What is the maximum compressive load that can be supported with a safety factor of 4? $E = 1.6 \times 10^6$ psi and $\sigma_{PL} = 5{,}400$ psi? Assume that the member is hinged at each end and is free to buckle in any direction.

10.16. What is the maximum compressive load F that can be supported by a hinge-ended structural steel column 2-m long made of an 89 × 76-mm angle section with a mass of 16.8 kg/m? Use a safety factor of 2, and assume that the member is free to buckle in any direction.

10.17. Four angle sections are riveted together to form a column with the cross section

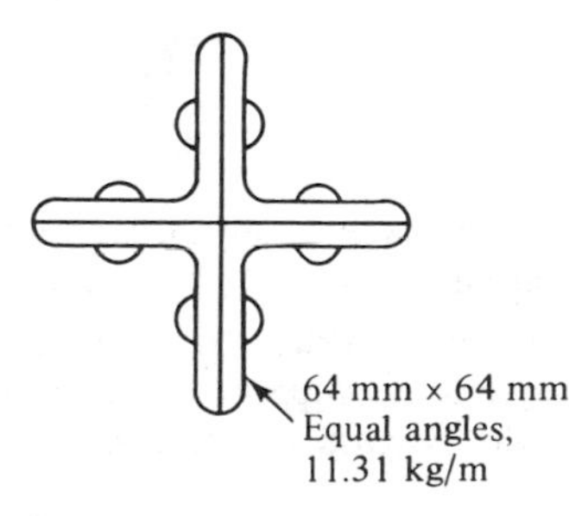

Problem 10.17

shown. Determine the ratio of the buckling load of the composite member and that of an individual angle. Assume the members are free to buckle in any direction. Note: The area moment of inertia of the composite member is the same about every transverse axis through the centroid of the cross section.

10.18. A 10-mm diameter rod 0.5-m long is placed between two immovable walls. What increase in temperature is required to buckle the rod if it is made of (a) structural steel and (b) 2024-T3 aluminum?

10.19. Member BD of the truss shown is made of a 1-in. (nominal) structural steel pipe. Is the member adequate to support an applied load of $F = 2$ kips? If so, what is the safety factor against failure by buckling?

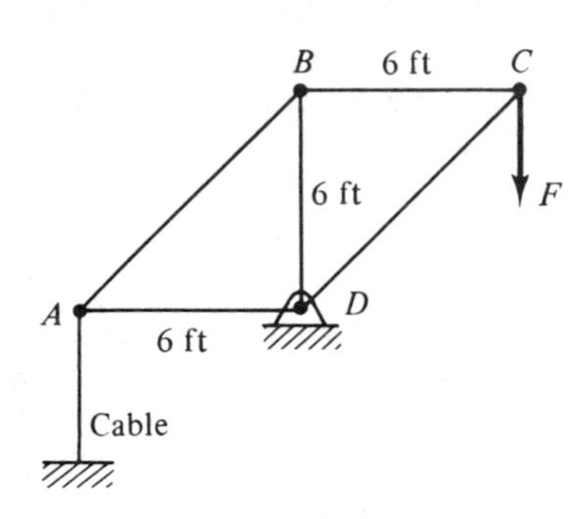

Problem 10.19

10.20. Same as Problem 10.19, except that the applied load is $F = 4$ kips.

10.21. What diameter structural steel rod is required to support a compressive load of 500 N with a safety factor of 4 if the member is 0.6 m long? Assume hinged ends.

10.22. What is the lightest weight Universal beam section that will support a compressive load of 12 kN? The member is 6 m long and is made of 6061-T6 aluminum. Assume hinged ends and freedom to buckle in any direction.

10.23. Brace BD of the structure shown is to be made of a steel ($\sigma_{YP} = 50$ ksi) channel section oriented with the flanges vertical. What is the lightest weight section that can be used if a safety factor of 3 against failure by buckling is required?

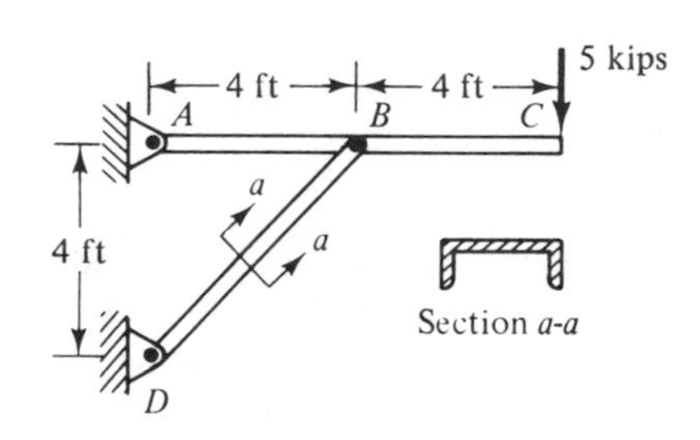

Problem 10.23

10.24. If the brace BD in Problem 10.23 is to be made of steel pipe ($\sigma_{YP} = 50$ ksi), what size pipe is required?

10.25. A beam-column with hinged ends supports an axial compressive load F and a uniformly distributed lateral load q. Derive an expression for the lateral deflection $v(x)$ of the member. Express the results in terms of the length L and flexural rigidity EI. Show that $v(x) \to \infty$ as $F \to F_{cr} = \pi^2 EI/L^2$.

10.4 EFFECT OF SUPPORT CONDITIONS

The manner in which a column is supported has a significant influence upon its buckling load. This will be demonstrated in this section by determining the critical load for elastic columns with various combinations of support conditions. The basic procedure will be the same as that used in the preceding section for hinge-ended columns.

Consider a column fixed at one end and hinged at the other and subjected to an axial compressive load F [Figure 10.9(a)]. The free-body diagram of the entire member when it is in a slightly deflected position is shown in Figure 10.9(b). Note that the problem is statically indeterminate because the reactions cannot all be determined from the equations of equilibrium. However, these equations do yield the relationships

$$A_x = F \qquad C_A = -A_y L \qquad A_y = -B_y$$

From the free-body diagram of Figure 10.9(c), the bending moment is found to be

$$M = A_y x + C_A - A_x v = A_y x - A_y L - Fv$$

Substituting this expression into Eq. (8.18) for the small deflection of linearly elastic beams, we obtain as the governing differential equation for the bending of the column

$$EI\frac{d^2v}{dx^2} = A_y x - A_y L - Fv$$

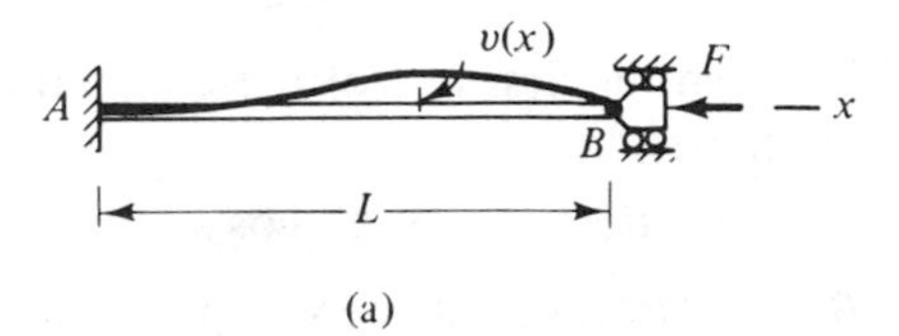

(a)

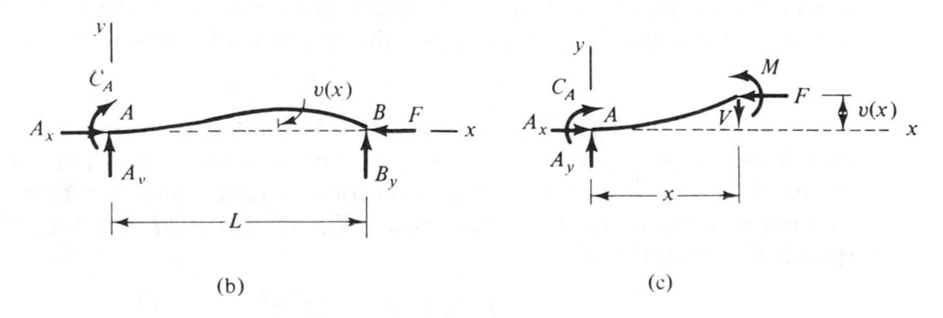

(b) (c)

Figure 10.9

It is convenient to rewrite this equation as

$$\frac{d^2v}{dx^2} + k^2v = \frac{A_y}{EI}(x - L); \qquad k^2 = \frac{F}{EI} \tag{10.20}$$

The general solution to Eq. (10.20) consists of the complementary solution, v_C, plus a particular solution, v_P. The complementary solution is obtained by setting the right-hand side of the equation equal to zero, in which case it reduces to Eq. (10.11). Thus, v_C is given by Eq. (10.13):

$$v_C = A \cos kx + B \sin kx$$

The particular solution is assumed to be of the same form as the terms on the right-hand side of the equation, or

$$v_P = Cx + D$$

where C and D are constants. Substituting this expression into Eq. (10.20), we find that it does satisfy the equation for $C = A_y/F$ and $D = -A_yL/F$. Thus, the complete solution is

$$v = v_C + v_P = A \cos kx + B \sin kx + \frac{A_y}{F}(x - L) \tag{10.21}$$

as can be readily confirmed by direct substitution.

Applying the boundary conditions $v(0) = 0$, $\theta(0) = 0$, and $v(L) = 0$, we have

$$v(0) = 0: \qquad 0 = A - \frac{A_yL}{F}$$

$$\theta(0) = 0: \qquad 0 = Bk + \frac{A_y}{F}$$

$$v(L) = 0: \qquad 0 = A \cos kL + B \sin kL \tag{10.22}$$

Elimination of A_y between the first two of these equations gives $A = -BkL$. Thus, the third condition becomes

$$B(\sin kL - kL \cos kL) = 0$$

One possibility is $B = 0$. However, this corresponds to the null solution $v(x) \equiv 0$, as can be seen from Eqs. (10.22) and (10.21). The nonzero solution is obtained when the terms in parentheses are zero. Thus, the critical load is defined by the condition

$$\tan kL = kL \tag{10.23}$$

The roots of Eq. (10.23) can be obtained by plotting both sides of the equation and determining the values of kL at the points of intersection. They can also be obtained by trial and error or other numerical schemes. Suffice it to say that the smallest nonzero value of kL that satisfies Eq. (10.23) is $kL = 1.43\pi$. This corresponds to a critical load of

$$F_{cr} = k^2EI = \frac{2\pi^2EI}{L^2} \tag{10.24}$$

where E, I, and L are as defined previously.

Although we shall not go through the details here, this same procedure applies for other columns with various combinations of fixed, free, or hinged ends. In every case, the critical load is found to differ from that for a hinge-ended column only by a constant factor. Thus, Euler's formula can be expressed in the general form

$$F_{\text{cr}} = \frac{C\pi^2 EI}{L^2} \tag{10.25}$$

where C is a constant, called the *end-fixity factor*, which depends upon the end conditions. The values of C for several common cases are given in Figure 10.10. Notice that the critical load increases as the degree of constraint increases. It is smallest for a column fixed at one end and free at the other end and it is largest for one fixed at both ends.

All of the results presented in the preceding section for hinge-ended columns also apply for columns with other end conditions if the general Euler's formula [Eq. (10.25)] is used in place of Eq. (10.15). Thus, the critical stress becomes

$$\sigma_{\text{cr}} = \frac{F_{\text{cr}}}{A} = \frac{C\pi^2 E}{(L/r)^2} \tag{10.26}$$

and the conditions for the validity of Euler's formula are

$$\sigma_{\text{cr}} = \frac{C\pi^2 E}{(L/r)^2} \leqslant \sigma_{PL} \tag{10.27}$$

or

$$(L/r)^2 \geqslant \frac{C\pi^2 E}{\sigma_{PL}} \tag{10.28}$$

In these equations, L/r is the slenderness ratio of the column and σ_{PL} is the proportional limit of the material.

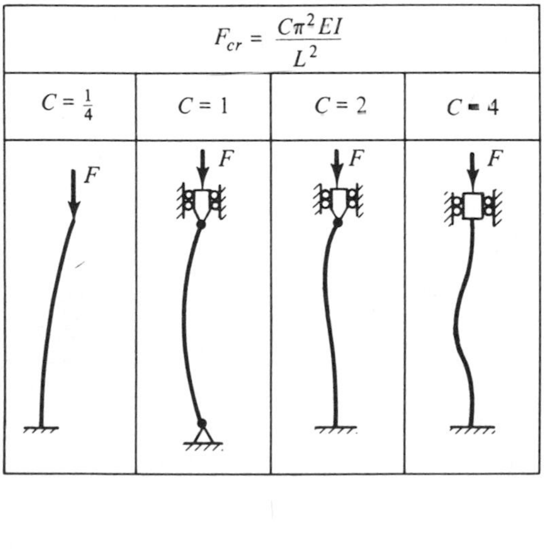

$F_{cr} = \frac{C\pi^2 EI}{L^2}$			
$C = \frac{1}{4}$	$C = 1$	$C = 2$	$C = 4$
F	F	F	F

Figure 10.10

Equation (10.28) indicates that the range of slenderness ratios for which Euler's formula applies, and for which a column can be considered to be long, depends upon the value of C. For example, we mentioned in the preceding section that a hinge-ended steel column ($C = 1$) with $E = 30 \times 10^6$ psi and $\sigma_{PL} = 30$ ksi can be considered to be long if $L/r \geqslant 100$. However, if one end is fixed and the other is free ($C = \frac{1}{4}$), it can be considered to be long if $L/r \geqslant 50$.

The end conditions for a column often depend upon the direction in which it buckles. For example, consider the column shown in Figure 10.11(a), which is fastened between massive cross members at each end by a single bolt. If the column buckles in the yz plane, as in Figure 10.11(b), the bolts act as hinges. However, if it buckles in the xz plane [Figure 10.11(c)], the supports act as fixed ends. The manner in which the column actually buckles can be determine only by computing the critical load for each case and seeing which is smallest (see Example 10.4). Note that the value of I is different in each case because the column bends about different axes.

Actually, the support conditions shown in Figure 10.10 are idealizations, as we have already discussed in Chapter 3. Few, if any, real columns duplicate these conditions exactly. In the preceding example, for instance, there will likely be some friction and resistance to rotation about the bolts. Thus, the support does not act exactly like a hinged end, but it certainly isn't fixed or free either. The hinged end is probably the closest approximation to reality for buckling in the xy plane.

Engineering judgment is usually required in the modeling of the support conditions for columns, just as it is for any other type of member. Since the buckling load decreases as the degree of constraint decreases, the computed critical load will be on the conservative side if we assume less support than actually exists.

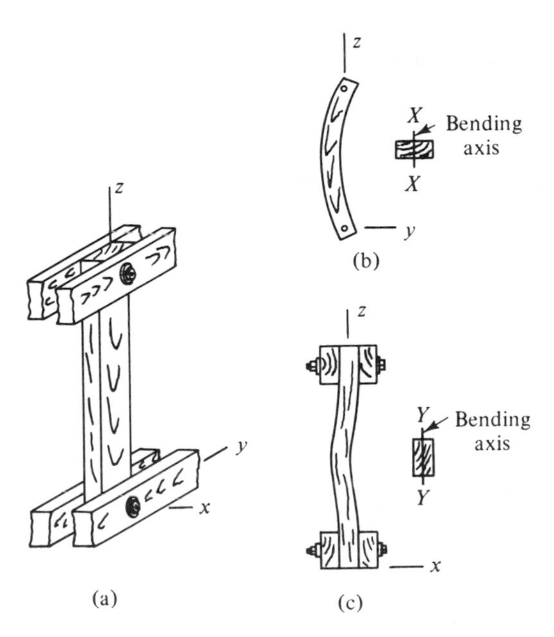

Figure 10.11

The wide range in the values of the constant C in the general Euler's formula ($\frac{1}{4} \leqslant C \leqslant 4$ for the cases shown in Figure 10.10) and the degree of uncertainty in determining its actual value are additional reasons why substantial safety factors are required in the design of columns.

The results presented in this and the preceding sections also apply to some columns with intermediate supports. The requirement is that the intermediate supports divide the member into identical sections. For example, each portion of the column shown in Figure 10.12(a) looks and behaves like a hinge-ended column with length $L/3$; therefore, the buckling load is equal to that for one of the segments: $F_{cr} = \pi^2 EI/(L/3)^2$. A second example to which our results apply is shown in Figure 10.12(b). Here, each segment acts like a fixed-hinged column with length $L/2$; therefore, the buckling load is $F_{cr} = 2\pi^2 EI/(L/2)^2$. Other problems involving intermediate supports are treated in texts on elastic stability.

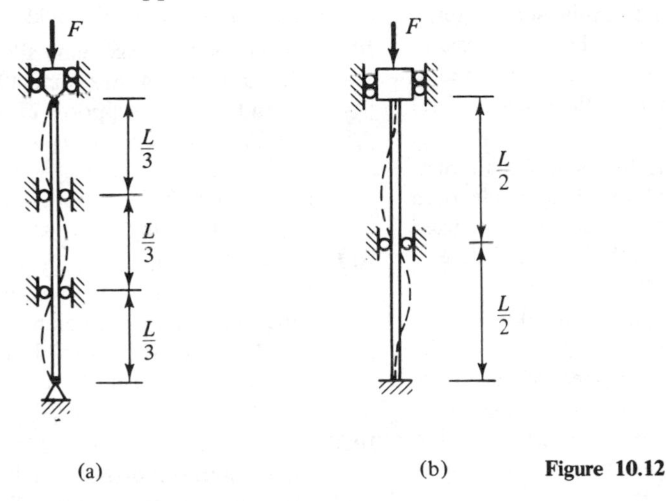

(a) (b) **Figure 10.12**

Example 10.4. Buckling Load of a Column. Determine the buckling load for the column shown in Figure 10.11(a) if it is 10 ft long and made of a 2-in. × 4-in. nominal timber with $E = 1.6 \times 10^6$ psi and $\sigma_{PL} = 5{,}400$ psi.

Solution. The geometric properties of the member obtained from Table A.12 are $A = 5.25$ in.,^{2}. $I_{XX} = 5.359$ in.^{4}, and $I_{YY} = 0.984$ in.^{4} Since the direction in which the column will buckle is not known beforehand, both possibilities must be checked. If it buckles in the yz plane, as in Figure 10.11(b), axis XX is the bending axis, and the supports act like hinged ends. The critical load for this case is

$$F_{cr} = \frac{\pi^2 EI}{L^2} = \pi^2 \frac{(1.6 \times 10^6 \text{ lb/in.}^2)(5.359 \text{ in.}^4)}{(120 \text{ in.})^2} = 5{,}880 \text{ lb}$$

If the member buckles in the xz plane [Figure 10.11(c)], the supports act like fixed ends, and YY is the bending axis. For this case, we have

$$F_{cr} = \frac{4\pi^2 EI}{L^2} = 4\pi^2 \frac{(1.6 \times 10^6 \text{ lb/in.}^2)(0.984 \text{ in.}^2)}{(120 \text{ in.})^2} = 4{,}320 \text{ lb}$$

Since the critical load is smallest in the second case, that is the manner in which the column will buckle. The buckling load is

$$F_{cr} = 4{,}320 \text{ lb} \qquad \textbf{Answer}$$

The critical stress is

$$\sigma_{cr} = \frac{F_{cr}}{A} = \frac{4{,}320 \text{ lb}}{5.25 \text{ in.}^2} = 823 \text{ psi}$$

which is well below the proportional limit.

PROBLEMS

10.26. What is the minimum slenderness ratio for which Euler's formula applies for a structural steel column that is (a) fixed at one end and free at the other, (b) fixed at one end and hinged at the other, and (c) fixed at both ends?

10.27. Determine the buckling load for a structural steel column 6-m long made of a 178-mm × 89-mm channel section fixed at each end. Assume the member is free to buckle in any direction.

10.28. A 1-in. diameter steel rod 10-ft long is fixed at the lower end and pinned at the upper end to a member of negligible weight that is free to expand into an opening. If the end of the bar is in the position shown at 60°F, what is the maximum temperature to which the bar can be heated before it will buckle?

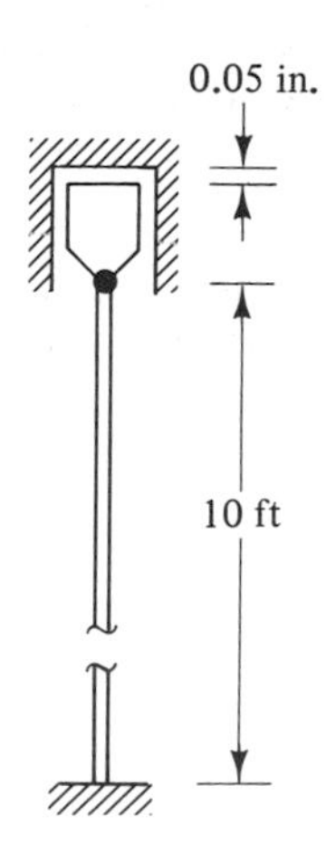

Problem 10.28

10.29. What is the maximum allowable length for a 4-in. × 4-in. (nominal) timber column that must support a compressive load of 10 kips with a safety factor of 3? $E = 1.6 \times 10^6$ psi and $\sigma_{PL} = 5{,}400$ psi. The member is fixed at both ends and is free to buckle in any direction.

10.30. What is the minimum required cross-sectional area of a square column 1-m long that must support a compressive load of 6 kN? The member is fixed at one end and hinged at the other; the material is 6061-T6 aluminum.

10.31. A 50-mm diameter rod can support a maximum compressive load of 4 kN when it is hinged at each end. What diameter rod is required to support this same load if both ends are fixed? Does the result depend upon the material used? Explain.

10.32. A cover over a sidewalk is supported by a central row of 10-ft tall columns made of high-strength ($\sigma_{YP} = 100$ ksi) steel pipe. The cover weighs 1,200 lb/ft of length, including estimated snow loads. What is the maximum allowable spacing between the columns if they are made of 3-in. (nominal) pipe? Use a safety factor of 5, and assume an end-fixity factor of $C = 3.5$.

10.33. If the columns in Problem 10.32 are to be spaced 20 ft apart, what size pipe is required?

10.34. Link BD of the structure shown is made of 2024-T3 aluminum. Determine the maximum load F that can be supported without buckling the link if $w = 40$ mm, $t = 15$ mm, and $L = 1$ m. Will the link buckle in the plane of the structure or out of plane?

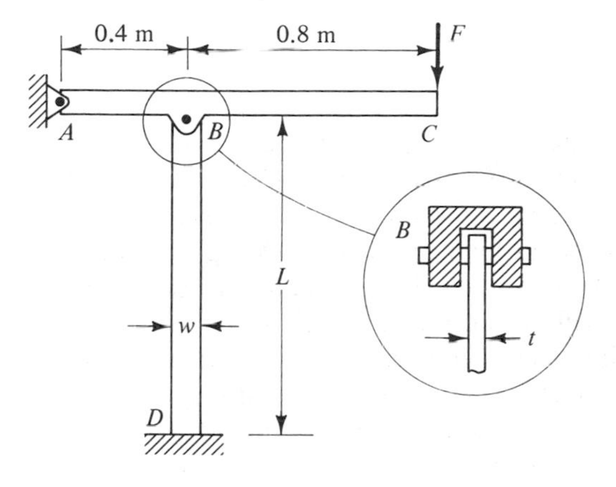

Problem 10.34

10.35. Determine the width to thickness ratio w/t for which link BD in Problem 10.34 will be equally strong for both in-plane and out-of-plane buckling. Assume elastic behavior.

10.36. The boom shown is made of a W 6 × 16 section oriented with the web horizontal. Determine the weight of the largest object that can be supported without causing the boom to buckle. Use a safety factor of 4; the material is 2024-T3 aluminum.

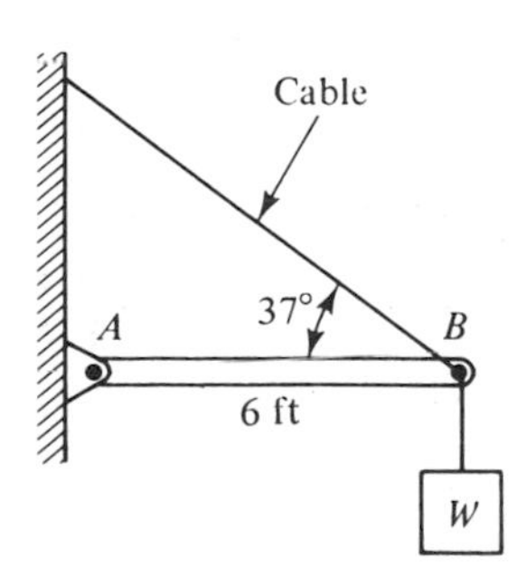

Problem 10.36

10.37. A 40-mm diameter acrylic rod supported by ball and socket joints at each end passes through a small clearance hole in a thin plate at its midpoint, as shown. Determine the buckling load for the member; $E = 2.8$ GN/m^2 and $\sigma_{YS} = 40$ MN/m^2.

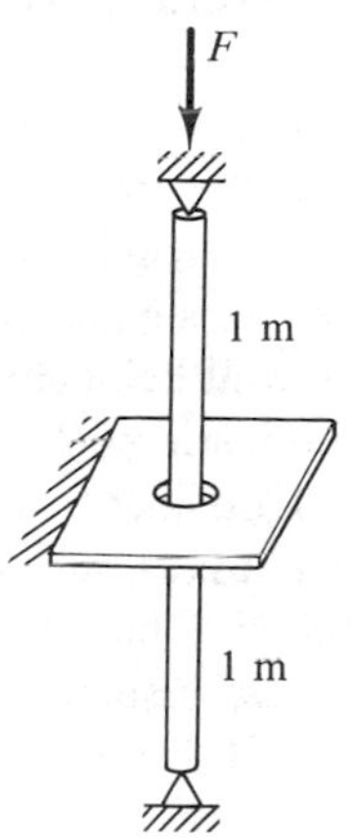

Problem 10.37

10.38. Same as Problem 10.37, except that the ends of the rod are fixed.

10.39. A hollow rectangular tube made of 2024-T3 aluminum is supported by ball and

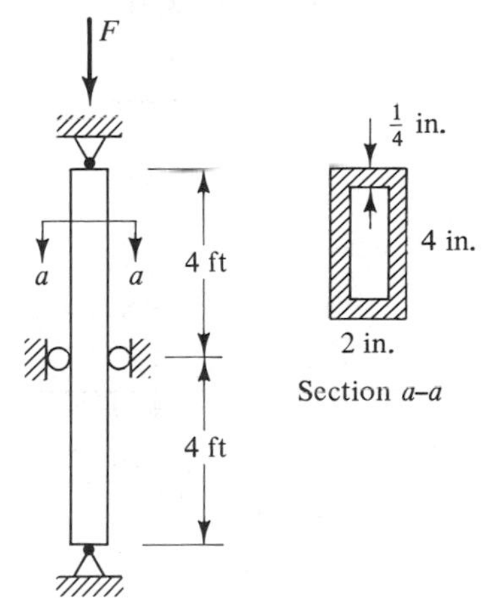

Problem 10.39

socket joints at each end and by two rollers at its midpoint, as shown. If the tube has a wall thickness of $\frac{1}{4}$ in., what is the buckling load for the member?

10.40. Same as Problem 10.39, except that the tube is made of titanium alloy.

10.41. Derive an expression for the buckling load of an elastic column fixed at one end and free at the other. Verify that the end-fixity factor is $C = \frac{1}{4}$ for this case.

10.5 INELASTIC BEHAVIOR

As indicated in Section 10.3, intermediate length columns buckle at stresses that exceed the proportional limit of the material. Consequently, Euler's formula does not apply, and a separate analysis is required. Since the stiffness of most materials decreases significantly above the proportional limit, inelastic buckling can occur at loads well below those predicted by Euler's formula. The procedure for determining the inelastic buckling load of columns will be considered in this section.

Consider a column of intermediate length subjected to an axial compressive load F that is just slightly less than the buckling load, whatever that may be [Figure 10.13(a)]. The stress-strain diagram for the material is shown in Figure 10.13(b). Since we assume that the column remains straight prior to buckling, the stress is $\sigma = F/A$.

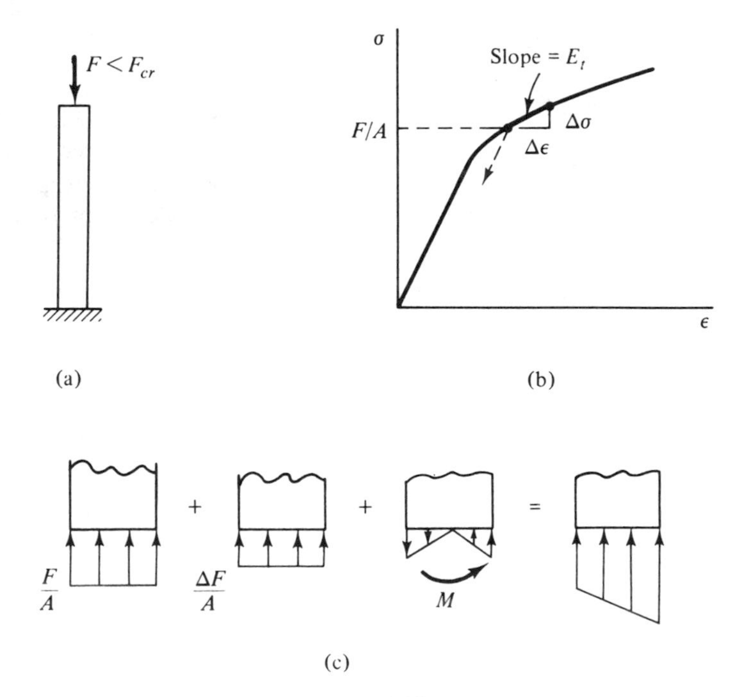

Figure 10.13

Now let us suppose that the applied load is increased by a small amount ΔF and that the column simultaneously starts to buckle. This increment in load produces an additional compressive stress of magnitude $\Delta F/A$ and a flexural stress due to the bending. The final stress distribution is as shown in Figure 10.13(c). It is assumed that the increment in compressive stress is larger than the bending stress, in which case the total stress increases at all points over the cross section and there is no unloading of any of the fibers.

As a result of the preceding assumptions, the increment of stress $\Delta\sigma$ on a typical fiber and the corresponding increment of strain $\Delta\varepsilon$, both of which are small for a small increment in load, must be as shown in Figure 10.13(b). The relationship between them is

$$\Delta\sigma = E_t\,\Delta\varepsilon \tag{10.29}$$

where E_t is the tangent modulus of the material at the stress $\sigma = F/A$. Equation (10.29) holds at all points over the cross section. This would not be true if the stress on some of the fibers decreased. In this case, the stresses would follow the path indicated by the dashed line in Figure 10.13(b), and the stress-strain relation for the fibers that experience unloading would be $\Delta\sigma = E\,\Delta\varepsilon$.

Now let us focus our attention upon the bending response, since it is directly associated with the buckling. The only difference between the bending encountered here and the linearly elastic bending considered in Chapter 8 is that the stress-strain relation is given by Eq. (10.29) instead of Hooke's law. Thus, all of our results for linearly elastic beams still apply if the elastic modulus, E, is replaced by the tangent modulus, E_t. In particular, Eq. (8.18) for the small deflections of beams still holds. Thus, the inelastic bending of the column is governed by the equation

$$E_t I \frac{d^2 v}{dx^2} = M \tag{10.30}$$

The expressions for the bending moment, M, and the boundary conditions are not affected by the material properties; therefore, the solutions of Eq. (10.30) will be the same as for the linearly elastic case, except that E is replaced by E_t. Consequently, we conclude that the buckling load and critical stress for inelastic columns are, respectively,

$$F_{\text{cr}} = \frac{C\pi^2 E_t I}{L^2} \tag{10.31}$$

and

$$\sigma_{\text{cr}} = \frac{F_{\text{cr}}}{A} = \frac{C\pi^2 E_t}{(L/r)^2} \tag{10.32}$$

In these equations, E_t is the tangent modulus of the material at the critical stress. All other quantities are as defined previously, including the end-fixity factor C, which has the same values as for elastic columns.

Equation (10.31) is called the *tangent modulus formula*. One may question the validity of the assumptions upon which this formula is based, but it gives values of the buckling load that are in satisfactory agreement with experimental results.

Although the tangent modulus formula is identical in form to Euler's formula, it is more difficult to apply because the tangent modulus, E_t, is not a constant, but rather depends upon the magnitude of the stress. The procedure to be followed depends upon what is known and what is to be determined, as we shall see in the following.

The stress-strain diagram for the material is usually available in the form of a graph. In this case, it is convenient to first determine the values of E_t at several points and plot its variation with the stress level, as in Figure 10.14. Plots of E_t for many common engineering materials are available in handbooks.

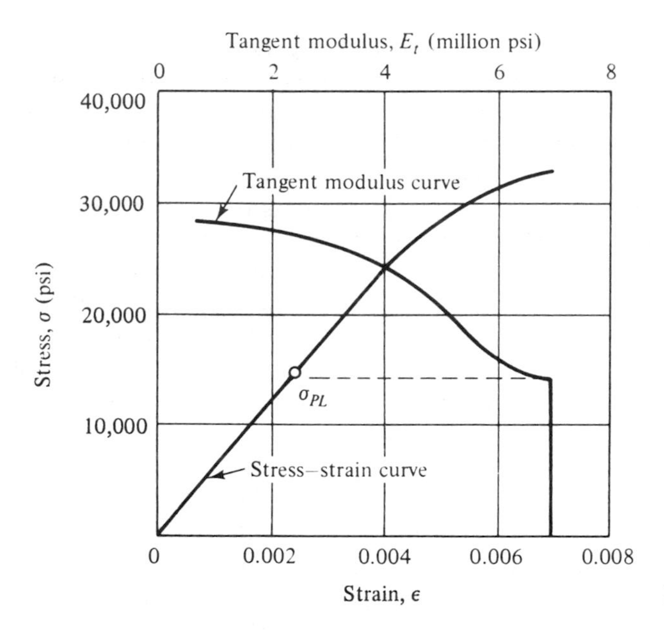

Figure 10.14

The simplest way to determine the buckling load for a given column is to plot Eq. (10.32) on the same graph as E_t (Figure 10.15). As indicated in the figure, this equation plots as a straight line whose slope depends upon the slenderness ratio. The point of intersection between this line and the curve of E_t determines the critical stress, σ_{cr}. Then, $F_{cr} = \sigma_{cr}A$, where A is the cross-sectional area of the member. Note that, for large values of the slenderness ratio, the value of the stress at the point of intersection is below the proportional limit. In this case, $E_t = E$, and the tangent modulus formula reduces to Euler's formula.

An alternate approach is to construct the curve showing the values of σ_{cr} at different values of the slenderness ratio for the material of interest. This is easily done by choosing several values of σ_{cr}, determining the corresponding values of E_t from the curve of E_t, and solving for L/r from Eq. (10.32). The central portion of

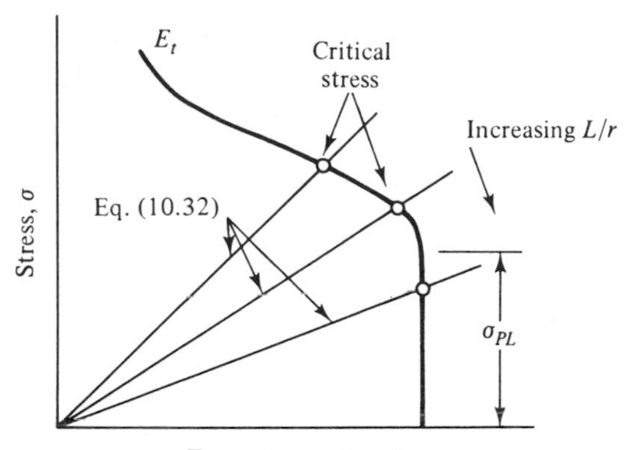

Figure 10.15

the curve in Figure 10.7 is an example of such a plot. Once these curves are available, the critical stress for a given column can be read directly from the plot.

This plot is also convenient for showing the effect of the end conditions in the inelastic range. Figure 10.16 shows the variation of the critical stress with the slenderness ratio for various values of C. Notice that the end conditions have less effect upon the critical stress for columns of intermediate length than they do for long columns. In particular, the critical stress, and hence the critical load, is not proportional to C, as in the elastic case. For example, the inelastic buckling load of a column with fixed ends is not four times that for one with hinged ends. The buckling load must be determined separately for each different set of end conditions considered.

A trial-and-error procedure (or other numerical or graphical scheme) is usually required to determine the size of a column needed to support a given load. Several possible approaches are illustrated in Example 10.5.

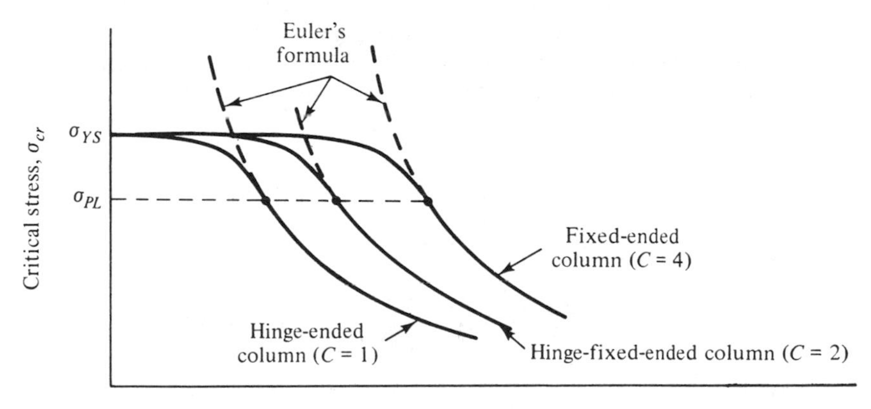

Figure 10.16

If the stress-strain curve for the material is known in equation form, the relation between the tangent modulus and the stress can be determined directly from the definition of E_t:

$$E_t = \frac{d\sigma}{d\varepsilon} \tag{10.33}$$

In this case, the critical stress can be expressed directly in terms of the slenderness ratio, and the design or analysis of the column can be carried out analytically. For example, if the equation for the stress-strain curve is

$$\sigma = k\sqrt{\varepsilon}$$

where k is a constant, we have

$$E_t = \frac{d\sigma}{d\varepsilon} = \frac{1}{2}\frac{k}{\sqrt{\varepsilon}} = \frac{k^2}{2\sigma}$$

Evaluating E_t at the critical stress and substituting it into Eq. (10.32), we get the desired relationship

$$\sigma_{cr}^2 = \frac{C\pi^2 k^2}{2(L/r)^2}$$

In addition to the tangent modulus formula, there are a number of empirical formulas available for predicting the critical stress for intermediate length columns. These formulas are obtained by fitting simple equations to plots of the experimentally determined values of the critical stress for members with different slenderness ratios. These equations are widely used in column design because they are often simpler to apply than the tangent modulus formula. However, since many of these equations hold only for specific materials, we will not consider them here. They are discussed in texts on structural design and stability.

Example 10.5. Determination of Inelastic Buckling Load. A column with a 2-in. × 2-in. square cross section is made of 6061-T6 aluminum alloy with the properties shown in Figure 10.17. The member is $3\frac{1}{2}$ ft long and is fixed at one end and hinged at the other. (a) What is the buckling load and (b) what size square cross section would be needed to support this same value of load if both ends were fixed?

Solution. (a) The moment of inertia of the cross section is

$$I = \frac{1}{12}bh^3 = \frac{1}{12}(2\text{ in.})^4 = 1.33\text{ in.}^4$$

From Eq. (10.31), we have

$$F_{cr} = \frac{C\pi^2 E_t I}{L^2} = \frac{2\pi^2(1.33\text{ in.}^4)}{(42\text{ in.})^2}E_t = (14.88 \times 10^{-3})\,E_t$$

therefore,

$$\sigma_{cr} = \frac{F_{cr}}{A} = \frac{(14.88 \times 10^{-3})}{4\text{ in.}^2}E_t = (3.72 \times 10^{-3})\,E_t$$

This expression for σ_{cr} is now plotted on the same graph as E_t (Figure 10.17). The two curves intersect at a stress of approximately 31,300 psi; therefore,

$$F_{cr} = \sigma_{cr} A = (31{,}300 \text{ lb/in.}^2)(4 \text{ in.}^2) = 125{,}200 \text{ lb} \qquad \textbf{Answer}$$

(b) Let a denote the length of each side of the cross section. Then

$$F_{cr} = \frac{4\pi^2 E_t (a^4/12)}{L^2} = 125{,}200 \text{ lb}$$

and

$$\sigma_{cr} = \frac{F_{cr}}{A} = \frac{125{,}200}{a^2} \text{ psi}$$

The problem is to determine a value of E_t and σ_{cr} from the tangent modulus curve such that both of these relations are satisfied. There are several ways to do this. One is trial and error.

We pick a value of E_t, solve for a from the expression for F_{cr}, and then compute σ_{cr}. If this value of σ_{cr} lies on the tangent modulus curve, it is the critical stress for the column, and the problem is solved. If not, the value of E_t is adjusted until this condition is met. Since there is always a possibility that Euler's equation will apply, we take $E_t = E$ as our first

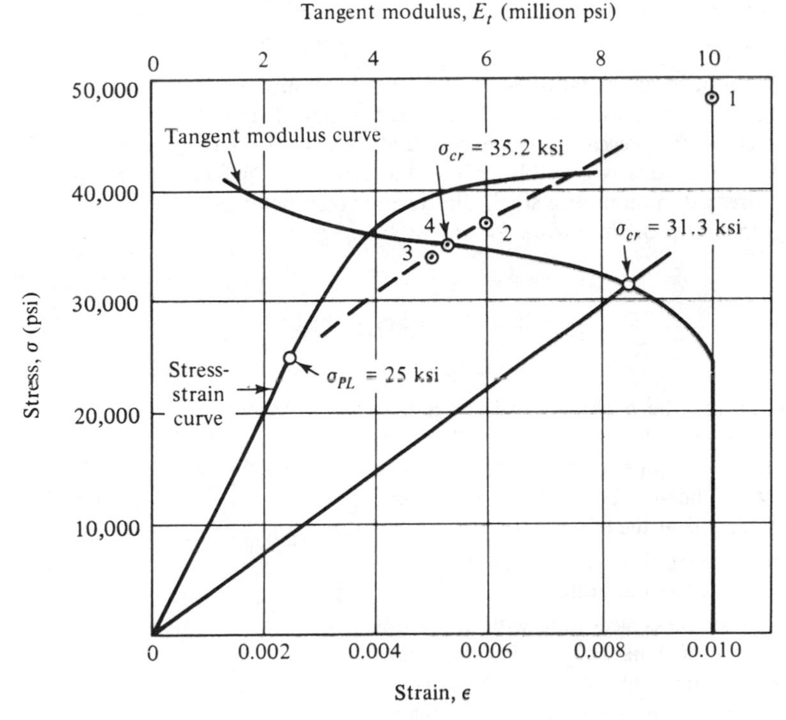

Figure 10.17

guess. The computations are summarized below:

1. $E_t = 10 \times 10^6$ psi $\quad \sigma_{cr} = 48{,}300$ psi (too high)
2. $E_t = 6 \times 10^6$ psi $\quad \sigma_{cr} = 37{,}400$ psi (too high)
3. $E_t = 5 \times 10^6$ psi $\quad \sigma_{cr} = 34{,}200$ psi (too low)
4. $E_t = 5.3 \times 10^6$ psi $\quad \sigma_{cr} = 35{,}200$ psi (close)

$$a^2 = \frac{F_{cr}}{\sigma_{cr}} = \frac{125{,}200 \text{ lb}}{35{,}200 \text{ lb/in.}^2} = 3.56 \text{ in.}^2$$

$$a = 1.9 \text{ in.} \qquad \textbf{Answer}$$

Notice that the required cross-sectional dimensions are very nearly the same as those in part (a). This is in keeping with our previous finding that the end conditions have a reduced effect upon the critical stress for inelastic buckling.

An alternate procedure is to plot the stresses obtained from the various trials on the same graph as E_t. These points are labeled 1, 2, 3, and 4, respectively, in Figure 10.17. The intersection of the curve through these points and the E_t curve defines the critical stress for the column. The advantage of this procedure is that we need consider only enough points to establish the curve through them.

Still another possibility is to obtain the equation relating σ_{cr} and E_t for the column by eliminating a between the expressions for F_{cr} and σ_{cr}. We have

$$\sigma_{cr}^2 = \frac{\pi^2 F_{cr} E_t}{3L^2} = \frac{\pi^2(125{,}200 \text{ lb})}{3(42 \text{ in.})^2} E_t = 233.5 E_t$$

The plot of this equation is indicated by the dashed curve in Figure 10.17. The intersection of this curve with the curve of E_t defines the critical stress for the column. This approach is the same as the preceding one, except that we actually obtain the equation relating σ_{cr} and E_t. Note that this is possible only if the cross-sectional area and moment of inertia can be expressed in terms of a single dimensional parameter. In contrast, the trial-and-error and first graphical procedure work for any problem.

PROBLEMS

10.42. Determine the buckling load for a 1-in. diameter stainless steel column 2 ft long with the compressive tangent modulus shown. The member is fixed at one end and hinged at the other.

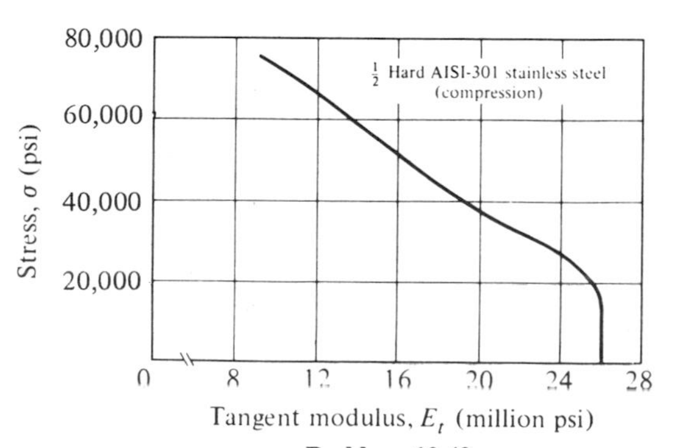

Problem 10.42

10.43. Same as Problem 10.42, except that the column is fixed at both ends.

10.44. A 4-ft stainless steel tube with an outside diameter of 2 in. and a $\frac{1}{4}$-in. wall thickness is hinged at each end. What is the maximum allowable length for the tube if it must support a compressive load of 20 kips with a safety factor of 2? Use the material properties given in Problem 10.42.

10.45. Same as Problem 10.44, except that the tube is made of 6061-T6 aluminum with the properties shown in Figure 10.17.

10.46. An 18-in. long magnesium column with the properties given in Figure 10.14 has a diameter of 1 in. and is hinged at each end. By what percentage can the buckling load be increased by fixing one end?

10.47. Determine the buckling load for a 6-ft long W 8 × 67 column hinged at each end. The material is 6061-T6 aluminum with the properties shown in Figure 10.17. Assume that the member is free to buckle in any direction.

10.48. What is the minimum required diameter of a column 20 in. long fixed at one end and hinged at the other that is to support a compressive load of 12 kips? The material is magnesium alloy with the properties given in Figure 10.14.

10.49. Same as Problem 10.48, except that the member is fixed at both ends.

10.50. What size pipe is required to support a compressive load of 125 kips if the member is 6 ft long and is hinged at each end? The material is stainless steel with the properties given in Problem 10.42.

10.51. A column with fixed ends and slenderness ratio $L/r = 30$ is made of a material whose compressive stress-strain diagram can be described by the equation $\sigma = 10^5\sqrt{\varepsilon}$ over the stress range of interest. What is the critical stress for the column?

10.52. Same as Problem 10.51, except that the stress-strain equation is $\sigma = 10^5\varepsilon^{1/3}$.

10.6 CLOSURE

In this chapter we have been concerned with the elastic and inelastic buckling of columns. There are several factors that distinguish buckling from other types of failures. First, the failure load does not depend upon the strength of the material. It depends only upon the material stiffness and the geometry of the member. Consequently, failure by buckling can occur at very small loads and stresses.

Second, since the deformations remain small until just prior to buckling, there is little warning of the impending failure. This is true even for members made of ductile materials. As a result, buckling failures tend to be catastrophic. In contrast, the other types of members considered in this text, when made of ductile materials, generally undergo large deformations prior to failure. In these cases, the deformations may provide a visual indication that a member is overloaded, thus making it possible to take corrective action.

Finally, the physical nature of the stability of an equilibrium position and buckling and the manner in which they are defined mathematically make it necessary to use the deformed geometry of the member in any force analysis. This is true even for small deformations. This is in contrast to the other types of members considered, for which the rigid-body assumption can be used when the deformations are small.

11

Experimental Strain and Stress Analysis

11.1 INTRODUCTION

In the preceding chapters we developed equations for determining the stresses and deformations in simple load-carrying members. More complicated members and loadings are considered in advanced texts, and, with the use of modern computerized methods of analysis, it is now possible to solve exceedingly complex problems. Nevertheless, there is a continuing need for experiments. Some problems are so complex that they still are not amenable to analysis, in which case the stresses and deformations can only be determined experimentally. Experimental results are also needed to assist in the development of new methods of analysis and to confirm the validity of existing ones.

Some of the common experimental techniques for determining stresses and strains will be considered in this chapter. Our discussions must necessarily be brief; for additional information, the reader is referred to texts on experimental stress analysis. Manufacturers of experimental equipment and supplies are also an excellent source of information.

Throughout our discussions it will be helpful to keep in mind that stresses and strains are mathematical quantities, and, as such, they cannot be measured directly. Instead they must be determined indirectly through their definitions and the measurement of physical quantities such as force and deformation.

11.2 ELECTRICAL RESISTANCE STRAIN GAGES

In Chapter 5 we briefly discussed instruments for determining the strains in materials tests. However, these devices are designed specifically for use on slender tension or compression specimens and would generally not be suitable for determining strains in members of more complicated geometry. One device commonly used to determine strains in such instances is the *electrical resistance strain gage*. This device operates on a physical principle discovered in 1854 by Lord Kelvin; namely, that the electrical resistance of a conductor changes when it is strained.

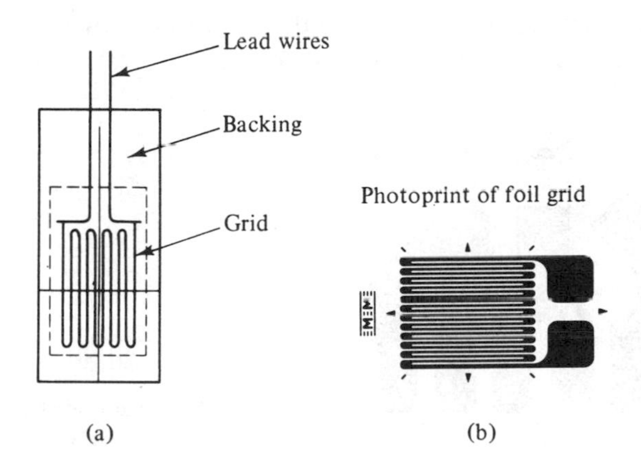

Figure 11.1. Wire Gage (a) and Etched-Foil Grid (b) (*Courtesy* Measurements Group, Inc.)

Electrical resistance strain gages typically consist of a grid attached to a piece of backing material and formed from a fine wire or a thin sheet of foil etched away to form the grid pattern (Figure 11.1) The entire unit is bonded to the surface of the test specimen, as illustrated in Figure 11.2, and a small electrical current is passed through the grid.

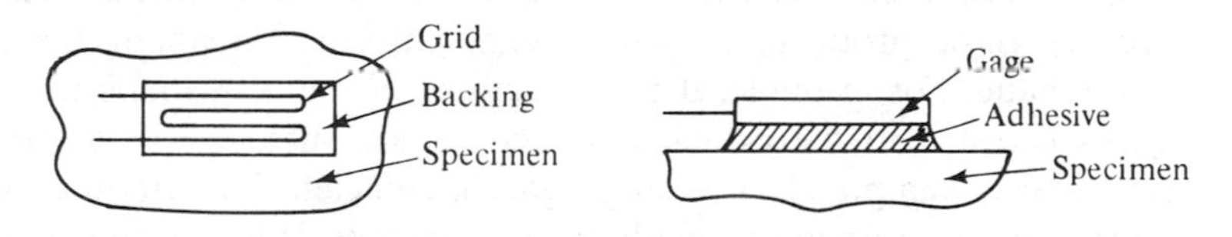

Figure 11.2

The adhesive transmits the deformations of the specimen to the grid, resulting in a change in its electrical resistance. This change in resistance is the physical quantity measured. When combined with the known strain-resistance characteristics of the gage, this information yields the normal strain in the direction along which the gage is aligned. The necessary conversion from change in resistance to strain is usually performed automatically by the instrumentation used. Both tensile and compressive strains can be measured because the gage is bonded over its entire length.

Figure 11.3 shows a typical *strain gage rosette* used to measure the strains in three different directions at a point. As explained in Section 9.8, this information is needed to determine completely the states of strain and stress in general two-dimensional problems. The rosettes consist of three separate grids with different orientations attached to the same piece of backing material (not shown in the figure) and are mounted to the specimen in the same way as a single gage. Each grid responds as an individual gage and provides the normal strain in the direction along which it is aligned. Two common rosette configurations are the *rectangular rosette* (shown in Figure 11.3), in which the grids are 45° apart, and the *delta rosette*, in which they are 60° apart.

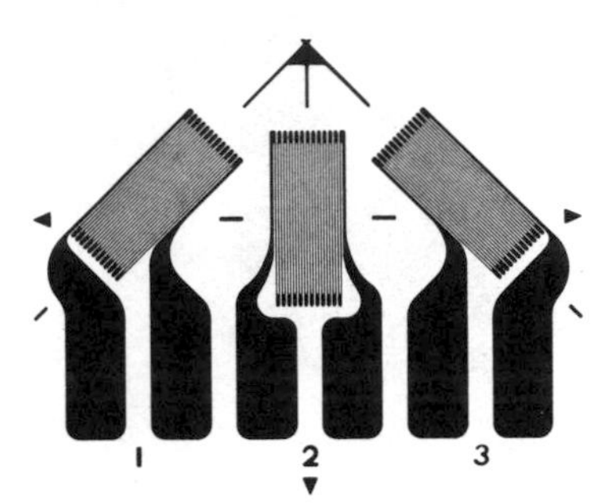

Figure 11.3. Strain Gage Rosette. (*Courtesy* Measurements Group, Inc.)

Three individual gages can be used instead of a rosette, but the rosette is more convenient to mount. Also, the angles between the grids are known to a higher degree of accuracy with a rosette, and the three grids are located more nearly at the same point.

The important thing to remember is that the strains generally must be measured in three different directions at each point of interest. The only exceptions are for those problems in which some information about the strains is known beforehand. For example, if the directions of the principal strains are known, two gages oriented in the principal directions are sufficient to determine the state of strain (see Example 11.1). A single gage is sufficient for determining the longitudinal or transverse strain in an axially loaded member. It is also sufficient for determining the shear strain in a shaft subjected to pure torsion (see Problem 11.8) and the bending strain in a beam subjected to pure bending.

The sensing element in electrical resistance strain gages is made of various metallic alloys and is formed into a grid pattern to provide a suitable length of conductor within a small overall distance. The backing material facilitates handling of the grid and provides the necessary electrical insulation between it and the specimen. Various materials are used for the backing, depending upon the environmental conditions to which the gage will be subjected. Paper and phenolic backings are common, but other materials, such as epoxies, polyesters, ceramics, and metals are also used.

Various kinds of adhesives are used to mount the gages, depending upon the loading and environmental conditions to be encountered and the type of backing material. Paper-backed gages are commonly mounted with cellulose nitrate cement (e.g., Duco household cement), and cyanoacrylate cement (e.g., Eastman 910) is used in a wide variety of cases. Phenolic, epoxy, and ceramic adhesives are also used. The manufacturers' recommendations and procedures should be followed closely to assure the integrity of the bond between the gage and the specimen. Obviously, a good bond is essential for proper performance of the gage.

All strain measuring devices give an average value of strain over their gage length, which, for an electrical resistance strain gage, is the length of the grid. Short gage lengths are required for accurate measurements at locations where the strains vary rapidly with position, such as near notches, holes, and other sources of strain concentration. Longer gage lengths are required whenever it is necessary to "average out" local variations in the material properties, such as occur in concrete and wood. The approximate ranges of gage lengths available are $\frac{1}{16}$ in. to 8 in. for wire gages and $\frac{1}{64}$ in. to 1 in. for foil gages.

The sensitivity of a strain gage is expressed in terms of the *gage factor F*, which is defined as

$$F = \frac{\Delta R/R}{\varepsilon} \tag{11.1}$$

Here, ε is the normal strain in the direction of the gage axis, R is the gage resistance, and ΔR is the change in resistance due to the strain. The larger the gage factor, the more sensitive the gage, i.e., the larger the change in resistance for a given strain. Most gages have a gage factor of approximately 2, but special gages with values of F exceeding 100 are available. The more common values for gage resistance are in the range 120–350 Ω.

Values of the gage factor are provided by the manufacturer and are determined experimentally by subjecting samples from each lot of gages manufactured to a known strain and measuring the resulting change in resistance. The resistance of the undeformed gage is also measured, and the value of F is obtained from Eq. (11.1). Careful quality control in the manufacturing process assures that the value of F obtained for the samples is an accurate indication of the gage factor for the entire lot.

Electrical resistance strain gages have the following attractive features:

1. They are easy to use and are relatively inexpensive. Standard single gages start at several dollars each, with rosettes costing approximately ten times that amount. The necessary instrumentation may be expensive, however, as are special purpose gages.
2. Their use is not restricted to the laboratory. Measurements can be made "in the field," and the gage can be read either on location or remotely.

3. They have high sensitivity and provide very good accuracy when properly installed and used. They also have a good range of response. Strains of the order of 1 $\mu\varepsilon$ can be detected, and an overall measurement system accuracy of ± 5 $\mu\varepsilon$ can be achieved in some instances. The largest strains that can be measured range from 1% to 3%, which is well into the inelastic range for ductile metals (mild steel yields at a strain of approximately 0.1%). Special gages are available with strain ranges up to 10%.
4. Their short gage lengths make it possible to obtain accurate results when the strains vary rapidly with position, such as at locations of strain concentration.
5. They are small and have no significant effect upon the response of the specimen, unless it is very thin or has a very low modulus of elasticity.
6. They can be used for both static and dynamic loadings.
7. They operate over a wide range of environmental conditions. Depending upon the type of backing material, the gages can be used at temperatures ranging from approximately -400 to $+1{,}800$°F. With proper protection, they can be used underwater and in other adverse environments.

The gages also have some unattractive features. For example, the change in resistance of the grid depends not only upon the strain in the direction of the gage axis, but, to some extent, upon the strain in the transverse direction. This effect, known as *transverse sensitivity*, can lead to appreciable errors for biaxial states of stress. Detailed consideration of this effect is beyond the scope of this text. Suffice it to say that a correction for this effect can be made in the data reduction using a *transverse sensitivity factor* for the gages provided by the manufacturer.

Changes in temperature during testing can also cause problems. Since the coefficient of thermal expansion of the gage metal is often different from that of the specimen, the two tend to expand or contract at different rates. However, since the two are bonded together, the gage will be stretched or compressed along with the specimen as it expands or contracts. This leads to an additional change in resistance of the grid and a false indication of the strain due to the applied loads.

Fortunately, problems due to temperature changes often can be circumvented by using the appropriate circuitry (see Section 11.3). Temperature effects also can be accounted for in the data reduction by using *apparent strain* data for the gages provided by the manufacturer. *Self-compensating gages* designed to expand or contract at the same rate as the metals and metallic alloys commonly used in design are available, but these gages are expensive and provide temperature compensation only over a limited temperature range.

The fact that strain gages can be used only on the surface of a member is not a serious disadvantage because the maximum stresses and strains usually occur at the surface. Recall, for example, that the maximum stresses in a twisted shaft occur at the outside surface and the maximum bending stresses in a beam occur at the top and bottom of the section. Since it is not feasible to place gages over the entire surface of a member, the critical areas of high strain must be identified before the gages are mounted. This often requires using some other form of strain analysis, such as brittle coatings (see Section 11.4). Most strain gages cannot be removed and reused, once they have been mounted.

Example 11.1. Strain Measurement in Principal Directions. Show that two strain gages, aligned in the directions of the principal strains, are sufficient for determining the state of strain at a point.

Solution. Let x and y be an arbitrary set of axes at the point and let 1 and 2 be the principal axes (Figure 11.4). Applying the strain transformation equations, Eqs. (9.13) and (9.15), in the principal directions and recognizing that the shear strain associated with the principal axes is zero, we obtain

$$\varepsilon_1 = \varepsilon_x \cos^2 \theta + \varepsilon_y \sin^2 \theta + \gamma_{xy} \sin \theta \cos \theta$$

$$\varepsilon_2 = \varepsilon_x \cos^2 (\theta + \pi/2) + \varepsilon_y \sin^2(\theta + \pi/2) + \gamma_{xy} \sin(\theta + \pi/2)\cos(\theta + \pi/2)$$

$$= \varepsilon_x \sin^2 \theta + \varepsilon_y \cos^2 \theta - \gamma_{xy} \sin \theta \cos \theta$$

$$\gamma_{12} = -2(\varepsilon_x - \varepsilon_y)\sin \theta \cos \theta + \gamma_{xy}(\cos^2 \theta - \sin^2 \theta) = 0$$

Thus, if the strains ε_1 and ε_2 are measured, we have three equations from which to determine the three unknowns ε_x, ε_y, and γ_{xy} for any choice of axes x and y. The angle θ is a known quantity because the principal axes are presumed known.

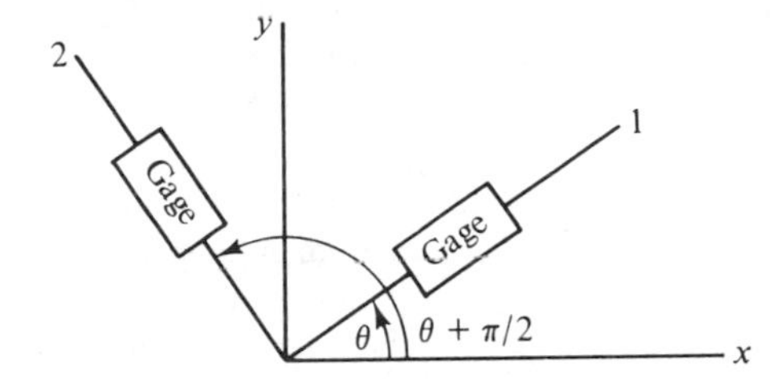

Figure 11.4

PROBLEMS

11.1. What is the change in resistance of a strain gage with gage factor $F = 1.96$ and resistance $R = 300\ \Omega$ when subjected to a strain of 200 $\mu\varepsilon$?

11.2. If the change in resistance of a strain gage with gage factor $F = 2.09$ and resistance $R = 120\ \Omega$ is 0.15 Ω, what is the normal strain in the direction of the gage axis?

11.3. A strain gage is calibrated by mounting it on a 1-mm thick strip that is then bent around a cylindrical surface with a radius of 1 m. If the gage resistance is $R = 350\ \Omega$ and the change in resistance is 0.38 Ω, what is the gage factor F?

11.4. A strain gage with resistance $R = 500\ \Omega$ is mounted on the top surface of a steel

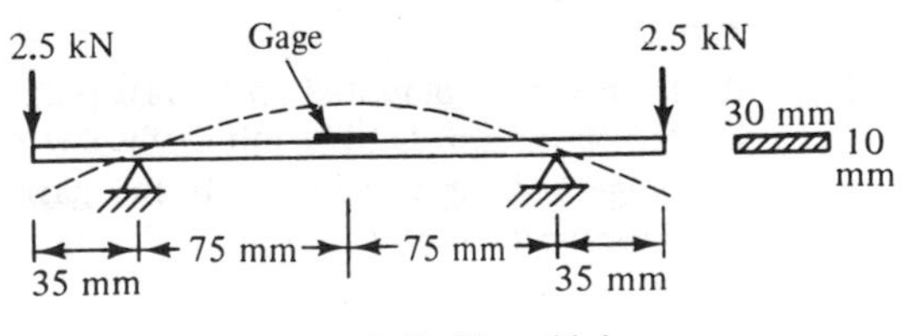

Problem 11.4

beam that is loaded as shown. If the change in gage resistance is 0.86 Ω, what is the gage factor F?

11.5. Two strain gages with $F = 2.09$ and $R = 240\ \Omega$ are mounted on a 0.505-in. diameter bar in the longitudinal and transverse directions. The changes in gage resistance are $\Delta R_L = 0.60\ \Omega$ and $\Delta R_T = -0.25\ \Omega$, respectively, for an applied axial load of 96 lb. Determine the modulus of elasticity and Poisson's ratio for the material. Assume linearly elastic behavior.

11.6. A strain gage with $F = 2.06$ and $R = 120\ \Omega$ is mounted on a 1-in. square 2024-T3 aluminum cantilever beam in the position shown. If the change in gage resistance is 0.68 Ω, what is the magnitude of the applied load Q?

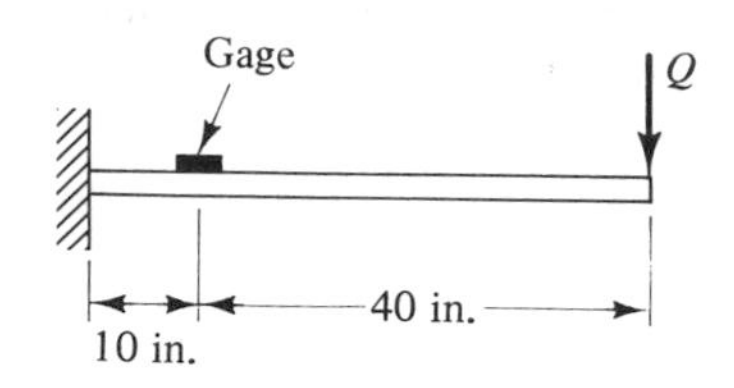

Problem 11.6

11.7. Same as Problem 11.6, except that the change in gage resistance is 0.85 Ω.

11.8. Show that a single strain gage oriented 45° from the longitudinal axis is sufficient to determine the state of strain in a shaft subjected to pure torsion. What is the anticipated change in resistance of such a gage when mounted on a 30-mm diameter shaft 2 m long subjected to an angle of twist of 5°? Assume $F = 2.0$ and $R = 300\ \Omega$.

11.9. A strain gage is mounted on a brass plate loaded as shown. During initial checkout the gage reading is 450 $\mu\varepsilon$. Is the gage functioning properly?

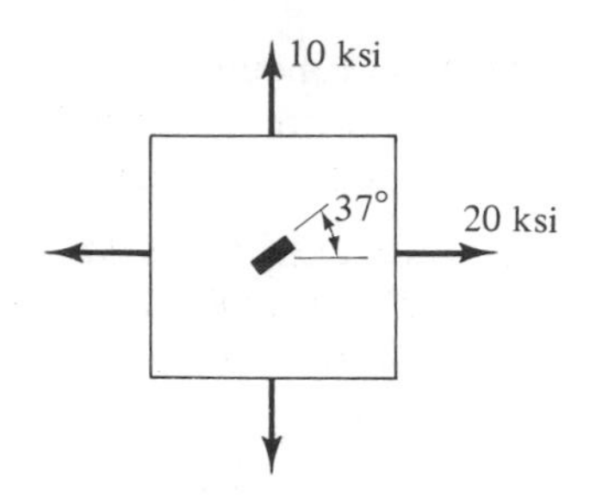

Problem 11.9

11.10. Same as Problem 11.9, except that the gage reading is 950 $\mu\varepsilon$.

11.11. A single strain gage is to be used to determine the longitudinal strain in an axially loaded bar with Poisson's ratio $\nu = 1/3$. During installation, the gage is misaligned with the axis of the member by a small angle θ. What is the percentage error made if the gage reading is taken to be the true value of the longitudinal strain? Plot the error versus θ for $-10° \leqslant \theta \leqslant 10°$.

11.12. Show that the shear strain γ_{xy} associated with axes xy can be determined using two strain gages placed symmetrically about the x axis, as shown. Determine the value of γ_{xy} for gage readings $\varepsilon_a = 500\ \mu\varepsilon$ and $\varepsilon_b = -200\ \mu\varepsilon$, with $\alpha = 15°$.

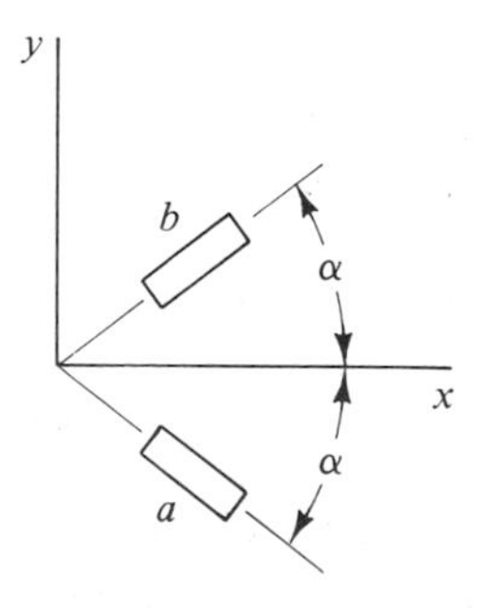

Problem 11.12

11.3 STRAIN-GAGE CIRCUITRY AND TRANSDUCERS

According to Eq. (11.1), an electrical resistance strain gage with a resistance of 120 Ω and a gage factor of 2 undergoes a change in resistance of $\Delta R = 0.24$ Ω for a strain of 1,000 $\mu\varepsilon$ (approximately the strain at yielding in mild steel). Smaller strains produce correspondingly smaller changes in resistance. Percentagewise, these changes in resistance are very small and can be measured accurately only by using the appropriate instrumentation. Most of the instruments used with strain gages are based on some form of Wheatstone-bridge circuit.

The basic Wheatstone-bridge circuit is shown in Figure 11.5, in which one or more of the arms is a strain gage and the others are resistors. When the resistances of the arms satisfy the condition

$$R_A R_D = R_B R_C \tag{11.2}$$

the bridge is balanced and there is zero output voltage.

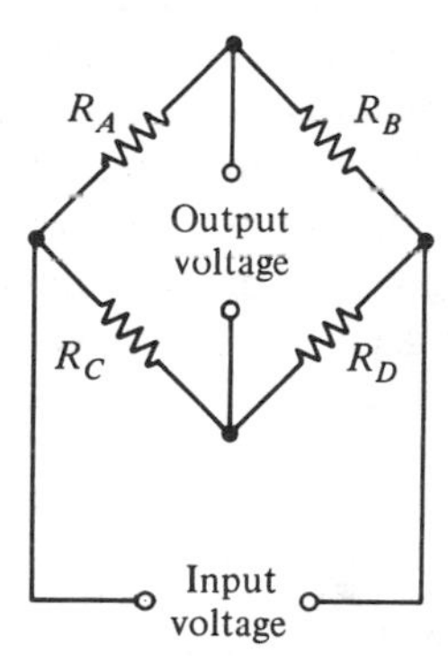

Figure 11.5

A change in resistance of one or more of the arms of the bridge creates an unbalanced condition and an output voltage proportional to the unbalance. If each arm undergoes a small change in resistance ΔR, the amount of unbalance is

$$U = (R_A + \Delta R_A)(R_D + \Delta R_D) - (R_B + \Delta R_B)(R_C + \Delta R_C)$$

Expanding this expression, making use of the balance condition of Eq. (11.2), and neglecting second-order terms involving the square and products of the ΔR's, we obtain

$$U = (R_A \Delta R_D + R_D \Delta R_A) - (R_B \Delta R_C + R_C \Delta R_B) \tag{11.3}$$

If each arm has the same initial resistance R, as is often the case, this expression reduces to

$$U = R[(\Delta R_A + \Delta R_D) - (\Delta R_B + \Delta R_C)] \tag{11.4}$$

Equation (11.4) indicates that changes in resistance of opposite arms of the bridge add while those of adjacent arms subtract. This fact makes it possible to connect the gage, or gages, into the bridge circuit in such a way as to achieve certain desired effects, one of which is to provide compensation for changes in temperature.

Temperature compensation can be achieved by using the *active gage* mounted to the specimen as one arm of the bridge, say arm A, and a *dummy gage* as the adjacent arm. The dummy gage is identical to the active gage and is mounted to a piece made of the same material as the test specimen. This piece is placed in the immediate vicinity of the active gage, but it is not loaded. The idea is to make the two gages experience the same changes in temperature. If they do, they will experience very nearly the same temperature-induced changes in resistance, ΔR_T. The active gage undergoes an additional change in resistance ΔR_L due to the applied loads. Since the resistances of the other two arms of the bridge don't change, we have from Eq. (11.4)

$$U = R[(\Delta R_T + \Delta R_L) - \Delta R_T] = R\Delta R_L$$

Thus, the temperature-induced change in resistance is canceled out by the circuitry, and the bridge output registers only the effect of the applied loads.

Other effects, such as increasing the bridge output for a given strain or canceling out strains due to certain types of loadings, can also be achieved by arranging the gages in an appropriate way. For example, suppose that we wish to determine only the bending strains in the member shown in Figure 11.6(a). This can be done by mounting one gage on top of the member and another on the bottom

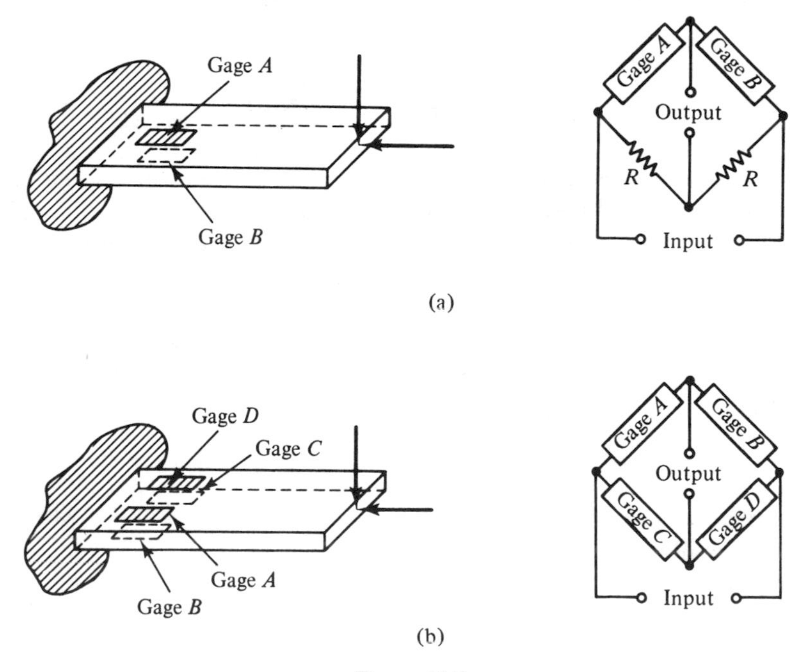

Figure 11.6

and connecting them into adjacent arms of a bridge circuit, as shown. This arrangement of the gages causes the strains due to the compressive load to cancel out.

To verify this, we note that gage A will experience a change in resistance ΔR_b due to the tensile bending strain and gage B will experience an equal and opposite change, $-\Delta R_b$, due to the compressive bending strain. Both gages experience the same changes $-\Delta R_c$ due to the compressive load. Let us also suppose that the gages undergo a change in resistance ΔR_T due to a temperature change. From (11.4), we obtain

$$U = R[(\Delta R_b - \Delta R_c + \Delta R_T) - (-\Delta R_b - \Delta R_c + \Delta R_T)] = 2R\Delta R_b$$

Thus, the output of the bridge registers only the bending strain, as desired. This arrangement of the gages also provides temperature compensation and gives twice the output of a single gage. The output can be doubled again by using four active gages arranged as shown in Figure 11.6(b).

If the gages A and B on the top and bottom of the beam shown in Figure 11.6(a) are connected into opposite arms of the bridge, the effects of the bending strains will cancel and only the strains due to the compressive load will be measured. This arrangement also gives twice the output of a single gage. However, it does not provide temperature compensation, unless dummy gages are used as the other two arms of the bridge.

Strain gages mounted to various types of members can also be used as *transducers* for measuring force, displacement, torque, and other physical quantities. For example, the cantilever beam shown in Figure 11.6(a) or 11.6(b) can be used as a transducer for measuring either lateral force or displacement. The system can be calibrated to measure force by hanging various known weights from the end of the beam and measuring the corresponding output of the bridge. It can be calibrated to measure displacement by deflecting the tip of the beam by various known amounts, say with a micrometer, and noting the corresponding bridge output.

Figure 11.7(a) shows a load cell for measuring tensile and compressive forces. It consists of a bar with strain gages mounted to the uniform central portion and arranged in a bridge circuit as shown. The cell can be calibrated by applying various known forces and measuring the corresponding bridge output. The arrangement shown in Figure 11.7(b) can be used to measure the torque in a shaft or its angle of twist. A transducer for measuring acceleration is shown in Figure 11.7(c). The acceleration of the mass results in bending of the beam element, which is detected by the strain gages.

The only basic requirement for a strain gage transducer is that the quantity to be measured produce a strain that can be detected by the gages. Thus, the number of possibilities is limited only by the imagination of the designer. Four active gages are used in most transducers in order to increase sensitivity. Whenever possible, the gages are arranged to provide temperature compensation. Most transducers can be designed so that they are linear (output proportional to input) over most, if not all, of their operating range.

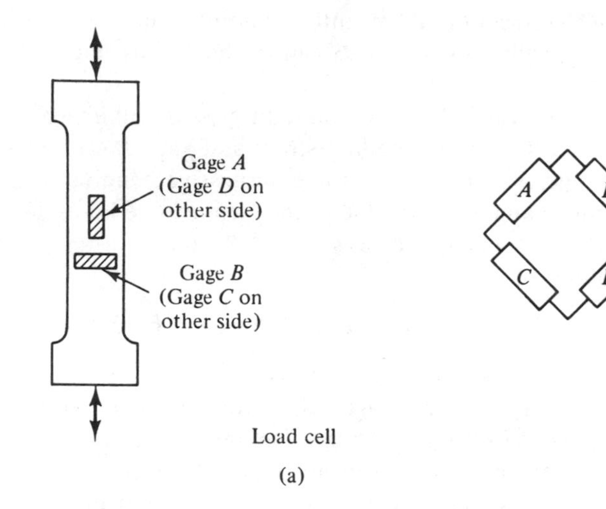

Load cell

(a)

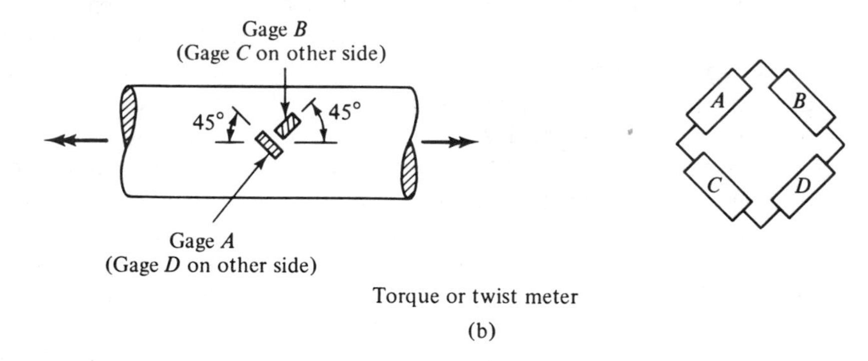

Torque or twist meter

(b)

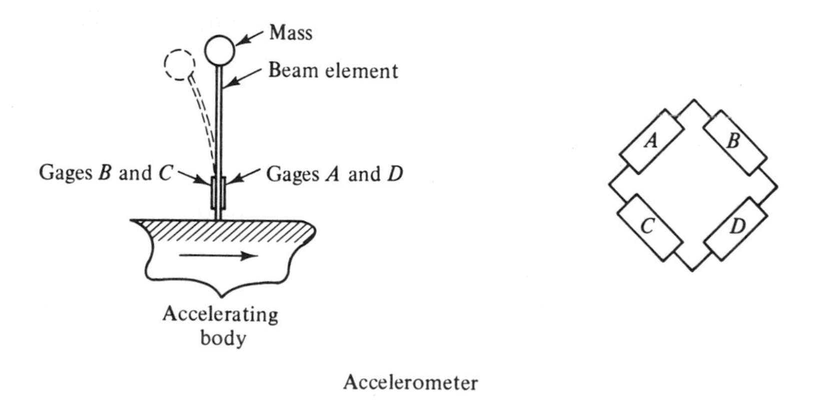

Accelerometer

(c)

Figure 11.7

Example 11.2. Measurement of Strains in a Shaft. The shear strain in a shaft is to be measured by using two electrical resistance strain gages aligned as shown in Figure 11.8(a). Should the gages be placed in adjacent or opposite arms of a bridge circuit? Does this arrangement provide temperature compensation?

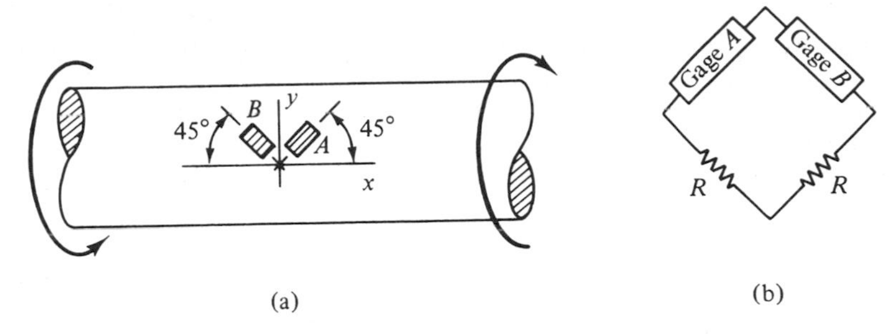

(a) (b)

Figure 11.8

Solution. According to the strain transformation equation, Eq. (9.13), the respective strains experienced by gages A and B are

$$\varepsilon_A = \frac{\gamma_{xy}}{2} \qquad \varepsilon_B = -\frac{\gamma_{xy}}{2}$$

Thus, the changes in resistance of the gages will be equal in magnitude and opposite in sign. Accordingly, the gages should be placed in adjacent arms of the bridge [Figure 11.8(b)] since the changes in resistance of these arms subtract. If placed in opposite arms, the outputs from the two gages add and would cancel out. Since a temperature change produces the same change in resistance in each gage, this arrangement does give temperature compensation.

To verify our conclusions, let ΔR_L be the change in gage resistance due to the applied loads (couples) and ΔR_T the change due to temperature effects. From Eq. (11.4), we obtain

$$U = R[(\Delta R_L + \Delta R_T) - (-\Delta R_L + \Delta R_T)] = 2R\Delta R_L$$

PROBLEMS

11.13. A load cell for measuring tensile and compressive forces is made by mounting two identical strain gages on a bar in the longitudinal and transverse directions and connecting them into a bridge circuit, as shown. What is the output of the load cell compared to one with only a single longitudinal gage? Is the cell temperature compensated?

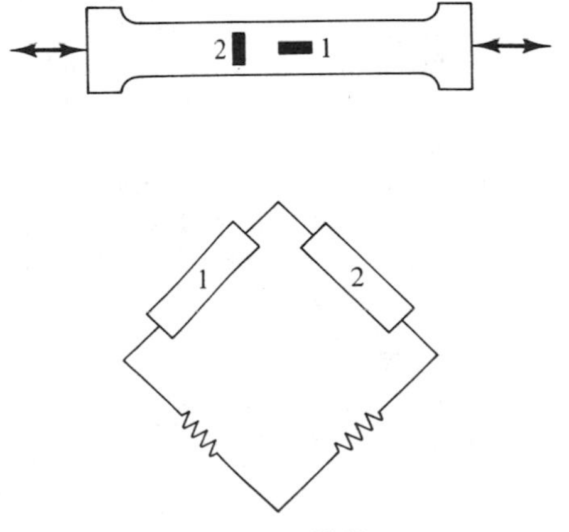

Problem 11.13

11.14. Determine the anticipated strain reading per kip of applied load for the load cell in Problem 11.13. The member is made of 2024-T3 aluminum and has a diameter of 0.505 in.

11.15. Four strain gages are mounted on a tensile specimen as shown. Show the gages arranged in a bridge circuit for measuring the longitudinal strain. Is the bridge output affected by any bending strains that may be present?

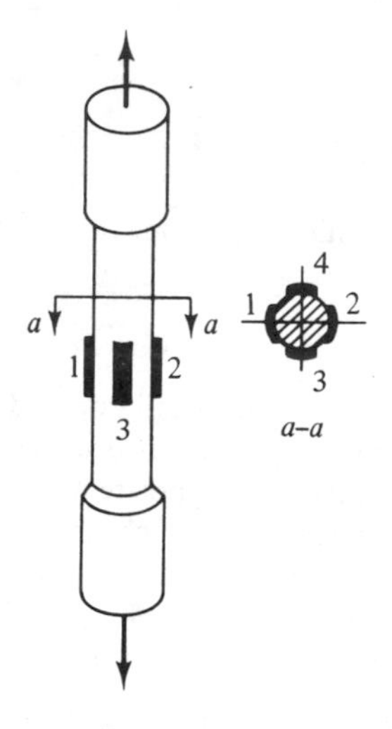

Problem 11.15

11.16. Determine the strain reading from the torque meter shown in Figure 11.7(b) per N · m of torque if the shaft is 2 m long, has a diameter of 20 mm, and is made of steel. What is the strain reading per degree of twist of the shaft?

11.17. Four strain gages are mounted as shown on a shaft subjected to combined torsion and axial loading. Show the gages arranged in a bridge circuit for measuring (a) the torque and (b) the axial load. Do these arrangements of the gages provide temperature compensation?

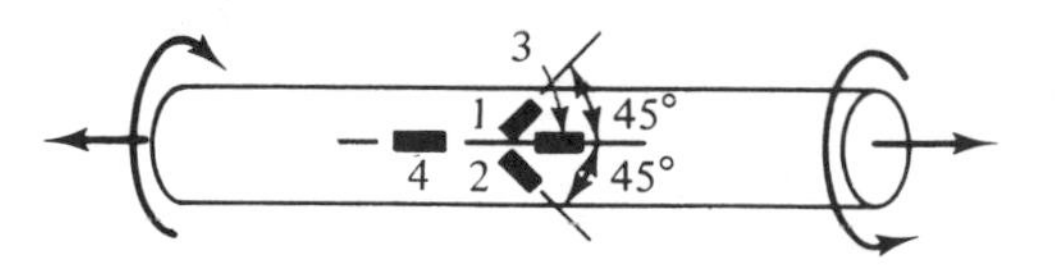

Problem 11.17

11.18. A proving ring for measuring large loads consists of an elastic ring with four strain gages mounted on the inner and outer surfaces, as shown. Show the gages arranged in a bridge circuit to give maximum output. Does this arrangement of the gages give temperature compensation?

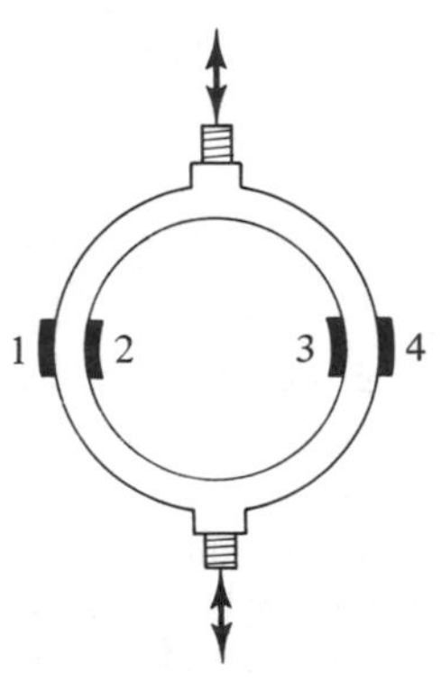

Problem 11.18

11.19. A pressure gage is fabricated from a thin-walled tube closed at one end and with two identical strain gages mounted in the circumferential direction. Show these gages arranged in a temperature-compensated bridge circuit for sensing the pressure. Use dummy gages as required.

11.20. If the tube in Problem 11.19 has a radius-to-thickness ratio of 30 and is made of a copper alloy with $E = 15 \times 10^6$ psi and $\nu = 0.35$, what is the strain reading from the bridge per psi of pressure?

11.21. Same as Problem 11.20, except that the tube is made of stainless steel with $E = 28 \times 10^6$ psi and $\nu = 0.3$.

11.22. A device for measuring torques consists of a rigid shaft attached to the end of an elastic strip with strain gages mounted on the top and bottom surfaces and arranged in a bridge circuit as shown. Determine an

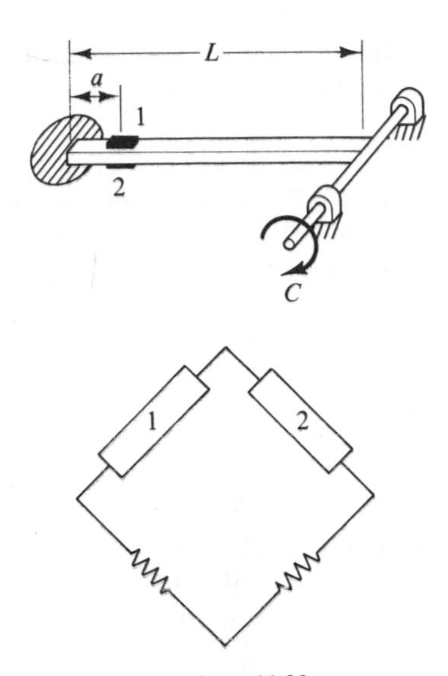

Problem 11.22

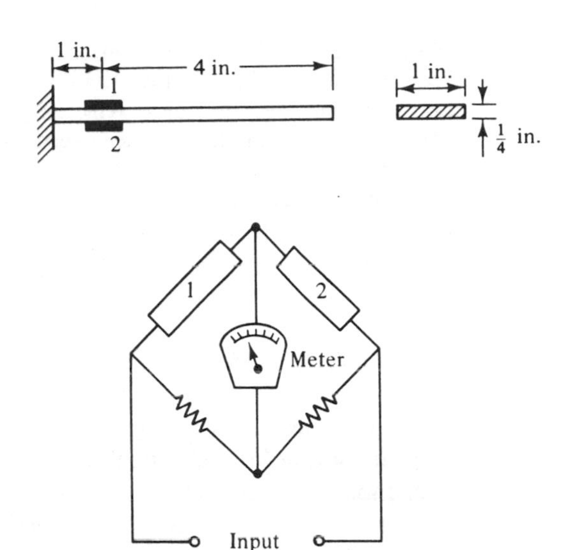

Problem 11.25

expression for the strain reading from the bridge in terms of the applied couple C, the dimensions a and L, and the flexural rigidity EI and thickness t of the strip.

11.23. Strain measuring systems are often calibrated by using simulated strains produced by connecting a resistor in parallel with the strain gage as shown. Given that the equivalent resistance of the parallel combination is $R_{eq} = RR_g/(R + R_g)$, determine the resistance R necessary to simulate a strain of 1,000 $\mu\varepsilon$ if $R_g = 300$ Ω and $F = 2$.

11.24. A 50,000-Ω resistor is placed in parallel with a strain gage with a resistance of 120 Ω and a gage factor of 2.1. What is the magnitude of the simulated strain? (See Problem 11.23.)

11.25. The cantilever beam shown is made of steel and is to be used as a deflection gage. When the beam is undeflected, the meter reading is zero. If a 100,000-Ω resistor produces a meter reading of 1,200 units when connected in parallel with one of the gages, what tip deflection of the beam will produce a meter reading of 58 units? (See Problem 11.23.) $R_g = 300$ Ω and $F = 2$.

11.26. Same as Problem 11.25, except that the calibration resistor has a resistance of 120,000 Ω.

11.27. Four identical strain gages ($R_g = 120$ Ω and $F = 2.0$) are mounted on a 203 ×

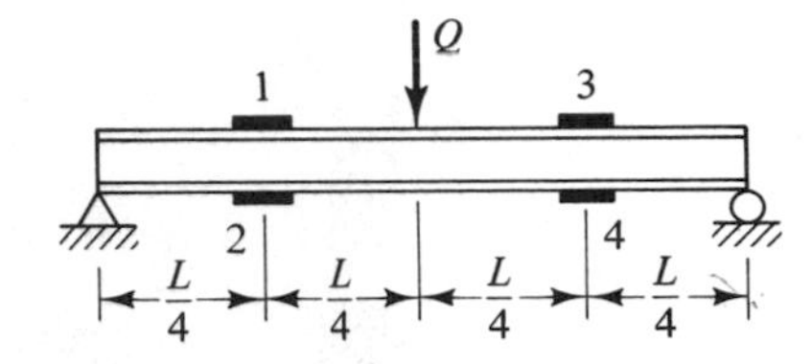

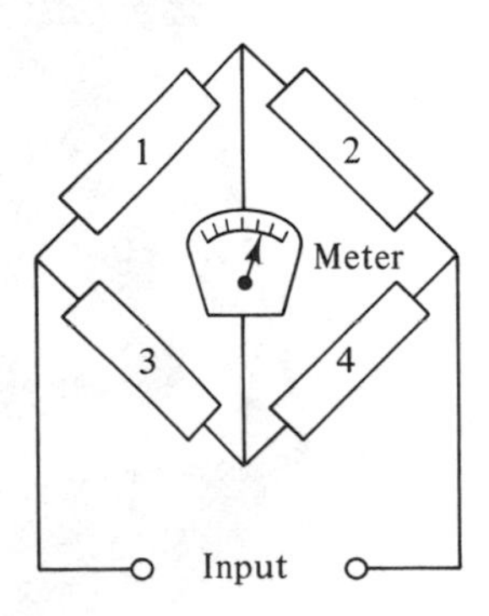

Problem 11.27

133-mm Universal beam (30 kg/m) at the locations shown. When the beam is unloaded, the meter reading is zero. If a 100,000-Ω resistor produces a meter reading of 60 units when connected in parallel with one of the gages, what is the applied load Q per unit of meter reading? The beam is 2 m long and is made of structural steel.

11.28. Same as Problem 10.27, except that the beam is 3 m long and is made of 2024-T3 aluminum.

11.4 STRAIN-GAGE INSTRUMENTATION

A wide variety of instrumentation is available for detecting, displaying, and recording the output from electrical resistance strain gages. The choice of instruments used depends on such factors as whether the loading is static or dynamic, the total number of gages involved, and whether or not a permanent record of the data is required.

Instruments called null-balance *strain indicators* (a typical example is shown in Figure 11.9) are commonly used for static tests. These instruments operate on AC line current transformed to DC or on an internal battery. Thus, they are completely portable and can be used in the field. They also have internal resistors that can be used in completing the strain-gage bridge circuit.

Strain indicators are very easy to operate. First, a knob is turned to adjust the instrument to the gage factor of the particular gages used. Next, the strain-gage

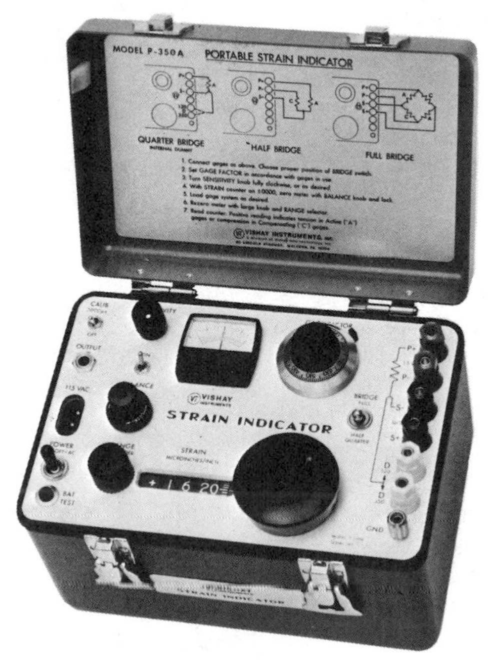

Figure 11.9. Vishay Instruments Model P-350A Portable Strain Indicator. (*Courtesy* Measurements Group, Inc.)

bridge is connected to the instrument and the circuit is balanced by rotating a second knob until a zero meter reading is obtained. The test specimen is then loaded. The resulting changes in resistance of the active gages cause an unbalance in the bridge, which is indicated by a nonzero meter reading. When it is adjusted so that the meter reading is again zero, the indicator provides a direct reading of the strain. The particular instrument shown in Figure 11.9 displays the value of strain in digital form; others display it in the form of a dial reading.

The indicators are designed for use with a single strain-gage bridge with one, two, or four active gages. If readings are to be taken from more than one set of gages, it is convenient to use a switch and balance unit. The unit shown in Figure 11.10 has provisions for taking readings from ten different sets of gages. By turning a switch, each different set can, in turn, be connected with the strain indicator. There are also provisions for initial balancing of each of the ten circuits. Systems capable of taking readings from hundreds of strain gages are used in the testing of large structures.

For dynamic loadings, it is necessary to use instruments that record the variation of the strain-gage signals with time. It is essential that the instrument have the proper frequency response, so that it doesn't distort the signals.

Oscillographs using galvanometers can be used to record dynamic strain signals involving components with frequencies in the range from 0 Hz to approximately 2,000 Hz. The output from the strain gages causes the galvanometers to rotate slightly. A record of the signal is provided by a light beam reflected from a

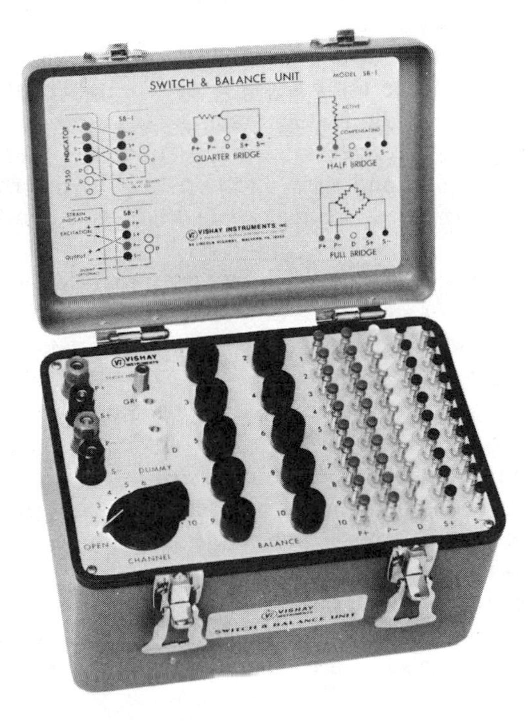

Figure 11.10. Vishay Instruments Model SB-1 Ten Channel Switch and Balance Unit. (*Courtesy* Measurements Group, Inc.)

small mirror attached to the galvanometer and falling onto a light-sensitive paper that passes through the instrument at a set, but changeable, speed. Oscillographs that record with a mechanical pen can be used for low-frequency signals (approximately 100 Hz and below). The frequency response of these instruments is limited by the mechanical inertia of the pen mechanism.

Cathode-ray oscilloscopes are commonly used for recording high-frequency strain signals, such as occur with impact loadings. A permanent record showing the variation of the strain with time can be obtained by photographing the signal trace on the screen of the scope with a still or moving-film camera. It is usually necessary to amplify the strain-gage signals before they are fed into the scope. Some strain indicators can be used for this purpose, and numerous other types of amplifiers are commercially available.

Computer-based systems are being increasingly used for the acquisition and reduction of strain gage data. Such systems greatly facilitate test programs and reduce the possibility of errors in the data reduction. Furthermore, the data can be output in a variety of forms. For example, inputs from strain-gage rosettes may be output as principal strains or principal stresses. The results also can be presented in various graphical formats.

11.5 BRITTLE COATINGS

There are several materials that form a thin, brittle coating when applied to the surface of a member and allowed to cure. When the member is loaded, the deformations in the plane of the surface are transmitted to the coating, which adheres tightly to the surface. Since the coating is brittle, it is relatively weak in tension and cracks when the maximum tensile strain reaches a certain critical value. By observing the resulting crack patterns in the coating, the maximum tensile surface strains within the member and their directions can be determined. Figure 11.11 shows the crack pattern in a coating applied to an automotive steering component.

The idea for this method of strain analysis was obtained by observing the cracking of glazes on old pottery and the flaking of mill scale from hot-rolled steel when it yields. Whitewash was one of the first coatings used, but it has low sensitivity and does not crack until the strains are very large. Most of the coatings in current use are a type of lacquer and are designed to crack at strains well below the yield strains for metals. They are available in different compositions for use at different temperatures and are sprayed onto the surface. Some skill is required to produce a uniform coating without runs and bubbles. The coatings must cure for approximately 24 hours at room temperature before they can be used. Curing at elevated temperatures is often desirable.

Before the specimen can be tested, the coating must be calibrated. This is done by using calibration strips in the form of rectangular bars that are coated at the same time as the specimen. These strips are loaded as cantilever beams subjected to

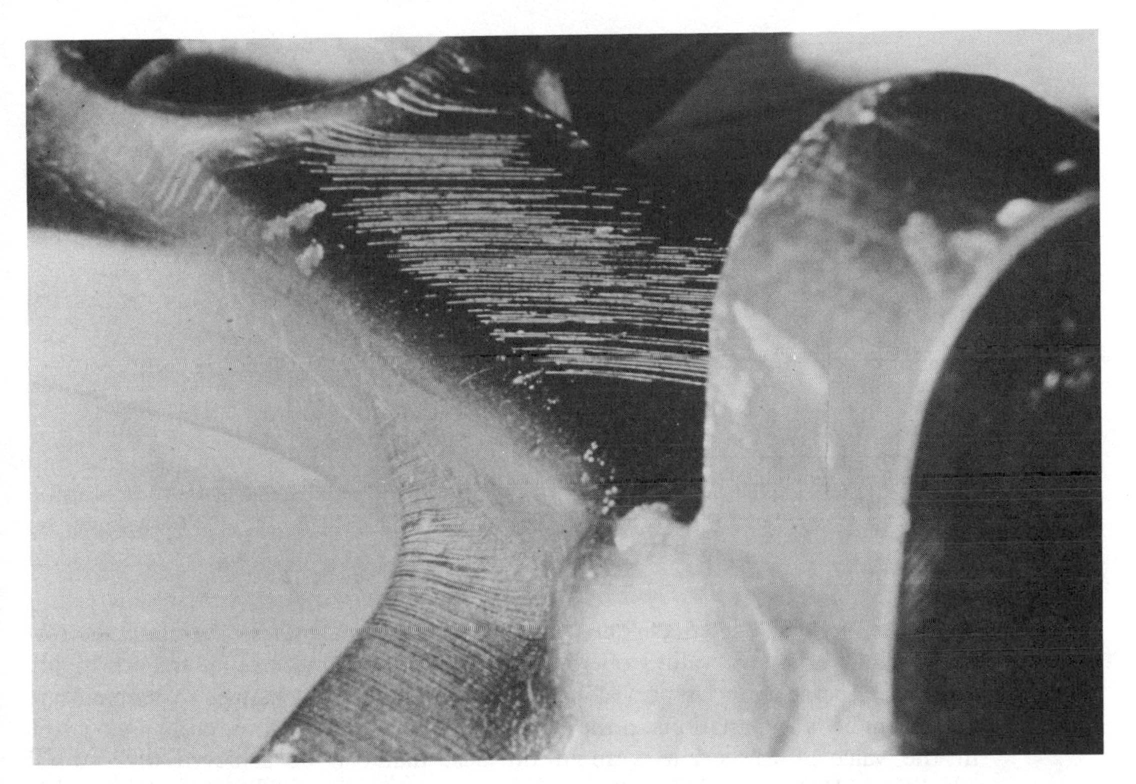

Figure 11.11. Crack Pattern in Brittle Coating on an Automotive Steering Component. (*Courtesy* Measurements Group, Inc.)

a prescribed tip displacement by a cam arrangement shown schematically in Figure 11.12(a). For this loading, the longitudinal normal strain is largest at the support and decreases linearly to a value of zero at the tip of the beam. The coating cracks near the support where the strain is large and remains uncracked near the tip where it is small [Figure 11.12(b)]. The strain at the location where the cracks stop is the threshold strain at which the coating fails. A scale is available for determining the value of this strain without computation.

Once the coating has been calibrated, the specimen can be tested. By increasing the loads on the specimen in increments and observing how the cracks form and grow, it is possible to map out the variation of the tensile strains over the surface of the member. Compressive strains can be determined by applying the coating to a loaded specimen and then removing these loads so that the coating is placed in tension.

The data analysis is simple. When the coating first cracks at a point, the maximum strain there in the direction perpendicular to the crack is taken to be equal to the calibration value for the coating. The maximum stress is then computed by multiplying the strain by the modulus of elasticity of the specimen. Since this

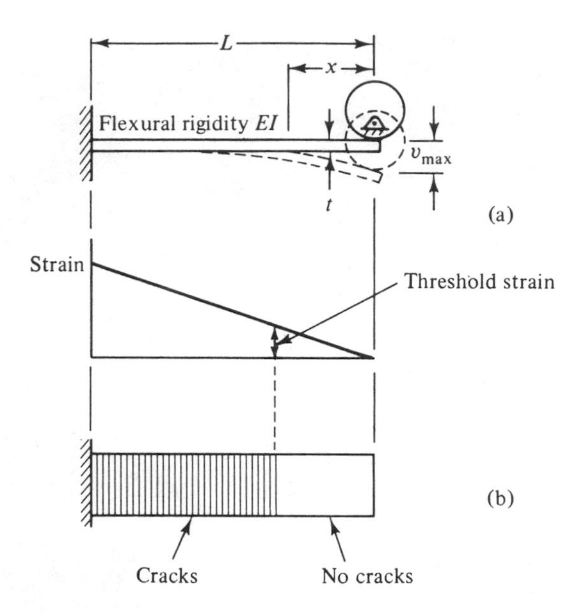

Figure 11.12

procedure neglects any effects due to the two-dimensional states of strain and stress that usually exist, the values obtained for the stresses and strains are not highly accurate. The accuracy is also affected by such factors as changes in temperature and humidity and variations in the thickness of the coating. An accuracy of $\pm 10\%$ in the value of the strain is about the best that can be achieved under ideal conditions. However, this is sufficient for many applications.

The coatings do give an accurate indication of the location of maximum strains and the directions of the principal strains. Thus, they are widely used as an aid in determining where and how to locate strain gages. The regions of high strain are distinguished by closely spaced cracks; the closer the cracks are together, the larger the strain. Since the coating fails due to the maximum tensile strain, the principal strains are oriented perpendicular and parallel to the cracks. This information makes it possible to use two strain gages at each point instead of three (see Example 11.1). Of course, the coating must be removed before the gages are mounted.

Brittle coatings have the advantage that they give an indication of the stresses and strains over the entire surface. Their use requires little or no special equipment, they can be used for both static and dynamic loadings, and they can be easily removed after testing. The common coatings can be used at temperatures up to approximately 100°F, and special ceramic coatings for use at higher temperatures are available.

Example 11.3. Determination of Pressure in a Cylinder. A thin-walled cylindrical vessel covered with a brittle lacquer is pressurized until longitudinal cracks form in the coating. Testing of a calibration strip indicates that the threshold value of strain for the coating is 600 $\mu\varepsilon$. Estimate the pressure in the cylinder if it is made of aluminum ($E = 10 \times 10^6$ psi and $\nu = 0.33$) and has a radius to thickness ratio of $R/t = 20$.

Solution. Taking the circumferential strain (perpendicular to the cracks) to be equal to the calibration value for the coating and multiplying by the modulus of elasticity of the cylinder, we obtain for the circumferential stress

$$\sigma_C = E\varepsilon_C = (10 \times 10^6 \text{ lb/in.}^2)(600 \times 10^{-6}) = 6{,}000 \text{ psi}$$

From Eq. (6.8), we have

$$\sigma_C = \frac{pR}{t} = 6{,}000 \text{ psi}$$

or

$$p = \frac{6{,}000 \text{ psi}}{20} = 300 \text{ psi} \qquad \textbf{Answer}$$

This value is only an estimate because the effects of the longitudinal stress are ignored in the analysis (see Problem 11.38).

PROBLEMS

11.29. Obtain an expression for the strain in the brittle coating calibration strip shown in Figure 11.12 as a function of the distance x from the tip. Thus, verify that the linear variation of strain indicated in Figure 11.12(a) is correct. Does the strain at a given location depend upon the material properties of the strip?

11.30. A brittle coating is applied to a 25-mm wide × 6-mm thick aluminum calibration strip with $E = 70 \text{ GN/m}^2$. The strip is 0.25 m long and is subjected to a tip displacement of 6 mm. If the last visible crack is 0.10 m from the tip, what is the threshold strain for the coating?

11.31. Same as Problem 11.30, except that the last visible crack is 0.12 m from the tip.

11.32. A brittle coating with a threshold strain of 800 $\mu\varepsilon$ is applied to a machine part made of grey cast iron with $E = 15 \times 10^6$ psi. What is the maximum tensile stress in the part when the coating first cracks? If strain gages are to be used to determine more accurate results, how many gages are required and how should they be oriented?

11.33. Same as Problem 11.32, except that the threshold strain for the coating is 650 $\mu\varepsilon$.

11.34. A 20-mm diameter steel shaft 2 m long is covered with a brittle coating and loaded in pure torsion until the coating cracks. The cracks form at 45° from the longitudinal axis. Calibration shows that the threshold strain for the coating is 400 $\mu\varepsilon$. Estimate the angle of twist of the shaft.

11.35. Same as Problem 11.34, except that the shaft is hollow and has an inside diameter equal to three-fourths the outside diameter.

11.36. A brittle coating is applied to a machine part at a point where the state of stress is

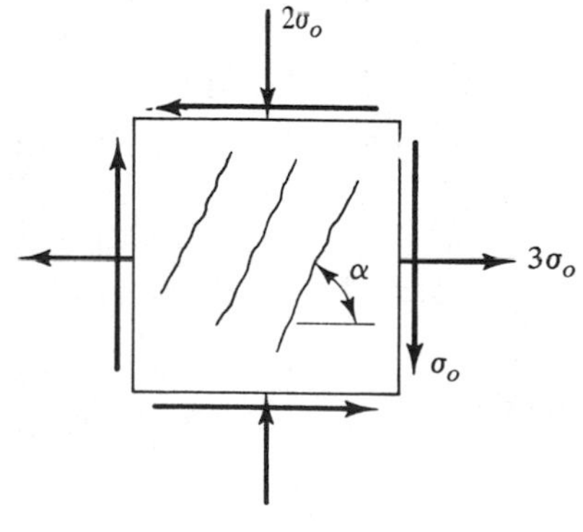

Problem 11.36

as shown. Estimate the angle α at which the coating will crack when the value of σ_0 is increased.

11.37. Proof testing of an aluminum pipe with reinforcing rings requires that the pipe be subjected to twice the normal operating pressure without showing evidence of yielding. Explain how a brittle coating could be used in this test. What threshold strain for the coating would be required?

11.38. Determine the percentage error in the value of the pressure obtained in Example 11.3 as a result of neglecting the effects of the longitudinal stress. Assume that the brittle coating gives an accurate indication of the circumferential strain.

11.6 MOIRÉ STRAIN ANALYSIS

When two grids consisting of equally spaced parallel lines on a transparent background are superposed and then displaced relative to one another, interference patterns are formed where the two sets of lines overlap. These patterns appear as dark bands, or fringes. This phenomenon is called the *moiré effect* and can be easily demonstrated by using two pieces of screen wire.

If one of the grids is applied to the surface of a member and the other is held stationary over it, the pattern formed when the member deforms can be used to

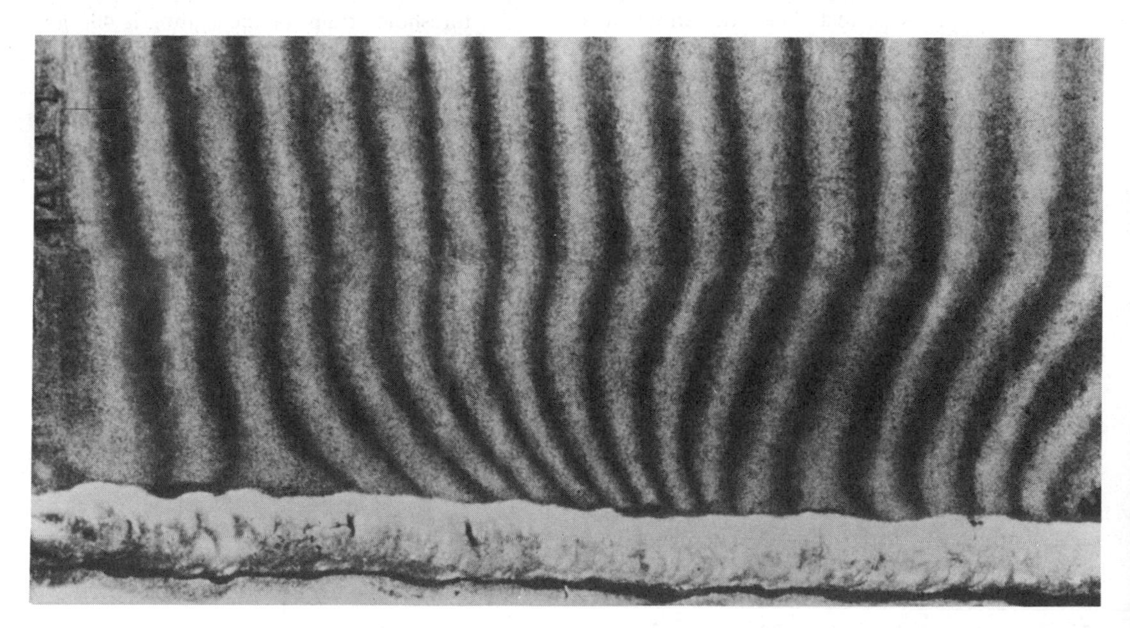

Figure 11.13. Moiré Fringe Pattern Resulting from Deformation of a Welded Plate. (*Courtesy* Measurements Group, Inc.)

determine the surface displacements and strains. The grid can be applied to the member photographically or it can be formed separately and attached with an adhesive. Figure 11.13 shows the moiré fringe pattern resulting from the deformation of a flat plate that has been welded to another plate. The distortions in the pattern are caused by residual stresses set up during the welding process.

Each fringe in the moiré pattern is a loci of points that undergo the same displacement δ in the direction perpendicular to the stationary grid lines. The change $\Delta\delta$ in the value of displacement from one fringe to the next is equal to the pitch p of the grid [Figure 11.14(a)]. Referring to the figure, we see that the average strain of a line element, such as AB, between the two fringes is

$$\varepsilon_{\text{avg}} = \frac{\Delta\delta}{\Delta L} = \frac{p}{d} \tag{11.5}$$

where $d = AB$ is the fringe spacing.

The actual strain at a point, $\varepsilon = d\delta/dL$, can be obtained by plotting the variation of the displacement with distance and then determining the slope of the resulting curve [Figure 11.14(b)]. Since only the change in δ is important in determining the strain, its value can be set equal to an arbitrary constant at one of the fringes. This procedure gives the variations of the normal strain along a line perpendicular to the stationary grid lines. By repeating the process along other such lines, the variation of the strain over the entire surface can be determined. The strains in other directions can be found by changing the orientation of the grids.

The complete state of strain at a point can be determined by measuring the normal strains in three different directions and applying the strain transformation

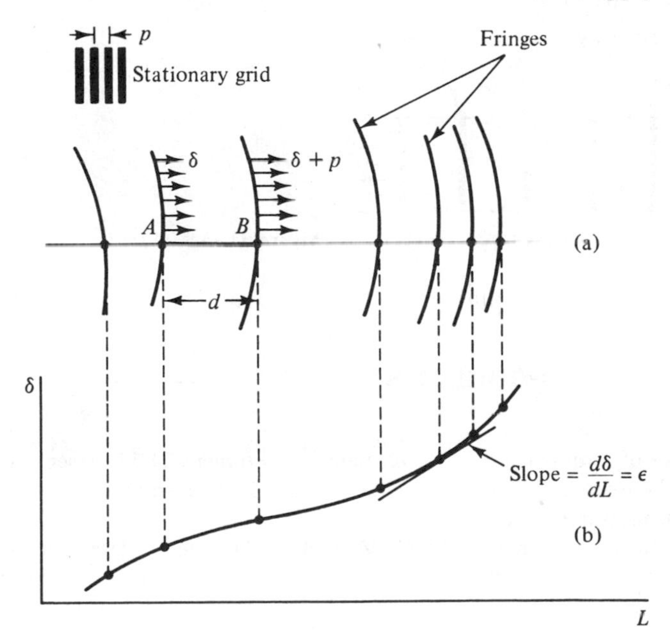

Figure 11.14

equations [Eqs. (9.13) and (9.15)], as in the case of strain gages. It is also possible to determine the state of strain by using grids aligned in two perpendicular directions, but we shall not go into that here.

Moiré strain analysis is useful primarily for problems involving large strains. If the strains are small, the fringes are far apart and there is not enough of a pattern to permit accurate analysis. The resolution can be enhanced by making the pitch p of the grid smaller, but there is a limit to how close the lines can be spaced. The most sensitive grids currently available have approximately 2,000 lines per inch (p = 0.0005 in.).

Like brittle coatings, moiré analysis gives an indication of the strains over the entire surface.

Example 11.4. Moiré Patterns for a Plate. A flat plate is loaded such that the longitudinal strain ε is uniform throughout. What would the resulting moiré fringe pattern look like if the grids used have 2,000 lines/in. and are oriented perpendicular to the plate axis? What is the spacing between the fringes for $\varepsilon = 1{,}000\ \mu\varepsilon$?

Solution. Since the strain does not vary along the length of the member, $\varepsilon = \varepsilon_{avg}$. From Eq. (11.5), we have

$$d = \frac{p}{\varepsilon} = \frac{0.0005 \text{ in.}}{0.001} = 0.5 \text{ in.} \qquad \textbf{Answer}$$

Since d is a constant, the moiré pattern will consist of a series of equally spaced fringes 0.5 in. apart (Figure 11.15).

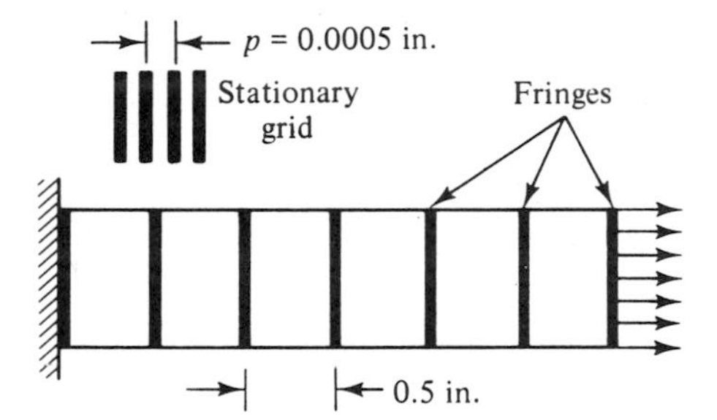

Figure 11.15

PROBLEMS

11.39. At a point on the surface of a rubber test specimen the spacing between adjacent moiré fringes is 10 mm, measured in the direction perpendicular to the stationary grid lines. If the grids have 10 lines/mm, what is the average strain at this point?

11.40. Same as Problem 11.39, except that the grids have 25 lines/mm.

11.41. A rectangular epoxy plate with $E = 0.6 \times 10^6$ psi and $\nu = 0.4$ is subjected to uniformly distributed tensile stresses σ_x

and σ_y as shown. Sketch the resulting moiré fringe pattern produced by the deformation of the plate and determine the fringe spacing if $\sigma_x = 1{,}000$ psi and $\sigma_y = 500$ psi. The grids have 2,000 lines/in. and are oriented with the grid lines parallel to the x axis.

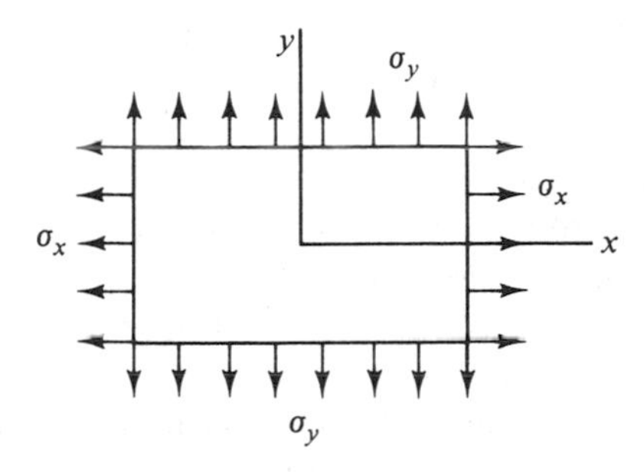

Problem 11.41

11.42. When the epoxy plate in Problem 11.41 is subjected to a stress $\sigma_y = 1{,}200$ psi, the resulting moiré fringes are found to be 0.25 in. apart. A stress σ_x is then applied, and the fringe spacing changes to 0.20 in. Determine σ_x.

11.43. Same as Problem 11.42, except that the fringe spacing changes to 0.30 in. after the stress σ_x is applied.

11.44. Same as Problem 11.41, except that the grid lines are parallel to the y axis.

11.45. A tracing of a portion of the moiré fringe pattern for a urethane rubber tensile specimen with a central hole is as shown. Determine the variation of the strain along line AB. The grid pitch is 200 lines/in.

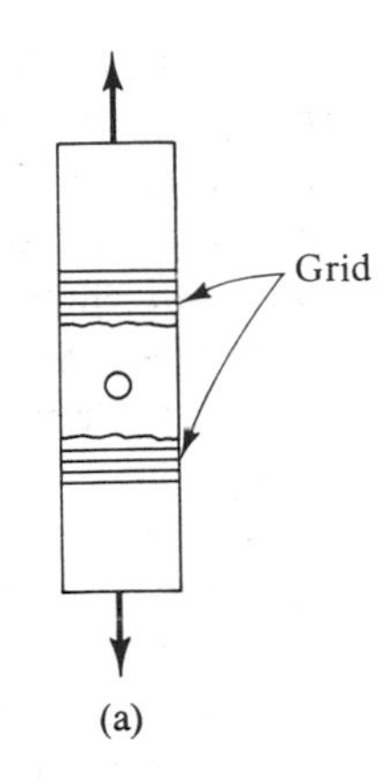

(a)

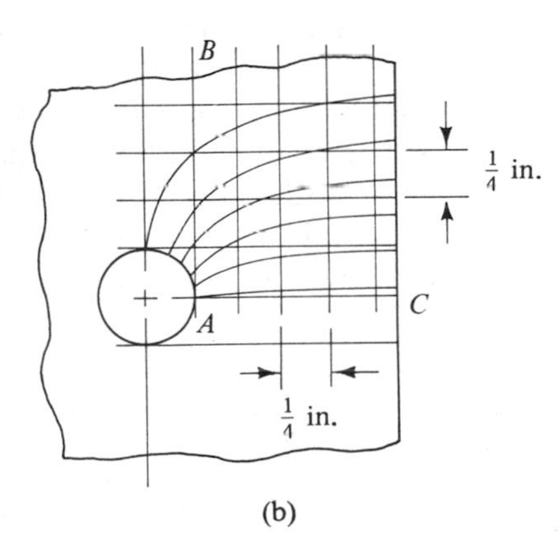

(b)

Problem 11.45

11.46. For the specimen in Problem 11.45, determine and sketch the variation in the strain along transverse axis AC. Compare the results with those shown in Figure 6.3 (Section 6.2).

11.7 PHOTOELASTICITY

The photoelastic method of stress analysis is based on the optical fringe patterns formed when a transparent model of a member is loaded in the presence of polarized light. An analysis of these fringe patterns yields the stresses in the model, which are then related to the prototype by using the appropriate model theory.

Before attempting to describe the photoelastic method in more detail, let us briefly review some of the relevant properties of light. Light is an electromagnetic radiation, which, for our purposes, can be thought of as consisting of waves vibrating in planes perpendicular to the direction of propagation. Color is associated with the frequency of the radiation. Monochromatic light is a radiation with a single frequency; white light is a mixture of radiations of different frequencies.

Unpolarized light consists of waves vibrating in an infinite number of planes. If unpolarized light is passed through a polarizing element, only the waves vibrating in the plane of polarization are transmitted. Thus, polarized light consists of waves vibrating in parallel planes.

Doubly refracting materials transmit light waves vibrating in two perpendicular planes. When polarized light passes through such a material, the light wave is broken into two perpendicular components. These components travel through the material at different speeds, and, as a result, appear to be shifted in space relative to one another through a distance called the *phase shift*.

Many materials that are not doubly refracting in their natural state become doubly refracting when they are stressed. The planes of light transmission at any one given point in the material are aligned with the principal axes of stress, and the phase shift between the transmitted waves is proportional to the difference of the principal stresses at that point. This *photoelastic effect* was discovered by Sir David Brewster in 1816, and it is the basis of the photoelastic method of stress analysis.

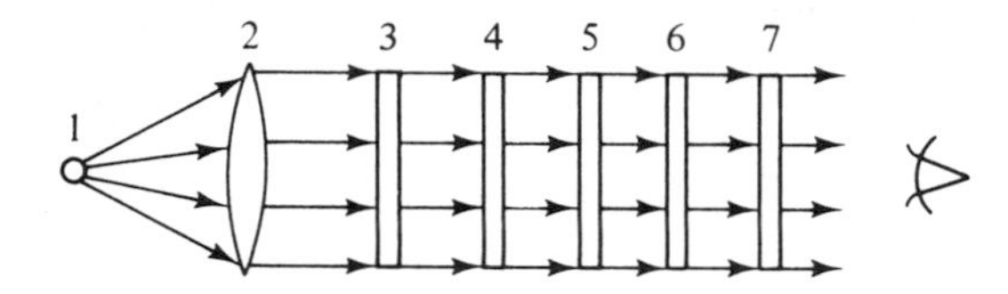

Figure 11.16

The first step in a photoelastic investigation is to make a model of the member of interest out of a suitable transparent material. This model is then placed in the optical field of an apparatus called a *polariscope*, which consists of a series of elements arranged as shown in Figure 11.16. In addition to the light source and collimating lens, there are two polarizing elements, the polarizer and analyzer, and two quarter-wave plates, which are discs of doubly refracting material whose thickness has been adjusted to give a phase shift equal to one-quarter of the wavelength of the particular monochromatic light used. Figure 11.17 is a photograph of a commercially available polariscope. The loaded specimen is placed at the center, between the two benches, and the camera at the right is used to record the optical patterns in the model. The light source is on the left.

When the stresses at a point in the model are such that the resulting phase shift is an integral multiple of the wavelength of the monochromatic light source, the light transmitted through the model will be the same as the background light.

Figure 11.17. Photoelastic Inc. Model 261 Collimated Light Split Bench Polariscope. (*Courtesy* Measurements Group, Inc.)

This condition can be expressed as

$$\sigma_1 - \sigma_2 = \frac{Kn}{t} \tag{11.6}$$

where σ_1 and σ_2 are the principal stresses, K is the experimentally determined *photoelastic constant* for the material, t is the model thickness, and n is an integer ($n = 0, 1, 2, \ldots$).

The loci of points in the model at which the condition in Eq. (11.6) is satisfied appear as a series of bands, called *integer isochromatic fringes*. The loci of points at which the phase shift is one-half the wavelength of the light source, or any odd multiple thereof, also form a series of bands. These bands are called *half-order isochromatic fringes* and correspond to the values $n = \frac{1}{2}, \frac{3}{2}, \frac{5}{2}, \ldots,$ in Eq. (11.6). If the polariscope is arranged to give a dark background, the whole-order fringes will be dark and the half-order fringes will be light. The reverse is true with a light background. Along any one given fringe, the quantity $\sigma_1 - \sigma_2$ is constant. Since

$\sigma_1 - \sigma_2$ is equal to twice the maximum in-plane shear stress [see Eq. (9.7)], the fringes can be thought of as lines of constant shear stress.

Figure 11.18 is a photograph of the isochromatic fringe pattern in a photoelastic model of a knee-frame, with the polariscope arranged to give a light background. Note the numbering of the fringes. The fringe order n can be determined either by counting from a known point of zero stress or by watching the formation of the fringes as the model is loaded. The maximum stress in the frame occurs at the fillet where $n = 8$. Also note that the fringes are narrowest where the stresses are highest. This fact can be used to locate regions of high stress in a member.

At the boundaries of the model, and at any other points where one of the principal stresses is known to be zero, the remaining principal stress can be determined directly from Eq. (11.6) and the observed fringe pattern. Consequently, the photoelastic method is particularly convenient for determining the stress concentration factor at holes, notches, and other geometric discontinuities. Many of the stress concentration factors given in handbooks were obtained by this method.

If both principal stresses are nonzero, they cannot be determined from the fringe pattern alone; additional information is required. This information can be obtained by removing the quarter-wave plates from the polariscope and using a white light source. With this setup, the isochromatic fringes appear as colored bands; hence the name isochromatics. Black bands, called *isoclinics*, will also be present. The isoclinics define the loci of points in the model at which the principal axes of stress are aligned parallel and perpendicular to the plane of polarization. By rotating the polarizer and analyzer together in increments, the directions of the

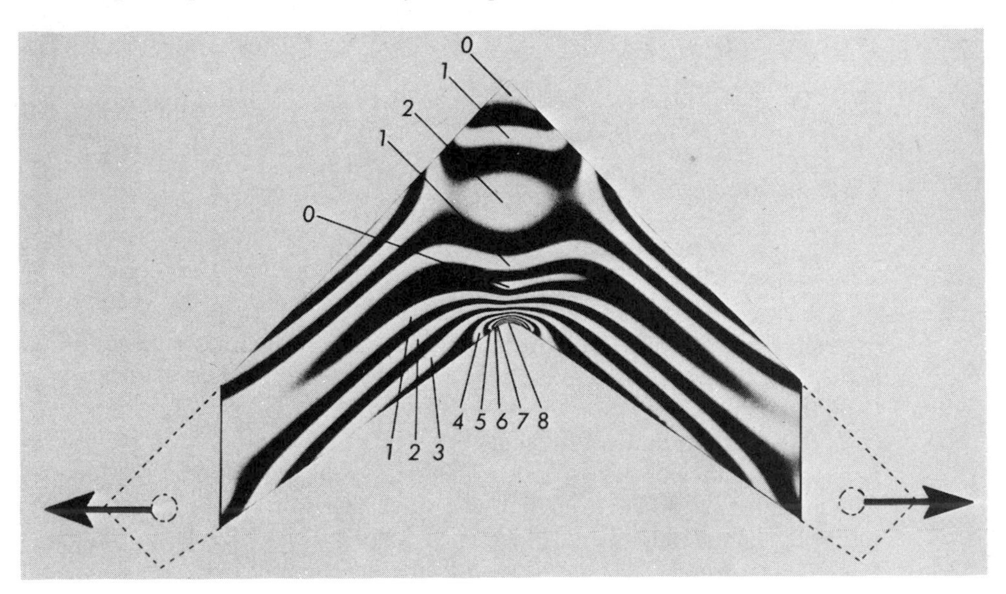

Figure 11.18. Isochromatic Fringe Pattern in a Photoelastic Model of a Knee-frame. (*Courtesy* Measurements Group, Inc.)

principal stresses can be determined at all points in the model. The individual stresses σ_1 and σ_2 can then be obtained by combining this information with that obtained from the isochromatic fringe pattern. However, the procedure is tedious and involves many computations. Computerized data processing greatly facilitates the process.

There are a number of different photoelastic materials that can be used for the model. Desirable properties are transparency, high optical sensitivity (low value of K), ease of machining, and linearly elastic response over a suitable range of stresses. The most commonly used materials are urethane rubbers, polyesters, epoxies, and types of phenolic.

The photoelastic constant K for the material is obtained from the fringe pattern in a specimen loaded in such a way that the stresses can be determined analytically. For example, if a strip of the material is loaded as a beam in pure bending, one principal stress will be equal to the bending stress and the other will be zero. The resulting fringe pattern is a series of equally spaced longitudinal bands, as illustrated in Figure 11.19. To determine K, we compute the stress at the location of each of the fringes from the bending stress formula and plot it versus the fringe order n. As can be seen from Eq. (11.6), the slope of the resulting straight line is equal to K/t (see also Example 11.5).

Photoelasticity gives an indication of the stresses throughout the entire member, and it can be used for both static and dynamic loadings. Three-dimensional problems can be analyzed by loading the model at an elevated temperature and letting it cool with the loads in place, thus causing the photoelastic patterns to be "frozen" into the material. The stresses over various cross sections can then be determined by examining thin slices cut from the model. Optical techniques for the direct analysis of three-dimensional problems have also been developed. Photoelasticity is particularly convenient for determining qualitative information about the stresses in a member. All that is required is a hand-loaded model held between two sheets of polaroid plastic.

The major disadvantage of the photoelastic method of stress analysis is that a model must be prepared and the stresses in the model related back to the prototype, a step that often leaves the final results open to question. This disadvantage has largely been overcome by the development of *photoelastic coatings* that can be bonded directly to the test specimen, in much the same way as a strain gage is

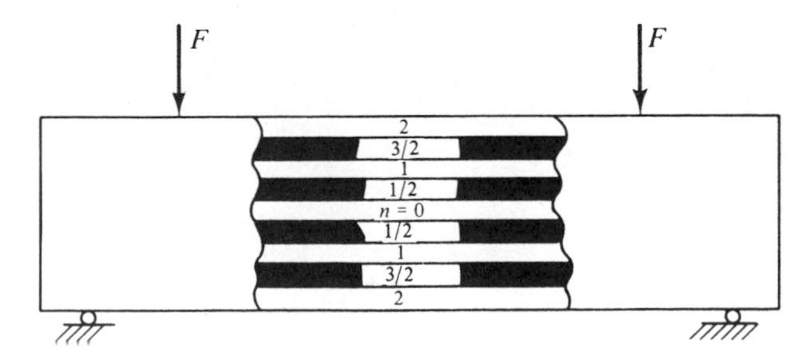

Figure 11.19

mounted. The surface deformations of the specimen are transferred to the coating by the adhesive; the resulting photoelastic patterns give an indication of the surface stresses in the member without use of a model. A special type of polariscope is required because the fringe patterns observed are formed by light reflected off the back surface of the coating. Otherwise, the basic procedure is the same as that for photoelastic models.

Example 11.5. Determination of Photoelastic Constant. If the beam shown in Figure 11.19 has a rectangular cross section 1.8 in. deep × 0.5 in. wide and the bending moment is $M = 200$ lb · in, what is the photoelastic constant K for the material? Careful measurement shows that the fringes are of equal width.

Solution. Since the fringes are of equal width, their centers are 0.2 in. apart. Computing the bending stress at the centers of each of the fringes, we obtain

$$n = 0 \qquad \sigma = 0$$
$$n = \tfrac{1}{2} \qquad \sigma = 167 \text{ psi}$$
$$n = 1 \qquad \sigma = 333 \text{ psi}$$
$$n = \tfrac{3}{2} \qquad \sigma = 500 \text{ psi}$$
$$n = 2 \qquad \sigma = 667 \text{ psi}$$

The stresses are plotted versus the fringe order n in Figure 11.20. The slope of the resulting straight line is K/t [see Eq. (11.6)]

$$\frac{K}{t} = 333 \frac{\text{lb/in.}^2}{\text{fringe}}$$

$$K = 167 \frac{\text{lb/in.}}{\text{fringe}} \qquad \textbf{Answer}$$

If the fringe order is known to be proportional to the stress, as it is in this example, the value of K can be obtained directly from Eq. (11.6) by using the value of the stress at any one of the fringes. However, it is usually best to work with a plot of the stress versus the fringe order because it gives a direct indication of the range of stress over which the material exhibits a linear optical response.

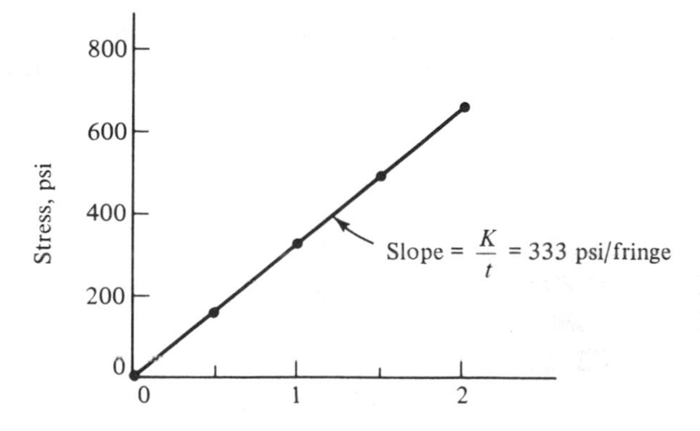

Figure 11.20

PROBLEMS

11.47. When a tensile specimen of a photoelastic material is tested in a polariscope, the loads at which the various isochromatic fringes appear are as follows:

Load (lb)	0	75	150	225	300	375	450
Fringe order	0	1	2	3	4	5	6

If the specimen is 1 in. wide and 1/4 in. thick, what is the photoelastic constant for the material?

11.48. Rework Example 11.5 for a specimen 50-mm deep × 20-mm wide.

11.49. Determine the stress at the fillet of the knee-frame model shown in Figure 11.18 if the photoelastic constant is $K = 15$ (kN/m)/fringe and the model is 8 mm thick.

11.50. Determine the stress at a point on the boundary of a photoelastic model where the fringe order is $n = 3$. The photoelastic constant for the material is $K = 80$ (lb/in.)/fringe and the model is 0.4 in. thick.

11.51. A photoelastic model of a rectangular bar with a circular hole in the middle is loaded in tension and the fringe order at the side of the hole is found to be $n = 5\frac{1}{2}$. At some distance away from the hole, the fringe order is $n = 2\frac{1}{2}$. What is the stress concentration factor due to the hole?

11.52. A photoelastic model of a rectangular bar with semicircular grooves is loaded in tension. The fringe order at the bottom of the grooves is $n = 6\frac{1}{2}$, while at some distance away it is $n = 4$. What is the stress concentration factor due to the grooves?

11.8 CLOSURE

The experimental methods discussed in this chapter are the ones commonly used in the solution of day-to-day engineering problems. Further information concerning their application can be obtained from texts on experimental stress analysis and from manufacturers and suppliers of experimental equipment and materials.

12

Appendix

TABLE A.1 GEOMETRIC PROPERTIES OF SOME COMMON SHAPES

Shape		$\bar{x}$	$\bar{y}$	Area, Length, or Volume	Moments of Inertia
Triangle	x', h, C, $\bar{y}$, x, $b/2$, $b/2$		$\frac{h}{3}$	$\frac{bh}{2}$	$\bar{I}_{x'} = \frac{1}{36} bh^3$ $I_x = \frac{1}{12} bh^3$
Circle	y, C, r, x, z	0	0	πr^2	$\bar{I}_x = \bar{I}_y = \frac{1}{4}\pi r^4$ $J_z = \frac{1}{2}\pi r^4$
Quarter circle	y, C, $\bar{y}$, $\bar{x}$, x	$\frac{4r}{3\pi}$	$\frac{4r}{3\pi}$	$\frac{\pi r^2}{4}$	$I_x = I_y = \frac{1}{16}\pi r^4$
Semi circle	y, C, r, x	0	$\frac{4r}{3\pi}$	$\frac{\pi r^2}{2}$	$I_x = I_y = \frac{1}{8}\pi r^4$
Rectangle	y, y', h, C, x', x, b	$\frac{b}{2}$	$\frac{h}{2}$	bh	$\bar{I}_{x'} = \frac{1}{12} bh^3$ $I_x = \frac{1}{3} bh^3$ $\bar{I}_{y'} = \frac{1}{12} hb^3$ $I_y = \frac{1}{3} hb^3$
Parabolic spandrel	y, b, $y = kx^2$, C, h, $\bar{y}$, x, $\bar{x}$	$\frac{3b}{4}$	$\frac{3h}{10}$	$\frac{bh}{3}$	
Arc of circle	y, r, α, C, x, α, $\bar{x}$	$\frac{r \sin \alpha}{\alpha}$	0	$2\alpha r$	
Solid hemisphere	r, C, $\bar{x}$	$\frac{3r}{8}$		$\frac{2}{3}\pi r^3$	
Solid cone	h, r, C, $\bar{x}$	$\frac{h}{4}$		$\frac{1}{3}\pi r^2 h$	

TABLE A.2 AVERAGE MECHANICAL PROPERTIES OF SOME COMMON ENGINEERING MATERIALS (VALUES IN SI ARE GIVEN IN PARENTHESES FOLLOWING THE CORRESPONDING VALUES IN US-BRITISH UNITS)

Material	Yield strength* 10^3 psi (MN/m^2)		Ultimate strength 10^3 psi (MN/m^2)			Elastic moduli 10^6 psi (GN/m^2)		Poisson's ratio	% Elong. in 2 in.	Coeff. ther. expansion 10^{-6} per °F (10^{-6} per °C)	Specific weight lb/ft^3 (kN/m^3)
	Tension	Shear	Tension	Comp.	Shear	Tension or comp.	Shear				
Structural steel (A36)	36 (248)	21 (145)	65 (448)	—	45 (310)	30 (207)	12 (83)	0.29	30	6.5 (11.7)	490 (77.0)
Alloy steel (4130) (heat tr.)	100 (689)	60 (414)	125 (862)	—	80 (552)	30 (207)	12 (83)	0.30	23	6.5 (11.7)	490 (77.0)
Aluminum alloy 2024-T3	50 (345)	30 (207)	68 (469)	—	40 (276)	10.5 (72)	4 (28)	0.33	15	12.6 (22.7)	173 (27.2)
Aluminum alloy 6061-T6	35 (241)	21 (145)	42 (290)	—	27 (186)	9.9 (68)	3.8 (26)	0.33	10	13.1 (23.6)	169 (26.5)
Titanium alloy 6M-4V	120 (827)	72 (496)	130 (896)	—	76 (524)	16 (110)	6.2 (43)	0.34	10	4.6 (8.3)	276 (43.4)
Brass	40 (276)	24 (165)	60 (414)	—	—	12 (83)	4.5 (31)	0.35	30	10.5 (18.9)	529 (83.1)
Grey cast iron	—	—	20 (138)	75 (517)	30 (207)	15 (103)	6.3 (43)	0.20	1	6.7 (12.1)	449 (70.5)
Concrete (high strength)	—	—	—	5 (34)	—	3 (21)	—	0.15	—	6.0 (10.8)	150 (23.6)
Douglas fir (par. to grain)	—	—	—	6.8 (47)	0.9 (6.2)	1.6 (11)	—	—	—	3.0 (5.4)	28 (4.4)

*At 0.2% permanent strain; tension and compression values approximately the same.

TABLE A.3* W SHAPES / PROPERTIES FOR DESIGNING

Designation	Area	Depth	Flange		Web thickness	Elastic properties					
						Axis X-X			Axis Y-Y		
	Area	Depth	Width	Thickness	thickness						
	A	d	b_f	t_f	t_w	I	S	r	I	S	r
	$In.^2$	In.	In.	In.	In.	$In.^4$	$In.^3$	In.	$In.^4$	$In.^3$	In.
W 10 × 29	8.54	10.22	5.799	0.500	0.289	158	30.8	4.30	16.3	5.61	1.38
× 25	7.36	10.08	5.762	0.430	0.252	133	26.5	4.26	13.7	4.76	1.37
× 21	6.20	9.90	5.750	0.340	0.240	107	21.5	4.15	10.8	3.75	1.32
W 10 × 19	5.61	10.25	4.020	0.394	0.250	96.3	18.8	4.14	4.28	2.13	0.874
× 17	4.99	10.12	4.010	0.329	0.240	81.9	16.2	4.05	3.55	1.77	0.844
× 15	4.41	10.00	4.000	0.269	0.230	68.9	13.8	3.95	2.88	1.44	0.809
× 11.5	3.39	9.87	3.950	0.204	0.180	52.0	10.5	3.92	2.10	1.06	0.787
W 8 × 67	19.7	9.00	8.287	0.933	0.575	272	60.4	3.71	88.6	21.4	2.12
× 58	17.1	8.75	8.222	0.808	0.510	227	52.0	3.65	74.9	18.2	2.10
× 48	14.1	8.50	8.117	0.683	0.405	184	43.2	3.61	60.9	15.0	2.08
× 40	11.8	8.25	8.077	0.558	0.365	146	35.5	3.53	49.0	12.1	2.04
× 35	10.3	8.12	8.027	0.493	0.315	126	31.1	3.50	42.5	10.6	2.03
× 31	9.12	8.00	8.000	0.433	0.288	110	27.4	3.47	37.0	9.24	2.01
W 8 × 28	8.23	8.06	6.540	0.463	0.285	97.8	24.3	3.45	21.6	6.61	1.62
× 24	7.06	7.93	6.500	0.398	0.245	82.5	20.8	3.42	18.2	5.61	1.61
W 8 × 20	5.89	8.14	5.268	0.378	0.248	69.4	17.0	3.43	9.22	3.50	1.25
× 17	5.01	8.00	5.250	0.308	0.230	56.6	14.1	3.36	7.44	2.83	1.22
W 8 × 15	4.43	8.12	4.015	0.314	0.245	48.1	11.8	3.29	3.40	1.69	0.876
× 13	3.83	8.00	4.000	0.254	0.230	39.6	9.90	3.21	2.72	1.36	0.842
× 10	2.96	7.90	3.940	0.204	0.170	30.8	7.80	3.23	2.08	1.06	0.839
W 6 × 25	7.35	6.37	6.080	0.456	0.320	53.3	16.7	2.69	17.1	5.62	1.53
× 20	5.88	6.20	6.018	0.367	0.258	41.5	13.4	2.66	13.3	4.43	1.51
× 15.5	4.56	6.00	5.995	0.269	0.235	30.1	10.0	2.57	9.67	3.23	1.46
W 6 × 16	4.72	6.25	4.030	0.404	0.260	31.7	10.2	2.59	4.42	2.19	0.967
× 12	3.54	6.00	4.000	0.279	0.230	21.7	7.25	2.48	2.98	1.49	0.918
× 8.5	2.51	5.83	3.940	0.194	0.170	14.8	5.08	2.43	1.98	1.01	0.889

*From *Manual of Steel Construction*, Seventh Edition, Courtesy of American Institute of Steel Construction, Inc.

TABLE A.4* CHANNELS / AMERICAN STANDARD / PROPERTIES FOR DESIGNING

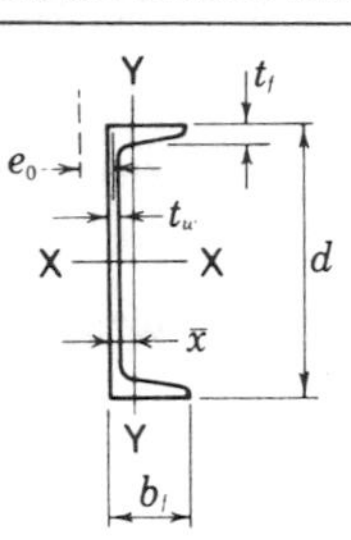

Designation	Area A	Depth d	Flange		Web thickness t_w	$\frac{d}{A_f}$	Axis X-X			Axis Y-Y			$\bar{x}$
			Width b_f	Average thickness t_f			I	S	r	I	S	r	
	In.^{2}	In.	In.	In.	In.		In.^{4}	In.^{3}	In.	In.^{4}	In.^{3}	In.	In.
C 9 × 20	5.88	9.00	2.648	0.413	0.448	8.22	60.9	13.5	3.22	2.42	1.17	0.642	0.583
× 15	4.41	9.00	2.485	0.413	0.285	8.76	51.0	11.3	3.40	1.93	1.01	0.661	0.586
× 13.4	3.94	9.00	2.433	0.413	0.233	8.95	47.9	10.6	3.48	1.76	0.962	0.668	0.601
C 8 × 18.75	5.51	8.00	2.527	0.390	0.487	8.12	44.0	11.0	2.82	1.98	1.01	0.599	0.565
× 13.75	4.04	8.00	2.343	0.390	0.303	8.75	36.1	9.03	2.99	1.53	0.853	0.615	0.553
× 11.5	3.38	8.00	2.260	0.390	0.220	9.08	32.6	8.14	3.11	1.32	0.781	0.625	0.571
C 7 × 14.75	4.33	7.00	2.299	0.366	0.419	8.31	27.2	7.78	2.51	1.38	0.779	0.564	0.532
× 12.25	3.60	7.00	2.194	0.366	0.314	8.71	24.2	6.93	2.60	1.17	0.702	0.571	0.525
× 9.8	2.87	7.00	2.090	0.366	0.210	9.14	21.3	6.08	2.72	0.968	0.625	0.581	0.541
C 6 × 13	3.83	6.00	2.157	0.343	0.437	8.10	17.4	5.80	2.13	1.05	0.642	0.525	0.514
× 10.5	3.09	6.00	2.034	0.343	0.314	8.59	15.2	5.06	2.22	0.865	0.564	0.529	0.500
× 8.2	2.40	6.00	1.920	0.343	0.200	9.10	13.1	4.38	2.34	0.692	0.492	0.537	0.512
C 5 × 9	2.64	5.00	1.885	0.320	0.325	8.29	8.90	3.56	1.83	0.632	0.449	0.489	0.478
× 6.7	1.97	5.00	1.750	0.320	0.190	8.93	7.49	3.00	1.95	0.478	0.378	0.493	0.484
C 4 × 7.25	2.13	4.00	1.721	0.296	0.321	7.84	4.59	2.29	1.47	0.432	0.343	0.450	0.459
× 5.4	1.59	4.00	1.584	0.296	0.184	8.52	3.85	1.93	1.56	0.319	0.283	0.449	0.458

*From *Manual of Steel Construction*, Seventh Edition, Courtesy of American Insitute of Steel Construction, Inc.

TABLE A.5* ANGLES/EQUAL LEGS / PROPERTIES FOR DESIGNING

Size and thickness	k	Weight per foot	Area	Axis X-X and axis Y-Y				Axis Z-Z
				I	*S*	*r*	*x or y*	*r*
In.	In.	Lb.	$In.^2$	$In.^4$	$In.^3$	In.	In.	In.
L 8 × 8 × 1 1/8	1 3/4	56.9	16.7	98.0	17.5	2.42	2.41	1.56
1	1 5/8	51.0	15.0	89.0	15.8	2.44	2.37	1.56
7/8	1 1/2	45.0	13.2	79.6	14.0	2.45	2.32	1.57
3/4	1 3/8	38.9	11.4	69.7	12.2	2.47	2.28	1.58
5/8	1 1/4	32.7	9.61	59.4	10.3	2.49	2.23	1.58
9/16	1 3/16	29.6	8.68	54.1	9.34	2.50	2.21	1.59
1/2	1 1/8	26.4	7.75	48.6	8.36	2.50	2.19	1.59
L 6 × 6 × 1	1 1/2	37.4	11.0	35.5	8.57	1.80	1.86	1.17
7/8	1 3/8	33.1	9.73	31.9	7.63	1.81	1.82	1.17
3/4	1 1/4	28.7	8.44	28.2	6.66	1.83	1.78	1.17
5/8	1 1/8	24.2	7.11	24.2	5.66	1.84	1.73	1.18
9/16	1 1/16	21.9	6.43	22.1	5.14	1.85	1.71	1.18
1/2	1	19.6	5.75	19.9	4.61	1.86	1.68	1.18
7/16	15/16	17.2	5.06	17.7	4.08	1.87	1.66	1.19
3/8	7/8	14.9	4.36	15.4	3.53	1.88	1.64	1.19
5/16	13/16	12.4	3.65	13.0	2.97	1.89	1.62	1.20
L 5 × 5 × 7/8	1 3/8	27.2	7.98	17.8	5.17	1.49	1.57	.973
3/4	1 1/4	23.6	6.94	15.7	4.53	1.51	1.52	.975
5/8	1 1/8	20.0	5.86	13.6	3.86	1.52	1.48	.978
1/2	1	16.2	4.75	11.3	3.16	1.54	1.43	.983
7/16	15/16	14.3	4.18	10.0	2.79	1.55	1.41	.986
3/8	7/8	12.3	3.61	8.74	2.42	1.56	1.39	.990
5/16	13/16	10.3	3.03	7.42	2.04	1.57	1.37	.994
L 4 × 4 × 3/4	1 1/8	18.5	5.44	7.67	2.81	1.19	1.27	.778
5/8	1	15.7	4.61	6.66	2.40	1.20	1.23	.779
1/2	7/8	12.8	3.75	5.56	1.97	1.22	1.18	.782
7/16	13/16	11.3	3.31	4.97	1.75	1.23	1.16	.785
3/8	3/4	9.8	2.86	4.36	1.52	1.23	1.14	.788
5/16	11/16	8.2	2.40	3.71	1.29	1.24	1.12	.791
1/4	5/8	6.6	1.94	3.04	1.05	1.25	1.09	.795

*From *Manual of Steel Construction*, Seventh Edition, Courtesy of American Institute of Steel Construction, Inc.

TABLE A.6* ANGLES/UNEQUAL LEGS / PROPERTIES FOR DESIGNING

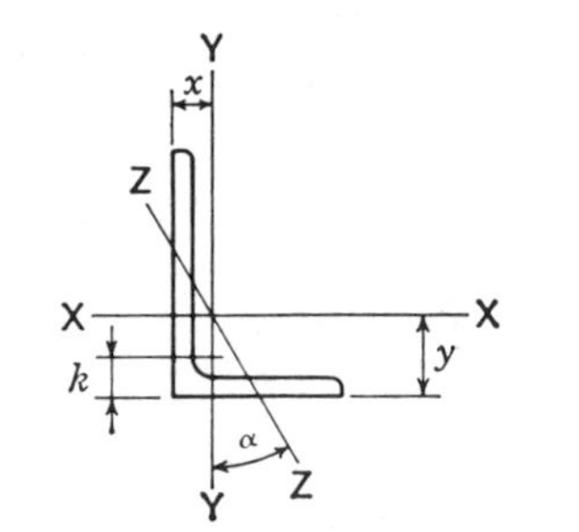

Size and thickness	k	Weight per foot	Area	Axis X-X				Axis Y-Y				Axis Z-Z	
				I	S	r	y	I	S	r	x	r	Tan
In.	In.	Lb.	$In.^2$	$In.^4$	$In.^3$	In.	In.	$In.^4$	$In.^3$	In.	In.	In.	α
L 4 × 3 1/2 × 5/8	1 1/16	14.7	4.30	6.37	2.35	1.22	1.29	4.52	1.84	1.03	1.04	.719	.745
1/2	15/16	11.9	3.50	5.32	1.94	1.23	1.25	3.79	1.52	1.04	1.00	.722	.750
7/16	7/8	10.6	3.09	4.76	1.72	1.24	1.23	3.40	1.35	1.05	.978	.724	.753
3/8	13/16	9.1	2.67	4.18	1.49	1.25	1.21	2.95	1.17	1.06	.955	.727	.755
5/16	3/4	7.7	2.25	3.56	1.26	1.26	1.18	2.55	.944	1.07	.932	.730	.757
1/4	11/16	6.2	1.81	2.91	1.03	1.27	1.16	2.09	.808	1.07	.909	.734	.759
L 4 × 3 × 5/8	1 1/16	13.6	3.98	6.03	2.30	1.23	1.37	2.87	1.35	.849	.871	.637	.534
1/2	15/16	11.1	3.25	5.05	1.89	1.25	1.33	2.42	1.12	.864	.827	.639	.543
7/16	7/8	9.8	2.87	4.52	1.68	1.25	1.30	2.18	.992	.871	.804	.641	.547
3/8	13/16	8.5	2.48	3.96	1.46	1.26	1.28	1.92	.866	.879	.782	.644	.551
5/16	3/4	7.2	2.09	3.38	1.23	1.27	1.26	1.65	.734	.887	.759	.647	.554
1/4	11/16	5.8	1.69	2.77	1.00	1.28	1.24	1.36	.599	.896	.736	.651	.558

L 3 1/2 × 3 × 1/2	15/16	10.2	3.00	3.45	1.45	1.07	1.13	2.33	1.10	.881	.875	.621	.714
7/16	7/8	9.1	2.65	3.10	1.29	1.08	1.10	2.09	.975	.889	.853	.622	.718
3/8	13/16	7.9	2.30	2.72	1.13	1.09	1.08	1.85	.851	.897	.830	.625	.721
5/16	3/4	6.6	1.93	2.33	.954	1.10	1.06	1.58	.722	.905	.808	.627	.724
1/4	11/16	5.4	1.56	1.91	.776	1.11	1.04	1.30	.589	.914	.785	.631	.727
L 3 1/2 × 2 1/2 × 1/2	15/16	9.4	2.75	3.24	1.41	1.09	1.20	1.36	.760	.704	.705	.534	.486
7/16	7/8	8.3	2.43	2.91	1.26	1.09	1.18	1.23	.677	.711	.682	.535	.491
3/8	13/16	7.2	2.11	2.56	1.09	1.10	1.16	1.09	.592	.719	.660	.537	.496
5/16	3/4	6.1	1.78	2.19	.927	1.11	1.14	.939	.504	.727	.637	.540	.501
1/4	11/16	4.9	1.44	1.80	.755	1.12	1.11	.777	.412	.735	.614	.544	.506
L 3 × 2 1/2 × 1/2	7/8	8.5	2.50	2.08	1.04	.913	1.00	1.30	.744	.722	.750	.520	.667
7/16	13/16	7.6	2.21	1.88	.928	.920	.978	1.18	.664	.729	.728	.521	.672
3/8	3/4	6.6	1.92	1.66	.810	.928	.956	1.04	.581	.736	.706	.522	.676
5/16	11/16	5.6	1.62	1.42	.688	.937	.933	.898	.494	.744	.683	.525	.680
1/4	5/8	4.5	1.31	1.17	.561	.945	.911	.743	.404	.753	.661	.528	.684
3/16	9/16	3.39	.996	.907	.430	.954	.888	.577	.310	.761	.638	.533	.688

*From *Manual of Steel Construction*, Seventh Edition, Courtesy of American Institute of Steel Construction, Inc.

TABLE A.7* PIPE / STANDARD WEIGHT / DIMENSIONS AND PROPERTIES

Dimension				*Weight per foot lbs. plain ends*	*Properties*			
Nominal diameter In.	*Outside diameter In.*	*Inside diameter In.*	*Wall thickness In.*		*A In.^{2}*	*I In.^{4}*	*S In.^{3}*	*r In.*
1/2	.840	.622	.109	.85	.250	.017	.041	.261
3/4	1.050	.824	.113	1.13	.333	.037	.071	.334
1	1.315	1.049	.133	1.68	.494	.087	.133	.421
1 1/4	1.660	1.380	.140	2.27	.669	.195	.235	.540
1 1/2	1.900	1.610	.145	2.72	.799	.310	.326	.623
2	2.375	2.067	.154	3.65	1.07	.666	.561	.787
2 1/2	2.875	2.469	.203	5.79	1.70	1.53	1.06	.947
3	3.500	3.068	.216	7.58	2.23	3.02	1.72	1.16
3 1/2	4.000	3.548	.226	9.11	2.68	4.79	2.39	1.34
4	4.500	4.026	.237	10.79	3.17	7.23	3.21	1.51
5	5.563	5.047	.258	14.62	4.30	15.2	5.45	1.88
6	6.625	6.065	.280	18.97	5.58	28.1	8.50	2.25
8	8.625	7.981	.322	28.55	8.40	72.5	16.8	2.94
10	10.750	10.020	.365	40.48	11.9	161	29.9	3.67
12	12.750	12.000	.375	49.56	14.6	279	43.8	4.38

*From *Manual of Steel Construction*, Seventh Edition, Courtesy of American Institute of Steel Construction, Inc.

TABLE A.8* UNIVERSAL BEAMS/DIMENSIONS AND PROPERTIES

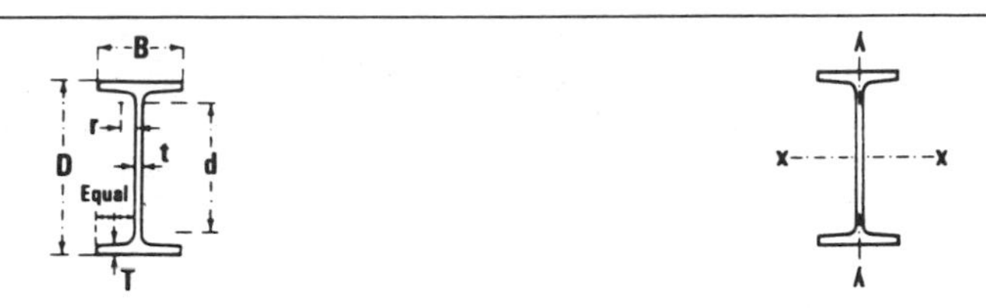

Serial size	Mass per meter	Depth of section D	Width of section B	Thickness		Area of section	Moment of inertia		Radius of gyration		Section modulus	
				Web t	Flange T		Axis x-x	Axis y-y	Axis x-x	Axis y-y	Axis x-x	Axis y-y
mm	kg	mm	mm	mm	mm	cm^2	cm^4	cm^4	cm	cm	cm^3	cm^3
381 × 152	67	388.6	154.3	9.7	16.3	85.4	21,276	947	15.8	3.33	1095	122.7
	60	384.8	153.4	8.7	14.4	75.9	18,632	814	15.7	3.27	968.4	106.2
	52	381.0	152.4	7.8	12.4	66.4	16,046	685	15.5	3.21	842.3	89.96
356 × 171	67	364.0	173.2	9.1	15.7	85.3	19,483	1278	15.1	3.87	1071	147.6
	57	358.6	172.1	8.0	13.0	72.1	16,038	1026	14.9	3.77	894.3	119.2
	51	355.6	171.5	7.3	11.5	64.5	14,118	885	14.8	3.71	794.0	103.3
	45	352.0	171.0	6.9	9.7	56.9	12,052	730	14.6	3.58	684.7	85.39
356 × 127	39	352.8	126.0	6.5	10.7	49.3	10,054	333	14.3	2.60	570.0	52.87
	33	348.5	125.4	5.9	8.5	41.7	8,167	257	14.0	2.48	468.7	40.99
305 × 165	54	310.9	166.8	7.7	13.7	68.3	11,686	988	13.1	3.80	751.8	118.5
	46	307.1	165.7	6.7	11.8	58.8	9,924	825	13.0	3.74	646.4	99.54
	40	303.8	165.1	6.1	10.2	51.4	8,500	691	12.9	3.67	559.6	83.71
305 × 127	48	310.4	125.2	8.9	14.0	60.8	9,485	438	12.5	2.68	611.1	69.94
	42	306.6	124.3	8.0	12.1	53.1	8,124	367	12.4	2.63	530.0	58.99
	37	303.8	123.5	7.2	10.7	47.4	7,143	316	12.3	2.58	470.3	51.11
305 × 102	33	312.7	102.4	6.6	10.8	41.8	6,482	189	12.5	2.13	414.6	37.00
	28	308.9	101.9	6.1	8.9	36.3	5,415	153	12.2	2.05	350.7	30.01
	25	304.8	101.6	5.8	6.8	31.4	4,381	116	11.8	1.92	287.5	22.85
254 × 146	43	259.6	147.3	7.3	12.7	55.0	6,546	633	10.9	3.39	504.3	85.97
	37	256.0	146.4	6.4	10.9	47.4	5,544	528	10.8	3.34	433.1	72.11
	31	251.5	146.1	6.1	8.6	39.9	4,427	406	10.5	3.19	352.1	55.53
254 × 102	28	260.4	102.1	6.4	10.0	36.2	4,004	174	10.5	2.19	307.6	34.13
	25	257.0	101.9	6.1	8.4	32.1	3,404	144	10.3	2.11	264.9	28.23
	22	254.0	101.6	5.8	6.8	28.4	2,863	116	10.0	2.02	225.4	22.84
203 × 133	30	206.8	133.8	6.3	9.6	38.0	2,880	354	8.71	3.05	278.5	52.85
	25	203.2	133.4	5.8	7.8	32.3	2,348	280	8.53	2.94	231.1	41.92

*From *Handbook on Structural Steelwork—Metric Properties and Safe Loads*, 1971, Courtesy of British Standards Institution, The Constructional Steel Research and Development Organization, and The British Constructional Steelwork Association, Ltd.

TABLE A.9* CHANNELS/DIMENSIONS AND PROPERTIES

Nominal size	Mass per meter	Depth of section D	Width of section B	Thickness		Area of section	Dimension p	Moment of inertia		Radius of gyration		Section modulus	
				Web t	Flange T			Axis x-x	Axis y-y	Axis x-x	Axis y-y	Axis x-x	Axis y-y
mm	kg	mm	mm	mm	mm	cm^2	cm	cm^4	cm^4	cm	cm	cm^3	cm^3
432 × 102	65.54	431.8	101.6	12.2	16.8	83.49	2.32	21,399	628.6	16.0	2.74	991.1	80.15
381 × 102	55.10	381.0	101.6	10.4	16.3	70.19	2.52	14,894	579.8	14.6	2.87	781.8	75.87
305 × 102	46.18	304.8	101.6	10.2	14.8	58.83	2.66	8,214	499.5	11.8	2.91	539.0	66.60
305 × 89	41.69	304.8	88.9	10.2	13.7	53.11	2.18	7,061	325.4	11.5	2.48	463.3	48.49
254 × 89	35.74	254.0	88.9	9.1	13.6	45.52	2.42	4,448	302.4	9.88	2.58	350.2	46.71
254 × 76	28.29	254.0	76.2	8.1	10.9	36.03	1.86	3,367	162.6	9.67	2.12	265.1	28.22
229 × 89	32.76	228.6	88.9	8.6	13.3	41.73	2.53	3,387	285.0	9.01	2.61	296.4	44.82
229 × 76	26.06	228.6	76.2	7.6	11.2	33.20	2.00	2,610	158.7	8.87	2.19	228.3	28.22
203 × 89	29.78	203.2	88.9	8.1	12.9	37.94	2.65	2,491	264.4	8.10	2.64	245.2	42.34
203 × 76	23.82	203.2	76.2	7.1	11.2	30.34	2.13	1,950	151.4	8.02	2.23	192.0	27.59
178 × 89	26.81	177.8	88.9	7.6	12.3	34.15	2.76	1,753	241.0	7.16	2.66	197.2	39.29
178 × 76	20.84	177.8	76.2	6.6	10.3	26.54	2.20	1,337	134.0	7.10	2.25	150.4	25.73
152 × 89	23.84	152.4	88.9	7.1	11.6	30.36	2.86	1,166	215.1	6.20	2.66	153.0	35.70
152 × 76	17.88	152.4	76.2	6.4	9.0	22.77	2.21	851.6	113.8	6.12	2.24	111.8	21.05
127 × 64	14.90	127.0	63.5	6.4	9.2	18.98	1.94	482.6	67.24	5.04	1.88	75.99	15.25
102 × 51	10.42	101.6	50.8	6.1	7.6	13.28	1.51	207.7	29.10	3.96	1.48	40.89	8.16
76 × 38	6.70	76.2	38.1	5.1	6.8	8.53	1.19	74.14	10.66	2.95	1.12	19.46	4.07

*From *Handbook on Structural Steelwork—Metric Properties and Safe Loads*, 1971, Courtesy of British Standards Institution, The Constructional Steel Research and Development Organization, and The British Constructional Steelwork Association, Ltd.

TABLE A.10* EQUAL ANGLES/DIMENSIONS AND PROPERTIES

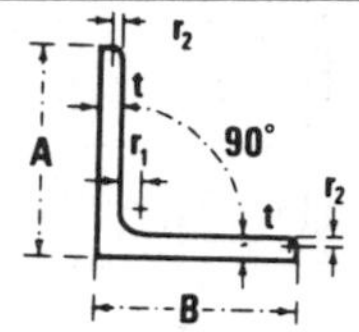

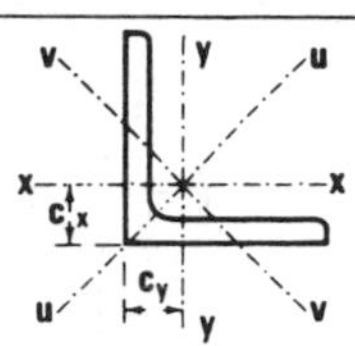

Nominal size	Leg lengths $A \times B$	Actual thickness	Mass per meter	Area of section	Center of gravity		Moment of inertia			Radius of gyration		
					Cx	Cy	Axis x-x or y-y	Axis u-u max.	Axis v-v min.	Axis x-x or y-y	Axis u-u max.	Axis v-v min.
mm	mm	mm	kg	cm^2	cm	cm	cm^4	cm^4	cm^4	cm	cm	cm
89 × 89	88.9 × 88.9	15.8	20.10	25.61	2.78	2.78	178	280	75.7	2.63	3.30	1.72
		14.2	18.31	23.32	2.72	2.72	164	259	69.1	2.65	3.33	1.72
		12.6	16.38	20.87	2.66	2.66	149	235	62.0	2.67	3.36	1.72
		11.0	14.44	18.40	2.60	2.60	133	211	55.0	2.69	3.38	1.73
		9.4	12.50	15.92	2.54	2.54	116	185	47.9	2.70	3.41	1.74
		7.9	10.58	13.47	2.48	2.48	99.8	159	41.0	2.72	3.43	1.74
		6.3	8.49	10.81	2.41	2.41	81.0	129	33.3	2.74	3.45	1.75
76 × 76	76.2 × 76.2	14.3	15.50	19.74	2.41	2.41	99.6	157	42.7	2.25	2.82	1.47
		12.6	13.85	17.64	2.35	2.35	90.4	143	38.2	2.26	2.84	1.47
		11.0	12.20	15.55	2.29	2.29	80.9	128	33.8	2.28	2.87	1.47
		9.4	10.57	13.47	2.23	2.23	71.1	113	29.5	2.30	2.89	1.48
		7.8	8.93	11.37	2.16	2.16	60.9	96.8	25.1	2.31	2.92	1.49
		6.2	7.16	9.12	2.10	2.10	49.6	78.8	20.3	2.33	2.94	1.49
64 × 64	63.5 × 63.5	12.5	11.31	14.41	2.03	2.03	50.4	78.9	21.8	1.87	2.34	1.23
		11.0	10.12	12.89	1.98	1.98	45.8	72.1	19.5	1.89	2.37	1.23
		9.4	8.78	11.18	1.92	1.92	40.5	64.0	17.0	1.90	2.39	1.23
		7.9	7.45	9.48	1.86	1.86	35.0	55.5	14.6	1.92	2.42	1.24
		6.2	5.96	7.59	1.80	1.80	28.6	45.4	11.8	1.94	2.45	1.25
57 × 57	57.2 × 57.2	9.3	7.74	9.86	1.76	1.76	28.6	45.0	12.1	1.70	2.14	1.11
		7.8	6.55	8.35	1.70	1.70	24.7	39.1	10.3	1.72	2.16	1.11
		6.2	5.35	6.82	1.64	1.64	20.6	32.6	8.53	1.74	2.19	1.12
		4.6	4.01	5.11	1.57	1.57	15.8	25.0	6.51	1.76	2.21	1.13
51 × 51	50.8 × 50.8	9.4	6.85	8.72	1.60	1.60	19.6	30.8	8.42	1.50	1.88	.98
		7.8	5.80	7.39	1.54	1.54	17.0	26.8	7.17	1.52	1.91	.98
		6.3	4.77	6.08	1.49	1.49	14.3	22.7	5.95	1.53	1.93	.99
		4.6	3.58	4.56	1.42	1.42	11.0	17.4	4.54	1.55	1.95	1.00

*From *Handbook on Structural Steelwork—Metric Properties and Safe Loads*, 1971, Courtesy of British Standards Institution, The Constructional Steel Research and Development Organization, and The British Constructional Steelwork Association, Ltd.

TABLE A.11* UNEQUAL ANGLES/DIMENSIONS AND PROPERTIES

Nominal size	Leg lengths A × B	Actual thickness	Mass per meter	Area of section	Center of Gravity Cx	Center of Gravity Cy	Moment of Inertia Axis x-x	Moment of Inertia Axis y-y	Moment of Inertia Axis u-u max.	Moment of Inertia Axis v-v min.	Angle Axis x-x to axis u-u
mm	mm	mm	kg	cm^2	cm	cm	cm^4	cm^4	cm^4	cm^4	tan α
102 × 64	101.6 × 63.5	11.0	13.40	17.07	3.51	1.62	174	51.9	194	31.4	.380
		9.5	11.61	14.79	3.45	1.56	152	45.8	171	27.4	.383
		7.8	9.69	12.35	3.38	1.49	129	38.9	145	23.2	.386
		6.3	7.89	10.05	3.31	1.43	106	32.2	119	19.1	.388
89 × 76	88.9 × 76.2	14.2	16.83	21.44	2.89	2.26	155	104	207	52.4	.710
		12.7	15.20	19.36	2.84	2.21	142	95.4	190	47.5	.713
		11.0	13.40	17.07	2.77	2.14	127	85.4	170	42.0	.715
		9.5	11.61	14.79	2.71	2.08	111	75.1	150	36.7	.718
		7.8	9.69	12.35	2.65	2.02	94.2	63.7	127	30.9	.720
		6.3	7.89	10.05	2.58	1.96	77.5	52.5	104	25.5	.721
89 × 64	88.9 × 63.5	11.0	12.20	15.55	2.97	1.71	118	49.8	140	28.2	.489
		9.4	10.57	13.47	2.91	1.65	104	43.9	123	24.6	.493
		7.8	8.93	11.37	2.85	1.59	88.8	37.7	106	21.0	.496
		6.2	7.16	9.12	2.78	1.53	72.1	30.7	85.8	17.0	.498
76 × 64	76.2 × 63.5	11.0	11.17	14.23	2.46	1.83	76.6	47.8	99.9	24.4	.669
		9.4	9.68	12.33	2.40	1.77	67.3	42.1	88.2	21.3	.673
		7.9	8.19	10.43	2.34	1.71	57.8	36.2	75.9	18.1	.676
		6.2	6.56	8.36	2.27	1.64	46.9	29.5	61.7	14.7	.678

*From *Handbook on Structural Steelwork—Metric Properties and Safe Loads*, 1971, Courtesy of British Standards Institution, The Constructional Steel Research and Development Organization, and The British Constructional Steelwork Association, Ltd.

TABLE A.12* PROPERTIES OF STRUCTURAL LUMBER — STANDARD DRESSED SIZES

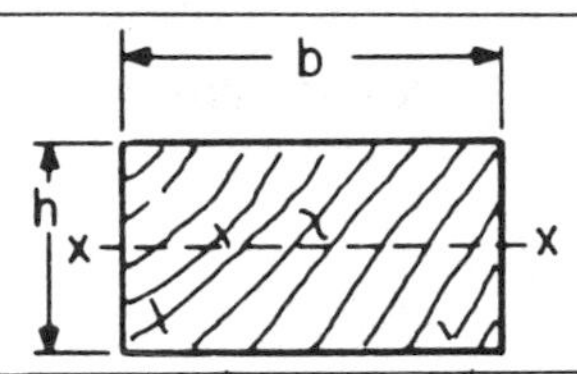

Nominal size b (inches) h	Standard dressed size (S4S) b (inches) h	Area of section A	Moment of inertia I_{xx}	Section modulus S	Weight in lb / ft when γ equals:	
					25 lb / ft^3	30 lb / ft^3
2 × 4	1 1/2 × 3 1/2	5.250	5.359	3.063	0.911	1.094
2 × 6	1 1/2 × 5 1/2	8.250	20.797	7.563	1.432	1.719
2 × 8	1 1/2 × 7 1/4	10.875	47.635	13.141	1.888	2.266
2 × 10	1 1/2 × 9 1/4	13.875	98.932	21.391	2.409	2.891
2 × 12	1 1/2 × 11 1/4	16.875	177.979	31.641	2.930	3.516
4 × 2	3 1/2 × 1 1/2	5.250	0.984	1.313	0.911	1.094
4 × 3	3 1/2 × 2 1/2	8.750	4.557	3.646	1.519	1.823
4 × 4	3 1/2 × 3 1/2	12.250	12.505	7.146	2.127	2.552
4 × 6	3 1/2 × 5 1/2	19.250	48.526	17.646	3.342	4.010
4 × 8	3 1/2 × 7 1/4	25.375	111.148	30.661	4.405	5.286
4 × 10	3 1/2 × 9 1/4	32.375	230.840	49.911	5.621	6.745
4 × 12	3 1/2 × 11 1/4	39.375	415.283	73.828	6.836	8.203
6 × 2	5 1/2 × 1 1/2	8.250	1.547	2.063	1.432	1.719
6 × 3	5 1/2 × 2 1/2	13.750	7.161	5.729	2.387	2.865
6 × 4	5 1/2 × 3 1/2	19.250	19.651	11.229	3.342	4.010
6 × 6	5 1/2 × 5 1/2	30.250	76.255	27.729	5.252	6.302
6 × 8	5 1/2 × 7 1/2	41.250	193.359	51.563	7.161	8.594
6 × 10	5 1/2 × 9 1/2	52.250	392.963	82.729	9.071	10.885
6 × 12	5 1/2 × 11 1/2	63.250	697.068	121.229	10.981	13.177
8 × 2	7 1/4 × 1 1/2	10.875	2.039	2.719	1.888	2.266
8 × 3	7 1/4 × 2 1/2	18.125	9.440	7.552	3.147	3.776
8 × 4	7 1/4 × 3 1/2	25.375	25.904	14.802	4.405	5.286
8 × 6	7 1/2 × 5 1/2	41.250	103.984	37.813	7.161	8.594
8 × 8	7 1/2 × 7 1/2	56.250	263.672	70.313	9.766	11.719
8 × 10	7 1/2 × 9 1/2	71.250	535.859	112.813	12.370	14.844
8 × 12	7 1/2 × 11 1/2	86.250	950.547	165.313	14.974	17.969
10 × 2	9 1/4 × 1 1/2	13.875	2.602	3.469	2.409	2.891
10 × 3	9 1/4 × 2 1/2	23.125	12.044	9.635	4.015	4.818
10 × 4	9 1/4 × 3 1/2	32.375	33.049	18.885	5.621	6.745
10 × 6	9 1/2 × 5 1/2	52.250	131.714	47.896	9.071	10.885
10 × 8	9 1/2 × 7 1/2	71.250	333.984	89.063	12.370	14.844
10 × 10	9 1/2 × 9 1/2	90.250	678.755	142.896	15.668	18.802
10 × 12	9 1/2 × 11 1/2	109.250	1204.026	209.396	18.967	22.760
12 × 2	11 1/4 × 1 1/2	16.875	3.164	4.219	2.930	3.516
12 × 3	11 1/4 × 2 1/2	28.125	14.648	11.719	4.883	5.859
12 × 4	11 1/4 × 3 1/2	39.375	40.195	22.969	6.836	8.203
12 × 6	11 1/2 × 5 1/2	63.250	159.443	57.979	10.981	13.177
12 × 8	11 1/2 × 7 1/2	86.250	404.297	107.813	14.974	17.969

*From *National Design Specification for Stress-Grade Lumber and its Fastenings*, 1973 Ed., Courtesy of National Forest Products Association.

TABLE A.13 SLOPES AND DEFLECTIONS OF COMMON BEAMS

Case	Deflection	Maximum deflection	Slope
① 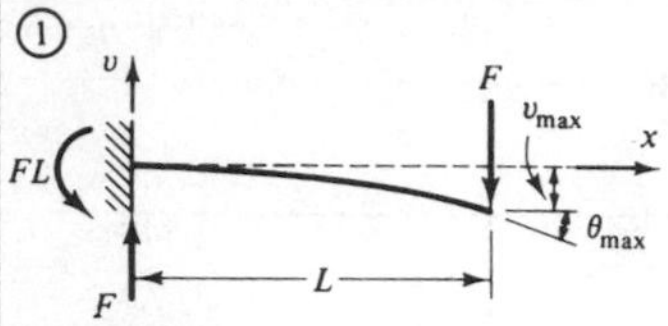	$v = \frac{Fx^2}{6EI}(x - 3L)$	$v_{max} = -\frac{FL^3}{3EI}$ at free end	$\theta_{max} = -\frac{FL^2}{2EI}$ at free end
② 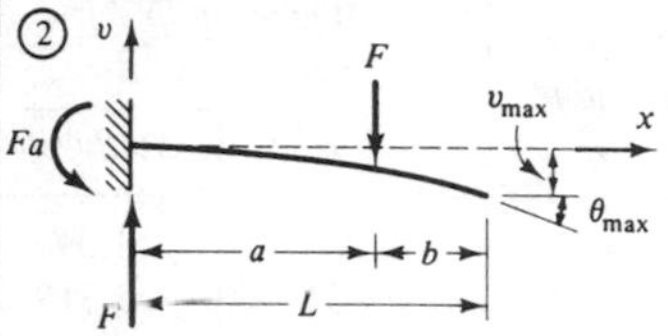	$v = \frac{Fx^2}{6EI}(x - 3a) \quad (x \leq a)$ $= \frac{F}{6EI}[x^3 - 3x^2a - (x - a)^3] \quad (x \geq a)$	$v_{max} = -\frac{Fa^2(3L - a)}{6EI}$ at free end	$\theta_{max} = -\frac{Fa^2}{2EI}$ at free end
③ 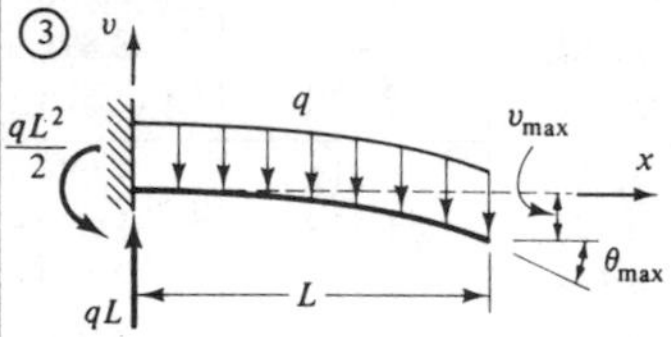	$v = \frac{qx^2}{24EI}(4Lx - x^2 - 6L^2)$	$v_{max} = -\frac{qL^4}{8EI}$ at free end	$\theta_{max} = -\frac{qL^3}{6EI}$ at free end
④ 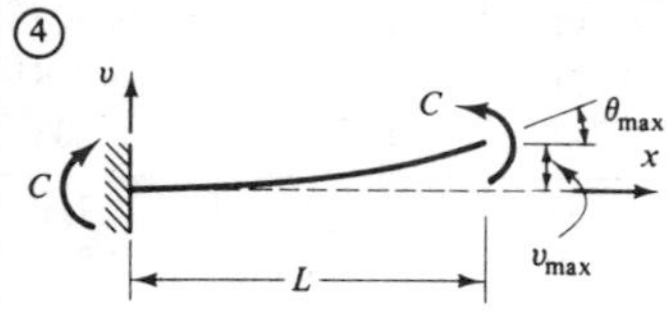	$v = \frac{Cx^2}{2EI}$	$v_{max} = \frac{CL^2}{2EI}$ at free end	$\theta_{max} = \frac{CL}{EI}$ at free end
⑤ 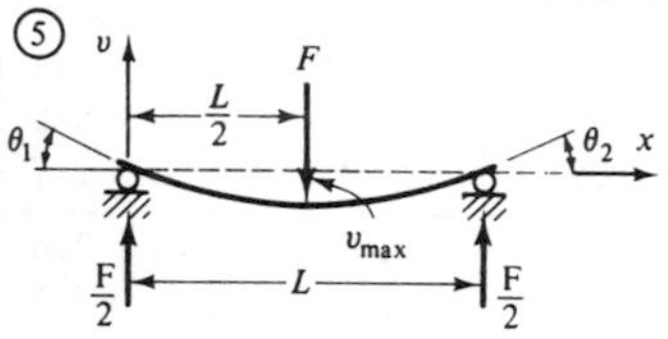	$v = \frac{Fx}{48EI}(4x^2 - 3L^2) \quad (x \leq \frac{L}{2})$	$v_{max} = -\frac{FL^3}{48EI}$ at center	$\theta_2 = -\theta_1 = \frac{FL^2}{16EI}$
⑥ 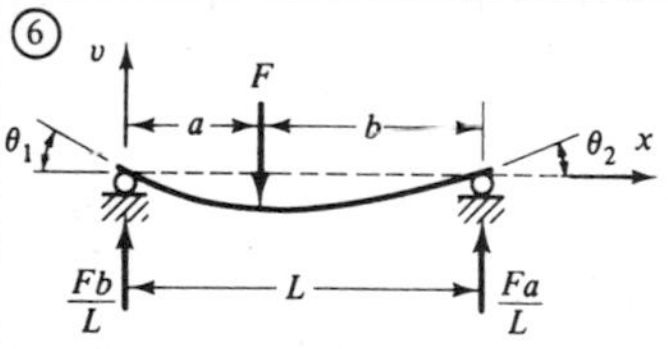	$v = \frac{Fbx}{6EIL}[x^2 - (L^2 - b^2)] \quad (x \leq a)$ $= \frac{Fb}{6EIL}[x^3 - (L^2 - b^2)x - \frac{L}{b}(x - a)^3] \quad (x \geq a)$	$v_{max} = -Fb(L^2 - b^2)^{3/2}$ at $x = \sqrt{\frac{L^2 - b^2}{3}}$ for $a > b$	$\theta_1 = \frac{-Fb}{6EIL}(L^2 - b^2)$ $\theta_2 = \frac{Fb}{6EIL}(L - b)(2L - b)$
⑦ 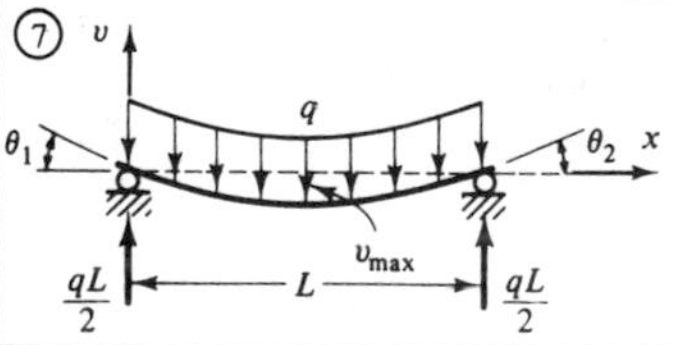	$v = -\frac{qx}{24EI}(2Lx^2 - x^3 - L^3)$	$v_{max} = -\frac{5qL^4}{384EI}$ at center	$\theta_2 = -\theta_1 = \frac{qL^3}{24EI}$

TABLE A.14 CONVERSION FACTORS: U.S.-BRITISH UNITS TO SI

To convert from	*To*	*Multiply by**
(Length)		
foot (ft)	meter (m)	3.048×10^{-1}
inch (in.)	meter (m)	2.540×10^{-2}
(Area)		
foot^2 (ft^2)	meter^2 (m^2)	9.290×10^{-2}
inch^2 (in.^{2})	meter^2 (m^2)	6.452×10^{-4}
(Volume)		
foot^3 (ft^3)	meter^3 (m^3)	2.832×10^{-2}
inch^3 (in.^{3})	meter^3 (m^3)	1.639×10^{-5}
(Force)		
pound (lb)	newton (N)	4.448
(Moment)		
pound · foot (lb · ft)	newton · meter (N · m)	1.356
pound · inch (lb · in.)	newton · meter (N · m)	1.130×10^{-1}
(Pressure, Stress)		
pound/foot^2 (lb/ft^2)	newton/meter^2 (N/m^2)	4.788×10
pound/inch^2 (lb/in.^{2})	newton/meter^2 (N/m^2)	6.895×10^3
(Specific Weight)		
pound/foot^3 (lb/ft^3)	newton/meter^3 (N/m^3)	1.571×10^2
pound/inch^3 (lb/in.^{3})	newton/meter^3 (N/m^3)	2.714×10^5

*To convert from SI to U.S.-British units, divide by the factor given.

Answers to Even-Numbered Problems

1

1.2. $5.3\ \mathrm{ft/s^2}$
1.4. \$2,360
1.6. sawdust
1.8. 30.4 ft/s, 9.27 m/s
1.10. city B
1.12. 6.894×10^3
1.14. 0.425

2

2.2. 86.7 lb ↗ 72°
2.4. 56.6 kN ↘ 45°
2.6. $\mathbf{A} - \mathbf{B}$ = 111.7 lb ↘ 1.3°, $\mathbf{B} - \mathbf{A}$ = 111.7 lb ↘ 1.3°
2.8. $P = 38.9$ kN, $F = 73.1$ kN
2.10. $T_1 = 333$ lb, $T_2 = 267$ lb
2.12. $\mathbf{A}$ = 42.4 lb ← +42.4 lb ↑, $\mathbf{B}$ = 69.3 lb → +40.0 lb ↑, $\mathbf{A}$ = 66.9 lb ↑ + 49.0 lb ↙ 30°
2.14. $F = 175.4$ kips, $T_1 = 93.3$ kips, $T_2 = 120.0$ kips

2.16. (a) -37 kips; (b) -25 kips; (c) $-3{,}700$ kips^2; (d) -58 kips, -80 kips (for positive direction to the right)

2.18. (a) 50 N; (b) 116 N; (c) 11,550 N^2; (d) 0, 200 N (for positive direction to the right)

2.20. $\mathbf{B} = 12.0 \nearrow$, $\mathbf{A}_{bb} = 4.3 \nearrow$ (both along bb)

2.22. $\mathbf{A} = 15.0$ m $\swarrow$ (along aa), $\mathbf{B} = 29.4$ m $\measuredangle$ 17.8°

2.24. $4\mathbf{i} - 5\mathbf{j} + 3\mathbf{k}$, $-3\mathbf{j} - 3\mathbf{k}$, $10\mathbf{i} - 11\mathbf{j} + 9\mathbf{k}$, 8; $\theta = 61.4°$

2.26. $10\mathbf{i} + 2\mathbf{j} - 6\mathbf{k}$, $6\mathbf{i} + 10\mathbf{j} - 2\mathbf{k}$, $22\mathbf{i} - 14\mathbf{k}$, 0; $\theta = 90°$

2.28. $-8/3$

2.30. $40\mathbf{i} - 40\mathbf{j}$ kN

2.32. $6\mathbf{i} + 2\mathbf{j} - 3\mathbf{k}$ m; $\theta_x = 31.0°$, $\theta_y = 73.4°$, $\theta_z = 115.4°$

2.34. $\mathbf{F} = 80\mathbf{i} + 34\mathbf{j} - 50\mathbf{k}$ lb; $\theta_x = 37.3°$, $\theta_y = 70.0°$, $\theta_z = 120.0°$

2.36. (a) $-3.9\mathbf{i} + 3.0\mathbf{j} + \mathbf{k}$ m; (b) -50.7 m^2; (c) 167.4°; $S = 5.0$ m, $\theta_x = 141.3°$, $\theta_y = 53.0°$, $\theta_z = 78.5°$

2.38. 30°; $T_1 = 10.4$ lb, $T_2 = 6.0$ lb

2.40. (a) $\mathbf{r}_{OP} = 4\mathbf{i} + 2\mathbf{k}$ ft, $\mathbf{r}_{OQ} = 6\mathbf{i} - 3\mathbf{j} + 2\mathbf{k}$ ft; (b) $\mathbf{r}_{PQ} = 2\mathbf{i} - 3\mathbf{j}$ ft; (c) $\mathbf{e}_{PQ} = (1/\sqrt{13})(2\mathbf{i} - 3\mathbf{j})$

2.42. $-1.0 \times 10^{-12}\mathbf{i} - 2.5 \times 10^{-12}\mathbf{j} + 2.5 \times 10^{-12}\mathbf{k}$ N

2.44. 198 lb, $124\mathbf{i} + 124\mathbf{j} + 93\mathbf{k}$ lb; 8.5°

2.46. $-110\mathbf{i} + 330\mathbf{j} - 50\mathbf{k}$ kN

2.48. $-4\mathbf{i} + 16\mathbf{j} + 20\mathbf{k}$

2.50. $-10\mathbf{i} + 48\mathbf{j} + 52\mathbf{k}$

2.52. $-6.0\mathbf{i} - 8.7\mathbf{j} + 2.7\mathbf{k}$ m^2, $\pm(-0.550\mathbf{i} - 0.798\mathbf{j} + 0.248\mathbf{k})$

2.54. $y = -2$, $z = 2$

2.56. $\mathbf{e}_N = \pm(1/\sqrt{5})(-\mathbf{i} + 2\mathbf{k})$

2.60. 340 lb · in. ↺

2.62. 125 N · m ↻

2.64. 880 N

2.66. 1.342 Fa ↻

2.68. (a) 11.96 N · m ↺; (b) 59.8 N ↓; (c) 47.8 N ↖ 37°

2.70. (a) $1{,}800\mathbf{i} + 400\mathbf{j} - 1{,}600\mathbf{k}$ lb · ft; (b) 3.13 ft

2.72. $\mathbf{M}_A = \mathbf{M}_B = \mathbf{0}$, $\mathbf{M}_C = -2{,}100\mathbf{j} + 2{,}100\mathbf{k}$ lb · ft, $\mathbf{M}_D = 1{,}800\mathbf{i} + 2{,}100\mathbf{j}$ lb · ft

2.74. $\mathbf{M}_A = -0.094\mathbf{i} - 0.227\mathbf{j} + 0.271\mathbf{k}$ N · m; $d = 73.2$ mm

2.76. $200\mathbf{i} - 40\mathbf{j} - 390\mathbf{k}$ N · m

2.78. 53

2.80. $4.8\mathbf{e}_{AB}$ N · m

2.82. $910\mathbf{e}_{DC}$ lb · ft; no

2.84. $750\mathbf{j}$ N · m

2.86. $-52\mathbf{j}$ kN · m; $90\mathbf{i}$ kN · m; $\mathbf{0}$

2.90 no; $-3.2\mathbf{i} - 28.6\mathbf{j}$ lb, $\mathbf{C} = 457$ lb · ft ↻

2.92. 1.35 kN · m ↻

2.94. 60 lb; smaller

2.96. 17.9 lb, at A and C

2.98. 746 lb · in. ↻

2.100. (a) $160\mathbf{i} + 172\mathbf{j} + 129\mathbf{k}$ N · m; (b) $53\mathbf{k}$ N · m

3

3.2. On log A: $\mathbf{F}_B$ and $\mathbf{F}_D$ exerted by logs B and D, $\mathbf{F}_G$ exerted by ground, $\mathbf{W}\downarrow$ due to earth's gravitational attraction. Countereffects: $-\mathbf{F}_B$ and $-\mathbf{F}_D$ exerted on logs B and D, $-\mathbf{F}_G$ exerted on ground, $\mathbf{W}\uparrow$ exerted on earth.

3.4. On child: $\mathbf{W}_C\downarrow$ due to earth's gravitational attraction, $\mathbf{N}\uparrow$ exerted by seat; countereffects: $\mathbf{W}_C\uparrow$ on earth, $\mathbf{N}\downarrow$ on seat. On seat: $\mathbf{N}\downarrow$ exerted by child, $\mathbf{W}_S\downarrow$ due to earth's gravitational attraction, $\mathbf{T}_L\uparrow$ and $\mathbf{T}_R\uparrow$ exerted by ropes; countereffects: $\mathbf{N}\uparrow$ on child, $\mathbf{W}_S\uparrow$ on earth, $\mathbf{T}_L\downarrow$ and $\mathbf{T}_R\downarrow$ on ropes. (Plus force interactions at hands.) $N = 314$ N

3.6. $\mathbf{T}_{AB}$ $\measuredangle$ 45° (at A), $\mathbf{C}_x\leftarrow$ and $\mathbf{C}_y\uparrow$ (at C), 30 lb↓ (at midpoint of bar)

3.8. 200 N $\measuredangle$ 30° (at hands), $\mathbf{N}\uparrow$ and $\mathbf{f}\leftarrow$ (on CD), 785 N↓ (at midpoint of crate)

3.10. 800 lb → and $\mathbf{T}$ $\measuredangle$ 30° (at B), $\mathbf{T}\uparrow$ (at D), $\mathbf{A}_x\leftarrow$ and $\mathbf{A}_y\uparrow$ (at A)

3.12. $A_x\mathbf{i}$ (at A), $\mathbf{T}_{CD}$ and $-490\mathbf{j}$ N (at C), $B_x\mathbf{i} + B_y\mathbf{j} + B_z\mathbf{k}$ (at B)

3.14. On 1: $\mathbf{N}_A\rightarrow$ (at A), $\mathbf{N}_B$ ↘ 30° (at B), $\mathbf{N}_D$ ↗ 30° (at D), 100 N↓ (at midpoint)
On 2: $\mathbf{N}_D$ $\measuredangle$ 30° (at D), $\mathbf{N}_E$ ↘ 30° (at E), 100 N↓ (at midpoint)

3.16. On 1: $\mathbf{A}_x\rightarrow$ and $\mathbf{A}_y\uparrow$ (at A), 100 lb↓ (at D), $\mathbf{B}_x\rightarrow$ and $\mathbf{B}_y\uparrow$ (at B)
On 2: $\mathbf{C}_x\rightarrow$ and $\mathbf{C}_y\uparrow$ (at C), $\mathbf{B}_x\leftarrow$ and $\mathbf{B}_y\downarrow$ (at B), 200 lb $\measuredangle$ 45° at E

3.18. 2.52 kips

3.20. $\mathbf{N}_A = 410$ N →, $\mathbf{N}_B = 640$ N ↘ 40°

3.22. 1,200 lb↓ and 4,800 lb · in. ↺

3.24. 0.77 m

3.26. $\mathbf{N}_A = 640$ lb↑, $\mathbf{N}_B = 160$ lb↑; $\mathbf{P} = 47$ lb ←

3.28. $T = \sqrt{3}\ W/2$, $\mathbf{A} = W/2$ ↘ 30°

3.30. $T = 223$ N; $\mathbf{O}_x = 106$ N ←, $\mathbf{O}_y = 278$ N↑

3.32. 35.4°

3.34. (a) $\mathbf{A} = 11.0$ kN↓, $\mathbf{B} = 13.5$ kN↑; (b) 2.7 m

3.36. $\sqrt{2}$ kN · m

3.38. 13.4 in.

3.40. $0.264\ W$

3.42. $T_{AB} = 4.81$ kips, $T_{AC} = 4.44$ kips, $T_{AD} = 1.62$ kips

3.44. $T = [(F^2 + W^2)/3]^{1/2}$, $h = 10 - 4[3W^2/(F^2 + W^2)]^{1/2}$ m; W = container weight

3.46. $T_A = 545$ lb, $T_B = 340$ lb, $T_C = 615$ lb

3.48. $\mathbf{A} = 150\mathbf{j} - 258\mathbf{k}$ N, $\mathbf{B} = -50\mathbf{j} + 86\mathbf{k}$ N

3.50. $\mathbf{O} = 3.8\mathbf{i} + 5.7\mathbf{k}$ kN, $T_{AB} = 6.6$ kN, $T_{BD} = 5.0$ kN

3.52. $\mathbf{A} = 2.6\mathbf{i} - 1.0\mathbf{k}$ kN and $\mathbf{C}_A = 4\mathbf{j}$ kN · m, $T_{AB} = 3.3$ kN

3.54. $\mathbf{O} = 160\mathbf{j} + 170\mathbf{k}$ lb, $\mathbf{C}_O = 1{,}530\mathbf{i} - 340\mathbf{j} + 80\mathbf{k}$ lb · ft

3.56. $\mathbf{O} = mg\,\mathbf{j}$, $\mathbf{C}_O = mgL_2 \sin 40°\mathbf{i} + (\sqrt{3}\ mg/2)(L_1 + L_2 \cos 40°)\mathbf{k}$

3.58. $M = 33.6$ kg; $\mathbf{A} = 460\mathbf{i} + 246\mathbf{j} - 123\mathbf{k}$ N (assuming A supports all thrust load), $\mathbf{B} = -276\mathbf{i} + 246\mathbf{j}$ N, $T_{CD} = 330$ N

3.60. yes; $\mathbf{A} = -200\mathbf{j}$ lb, $\mathbf{C}_A = -1{,}600\mathbf{i}$ lb · ft, $T_B = 200$ lb

3.62. $P = 654$ N, $Q = 286$ N, $T_{BC} = 817$ N

3.64. 1 kip

3.66. $0.1\ W$

3.68. $\mathbf{A}_x = 118$ N $\leftarrow$, $\mathbf{A}_y = 245$ N$\uparrow$, $\mathbf{B}_x = 118$ N $\rightarrow$, $\mathbf{B}_y = 245$ N$\uparrow$

3.70. $P = 69$ lb; $P/F = 8.64$

3.72. $\mathbf{A}_x = 1.37\ W \leftarrow$, $\mathbf{A}_y = -\mathbf{C}_y = 0.5W\uparrow$, $\mathbf{B}_x = 5.24\ W \rightarrow$, $\mathbf{C}_x = 3.87\ W \leftarrow$

3.74. $\mathbf{A}_x = 1.30$ kN $\leftarrow$, $\mathbf{A}_y = 0.75$ kN$\uparrow$, $\mathbf{C}_x = 1.30$ kN $\rightarrow$, $\mathbf{C}_y = 5.25$ kN$\uparrow$

3.76. $\mathbf{A}_x = 2{,}500$ lb $\leftarrow$, $\mathbf{A}_y = 800$ lb$\downarrow$, $\mathbf{B}_x = 3{,}500$ lb $\rightarrow$, $\mathbf{B}_y = 1{,}800$ lb$\uparrow$, $\mathbf{C}_x = 1{,}000$ lb $\leftarrow$, $\mathbf{C}_y = 1{,}000$ lb$\downarrow$

3.78. On ABC: $\mathbf{A}_y = 200$ lb$\uparrow$, $\mathbf{B}_x = 150$ lb $\rightarrow$, $\mathbf{B}_y = 200$ lb$\downarrow$, $\mathbf{C}_x = 150$ lb $\leftarrow$
On CDE: $\mathbf{C}_x = 150$ lb $\rightarrow$, $\mathbf{B}_x = 150$ lb $\leftarrow$, $\mathbf{B}_y = 200$ lb$\downarrow$, $\mathbf{E}_y = 200$ lb$\uparrow$
On BD: $\mathbf{B}_x = 150$ lb $\leftarrow$, $\mathbf{B}_y = 200$ lb$\uparrow$, $\mathbf{D}_x = 150$ lb $\rightarrow$, $\mathbf{D}_y = 200$ lb$\uparrow$

3.80. $\mathbf{A} = 7.00$ kips$\uparrow$, $\mathbf{B} = 7.54$ kips$\uparrow$, $\mathbf{C} = \mathbf{D} = 16.73$ kips$\uparrow$

3.82. $\mathbf{C} = 76$ N · m$\circlearrowright$, $\mathbf{F}_{AB} = 2.9$ kN ↖ 48°

3.84. $2\sqrt{3}\ W/9$

3.86. $F_{AB} = F_{BC} = 2.31$ kN (T), $F_{AD} = 1.16$ kN (C), $F_{BD} = 2.31$ kN (C), $F_{CD} = 4.62$ kN (C)

3.88. $F_{AB} = 542$ lb (C), $F_{AD} = F_{CD} = 1{,}469$ lb (T), $F_{BD} = 0$, $F_{BC} = 1{,}696$ lb (C); $e_{AB} = -21.8\%$, $e_{AD} = e_{CD} = e_{BC} = -6.9\%$

3.90. $F_{DF} = 20.0$ kN (T), $F_{EC} = 37.7$ kN (C), $F_{EG} = 21.5$ kN (C)

3.92. $F_{FG} = 1.2$ kips (T), $F_{CF} = 11.6$ kips (C), $F_{DF} = 4.6$ kips (C), $F_{EF} = 2.3$ kips (C)

3.94. Zero force members: GH, IF, EJ; $F_{FH} = 25.5$ kN (T), $F_{IJ} = 18.0$ kN (C)

3.96. $F_{DC} = 10.0$ kN (C), $F_{DE} = 0$, $F_{EC} = 6.0$ kN (C), $F_{EF} = 8.0$ kN (C), $F_{CF} = 9.3$ kN (T), $F_{BF} = 11.6$ kN (C), $F_{FG} = 8.6$ kN (C)

3.98. $F_{CD} = 1{,}620$ lb (C), $F_{CJ} = F_{DJ} = 0$, $F_{KJ} = 1{,}620$ (T)

3.100. $F_{FG} = F_{GH} = F_{GJ} = 0$, $F_{CG} = F_{GK} = 3{,}080$ lb (C)

3.102. (a) 14.46 kN; (b) 13.10 kN

3.104. $100 \text{ lb} \leqslant W \leqslant 400$ lb

3.106. $F = \mu_s W/\sin\alpha$

3.108. 760 N · m

3.110. (a) 23°; (b) 25°; (c) 42°

3.112. 0.47

3.114. equilibrium; $\mathbf{f}_B = 6.01$ kips $\leftarrow$

3.116. 485 N; A and B move as a unit

3.118. 1.06

3.120. $L[1 + 6\mu_s(1 + 9\mu_s^2)^{-1/2}]$

4

4.2. not equivalent: add $\mathbf{C} = 15a$ ↺

4.4. $\mathbf{F} = \mathbf{0}$, $\mathbf{C} = 6$ N · m ↺

4.6. $\mathbf{F} = -2\mathbf{k}$ kips, $\mathbf{C} = -1.00\mathbf{i} - 3.46\mathbf{j}$ kip · in.; (a) $-1.00\mathbf{i}$ kip · in.; (b) $-3.46\mathbf{j}$ kip · in.; no

4.8. $\mathbf{F} = 2.684T\mathbf{i} + 0.025T\mathbf{j} - 0.043T\mathbf{k}$, $\mathbf{C} = -0.283T\mathbf{j} - 0.163T\mathbf{k}$; yaw and pitch, no roll

4.10. $\mathbf{F} = 80$ lb →, 3 ft above AD

4.12. $\mathbf{R} = 2\mathbf{j}$ kN, $\mathbf{C}^R = 2.0\mathbf{i} + 1.2\mathbf{k}$ kN · m; yes

4.14. $\mathbf{R}_x = 76.6$ N →, $\mathbf{R}_y = 18.7$ N↓ (at intersection of forces)

4.16. $\mathbf{R} = 6$ kN↓, 4.67 m from left end

4.18. $\mathbf{C}^R = 180$ lb · in. ↻

4.20. $F = 180$N, $\theta = 56.3°$; $\mathbf{C}^R = 108$ N · m ↻

4.22. $F_1 = 10.0$ kips, $F_2 = 12.5$ kips, $F_3 = 1.5$ kips; $\mathbf{C}^R = 22$ kip · ft ↻

4.24. $\mathbf{R} = 300$ lb↓, at $x = 8.33$ ft, $y = 3.67$ ft

4.26. (a) 32.12 ft/s^2; (b) 29.11 ft/s^2

4.28. $F = 146 \times 10^{-9}$ N, $F/W = 3.17 \times 10^{-9}$

4.30. $\bar{x} = 0.500\ L$, $\bar{y} = 0.289L$

4.32. $\bar{x} = 3a$, $\bar{y} = 0$

4.34. 60°

4.36. 7.34 ton, 2.0 m

4.38. $\bar{x} = 5.22$ ft, $\bar{y} = 1.38$ ft

4.40. $\bar{x} = 0.675$ m, $\bar{y} = 1.200$ m, 190 N↑

4.42. $\bar{x} = 3.0$ in., $\bar{y} = 1.2$ in.

4.44. $\bar{x} = L/2$, $\bar{y} = \pi A/8$

4.46. $\bar{x} = 0.657$ m, $\bar{y} = 0.560$ m

4.48. $\bar{x} = 2.00$, $\bar{y} = 1.60$

4.50. $\bar{x} = 0.630$ in., $\bar{y} = 0.410$ in.

4.52. $h/3$ from base

4.54. $\bar{x} = 3.33$ in., $\bar{y} = \bar{z} = 0$

4.56. $\bar{x} = 0.637\ R$, $\bar{y} = 0.171\ R$, $\bar{z} = L/2$

4.58. $A = \pi R(R^2 + h^2)^{1/2}$, $V = \pi R^2 h/3$

4.60. 5.08 ft^2

4.62. $\bar{x} = 0$, $\bar{y} = 4.21$ in.

4.64. $\bar{x} = 40.0$ mm, $\bar{y} = 19.3$ mm

4.66. $\bar{x} = 0.562$ m, $\bar{y} = -0.300$ m

4.68. $\bar{x} = 0$, $\bar{y} = 2.25$ in.

4.70. $\bar{x} = 0$, $\bar{y} = 0.097$ in., $\bar{z} = -0.162$ in.

4.72. 1.65 in.

4.74. $\bar{x} = 7.12$ in., $\bar{y} = 2.37$ in.

4.76. $\bar{x} = -22.0$ mm, $\bar{y} = 77.8$ mm, $\bar{z} = 75.0$ mm

4.78. G: ($\bar{x} = 0$, $\bar{y} = -0.57$ in., $\bar{z} = -4.50$ in.), C: ($\bar{x} = 0$, $\bar{y} = -1.25$ in., $\bar{z} = -3.91$ in.)

4.80. 3 kg, 3 m

4.82. $0.156L$

4.84. 39 gal.

4.86. $\mathbf{R}_A = 1.33$ kN ↑, $\mathbf{R}_B = 1.67$ kN ↑

4.88. $\mathbf{R}_B = 24$ kN ↑, $\mathbf{M}_B = 12$ kN · m ↻

4.90. $\mathbf{A}_x = 600$ lb →, $\mathbf{A}_y = 450$ lb ↑, $\mathbf{D}_x = 300$ lb →, $\mathbf{D}_y = 450$ lb ↓, $F_{AC} = 750$ lb (C)

4.92. 24.4 kN/m^2

4.94. overflow (assuming dam doesn't slide)

4.96. 13.2 lb/ft, 2.62 in. above bottom

4.98. $R = \gamma abh$, $d = d_c + b^2/(12d_c)$

4.100. 111 kN/m, 1.88 m above bottom

4.102. 14.84 kips

4.104. $R = (4/15)a\gamma h^2$, at $x = 0$, $y = 3h/7$

4.106. 55.5 mN

4.108. $T/W = 0$ for $(0 \leqslant d/a \leqslant 0.65)$; $T/W = (2/3)[3(d/a)^2 - (d/a)^3 - 1]$ for $(0.65 \leqslant d/a \leqslant 1)$; $T/W = 2/3$ for $(d/a \geqslant 1)$

4.110. $I_x = 4.27$, $I_y = 19.50$, $J_z = 23.77$

4.112. $I_x = I_y = \pi R^4/16$, $J_z = \pi R^4/8$

4.114. $I_a = 72$ mm^4, $b = 4$ mm

4.116. $I_x = I_y = 0.137a^4$

4.118. $\bar{y} = 6$ in., $\bar{I}_x = 136$ in.^{4}

4.120. 288×10^4 mm^4

4.122. $\bar{I}_x = 370$ in.^{4}, $\bar{I}_y = 106$ in.^{4}

4.124. $(8.36 \times 10^4)t$ mm^4 (t in mm)

5

5.2. On a–D: **P** = 2 kips → ; On b–D: **P** = 6 kips → ; On c–D: **P** = 2 kips ←

5.4. On A–a: **P** = 4 kN ←, **V** = 2 kN ↑, **M** = 0.4 kN · m ↻;
Section b–b: $P = 5.66$ kN (T)

5.6. On a–B: **T** = 300 lb · in. ↠ ; On b–B: **T** = 700 lb · in. ↞

5.8. On A–a: **V** = 4 kN ↓, **M** = 0.8 kN · m ↻; On b–B: **V** = **M** = **O**

5.10. On a–B: **P** = **O**, **V** = 150**j** + 100**k** N, **T** = −30**i** N · m, **M** = −17.0**j** + 25.5**k** N · m; On b–B: **P** = **0**, **V** = 150**j** + 100**k** N, **T** = −15**i** N · m, **M** = −6**j** + 9**k** N · m

5.12. 45.84 MN/m^2

5.14. $d_{AD} = 0.296$ in., $d_{CD} = 0.197$ in.

5.16. (a) $F = 22{,}050$ lb; (b) $F = 22{,}035$ lb

5.18. $F = 9.3$ kN

5.20. 63.4°

5.22. $A_{CG} = 337\ mm^2$, $A_{GH} = 96\ mm^2$

5.24. 1.02 $in.^2$; 2 in.

5.26. 600 N or 1 kN

5.28. $\sigma = 55(y + 20) - 407{,}000/(y + 20)^2\ lb/ft^2$

5.30. 39 MN/m^2

5.32. $d = 0.142$ in.

5.34. $a = 20$ mm, $b = 100$ mm

5.36. $\tau_{avg} = 11.3$ ksi, $\sigma_{bearing} = 13.3$ ksi; 6 years

5.38. (a) 12.73 ksi; (b) 8 ksi; (c) 20 ksi

5.40. $P = 34.2$ kN

5.42. 33%

5.44. 4

5.46. 0.57

5.48. 0.45%

5.50. 0.152

5.52. $\varepsilon_{AB} = 333\ \mu\varepsilon$, $\varepsilon_{BC} = -500\ \mu\varepsilon$, $\varepsilon_{AC} = 0$

5.54. $\gamma_{xy} = 0.0524$

5.56. (a) $3ay^2/h^3$; (b) 0; (c) $3a/(4h)$

5.58. $\varepsilon_x = 1{,}000\ \mu\varepsilon$, $\varepsilon_y = 2{,}000\ \mu\varepsilon$, $\gamma_{xy} = -1{,}000\ \mu\varepsilon$

5.60. $\varepsilon_n = \varepsilon_t = 0.010$, $\gamma_{nt} = 0.023$

5.62. $E = 5.0$ ksi, $\nu = 0.33$

5.64. 18 ksi

5.66. 27.07 GN/m^2

5.68. Cast iron: 0.9%; Steel: 26%

5.70. (a) 400 MN/m^2; (b) 100 GN/m^2; (c) 540 MN/m^2; (d) 520 MN/m^2; 253.5 mm; 157 kN

5.72. 24.06 in.

5.74. $K = 77.2$ ksi, $n = 0.286$

5.76. $\varepsilon_x = -590\ \mu\varepsilon$, $\varepsilon_y = 425\ \mu\varepsilon$, $\gamma_{xy} = 1{,}333\ \mu\varepsilon$

5.78. 7.5×10^{-5} in.

5.80. $\sigma_x/\sigma_y = (1 + 1.5\nu)/(1.5 + \nu)$

5.82. 7.5 kips; $\delta_y = -3.28 \times 10^{-3}$ in., $\delta_z = 1.56 \times 10^{-3}$ in.

5.84. $\sigma_y = -\nu\sigma$, $\delta_z = \nu(1 + \nu)\sigma t/E$

5.86. $2E_A\nu_B p/(E_A + E_B)$ ↳ $\varepsilon_z = -\frac{\nu(\nu-1)}{E}\sigma$

6

6.2. 30 ft

6.4. 0.821 mm

6.6. 1.67 in.^{2}

6.8. $\delta_{AD} = 4.72$ mm; $\sigma_{max} = 173.3$ MN/m^2 (T) (section BC)

6.10. 84 kN; -2.90 mm

6.12. 0.064 in.

6.14. (a) 0.152 mm; (b) 0.188 mm

6.16. W 8 $\times$ 48

6.18. $F = 2.4$ kN, $\delta = 0.980$ mm

6.20. 16.5 kN

6.22. $A = 0.2$ in.^{2}; $d_{pin} = 1.0$ in.

6.24. $F_{st} = 58.9$ kN, $F_{al} = 29.4$ kN

6.26. $\sigma_1 = 20.4$ ksi, $\sigma_2 = 24.4$ ksi, $\sigma_3 = 16.4$ ksi; $\delta = 0.491$ in.

6.28. 172.5 kips

6.30. $\sigma_{br} = 26.57$ MN/m^2 (C), $\sigma_{st} = 41.52$ MN/m^2 (T)

6.32. 114.4 MN/m^2

6.34. $\sigma_{AB} = 168$ MN/m^2 (T), $\sigma_{BC} = 207$ MN/m^2 (C)

6.36. No

6.38. 17.3 ft

6.40. $\delta_L = 0.037$ mm, $\delta_D = -0.023$ mm; $\sigma_{max} = 102$ MN/m^2

6.42. $\sigma_{max} = 52.7$ MN/m^2 (C); 0.098 mm (to the right)

6.44. 1.23 in. $\leqslant d \leqslant$ 2.41 in.

6.46. $\sigma_{st} = 666$ psi (T), $\sigma_{zinc} = 7{,}200$ psi (T)

6.48. (a) $F_{wood} = 7.24$ kN, $F_{con} = 12.08$ kN; (b) $F_{wood} = 0$, $F_{con} = 15.70$ kN

6.50. $\sigma_{st} = 13.50$ MN/m^2 (T), $\sigma_{al} = 27.00$ MN/m^2 (C); $\tau_{rivet} = 35.79$ MN/m^2

6.52. (a) 1.60; (b) 1.60; (c) 1.26

6.54. (a) 13.4 kips; (b) 17.5 kips; (c) 10.0 kips

6.56. 50.7 mm

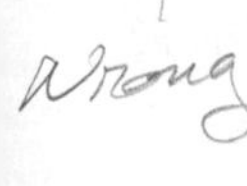

6.58. $d_{bolt} = 1\frac{1}{4}$ in., $t = \frac{1}{2}$ in., $w = 2\frac{1}{2}$ in., $s \geqslant 1$ in.

6.60. 6 mm

6.62. $F_{limit} = 40.8$ kN

6.64. $F_{limit} = 17.2$ kips

6.66. $F_{limit} = 800$ kips

6.68. $F_{limit} = 378$ kN

6.70. $F_{limit} = 1.4\sigma_{YS}A$

6.72. 47.1 kips

6.74. 26.7 mm

6.76. 32.3°C; 583 kN/m^2
6.78. $\sigma = 26.25$ ksi, $p = 3.28$ ksi
6.80. (a) 37.2 N; (b) 3.1 kN/m^2
6.82. $F = 880$ N; $\sigma = 6.67$ MN/m^2
6.84. $\delta_L = (\nu^2 - 1)\sigma L/E$, $p = \nu\sigma t/R$

7

7.2. 70 hp
7.4. 7.54 MN/m^2
7.6. (a) 306 N · m; (b) 200 N · m
7.8. 5.1 ksi (section BC)
7.10. 2.3 MW
7.12. 10.18 kip · in.
7.14. $d \leqslant 5.27$ mm
7.16. wood shaft
7.18. $\sigma_\theta = 5.2$ ksi, $\tau_\theta = 3.0$ ksi
7.20. 22.8°
7.22. 14.4°
7.24. $d_i = 65.7$ mm; $W_H/W_S = 0.36$, $\tau_H/\tau_S = 1.25$
7.26. (a) $L_1 = 8.6$ in., $L_2 = 51.4$ in.; (b) 15.3 ksi
7.28. (a) 0.2° CCW; (b) 6.8° CW (as viewed from right end)
7.30. 0.983 m; no
7.32. $C/2$ ↞ (at each end)
7.34. 34.15 kip · in.
7.36. $F_{max} = 2.18$ kN; $\theta_{AB} = 8.3°$ CCW, $\theta_{CD} = 6.2°$ CW
7.38. 14.3%; 21.4% reduction in AB, none in BC
7.40. $\tau_A = T\rho G_A/(G_A J_A + G_B J_B)$, $\tau_B = T\rho G_B/(G_A J_A + G_B J_B)$, $\theta = TL/(G_A J_A + G_B J_B)$
7.42. $(T_{fp})_H/(T_{fp})_S = (T_U)_H/(T_U)_S = 0.875$
7.44. (a) 3.51 kip · in.; (b) 13.09 kip · in.; (c) 19.64 kip · in.
7.46. $\theta = 11.46°$, $T = 23.38$ kN · m
7.48. 204.5 kip · in.
7.50. $\rho_E = 0.508$ in., $\theta = 28.9°$
7.52. 15.31 kN · m
7.54. $C_{limit} = 20.32$ kN · m
7.56. $C_{limit} = 55.76$ kip · in.
7.58. 3.26 kN

8

8.2. $V = -5x$ kN, $M = -2.5x^2$ kN · m

8.4. $(0 \leqslant x \leqslant a)$: $V = 4F/3$, $M = 4Fx/3$; $(a \leqslant x \leqslant 2a)$: $V = F/3$, $M = F(x/3 + a)$; $(2a \leqslant x \leqslant 3a)$: $V = -5F/3$, $M = 5F(a - x/3)$

8.6. $V = 2/3 - x^2/2$ kN, $M = 2x/3 - x^3/6$ kN · m

8.8. $(0 \leqslant x \leqslant 2a)$: $V = -F/2$, $M = -Fx/2$; $(2a \leqslant x \leqslant 3a)$: $V = F$, $M = F(x - 3a)$

8.14. $V_{max} = -5qa/4$ (at left of right support), $M_{max} = -qa^2/2$ (at right support)

8.16. $V_{max} = 4.8$ kips (at left support), $M_{max} = 10.8$ kip · ft (at right of applied couple)

8.18. $V_{max} = 6$ kN (at left end), $M_{max} = 1.6$ kN · m (1.2 m from left end)

8.20. $V_{max} = -2.4$ kips (at left of right support), $M_{max} = 2.8$ kip · ft (3.46 ft from left end)

8.22. -6.7%

8.24. $(\sigma_t)_{max} = 140$ MN/m^2, $(\sigma_c)_{max} = 100$ MN/m^2

8.26. vertically

8.28. $L = 4$ m; $\rho = \infty$

8.30. (a) $\sigma_{max} = 4.35$ ksi, $\rho = 84.2$ ft; (b) $\sigma_{max} = 4.53$ ksi, $\rho = 81.0$ ft

8.32. $C = 2kbh^{n+2}R^{-n}/(n + 2)$; $\sigma_{max} = k(h/R)^n$

8.34. $(\sigma_t)_{max} = 6.37$ ksi (4.2 ft from left end—bottom of section), $(\sigma_c)_{max} = 5.78$ ksi (at right support—bottom of section)

8.36. $(\sigma_t)_{max} = 21.3$ MN/m^2 (at left support—top of section), $(\sigma_c)_{max} = 26.7$ MN/m^2 (2.5 m from left end—top of section)

8.38. W 10 × 25

8.40. $\sigma_{max} = (6qL^2/ta^2)(2X - X^2 - 1)/(2 - X)^2$, $X = x/L$

8.42. 3/16 in.

8.44. (a) 360 MN/m^2; (b) 240 MN/m^2

8.46. $(\sigma_S)_{max} = 24.48$ ksi, $(\sigma_A)_{max} = 4.28$ ksi

8.48. 6 in.^{2}

8.50. $\tau_{aa} = 1.45$ MN/m^2, $\tau_{bb} = 5.82$ MN/m^2, $\tau_{cc} = 5.96$ MN/m^2

8.52. 750 psi

8.54. $\sigma_{max}/\tau_{max} = 5.14$

8.56. $d \geqslant 69$ mm

8.58. $\sigma_{max} = 912$ psi, $\tau_{max} = 99$ psi

8.60. 44 mm

8.62. 11.1%; 2.97 in.

8.64. $EIv(x) = (q_0/120L)(x^5 - 5xL^4 + 4L^5)$

8.66. $EIv(x) = (F/144)(-72ax^2 + 38x^3 - x^5/a^2 - 45a^2x)$

8.68. $EIv(x) = (Cx/24L)(4x^2 - L^2) \quad (0 \leqslant x \leqslant L/2)$
$= (C/24L)[4x^3 - 12L(x - L/2)^2 - L^2x] \quad (L/2 \leqslant x \leqslant L)$

8.70. $v_{max} = -7FL^3/(12EI)$ (at $x = 0$)

8.72. $v_{max} = \mp 0.008\ CL^2/EI$ (at $x = 0.289L$ and $0.711L$)
8.74. $d \leqslant 0.607$ mm
8.76. 305×102 (28 kg/m)
8.78. $M_{max} = 8.4$ kip · ft (at wall)
8.80. $EIv(x) = (C/24L)[4x^3 - 12L\langle x - L/2\rangle^2 - L^2x]$
8.82. $EIv(x) = (q/384)[16x^3L - 16\langle x - L/4\rangle^4 + 16\langle x - 3L/4\rangle^4 - 11L^3x]$; $-0.148qL^4/EI$
8.84. $EIv(x) = (q/192)[-8Lx^3 + 8x^4 - 8\langle x - L/2\rangle^4 + L^3x]$; $v_{max} = \pm 5qL^4/(6{,}144EI)$ (at $x = L/4, 3L/4$); $\theta(L/2) = -qL^3/(192\ EI)$
8.86. $v = -1.13$ in., $\theta = -1.24°$
8.88. 4×6 in.; -4.9 in.
8.90. Left end: $3C/2L \downarrow$, $C/4$ ↻; Right end: $3C/2L \uparrow$, $C/4$ ↻
8.92. 1.77; 2.04
8.94. 305×102 (25 kg/m)
8.96. $-$ 6.7 mm
8.98. $\sigma_{beam} = 3.94$ ksi; $\sigma_{wire} = 1.86$ ksi
8.100. $F = qL/4$, $v = -qL^4/(24EI)$
8.102. 17.2 kips
8.104. $\sigma_{YS}d^3/6$
8.106. $0.052\sigma_{YS}b^3$
8.108. 22.5 kN
8.110. $d \geqslant 1.2$ in.
8.112. 1.77 in.
8.114. 66.8 N · m
8.116. $F_{limit} = 6M_{fp}/L$
8.118. $q_{limit} = 28.13$ kip/ft
8.120. $(q_0)_{limit} = 24M_{fp}/L^2$
8.122. $F_{limit} = 2M_{fp}/L$

9

9.2. 0.56 kips
9.4. 15 mm
9.6. 3.2 in.
9.8. $\sigma_A = 9.33\ MN/m^2\ (C)$, $\sigma_B = \sigma_D = 1.33\ MN/m^2\ (C)$, $\sigma_C = 6.67\ MN/m^2\ (T)$
9.10. W 8×31
9.12. Member *ACD*: 203×133 (25 kg/m) or 254×102 (25 kg/m), Member *BEC*: 254×102 (22 kg/m)
9.14. 23.2 ft

9.16. A: 6.8 ksi; B: 0.8 ksi, 2.8 ksi; C: 0.9 ksi, 0.8 ksi

9.18. $\sigma_n = -5.00\ \text{MN/m}^2$, $\sigma_t = 5.00\ \text{MN/m}^2$, $\tau_{nt} = 8.66\ \text{MN/m}^2$

9.20. $\sigma_n = -40\ \text{MN/m}^2$, $\sigma_t = 40\ \text{MN/m}^2$, $\tau_{nt} = 0$

9.22. $\sigma = 10$ ksi, $\tau = 0$

9.24. $12\ \text{MN/m}^2$

9.26. 70 ksi

9.28. $\sigma = -57$ psi, $\tau = 833$ psi

9.30. $\sigma_n = -4$ ksi, $\sigma_t = 4$ ksi, $\tau_{nt} = 0$

9.32. $\sigma_n = -4.10\ \text{MN/m}^2$;, $\sigma_t = -5.90\ \text{MN/m}^2$, $\tau_{nt} = -7.57\ \text{MN/m}^2$

9.34. $\sigma = 5$ ksi, $\tau = 3$ ksi

9.36. $\sigma_x = -2\ \text{MN/m}^2$, $\sigma_y = 14\ \text{MN/m}^2$, $\tau_{xy} = 6\ \text{MN/m}^2$

9.38. $\sigma_x = 23.3$ ksi, $\sigma_y = -17.7$ ksi, $\tau_{xy} = 0$

9.40. $12\ \text{MN/m}^2$

9.42. $p = 50$ psi, $F = 157.9$ kips; $\tau_{nt} = -8.67$ ksi

9.44. $\sigma_1 = -\sigma_2 = |(\tau_{nt})_{\max}| = 44.7\ \text{MN/m}^2$

9.46. $\sigma_1 = 4.16$ ksi, $\sigma_2 = -2.16$ ksi, $|(\tau_{nt})_{\max}| = 3.16$ ksi

9.48. $\sigma_1 = 100\ \text{MN/m}^2$, $\sigma_2 = 0$, $|(\tau_{nt})_{\max}| = 50\ \text{MN/m}^2$

9.50. 1,073 psi

9.52. (a) $\sigma_x = 10\sigma/3$, $\tau_{xy} = 0$, $\tau = 2.31\sigma$; (b) $\sigma_1 = 10\sigma/3$, $\sigma_2 = -2\sigma$

9.54. $9.64\ \text{MN/m}^2$; $10.60\ \text{MN/m}^2$

9.56. $M = 12.57$ kip · in., $T = 9.42$ kip · in.

9.58. Point A: $\sigma_1 = 43.2\ \text{MN/m}^2$, $\sigma_2 = -2.4\ \text{MN/m}^2$, $|(\tau_{nt})_{\max}| = 22.8\ \text{MN/m}^2$; Point B: $\sigma_1 = -\sigma_2 = |(\tau_{nt})_{\max}| = 10.9\ \text{MN/m}^2$

9.60. 17.5 hp

9.62. (a) 27 mm; (b) 20 mm; $W_T/W_A = 0.88$

9.64. 17.5%

9.66. 2.2 mm

9.68. $1\frac{1}{4}$ in.

9.70. fracture; 51 N

9.72. $(P/P_{YS})^2 + (T/T_{YS})^2 = 1$

9.74. $\varepsilon_1 = 600\ \mu\varepsilon$, $\varepsilon_2 = 100\ \mu\varepsilon$, $|(\gamma_{nt})_{\max}| = 500\ \mu\varepsilon$

9.76. $\varepsilon_1 = 800\ \mu\varepsilon$, $\varepsilon_2 = -200\ \mu\varepsilon$, $|(\gamma_{nt})_{\max}| = 1{,}000\ \mu\varepsilon$

9.78. $\varepsilon_1 = 566\ \mu\varepsilon$, $\varepsilon_2 = -866\ \mu\varepsilon$, $|(\gamma_{nt})_{\max}| = 1{,}432\ \mu\varepsilon$

9.80. $\varepsilon_n = 102\ \mu\varepsilon$, $\varepsilon_t = 598\ \mu\varepsilon$, $\gamma_{nt} = 60\ \mu\varepsilon$

9.82. $\theta = 27.0°$, with $\varepsilon_t = -450\ \mu\varepsilon$ and $\gamma_{nt} = 630\ \mu\varepsilon$ or $\theta = -49.6°$, with $\varepsilon_t = -450\ \mu\varepsilon$ and $\gamma_{nt} = -630\ \mu\varepsilon$

9.84. $\varepsilon_x = -0.072$, $\varepsilon_y = 0.032$, $\gamma_{xy} = 0.060$; $|(\gamma_{nt})_{\max}| = 0.120$

9.86. $\sigma_1 = -4.27$ ksi, $\sigma_2 = -10.66$ ksi

9.88. 182 $\mu\varepsilon$

9.90. 2.19 kN · m

9.92. $\varepsilon_x = -410\ \mu\varepsilon$, $\varepsilon_y = 1{,}940\ \mu\varepsilon$, $\gamma_{xy} = 1{,}880\ \mu\varepsilon$

9.94. $\varepsilon_x = -410\ \mu\varepsilon$, $\varepsilon_y = 2{,}170\ \mu\varepsilon$, $\gamma_{xy} = -680\ \mu\varepsilon$

9.96. $\varepsilon_1 = 877\ \mu\varepsilon$, $\varepsilon_2 = -341\ \mu\varepsilon$

9.98. $\sigma_1 = 57.2$ MN/m^2, $\sigma_2 = 7.3$ MN/m^2, $|(\tau_{nt})_{max}| = 25.0$ MN/m^2; $|\tau_{max}| = 28.6$ MN/m^2

9.100. $\sigma_1 = 23.43$ ksi, $\sigma_2 = 3.57$ ksi, $|(\tau_{nt})_{max}| = 9.93$ ksi; $|\tau_{max}| = 11.71$ ksi

9.102. $\sigma_{1,2} = \dfrac{E}{3(1-\nu)}(\varepsilon_a + \varepsilon_b + \varepsilon_c) \pm \dfrac{\sqrt{2}\,E}{3(1+\nu)}[(\varepsilon_a - \varepsilon_b)^2 + (\varepsilon_a - \varepsilon_c)^2 + (\varepsilon_b - \varepsilon_c)^2]^{1/2}$

9.104. $p = 74.3$ kN/m^2, $F = 494$ N (C)

10

10.2. $k_T/(2L)$

10.4. $2kL - W/2$

10.6. (a) $2kL$; (b) $2kL - W$

10.8. $k > W/(4L)$

10.10. (a) 90.7; (b) 45.5; (c) 36.3; (d) 54.4

10.12. $F_{cr} = 22.9$ kN; $L/r = 120$

10.14. 70 in.

10.16. 134 kN

10.18. (a) 21.1°C; (b) 10.9°C

10.20. $SF = 1.24$

10.22. 305 × 165 (40 kg/m)

10.24. 2 in.

10.26. (a) 45.3; (b) 128; (c) 181

10.28. 137°F

10.30. 232 mm^2

10.32. 36 ft

10.34. 10.7 kN; out of plane

10.36. 16.6 kips

10.38. 6.94 kN

10.40. 91 kips

10.42. 31.4 kips

10.44. 56 in.

10.46. 53%

10.48. 0.96 in.

10.50. $3\frac{1}{2}$ in.

10.52. 24.5 ksi

11

11.2. 600 $\mu\varepsilon$

11.4. 2.04

11.6. 120 lb

11.8. 1.96 ohm

11.10. yes

11.12. $-$ 1,400 $\mu\varepsilon$

11.14. 633 $\mu\varepsilon$/kip

11.16. 15.3 $\mu\varepsilon/(\mathrm{N}\cdot\mathrm{m})$; 174 $\mu\varepsilon$/deg

11.18. gages 1 and 2 and gages 3 and 4 in adjacent arms; yes

11.20. 2.6 $\mu\varepsilon$/psi

11.22. $(Ct/2EI)(1 - 3a/L)$

11.24. 1,140 $\mu\varepsilon$

11.26. 0.0025 in.

11.28. 134 N/unit

11.30. 346 $\mu\varepsilon$

11.32. 12 ksi; two gages aligned parallel and perpendicular to crack

11.34. 9.2°

11.36. 79.1°

11.38. $-$ 16.7%

11.40. 4,000 $\mu\varepsilon$

11.42. 500 psi (T)

11.44. 1,980 psi (T)

11.46. $\varepsilon_A = 0.067$, $\varepsilon_C = 0.024$

11.48. 24.1 (kN/m)/fringe

11.50. 600 psi

11.52. 1.6

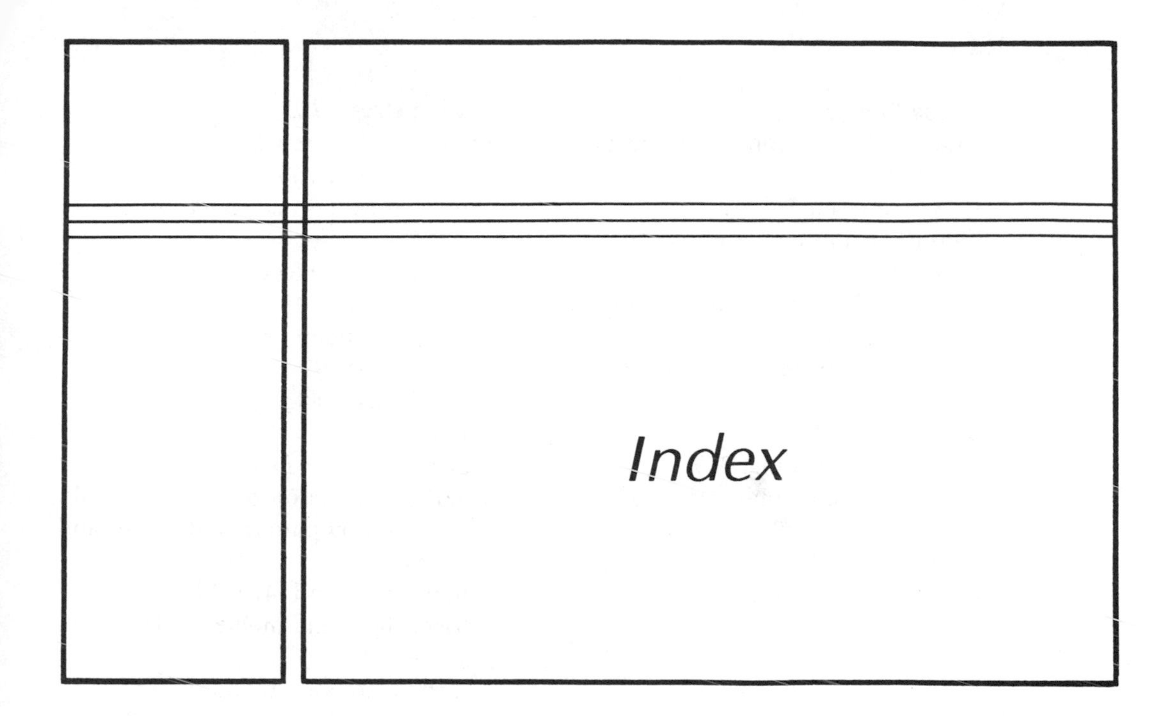

Index

B

C

G

H

I

J

K

L

M

N

O

P

R

S

T

U

V

W

Y

Z

REFERENCES

1. **Rosenberg, E. W. and Fischer, R. W.,** DNCB allergy in the guinea-pig colon, *Arch. Dermatol.*, 89, 159, 1964.
2. **Boughton-Smith, N. K. and Whittle, B. J. R.,** The role of eicosanoids in animal models of inflammatory bowel disease, in *Inflammatory Bowel Disease, Basic Research and Clinical Implications,* Goebell, H., Peskar, B. M., and Malchow, H., Eds., MTP Press, Lancaster, England, 1988, 175.
3. **Bicks, R. O., Azar, M. M., Rosenberg, E. W., Dunham, W. G., and Luther, J.,** Delayed hypersensitivity reactions in the intestinal tract. I. Studies of 2,4-dinitrochlorobenzene caused guinea-pigs and swine colon lesions, *Gastroenterology,* 53, 422, 1967.
4. **Norris, A. A., Lewis, A. J., and Zeitlin, I. J.,** Actions of anticolitic drugs on a guinea-pig model of immune colitis, *Agents Actions,* 12, 239, 1982.
5. **Morris, G. P., Beck, P. L., Herridge, M. S., Depew, W. T., Szewczuk, M. R., and Wallace, J. L.,** Hapten-induced model of chronic inflammation and ulceration in the rat colon, *Gastroenterology,* 96, 795, 1989.
6. **Boughton-Smith, N. K., Wallace, J. L., and Whittle, B. J. R.,** Relationship between arachidonic acid metabolism, myeloperoxide activity and leukocyte infiltration in a rat model of inflammatory bowel disease, *Agents Actions,* 25, 115, 1988.
7. **Pfeiffer, C. J.,** Animal models of colitis, in *Animal Models of Intestinal Disease,* Pfeiffer, C. S., Ed., CRC Press, Boca Raton, FL, 1985, 148.
8. **Wallace, J. L., Whittle, B. J. R., and Boughton-Smith, N. K.,** Prostaglandin protection of rat colonic mucosa from damage induced by ethanol, *Dig. Dis. Sci.,* 30, 866, 1985.
9. **Kirsner, J. B., Elahlepp, J. G., Goldgraber, M. B., Ablaza, J., and Fort, H.,** Production of experimental ulcerative 'colitis' in rabbits, *Arch. Pathol.,* 68, 392, 1959.
10. **Hodgson, H. J. F., Potter, B. J., Skinner, J., and Jewell, D. P.,** Immune-complex mediated colitis in rabbits, *Gut,* 19, 225, 1978.
11. **Mee, A. S., McLaughlin, J. E., Hodgson, H. J. F., and Jewell, D. P.,** Chronic immune colitis in rabbits, *Gut,* 20, 1, 1979.
12. **Marcus, R. and Watt, J.,** Seaweeds and ulcerative colitis in laboratory animals, *Lancet,* 2, 489, 1969.
13. **Watt, J. and Marcus, R.,** Carrageenan induced ulceration of the large intestine in the guinea-pig, *Gut,* 20, 1, 1971.
14. **Abraham, F., Fablan, R. J., Goldberg, L., and Coulston, F.,** Role of lysosomes in carrageenan induced cecal ulceration, *Gastroenterology,* 67, 1169, 1974.
15. **Onderdonk, A. B., Hermos, J. A., Dzink, J. L., and Bartlett, J. G.,** Protective effect of metronidazole in experimental ulcerative colitis, *Gastroenterology,* 74, 521, 1978.
16. **Onderdonk, A. B. and Bartlett, J. G.,** Bacteriological studies of experimental ulcerative colitis, *Am. J. Clin. Nutr.,* 32, 258, 1979.
17. **Watt, J., McLean, C., and Marcus, R.,** Degradation of carrageenan for the experimental production of ulcers in the colon, *J. Pharm. Pharmacol.,* 31, 645, 1979.
18. **Benitz, K. F., Goldberg, L., and Coulston, F.,** Intestinal effects of carrageenans in the rhesus monkey, *Food Cosmetol. Toxicol.,* 11, 565, 1973.
19. **Grasso, P., Sharratt, M., Carpanini, F. M. B., and Gangolli, S. D.,** Studies on carrageenan and large-bowel ulceration in mammals, *Food Cosmetol. Toxicol.,* 11, 555, 1973.
20. **Norris, A. A., Levis, A. J., and Zeitlin, I. J.,** Inability of degraded carrageenan fractions to induce inflammatory bowel ulceration in the guinea-pig, *J. Pharm. Pharmacol.,* 33, 612, 1981.
21. **Dupont, H. L. et al.,** Pathogenesis of *Escherichia coli* diarrhoea, *N. Engl. J. Med.,* 285, 1, 1971.
22. **Halpern, B., Zlnebaum, A., Orial Palou, R., and Morard, J. C.,** Experimental immune ulcerative colitis, in *Immunopathology, 5th Int. Symp. Mechanisms of Inflammation Induced by Immune Reactions,* Grune & Stratton, New York, 1967, 161.
23. **Cooke, E. M., Filipe, M. T., and Dawson, I. M. P.,** The production of colonic autoantibodies in rabbits by immunization with *Escherichia coli, J. Pathol. Bacteriol.,* 96, 125, 1968.
24. **Victor, R. G., Kirsner, J. B., and Palmer, W. L.,** Failure to induce ulcerative colitis experimentally with infiltrates of faeces and rectal mucosa, *Gastroenterology,* 14, 398, 1950.
25. **Kraft, S. C. and Kirsner, J. B.,** Immunological apparatus of the gut and inflammatory bowel disease, *Gastroenterology,* 60, 922, 1971.
26. **Humphrey, C. D., Lushbaugh, W. B., Condon, C. W., Pittman, J. C., and Pittman, F. E.,** Light and electron microscopic studies of antibiotic associated colitis in the hamster, *Gut,* 20, 6, 1979.
27. **Rothman, S. W.,** Presence of *Clostridium difficile* toxin in guinea-pigs with penicillin-associated colitis, *Med. Microbiol. Immunol.,* 169, 187, 1981.
28. **Virchov, R.,** Historisches, Kritisches und Positives zur Lehre der Unterleibsaffektionen, *Virchows Arch. Patt. Anat. Physiol.,* 5, 632, 1953.

29. **Kirsner, J. B.,** Experimental 'colitis' with particular reference to hypersensitivity reactions in the colon, *Gastroenterology,* 40, 307, 1961.
30. **Marston, A., Marcuson, R. W., Chapman, M., and Arthur, J. F.,** Experimental study of devascularization of the colon, *Gut,* 10, 121, 1969.
31. **Madara, J. L., Podolsky, D. K., King, N. W., Sehgan, P. K., Moore, R., and Winter, M. S.,** Characterization of spontaneous colitis in cotton-top tamarins (*Saguinus oedipus*) and its response to sulfasalazine, *Gastroenterology,* 88, 13, 1985.
32. **Kennedy, P. C. and Cello, R. M.,** Colitis of boxer dogs, *Gastroenterology,* 51, 926, 1966.
33. **von Kruiningen, H. J.,** Canine colitis comparable to regional enteritis and mucosal colitis of man, *Gastroenterology,* 62, 1128, 1972.
34. **Flower, R. J.,** Steroidal anti-inflammatory drugs as inhibitors of phospholipase A_2, Samuelsson, B. and Paoletti, R., Eds., as cited in *Adv. Prost. Thrombox. Res.,* 3, 105, Raven Press, New York.
35. **Higgs, G. A., Moncada, S., and Vane, J. R.,** Eicosanoids in inflammation, *Ann. Clin. Res.,* 16, 287, 1984.
36. **Higgs, G. A., Palmer, R. M. J., Eakins, K. E., and Moncada, S.,** Arachidonic acid metabolism as a source of inflammatory mediators and its inhibition as a mechanism of action for anti-inflammatory drugs, *Mol. Aspects Med.,* 4, 275, 1981.
37. **Salmon, J. A. and Flower, R. J.,** Prostaglandins and related compounds, in *Hormones in the Blood,* Vol. 5, Gray, G. H. and James, V. H. T., Eds., Academic Press, London, 1983, 138.
38. **Whittle, B. J. R. and Vane, J. R.,** Prostacyclin, thromboxanes and prostaglandins—actions and roles in the gastrointestinal tract, in Progress in Gastroenterology, Jerzy-Glass, G. B. and Sherlock, P., Eds., Grune & Stratton, New York.
39. **Rask Madsen, J. and Bukhave, K.,** Prostaglandins and chronic diarrhoea: clinical aspects, *Scand. J. Gastroenterol.,* 14(Suppl. 53), 73, 1979.
40. **Rachmilewitz, D.,** Prostaglandins and diarrhoea, *Dig. Dis. Sci.,* 25, 897, 1980.
41. **Samuelsson, B., Borgeat, P., Hammarstrom, S., and Murphy, R. C.,** Leukotrienes: a new group of biologically active compounds, in *Adv. Prost. Thrombox. Res.,* Vol. 6, Samuelsson, B., Ramwell, P. and Paoletti, R., Eds., Raven Press, New York, 1980, 1.
42. **Samuelsson, B.,** Leukotrienes: mediators of immediate hypersensitivity reactions and inflammation, *Science,* 220, 568, 1983.
43. **Rampton, D. S. and Hawkey, C. J.,** Prostaglandins and ulcerative colitis, *Gut,* 25, 1399, 1984.
44. **Norris, A. A., Lewis, A. J., and Zeitlin, I. J.,** Changes in colonic tissue levels of inflammatory mediators in a guinea-pig model of immune colitis, *Agents Actions,* 12, 243, 1982.
45. **Boughton-Smith, N. K. and Whittle, B. J. R.,** Increased metabolism of arachidonic acid in an immune model of colitis in guinea-pigs, *Br. J. Pharmacol.,* 86, 439, 1985.
46. **Hoult, J. R. S., Moore, P. K., Marcus, A. J., and Watt, J.,** On the effect of sulphasalazine on the prostaglandin system and the defective inactivation observed in experimental ulcerative colitis, *Agents Actions,* 4, 232, 1979.
47. **Boughton-Smith, N. K., Wallace, J. L., Morris, G. P., and Whittle, B. J. R.,** The effect of anti-inflammatory drugs on eicosanoid formation in a chronic model of inflammatory bowel disease in the rat, *Br. J. Pharmacol.,* 94, 65, 1988.
48. **Wallace, J. L., Morris, G. P., and MacNaughton, W. K.,** Evaluation of the role of leukotrienes as mediators of colonic inflammation and ulceration in an animal model, in *Inflammatory Bowel Disease: Current Status and Future Approach,* MacDermot, R. P., Ed., Elsevier, New York, 1988, 279.
49. **Wallace, J. L., MacNaughton, W. K., Morris, G. P., and Beck, P. L.,** Inhibition of leukotriene synthesis markedly accelerates healing in a rat model of inflammatory bowel disease, *Gastroenterology,* 96, 29, 1989.
50. **Allgayer, H., Deschryver, K., and Stenson, W. F.,** Treatment with 16,16 dimethyl prostaglandin E_2 before and after induction of colitis with trinitrobenzene sulfonic acid in rats decreases inflammation, *Gastroenterology,* 96, 1290, 1989.
51. **Sharon, P. and Stenson, W. F.,** Metabolism of arachidonic acid in acetic acid colitis in rats. Similarity to human inflammatory bowel disease, *Gastroenterology,* 88, 55, 1985.
52. **Moqbel, R., King, S. J., MacDonald, A. J., Miller, H. R. P., Cromwell, O., Shaw, R. J., and Kay, B.,** Enteral and systemic release of leukotrienes during anaphylaxis of (Holies) *Nippostrongylus brasiliensis* primed rats, *J. Immunol.,* 137, 296, 1986.
53. **Wolbling, R. H., Aehringhaus, U., Peskar, B. A., Morgenroth, K., and Peskar, B. M.,** Release of slow-reacting substance of anaphylaxis and leukotriene C_4-like immunoreactivity from guinea-pig colonic tissue, *Prostaglandins,* 25, 809, 1983.
54. **Zipser, R. D., Patterson, J. B., Kao, H. W., Hauser, C. J., and Lecke, R.,** Hypersensitive prostaglandin and thromboxane response to hormones in rabbit colitis, *Am. J. Physiol.,* 249, G457, 1985.
55. **Zipser, R. D., Nast, C. C., Lee, M., Kao, H. W., and Duke, R.,** *In vivo* production of leukotriene B_4 and leukotriene C_4 in rabbit colitis. Relationship to inflammation, *Gastroenterology,* 92, 33, 1987.

Chapter 9

THE PATHOLOGY OF ULCERATIVE COLITIS

A. Brian West and Jay A. Gates

TABLE OF CONTENTS

I. INTRODUCTION

While the diagnosis of ulcerative colitis is usually made clinically and takes into account history, symptoms, clinical findings, endoscopic appearances, and radiologic studies, the pathologic features of this disease are discrete. It is an inflammatory process directed against the colonic mucosa, in which the crypts show epithelial injury and inflammation, leading to crypt abscess formation when neutrophils aggregate in the lumen. This injury, which is focused mainly on the lower one third of the crypts, is associated with histologic exhaustion of the mucus-secreting cells and focal ulceration and results in severe protracted blood-stained mucoid diarrhea, which is the common presenting feature of ulcerative colitis. It is a process with a remarkably variable natural history: at one extreme it may present as an acute fulminant colitis requiring colectomy; however, much more commonly it is a relatively mild, protracted process characterized by exacerbations and remissions over decades. Whether or not true complete recovery is ever seen is an open question.[1]

The inflammatory process is virtually exclusive to the colorectum, rarely extending into the terminal ileum and never further than approximately 25 cm, despite the many similarities of small and large bowel mucosa and regardless of the severity of colonic involvement. The affected area is continuous without intervening normal mucosa. Usually it starts in the rectum and is most severe there, extending a variable distance proximally toward the ileocecal valve. Treatment, however, may alter this; rectal suppositories and antiinflammatory enemas may down regulate the inflammatory process in the distal large bowel so that the disease appears more active proximally. Moreover, in a small proportion of cases, probably less than 5%, there is apparent (endoscopic or radiologic) rectal sparing; however, biopsies from the "normal" rectal mucosa are usually abnormal and compatible with ulcerative colitis.[2] When the disease process is confined to the rectum, as happens in about 30 to 40% of patients,[3] it is often referred to as "ulcerative proctitis"; however, the histologic features are no different from those seen in cases with more extensive bowel involvement.

The risk of colorectal carcinoma in patients with long standing ulcerative colitis is significantly higher than in noncolitics, and is related to the duration and extent of disease. In a series of 396 children, cancer developed in 3% by 10 years, 23% by 20 years, and 43% by 35 years after onset of ulcerative colitis.[4] In a more recent study of a group of 99 patients, the annual probability that a patient free of cancer or high-grade dysplasia at the beginning of a year would develop such a lesion during that year (the hazard rate) was 2.5% at 20 years after the onset of colitis, 7% at 30 years, and 20% at 40 years.[5] The risk of developing cancer is not obviously related to disease activity. When this risk was first recognized, prophylactic colectomy for patients with long-standing ulcerative colitis became commonplace and provided abundant material for study from colons with mild active and quiescent colitis in addition to those resected for severe and fulminant disease and for cancer. Fiberoptic endoscopes have since extended the opportunities for obtaining biopsies from within the sigmoid and rectum to the entire large bowel. The development of endoscopic surveillance with multiple biopsies as a means of detecting early neoplastic changes, in addition to monitoring severity of disease and response to therapy, has resulted in a need for general as well as specialist anatomic pathologists to be familiar with the mucosal changes seen in ulcerative colitis.

Ulcerative colitis affects the sexes equally. It presents most commonly about the third decade, with a second peak in the seventh or eighth. It may also present in childhood: in parts of the U.S. it is more common than Crohn's disease in children under 5 years old;[6] and in Scottish 6 to 16 year olds the incidence is about 15 to 20 per million.[7] The pathologic features in children are similar to those seen in adults, though fulminant colitis appears to be less frequent.

The main extraintestinal manifestations and complications of ulcerative colitis are listed in Table 1.[8]

TABLE 1
Complications and Extraintestinal Manifestations of Ulcerative Colitis

Local
Anal and perianal excoriation, superficial fissures, abscesses
Toxic megacolon, perforation, peritonitis
Pseudopolyposis
Colorectal carcinoma
Carcinoid tumor, malignant lymphoma
Distant
Primary sclerosing cholangitis and cirrhosis
Cholangiocarcinoma
Hepatic steatosis
Thromboembolism
Erythema nodosum, pyoderma gangrenosum
Conjunctivitis, iritis, episcleritis
Arthritis, sacroileitis, ankylosing spondylitis
Peripheral neuropathy
Wegener's granulomatosis

II. PATHOLOGIC FEATURES

A. ACTIVE CHRONIC ULCERATIVE COLITIS

The large bowel affected by active chronic ulcerative colitis is typically shortened, with a soft, pliable, mildly thickened wall due in part to hypertrophy of the muscularis propria. Mural fibrosis, strictures, and fistulas are not features of this disease, nor is "creeping fat" (extension of adipose tissue around the serosa from the mesenteric toward the antimesenteric border). The serosa usually is smooth and without inflammation and adhesions. The mucosa is boggy and congested with loss of haustral folds and often a marked nodularity (Figure 1). Erosions and shallow ulcers may be covered with inflammatory exudate or reepithelialized. Crypt abscesses rupture and become confluent at the level of the crypt bases, dissecting off the superficial mucosa from just above the muscularis mucosa. The resulting large, flat, deep-mucosal ulcers are separated by the intervening inflamed nonulcerated areas which stand proud of them, appearing as inflamed pseudopolyps (Figure 2). These structures are usually nodular or polyploid and may be markedly inflamed. Occasionally, in long-standing disease, they are filiform, 2 to 3 mm in diameter and up to 3 cm long, simple or branched. Mucosal bridges may form by fusion of polyps to one another or to adjacent mucosa.[9,10] Depending upon the extent and severity of the disease in the past there may be varying mucosal destruction, the final result being flat, finely granular, atrophic mucosa. Occasionally, regeneration leads to the development of coarse villi which have a velvety gross appearance and on microscopic examination resemble a villous adenoma, especially when dysplasia is present.[11]

Ulcerative colitis tends to be most severe distally, and in a resected specimen the full range of activity may be seen in a continuous gradation from severe active disease to normal mucosa. The appendix is involved in about half the cases of pan-colitis, but the inflammation, as elsewhere in the bowel, is confined to the mucosa.[12] Even when the entire colorectum is involved the disease usually ends abruptly at the ileocecal valve. If the terminal ileum is affected, it is in continuity with the ileocecal valve, extending proximally for up to 25 cm. This is usually attributed to reflux of cecal contents through an inflamed, patulous, incompetent ileocecal valve, and is termed "back-wash ileitis".

Under low-power microscopy the characteristic features of ulcerative colitis are markedly inflamed, congested mucosa with expansion of the lamina propria, shallow ulcers which infrequently extend into the submucosa, inflammation of the submucosa beneath ulcers, and mild hypertrophy of the muscularis propria (Figure 2). Lymphoglandular complexes are reactive and the number per unit area of mucosa is increased.[13] Where there has been ulceration, fibrosis may be present in the submucosa; however, transmural inflammation and

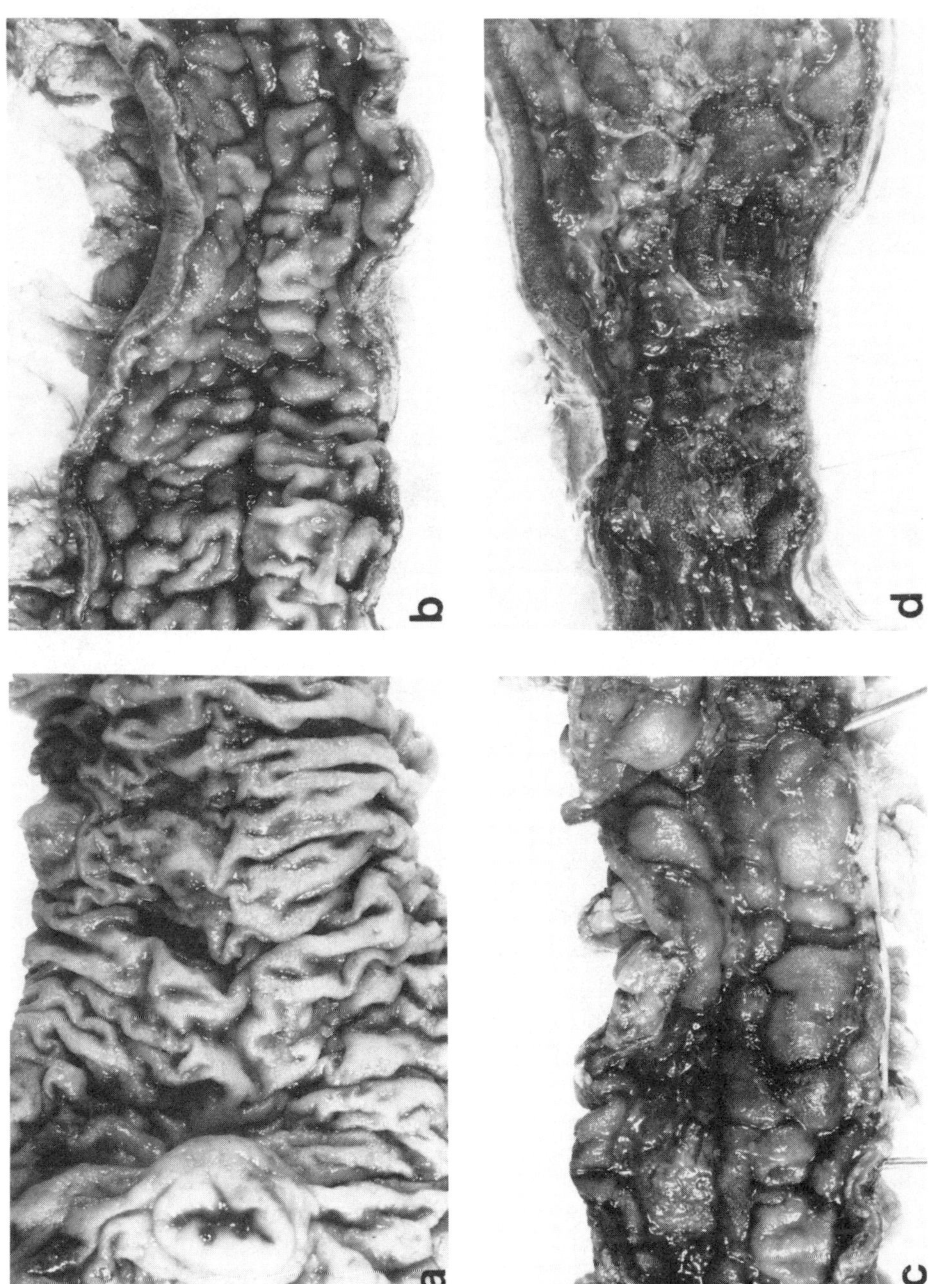

FIGURE 1. Total proctocolectomy specimen resected because of long-standing active ulcerative colitis. In the four segments of the specimen illustrated, mucosal abnormalities increase from minimal in the cecum (a; ileocecal valve on left), through swollen, inflamed, congested, nodularity in the transverse (b) and descending colon (c) to severe ulceration and loss of normal contours in the rectum (d; anus on right).

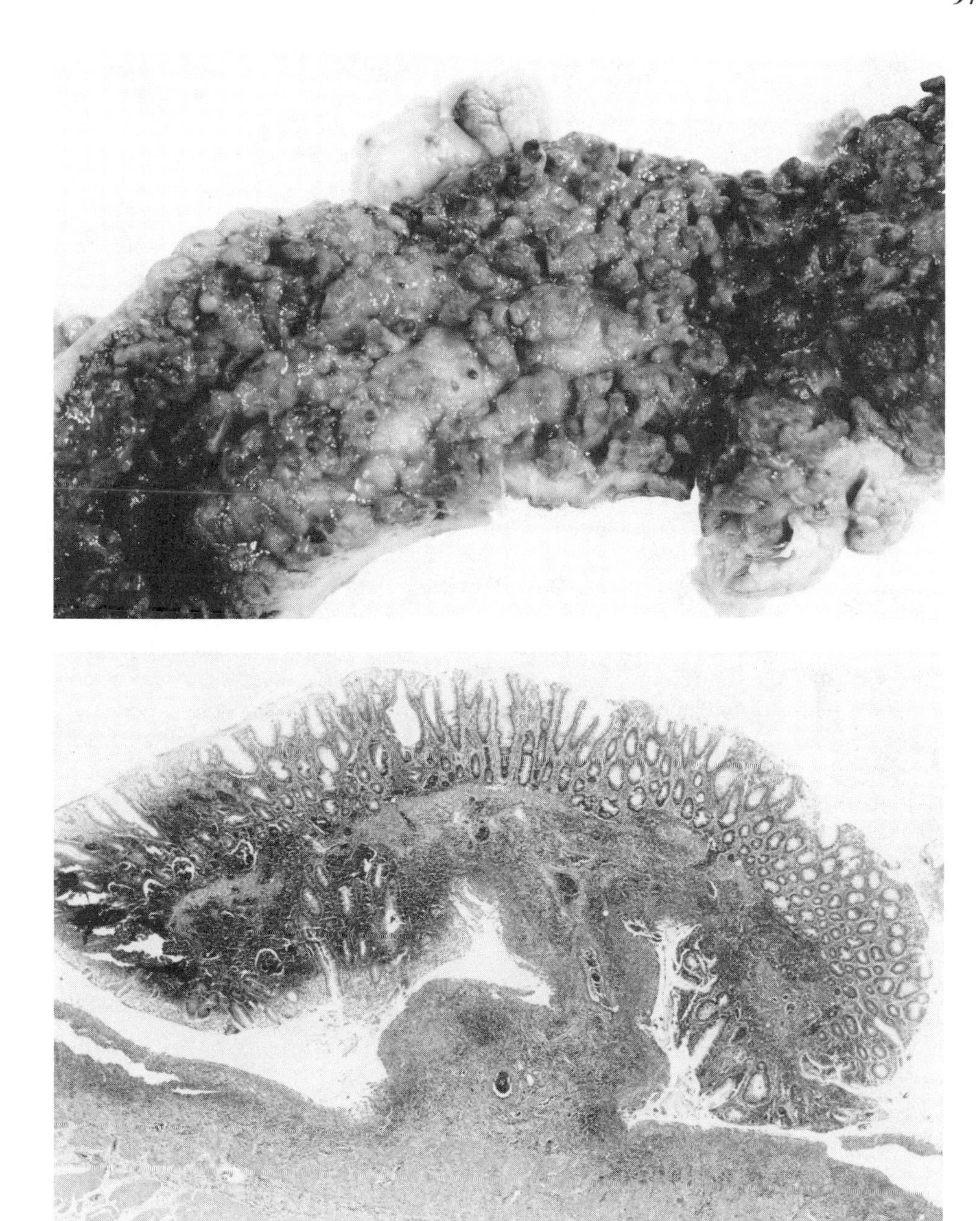

FIGURE 2. Severe active ulcerative colitis with inflamed pseudopolyps. The polypoid, inflamed mucosa stands proud of the surrounding flat ulcer bed which formed by coalescence of deep crypt abscesses and mucosal sloughing virtually down to the level of the flat investing sheet of muscularis mucosa. Inflammation extends only into the superficial submucosa.

fibrosis are not seen, nor are granulomas. These features underline the mucosal nature of this disease process and distinguish it from Crohn's disease.

Other characteristics of ulcerative colitis are evident upon close examination of the mucosa; notably, the chronicity of the process and the cryptocentric nature of the injury. The hallmark of chronicity is inflammation (usually lymphoplasmacytic) in the presence of mucosal architectural distortion (Figure 3). This is manifested by loss of glands (irregular spacing), branching (bifid and trifid forms), bizarre shapes, and nonparallel arrangement of glands, increased space between the crypt bases and the muscularis mucosa, and Paneth cell metaplasia. (Paneth cells are present in small numbers in the normal cecum and ascending colon, but absent distally.[14]) A lymphoplasmacytic infiltrate (so called "chronic inflammation") alone does not warrant a diagnosis of chronic colitis, at least in the sense of ulcerative colitis. Increased plasma cells, including secretors of IgA, IgG, and IgM,[15] expand the deep lamina propria ("basal plasmacytosis").[16] Eosinophils, macrophages, and mast cells may be present in increased numbers. "Cryptocentricity" refers to the main site of histologic injury in the mucosa, the crypt epithelium, which is also the focal point for neutrophilic inflammation. Neutrophils appear to home preferentially to the basolateral part of the crypts where the epithelium undergoes marked changes, though these features may extend to involve the entire crypt and the surface epithelium. Such cryptitis leads to the development of crypt abscesses when neutrophils extend into the lumen of the damaged crypts. Some neutrophils may also be present in the lamina propria where there is capillary dilatation and congestion and, in active areas, red-cell extravasation and hemorrhage may be seen. There is marked mucin depletion in the epithelial cells, both in degenerating (injured) and regenerating (reparative) forms.

Destruction of crypts may progress to more extensive epithelial injury and, if confluence occurs (as described above), to macroulceration. Because injury is focused on the lower one third of the crypts, ulceration generally occurs at this level, with sloughing of almost the entire thickness of the mucosa (Figure 2). Ulcers are predominantly mucosal, but occasionally extend into the superficial submucosa. Fissure ulcers, common in Crohn's disease, are rare and virtually confined to fulminant colitis. Concomitant with injury there is regeneration and repair; granulation tissue forms the base of large ulcers, which reepithelialize in time, and there is marked regenerative and reparative activity in injured crypts. During this process, fragments of epithelium may become trapped in granulation or fibrous tissue deep in the regenerating mucosa or in the submucosa, resulting in the formation of glands, mucus cysts, and epithelial cell clusters which may be mistaken for carcinoma.[17]

B. FULMINANT COLITIS

Fulminant colitis may occur at any stage during the disease but is relatively uncommon. It is an acute emergency requiring active intervention, often colectomy. There is pan-colitis with extensive ulceration and bleeding which may progress to toxic megacolon, systemic sepsis, and perforation. Histologic examination may reveal transmural inflammation (often acute) and fissure ulcers and confirms the extensive mucosal loss and ulceration seen on gross examination. In toxic megacolon the bowel is dilated and the wall attenuated and mural necrosis or perforation may be seen.

The factors which trigger the development of fulminant colitis in patients with quiescent or mildly active ulcerative colitis are not understood. In some cases, however, bacterial superinfection has been implicated.[18] Attempts to culture pathogens should therefore be made. Superinfection may also be responsible for some exacerbations in ulcerative colitis; however, there is little evidence to support this as a general phenomenon.

C. QUIESCENT COLITIS

Quiescent ulcerative colitis is seen in colectomy specimens resected prophylactically or because of carcinoma and in biopsies performed for surveillance or to monitor disease activity. It is characterized by features of chronicity without activity (Figure 4). There is architectural distortion of the mucosa as described above, with Paneth cell metaplasia, enteroendocrine cell hyperplasia, and a low-grade lymphoplasmacytic infiltrate (often in the deep lamina propria), but without neutrophilic infiltration, cryptitis, crypt abscesses, and epithelial injury. Hyperplasia of the muscularis mucosa may be prominent. These features in themselves are not pathognomonic of ulcerative colitis; however, they are suggestive of this diagnosis.

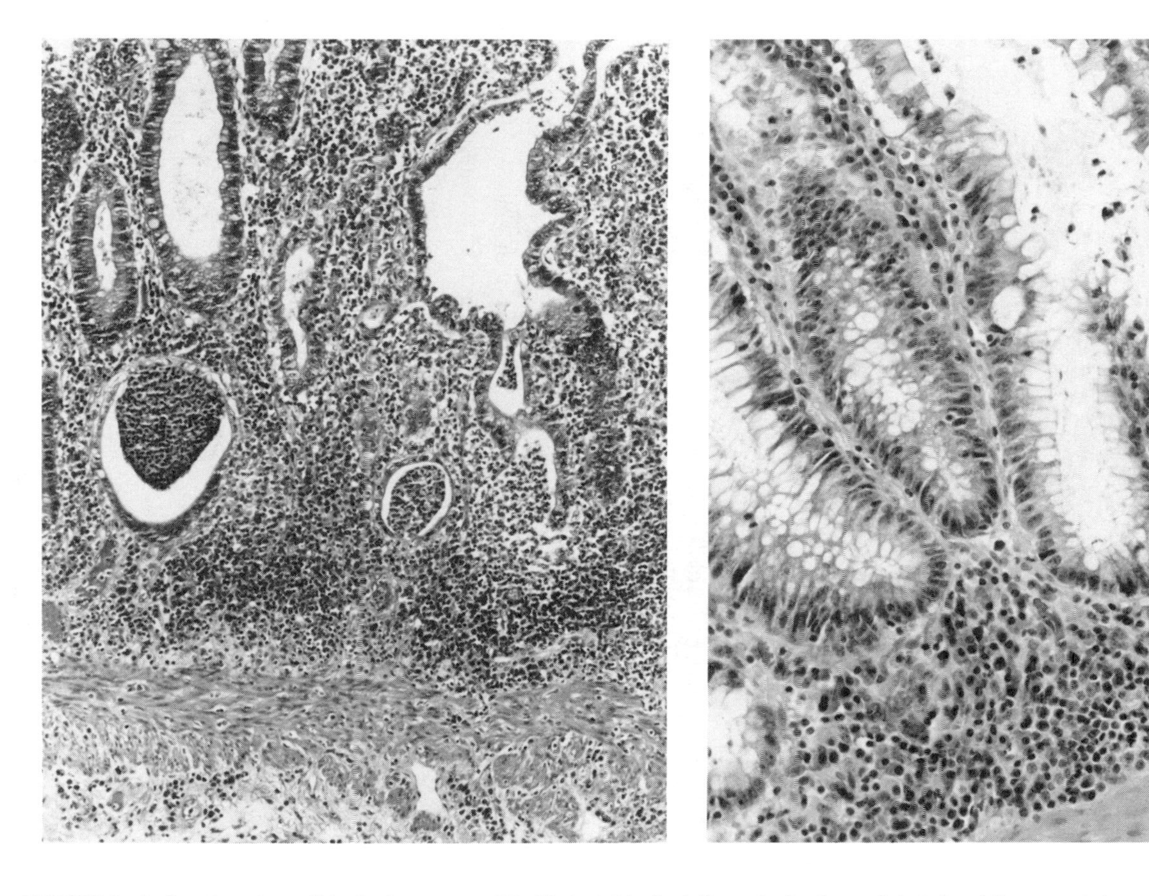

FIGURE 3. Active ulcerative colitis. In the severe colitis illustrated in the left panel, glands are distorted and three crypt abscesses are present, one of which has ruptured. The epithelium in the gland necks is reactive, but the most conspicuous injury is in the lower one third of the glands. Mucin depletion is severe. Chronic inflammatory cells expand the lamina propria, with a concentration of plasma cells in the enlarged space between crypt bases and the muscularis mucosa. In the mild colitis illustrated in the right panel, glands are separated from the muscularis mucosa and show regenerative features with slight mucin depletion; one is branched. A predominantly plasmacytic infiltrate fills the lamina propria beneath and between the glands. No neutrophils are seen.

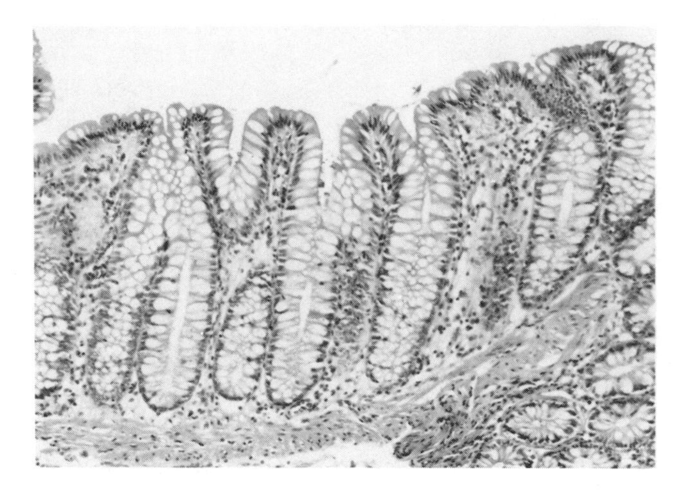

FIGURE 4. Quiescent ulcerative colitis. There is subtle distortion of the mucosal architecture: glands are irregularly spaced and a few are branched. There is no significant inflammation or mucin depletion. The patient had a long history of ulcerative colitis, which was active in the rectum at the time this biopsy was taken from the transverse colon.

D. ULTRASTRUCTURAL FEATURES

The ultrastructural abnormalities of the mucosa in ulcerative colitis have been described in a number of early[19–22] and recent studies.[1,23–25] The epithelial cells in both active and quiescent disease show abnormalities which include cisternal dilatation of the endoplasmic reticulum, reduction in number and length of microvilli, and enlargement of the intercellular spaces. Swollen mitochondria, glycocalyceal and "R" bodies, protrusions of apical cytoplasm, and necrosis are seen in active disease. Abnormal epithelial cells are present both in crypts and on the mucosal surface.[21] The epithelial cell changes include features suggestive of impaired maturation, consistent with the increased rate of turnover observed in kinetic studies of both active and quiescent ulcerative colitis[26] and of epithelial cell injury which is present even in the absence of inflammation.[25] Curiously, the ultrastructural features of ulcerative colitis have received much less attention than those of Crohn's disease, and the reported abnormalities of the epithelial cells warrant more intensive study.

E. DIFFERENTIAL DIAGNOSIS

The differential diagnosis to be considered will depend upon the clinical circumstances and upon whether a resection or biopsies have been submitted for examination. Patients undergoing resection usually have the diagnosis already established by biopsy and clinicopathologic correlation, the major exception being those who present with acute fulminant colitis. The gross features of the large bowel in ulcerative colitis have been described above and are best seen in specimens examined both fresh and after formalin fixation of the opened pinned-out bowel. Distinction from Crohn's disease and other chronic conditions such as ischemic and radiation colitis are made easily on the basis of the location of the lesions and the presence or absence of fibrosis, strictures, fissures and fistulas, adhesions, and creeping fat. Microscopy is usually confirmatory; however, some cases of chronic inflammatory bowel disease (perhaps 5 to 10%) have mixed features which preclude a definitive diagnosis of Crohn's disease or ulcerative colitis and are termed "indeterminate colitis". The true nature of the disease may declare itself in time (e.g., by development of small bowel involvement), but many cases remain indeterminate.[27,28]

The biopsy diagnosis of intestinal inflammatory diseases presents problems quite different from those discussed above. Virtually all inflammatory diseases of the large bowel affect the mucosa (Table 2), which is the only tissue sampled adequately in most endoscopically retrieved biopsies. Thus, the submucosa and deeper layers are usually unavailable for examination. In order to determine the distribution and activity of the disease process along the length of the bowel, serial biopsies taken at 10-cm intervals from cecum to rectum should be examined. These should include biopsies from endoscopically normal mucosa, if present. Quiescent chronic colitis may appear normal on endoscopy, and its recognition in some parts of the bowel may be highly significant. Determining the distribution of disease along the length of the colorectum is particularly helpful in distinguishing processes which are segmental from those which extend in continuity (Table 3). A minimum of three sections from at least three levels of each biopsy are usually examined and more are often recommended.[29] In evaluating the biopsies, the following are critically reviewed: mucosal architecture (gland branching, distortion, orientation, and drop out; expansion of the area between crypts and muscularis mucosa, and fibrosis), epithelium (injury, cryptitis, crypt abscesses, ulceration, regeneration, Paneth cell metaplasia, and dysplasia), and lamina propria (quantity and composition of inflammatory infiltrates, edema, congestion, and granulomas). Lymphoglandular complexes, which may be prominent in acute as well as chronic inflammation and especially in childhood, must not be misinterpreted as chronic inflammation. Evidence of acute self-limited colitis (superficial edema, neutrophils in the lamina propria, cryptitis, crypt ulcers and crypt abscesses, surface ulceration, and mucus depletion in the absence of architectural distortion and basal plasmacytosis,[16,30–34] of ischemic and pseudomembranous colitis, and of specific pathogens such as cytomegalovirus and *Entamoeba histolytica* should be sought. Increasingly, unusual pathogens are being seen in the immunosuppressed, and it is prudent to be alert to the possibility of infection with organisms such as atypical mycobacteria, *Cryptosporidium*, *Strongyloides*, and *Histoplasma*.[35] Other important considerations are the microscopic and collagenous colitis spectrum, solitary rectal ulcer syndrome (SRUS), and diversion colitis. No architectural distortion is seen in microscopic or collagenous colitis, in which epithelial inflammation is predominantly lymphocytic and centered on the luminal surface rather than in crypts.[36] In SRUS, crypt hyperplasia with strands of smooth muscle extending from the muscularis mucosa high into the lamina propria are characteristic, and crypt distortion, inflammation, and ulceration may be present.[37,38] Diversion colitis, occurring in a diverted segment of large bowel, often a Hartmann pouch or mucous fistula, may cause particular difficulty, especially in a patient who has had surgery previously for inflammatory bowel disease. Intense chronic inflammation in the lamina propria with prominent nodular lymphoglandular complexes and occasional crypt abscesses are seen, and there may be superficial ulceration. However, architectural distortion is not a feature and the acute inflammation is less obviously cryptocentric than in ulcerative colitis.[39] Other entities to be considered are listed in Table 2.

Distinguishing Crohn's disease from ulcerative colitis is a major challenge in evaluating biopsies from patients with known chronic inflammatory bowel disease. For many years Crohn's colitis was poorly recognized,[40] and our experience reviewing material in consultation leads us to believe that it is still not infrequently misdiagnosed as ulcerative colitis (e.g., see Table 4). While both conditions exhibit histologic features of chronic inflammatory bowel disease, a pattern of distribution showing rectal sparing and skip zones, the presence of mucosal noncaseating granulomas, and only modest amounts of cryptitis and mucin depletion relative to the extent of the inflammation all point toward Crohn's disease (Table 3). There are difficulties with all these criteria; however, ulcerative colitis patients treated with suppositories and enemas may have apparent sparing of the distal bowel, and a few have no gross rectal involvement even without treatment (though microscopic features compatible with ulcerative colitis are usually present);[2] granulomas occur in ulcerative colitis in response to the presence of foreign material in the lamina propria or submucosa, most commonly extravasated

TABLE 2
Differential Diagnosis of Colitis/Proctitis in Biopsies

Inflammatory bowel disease
- Crohn's disease
- Ulcerative colitis
- Indeterminate colitis

Infectious
- Bacterial: acute self limited,
  - hemorrhagic,
  - pseudomembranous,
  - yersinial,
  - tuberculous,
  - lymphogranuloma venereum,
  - chlamydial
- Viral: herpetic,
  - cytomegaloviral,
  - AIDS[a] related
- Fungal: histoplasma
- Parasitic: amoebic,
  - cryptosporidial,
  - schistosomal,
  - *Strongyloides*

Ischemic
- Occlusive: atherosclerotic,
  - microembolic,
  - thrombotic,
  - vasculitic
- Hypotensive

Iatrogenic
- Drugs: enemas,
  - antibiotics,
  - gold,
  - NSAIDS[b]
- Diversion
- Radiation
- GVHD[c]

Others
- Eosinophilic
- Uremic
- Microscopic/collagenous
- Diverticulitis
- Solitary rectal ulcer
- Behcet's disease

[a] AIDS = acquired immunodeficiency syndrome.
[b] NSAIDS = nonsteroidal antiinflammatory drugs.
[c] GVHD = graft vs. host disease.

mucin from injured epithelium; and evaluation of the "appropriate" amount of cryptitis and mucin depletion is subjective. These factors probably lead to some overdiagnosis of ulcerative colitis and underdiagnosis of Crohn's disease of the colon.

III. DYSPLASIA AND CARCINOMA

A. INTRODUCTION

The colorectal carcinomas which arise in the background of ulcerative colitis are adenocarcinomas. There is a tendency toward mucinous differentiation, and growth is more

TABLE 3
Comparison of Biopsy Findings in Active Ulcerative, Crohn's, and Acute Self-Limited Colitis

	Ulcerative colitis	Crohn's colitis	Acute self-limited colitis
Distribution	Continuous	Segmental from rectum	Diffuse
Architecture	Distorted	Distorted	Normal
Crypt abscesses	+++	++	+++
Paneth cell metaplasia	++	+	–
Mucin depletion	+++	+	++
Plasma cell infiltrates	+++	++	+
Granulomas	–	+	–
Neutrophil infiltrates between crypts	+	+	+++
Submucosal inflammation	+	+++	–

TABLE 4
Overdiagnosis of Ulcerative Colitis: Example of a Case

Endoscopies with biopsy	14
Biopsies	107
Sections examined	454
Granulomas	7
Sections containing granulomas	7

Note: This patient carried a clinical diagnosis of ulcerative colitis since 1972, when he had his first sigmoidoscopy. Long colonoscopy biopsy series from 1978 on showed chronic IBD with skip lesions and rectal sparing; cryptitis and mucin depletion were relatively slight. On review in 1989, subtle (undetected) granulomas were present, though only in 7 of 454 sections examined (in biopsies from 1976, 1978, 1983, and 1988). In 1989 a small bowel stricture developed and the clinical and pathologic features were those of Crohn's disease.

infiltrative and less exophytic than in the usual colorectal adenocarcinomas, which generally arise in an older age group. The tumors are typically flat or slightly elevated lesions with extensive mural invasion and lymphatic involvement in relation to their size. They occur slightly more often in the rectum and sigmoid than in other parts; however, as many as 40% may occur proximal to the splenic flexure. In a series of 82 patients who developed colorectal cancer in a background of ulcerative colitis, two thirds were male, 90% had had extensive colitis for more than 10 years, and the average age at diagnosis of carcinoma was 43 years.[41] Of particular interest was the presence of multifocal carcinoma in 13% of these patients, tumor proximal to the splenic flexure in 44%, and poorly differentiated adenocarcinoma (including mucinous tumors) in one third. In another series, 18% of cancer patients with ulcerative colitis had multiple synchronous colorectal cancers, as opposed to 2.5% of those with nonpolyposis, noncolitis large bowel adenocarcinomas.[42]

The recognition of dysplasia in association with cancer in the rectal mucosa in ulcerative colitis led to the hypothesis that detection of dysplasia in endoscopic biopsies could be used to determine when prophylactic colectomy should be performed.[43] Detailed studies of resection specimens confirmed the frequent association of dysplasia and carcinoma, though in some cases carcinoma appears to develop directly.[44,45] Colonoscopic surveillance of patients with long-standing ulcerative colitis followed, and serial biopsies of diseased mucosa from cecum to rectum, often with a request to the pathologist to "rule out dysplasia", became commonplace. Since objective criteria for diagnosing and grading dysplasia were not established at that time, and agreed strategies for managing ulcerative colitis patients with epithelial dysplasia of different grades and extent did not exist, a number of gastrointestinal pathologists from several countries came together and formed the Inflammatory Bowel Disease — Dysplasia Morphology Study Group to address these issues. The Group presented a comprehensive report in 1983 which set the standards for diagnosis and management of dysplasia in ulcerative colitis.[46]

B. DEFINITION, DIAGNOSIS, AND GRADING OF DYSPLASIA

Dysplasia is defined as an unequivocal neoplastic alteration of the colorectal epithelium. It is a morphologic concept and implies a genomic alteration (or more probably a series of alterations) leading to changes in cell shape, size, and staining characteristics which pathologists recognize as indicative of neoplastic transformation. These changes are probably incidental to the critical (genomic) events in development of neoplasia; indeed, it is quite possible that the morphologic changes may not take place in some instances of neoplastic transformation, though fortunately this seems to be relatively uncommon. For the pathologist, attention has been focused on defining and recognizing those morphologic changes which are characteristic and diagnostic of neoplasia and which enable it to be differentiated from other alterations in colorectal epithelium seen in ulcerative colitis during the processes of injury and repair.

Dysplastic mucosa exhibits various abnormal architectural arrangements, including a villous surface and enlarged crowded glands similar to those seen in adenomatous polyps in noncolitics (Figure 5). A back-to-back pattern of glands is sometimes seen. The glands are lined by epithelial cells with enlarged, hyperchromatic, pleomorphic nuclei which may be crowded and stratified, exhibiting loss of polarity and occurring in the middle or apical portions of the cells. The nuclear:cytoplasmic ratio is increased. The glands may become distorted and complex in form. Dystrophic goblet cells and columnar mucous cells are also seen, but may be less easily evaluated.[11,47] Inflammation is often severe in the surrounding mucosa; however, the inflammatory cells tend to avoid the neoplastic epithelium. This is in contrast to what is seen in inflammatory or reactive atypia, in which the epithelium is often infiltrated by inflammatory cells and the colonocytes show degenerative or regenerative changes but with less nuclear pleomorphism and stratification. The distinction between dysplasia and inflammatory atypia, however, is far from clear cut; for extensive treatment of this topic the report of the Inflammatory Bowel Disease Study Group should be consulted.[46]

The Group proposed the schema outlined in Table 5 for reporting on dysplasia in colonic biopsies from patients with ulcerative colitis. The criteria for distinguishing between low- and high-grade dysplasia are described in detail; they depend largely upon the degrees of glandular distortion and of hyperchromatism and pleomorphism of the nuclei. In most cases the extent of stratification can be used to make this distinction: if more than 50% of the nuclei lie basal to the midline of the cell it is called low grade, if they lie on the apical side it is called high grade. Dysplasia in a biopsy from an endoscopically recognizable lesion (the so-called dysplasia-associated lesion or mass, DALM) is of particular importance and must be distinguished from that arising in endoscopically normal mucosa, as it has a significantly greater risk of being associated with invasive carcinoma, especially if the lesion is polypoid.[48,49] High-grade dysplasia is considered synonymous with intraepithelial (*in situ*) carcinoma by many pathologists, though others prefer to make a distinction. Carcinoma invasive only in the lamina propria is best termed intramucosal, to distinguish it from invasive carcinoma which has penetrated the muscularis mucosa and gained access to the rich lymphatic plexus of the submucosa.

The diagnosis and grading of dysplasia, as discussed above, are usually determined on the basis of histologic sections of mucosal biopsies, stained conventionally with hematoxylin and

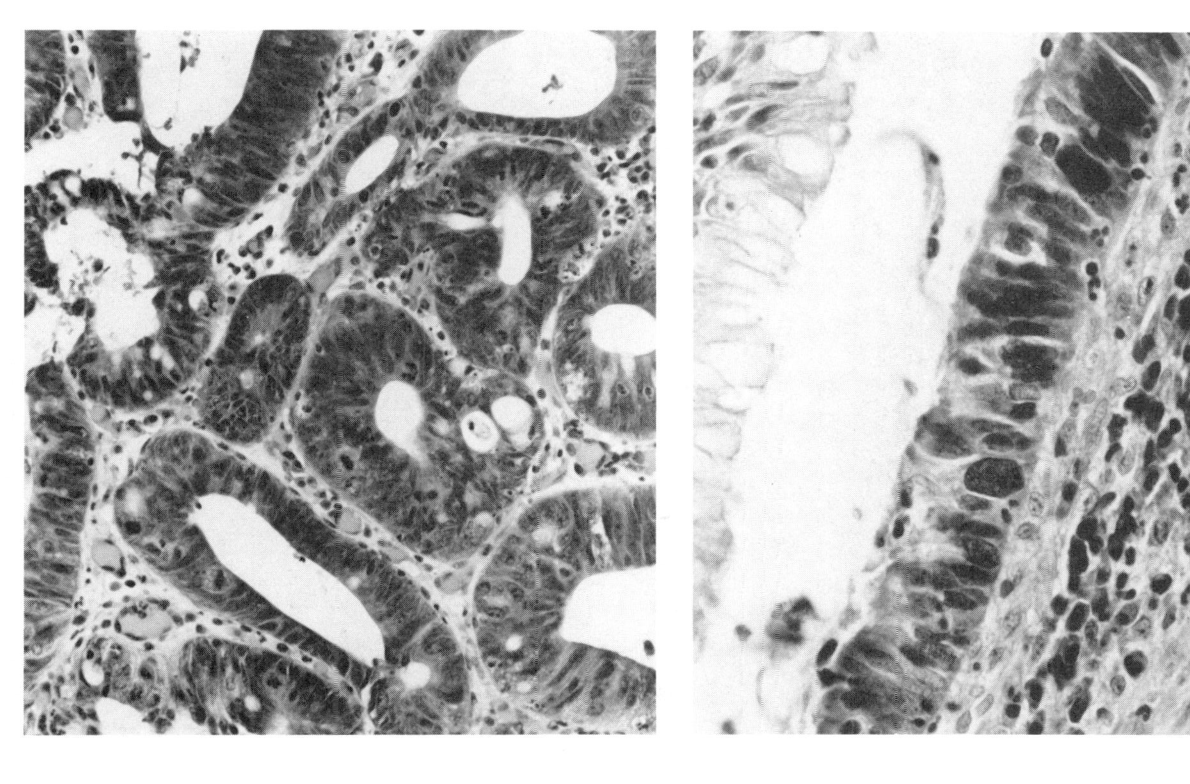

FIGURE 5. High-grade dysplasia. In the left panel distorted glands are lined by epithelial cells with crowded, stratified, pleomorphic, hyperchromatic nuclei, high nuclear:cytoplasmic ratio and abundant mitoses. There is little inflammation. The features are in marked contrast to those seen in Figures 3 and 4. In the right panel, high-grade dysplasia in an area of villous regeneration is shown. Highly atypical, crowded, stratified pleomorphic nuclei line a villus which contains a lymphoplasmacytic infiltrate in the lamina propria. The villus to the left is lined by nondysplastic epithelium.

TABLE 5
Scheme for Classification of Dysplasia in Colonic Biopsies from Patients with Ulcerative Colitis and Recommended Management

Classification	Management
Negative for dysplasia	
Normal	Surveilance
Colitis, inactive	Surveillance
Colitis, active	Surveillance (+/– treatment)
Indefinite for dysplasia	
Probably negative (probably inflammatory)	Surveillance
Unknown	Early colonoscopy and biopsy
Probably positive (probably dysplastic)	Early colonoscopy and biopsy
Positive for dysplasia	
Low grade	If endoscopically normal, recolonoscope and biopsy; if persists or progresses confirmed, advise proctocolectomy If from lesion or mass (DALM) and confirmed, advise proctocolectomy
High grade	If confirmed, advise proctocolectomy

From Riddell, R. H. et al., *Hum. Pathol.*, 14, 931, 1983.

eosin. Brush cytology has one theoretical advantage over biopsies in that this technique preferentially samples the less cohesive epithelial cells, which include those which are dysplastic and malignant. In experienced hands it may prove to be of value in screening for dysplasia and carcinoma.[50,51]

C. PRACTICAL PROBLEMS IN DIAGNOSING DYSPLASIA

In practice, the diagnosis of dysplasia in this setting is of immense significance as a firm diagnosis generally implies a recommendation of proctocolectomy (Table 5). Sphincter-sparing procedures may be used; however, all the colorectal mucosa is removed as any that is left is at risk of developing carcinoma.[45] For this reason it is recommended that the diagnosis of dysplasia should be confirmed, preferably in more than one biopsy and by more than one pathologist, ideally by a pathologist with specialist expertise in this field.[46,52] However, even among experts there is considerable disagreement about the diagnosis of dysplasia. In the Inflammatory Bowel Disease — Dysplasia Morphology Study Group report, for example, in which high-grade lesions were classified as Grade 3, low-grade lesions as Grade 2, indefinite, unknown, and probably positive as Grade 1, and negative as Grade 0, 33% of observations by the 12 expert gastrointestinal pathologists were assigned to the "wrong" grades; and of 77 observations of high-grade dysplastic (Grade 3) lesions, 17% were classified as low grade (Grade 2), 5% were called Grade 1, and 1 was considered negative (Grade 0).[53] Only minor interobserver and intraobserver disagreements in assessing dysplasia in biopsies of ulcerative colitis were found by Dundas et al., but Dixon et al. and Melville et al. reported significant interobserver disagreement, which has also been our experience.[47,52,54]

In view of these difficulties, it is not surprising that attempts have been made to develop more objective tests for neoplastic change in ulcerative colitis. Mucin histochemistry has been found by some authors to be of possible value in detecting precancerous changes,[55–60] although others have been unable to distinguish between reactive and dysplastic changes any better than with hematoxylin and eosin stains.[61,62] The promise of initial studies using lectin histochemistry[63] has not been fulfilled by subsequent work,[57,62,64] and attempts to use other markers, including carcinoembryonic antigen,[62] tumor-associated glycoprotein (TAG)-72,[65] and c-*myc* and *ras* oncogene products,[66,67] have been unsuccessful in specifically detecting premalignant change in colonic mucosa. Flow cytometry has been used to measure nuclear DNA content;[68–73] however, its value in diagnosing dysplasia appears to be limited. Fozard et al.[69] reported identical prevalences of aneuploidy in long-standing colitis with and without dysplasia,[69] and only 4 of 14 carcinomas were aneuploid; Melville et al.[72] found aneuploidy in 74% of carcinomas but only in 39% of examples of dysplasia, while 5% of morphologically nondysplastic samples were aneuploid; finally, Rutegard et al.[73] studied 60 long-standing colitics, including 5 with dysplasia without aneuploidy (1 developed carcinoma, 1 was high grade), and 6 with persistent aneuploidy but without dysplasia.[73] Clearly, the relationship between aneuploidy and dysplasia is not simple; the significance of findings of aneuploidy, with and without dysplasia, in mucosal biopsies from patients with ulcerative colitis requires elucidation. Morphometric analysis has also been attempted.[74] So far, none of these or other innovative tests has found wide acceptance and morphologic evaluation of dysplasia, with all its shortcomings, remains the procedure of choice.

Dysplasia occurring in flat mucosa (i.e., not associated with a lesion or mass) is undetectable endoscopically. In resection specimens scrutinized with a hand lens dysplastic and nondysplastic mucosa cannot be distinguished reliably. In addition, dysplasia is often patchy and multifocal. Therefore, a significant sampling problem is encountered when tiny endoscopic biopsies taken randomly from a series of sites in the relatively vast area of colorectal mucosa are used in attempts to detect dysplasia. This has obvious implications for endoscopic surveillance programs: visible lesions or masses can readily be detected and biopsied; however, the probability of sampling dysplasia in flat mucosa in protocol biopsies would appear low unless the dysplasia is very extensive. The utility of surveillance as currently practiced has recently been questioned,[75] and Romano et al.[53] have suggested that colonoscopic surveillance with biopsy only of visible lesions may be the most rational approach in the present state of knowledge.[53,75] In addition, there is uncertainty about the significance of dysplasia arising during surveillance, as opposed to that found on initial examination, and its association with carcinoma may be less clear cut than was previously thought.[76,77] We can anticipate changes in the protocols for surveillance, biopsy, and surgery as we learn more about the natural history of dysplasia in ulcerative colitis.

IV. THE PATHOLOGIST'S ROLE IN ULCERATIVE COLITIS

In evaluating resection specimens the pathologist determines the type, extent, and activity of inflammatory bowel disease, the adequacy of excision of the colorectal mucosa (which should be completely removed), and the presence of complications. The specimen should be sufficiently sampled to determine if dysplasia or carcinoma is present, with special attention being paid to mass lesions. Dysplasia is graded and the extent and multifocality are noted. Carcinoma is graded and staged in the usual way, with evaluation of the gross and microscopic growth pattern, inflammatory response, vascular and lymphatic invasion, and adequacy of resection.

Serial endoscopic biopsies are taken for diagnosis to monitor disease activity or for surveillance. They provide information about the distribution, activity, and type of intestinal inflammatory disease and about the presence and grade of dysplasia or carcinoma. When previous biopsies have been taken it is important to obtain them and to review all available sections, especially if the diagnosis is in doubt (as illustrated by the example shown in Table

4), or if dysplasia is present or was recorded previously. Particular attention should be paid to biopsies of mass lesions. We believe that the history, clinical and endoscopic findings, and results of laboratory tests and radiologic studies should be reviewed with the histology before a final clinical diagnosis is reached. This requires close clinicopathologic cooperation. The pathologist should not be overzealous in making a firm diagnosis of ulcerative colitis or of Crohn's colitis on the basis of biopsies alone, although, if either one is favored, this should be clearly stated in the report. Application of the standardized classification of dysplasia (Table 5) in reporting all biopsies leads to uniformity of style and good communication, and if dysplasia is detected or suspected, review of the slides by an experienced gastrointestinal pathologist is prudent in view of the clinical implications of this diagnosis.

REFERENCES

1. **Rubio, C. A., May, I., and Slezak, P.,** Ulcerative colitis in protracted remission — a quantitative scanning electron microscopic study, *Dis. Colon Rectum,* 31, 939, 1988.
2. **Spiliadis, C. A., Spiliadis, C. A., and Lennard-Jones, J. E.,** Ulcerative colitis with relative sparing of the rectum, *Dis. Colon Rectum,* 30, 334, 1987.
3. **Both, H., Torp-Pedersen, K., Kreiner, S., and Hendriksen, C.,** Clinical appearance at diagnosis of ulcerative colitis and Crohn's disease in a regional patient group, *Scand. J. Gastroenterol.,* 18, 987, 1983.
4. **Devroede, G. J., Taylor, W. F., Sauer, W. G., Jackman, R. J., and Stickler, G. B.,** Cancer risk and life expectancy of children with ulcerative colitis, *N. Engl. J. Med.,* 285, 17, 1971.
5. **Lashner, B. A., Silverstein, M. D., and Hanauer, S. B.,** Hazard rates for dysplasia and cancer in ulcerative colitis: results from a surveillance program, *Dig. Dis. Sci.,* 34, 1536, 1989.
6. **Gryboski, J. D.,** Diseases of the colon, rectum and anus in children, in *Diseases of the Colon, Rectum, and Anal Canal,* Kirsner, J. B. and Shorter, R. G., Eds., Williams & Wilkins, Baltimore, 1988, 595.
7. **Barton, J. R., Gillon, S., and Ferguson, A.,** Incidence of inflammatory bowel disease in Scottish children between 1968 and 1983; marginal fall in ulcerative colitis, three-fold rise in Crohn's disease, *Gut,* 30, 618, 1989.
8. **Danzi, J. T.,** Extraintestinal manifestations of idiopathic inflammatory bowel disease, *Arch. Int. Med.,* 148, 297, 1988.
9. **Brozna, J. P., Fisher, R. L., and Barwick, K. W.,** Filiform polyposis: an unusual complication of inflammatory bowel disease, *J. Clin. Gastroenterol.,* 7, 451, 1985.
10. **Goldenberg, B., Mori, K., Friedman, I. H., Shinya, H., and Buchwald, R. P.,** Fused inflammatory polyps simulating carcinoma in ulcerative colitis, *Am. J. Gastroenterol.,* 73, 441, 1980.
11. **Lee, R. G.,** Villous regeneration in ulcerative colitis, *Arch. Pathol. Lab. Med.,* 111, 276, 1987.
12. **Lumb, G. and Protheroe, R. H. B.,** Ulcerative colitis: a pathological study of 152 surgical specimens, *Gastroenterology,* 34, 381, 1958.
13. **O'Leary, A. D. and Sweeney, E. C.,** Lymphoglandular complexes of the colon: structure and distribution, *Histopathology,* 10, 267, 1986.
14. **Symonds, D. A.,** Paneth cell metaplasia in diseases of the colon and rectum, *Arch. Pathol. Lab. Med.,* 97, 343, 1974.
15. **Keren, D. F., Appelman, H. D., Dobbins, W. O., Wells, J. J., Whisenant, B., Foley, J., Dieterle, R., and Geisinger, K.,** Correlation of histopathologic evidence of disease activity with the presence of immunoglobulin-containing cells in the colons of patients with inflammatory bowel disease, *Hum. Pathol.,* 15, 757, 1984.
16. **Nostrant, T. T., Kumar, N. B., and Appelman, H. D.,** Histopathology differentiates acute self-limited colitis from ulcerative colitis, *Gastroenterology,* 92, 318, 1987.
17. **Allen, D. C. and Biggart, J. D.,** Misplaced epithelium in ulcerative colitis and Crohn's disease of the colon and its relationship to malignant mucosal changes, *Histopathology,* 10, 37, 1986.
18. **Schmidt, T., Pfeiffer, A., Ehret, W., Keiditsch, E., Ruckdeschel, G., and Kaess, H.,** Legionella infection of the colon presenting as acute attack of ulcerative colitis, *Gastroenterology,* 97, 751, 1989.
19. **Gonzalez-Licea, A. and Yardley, J. H.,** Nature of the tissue reaction in ulcerative colitis: light and electron microscopic findings, *Gastroenterology,* 51, 825, 1966.
20. **Donellan, W. L.,** Early histological changes in ulcerative colitis: a light and electron microscopic study, *Gastroenterology,* 50, 519, 1966.

21. **Nagle, G. J. and Kurtz, S. M.,** Electron microscopy of the human rectal mucosa. A comparison of idiopathic ulcerative colitis with inflammation of known etiologies, *Am. J. Dig. Dis.*, 12, 541, 1967.
22. **O'Connor, J. J.,** An electron microscopic study of inflammatory colonic disease, *Dis. Colon Rectum*, 15, 265, 1972.
23. **Shields, H. M., Bates, M. L., Goldman, H., Zuckerman, G. R., Mills, B. A., Best, C. J., Bair, F. A., Goran, D. A., and deSchryver-Kecskemeti, K.,** Scanning electron microscopic appearance of chronic ulcerative colitis with and without dysplasia, *Gastroenterology*, 89, 62, 1985.
24. **Balazs, M. and Kovacs, A.,** Ulcerative colitis: electron microscopic studies with special reference to development of crypt abscesses, *Dis. Colon Rectum*, 32, 327, 1989.
25. **Delpre, G., Avidor, I., Steinherz, R., Kadish, U., and Ben-Bassat, M.,** Ultrastructural abnormalities in endoscopically and histologically normal and involved colon in ulcerative colitis, *Am. J. Gastroenterol.*, 84, 1038, 1989.
26. **Allan, A., Bristol, J. B., and Williamson, R. C. N.,** Crypt cell production rate in ulcerative proctocolitis: differential increments in remission and relapse, *Gut*, 26, 999, 1985.
27. **Price, A. B.,** Overlap in the spectrum of non-specific inflammatory bowel disease — colitis indeterminate, *J. Clin. Pathol.*, 31, 567, 1978.
28. **Pezim, M. E., Pemberton, J. H., Beart, R. W., Wolff, B. G., Dozois, R. R., Nivatvongs, S., Devine, R., and Ilstrup, D. M.,** Outcome of "indeterminant" colitis following ileal pouch-anal anastomosis, *Dis. Colon Rectum*, 32, 653, 1989.
29. **Surawicz, C. M., Meisel, J. H., Ylvisaker, T., Saunders, D. R., and Rubin, C. E.,** Rectal biopsy in the diagnosis of Crohn's disease: value of multiple biopsies and serial sectioning, *Gastroenterology*, 81, 66, 1981.
30. **Price, A. B., Jewkes, J., and Sanderson, P. J.,** Acute diarrhoea: Campylobacter colitis and the role of the rectal biopsy, *J. Clin. Pathol.*, 32, 990, 1979.
31. **Kumar, N. B., Nostrant, T. T., and Appelman, H. D.,** The histopathologic spectrum of acute self-limited colitis (acute infectious-type colitis), *Am. J. Surg. Pathol.*, 6, 523, 1982.
32. **Surawicz, C. M. and Belic, L.,** Rectal biopsy helps to distinguish acute self-limited colitis from idiopathic inflammatory bowel disease, *Gastroenterology*, 86, 104, 1984.
33. **Allison, M. C., Hamilton-Dutoit, S. J., Dhillon, A. P., and Pounder, R. E.,** The value of rectal biopsy in distinguishing self-limited colitis from early inflammatory bowel disease, *Q. J. Med.*, 65, 985, 1987.
34. **Surawicz, C. M.,** The role of rectal biopsy in infectious colitis, *Am. J. Surg. Pathol.*, 12(Suppl. 1), 82, 1988.
35. **Rotterdam, H.,** The pathology of the gastrointestinal tract in AIDS, in *Digestive Disease Pathology*, Vol. 2, Watanabe, S., Wolff, M., Sommers, S. C., Eds., Field & Wood, New York, 1989, 21.
36. **Lazenby, A. J., Yardley, J. H., Giardiello, F. M., Jessurun, J., and Bayliss, T. M.,** Lymphocytic ("microscopic") colitis: a comparative histopathologic study with particular reference to collagenous colitis, *Hum. Pathol.*, 20, 18, 1989.
37. **Saul, S. H. and Sollenberger, L. C.,** Solitary rectal ulcer syndrome: its clinical and pathological underdiagnosis, *Am. J. Surg. Pathol.*, 9, 411, 1985.
38. **Levine, D. S.,** "Solitary" rectal ulcer syndrome. Are "solitary" rectal ulcer syndrome and "localized" colitis cystica profunda analagous syndromes caused by rectal prolapse?, *Gastroenterology*, 92, 243, 1987.
39. **Murray, F. E., O'Brien, M. J., Birkett, D. H., Kennedy, S. M., LaMont, J. T.,** Diversion colitis: pathologic findings in a resected sigmoid colon and rectum, *Gastroenterology*, 93, 1404, 1987.
40. **Levine, J.,** Where was Crohn's colitis in 1932, *J. Clin. Gastroenterol.*, 11, 187, 1989.
41. **Mir-Madjlessi, S. H., Farmer, R. G., Easley, K. A., and Beck, G. J.,** Colorectal and extracolonic malignancy in ulcerative colitis, *Cancer*, 58, 1569, 1986.
42. **Greenstein, A. J., Heimann, T. M., Sachar, D. B., Slater, G., and Aufses, A. H.,** A comparison of multiple synchronous colorectal cancer in ulcerative colitis, familial polyposis coli, and *de novo* cancer, *Ann. Surg.*, 203, 123, 1986.
43. **Morson, B. C. and Pang, L. S. C.,** Rectal biopsy as an aid to cancer control in ulcerative colitis, *Gut*, 8, 423, 1967.
44. **Allen, D. C., Biggart, J. G., and Pyper, P. C.,** Large bowel mucosal dysplasia and carcinoma in ulcerative colitis, *J. Clin. Pathol.*, 38, 30, 1985.
45. **Thomas, D. M., Filipe, M. I., and Smedley, F. H.,** Dysplasia and carcinoma in the rectal stump of total colitics who have undergone colectomy and ileo-rectal anastomosis, *Histopathology*, 14, 289, 1989.
46. **Riddell, R. H., Goldman, H., Ransohoff, D. F., Appelman, H. D., Fenoglio, C. M., Haggitt, R. C., Ahren, C., Correa, P., Hamilton, S. R., Morson, B. C., Sommers, S. C., and Yardley, J. H.,** Dysplasia in inflammatory bowel disease: standardized classification with provisional clinical applications, *Hum. Pathol.*, 14, 931, 1983.
47. **Dundas, S. A. C., Kay, R., Beck, S., Cotton, D. W. K., Coup, A. J., Slater, D. N., and Underwood, J. C. E.,** Can histopathologists reliably assess dysplasia in chronic inflammatory bowel disease?, *J. Clin. Pathol.*, 40, 1282, 1987.

48. **Blackstone, M. O., Riddell, R. H., Rogers, B. H. G., and Levin, B.,** Dysplasia-Associated Lesion or Mass (DALM) detected by colonoscopy in long-standing ulcerative colitis: an indication for colectomy, *Gastroenterology,* 80, 366, 1981.
49. **Butt, J. H., Konishi, F., Morson, B. C., Lennard-Jones, J. E., and Ritchie, J. K.,** Macroscopic lesions in dysplasia and carcinoma complicating ulcerative colitis, *Dig. Dis. Sci.,* 28, 18, 1983.
50. **Festa, V. I., Hajdu, S. I., and Winawer, S. J.,** Colorectal cytology in chronic ulcerative colitis, *Acta Cytol.,* 29, 262, 1985.
51. **Melville, D. M., Richman, P. I., Shepherd, N. A., Williams, C. B., and Lennard-Jones, J. E.,** Brush cytology of the colon and rectum in ulcerative colitis: an aid to cancer diagnosis, *J. Clin. Pathol.,* 41, 388, 1988.
52. **Dixon, M. F., Brown, L. J. R., Gilmour, H. M., Price, A. B., Smeeton, N. C., Talbot, I. C., and Williams, G. T.,** Observer variation in the assessment of dysplasia in ulcerative colitis, *Histopathology,* 13, 385, 1988.
53. **Romano, T. J., Conn, A., West, A. B., and Spiro, H. M.,** Dysplasia and cancer in ulcerative colitis, in preparation.
54. **Melville, D. M., Jass, J. R., Morson, B. C., Pollock, D. J., Richman, P. I., Shepherd, N. A., Ritchie, J. K., Love, S. B., and Lennard-Jones, J. E.,** Observer study of the grading of dysplasia in ulcerative colitis: comparison with clinical outcome, *Hum. Pathol.,* 20, 1008, 1989.
55. **Ehsanullah, M., Filipe, M. I., and Gazzard, B.,** Mucin secretion in inflammatory bowel disease: correlation with disease activity and dysplasia, *Gut,* 23, 485, 1982.
56. **Ehsanullah, M., Morgan, M. N., Filipe, M. I., and Gazzard, B.,** Sialomucins in the assessment of dysplasia and cancer risk in patients with ulcerative colitis treated with colectomy and ileorectal anastomosis, *Histopathology,* 9, 223, 1985.
57. **Fozard, J. B. J., Dixon, M. F., Axon, A. T. R., and Giles, G. R.,** Lectin and mucin histochemistry as an aid to cancer surveillance in ulcerative colitis, *Histopathology,* 11, 385, 1987.
58. **Allen, D. C., Connolly, N. S., and Biggart, J. D.,** Mucin profiles in ulcerative colitis with dysplasia and carcinoma, *Histopathology,* 13, 413, 1988.
59. **Agawa, S., Muto, T., and Morioka, Y.,** Mucin abnormality of colonic mucosa in ulcerative colitis associated with carcinoma and/or dysplasia, *Dis. Colon Rectum,* 31, 387, 1988.
60. **Filipe, M. I., Sandey, A., and Ma, J.,** Intestinal mucin antigens in ulcerative colitis and their relationship with malignancy, *Hum. Pathol.,* 19, 671, 1988.
61. **Jass, J. R., England, J., and Miller, K.,** Value of mucin histochemistry in follow up surveillance of patients with long-standing ulcerative colitis, *J. Clin. Pathol.,* 39, 393, 1986.
62. **Ahnen, D. J., Warren, G. H., Greene, L. J., Singleton, J. W., and Brown, W. R.,** Search for a specific marker of mucosal dysplasia in chronic ulcerative colitis, *Gastroenterology,* 93, 1346, 1987.
63. **Pihl, E., Peura, A., Johnson, W. R., McDermott, F. T., and Hughes, E. S. R.,** T-antigen expression by peanut agglutinin staining relates to mucosal dysplasia in ulcerative colitis, *Dis. Colon Rectum,* 28, 11, 1985.
64. **Cooper, H. S., Farano, P., and Coapman, R. A.,** Peanut lectin binding sites in colons of patients with ulcerative colitis, *Arch. Pathol. Lab. Med.,* 111, 270, 1987.
65. **Thor, A., Itzkowitz, S. H., Schlom, J., Kim, Y. S., and Hanauer, S.,** Tumor-associated glycoprotein (TAG-72) expression in ulcerative colitis, *Int. J. Cancer,* 43, 810, 1989.
66. **Ciclitira, P. J., Macartney, J. C., and Evan, G.,** Expression of c-myc in non-malignant and pre-malignant gastrointestinal disorders, *J. Pathol.,* 151, 293, 1987.
67. **McKenzie, J. K., Purnell, D. S., and Shamsuddin, A. M.,** Expression of carcinoembryonic antigen, T-antigen, and oncogene products as markers of neoplastic and preneoplastic colonic mucosa, *Hum. Pathol.,* 18, 1282, 1987.
68. **Hammarberg, C., Slezack, P., and Tribukait, B.,** Early detection of malignancy in ulcerative colitis. A flow-cytometric DNA study, *Cancer,* 53, 291, 1984.
69. **Fozard, J. B. J., Quirke, P., Dixon, M. F., Giles, G. R., and Bird, C. C.,** DNA aneuploidy in ulcerative colitis, *Gut,* 27, 1414, 1986.
70. **Borkje, B., Hostmark, J., Skagen, D. W., Schrumpf, E., and Laerum, O. D.,** Flow cytometry of biopsy specimens from ulcerative colitis, colorectal adenomas, and carcinomas, *Scand. J. Gastroenterol.,* 22, 1231, 1987.
71. **Lofberg, R., Tribukait, B., Ost, A., Brostrom, O., and Reichard, H.,** Flow cytometric DNA analysis in longstanding ulcerative colitis: a method of prediction of dysplasia and carcinoma development?, *Gut,* 28, 1100, 1987.
72. **Melville, D. M., Jass, J. R., Shepherd, N. A., Northover, J. M. A., Capellaro, D., Richman, P. I., Lennard-Jones, J. E., Ritchie, J. K., and Andersen, S. N.,** Dysplasia and deoxyribonucleic acid aneuploidy in the assessment of precancerous changes in chronic ulcerative colitis: observer variation and correlations, *Gastroenterology,* 95, 668, 1988.
73. **Rutegard, J., Ahsgren, L., Stenling, R., and Roos, G.,** DNA content in ulcerative colitis: flow cytometric analysis in a patient series from a defined area, *Dis. Colon Rectum,* 31, 710, 1988.

74. **Allen, D. C., Hamilton, P. W., Watt, P. C. H., and Biggart, J. D.,** Morphometrical analysis in ulcerative colitis with dysplasia and carcinoma, *Histopathology,* 11, 913, 1987.
75. **Collins, R. H., Feldman, M., and Fordtran, J. S.,** Colon cancer, dysplasia and surveillance in patients with ulcerative colitis: a critical review, *N. Engl. J. Med.,* 316, 1654, 1987.
76. **Fuson, J. A., Farmer, R. G., Hawk, W. A., and Sullivan, B. H.,** Endoscopic surveillance for cancer in chronic ulcerative colitis, *Am. J. Gastroenterol.,* 73, 120, 1980.
77. **Nugent, F. W. and Haggitt, R. C.,** Results of a longterm prospective surveillance program for dysplasia in ulcerative colitis, *Gastroenterology,* 86(Abstr.), 1197, 1984.

Chapter 10

ULCERATIVE COLITIS — CLINICAL ASPECTS AND COMPLICATIONS

Humphrey J. O'Connor

TABLE OF CONTENTS

I. INTRODUCTION

On the basis of the patient's history alone, one can often suspect ulcerative colitis even before sigmoidoscopy or a barium enema has been performed. Most physical signs in ulcerative colitis result from the complications of the disease. This chapter outlines the symptoms and signs of ulcerative colitis, their value in the assessment of disease severity, and the local and systemic complications of the condition.

II. SYMPTOMS

The most common symptoms of ulcerative colitis are rectal bleeding, diarrhea, abdominal pain, and weight loss.

A. RECTAL BLEEDING

The passage of blood from the rectum is the most frequent, and often the first, symptom of ulcerative colitis. In disease limited to the rectum, blood and mucus are often passed without stool of any kind, blood may be mixed in a liquid stool, or the patient passes formed stool coated or streaked with blood. In more extensive disease, the amount of rectal bleeding may be considerable, occasionally amounting to torrential hemorrhage in the acutely ill patient.

B. DIARRHEA

Diarrhea and urgency to defecate are often the most disabling symptoms of ulcerative colitis and occur in about 50% of patients as initial symptoms. The severity of diarrhea is in general related to the extent of large bowel involved. Patients with proctitis, for instance, may not suffer at all from diarrhea and constipation may predominate with stool impacted in the uninvolved colon proximal to the affected area. In contrast, patients with acute total colitis may have urgent bloody diarrhea more than 20 times in 24 h which persists through the night and is associated with episodic and embarrassing incontinence. Nocturnal diarrhea usually implies severe disease. In the less severe case, diarrhea tends to occur in the morning and after meals.

Some patients with ulcerative colitis complain of the passage of either purulent material or clear to colored mucus which usually accompanies diarrhea or bleeding. A heavy flow of pus suggests the presence of a perianal fistula or abscess draining via the rectum.

C. ABDOMINAL PAIN

Ulcerative colitis is generally not a painful condition and in mild disease anything more than lower abdominal cramping, often relieved by defecation, is rare. In the acutely ill patient with severe ulcerative colitis, however, intense abdominal pain may be a prominent initial symptom and should alert the clinician to the possibility of serosal inflammation with impending toxic dilatation or perforation of the colon.

Some patients with ulcerative colitis experience pain in the rectum in the form of tenesmus — a severe bearing-down pain lasting for many minutes and usually accompanied by an inability to defecate.

D. WEIGHT LOSS

In a severe attack of ulcerative colitis, the combination of an intense catabolic state, diminished intake, and excessive loss of nutrients may result in profound weight loss. In the long term, patients with ulcerative colitis, given proper medical management, maintain normal or near-normal body weight.

E. OTHER SYMPTOMS

When ulcerative colitis is active, patients often complain of profound generalized weakness and fatigue and some will admit to feeling depressed. Some patients experience nausea, though vomiting is unusual in ulcerative colitis.

An important long-term effect of ulcerative colitis in childhood is growth retardation. This clinical feature is associated with the presence of active disease and growth resumes when the disease remits or following total colectomy.

III. PHYSICAL SIGNS

Physical examination of the patient with a mild or moderate attack of ulcerative colitis may be essentially normal. In contrast, the patient with the most severe form of colitis, so-called fulminant colitis, appears distinctly unwell, obtunded, pale, dehydrated, and commonly has a high fever. Muscle wasting in fulminant colitis is often striking and may develop within a relatively short time. The skin, eyes, and joints should be examined carefully for evidence of the systemic complications of ulcerative colitis (see below). Recurrent aphthous ulcers of the mouth occur commonly in ulcerative colitis and may be particularly severe during an acute attack.

A. ABDOMINAL EXAMINATION

Examination of the abdomen is unremarkable in most patients with ulcerative colitis. In the severely ill patient with fulminant colitis, the onset of abdominal tenderness particularly with rebound and distension suggests that the inflammatory process is extending through the colonic wall with the consequent risk of toxic dilatation and/or perforation. Particular attention should be paid to bowel sounds and percussion; the finding of a tympanitic abdomen with absent bowel sounds should be considered indicative of toxic dilatation or perforation until proved otherwise by abdominal X-rays. Clearly, abdominal examination should be performed several times each day in patients with fulminant colitis.

The finding of a mass in the abdomen is suspicious of a tumor or an abscess.

B. RECTAL EXAMINATION

Visual examination of the perianal area should precede digital examination which in turn should always precede sigmoidoscopy. Perianal fistulae can be detected without difficulty on inspection. Digital examination which produces pain suggests the presence of an anal fissure or perianal abscess.

In severe acute colitis, fresh blood and mucopus usually coat the examining finger and spill from the anal verge immediately after the examination. In chronic long-standing disease, pseudopolyps or fibrous strictures can often be felt on rectal examination and carcinoma complicating ulcerative colitis can, in up to 25% of cases, be detected by a diligent rectal examination.

IV. CLINICAL FEATURES AND SEVERITY OF DISEASE

The importance of careful assessment of clinical features in ulcerative colitis is underlined by the fact that the severity of symptoms correlates with both the extent of colitis and the severity of the inflammation. Hence, the severity of attacks of ulcerative colitis can be usefully classified according to simple clinical criteria as first described by Truelove and Witts.[1]

Severe disease — Characterized by severe diarrhea (six or more motions a day) with macroscopic blood in the stool, fever, tachycardia, anemia, and an E.S.R. more than 30 mm in 1 h.

Mild disease — Characterized by mild diarrhea (four or less motions a day) with no more than small amounts of macroscopic blood in the stools, absence of both fever and tachycardia, a minor degree of anemia, and an E.S.R. of 30 mm in 1 h or less.

Moderate disease — Intermediate between severe and mild. In their unique study of 624 patients with ulcerative colitis, Edwards and Truelove found that about 60% of the patients had mild attacks,[2] 25% moderate, and 15% severe. Furthermore, the clinical severity of an attack of colitis had a profound influence on the short-term prognosis; mild attacks carried a very low risk of death whereas severe attacks carried a grave short-term prognosis with nearly one third dying. A clinically severe attack almost always implied substantial or entire colonic involvement whereas total colitis was found in only 20% of the mild cases.

Lennard-Jones et al.[3] have emphasized the importance of early recognition of certain features of the acute attack which carry a bad prognosis, namely, the maximum daily temperature, the maximum daily pulse rate, the frequency of bowel actions, and the plasma albumin concentration. Thus, patients who pass more than eight stools in the first 24 h of hospitalization and have a pyrexia greater than 38°C should be regarded as possible candidates for surgical intervention during the attack. Furthermore, the failure of therapy over the first 4 d to reduce the pulse rate, temperature, and stool frequency and the appearance of colonic dilatation are strong indications for operation. Between 5 and 15% of patients require colectomy for their first attack of ulcerative colitis.

For most of the patients who survive their first attack, ulcerative colitis in the long-term pursues a chronic intermittent course with periods of complete freedom from symptoms interspersed with attacks of the disease. A minority of patients continue to suffer from continuous symptoms of variable severity, the so-called chronic continuous form of the disease. Some patients seem never to have a second attack, though the longer the period of follow-up, the smaller is the proportion of patients who have had only one attack. In the Oxford series[2], only 4% had escaped a second attack after 15 years from the onset. Hence, ulcerative colitis is a chronic disease which is always likely to recur in a symptomatic form. It is of interest that the clinical severity of the first attack has little influence on the pattern of the long-term course of ulcerative colitis.

V. COMPLICATIONS

The complications of ulcerative colitis fall logically into two groups: local complications in and around the colon and remote or systemic complications.

A. LOCAL COMPLICATIONS

The local complications may be grouped into those associated with the acute attack and those associated with chronic long-standing disease.

1. Acute Ulcerative Colitis

In an acute attack of ulcerative colitis, the likelihood of complications increases with the severity of the attack. Thus, patients with fulminant colitis are especially prone to one of the following complications: toxic megacolon, perforation, or hemorrhage.

a. Toxic Megacolon

Acute dilatation of the colon may occur spontaneously in severe ulcerative colitis or may be iatrogenically precipitated by the inadvertent use of drugs which paralyze the colonic musculature including opiates, anticholinergic drugs, and antidiarrheal agents. Other precipitants of toxic megacolon are hypokalemia and the performance of barium enema examinations, particularly air-contrast examinations, during a fulminant attack of colitis. Toxic megacolon is as likely to occur during relapses as first attacks of the disease.[4]

Megacolon has been arbitrarily defined as dilatation of the transverse colon to a diameter greater than 5.5 cm[5,6] though this definition does not take into account "toxicity" and the clinical state of the patient. Some patients may be severely ill with a 5-cm diameter colon, while others may be relatively well with a 6-cm diameter colon. Despite this, careful attention should be paid to any dilatation of the colon in severe colitis as it usually indicates that inflammation is extending into the muscle coats of the colon with an attendent high risk of perforation.

An early symptom of incipient toxic megacolon is often a sudden decrease in diarrhea which at first glance appears to be an improvement but is usually accompanied by a marked deterioration in the patient's general condition. Vomiting, an unusual symptom in ulcerative colitis per se, may occur with toxic megacolon. On examination, the abdomen is tympanitic and bowel sounds are reduced or absent. Abdominal distension is the next physical sign often occurring within several hours after the initial symptoms.

Perforation occurs in approximately 20% of patients with toxic megacolon and carries a mortality of over 50%.[7] The overall mortality from toxic megacolon in a recent series was 30% despite intensive medical therapy and a policy of early surgery.[8] Even after successful medical therapy, toxic megacolon has a predeliction to recur in the same patient and nearly one half eventually require colectomy.[9]

b. Perforation

Perforation of the colon is the most dangerous local complication of ulcerative colitis, resulting in a generalized peritonitis which is fatal in up to 75% of cases.[4] Though the majority of perforations occur in patients with fulminant colitis or toxic megacolon, this complication can also occur in moderately severe disease and is especially likely to occur during a first attack of colitis.[4] Because of the paper-thin bowel wall in severe colitis, perforation may occur at more than one site.

The presenting features of perforation include generalized abdominal pain, distension, rebound tenderness, and absent bowel sounds. However, these classic signs of perforation may be masked by corticosteroid therapy and the only indication that some disaster has occurred may be a marked deterioration in the general condition of the patient. Under these circumstances, an immediate plain abdominal X-ray to detect free gas is mandatory. Both delayed recognition and delay in performing emergency colectomy contribute to the excessively high mortality from perforation in ulcerative colitis.

c. Hemorrhage

Massive hemorrhage is a rare complication of ulcerative colitis and is almost always associated with severe disease.[4] It can usually be treated successfully by blood transfusion; however, exceptionally, it is an indication for emergency colectomy.

2. Chronic Ulcerative Colitis

a. Pseudopolyposis

The development of pseudopolyps is the result of epithelialization of nodules of exuberant granulation tissue. This complication is associated with severe total ulcerative colitis and patients may have survived an episode of fulminant colitis or toxic megacolon. Pseudopolyps can develop rapidly and not infrequently during the first attack of the disease. Once developed, pseudopolyps usually persist though this is not invariable and regression can occur.[4] It is sometimes supposed that pseudopolyps may be precancerous. There is no good evidence to support this idea; however, since the patient with pseudopolyps usually has extensive, chronic, and often long-standing disease, such a patient may also be in a relatively high-risk category for the development of a neoplasm. Pseudopolyposis in itself is not a firm indication for colectomy.

b. Stricture

The development of a fibrous stricture may complicate chronic ulcerative colitis at any time though approximately one third will be diagnosed within 5 years of disease onset.[4] Strictures are most frequent in the rectum and sigmoid colon but may occur in any part of the colon; they are occasionally multiple. Patients with the chronic continuous form of the disease may be more liable to develop stricture than those with the chronic intermittent form. The finding of a stricture automatically raises the possibility of carcinoma complicating ulcerative colitis and strictures should be examined by colonoscopy and biopsy specimens and brushings taken to exclude malignancy. Benign strictures rarely cause symptoms per se and a stricture producing obstruction is more likely to be due to malignancy or to a misdiagnosis of Crohn's disease.

c. Carcinoma of the Colon

Carcinoma of the colon is the most feared complication of chronic ulcerative colitis. Of the 624 patients followed up by Edwards and Truelove,[22] 3.5% developed carcinoma of the colon of whom 17 died as a direct consequence. Patients with ulcerative colitis have at least an eightfold increased risk of cancer of the colon relative to the risk in the general population.[10] Certain clinical features are especially associated with a high risk of carcinoma including extensive colitis, increasing duration of disease, a chronic continuous clinical course, and young age at the time of diagnosis.

In a comprehensive survey of 823 patients with ulcerative colitis,[10] Gyde et al. showed that patients with extensive colitis had a 19-fold increase in risk compared with a 4-fold increased risk in patients with either left-sided colitis or proctitis. Life table analysis in their patients with extensive colitis gave a cumulative risk of 7% at 20 years and 17% at 30 years from onset of disease. Very few cases develop when the disease has been present for less than 10 years.

Patients whose clinical course has been a chronic continuous one have a much higher risk of cancer than those with the chronic intermittent form and deDombal et al.[11] relate cancer risk to "patient-years" of activity. Young age at the time of disease onset is an additional risk factor but may only exert a substantial effect when the onset is under the age of ten.[4,10]

Carcinoma complicating ulcerative colitis is difficult to diagnose. The symptoms of colitis may mimic those of colonic cancer and the tumors themselves tend to be flat, infiltrating, and poorly differentiated leading to delayed presentation. These considerations and the increasing availability of colonoscopy in recent years have sparked intense interest in the question of long-term cancer surveillance in patients with ulcerative colitis and will be dealt with in another chapter.

d. Perianal Complications

The perianal complications seen in 1089 patients followed with ulcerative colitis included hemorrhoids (20%), anal fissures (12%), anal fistulae (4 to 5%), perianal or ischiorectal abscesses (4 to 6%), rectal prolapse (2%), and rectovaginal fistulae (2 to 3%).[12] Most of these complications occur during periods of disease activity and are more frequent in patients with total colitis than distal disease. Patients with perianal complications complain of pain and discomfort around the anus, especially on defecation, and the passage of purulent material or bright red blood per rectum. The diagnosis is usually apparent on visual and digital examination.

B. SYSTEMIC COMPLICATIONS

The extraintestinal complications of ulcerative colitis usually present after the colonic symptoms but may be present when the patient is first seen and occasionally may precede the development of colonic symptoms. Sometimes, the extracolonic complications of ulcerative colitis can be more devastating than the disease itself and carry a prognosis independent of that of the colonic disease.

1. Liver and Biliary Tract

The incidence of liver disease in ulcerative colitis is uncertain. Up to 15% of patients show evidence of biochemical dysfunction; however, clinically overt liver disease is uncommon (1 to 3%).[4,13] Histological assessment shows that several liver and biliary lesions occur more frequently in patients with ulcerative colitis than in control populations and include fatty change, pericholangitis, chronic active hepatitis, amyloidosis, sclerosing cholangitis, and bile duct carcinoma.

Fatty change — The most common pathological change occurring alone or associated with other histological changes.[13] Fat accumulation is heaviest at the periphery of the hepatic lobule and serial biopsies have shown the lesion is reversible. The etiology of fatty infiltration is unknown though malnutrition and protein depletion have been implicated. Minimal enlargement of the liver may be the only physical sign and liver function tests may be normal or mildly abnormal.

Pericholangitis — Characterized by a periportal inflammatory infiltrate of predominantly mononuclear cells, degenerative changes in bile ductules, and varying degrees of periportal fibrosis. In an unselected series of 300 patients with ulcerative colitis,[13] pericholangitis was found in 15 (5%). The etiology and natural history of pericholangitis are unclear, although serial biopsies have shown evolution to cirrhosis in a few patients.[14] There are no symptoms associated with pericholangitis. Serum alkaline phosphatase is elevated in contrast to almost normal values for transaminases. Jaundice is rare in patients with pericholangitis uncomplicated by cirrhosis.

Chronic active hepatitis — Associated with ulcerative colitis and either condition can precede the other. Physical signs such as splenomegaly are usually the result of a complicating cirrhosis. Hepatic amyloidosis is a rare complication of ulcerative colitis comprising less than 1% of lesions seen on liver biopsy.

Sclerosing cholangitis — Increasingly recognized as a rare but serious complication of ulcerative colitis.[15] A chronic inflammatory process directed at the extrahepatic and/or intrahepatic bile ducts results in fibrosis and irregular narrowing and beading of the biliary tree. This process leads, over a variable period of time, to biliary obstruction and eventually to cirrhosis. Patients present with recurrent attacks of jaundice, right upper-quadrant pain, and fever though some may present initially with variceal bleeding. Compared with other forms of cirrhosis, the classic peripheral stigmata of chronic liver disease — clubbing, palmar erythema, etc., are relatively uncommon in sclerosing cholangitis. Liver function tests indicate extrahepatic biliary obstruction. The diagnosis of sclerosing cholangitis is made using the combination of liver biopsy and direct contrast cholangiography by either the percutaneous transhepatic route or ERCP (Figure 1).

Biliary tract carcinoma — Approximately ten times more frequent in patients with ulcerative colitis than in the general population. Most patients with this complication have both a long history of colitis and total involvement of the colon. The tumor may appear several years after total proctocolectomy.[16] Recent evidence suggests that cholangiocarcinoma may complicate sclerosing cholangitis,[17] and accurate differentiation of the two conditions in patients with ulcerative colitis is often difficult despite the use of sophisticated imaging techniques and ultrasound- or computerized axial tomography-guided biopsy may be required.

2. Skin

Skin lesions have been reported in up to 10% of patients with ulcerative colitis.[18] Nonspecific eruptions are seen, including urticaria, angioedema, erythema, and purpura; however, the distinctive skin lesions associated with ulcerative colitis are pyoderma gangrenosum and erythema nodosum.

Pyoderma gangrenosum — Characterized by papules and pustules which enlarge and coalesce to form large areas of painful, necrotic ulceration surrounded by a violaceous

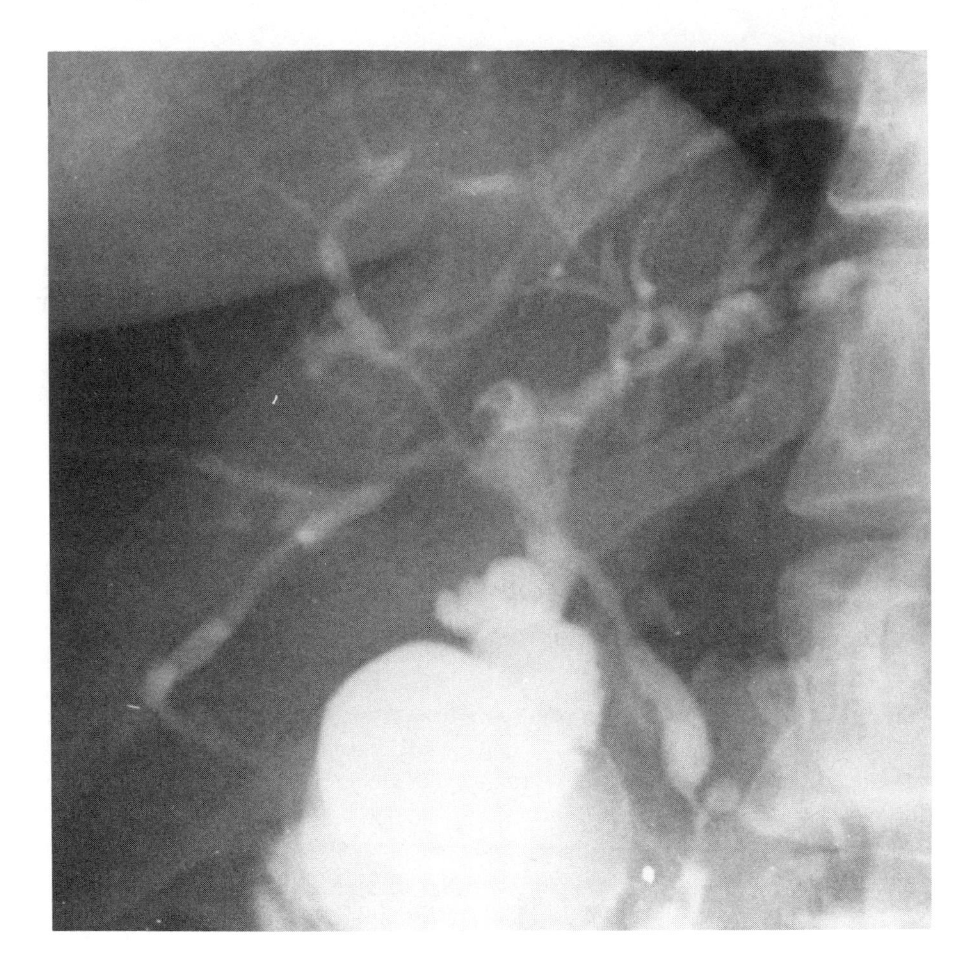

FIGURE 1. Endoscopic retrograde cholangiogram showing the characteristic appearances of sclerosing cholangitis with beading and strictures of the extrahepatic and intrahepatic bile ducts.

undermined border (Figure 2). Though the condition may appear unrelated to the activity of the colitis, pyoderma gangrenosum usually complicates a severe attack of pancolitis and is associated with marked systemic symptoms.[4] Pyoderma can occur anywhere on the body but is typically found on the shin, often on the arms, and may even involve the face. Successful treatment of the colitis usually results in healing of the skin. Persistent severe pyoderma gangrenosum is itself an indication for colectomy.

Erythema nodosum — Characterized by raised, tender erythematous swellings most often found on the extensor surface of the legs and occasionally the arms. About 3% of patients will develop erythema nodosum usually during an exacerbation of their colitis. The lesion is more common in women than men and frequently appears in conjunction with the arthritis of ulcerative colitis.

Necrotizing leukocytoclastic vasculitis has been reported in ulcerative colitis.[19]

3. Joints

Specific arthritis — The specific arthritis of ulcerative colitis usually coincides with a first attack or a relapse of the colitis though joint symptoms may occasionally antedate bowel symptoms.[20] The arthritis is typically a monoarthritis or an oligoarthritis affecting large weight-bearing joints and tends to be migratory. In some patients, the smaller joints of the

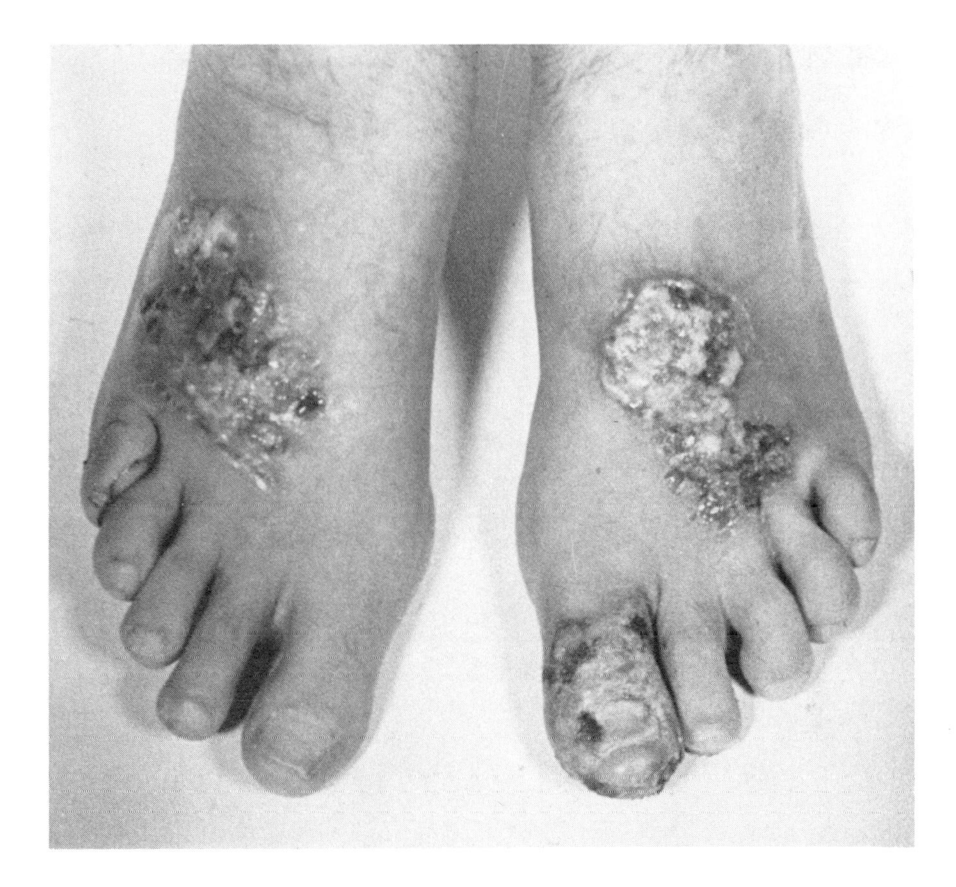

FIGURE 2. Pyoderma gangrenosum in a patient with ulcerative colitis. Note the large area of necrotic ulceration surrounded by an undermined border.

hands and feet are affected. There is synovitis, similar to that in rheumatoid arthritis with effusion, and the joint is swollen, tender, and reddened. In contrast to rheumatoid disease, however, there is no erosion of the juxtaarticular bone. Though the onset of bowel and joint symptoms often coincides, there may be disparity between the severity of the arthritis and the colitis: the colitis may be mild, yet the disability from the arthritis may be severe. When the colitis is treated effectively, the arthritis usually subsides without residual joint damage.[4]

Ankylosing spondylitis — This condition is about 10 to 20 times more frequent in patients with ulcerative colitis than in the general population and, conversely, the incidence of ulcerative colitis is high in patients with ankylosing spondylitis.[21] Men are only about twice as frequently affected as women in contrast to ankylosing spondylitis occurring in the absence of colitis where the male:female ratio is about 10:1. A clue to the etiology comes from the strong association between histocompatibility antigens and ankylosing spondylitis. In ankylosing spondylitis unassociated with ulcerative colitis, HLA-B27 is found in about 90% of patients, whereas only about 50% of patients with ankylosing spondylitis who also suffer from ulcerative colitis have this antigen.[22] The course of ankylosing spondylitis is the same whether or not there is associated ulcerative colitis with chronic, gradually progressive, painful deformity of the spine. Ankylosing spondylitis progresses despite quiescent colitis. The incidence of sacroileitis is higher than that of ankylosing spondylitis in patients with ulcerative colitis, although many of these patients are asymptomatic.[23]

Rheumatoid arthritis — There is an increased incidence of rheumatoid arthritis in patients with ulcerative colitis but the two conditions usually run separate and unrelated courses in the individual patient.

4. Eye Lesions

Uveitis — This condition, affecting the anterior and/or posterior uveal tract, is the most common ocular lesion seen with ulcerative colitis, affecting some 5 to 10% of patients.[24] Anterior uveitis (iritis) presents as a painful red eye and the patient's marked discomfort distinguishes the condition from conjunctivitis. Physical examination will reveal erythema concentrated around the iris, inflammatory exudate in the anterior chamber, and corneal precipitates. Any patient with ulcerative colitis who has a sudden change in vision should be promptly examined for evidence of posterior uveitis which, if left untreated, can lead to retinal destruction and blindness. In almost one half the cases, uveitis will be bilateral. Attacks of uveitis usually coincide with frank attacks of colitis and not uncommonly with other systemic complications, notably arthritis and skin lesions.

Episcleritis — Also more common in ulcerative colitis and is characterized by inflammation of episcleral tissues with associated discomfort and localized reddening.

Other eye lesions seen in association with ulcerative colitis are superficial or interstitial keratitis, retrobulbar neuritis, and Sjogren's syndrome.

5. Blood Disorders

Iron-deficiency anemia — The most common blood disorder seen in ulcerative colitis is iron-deficiency anemia, the result of persistent or recurrent loss of blood from the colon. Of the 624 patients in the Oxford series,[4] 167 (20%) had a hemoglobin level <9.0 g/dl at some time during the course of their illness. Even patients who are clinically asymptomatic may continue to have increased blood loss from the colon[25] and a persistent hypochromic anemia may indicate continuing disease activity. During an acute attack of colitis, the combination of blood loss and marrow suppression may produce a severe degree of anemia in a surprisingly short time requiring blood transfusion for correction.

Hemolytic anemia — This condition with Heinz bodies in the peripheral blood film is seen in some patients receiving sulfasalazine.[26] A more severe autoimmune Coombs-positive hemolytic anemia may also complicate ulcerative colitis[27] as can a microangiopathic hemolytic anemia with or without disseminated intravascular coagulopathy.

Hypoplasia and aplasia — These conditions of the bone marrow may occur in patients receiving azathioprine and, rarely, sulfasalazine. Azathioprine-induced marrow suppression is usually reversible, whereas that secondary to sulfasalazine may be irreversible and fatal.

Hypoprothrombinemia — This condition with a prolonged prothrombin time is sometimes seen in moderate or severe ulcerative colitis even in the absence of complicating chronic liver disease. Conversely, a *hypercoagulable state* may complicate ulcerative colitis with thrombocytosis, increased levels of factors V and VIII and fibrinogen and reduced levels of antithrombin.[28] Deep-vein thrombosis and pulmonary embolism are known complications of ulcerative colitis which usually coincide with an acute attack of the disease.[4]

6. Other Complications

For reasons that remain unclear, the incidence of *nephrolithiasis* and *urolithiasis* is about twice that in the normal population with a particular predeliction for uric acid stones.[18] Interstitial pulmonary fibrosis and pulmonary vasculitis have been reported in association with ulcerative colitis,[29] as well as an increased incidence of hay fever and asthma.[30] There are reported cases of pericarditis and myocarditis occurring in association with active ulcerative colitis.[31] Finally, there is an increased incidence of goiter and thyrotoxicosis in patients with ulcerative colitis.[32]

REFERENCES

1. **Truelove, S. C. and Witts, L. J.,** Cortisone in ulcerative colitis: final report on a therapeutic trial, *Br. Med. J.*, 2, 1041, 1955.
2. **Edwards, F. C. and Truelove, S. C.,** The course and prognosis of ulcerative colitis, *Gut,* 4, 299, 1963.
3. **Lennard-Jones, J. E., Cooper, G. W., Newell, A. C., Wilson, C. W. E., and Avery Jones, F.,** Observations on idiopathic proctitis, *Gut,* 3, 201, 1962.
4. **Edwards, F. C. and Truelove, S. C.,** The course and prognosis of ulcerative colitis. Part III. Complications, *Gut,* 5, 1, 1964.
5. **Hywel-Jones, J. and Chapman, M.,** Definition of megacolon in colitis, *Gut,* 10, 562, 1969.
6. **Bartram, C. I.,** Plain abdominal X-ray in acute colitis, *Proc. R. Soc. Med.,* 69, 617, 1976.
7. **Binder, S. C., Patterson, J. F., and Glotzer, D. J.,** Toxic megacolon in ulcerative colitis, *Gastroenterology,* 66, 909, 1974.
8. **Muscroft, T. J., Warren, P. M., Montgomery, R. D., Asquith, P., and Sokhi, G. S.,** Toxic megacolon in ulcerative colitis: a medico-surgical audit, *Gut,* 21, A454, 1980.
9. **Grant, C. S. and Dozois, R. R.,** Toxic megacolon: ultimate fate of patients after successful medical management, *Am. J. Surg.,* 147, 106, 1984.
10. **Gyde, S. N., Prior, P., Allan, R. N., Stevens, A., Jewell, D. P., Truelove, S. C., Lofberg, R., Brostrom, O., and Hellers, G.,** Colorectal cancer in ulcerative colitis: a cohort study of primary referrals from three centres, *Gut,* 29, 206, 1988.
11. **deDombal, F. T., Watts, J. McK., Watkinson, G., and Goligher, J .C.,** Local complications of ulcerative colitis: stricture, pseudopolyposis and carcinoma of colon and rectum, *Br. Med. J.,* 1, 1442, 1966.
12. **deDombal, F. T., Watts, M. B., Watkinson, G., and Goligher, J. C.,** Incidence and management of anorectal abscess, fistula and fissure in patients with ulcerative colitis, *Dis. Colon Rectum,* 9, 201, 1966.
13. **Perrett, A. D., Higgins, G., Johnston, H. H., Massarella, G. R., Truelove, S. C., and Wright, R.,** The liver in ulcerative colitis, *Q. J. Med.,* 40, 211, 1971.
14. **Mistilis, S. P.,** Pericholangitis and ulcerative colitis. I. Pathology, etiology, and pathogenesis, *Ann. Intern. Med.,* 63, 1, 1965.
15. **Chapman, R. W. G., Marborgh, B. A., Rhodes, J. M., Summerfield, J. A., Dick, P., Scheuer, P. J., and Sherlock, S.,** Primary sclerosing cholangitis: a review of its clinical features, cholangiography, and hepatic histology, *Gut,* 21, 870, 1980.
16. **Ritchie, J. K., Allan, R. N., Macartney, J., Thompson, H., Hawley, P. R., and Cooke, W. T.,** Biliary tract carcinoma associated with ulcerative colitis, *Q. J. Med.,* 43, 263, 1974.
17. **MacCarty, R. C., LaRusso, N. F., May, G. R., Bender, C. E., Wiesner, R. H., King, J. E., and Coffey, R. J.,** Cholangiocarcinoma complicating primary sclerosing cholangitis: cholangiographic appearances, *Radiology,* 156, 43, 1985.
18. **Greenstein, A. J., Janowitz, H. D., and Sachar, D. B.,** The extraintestinal complications of Crohn's disease and ulcerative colitis: a study of 700 patients, *Medicine (Baltimore),* 55, 401, 1976.
19. **Wackers, F. J. Th., Tytgat, G. N., and Vreeken, J.,** Necrotizing vasculitis and ulcerative colitis, *Br. Med. J.,* 4, 83, 1974.
20. **McEwen, C., Lingg, C., and Kirsner, J. B.,** Arthritis accompanying ulcerative colitis, *Am. J. Med.,* 33, 923, 1962.
21. **Acheson, E. D.,** An association between ulcerative colitis, regional enteritis, and ankylosing spondylitis, *Q. J. Med.,* 29, 489, 1960.
22. **Dekker-Saeys, B. J., Meuwissen, S. G. M., Van Den Berg-Loonen, E. M., Dehaas, W. H. D., Meijers, K. A. F., and Tytgat, G. N. J.,** Clinical characteristics and results of histocompatibility typing (HLA B27) in 50 patients with both ankylosing spondylitis and inflammatory bowel disease, *Ann. Rheum. Dis.,* 37, 36, 1978.
23. **Wright, R., Lumsden, K., Luntz, M. H., Sevel, D., and Truelove, S. C.,** Abnormalities of the sacro-iliac joints and uveitis in ulcerative colitis, *Q. J. Med.,* 34, 229, 1965.
24. **Korelitz, B. I. and Coles, R. S.,** Uveitis (iritis) associated with ulcerative and granulomatous colitis, *Gastroenterology,* 52, 78, 1967.
25. **Beal, R. W., Skyring, A. P., McRae, J., and Firkin, B. G.,** The anemia of ulcerative colitis, *Gastroenterology,* 45, 589, 1963.
26. **Pounder, R. E., Craven, F. R., and Henthern, J. S.,** Red cell abnormalities associated with sulphasalazine therapy for ulcerative colitis, *Gut,* 16, 1975, 1981.
27. **Goldstone, A. H.,** Autoimmune haemolytic anaemia in ulcerative colitis, *Br. Med. J.,* 2, 556, 1974.
28. **Lam, A., Borda, I., and Inwood, M.,** Coagulation studies in ulcerative colitis and Crohn's disease, *Gastroenterology,* 68, 245, 1975.
29. **Craft, S. C., Earle, R. H., and Roesler, M.,** Unexplained bronchopulmonary disease with inflammatory bowel disease, *Arch. Intern. Med.,* 136, 454, 1976.

30. **Jewell, D. P. and Truelove, S. C.,** Reaginic hypersensitivity in ulcerative colitis, *Gut,* 13, 903, 1972.
31. **Goodman, M. J., Moir, D. J., and Holt, J. M.,** Pericarditis associated with ulcerative colitis and Crohn's disease, *Am. J. Dig. Dis.,* 21, 98, 1976.
32. **Jarnerot, G., Azad Khan, A. K., and Truelove, S. C.,** The thyroid in ulcerative colitis and Crohn's disease, *Acta Med. Scand.,* 197, 83, 1975.

Chapter 11

PSYCHOLOGICAL ASPECTS OF ULCERATIVE COLITIS

P. J. Shoenberg

George Engel observed that the sickness of ulcerative colitis and its consequent disabling effects makes the person with this condition much less able to cope with his environment and his personal problems. Much of his energy is bound to be concentrated on his illness. This places a limitation on his psychological resources and the loss of energy renders him helpless at the times of exacerbation and so makes him more dependent on others.[1,2] In this chapter I will explore this relationship between the person with ulcerative colitis and his disease and examine his dependency needs.

It is now 59 years since Cecil Murray published his observations on the psychogenic factors in the etiology of ulcerative colitis.[3] Since then clinicians have been intrigued by the possibility that there might be psychosomatic aspects to this condition. Murray noticed that his 12 patients were emotionally immature, fearful, and had too close a tie with their mothers. He also noticed that there was a close association in time between the emergence of a difficult psychological situation and the onset of the physical symptoms of ulcerative colitis. He concluded that psychological factors must play a significant role in the causation of ulcerative colitis. How far have his observations been born out?

As Drossman has pointed out,[4] research in this area has been dogged by a variety of problems of methodology: (1) because there has been a lack of agreement as to the terminology, conceptualization, and research design among behavioral disciplines, (2) because there have been difficulties in obtaining relevant psychological data, (3) because it has been difficult to control for the numerous psychosocial variables, (4) because much of the data have been retrospective, (5) because the early studies could not distinguish between ulcerative colitis and Crohn's disease, (6) because the assessments were made without healthy or medical comparison groups, (7) because the symptom severity or illness chronicity were not considered, and (8) because conclusions were often drawn from studies using unskilled interviewers or questionnaires which would not discriminate for sensitive or deep-seated psychological data.

Evidence from two separate, adequately controlled studies[5,6] suggests that there may be a close association between the emergence of a difficult psychological situation and the onset of the physical symptoms of ulcerative colitis. These studies show that significantly stressful life events, such as a bereavement, a terminal illness in a partner, a marriage or a divorce, a pregnancy or a childbirth, or immigration,[5] are associated with ulcerative colitis and that this association is statistically significant. George Engel,[1,2] the physician and psychoanalyst, drawing on his own work as a psychotherapist with ulcerative colitis, has argued that each of these stressful life events represents for the patient a loss of, or the threat of the loss of, a key personal relationship to another person. Why should such a threat have such severe effects on the patient prone to colitis?

The early psychoanalytic studies of the personality of patients with ulcerative colitis suggested that they were, as Murray had observed,[3] often immature and emotionally vulnerable people. However, the first adequately controlled study by Feldman and colleagues tended to contradict these findings,[7] they showed that when a consecutive series of ulcerative colitis patients was compared with a control group they showed no evidence of significantly increased psychiatric morbidity. However, Feldman's study did not use particularly sensitive methods to explore the emotional difficulties of these patients. In the last 2 decades psychoanalysts working in Paris and Boston[8,9] published new findings on patients with a variety of psychosomatic disorders. These suggested that, although in many respects these patients appeared to be outwardly psychologically adjusted, in psychoanalytic interviews they emerged as being crippled by a particular *inability to find words for their feelings* and as suffering with a poverty of emotional life in which their imaginative capacity was limited. Sifneos called this hidden disorder of the personality alexithymia (which means the incapacity to find words for feelings). A more recent study by Fava and Pavan[6] compared a consecutive series of three groups of patients with ulcerative colitis, irritable bowel syndrome, and appendicitis looking at psychiatric morbidity and alexithymia as separate variables. They showed that while the patients with ulcerative colitis had a low score for psychiatric morbidity in comparison with those with irritable bowel syndrome (who had a high score), the colitis group had a significantly high score (compared with the other two groups) for the personality trait alexithymia. This finding may go some way towards explaining the earlier negative results of Feldman's study[7] and supporting the possibility that people with colitis really are emotionally vulnerable. Engel has argued that the emotional vulnerability of the patient with colitis is derived from his childhood experiences which have made it difficult for him to grow into a fully mature and independent person; he has argued that the colitic's tendency is to make clinging and demanding relationships to others and to feel easily humiliated by slights and rejections, without being able to express the anger felt (which remains hidden and bottled up).[1,2] In psychotherapy these problems of dependency and emotionality are reenacted with the therapist, as they so often are also reenacted in the relationship to the physician. These difficulties in relating to others may stem from very complex early and later relationships of the child to the mother, who herself may have been depressed or vulnerable in some other way during the early phases of the child's life. The child has been over-sensitive to the mother's needs and has felt too easily aware of her moods and fearful of her disapproval. It is for these reasons he finds it hard to find an independence from her. In adult life this way of relating may be repeated with other key persons such as a husband or a wife. When there is any disruption or threatened disruption of this key relationship the patient suffers with feelings of rejection, fear, and hurt more easily than someone else might do. It is then that he feels as he did in childhood, helpless and hopeless, and it is then that an attack of colitis may occur. Examples of this follow.

Caroline, a 30-year-old single secretary with a 10-year history of proctocolitis, had been seeing me for once-weekly individual psychoanalytic psychotherapy for some months. She came from a family (with whom she was currently living) where there was a rift between her parents. Her father was a rather aggressive greengrocer who frequently quarrelled with her mother, a somewhat withdrawn and possibly depressed housewife. Caroline and her younger brother often acted as go betweens in these parental conflicts. She still felt an over dependence on her mother, always wishing for her approval. At Christmas time, during a break in her psychotherapy, her father quarrelled with his mother-in-law, who was visiting their home. He refused in the end to speak to the mother-in-law. Shortly after this row, Caroline had an attack of rectal bleeding. Some months further on into her psychotherapy she decided to move out of her family home into a flat of her own. She was worried that her mother would not approve of the flat she had purchased. Within a few weeks of moving into this flat she again began to bleed from the rectum.

Another patient, Janet, a single housing manager, also 30 years old but with a far more extensive colitis which eventually involved the whole of her colon, developed her second attack of colitis when she was 15 years old, shortly after her mother, with whom she had a very difficult relationship, had to go into the hospital for a hysterectomy. Some years later, shortly after the breakup of a love affair, her father died and Janet had a further attack of colitis. This time she became quite severely depressed as well. In both Caroline's and Janet's cases the onset of their symptoms occurred when they were preparing for school exams, ("O" levels in the case of Caroline and the 11+ exam in the case of Janet).

We do not yet know how it is that such stress acting on such vulnerable personalities can affect the brain to then act on a vulnerable target organ such as the colon to start the colitis. Animal studies with the Siamang gibbon and the cotton topped tamarin,[10,11] however, do confirm that under experimental conditions stressful situations may induce colitis-like lesions. Work in the new field of psychoneuroimmunology has shown that significant losses such as bereavements dramatically affect the immune responses; however, no specific studies have yet been done on the immune system and stress in ulcerative colitis.

There is some evidence from studies on a controlled series of patients with colitis that, in some cases of colitis, psychoanalytic psychotherapy may very much help the patient not only with their emotional conflicts but also with their physical prognosis as compared with those who have received medical treatment alone.[12] My own experience as a medical psychotherapist working alongside a team of gastroenterologists and surgeons has been that these patients with ulcerative colitis are sensitive and vulnerable personalities who feel too easily at the mercy of others and long for approval and love, yet have difficulty in trusting they can find it. Outwardly, they appear independent, even unwilling to get close, but inwardly they are more childlike and dependent, easily feeling hurt and rejected by a world with which they try too hard to comply. Sometimes they have a whole scatter of helpers, including a nutritionist, psychotherapist, and a relaxation therapist, in addition to their physician and general practitioner.[13] They feel intensely how different they are from other people and sense the stigma of their illness, fearing its consequences, especially hospitalizations. They are also often very lonely people who have great difficulty in making relationships to others and may continue in an overdependent one to their family. Only some can openly acknowledge a link between stress and their relapses. Others are more likely to see things in terms of purely physical things, such as their diet.

Caroline, although outwardly independent, was very vulnerable inside and particularly sensitive to the stigma of her illness. One day, well into the second year of her psychotherapy, she reported a dream in which she was at a dance where all the men were in evening dress. She noticed that for some reason in the dream their legs were floppy like those of puppets. One of these men came up to her to ask her for a dance. He looked like her lover but also like me. After she had told me the dream she recalled that it had occurred after a conversation with her lover in which he had suggested that she take out a life insurance policy because she suffered with colitis. She said she had felt humiliated by this suggestion of his but had said nothing, preferring to hide her feelings. Also she now recalled in this discussion about the dream that as a teenager she used to enjoy helping out at dances that were held in her neighborhood for physically handicapped children. I commented that in the dream in her rage against her lover who was physically fit she had turned tables on him and also on me, giving us the physical handicap (i.e., the floppy legs). She agreed with this interpretation of her dream.

In practice the character trait of alexithymia is more elusive than the results of Fava and Pavan's study might suggest.[6] Caroline and Janet certainly exhibited some aspects of alexithymia in their rather restricted fantasy lives and perhaps in the somewhat concrete way their narratives emerged in their respective psychotherapy sessions. Yet both were able to bring dreams for discussion in their psychotherapies and both could free associate to these dreams so that their symbolic meaning could be shared. At the same time, although often they did seem to find it difficult to find words to express their feelings, there were certainly plenty of times when they expressed their feelings as directly as any neurotic patient might in such a psychotherapy. What I noticed was a marked tendency for each of them to hide, suppress, and repress their aggressive feelings which were often readily translated into increased bowel action with diarrhea and bleeding per rectum. Both patients exhibited a marked tendency to form clinging, overdependent relationships to other key persons. In addition, Janet could very easily experience a sense of loss of self which was painful and threatening to her. Sometimes she came to psychotherapy sessions with a ghost-like presence. In childhood she had suffered from a very intrusive mother who was then, but also now, too emotionally demanding of Janet.

As a small girl she could remember her mother forcing her to eat special diets and take plenty of fresh air — she responded to these demands with deviousness.

Caroline, who came to psychotherapy by choice, was with me in once- and later twice-weekly psychoanalytic psychotherapy for 3 years. She certainly benefited from this experience both emotionally and physically. Her episodes of diarrhea diminished in frequency and she was able to mature as a person, leaving home and moving into a flat of her own and eventually marrying her lover. Janet, on the other hand, was referred to me because of a severe depression with episodes of depersonalization which had followed being told there was a prospect of her having to have radical surgery which might leave her with an ileostomy. At this time I arranged for her to see a psychotherapist in our hospital on a weekly basis. In this therapy Janet was dismissive of her therapist's attempts to relate to her. At the end of 6 months she broke off the psychotherapy, but a year later a sigmoidoscopy showed no recurrence of ulcers. When some years later her colitis, now involving the whole of her colon, flared up, her surgeon told her that he planned to do a colectomy with an ileoanal anastomosis. She became very distressed and depressed and complained of nightmares in which she saw witches coming towards her. Each of the witches had unsightly ileostomies. After this operation she developed an inflammation in the ileal pouch the surgeon had fashioned and she was again very depressed. I tried to persuade Janet to come into our inpatient psychosomatic unit. She initially agreed but left hospital to recuperate with her mother in Wales. When she returned, her "pouchitis" was still troubling her. I arranged for her to see a more experienced psychotherapist this time, now on a twice-weekly basis. By the end of a year of psychotherapy she was less depressed and her pouchitis had settled, but she herself remained an intensely vulnerable and immature personality.

Only very few patients are likely to come into psychotherapy, either by virtue of a true personal motivation and insight or else by virtue of a major psychological crisis such as a severe depressive reaction to surgery. How can the insights of psychoanalysis into the psychology of these few patients help the physician and the surgeon in their own care of their many other colitic patients? As Engel has said, when the patient is physically ill, he or she feels relatively helpless and correspondingly dependent on others. As Engel has also pointed out,[1,2] this may be to the advantage and to the disadvantage of the patient; this is because on the one hand it helps to remove him from the immediate life situation in which his attack has developed, but on the other hand it may lead to a revival of other conflicts of early life when the patient was also so helpless and dependent. For example, if, when as a small child, he experienced problems of cleanliness and soiling, then these are now intensified in his present experience of a colitic attack. Further physical illness reduces greatly the patient's means of coping with the stresses and frustrations of his life. When a colitic is very physically ill it may be inadvisable to attempt too active a psychological exploration until an effective relationship is established with the patient. These patients are acutely sensitive to rejection and so they need, but they may not be able to demand, extra time from the physician, whom they seek out but fear to become dependent on. Their dependency is one in which they hope the physician cannot only take over their physical but also their psychological care and so the physician may become an object of extreme dependency. If then the physician can recognize and also tolerate this demand, he may help such complex persons to find emotional as well as physical safety in the midst of the uncertainties of their illness. Often, other members of the staff find the colitic very difficult to cope with, especially the nurses who have to spend long periods of time at a stretch with the patient. They very much need to share their anxieties and apprehensions as a team with the physician. These patients above all need real continuity of care as relapses may well occur when the doctor, with whom they have established a good working relationship, leaves and is replaced by another doctor with whom they have to start all over again, and they naturally feel rejected.

When a decision for surgery is made it is important that much time is given to discuss this with the patient so that his emotional reactions to this prospect can be allowed for, especially when an ileostomy is likely. The patient who becomes clearly very depressed will need additional psychotherapeutic help. Many however tend to hide their feelings, fearing humiliation if these feelings are exposed or worse still not adequately understood. A very few can make links between the pattern of their emotional life and the pattern of their relapses of colitis. In these cases a referral to a psychotherapist should be considered.

In conclusion, I want to emphasize how difficult it can be to make effective psychological contact with these people who feel so physically damaged and handicapped by their condition. If such contact can be achieved, its results may be rewarding.

REFERENCES

1. **Engel, G. L.,** Psychological aspects of the management of patients with ulcerative colitis, *N.Y. State J. Med.*, September 15, 2255, 1952.
2. **Engel, G. L.,** Studies of ulcerative colitis. V. Psychosocial aspects and their implications for treatment, *Am. J. Dig. Dis.*, 3, 315, 1958.
3. **Murray, C. D.,** Psychogenic factors in the etiology of ulcerative colitis and bloody diarrhoea, *Am. J. Med. Sci.*, 180, 239, 1930.
4. **Drossman, D. A.,** Psychosocial aspects of ulcerative colitis and Crohn's disease, in *Inflammatory Bowel Disease*, 3rd ed., Kirsner, J. B. and Shorter, R. G., Eds., Lea & Febiger, Philadelphia, 1988, 209.
5. **Hislop, I. G.,** Onset setting in inflammatory bowel disease, *Med. J. Aust.*, 1, 981, 1974.
6. **Fava, G. A. and Pavan, L.,** Large bowel disorders. Illness configuration and life events; Psychopathology and alexithymia, *Psychother. Psychosom.*, 27, 93, 1976–1977.
7. **Feldman, F., Cantor, D., Soll, S., and Bachrach, W.,** Psychiatric study of a consecutive series of 34 patients with ulcerative colitis, *Br. Med. J.*, 3, 14, 1967.
8. **Marty, P. and de M'Uzan, M.,** La pensée operatoire, *Rev. Fr. Psychoanal. (Suppl.)*, 27, 1345, 1963.
9. **Sifneos, P. E.,** The prevalence of "Alexithymic" characteristics in psychosomatic patients, *Psychother. Psychosom.*, 22, 255, 1973.
10. **Stout, C. and Snyder, R. L.,** Ulcerative colitis-like lesions in Siamang gibbon, *Gastroenterology*, 57, 256, 1969.
11. **Chalifoux, L. V. and Bronson, R. T.,** Colonic adenocarcinoma associated with chronic colitis in cotton top marmosets, *Sanguinus oedipus*, *Gastroenterology*, 80, 942, 1981.
12. **Karush, A., Ed.,** *Psychotherapy in Chronic Ulcerative Colitis*, W. B. Saunders, Philadelphia, 1977.
13. **Winnicott, D. W.,** Psychosomatic illness in its positive and negative aspects, *Int. J. Psycho-Anal.*, 47, 510, 1966.

Chapter 12

DIET AND ULCERATIVE COLITIS

Luke J. D. O'Donnell

TABLE OF CONTENTS

I. INTRODUCTION

The simple logic that the food to which the gut is exposed may produce disordered physiology and disease has proved attractive to both physicians and laymen. Nevertheless, despite the attraction of this logic, no consensus exists that diet is of prime etiological importance in the vast majority of the major gastrointestinal diseases. This is particularly true in ulcerative colitis. Assessment of causation by epidemiological means presents many difficulties. The lack of simple methods to accurately define the exact prevalence and incidence of these diseases, the relative rarity of the diseases, and the inability of present day techniques of dietary assessment to portray the complexity of dietary intake all contribute to the difficulties.

Investigation of the relationship between dietary intake and gastrointestinal disease often is the domain of nutritionists rather than clinical gastroenterologists. This lack of enthusiasm on the part of gastroenterologists is surprising when compared to their large output of pharmacological, physiological, biochemical, and immunological data on all aspects of gastrointestinal disease. Nearly all patients with diseases of the gut ask their doctor "Could it be due to something I ate?" or "Do I need to change my diet?". The inadequacy of the reply however does not seem to have concentrated the minds and efforts of clinicians on the problem.

The fact that diet is often an emotive issue which attracts more than its fair share of evangelists and cynics does not help matters either. Furthermore, although there is no agreement regarding the role of diet in the causation of many gastrointestinal diseases, it has not hindered those who advocate a myriad of exotic and useless dietary treatments. The investigation of a dietary etiology in ulcerative colitis has not been a particularly fruitful field of investigation. This may reflect the possibility that diet has no etiological role; however, it is also possible that these investigations have failed to deliver an answer due to difficulties and inadequacies in their design and execution.

Dietary components may give rise to gut diseases due to excessive or decreased intake of a particular foodstuff over a prolonged period of time resulting in a maladaptive physiological response or in toxicity. Alternatively, an idiosyncratic reaction to a food component may occur. In the following sections the evidence regarding both of these etiological possibilities will be discussed. The nutritional management of patients with ulcerative colitis is also outlined.

II. CAUSATION AND OVERALL DIETARY INTAKE

There are seven reports in the literature which have investigated the relationship between overall dietary intake and ulcerative colitis. Most of these studies were designed to look at the relationship between Crohn's disease and diet and only three of them were designed specifically at the outset to study dietary factors in ulcerative colitis. All were case-control studies which used hospital-based patients not suffering from gastrointestinal diseases as the control group (mostly fracture and gynecological clinic patients). No data, therefore, are available comparing the dietary intake of ulcerative colitis patients and randomly selected population controls.

Of the four studies reporting on ulcerative colitis and diet, but which were primarily set up to look at the relationship between Crohn's disease and diet, two contain less than 20 patients with ulcerative colitis,[1,2] one reports only on intake of breakfast nutrients,[3] and the fourth only assessed the amount of sugar added to beverages and cereals.[4] Not surprisingly, none of these studies found that there was a relationship between diet and ulcerative colitis. In the largest of these studies, current sugar intake in 100 patients with ulcerative colitis was not significantly different from that of controls as assessed by postal questionnaire.[4] No comparative data on premorbid sugar intake in ulcerative colitis patients and controls were available. Moreover, only visible (added) sugar intake, rather than total sugar intake, was assessed in this study.

Two studies which specifically assessed the intake of sugar and unrefined carbohydrates as well as all other dietary constituents have been reported. Thornton et al. found no difference in dietary intake between patients and controls.[5] Panza et al., however, found a strong positive association between consumption of highly refined food (sugar, bread, sweets) and ulcerative colitis whereas negative association existed for consumption of vegetables and fruit.[6] Although Thornton et al. questioned patients about their premorbid dietary intake within 5 months of the onset of symptoms, their study suffered from the disadvantage that only 30 subjects were studied in each group and the method of relative risks was not used to analyze the data. Panza et al. had 124 patients and 250 controls in their study and analyzed their results using relative risks. However, in the study of Panza et al. patients were questioned about their premorbid diet up to 7 or more years after the onset of their illness.

An earlier study from Germany investigated the premorbid nutritional habits of 114 patients with ulcerative colitis by questionnaire administered a mean of 8.6 years after diagnosis. The results indicated that patients consumed more cereal products and potatoes and less dairy products, coffee, tea, fruit juice, lemonade, vegetables, fruit, and alcohol than controls.[7] No difference in sugar intake was noted between the two groups. The observation that ulcerative colitis patients consumed less milk than controls has not been observed in other studies and would indicate that therapeutic advice to reduce milk intake may have contaminated the results.

All of the studies investigating the relationship between diet and ulcerative colitis therefore are less than perfect. These imperfections reflect the temporal difficulties encountered in this type of work. A long-term cohort study of initially healthy people which assessed dietary intake and documented the incidence of ulcerative colitis would be the ultimate method of investigating the problem. However, as the incidence of the disease is so low, the logistics involved are daunting; 10,000 people would need to be followed up for 10 years to enroll approximately 30 cases — this is not a practical option.

Furthermore, because of the inadequacies of present-day methods of assessing the complexity of dietary intake, a dietary cause of ulcerative colitis remains a possibility. For instance, there have been no systemic studies of food additive intake in colitis. These additives have been increasingly used in food processing during this century in Western countries where the incidence of colitis has increased. Similarly, contemporary methods of dietary assessment do not indicate the physical format of the ingested foodstuffs.[8,9] The significance of the physical format of foods becomes apparent when one considers starch digestion. More bread starch is delivered to the colon when bread is baked with course flour rather than with fine flour.[10] Delivery of low amounts of starch to the colon has been associated with colonic diseases,[11,12] and fermentation of starch produces volatile fatty acids, a deficiency of which causes diversion colitis.[13] Hence, the ingestion of similar amounts of starch in either large or small particle size could have different metabolic consequences for the colon and conceivably ingestion of starch foodstuffs composed of fine particles could contribute to the occurrence of ulcerative colitis. The discovery of a dietary factor in the etiology of ulcerative colitis may have to await advances in nutritional science.

III. FOOD INTOLERANCE AND ULCERATIVE COLITIS

Studies examining overall dietary intake in groups of patients and controls cannot easily detect idiosyncratic reactions to foodstuffs as a cause of disease. The presence of idiosyncrasies are usually discovered by the patient and/or doctor in a clinical context. Objective testing is then required to confirm the diagnosis. Idiosyncracies to food may have a metabolic basis related for instance to an enzyme defect or alternatively there may be an allergic/immunological basis.

Coombs and Gell have described four types of hypersensitivity in which there is an exaggerated or inappropriate immune response leading to tissue damage. The immunological evidence for food allergy in ulcerative colitis can be reviewed according to this classification.

A. TYPE I REACTIONS

No definite evidence exists to suggest that adult ulcerative colitis is a disease due to Type I hypersensitivity to a food allergen. In ulcerative colitis rectal mucosa mast cells and eosinophils may be increased in number during acute exacerbations,[14,15] but this is likely to be a nonspecific reaction. Earlier reports suggested that eosinophilia was also a feature of active disease;[16,17] however, more recently this has been ascribed to the use of sulfasalazine.[18] The rectal mucosa histamine content, one of the principal mediators released during mast cell degranulation, is similar in active and inactive ulcerative colitis and in normal controls.[19] Positive intradermal tests and specific serum IgE titers to milk proteins, the most commonly suspected allergen, are no more common in patients with ulcerative colitis than controls.[20] The prevalence of atopic disorders is also similar in ulcerative colitis patients and controls.[21] Oral or topical administration of disodium cromoglycate was reported to be beneficial in the treatment of ulcerative colitis;[22,23] however, later reports refuted these findings.[24,26]

In children under 2 years, however, food allergy is the major cause of colitis.[27] Cows milk protein is the most commonly occurring allergen but soya and beef may also be implicated. These patients have blood eosinophilia and high titers of antigen-specific serum IgE antibodies. Exclusion of the offending food from the diet results in clinical remission.

B. TYPE II REACTIONS

Circulating antibodies to food proteins occur in ulcerative colitis; however, these are thought not to have any etiological significance as they occur to a large number of different food antigens. They are found also in other inflammatory diseases of the gut and in healthy individuals.[28,30]

C. TYPE III REACTIONS

The systemic accompaniments of ulcerative colitis such as iritis, arthritis, and erythema nodosum have been likened to a serum sickness-type reaction due to circulating immune complexes.[31] Immune complexes have also been used in the production of an experimental animal model of colitis.[32] Apart from this indirect evidence, which does not specifically incriminate food antigens, Type III reactions have not been implicated in the pathogenesis of ulcerative colitis.

D. TYPE IV REACTIONS

Intestinal mononuclear cells which give rise to antibody-dependent and spontaneous cell-mediated cytotoxicity to colonic epithelium have been demonstrated in patients with ulcerative colitis despite the fact that there is decreased general cell-mediated cytotoxic activity.[33,37] However, there is no evidence to suggest that food antigens are involved in either initiating or perpetuating these phenomena.

E. MILK INTOLERANCE

Of all the possible food intolerances which might contribute to the etiology of ulcerative colitis, milk and dairy products have been most frequently indicted and therefore deserve special consideration. In 1961 a series of five patients with ulcerative colitis whose disease went into remission when milk or cheese was excluded from their diet and who developed exacerbations of their disease on reintroduction of these foods was described.[38] Flare ups usually occurred within a few days of milk challenge, but in some instances took up to 6 weeks to develop. Once exacerbations became established in these patients, withdrawal of milk and cheese did not produce improvement and treatment with corticosteroids was necessary.

Following the description of these patients, the same Oxford group reported the presence of circulating antibodies to milk proteins and that early weaning occurred in ulcerative colitis patients.[28,39] High titers of antibodies to casein and lactoglobulin occurred in 5 to 28% of ulcerative colitis patients. However, up to 8% of healthy subjects had similar antibody titers. There was also a higher prevalence of circulating ovalbumin antibodies in ulcerative colitis patients compared to controls. This latter finding was confirmed in a later study.[29] These findings indicated that the presence of antibodies to milk proteins do not have a specific etiological role. The positive correlation between milk protein antibodies and disease chronicity supports this contention and suggests that the antibodies arise as a consequence of the disease rather than vice versa.

Regarding weaning practices, 18 to 30% of ulcerative colitis patients and 7 to 12% of controls have never been breast fed.[39,40] Although the difference in weaning practice is statistically significant, it is difficult to see how this can have an important etiological role as the absolute difference between these two groups is small. Furthermore, at 1 month of age, weaning in patients and controls is similar, due to a higher failure rate with breast feeding in the controls. No reports of weaning practices in ulcerative colitis patients with supposed specific milk intolerance have been published.

In 1965, a controlled trial examined the relapse rate in a group of patients with ulcerative colitis who were advised to consume milk liberally, compared to a group with milk excluded from their diet.[41] During the trial period of 1 year, 16 of 26 and 19 of 24 patients relapsed on the milk-free and milk-containing diets, respectively. The difference achieved borderline statistical significance. However, the application of these findings to the everyday care of patients needs to be approached with caution as there was a very high relapse rate in both trial groups. Patients on the "normal" diet were advised to consume liberal quantities of milk and patient use of sulfasalazine was not studied.

Minor but definite morphological abnormalities of the small intestine have been described in ulcerative colitis.[42] The possibility that this may predispose to milk intolerance due to lactase deficiency was raised in a report of three patients with high fecal lactic acid secretion.[43] In a larger randomized controlled crossover study, however, there was no difference in symptom scores between ulcerative colitis patients and controls after 4 to 8 weeks of feeding with 20 to 100 g of either glucose or lactose.[44]

Hence, the evidence for milk intolerance in ulcerative colitis is weak and patients should not lightly be denied the use of this nutritious food. Likewise, the review of possible hypersensitivity immune reactions does not implicate an allergy to milk or other food allergens in the pathogenesis of ulcerative colitis. The fact that substitution of oral food intake by total parenteral nutrition does not produce any benefit in acute colitis further supports the contention that food intolerance is not involved in the etiology.[45,46]

IV. DIETARY MANAGEMENT AND NUTRITION

Apart possibly from iron or folate deficiency, patients with mild to moderate ulcerative colitis have a normal nutritional status; however, patients sick enough to require hospital admission for their disease often are undernourished. Of 33 patients with ulcerative colitis requiring admission to St. Mark's Hospital, London in 1977 and 1978, 29 had weight loss ranging from 0.25 to 17 kg.[47] In a representative subgroup of these patients mean fat content was significantly less than that of normal age and sex-matched lean controls (13.8 vs. 18.1 kg; $p < 0.05$). Total body protein, as assessed by mid-arm muscle circumference, was the same in both patients and controls. In patients with severe exacerbations of colitis, however, daily fecal losses of nitrogen can be increased to 6.5 g from a normal level of about 0.75 g.[48] As urinary losses of nitrogen are at least 4 to 5 g/d, it can be predicted that with a normal daily intake of about 10 g of nitrogen, a negative nitrogen balance may occur in severe colitis. Thus, patients with severe colitis with loss of appetite and unable to maintain a normal nitrogen intake may easily slip into negative nitrogen balance. Whole body protein breakdown may be increased in severe colitis but this is compensated for by increased synthesis.[49] This increased turnover of protein throughout the body correlates with inflammatory activity as assessed by the ESR.

Many patients with ulcerative colitis have chronic gastrointestinal blood loss so it is therefore not surprising that up to 80% of patients may have iron deficiency which often is not associated with anemia.[50] Serum folic acid concentrations less than 2 SD below the population mean have been described in about 50% of a Danish series of patients.[51] This was ascribed to a low-grade hemolytic process associated with sulfasalazine usage. Clinical anemia was not a problem although up to 14% of the patients had megaloblastic changes in their bone marrow. Deficiencies of other vitamins and minerals in ulcerative colitis do not constitute a clinical problem.

Although growth retardation is the most common physical abnormality in children with Crohn's disease, in children with ulcerative colitis that is not a frequent problem. In a series of 130 young Americans with ulcerative colitis, only 3 had severe growth retardation before the initiation of steroid therapy.[52] As severe growth retardation was defined as height below the third percentile this figure does not differ from that occurring in the general population. However, apart from their ulcerative colitis no other cause for their growth retardation was pinpointed in these children.

Two controlled trials have examined the role of nutritional therapy for acute colitis using i.v. hyperalimentation.[45,46] The rationale behind these studies was to promote "bowel rest" and to improve nutritional intake. In both of these trials conventional management with steroids was compared with that of combined steroids and i.v. nutrition. Neither of these studies demonstrated any benefit from using parenteral nutrition, confirming the uncontrolled data of Truelove et al.[53]

As regards nutritional advise, specific recommendations are usually not necessary for patients with mild to moderate ulcerative colitis. The possibility of iron deficiency should be explored in patients with chronic activity and iron supplementation prescribed as necessary. Detection of folate deficiency should prompt the discontinuation of sulfasalazine and its substitution by mesalazine, olsalazine, or balsalazide. In patients with severe disease, attention to anthropometric data will identify protein/calorie malnutrition for which the patient should be encouraged to take high calorie, high protein oral supplements. Anorexia is the most important cause of weight loss in these patients and overall calorific intake can only be increased by careful supervision and encouragement. Parenteral nutrition cannot be recommended for acute colitis as it is not of proven benefit, it exposes patients to the risks of central venous lines, and it may dangerously postpone the decision in favor of total colectomy in severely ill patients. It may however be occasionally useful in nutritionally depleted patients who definitely require a colectomy and in whom postoperative return of small intestinal function may be slow.

V. CONCLUSION

No dietary factor can be definitely incriminated in the etiology of ulcerative colitis. Ulcerative colitis patients may have a relatively high premorbid intake of sugar but the vast majority of people with high sugar intake do not develop colitis. There is no good evidence that intolerance to milk or any other food causes ulcerative colitis. Folate and iron deficiencies may occur in colitis, necessitating either supplementation or dietary advice. Protein/calorie malnutrition only occurs in severe colitis associated with anorexia; patient encouragement and occasionally simple supplementation are all that is necessary in this situation.

REFERENCES

1. **Brauer, P. M., Gee, M. I., Grace, M., and Thomson, A. B. R.,** Diet of women with Crohn's disease and other gastrointestinal diseases, *J. Am. Diet. Assoc.,* 82, 659, 1983.
2. **Gee, M. I., Grace, M. G. A., Wensel, R. M., Sherbaniuk, R. W., and Thomson, A. B. R.,** Nutritional status of gastroenterology outpatients: comparison of inflammatory bowel disease with functional disorders, *J. Am. Diet. Assoc.,* 85, 1591, 1985.
3. **Archer, L. N. J. and Harvey, R. F.,** Breakfast and Crohn's disease. II, *Br. Med. J.,* ii, 540, 1978.
4. **Maybetty, J. F., Rhodes, J., and Newcombe, R. G.,** Increased sugar consumption in Crohn's disease, *Digestion,* 20, 323, 1980.
5. **Thornton, J. R., Emmett, P. M., and Heaton, K. W.,** Diet and ulcerative colitis, *Br. Med. J.,* 1, 293, 1980.
6. **Panza, E., Franceschi, S., and La Vecchia, C., et al.,** Dietary factors in the aetiology of inflammatory bowel disease, *Ital. J. Gastroenterol.,* 19, 205, 1987.
7. **Brandes, V. J. W., Stenner, A., and Martini, G. A.,** Ernahrungsgewohnheiten der Patienten mit Colitis ulcerosa, *Z. Gastroenterol.,* 12, 834, 1979.
8. **O'Dea, K., Nestel, P. J., and Antonoff, L.,** Physical factors influencing postprandial glucose and insulin responses to starch, *Am. J. Clin. Nutr.,* 33, 760, 1980.
9. **Wursch, P., Del Vedovo, S., and Koellreutter, B.,** Cell structure and starch nature as key determinants of the digestion rate of starch in legume, *Am. J. Clin. Nutr.,* 43, 25, 1986.
10. **O'Donnell, L. J. D., Emmett, P. M., and Heaton, K. W.,** Size of flour particles and its relation to glycaemia, insulinaemia and colonic disease, *Br. Med. J.,* 298, 1616, 1989.
11. **Thornton, J. R., Dryden, A., Kelleher, J., and Losowsky, M. S.,** Does supperefficient starch absorbtion promote diverticular disease?, *Br. Med. J.,* 292, 1708, 1987.
12. **Thornton, J. R., Dryden, A., Kelleher, J., and Losowsky, M. S.,** Superefficient starch absorbtion. A risk factor for colonic neoplasia?, *Dig. Dis. Sci.,* 32, 1088, 1987.
13. **Harig, J. M., Soegel, K. H., Komorowski, R. A., and Wood, C. M.,** Treatment of diversion colitis with short-chain fatty acid irrigation, *N. Engl. J. Med.,* 320, 23, 1989.
14. **Lloyd, G., Green, F. H. Y., Fox, H., Mani, V., and Turnberg, L. A.,** Mast cells and immunoglobulin E in inflammatory bowel disease, *Gut,* 16, 861, 1975.
15. **Heatley, R. V. and James, P. D.,** Eosinophils in the rectal mucosa. A simple method of predicting the outcome of ulcerative proctocolitis?, *Gut,* 20, 787, 1978.
16. **Riisager, P. M.,** Eosinophil leucocytes in ulcerative colitis, *Lancet,* ii, 1008, 1059.
17. **Juhlin, L.,** Basophil and eosinophil leucocytes in various internal disorders, *Acta Med. Scand.,* 174, 249, 1963.
18. **Benfield, G. F. A. and Asquith, P.,** Blood eosinophilia and ulcerative colitis — influence of ethnic origin, *Postgrad. Med. J.,* 62, 1101, 1986.
19. **Binder, V. and Hyidberg, E.,** Histamine content of rectal mucosa in ulcerative colitis, *Gut,* 8, 24, 1967.
20. **Jewell, D. P. and Truelove, S. C.,** Reaginic hypersensitivity in ulcerative colitis, *Gut,* 13, 903, 1972.
21. **Mee, A. S., Brown, D., and Jewell, D. P.,** Atopy of inflammatory bowel disease, *Scand. J. Gastroenterol.,* 14, 743, 1979.
22. **Heatley, R. V., Calcraft, B. J., Rhodes, J., Owem, E., and Evans, B. K.,** Disodium cromoglycate in the treatment of chronic proctitis, *Gut,* 16, 559, 1975.
23. **Mani, V., Green, F. H. Y., and Lloyd, G. et al.,** Treatment of ulcrative colitis with oral disodium cromoglycate. A double-blind controlled trial, *Lancet,* i, 439, 1976.
24. **Dronfield, M. W. and Langman, M. J. S.,** Comparative trial of sulphasalazine and oral sodium cromoglycate in the maintenance of remission in ulcerative colitis, *Gut,* 19, 1136, 1978.
25. **Buckell, N. A., Gould, S. R., Day, D. W., Lennard-Jones, J. E., and Edwards, A. M.,** Controlled trial of disodium cromoglycate in chronic persistent ulcerative colitis, *Gut,* 19, 1140, 1978.
26. **Binder, V., Elsborg, L., and Griebe, J. et al.,** Disodium cromoglycate in the treatment of ulcerative colitis and Crohn's disease, *Gut,* 22, 55, 1981.
27. **Jenkins, H. R., Pincott, J. R., Soothill, J. F., Milla, P. J., and Harries, J. T.,** Food allergy: the major cause of infantile colitis, *Arch. Dis. Child.,* 59, 326, 1984.
28. **Taylor, K. B. and Truelove, S. C.,** Circulating antibodies to milk proteins in ulcerative colitis, *Br. Med. J.,* 2, 924, 1961.
29. **Wright, R. and Truelove, S. C.,** Circulating antibodies to dietary proteins in ulcerative colitis, *Br. Med. J.,* 2, 142, 1965.
30. **Falchuck, K. R. and Isselbacher, K. J.,** Circulating antibodies to bovine albumin in ulcerative colitis and Crohn's disease, *Gastroenterology,* 70, 5, 1976.
31. **Hodgson, H. J. F., Potter, B. J., and Jewell, D. P.,** Immune complexes in ulcerative colitis and Crohn's disease, *Clin. Exp. Immunol.,* 29, 187, 1977.

32. **Mee, A. S., McLaughlin, J. E., Hodgson, H. J. F., and Jewell, D. P.,** Chronic immune colitis in rabbits, *Gut,* 20, 1, 1979.
33. **Targan, S., Britvan, L., Kendal, R., Vimadalal, S., and Soll, A.,** Isolation of spontaneous and interferon inducible natural killer-like cells from human colonic mucosa: lysis of lymphoid and autologues epithelial target cells, *Clin. Exp. Immunol.,* 54, 14, 1983.
34. **Shorter, R. G., McGill, D. B., and Bahn, R. C.,** Cytotoxicity of mononuclear cells for autologous colonic epithelial cells in colonic disease, *Gastroenterology,* 86, 13, 1984.
35. **Roche, J. K., Fiocchi, C., and Youngman, K.,** Sensitization to epithelial antigens in chronic mucosal inflammatory disease, *J. Clin. Invest.,* 75, 522, 1985.
36. **Falchuk, Z. M., Barnhard, E., and Machado, I.,** Human colonic mononuclear cells: studies of cytotoxic functions, *Gut,* 22, 290, 1981.
37. **McDermott, R. P., Bragdon, M. J., Kodner, I. J., and Bertovic, H.,** Deficient cell-mediated cytotoxicity and hyporesponsiveness to interferon and mitogenic lectin activation by inflammatory bowel disease peripheral blood and intestinal mononuclear cells, *Gastroenterology,* 90, 6, 1986.
38. **Truelove, S. C.,** Ulcerative colitis provoked by milk, *Br. Med. J.,* 1, 154, 1961.
39. **Acheson, E. D. and Truelove, S. C.,** Early weaning in the aetiology of ulcerative colitis. A study of feeding in infancy in cases and controls, *Br. Med. J.,* 2, 929, 1961.
40. **Whorwell, P. J., Holdstock, G., Whorwell, G. M., and Wright, R.,** Bottle feeding, early gastroenteritis and inflammatory bowel disease, *Br. Med. J.,* 1, 382, 1979.
41. **Wright, R. and Truelove, S. C.,** A controlled therapeutic trial of various diets in ulcerative colitis, *Br. Med. J.,* 2, 139, 1965.
42. **Salem, S. N. and Truelove, S. C.,** Small intestinal and gastric abnormalities in ulcerative colitis, *Br. Med. J.,* 1, 827, 1965.
43. **Frazer, A. C., Hood, C., and Davies, A. G.,** Carbohydrate intolerance in ulcerative colitis, *Lancet,* i, 503, 1966.
44. **Cady, A. B., Rhodes, J. B., Littman, A., and Crane, R. K.,** Significance of lactase deficiency in ulcerative colitis, *J. Lab. Clin. Med.,* 70, 279, 1967.
45. **Dickinson, R. J., Ashton, M. G., Axon, A. T. R., Smith, R. C., Yeung, C. K., and Hill, G. L.,** Controlled trial of intravenous hyperalimentation and total bowel rest as an adjunct to the routing therapy of acute colitis, *Gastroenterology,* 79, 1199, 1980.
46. **McIntyre, P. B., Powell-Tuck, J., and Wood, S. R., et al.,** Controlled trial of bowel rest in the treatment of severe acute colitis, *Gut,* 27, 481, 1986.
47. **Powell-Tuck, J.,** Protein metabolism in inflammatory bowel disease, *Gut,* S1, 67, 1986.
48. **Buckell, N. A., Gould, S. R., and Hernandez, N. A. et al.,** Direct method of measuring faecal protein loss in patients with ulcerative colitis, *Gut,* 18, A971, 1977.
49. **Powell-Tuck, J., Garlick, R. J., Lennard-Jones, J. E., and Waterlow, J. C.,** Rates of whole body protein synthesis and breakdown increase with the severity of inflammatory bowel disease, *Gut,* 25, 460, 1984.
50. **Driscoll, R. H. and Rosenberg, I. H.,** Total parenteral nutrition in inflammatory bowel disease, *Med. Clin. North Am.,* 62, 185, 1978.
51. **Elsborg, L. and Larsen, L.,** Folate deficiency in chronic inflammatory bowel disease, *Scand. J. Gastroenterol.,* 14, 1019, 1979.
52. **McCaffrey, T. D., Nasr, K., Lawrence, A. M., and Kirsner, J. B.,** Severe growth retardation in children with inflammatory bowel disease, *Pediatrics,* 45, 386, 1970.
53. **Truelove, S. C., Willoughby, C. P., Lee, E. G., and Kettlewell, M. G. W.,** Further experience in the treatment of severe attacks of ulcerative colitis, *Lancet,* ii, 1086, 1978.

Chapter 13

THE MEDICAL TREATMENT OF ULCERATIVE COLITIS

M. G. Courtney

TABLE OF CONTENTS

I. INTRODUCTION

Ulcerative colitis (UC) is a chronic relapsing inflammatory disease of unknown origin primarily affecting the colonic mucosa in young adults of either sex.

Rarely, extracolonic manifestations of obscure etiology involve the skin, eyes, joints, biliary tract, and liver. However, the predominant symptoms and signs arise from the mucosal lesion which has a characteristic distribution and histology allowing, in most cases, for clear differentiation from other colonic diseases, most particularly Crohn's disease. Infectious diseases which can mimic many aspects of UC should be excluded by repeated fecal culture.

Treatment is generally directed toward alleviating symptoms and signs but is influenced in long-standing disease by the risk of malignant transformation in the colonic mucosa.

The main objective in therapy is to maintain health with the important ramifications of preserving work and income in this age group. Herein lie some of the many contradictions of this perplexing disease. As there is no known cause, no cure exists. The natural history is extraordinarily varied and impossible to predict in any one individual. Furthermore, the physician cannot always rely on reported symptoms as patients rapidly learn that the consequences of full disclosure may mean enforced bed rest or hospitalization with loss of income, promotion prospects, and social activities. Indeed, no two patients react in similar fashion to similar disease extent and activity and also adapt to the disease in a very dissimilar manner.

The attending physician must therefore take many factors into consideration when deciding on a treatment plan. In particular the full cooperation of the patient and family must be enlisted which allows for decreased stress, better compliance with therapy, and earlier presentation during subsequent flare ups in disease activity. In this regard patient/family education and self-help groups are of paramount importance, allowing trust to develop in the doctor-patient relationship.

Treatment to prevent relapse or preserve remission is directed toward decreasing stress, improving nutrition, and developing awareness of early signs of exacerbation. Known initiating factors in susceptible individuals, such as antibiotics and nonsteroidal antiinflammatory drugs (NSAIDs), should be avoided unless absolutely necessary.

Treatment can be divided into several categories: maintenance of remission, control of exacerbations, and treatment of complications.

Any one individual will generally settle into patterns ranging from prolonged remission with infrequent mild relapses to unremitting disease activity, though random and entirely unpredictable variations may supervene.

Another confounding variable is response to disease, with some patients seemingly unaffected by quite marked disease activity while others are unable to cope, psychologically or otherwise, with mild relapses.

II. MAINTENANCE OF REMISSION

Most patients with UC are healthy most of the time and rightly do not wish to dwell on the disease and may in fact deny the problem. Though it is important to reinforce the natural desire to get on with their lives, it is also important to ensure compliance with their physicians' remedies. For over 4 decades the mainstay of treatment has been sulfasalazine (Salazopyrin®, Azulfidine®) which was originally developed by Nanna Svartz for treatment of rheumatoid arthritis but which was also noted, serendipitously, to be active in UC.[1]

Sulfasalazine (2 to 4 g/d in divided doses) taken orally for prolonged periods undoubtedly reduces the incidence of relapse.[2] Sulfasalazine is thought to be cleaved at the azo bond between the 5-aminosalicylate (5-ASA) moiety and sulfapyridine by an azoreductase enzyme in colonic bacteria and/or mucosal cells.[3] The 5-ASA component (the generic name is mesalazine in Europe and mesalamine in the U.S.) is thought to be the active agent in reducing

TABLE 1
Adverse Effects of Sulfasalazine

General:
Anorexia, nausea, vomiting, diarrhea, abdominal pain, anaphylaxis

Skin
Pruritic maculopapular rash, Stevens-Johnson syndrome

Blood
Leucopenia, thrombocytopenia, autoimmune hemolytic anemia, megaloblastic anemia

Fertility
Oligospermia, abnormal forms, decreased motility

Gut
Decreased folate/digoxin absorption

Neurological
Depression, irritability, mood disturbance

Miscellaneous
Yellow/orange discoloration of skin, urine, tears (may stain contact lenses), hair loss, systemic lupus erythematosus

mucosal inflammation, though intact sulfasalazine may contribute in some degree to the antiinflammatory effect.[4] These products probably act by interruption of the generation pathways of inflammatory mediators such as leukotrienes, hydroxyeicosotetraenoic acid, platelet activating factor, and bradykinin, or possibly by scavenging the excess tissue-damaging free radicals found in active disease.[5,6]

The well-absorbed sulfapyridine component is responsible for most of the adverse effects, which are more likely in those "slow" acetylators of sulfapyridine, which by genetic design clear the offending agent less quickly than normal.[7] Though side effects occur to some degree in up to 30% of those receiving the medication,[8] they are rarely serious (Table 1). Most minor side effects can be overcome by decreasing dosage followed by gradual increase or by using the enteric coated (EN-tab) formulation.[9] As most severe adverse events occur early, increased vigilance is to be recommended in the initial 3 to 6 months of therapy.

Unfortunately, 5-ASA alone cannot be taken by mouth,[10] as it is unstable in gastric acid, rapidly absorbed in the upper gut, metabolized in the liver, and may, it is suggested, in high dosage have nephrotoxic effects.[11] The poorly absorbed sulfasalazine allows most 5-ASA to be "carried" to the colon and then to be liberated at the inflamed site. Most 5-ASA is excreted unchanged in feces but small amounts are absorbed, N-acetylated in the liver, and quantitatively excreted in the urine.[12] Some 5-ASA is also acetylated in the colon, either by the action of bacteria or by colonocytes.[13] The rate of acetylation of 5-ASA, unlike sulfapyridine, is not genetically determined.[14] Acetyl-5-ASA however does not appear to possess intrinsic anti-inflammatory activity.[15]

In an attempt to reduce the incidence of intolerance to sulfapyridine, alternative "carrier" molecules have been fused with 5-ASA and also alternative methods of delivery have been designed. These have included 5-ASA suppositories (0.5 to 1.0 g b.d.) and enemata (1 to 4 g nocte or b.d.) which are designed for proctitis or distal colonic disease, respectively, but which were hampered initially by difficulty in maintaining the 5-ASA in active, unoxidized form during prolonged storage. Recently, however, newer preparations (Asacol®/Asacolon® suppositories/enema) have overcome these problems. Systemic absorption of 5-ASA from these rectal preparations is low, significant toxicity rare, and efficacy appears at least equal to currently available rectal steroid preparations.[16] Ulcerative proctitis, refractory to oral medications, may respond to rectal 5-ASA administration.[17] With respect to enemata, further clarification is required concerning the efficacy advantages of high vs. low dosage, suspension vs. solution, neutral vs. acidic pH, and high vs. low volume of administration.

TABLE 2
Slow- and Delayed-Release 5-ASA Compounds

Generic name	Trade name	Coating	Solubility	Delivery sites
Mesalazine	Asacol® Asacolon®	Eudragits S[a]	pH > 7	Terminal ileum Colon
Mesalazine	Mesasal® Claversal Salofalk®	Eudragit L[a]	pH > 6	Ileum Colon
Mesalazine	Pentasa®	Ethylcellulose-coated microgranules (multiple)	pH not yet defined	Duodenum, jejunum, ileum, colon

[a] Acrylic polymer coating.

4-ASA, a stable isomer and (intriguingly) an antituberculosis drug (PAS), may possess similar antiinflammatory effects to 5-ASA which, if confirmed in further clinical efficacy studies, will be an important (and cheap!) addition to our armamentarium in the treatment of UC.[18] The major problem, however, of all rectally administered medications remains lack of compliance in long-term maintenance therapy. Perhaps a compromise of prolonged low-dose oral 5-ASA therapy combined with intermittent (e.g., once weekly) high-dose rectal 5-ASA therapy may extend remission and reduce toxicity with acceptable compliance and affordable cost.

Alternative oral delivery systems have been developed to overcome the problems of 5-ASA absorption in the upper gut and to remove the sulfapyridine-associated side effects of sulfasalazine.[19] These new products consist of slow- and delayed-release* formulations of 5-ASA alone (Table 2) or 5-ASA azo linked to novel sulfa-free carriers (Table 3). Obviously, these new drugs have differing bioavailabilities[20] and, although experience is still limited, early indications suggest that tolerance is much better and efficacy is preserved but not improved over sulfasalazine. It may well be possible to better tailor therapy with these novel compounds to the individual needs of patients with UC (and also Crohn's disease).

One obvious drawback of the concept of slow- and delayed-release formulations is the variability of gastrointestinal pH in health and in disease which may cause unpredictable early or late release of contents with consequent reduction in efficacy. Another problem with such pH- and time-dependent systems is related to alterations in mouth-to-cecum transit times consequent to disease activity, which may occur as an adaptive response of the upper/mid gut to hind gut disease. Recent evidence also suggests that abnormal colonic motility at sites of active disease is secondary to excess bradykinin produced locally in the inflamed mucosa. Medications, such as lactulose which lowers colonic pH, will substantially reduce or prevent release of active agent and therefore should be avoided.

Alternative carrier molecules fused to 5-ASA have been developed (Table 3) but most have not gained widespread usage as yet. The best characterized prodrug of this type, olsalazine (Dipentum®), utilizes the elegant concept of back-to-back azo linkage of two 5-ASA molecules, which on exposure to the colonic milieu will release double molar quantities of active agent.[21] This product, in encapsulated powder form, has shown equal efficacy to sulfasalazine but with much fewer side effects due to the absent sulfapyridine.[22] 5-ASA-dependent skin rashes, diaphoresis, pyrexia, nausea, and abdominal pain may occur and early indications suggest that approximately 10% of patients may be intolerant to the newer products. In such cases switching to another new formulation may be successful.

With all such pharmaceuticals, a careful watch should be maintained for rare but potentially

*Also known as positioned-release formulations.

TABLE 3
Sulfa-Free Azo Linked 5-ASA "Carrier" Drugs (Oral)

Generic name	Trade name	"Carrier"
Olsalazine	Dipentum®	–5-ASA
Balsalazide	Colazide	–4-aminobenzoylalanine
Ipsalazide	?	–4-aminobenzoylglycine
Benzalazine	?	–*p*-aminobenzoic acid
Poly-ASA	?	-sulfanilamido-ethylene polymer

dangerous episodes of 5-ASA-induced exacerbations of colitis and, in patients who are not well controlled with sulfasalazine or other 5-ASA-based products, a trial of medication withdrawal should be instituted.[23] Initially, after introduction of all 5-ASA-containing agents, but most particularly with olsalazine, some loosening of the stools may be observed but this generally disappears with continued treatment or temporary reduction in dosage, as colonic adaptation occurs.[24] The imminent introduction of a tablet formulation of Dipentum® may obviate this problem by ensuring slower disintegration of tablet than capsule, thus achieving a more gradual rise in gut concentration of olsalazine and 5-ASA species.

Optimal dosages of these newer products have not yet been clearly defined, but generally aim to deliver equal amounts of 5-ASA to the colon when compared to sulfasalazine therapy. Thus, daily dosage of 2 to 4 g of sulfasalazine could be replaced with 1 to 2 g of Dipentum® or 0.8 to 1.6 g of Asacol®/Asacolon®. With these new products, 5-ASA dosage is not limited by sulfapyridine-associated side effects; however, as yet no improvement in efficacy has been demonstrated by two- and threefold increases in dosage.[25] Unfortunately, the most troublesome side effect of 5-ASA, i.e., diarrhea, is more likely with the high dosages used in extensive disease. Several clinical trials suggest that all of the new 5-ASA products are of equal efficacy to sulfasalazine in maintaining/inducing remission in UC but with a better immediate/short-term side-effect profile.[22,26–28] However, information on delayed toxicity after several years therapy is not yet available and is awaited with interest.

In view of the possibility of nephrotoxic effects of absorbed 5-ASA during long-term dosage,[29] the product presenting least exposure of 5-ASA to the small intestine would be most suitable for maintenance therapy and would have the additional benefit of delivering maximal 5-ASA to the colon, where it is needed. Pentasa® releases approximately 50% of its 5-ASA content in the small intestine and would appear unsuited to the long-term therapy of UC,[14] but may have a role in the treatment of small intestinal Crohn's disease.

Asacol®, Asacolon®, Claversal®, Mesasal®, and Salofalk® deliver approximately 30% of the 5-ASA load in the lower small intestine and may be more suited to the treatment of ileocolonic Crohn's disease, ileorectal anastamosis, or UC exacerbations. Olsalazine requires the action of colonic bacteria to cleave the 5-ASA dimer, so consequently very little 5-ASA is presented to the small intestine and therefore Dipentum® would appear the most appropriate agent at present for long-term maintenance therapy in UC.[30] Olsalazine is not suitable therapy after ileorectal anastamosis as insufficient bacteria are present to ensure cleavage to active 5-ASA.

At present the licensing arrangements for various 5-ASA compounds are complicated and vary in different countries, from sulfasalazine-intolerant use only to free license. Generally, sulfasalazine still remains the first-line agent in the treatment of UC[31] and is best prescribed in enteric-coated (EN-tab) form and in increments initially to decrease gastrointestinal intolerance. It is especially useful in those cases complicated by arthritis and/or ankylosing spondylitis — reflecting the historical origins of the drug and its more recent success as a second-line agent in rheumatoid arthritis,[32] which is at least one example of a beneficial effect of sulfapyridine! In sulfasalazine-intolerant individuals very gradual incremental desensitization[9] may be attempted or a nonsulfapyridine-containing 5-ASA compound may be prescribed.

In countries with an open license it is to be expected that alternative 5-ASA compounds will be prescribed initially in preference to sulfasalazine to reduce side effects and ensure better compliance. This changeover may be hastened by emerging evidence that sulfasalazine and sulfapyridine, but not 5-ASA, may be responsible for chromosomal damage in UC.[33] Some public resistance to sulfasalazine is developing also in view of the associated oligomegalospermia with excess abnormal forms and poor motility.[34] The consequent decreased fertility can be slowly reversed over 6 months to 1 year by substitution of sulfasalazine with 5-ASA alone, suggesting that sulfapyridine is responsible.[35] Potency is not affected by either drug, but is frequently reduced in disease exacerbations.

During and after this period of product substitution, postmarketing surveillance should be emphasized to rapidly identify any previously unrecognized complications of stand-alone 5-ASA products. Caution should be exercised in prescribing for patients in renal failure and in the elderly. Carefully designed long-term maintenance studies of these products are required, not only from a toxicological point of view but also to determine the optimal maintenance dosage, which may vary with each product. It may well take several years for the various new pharmaceuticals to establish their own particular "niche" in the market and, until the present fluid situation has crystalized, therapeutic recommendations are still provisional.

It should be emphasized that oral steroid therapy has almost no role in the maintenance therapy of UC and of course systemic steroids are associated with significant adverse effects. In a few "steroid-dependent" patients, 5 to 15 mg of prednisolone per day orally are required to prevent immediate relapse following complete withdrawal. Attempts to reduce steroid side effects include alternate-day dosing, ultra-slow steroid reduction, and the use of azathioprine (2 mg/kg/d) orally for its steroid-sparing effect to cover withdrawal. Sometimes colectomy is the preferred option.

New, potent, topically active steroids (fluticasone propionate sp.) with low systemic bioavailability due to poor absorption after oral dosage and high "first-pass" metabolism are at present undergoing clinical trials for UC and Crohn's disease. If the beneficial effects of steroids are indeed independent of absorption and systemic activity, these products will constitute such a major advance in therapy as to require a complete reappraisal of present management strategies in chronic inflammatory bowel disease.

Repeated short courses of rectally administered steroids in solution (prednisolone phosphate/prednisolone metasulfobenzoate, 20 to 40 mg nocte/b.d.), or in foam (hydrocortisone acetate 10%, 1 to 2 g nocte/b.d.) may be useful in controlling symptoms in those individuals with chronic distal disease and low-grade activity (proctosigmoiditis) or after ileorectal anastamosis. Though sufficient systemic absorption of rectal steroid preparations does occur to affect the measured hypothalamic-pituitary-adrenal axis, it is unlikely that such effects are of great clinical importance. In any case, the recent development of low-absorption, rapidly metabolized rectal steroids (beclomethasone dipropionate, Tixocortol pivalate, prednisolone metasulfobenzoate) should obviate this problem — theoretical or otherwise.[36]

Rectally administered 5-ASA preparations are being increasingly used but suffer the same disadvantages as rectally administered steroid preparations, i.e., they are inconvenient, awkward, embarrassing, time consuming, and expensive. In addition, long-continued use of rectal 5-ASA enemata may lead to perianal irritation, though the provision of additional lubrication for tube insertion may reduce this problem. When used as monotherapy in active disease, both rectal products take a long period to induce remission and have high relapse rates following withdrawal.

A. USE IN PREGNANCY

It is generally accepted that uncontrolled disease activity is a more significant risk to mother and fetus than sulfasalazine or steroids used to maintain remission.[37] Experience with the newer 5-ASA agents is limited and caution should be exercised with their use in pregnancy. However, it is known that in sulfasalazine-treated pregnant women, intact sulfasalazine, sulfapyridine, and 5-ASA cross the placenta and penetrate the fetus.[38] Thus, there are grounds to suppose that 5-ASA use in pregnancy may be no more hazardous than the current use of sulfasalazine. However, predictions in this area are fraught with risk; in particular the fetal exposure to 5-ASA may vary with different drugs which may have totally distinct embryotoxic effects.

UC is not a contraindication to pregnancy and generally does not adversely affect the pregnancy, though disease presenting in pregnancy is often severe and should be treated aggressively. Folate deficiency is independently associated with UC, sulfasalazine therapy, and pregnancy itself and therefore a folate supplement should be routinely prescribed during (and if possible, before a planned) pregnancy. Sulfasalazine, sulfapyridine, but not 5-ASA are secreted into breast milk,[38] but in such low levels that risk to the infant is extremely unlikely.

Despite good compliance to a 5-ASA-based product, approximately 25% of patients with UC will relapse within 1 year.[39]

B. WITHDRAWAL OF MEDICATION

In view of the extremely variable natural history of UC and wide interpatient variations, it is not possible to generalize about eventual elimination of therapy in all patients. However, many patients only ever have one or two initial attacks while many others have clinical remission lasting several years at a time; as the majority of patients are poorly compliant in the long term, it is not unreasonable to attempt withdrawal of an expensive and potentially dangerous remedy given a suitable opportunity.

Sensible guidelines could include disease remission for at least 2 years after initial attack and at least 4 years after subsequent attacks, absence of significant complications, and continued low-intensity medical review for 2 or 3 years after withdrawal. Patients should be encouraged to seek medical attention early during a subsequent relapse. The question of whether there is an increased cancer risk in inactive, untreated colitis is unanswered, but would probably reflect known risk factors operative during the period of disease activity.

C. OTHER IMMUNOSUPPRESSIVE AGENTS

Azathioprine, its active metabolite, 6-mercaptopurine, and methotrexate are toxic and should not be routinely employed in the treatment of UC. Such drugs are used in specialist units as part of the therapy of unusually complicated or refractory disease, where it is considered that potential benefit outweighs potential risk. Such decisions should only be made after failure of other medical remedies and in units capable of detection and treatment of possible adverse effects, such as myelosuppression and/or infection. Another problem with such therapy is that some side effects mimic disease exacerbation and may trigger potentially dangerous increases in the offending medication by the unwary physician.

In refractory/complicated Crohn's disease, there have been encouraging early reports concerning the use of the antirejection agent cyclosporin, but as yet no recommendations can be made concerning its use in UC.

III. CONTROL OF EXACERBATIONS

A close doctor/patient relationship will result in early, accurate reporting of worsening symptoms and prompt institution of appropriate corrective measures. However, when the patient and doctor are unfamiliar or if the patient presents for the first time acutely ill, then the greatest care must be taken in evaluation of disease severity and monitoring of treatment efficacy, as any laxity in such cases may have catastrophic consequences.

Every physician handling large numbers of UC cases will rapidly develop his/her own particular scheme for assessment and treatment which may or may not involve reference to "activity indices". Certain physicians find these of great use, but while frequently employed in clinical trials, there has not been a simple index to command wide acceptance among practicing clinicians. Having decided on which category the relapse falls into, such as mild, moderate, severe, or fulminating, a rapid response should be instituted in order to prevent deterioration of disease.

Generally, disease in childhood should be treated actively and good nutrition stressed to prevent growth arrest and/or failure of educational achievement. Education about the disease should extend beyond the child and family to include schoolfriends and teachers. Children (and some adults) are extraordinarily embarrassed by the symptom complex in UC and very sensitive to peer opinion of their malady.

A. MILD DISEASE

This may respond to increase in oral sulfasalazine/5-ASA dosage alone, though as UC tends to be worse distally and spread proximally, it is often wise to institute a short course of rectal steroids/5-ASA for additional control. Though patients may continue work, and moderation of social activities is advised in a sensible fashion to ensure adequate rest and nutrition. An attempt is made to identify and remove existing and particularly new stresses. Potential initiating factors such as antibiotic or NSAID usage should be discontinued if feasible.[40] Milk-product withdrawal may help to reduce diarrhea if there is an associated lactose intolerance. Antidiarrheal agents may be cautiously prescribed to control troublesome symptoms but should be discontinued after a short period.

It should be appreciated that patients may self medicate with such agents (and also steroids) without informing the doctor. Especially in severe disease, antidiarrheals may be associated with toxic dilatation of the colon.

B. MODERATE DISEASE

In addition to increased oral sulfasalazine or 5-ASA dosage, a short course of oral steroids should be prescribed. Bed rest at home and a low residue diet are sensible, simple ways to rapidly relieve symptoms. Prednisolone (20 to 40 mg/d orally in single or divided doses) is usually employed as no advantage accrues from adrenocorticotrophic hormone (ACTH) (40 to 80 U/d), which carries the additional disadvantage of parenteral administration. Oral steroids are tailed over 3 to 6 weeks, depending on response, and withdrawal may be overlapped by a rectally administered steroid/5-ASA course to prevent a "flare". Depending on the result, the systemic steroids may need to be increased and/or prolonged. If rapid amelioration of symptoms does not occur, i.e., within 2 or 3 d or if there is any deterioration, hospitalization is advisable.

Fecal cultures should be routinely employed as part of the investigation protocol in all cases of "so-called" relapse/exacerbations of UC as infectious agents may be responsible for up to 20% of such events.

C. SEVERE DISEASE

Though this can develop rapidly, it is more likely that symptoms have been prolonged and have not responded to at least some form of therapy, usually given in inadequate degree. Increased dosage of sulfasalazine/5-ASA will not control severe disease and is not well tolerated by ill patients. Apart entirely from the serious colonic inflammation which may be present, there may be in addition significant systemic disorders consequent on electrolyte/fluid disturbance which may require urgent treatment in their own right.

The appropriate course is urgent hospital admission for bed rest and restitution of fluid and electrolytes (especially K^+ and Mg^{2+}). Blood transfusion may be required, as iron therapy is usually insufficient to correct anemia in active disease. Low-residue diet and oral or i.v. high-dose steroid therapy should be instituted without delay. The usual steroid dosage is 40 to 60 mg prednisolone per day orally (in single or divided doses) or hydrocortisone 400 mg/d i.v. in divided doses. This is gradually tailed as improvement occurs over 7 to 10 d and, after discharge, complete withdrawal is accomplished over 6 to 12 weeks. If symptoms deteriorate during steroid reduction, the dosage should be increased by the previous decrement and, after stabilization, an even more gradual reduction should be recommenced.

Broad-spectrum antibiotic cover may be instituted by the i.v. route and is especially indicated if there is the least suspicion of colonic perforation. The possibility of an associated antibiotic-induced pseudomembranous colitis should be borne in mind if an ill, debilitated patient on maximal therapy suddenly deteriorates. Appropriate measures to detect and treat *Clostridium difficile* are recommended in such circumstances. Nasogastric suction is employed to prevent aerophagy, reduce bowel gas, and decrease abdominal girth which provides relief to the patient.

The overall situation should be carefully and repeatedly monitored with particular emphasis on clinical state, pulse, temperature, and abdominal signs, including girth. Systemic steroid therapy may mask the evidence of free perforation which can occur without intervening colonic dilatation. Thus, a high index of suspicion must be maintained and combined with frequent regular clinical assessment. Factors known to be associated with decreased colonic peristalsis such as low serum potassium, magnesium, and albumin should be corrected promptly and opiates or anticholinergics discontinued immediately. Procedures which increase intraluminal pressure such as barium enema or colonoscopy are best avoided. If the diagnosis is in doubt in the first presentation of colitis, then proctoscopy and rectal biopsy alone are adequate until the situation improves. If objective evidence of improvement is not forthcoming, repeated plain X-ray film of abdomen and surgical consultation are mandatory to detect and treat toxic dilatation of the colon.

Dietary therapy of severe relapse is important in the short-term management of relapse but has not been shown convincingly to influence long-term outcome. However, the bowel "rest" certainly reduces stool frequency which is greatly appreciated by the patient. Diet therapy can be adjusted over a wide spectrum from nil orally, low-residue oral diet, enteral oligopeptide diet, enteral elemental diet, to total parenteral nutrition by central venous access. Peripheral i.v. alimentation has a small role to play in nutritional therapy but has the great advantage of avoiding the potential significant hazards of total parenteral nutrition (TPN). No convincing benefit in relapse prevention has been demonstrated in patients treated in remission with high-fiber diets, though they are frequently recommended in any case. If folate deficiency is detected despite adequate diet, folate supplementation should be introduced.

D. FULMINANT DISEASE

A gravely ill, often collapsed patient in imminent danger of colonic perforation requires emergency maximal medical intervention in an intensive care unit. If no discernible beneficial response is observed within 48 to 72 h, or if signs of deterioration supervene, then urgent colectomy should be considered.

The actual timing and nature of surgery is obviously a decision for the consulting surgeon, made in the light of prevailing clinical circumstances, and is beyond the scope of this work. However, the management of severely ill patients with UC calls for the closest cooperation between physician and surgeon which ensures optimal management in such critical cases. Only 10 to 20% of fulminant cases respond to medical means alone and most of these will eventually require colectomy due to functional failure or a subsequent episode of toxic dilatation.

E. GENERAL

Recovery from acute illness is greatly accelerated by proper rest and nutrition and encour-

TABLE 4
Extracolonic Manifestations of Ulcerative Colitis

Skin
Erythema nodosum, pyoderma gangrenosum, aphthous stomatitis

Musculoskeletal
Arthritis, sacroileitis, ankylosing spondylitis, clubbing, hypertrophic osteoarthropathy

Ocular
Episcleritis, iritis, uveitis

Hepatic
Abnormal liver blood tests, fatty change, pericholangitis, sclerosing cholangitis, chronic hepatitis, cirrhosis, cholangiocarcinoma

Blood
Anemia
(iron deficiency, megaloblastic, autoimmune hemolytic), hypercoagulable state (thrombosis, embolism)

Cardiovascular
Pericarditis, vasculitis, (?)amyloid

Pulmonary
Fibrosing alveolitis,[a] eosinophilic pneumonia[a]

Endocrine
Thyroid disease, amenorrhea

Metabolic
Malnutrition, growth retardation

[a] Possibly sulfasalazine related.

agement by the attending physician. Of particular benefit, the doctor can stand between the frequent demands placed on patients for premature return to work or household duties.

It must be understood by the doctor that colitis is a frequent cause of marital disharmony and a sensitive and sympathetic approach can defuse potential conflicts. Again, these delicate aspects are best approached from the direction of patient education to obviate such problems before they became ingrained. Similarly, many patients have deep and unexpressed anxieties about the consequences of surgery, including impotence and ileostomy. These questions are best confronted directly and frankly ventilated, initially in one-to-one and later in group discussion. An informal meeting with another patient with an ileostomy and who is leading a successful, active life can provide enormous reassurance. Undoubtedly, a vast, untapped amount of despair, depression, and disharmony surrounds this disease and is all too often avoided by both clinician and patient alike.

Chronic ill health in elderly patients without community support may lead to suicide. In such a situation, appropriate liaison with the general practitioner, psychiatrist, public-health nurse, and home-help and self-help groups is to be strongly recommended.

IV. TREATMENT OF COMPLICATIONS

The complications of UC can be immediate or delayed, local or systemic, trivial or serious, disease or drug related, or any combination of the foregoing (Table 4).

The treatment of the disease itself with sulfasalazine and immunosuppressives can lead to some severe adverse effects such as anaphylaxis, agranulocytosis, Stevens-Johnson syndrome, and hemolytic anemia, which generally respond to prompt discontinuation of the causative agent (usually sulfasalazine) and institution of appropriate therapy. A multitude of steroid-associated adverse effects can also occur either from patient self medication or iatrogenic routes.

Fortunately, most disease-related complications are not serious and are usually reversible. Even potentially grave complications such as iritis, arthritis, and chronic active hepatitis usually respond to steroid therapy employed against the primary lesion, though rarely may progress even in the face of quiescent or chronic low-grade colitis. Colectomy, performed for other indications, may greatly ameliorate complications such as uveitis but complications per se should rarely provide the sole indication for surgery. Sclerosing cholangitis is an instance where the complication and not the colitis may eventually dominate the clinical picture and where neither colectomy nor steroid therapy have proved useful in arresting the situation. Generally, supportive therapy of coagulopathy and bile-salt deficiency are required for a variable and unpredictable period. Ultimately, liver transplantation may be required as secondary biliary cirrhosis and liver failure supervene. Recurrence of sclerosing cholangitis in the grafted liver is distressing news for the patient but provides a fascinating insight into the pathogenesis of both diseases.

Certain complications are immediately life threatening such as colonic perforation (with or without toxic dilation), massive hemorrhage, overwhelming septicemia, or pulmonary embolism. Other complications such as carcinoma may be ultimately fatal, but if so would represent a failure (or worse still, an absence) of an effective surveillance program. In view of the significantly increased risk of colonic carcinoma in long-standing, chronic pancolitis, such patients should have full colonoscopy with serial biopsies every year. Colonic cancer in UC patients may be atypical, presenting with multifocal, flat lesions and of course symptoms may be easily confused with those of UC itself, resulting in diagnostic delay and decreased likelihood of surgical cure.[41] If colonoscopy is not possible or subtotal in extent, barium enema may be employed but does carry the risk of substantial radiation exposure and, of course, cannot evaluate borderline lesions such as pseudopolyps or benign stricture accurately or detect the presence of dysplasia.

However, even with careful histological assessment by an experienced pathologist, it is sometimes difficult to be specific about the degree of dysplasia or even be certain of its significance. In such difficult cases, regular review is essential. Joint conferences involving gastroenterologists, surgeons, radiologists, and pathologists are extremely useful in establishing a clear, common, and sensible management policy which is essential in such complex cases.

V. PROGNOSIS

Most patients with UC are able to lead productive and enjoyable lives with only occasional disruptions caused by disease exacerbations. Even in hospitalized patients with moderate to severe disease, remission rates of approximately 90% are attainable. Response rates decrease with greater disease extent/activity, repeated relapses, and with age greater than 60 years. Usually, most patients fall into a set pattern after a few years, which can vary along a continuum from well-controlled, limited disease of low activity to poorly controlled disease of maximal extent and marked activity. Overall, 25% of patients will require colectomy at some time in the course of the disease, usually for chronic ill health, fulminating disease, dysplasia/neoplasia, or because of drug toxicity. Colectomy is almost always curative, except in rare instances where extracolonic complications persist. Recent reports of carcinoma developing in long-standing ileostomies, possibly due to colonic metaplasia, merit attention and would suggest long-term follow-up of ileostomy cases.

VI. ALTERNATIVE THERAPIES

A considerable and growing number of patients volunteer the information that they are attending "alternative" therapists. It is likely that many more patients may hide this fact in order to avoid any opprobrium from their medical attendants. Faced with this situation, the doctor should eschew any antagonism lest it damage the support between physician and patient, but should clearly point out that none of these alternative strategies, ranging from simple herbal concoctions to acupuncture, have ever been shown to be beneficial in an objective way. If the treatment seems harmless and the patient is fixed upon trying it, then it is counterproductive to discourage this trial and, upon failure, the patient may perhaps be less disposed to further adventure in the area. However, if any potential harm should occur then the physician should inform the patient of his concerns at once.

On occasion, unorthodox remedies such as hypnotherapy may in fact be associated with a beneficial response in a particular patient and can be easily incorporated into a regular treatment regimen providing a pointer to the alert physician for an increased need for psychotherapy or counseling in that patient. It is never correct to withdraw medical care from a patient who dabbles with "alternative" strategies and the doctor should retain a responsible, caring, and professional attitude at all times.

VII. FUTURE CONSIDERATIONS

Research into UC is proceeding at a rapid pace which augurs well for exciting developments within a decade.

Further progress in mapping out the complex, interrelated pathways of the various inflammatory mediators will allow better design of specific antiinflammatory agents. Greater refinement of delivery systems will allow more site-specific targeting of medication. Decreased toxicity of drugs will enable higher dosages to be employed to better control disease and maintain longer remission. Reliable, early means of detection of neoplasia will allow earlier, safer surgery. Means to prevent, and even reverse, the development of dysplasia will be developed (with folate supplementation a prime candidate already[42]). Reliable indicators of eventual outcome will be invaluable in the long-term management of this chronic disease.

However, until the definite etiopathogenesis of UC is determined, cure will remain frustratingly elusive. Despite this, we must strive to understand, to counsel, to comfort, and to relieve, always.

REFERENCES

1. **Svartz, N.,** Salazopyrin, a new sulphanilamide preparation, *Acta Med. Scand.,* 110, 577, 1942.
2. **Misiewicz, J. J. et al.,** Controlled trial of sulfasalazine in maintenance therapy for ulcerative colitis, *Lancet,* i, 185, 1965.
3. **Azad Khan, A. K. et al.,** An experiment to determine the active therapeutic moiety of sulphasalazine, *Lancet,* ii, 892, 1977.
4. **Hawkey, C. J.,** Sulfasalazine: drug or pro-drug? (editorial), *Lancet,* i, 1299, 1987.
5. **Sharon, P. et al.,** Enhanced synthesis of leukotrienes by colonic mucosa in inflammatory bowel disease, *Gastroenterology,* 86, 453, 1984.
6. **Stenson, W. F.,** Platelet activating factor and inflammatory bowel disease (editorial), *Gastroenterology,* 95, 1416, 1988.
7. **Schroder, H. et al.,** Acetylator phenotype and adverse effects of SASP in healthy subjects, *Gut,* 13, 278, 1972.
8. **Nielsen, O. H.,** Sulfasalazine intolerance, *Scand. J. Gastroenterol.,* 17, 389, 1982.
9. **Holdsworth, C. D.,** Sulfasalazine desensitization, *Br. Med. J.,* 282, 110, 1981.

10. **Nielsen, O. H. et al.,** Kinetics of 5-aminosalicylic acid after jejunal instillation in man, *Br. J. Clin. Pharmacol.,* 16, 738, 1983.
11. **Calder, I. C. et al.,** Nephrotoxic lesions from 5-aminosalicylic acid, *Br. Med. J.,* i, 157, 1972.
12. **Jarnerot, G.,** 5-ASA drugs in inflammatory bowel disease, *Drugs,* 37, 73, 1989.
13. **Ireland, A. et al.,** Acetylation of 5-aminosalicylic acid by human colon epithelial cells, *Gastroenterology,* 90, 1471, 1986.
14. **Rasmussen, S. N. et al.,** 5-aminosalicylic acid in a slow-release preparation: bioavailability plasma level and excretion in humans, *Gastroenterology,* 83, 1062, 1982.
15. **Van Hogezand, R. A. et al.,** Double-blind comparison of 5-aminosalicylic acid and acetyl 5-aminosalicylic acid suppositories in patients with idiopathic proctitis, *Aliment. Pharmacol. Ther.,* 2, 33, 1988.
16. **Danish 5-ASA Group,** Topical 5-aminosalicylic acid vs. prednisolone in ulcerative proctosigmoiditis: a randomised, double-blind, multicentre trial, *Dig. Dis. Sci.,* 32, 598, 1987.
17. **Guarino, J. et al.,** 5-aminosalicylic acid enemas in refractory distal ulcerative colitis: long-term results, *Am. J. Gastroenterol.,* 82, 732, 1987.
18. **Campieri, M. et al.,** A double-blind clinical trial to compare the effects of 4-aminosalicylic acid to 5-aminosalicylic acid in topical treatment of ulcerative colitis, *Digestion,* 29, 204, 1984.
19. **Peppercorn, M. A.,** Sulfasalazine: pharmacology, clinical use, toxicity and related new drug development, *Ann. Intern. Med.,* 3, 377, 1984.
20. **Dew, M. J. et al.,** Comparison of the absorption and metabolism of sulfasalazine and acrylic-coated 5-aminosalicylic acid in normal subjects and patients with colitis, *Br. J. Clin. Pharm.,* 17, 474, 1984.
21. **Truelove, S. C.,** Evolution of olsalazine, *Scand. J. Gastroenterol.,* 23(Suppl. 148), 3, 1988.
22. **Sandberg-Gertzen, H. et al.,** Azodiasal sodium in the treatment of ulcerative colitis: a study of tolerance and relapse prevention properties, *Gastroenterology,* 90, 1024, 1986.
23. **Chakraborty, T. K. et al.,** Salicylate induced exacerbation of ulcerative colitis, *Gut,* 28, 613, 1987.
24. **Sandberg-Gertzen, H. et al.,** Long term treatment with olsalazine for ulcerative colitis: safety and relapse prevention, a follow-up study, *Scand. J. Gastroenterol.,* 23(Suppl. 148), 48, 1988.
25. **Dew, M. J. et al.,** Maintenance of remission in ulcerative colitis with 5-aminosalicylic acid in high doses by mouth, *Br. Med. J.,* 287, 23, 1983.
26. **Reilly, S. A. et al.,** Comparison of delayed-release 5-aminosalicylic acid and sulphasalazine as maintenance treatment for patients with ulcerative colitis, *Gastroenterology,* 94, 1383, 1988.
27. **Meyers, S. et al.,** Olsalazine sodium in the treatment of ulcerative colitis among patients intolerant of sulfasalazine, *Gastroenterology,* 93, 1255, 1987.
28. **Dew, M. J. et al.,** Maintenance of remission in ulcerative colitis with oral preparation of 5-aminosalicylic acid, *Br. Med. J.,* 285, 1012, 1982.
29. **Tremaine, W. J. et al.,** Urinary sediment abnormalities in patients on long-term oral 5-aminosalicylic acid for chronic ulcerative colitis, *Gastroenterology,* 94, A465, 1988.
30. **Sandberg-Gertzen, H. et al.,** Absorption and excretion of a single 1 g dose of azodiasal sodium in subjects with ileostomy, *Scand. J. Gastroenterol.,* 18, 107, 1983.
31. **Peppercorn, M. A.,** Update on the aminosalicylates: a promise fulfilled (editorial), *Gastroenterology,* 95, 1677, 1988.
32. **Neuman, V. C. et al.,** Comparison between penicillamine and sulphasalazine in rheumatoid arthritis: Leeds-Bermingham trial, *Br. Med. J.,* 287, 1099, 1983.
33. **Konstantinova, B. et al.,** Chromosomal aberrations in patients with ulcerohaemorrhagic colitis, *Digestion,* 2, 329, 1969.
34. **Levi, A. J. et al.,** Male infertility due to sulphasalazine, *Lancet,* ii, 276, 1979.
35. **Cann, P. A. et al.,** Reversal of male infertility on changing treatment from sulphasalazine to 5-aminosalicylic acid, *Lancet,* i, 1119, 1984.
36. **Kumana, C. R. et al.,** Beclomethasone dipropionate enemas for treating inflammatory bowel disease without producing Cushing's syndrome or hypothalamic pituitary adrenal suppression, *Lancet,* i, 579, 1982.
37. **Willoughby, C. P.,** Ulcerative colitis and pregnancy, *Gut,* 21, 469, 1980.
38. **Azad Kahn, A. K. et al.,** Placental and mammary transfer of sulfasalazine, *Br. Med. J.,* ii, 1553, 1979.
39. **Misiewicz, J. J. et al.,** Controlled trial of sulfasalazine in maintenance therapy for ulcerative colitis, *Lancet,* i, 185, 1965.
40. **Rampton, D. S. et al.,** Analgesic ingestion and other factors preceding relapse in ulcerative colitis, *Gut,* 24, 187, 1983.
41. **Lennard-Jones, J. E. et al.,** Cancer in colitis: assessment of the individual risk by clinical and histological criteria, *Gastroenterology,* 73, 1280, 1977.
42. **Lashner, B. A. et al.,** Effect of folate supplementation on the incidence of dysplasia and cancer in chronic ulcerative colitis, *Gastroenterology,* 97, 255, 1989; **Rosenberg, I. H. and Mason, J. B.,** (editorial) *Gastroenterology,* 97, 502, 1989.

Chapter 14

SURGERY FOR ULCERATIVE COLITIS

P. Ronan O'Connell and Francis B. V. Keane

TABLE OF CONTENTS

I. INTRODUCTION

Surgical treatment of ulcerative colitis has, for almost 100 years, been the final remedy for patients with acute complicated or chronic disease which has failed to respond to medical treatment. For 40 years, panproctocolectomy with permanent ileostomy has been the standard surgical treatment. No patient faces the prospect of major surgery and permanent ileostomy gladly. Yet, panproctocolectomy and ileostomy can be life saving and may transform a life style, although there is a substantial psychological and social price to pay. Surgical consultation often takes place in an atmosphere of foreboding on the part of the patient and a sense of failure on the part of the physician. It is against this background that recent developments in continence preservation may herald a revolutionary change in attitudes to surgical treatment. In this chapter we review the development of, the indications for, and the clinical results of panproctocolectomy. We document the evolution and results of surgical alternatives to permanent ileostomy. Finally, we discuss the current areas of controversy in surgical treatment of ulcerative colitis.

II. DEVELOPMENT OF PROCTOCOLECTOMY AND ILEOSTOMY

Surgical intervention in ulcerative colitis was first recorded by Mayo-Robson of Leeds who used a sigmoid colostomy to irrigate the distal colon of a woman with ulcerative proctitis.[1] The patient improved and the colostomy was later closed. The practice of colonic irrigation with antiseptic or astringent solutions gradually became established in treatment of fulminant colitis. Appendicostomy became a popular access route for irrigation of the inflamed colon and was in use up to the late 1930s.[2,3] Appendicostomy was simple to perform and fecal leakage was minimal. The operation was reversible by simple appendicectomy when the colitis was in remission. The major drawback of colonic irrigation was failure in most cases to achieve a sustained remission.

Complete rest of the large bowel with diversion of fecal flow in addition to irrigation was introduced by Brown of St. Louis.[4] Ten patients were described in whom the ileum was divided, the distal end closed over a catheter, and the proximal end brought out flush with the skin. The mortality associated with ileostomy diversion was high, even in the best centers. Some 32% of patients died following ileostomy at the Mayo Clinic between 1921 and 1930.[5] The majority of postoperative deaths were due to progression of the colonic disease and not to complications of the ileostomy.[6] Colectomy was only performed as a last resort and then usually as a staged procedure. For those who did survive, the difficulty of protecting the peristomal skin was a constant problem, and some were reduced to a state of chronic invalidism because of it. The ileostomy stoma proved very difficult to manage as appliances used for colostomy control were unsatisfactory for collection of the liquid and the highly irritating ileostomy effluent. Nevertheless, some patients, especially those in whom the colon and rectum had been removed, were able to return to reasonably productive lives.[7]

Development of an efficient stoma effluent collection bag, the Koenig-Rutzen appliance, represented a major advance in surgical management of ulcerative colitis.[8] The problem of leakage was overcome and the appliance did not have to be removed and reapplied each time it was emptied. Yet, simple fecal diversion by means of a loop ileostomy was not enough to ensure good health. It became apparent the proctocolectomy was needed for cure. At first, colectomy was performed at a second operation after initial loop ileostomy, but rapidly panproctocolectomy and end ileostomy was adopted as conventional surgical treatment for ulcerative colitis.

To minimize peristomal skin inflammation, it had been the practice since the 1930s to construct an ileostomy spout by exteriorization of at least 3 cm of ileum. In the weeks that followed the exposed ileal serosa became inflamed and later contracted with scar tissue. Ileostomy stenosis and prestomal ileitis were frequent problems.[9] To overcome this, Dragstedt described a technique of skin grafting the exposed serosa.[10] Others used abdominal skin flaps[11] or ileal mucosal grafts to cover the exposed ileal serosa.[12] None of these techniques were effective. In 1952, Brook described a technique of partial intussusception of the distal 5 cm of ileum to form an everted stoma covered with ileal mucosa.[13] Development of the everted spout ileostomy was the final technical innovation in development of modern panproctocolectomy. By the mid 1950s surgical treatment for ulcerative colitis had ceased to be a means of last resort, undertaken in a septic and malnourished patient. Panproctocolectomy and ileostomy was firmly established as the standard surgical treatment for ulcerative colitis.

III. INDICATIONS FOR SURGERY

A. URGENT

The great majority of patients with ulcerative colitis have mild to moderate disease which is intermittently active and can be adequately managed with medical therapy alone. Only a minority of patients, perhaps 5%, will require acute surgical intervention, either because of acute colitis which is unresponsive to medication or because of a life-threatening complication of the disease. All patients with a severe attack of ulcerative colitis should be admitted to hospital. A severe attack has been defined as "passage of more than six bloody stools per day in association with tachycardia (>90/min), fever (>37.5°C), anemia (Hb < 10.0 g/dL), ESR > 30 mm/hr, or serum albumin < 35 g/L."[14] Such patients should be seen jointly by a physician and a surgeon and initially undergo intensive medical treatment. Serial plain abdominal X-rays are of value in identification of acute toxic dilatation and free perforation. The presence of mucosal islands on the plain X-ray predicts a poor response to medical treatment.[15] Patients over the age of 60, those with pancolitis, and those suffering a first episode of colitis are more likely to require surgical intervention.[16] Surgical intervention is indicated in patients undergoing medical treatment who deteriorate within 48 h, who fail to improve with 7 d, or who have evidence of a supervening complication.[17]

Toxic dilatation of the colon usually affects the transverse colon and is diagnosed when the diameter of the colon exceeds 6.5 cm on plain X-ray. Opinion is divided on whether a diagnosis of toxic dilation is an absolute indication for operation.[16,17] Undoubtedly, some will respond to medical therapy, but failure with perforation can occur in 20 to 30%.[16,18] Furthermore, 90% of patients who have an episode of toxic dilation treated medically ultimately require operation.[19] A reasonable consensus would be that once a diagnosis of toxic dilatation has been made, intensive medical treatment should be undertaken with a view to preparing the patient for urgent laparotomy and that any deterioration in the patient's condition be an indication for emergency exploration.

The question of which operation to perform must be individualized for both the patient and the surgeon. In general, it may be said that colectomy, ileostomy, and mucus fistula is the quickest and safest operation for acute fulminating colitis. Proctectomy should not be performed except where there is massive hemorrhage from the rectum or where a rectal perforation below the peritoneal reflection has occurred. In the latter case, a Hartmann's procedure with drainage of the presacral space may be preferable to proctectomy. Apart from the speed and added safety of preserving the rectum in acute colitis, a further advantage is the possibility of a future continence-preserving operation such as ileorectal or ileal pouch-anal anastomosis.

The surgical management of toxic megacolon is controversial because of the great technical difficulty of performing colectomy in this situation without intraoperative perforation. In addition to the fragile nature of the inflamed colon, sealed perforations are common. In these circumstances many would favor the approach of controlled decompression through a loop ileostomy, transversostomy, and if necessary sigmoidostomy as proposed by Turnbull.[20] The technique can be life saving in the gravely ill patient and Fazio at the Cleveland Clinic has reported 83 patients treated in this way, with 3 postoperative deaths.[16] Nevertheless, the majority of surgeons would favor colectomy, ileostomy, and mucus fistula as the operation of choice in toxic megacolon and would reserve the Turnbull technique for the gravely ill patient.

B. ELECTIVE

The indications for elective operation on patients with ulcerative colitis are less well defined than for urgent operation. In the majority of patients, ulcerative colitis pursues an intermittent course often with prolonged periods of remission. Operation should be considered for patients with chronic ill health and/or steroid dependence. The timing of a surgical consultation will depend greatly on the attitude of the patient and the surgical bias of the physician. Probably, the most controversial indication for operation is chronicity of disease and potential for malignancy. This subject will be dealt with in more detail elsewhere. Suffice it to say that there is an established association between ulcerative colitis and colon cancer and that the risk increases with the extent and the duration of disease. At present, yearly colonoscopic screening for dysplasia and early carcinoma is indicated in those with disease of 8 to 10 years' duration.[21] The presence of confirmed high-grade dysplasia or carcinoma *in situ* is an indication for surgical excision.

Within the past decade, operative technique in elective surgical treatment of ulcerative colitis has become one of the most hotly debated issues in general surgery. Panproctocolectomy with permanent ileostomy, so long the gold standard of treatment, has come under challenge by renewed interest in techniques of continence preservation. In view of this, the advantages and disadvantages of operative techniques available for elective surgical treatment of ulcerative colitis are discussed further.

IV. PROCTOCOLECTOMY AND ILEOSTOMY — CLINICAL RESULTS

Since the 1960s ileostomy appliance technology has improved and endostomal therapy has developed as a nursing specialty. The life style of an "ileostomate" has also been made easier by the formation of self-help groups in most major cities. The development of perirectal and intersphincteric dissection techniques has reduced injury to the pelvic sympathetic nerves and thereby lowered the incidence of postoperative male sexual dysfunction and bladder dysfunction.[22]

The majority of patients come to accept their stoma and to enjoy their new-found good health.[23] Ileostomy does not appear to prevent patients returning to previous employment as this is accomplished by more than 90% of patients without major difficulty.[23,24] However, as many as 20% of patients with an ileostomy feel in some way restricted in their social life by the presence of an ileostomy and more than 25% of patients have difficulties with stoma care.[23] In a survey of members of the Ileostomy Association of Great Britain and Ireland, nearly one third of patients reported sexual dysfunction after proctocolectomy and 12% felt that the ileostomy contributed to marital tension and unhappiness, and even divorce in 2%.[25] All of these findings are supported by similar data from a review of 675 patients who had an ileostomy constructed at the Mayo Clinic between 1966 and 1980.[26]

Eventually, 90 to 95% of patients are satisfied with an ileostomy. On the other hand, as many as 40% of patients with a permanent ileostomy either definitely desired or would, if the opportunity presented itself, seriously consider a change to a continent ileostomy.[26] The search for ways to improve the physical and mental well being of patients who have had or who face proctocolectomy has been a priority among gastrointestinal surgeons for many years.

V. ALTERNATIVES TO ILEOSTOMY

A. ILEORECTOSTOMY

Originally described as ileosigmoidostomy by Lilienthal and popularized by Devine,[27,28] the role of ileorectostomy as an alternative to proctocolectomy has remained controversial. Proponents view the operation as a definitive method of treatment,[29,30] whereas others either feel that it is an acceptable alternative in selected cases.[31,32] The controversy stems from preservation of diseased rectal mucosa and the long-term malignant potential of so doing. In addition, there is continuing proctitis which results in long-term failure in approximately 30% of cases.[32,33]

It is difficult to predict at the time of operation those patients who will have a good functional result following ileorectostomy, as in some cases severe proctitis will improve following colectomy, whereas other cases with mild proctitis will progressively deteriorate. The functional results following the operation are similar to, and some would say better than, those reported following the ileal pouch-anal procedure, most patients ending up with six or fewer semiformed stools per day.[34] It is generally agreed that a patient with a contracted or stenosed rectum from ulcerative colitis is unlikely to obtain a satisfactory functional result and should not be offered the procedure.

The risk of carcinoma developing in the rectal stump following ileorectostomy for ulcerative colitis is itself the subject of controversy. It appears that like carcinoma of the colon in ulcerative colitis the risk of cancer in the rectal stump increases with time and that the cumulative risk reaches 13 to 17%, 25 to 30 years after the onset of the ulcerative colitis.[35,36] The introduction of regular lifetime proctoscopy and screening for epithelial dysplasia may help identify those patients who would develop cancer allowing a safe interval proctectomy.[33]

A fair comparison between ileorectal anastomosis and ileal pouch anal anastomosis is difficult to make. Proponents of ileorectal anastomosis point to the technical ease of the operation, absence of a second stage ileostomy closure, and a much lower risk of perioperative autonomic nerve damage.[32] Against this is the argument that ileoanal anastomosis removes all diseased mucosa, avoids an eventual proctectomy rate of 30%, and that the functional results of the two procedures are similar. The logic of these arguments is persuasive and more surgeons are tending toward ileoanal anastomosis, but the final answer is yet to be determined.

B. THE KOCK POUCH

The continent ileostomy was the first realistic alternative to conventional ileostomy that also permitted complete excision of the diseased colonic and rectal mucosa.[37] Originally interested in developing a continent ileal urinary conduit, Kock realized the possibility of applying the same principles to construction of a continent ileostomy reservoir. He found that reliable continence could be achieved by construction of a pouch reservoir into which the efferent limb of ileum was intussuscepted to produce a continent nipple valve. The ileal pouch reservoir was designed to overcome the peristaltic response of the ileum to distension, which had been the principal cause of failure in previously designed intestinal urinary bladder substitutes.[37] The operation was received with considerable interest and during the 9 years that followed more than 1100 cases were reported.[38] The operative mortality rate was less than 2% and the anastomotic leak rate was as low as 3%.[38,39] A major long-term drawback with the operation was the tendency in 30 to 40% of patients for the nipple valve to prolapse and become incontinent.[40] With refinements in operative technique the reoperation rate has been reduced to 20 to 30%.[41,42]

Since the advent of the ileal pouch-anal anastomosis, the indications for Kock pouch ileostomy are diminishing. The reasons for this are that the operation is technically more difficult than ileal pouch-anal anastomosis, it does not avoid a stoma, and it has a substantial failure rate. The indications for a Kock pouch now appear to be limited to those patients already with a Brooke ileostomy who wish to convert to a continent stoma, patients with a failed ileal pouch-anal anastomosis who wish to avoid an incontinent ileostomy, and patients in whom poor anal sphincter function precludes an ileoanal procedure.[41]

C. STOMAL OCCLUSION DEVICE

An interesting alternative to the Kock pouch is a method of producing a continent ileostomy using a prosthetic valve. An early prototype was the Erlanger magnetic closure system. Infection around the implanted metalic ring prevented widespread application of the device. Several investigators have explored the possibility of using intermittent stoma occlusion with a balloon catheter to produce continence.[43,44] Early experimental work in animals, using a program of ileostomy occlusion for increasing periods of time, demonstrated that the distal ileum could dilate to form a compliant reservoir. Dilatation was not found to interfere with ileal absorption or secretion.[43] A study of four patients with a Brook ileostomy concluded that chronic intermittent ileostomy occlusion achieved enteric continence without impairing intestinal function.[45] However, 2 years later only 2 of the 4 patients continued to use the device.[46]

The efficiency of the stomal occlusion device was studied in a second group of seven patients in whom a prestomal ileal pouch was constructed following an initial favorable study in animals.[45] Four of the patients developed complications related to use of the occlusion catheter.[46] Only 3 of the 7 patients continued to use the device at 1 year or more after the operation.

The only useful role that has been found to date for the stomal occlusion device is in patients with an established Kock pouch which has failed and who elect not to undergo nipple valve reconstruction. Of 50 such patients at the Mayo Clinic, 42 continue to use the device and have achieved complete fecal continence. The other patients objected to the bulkiness and occasional discomfort of the device, and chose either to have reoperation on the Kock pouch or to wear a permanent external ileostomy bag.[46]

D. MUCOSAL PROCTECTOMY AND ILEOANAL ANASTOMOSIS

1. Development of Ileoanal Anastomosis

The concept that ulcerative colitis is a disease of colonic mucosa and therefore could be treated by colectomy, mucosal proctectomy, and ileoanal anastomosis has been widely attributed to Ravitch and Sabiston.[47] Although seven cases of "anal ileostomy" after proctocolectomy had previously been described,[48–51] Ravitch and Sabiston were the first to study in a systematic fashion the hypothesis that continence could be maintained after colectomy, mucosal proctectomy, and ileoanal anastomosis. A series of animal experiments were performed to perfect the technique before application in humans.[47] Two patients with ulcerative colitis were then operated upon[52] and, following initial success, nine further patients were added to the series, five with ulcerative colitis and four with polyposis coli.[53] Three patients required excision of the ileoanal anastomosis and conversion to an abdominal stoma because of pelvic sepsis. The functional results of the remaining patients were not satisfactory. All had excessive stool frequency and troublesome perianal skin excoriation. Even Ravitch himself became disillusioned with the outcome and eventually abandoned the technique.

Ileoanal anastomosis was applied by several other surgeons in small numbers of patients, with almost universally poor results. Goligher,[54] Devine and Webb,[55] Best,[56] Schneider,[57] and Casanova-Diaz[58] documented numerous complications of the operation which included pelvic sepsis, pelvic fistula, anastomotic dehiscence, anastomotic stricture, and small bowel obstruction. In addition, most patients had intractable stool frequency, poor continence, and perineal excoriation. Of the 41 patients operated on before 1960, only 22 patients (54%) were continent and 15 (37%) eventually required a permanent abdominal ileostomy.[59]

Between 1960 and 1976, a total of 45 patients were reported to have undergone ileoanal anastomosis. Of these, 35 (77%) were grossly continent and only 3 had required conversion to a permanent ileostomy.[59] The improved success rate owed as much to advances in perioperative nutritional and medical support as to improvements in the technique of mucrosal proctectomy introduced by Soave.[60] While most patients achieved gross fecal continence, stool frequency, urgency, and occasional minor incontinence were constant problems. As a result, ileoanal anastomosis remained principally a research interest.

In 1977, Martin reported the outcome of ileoanal anastomosis in 17 young patients. There were only two failures, both due to pelvic sepsis, the remainder obtained acceptable continence.[61] With assiduous attention to preoperative nutrition, preoperative hemostasis, and postoperative drainage of the space between the ileum and the cuff of rectal muscle, the incidence of postoperative sepsis was reduced.[62] Like Soave, Martin preserved the most distal 1 cm of rectal mucosa, believing that in doing so anorectal sensation was maintained and postoperative continence improved. While all his patients did encounter early postoperative difficulty with stool frequency and perineal skin excoriation, improvement occurred over a 5- to 12-month period following closure of a temporary protective ileostomy.

A most important aspect of Martin's paper was timing. By 1977 there was widespread interest in continence preservation for patients with ulcerative colitis and polyposis coli. The Kock pouch technique had raised expectations, but it was plagued by technical difficulty and high revision rates. Surgeons with an interest in continence preservation were looking for an alternative. Prompted by Martin's work, Telander and others at the Mayo Clinic performed ileoanal anastomosis on 12 young patients between 1977 and 1979. Two cases failed but the remainder had daytime continence. All occasionally experienced fecal soiling at night. Stool frequency and an improvement in continence was seen up to 1 year following operation. The improvement in stool frequency and continence was associated with an increase in the reservoir capacity of the ileum proximal to the ileoanal anastomosis.[63] Clinical improvement in stool frequency and continence was hastened by daily balloon dilatation of the "neorectum".[63]

Clinical results following ileoanal anastomosis in adults were not satisfactory. Of 50 adults undergoing ileoanal anastomosis at the Mayo Clinic between 1978 and 1981, 33% required conversion to a Brooke ileostomy because of excessive stool frequency and incontinence.[64] A detailed physiological study of these patients found a direct relationship between the maximum capacity of the "neorectum" and the postoperative stool frequency.[65] It was found that the anal canal resting and squeeze pressures were similar to those of healthy controls indicating that endorectal dissection did not damage anal sphincter function. Loss of reservoir capacity of the rectum was the principal cause of the poor functional outcome in adults.[64]

2. Development of the Ileal Reservoir

Gaston, in 1951, had studied the effect of colectomy and mucosal protectomy on stool frequency in one of the first patients to have an ileoanal anastomosis. He concluded that intractable stool frequency would result from the loss of "reservoir" continence provided by the left colon and rectum whether or not the anal sphincter mechanism was preserved.[66] The search for a means of replacing the rectal reservoir with an ileal pouch reservoir proximal to the anastomosis began in 1955 with an investigation of an "S" shaped ileal pouch constructed in dogs during colectomy, mucosal proctectomy, and ileoanal anastomosis.[67] Animals with an ileal pouch had better results than a control group with a straight ileoanal anastomosis. As a result, a 26-year-old woman with familial polyposis coli underwent colectomy, mucosal proctectomy, and ileal pouch-anal anastomosis with construction of a double lumen isoperistaltic ileal pouch proximal to the ileoanal anastomosis. The clinical outcome was "equivocal" and the authors remarked that the technical difficulty of the procedure far exceeded that experienced in the dogs. It was hard to imagine that ileal pouch-anal anastomosis would ever be of clinical use.[68]

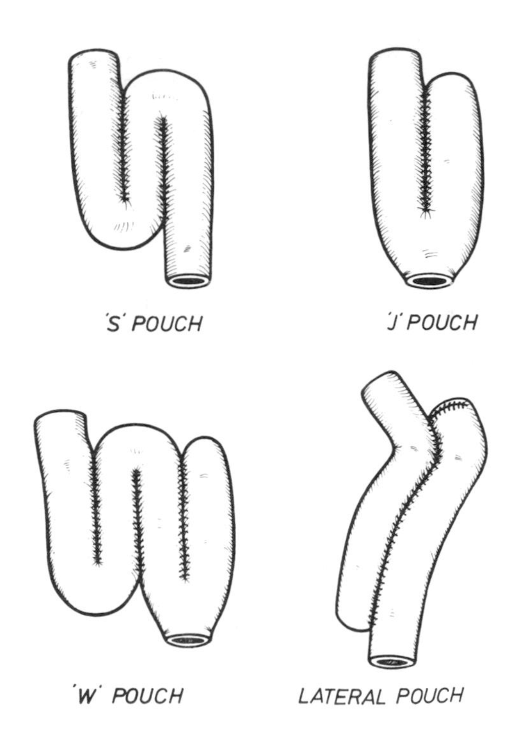

FIGURE 1. Figure showing four ileal pouch designs: "S" triple lumen pouch, "J" double lumen pouch, "lateral" isoperistaltic double lumen pouch, and "W" quadruple lumen pouch.

It was the initial success and apparent safety of the Kock pouch technique that led Parks and Nicholls to again combine an ileal pouch with the ileoanal anastomosis technique.[69] The pouch used was a triple loop "S" pouch similar to that described by Valiente and Bacon (Figure 1). The outcome was encouraging, since four of the initial five patients were fully continent, but three of the five had difficulty evacuating the pouch and needed intermittent intubation. To overcome this difficulty, double lumen "J" pouch and isoperistaltic double lumen pouch designs were devised (Figure 1).[70,71] The "J" pouch appeared to have advantages over the "S" type pouch as it was technically easier to construct and did not possess an efferent limb which caused impaired evacuation in some patients.[72,73] At the Mayo Clinic, the results of 74 patients with colectomy, mucosal proctectomy, and ileal "J" pouch-anal anastomosis were compared to the clinical results in 50 patients with "straight" ileoanal anastomosis.[74] Continence and stool frequency were both significantly better in the patients with pouches because of their larger neorectal capacity and compliance.

3. Clinical Results of Ileal Pouch-Anal Anastomosis

In the 10 years since Parks reported the outcome of ileal pouch-anal anastomosis in 5 patients there has been an exponential increase in the numbers of ileal pouch operations performed. By 1982 there had been 25 reports of ileoanal anastomosis in the literature.[59] After 5 years more, a further 66 reports have appeared.[75] Reported experience with ileal pouch-anal anastomosis now approaches 2000 cases. Ileoanal anastomosis can be performed safely with a postoperative mortality of 0.3% (Table 1) which is comparable to that following standard proctocolectomy. Postoperative morbidity is, however, substantial and a significant proportion of the morbidity occurs following ileostomy closure. The overall failure rate requiring ileal pouch excision is approximately 6% (Table 1). Early postoperative failure is most commonly due to pelvic sepsis. Unacceptable functional results and recurrence of previously

TABLE 1
Morbidity and Mortality Following Ileoanal Anastomosis

Author, year (ref.)	No.	Pouch type	Mor-tality	Failed (%)	Sepsis (%)	Obstruction (%)	Stricture (%)
Taylor and Dozois, 1987 (87)	542	J, S	1	5	8	15	12
Fonkalsrud, 1987 (96)	145	L, S, J	1	5	15	8	16
Nicholls, 1987 (84)	119	S, J, W	1	6	18	14	9
Cohen, 1986 (98)	125	S, J	0	2	8	4	NS
Becker, 1986 (99)	100	J	0	NS	0	15	NS
Goldberg and Rothenberger, 1986 (100)	91	S, J	0	6	10	18	10
Schoetz, 1986 (101)	91	S, J	0	3	16	29	3
Keighley et al., 1987 (95)	40	S, J, W	0	23	35	15	23
Nasmyth et al., 1986 (88)	39	S, J, L	0	18	21	10	8
TOTAL	1292		0.3%	5%	10%	14%	12%

Note: NS = not stated, J = "J" double lumen pouch, S = "S" triple lumen pouch, W = "W" quadruple lumen pouch, and L = "lateral" isoperistaltic double lumen pouch.

unsuspected Crohn's disease are the most frequent causes of late failure.[76] Approximately 35% of patients experience a postoperative complication. Pelvic sepsis (11%) is potentially the most serious. The incidence of postoperative sepsis has been reduced in most large series by use of a shorter cuff of denuded rectum, suction drainage of the presacral space, and assiduous aseptic technique. Particular care is needed in ileostomy closure as a 7% leakage rate with peritonitis has been reported.[77]

Overall, 13% of patients experience postoperative intestinal obstruction, and approximately one half of these require laparotomy (Table 1). The incidence of intestinal obstruction is comparable to that reported after panproctocolectomy and after ileorectostomy.[78,79] Stricture of the ileoanal anastomosis requiring dilatation on at least one occasion is common, occurring in approximately 12% of patients. Recurrent strictures are uncommon and are usually the result of excess tension on the ileoanal anastomosis with partial dehiscence.

The functional outcome of ileal pouch-anal anastomosis following ileostomy closure has been documented in several centers (Table 2). Almost all patients are continent, but about one quarter experience intermittent minor soiling. Soiling particularly occurs at night and during times of stress. Minor continence difficulties are more common in women and patients over 50 years of age. Continence improves with time and is better in those who have had the operation for polyposis coli rather than ulcerative colitis.[75]

Stool frequency following ileal pouch anal anastomosis decreases in the first year after operation[78] and remains constant thereafter.[75] Stool frequency tends to be lower in patients with a large ileal pouch (Table 2);[79,80] thus, it is less in patients with a triple lumen "S" or a quadruple lumen "W" pouch (Figure 1) compared to those with the smaller double lumen "J" or "lateral" pouches. In the first year, approximately one half the patients regularly use antidiarrheal medication or psyllium colloid bulking agents. This falls to approximately one third at 5 years.[75]

TABLE 2
Functional Results After Ileal Pouch-Anal Anastomosis

Author, year (ref.)	No.	Follow up	Stool frequency			Major	Minor
			Day	Night	Total		
Pemberton et al., 1987 (75)	390 (J)	28	6	2	8	4	22
Nicholls, 1987 (84)	58 (S)	24	3	1	4	5	28
	32 (W)	7	3	1	4	3	6
	13 (J)	13	5	1	6	0	25
McHugh et al., 1987 (86)	51 (J)	22	6	1	7	0	22
	19 (S)	11	5	1	6	0	31
Becker, 1986 (99)	100 (J)	12	5	1	6	0	25
Goldberg and Rothenberger, 1986 (100)	52 (S)	19	4	2	6	0	27
Nasmyth et al., 1986 (88)	9 (S)	>6	5	1	6	0	50
	12 (J)	>6	6	1	7	0	83
	4 (L)	>6	6	1	7	0	100

Note: *J* = "J" double lumen pouch, S = "S" triple lumen pouch, W = "W" quadruple lumen pouch, and L = "lateral" isoperistaltic double lumen pouch.

An important cause of postoperative morbidity is recurrent pouch ileitis. This syndrome occurs in approximately 15% of patients and features malaise, ileal pouch inflammation, and bloody diarrhea. The syndrome was originally recognized in patients with Kock pouch.[81] The symptoms respond to antibiotic therapy, particularly if directed against anaerobic organisms. This has led to the hypothesis that pouch ileitis is primarily due to stasis within an ileal reservoir, leading to bacterial overgrowth and mucosal inflammation.[39,82] Nevertheless, investigations have failed to show quantitative bacterial overgrowth or impaired ileal pouch emptying as the cause. Pouch ileitis occurs almost exclusively in patients who have had ulcerative colitis and not those who have had polyposis coli. This observation has led to speculation regarding an interaction between luminal contents and the intestinal wall which is exclusive to patients with ulcerative colitis.[83]

The incidence of sexual dysfunction is low after ileal pouch-anal anastomosis. In men impotence is rare (1 to 5%) but retrograde ejaculation occurs in 10 to 12%.[84,85] Dyspareunia was less in women following the operation than before, although 5 to 10% of women reported that fear of incontinence adversely affected their sexual activities.[85,86] A number of pregnancies with successful vaginal delivery have been reported after ileal pouch-anal anastomosis.[87]

The quality of life after ileal pouch-anal anastomosis is difficult to assess. The great majority of patients are satisfied with the outcome of their operation. Approximately one third of patients feel that stool frequency is a significant problem and one quarter feel that their life style is restricted by stool frequency, urgency, or continence difficulties.[85,86,88] Still, there was almost universal preference for the life style experienced with an ileal pouch than with a defunctioning ileostomy.[26,89]

VI. AREAS OF CONTROVERSY

A. POUCH DESIGN

The major controversy that continues among surgeons with an interest in ileal pouch-anal anastomosis concern the optimum pouch design. Long-term continence rates are similar in most series; however, stool frequency is greater with double lumen "J" and "Lateral" pouches compared to larger "S" and "W" pouches (Figure 1). Theoretically, the "W" pouch has a number of advantages as it combines the capacity of an "S" pouch with the efficiency of emptying of the "J" pouch.[79] Early experience with the "W" pouch has been satisfactory; however, it is technically more difficult than the "J" pouch to construct and, because of its bulk, can be difficult to pull into the pelvis. Moreover, the volume of stool passed per 24 h and the motility of the ileal pouch are more important determinants of stool frequency than ileal pouch capacity.[80] Therefore, efforts to reduce postoperative stool frequency should be directed at reduction of stool volume and modification of pouch motility as well as the design of large reservoirs.

The final determination on which ileal pouch type provides the most physiological reservoir for ileal stool has yet to be made. Present evidence suggests that once the ileal pouch is of "adequate" capacity and has an unobstructed channel for evacuation, there is little to choose between the various techniques. It appears likely, however, that the "J" pouch type will continue to be the most widely adopted, simply because it is technically the easiest to construct.

B. MUCOSAL PROCTECTOMY

Complete excision of diseased rectal mucosa with preservation of the anal sphincteric mechanism is fundamental to cure in patients with ulcerative colitis undergoing ileoanal anastomosis. It is now clear that a 2- to 3-cm cuff of rectal muscle is sufficient to preserve continence. Furthermore, it is possible to perform a complete rectal mucosectomy without damage to the sphincter mechanism. However, injury to both the internal and external sphincters may occur during mucosectomy and these patients are more likely to have postoperative difficulties with continence.[90] To avoid injury to the anal sphincter mechanism, some advocate ileoanal anastomosis to the top of the anal canal, not to the dentate line.[91,92] In doing so, 1 to 2 cm of anal transitional zone mucosa is preserved. Manometric studies have confirmed that this approach is less likely to damage the sphincter mechanism. Moreover, anal sensation and discriminatory function are significantly better after end-to-end ileoanal anastomosis, preserving the proximal anal mucosa, than after mucosal proctectomy with endoanal anastomosis.[93] Histological studies show that the anal transitional zone is not inflamed in ulcerative colitis.[94] Nevertheless, the long-term safety of preservation of the anal transitional zone is uncertain and most proponants of ileoanal anastomosis continue to perform mucosal proctectomy.

C. PATIENT SELECTION

Ileal pouch-anal anastomosis is not suitable in obese patients because of the difficulty in obtaining sufficient length in the small bowel mesentery to allow safe ileal-pouch anal anastomosis. The operation is not indicated in patients with impaired anal sphincter function as the increased stool volume following the operation places the sphincter mechanism under stress.[90] For similar reasons, the operation is rarely performed on patients over the age of 55 years. If there is any doubt concerning the integrity of the sphincter mechanism, preoperative anal manometry should be performed.

A diagnosis of Crohn's disease is a contraindication to ileal pouch-anal anastomosis. Where there is doubt about the preoperative diagnosis, it may be better to perform a staged procedure with initial colectomy, ileostomy, and mucus fistula. This will allow histological examination of the entire colon before a decision to proceed is made. In approximately 10% of cases the histological diagnosis is "indeterminate colitis". It appears that the great majority of these cases behave like ulcerative colitis and it may be reasonable to proceed with a pouch procedure.[89] Ileal pouch-anal anastomosis should be performed as an elective operation and is rarely suitable as an urgent or emergency operation. Nevertheless, steroid dependence does not seem to adversely affect the outcome in elective operations.

VII. CONCLUSIONS

Despite the difficulties patients with ileostomy experience, proctocolectomy and ileostomy is a good operation which cures the underlying disease and rapidly restores the patient to good health. For almost 40 years proctocolectomy and ileostomy have been the "gold standard" against which alternative treatment options must be compared. The operation can be performed safely by most general surgeons. It does not require patient preselection on the grounds of age, obesity, or severity of disease. Furthermore, the operation can be performed in one stage and does not require a defunctioning ileostomy which appears to be necessary with the more recent continence-preserving techniques.

Most surgeons will perform ileorectostomy in selected cases with very good results. A few surgeons continue to advocate the almost exclusive use of ileorectostomy. Early enthusiasm for Kock pouch ileostomy has been tempered by experience and the operation is now used only by those with particular expertise and, even then, only in selected cases which are unsuitable for ileal pouch-anal anastomosis. The results of ileoanal anastomosis have improved beyond recognition when compared with those early results obtained by Ravitch.[53] Ileal pouch-anal anastomosis is now the surgical procedure routinely performed for chronic ulcerative colitis and polyposis coli in many centers throughout the world. However, proctocolectomy and ileostomy continues to be the operation performed by the great majority of surgeons treating ulcerative colitis.

In the past, surgeons have been reluctant to change from the tried and trusted proctocolectomy and ileostomy because of the poor fecal continence reported following early attempts at ileoanal anastomosis. Whether ileal pouch-anal anastomosis will become more widely practiced in the future depends to a great extent on critical appraisal of the clinical results obtained by surgeons in specialist centers. Much of these data are now available. Ileal pouch-anal anastomosis can be performed with a mortality equivalent to that of proctocolectomy and ileostomy. Postoperative morbidity is higher after ileoanal anastomosis, and two, sometimes three, operations may be required. The functional results and the quality of life eventually achieved by most patients after ileal pouch-anal anastomosis appear to justify a more widespread introduction of this procedure.

However, here lies a problem. To date the outcome of ileal pouch procedures has only been reported by surgeons in specialist centers operating on young, informed, and highly motivated patients. Ileal pouch-anal anastomosis is a technically demanding operation and there is an undoubted learning curve reflected in a substantial decrease in postoperative morbidity with experience.[85,95,96] Should ileal pouch procedures be performed exclusively in specialist centers or can they be safely undertaken by general surgeons with an interest in colorectal disease who previously would have performed a proctocolectomy? The results of a recent multicenter review from Italy are helpful in answering this question. The morbidity and clinical outcome in 84 patients operated upon in 21 separate surgical departments were comparable to those of any large series from a single center.[97] Thus, it is reasonable to anticipate that in the near future, a patient with ulcerative colitis with an indication for operation will be offered the choice between panproctocolectomy with permanent ileostomy and some form of continence-preserving alternative.

Life with an ileostomy is not easy, but continence after ileal pouch-anal anastomosis is not normal. The patient and the gastroenterologist as well as the surgeon must be fully aware of the advantages as well as the disadvantages of an ileal pouch procedure. Careful patient selection and assiduous attention to surgical technique and postoperative care are critical to a satisfactory outcome. Under these conditions, it is hard not to offer a young, otherwise healthy, patient the opportunity of life without an ileostomy.

REFERENCES

1. **Mayo Robson, A. R.,** Case of colitis with ulceration treated by inguinal ileostomy, *Trans. Clin. Soc. London,* 26, 213, 1893.
2. **Corbett, R. S.,** A review of the surgical treatment of chronic ulcerative colitis, *Proc. R. Soc. Med.,* 38, 277, 1945.
3. **Brook, B. N.,** Conventional ileostomy: historical perspectives, in *Alternatives to Conventional Ileostomy,* Dozois, R. R., Ed., Year Book Medical Publishers, Chicago, 1985.
4. **Brown, J. Y.,** The value of complete physiological rest of the large bowel in the treatment of certain ulcerative and obstructive conditions of that organ, *Surg. Gynecol. Obstet.,* 7, 610, 1913.
5. **Bargan, J. A., Brown, P. W., and Rankin, F. W.,** Indication for and technique of ileostomy in chronic ulcerative colitis, *Surg. Gynecol. Obstet.,* 55, 196, 1932.
6. **McKittrick, L. S. and Moore, F. D.,** Ulcerative colitis. Ileostomy: problem or solution?, *JAMA,* 139, 201, 1949.
7. **Rankin, F. W.,** Total colectomy; its indications and technique, *Trans. Am. Surg. Assoc.,* 49, 263, 1931.
8. **Strauss, A. A. and Strauss, S. F.,** Surgical treatment of ulcerative colitis, *Surg. Clin. North Am.,* 24, 211, 1944.
9. **Warren, R. and McKittrick, L. S.,** Ileostomy for ulcerative colitis; technique, complications, management, *Surg. Gynecol. Obstet.,* 93, 555, 1951.
10. **Dragstedt, L. R., Dack, G. M., and Kirsner, J. B.,** Chronic ulcerative colitis: summary of evidence implicating *Bacterium necrophorum necrophorum* as an etiologic agent, *Ann. Surg.,* 114, 653, 1941.
11. **Monroe, C. W. and Olwin, J. H.,** Use of an abdominal flap graft in construction of a permanent ileostomy, *Arch. Surg.,* 59, 565, 1949.
12. **Crile, G. and Turnbull, R. B.,** Mechanism and prevention of ileostomy dysfunction, *Ann. Surg.,* 140, 459, 1954.
13. **Brooke, B. N.,** Management of ileostomy including its complications, *Lancet,* ii, 102, 1952.
14. **Truelove, S. C. and Witts, L. J.,** Cortisone in ulcerative colitis: final report on a therapeutic trial, *Br. Med. J.,* 2, 1041, 1955.
15. **Lennard-Jones, J. E., Ritchie, J. K., Hilder, W., and Spicer, C. C.,** Assessment of severity in colitis: a preliminary study, *Gut,* 16, 579, 1975.
16. **Goligher, J. C.,** Ulcerative Colitis, in *Surgery of the Anus, Rectum and Colon,* Goligher, J. C., Ed., Bailliere Tindal, London, 1984.
17. **Jewell, D. P.,** Ulcerative colitis: indications for surgery, in *Surgery of Inflammatory Bowel Disorders,* Lee, E. C. G. and Nolan, D. J., Eds., Churchill Livingstone, Edinburgh, 1987, 33.
18. **Wolff, B. G., Culp, C. E., and Dozois, R. R.,** Ulcerative colitis: diagnosis and indications for ileostomy, in *Alternatives to Conventional Ileostomy,* Dozois, R. R., Ed., Year Book Medical Publishers, Chicago, 1985, 3.
19. **Grant, C. S. and Dozois, R. R.,** Toxic megacolon: ultimate fate of patients after "successful" medical management, *Am. J. Surg.,* 147, 106, 1984.
20. **Turnbull, R. B., Weakley, Hawk, W. A., and Schofield, P.,** Choice of operation for the toxic megacolon phase of non-specific ulcerative colitis, *Surg. Clin. North Am.,* 50, 1151, 1970.
21. **Yardley, J. H., Ransohoff, D., Riddell, R., and Goldman, H.,** Cancer in inflammatory bowel disease: how serious is the problem and what should be done about it?, *Gastroenterology,* 85, 197, 1983.
22. **Lee, E. C. G. and Dowling, B. L.,** Perimuscular excision of the rectum for Crohn's disease and ulcerative colitis, *Br. J. Surg.,* 59, 29, 1972.
23. **Roy, P. H., Sauer, W. G., Bearhs, O. H., and Farrow, G. M.,** Experience with ileostomies: evaluation of long term rehabilitation, *Am. J. Surg.,* 91, 459, 1970.
24. **Whates, P. D. and Irving, M.,** Return to work following ileostomy, *Br. J. Surg.,* 71, 619, 1984.
25. **Burnham, W. R., Lennard-Jones, J. E., and Brooke, B. N.,** Sexual problems among married ileostomies, *Gut,* 18, 673, 1977.
26. **Pemberton, J. H., Phillips, S. F., Dozois, R. R., and Wendorf, L. J.,** Conventional ileostomy: current clinical results, in *Alternatives to Conventional Ileostomy,* Dozois, R. R., Ed., Year Book Medical Publishers, Chicago, 1985, 40.
27. **Lilienthal, H.,** Extirpation of the entire colon, the upper portion of the sigmoid flexure and four inches of the ileum for hyperplastic colitis, *Ann. Surg.,* 37, 616, 1903.
28. **Devine, H.,** A method of colectomy for desperate cases of ulcerative colitis, *Surg. Gynecol. Obstet.,* 76, 136, 1943.
29. **Aylett, S. O.,** Total colectomy and ileorectal anastomosis, *Proc. R. Coll. Surg. Edinburgh,* 26, 28, 1981.
30. **Parc, R., Levy, E., Frileux, P., and Loygue, J.,** Ileorectal anastomosis after total abdominal colectomy for ulcerative colitis, in *Alternative to Conventional Ileostomy,* Dozois, R. R., Ed., Year Book Medical Publishers, Chicago, 1985.

31. **Adson, M. A., Cooperman, A. M., and Farrow, G. M.,** Ileorectostomy for ulcerative colitis of the colon, *Arch. Surg.,* 104, 424, 1972.
32. **Hawley, P. R.,** Ileorectal anastomosis, *Br. J. Surg.,* 72, S75, 1985.
33. **Farnell, M. B. and Adson, M. A.,** Ileorectostomy; current results, the Mayo Clinic experience, in *Alternatives to Conventional Ileostomy,* Dozois, R. R., Ed., Year Book Medical Publishers, Chicago, 1985.
34. **Jagelman, D. G.,** Colectomy with ileorectal anastomosis for ulcerative colitis: a historical perspective, in *Alternatives to Conventional Ileostomy,* Dozois, R. R., Ed., Year Book Medical Publishers, Chicago, 1985, 55.
35. **Grundfest, S. F., Fazio, V., and Weiss, R. A., et al.,** The risk of cancer following colectomy and ileorectal anastomosis, *Ann. Surg.,* 193, 9, 1981.
36. **Johnson, W. R., McDermott, F. T., and Hughes, E. S. R., et al.,** The risk of rectal carcinoma following colectomy in ulcerative colitis, *Dis. Colon Rectum,* 26, 44, 1983.
37. **Kock, N. G.,** Intra-abdominal "reservoir" in patients with permanent ileostomy: preliminary observation on a procedure resulting in faecal "continence" in 5 ileostomy patients, *Arch. Surg.,* 99, 223, 1969.
38. **Myrvoid, H. E.,** The continent ileostomy, *World J. Surg.,* 11, 720, 1987.
39. **Kock, N. G., Myrvoid, H., Nilsson, L., and Philipson, B.,** Continent ileostomy: an account of 314 patients, *Acta Chir. Scand.,* 147, 67, 1981.
40. **Dozois, R. R., Kelly, K. A., Beart, R. W., and Beahrs, O. J.,** Improved results with continent ileostomy, *Ann. Surg.,* 192, 319, 1980.
41. **Dozois, R. R., Kelly, K. A., Beart, R. W., and Bearhs, O. H.,** Continent ileostomy, the Mayo Clinic experience, in *Alternatives to Conventional Ileostomy,* Dozois, R. R., Ed., Year Book Medical Publishers, Chicago, 1985.
42. **Kock, N. G., Myrvoid, H. E., Nilsson, L. O., and Philipson, B. M.,** Continent ileostomy: the Swedish experience, in *Alternatives to Conventional Ileostomy,* Dozois, R. R., Ed., Year Book Medical Publishers, Chicago, 1985, 163.
43. **Pemberton, J. H., Kelly, K. A., and Phillips, S. F.,** Achieving ileostomy continence with an indwelling stomal device, *Surgery,* 90, 336, 1981.
44. **Sandra, Y., Fonkalsrud, E. W., and Kojima, Y.,** Intermittent ileostomy occlusion for fecal storage using balloon catheterization, *Surgery,* 91, 459, 1982.
45. **Pemberton, J. H., Kelly, K. A., and Beart, R. W.,** Achieving continence with a prestomal ileal pouch and occlusive device, *Surgery,* 94, 72, 1983.
46. **Kelly, K. A., van Heerden, J., Beart, R. W., and Pemberton, J. H.,** The continent ostomy valve: early clinical results, in *Alternatives to Conventional Ileostomy,* Dozois, R. R., Ed., Year Book Medical Publishers, Chicago, 1985, 226.
47. **Ravitch, M. M. and Sabiston, D. C.,** Anal ileostomy with preservation of the sphincter, *Surg. Gynecol. Obstet.,* 84, 1095, 1947.
48. **Nissen, R.,** Demonstratonen aus operativen chirurgie zunachst einige beobachtungen, *Zentralbl. Chir.,* 60, 883, 1933.
49. **Wangensteen, O. H.,** Primary resection of the colon and rectosigmoid, *Surgery,* 14, 403, 1943.
50. **Wangensteen, O. H. and Toon, R. W.,** Primary resection of the colon with reference to cancer and colitis, *Am. J. Surg.,* 75, 384, 1948.
51. **Best, R. R.,** Anastomosis of the ileum to the lower part of the rectum and anus, *Arch. Surg.,* 57, 276, 1948.
52. **Ravitch, M. M.,** Anal ileostomy with preservation of the sphincter, *Surg. Gynecol. Obstet.,* 84, 1095, 1948.
53. **Ravitch, M. M. and Handelsman, J. C.,** One stage resection of the colon and rectum for colitis and polyposis, *Bull. Johns Hopkins Hosp.,* 88, 59, 1951.
54. **Goligher, J. C.,** The functional results after sphincter saving resections of the rectum, *Ann. R. Coll. Surg. Engl.,* 8, 421, 1951.
55. **Devine, J. and Webb, R.,** Resection of the rectal mucosa, colectomy, and anal ileostomy, *Surg. Gynecol. Obstet.,* 92, 437, 1951.
56. **Best, R. R.,** Evaluation of ileoproctostomy to avoid ileostomy in various colonic lesions, *JAMA,* 150, 637, 1952.
57. **Schneider, S.,** Anal ileostomy; experiences with a new three stage procedure, *Arch. Surg.,* 70, 539, 1955.
58. **Casanova-Diaz, A. S.,** as quoted by Valiente, M. A. and Bacon, H. E., Construction of pouch using 'pantaloon' technique for pull-through of ileum following total colectomy, *Am. J. Surg.,* 90, 742, 1955.
59. **Pemberton, J. H., Hepell, J., Beart, R. W., Dozois, R. R., and Telander, R. L.,** Endorectal ileoanal anastomosis, *Surg. Gynecol. Obstet.,* 155, 417, 1982.
60. **Soave, F.,** A new surgical technique for treatment of Hirschsprung's disease, *Surgery,* 7, 416, 1964.
61. **Martin, L. W., LeCoultre, C., and Schubert, W. K.,** Total colectomy and mucosal proctectomy with preservation of continence, *Ann. Surg.,* 186, 477, 1977.
62. **Martin, L. W. and LeCoultre, C.,** Technical considerations in performing total colectomy and Soave endorectal anastomosis for ulcerative colitis, *J. Pediatr. Surg.,* 13, 762, 1978.

63. **Telander, R. L. and Perrault, J.,** Colectomy with rectal mucosectomy and ileoanal anastomosis in young adults: its use for ulcerative colitis and familial polyposis, *Arch. Surg.,* 116, 623, 1981.
64. **Beart, R. W., Dozois, R. R., and Kelly, K. A.,** Ileoanal anastomosis in the adult, *Surg. Gynecol. Obstet.,* 154, 826, 1982.
65. **Heppell, J., Kelly, K. A., and Phillips, S. F., et al.,** Physiologic aspects of continence after endorectal ileo-anal anastomosis, *Ann. Surg.,* 195, 435, 1982.
66. **Gaston, E. A.,** Physiological basis for preservation of fecal continence, *JAMA,* 146, 1486, 1951.
67. **Valiente, M. A. and Bacon, H. E.,** Construction of pouch using "pantaloon" technique for pull-through of ileum following total colectomy, *Am. J. Surg.,* 90, 742, 1955.
68. **Karlan, M., McPherson, R. C., and Watman, R. N.,** An experimental evaluation of fecal continence in the dog, *Surg. Gynecol. Obstet.,* 100, 469, 1959.
69. **Parks, A. G. and Nicholls, R. J.,** Proctocolectomy without ileostomy for ulcerative colitis, *Br. Med. J.,* 2, 85, 1978.
70. **Utsunomiya, J., Iwama, T., and Imajo, M., et al.,** Total colocetomy, mucosal proctectomy and ileoanal anastomosis, *Dis. Colon Rectum,* 23, 459, 1980.
71. **Fonkalsrud, E. W.,** Endorectal ileal pullthrough with lateral ileal reservoir for benign colorectal diseases, *Ann. Surg.,* 194, 761, 1981.
72. **Schraut, W. H., Rosemurgy, A. S., Wang, C. H., and Block, G. E.,** Determinants of optimal results after ileoanal anastomosis: anal proximity and motility patterns of the ileal reservoir, *World J. Surg.,* 7, 400, 1983.
73. **Templeton, J. L. and McKelvery, S. T. D.,** An experimental comparison of 3-loop and 2-loop pelvic ileal reservoirs, *Br. J. Surg.,* 72(Abstr.), 397, 1985.
74. **Telander, R. L. and Dozois, R. R.,** The endorectal ileoanal anastomosis, *Prob. Gen. Surg.,* 1, 39, 1984.
75. **Pemberton, J. H., Kelly, K. A., Beart, R. W., Dozois, R. R., Wolff, B. G., and Ilstrup, D. W.,** Ileal pouch-anal anastomosis for chronic ulcerative colitis, *Ann. Surg.,* 206, 504, 1987.
76. **O'Connell, P. R., Pemberton, J. H., and Weiland, L. H., et al.,** Does rectal mucosa regenerate after ileoanal anastomosis?, *Dis. Colon Rectum,* 30, 1, 1987.
77. **Metcalf, A. M., Dozois, R. R., Beart, R. W., Kelly, K. A., and Wolff, B. G.,** Temporary ileostomy for ileal pouch-anal anastomosis: function and complications, *Dis. Colon Rectum,* 29, 300, 1986.
78. **Metcalf, A., Beart, R. W., Dozois, R. R., Kelly, K. A., and Wolff, B. G.,** Ileal J pouch-anal anastomosis: clinical outcome, *Ann. Surg.,* 202, 735, 1985.
79. **Nicholls, R. J. and Pezim, M. E.,** Restorative proctocolectomy with ileal reservoir: a comparison of 3 types of reservoir, *Br. J. Surg.,* 72, 470, 1985.
80. **O'Connell, P. R., Pemberton, J. H., and Kelly, K. A.,** Determinants of stool frequency after ileal pouch-anal anastomosis, *Am. J. Surg.,* 153, 157, 1987.
81. **King, S. A.,** Enteritis and the continent ileostomy, *Conn. Med.,* 41, 477, 1977.
82. **Bonello, J. C., Thow, G. B., and Manson, R. R.,** Mucosal enteritis, a complication of the continent ileostomy, *Dis. Colon Rectum,* 24, 37, 1981.
83. **O'Connell, P. R., Rankin, D. R., Weiland, L. H., and Kelly, K. A.,** Enteric bacteriology, absorption, morphology and emptying after ileal pouch-anal anastomosis, *Br. J. Surg.,* 73, 909, 1986.
84. **Nicholls, R. J.,** Restorative proctocolectomy — the St. Mark's experience, in *Surgery of Inflammatory Bowel Disorders,* Lee, E. C. G. and Nolan, D. J., Eds., Churchill Livingstone, Edinburgh, 1987, 105.
85. **Taylor, B. A. and Dozois, R. R.,** The J ileal pouch-anal anastomosis, *World J. Surg.,* 11, 727, 1987.
86. **McHugh, S. M., Diamant, N. E., McLeod, R., and Cohen, Z.,** S-Pouches vs. J-Pouches: a comparison of functional outcomes, *Dis. Colon Rectum,* 30, 671, 1987.
87. **Metcalf, A. M., Dozois, R. R., Beart, R. W., and Wolff, B. G.,** Pregnancy following ileal pouch-anal anastomosis, *Dis. Colon Rectum,* 28, 859, 1985.
88. **Nasmyth, D. G., Williams, N. S., and Johnston, D.,** Comparison of the function of triplicated ileal reservoirs after mucosal proctectomy and ileoanal anastomosis for ulcerative colitis and polyposis coli, *Br. J. Surg.,* 73, 361, 1986.
89. **Pezim, M. E., Pemberton, J. H., Beart, R. W., et al.,** Outcome of indeterminant colitis following ileal pouch-anal anastomosis, *Dis. Colon Rectum,* 32, 653, 1989.
90. **O'Connell, P. R., Stryker, S. J., Metcalf, M. A., Pemberton, J. H., and Kelly, K. A.,** Anal manometry following ileoanal anastomosis, *Surg. Gynecol. Obstet.,* 166, 47, 1988.
91. **Martin, L. W., Torres, A. M., Fischer, J. F., and Alexander, F.,** The critical level for preservation of continence in the ileoanal anastomosis, *J. Pediatr. Surg.,* 20, 664, 1985.
92. **Johnston, D., Holdsworth, P. J., and Nasmyth, D. G., et al.,** Preservation of the entire anal canal in conservative proctocolectomy for ulcerative colitis, *Br. J. Surg.,* 74, 940, 1987.
93. **Holdsworth, P. J. and Johnston, D.,** Anal sensation after restorative proctocolectomy for ulcerative colitis, *Br. J. Surg.,* 75, 993, 1988.
94. **Deasy, J. M., Quirke, P., Dixon, M. F., Lagopoulos, M., and Johnston, D.,** The surgical importance of the anal transitional zone in ulcerative colitis, *Br. J. Surg.,* 74(Abstr.), 533, 1987.

95. **Keighley, M. R. B., Winslet, M. C., Pringle, W., and Allan, R. N.,** The pouch as an alternative to permanent ileostomy, *Br. J. Hosp. Med.,* 4, 286, 1987.
96. **Fonkalsrud, E. W.,** Update on clinical experience with different surgical techniques of the endorectal pull-through operation for colitis and polyposis, *Surg. Gynecol. Obstet.,* 165, 309, 1987.
97. **Pescatori, M., Mattana, C., and Castagneto, M.,** Clinical and functional results after restorative proctocolectomy, *Br. J. Surg.,* 75, 321, 1988.
98. **Cohen, Z.,** Restorative proctocolectomy with ileal reservoire, *Int. J. Colorect. Dis.,* 1, 2, 1986.
99. **Becker, M. J. and Raymond, J. L.,** Ileal pouch-anal anastomosis: a single surgeon's experience, *Ann. Surg.,* 204, 375, 1986.
100. **Goldberg, S. M. and Rothenberger, D. A.,** Symposium: Restorative proctocolectomy with ileal reservoire, *Int. J. Colorect. Dis.,* 1, 2, 1986.
101. **Schoetz, D. J., Coller, J. A., and Veidonheimer, M. C.,** Ileoanal reservoir for ulcerative colitis and familial polyposis, *Arch. Surg.,* 121, 404, 1986.

Chapter 15

CANCER IN ULCERATIVE COLITIS: WHAT IS THE RISK AND IS SCREENING EFFECTIVE IN SAVING LIVES?

Sylvia Gyde

TABLE OF CONTENTS

I. HISTORICAL PERSPECTIVE

Although ulcerative colitis was first distinguished by Wilks in 1879[1,2] as distinct from bloody diarrheas of infective origin, the disease received little attention until further detailed descriptions were made in 1921 by Hurst[3] in a Guy's Hospital Report. By 1928,[4] Bargen at the Mayo Clinic was describing colon cancer complicating the disease. Over the next 35 years there followed numerous reports describing cases of colorectal cancer occurring in association with ulcerative colitis. These reports were summarized in a review article in 1964 by Goldgraber and Kirshner.[5]

II. CASE CONTROL STUDY

Nefzger and Acheson,[6] reporting in the 1960s, described an investigation of 525 men admitted to the U.S. Army in 1944 having ulcerative colitis. They found the mortality in these men was double that in a group of matched controls, the excess mortality in the early years being due to the disease process itself, but in the later years (17-year follow-up) due solely to cancer of the colon and rectum.

The authors recognized that the wide fluctuations in the proportion of cancer cases in reported series probably reflected selection biases in patients recruited to the series and also to lack of complete follow-up of patients. They also noted that estimates of the cancer risk (which up to this time had been expressed as crude percentages) were misleading in that they did not take account of the number of patient years of exposure to the disease from onset or diagnosis.

From this point in the mid 1960s, most investigators used actuarial methods to calculate the cumulative cancer risk over time which involved calculating the number of person years at risk in the group under review.

III. AGE OF CANCER ONSET

In 1964, Edwards and Truelove,[7] in a classic series of reports ("The Course and Prognosis of Ulcerative Colitis") based on the long-term follow-up of a large series of patients at Oxford, described the cancer risk and recognized that cancer was occurring in these patients at a "much younger age than when it occurs in a previously healthy person." This was subsequently confirmed by many studies[8–12] including a large study from St. Mark's Hospital, London.[13] In the St. Mark's study the average age at diagnosis of cancer was calculated for 4817 noncolitis and 67 colitic colorectal cancers. The average age at diagnosis of cancer was 63 years in the noncolitic group and 49 years in the colitic group. Ulcerative colitic patients were therefore tending to develop colorectal cancer at least 10 years younger than the general population.

As the number of reports on large series of ulcerative patients increased, certain findings were repeatedly made concerning the cancer risk.

IV. HIGH-RISK PATIENTS

In all studies, patients at the highest risk of developing cancer were those in whom, at some time during the course of their disease, the inflammation has involved most or the whole of the colonic mucosa (extensive colitis and total colitis).[14–23]

Cancer has been rarely observed in extensive colitis patients before 10 years from onset of disease. Patients with left-sided disease or proctitis appear to be slightly more at risk of developing cancer than the general population, even though they experience the same chronic

TABLE 1
Crude 5-Year Survival in Patients with Colorectal Cancer Complicating Ulcerative Colitis—Summary of Reported Studies

UC patients with colorectal cancer (no.)	Mean interval symptoms of UC to diagnosis of cancer (years)	Crude 5-year survival (%)	Patients with distant metastases on referral		Incidental with multiple cancer				
			No.	%	No.	%	No.	%	Ref.
29	19.6	55.1	NK	—	4	14	NK	—	8
79	17	41.0	28	35	3	4	19	24	9
70	17.1	41.7	17	24	16	23	14	20	12
67	10	65.1	19	29	15	26	NK	—	13
35	19.7	33.5	8	28*	4	11.4	10	29	27

Note: Data not available in 6 of 35 cancers, UC = ulcerative colitis, and NK = not known.

inflammatory changes in the mucosa involved over long time periods.[10,17,23] This is somewhat surprising, given that in the "high-risk", extensive colitis group we associate the cancer risk with long-standing inflammation. Other factors besides inflammation must therefore be involved in producing the observed increased cancer incidence. It is possible that the increased cancer risk is not related to the inflammatory process. Genetic studies suggest that ulcerative colitis patients have a genetic predisposition to develop the disease.[24,25] It is possible that there is concurrently a genetic predisposition to develop colon cancer at a young age, independent of the inflammatory process, in patients with the "full-blown" condition (extensive and total colitis).

V. DISTRIBUTION OF CANCERS IN THE COLON

The distribution of colorectal cancer throughout the colon in ulcerative colitis differs from the general population.[10,26] In ulcerative colitis the cancers tend to occur relatively evenly distributed throughout the colon whereas, in the general population, over two thirds of the cancers occur distally in the sigmoid colon and rectum. This fact carries implications for early diagnosis of cancer in that approximately 75% of colorectal cancers in the general population will be within reach of the sigmoidoscope as compared to only 47% in ulcerative colitis patients.

Multiple (synchronous) cancers occur more commonly in ulcerative colitis than in the general population.

VI. SURVIVAL

Early analyses of survival from colorectal cancer based upon individual case reports in the literature suggested that colorectal cancer in ulcerative colitis carried a much poorer prognosis than colorectal cancer in the general population[11] (see also References 8, 9, 12, 13, 27, and 28, Tables 1 and 2, and Figure 1). However, several more recent careful analyses of survival in large series, taking into account the Dukes stage of the cancer at diagnosis, have shown that if patients are matched for stage at diagnosis of cancer, survival of ulcerative colitis patients developing colorectal cancer is similar to the survival of patients developing colorectal cancer in the general population.

Late diagnosis therefore appears to be the cause of the observed poor survival from cancer in ulcerative colitis. Late diagnosis may in part be explained by the symptoms of cancer

TABLE 2
Actuarial 5-Year Survival of 35 Patients with Colorectal Cancer Complicating Ulcerative Colitis Compared with the Survival of Colorectal Cancer in Patients with the Same Median Age in the Relevant General Population (West Midlands Region)

Source	Patients (no.)	Sex	Site	Mean age at diagnosis of cancer	Actuarial 5-year survival (%)	95% Confidence limits
Ulcerative colitis series	35	M and F	Colon, rectum	47.5[a]	33.5	16.9—50.1
West Midlands region (1962 to 1967)	462	M and F	Colon, rectum	47.5[b]	32.6	28.2—37.0

[a] Age range 33 to 70, but median fell in age group 45 to 49 years.
[b] All patients aged 45 to 49 years.

From Gyde, S. N., et al., *Gut,* 1983. With permission.

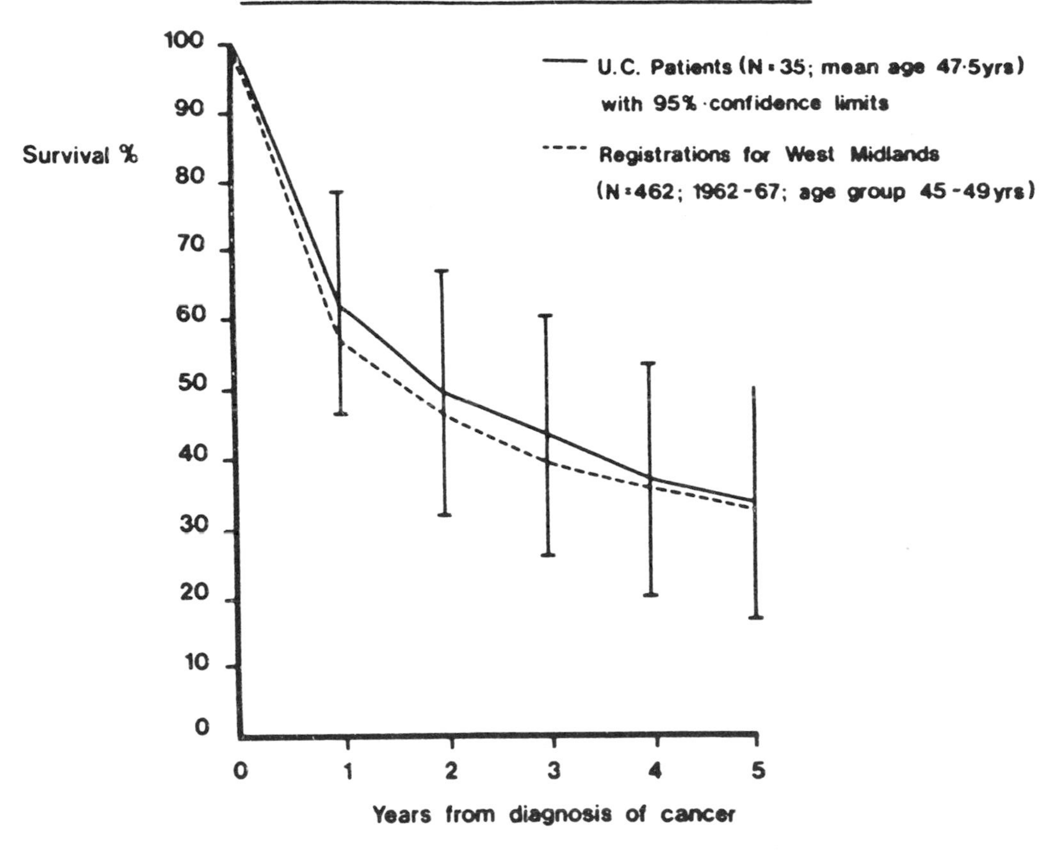

FIGURE 1.

mimicking the symptoms of the disease itself and therefore not leading to the intensive investigation which the same symptoms would cause in a healthy member of the general population.

VII. CANCER RISK OVER TIME

A. EARLY REPORTS

From the 1960s onwards, there were many reports in the literature of the cancer incidence over time based on hospital series of patients using actuarial methods to calculate the cumulative probability of developing cancer from onset of disease or diagnosis (Table 3). The varying results may be a reflection of the different selection biases in the composition of the series. Hospitals specializing in the treatment of the condition tend to collect large series of patients ideal for research but comprising high proportions of patients with severe disease or difficult cases. These selection biases can be minimized if analysis is done by extent of disease so that the proportion of extensive colitis patients (selection bias for severity of disease) will not affect the results. Patients with cancer at first referral or developing cancer within the first year of follow-up should be excluded, and patients should be drawn from a defined geographical area.

In these early reports, methods of analysis showed considerable variation making comparison between series difficult. Also, the series upon which estimates of the risk were based were small.

TABLE 3
Cancer Incidence in Ulcerative Colitis: Hospital Series

Authors and hospital series	Review period	Cases		Cancers		Results
		Whole series	Extensive colitis	Whole series	Extensive colitis	Cumulative cancer incidence
de Dombal et al.[14] General Infirmary, Leeds	1952—1963	428	210	8	8	10 years 5% 20 years 21% 25 years 42%
McDougall[15] Gordon Hospital, London	1947—1963	637	196	15	9	Extensive colitis Observed = 9, O/E = 30 Expected = 0.3
Edwards and Truelove Radcliffe Infirmary, Oxford	1938—1962	624	236	22	17	20 years Whole series 5.5% 1st attack (n = 250) 12.6%
Greenstein et al.[16, 17] Mount Sinai Hospital, New York	1960—1976	267	NS	26	21	Left sided colitis O/E = 8.6 Extensive colitis O/E = 26.5
Lennard-Jones et al.[18] St Marks Hospital, London (Cancer surveillance in extensive colitis)	1966—1980	—	303	—	13	Surveillance and prophylactic PPC for dysplasia possibly prevented development of carcinoma in an additional 8 patients
Prior et al.[19] Queen Elizabeth and General Hospitals, Birmingham	1944—1976	676	462	35	35	Whole series (68% with extensive colitis) 25 years 8% 30 years 20%
Katzka et al.[20] Private practice, New York, USA	1955—1980	—	106	—	4	Extensive colitis >10 years from onset risk at 11 years 7.2%

Note: O = observed; E = expected; NS = not stated.

B. RECENT REPORTS

Two recent reports of large series with long follow up show a very similar cumulative cancer incidence over time (Table 4). A report from the Cleveland Clinic described an analysis in a series of 1248 cases of ulcerative colitis and found a cancer incidence in extensive colitis patients of just over 10% at 20 years and 25% at 30 years — approximately a 1% increase in risk per year from 10 years after diagnosis of disease.[10]

Very similar results were reported in a recent large cohort study, recruiting patients from three hospital series in Stockholm, Oxford, and Birmingham. Extensive colitis patients showed a cumulative cancer risk of just less than 10% at 20 years and just less than 20% at 30 years. Again, an incremental risk 10 years from diagnosis of 1% per year.[23]

In both studies there was a small increased risk in left-sided disease of around 4% at 30 years from diagnosis.

The log-rank method was used in the cohort study[23] to examine whether there was a significant difference in the cumulative cancer incidence between various groups.

VIII. CUMULATIVE CANCER INCIDENCE

A. COMPARISON OF EXPECTED RISK BETWEEN ULCERATIVE COLITIS PATIENTS AND THE GENERAL POPULATION

Figure 2 shows the cumulative cancer incidence in the extensive colitis group (with the 95% confidence limits defined) showing a steep rise compared with that expected in a general population sample of similar age and sex. The numbers of cancers "expected" are small due to the young age of patients in the series under review. When these results were plotted on a log scale (Figure 2A) the rate of rise of the cancer incidence in the general population was demonstrated to rise at the same rate as in the extensive colitis patients. It appears that in ulcerative colitis the whole incidence has a similar pattern but is simply set at a higher level.

1. Males and Females

There was no significant difference at any point in time in the cancer incidence between males and females (Figure 3).

2. By Center

There was no significant difference in the cumulative cancer incidence at any point in time by center (Figure 4). This suggests that the entry criteria used to select patients from the three hospital series (Birmingham, Oxford, and Stockholm) was successful in eliminating selection biases in the hospital series under review. Given that the cumulative cancer incidence over time by each separate center in the cohort study is so similar to that found in the Cleveland Clinic study, we are probably defining a cumulative cancer incidence over time which is generally applicable and not a biased result.

3. By Age at Onset of Disease

In earlier studies it was suggested that early age at onset of disease increased the risk of cancer. Figure 5 demonstrates that in the three broad age bands examined at onset of disease. In this study, early age at onset of disease did not produce a cancer incidence significantly different from onset of disease later in life.

4. By Extent of Disease

Figure 6 demonstrates the marked difference in the cumulative cancer incidence over time in extensive colitis patients compared with all others in the series (left-sided disease and proctitis).

TABLE 4
Colorectal Cancer in Ulcerative Colitis: Recent Studies

		Patients		% Cancers		Cumulative cancer incidence		
Center	Review period	Whole series	Extensive colitis	Whole series	Extensive colitis	Extensive colitis (%)	C.I.[a]	Left-sided colitis (%)
Oxford, Birmingham, and Stockholm	1945—1982	823	486	35	29	20 years = 7.2 30 years = 16.5	3.6—10.8 9.0—24.0	20 years = 1 30 years = 3.4
Cleveland Clinic	All cases to 1984	1248	—	82	74	20 years = 11.9 30 years = 25.3		20 years = 1.8 30 years = 3.7

[a]C)I. = 95% confidence interval.

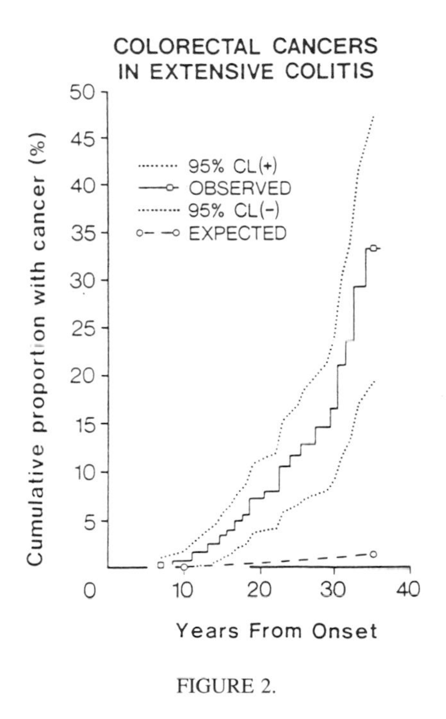

FIGURE 2.

B. THE EFFECT OF PANPROCTOCOLECTOMY ON THE OBSERVED CANCER INCIDENCE

If all extensive colitis patients retained their colons and rectums throughout their lifetime, the cancer incidence observed would undoubtedly be much higher. If earlier policies (such as were pursued in some Scandinavian countries) of prophylactic panproctocolectomy at 10 years from onset of disease in extensive colitis patients were operative, the cancer risk would be virtually nil. What is being estimated in all cancer incidence studies is the cumulative cancer risk in extensive colitis patients with relatively mild disease. The policy for panproctocolectomy in any given hospital series will affect the risk found. The fact that no cancers were observed in the early years of the Copenhagen County population study must have been due in part to the policy at that time of performing prophylactic panproctocolectomy for the cancer risk in extensive colitis patients 10 years from onset of disease.

C. POPULATION-BASED STUDIES

Ideally, one would determine the cancer incidence in a disease by following all patients developing the disease in a defined geographical area for long enough to determine which patients developed cancer (Table 5). However, the practical problems which this involves in the instance of a disease such as ulcerative colitis are almost insurmountable. The first problem is that ulcerative colitis is an uncommon disease (incidence 5 to 10/100,000 population)[29–34] so that recruitment to build up a series of any size is a slow process taking many years. Second, having accrued patients, the cancer risk does not start to rise until at least 10 years from onset of disease, and even then the risk is concentrated only in a small proportion of the total series (the high-risk extensive colitis group). It would not be possible to obtain meaningful results in one investigator's lifetime unless extremely large numbers of patients were recruited in a multicenter trial.

The problems (and also strengths) of population-based studies are well illustrated by an excellently designed and executed study ongoing in Denmark.[21] This study involves the ascertainment of all cases of ulcerative colitis diagnosed in Copenhagan County (population 500,000). Patients were recruited into the series between the years 1960 and 1978, during

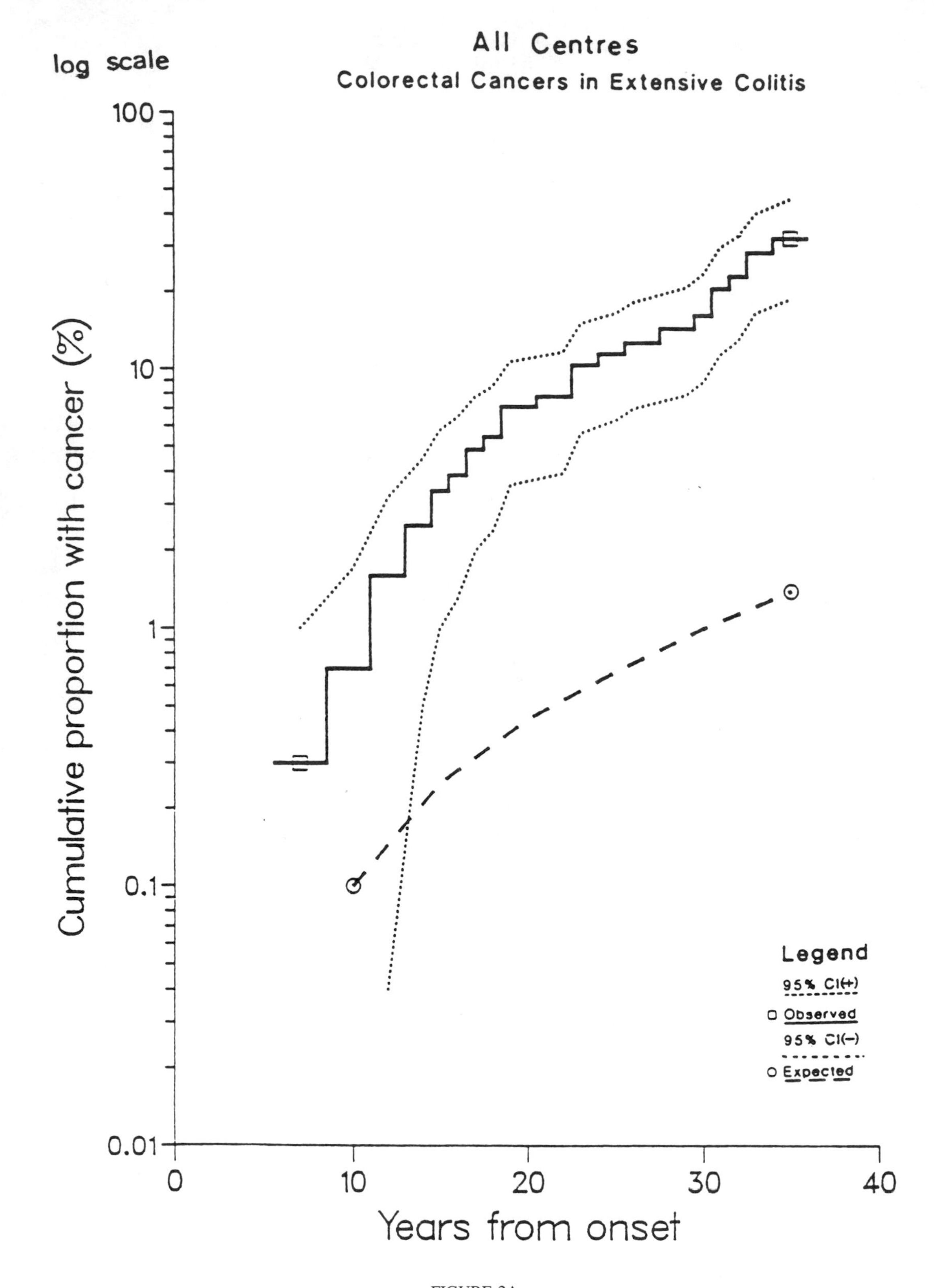

All Centres
Colorectal Cancers in Extensive Colitis
log scale
Cumulative proportion with cancer (%)
100
10
1
0.1
0.01
0
10
20
30
40
Years from onset
Legend
95% CI(+)
Observed
95% CI(−)
Expected

FIGURE 2A.

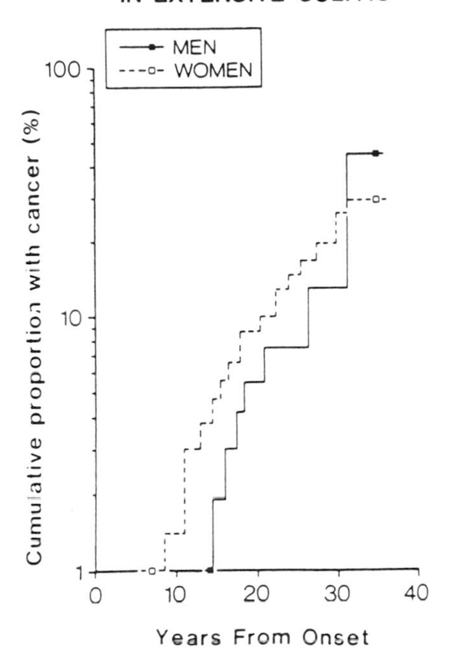

FIGURE 3.

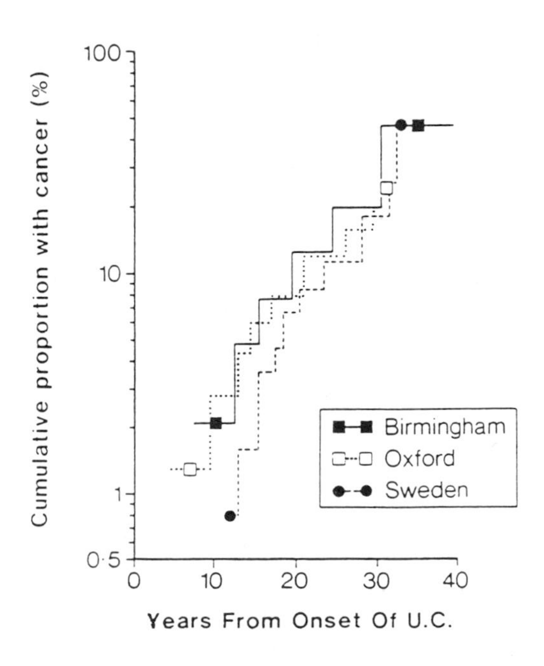

FIGURE 4.

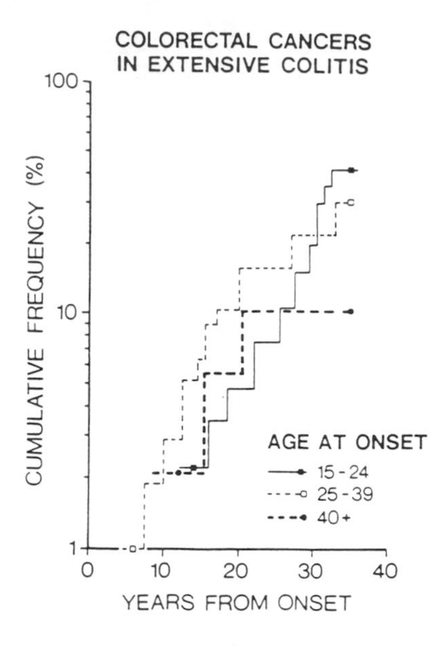

FIGURE 5.

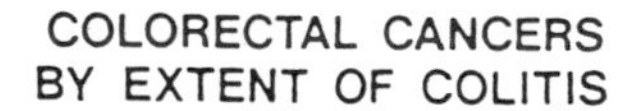

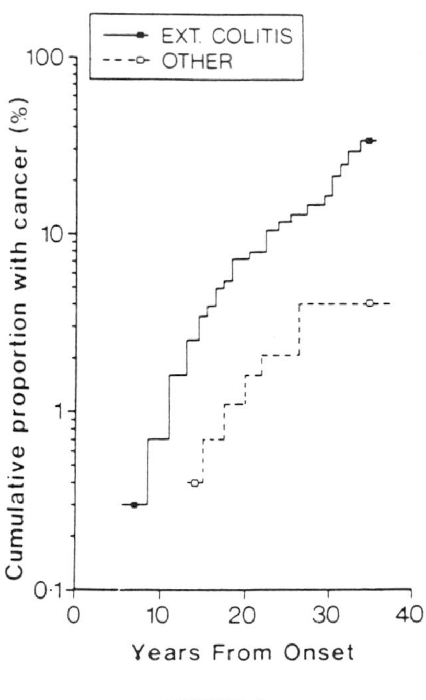

FIGURE 6.

which time 783 patients were diagnosed with the disease, 124 of whom (16%) had extensive colitis.

In the most recent report in 1985, with a median follow up of 7 years, 7 colorectal cancers had been diagnosed in the series, 2 of which were in extensive colitis patients. The cancer risk will presumably become increasingly apparent in this series as more and more patients in the extensive colitis group have been followed for more than 10 years from onset of disease.

TABLE 5
Cancer Incidence in Ulcerative Colitis: Population-Based Series

	Cases		Cancers		Cummulative cancer incidence
Review period	Whole series	Extensive colitis	Whole series	Extensive colitis	(whole series, %)
1960—1978	783	124 (total colitis)	7	2 (total colitis)	10 years = 0.8 15 years = 1.1 18 years = 1.4 (CI = 0.7—2.8)

Note: Continuation of the Bonnevie study in Copenhagan County. All cases of ulcerative colitis in Copenhagan County. Median follow up = 6.7 years. CI = 95% confidence interval.

From Hendriksen, C., Kreiner, S., and Binder, V., *Gut,* 26, 158, 1985. With permission.

The cumulative cancer risk in the series as a whole was less than 1% at 15 years and less than 2% at 18 years. It was not possible to calculate a meaningful risk for the extensive colitis group over time given only two cancers had been observed.

IX. THE CANCER RISK: ILEORECTAL ANASTOMOSIS

Ileorectal anastomosis has been a highly successful operation in symptomatic patients with extensive disease and in restoring health while retaining bowel continuity[35–38] (Table 6). The retained rectum, however, carries the same high cancer risk of the extensive colitis patient. Although clinicians caring for these patients are very aware of this fact, cancers in the retained rectum are still missed. In early years, patients with ileorectal anastomosis were restored to health but not followed up or reviewed on a regular basis. Many returned in later years with advanced cancer of the rectum. More recently, since the late 1960s with surveillance for dysplasia by rectal biopsy, follow-up has been much more rigorous and although cancers are still missed, more are being diagnosed at an early stage where cure by proctectomy is likely. The cancer risk following ileorectal anastomosis was recently reviewed by Madjlessi and Farmer.[38]

X. BILE DUCT CARCINOMA IN ULCERATIVE COLITIS

The only site apart from the colon and rectum showing an increase in cancer risk in ulcerative colitis is bile duct carcinoma.[39–42] Bile duct cancer is usually found in patients with extensive colitis with long-standing disease and the diagnosis is often only made at post-mortem (25% of cases).

XI. MANAGEMENT OF PATIENTS AT HIGH RISK OF DEVELOPING COLORECTAL CANCER

There is no doubt from the literature that patients at high risk of developing colorectal cancer are extensive colitis patients at least 10 years from onset of disease. Patients with left-sided colitis or proctitis appear to have a slightly increased risk but this may be due to some extensive colitis patients remaining undiagnosed in the left-sided disease group.

We know from the studies on cancer incidence in extensive colitis patients that, given 100 patients 20 years from onset of disease, approximately 10 of them will be likely to have developed colorectal cancer by that time and that in each subsequent year one more patient will develop cancer. Survival studies indicate that if the cancer could be diagnosed at a

TABLE 6
Rectal Cancer in Ileorectal Anastomosis

Author (Location)	Review period	Extensive disease	Cancer	Dukes stage
Baker[35] (Gordon Hospital)	1952—1976	362	21	12 Dukes C 4 distant metastases
Grundfest[36] (Cleveland Clinic)	1957—1977	89	4	2 Dukes A
Oakley[37] (Cleveland Clinic)	1960—1982	145 (Whole series)	5	3 Dukes A

precancerous stage or Dukes stage A and surgery carried out, a complete cure is likely. How then should we best manage these patients?

A. PROPHYLACTIC SURGERY

One option is to perform prophylactic surgery at some specified point, e.g., in extensive colitis patients 10 years from onset of disease. Although this option might be preferred in patients who are continuously symptomatic and unwell with the disease itself, it has many drawbacks. Although young patients, for the most part, adapt well to life with a permanent stoma, panproctocolectomy still carries a small mortality.

There is a distinct possibility of morbidity in terms of obstruction and further surgery if stoma refashioning is required. There is a small risk of impotence in males following damage to the pelvic nerves at panproctocolectomy. Ileorectal anastomosis is often a more acceptable alternative to the patient. However, unless extremely vigilant follow-up is maintained after surgery, these patients tend to become lost to follow-up, retaining a rectum with the high cancer risk of the extensive colitis patients. Many such patients are only seen again in a clinic, years later, returning with symptomatic rectal cancer.

B. SCREENING FOR CANCER

The other option is to delay surgery in high risk patients until some "marker" indicates that the development of cancer is imminent.

1. Dysplasia as a Marker

In 1967 Morson and Pang[43] described the morphology of precancerous changes which occurred in association with cancer in ulcerative colitis. They examined 27 surgical specimens with frank carcinoma from ulcerative colitis patients and found extensive precancerous "dysplastic" changes in 23 of these specimens. They then examined 172 consecutive colectomy specimens from ulcerative colitis patients where surgery had been performed for symptomatic disease. In the 134 specimens removed from patients with extensive colitis, they found precancerous changes present in 12 of these specimens (9%) and no precancerous changes in the 38 specimens with left-sided disease.

The detection of dysplastic changes in the colon and rectum therefore seemed to be a marker for cancer. Many subsequent publications have been concerned in assessing the reliability of dysplasia as a marker,[44–54] and this literature has been reviewed.[55–61]

Theoretically, there would be problems to consider if dysplasia were used as a marker for cancer.

Dysplastic changes may be precancerous in certain instances, but in others some of these lesions will regress, some will stay the same, and some will go on to become invasive carcinoma.[59–62] If you remove all colons and rectums with dysplasia you might be removing many colons and rectums which would never have gone on to develop cancer (false positives of the screen).

2. *De Novo* Cancers

The literature subsequent to Morson and Pang's original paper has shown that not all cancers in ulcerative colitis occur in association with dysplastic change. Approximately 30% of cancers appear to arise *de novo* from morphologically normal mucosa.[62] Unless biopsies happened to be taken directly from the cancer, these cancers would be missed by the screen (false negatives of the screen).

3. Inter- and Intraobserver Error

It has been well documented that histopathologists, given that they view the same specimen slides, do not give a high level of concordance concerning the presence or absence or grading of dysplastic changes.[51] They do not always agree with themselves, if they are shown the same slides on more than one occasion. Recently, attempts have been made by groups of interested pathologists to standardize morphological changes in relation to the grading of dysplasia to minimize these errors.

4. Patchy Distribution of Dysplasia

Usual practice at colonoscopy is to take multiple random biopsies at specified intervals along the colon. Due to its patchy distribution, dysplasia can still be missed by this method, and even when dysplasia has been found the *repeatability* of the dysplasia at a second series of biopsies is not 100%.

Although, therefore, we have a marker for cancer in the form of dysplasia, there are good theoretical grounds for thinking that it might not be a very effective marker.

C. PRELIMINARY TRIALS BEFORE INTRODUCING ANY SCREENING TEST FOR CANCER

Before any screening test for cancer is introduced at a population level, it should have been proved to be beneficial in a clinical trial situation. Survival in the group screened must be shown to be increased in relation to a similar unscreened group. This can be best done by carrying out a randomized controlled trial.

In the case of breast cancer screening, a large randomized trial showed that survival was significantly increased in the screened group in women over the age of 50 years. Case control studies showed similar results and breast cancer screening is following the positive results of these trials being introduced in England at a population level.

Even when a screening procedure has proved successful in a trial situation in prolonging life it may still not prove to be successful when introduced on a population basis. Screening for cervical cancer has signally failed to produce a benefit in terms of reduction in cancer incidence or increase in survival in England, not because the screen is ineffective per se if carried out at near 100% efficiency (i.e., under trial conditions), but because having no accurate registers of women upon which to base call and recall for the screen and also because the compliance rate is only 50 to 60%, no benefit accrues from screening.

D. THE PRESENT POSITION OF SCREENING FOR CANCER IN ULCERATIVE COLITIS

Following Morson and Pang's paper in 1967, screening for colorectal cancer using dysplasia as a marker has gradually been introduced into clinical practice without any prior clinical trials to test the effectiveness of the procedure in terms of reducing the cancer incidence or increasing life expectancy in the high risk group. Some prospective studies have been running for many years and have recently reported. None of these studies, however, randomized high risk patients eligible for the screen into "screened" and "unscreened" groups so that there is no comparison group in any of these studies when considering long-term survival.

In a prospective trial of screening for cancer in extensive colitis patients at St. Mark's

TABLE 7
Recent Reports: Screening for Cancer in Ulcerative Colitis

Center	No. of patients	No. of cancers	Dukes stage A	Dukes stage B	Dukes stage C	Dukes stage D	Ref.
St. Mark's Hospital, London, 1983	303	16 (13 patients)	11	3	1	1	63
Lakey Clinic Medical Center, U.S. (1984)	151	6	1	1	4		60, 61
Cleveland Clinic U.S. (1985)	248	7	1	3	3		65

Hospital,[63] London, which has been running for over 15 years, the authors state, "Despite regular surveillance, carcinoma developed in 13 of the 303 extensive colitis patients" (16 cancers in all). However, only 3 of these 13 patients were diagnosed as having cancer by the screening procedure per se and "The rest of the cancers were diagnosed either preoperatively by barium enema or sigmoidoscopy or at operation performed for other reasons." Therefore, although in this major study a high proportion of the cancers (11/16) were diagnosed at an early stage (Dukes stage A), the majority were diagnosed by good routine clinical care rather than by the screening procedure itself. These conclusions were reinforced by recent results from another surveillance study recently reporting from Leeds.[64] Table 7 summarizes the results of this study with two other recent reports of American studies of screening.[61,65] It can be seen in the American studies that the majority of cancers diagnosed were at a late stage where survival would be unlikely to be increased.

The results of these studies give cause for concern, regarding whether this time-consuming, expensive, and unpleasant screening procedure is actually saving lives.[66,67] The way forward would be for all centers presently screening to adopt a common protocol and enter their patients into a multicenter, randomized control trial. Given sufficient numbers it would rapidly become apparent whether the screening procedure was any more effective than ordinary clinical care at increasing survival in patients at high risk of developing cancer.

If the screen was proved not to be effective, the large amount of resources being presently poured into this procedure could be better used in other ways, perhaps in developing a more effective marker than dysplasia for a new screening procedure.

REFERENCES

1. **Wilks, S.,** *Lectures of Pathological Anatomy,* 1st ed., Langman and Roberts, London, 1859.
2. **Wilks, S.,** Morbid appearances in the intestines of Miss Banks, *Med. Times Gaz.,* 19, 264, 1859.
3. **Hurst, A. F.,** *Guy's Hosp. Rep.,* 71, 24, 1921.
4. **Bargen, J. A.,** Chronic ulcerative colitis complicated by malignant neoplasia, *Arch. Surg.,* 17, 561, 1928.
5. **Goldgraber, M. B. and Kirsner, J. B.,** Carcinoma of the colon in ulcerative colitis, *Cancer,* 17, 657, 1964.
6. **Nefzger, M. D. and Acheson, E. D.,** Ulcerative colitis in the United States Army in 1944, *Gut,* 4, 183, 1963.
7. **Edwards, F. C. and Truelove, S. C.,** The course and prognosis of ulcerative colitis, Parts I, III, and IV. Carcinoma of the colon, *Gut,* 4, 299, 1964.
8. **Hughes, R., Hall, T., and Block, G., et al.,** The prognosis of carcinoma of the colon and rectum complicating ulcerative colitis, *Surg. Gynecol. Obstet.,* 136, 46, 1978.
9. **Lavery, I., Chiulli, R., and Jagelman, D., et al.,** Survival with carcinoma arising in mucosal ulcerative colitis, *Ann. Surg.,* 195, 508, 1978.
10. **Madjlessi, S. H., Farmer, R. G., Easley, K. A., and Beck, G. J.,** Colorectal and extracolonic malignancy in ulcerative colitis, *Cancer,* 58(7), 1569, 1986.
11. **Slaney, G. and Brooke, B. N.,** Cancer in ulcerative colitis, *Lancet,* ii, 694, 1959.

12. **Van Heerden, J. A., Judd, E. S., and Dockerty, M. B.,** Carcinoma of the extrahepatic bile ducts: a clinicopathologic study, *Am. J. Surg.*, 113, 49, 1967.
13. **Ritchie, J., Hawley, P., and Lennard-Jones, J.,** Prognosis of carcinoma in ulcerative colitis, *Gut,* 22, 752, 1981.
14. **de Dombal, F. T., Watts, J., and Watkinson, G., et al.,** Local complications of ulcerative colitis: stricture, pseudopolyposis and carcinoma of the colon and rectum, *Br. Med. J.*, i, 1442, 1966.
15. **McDougall, I.,** The cancer risk in ulcerative colitis, *Lancet,* ii, 655, 1964.
16. **Greenstein, A. J., Sachar, D. B., and Smith, H., et al.,** A comparison of the cancer risk in Crohn's disease and ulcerative colitis, *Cancer,* 48, 2742, 1981.
17. **Greenstein, A. J., Sachar, D. B., and Smith, H., et al.,** Cancer risk in universal and left sided ulcerative colitis: factors determining risk, *Gastroenterology,* 77, 290, 1979.
18. **Lennard-Jones, J., Morson, B., and Ritchie, J., et al.,** Cancer in colitis: assessment of the individual risk by clinical and histological criteria, *Gastroenterology,* 73, 1280, 1977.
19. **Prior, P., Gyde, S. N., and Macartney, J. C., et al.,** Cancer morbidity in ulcerative colitis, *Gut,* 23, 490, 1982.
20. **Katza, I., Brody, R., Morris, E., and Katz, S.,** Assessment of colorectal cancer risk in patients with ulcerative colitis. Experience from a private practice, *Gastroenterology,* 85, 22, 1983.
21. **Hendriksen, C., Kreiner, S., and Binder, V.,** Long term prognosis in ulcerative colitis — based on results from a regional patient group from the County of Copenhagan, *Gut,* 26, 158, 1985.
22. **Kewenter, J., Ahlman, H., and Hulten, L.,** Cancer risk in extensive colitis, *Ann. Surg.,* 188, 824, 1978.
23. **Gyde, S. N., Prior, P., and Allan, R. N., et al.,** Colorectal cancer in ulcerative colitis: a cohort study of primary referrals from three centres, *Gut,* 29, 206, 1988.
24. **Mayberry, J. F., Rhodes, J., and Newcombe, R. G.,** Familial presence of inflammatory bowel disease in relatives of patients with Crohn's disease, *Br. Med. J.,* 1, 84, 1980.
25. **McConnell, R. B.,** Inflammatory bowel disease: newer views of genetic influences, in *Developments in Digestive Disease,* Vol. 3, Berk, J. E., Ed., Lea & Febiger, Philadelphia, 1980, 271.
26. **Langman, M. J.,** Epidemiology of cancer of the large intestine, *Proc. R. Soc. Med.,* 59, 132, 1966.
27. **Gyde, S. N., Prior, P., and Thompson, H., et al.,** Survival of patients with colorectal cancer complicating ulcerative colitis, *Gut,* 1983.
28. **Hulten, L., Kewenter, J., and Ahren, C.,** Precancer and carcinoma in chronic ulcerative colitis: a histopathological and clinical investigation, *Scand. J. Gastroenterol.,* 7, 6634, 1971.
29. **Bonnevie, O., Riis, P., and Anthonisen, P.,** An epidemiological study of ulcerative colitis in Copenhagan County, *Scand. J. Gastroenterol.,* 3, 432, 1963.
30. **Evans, J. G. and Acheson, E. D.,** An epidemiological study of ulcerative colitis and regional enteritis in the Oxford area, *Gut,* 6, 311, 1965.
31. **Linden, G. and Noller, C.,** Ulcerative colitis in Finland. II. One year incidence in all hospitals, *Dis. Colon Rectum,* 14, 264, 1971.
32. **Wigley, R. D. and MacLaurin, B. P.,** A study of ulcerative colitis in New Zealand showing a low incidence in Maoris, *Br. Med. J.,* ii, 228, 1962.
33. **Monk, M., Mendeloff, A. L., Siegel, C. I., and Lilienfield, A.,** An epidemiological study of ulcerative colitis and regional enteritis among adults in Baltimore. I. Hospital incidence and prevalence 1960–1963, *Gastroenterology,* 53, 198, 1967.
34. **Sinclair, T. S., Brunt, P. W., Ashley, N., and Mowat, G.,** Non-specific proctocolitis in North Eastern Scotland: a community study, *Gastroenterology,* 85, 1, 1983.
35. **Baker, W., Glass, R., Ritchie, J., and Aylett, S.,** Cancer of the rectum following colectomy and ileorectal anastomosis for ulcerative colitis, *Br. J. Surg.,* 65, 862, 1978.
36. **Grundfest, S. F., Fazio, V., and Weiss, R., et al.,** The cancer risk following colectomy and ileorectal anastomosis for extensive mucosal ulcerative colitis, *Ann. Surg.,* 193, 9, 1981.
37. **Oakley, J. R., Jagelman, D. G., and Fazio, V. W., et al.,** Complications and quality of life after ileorectal anastomosis for ulcerative colitis, *Am. J. Surg.,* 149, 24, 1985.
38. **Madjlessi, S. H. and Farmer, R. G.,** Squamous cell carcinoma of the rectal stump in a patient with ulcerative colitis. Report of a case and review of the literature, *Cleveland Clin. Q.,* 22(2), 257, 1985.
39. **Akwari, O., Van Heerden, J., Foulk, W., and Baggenstoss, A.,** Cancer of the bile ducts associated with ulcerative colitis, *Ann. Surg.,* 181, 303, 1975.
40. **Converse, C., Reagan, J., and Decosse, J.,** Ulcerative colitis and carcinoma of the bile ducts, *Am. J. Surg.,* 121, 39, 1971.
41. **Madjlessi, S. H., Farmer, R. G., and Sivak, H. V., Jr.,** Bile duct carcinoma in patients with ulcerative colitis. Relationship to sclerosing cholangitis: report of six cases and review of the literature, 1987.
42. **Ritchie, J., Allan, R. N., and Macartney, J., et al.,** Biliary tract carcinoma associated with ulcerative colitis, *Q. J. Med.,* 170, 263, 1974.
43. **Morson, B. and Pang, L.,** Rectal biopsy as an aid to cancer control in ulcerative colitis, *Gut,* 8, 423, 1967.

44. **Cook, M. G., Path, M. R. C., and Goligher, J. C.,** Carcinoma and epithelial dysplasia complicating ulcerative colitis, *Gastroenterology,* 68, 1127, 1975.
45. **Evans, D. J. and Pollack, D. J.,** *In situ* and invasive carcinoma of the colon patients with ulcerative colitis, *Gut,* 13, 566, 1952.
46. **Fenoglio, C. M. and Pascal, R. R.,** Adenomatous epithelium, intraepithelial anaplasia, and invasive carcinoma in ulcerative colitis, *Am. J. Dig. Dis.,* 18, 556, 1973.
47. **Granqvist, S., Gabrielson, N., Sundelin, P., and Thorgeirsson, T.,** Precancerous lesions in the mucosa in ulcerative colitis, *Scand. J. Gastroenterol.,* 15, 289, 1980.
48. **Hulten, L., Kewenter, J., Ahren, C., and Ojewkog, B.,** Clinical and morphological characteristics of colitis carcinoma and colorectal carcinoma in young people, *Scand. J. Gastroenterol.,* 14, 673, 1979.
49. **Myrvoid, H. E., Kock, N. G., and Ahren, C.,** Rectal biopsy and precancer in ulcerative colitis, *Gut,* 15, 301, 1974.
50. **Ransohoff, D. F., Riddell, R. H., and Levin, B.,** Ulcerative colitis colonic cancer: problems in assessing the diagnostic usefulness of mucosal dysplasia, *Dis. Colon Rectum,* 28, 383, 1985.
51. **Riddell, R. H., Goldman, H., and Ransohoff, D. F., et al.,** Dysplasia in inflammatory bowel disease: standardized classification with provisional clinical applications, *Hum. Pathol.,* 14, 931, 1985.
52. **Teague, R. H. and Read, A. E.,** Polyposis in ulcerative colitis, *Gut,* 16, 792, 1975.
53. **Yardley, J. H. and Keren, D. F.,** 'Precancer' lesions in ulcerative colitis; a retrospective study of rectal biopsy and colectomy specimens, *Cancer,* 34, 835, 1974.
54. **Blackstone, M., Riddell, R., Rogers, G., and Levin, B.,** Dysplasia-associated lesion or mass (DALM) detected by colonoscopy in longstanding ulcerative colitis: an indication for colectomy, *Gastroenterology,* 80, 366, 1981.
55. **Butt, J. H., Price, A., and Williams, C. B.,** Dysplasia and cancer in ulcerative colitis, in *Inflammatory Bowel Diseases,* 1st ed., Allan, R. N., Keighley, M. R. B., Alexander-Williams, J., Hawkins, C. F., Eds., Churchill Livingstone, Edinburgh, 1983, 140.
56. **Butt, J. H., Lennard-Jones, J. E., and Ritchie, J. K.,** A practical approach to the risk of cancer in inflammatory bowel disease: research, watch, or act?, *Med. Clin. North Am.,* 1980.
57. **Dobbins, W. O.,** Current status of the pre-cancer lesions in ulcerative colitis, *Gastroenterology,* 73, 1431, 1977.
58. **Lennard-Jones, J. E.,** Cancer risk in ulcerative colitis: surveillance or surgery, *Br. J. Surg.,* Suppl, 584, 1985.
59. **Nugent, F. W., Haggitt, R. C., Colcher, H., and Kutteruf, G. C.,** Malignant potential of chronic ulcerative colitis: preliminary report, *Gastroenterology,* 76, 1, 1979.
60. **Nugent, F. W.,** Surveillance of patients with ulcerative colitis: Lakey Clinic results, in *Colorectal Cancer: Prevention, Epidemiology, and Screening,* Winawer, S. J., Schottenfield, D., Sherlock, P., Eds., Raven Press, New York, 1980, 375.
61. **Nugent, F. W. and Haggitt, R. C.,** Results of a longterm prospective surveillance program for dysplasia in ulcerative colitis, *Gastroenterology,* 86, 1197, 1984.
62. **Rosenstock, E., Farmer, R. G., and Petras, R., et al.,** Surveillance for colonic carcinoma in ulcerative colitis, *Gastroenterology,* 89, 1342, 1985.
63. **Lennard-Jones, J. E., Morson, B. C., and Ritchie, J. K., et al.,** Cancer surveillance in ulcerative colitis. Experience over 15 years, *Lancet,* ii, 149, 1983.
64. **Manning, A. P., Bulgim, O. R., Dixon, N. F., and Axon, A. T. R.,** Screening by colonoscopy for colonic epithelial dysplasia in inflammatory bowel disease, *Gut,* 28, 11, 1489, 1987.
65. **Rosenstock, E., Farmer, R. G., Petras, R., Sivak, M. V., Jr., Rankin, G. B., and Sullivan, B. H.,** Surveillance for colonic carcinoma in ulcerative colitis, *Gastroenterology,* 89, 1342, 1985.
66. **Fozard, J. B. J. and Dixon, M. F.,** Colonic surveillance in ulcerative colitis — dysplasia through the looking glass, *Gut,* 30, 285, 1989.
67. **Collins, R. H., Feldman, M., and Fortran, J. S.,** Colon cancer, dysplasia and surveillance in patients with ulcerative colitis: a critical review, *N. Engl. J. Med.,* 316, 1654, 1987.

INDEX

A

B

C

D

E

F

G

H

I

J

K

L

M

N

O

P

V

W